Soil Components and Human Health

Coltan miner in North Burundi exposed to potentially toxic mining material (Photo: Andreas Bürkert, with kind permission)

Rolf Nieder • Dinesh K. Benbi • Franz X. Reichl

Soil Components and Human Health

Springer

Rolf Nieder
Institute of Geoecology
Technische Universität Braunschweig
Braunschweig, Germany

Dinesh K. Benbi
Department of Soil Science
Punjab Agricultural University Ludhiana
Ludhiana, India

Franz X. Reichl
Walther-Straub Institute of Pharmacology
and Toxicology
LMU
Munich, Germany

ISBN 978-94-024-1221-5 ISBN 978-94-024-1222-2 (eBook)
https://doi.org/10.1007/978-94-024-1222-2

Library of Congress Control Number: 2017958956

Printed on acid-free paper

This Springer imprint is published by Springer Nature
The registered company is Springer Science+Business Media B.V.
The registered company address is: Van Godewijckstraat 30, 3311 GX Dordrecht, The Netherlands

Dedicated to Dr. Karl Nieder, MD,
and Lakshmi Benbi

Preface

Soil behaves as a medium of biogeochemical transformations and fluxes and comprises mineral, gaseous, liquid and biological components that have direct impacts on human health. Soil can also affect public health indirectly, e.g. through emission of greenhouse gases, viz. carbon dioxide (CO_2), methane (CH_4) and nitrous oxide (N_2O), and their feedback to climate. Global climate change is predicted to result in catastrophic events such as droughts, flooding and soil degradation. Soils constitute the largest terrestrial carbon pool and function both as sink and source of CO_2. Optimizing soil management to increase soil organic carbon (SOC) stocks and remove CO_2 from the atmosphere is an important option for mitigating greenhouse gas emissions. Nitrous oxide is considered the largest ozone-depleting substance in the Earth's stratosphere emitted through human activities. Because the ozone layer prevents most harmful ultraviolet (UV) rays from passing through the Earth's atmosphere, depletion of ozone has generated worldwide concern.

Besides natural or soil-borne compounds, anthropogenic chemical and biological materials in soils are health relevant as their transfer to humans could occur by means of drinking water/beverages, consumption of plant and animal products, air (inhalation) or skin exposure. Factors and components that directly affect human health include (i) inhalation of soil-borne fibrogeneous or carcinogenic dusts; (ii) inhalation of toxic or carcinogenic elements or gases; (iii) nutrient deficiencies in soils, food crops and water; (iv) surplus of essential, toxic or carcinogenic elements as well as organic substances in soils, food crops, water and air; and (v) uptake of pathogens from soil materials, food crops and water.

Currently, some books are available that deal with the impact of soil components and chemical materials on environmental compartments, such as pedosphere, hydrosphere and atmosphere, and/or on ecosystems (environmental sciences). These books have little or limited focus on human health (environmental medicine). Recently developed branches of science, such as medical geology, medical mineralogy and medical geochemistry, concentrate on the relationship between natural geological factors (mainly solid earth materials and toxic elements from the lithosphere) and human health. This unique, interdisciplinary volume presents comprehensive knowledge on the behaviour and mobility of elements, chemical and biological materials in the rock–soil–hydrosphere–atmosphere continuum and their interactions with the human body. It is the first publication of its kind that integrates soil and environmental sciences with medical sciences demonstrating the influence of soil components on the environment and humans.

A wide spectrum of components released from soils impact the environment and human health. Some of the components have been discussed as potential sources for diseases. A number of factors have an influence on the health effects of soil materials, such as physical and chemical characteristics; bioavailability; intensity of exposure; exposure route (e.g. inhalation, ingestion, skin or wound uptake); human immune response; processes that control absorption, distribution, metabolization, decomposition and excretion; and human health state.

Following an introductory section, Chap. 1 deals with the relationship between soil quality and human health. The chapter illustrates the impacts of land management and climate change on soil quality and the influence of soil properties on plant and animal products for human nutrition. Component-specific chapters (Chaps. 2, 3, 4, 5, 6, 7, 8, 9, 10, 11, 12, 13, and 14) summarize and discuss potential sources, behaviour and impacts of different substances on the environment and on human health. For example, clay, peat and mud are used for certain therapeutic applications. Fangotherapy is a simple therapy used in treatment of several disorders. Soil is a habitat of numerous fungi and bacteria, many of which produce medicinal agents called antibiotics. They can be used against a variety of infections. Humans can ingest soil either involuntarily or deliberately. The latter is known as geophagia. Soil ingestion is a source of mineral nutrients and potentially harmful elements. Irritant soil materials are responsible for a geochemical, non-infectious disease termed podoconiosis or endemic non-filarial elephantiasis. Nanoparticles are present in nature and are successfully used in many products of daily life. They are also embedded in dental materials. All these aspects are discussed in Chap. 2.

Chapter 3 summarizes risks due to human contact with potentially toxic natural and anthropogenic dusts. Potentially toxic natural particulate dusts include mineral dusts (silica, asbestos and others) and aerosols from rocks and soils generated by natural weathering and erosion as well as volcanic ash particles, which hold transition metals and other toxic elements on their surfaces. Anthropogenic sources of particulate matter include dusts from mining and quarrying, agricultural soils and combustion of fossil fuel for energy generation and heating. Petrol- and diesel-powered vehicles are an important source of particulate and gaseous atmospheric pollution. Particulate matter released by biomass burning from forest clearance and agricultural practices continues to be important.

Chapter 4 presents the major natural and anthropogenic sources of the soil-borne gases CO_2, carbon monoxide (CO), methane (CH_4), nitrous oxide (N_2O) and ammonia (NH_3) and their effects on climate and health. Chapter 5 provides an overview of reactive water-soluble forms of nitrogen (N) and phosphorus (P) such as nitrite (NO_2^-), nitrate (NO_3^-) and phosphate (HPO_4^{2-}, $H_2PO_4^-$) and their origin and impacts on surface and groundwater, ecosystems and humans.

While Chap. 6 gives information on soil-borne and added macroelements such as N, P, potassium (K), calcium (Ca), magnesium (Mg) and sulphur (S), Chap. 7 discusses the role of the microelements, viz. zinc (Zn), iron (Fe), manganese (Mn), copper (Cu), molybdenum (Mo), boron (B), chlorine (Cl), nickel (Ni), chromium (Cr), iodine (I) and selenium (Se), including their bioavailability, impacts on ecosystems and role for human nutrition.

Chapter 8 discusses elements like arsenic (As), beryllium (Be), cadmium (Cd), chromium (Cr), mercury (Hg), lead (Pb), tin (Sn), tungsten (W) and zirconium (Zr). These elements may have toxic effects on living organisms. Metals such as Cu, Ni and Zn are also known as micronutrients (Chap. 7) which may be toxic to living organisms if present in excessive concentrations.

Chapter 9 focuses on radionuclides in soils from natural and man-made sources that constitute a direct route of exposure to humans. Soil- or rock-borne radionuclides such as thorium (^{232}Th), uranium (^{238}U, ^{235}U) and potassium (^{40}K) generate a significant component of the background radiation people are exposed to. The most important man-made radionuclide is cesium (^{137}Cs) which is released from nuclear fission and activation processes. Large amounts of ^{137}Cs were released into the atmosphere during the nuclear weapons tests in the 1950s and 1960s and during the nuclear reactor accidents of Chernobyl (1986) and Fukushima Daiichi (2011) power stations. Radiation is harmful to life. It can cause cancer, genetic and organ damages, cell killing as well as rapid death.

Pesticides are organic chemicals used to protect plants against weeds, insects and diseases (e.g. herbicides, insecticides, fungicides and nematicides). Their behaviour in soils, role in the food chain and impact on humans are discussed in Chap. 10. There are also numerous industrial and unintentionally formed organic substances, which can reach the soil by direct or indirect means and partly have considerable damaging impact to the environment and to human health. Examples are dioxins, furans, polycyclic aromatic hydrocarbons (PAHs) and polychlorinated biphenyls (PCBs) which are presented in Chap. 11. Among the organic chemicals (Chaps. 10 and 11), the persistent organic pollutants (POPs) play a key role.

Human and veterinary medicinal products are a class of emerging environmental contaminants worldwide. Sources of human medicinal products include release from industrial production of pharmaceuticals, discharge of pharmaceuticals from wastewater treatment plants into rivers, field application of sewage sludge as organic fertilizer and use of treated wastewater for irrigation. Sources of veterinary pharmaceuticals include medical treatment of livestock and medicines from surface-applied liquid or farmyard manure. Medicinal products entering the terrestrial environment can reach surface water and groundwater. The organic-based pollutants have the potential to alter significantly the content and diversity of organisms in the soil and to disrupt hormonal systems and modify the natural growth of humans and animals. In Chap. 12, these health risks are presented.

Chapter 13 treats the soil as a vector and source of human pathogens. Soil is a substantial reservoir for viruses, bacteria, fungi, protozoa and multicellular organisms, some of which are pathogenic. Numerous pathogens are applied to soils with human wastewater or animal manures. Viruses are the most hazardous and have the lowest infectious doses of any of the enteric pathogens. Hantavirus, hepatitis A and E, enteric adenoviruses, poliovirus, multiple strains of echoviruses and coxsackievirus are enteric viruses associated with human wastewater. Hazardous pathogens also include bacteria such as *Bacillus anthracis* (causing anthrax) and *Clostridium tetani* (causing tetanus) and fungi such as *Coccidioides immitis* (agent for valley fever). Gastrointestinal infections are the most common diseases caused by enteric bacteria. Some examples are salmonellosis (*Salmonella* ssp.), dysentery (*Shigella* ssp.) and other infections caused by *Campylobacter jejuni*, *Yersinia* ssp., *Escherichia coli* O157:H7 and many others. The most commonly detected protozoa in sewage are *Entamoeba histolytica*, *Giardia intestinalis* and *Cryptosporidium parvum*. Helminths can be grouped into the nematodes (including hookworms, roundworms, whipworms and pinworms), the trematodes (flukes) and the cestodes (tapeworms). Human pathogenic helminths inhabit the human intestine at some point in their life cycle.

Chapter 14 presents the role of prions (proteinaceous infectious disease-causing agents) in soils including risk assessment of prion exposure through environmental pathways. Prions cause transmissible spongioform encephalopathies which are fatal neurodegenerative diseases impacting various mammalian species such as cattle (bovine spongiform encephalopathy (BSE)); goats and sheep (scrapie); moose, deer and elk (chronic wasting disease (CWD)); and humans (Creutzfeldt–Jakob disease (CJD)). As they are horizontally transmissible and remain infectious in soils for many years, CWD and scrapie are of particular environmental concern.

We are thankful to our families – Alexandra, Raphaela and Petra (R. Nieder) and Adwitheya and Meenu (D.K. Benbi) – for their understanding and patience during the preparation of this book. We are grateful to Mr. Marcus Schiedung, Ms. Ramandeep Kaur and Ms. Sunita Siag for their help in the preparation of illustrations. Dr. Christof Högg is acknowledged for the revision of Chap. 2. We would like to express our great appreciation to Ms. Judith Terpos (Springer Life Science) for her continuous encouragement and patience throughout the time of writing this book, and her team (Ms. Corina van der Giesen, Ms. Rajeswari Sathiamoorthy and Mr. Varun Ilangovan) in getting it ready for publication.

Braunschweig, Germany — Rolf Nieder
Ludhiana, India — Dinesh K. Benbi
Munich, Germany — Franz X. Reichl

Introduction

A wide spectrum of interactions exists between soil materials and humans. Soil materials may have deleterious, beneficial or no impacts on human health. In several cases, the relationship between soil materials and public health is well known, for instance, between asbestos and asbestosis, radionuclides and cancer, soil pathogens and gastrointestinal infections and prions and Creutzfeldt–Jakob disease. Understanding the complex relationships between soil materials and human health requires creative cooperation between soil and environmental sciences and environmental medicine. These disciplines are important components of cross-disciplinary research aimed at understanding interactions between biotic and abiotic soil components and the human body. Despite their importance, these interdisciplinary cross-linkages up to now have received only little attention by researchers, policymakers and the wider public, and comprehensive reviews on this topic are rare (Oliver 1997; Abrahams 2002, Pepper et al. 2009). Due to the limited interdisciplinary collaboration, solutions related to complex health problems that result from soil materials for many cases are yet to come.

Soils or the soil zone can be associated with the pedosphere which exists at the interface of the atmosphere, the lithosphere, the hydrosphere and the biosphere. The pedosphere acts as the mediator of biogeochemical transformations and fluxes into and out of the adjacent spheres and is composed of mineral, fluid, gaseous and biological components. Soils are the most complicated biomaterial on the planet and are most important for human life and health since they influence the availability and quality of food, filter and store water, detoxify pollutants and induce human contact with minerals, chemicals and pathogens. Soil is also the habitat of a wide range of organisms necessary for the cycling of elements and for maintaining a healthy environment for human beings (Abrahams 2002).

Several major exposure routes exist through which humans can get into contact with potential toxicants or pathogens. These include ingestion, inhalation and skin, eye or wound contact. Depending on the exposure route, toxicants or pathogens can have different effects due to differences in physical and physiological processes

(Plumlee et al. 2006). Studies investigating the health effects of soil materials originally focused on workplaces where asbestos or silica was commercially produced or on earth materials containing toxic elements (e.g. arsenic, lead, mercury). During the last few decades, studies were also concentrated on exposures to substrates (organic or inorganic) which are released into the environment by anthropogenic or natural processes and reach the soil.

Soils can have both direct and indirect influences on human health, and these effects can be beneficial or harmful (Pepper et al. 2009). Direct effects can result from soil materials themselves (e.g. mineral dusts), contaminants or pathogens contained within or carried by the soil materials. Food and animal feedstuff are increasingly being transported around the world. Pathogens have, therefore, enhanced potential for global spread. Soil particles that are involuntarily ingested are commonly small and have a larger reactive surface compared to their mass and contain larger concentrations of contaminants than the soil from which they derive (Sheppard 1995).

Different soil components can exhibit various physical and chemical characteristics. Physical characteristics particularly include its form, i.e. gaseous, solid, aqueous or non-aqueous liquid. For solid components it is also important whether the material is a glass or crystalline, its size, density, mineralogy and surface charge (Plumlee et al., 2006). Chemical characteristics include type (inorganic, organic), amount and speciation of potentially toxic substances in soil materials. The speciation includes how the toxicant is fixed within the soil material (e.g. adsorbed on the surface or tied within the crystal structure) and the oxidation state. Soils, dusts and other earth materials are also hosts of various microbes, prions and multicellular organisms (Bultman et al. 2005). These pathogens can be permanent, periodic, transient or incidental (e.g. through soil management) soil inhabitants. Human exposure can result from dust inhalation, ingestion of soil, skin contact with soil or ingestion of water which had contact with soils containing pollutants or pathogens.

Potential health concerns can include cancers, respiratory diseases, neurological disorders, diseases of the excretory system, skin diseases and secondary diseases like heart failure, increased susceptibility to infections and others. For example, dusts and aerosols from soils cause pulmonary diseases; irritant soil materials cause dermatosis; pathogens (viruses, bacteria, fungi, protozoa and multicellular pathogens) and prions cause various types of partly lethal infections; pollutants (toxic elements, agrochemicals, organic pollutants, radionuclides) cause damage to digestive organs, neurological diseases and cancer; and uptake of human or veterinary medicinal products from different sources causes decreased effectiveness or even ineffectiveness of antibiotics in the human body due to boosting of bacterial resistance against antibiotic substances. Soil itself is also a source of antibiotics and other natural products that commonly have a positive effect on human health. Soil-borne antibiotics are used to treat human infections. Natural products isolated from soil resulted in 60% of new cancer drugs between the period 1983 and 1994 (Pepper et al. 2009).

Soils can also have an indirect impact on human health through dynamic interaction between pedosphere, biosphere, atmosphere and hydrosphere. Indirect effects are from consumption of plants and water or organisms that can take up contaminants by interacting with soil materials. Soil is the most important source of nutritive elements (macro-, secondary and microelements) that reach humans through plant and animal products. High crop yields are encouraged by fertilizer additions to soil, and synthetic pesticides are frequently utilized to ensure successful yield outcomes. The link between a rising world population that in this century may increase to nine billion before it stabilizes and the ability of soils to support a drastically increasing food supply is causing increasingly serious concern. Even though soils are not the only resource enabling food production, their efficient and at the same time sustainable use and protection will be some of humankind's most complex challenges in the future.

There is also serious concern about a wide range of potentially hazardous compounds that are leached from soils and accumulate in the hydrosphere. Increasing development worldwide has created an increasing need for water of acceptable quality. Though there are numerous sources of water in the Earth's hydrosphere, the availability of water suitable for human consumption and use is limited. The main sources of water pollution are industrial and domestic wastewater and agricultural activities (excessive mineral and organic fertilizer application, use of pesticides). As the hydrosphere is in contact with other components of the environment, this may be dangerous for human health and the environment.

Through their interaction with the atmosphere, soils have also effects on climate change, caused by enhanced emissions of greenhouse gases (GHGs, namely, CO_2, CH_4 and N_2O) into the atmosphere. Among these effects are changes of weather patterns, which may result in extreme meteorological events; rising sea levels; water and food shortages; accelerated destruction of the ozone layer, which will lead to increased UV-B radiation; shifts in the regional distribution of infectious diseases such as malaria, schistosomiasis and dengue; and increased thermal stress to people resulting in increased cardiovascular and respiratory mortality (Martens 1998; Abrahams 2002). Even though fossil fuel combustion is the main cause of the problem, clearing of forests, wetland drainage and cultivation of prairie and forest soils for agriculture lead to significant increases in atmospheric CO_2 as soil organic carbon (SOC) and aboveground biomass were mineralized (Nieder and Benbi 2008). The SOC content of most agricultural soils is now about one third less than that in its native condition as either forest or grassland. Increase of SOC storage and removal of CO_2 from the atmosphere provides a low-cost, immediately available option for mitigating GHG emissions. From a global perspective, soils are vital to the future well-being of nations through their impact on climate change and global warming.

Overexploitation of soils attributable to a desire to increase food production led to serious soil degradation. The area of soils disturbed by human activities was estimated to occupy worldwide 1216 Mha (Lal 2004). Soil erosion by wind and water appears to be the most widespread process of soil degradation. Other important soil degradation processes include loss of soil organic matter (SOM) and nutrients, salinization, acidification, pollution, compaction and subsidence. Degraded soils possess physical and chemical constraints for crop growth like calcareous or acidic chemical condition, low SOM and nutrient contents, shallow solum, presence of toxic substances and compaction (Oldeman 1994). Degraded lands in several regions of the world can be seen as a major contributor to the occurrence of disastrous floods and to poor economic and health conditions of the rural population. Appropriate and sustainable soil management strategies are therefore needed that in turn require detailed knowledge of the properties and limitations of degraded soils which up to now is limited (Asio et al. 2009).

The objectives of this volume are to highlight important links existing between soils and human health and to stress the significance of soils for humans. Moreover, some of the actual research opportunities and challenges will encourage the exchange of ideas and concepts between soil and environmental scientists on the one hand and medical scientists on the other hand. The topics covered in this book will be of immense value to a wide range of readers, including soil scientists, medical scientists and practitioners, nursing scientists and staff, ecologists, agronomists, geologists, geochemists, planners, public health professionals and several others dedicated to the quality and protection of the environment and human health.

References

Abrahams PW (2002) Soils: their implications to human health. Sci Tot Environ 291:1–32

Asio VB, Jahn R, Perez FO, Navarrete IA, Abit SM (2009) A review of soil degradation in the Philippines. Ann Trop Res 31:69–94

Bultman MW, Fisher FS, Pappagianis D (2005) The ecology of soil-borne human pathogens. In: Selenius O, Alloway B, Centeno J, Finkelman R, Fuge R, Lindh U, Smedley P (eds) Essentials of medical geology. Elsevier, pp 481–512

Lal R (2004) Soil carbon sequestration to mitigate climate change. Geoderma 123:1–22

Martens WJM (1998) Health and climate change: modelling the impacts of global warming and ozone depletion. Earthscan, London, 176 pp

Nieder R, Benbi DK (2008) Carbon and nitrogen in the terrestrial environment. Springer, Heidelberg, 430 pp

Oldeman LR (1994) Global extent of soil degradation. In: Oldeman LR (ed) Soil resilience and sustainable use. CAB International, UK, pp 99–118

Oliver MA (1997) Soil and human health: a review. Eur J Soil Sci 48:573–592

Pepper I.L, Gerba CP, Newby DT, Rice CW (2009) Soil: a public health threat or saviour? Crit Rev Environ Sci Technol 39:416–432

Plumlee GS, Morman SA, Ziegler TL (2006) The toxicological geochemistry of earth materials: An overview of processes and the interdisciplinary methods used to understand them. Rev Min Geochem 64:5–57

Paustian K, Antle JM, Sheenan J, Paul EA (2006) Agriculture's role in greenhouse gas mitigation. Pew Center on Global Climate Change. Arlington

Sheppard SC (1995) A model to predict concentration enrichment of contaminants on soil adhering to plants and skin. Environ Geochem Health 17:13–20

Contents

Acronyms and Abbreviations

Ac	Actinium
ACE	Acenaphthene
ACY	Acenaphthylene
AEC	Anion exchange capacity
Ag	Silver
Al	Aluminium
AI	Adequate intake
ALA-D	Aminolevulinic acid dehydratase
Am	Americium
AMDOPH	1-Acetyl-1-methyl-2-dimethyl-oxamoyl-2-phenylhydrazide
ANT	Anthracene
Ar	Argon
As	Arsenic
ATP	Adenosine triphosphate
ATSDR	Agency for Toxic Substances and Disease Registry
Au	Gold
B	Boron
Ba	Barium
BaA	Benz[a]anthracene
BAL	British antilewisite
BaP	Benzo[a]pyrene
BbF	Benzo[b]fluoranthene
BCF	Bioconcentration factor
Be	Beryllium
BGP	Benzo[ghi]perylene
BkF	Benzo[k]fluoranthene
BNF	Biological nitrogen fixation
Bq	Becquerel
BSE	Bovine spongiform encephalopathy
C	Carbon

Ca	Calcium
CA	Conservation agriculture
CCME	Canadian Council of Ministers of the Environment
Cd	Cadmium
Ce	Cerium
CEC	Cation exchange capacity
CGIAR	Consultative Group on International Agricultural Research
CH_4	Methane
CHR	Chrysene
Ci	Curie
CJD	Creutzfeldt–Jakob disease
fCJD	Familial Creutzfeldt–Jakob disease
iCJD	Iatrogenic Creutzfeldt–Jakob disease
vCJD	Variant Creutzfeldt–Jakob disease
Cl	Chlorine
Cm	Curium
CNS	Central nervous system
Co	Cobalt
CO	Carbon monoxide
CO_2	Carbon dioxide
COX	Cyclooxygenase enzyme
Cr	Chromium
Cs	Cesium
CT	Conventional tillage
Cu	Copper
CWD	Chronic wasting disease
2,4-D	2,4-Dichlorophenoxyacetic acid
DAP	Diammonium phosphate
DBA	Dibenz[a,h]anthracene
DCD	Dicyandiamide
DCP	Dicalcium phosphate
DDE	Dichlorodiphenyldichloroethylene
DDT	Dichlorodiphenyltrichloroethane
DDTC	Diethyldithiocarbamate trihydrate
DMA	Dimethylarsinic acid
DMPP	3,4-Dimethylpyrazole phosphate
DMPS	Dimercaptopropanesulphonate
DMSA	Dimercaptosuccinic acid
DNA	Deoxyribonucleic acid
DON	Dissolved organic nitrogen
2,4-DP	Butoxyethyl ester of ($\pm$) 2-(2,4-dichlorophenoxy)propanoic acid
DRI	Dietary reference intake
DT_{50}	Half-life time
DTPA	Diethylenetriamine pentaacetic acid

EAR	Estimated average requirement
EC	Electrical conductivity
EC_{50}	Half maximal effective concentration
EDDHA	Ethylenediamine di(o-hydroxy) phenyl acetic acid
EDDS	Ethylenediaminedisuccinic acid
EDTA	Ethylenediamine tetraacetic acid
EE2	17-alpha-Ethinylestradiol
EEA	European Environment Agency
EFSA	European Food Safety Authority
Eh	Redox potential (tendency of a chemical species to be reduced)
EHEC	Enterohaemorrhagic *E. coli*
EIEC	Enteroinvasive *E. coli*
EPEC	Enteropathogenic *E. coli*
EPTC	S-Ethyl-N,N-dipropylthiocarbamate
ESBL	Extended-spectrum β-lactamase
ETEC	Enterotoxigenic *E. coli*
EUE	Exotic ungulate encephalopathy
ExPEC	Extraintestinal pathogenic *E. coli*
F	Fluorine
FAO	Food and Agriculture Organization
FDC	Follicular dendritic cell
Fe	Iron
FFI	Fatal familial insomnia
FLT	Fluoranthene
FLU	Fluorine
FSE	Feline spongiform encephalopathy
GHG	Greenhouse gas
GJ	Gigajoule
GLASOD	Global Assessment of Human-Induced Soil Degradation
GSSS	Gerstmann–Sträussler–Scheinker syndrome
Gy	Gray
H	Hydrogen
H_2	Molecular hydrogen
H_c	Henry's law constant
HAB	Harmful algal bloom
HAV	Hepatitis A virus
Hb	Haemoglobin
HBB	Hexabromobiphenyl
HBCD	Hexabromocyclodecane
HCB	Hexachlorobenzene
HCBD	Hexachlorobutadiene
HCH	Hexachlorocyclohexane
HEDTA	Hydroxyethyl ethylenediamine tetraacetic acid
HEV	Hepatitis E virus

HFRS	Haemorrhagic fever with renal syndrome
Hg	Mercury
HNO_3	Nitric acid
HPS	Hantavirus pulmonary syndrome
I	Iodine
IAA	Indole acetic acid
IAEA	International Atomic Energy Agency
ICRP	International Commission on Radiological Protection
IEA	International Energy Agency
IEO	International Energy Outlook
IFA	International Fertilizer Industry Association
IND	Indeno[1,2,3-cd]pyrene
IOM	Institute of Medicine
IPCC	Intergovernmental Panel on Climate Change
IPCS	International Programme on Chemical Safety
IUSS	International Union of Soil Sciences
IZiNCG	International Zinc Nutrition Consultative Group
K	Potassium
Kd	Coefficient related to the partitioning between solid and aqueous phases (sorption constant)
K_{aw}	Air–water partitioning coefficient
K_{oa}	Octanol–air partitioning coefficient
K_{oc}	Organic carbon–water partitioning coefficient
K_{ow}	Octanol–water partitioning coefficient
LC_{50}	Lethal concentration 50% (concentration of a chemical required to kill 50% of a test population)
LD_{50}	Lethal dose 50% (amount of a substance required to kill 50% of a test population)
LDL	Low-density lipoprotein
MAFF	Ministry of Agriculture, Fisheries and Food
MAP	Monoammonium phosphate
MCP	Monocalcium phosphate
MetHb	Methaemoglobin
Mg	Magnesium
Mg	Megagram (unit of mass; 1 Mg is equal to 1,000,000 grams = 1 tonne)
MHG	Methaemoglobinaemia
MIT	Mineralization–immobilization turnover
MKP	Monopotassium phosphate
MMA	Monomethylarsonic acid
Mn	Manganese
Mo	Molybdenum
MPa	Megapascal (1 MPa = 1 million Pa)
MRSA	Methicillin-resistant *Staphylococcus aureus*

MSM	Methylsulphonylmethane
N	Nitrogen
N_2	Molecular nitrogen
N_r	Reactive nitrogen
N_2O	Nitrous oxide
Na	Sodium
Nap	Naphthalene
Nb	Niobium
NH	Imidogen
NH_2	Amidogen
NH_3	Ammonia
NH_4^+	Ammonium
NH_y	Collective term for NH_3 and NH_4^+
Ni	Nickel
NO	Nitric oxide
NO_2	Nitrogen dioxide
NO_2^-	Nitrite
NO_3^-	Nitrate
NO_x	Nitrogen oxides (sum of NO and NO_2)
NORM	Naturally occurring radioactive materials
Np	Neptunium
NRC	National Research Council
NT	No tillage
NTA	Nitrilotriacetic acid
NTS	Non-typhoid *Salmonella*
O	Oxygen
O_2	Molecular oxygen
O_3	Ozone
OECD	Organisation for Economic Co-operation and Development
P	Phosphorus
Pa	Protactinium
PAH	Polycyclic aromatic hydrocarbon
Pb	Lead
PBB	Polybrominated biphenyl
PBDE	Polybrominated diphenyl ether
PCB	Polychlorinated biphenyl
PCDD	Polychlorinated dibenzo-p-dioxin
PCDF	Polychlorinated dibenzofuran
PCN	Polychlorinated naphthalene
Pd	Palladium
PeCB	Pentachlorobenzene
PFOS	Perfluorooctane sulphonate
Pg	Petagram (unit of mass; 1 Pg = 15^{15} g = 1 billion tonnes)
pH	Potential of hydrogen

PHE	Phenanthrene
pK_a	Acid dissociation constant
PM	Particulate matter
$PM_{0.1}$	Particles with an aerodynamic equivalent diameter <0.1 μm
$PM_{2.5}$	Particles with an aerodynamic equivalent diameter <2.5 μm
PM_{10}	Particles with an aerodynamic equivalent diameter <10 μm
Po	Polonium
POP	Persistent organic pollutant
PP	Peyer's patches
ppb	Parts per billion
ppm	Parts per million
PrP	Prion protein
Pu	Plutonium
PWP	Permanent wilting point
PYR	Pyrene
Ra	Radium
RDA	Recommended dietary allowance
RDI	Recommended daily intake
Rem	REM
Rn	Radon
RNA	Ribonucleic acid
RSG	Reference soil group
RT	Reduced tillage
Ru	Ruthenium
RUM	Returning unwanted medicine
S	Sulphur
S^o	Elemental sulphur
Sb	Antimony
SCCP	Short-chained chlorinated paraffin
Se	Selenium
SECIS	Swedish Environmental Classification and Information System
Si	Silicon
Sn	Tin
SO_2	Sulphur dioxide
SOC	Soil organic carbon
SOM	Soil organic matter
SO_x	SO_2 and SO_4^{2-}
SQI	Soil quality index
Sr	Strontium
SSP	Single superphosphate
STEC	Shiga toxin-producing *E. coli*
Sv	Sievert
2,4,5-T	2,4,5-Trichlorophenoxyacetic acid
Ta	Tantalum

Tc	Technetium
Te	Tellurium
TE	Toxic element
Tg	Teragram (unit of mass; 1 TG = 10^{12} g = 1 million tonnes)
Th	Thorium
Ti	Titanium
TI	Thallium
TME	Transmissible mink encephalopathy
TOMS	Total Ozone Mapping Spectrometer
TRIEN	Triethylene tetramine
TSE	Transmissible spongiform encephalopathy
TSP	Triple superphosphate
TUL	Tolerable upper limit
U	Uranium
UNCCD	United Nations Convention to Combat Desertification
UNEP	United Nations Environment Programme
UNSCEAR	United Nations Scientific Committee on the Effects of Atomic Radiation
UNSD	United Nations Statistics Division
US ATSDR	United States Agency for Toxic Substances and Disease Registry
US EPA	United States Environmental Protection Agency
USGS	United States Geological Survey
V	Vanadium
VOC	Volatile organic compound
WCA	World Coal Association
WHO	World Health Organization
WMO	World Meteorological Organization
WRB	World Reference Base for Soil Resources
WRI	World Resources Institute
Y	Yttrium
Zn	Zinc
Zr	Zirconium

Chapter 1
Soil Quality and Human Health

Abstract Protecting soil and preserving its overall quality has become a key international goal. Early concepts of soil quality dealt mainly with soil properties that contribute to soil productivity, with little consideration for environment regulation and human health. It is only recently that studies integrating soil and human health have been initiated. While soil performs several important functions related to ecosystem services, the most significant functions for human health are production of safe and nutritious food and protecting from environmental pollution. Human health is greatly dependent on the soil-water-air continuum which is strongly moderated by processes in the soil. The functions of soil such as filtering, buffering and transformation help in protecting the environment, including human beings against the contamination of groundwater and the food chain. Human activities impact several processes in soil that could lead to physical (accelerated erosion, deterioration of soil structure, crusting, compaction, hard-setting), chemical (nutrient depletion and imbalance, acidification, salinization) and biological (depletion of soil organic matter, loss of biodiversity) degradation of soil. Soil degradation directly affects food security through reduction in crop yields, decline in their nutritional quality and reduced input use efficiency. Plant availability of mineral nutrients in the soil is the main source of mineral supply to human beings. Plants, which absorb minerals from soil, are either eaten directly by humans or fed to animals that are then included in human diet. Therefore, any deficiency in plant products could manifest in human beings. Global warming associated with altered rainfall pattern could subject soils to significant risk of climate induced physical and chemical degradation. Therefore, it is imperative to manage soils to minimize soil degradation and to derive benefits for human health. In this chapter, land as a resource for supporting global population and the role it plays in performing ecosystem functions vis-à-vis soil quality, and the impact of anthropogenic activities on soil quality, plant and animal products and human health are discussed.

Keywords Land resources • Soil quality indicators • Soil degradation • Organic agriculture • Soil and human health • Land suitability classes

R. Nieder et al., *Soil Components and Human Health*,
https://doi.org/10.1007/978-94-024-1222-2_1

1.1 Public Awareness for Soil Quality and Protection

During the late twentieth century, there has been increased recognition about the key involvement of the soil in crop production and in water and atmospheric purification, emphasizing the role of soil both for crop production and for environmental quality. Protecting soil and preserving its overall quality has become a key international goal. Early concepts of soil quality dealt mainly with soil properties that contribute to soil productivity, with little consideration for environment regulation and human health. During the last two decades, the scope of soil quality has been considerably widened to represent compositional structure and multiple functions of soil. Though it is known since millennia that soils play an important role in human health but studies integrating soil and human health have been initiated only recently. One example is the International Union of Soil Sciences 18th World Congress of Soil Science held at Philadelphia, USA (2006) during which a symposium on "Soil and human health" was organized. Recognizing the role of soil in food production, environment moderation and human health maintenance, several definitions of soil quality with similar elements have been proposed (e.g. Carter et al. 1997; Karlen et al. 1997; Arshad and Martin 2002; Filip 2002). The following definition proposed by the Soil Science Society of America: "The fitness of a specific kind of soil, to function within its capacity and within natural or managed ecosystem boundaries, to sustain plant and animal productivity, maintain or enhance water and air quality, and support human health and habitation" is one of the most comprehensive definitions that integrates scientific knowledge and practical approach. For the soil to perform its functions at optimal level, it is imperative to manage it properly so as to maintain its quality. Soil may be lost in a relatively short period of time if used inappropriately or mismanaged with very limited opportunity for regeneration or replacement, indicating that the soil is not an inexhaustible resource. Changes in soil quality can be assessed by measuring appropriate properties and comparing them with desired values (critical limits or threshold level), at different time intervals, for a specific use in a selected agro-ecosystem. Such a monitoring system provides information on the effectiveness of the selected farming system, land use practices, technologies and policies that influence soil quality. A farming system or policy that contributes negatively to any of the selected indicators could be considered potentially unsustainable and thus, discouraged or modified. Systems that improve performance of the indicators can be promoted and advanced to ensure sustainability. Present approaches to quantify soil quality are concerned with either directly characterizing soil properties or identifying specific functions that can represent the attribute in question (Gregorich et al. 1994). Soil quality influences the quality of plant and animal products, which impact human health. However, soil quality cannot be evaluated independent of other elements of physical environment such as climate, topography, relief, vegetation, and hydrology. In this context, soil is one of the several compartments of the physical environment (land) having two-way interaction with other elements (Tóth 2008). Therefore, characteristics of soil are determined by other land forming factors that in-turn are influenced by soil's effect on environmental processes.

1.2 Land Resources and Land Quality

Globally, there is 130.8 million km^2 of ice-free land out of which about 14.5 million km^2 (11.1%) is arable land used for cultivation of agricultural crops and an additional 2.4 million km^2 of land, largely in the arid parts of the world, is irrigated (Eswaran et al. 1999). About 35% of total land is under grasslands and woodlands and 28% is forest ecosystems. Not all the available land is suitable for food production as the quality of land depends on other elements of physical environment. Land quality, which represents the condition or health of land relative to its capacity for sustainable land use and environmental management, integrates resources of soil, water, vegetation and terrain. Collectively, these sources provide the basis for land use. Based on properties of the soils, the major stresses they experience and their resilience to stress, nine land quality classes have been created (Eswaran et al. 1999). The properties of different land quality classes, and the area and population associated with each land quality class are given in Tables 1.1 and 1.2, respectively. Class I land, which is considered ideal for crop production

Table 1.1 Properties of inherent land quality classes obtained by a combination of the performance and resilience attributes of soils

Land quality class	Properties
I	Prime land; highly productive soils with few management-related constraints; ideal temperature and moisture conditions for annual crops; soil management involves conservation practices to minimize erosion, optimum fertilization, and use of best available plant materials
II and III	Soils are good with very high productivity and response to management; have few problems for sustainable production, particularly for Class II soils, care must be taken to reduce degradation; lower resilience characteristics of Class II soils make them more risky; particular for low-input grain crop production. Conservation tillage is essential, buffer strips are generally required and fertilizer use must be carefully managed; the land is suitable for national parks and biodiversity zones
IV, V, and VI	These soils may not be used for grain crop production, particularly those belonging to class IV; productivity is not high; lack of plant nutrients is a major constraint necessitating adoption of a good fertilizer use plan; require important inputs of conservation management; need continuous monitoring of soil degradation; can be set aside for national parks or as biodiversity zones; in the semi-arid areas these can be managed for range
VII	May only be used for grain crop production if there is a real pressure on land; definitely not suitable for low-input grain crop production; their low resilience makes them easily prone to degradation; should be retained under natural forests or range and some localized areas can be used for recreational purposes; biodiversity management is crucial in these areas
VIII and IX	Belong to very fragile ecosystems and are very uneconomical to use for grain crop production; should be retained under their natural state. Class IX soils are mainly confined to the Boreal area, timber harvesting must be done carefully to avoid ecosystem damage. Class VIII is mainly the deserts

Adapted from Eswaran et al. (1999)

Table 1.2 Land area (million km^2) in different land quality classes and the estimates of population in each class

Land quality class	Land area (million km^2)	Land area (%)	Population %	Soil performance	Soil resilience	Risk for sustainable grain production (%)
I	3.11	2.38	6.1	H	H	<20
II	6.51	4.98	14.2	H	M	20–40
III	5.95	4.55	4.8	M	H	20–40
IV	5.17	3.95	11.8	H	L	40–60
V	21.60	16.51	29.7	M	M	40–60
VI	17.42	13.32	12.1	L	H	40–60
VII	11.79	9.01	11.5	M	L	60–80
VIII	21.83	16.69	1.9	L	M	>80
IX	36.96	28.59	11.2	L	L	>80

Adapted from Eswaran et al. (1999)
L low, *M* medium, *H* high

occupies only 2.4% of the world's land surface but contributes more than 40% of global food and feed production. Over 90% of Class I soils are used for grain production. By definition, Class I lands are absent in the tropics, which could be the reason for low productivity of tropical soils. Land quality classes I, II and III together representing 12% of the total land area support 25% of the world population. These lands are generally free of constraints for most agricultural uses. They are, however, unequally spread around the globe with a larger portion in the temperate countries. Class II and III lands are extensive in the tropical and temperate regions, and are under agriculture. Class IV, V, and VI lands occupy a significant part of the earth's surface (34%) and also support about 54% of the world's population. A major part of Class IV, V and VI lands occur in the tropics and low-input agriculture promotes land degradation. Class VII lands, supporting 11.5% of the population, comprise shallow soils, with high salt concentrations and high organic matter content. These lands are not considered suitable for agriculture. Peat lands, which are very fragile, are also included in this group but they have tremendous capacity to sequester carbon. Class VIII lands occur at low temperatures and/or steep slopes. These lands include highly fragile peat lands of the high latitudes. Land quality class IX soils have inadequate moisture to support most annual crops, and are characterized by very low net primary productivity. This class also includes deep soils that are highly productive under irrigation but efficient use of water is crucial for management of these soils. From a sustainable agriculture point of view, Classes IX, VIII, and VII are not recommended for agriculture, though small areas may currently be used. The global land area that is generally free of constraints for most agriciultural uses is unequally spread around the globe with a larger portion in the temperate countries of the world. In addition to poorer land quality in tropical regions, land degradation is also well-entrenched, which aggravates food insecurity and thus influences human health. Currently, about 4.93 billion ha area is under agriculture showing an increase of about 10% during the

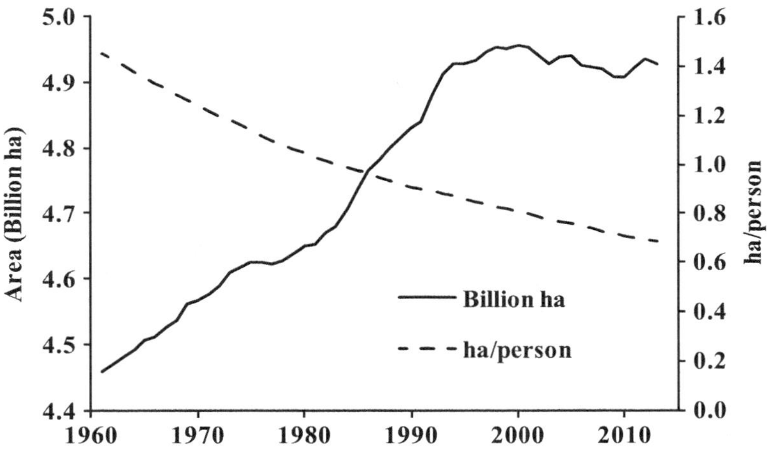

Fig. 1.1 Agricultural area and per person availability of agricultural area worldwide during 1961 to 2013 (Data source: FAOSTAT Database n.d.)

last over 50 years (Fig. 1.1). The increase in area has occurred because of conversion of forest, wetlands and grasslands. In the same time period, per capita availability of agricultural area has declined by about 52% because of increase in population. This could severely influence food and feed security. Cultivation is a major land use (20% or more of the land area) in South and Southeast Asia, Western and Central Europe, and Central America and the Caribbean, but is less prevalent in sub-Saharan and Northern Africa, where cultivated land covers less than 10% of the total area (Fig. 1.2). Land suitability for agriculture has also been defined as prime, good and marginal in terms of their capacity to produce potentially attainable crop yields (FAO 2011). Currently, 28% of cultivated land is of prime quality and 53% of good quality. The soils are often marginal in low-income countries and only 28% of total cultivated land is classified as prime. In high-income countries, 32% of the currently cultivated land is of prime quality.

1.3 Soil Quality

Soil quality is generally distinguished into inherent soil quality and dynamic soil quality depending on its response to management (Carter et al. 1997). Inherent soil quality depends on the soil's mineral composition, soil texture and depth and shows little change over time unless degraded by human activities. Inherent soil quality is a measure of the soil to perform specific functions (Beinroth et al. 2001), which include production of biomass, environmental conservation, climate regulation, protection of food chain from pollution, preservation of gene reserves and geogenic/cultural heritage, and as source of raw material and physical basis for

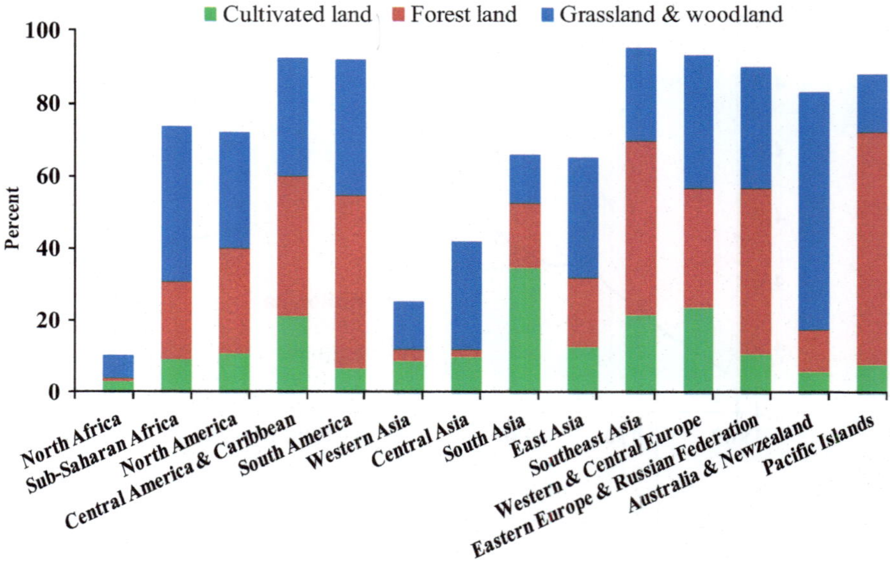

Fig. 1.2 Regional distribution of cultivated, forest and grassland and woodland ecosystems (Adapted from Fischer et al. 2010)

infrastructural development (Blum and Eswaran 2004; FAO 1995). While all the functions are important, the most significant functions for human health are production of safe and nutritious food and protecting from environmental pollution. Human health is greatly dependent on the soil-water-air continuum, which is strongly moderated by processes in the soil. The functions of soil such as filtering, buffering and transformation help in protecting the environment, including human beings against the contamination of ground water and food. Because of rapid industrialization and urbanization in several regions of the world, large amounts of solid and liquid by-products including potentially toxic organic and inorganic compounds and pollutants are produced and deposited/dumped in soils. The soils respond to these depositions through mechanical filtration, physical or physico-chemical adsorption and precipitation reactions and microbial and biochemical transformation processes to minimize the bioavailability and concentration of toxic elements in soil solution and their movement to the groundwater (Blum 2008). However, capacities of soils to perform these functions are limited and vary according to specific soil conditions or properties, which are a consequence of soil forming processes and anthropogenic activities. Any disturbance of soil functions or modification of soil forming processes has the potential for adverse human health effects.

The soil-forming factors viz. climate and living organisms acting on parent materials over time, as conditioned by relief have geographic order. There are currently two major international soil classification systems for naming soils and creating legends for soil maps. According to US Soil Taxonomy (Soil Survey Staff

1999) the soils are grouped into 12 orders that include numerous suborders. The World Reference Base (WRB) for Soil Resources (IUSS Working Group WRB 2015) classifies soils using 32 Reference Soil Groups (RSGs). In both systems the soils are classified based on the presence or absence of features or horizons (diagnostic properties and horizons) that reflect soil forming processes. The soil orders Ultisol (e.g. Lixisol, Alisol, Acrisol and Nitisol according to WRB), Alfisol (e.g. Luvisol and Albeluvisol according to WRB), Inceptisol (e.g. Leptosol, Cambisol and Anthrosol according to WRB), and Entisol (e.g. Fluvisol, Gleysol and Arenosol according to WRB) occupy about 44% of the global land area and together support about 71% of the world population (Blum and Eswaran 2004) (Fig. 1.3).

In the tropics, much of the population is associated with Entisols, Inceptisols and Ultisols, and in the temperate regions Alfisols and Mollisols have high density of people. Mollisols, which occupy about 6.9% of the land surface are associated with about 6.7% of the population. Soils of the suborder Udoll of the Mollisol order (Chernozems according to WRB; Fig. 1.4a) are considered the most fertile soils. Ultisols (e.g. Lixisols according to WRB; Fig. 1.4b) and Vertisols (same term in the WRB system; Fig. 1.4c) dominate in the tropics and together occupy 10.9% of the land surface and are associated with 23.5% of the population. Ultisols and Oxisols (Ferralsols according to WRB; Fig. 1.4d) are problem soils for low-input agriculture. Gelisols (Cryosols according to WRB; Fig. 1.4e), occupying 8.6% of the land surface, have the lowest population density with about two persons km^{-2}. Andisols (Andosols according to WRB; Fig. 1.4f), which occupy only 0.7% of the land area, are associated with 1.7% of the world population. Fragile soils such as Histosols

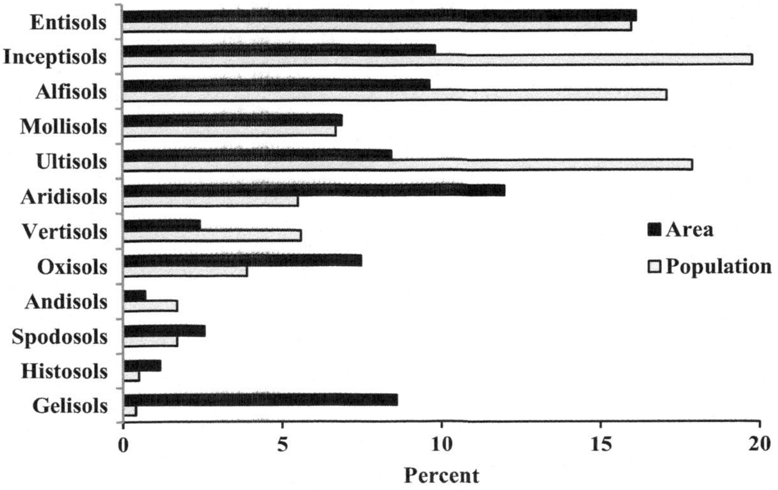

Fig. 1.3 Global land area under different soil orders (Soil Survey Staff 1999) and estimated percentage of population living on such soils. Total ice-free land area estimated as 130.8 million km^2 and total population equals 6.4 billion in 2002 (Drawn from Blum and Eswaran 2004)

Fig. 1.4 (**a**) Profile of a Haplic Chernozem (US Soil Tax.: Mollisol) in loess near Halle (Germany) showing a mollic Ah horizon (earthworm mull), high soil organic matter and nutrient contents, neutral pH, high activity clays, high cation exchange capacity (CEC) and high water storage capacity (Photo: Rolf Nieder). (**b**) Profile of a Haplic Lixisol (US Soil Tax.: Ultisol) in western

Fig. 1.4 (continued) Rwanda (near Gatumba) on strongly weathered schist with poor soil organic matter and nutrient content, slightly acidic pH, low-activity clays, low CEC and low water storage capacity (Photo: Rolf Nieder). (**c**) Profile of a Haplic Vertisol in North Morocco with dark clay (stable organic complexes with smectites (i.e., high-activity clays)), a crumby to polyedric structure in the surface soil horizon and a polyedric to prismatic structure in the vertic subsurface horizon (Photo: Rolf Nieder). (**d**) Profile of a Rhodic Ferralsol (US Soil Tax.: Oxisol) in southern Rwanda (near Butare) with a deep solum (usually several meters thick). Deep and intensive

Fig. 1.4 (continued) weathering has resulted in a high concentration of sesquioxides and well-crystallized kaolinite (low-activity clay). This mineralogy and the low pH explain the stable microstructure (pseudo-sand) and reddish (hematite) soil colours. Ferralsols are chemically poor soils, and generally have low nutrient contents and low CEC. Anion exchange capacity often equals or exceeds the CEC. Ferralsols mostly exhibit an excellent porosity and thus good permeability and favorable infiltration rates (Photo: Rolf Nieder). (**e**) Profile of a Gleyi-Turbic Cryosol (US Soil Tax.: Gelisol) in North Siberia characterized by permanently frozen soil within 1 m of the land surface and by waterlogging during periods of thaw. Cryoturbation is the most

and Aridisols (Fig. 1.4g) together cover 13.2% of the land surface and are associated with 6% of the world population. In the developing countries the population pressure has forced the poor landless people to move into fragile ecosystems resulting in deterioration of soil and environmental quality. Drainage of wetlands is one example (Fig. 1.4h).

In contrast to inherent soil quality, the dynamic soil quality is subject to change over relatively short time periods and its attributes are strongly influenced by management. Several soil properties generally referred to as soil quality indicators are used to describe quality of a soil. Attributes that are sensitive to management are most desirable as indicators. A minimum number of indicators need to be measured to evaluate changes in soil quality resulting from various management practices. The commonly used indicators include soil texture, structure, pH, organic matter, microbial biomass C, and respiration or enzyme activities. The indicators are generally identified based on subjective judgement or through statistical data reduction techniques. While the indicators (Table 1.3) have generally been proposed with respect to soil's functions for agricultural production and sustainability, these directly and indirectly affect the food chain and human health.

The soil quality indicators that could directly influence human health include plant availability of essential nutrients, and contamination of soil with organic chemicals, concentration of toxic or pollutant elements, and existence of pathogens. However, in any soil quality assessment program soil organic matter (SOM) is a key attribute that exerts a profound influence on several physical, chemical and biological properties of soil. The effects of SOM on soil quality often involve the

Fig. 1.4 (continued) important soil-forming process acting in polar areas as well as other regions with permafrost. The process can be noted as discontinuous soil horizons, incorporation of the surface horizon into the subsoil and vice versa (so-called involutions, intrusions or injections), and occurrence of sorted circles, polygons or stripes on the soil surface. Cryoturbation has a strong influence on agriculture, forestry, and the construction of roads and buildings (Photo: Pavel Barsukov; with kind permission). (**f**) Mollic Andosol (US Soil Tax.: Andisiol) developed from volcanic pumice in the Chimborazo area, Ecuador. The organic matter (up to 8%) of the dark surface horizon is protected by allophane. Andosols are fertile soils, particularly those in intermediate or basic volcanic ash. Most Andosols have a good internal drainage, high porosity, high water infiltration and good aggregate stability (Photo: Rolf Nieder). (**g**) Landscape with Aridisols (mainly Leptosols and Arenosols according to WRB) in the Tabernas Desert, Province of Almería, Spain. Almería is the driest region of Europe, with the continent's only true desert climate. The annual rainfall reaches levels as low as 156 mm in coastal areas. The desertlike soils have low soil organic matter content and show accumulation of translocated soluble salts or sodium ions. They are either not or only sparsely vegetated by drought- or salt-tolerant plants. Dry climate and low humus content limit their arability without irrigation (Photo: Rolf Nieder). (**h**) Drainage of wetlands for cropping near Butare, southern Rwanda. African wetlands are a very important source of natural resources. Despite their economic (people depend on wetlands for water, medicine, food and building materials) and ecological (habitat for numerous plants and animals) importance, large wetland areas throughout Africa are being reclaimed. Decision-makers often have insufficient understanding of these values, so the protection of wetlands is not a serious alternative. Drainage of wetlands causes organic matter decomposition and release of greenhouse gases such as CO_2 and N_2O (Photo: Rolf Nieder)

Table 1.3 Commonly used soil quality indicators

Indicator	Soil property
Physical	Soil texture, bulk density, aggregate stability and distribution, rooting depth, penetration resistance, porosity, hydraulic conductivity, infiltration rate, water holding capacity, mineralogy
Chemical	Organic C, labile C fractions of different oxidizability, total N, mineral N, pH, electrical conductivity, available nutrient status, cation exchange capacity, potentially toxic elements, organic chemical contamination
Biological	Microbial biomass C and N, potentially mineralizable N, soil respiration, metabolic quotient, respiratory quotient, enzyme activities, phospholipid, fatty acid, DNA

mineral fraction. Thus, variations across different soils are also a consequence of variations in the soil inorganic component. The physical properties favourably influenced by SOM include bulk density, aggregate stability, water retention and transmission. In general, soil bulk density decreases as the rate of organic residue application and the SOM content increases regardless of the soil texture or the residue type used. Aggregation, or the binding together of individual soil particles, gives rise to what is known as soil structure. Typically, a well-structured soil has greater resistance to the forces of erosion and has improved air-water relationships. Generally, hydraulic conductivity, infiltration rate, air diffusivity, surface drainage, and ease of root penetration increase with increasing aggregation. Soil organic matter favourably impacts water retention and transmission properties of soils. Increase in SOM content results in increased water retention at both field capacity (FC: −33 kPa) and permanent wilting point (PWP: −1500 kPa), which is due to the higher water absorption capacity of SOM. Soil organic matter can absorb and hold substantial quantities of water, up to twenty times its mass (Stevenson 1994). The soil chemical properties influenced by SOM include nutrient availability, exchange capacity, interaction with metals and organics, and buffering capacity. Organic matter provides a large pool of macronutrients in the soil, which become available to the plant through mineralization. The biological and biochemical properties mainly influenced by SOM include supply of energy for biological processes, microbial biomass activity, enzyme activities, and ecosystems resilience. In the last two decades there is increasing realization that it is not only the quantity of SOM but also its composition or turnover rate that is important for evaluating management-induced changes in soil quality. Physical fractions of SOM separated according to size or density have been used as indicators of active or labile and recalcitrant pools of SOM (Benbi et al. 2012). The active C pool includes soil microbes and microbial products, with short turnover times, and the passive C pool includes physically and chemically stabilized organic C that is very resistant to decomposition.

A soil quality assessment program generally includes one or more of the three basic components viz.: (1) the ability of soil to enhance crop production (productivity component); (2) the ability of soil to function in attenuation of environmental contaminants, pathogens, and offsite damage (environment component); and (3) the linkage between soil quality and plant, animal and human health (health component). Several formulations combining specific soil quality elements have been proposed for developing soil quality indices (Doran and Parkin 1994; Karlen and Stott 1994). Parr et al. (1992) proposed a soil quality index (SQI) as a function of soil properties (SP), potential productivity (P), environmental factors (E), human and animal health (H), erodibility (ER), biological diversity (BD), food quality/ safety (FQ), and management inputs (MI) (Eq. 1.1).

$$\mathrm{SQI} = f\ (\mathrm{SP, P, E, H, ER, BD, FQ, MI}) \tag{1.1}$$

Doran and Parkin (1994) described a performance based SQI that consists of six elements (Eq. 1.2) viz. food and fiber production (SQE1), erosivity (SQE2), ground water quality (SQE3), surface water quality (SQE4), air quality (SQE5), and food quality (SQE6).

$$\mathrm{SQI} = f\ (\mathrm{SQE1, SQE2, SQE3, SQE4, SQE5, SQE6}) \tag{1.2}$$

Andrews et al. (2004) proposed a framework for soil quality assessment with emphasis on environment protection based on soil functions including microbial biodiversity and habitat, filtration of contaminants, nutrient cycling, and resistance to degradation and resilience, and water relations. One or more indicators were used for each function. Advantage of soil quality index approach is that soil functions can be assessed based on specific performance criteria established for each element, for a given ecosystem. While several soil quality indices have been used for soil quality assessment, but these do not provide guidelines or threshold values for characterizing soils with respect to specific function such as human health and quality of food. The soil quality indices may only be used for relative rating of quality of various soils. The Canadian Council of the Ministers of the Environment (CCME 2006) presented a protocol that provides the rationale and guidance for developing environmental and human health soil quality guidelines for contaminated sites in Canada. The protocol considers the effects of contaminated soil exposure on human and ecological receptors for given land uses. The pathways and receptors of contaminated soil considered in the derivation of soil quality guidelines were selected based on exposure scenarios for agricultural, residential/ parkland, commercial, and industrial land uses.

1.4 Influence of Land Management on Soil Quality

For decades, soils have been subjected to degradation and pollution. Soil degradation results in long-term loss of ecosystem function and productivity from which land cannot recover unaided (Bai et al. 2008). Human activities impact several processes in soil that could lead to deterioration of soil quality. Principal soil degradation processes include physical (accelerated erosion, deterioration of soil structure, crusting, compaction, hard-setting), chemical (nutrient depletion and imbalance, acidification, salinization, decrease in cation retention capacity) and biological degradation (depletion of soil organic matter, reduction in the activity and species diversity of soil microorganisms). Land quality together with climate, terrain and landscape position, vegetation and soil biodiversity determine the kind of land degradation process. Several studies have shown that a significant decline in soil quality has occurred worldwide through adverse changes in its physical, chemical and biological properties and contamination by inorganic and organic chemicals. Soil erosion by wind (Chap. 3) and water (Fig. 1.5a), soil organic matter and nutrient depletion (Fig. 1.5b), salinization (Fig. 1.5c), acidification (Fig. 1.5d) and soil sealing by housebuilding, roads and other infrastructure (Fig. 1.5e, f) are considered the most important forms of soil degradation.

Global Assessment of Human-induced Soil Degradation (GLASOD) showed that worldwide 15% of the land covering an area of 1964 million ha is affected by human-induced soil degradation out of which 1642 million ha (84%) is affected by water and wind erosion (Table 1.4). Total annual production of dust by deflation of soils and sediments has been estimated to be 61–366 million Mg. The amount of dust arising from the Sahel zone has been reported to be around 270 million Mg per annum, which corresponds to a loss of a layer of 20 mm over the entire area (WMO 2005). Besides leading to loss of organic matter and plant productivity, wind erosion can cause serious health problems by blowing soil particles, pollutant and microbes into the air, aggravating allergies, asthma and opportunistic infection of the lungs (Korenyi-Both et al. 1992; Peters et al. 2001; Prahalad et al. 2001; Griffin and Kellogg 2004). Airborne bacterial and fungal spores and microbial molecules such as endotoxins and fungal mycotoxins can cause allergic reactions and respiratory stress in children (Braun-Fahrlander et al. 2002). Desert dust in Kuwait during the 1990s was reported to cause cellular membrane and DNA damage (Athar et al. 1998).

Chemical soil degradation covers about 239 million ha (12%) and physical soil deterioration, which includes compaction and water logging, occupies around 83 million ha (4%). Soils affected by pollution occupy an area of 22 million ha worldwide. Recent estimates made by Bai et al. (2008) show that 24% of the global land, often in very productive areas, has degraded during the period 1981–2003. Comparison of degrading areas with global land cover revealed that 19% of degrading land is cropland, 24% is broad-leaved forest and 19% needle-based forests. Comparison, of the new analysis with the previous GLASOD estimates show that much of the area estimated by two approaches does not overlap.

Loss of nutrients is the major sub-type of chemical deterioration of the soils followed by salinization. More than two-thirds of area affected by salinization (76 million ha) is located in Asia. Salinization leads to an excessive accumulation of water-soluble salts such as sodium, potassium, calcium, magnesium, chloride, sulphate, carbonate and bicarbonate in the soil and soil solution. The salinization could be caused through natural processes (primary salinization) or human interventions (secondary salinization). Primary salinization is caused due to high salt contents in parent material or groundwater particularly in arid regions. Secondary

Fig. 1.5 (**a**) Severe soil erosion by water (gully erosion) on a loess soil (Luvisol) cropped with sugar beets (*Beta vulgaris* ssp. *vulgaris*, Altissima) near Bad Gandersheim, Germany. Until about

Fig. 1.5 (continued) June, sugar beets do not provide a complete vegetation coverage and the soil on its surface is thus vulnerable against rainfall. Soil erosion by water is a result of rain detaching and transporting soil material (Photo: Rolf Nieder). (**b**) Cultivation of cassava *(Manihot esculenta)* on an eroded, soil organic matter and nutrient depleted Ferric Acrisol with low yield level near Gatumba, Rwanda (Photo: Rolf Nieder). (**c**) Landscape in the Tarim Basin of Xinjiang Province,

Fig. 1.5 (continued) northwest China. Soil salinization is one of the most important eco-environmental problems in arid regions. It can induce land degradation, inhibit vegetation

salinization develops due to inappropriate irrigation practices such as application of salt-rich irrigation water and/or insufficient drainage. Salinization is considered a major threat in the irrigation systems of the Indus, Tigris, and Euphrates River basins, in north eastern Thailand and China, in the Nile delta, in northern Mexico, and in the Andean highlands (Bai et al. 2008). Dryland salinity, which currently affects about 1 million hectare area in southwest Australia, has been linked to serious human health problems. Jardine et al. (2007) identified several potential human health impacts resulting from dryland salinity viz. wind-borne dust and respiratory health including altered ecology of the mosquito-borne disease Ross River virus and mental health consequences of salinity-induced environmental degradation.

Of the 135 million ha influenced by nutrient depletion worldwide, 68 million ha is located in South America followed by Africa (Fig. 1.6). The official report of the Earth Summit (1992) expressed concern over major declines in the mineral values in farm and range soils throughout the world. This concern was based on data showing that during the previous 100 years, average mineral levels in agricultural soils had declined worldwide, by 72% in Europe, 76% in Asia, 74% in Africa, 55% in Australia, and 85% in North America.

Nutrient depletion can be attributed to soil mining because of insufficient and imbalanced fertilizer use, soil erosion, and leaching. Nutrient depletion is predicted to cause serious problems in the mid-altitude hills of Nepal; in poor soil quality areas of north-eastern India and Myanmar, now undergoing transition to permanent agriculture, and in areas in north eastern Thailand. It is also expected to cause major problems in large areas of Africa under transition to short fallow or permanent cropping, in areas of reduced silt deposits in the Nile delta, in the sub-humid Mesoamerican hill sides, and in the semi-arid Andean valleys, north eastern Brazil, and the Caribbean Basin lowlands, where agriculture is undergoing intensification (Bai et al. 2008). Given the lack of alternatives available to smallholders and their limited resources, soil mining tends to be associated with poverty. In contrast, soils in many developed countries have excess nutrients. For example Western Europe has considerable surpluses of nitrogen, phosphorus and potassium (Bach and Frede 1998). These surpluses are the result of excessive mineral fertilizer input as well as due to addition of nutrients through imported food products that enter the nutrient cycle via animal dung or liquid manure. During the years 2008 and 2010, about

Fig. 1.5 (continued) growth and thus impede regional agricultural production (Photo: Andreas Bürkert, with kind permission). (**d**) Forest (spruce: *Picea abies*) decline (German expression: "Waldsterben") as a visible issue of long-term atmospheric acidic deposition and soil acidification in the Harz National Park, Germany. Forest decline is not caused solely by acidic deposition but also by other factors such as bark beetle infestation (Photo: Rolf Nieder). (**e**) Small-scale soil sealing caused by construction of family houses and roads in Wolfenbüttel, Germany. Funding of single family houses at urban fringes is a counterproductive policy with respect to limiting soil sealing (Photo: Rolf Nieder). (**f**) Large-scale soil sealing caused by building houses, roads, parking areas and other infrastructure in Melbourne, Australia. Melbourne is Australia's fastest growing capital city (Photo: Rolf Nieder)

Table 1.4 Global Assessment of Human-induced Soil Degradation (GLASOD) for different regions (million ha). Numbers in parenthesis indicate percent of total degraded area (Adapted from Oldeman et al. 1991)

Kind of degradation	World	Asia	Africa	South America	North and Central America	Australasia	Europe
Water erosion	1094 (56)	441 (59)	227 (46)	123 (51)	106 (67)	83 (81)	115 (52)
Wind erosion	548 (28)	222 (30)	187 (38)	42 (17)	39 (25)	16 (16)	42 (19)
Nutrient depletion	135 (7)	15 (2)	45 (9)	68 (28)	4 (3)	1 (–)	3 (2)
Salinization and acidification	82 (4)	57 (8)	16 (3)	2 (1)	2 (2)	1 (1)	4 (2)
Pollution	22 (1)	2 (–)	0 (–)	–	1 (–)	–	19 (9)
Physical[a]	83 (4)	12 (2)	19 (4)	8 (3)	6 (4)	2 (2)	36 (17)
Total	1,964 (100)	747 (100)[b]	494 (100)	243 (100)	158 (100)[a]	103 (100)	219 (100)[a]

[a]Physical degradation includes compaction, water logging, and subsidence organic soils

[b]Summation of the percentages may not equal 100 because of rounding off

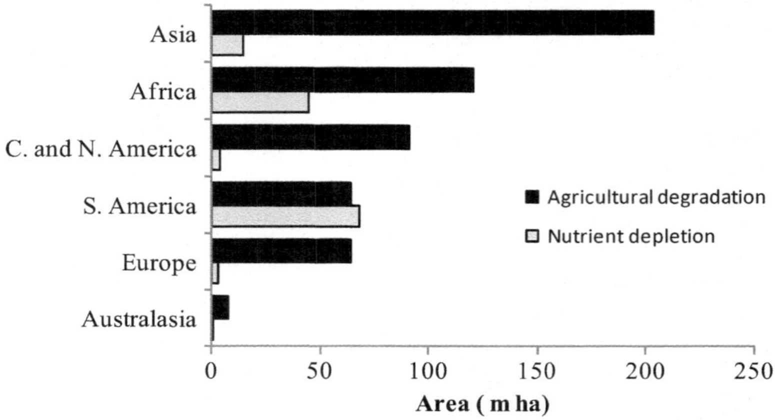

Fig. 1.6 Depletion of agricultural land and nutrient depletion (GLASOD) (Drawn from Oldeman et al. 1991)

35 million Mg of soy and soybean products were imported into the EU. Soybeans are processed into soybean oil and soy flour. Virtually all the soy flour goes into animal feed (Kotschi 2013). In Asia, high nutrient surpluses occur as a consequence of excessive nitrogen and phosphorus fertilization such as in China, South Korea, and Malaysia (Lin et al. 1996; Tan et al. 2005). Globally, soil nutrient deficits for cereal production (wheat, rice, maize and barley) were estimated at an average rate (kg ha^{-1} $year^{-1}$) of 18.7 N, 5.1 P, and 38.8 K, covering respectively 59%, 85%, and 90% of harvested area in the year 2000 (Tan et al. 2005).

Current fertilization practices are skewed towards greater application of nitrogen fertilizers. The application of ammonium and urea form of fertilizers leads to soil acidification, SOM depletion and emission of greenhouse gases. Urea, which comprises 67% of the global fertilizer N consumption, has an acidity index of 0.71 meaning thereby that 0.71 kg of lime is required to neutralize 1 kg of fertilizer N. Soil acidity impacts soil health through its adverse effect on nutrient availability especially phosphorus which is readily fixed in acidic soils, the increase in concentration and bioavailability of toxic metals in the soil solution, and impairment of life and activities of soil microorganisms. In alkaline soils, most of the heavy metals are bound and relatively immobile and not taken up by plants. In strongly acidic soils, the availability and mobility of some of the heavy metals is increased and plants can take up these metals, which may exceed the maximum permissible limit for human consumption. Soil pH is also crucial in affecting the transport of pollutants and toxic elements and can also influence the transport of viruses. All these effects, besides leading to soil degradation influence the quality of food and human health.

There is a strong link between soil and environmental quality and human health (Fig. 1.7). According to United Nations Convention to Combat Desertification (UNCCD), over 250 million people are directly affected by land degradation. In addition, some one billion people in over 100 countries are at risk. These people include many of the world's poorest and most marginalized people. Soil

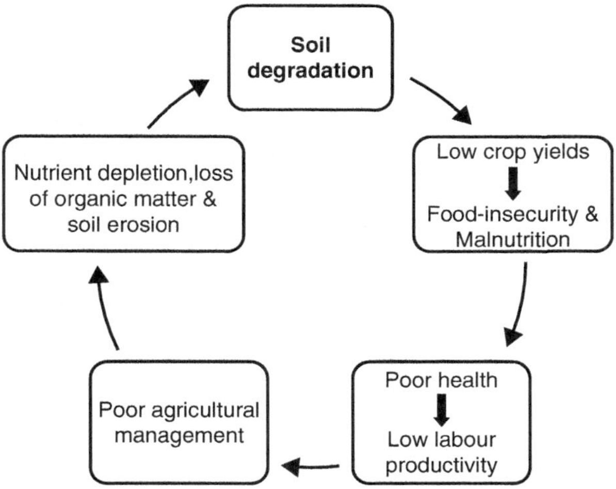

Fig. 1.7 Link between soil degradation and human health (Adapted from Deckelbaum et al. 2006)

degradation directly affects food insecurity through reduction in crop yields, decline in their nutritional quality (protein content, micronutrients etc.), and reduced input (e.g. fertilizer, irrigation water) use efficiency (Lal 2009).

Farmers may need to use greater amount of inputs such as fertilizer or manure in order to maintain yields. Indirect effects of soil degradation are related to pollution of soil, air, and water with severe impact on human health (Pimentel et al. 2007). Soil degradation can also have adverse effects off the farm, such as deposition of eroded soil in streams or behind dams, contamination of drinking water by agrochemicals, and loss of habitat. Agrochemical pollution is expected to be critical in cotton-producing areas in Turkey, in high-density and coastal areas in East and Southeast Asia, on banana plantations in Central America, in areas of intensive agriculture in Bolivia, and in peri-urban agriculture in Southeast Asia and Mexico (Bai et al. 2008). Soil erosion will create serious production problems in southeast Nigeria, in Haiti, and on the sloping lands of the Himalayan foothills, southern China, Southeast Asia, and Central America. These effects are exacerbated by environmental change because the positive feedback between soil degradation and the projected global warming may also adversely impact food security. An important indirect effect of the projected global warming on food security is through increase in risks of soil degradation with associated increase in losses of water and nutrients. In the next few decades, land degradation may pose a serious threat to food production, particularly in poor and densely populated areas of the developing world. Appropriate policies are required to encourage land-improving investments and better land management if developing countries are to sustainably meet the food needs of their populations.

Globally, there are few studies on the impact of degradation on agricultural production but it is estimated to cause reduced yields on approximately 16% of the agricultural land, especially cropland in Africa, Central America and pastures in Africa. Losses in productivity of cropping land in sub-Saharan Africa are estimated to be 0.5–1% annually, suggesting productivity loss of at least 20% over the last 40 years (WMO 2005). Globally, the cumulative productivity loss for cropland from soil degradation over the past 50 years is estimated to be about 13%, and for pasture lands 4%. Crop yield losses in Africa from 1970 to 1990 due to water erosion alone are estimated to be 8% (Scherr and Yadav 1997). Sub-regional studies have documented large aggregate declines in crop yields due to degradation in many parts of Africa, China, South Asia, and Central America. Although some types of degradation are irreversible, most can be prevented or reversed by, for example, adding nutrients to nutrient-depleted soil, restoring topsoil through soil amendments, establishing vegetation, or buffering soil acidity. Sustainable land management practices can avoid land degradation.

1.5 Impacts of Climate Change on Soil Quality

Global warming associated with altered rainfall pattern due to climate change are expected to subject soils to significant risk of climate induced physical and chemical degradation. Climate change might exacerbate land degradation through changes in spatial and temporal patterns of temperature, rainfall, solar radiation, and winds. In most parts of Asia, forest cover is shrinking, agriculture is gradually expanding to marginal lands and land degradation is accelerating through nutrient leaching and soil erosion. Unsustainable irrigation and management practices in conjunction with climate change have led to deterioration of soil health due to increased salinization, nutrient depletion and erosion. While several reports are available relating soil quality to land use and management, studies describing the effect of climate change on soil quality are rare. This is mainly because the changes in soil properties and the climate take place over a long-term. A change in soil quality can only be perceived when all the effects are combined over a period of time. Generally, biological processes in soil such as decomposition and storage of organic matter, C and N cycling, microbial and metabolic quotients are likely to be influenced greatly by climate change and have thus high relevance to assess climate change impacts (Table 1.5). Physical indicators of soil quality such as porosity and available water capacity have also high relevance and are occasionally used to assess climate change impacts. Chemical indicators of soil quality such as pH, electrical conductivity (EC) and availability of nutrients have medium relevance and are frequently used to assess climate change impacts (Table 1.6). Simulation models are increasingly used to generate scenarios and predict the possible effects of climate change on soil health. Soil organic matter turnover models are being used for regional and global analysis of soil C dynamics (Allen et al. 2011).

Changes in climate are likely to influence the rates of accumulation and decomposition of SOM, both directly through changes in temperature and water balance and indirectly through changes in primary productivity and rhizodepositions (Fig. 1.8). Atmospheric CO_2 concentration influences SOM storage through its effect on primary production. Generally, it is expected that increase in temperature

Table 1.5 Soil quality indicators and soil processes with high relevance to assess climate change impacts

Indicator	Soil processes affected
Soil organic matter fractions	Residue decomposition, organic matter storage and quality
Mineralizable C and N	Metabolic activity of soil organisms, mineralization-immobilization turnover
Total C and N	C and N mass and balance
Soil respiration, soil microbial biomass	Microbial activity
Microbial quotients	Substrate use efficiency
Microbial diversity	Nutrient cycling and availability
Porosity	Air capacity, plant available water capacity
Available water	Field capacity, permanent wilting pointing, water flow

Adapted from Allen et al. (2011)

Table 1.6 Soil quality indicators and soil processes with medium relevance to assess climate change impacts

Indicator	Soil processes affected
Soil structure	Aggregate stability, soil organic matter turnover
pH	Biological and chemical activity thresholds
EC	Plant and microbial activity thresholds
Available N, P and K	Plant available nutrient and potential for loss

Adapted from Allen et al. (2011)

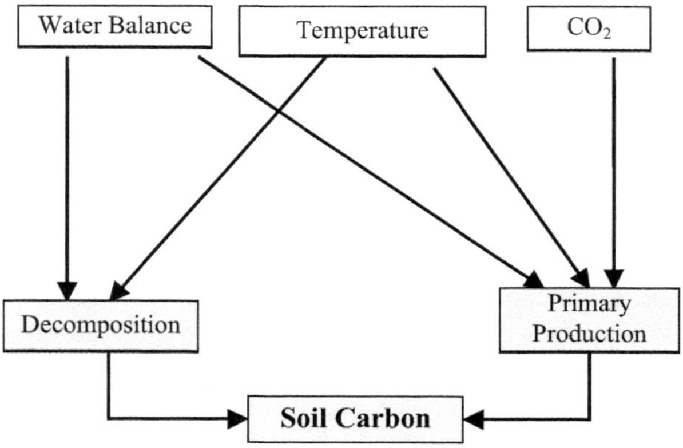

Fig. 1.8 Climatic factors affecting soil carbon pool (Adapted from Benbi 2012)

will enhance the rate of SOM decomposition, which decreases the SOC content. Increased temperature together with elevated CO_2 concentration may lead to increase in net primary productivity, which provides input to soil organic C (SOC). The change in soil C storage represents the net effect of organic matter decomposition and primary production. Studies in the past have shown that soil organic C and N pools are positively correlated with precipitation and negatively correlated with temperature. Accordingly, the soils of arid and hot eco-regions have lower SOC stocks compared to soils in the temperate regions. There are contradictory reports on the effect of temperature increase on SOC. Simulation studies of Kirschbaum (1993) showed that temperature increase could result in loss of SOC due to increased decomposition. On the contrary, Gifford (1992) predicted no loss of SOC due to temperature increase. Since SOC pools is influenced by rate of organic matter decomposition and net primary productivity, the future trend in SOC pool will depend on the relative temperature sensitivities of the two processes. The temperature sensitivity of organic matter decomposition decreases with increasing temperature and particulate organic matter is considered to be more sensitive to temperature than the mineral associated organic matter (Benbi et al. 2014). Using a simple productivity model coupled to an SOM model, Kirschbaum (1995) showed that the SOC content may decrease greatly at low temperature and the loss of C per degree warming may decrease with increasing temperature. However, there is no consensus on the effect of global warming on SOC stocks. In order to predict the fate of SOC stocks in relation to global warming it is essential to understand the temperature response of the processes that control substrate availability, depolymerization, microbial efficiency and enzyme production (Conant et al. 2011).

Climate change can greatly influence N cycling processes, especially atmospheric N deposition in terrestrial ecosystems. Global estimates of total atmospheric N deposition show great increase during the last over 150 years and these are projected to increase further by ~2.5 times by the year 2100 (Lamarque et al. 2005). Total annual N deposition has increased from 12.8 Tg N in 1860 to 45.8 Tg N in the early 1990s (Galloway et al. 2004). Compared to an estimated input of 1–3 kg N ha^{-1} $year^{-1}$ in the early 1900s (Galloway 1995; Asman et al. 1998), the atmospheric N deposition rates of 20–60 kg N ha^{-1} $year^{-1}$ in non-forest ecosystems, and up to 100 kg N ha^{-1} $year^{-1}$ in forest stands in Europe or the USA have been reported (Bobbink et al. 2003). The high deposition rates could be attributed to biomass burning, soil emissions of NO_x and NH_3 as well as lightning production of NO_x. The deposited N besides impacting a number of processes in soil significantly modifies the global C and N cycles. Nitrogen deposited to agricultural or croplands could serve as a source of nutrient, but it could also adversely affect several processes in the soil. Increased N deposition may lead to reduction in biodiversity, soil acidification, increased N_2O emissions from denitrification and nitrification in soils, altered balance of nitrification and mineralization/immobilization, increased nitrate leaching and eutrophication. Because a number of factors determine the severity of an effect of N deposition, there is high variation in response of different ecosystems to atmospheric N deposition. In N-limited temperate ecosystems, N deposition can enhance C storage, which could impact atmospheric CO_2 concentration (Townsend et al. 1996).

Climate changes could influence soil organisms through changes in quantity and quality of plant-mediated soil C inputs. Elevated atmospheric CO_2 could stimulate flow of organic C into the soil system, increase root production and exudation, and enhance mycorrhizal and N_2-fixing relationships (Pritchard 2011). It has been suggested that global warming and enhanced atmospheric CO_2 concentration may enhance energy flow through fungal pathways compared to bacterial pathways.

However, it is not known whether the shift toward fungal domination of soils will increase occurrence of soil-borne fungal diseases (Pritchard 2011).

Severe weather can affect the resilience of the food chain by affecting soil (e.g. erosion caused by heavy rainfall), growing conditions and yield amount and quality (by affecting temperature and water availability), harvesting and planting conditions (via dryness, wetness or snow, or by lack of seed availability from a previous poor year), storage and transport logistics, and the collective impacts working on price through the market and therefore affecting access to food (including animal feed) as well as availability. To mitigate climate change effects, it is imperative that soil health is maintained so that it can sustain physical, chemical and biological functions and provide ecosystem resilience.

1.6 Relationship Between Soil Quality and Agricultural Products

Soil quality affects human nutrition and health through its impacts on quantity and quality of food production. Decline in crop yields and agronomic production aggravate food-insecurity that currently affects 796 million people globally (Table 1.7), and low concentration of protein and micronutrients (e.g., Zn, Fe, Se, B, I) exacerbate malnutrition and hidden hunger that affects 3.7 billion people, especially children (Lal 2009). The highest number (488 million) of undernourished people lives in low-income food-deficit countries.

Plant products are a source of mineral nutrients, proteins, vitamins, carbohydrates, and fats in the human diet and more than half (52%) of the total dietary energy supply is derived from cereals, roots and tubers. Therefore, any deficiency in the plant products could manifest in the human beings. As discussed in the preceding section, minerals are depleting from agricultural soils mainly because of increased crop yield per unit area coupled with imbalanced use of fertilizers skewed towards nitrogen. Intensive cultivation of high yielding crop varieties have resulted in the plants using 17 essential elements with a return of mainly three macronutrients viz. nitrogen, phosphorus and potassium. The essential micronutrients are generally not applied or supplied through foliar sprays. Plant availability of mineral nutrients in the soil is the main source of mineral supply to

Table 1.7 Number and percentage of undernourished persons in the world

Year	Number (million)	Percent
1990–1992	1011	18.6
1995–1997	966	16.6
2000–2002	930	14.9
2005–2007	942	14.3
2010–2012	821	11.8
2013–2015	796	11.0

Source: FAOSTAT Database (n.d.)

human beings. Plants which absorb nutrients from soil are either eaten directly by humans or fed to animals that are then included in human diet. Hence soil health indicators that affect plant nutrient availability and uptake like SOM, texture, pH, drainage and management practices such as manure and fertilizer application are crucial to the availability of mineral nutrients to human beings. Parent material, pedogenic factors and climate also play an important role (Deckers and Steinnes 2004). Unfortunately, the link between the quality of food and the soil that produces it has not received the desired attention. A healthy soil produces a healthy plant by supplying essential nutrients and protecting it from diseases and pests. Human beings and animals eating such plants are obviously healthy. Results of several studies have shown that soil fertility and nutrient status besides influencing crop yields impact nutrient uptake and mineral composition of the plant products.

Since the green revolution period (1960s) there has been tremendous improvement in the yield levels of cultivated crops including cereals, pulses, fruits and vegetables. The global rate of increase in agronomic yield (kg ha^{-1} $year^{-1}$) for the 50 year period between 1961 and 2011 was 65 for corn, 54 for rice, 41 for wheat and 27 for soybean (Lal 2013). However, there are reports of decline in the mineral content of plant products. A comparison of 56 historical cultivars and 7 modern cultivars of spring wheat showed that while the modern cultivars yielded 0.83 Mg ha^{-1} higher grains than the historical cultivars, the concentration of essential elements in the grain decreased by 6–50% (Table 1.8).

Though the increased yield of modern crop varieties could compensate for lower concentration of mineral elements in the food product, the mineral intake by the humans per unit weight of cereals/pulses/fruits/vegetables consumed is reduced. This has a cascading effect on meeting the recommended daily allowance (RDA) of minerals and vitamins and human health. Murphy et al. (2008) compared historical and modern wheat cultivars and showed that considerably more bread was required for modern cultivars than the historic cultivars to meet the RDA of eight minerals. As an example, they (Murphy et al. 2008) showed that females aged between 19 and 30 would have to eat 15.2 slices of bread made from modern cultivars compared to 10.6 slices of whole wheat bread made from historical cultivars (high in Zn) to meet the RDA of Zn. Similarly, modern varieties of fruits and vegetables have also been

Table 1.8 Mineral concentration (mg kg^{-1}) in modern and historical wheat cultivars

Mineral	Historical cultivar	Modern cultivar	% change
Ca	421.6 ± 10.9	398.5 ± 16.1	−6
Cu	4.8 ± 0.1	4.1 ± 0.2	−16
Fe	35.7 ± 1.0	32.3 ± 1.8	−11
Mn	50.0 ± 1.2	46.8 ± 3.1	−7
Mg	1402.6 ± 21.0	1307.6 ± 25.6	−7
P	3797.1 ± 55.7	3492.7 ± 119.3	−9
Se (μg kg^{-1})	16.2 ± 1.7	10.8 ± 2.7	−50
Zn	33.9 ± 0.9	27.2 ± 1.9	−25

Adapted from Murphy et al. (2008)

Table 1.9 Changes (%) in mineral content of different types of vegetables (27 varieties), fruit (17 types) and meat (10 cuts) measured between 1940 and 1991

Mineral	Vegetable	Fruit	Meat
Sodium (Na)	−49	−29	−30
Potassium (K)	−16	−19	−16
Phosphorus (P)	+9	+2	−28
Magnesium (Mg)	−24	−16	−10
Calcium (Ca)	−46	−16	−41
Iron (Fe)	−27	−24	−54
Copper (Cu)	−76	−20	−24

Adapted from Thomas (2003)

found to be lower in a range of minerals and trace elements compared to those grown in 1940s (Table 1.9). Similar findings were reported for animal derived foodstuffs, including meat and dairy produce. Since fruits, vegetables, and cereals form bulk of our diet, the decreased level of mineral nutrients in them could greatly influence human health and nutrition. In addition to the overall mineral depletion, significant changes have also taken place in mineral ratios, which can significantly influence biochemical processes in human body.

1.7 Organic vs. Conventional Farming

In recent years, greater attention is being paid to organic production as an alternative form of agriculture, to obtain high-quality food in an environment-friendly manner. Organic production methods aim to avoid the use of synthetic fertilizers and harmful pesticides, growth regulators and livestock feed additives. They rely on ecological processes, biodiversity and biological cycles adapted to local conditions. The organic farming practices which are based on minimal use of off-farm inputs resort to addition of animal manures and green manure as source of nutrients, and integrated pest management to control pest and weeds. Applications of chemical fertilizers and pesticides are known to pollute soil, water, and air, harming both the environment and human health. The primary goal of organic agriculture is to optimize soil, plant, human and animal health. A number of studies comparing the nutritional composition of foods produced through organic and conventional methods of farming indicated either minor or no significant differences (Table 1.10). A review of 41 studies showed that organically produced crops contained 27% more vitamin C, 21% more iron, 29% more magnesium, and 14% more phosphorus than did conventional crops (Worthington 2001). Some studies have suggested positive effects of organic food production on organic acids and polyphenolic compounds, which are considered to have potential human health benefits as antioxidants. The increase in organic acids and polyphenolics in organic foods has been attributed to the differences in plant metabolism and the response of plants to stressful environments in the absence of mineral fertilizers and pesticide application. In conventional agriculture, ready availability of mineral fertilizers

Table 1.10 Summary of studies comparing organic and conventional foods with respect to some quality attributes

Chemical studied	Food	Results	References
Flavonols	Strawberries, blueberries, black currants	No consistent effect	Hakkinen and Torronen (2000); Mikkonen et al. (2001)
	Chinese cabbage, spinach, welsh onion, green pepper	Organic foods generally had higher levels of flavonoids	Ren et al. (2001)
Polyphenol oxidase enzyme activity, total phenolics, organic acids	Peach, pear	Organic peaches and pears had higher phenolic and polyphenol oxidase levels; organic peaches had higher levels of ascorbic acid and citric acid	Carbonaro and Mattera (2001); Carbonaro et al. (2002)
Polyphenol oxidase and diphenolase enzymes	Grapes	No difference in polyphenol oxidase enzyme levels; diphenolase activity higher in organic grapes	Nunez-Delicado et al. (2005)
Phenolics and ascorbic acid	Marionberries, corn, strawberries	Phenolics and ascorbic acid higher in organics	Asami et al. (2003)
Phenolics	Lettuce, collards, pac choi	No difference in lettuce and collards; phenolics higher in organic pac choi	Young et al. (2005)
Phenolics	Apples	Higher in organic apple pulp; no difference in apple peels	Veberic et al. (2005)
Phenolics	Oats grain	No difference	Dimberg et al. (2005)
Vitamin C, carotenoids, polyphenols	Tomatoes	Higher levels of Vitamin C, carotenoids, and polyphenols in organic tomatoes	Caris-Veyrat et al. (2004)

Adapted from Winter and Davis (2006)

such as nitrogen may accelerate plant growth and development leading to preferential allocation of plant resources for growth purposes. This could result in a decrease in the production of plant secondary metabolites such as organic acids, polyphenolics, chlorophyll, and amino acids under conventional system of production. Secondly, organic production methods, which do not involve the use of agrochemicals for pest control, may put more stress on plants necessitating greater allocation of resources toward the synthesis of their own chemical defense mechanisms. Increases in antioxidants such as plant polyphenolics have been attributed to their production in plant defense (Asami et al. 2003).

However, the same defense mechanisms may result in the elevations of other plant secondary metabolites that could be of toxicological rather than nutritional significance. Several studies and monitoring programs have shown much less

Table 1.11 Results of different monitoring programs for detection of pesticide residue in conventional and organic produce

	Percentage detected		
Monitoring agency/country	Conventional	Organic	Conventional/organic ratio
USDA pesticide data program	73	23	3.2
CDPR marketplace surveillance program	31	6.5	4.8
Consumer union	79	27	2.9
Belgium	49	12	4.1

Sources: Baker et al. (2002) and Pussemier et al. (2006)
Adapted from Winter and Davis (2006)

occurrence of pesticide residues in organic produce compared to the conventional produce (Rembiałkowska 2007; Table 1.11). The pesticides used in conventional agriculture are associated with elevated cancer risks for workers and consumers and are being investigated for their links to endocrine disruption and reproductive dysfunction (Horrigan et al. 2002). Besides pesticides, organically produced foods are reported to have lower content of nitrates and nitrites than in conventionally grown crops. High nitrates, which can easily be converted to nitrites, can cause serious human health risks such as methaemoglobinaemia in babies and infants. Nitrate accumulation in plant products and its effect on human health is further discussed in Chap. 5.

Contrary to above studies, several studies did not demonstrate any significant differences in nutritional composition of foods produced by organic and conventional methods of production. Based on a review of 150 comparative studies published between 1926 and 1994 involving a wide range of foods (cereals, potatoes, vegetables, fruits, wine, beer, bread, milk and other dairy products, meat and meat products, eggs, and honey), Woese et al. (1997) concluded that there were no major differences in nutrient levels between the different production methods in some cases and contradictory findings in other cases did not permit definitive conclusions about the influence of production methods on nutrient levels. Improper experimental designs, too variable results, and lack of statistical analyses generally pose problems in making definitive conclusions about the impact of two production systems on the nutritional value of foods (Bourn and Prescott 2002). Rosen (2010) also argued that claims made about the higher nutritional value of organic foods are based either on non-peer reviewed publications or on data that are not statistically significant. The evidence available so far suggests that the intake of organic foods may lead to some advantages, such as the ingestion of a higher content of phenolic compounds and some vitamins, such as vitamin C, and a lower content of nitrates and pesticides (Lima and Vianello 2011). However, it has been suggested that care should be taken about the ingestion of foods in relation to the content of some substances, e.g. polyamines, substances stimulating cellular division, because some foods coming from organic origin present a higher content of these compounds (Lima and Vianello 2011). There are apprehensions that the bacterial contamination (mainly *Salmonella* ssp. and *Campylobacter* ssp.) may sometimes be higher in organic produce, but scientific evidence of this is still not clear. In spite of some positive indications, the impact of organic food consumption on human health remains essentially unknown and needs to be further investigated. Finally, the organic method of cultivation generally results in significantly lower yields than the conventional method of cultivation. This apart from increasing the prices and decreasing accessibility of food to poor people could influence food insecurity.

References

Allen DE, Singh BP, Dalal RC (2011) Soil health indicators under climate change: a review of current knowledge. In: Singh BP, Cowie L, Yin Chan K (eds) Soil health and climate change. Springer, Berlin, pp 25–45

Andrews SS, Karlen DL, Cambardella CA (2004) The soil management assessment framework: a quantitative soil quality evaluation method. Soil Sci Soc Am J 68:1945–1962

Arshad MA, Martin S (2002) Identifying critical limits for soil quality indicators in agro-ecosystems. Agric Ecosyst Environ 88:153–160

Asami DK, Hong YJ, Barrett DM, Mitchell AE (2003) Comparison of the total phenolic and ascorbic acid content of freeze-dried and air-dried marionberry, strawberry, and corn grown using conventional, organic, and sustainable agricultural practices. J Agric Food Chem 51:1237–1241

Asman WAH, Sutton MA, Schjørring JK (1998) Ammonia: emission, atmospheric transport and deposition. New Phytol 139:27–48

Athar M, Iqbal M, Beg MU, Al-Ajmi D, Al-Muzaini S (1998) Airborne dust collected from Kuwait in 1991–1992 augments peroxidation of cellular membrane lipids and enhances DNA damage. Environ Int 24:205–212

Bach M, Frede HG (1998) Agricultural N, P, and K balances in Germany 1970 to 1995. Z Pflanzenernaehr Bodenkd 161:385–393

Bai ZG, Dent DL, Olsson L, Schaepman ME (2008) Proxy global assessment of land degradation. Soil Use Manag 24:223–234

Baker BP, Benbrook CM, Groth E, Benbrook KL (2002) Pesticide residues in conventional, integrated pest management (IPM)-grown and organic foods: insights from three U.S. data sets. Food Addit Contam 19:427–446

Beinroth FH, Eswaran H, Reich PF (2001) Land quality and food security in Asia. In: Bridges EM, Hannam ID, Oldeman LR, de Pening VFWT, Scherr SJ, Sompatpanit S (eds) Responses to land degradation. Proc. 2nd. International Conference on Land Degradation and Desertification, Khon Kaen, Thailand. Oxford Press, New Delhi

Benbi DK (2012) Impact of climate change on soil health. Ann Agric Res New Ser 33:204–213

Benbi DK, Toor AS, Kumar S (2012) Management of organic amendments in rice-wheat cropping system determines the pool where carbon is sequestered. Plant Soil 360(1–2):145–162

Benbi DK, Boparai AK, Brar K (2014) Decomposition of particulate organic matter is more sensitive to temperature than the mineral associated organic matter. Soil Biol Biochem 70:183–192

Blum WEH (2008) Characterisation of soil degradation risk: an overview. In: Tóth G, Montanarella L, Rusco E (eds) Threats to soil quality in Europe. European Commission Joint Research Centre, Institute for Environment and Sustainability, Italy, pp 5–10

Blum WEH, Eswaran H (2004) Soils for sustaining global food production. J Food Sci 69:CRH37–CRH42

Bobbink R, Ashmore M, Braun S, Flückiger W, van den Wyngaert IJJ (2003) Empirical nitrogen critical loads for natural and semi-natural ecosystems: 2002 update. In: Achermann B, Bobbink R (eds) Empirical critical loads for nitrogen. Environmental Documentation No. 164. Swiss Agency for Environment, Forest and Landscape SAEFL, Berne, pp 43–170

Bourn D, Prescott J (2002) A comparison of the nutritional value, sensory qualities, and food safety of organically and conventionally produced foods. Crit Rev Food Sci Nutr 42:1–34

Braun-Fahrlander C, Riedler J, Herz U, Eder W, Waser M, Grize L, Maisch S, Carr D, Gerlach F, Bufe A, Lauener RP, Schierl R, Renz H, Nowak D, von Mutius E (2002) Environmental exposure to endotoxin and its relation to asthma in school-age children. N Engl J Med 347:869–877

Carbonaro M, Mattera M (2001) Polyphenoloxidase activity and polyphenol levels in organically and conventionally grown peach (*Prunus persica* L., cv. Regina bianca) and pear (*Pyrus communis* L., cv. Williams). Food Chem 72:419–424

Carbonaro M, Mattera M, Nicoli S, Bergamo P, Cappelloni M (2002) Modulation of antioxidant compounds in organic vs. conventional fruit (peach, Prunus persica L., and pear, *Pyrus communis* L.) J Agric Food Chem 50:5458–5462

Caris-Veyrat C, Amiot MJ, Tyssandier V, Grasselly D, Buret M, Mikolajczak M, Guilland JC, Bouteloup-Demange C, Borel P (2004) Influence of organic versus conventional agricultural practice on the antioxidant microconstituent content of tomatoes and derived purees; consequences on antioxidant plasma status in humans. J Agric Food Chem 52:6503–6509

Carter MR, Gregorich EG, Anderson DW, Doran JW, Janzen HH, Pierce FJ (1997) Concepts of soil quality and their significance. In: Gregorich EG, Carter MR (eds) Soil quality for crop production and ecosystem health. Elsevier, Amsterdam, pp 1–19

CCME (Canadian Council of Ministers of the Environment) (2006) A protocol for the derivation of environmental and human health soil quality guidelines. Canadian Council of Ministers of the Environment, Winnipeg. 186 pp

Conant R, Ryan MC, Ågren GI, Birge HE, Davidson EA, Eliasson PE, Evans SE, Frey SD, Giardina CP, Hopkins F, Hyvonen R, Kirschbaum MUF, Lavallee JM, Leifeld J, Parton WJ, Steinweg JM, Wallenstein MD, Wetterstedt JAM, Bradford MA (2011) Temperature and soil organic matter decomposition rates- synthesis of current knowledge and a way forward. Glob Chang Biol 17:3392–3404

Deckelbaum RJ, Pam C, Mutuo P, DeClerck F (2006) Econutrition: implementation models from the Millennium Villages Project in Africa. Food Nutr Bull 27:335–342

Deckers J, Steinnes E (2004) State of the art on soil-related geo-medical issues in the world. Adv Agron 84:1–35

Dimberg LH, Gissén C, Nilsson J (2005) Phenolic compounds in oat grains (*Avena sativa* L.) grown in conventional and organic systems. AMBIO: A J Hum Environ 34:331–337

Doran JW, Parkin TB (1994) Defining and assessing soil quality. In: Doran JW, Coleman DC, Bezdicek DF, Stewart BA (eds) Defining soil quality for a sustainable environment, vol 35. Soil Sci Soc Am, Madison, pp 3–21 (special publication)

Earth Summit (1992) Rio earth summit declaration. Development Rio de Janerio, Brazil, 3 to 14 June 1992 Agenda 21

Eswaran H, Beinroth F, Reich P (1999) Global land resources and population supporting capacity. Am J Altern Agric 14:129–136

FAO (Food and Agriculture Organisation) (1995) Planning for sustainable use of land resources: towards a new approach. In: Sombroek WG, Sims D (eds) Land and water bulletin no. 2. FAO, Rome

FAO (Food and Agriculture Organisation) (2011) The state of the world's land and water resources for food and agriculture (SOLAW) – managing systems at risk. Food and Agriculture Organization of the United Nations, Rome and Earthscan, London

FAOSTAT Database (n.d.) Food and Agriculture organization (FAO), Rome

Filip Z (2002) International approach to assessing soil quality by ecologically-related biological parameters. Agric Ecosyst Environ 88:169–174

Fischer G, Hizsnyik E, Prieler S, Wiberg D (2010) Scarcity and abundance of land resources: competing uses and the shrinking land resource base. SOLAW Background Thematic Report TR02. FAO, Rome. Available at http://www.fao.org/nr/solaw/

Galloway JN (1995) Acid deposition: perspective in time and space. Water Air Soil Pollut 85:15–24

Galloway JN, Dentener FJ, Capone DG, Boyer EW, Howarth RW, Seitzinger SP, Asner GP, Cleveland CC, Green PA, Holland EA, Karl DM, Michaels AF, Porter JH, Townsend AR, Vörösmarty CJ (2004) Nitrogen cycles: past, present, and future. Biogeochemical 70:153–226

Gifford RM (1992) Implications of the globally increasing atmospheric CO_2 concentration and temperature for the Australian terrestrial carbon budget: integration using a simple model. Austral J Bot 40:527–543

Gregorich EG, Carter MR, Angers DA, Monreal CM, Ellert BH (1994) Towards a minimum data set to assess soil organic matter quality in agricultural soils. Can J Soil Sci 74:367–385

Griffin DW, Kellogg CA (2004) Dust storms and their impact on ocean and human health: dust in earth's atmosphere. EcoHealth 1:284–295

Hakkinen SH, Torronen AR (2000) Content of flavonols and selected phenolic acids in strawberries and vaccinium species: influence of cultivar, cultivation site and technique. Food Res Int 33:517–524

Horrigan L, Lawrence RS, Walker P (2002) How sustainable agriculture can address the environmental and human health harms of industrial agriculture. Environ Health Perspect 110:445–456

IUSS Working Group WRB (2015) World reference base for soil resources 2014, update 2015 International soil classification system for naming soils and creating legends for soil maps. World Soil Resources Reports No. 106. FAO, Rome

Jardine A, Speldewinde P, Carver S, Weinstein P (2007) Dryland salinity ecosystem distress syndrome: human health implications. EcoHealth 4:10–17

Karlen DL, Stott DE (1994) A framework for evaluating physical and chemical indicators of soil quality. In: Doran JW, Coleman DC, Bezdicek DF, Stewart BA (eds) Defining soil quality for a sustainable environment, vol 35. Soil Sci Soc Am J, Madison, pp 53–72 (special publication)

Karlen DL, Mausbach MJ, Doran JW, Kline RG, Harris RF, Schuman GE (1997) Soil quality: a concept, definition, and framework for evaluation. Soil Sci Soc Am J 61:4–10

Kirschbaum MUF (1993) A modeling study of the effects of changes in atmospheric CO_2 concentration, temperature and atmospheric nitrogen input on soil organic carbon storage. Tellus 45B:321–334

Kirschbaum MUF (1995) The temperature dependence of soil organic matter decomposition, and the effect of global warming on soil organic C storage. Soil Biol Biochem 27:753–760

Korenyi-Both AL, Kornyi-Both AL, Molnar AC, Fidelus-Gort R (1992) Al Eskan disease: desert storm pneumonitis. Mil Med 157:452–462

Kotschi J (2013) A soiled reputation: adverse impacts of mineral fertilizers in tropical agriculture. Heinrich Böll Stiftung (Heinrich Böll Foundation), WWF Germany

Lal R (2009) Soil degradation as a reason for inadequate human nutrition. Food Sci 1:45–57

Lal R (2013) Climate-strategic agriculture and the water-soil-waste nexus. J Plant Nutr Soil Sci 176:479–493

Lamarque JF, Kiehl JT, Brasseur G, Butler T, Cameron-Smith P, Collins WD, Collins WJ, Granier C, Hauglustaine D, Hess P, Holland E, Horowitz L, Lawrence M, McKenna D, Merilees P, Prather M, Rasch P, Rotman D, Shindell D, Thornton P (2005) Assessing future nitrogen deposition and carbon cycle feedback using a multimodel approach: analysis of nitrogen deposition. J Geophys Res 110:D19303. https://doi.org/10.1029/2005JD005825

Lima GPP, Vianello F (2011) Review on the main differences between organic and conventional plant-based foods. Int J Food Sci Technol 46:1–13

Lin X, Yin C, Xu D (1996) Input and output of soil nutrients in high-yield paddy fields in South China. Proceedings of the international symposium on maximizing rice yields through improved soil and environmental management. Khon Kaen, Thailand, pp 93–97

Mikkonen TP, Maatta KR, Hukkanen AT, Kokko HI, Torronen AR, Karenlampi SO, Karjalainen RO (2001) Flavonol content varies among black currant cultivars. J Agric Food Chem 49:3274–3277
Murphy KM, Reeves PG, Jones SS (2008) Relationship between yield and mineral nutrient concentrations in historical and modern spring wheat cultivars. Euphytica 163:381–390
Nunez-Delicado E, Sanchez-Ferrer A, Garcia-Carmona FF, Lopez-Nicolas JM (2005) Effect of organic farming practices on the level of latent polyphenol oxidase in grapes. J Food Sci 70: C74–C78
Oldeman LR, Hakkeling RTA, Sombroek WG (1991) World map of the status of human-induced soil degradation, 2nd edn. ISRIC, Wageningen
Parr JF, Papendick RI, Hornick SB, Meyer RE (1992) Soil quality: attributes and relationship to alternative and sustainable agriculture. Am J Altern Agric 7:5–11
Peters A, Fröhlich M, Döring A, Immervoll T, Wichmann HE, Hutchinson WL, Pepys MB, Koenig W (2001) Particulate air pollution is associated with an acute phase response in men. Eur Heart J 22(14):1198–1204
Pimentel D, Cooperstein S, Randell H, Filiberto D, Sorrentino S, Kaye B, Nicklin C, Yagi J, Brian J, O'Hern J, Habas A, Weinstein C (2007) Ecology of increasing diseases; population growth and environmental degradation. Hum Ecol 35:653–668
Prahalad AK, Inmon J, Dailey LA, Madden MC, Ghio AJ, Gallagher JE (2001) Air pollution particles mediated oxidative DNA base damage in a cell free system and in human airway epithelial cells in relation to particulate metal content and bioreactivity. Chem Res Toxicol 14:879–887
Pritchard SG (2011) Soil organisms and global climate change. Plant Pathol 60:82–99
Pussemier L, Larondelle Y, Van Peteghem C, Huyghebaert A (2006) Chemical safety of conventionally and organically produced foodstuffs: a tentative comparison under Belgian conditions. Food Control 17:14–21
Rembiałkowska E (2007) Quality of plant products from organic agriculture. J Sci Food Agric 87:2757–2762
Ren H, Endo H, Hayashi T (2001) Antioxidative and antimutagenic activities and polyphenol content of pesticide-free and organically cultivated green vegetables using water-soluble chitosan as a soil modifier and leaf surface spray. J Sci Food Agric 81:1426–1432
Rosen JD (2010) A review of the nutrition claims made by proponents of organic food. Compr Rev Food Sci Food Saf 9:270–277
Scherr SJ, Yadav S (1997) Land degradation in the developing world- issues and policy options for 2020. Food, Agriculture and Environment Discussion paper 14. International Food Policy Research Institute, Washington, DC
Soil Survey Staff (1999) Soil taxonomy: a basic system of soil classification for making and interpreting soil surveys, 2nd edn. Natural Resources Conservation Service. U.S. Department of Agriculture Handbook 436
Stevenson FJ (1994) Humus chemistry. Genesis, composition, reactions, 2nd edn. Wiley, New York
Tan ZX, Lal R, Wiebe KD (2005) Global soil nutrient depletion and yield reduction. J Sustain Agric 26:123–146
Thomas D (2003) A study on the mineral depletion of the foods available to us as a nation over the period 1940 to 1991. Nutr Health 17:85–115
Tóth G (2008) Soil quality in the European Union. In: Tóth G, Montanarella L, Rusco E (eds) Threats to soil quality in Europe. European Commission Joint Research Centre, Institute for Environment and Sustainability, Italy, pp 11–19
Townsend AR, Braswell BH, Holland EA, Penner JE (1996) Spatial and temporal patterns in terrestrial carbon storage due to deposition of fossil fuel nitrogen. Ecol Appl 6:806–814
Veberic R, Trobec M, Herbinger K, Hofer M, Grill D, Stampar F (2005) Phenolic compounds in some apple (*Malus domestica* Borkh) cultivars of organic and integrated production. J Sci Food Agric 85:1687–1694

Winter CK, Davis SF (2006) Organic foods. J Food Sci 71:R117–R124
WMO (World Meteorological Organization) (2005) Climate and land degradation. WMO-report No. 989. World Meteorological Organization, Geneva
Woese K, Lange D, Boess C, Bögl KW (1997) A comparison of organically and conventionally grown food- Results of a review of the relevant literature. J Sci Food Agric 74:281–293
Worthington V (2001) Nutritional quality of organic versus conventional fruits, vegetables, and grains. J Altern Complement Med 7:161–173
Young JE, Zhao X, Carey EE, Welti R, Yang S-S, Wang W (2005) Phytochemical phenolics in organically grown vegetables. Mol Nutr Food Res 49:1136–1142

Chapter 2
Medicinal Uses of Soil Components, Geophagia and Podoconiosis

Abstract Soils have an impact on human health in many ways. The link between soils and human health has been recognized for thousands of years. Examples of how soils influence human health include the transfer of nutrients from soil to people through plant and animal sources as well as through direct ingestion. The principal source materials for soil components that are beneficial to human health or that are to be used as medicine can be grouped into soil minerals as a source of elements essential to the human body or used in healthcare products and mud, peat, and clay for fangotherapy for healing purposes, as well as soil microorganisms that produce drugs. Involuntarily or deliberately ingested components of soils can be beneficial to humans. However, minerals can also have an adverse effect on human health when they are inhaled over a very long period. Podoconiosis, a non-infectious disease, is associated with chronic barefoot exposure to red volcanic soil, with greater prevalence in high-altitude, impoverished areas of the tropics.

Minerals can be beneficial to human health by serving as active principles or excipients in pharmaceutical preparations, in spas, and in beauty therapy medicine. In some cases, however, these minerals can be harmful to human health. Because of their capacity for adsorption and absorption, phyllosilicate clays have a long history of medicinal use. Clays are commonly used in the pharmaceutical industry as active substances or excipients. Clay minerals (including smectites, kaolinite and fibrous clay minerals) have been extensively applied for prolonged release, especially of basic drugs. Smectites in particular have been the most commonly used substrates, as they can retain large amounts of drug due to their high cation exchange capacity. Clays can act alone or in formulations containing polymers, forming composite materials which show an improved and/or synergic effect when acting together with drugs. When clays are administered concurrently, they may interact with drugs reducing their absorption. Therefore, such interactions can be used to achieve technological and biopharmaceutical advantages, regarding the control of release. Natural clay mixtures are able to kill *Escherichia coli* and methicillin-resistant *Staphylococcus aureus* in vitro. Smectites adsorb aflatoxins, which are naturally occurring carcinogenic mycotoxins produced by several Aspergillus species. Iron-rich smectite and illite clays showed natural antibacterial properties. Kaolin has long been used in diarrhoea medicines and is also used in ointments as an emollient. In addition, various clays are used in antacids. Other minerals, such as kaopectate

R. Nieder et al., *Soil Components and Human Health*,
https://doi.org/10.1007/978-94-024-1222-2_2

(kaolin), tums (calcite), milk of magnesia (magnesite), talcum powder (talc), toothpastes (quartz, rutile), anti-perspirants (bauxite), calcium supplement (ground up coral) are used in a wide range of everyday healthcare products.

The use of mud, peat, and clay for healing purposes can be summarized by the term "fangotherapy". Each of these materials (mud, clay and peat) has its own special properties but in general they hold heat and are useful as a thermal application for chronic conditions.They also stimulate circulation and lymph flow, support detoxification and help the body to relax. Some types of fango have anti-inflammatory and pain relieving properties that make them useful for soft-tissue injury. Fangotherapy is a simple therapy used in treatment of neurological, rheumatological, cardiovascular, gynecological, inflammatory and menstrual cycle disorders.

Soils presumably harbor the most diverse populations of microorganisms of any environment. For example, estimates for the number of bacterial species per gram of soil range from 2000 to several millions. Soils are an important source of medicines such as antibiotics. Numerous bacterial genera and species that produce antibiotics in vitro have been isolated from different soils. Actinomycetes, in particular *Streptomyces* species, have been the primary resource of clinical antibiotics and other therapeutics. To date, more than 7000 different secondary metabolites have been discovered in *Streptomyces* isolates, many of them being antibiotics. Soil fungi, for example the cephalosporin group, are also major sources of antibiotics. Although penicillin (isolated by Sir Alexander Fleming in 1929) was not discovered from the soil itself, it forms from a soil-borne fungus. Beyond antibacterial agents, numerous other drugs have their origin in soil. For example, a wide spectrum of cancer drugs have been developed from natural products in soil.

Soil ingestion occurs either involuntarily or deliberately. For the former, exposed population may ingest at least small quantities of soil. One reason for this is that any soil adhering to fingers may be inadvertently ingested by so-called hand-to-mouth activity. Any outdoor activity may result in an increase in ingestion. Particularly young children are vulnerable to soil ingestion, as they have a predilection for eating non-food items such as soil. The term pica is used in context with any form of abnormal ingestion which involves substances that are not normally regarded as edible. The terms geophagy or geophagia relate to the deliberate ingestion of soil which is especially associated with certain geographic areas mainly in the tropics. For many people, however, geophagia is difficult to comprehend. The reasons why people indulge in eating soil are manifold. One prevailing theory about the practice of geophagia is that the consumed soil acts as a nutrient source. Mineral nutrients such as Ca, Cu, Fe, Mn, Mg and Zn may be supplied directly to geophagists via soil ingestion. Ingested mineral, chemical and biological components of soils can also be detrimental to human health. One negative consequence of soil ingestion is that the amounts and balance of mineral nutrients within the individual may be affected. In terms of toxicity of elements via soil ingestion, most concern to date has been concentrated on Pb. A number of infectious diseases can also result from the ingestion of soil.

Irritant red clay soil derived from basalt volcanic rocks is responsible for a geochemical, non-infectious disease termed podoconiosis or endemic non-filarial elephantiasis. The disease affects barefooted farmers in at least ten countries across tropical Africa, central America and North India, and is characterized by ascending bilateral and asymmetrical lymphedema of the feet and legs, generally occurring below the knees.

Nanoparticles having a size from 1 to 100 nm are present in nature and are successfully used in many products of daily life. They are also embedded per se or as byproducts from the milling process of larger filler particles in many dental materials. Recently, possible adverse effects of nanoparticles have gained increased interest and the main target organs are considered to be the lungs. Exposure to nanoparticles in dentistry may occur in the dental laboratory, where dust is produced e.g. by processing gypsum type products or by grinding and polishing materials. In the dental practice virtually no exposure to nanoparticles occurs when handling unset materials. However, nanoparticles are produced by intraoral adjustment of the set materials through grinding and polishing regardless whether they contain nanoparticles or not. They may also be produced through wear of restorations or released from dental implants and enter the environment when removing restorations. The risk for dental technicians is known and taken care of by legal regulations. Based on model worst case calculations, the risk for dental practice personnel and for patients due to nanoparticles exposure through intraoral grinding/polishing and wear is rated low to negligible. Measures to reduce exposure to nanoparticles include intraorally grinding/polishing using water coolants, proper sculpturing to reduce the need for grinding and sufficient ventilation of treatment areas.

Keywords Minerals for therapeutic purposes • Cosmetic products • Fangotherapy • Microorganisms producing drugs • Geophagia • Podoconiosis • Nano-particles released from dental materials • Titan nanoparticles • Silver nanoparticles

2.1 Minerals as a Source of Essential Elements

Humans and minerals (natural inorganic solids, generally crystalline) are chemical systems having in common a number of elements. Soil can be a significant source of minerals providing elements which are essential to the body. These elements are commonly provided as ions through weathering of minerals, defixation from the interlayer space of clay minerals, desorption from mineral surfaces or solution processes. Except for carbon (C), hydrogen (H) and oxygen (O), which are obtained from air and water, plants take up other nutrients from soil. The elements C, H, O, nitrogen (N), sodium (Na), magnesium (Mg), calcium (Ca), phosphorus (P), sulfur (S), chlorine (Cl), potassium (K), manganese (Mn), iron (Fe), copper (Cu), zinc (Zn), and selenium (Se) are considered essential elements to all animals and vegetation (Gomes and Silva 2007). H, O, C, and N make up just over 96% of the human body mass, while Na, K, Ca, Mg, P, S, and Cl make up 3.78% of the body. All these elements are called macronutrients and their concentration is expressed in mg g^{-1}. The remaining elements and others (about 70) are called trace elements, their concentration being expressed in μg g^{-1}.

Considering their physical and chemical properties, elements can be essential to keep human health in good shape, but in certain circumstances, deficiency or excess of elements can be factors of human disease generation. Although being essential to human health, some elements can be good, toxic or lethal, depending on the individual dosage. Non-essential elements can also be tolerable, toxic or lethal, depending on the dosage too. This principle was established by Paracelsus (1493–1541 AD) "all substances are poisons; there is none which is not a poison; the right dose differentiates a poison and a remedy". The Periodic Table representing essential and toxic elements as well as those considered as essential and toxic is shown in Fig. 2.1.

The intake of minerals and non-toxic or toxic elements by food, water and/or soil, e.g. through geophagia (see Sect. 2.7.4 in this chapter) or dust (Chap. 3) takes place either by ingestion, inhalation or dermal absorption (Fergusson 1990). Ingestion is the most common pathway of exposure for the general people, whereas inhalation and dermal sorption are significant in certain occupational settings (Adriano 2001). The sources and intake pathways of minerals and non-toxic or toxic chemical elements are presented in Fig. 2.2.

Many health problems are related to dietary deficiencies of certain essential chemical elements. There is an old aphorism "We are what we eat" that expresses this fact very well. Dietary deficiencies and excesses with repercussion on human health involve chemical elements characterized by specific functions such as Ca, Mg, K, Na, Fe, Zn, Cu, F, and Se. The deficiency of these elements can cause severe health problems (Gomes and Silva 2007). Table 2.1 summarizes the specific functions of most macro and secondary elements and micronutrients.

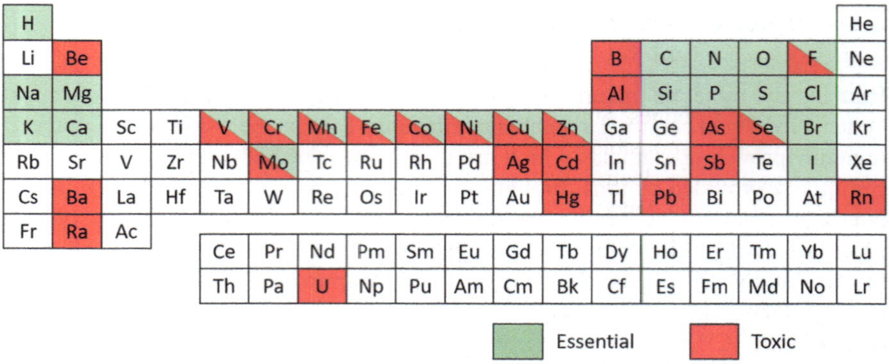

Fig. 2.1 Periodic table of elements distinguishing in essential, toxic and essential/toxic chemical elements (Adapted from Gomes and Silva 2007)

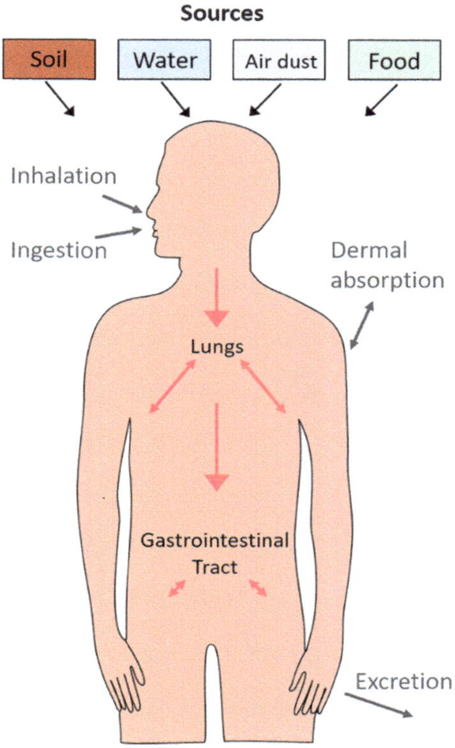

Fig. 2.2 Sources, intake pathways, and uptake of minerals by humans (Adapted from Gomes and Silva 2007)

Calcium is the most abundant element in the human body with 99% being found in teeth and bones whereas the rest is distributed in ionic form throughout our system in blood, cells and tissues. It is probably the most important essential mineral needed by the body because it helps maintain the strength of bones, the internal support structure of the whole body (Seeman 2009). Calcium deficiency in children results in short bones, especially in the arms and legs, but also can have a great impact on the bones in the ribcage. In adults, Ca deficiency leads to brittle bones that break easily. Besides Ca, P and Mg are fundamental for a good state and performance of bones. Sulfur helps to stabilize protein structures. The elements K, Na and Cl are responsible for maintaining the adequate equilibrium of intra-cellular and extra-cellular electrolytes in the human body. Magnesium works to maintain the acid/alkaline balance in the body which is critical for the healthy functioning of nerves and muscles and also helps regulate the heart rhythm to keep it steady, normalizes blood pressure and helps regulate blood sugar levels (Fox et al. 2004). Magnesium is also important for proper metabolism of calcium, phosphorus, sodium, potassium, and vitamin D. Common conditions such as migraines, attention deficit disorder, mitral valve prolapse, fibromyalgia, asthma and allergies have been linked to Mg deficiency. Manganese helps the body convert protein and fat to energy. It also promotes normal bone growth, helps maintain healthy reproductive,

Table 2.1 Functions of essential elements in the human body

Element	Major function in the human body
Ca	Development and maintainance of healthy bones and teeth; supports blood clotting, muscle contraction and nerve transmission; reduces risk of osteoporosis
K	Regulation of heartbeat, maintainance of fluid balance and support of muscles contraction
Mg	Activation of over 100 enzymes and support of nerves and muscles function
P	Together with Ca develops and maintains strong bones and teeth; enhances the use of other nutrients
S	Necessary for muscle protein and hair; deficiency can result in degeneration of collagen, cartilage, ligaments, and tendons
Fe	Essential for red blood cells formation and function; important for brain function
Cr	Needed for glucose metabolism and regulation of blood sugar
Co	Promotion the of red blood cell formation
Cu	Promotion of the formation of red blood cells and connective tissue; function as a catalyst to store and release iron to help form haemoglobin; contribution to central nervous system function
I	Necessary for the thyroid hormone to support metabolism
Se	Component of a key anti-oxidant enzyme, needed for normal growth and development
Zn	Component of more than 200 enzymes involved in digestion, metabolism, reproduction and wound healing
Mo	Contribution to normal growth and development
F	Needed for bones and teeth; Ca by itself cannot build bone
Si	Needed for normal bone growth, and for proper integrity of the skin

Adapted from Gomes and Silva (2007)

nervous, and immune systems, and is involved in blood sugar regulation. In addition, manganese is involved in blood clotting and the formation of cartilage and lubricating fluid in the joints (Fraga et al. 2013). Manganese in excess can inhibit proper functioning of the human body resulting in deleterious effects. Elevated levels of manganese in the body may result in multiple neurologic problems (Keen et al. 1999). Zinc is an essential mineral and is found in just about every cell in our bodies. It is important for the production of enzymes that help our immune system function effectively and in regulating cell division and growth and wound healing effectively (Whittaker 1998). Long-term higher Zn intakes than the requirements could, however, interact with the metabolism of other trace elements. Zinc in high concentration can result in acute Zn poisoning. Iron in the human body is necessary for oxygen transport in the blood since it is the central atom of the heme group, a metal complex that binds molecular oxygen (O_2) in the lungs and carries it to all of the other cells in the body that need oxygen to perform their activities. Without iron in the heme group, there would be no site for the oxygen to bind, and thus no oxygen would be delivered to the cells resulting in cells dying (Casiday and Frey 2010; Whittaker 1998). Copper contributes to the formation of diverse enzymes required for several functions, production of energy, anti-oxidation, synthesis of a hormone named adrenaline, formation of conjunctive tissue, and Fe metabolism and formation of haemoglobin, the pigment of erythrocytes that transports the oxygen (Lindh 2005). More details on the role of nutrients for human health are given in Chaps. 6 and 7.

2.2 Role of Minerals for Therapeutic Purposes

A large spectrum of minerals are used for pharmaceutical, medical and cosmetic purposes (reviewed by Carretero and Pozo 2010). These minerals may be taken up orally or parenterally, or may be used topically. The therapeutic activity of the minerals is controlled by their chemical composition as well as their physical and physico-chemical properties. The important properties include large specific surface area, high sorption capacity, solubility in water, reactivity toward acids, high refractive index, high heat retention capacity, opacity, low hardness, astringency, and high reflectance (Carretero and Pozo 2010). In the case of clay minerals surface areas reaches values of around 800 $m^2\ g^{-1}$ (Gomes and Silva 2007).

2.2.1 Minerals Used for Pharmaceutical Preparations and Cosmetic Products

A large number of minerals are used as active ingredients in pharmaceutical preparations as well as in cosmetic products. The minerals in use include elements, oxides, hydroxides, carbonates, sulphates, chlorides, phyllosilicates, sulphides, phosphates, nitrates and borates (Table 2.2). The therapeutic activity of these minerals is controlled by their chemical composition and their physico-chemical properties. Minerals that are capable of reacting with acids can serve as antacids. They are also effective as antidiarrhoeaics, osmotic oral laxatives, and mineral supplements because in reacting with hydrochloric acid of the stomach, cations disposable for absorption or capable to be effective in the bowel are released. Minerals such as phyllosilicates displaying a high sorption capacity and a large specific surface area, can also function as gastrointestinal and dermatological protectors, and antiinflammatories and local anesthetics (Carretero and Pozo 2010). Water-soluble minerals can be used as homeostatics, antianemics and decongestive eye drops. Likewise, minerals with a high heat retention capacity can serve as anti-inflammatories and local anesthetics and minerals with high astringency are used as antiseptics and disinfectants. Such minerals which react with cysteine can serve as keratolytic reducers.

Occasionally, it is not essential that the mineral is composed of nontoxic ions in order that it may serve as an active ingredient in pharmaceutical preparations. This is true for minerals that induce vomiting since they are quickly eliminated. These water-soluble minerals, composed of Cu^{2+} and Zn^{2+} ions, are used as direct emetics

Table 2.2 Overview of minerals used as active substances in pharmaceutical preparations and cosmetic products

Mineral group/ mineral	Chemical formula	Administration	Therapeutic effect or cosmetic action
Oxides/			
Rutile	TiO_2	Topical	DermProt[a], SolProt[b]
Periclase	MgO	Oral	Antacid, OsOrLax[c], MinSup[d]
Zincite	ZnO	Topical	Disinfectant, DermProt[a], SolProt[b]
Hydroxides/			
Brucite	$Mg(OH)_2$	Oral	Antacid, OsOrLax[c], MinSup[d]
Gibbsite	$Al(OH)_3$	Oral	Antacid, GastProt[e], Antidiarr[f]
Hydrotalcite	$Mg_6Al_2(CO_3)(OH)_{16}\cdot 4H_2O$	Oral	Antacid
Carbonates/			
Calcite	$CaCO_3$	Oral, topical	Antacid, Antidiarr[f], MinSup[d], Tooth[g]
Magnesite	$MgCO_3$	Oral	Antacid, OsOrLax[c], MinSup[d]
Hydrozincite	$Zn_5(CO_3)_2(OH)_6$	Topical	DermProt[a]
Smithsonite	$ZnCO_3$	Topical	DermProt[a]
Sulphates/			
Chalcanthite	$CuSO_4\cdot 5H_2O$	Oral, topical	Direct emetic, disinfectant
Epsomite	$MgSO_4\cdot 7H_2O$	Oral, topical	OsOrLax[c], MinSup[d], BathSal[h]
Mirabilite	$Na_2SO_4\cdot 10H_2O$	Oral, topical	OsOrLax[c], bathroom salts
Melanterite	$FeSO_4\cdot 7H_2O$	Oral, topical	Antianemic, MinSup[d]
Zincosite	$ZnSO_4$	Oral, topical	Direct emetic, disinfectant
Goslarite	$ZnSO_4\cdot 7H_2O$	Oral, topical	Direct emetic, disinfectant
Alum	$KAl(SO_4)2\cdot 12H_2O$	Topical	Disinfectant, deodorant
Chlorides/			
Halite	$NaCl$	Oral, parenteral, topical	Homeostatic, MinSup[d], eye drops, BathSal[h]
Sylvite	KCl	Oral, parenteral, topical	Homeostatic, MinSup[d], eye drops, BathSal[h]
Phyllosilicates[i]/			
Montmorillonite	$(Al_{1,67}Mg_{0,33})Si_4O_{10}(OH)_2M^+_{0,33}$	Oral, topical	Antacid, GastProt[e], Antidiarr[f], DermProt[a]
Saponite	$Mg_3(Si_{3,67}Al_{0,33})O_{10}(OH)_2M^+_{0,33}$	Oral, topical	Antacid, GastProt[e], Antidiarr[f], DermProt[a]
Hectorite	$(Mg_{2.67}Li_{0,33})Si_4O_{10}(OH)_2M^+_{0,33}$	Oral, topical	Antacid, GastProt[e], Antidiarr[f], DermProt[a]

(continued)

Table 2.2 (continued)

Mineral group/ mineral	Chemical formula	Administration	Therapeutic effect or cosmetic action
Sepiolite	$Mg_8Si_{12}O_{30}(OH)_4(OH_2)_4{\cdot}8H_2O$	Oral, topical	Antacid, GastProt[e], Antidiarr[f], DermProt[a]
Kaolinite	$Al_2Si_2O_5(OH)_4$	Oral, topical	Antacid, GastProt[e], Antidiarr[f], DermProt[a]
Talc	$Mg_3Si_4O_{10}(OH)_2$	Topical	DermProt[a]
Muscovite	$KAl_2(Si_3Al)O_{10}(OH)_2$	Topical	
Others			
Sulphur	S	Topical	Disinfectant, keratolytic reducer
Greenokite	CdS	Topical	Keratolytic reducer
Borax	$Na_2B_4O_7{\cdot}10H_2O$	Topical	Disinfectant
Hydroxiapatite	$Ca_5(PO_4)_3(OH)$	Oral	MinSup[d]
Niter	KNO_3	Topical	Anaesthetizer in toothpastes

Adapted from Carretero and Pozo (2010)
[a]DermProt: Dermatological protector
[b]SolProt: Solar protector
[c]OsOrLax: Osmotic oral laxative
[d]MinSup: Mineral supplement
[e]GastProt: Gastrointestinal protector
[f]Antidiarr: Antidiarrhoeaic
[g]Tooth: Agent in toothpaste
[h]BathSal: Bathroom salts
[i]Phyllosilicates listed in Table 2.2 are also present in cosmetic creams, powders and emulsions

(Carretero and Pozo 2010). Minerals are also used in cosmetic products for their physical and physico-chemical properties as well as their chemical composition. Those minerals exhibiting a high refraction index can be used as solar protectors. Water-soluble species can be utilized as ingredients in toothpastes and bathroom salts, while minerals with a high sorption capacity and a large specific surface area can function as creams, powders and emulsions. Minerals with proper hardness can act as abrasives in toothpastes. Highly opaque minerals and minerals of high reflectance are used in creams, powders and emulsions. Likewise, minerals with high astringency are included in deodorants.

2.2.2 Therapeutic Effects of Minerals Administered Orally

Minerals that are administered orally to the patient may act as mineral supplements, antacids, gastrointestinal protectors, antidiarrhoeaics, oral laxatives, direct emetics, antianemics or homeostatics.

2.2.2.1 Mineral Supplements

Mineral supplements are usually administered orally as tablets. They are administered in situations of physical weakness, convalescence, or asthenia in order to remedy slight deficiencies of essential elements, such as P, Ca, Mg, Na, K and Fe (Carretero and Pozo 2010). The minerals used for this purpose are calcite, magnesite, hydroxyapatite, epsomite, periclase, brucite, halite, sylvite, and melanterite (Table 2.2). These minerals are soluble in water and hydrochloric acid, the latter being constituent of the gastric fluid.

2.2.2.2 Antiacids

As antiacids, minerals can be administered to the patient orally in the form of pills, powders, suspensions, and emulsions. Acidity in the stomach, caused by excessive production of hydrochloric acid, can be effectively reduced by oral administration of non-systemic antacids. Minerals can function as antacids through two mechanisms: (i) neutralization of gastric acid and (ii) adsorption of H^+ ions to mineral surface sites. Depending on the mineral type, acid neutralization may increase the pH of the gastric fluid from pH 1.5–2.0 to pH $\geq$7. An effective antacid is one that elevates the pH by 3–4 units, and causes the disappearance of "free acidity" (Carretero and Pozo 2010). Carbonates (calcite, magnesite), oxides (periclase), and hydroxides (brucite, gibbsite, hydrotalcite) are widely used as antacids. Since calcite tends to cause 'acid rebound', it should only be administered in treatments of short duration. Examples of neutralization of hydrochloric acid in the stomach by minerals are represented by the following equations:

$$CaCO_3 + 2H^+ \rightarrow Ca^{2+} + CO_2 + H_2O \tag{2.1}$$

$$MgO + 2H^+ \rightarrow Mg^{2+} + H_2O \tag{2.2}$$

$$Mg(OH)_2 + 2H^+ \rightarrow Mg^{2+} + 2H_2O \tag{2.3}$$

$$Al(OH)_3 + 3H^+ \rightarrow Al^{3+} + 3H_2O \tag{2.4}$$

The main reaction products are CO_2 (gas), Mg^{2+}, Ca^{2+} and Al^{3+} ions. Gastric acidity may also be reduced through adsorption of H^+ ions (protons) to the mineral surface. Clay minerals as cation exchangers (palygorskite, sepiolite, montmorillonite, saponite, or mixtures of any two) can act as antacids through this mechanism. Kaolinite is not suitable as an antacid because it has a low capacity for adsorbing protons.

2.2.2.3 Gastrointestinal Protectors

Gastrointestinal protectors can be administered orally as pills, powders, suspensions and emulsions. When the thickness of the mucous membrane diminishes, the mucolytic activity of gastric enzymes increases, requiring the application of these minerals which have a large specific surface and a high sorption capacity. They should also be non-toxic to organisms. Minerals used as gastrointestinal protectors include metal hydroxides (e.g., gibbsite) and clay minerals (e.g., palygorskite, sepiolite, kaolinite, smectites) (Table 2.2). These minerals, by adhering to the gastric and intestinal mucous membrane, diminish irritation and gastric secretion, take up gases, toxins, bacteria, and even viruses (Carretero and Pozo 2010). The mode of action involves increasing the viscosity and stability of the gastric mucus, and decreasing the degradation of glycoproteins in the mucus (López Galindo and Viseras 2004). Since these minerals can also remove enzymes, vitamins, and other vital substances, their continual and prolonged administration is not recommended. Except for acid-stable kaolinite, all the minerals used as gastrointestinal protectors will dissolve to varying extents, releasing Mg^{2+}, Al^{3+}, and silica gel.

2.2.2.4 Antidiarrhoeaics

Diarrhoea is either an acute or chronic pathological state characterised by an increase in the fluidity of the faeces and the frequency of their evacuation. It can be caused by bacterial infections, food poisoning, defective intestinal absorption or allergic states (Carretero 2002). Treatment should particularly focus on eliminating its cause although the symptoms of diarrhoea may be eliminated using drugs that act by antidiarrhoeaics (Carretero and Pozo 2010). Most drugs against diarrhoea act by reducing the quantity of liquid that arrives in the colon from the thin bowel. Minerals used as antidiarrhoeaics should have a high specific surface and sorption capacity, and be non-toxic to organisms (Table 2.2). Besides removing excess water and compacting the faeces, effective antidiarrhoeaics can adsorb excess gases from the digestive tract. Clay minerals, such as kaolinite, palygorskite, sepiolite, and smectites (e.g., montmorillonite) are widely used as antidiarrhoeaics. Prior to application, palygorskite, sepiolite and smectites are normally "activated" by heating or acid treatment in order to enhance their sorption capacity (Christidis et al. 1997). Except for kaolinite, these clay minerals can undergo partial degradation in the acidic medium of the stomach (Vicente Rodríguez et al. 1994). The silica gel produced has a high sorption capacity and can contribute to the antidiarrhoeaic action. The long-term use of clay minerals antidiarrhoeaics is not recommended because of the risk of kidney stone formation (Levison et al. 1982). Calcite and gibbsite (Table 2.2) also feature as antidiarrhoeaics. In this case, the therapeutic action may be ascribed to the release of Ca^{2+} ions (from calcite) and Al^{3+} ions (from gibbsite), followed by the formation of insoluble salts in the bowel.

2.2.2.5 Osmotic Oral Laxatives

Orally administered laxatives are those laxatives whose active principles' main effect is to encourage defecation. They can act by osmosis, by irritating the small bowel or the colon–rectum (Carretero 2002). Minerals used as oral laxatives contain Na^+ or Mg^{2+} ions and nontoxic anions, and have a high solubility in water or hydrochloric acid (Table 2.2). Examples are mirabilite, epsomite, periclase, brucite and magnesite. Sodium and magnesium ions exert their therapeutic action as they spread through the stomach's fluids and reach the small bowel where they increase the osmotic pressure of the intestinal contents which induces water to pass from the blood plasma through the bowel wall in order to re-establish osmotic balance (Carretero and Pozo 2010). As a result, there is a considerable increase in the volume of the bowel's contents which, in turn, stimulates the propulsive motor activity of the smooth intestinal muscle. This effect continues in the colon-rectum, producing liquid faeces. The laxative activity of minerals decreases in the order: mirabilite > epsomite > periclase > brucite > magnesite. The highly water-soluble mirabilite and epsomite are orally administered as a solution, or as dissolved granules. Sparingly soluble minerals are usually administered as suspensions, and dissolve in the acidic medium of the stomach.

2.2.2.6 Direct Emetics

Mineral emetics are administered orally as aqueous solutions. Emetics are substances that cause vomiting. Emetics, being absorbed by and distributed through the blood plasma, can act directly on the nerve endings of the stomach, or indirectly on the vomit centre in the medulla oblongata (Carretero and Pozo 2010). Chalcanthite, goslarite and zincosite have been used as emetics. They contain Cu^{2+} and Zn^{2+} ions, are highly soluble in water (Table 2.2), and act directly. These cations irritate the gastric mucosa, and stimulate the vomit centre. When vomiting is delayed, however, the ions can move to the bowel causing intestinal colic and diarrhoea. Chalcanthite is also useful as an antidote in phosphorus poisoning.

2.2.2.7 Antianemics

Anemia is a blood dysfunction caused by a deficiency of red blood cells, or a low concentration of hemoglobin (Carretero and Pozo 2010). The body needs iron to manufacture hemoglobin. If there is not enough available iron, hemoglobin production is limited and this affects the production of red blood cells. Ferrous ions are much less irritating to, and more quickly absorbed by, the organism than are ferric ions. Melanterite is widely used as an antianemic because this mineral is highly soluble in water, and contains Fe^{2+} ions in its structure. Released Fe^{2+} diffuse to the small bowel where absorption of Fe^{2+} ions occurs. These ions then pass to the blood plasma where they bind to transferrin, a glycoprotein (globulin β), and convert into the ferric form which is then transported to the bone marrow for incorporation into hemoglobin.

2.2.2.8 Homoeostasis

Homoeostasis refers to the tendency toward electrostatic equilibrium between body liquids (Carretero and Pozo 2010). Water is the main liquid constituent of living organisms, making up about 60% of body weight. Mineral substances are also indispensable for the maintenance of life, since they are actively involved in regulating osmotic pressure, acid-base equilibrium, and diverse organic functions. During dehydration the loss of water is always accompanied by the loss of electrolytes of which Na^+ and K^+ ions are the most important. Water-soluble minerals containing Na^+ or K^+ ions (and their complementary anions) act as homoeostatics (Table 2.2). The most commonly used minerals are halite and sylvite. These minerals are administered as saline solutions, either orally in mild cases or parenterally in very severe cases.

2.2.3 Therapeutic Effects of Minerals Administered Topically

Minerals can be administered to the patient topically as decongestive eye drops, anti-inflammatories and local anesthetics, antiseptics and disinfectants, dermatological protectors and keratolytic reducers.

2.2.3.1 Decongestive Eye Drops

For its high solubility in water, halite is the principal mineral used in decongestive eyedrops. The mineral, previously dissolved, is administered in the form of an isotonic collyrium to treat eye dryness, stinging and weeping induced by smoke, dust, or air-borne particles, light ocular irritation, and problems of visual fixation.

2.2.3.2 Anti-inflammatories

Inflammation is a reaction of the organism against an irritating or infectious agent. It can produce pain, swelling, reddening, rigidity or loss of mobility, and heat in the affected area. Inflammation is the body's attempt at self-protection with the aim to remove harmful stimuli, including damaged cells, irritants, or pathogens. Pain can be alleviated by applying some cold compresses, relaxation, and the complete immobilization of the affected area. Sometimes, in the treatment of musculoskeletal disorders, rheumatism, trauma or stress, hot poultices are also used, because the heat is also a therapeutic agent. These therapeutic actions may be enhanced by using minerals with adsorbent and heat retention properties. Kaolinite, being the most common, is topically applied as a poultice to reduce inflammation and alleviate pain.

2.2.3.3 Antiseptics and Disinfectants

Substances that destroy microorganisms or inhibit microbial growth in living tissues are called antiseptics. In contrast, disinfectants are applied to surfaces of inanimate objects. Some substances can be used as both antiseptic and disinfectant. Mineral antiseptics and disinfectants have a high astringent capacity (Table 2.2). Sulphur, borax, chalcanthite, zincite, goslarite, zincosite, and alum are widely used in liquid forms (lotions, drops) or as powders (Carretero and Pozo 2010). The antiseptic or disinfectant activity of these minerals is mainly controlled by their concentration. However, at high concentrations these substances are corrosive and may be toxic to organisms because they can be readily absorbed by the skin. For this reason, they should not be applied continually over extensive areas of skin, or spread over broken skin. Sulphur is used as a soft antiseptic and disinfectant. It is also used as lotions for colutories and gargles for aphthous ulcers and stomatitis. Chalcanthite is used as an astringent and fungicide solution. The oxides and sulphates of zinc are used as astringent lotions for painless ulcers, and as gargles and astringent colutories. They are also used as ophthalmic drops (0.25% solutions) to treat conjunctivitis and chronic cornea inflammation. Alum (1–4% solutions) is used in gargles and colutories for stomatitis and faringitis. On the other hand, in solid form (powder), alum is also applied as a hemostatic agent (to inhibit hemorrhages) for cuts or superficial skin damage. Borax acts as a weak bacteriostatic and soft astringent agent. Historically, borax was used as a disinfectant in the form of skin lotions, and eye-, nose-, and mouth-washes.

2.2.3.4 Dermatological Protectors

Dermatological protectors are creams, ointments, and powders that protect the skin against external agents and, occasionally, against exudations and liquid excretions (Carretero and Pozo 2010). Minerals with a high sorption capacity are used as dermatological protectors (Table 2.2). Fibrous minerals (palygorskite, sepiolite), however, are excluded due to their possible carcinogenic effect when inhaled (Carretero et al. 2006). Suitable minerals are kaolinite, talc, smectites (e.g., montmorillonite), zincite, hydrozincite, smithsonite and rutile. These minerals can adhere to skin to form a film, which provides (mechanical) protection against external physical and chemical agents. By taking up skin exudations and creating a surface for evaporation, such minerals also have a refreshing action. In addition, they exert a soft antiseptic action by producing a water-poor medium that is unfavourable to bacterial growth. Indeed, some of these minerals (e.g., kaolinite, talc, smectites) are good sorbents of dissolved or suspended substances (greases, toxins), bacteria and viruses.

2.2.3.5 Keratolytic Reducers

Keratolytic reducers act on the corneous, superficial layer of the epidermis, reducing its thickness, or causing its peeling (Carretero and Pozo 2010). They are used to treat cutaneous affections such as seborrheoic dermatitis, psoriasis, chronic eczemas or acne. Sulphur and greenockite are effective keratolytic reducers. Sulphur has been used extensively in dermatology for its keratolytic effect and its supposed antimicrobial effect. The keratolytic effect of sulphur is probably due to the reaction between the sulphur and the cysteine in keratinocytes forming hydrogen sulphide. Hydrogen sulphide can break down keratin, thus demonstrating sulphur's keratolytic activity. The smaller the particle size, the greater is the degree of such interaction and therefore the therapeutic efficacy. The supposed antimicrobial effect depends on the conversion of sulphur to pentathionic acid, which is toxic to fungi, by the normal skin flora or the keratinocytes. The keratolytic properties may promote fungal shedding from the stratum corneum. Sulphur has an inhibitory effect on the growth of *Propionibacterium acnes* as well as *Sarcoptes scabiei*, some *Streptococci*, and *Staphylococcus aureus*. This suggested antibacterial activity purportedly results from the inactivation of sulfhydryl groups contained in bacterial enzyme systems. The precise mechanism of action is still unknown (Carretero and Pozo 2010). Cadmium sulphide, applied as shampoo or a suspension in an alkaline detergent base, is effective against dandruff and seborrhoea. Although it is not absorbed by the scalp, cadmium sulphide is very toxic when ingested, and hence has fallen into disuse. Sulphur has traditionally been applied as lotions, ointments, and creams to treat acne. Since these minerals are applied as shampoo, suspensions, lotions, ointments and creams, they are practically not absorbed. In fact about only 1% of topically applied sulphur is systemically absorbed. Adverse effects from topically applied sulphur are uncommon and are mainly limited to the skin (Lin et al. 1988).

2.2.4 Minerals Administered Parenterally

Parenteral therapy is a route of administration for drugs which are poorly absorbed via the oral or the topical route and it can provide a rapid response during an emergency. In severe cases, minerals used as antianemics (melanterite) and homoeostatics (halite and sylvite) are administered parenterally in a dissolved form.

2.3 Minerals Used as Excipients

Excipients are substances introduced into certain pharmacals in order to: (i) improve its organoleptic characteristics such as taste, smell and colour, or its physical–chemical properties such as viscosity, (ii) facilitate the pharmaceutical formulation's preparation, and (iii) promote the pharmaceutical formulation's disintegration when it is orally administered in the form of pills, capsules, etc. (Carretero 2002). Phyllosilicates, particularly clays, are commonly used as excipients. The technological properties of clays are directly related to their colloidal size and crystalline structure in layers, meaning a high specific surface area, optimum rheological characteristics and/or excellent sorptive ability (Aguzzi et al. 2007). The use of a clay mineral for any specific application depends on both its type of structure (1:1 or 2:1 layer type) and on its chemical composition. The different types of cations in the octahedral sheet and isomorphic substitutions in the octahedral and tetrahedral sheets result in net charge deficits, varying according to the sheet unit, and, ultimately, in different mineral phases giving rise to very varied technical behaviour. Clay minerals are cation exchangers and so they may undergo ion exchange with basic drugs in solution.

Palygorskite, smectites, kaolinite and talc are used as (i) lubricants to ease the manufacture of pills (talc), (ii) agents to aid disintegration, due to their ability to increase in volume in the presence of water (smectites), or the dispersion of fibres (palygorskite), which favour the liberation of the drug of the pharmaceutical formulation when it arrives in the stomach, (iii) inert bases for cosmetics (palygorskite, kaolinite, smectites, talc), and (iv) as emulsifying, polar gel and thickening agents because of their colloidal characteristics (palygorskite, smectites) and to avoid the segregation of the pharmaceutical formulation's components (Carretero 2002).

Smectites, especially montmorillonite and saponite, due to their higher cation exchange capacity compared to other pharmaceutical silicates (such as talc, kaolin and fibrous clay minerals) have also been extensively applied for prolonged release, especially of basic drugs (Aguzzi et al. 2007). The main concept in prolonged release technology is that any pharmaceutical dosage form should be designed to provide therapeutic levels of drug to the site of action and maintain them throughout the treatment. These goals may be achieved by modifying the rate and/or time and/or site of drug release in comparison with conventional formulations. Such modification in release of active substances is provided to reduce toxic effects or for some other therapeutic purpose.

2.4 Minerals Used in Cosmetic Products

2.4.1 Toothpaste

Minerals are used in toothpaste for (i) formulation for sensitive teeth, and (ii) as abrasives or polishing agents. For sensitive teeth, minerals comprising potassium and non-toxic anions with a high solubility in water are used in toothpaste. Niter is a prime example because it contributes K^+ ions when the mineral dissolves on contact with saliva (Table 2.2). These ions act on nerve endings inside the dentine, inhibiting transmission of painful stimuli (Wara-aswapati et al. 2005). Calcite is used as an abrasive/polishing agent because it is nontoxic to organisms, while its hardness of 3 (in the Mohs scale) is less than that of tooth enamel (5 in the Mohs scale).

2.4.2 Sun Protection Products

Sun protection products minimize skin damage by preventing solar ultraviolet radiation from penetrating the skin. Minerals used as sun protection products must have a high refractive index and good light-scattering properties. Generally, rutile and zincite fulfil these requirements (Carretero and Pozo 2010). Natural rutile, however, is not used. Instead, the synthetic analogous obtained of rutile or ilmenite feature as solar protectors. Synthetic titaniumdioxide is white and has a very high refractive effect. This mineral reflects ultraviolet radiation when applied to skin as a fine layer. It is also more effective than other photosensitive organic compounds because of its high stability against photodegradation. The light-scattering ability of the mineral is controlled by the particle size. For example, titanium dioxide with an average particle size of 230 nm scatters visible light, while its counterpart with an average particle size of 60 nm scatters ultraviolet light and reflects visible light (Hewitt 1992). The disadvantage of using (synthetic) titanium dioxide is that it gives a white appearance to the skin. Transparent sunscreen preparations use TiO_2 with a very small particle size of 50 nm which is considered optimum. It provides good protection against ultraviolet radiation while the dispersion of visible light is such that the cream is not white on the skin. Nanosize TiO_2 is extensively used in sunscreen lotions (Jaroenworaluck et al. 2006). Zincite is also used as a solar protector because its physical properties are similar to those of TiO_2.

2.4.3 Creams, Powders and Emulsions

As cosmetic products, creams, powders and emulsions are applied to external parts of the body in order to embellish or modify physical appearance, or preserve the physico-chemical conditions of skin. Opaque minerals with a high sorption capacity are used in cosmetic formulations as creams, powders, emulsions in order to give opacity, remove shine, and cover blemishes. Besides adhering to skin and forming a protective film, these minerals can take up grease and toxins. Palygorskite and sepiolite (in liquid preparations) together with kaolinite, smectites, and talc are the principal minerals used for this purpose. On the other hand, micas have long been used in eye shades and lipsticks for their high reflectance and iridescence. More recently, muscovite mica are added to moisturizing creams in order to produce a luminous effect on skin.

2.4.4 Bathroom Salts and Deodorants

Because of their high solubility in water and their specific composition, halite, sylvite, epsomite and mirabilite are used in bathroom salts. Alum is widely used as a

Fig. 2.3 The salt lake Baskunchak is located in Astrakhan region, about 270 km north of the Caspian Sea, and 53 km to the east of the Volga River. The lake is the place of salt mining from the eighth century. Today, pure salt (99.8% NaCl) extracted makes up to 80% of the total salt production in Russia. Depending on demand, 1.5–5 million Mg of salt is mined per year (Photo: Yakov Kuzyakov, with kind permission)

deodorant because of its high astringence. There are numerous sources for high-quality salts worldwide. Lake Baskunchak is one example (Fig. 2.3).

2.5 Soil Materials for Fangotherapy

Fangotherapy is the use of peat, mud and clay for healing purposes (Ekmekcioglu 2002). The word fango is the Italian word for mud and so strictly speaking peat and clay should not be labeled as fango treatments (Patel et al. 2015). While fango treatments are mainly used for skin care, massage therapists will find fango useful for spa treatments aimed at the reduction of soft tissue pain and dysfunction, and to relax and revitalize the body. The therapeutic substances used in fango spa treatments have different characteristics that affect their therapeutic properties and uses. Each of these materials (mud, clay and peat) has its own special properties but in general they hold heat and are useful as a thermal application for chronic conditions (Sukenik et al. 1992). They also stimulate circulation and lymph flow, support detoxification and help the body to relax. Some types of fango have anti-inflammatory and pain relieving properties that make them useful for soft-tissue injury (Patel et al. 2015). Clay as a mineral (phyllosilicate), is the most absorbent of

Table 2.3 Composition and properties of fangotherapy

Composition	Category	Main types	Commercial name	Properties	Applications
Mainly organic	High moor peat	Mainly mosses	Many European types	Immune-boosting, endocrine balancing, thermal, relaxing, others	Relaxation, revitalization, esthetics, others
	Low moor peat	Mosses and other plants	Moor mud, many others	Anti-inflammatory, circulatory stimulant, antiviral, antiseptic	Arthritis, muscle pain or soreness, joint pain, inflammation
Mainly mineral	Clay	Kaolinite (1:1 crystalline structure)	Kaolin, China white	Thermal, relaxing, circulatory stimulant, absorbs excess oil and draws out impurities, suspends to form an emulsion to hold cosmetic substances together, acts as a carrier for other therapeutic products	Thermal agents to warm and relax the body, base for treatment products, esthetics,
		Illite (2:1 crystalline structure)	French green		
		Smectite (2:1 crystalline structure)	Bentonite, Fuller's earth, MAS		Cosmetic emulsion
	Mud	Sulfur containing and 'matured'	Dead sea, Euganaean, Piestany, many others	Anti-inflammatory	Arthritis, muscle pain or soreness, imflammation, joint pain, revitalization, esthetics, others

Adapted from Patel et al. (2015)

the fango substances. It is used to pull impurities from the skin and to stimulate circulation. Mud is also predominantly mineral but has low contents of organic matter that give it a wider range of properties. Mud may be anti-inflammatory, antiviral and immune boosting. Peat is therapeutically the most active substance of the three as it is mainly organic and derived from the break down of plant material over thousands of years. Table 2.3 gives an overview of composition and properties of fangotherapy.

2.5.1 Peat

Peat is an easily available natural material and a source of biologically active substances widely used, not only in horticulture but also in human medicine

(Trckova et al. 2005). Peatlands are characteristically waterlogged locations usually found on relatively flat landscapes. Cold and anaerobic conditions up to tens of centimetres beneath the surface cause organic residues to accumulate, to depths of at least 30 cm and often up to several metres (Gorham 1991). A series of complex decomposition and accumulation processes characteristic of this kind of environment lead to the formation of peat. Peat bogs form where the presence of excess of water for most or all the year prevents the complete degradation of organic substances, leading to an accumulation of sediments forming peat (Bozkurt et al. 2001). The general classification of peatland ecosystems into ombrotrophic bogs, minerotrophic fens and transitional mires is based on the origin of the mire water. In ombrotrophic peatlands the water is supplied exclusively by precipitation, while minerotrophic mires are additionally supplied with water from mineral soil (Andriesse 1988; Bozkurt et al. 2001).

High-moor (ombrotrophic) bog is formed in beds with poor drainage ability as a consequence of heavy rainfalls, and a lack of minerals, particularly in calcium deficient acidic rocks (Trckova et al. 2005). High levels of aerial humidity can be observed at higher altitudes so bogs are found mainly in mountains and foothills. Extreme humidity causes delayed decay and accumulation of organic material. Under such conditions; only undemanding vegetation can survive, such as sphagnum moss (*Sphagnum* sp.), haircap moss (*Polytrichum* sp.), cotton grass (*Eriophorum* sp.), dendriform plants, stunted trees (particularly spruce (*Picea* sp.), and pine trees (*Pinus* sp.)). Growth of higher plants is curbed by the peat mosses themselves as they bind all available nutrients and render the bog water acidic. Oligotrophic peat of high-moor bogs contains very low concentrations of inorganic substances and ashes (2–4%); accordingly, it is deficient in mineral nutrients and the level of acidity is high (pH 3–4). It is light, crumbly and yellowbrown to brown in colour (Bozkurt et al. 2001; Pereverzev 2005). Figure 2.4 shows the profile of an ombrogenic bog in North Germany.

Low-moor (minerotrophic) fens occur near eutrophic soils or are formed when nutrient rich waters are filled with sediment (Trckova et al. 2005). The abundance of plant nutrients allows fens to support luxuriant vegetation. Arising in basins rich in minerals (such as limestone clay, marl, pond mud); they contain rather high proportions of inorganic substances, have an elevated ash content (6–18%) and a neutral to slightly alkalic pH is (6–8). Eutrophic peat of low moors is dark brown to black in colour, heavy, slimy in wet weather, and may be powdered when dried (Bozkurt et al. 2001; Pereverzev 2005). In particular higher plants such as sedge (*Carex* sp.), reed (*Phragmites* sp.), cane (*Calamagrostis* sp.), and horsetail (*Equisetum* sp.), birch trees (*Betula* sp.), alder (*Alnus* sp.), pine (*Pinus* sp.), spruce (*Picea* sp.), and some species of moss (*Sphagnum* sp., *Musci* sp. and *Hypnum* sp.) contribute to minerotrophic fen formation (Bozkurt et al. 2001). Figure 2.5 shows the profile of a low-moor fen in North Germany. In Fig. 2.6, the traditional way of transport of peat material for the use of fango treatment is presented.

Fig. 2.4 Profile of an ombrogenous sphagnum moss peat Histosol (Fibric Histotsol) in the Teufelsmoor near Bremen, North Germany. The peat consists mainly of *Sphagnum* moss (Photo: Rolf Nieder)

2.5.2 Clay

Clays have been defined as finely grained natural mineral materials, which can act plastically and will harden on drying (Guggenheim and Martin 1995). However, more relative to this review is the definition of a clay mineral, which refers to the specific minerals as phyllosilicate (layered silicate) molecules that give the plastic qualities to clays. These microcrystalline-layered minerals are generally based on hydrous metal silicates with sheetlike structures (Gaskell and Hamilton 2014). The properties of clay minerals are directly affected by their structure and influence their subsequent application. There are two different sheet structures, tetrahedral (Fig. 2.7a, b) and octahedral (Fig. 2.7c, d), that build the clay mineral layers, and it is the proportion and organization of these sheets in the layers that distinguish between different clay types and impart some of the physicochemical characteristics they possess (Brigatti et al. 2013).

Tetrahedral and octahedral sheets are linked via sharing oxygen atoms at the apical point of the tetrahedral sheets to form continuous layers that then further associate to produce planar particles (Bergaya and Lagaly 2013). An octahedral

Fig. 2.5 Profile of a topogenous low-moor peat Histosol (Eutric Histosol) in the Ochsenmoor near Osnabrück, North Germany (Photo: Rolf Nieder)

sheet can share oxygen atoms with one or two tetrahedral sheets, forming 1:1 (e.g. kaolinite) or 2:1 (e.g. illite and smectite) crystalline structures respectively, as shown in Fig. 2.7a, b. The formed layers can be either electrically neutral or display a negative charge. The negative charge arises if there are deviations in the cations present in the tetrahedral and/or octahedral sheets (Gaskell and Hamilton 2014). The space between the layers in a clay particle is referred to as the interlayer space. In 1:1 type clay minerals, due to the layer structures, strong bonds are formed between the octahedral sheet of one layer with the tetrahedral sheet of the other. These clay minerals are electrically neutral thus do not need counterbalancing cations to occupy the interlayer space (Fig. 2.8).

As a result, these clay minerals do not hydrate and are considered non-swelling (Gaskell and Hamilton 2014). On the other hand, 2:1 clay minerals have weaker bonds between the tetrahedral sheets of two layers. The layer charge varies across different 2:1 clay mineral classes and as such the interlayer space is either unoccupied (for neutral clays) or occupied by counterbalancing cations (Fig. 2.8b). As such, some of the clay minerals that belong within this category display the ability to swell, by absorbing water into their interlayer space, causing the expansion of the mineral structure (Hensen and Smit 2002). Properties such as these are suited to applications where excess liquid needs to be removed (Choy et al. 2007), including

Fig. 2.6 Transport of peat material with a diesel locomotive to the specialist clinic Teufelsbad in Blankenburg, Germany (Pawel et al. 2013, p. 37; Photo: Fundus Teufelsbad, with kind permission from Bergverein zu Hüttenrode e. V.)

wound exudates aiding healing and controlling odour. The cation-exchange capacity (CEC) of clay minerals is directly linked to the extent and type of their counterbalancing cations. This feature allows for the exchange of the counterbalancing cations with guest species, changing the properties of the clay minerals and facilitating their applicability in the biomedical field. The exchanged molecules can range from alternative metal cations, therapeutic molecules, polymers and toxins (Ladewig et al. 2009), some of which are covered in more detail below. The classification of clay minerals is primarily based on the ratio of the tetrahedral to octahedral sheets within the clay layers, as well as the layer charge that may arise throughout the structure (Gaskell and Hamilton 2014).

2.5.3 *Mud*

Like clay, mud is mainly mineral in origin, It contains 2–4% organic substances which play an important role in its therapeutic use (Hendrickson 2003). Therapeutic mud is "matured" or "ripened" in natural mineral water (Patel et al. 2015). The maturing process for each mud may be slightly different, but generally it involves the oxidation and reduction of the mud over a period of up to 12 months. The

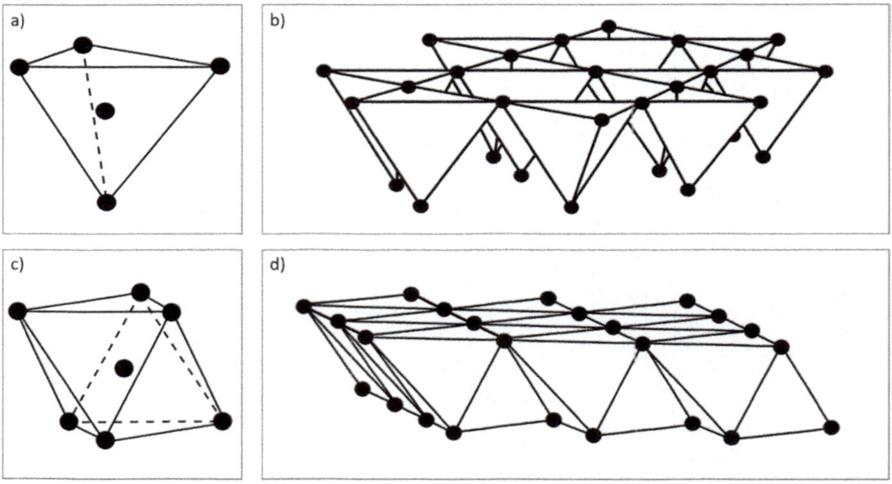

Fig. 2.7 Clay mineral sheet structures. Single tetrahedron unit (**a**) illustrating the tetrahedral ion (*small circle*), usually Si^{4+}, surrounded by four oxygen atoms (larger circles) and tetrahedral sheet (**b**) showing the corner sharing of tetrahedral basal oxygen atoms. Single octahedron unit (**c**) illustrating the octahedral ion (small circle), commonly Al^{3+}, Fe^{3+}, Mg^{2+} or Fe^{2+}, surrounded by six oxygen atoms (larger circles) and octahedral sheet (**d**) showing the sharing of octahedral edges (Adapted from Gaskell and Hamilton 2014)

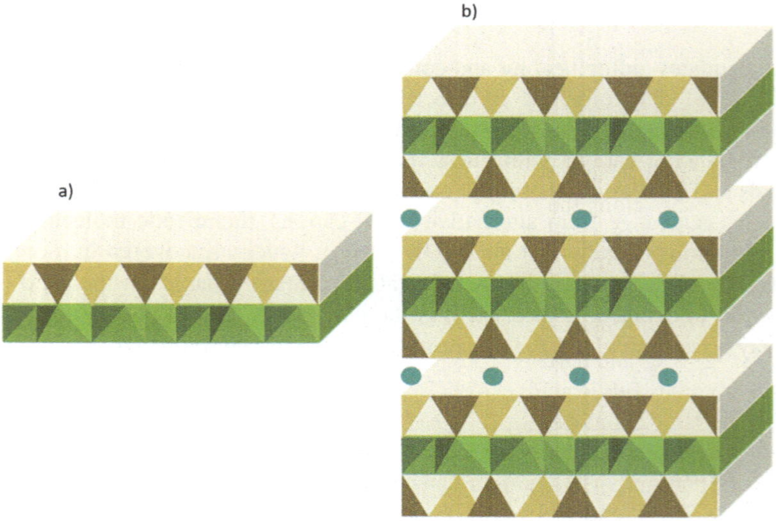

Fig. 2.8 The arrangement of clay mineral layers into particles. The 1:1 crystalline arrangement, representative for kaolinite (**a**), and the 2:1 crystalline arrangement, representative for illite and smectite (**b**), with the *circles* illustrating counterbalancing cations. The continuous *triangular* and *rectangular shapes* represent tetrahedral and octahedral sheets, respectively (Adapted from Gaskell and Hamilton 2014)

process of maturing the mud is characterized by changes in the chemical composition of the mud, and changes in its appearance. Sulfur is perhaps the most important component in the different kinds of therapeutic mud and occurs naturally in the vicinity of volcanoes and hot springs (Patel et al. 2015). Sulfur baths have been researched as a viable means of reducing oxidative stress on the body and decreasing inflammation in muscles and joints. Sulfur rich mineral and mud baths are useful in the treatment of osteoarthritis, rheumatoid arthritis and other inflammatory conditions (Patel et al. 2015). Individuals report that they experience increased strength, decreased morning stiffness, better walking ability and decreased pain after a course of sulfur mud treatment. Therapeutic mud is also used successfully for bursitis, tendonitis, sprains, strains and other musculoskeletal injuries and disorders.

Mud found in different parts of the world has different properties. Mud composition varies with the place of origin. Firstly, mineral constituents of mud vary with the kind of rocks found in the region and the process of soil formation. Secondly, mud property is influenced by kind of flora and fauna of the region. Therefore, it is essential to learn about properties of mud before utilizing it benefits (Rastogi 2012). It is important to note that before using any type of mud it should be dried, powdered and sieved to remove any type of impurities such as stones, grass, etc. Black mud ("dark cotton soil") is suitable for mud therapy as it is rich in minerals and also retains water for long time. It should always be free from contamination and any kind of pollution (Patel et al. 2015). Mud from the Dead Sea was already used by Cleopatra and Queen Sheeba for enhancing beauty; black mud of Dead Sea has beautifying and therapeutic powers. It contains more than 20 kinds of salts and minerals including Magnesium, Calcium, Potassium Bromide, Silicates, Natural Tar and organic elements. While these beneficial minerals are useful for healing any kind of skin disorders, the presence of silicates make its masks very beneficial for softening and cleansing skin. The mud enhances blood circulation and leaves the skin with a glow (Patel et al. 2015). Hundreds of spa centers are active in Germany. Seaside resorts along the Baltic and North Sea are common. They are under specific standards of governmental regulation (Titzmann and Balda 1996). Mud from the North Sea (Fig. 2.9) is used for the treatment of both dermatological and musculoskeletal disorders.

Moor Mud is mud produced over thousands of years from organic residue of flowers, grasses and herbs. This residue transformed over several years to fine paste which contains fulvic acids, vitamins, amino acids, plant hormones, humic acids in a form which could be easily absorbed by human body. The mud has chelatic properties which enables its top layer to filter out impurities/pollutants and preserve purity of the mud. This mud has therapeutic properties and is useful in detoxification, healing, beautification, nourishing human body. The mud has anti-inflammatory and anti-aging effects. It is also useful in conditions such as Arthritis and recovery from injury in sports (Evcik et al. 2007).

Fig. 2.9 Mud from the North Sea near Cuxhaven, Germany. The mud originates from the Wadden Sea which is an intertidal zone in the southeastern part of the North Sea (Photo: Rolf Nieder)

2.5.4 Therapies

2.5.4.1 Mud Packs

Mud packs are typically used for local application. Frequent application of packs of mud helps in improving skin complexion and acts on skin spots and patches to reduce them (Patel et al. 2015). The construct and usage of a mudpack is similar for all applications though the thickness and the size varies as per the usage. To make a mud pack first soak mud in water for 30 min. Now take a thin wet muslin cloth and apply mud evenly on the muslin cloth to form a uniform thin layer of half to 1 in. in thickness. Fold all the sides to make it a compact pack (Chadzopulu et al. 2011).

For mudpack for eyes, mud soaked in water is spread to make a half inch thick layer. The pack is placed on the eyes for 20–30 min. An eye mud pack helps in relaxing the eyes; this is especially good for those who require to sit in front of a computer for long hours. Therapeutically, it reduces irritation, itching or other allergic conditions such as conjunctivitis and hemorrhage of the eye ball. It also helps in correcting refractive errors like short/long sightedness. It is effective in Glaucoma, where it works to reduce the eye ball tension (Patel et al. 2015). For mudpack for head, a mud pack is normally a thick narrow band. It is applied over the fore head and helps to heal congestive headache relieves pain immediately (Patel et al. 2015). For mudpack for face, fine mud is used and a smooth paste is first made using cold water. This paste is evenly applied on the face and left to dry for 30 min. After 30 min one must thoroughly wash the face with cold water. It helps in improving the complexion of the skin. In cases of acne it helps by absorbing excess oils and toxins from the skin. It also helps in reducing dark circles around the eyes

(Patel et al. 2015). Mudpack for abdomen should be applied for 20–30 min. The body and the mud pack should be covered with a blanket, if applied during the cold weather. An abdomen mud pack helps in all kinds of indigestion. It is very helpful in decreasing intestinal heat and stimulates peristalsis (Patel et al. 2015).

2.5.4.2 Mud Bath

Mud bath involves application of special kind of mud rich in natural salts and mineral over the entire body (except head). Mud baths are useful in many skin diseases such as Psoriasis, Urticaria, and Leucoderma (Riyaz and Arakkal 2011). First the mud is prepared by soaking it in water. The mud is then applied to the full body either in sitting or lying down position (Fig. 2.10).

Mud is kept for 45–60 min and ideally be exposed to sun light, at least intermittently. Remember that the head should always be covered when exposing the body to sunlight (Patel et al. 2015). Afterwards, the person should be thoroughly washed with cold to luke warm water. Dry the person quickly and transfer to a warm bed. A mud bath helps in increasing the blood circulation and energizing the skin tissues. It thus helps in cleansing and improving the skin condition generally. Regular mud baths may be considered as natural beauty treatment procedure as it also helps in improving skin complexion and reducing spots and patches, possibly the result of some skin disorder like chickenpox or small pox. Mud baths are useful in many skin diseases such as Psoriasis, Urticaria, leucoderma, Leprosy and other skin allergic conditions (Patel et al. 2015).

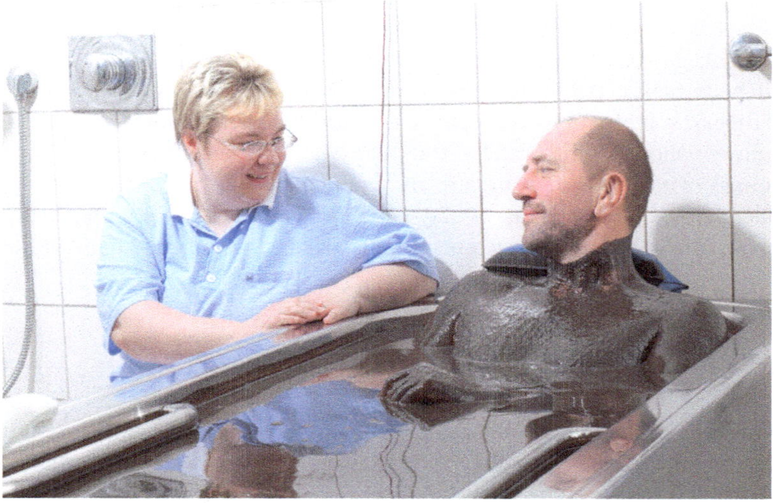

Fig. 2.10 Patient in a mud bath of the specialist clinic Teufelsbad in Blankenburg, North Germany (Pawel et al. 2013, p. 41; Photo: Fundus Teufelsbad, with kind permission from Bergverein zu Hüttenrode e. V.)

Despite therapeutic properties of mud which are very useful in maintaining good health and glowing skin, the over-use of mask applications may lead to excessive drying of the skin. Therefore it is essential to keep the pack moistened to avoid excessive dryness and stretching. Also, the mud must be carefully chosen for obtaining desired results, and its quality must be ensured to avoid any complications (Bellometti et al. 1997).

2.5.4.3 Therapeutic Benefits of Mudpack and Claypack

Mud pack is broadly made use of clay treatment due to its several wellbeing many advantages. The character remedy therapists are adopting mud therapy to body in wounds, boils, bruises and other disorders along with acceptable food regimen and therapies (Patel et al. 2015). It is much more useful than the cold packs, as it retains the coolness and moisture in the entire body for very long time. Other added benefits of mudpack are uptake of toxic compounds. It relaxes the skin pores, and draws the blood to the pores and skin, and relieves the congestion and soreness and agony and eradicates the morbid make any difference. It will increase the high temperature radiation. It is a good hair conditioner and is good for skin. It relaxes muscles and improves blood circulation. It maintains metabolism rendering positive impact on digestion. It is useful in conditions of inflammation/swelling and relieves pain. It is useful in condition of stiff joints (Flusser et al. 2002). The mud pack is organized upon acquiring clay from 10 cm below the area of earth. It has to not comprise dangerous substances or chemical substances. The clay is made as a sleek paste with warm drinking water. The length is 10–30 min (Patel et al. 2015). Clay pack or mud pack is extremely worthwhile therapy in nervous diseases, weak point, fever, scarlet fever, influenza, measles, swellings, ear and eye troubles, rheumatism, gout, liver and kidney problems, tooth ache, head ache, and typical overall body aches. It is also helpful in abdominal disorders as indigestion and bowel illnesses. It promotes labor pains. It is also used subsequent to fomentation to the figure for about 15 min (Patel et al. 2015). Mud pack reduces the swelling of any component of the physique. In case of significant fever, the temperature can be introduced down with the help of clay pack. It has to be modified repeatedly until the sought after decreasing of temperature is affected (Galzigna et al. 1998).

2.6 Soil Microorganisms Producing Drugs

Soils presumably harbor the most diverse populations of bacteria of any environment on this planet because of the extensive heterogeneity in soil properties and the large spatiotemporal variations in biotic and abiotic conditions (Raaijmakers and Mazolli 2012). Considerable efforts have been made in the past to assess the number and diversity of bacterial species in the soil. With the development of culture-independent methods, such as ribosomal RNA (rRNA) sequencing, it has become evident that only a minority (~1% to 5%) of soil bacteria can be cultured by conventional methods and that approximately one third of the bacterial divisions have so far no cultured representatives (Clardy et al. 2006). Estimates of the number of bacterial species per gram of soil range between 2000 and 8.3 million (Gans et al. 2005; Schloss and Handelsman 2006). Torsvik et al. (2002), using extrapolations of DNA:DNA hybridizations, estimated the number of bacterial species in a gram of boreal forest soil to be approximately 10,000. Reevaluation of this approach by Gans et al. (2005) led to the prediction of 8.3 million microbial species per gram of soil. By sequencing two clone libraries from different sites, subsequent work by Schloss and Handelsman (2006) led to estimates of 2000–5000 operational taxonomic units in a gram of soil. To proliferate and survive in soil microhabitats, bacteria have developed multiple strategies to ward off competitors and predators.

Fungi are another rich source of antibiotics (Peláez 2006). Among the potential tactics, antibiotics are assumed to act as "weapons in the numerous conflicts involved in microbial empire building" (Davies 1990). These anthropogenic views of antibiotics as weapons or shields originated, in part, from the Golden Age of antibiotic discovery and from the exploitation of antibiotics to combat bacterial infections of humans and animals. The microbial drug era began with Alexander Fleming (1929) when he discovered in a Petri dish seeded with *Staphylococcus aureus* that a compound produced by a mold killed the bacteria. The mold, identified as the fungus *Penicillium notatum*, produced an active agent that was named penicillin. Later, penicillin was isolated as a yellow powder and used as a potent antibacterial compound during World War II. By using Fleming's method, other naturally occurring substances, such as chloramphenicol and streptomycin, were isolated. Naturally occurring antibiotics are produced by fermentation, an old technique that can be traced back almost 8000 years, initially for beverages and food production. Beer is one of the world's oldest beverages, produced from barley by fermentation, possibly dating back to the sixth millennium BC and recorded in the written history of ancient Egypt and Mesopotamia. Another old fermentation, used to initiate the koji process, was that of rice by Aspergillus oryzae. During the past 4000 years, *Penicillium roqueforti* has been utilized for cheese production, and for the past 3000 years soy sauce in Asia and bread in Egypt has represented examples of traditional fermentations (Hölker et al. 2004).

To date only a small proportion of the antimicrobial compounds produced by soil bacteria has been studied (Raaijmakers and Mazolli 2012). In fact, there is still limited evidence that antibiotic compounds are actually produced by soil bacteria in situ and, if so, at concentrations that are sufficient to exert their presumed weapon/ shield functions (Monier et al. 2011). Hence, Davies (1990) postulated that the term antibiotic is more a pharmaceutical description and that antibiotics may have completely different natural roles. Recent studies have demonstrated that antibiotics, at subinhibitory concentrations, can indeed exert other effects on microorganisms, including changes in transcription, virulence, motility, and biofilm formation (Romero et al. 2011).

Table 2.4 Examples of antibiotics originating from microbial natural products

Original substance	Trade name (generic name)	Producing microbe
Cephalosporins	Mefoxin (cefoxitin); ceclor (cefaclor);	*Acremonium* spp.; *Emericellopsis* spp.;
	Claforan (cefotaxime);	*Amycolatopsis lactamdurans*;
	Rocephin (ceftriaxone);	*Streptomyces clavuligerus*
	Ceftin (cefuroxime)	
Daptomycin	Cubicin (daptomycin)	*Streptomyces roseosporus*
Erythromycin	Erythrocin;	*Saccharopolyspora erythraea*
	Zithromax (azithromycin);	
	Biaxin (clarithromycin);	
	Ketek (telithromycin)	
Fosfomycin	Monuril	*Streptomyces fradiae*
Mupirocin (pseudomonic acid)	Bactroban (mupirocin)	*Pseudomonas fluorescens*
Penicillins	Penicillin G, V; ampicillin; methicillin;	*Penicillium* spp.;
	Amoxicillin; carbenicillin	*Aspergillus* spp.
Streptogramins	Synercid (dalfopristin/ quinupristin)	*Streptomyces pristinaespiralis*
Thienamycin	Primaxin (imipenem);	*Streptomyces cattleya*
	Invanz (ertapenem)	
Vancomycin	VANCOCIN	*Streptomyces orientalis*

Commercial name is in capitals. Names in parentheses refer to marketed semisynthetic derivatives from the original natural compound
Adapted from Peláez (2006)

2.6.1 Antibiotics from Natural Products

Antibiotics encompass a chemically heterogeneous group of organic, low-molecular weight compounds produced by microorganisms that are deleterious to the growth or metabolic activities of other microorganisms (Raaijmakers and Mazolli 2012). Antibiotics generally act on several vital processes in microbes, including cell wall biosynthesis and DNA, RNA, and protein synthesis. Drugs of natural origin have been classified as (i) original natural products, (ii) products derived or chemically synthesized from natural products or (iii) synthetic products based on natural product structures. However, microbial natural products are the origin of most antibiotics present in the market (Peláez 2006). Out of about 90 antibiotics marketed in the years 1982–2002, 70 originated in natural products (Newman et al. 2003). Most of the remaining products belonged to the fluoroquinolones class, and it could even be argued that this category has its origin in nature as well, since they were based on nalidixic acid, a compound discovered during the attempts to synthesize the natural anti-malarial agent chloroquin (Peláez 2006). Many antibiotics discovered until the 1970s reached the market, and their chemical scaffolds were later used as leads to generate new generations of clinically useful antibiotics by chemical modification (Table 2.4).

During the 1990s the streptogramins dalfopristin/quinupristin (Synercid1), linezolid (Zyvox1), and the lipopeptide daptomycin (Cubicin1) were introduced as new antibiotic classes (Bush and Macielag 2000; Coates et al. 2002). While linezolid is a purely synthetic drug, the other two are natural products. Tigecylcine, a derivative of tetracycline, is also an old class of natural antibiotics (Nathwani 2005). Other older class antibiotics include β-lactams and macrolides, that represent significant advantages over the original products in spectrum, potency or pharmacokinetic properties (Coates et al. 2002). Compounds cited in the literature "as under development" belong to the classes of antibiotics mentioned above or to others equally old (Peláez 2006). A comprehensive overview of current human and veterinary antibiotics is given in Chap. 12.

2.6.2 The Need for New Antibiotics

During recent decades, the number of reports on the progressive development of bacterial resistance to almost all available antimicrobial agents has increased. In the 1970s, the major problem was the multidrug resistance of Gram-negative bacteria, but later in the 1980s the Gram-positive bacteria became important, including methicillin-resistant staphylococci, penicillin-resistant pneumococci and vancomycin-resistant enterococci (Moellering 1998). In a quantitative study of changes in antibiotic resistance from soil over time, Knapp et al. (2010) isolated DNA from agricultural soils collected from the Netherlands from 1940 to 2008 and quantified antibiotic resistance genes. The authors found that levels of all resistance genes investigated rose over time from the pre-antibiotic era (before 1940) to the present, which was especially dramatic for tetracycline and β-lactam resistance elements, likely reflecting changes in agricultural practice and demonstrating the enrichment of resistant organisms with the modern use of antibiotics. New methods have revolutionized the screening of natural products and offer a unique opportunity to re-establish natural products as major source of drug leads. Examples of recent advances in the application of these technologies that have immediate impact on the discovery of novel drugs are: (i) development of a streamlined screening process for natural products; (ii) improved natural product sourcing, and advances in (iii) organic synthetic methodologies, (iv) combinatorial biosynthesis, and (v) microbial genomics (Lam 2007).

Microbial natural products still remain the most promising source of novel antibiotics, although new approaches are required to improve the efficiency of the discovery process (Peláez 2006). The advent of resistant Gram-positive bacteria has been noticed by the pharmaceutical, biotechnology and academic communities and efforts have been made to find novel antimicrobial agents to meet this need. For example, a new glycopeptide antibiotic, teicoplanin, was developed against infections with resistant Gram-positive bacteria, especially bacteria resistant to the glycopeptide vancomycin (Demain and Sanchez 2009). In another instance, the approach involved the redesign of a mixture of two compounds, called streptogramin, into a new mixture, called pristinamycin, to allow administration of the drug parenterally and in higher doses than the earlier oral preparation (Bacque et al. 2005). Quinupristin and dalfopristin, the two components of streptogramin, were chemically modified to allow intravenous administration. The new combination, pristinamycin, was approved for use against infections caused by vancomycin-resistant *Enterococcus faecium*. Additional moves against resistant microorganisms are the glycylcyclines developed to treat tetracycline-resistant bacteria. These modified tetracyclines show potent activity against a broad spectrum of Gram-positive and Gram-negative bacteria, including strains that carry the two major tetracycline-resistance determinants, involving efflux and ribosomal protection (Demain and Sanchez 2009). Two glycylcyline derivatives (DMG-MINO and DMG-DMDOT) have been tested against a large number of clinical pathogens isolated from various sources. The spectrum of activity of these compounds includes organisms with resistance to antibiotics other than tetracyclines, for example, methicillin-resistant staphylococci, penicillin-resistant *S. pneumoniae* and vancomycin-resistant enterococci (Sum et al. 1998). Tigecycline was approved in 2005 as an injectable antibiotic (Sum 2006). The novel class of antimicrobial agents used in treating resistance to Gram-positive infections also includes the cyclic lipopeptide antibiotic daptomycin produced by *Streptomyces roseosporus*. This compound was approved in for skin infections resulting from complications following surgery, diabetic foot ulcers and burns (Demain and Sanchez 2009). It acts against most clinically relevant Gram-positive bacteria (*Staphylococcus aureus, Streptococcus pyogenes, Streptococcus agalactiae, Streptococcus dysgalactiae* ssp. equisimilis and *Enterococcus faecalis*), and retains in vitro potency against isolates resistant to methicillin, vancomycin and linezolid. Traditionally, these infections were treated with penicillin and cephalosporins, but resistance to these agents became widespread (Demain and Sanchez 2009). Daptomycin seems to have a favorable side effect profile, and it might be used to treat patients who cannot tolerate other antibiotics (Demain and Sanchez 2009). Telithromycin, a macrolide antibiotic, is the first orally active compound of a new family of antibacterials named the ketolides. It shows potent activity against pathogens implicated in communityacquired respiratory tract infections, irrespective of their β-lactam, macrolide or fluoroquinolone susceptibility. Some of the microorganisms susceptible to this antibiotic are pneumococci, *Haemophilus influenzae* and *Moraxella catarrhalis*, including β-lactamase-positive strains. In addition, telithromycin has a very low potential for selection of resistant isolates or induction of cross-resistance found with other macrolides (Leclercq 2001).

The use of antineoplastic and broad-spectrum antibiotics, prosthetic devices and grafts, and more aggressive surgery has increased invasive fungal infections.The rising incidence of invasive fungal infections and the emergence of broader fungal resistance have led to the need for novel antifungal agents. Amphotericin B is the first-line therapy for systemic infection because of its broad spectrum and fungicidal activity (Demain and Sanchez 2009). However, considerable side effects limit its clinical utility. Echinocandins are large lipopeptide molecules that inhibit the synthesis of 1,3-β-D-glucan, a key component of the fungal cell wall. The echinocandins caspofungin and micafungin have reached the market. Caspofungin is also known as pneumocandin. This compound was the first cell-wall-active antifungal approved as a new injectable antifungal (Hoang 2001). It irreversibly inhibits 1,3-β-D-glucan synthase, preventing the formation of glucan polymers and disrupting the integrity of fungal cell walls. Caspofungin is more active and less toxic than amphotericin B and shows a broad spectrum of activity against *Candida* (including fluconozole resistance), *Aspergillus*, *Histoplasma* and *P. carinii*, the major cause of HIV death. Micafungin is licensed for clinical use in Asian countries and in the USA. This compound is extremely potent against clinically important fungi, including Aspergillus and azole-resistant strains of Candida. Although several new antifungal drugs have been developed, some patients remain resistant to treatments. The main reasons for this include intrinsic or acquired antifungal resistance, organ dysfunction preventing the use of some agents and drug interactions (Demain and Sanchez 2009). Posaconazole is a new member of the triazole class of antifungals. It has shown clinical efficacy in the treatment of oropharyngeal candidiasis and has potential as a salvage therapy for invasive aspergillosis, zygomycosis, cryptococcal meningitis and a variety of other fungal infections (Demain and Sanchez 2009). It is available as an oral suspension and has a favorable toxicity profile.

In addition to the screening programs for antibacterial and antifungal activity, the pharmaceutical industry has extended these programs to other disease areas. Microorganisms are a potent source of structurally diverse bioactive metabolites and have yielded some of the most important products of the pharmaceutical industry. Microbial secondary metabolites are now also being used for applications other than antibacterial and antifungal infections.

2.6.3 Other Drugs from Natural Products

2.6.3.1 Anticancer Drugs

Natural products have also had a major impact on cancer chemotherapy. Anti-tumor antibiotics are amongst the most important of the cancer chemotherapeutic agents that started to appear around 1940 with the discovery of actinomycin. Since then many compounds with anti-cancer properties have been isolated from natural sources. More than 60% of the compounds with antineoplasic activity were originally isolated as natural products or are their derivatives. Among the approved products deserving special attention are actinomycin D, anthracyclines (daunorubicin, doxorubicin, epirubicin, pirirubicin and valrubicin, bleomycin), mitosanes (mitomycin C), anthracenones (mithramycin, streptozotocin and pentostatin), enediynes (calcheamycin), taxol and epothilones (reviewed by Demain and Sanchez 2009). Actinomycin D is the oldest microbial metabolite used in cancer therapy. Its relative, actinomycin A, was the first antibiotic isolated from actinomycetes. The latter was obtained from *Actinomyces antibioticus* (now *Streptomyces antibioticus*) by Waksman and Woodruff (1941). It prevents elongation by RNA polymerase as it binds DNA at the transcription initiation complex. This property confers some human toxicity and it has been used primarily as an investigative tool in the development of molecular biology. Despite the toxicity, it has served well against Wilms tumor in children. Anthracyclines are used to treat a wide range of cancers, including leukemias, lymphomas, and breast, uterine, ovarian and lung cancers (Minotti et al. 2004). They act by intercalating DNA strands, which result in a complex formation that inhibits the synthesis of DNA and RNA.

Mitosanes are composed of several mitomycins that are formed during the cultivation of *Streptomyces caespitosus*. Although the mitosanes are excellent antitumor agents, they have limited utility owing to their toxicity. Mitomycin C shows activity against several types of cancer (lung, breast, bladder, anal, colorectal, head and neck), including melanomas and gastric or pancreatic neoplasms. Mithramycin (plicamycin) is an antitumor aromatic polyketide produced by *Streptomyces argillaceous* that shows antibacterial and antitumor activity (Fernández et al. 1998). It is used in the treatment of testicular cancer, disseminated neoplasms and hypercalcemia. With repeated use, organotoxicity (kidney, liver and hematopoietic system) can become a problem. Streptozotocin is a microbial metabolite produced by *Streptomyces achromogenes*. Chemically, it is a glucosamine-nitroso-urea compound. As with other alkylating agents in the nitroso-urea class, it is toxic to cells by causing damage to DNA, although other mechanisms may also contribute. Pentostatin (deoxycoformycin) is an anticancer chemotherapeutic drug produced by *Streptomyces antibioticus*. It interferes with the cell's ability to process DNA (Showalter et al. 1992) and is commonly used to treat leukemia. However, it can cause kidney, liver, lung and neurological toxicity. Calicheamycins are highly potent antitumor microbial metabolites of the enediyne family produced by *Micromonospora echinospora*. Their antitumor activity is apparently due to the cleavage of double-stranded DNA (Walker et al. 1992). They are highly toxic, but it was possible to introduce one such compound into the clinic by attaching it to an antibody that delivered it to certain cancer types selectively.

Taxol (paclitaxel) is a successful non-actinomycete molecule, which was first isolated from the Pacific yew tree, *Taxus brevifolia*, but is also produced by the endophytic fungi *Taxomyces andreanae* and *Nodulisporium sylviforme* (Zhao et al. 2004). This compound inhibits rapidly dividing mammalian cancer cells by promoting tubulin polymerization and interfering with normal microtubule breakdown during cell division. The drug also inhibits several fungi (*Pythium*, *Phytophthora* and *Aphanomyces*) by the same mechanism. Taxol was first approved for refractory ovarian cancer, and today it is used against breast and advanced forms of Kaposi's sarcoma (Newman and Cragg 2007). Epothilones are macrolides originally isolated from the broth of the soil myxobacterium *Sorangium cellulosum* as weak agents against rust fungi (Gerth et al. 1996). They were identified as microtubulestabilizing drugs, acting in a similar manner to taxol. However, they are generally 5–25 times more potent than taxol in inhibiting cell growth in cultures (Demain and Sanchez 2009).

In men between the ages of 15 and 35 years, testicular cancer is the most common type of cancer (Jemal et al. 2007). The majority (95%) of testicular neoplasms are germ cell tumors, which are relatively uncommon carcinomas, accounting for only 1% of all male malignancies. Significant progress has been made in the medical treatment of advanced testicular cancer, with a substantial increase in cure rates from approximately 5% in the early 1970s to almost 90% today (Einhorn 2002). A major advance in chemotherapy for testicular germ cell tumors was the introduction of cisplatin in the mid-1970s. Two chemotherapy regimens are effective for patients with a good testicular germ cell tumor prognosis: three cycles of bleomycin, etoposide and cisplatin or four cycles of etoposide and cisplatin (Culine et al. 2007). Of the latter three agents, bleomycin and etoposide are natural products.

2.6.3.2 Enzyme Inhibitors

Enzyme inhibitors have received increasing importance for potential utilization in medicine. Enzyme inhibitors with various industrial uses have been isolated from microbes (reviewed by Demain and Sanchez 2009). The most important are clavulanic acid, the inhibitor of β-lactamases and the statins, which are hypocholesterolemic drugs. Some of the common targets for other inhibitors are glucosidases, amylases, lipases, proteases and xanthine oxidase. Acarbose is a pseudotetrasaccharide made by *Actinoplanes* sp. SE50. It contains an aminocyclitol moiety, valienamine, which inhibits intestinal α-glucosidase and sucrase. This results in a decrease in starch breakdown in the intestine, which is useful in combating diabetes in humans (Truscheit et al. 1981).

Amylase inhibitors are useful for the control of carbohydrate-dependent diseases, such as diabetes, obesity and hyperlipemia (Boivin et al. 1988). Amylase inhibitors are also known as starch blockers because they contain substances that prevent dietary starches from being absorbed by the body. The inhibitors may thus be useful for weight loss, as some amylase inhibitors show potential for reducing carbohydrate absorption in humans (Díaz et al. 2004). The use of amylase inhibitors for the treatment of rumen acidosis has also been reported. Examples of microbial α-amylase inhibitors are paim, obtained from culture filtrates of Streptomyces corchorushii (Arai et al. 1985), and TAI-A, TAI-B, oligosaccharide compounds from *Streptomyces calvus* TM-521 (Demain and Sanchez 2009). Lipstatin is a pancreatic lipase inhibitor produced by *Streptomyces toxytricini* that is used to combat obesity and diabetes. It interferes with the gastrointestinal absorption of fat (Weibel et al. 1987). In the pathogenic processes of emphysema, arthritis, pancreatitis, cancer and AIDS, protease inhibitors are potentially powerful tools for inactivating target proteases. Examples of microbial products include antipain, produced by *Streptomyces yokosukaensis*, leupeptin from *Streptomyces roseochromogenes* and chymostatin from *Streptomyces hygroscopicus*. Leupeptin is produced by more than 17 species of actinomycetes (Demain and Sanchez 2009). Fungal products are also used as enzyme inhibitors against cancer, diabetes, poisonings, Alzheimer's disease, etc. The enzymes inhibited include glycosidases, acetylcholinesterase, protein kinase, tyrosine kinase, and others (Paterson 2008).

2.6.3.3 Immunosuppressants

Suppressor cells are critical in the regulation of the normal immune response. An individual's immune system is capable of distinguishing between native and foreign antigens and of mounting a response only against foreign antigens. A major role has been established for suppressor T lymphocytes. Suppressor cells also play a role in regulating the magnitude and duration of the specific antibody response to an antigenic challenge. Suppression of the immune response either by drugs or by radiation, to prevent the rejection of grafts or transplants or to control autoimmune diseases, is called immunosuppression. A number of microbial compounds capable of suppressing the immune response have been discovered (Demain and Sanchez 2009). Cyclosporin A was originally introduced as a narrow-spectrum antifungal peptide produced by the mold *Tolypocladium nivenum* (originally classified as *Trichoderma polysporum*), by aerobic fermentation. Cyclosporins are a family of neutral, highly lipophilic, cyclic undecapeptides containing some unusual amino acids, synthesized by a non-ribosomal peptide synthetase, cyclosporin synthetase. Discovery of the immunosuppressive activity led to its use in heart, liver and kidney transplants and to the overwhelming success of the organ transplant field (Borel 2002). Cyclosporin is thought to bind to the cytosolic protein cyclophilin (immunophilin) of immunocompetent lymphocytes, especially T lymphocytes. This complex of cyclosporin and cyclophilin inhibits calcineurin, which under normal circumstances is responsible for activating the transcription of interleukin-2. It also inhibits lymphokine production and interleukin release and therefore leads to a reduced function of effector T cells.

Other important transplant agents include sirolimus (rapamycin) and tacrolimus (FK506), which are produced by actinomycetes. The macrolide rapamycin, produced by *Streptomyces hygroscopicus,* is especially useful in kidney transplants as it lacks the nephrotoxicity seen with cyclosporin A and tacrolimus. This compound binds to the immunophilin FK506-binding protein (FKBP12), and this binary complex interacts with the rapamycin-binding domain and inactivates a serine-threonine kinase termed the mammalian target of rapamycin. The latter is known to control proteins that regulate mRNA translation initiation and G1 progression (Aggarwala et al. 2006). The antiproliferative effect of rapamycin has also been used in conjunction with coronary stents to prevent restenosis, which usually occurs after the treatment of coronary artery disease by balloon angioplasty. Rapamycin also shows promise in treating tuberous sclerosis complex, a congenital disorder that leaves sufferers prone to benign tumor growth in the brain, heart, kidneys, skin and other organs. As rapamycin has poor aqueous solubility, some of its analogs, RAD001 (everolimus), CCI-799 (tensirolimus) and AP23573 (ARIAD), have been developed with improved pharmaceutical properties (Demain and Sanchez 2009).

Tacrolimus (FK 506), produced by *Streptomyces tsukubaensis,* is a macrolide immunosuppressant which possesses similar but more potent immunosuppressant properties compared with cyclosporin, inhibiting cell-mediated and humoral immune responses (Peters et al. 1993). Like cyclosporin, tacrolimus demonstrates considerable interindividual variation in its pharmacokinetic profile. This has caused difficulty in defining the optimum dosage regimen and has highlighted the usefulness of therapeutic drug monitoring. Most clinical studies with tacrolimus have neither been published in their entirety nor subjected to extensive peer review; there is also a paucity of published randomised investigations of tacrolimus versus cyclosporin, particularly in renal transplantation. Despite these drawbacks, tacrolimus has shown notable efficacy as a rescue or primary immunosuppressant therapy when combined with corticosteroids in adult and paediatric recipients following liver or kidney transplantation (Peters et al. 1993).

2.6.3.4 Hypocholesterolemic Drugs

Atherosclerosis is generally viewed as a chronic, progressive disease characterized by the continuous accumulation of atheromatous plaque within the arterial wall. Statins form a class of hypolipidemic drugs used to lower cholesterol by inhibiting the enzyme HMG-CoA reductase, the rate-limiting enzyme of the mevalonate pathway of cholesterol biosynthesis. Inhibition of this enzyme in the liver stimulates low-density lipoprotein (LDL) receptors, resulting in an increased clearance of LDL from the bloodstream and a decrease in blood cholesterol levels. Through their cholesterol-lowering effect, they reduce the risk of cardiovascular disease, prevent stroke and reduce the development of peripheral vascular disease (Nicholls et al. 2007). Moreover, they are anti-thrombotic and antiinflammatory.

The first statin (compactin; mevastatin) was isolated as an antibiotic product of *Penicillium brevicompactum* and later from *Penicillium citrinum.* Pravastatin, another statin, is made through different biotransformation processes from compactin by *Streptomyces carbophilus* (Serizawa and Matsuoka 1991) and *Actinomadura sp.* (Peng and Demain 1998). Other genera involved in the production of statins are *Doratomyces*, *Eupenicillium*, *Gymnoascus*, *Hypomyces*, *Paecilomyces*, *Phoma*, *Pleurotus* and *Trichoderma* (Alarcon et al. 2003).

2.7 Geophagia

The practice of eating soil (geophagia or geophagy) which is a special type of pica referring to the deliberate consumption of soil/clay was first documented by Hippocrates (460–380 BC) more than 2000 years ago (Wilson 2003). Geophagia cuts across socio-economic, ethnic, religious and racial divides and has been observed in hundreds of cultures on all inhabited continents (Young et al. 2007, 2011). Geophagic tendencies have been postulated to arise from cultural, medicinal, physiological and nutritional factors (Callahan 2003). Young children are particularly vulnerable to the habit of soil-eating (Bisi-Johnson et al. 2010). Children under the age of 18–20 months normally explore and acquaint themselves with the environment by mouthing everything they come across (Fessler and Abrams 2004). Issues around geophagia are conflicting. While some beneficial roles of geophagia have been described, others have found it detrimental. Beneficial aspects include the use of kaolin to treat diarrhea, gastritis, colitis, enhancement of bioactivities and maintenance of normal intestinal flora by commensal flora found in soil. Clay or soil containing special constituents are valuable oral and topical antimicrobials as well as adsorbents of toxins. Soils and clay deposits have also been reported to contain toxic elements such as lead, chromium, arsenic and aluminium that can lead to metal toxicity even in low levels (Ekosse and Jumbam 2010). Since geophagic materials have been reported to contain both essential and toxic minerals, studies on the materials may provide clues on the relationship between soil's mineralogical properties and health effect on geophagic individuals. Determining mineral composition may help identifying the motivation behind geophagic practice (Wilson 2003).

The practice of geophagy still occurs in many parts of the world and has been reported to be practiced in Africa, America including USA, Asia including India and China, Australia and Europe (Boyle and Mackay 1999; Reilly and Henry 2000; Woywodt and Kiss 2002). Previously, geophagy was believed to be common among communities of low social status; however, recent studies have indicated that pregnant females in affluent societies also indulge in geophagic practice (Ngole and Ekosse 2012). Although the practice is common among pregnant women, women of all ages, irrespective of their educational level and social status engage in the habit (Geissler et al. 1999; Prince et al. 1999; Woywodt and Kiss 2002; Callahan 2003).

In sub-Saharan Africa, geophagia is more prevalent among pregnant women, lactating mothers and children under the age of 5 especially those living in rural areas where culture and indigenous practices remain prevalent and well-established (Kutalek et al. 2010; Ngole and Ekosse 2012). In African regions where geophagy occurs, between 46% and 73% of pregnant or breast-feeding women consume soil regularly but the amounts consumed differ considerably, with values ranging mostly from 1 to 100 g day^{-1} (Luoba et al. 2004, 2005). Clay eating is most widespread among women in Malawi, Zambia, Zimbabwe, Swaziland and South Africa, where an estimated prevalence level in the rural areas is put at 90% (Walker et al. 1997). In Malawi, it was reported that it is surprising for a pregnant woman not to practice geophagia since it is their way of identifying if she is really pregnant. The taste of clay is claimed to diminish the nausea, discomfort and vomiting in "morning sickness" during pregnancy. Clay eating in this case is seen as normal during pregnancy (Hunter 1993). Studies have also established that geophagia is common in Tanzania and Kenya (Young et al. 2007). A study by Luoba et al. (2004) in western Kenya showed that among 827 pregnant women during and after pregnancy 65% practiced eartheating before pregnancy. The prevalence remained high during pregnancy, and then declined to 34.5% and

29.6% at 3 and 6 months post-partum, respectively. Geophagia has also been reported in Guinea and Nigeria (Bisi-Johnson et al. 2010; Glickman et al. 1999). In South Africa, the prevalence of geophagia among urban and rural black South African women was reported to be 38.3% and 44.0% respectively as compared to the prevalence among the Indian, coloured and white women put at 2.2%, 4.4% and 1.6%, respectively (Walker et al. 1997). Poverty, starvation and famine have also been associated with geophagia (Hawass et al. 1987), in which case the substances function as a bulking agent to supplement insufficient food or a poor diet (Mcloughlin 1987).

Soil materials consumed by geophagic individuals include red, white, yellow and brown clay types, termite mounds and various other types of soil. In Australia, some aborigines eat white clay found mostly in the billabounds of the coastal areas of the North territory, fresh water springs and riverbeds mainly for medicinal purposes (Beteson and Lebroy 1978). The preferred type of earth eaten by Kenyan women according to Luoba et al. (2004) was soft stone, known locally as odowa and earth from termite mounds.

2.7.1 Nutritional Aspects of Geophagia

Whilst it is recognised that geophagia is a multi-causal behaviour, a prevalent explanation of the practice has been that the deliberate consumption of soil is a response to a mineral nutrient deficiency. Living organisms require certain naturally occurring elements (essential nutrients) for their metabolism to function efficiently. For humans there are a nuber of elements that are believed to be essential. Humans have obtained the essential nutrients by consuming plants that accumulated these elements from the soil, by eating meat from animals that accumulated the elements from plants or from animals that had obtained the elements from plant eating organisms lower in the food chain, by drinking water in which the nutrients are suspended or dissolved or by directly ingesting soil (Finkelman 2006).

Since different regions of the earth's crust vary in levels of minerals because of differences in their physical, mineralogical and chemical properties, there are positive and negative health effects of ingesting soils. Such effects vary depending on the mineralogy and chemical composition of the ingested soils which are largely influenced by the soils pedogenetic development, the quantity of soil ingested and the rate of consumption (Ngole and Ekosse 2012). Many sites are available from which soil materials can be obtained including the walls of mud houses, termite moulds, quarry mines and hills in rural areas (Luoba et al. 2004; Young et al. 2011). Several studies have shown that the percentage of nutrients supplied by clay samples to geophagic individuals vary with the source and the variation is as a result of physico-chemistry, mineralogy and geochemistry of the ingested soils which is mainly influenced by the soils pedogenetic development and the quantity of soil/clay ingested (Ngole and Ekosse 2012).

For humans, Ca, Mg, Mn, Zn and Fe are the most prominent minerals in context with geophagia. Recommended dietary intake levels are used to identify average requirements needed by individuals to adequately meet nutritional needs (Table 2.5).

In pregnant women, extra Ca is required for skeletal and bone development of the fetus. In lactating mothers, significant demineralization of maternal bone occurs to meet the increased calcium demand of the neonate. Calcium-rich clays were reported to have diverse benefits among pregnant and lactating women (Wiley and Katz 1998; Lakudzula and Khonje 2011). Wiley and Katz (1998) reported that clay may serve as an important Ca supplement as it reduces risk of skeletal problems in the fetus. Aufreiter et al. (1997) found that geophagic soils from China had high content of more than 14 mg g^{-1} Ca mineral and concluded that such soil could be a significant source of Ca. Ngole and Ekosse (2012) found levels of Ca to range between 0.1 and 18.1 mg g^{-1} in geophagic soils from Eastern Cape-South Africa. However, when assuming a 50% bioaccessibility rate, they deduced that geophagic individuals consuming 5–30 g of soil per day would contribute to <1% of RDI. Ca present in these soils would thus negligibly contribute to the nutrient demands of the geophagic individuals irrespective of their age. Contribution of 0.2% of recommended daily minimum intake of dietary Ca based on an average ingestion of 50 g of Blantyre and Indian 'edible' soils in Malawi has been reported by Lakudzula and Khonje (2011). Similarly, Gichumbi et al. (2012) reported that samples obtained from open-air markets in Kiambu exhibited Ca contents between 2.2 and 5.4 mg g^{-1} and concluded that the samples were not a significant source of Ca.

Ngole and Ekosse (2012) analyzed geophagic soils from Eastern Cape-South Africa for the Mg content. They found the levels of Mg to be in the range from 2.1 to 9.1 mg g^{-1}. Assuming a 50% bioaccessibility rate, they reported that geophagic individuals consuming 5–30 g of soil per day would be supplied with less than 1% of RDI and concluded that the bioaccessible fraction of Mg in the studied geophagic samples makes little contribution to the nutrient demand of a geophagic adult. Lakudzula and Khonje (2011) reported similar observations after analyzing Blantyre and Indian 'edible' soils from Malawi. Based on an average ingestion of 50 g soil, they

Table 2.5 Recommended daily intake (RDI) for Ca, Mg, Mn, Zn and Fe

		RDI in mg day^{-1}				
Life stage group	Range in years	Ca	Mg	Mn	Zn	Fe
Infants	0–6 months	200	30	0.003	2	0.27
	6–12 months	260	75	0.6	3	11
Children	1–3 years	700	80	1.2	3	7
	4–8 years	1000	130	1.5	5	10
Males	9 to >70 years	1000–1300	240–420	1.9–2.3	8–11	8–11
Females	9 to >70 years	1000–1300	240–320	1.6–1.8	8–9	8–18
Pregnancy	14–50 years	1000–1300	350–400	2.0	11–12	27
Lactation	14–50 years	1000–1300	310–360	2.6	12–13	9–10

Adapted from Food and Nutrition Board (2011)

found out that such soils contributed only 0.006% of recommended daily minimum intake of dietary Mg and concluded that ingesting such soils could not sufficiently contribute to dietary Mg needed by geophagic individuals. Gichumbi et al. (2012) reported samples obtained from Kiambu County could not be a significant source of the mineral since the Mg range was 1.7–3.2 mg g^{-1}. Data from Mg elemental analyses (0.9 mg g^{-1}) of Rwandan soil samples eaten by gorillas led Mahaney et al. (2000) to the conclusion that mineral supplementation is not an explanation for geophagy.

Ekosse and Anyangwe (2012) found that geophagic materials obtained from Botswana may have the potential to provide Mn needed by humans. In contrast, Odewumi (2013) reported Mn to be 0.2–0.4 mg g^{-1} in geophagic materials obtained from Share area, Nigeria and concluded that these materials could not sufficiently contribute to dietary Mn necessary for humans. Ngole and Ekosse (2012) reported Mn to be less than 1.0% present in geophagic materials obtained from Eastern Cape, South Africa and based on nutrient biaccessibility calculations concluded that the soils could negligibly contribute to the nutrient demands of the geophagic individual.

Okunlola and Owoyemi (2011) concluded that the geophagic materials' content for zinc whose range was 43.17–200.13 mg kg^{-1} may be suitable to act as a source of zinc to geophagic individuals. Different findings have been reported by Ngole and Ekosse (2012). They found geophagic materials to contain Zn in the range 23.0–127.0 mg kg^{-1}. Based on calculations of nutrient bioaccessibility they concluded that the minerals in geophagic materials could not sufficiently contribute to the nutrient demands of geophagic individuals. Hooda et al. (2004) reported the levels of Zn present in geophagic materials to be in the range 24–88 mg kg^{-1}. Although the Zn present was in appreciable amounts, they concluded that the ingestion of these relatively Zn-rich soils Zn dissolution and adsorption in the gastrointestinal tract may be strongly limited.

Various studies have shown that soils in large areas of the tropics contain beneficial amounts of Fe. However, informations on the bioaccessibility fraction of Fe are contradictory for soils from different regions. For example, analyses of geophagic samples obtained from Embu, Meru and Chuka town of Eastern Province, Kenya, yielded levels of Fe in a range between 37.0 and 68.9 mg g^{-1} (Mwangi and Ombaka 2010) and it was concluded that geophagic soil obtained from these regions could be an important source of Fe supplement. A study of geophagic materials from Eastern Cape Province in South Africa yielded levels of Fe to be 9.0–55.2 mg g^{-1} (Ngole and Ekosse 2012). Based on calculations of nutrient bioaccessibility, the authors concluded that Fe in the soils could not sufficiently contribute to the nutrient demands of the geophagic individuals. Similarly, Ngole et al. (2010) reported Fe to be in the range 5.9–145.7 mg g^{-1} in Eastern Cape and Swasiland soils and concluded that geophagic materials could negligibly supply the nutrient demands of geophagic individuals.

In summary, on the basis of the above results, it cannot be generally assumed that geophagia may help supplement mineral nutrients. Moreover, Hooda et al. (2004) on the basis of an in vitro soil ingestion simulation test concluded that soil ingestion through geophagia can indeed reduce the absorption of already bioavailable nutrients, particularly micronutrients such as Fe, Cu and Zn. It was found that a large fraction of mineral nutrients will not be soluble in environments like the stomach (pH 2) and their solubility is expected to further decrease as the pH becomes alkaline when ingested soil enters the intestine (Hooda et al. 2002). This has been clearly demonstrated by the use of 'physiologically based extractions' which showed that only a small fraction of the total mineral nutrients/metals in the soils was potentially available for absorption in the intestine. These findings, while disagreeing with the commonly held view of geophagia as a source of nutrient supplementation, were consistent with micronutrient deficiency problems observed in clinical nutrition studies conducted amongst geophagic populations. The work also showed, however, that in some cases, the ingested soils may become a source of Ca, Mg and Mn, although it is not clear why other similar soils should not release any of these mineral nutrients (Hooda et al. 2004).

2.7.2 Microbiological Benefits of Geophagia

Soil materials rich in clay have been reported to adsorb intestinal unwanted substances (Bisi-Johnson et al. 2010). Microbial agents such as *Yersinia enterocolitica*, *Escherichia coli*, *Streptococcus faecalis*, *Helicobacter pylori*, and mycobacteria have been postulated to play a role in the aetiology of Crohn's disease which is characterized by a severe, non-specific, chronic inflammation of the intestinal wall (Liu et al. 1995; Lamps et al. 2003). Alterations of the flora with probiotics and antibiotic strategies have putative beneficial effects in humans. Several studies have emphasized the role of probiotic bacteria to favorably alter the intestinal microflora balance, inhibit the growth of harmful bacteria, promote digestion, boost immune function, and increase resistance to infection (reviewed by Bisi-Johnson et al. 2010). A food supplement made up of homeostatic soil organisms including *Lactobacillus acidophilus, Lactobacillus bulgaricus, Lactobacillus delbreukii, Lactobacillus caseii, Lactobacillus plantarum, Lactobacillus brevis, Lactobacillus lactis, Bacillus licheniformus, Bacillus subtilis, Bifido bifidus, Sacchromyces* helps to maintain a healthy balance of the intestinal flora which is done by producing organic compounds such as lactic acid, hydrogen peroxide and acetic acid, thus increasing the acidity of the intestine and inhibiting the reproduction of massive amounts of harmful microorganisms (Bisi-Johnson et al. 2010). There is therefore evidence that supports the usefulness of the commensal flora found in soil as vital in the establishment of healthy bacteria within the digestive tract, addressing the problems presented by Crohn's disease and leaky gut syndrome.

Ingested clay in pregnant women may improve digestive efficiency and also reduce fetal exposure to toxins tolerated by the mother (Bisi-Johnson et al. 2010). Aflatoxin is a known toxin produced by fungi. There have been reports of the amelioration of the toxic effect of aflatoxin (Rosa et al. 2001), afflatoxin and fumonisin (Miazzo et al. 2005) in animal feeds by the addition of sodium bentonite to the broiler chick diets. Some specific beneficial pharmaceutical usage of edible soil includes that of white clay (kaolin) which is used as a diarrhea remedy (Bisi-Johnson et al. 2010). Several studies have reported that minerals from clay could provide inexpensive, highly-effective antimicrobials to fight numerous human bacteria infections including MRSA (Methicillin-resistant *Staphylococcus aureus*) (Haydel et al. 2008; Williams et al. 2008) and buruli ulcer, an infection caused by *Mycobacterium ulcerans* (Bisi-Johnson et al. 2010). The bacterium produces a potent immunosuppressant toxin that causes necrotic lesions and destroys the fatty tissues under the skin. Specific clay minerals may prove valuable in the treatment of Buruli ulcer and other bacterial diseases, for which there are no effective antibiotics (Haydel et al. 2008). Clay minerals, in particular montmorillonite, have also been reported to provide protection to DNA against degradation (Cai et al. 2006).

2.7.3 Hazards Resulting from Soil Pathogens

Soil as a reservoir for many microbes has been implicated as one of the sources and vehicles by which many human pathogens are transmitted. For example, *Mycobacterium avium* subspecies *paratuberculosis* which is present in many dairy herds could be transmitted to humans via foodstuff and via direct contact with contaminated soil (Bisi-Johnson et al. 2010). Soil is a reservoir for *Pseudomonas aeruginosa* and the bacterium has the capacity to colonize plants during favorable conditions of temperature and moisture (Green et al. 1974). Soils also harbour pathogenic microorganism such as *Sporotrichum schenckii*, *Histoplasma capsulatum*, *Cryptococcus neoformans* and medically important actionmycetes (Bisi-Johnson et al. 2010). A recent molecular characterization of a novel clinical isolate of *Francisella* spp. (the causative pathogen of tularemia) showed genetic relatedness to species of the pathogen cloned from soil sample (Kugeler and Mead 2008). Therefore, a lot of pathogenic microorganism which are resident in soil can be detrimental to health if such soil is consumed. The spore-bearing *Clostridium perfringens* and *C. tetani* are often encountered in the surface layers of soil (and in human and animal excreta). They are particularly abundant in cultivated and manured fields, especially in the tropics (Abrahams 2002). These bacilli belonging to the family of bacteria which produce one of the deadliest toxins often cause infections by contamination of wounds exposed to the spore-bearing soil. *C. tetani* is the causative organism of tetanus while *C. perfringens* is the etiology of gas gangrene. Occupations may be a predisposing factor to tetanus, as in the case of soldiers and farmers hence, more tetanus can be expected in rural areas where manual labour is intense. However, people living in rural communities who probably ingest regular quantities of *C. tetani,* develop natural immunity through mouth and gut absorption and thus get protected from tetanus (Sanders 1996). Anemia resulting from geophagia, to which the craving for soil is attributed, is believed in some cases to have actually resulted from microbial or worm infection encountered by ingestion of soil. The global burden of geohelminth infections is of great enormity. Worldwide, there are approximately 3.5 billion infections with parasitic geohelminths: *Ascaris, Trichuris*, and hookworm (Chan 1997). Eggs of parasitic worms (geohelminths) as well can be consumed with ingestion of soil. Ascariasis (characterized by abdominal pain and nausea with disturbed functioning of the alimentary tract) and trichiuriasis are caused by the ingestion of *Ascaris lumbricoides* and *Trichuris trichiura* eggs, respectively (Abrahams 2002). The role of plasmid DNA in genetic material transfer, particularly the transfer of antibiotic resistance genes among species of microorganism, humans and animals cannot be overemphasized. This means that the use and consumption of soils or clays which contain antibiotic resistance plasmid may significantly contribute to the problem of emerging antibiotic resistance. Detailed information on the role of soil as a transmitter of human pathogens is given in Chap. 13.

Also of important microbiological consideration is the association and survival of prions in soil (Bisi-Johnson et al. 2010). Prions are incriminated in transmissible spongiform encephalopathies (TSEs) in humans and animals, primarily affecting the central nervous system (Prusiner 1998). Prion protein released into soil environments may be preserved in a bioavailable form, exposing humans to the infectious agent. The act of processing clay-rich soil for consumption by heat treatment may probably render the soil safe of pathogens. However, the effectiveness of the processing remains a question to be unraveled. This will depend on the type of microorganism present as sporeformers tend to be resistant to heat treatment. In an experiment conducted by Pedersen et al. (2000) where by bentonite was exposed to low (20–30 °C) and high (50–70 °C) temperatures, the spore-forming *Desulfotomaculum nigrificans* and *B. subtilis* were the only surviving bacteria. However, too much processing may reduce the clay's therapeutic potential. Detailed information on the role of soil as an environmental reservoir of prion diseases is given in Chap. 14.

2.7.4 Hazards Resulting from Toxic Elements

Geophagic materials can contain toxic elements such as lead, mercury and cadmium which are ubiquitous and persistent environmental pollutants. However, in terms of toxicity of elements via soil ingestion, most concern to date has been concentrated on lead (Pb). Lead poisoning is caused by increased levels of the heavy metal Pb in the body which is toxic to many organs and tissues including the heart, bones, intestines, kidneys, and reproductive and nervous systems (for more details see Chap. 8). High Pb levels interfere with the production of blood cells and subsequently, the absorption of calcium which is required for healthy and strong bones. If circulating levels of the hormones in the body are within normal range, Pb alters the bone cell function by interfering with bone cells' ability to respond to hormonal stimuli (Patrick 2006). The World Health Organization (1996) limits soil lead levels to 1 mg kg^{-1}.

The potential hazard of Pb-contaminated soil to geophagists has been emphasised not only in children, but also in adults. However, children are of most concern, with Pb as a neurotoxin being especially harmful to the developing brains and nervous systems of young people. Levels of Pb in geophagic materials in Kenya have been reported to be within ranges of 0.14–0.96 mg kg^{-1} (Mwangi and Ombaka 2010) and 0.68–1.54 mg kg^{-1} (Gichumbi et al. 2012). These levels partly reach or even exceed the WHO limit (WHO 1996) and thus may have deleterious effects on the consumer. Ekosse and Jumbam (2010) concluded that the high levels of Pb could pose a health risk particularly as a result of long-term continued use of such contaminated soils and that their consumption should be discouraged.

Pb poisoning is still an important health issue in the USA that has been described as the silent epidemic. However, there has been a significant decline in blood Pb concentrations during the 1990s, attributed to removal of Pb from petrol, as well as reducing Pb in the food canning (Abrahams 2002). Urban soils in large US cities, especially those found in the central districts where the highway networks have concentrated traffic, are a giant reservoir of Pb (and other elements such as Cd and Zn) because of pollutants such as leaded petrol and paint (Mielke 1999). Many children in such areas face a significant risk of Pb poisoning from the deliberate or otherwise ingestion of soil from their local yards, school playgrounds (Higgs et al. 1999) and, to a lesser extent, in the open spaces around their homes. For this reason, it is important that soil ingestion be considered in any risk assessments involving not only Pb, but also other potentially harmful elements and organic contaminants (Abrahams 2002).

Soil ingestion can be detected by dental inspection which can reveal excessive tooth wear (Abbey and Lombard 1973), and by radiological examination of the abdomen that will reveal opaque masses of soil in the colon (Mengel and Carter 1964). Constipation, the reduction of the power of absorption of food materials by the body, severe abdominal pain, and obstruction and perforation of the colon may result following the internal accumulation of soil (Amerson and Jones 1967; Bateson and Lebroy 1978). In pregnant women, this can lead to dysfunctional labour and maternal death (Horner et al. 1991).

2.8 Podoconiosis

Podoconiosis has been documented in tropical highland areas of Africa, northwest India and Central America. Areas of high prevalence are found in Uganda, Tanzania, Kenya, Rwanda, Burundi, Sudan, Ethiopia, Equatorial Guinea, Cameroon, the Central American highlands of Mexico and Guatemala, Ecuador, Brazil, northwest India, Sri Lanka and Indonesia (reviewed by Davey et al. 2007). Men and women are equally affected in most communities. In the past, podoconiosis was also common in North Africa (Algeria, Tunisia, Morocco and the Canary Islands) and

Europe (France, Ireland and Scotland), but since the use of footwear has become standard is no longer found in these areas (Price 1990). Price (1976) and Price and Bailey (1984) superimposed maps of disease occurrence onto geological surveys that showed evidence of a link with red clays derived from basaltic rocks rich in alkali metals such as sodium and potassium. Increased prevalence has been noted at altitudes of 1500 m above sea level or higher, and in areas that receive an annual rainfall ≥1500 mm with an average annual temperature between 19 and 21 °C (Dwek et al. 2015). These conditions contribute to the steady disintegration of basaltic lava and the reconstitution of the mineral components into clays (Price and Henderson 1978). Development of podoconiosis is closely related to living and working barefoot on these soils. Population-based surveys suggest a prevalence of 5–10% in barefoot populations living on irritant soil. Farmers are at high risk, but the risk extends to any occupation with prolonged contact with the soil, and the condition has been noted among goldmine workers and weavers who sit at a ground level loom.

2.8.1 Clinical Effects

Most evidence suggests an important role for mineral particles on a background of genetic susceptibility (Davey et al. 2007). Colloid-sized particles dominated by elements common in irritant clays (aluminium, silicon, magnesium, iron) are absorbed through the foot and have been demonstrated in the lower limb lymph node macrophages of those living barefoot on the clays (Price and Henderson 1978). Electron microscopy shows local macrophage phagosomes to contain particles of stacked kaolinite ($Al_2Si_2O_5(OH)_4$), while light microscopy shows subendothelial oedema and subsequent collagenization of afferent lymphatics reducing and finally obliterating the lumen (Price 1977).

Podoconiosis is characterized by a prodromal phase before elephantiasis sets in. Early symptoms commonly include itching of the skin of the forefoot and a burning sensation in the foot and lower leg. Early changes that may be observed are splaying of the forefoot, plantar oedema with lymph ooze, increased skin markings, hyperkeratosis with the formation of moss-like papillomata, and 'block' (rigid) toes. Later, the swelling may be one of two types: hard and fibrotic ('leathery' type) and soft and fluid ('water-bag' type) (Price 1990). Figure 2.11 shows podoconiosis at different stages.

Very long-standing disease is associated with fusion of the interdigital spaces and ankylosis of the interphalangeal or ankle joints. Acute episodes occur in which the patients is pyrexial and the limb warm and painful. These episodes appear to be related to progression to the hard, fibrotic leg. A serious complication of podoconiosis is acute adenolymphangitis, which represents a warm, painful sensation in the limbs, accompanied by fever (Dwek et al. 2015).

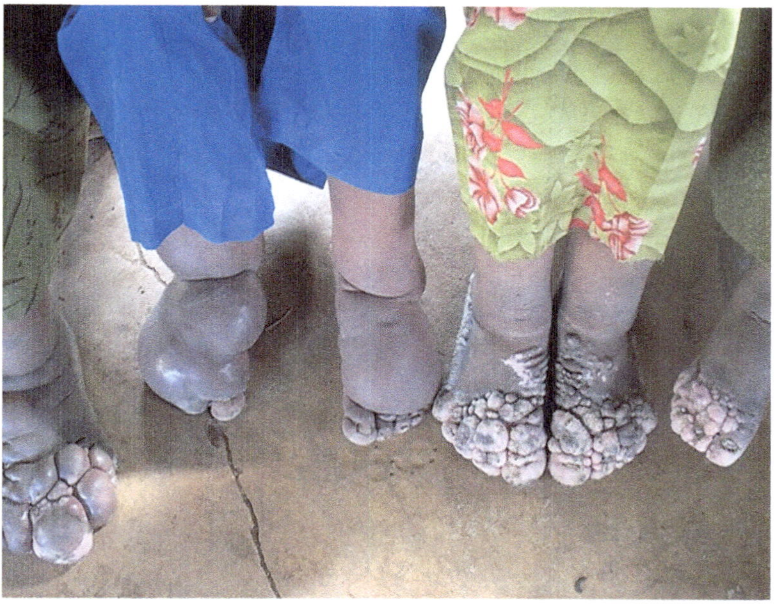

Fig. 2.11 Podoconiosis at different stages in four approximately 50 year old women from Ethiopia (Deribe et al. 2013, Fig. S1; unrestricted reproduction permitted by PLOS ONE)

2.8.2 Prevention and Therapy

Podoconiosis is a preventable disease that is treatable in the early stages if shoes are worn consistently and if feet are washed regularly (Dwek et al. 2015). Therefore, podoconiosis has a high potential for elimination. Footwear, however, remains unaffordable for numerous residents of most affected areas in the tropics. Prevention of the progression of early symptoms and signs to overt elephantiasis includes thorough foot hygiene (washing with water and soap, using antiseptics and emollients), and the use of socks and shoes. Compression bandaging is effective in reducing the size of the soft type of swelling. Progression can be completely averted if these measures are strictly adhered to, but compliance must be life-long. Relocation from an area of irritant soil or adoption of a non-agricultural occupation are also effective, but may not be feasible for many patients (Davey et al. 2007). Management of patients with advanced elephantiasis additionally encompasses elevation and compression of the affected leg. In selected cases, removal of prominent nodules are recommended. For elevation to be successful, at least 18 h with the legs at or above the level of the heart are needed each day. Where electricity is available, intermittent compression machines may reduce swelling rapidly (Price 1975). The aim of elevation and compression is to reduce the swelling of the leg so that tall compression footwear can be worn, or nodulectomy undertaken. Patients unable to strictly avoid contact with soil experience recurrent swelling which is more painful than the original disease because of scarring. Social rehabilitation is vital, and includes training treated patients in skills that enable them to generate income without contact with irritant soil (Davey et al. 2007).

2.9 Biocompatibility of (Nano-)particles Released from Dental Materials

2.9.1 Inhaled Nanoparticles

Dental personnel often come into contact with composite dust upon polishing or grinding composites. Contemporary composites typically contain high amounts of (silica) nano-filler that are functionalized with a silane coupling agent to attach them to the resin matrix (Albers 2002). The size of the filler particles varies widely, but contemporary composites may contain large quantities of submicron and even nanoparticles. Typically, 40-nm-sized pyrogenic silica ('microfill') is very often used (Chen 2010). The most recent class of composites is coined 'nano-composites', as they contain larger amounts of amorphous nano-particles (<100 nm) than traditional 'hybrid composites'. Beside pyrogenic silica, they also contain filler produced in a solgel process, which are typically composed of mixed oxides, like ZrO_2–SiO_2, and which can be added to a greater extent without overly increasing the viscosity of the unpolymerized composite. Like in other fields of industry involving nano-technology (Elsaesser and Howard 2012), some dental researchers and practitioners have expressed concerns that there may be health risks connected to the use of these nano-particles (Jandt and Sigusch 2009). Dental personnel may inhale aerosolized composite dust while shaping and polishing new composite restorations, and upon removal of old composite restorations (Leggat et al. 2007). Especially in case of inadequate aspiration and/or lack of assistance, dental professionals (and patients) may be exposed to composite dust. In general, dust smaller than 5 μm and larger than 0.01 μm may be respired and can penetrate deep into the alveolar region of the lungs, beyond the body's natural mechanisms of cilia and mucous clearing (Klaassen 2013). Chronic inhalation of respirable dust (<5 μm) and nano-particles may provoke both local and systemic toxicity (Klaassen 2013; Kunzmann et al. 2011). Excessive and long-term exposure to respirable dusts can induce pneumoconiosis (Choudat et al. 1993), a chronic lung condition marked by nodular fibrosis. Nano-particles (<100 nm) may also be absorbed in the blood or the lymphatic system resulting in systemic toxicity (Oberdorster et al. 2005; Delfino et al. 2005; Napierska et al. 2010). It was shown that respirable crystalline silica could be released upon abrading traditional quartz-containing composites (which are now no longer commonly used) (Collard et al. 1991). Composites released respirable dust (<5 μm) in vitro (Van Landuyt et al. 2012). These observations were corroborated by the clinical measurements; however only short episodes of high concentrations of respirable dust upon polishing composites could be observed (Van Landuyt et al. 2012). Electron microscopic analysis showed that the size of the dust varied widely with particles larger than 10 μm, but submicron and even nano-sized particles could also be observed (Van Landuyt et al. 2012). The dust particles often consisted of multiple filler particles contained in resin, but single nano-filler particles could also frequently be distinguished. Therefore inhalation of composite dust is better avoided. Furthermore, it is recommended to always use water-cooling upon polishing or removing composites, to use good aspiration, to frequently ventilate the dental operatory and to wear masks with high particle-filtration efficiency for small particle sizes (Van Landuyt et al. 2012).

2.9.2 *Risk Assessment for the Uptake of Nanoparticles Abraded from Dental Materials*

During the chewing situation only low amounts of nanoparticles (<100 nm) were abraded in vitro from nano(hybride-)composites (Reichl et al. 2014). Following calculation can be given for the worst case situation in patients: In a clinical study a composite abrasion was found of 0.25 mm within 3 years (Lewis 1991). It is assumed that all abraded particles are nanoparticles. It is further assumed that 32 teeth are filled with composites, each with an area of 3 mm^2 (Lewis 1991). Then a total area can be calculated of about 100 mm^2. This area multiplicated by 0.25 mm results in a total wear of about 25 mm^3 within 3 years or 50 μg day^{-1}. Assuming a composite density of 2 g cm^{-3}, a total wear can be calculated of 50 mg within 3 years or 50 μg day^{-1}. The normal daily uptake of nanoparticles in human beings (western hemisphere) is about 400 μg day^{-1} (Terzano et al. 2010). Therefore in the normal physiological situation the uptake of nanoparticles abraded from composites and the health risk in patients are only of minor relevance.

2.9.3 *Toxicity of Titan (Nano-)particles In Vitro*

Titanium (Ti) and its alloys have been used as source materials for biomedical applications especially in dentistry, since previous studies showed that these materials stand out for good mechanical properties, excellent corrosion resistance and high biocompatibility (Okabe and Hero 1995; Lautenschlager and Monaghan 1993; Watanabe et al. 2003; Özcan and Hämmerle 2012). For example, it was found that Ti is one of the most biocompatible metallic material because of its ability to form a stable and insoluble protective oxide layer (TiO_2) on its surface (Castilho et al. 2006; Elias et al. 2008). Ti is preferentially used for endosseous dental implant material (Castilho et al. 2006). Besides Ti, some Titanium alloys are also used for dental applications, such as Nickel Titanium (NiTi). The alloy NiTi is used for castings of crowns and denture construction, orthodontic archwires and brackets (Thompson 2000; Elias et al. 2008; Setcos et al. 2006). It has been found that the properties of Ti implants can be improved by using nanostructured Ti consisting of Ti-nanoparticles (Ti-NPs) (Valiev et al. 2008). Even though Ti based implants are considered to be biocompatible, their induced side effects such as hypersensitivity and allergic reactions have been reported (Lalor et al. 1991; Sicilia et al. 2008; Egusa et al. 2008). It has also been found that Ti based materials can cause immuno-inflammatory reactions (Voggenreiter et al. 2003). These side effects might have been caused by the interaction between tissues and implants (Hansen 2008; Yang and Merritt 1994). In vitro and in vivo studies showed that Ti ions can be released from Ti based implants, for example by corrosion, wear and electrochemical processes (Okazaki and Gotoh 2005; Jacobs et al. 1999; Browne and Gregson 2000; Bianco et al. 1997). The release of Ni ions from NiTi alloy also has been reported. Previous studies indicated that Ti ions and Ni ions induced cytotoxicity and DNA damage in human cells (Soto-Alvaredo et al. 2014; Faccioni et al. 2003). Furthermore, Ti-particles/debris (3–250 μm) was found in the peri-implant animal tissues after application of Ti based implants (Martini et al. 2003; Franchi et al. 2004). Clinical studies also showed that Ti particles in nanometer- and micrometer-size could be released into human tissues/organs of the patients with Ti based implants or replacements (Soto-Alvaredo et al. 2014; Faccioni et al. 2003; Martini et al. 2003; Franchi et al. 2004; Urban et al. 1996, 2000). The toxic effect of Ti-particles has been described: phagocytosis of Ti-particles could induce cytotoxicity in rat calvarial osteoblasts and MG63 cells (Pioletti et al. 1999; Lohmann et al. 2000). Genotoxic effect of Ti-particles has also been detected, which induced apoptosis in mesenchym stem cells (Wang et al. 2002, 2003). It was found that the particles size can influence the toxicity of metal particles (Karlsson et al. 2009; Hackenberg et al. 2011). The ability of different particles entering cells may also affect the toxicity (Karlsson et al. 2009; Hackenberg et al. 2011; Oh et al. 2010), and it is reported that particle size can impact the cellular uptake efficiency and pathway (Lee et al. 1993; Zhu et al. 2013; Jiang et al. 2008).

A recent in vitro study (He et al. 2015a) demonstrated cytotoxicity and DNA damage of size dependent Ti particles in periodontal ligament (PDL)-hTERT cells: Ti microparticles (Ti-MPs, <44 μm), NiTi microparticles (NiTi-MPs, <44 μm), and Ti nanoparticles (Ti-NPs, <100 nm). The EC_{50} values (representative for cytotoxicity) of investigated particles were: 2.8 mg/ml (Ti-NPs), 41.8 mg/ml (NiTi-MPs) and >999 mg/ml (Ti-MPs). Genotoxicity were described for Ti particles in the following range (Ti-NPs highest toxicity): Ti-MPs $<$ NiTi-MPs $<$ Ti-NPs (He et al. 2015a). The highest cellular uptake efficiency was observed with Ti-NPs, followed by Ti-MPs and NiTi-MPs. Only Ti-NPs were found in the nucleus. Compared to Ti-MPs and NiTi-MPs, Ti-NPs induced higher cellular uptake efficiency and higher toxic potential in PDL-hTERT cells (He et al. 2015a).

2.9.4 *Titan (Nano-)particles in Human Jawbones with Ti Implants*

Although Ti based implants have been considered to be biological inert, it has been found that implants in the body can undergo corrosion and release particulate debris over time (Meachim and Williams 1973; Galante et al. 1991). Metallic debris from Ti based implants might exist in several forms including particles (micrometre to nanometre size), colloidal and ionic forms (e.g. specific/unspecific protein binding) (Hallab et al. 2000, 2001), organic storage forms (e.g. hemosiderin, as an iron-storage complex), inorganic metal oxides and salts (Hallab et al. 2001). According to a previous study (Jacobs et al. 1998), the degradation of implants in the human body is primarily induced by wear and corrosion: wear is the mechanical/physical form of implant degradation which produces (nano)particles; while corrosion refers to the chemical/electrochemical form of degradation that mainly produces soluble metal ions (Jacobs et al. 1998). Ti particles released from implants have been found in the regenerated bone and peri-implant tissues in animals (Martini et al. 2003; Franchi et al. 2004). It has been shown that also Ti ions can be released from embedded implants in animals (Edwin 1962; Ducheyne et al. 1984). Ti particles/ions are able to enter the circulation of blood and lymph (Hackenberg et al. 2011; Hallab et al. 2001; Brien et al. 1992). Ti particles detached from hip, knee and mandible implants have been detected in organs such as liver, spleen, lung, and lymph nodes (Hackenberg et al. 2011; Schliephake et al. 1993). Increased levels of elementary Ti have also been detected in the blood of patients with poorly functioning implants (Hallab et al. 2001). Increased concentrations of metals (e.g. Ti, Cr, Co and Al) derived from implants in body fluids might induce acute or chronic toxicological effects (Hallab et al. 2001). The long-term effects of Ti derived from implants are still not fully understood, but associated hypersensitivity and allergic reactions in patients have been reported (Lalor et al. 1991; Sicilia et al. 2008). In a clinical study, 0.6% of 1500 patients were found to exhibit Ti allergic reactions (He et al. 2015b). Additionally, it has been found that detached metal debris from implants might cause marrow fibrosis, necrosis, and granulomatosis (Case et al. 1994; Amstutz et al. 1992; Dannenmaier et al. 1985).

In a recent study (He et al. 2015b) the release of Ti and other metallic elements was measured from dental jawbone implants through detailed post-mortem studies of human subjects. The highest Ti content detected in human mandibular bone was 37,700 μg kg^{-1} bone weight at a distance of 556–1587 μm from implants, and the intensity increased with decreasing distance from implants. Particles with sizes of 0.5–40 μm were found in human jawbone marrow tissues at distances of 60–700 μm from implants (He et al. 2015b).

2.9.5 Histological Analyses

Ti particles could not only be observed in peri-implant tissues but also in newly formed bone (Martini et al. 2003; Franchi et al. 2004). It was suggested that those Ti particles might be detached during the insertion of the implant (Jacobs et al. 1998; Flatebo et al. 2011; Franchi et al. 2004) or released after insertion (Franchi et al. 2004). Previous studies reported that particles <1 μm could be taken up by non-phagocytic eukaryotic cells via endocytosis (Rejman et al. 2004) and particles with diameters exceeding 0.75 μm can be taken up by macrophages, neutrophils and monocytes via phagocytosis or through macropinocytosis (>1 μm) by all cell types (Conner and Schmid 2003). Therefore, transportation of Ti particles with size of 0.5–5 μm that were observed in human bone marrow tissues described in present study into cells is possible (He et al. 2015b). In this study, bone marrow fibrosis, avital bone tissues and multinucleated cells were observed near the implant (He et al. 2015b). Similarly, a previous post-mortem study about metal particles released from implants showed bone marrow fibrosis in lymph-node tissue (Case et al. 1994). Furthermore, multinucleated cells have been observed after hydroxy-apatite implantation in rat periodontal tissues (Kawaguchi et al. 1992). Bone marrow fibrosis might be induced by marrow injury and inflammation (Travlos 2006). These effects are reported to be associated with surgical trauma during insertion of implants (Piattelli et al. 1998). Multinucleated giant cells can be elicited by wear particles in periprosthetic tissues in human specimens and these cells might contain phagocytized wear particles (Anazawa et al. 2004). For the worst case situation for patients with Ti-implants following calculation can be given: The highest Ti-content detected in human mandibular bone was 37,700 μg kg^{-1} bone (He et al. 2015b). It is assumed that all Ti in the bone is Nano-Ti and 1 kg bone is 1 L fluid, a Nano-Ti-concentration can be calculated of 37 μg ml^{-1}. The EC_{50} value for Nano-Ti in human cells is 2800 μg ml^{-1} (He et al. 2015a). Therefore it is suggested that Ti dental implants might have no adverse clinical effects. However in the human jawbone with high titan content some histological changes can be observed, compared to control jawbone (He et al. 2015b).

2.9.6 Silver Nanoparticles

Bacterial biofilms are responsible for dental diseases, such as caries and periodontitis (Whittaker et al. 1996). Due to the high frequency of recurrent caries after restorative treatment, much attention has been paid to the therapeutic effects revealed by direct filling materials (Gama-Teixeira et al. 2007; Mjor 1996; Wilson et al. 1997). Resin composites containing silver ion-implanted fillers that release silver ions have been found to have antibacterial effects on oral bacteria e.g. streptococcus mutans (Yamamoto et al. 1996). Most studies available on the antimicrobial effect from silver containing composite describe the effect of the silver particles to different species of cariogenic bacteria or these studies deal with modified material properties belonging to the addition of silver particles. The less the degree of conversion of monomers the higher are the amounts of elutable residual monomers from cured composite (Miletic and Santini 2008). The amount of elutable substances has an important part in the biocompatibility and toxicity of the material, because from some monomers or compounds eluted from composites it is known that they can cause allergic reactions or may be metabolized to reactive oxygen species (Goon et al. 2006; Durner et al. 2009). From the (co)monomers bisphenol-A-glycidyldimethacrylate (BisGMA), 2-hydroxyethyl methacrylate (HEMA) and triethyleneglycol dimethacrylate (TEGDMA) it is known that they can cause DNA strand breakage (Durner et al. 2010; Urcan et al. 2010). From camphorquinone (CQ) it is known that it can cause oxidative stress (induction of ROS) and DNA damage (Volk et al. 2009). The identification and quantification of eluatable substances is important for toxicological risk assessments. It could be demonstrated that silver nanoparticles in dental composites can influence the chemical polymerization process in dental composites measured by the increase of elutable residual substances (e.g. TEGDMA; BisGMA, CQ) from the light hardened specimen (Durner et al. 2011).

References

Abbey LM, Lombard JA (1973) The etiological factors and clinical implications of pica: report of a case. J Am Dent Assoc 87:885–887

Abrahams PW (2002) Soils: their implications to human health. Sci Total Environ 291:1–32

Adriano DC (2001) Trace elements in terrestrial environments: biochemistry, bioavailability and risk of metals. Springer, Berlin

Aggarwala D, Fernandez ML, Solimanb GA (2006) Rapamycin, an mTOR inhibitor, disrupts triglyceride metabolism in guinea pigs. Metabolism 55:794–802

Aguzzi C, Cerezo P, Viseras C, Caramella V (2007) Use of clays as drug delivery systems: possibilities and limitations. Appl Clay Sci 36:22–36

Alarcon J, Aguila S, Arancibia-Avila P, Fuentes O, Zamorano-Ponce E, Hernandez M (2003) Production and purification of statins from *Pleurotus ostreatus* (Basidiomycetes) strains. Z Naturforsch 58:62–64

Albers HF (2002) Tooth-colored restoratives. Principles and techniques, 9th edn. BC Decker, Hamilton

Amerson JR, Jones HQ (1967) Prolonged kaolin (clay) ingestion: a cause of colon perforation and peritonitis. Bull Emory Univ Clin 5:11–15

Amstutz HC, Campbell P, Kossovsky N, Clarke IC (1992) Mechanism and clinical significance of wear debris-induced osteolysis. Clin Orthop Relat Res 276:7–18

Anazawa U, Hanaoka H, Morioka H, Morii T, Toyama Y (2004) Ultrastructural cytochemical and ultrastructural morphological differences between human multinucleated giant cells elicited by wear particles from hip prostheses and artificial ligaments at the knee. Ultrastruct Pathol 28:353–359

Andriesse JP (1988) Nature and management of tropical peat soils. FAO Soils Bull 59:165

Arai M, Oouchi N, Murao S (1985) Inhibitory properties of an α-amylase inhibitor, paim, from Streptomyces corchorushii. Agric Biol Chem 49:987–991

Aufreiter S, Hancock RGV, Mahaney WC, Stambolic-Robb A, Sanmugadas K (1997) Geochemistry and mineralogy of soils eaten by humans. Int J Food Sci Nutr 48:293–305

Bacque E, Barriere JC, Berthand N (2005) Recent progress in the field of antibacterial pristinamycins. Curr Med Chem Anti-Infect Agents 4:185–217

Bateson EM, Lebroy T (1978) Clay eating by aboriginals of the Northern Territory. Med J Aust Spec Suppl Aborig Health 10:1–3

Bellometti S, Giannini S, Sartori L, Crepaldi G (1997) Cytokine levels in osteoarthrosis patients undergoing mud bath therapy. Int J Clin Pharmacol Res 17:149–153

Bergaya F, Lagaly G (2013) General introduction: clays, clay minerals, and clay science. In: Bergaya F, Lagaly G (eds) Handbook of clay science part A: fundamentals. Elsevier, Oxford, pp 1–19

Beteson EM, Lebroy T (1978) Clay eating by the aboriginals of the northern territory. Med J Aust 1:51–53

Bianco PD, Ducheyne P, Cuckler JM (1997) Systemic titanium levels in rabbits with a titanium implant in the absence of wear. J Mater Sci Mater Med 8:525–529

Bisi-Johnson MA, Obi CL, Ekosse GE (2010) Microbiological and health related perspectives of geophagia: an overview. Afr J Biotechnol 9(19):5784–5791

Boivin M, Flourie B, Rizza RA, Go VL, DiMagno EP (1988) Gastrointestinal and metabolic effects of amylase inhibition in diabetics. Gastroenterology 94:387–394

Borel JF (2002) History of the discovery of cyclosporin and of its early pharmacological development. Wien Klin Wochenschr 114:433–437

Boyle JS, Mackay MC (1999) Pica: sorting it out. J Transcult Nurs 10:65–68

Bozkurt S, Lucisano M, Moreno L, Neretnieks I (2001) Peat as a potential analogue for the long-term evolution in landfills. Earth Sci Rev 53:95–147

Brien WW, Salvati EA, Betts F, Bullough P, Wright T, Rimnac C, Buly R, Garvin K (1992) Metal levels in cemented total hip arthroplasty. A comparison of well-fixed and loose implants. Clin Orthop Relat Res 276:66–74

Brigatti MF, Galan E, Theng BKG (2013) Structure and mineralogy of clay minerals. In: Bergaya F, Lagaly G (eds) Handbook of clay science part A: fundamentals. Elsevier, Oxford, pp 21–81

Browne M, Gregson PJ (2000) Effect of mechanical surface pretreatment on metal ion release. Biomaterials 21:385–392

Bush K, Macielag M (2000) New approaches in the treatment of bacterial infections. Curr Opin Chem Biol 4:433–439

Cai P, Huang QY, Zhang XW (2006) Interactions of DNA with clay minerals and soil colloidal particles and protection against degradation by DNase. Environ Sci Technol 40(9):2971–2976

Callahan GN (2003) Eating dirt. Emerg Infect Dis J 9:1016–1021

Carretero MI (2002) Clay minerals and their beneficial effects upon human health. A review. Appl Clay Sci 21:155–163

Carretero MI, Pozo M (2010) Clay and non-clay minerals in the pharmaceutical and cosmetic industries Part II. Active ingredients. Appl Clay Sci 47:171–181

Carretero MI, Gomes C, Tateo F (2006) Clays and human health. In: Bergaya F, Theng BKG, Lagaly G (eds) Handbook of clay science. Elsevier, Amsterdam, pp 717–741

Case CP, Langkamer VG, James C, Palmer MR, Kemp AJ, Heap PF, Solomon L (1994) Widespread dissemination of metal debris from implants. J Bone Joint Surg (Br) 76:701–712

Casiday R, Frey R (2010) Iron use and storage in the body: ferritin and molecular representations. Department of Chemistry, Washington University, St. Louis

Castilho GA, Martins MD, Macedo WA (2006) Surface characterization of titanium based dental implants. Braz J Phys 36:1004–1008

Chadzopulu A, Adraniotis J, Theodosopoulou E (2011) The therapeutic effects of mud. Prog Health Sci 1(2):132–136

Chan MS (1997) The global burden of intestinal nematode infections – fifty years on. Parasitol Today 13:438–443

Chen MH (2010) Update on dental nanocomposites. J Dent Res 89:549–560

Choudat D, Triem S, Weill B, Vicrey C, Ameille J, Brochard P, Letourneux M, Rossignol C (1993) Respiratory symptoms lung function pneumoconiosis among self employed dental technicians. Br J Ind Med 50:443–449
Choy J, Choi S, Oh J, Park T (2007) Clay minerals and layered double hydroxides for novel biological applications. Appl Clay Sci 36(1–3):122–132
Christidis GE, Scott PW, Dunham AC (1997) Acid activation and bleaching capacity of bentonites from the islands of Milos and Chios, Aegean, Greece. Appl Clay Sci 12:329–347
Clardy J, Fischbach MA, Walsh CT (2006) New antibiotics from bacterial natural products. Nat Biotechnol 24:1541–1550
Coates A, Hu Y, Bax R, Page C (2002) The future challenges facing the development of new antimicrobial drugs. Nat Rev Drug Discov 1:895–910
Collard SM, Vogel JJ, Ladd GD (1991) Respirability, microstructure and filler content of composite dusts. Am J Dent 4:143–151
Conner SD, Schmid SL (2003) Regulated portals of entry into the cell. Nature 422:37–44
Culine S, Kerbrat P, Kramar A, Théodore C, Chevreau C, Geoffrois L, Bui NB, Pény J, Caty A, Delva R, Biron P, Fizazi K, Bouzy J, Droz JP (2007) Refining the optimal chemotherapy regimen for good-risk metastatic nonseminomatous germ-cell tumors: a randomized trial of the Genito-Urinary Group of the French Federation of Cancer Centers (GETUG T93BP). Ann Oncol 18:917–924
Dannenmaier WC, Haynes DW, Nelson CL (1985) Granulomatous reaction and cystic bony destruction associated with high wear rate in a total knee prosthesis. Clin Orthop Relat Res 198:224–230
Davey G, Fasil Tekola F, Newport MJ (2007) Podoconiosis: non-infectious geochemical elephantiasis. Trans R Soc Trop Med Hyg 101:1175–1180
Davies J (1990) What are antibiotics? Archaic functions for modern activities. Mol Microbiol 4:1227–1232
Delfino RJ, Sioutas C, Malik S (2005) Potential role of ultrafine particles in associations between airborne particle mass and cardiovascular health. Environ Health Perspect 113:934–946
Demain AL, Sanchez S (2009) Microbial drug discovery: 80 years of progress. J Antibiot 62:5–16
Deribe K, Brooker SJ, Pullan RL, Hailu A, Enquselassi F, Reithinger R, Newport M, Davey G (2013) Spatial distribution of podoconiosis in relation to environmental factors in Ethiopia: a historical review. PLoS One 8(7):e68330. https://doi.org/10.1371/journal.phone.0068330
Díaz E, Aguirre C, Gotteland M (2004) Effect of an α-amylase inhibitor on body weight reduction in obese women. Rev Chil Nutr 31:306–317
Ducheyne P, Willems G, Martens M, Helsen J (1984) In vivo metal-ion release from porous titanium-fiber material. J Biomed Mater Res 18:293–308
Durner J, Kreppel H, Zaspel J, Schweikl H, Hickel R, Reichl FX (2009) The toxicokinetics and distribution of 2-hydroxyethyl methacrylate in mice. Biomaterials 30:2066–2071
Durner J, Debiak M, Burkle A, Hickel R, Reichl FX (2010) Induction of DNA strand breaks by dental composite components compared to X-ray exposure in human gingival fibroblasts. Arch Toxicol 85(2):143–148
Durner J, Stojanovic M, Urcan E, Hickel R, Reichl FX (2011) Influence of silver nano-particles on monomer elution from light-cured composites. Dent Mat 27:631–636
Dwek P, Kong LY, Wafer M, Cherniak W, Pace R, Malhamé I, Simonsky D, Anguyo G, Stern E, Silverman M (2015) Case report and literature review: podoconiosis in Southwestern Uganda. Int J Trop Dis Health 9(3):1–7
Edwin SH (1962) Characteristics of trace ions released from embedded metal implants in the rabbit. J Bone Joint Surg 44(2):323–336
Egusa H, Ko N, Shimazu T, Yatani H (2008) Suspected association of an allergic reaction with titanium dental implants: a clinical report. J Prosthet Dent 100:344–347
Einhorn LH (2002) Curing metastatic testicular cancer. Proc Natl Acad Sci U S A 99:4592–4595
Ekmekcioglu C (2002) Effect of sulfur baths on antioxidative defense systems, peroxide concentrations and lipid levels in patients with degenerative osteoarthritis. Res Complement Class Nat Med 9:216–220

Ekosse GIE, Anyangwe S (2012) Mineralogical and particulate characterization of geophagic clayey soils from Botswana. Bull Chem Soc Ethiop 26:373–382

Ekosse GE, Jumbam DN (2010) Geophagic clays: their mineralogy, chemistry and possible human health effects. Afr J Biotechnol 9:6755–6767

Elias C, Lima J, Valiev R, Meyers M (2008) Biomedical applications of titanium and its alloys. JOM 60:46–49

Elsaesser A, Howard CV (2012) Toxicology of nanoparticles. Adv Drug Deliv Rev 64:129–137

Evcik D, Kavuncu V, Yeter A, Yigit I (2007) The efficacy of balneotherapy and mud-pack therapy in patients with knee osteoarthritis. Joint Bone Spine 74(1):60–65

Faccioni F, Franceschetti P, Cerpelloni M, Fracasso ME (2003) In vivo study on metal release from fixed orthodontic appliances and DNA damage in oral mucosa cells. Am J Orthod Dentofac Orthop 124:687–693

Fergusson JE (1990) The heavy metals: chemistry, environmental impact and health effects. Pergamon Press, Oxford. 614 pp

Fernández E, Weißbach U, Sánchez Reillo C, Braña AF, Méndez C, Rohr J, Salas JA (1998) Identification of two genes from Streptomyces argillaceus encoding glycosyltransferases involved in transfer of a disaccharide during biosynthesis of the antitumor drug mithramycin. J Bacteriol 180:4929–4937

Fessler DMT, Abrams ET (2004) Infant mouthing behavior: the immunocalibration hypothesis. Med Hypotheses 63:925–932

Finkelman RB (2006) Health benefits of geologic materials and geologic processes. Int J Environ Res Public Health 3(4):338–342

Flatebo RS, Hol PJ, Leknes KN, Kosler J, Lie SA, Gjerdet NR (2011) Mapping of titanium particles in peri-implant oral mucosa by laser ablation inductively coupled plasma mass spectrometry and high-resolution optical darkfield microscopy. J Oral Pathol Med 40:412–420

Fleming A (1929) On the antibacterial action of cultures of Penicillium, with special reference to their use in the isolation of B. influenzae. Br J Exp Pathol 10:226–236

Flusser D, Abu-Shakra M, Friger M, Codish S, Sukenik S (2002) Therapy with mud compresses for knee osteoarthritis: comparison of natural mud preparation with mineral-depleted mud. J Clin Rheumatol 8:197–203

Food and Nutrition Board (2011) Institute of Medicine, National Academies. http://www.nationalacademies.org/hmd/~/media/Files/Activity%20Files/Nutrition/DRI-Tables/2_%20RDA%20and%20AI%20Values_Vitamin%20and%20Elements.pdf?la=en. Accessed 4 Jan 2017

Fox C, Ramsoomair D, Carter C (2004) Magnesium: its proven and potential clinical significance. South Med J 94:12

Fraga CG, Oteiza PI, Keen CL (2013) Trace elements and human health. Mol Asp Med 26 (4):233–234

Franchi M, Bacchelli B, Martini D, Pasquale VD, Orsini E, Ottani V, Fini M, Giavaresi G, Giardino R, Ruggeri A (2004) Early detachment of titanium particles from various different surfaces of endosseous dental implants. Biomaterials 25:2239–2246

Galante JO, Lemons J, Spector M, Wilson PD Jr, Wright TM (1991) The biologic effects of implant materials. J Orthop Res 9:760–775

Galzigna L, Ceschi-Berrini C, Moschin E, Tolomio C (1998) Thermal mud-pack as an anti-inflammatory treatment. Biomed Pharmacother 52:408–409

Gama-Teixeira A, Simionato MR, Elian SN, Sobral MA, Luz MA (2007) Streptococcus mutans-induced secondary caries adjacent to glass ionomer cement, composite resin and amalgam restorations in vitro. Braz Oral Res 21:368–374

Gans J, Wolinsky M, Dunbar J (2005) Computational improvements reveal great bacterial diversity and high metal toxicity in soil. Science 309:1387–1390

Gaskell EE, Hamilton AR (2014) Antimicrobial clay-based materials for wound care. Future Med Chem 6(6):641–655

Geissler PW, Prince RJ, Levene M, Poda C, Beckerleg SE, Mutemi W, Shulman CE (1999) Perceptions of soil-eating and anaemia among pregnant women on the Kenyan Coast. Soc Sci Med J 48:1069–1079

Gerth K, Bedorf N, Hofle G (1996) Epothilons A and B: antifungal and cytotoxic compounds from Sorangium cellulosum (Myxobacteria): production, physico-chemical and biological properties. J Antibiot 49:560–563

Gichumbi JM, Ombaka O, Gichuki JG (2012) Geochemical and mineralogical characteristics of geophagic materials from Kiambu, Kenya. Int J Mod Chem 2:108–116

Glickman LT, Camara AO, Glickman NW, McCabe GP (1999) Nematode intestinal parasites of children in rural Guinea, Africa: prevalence and relationship to geophagia. Int J Epidemiol 28:169–174

Gomes CSF, Silva JBP (2007) Minerals and clay minerals in medical geology. Appl Clay Sci 36 (1–3):4–21

Goon AT, Isaksson M, Zimerson E, Goh CL, Bruze M (2006) Contact allergy to (meth)acrylates in the dental series in southern Sweden: simultaneous positive patch test reaction patterns and possible screening allergens. Contact Dermatitis 55:219–226

Gorham E (1991) Northern peatlands – role in the carbon cycle and probable responses to climatic warming. Ecol Appl 1:182–195

Green SK, Schroth MN, CHO JJ, Kominos SD, Vitanza-Jack VB (1974) Agricultural plants and soil as a reservoir for *Pseudomonas aeruginosa*. Appl Microbiol 28(6):987–991

Guggenheim S, Martin RT (1995) Report definition of clay and clay mineral: joint report of the AIPEA nomenclature and CMS nomenclature committees. Clay Clay Miner 43(2):255–256

Hackenberg S, Scherzed A, Technau A, Kessler M, Froelich K, Ginzkey C, Koehler C, Burghartz M, Hagen R, Kleinsasser N (2011) Cytotoxic, genotoxic and pro-inflammatory effects of zinc oxide nanoparticles in human nasal mucosa cells in vitro. Toxicol in Vitro 25:657–663

Hallab NJ, Jacobs JJ, Skipor A, Black J, Mikecz K, Galante JO (2000) Systemic metal-protein binding associated with total joint replacement arthroplasty. J Biomed Mater Res 49:353–361

Hallab NJ, Mikecz K, Vermes C, Skipor A, Jacobs JJ (2001) Orthopaedic implant related metal toxicity in terms of human lymphocyte reactivity to metal-protein complexes produced from cobalt-base and titanium-base implant alloy degradation. Mol Cell Biochem 222:127–136

Hansen DC (2008) Metal corrosion in the human body: the ultimate bio-corrosion scenario. Electrochem Soc Interf 17:31–34

Hawass SED, Alnozha MM, Kalowole T (1987) Adult geophagia: a report of three cases with review of pregnant women. J Am Diet Assoc 91:34–38

Haydel SE, Remenih CM, Williams LB (2008) Broad-spectrum in vitro antibacterial activities of clay minerals against antibiotic-susceptible and antibiotic-resistant bacterial. Pathog J Antimicrob Chem 61:353–361

He X, Hartlieb E, Rothmund L, Waschke J, Wu X, Van Landuyt KL, Milz S, Michalke B, Hickel R, Reichl FX, Högg C (2015a) Intracellular uptake and toxicity of three different Titanium particles. Dent Mater 31(6):734–744

He X, Reichl FX, Wang Y, Michalke B, Milz S, Yang Y, Stolper P, Lindemaier G, Graw M, Hickel R, Högg C (2015b) Analysis of Titanium and other metals in human jawbones with dental implants. Dent Mater 31(6):745–757

Hendrickson T (2003) Massage for orthopedic conditions. Lippincott Williams & Wilkins, Balitmore

Hensen EJM, Smit B (2002) Why clays swell. J Phys Chem B 106(49):12664–12667

Hewitt JP (1992) Titanium dioxide: a different kind of sunshield. Drug Cosmet Ind 151(3):26–32

Higgs FJ, Mielke HW, Brisco M (1999) Soil lead at elementary public schools: comparison between school properties and residential neighbourhoods of New Orleans. Environ Geochem Health 21:27–36

Hoang A (2001) Caspofungin acetate: an antifungal agent. Am J Health Syst Pharm 58:1206–1217

Hölker U, Höfer M, Lenz J (2004) Biotechnological advantages of laboratory-scale solidstate fermentation with fungi. Appl Microbiol Biotechnol 64:175–186

Hooda PS, Henry CJK, Seyoum TA, Armstrong LDM, Fowler MB (2002) The potential impact of geophagia on the bioavailability of iron, zinc and calcium in human nutrition. Environ Geochem Health 24:305–319

Hooda PS, Henry CJ, Seyoum TA, Arnstrong LD, Fowler MB (2004) The potential impact of soil ingestion in human mineral nutrition. Sci Tot Environ J 333:75–87
Horner RD, Lackey CJ, Kolasa K, Warren K (1991) Pica practices of pregnant women. J Am Diet Assoc 91:34–38
Hunter JM (1993) Macroterm geophagy and pregnancy clays in Southern Africa. J Cult Geogr 14:69–92
Jacobs JJ, Gilbert JL, Urban RM (1998) Current concepts review-corrosion of metal orthopaedic implants. J Bone Joint Surg 80:268–282
Jacobs JJ, Silverton C, Hallab NJ, Skipor AK, Patterson L, Black J, Galante JO (1999) Metal release and excretion from cementless titanium alloy total knee replacements. Clin Orthop Relat Res 358:173–180
Jandt KD, Sigusch BW (2009) Future perspectives of resin-based dental materials. Dent Mat 25:1001–1006
Jaroenworaluck A, Sunsaneeyametha W, Kosachan N, Stevens R (2006) Characteristics of silica-coated TiO_2 and its UV absorption for sunscreen cosmetic applications. Surf Interface Anal 38:473–477
Jemal A, Siegel R, Ward E, Murray T, Xu J, Thun MJ (2007) Cancer statistics. CA Cancer J Clin 57:43–66
Jiang W, Kim BY, Rutka JT, Chan WC (2008) Nanoparticle-mediated cellular response is size dependent. Nat Nanotechnol 3:145–150
Karlsson HL, Gustafsson J, Cronholm P, Moller L (2009) Size-dependent toxicity of metal oxide particles – a comparison between nano- and micrometer size. Toxicol Lett 188:112–118
Kawaguchi H, Ogawa T, Shirakawa M, Okamoto H, Akisaka T (1992) Ultrastructural and ultracytochemical characteristics of multinucleated cells after hydroxyapatite implantation into rat periodontal tissue. J Periodontal Res 27:48–54
Keen CL, Ensunsa JL, Watson MH (1999) Nutritional aspects of manganese from experimental studies. J Neuro-Oncol 20:213–223
Klaassen CD (2013) Casarett and Doull's toxicology: the basic science of poisons, 8th edn. McGraw Hill Education Medical Hamilton, London
Knapp CW, Dolfing J, Ehlert PA, Graham DW (2010) Evidence of increasing antibiotic resistance gene abundances in archived soils since 1940. Environ Sci Technol 44:580–587
Kugeler KJ, Mead PS (2008) Isolation and characterization of a novel *Francisella* sp. from human cerebrospinal fluid and blood. J Clin Microbiol 46(7):2428–2431
Kunzmann A, Andersson B, Thurnherr T, Krug H, Scheynius A, Fadeel B (2011) Toxicology of engineered nanomaterials: focus on biocompatibility, biodistribution and biodegradation. Biochim Biophys Acta 1810:361–373
Kutalek RG, Wewalka C, Gundacker H, Auer J, Wilson D, Haluza D, Prinz A (2010) Geophagy and potential health implications: geohelminths, microbes and heavy metals. Trans R Soc Trop Med Hyg 104:787–795
Lakudzula DD, Khonje JJ (2011) Nutritive potential of some 'edible' soils in Blantyre city, Malawi. Malawi Med J 23:38–42
Lalor PA, Revell PA, Gray AB, Wright S, Railton GT, Freeman MA (1991) Sensitivity to titanium. A cause of implant failure? J Bone Joint Surg (Br) 73:25–28
Lam KS (2007) New aspects of natural products in drug discovery. Trends Microbiol 15 (6):279–289
Lamps LW, Madhusudhan KT, Havens JM, Greenson JK, Bronner MP, Chiles MC, Dean PJ, Scott MA (2003) Pathogenic Yersinia DNA is detected in bowel and mesenteric lymph nodes from patients with Crohn's disease. Am J Surg Pathol 27:220–227
Lautenschlager EP, Monaghan P (1993) Titanium and titanium alloys as dental materials. Int Dent J 43:245–253
Ladewig K, ZP X, GQ L (2009) Layered double hydroxide nanoparticles in gene and drug delivery. Expert Opin Drug Deliv 6(9):907–922
Leclercq R (2001) Overcoming antimicrobial resistance: profile of a new ketolide antibacterial, telithromycin. J Antimicrob Chemother 48:9–23

Lee KD, Nir S, Papahadjopoulos D (1993) Quantitative analysis of liposome-cell interactions in vitro: rate constants of binding and endocytosis with suspension and adherent J774 cells and human monocytes. Biochemistry 32:889–899

Leggat PA, Kedjarune U, Smith DR (2007) Occupational health problems in modern dentistry: a review. Ind Health 45:611–621

Levison DA, Crocker PR, Banim S, Wallace DMA (1982) Silica stones in the urinary bladder. Lancet 27:704–705

Lewis G (1991) In vivo occlusal wear of posterior compostite restorations. Oper Dent 16:61–69

Lin AN, Reimer RJ, Carter DM (1988) Sulphur revisited. J Am Acad Dermatol 18(3):553–558

Lindh U (2005) Biological functions of the elements. In: Selinus O, Alloway B, Centeno JA, Finkelman RB, Fuge R, Lindh U, Smedley P (eds) Essentials of medical geology: impacts of the natural environment on public health. Elsevier Academic, Amsterdam, pp 115–160

Liu Y, Van Kruiningen HJ, West AB, Cartun RW, Cortot A, Colombel JF (1995) Immunocytochemical evidence of Listeria, *Escherichia coli*, and Streptococcus antigens in Crohn's disease. Yeast 108:1396–1404

Lohmann CH, Schwartz Z, Koster G, Jahn U, Buchhorn GH, MacDougall MJ, Casasola D, Liu Y, Sylvia VL, Dean DD, Boyan BD (2000) Phagocytosis of wear debris by osteoblasts affects differentiation and local factor production in a manner dependent on particle composition. Biomaterials 21:551–561

López Galindo A, Viseras C (2004) Pharmaceutical and cosmetic applications of clays. In: Wypych F, Satyanarayana KG (eds) Clay surfaces. Fundamentals and applications. Elsevier, Amsterdam, pp 267–289

Luoba AI, Geissler PW, Estambale B, Ouma JH, Magnussen P, Alusala D, Ayah R, Mwaniki D, Friis H (2004) Geophagy among pregnant and lactating women in Bondo District, western Kenya. Trans R Soc Trop Med Hyg 98:734–741

Luoba AI, Geissler PW, Estambale B, Ouma JH, Alusala D, Ayah R, Mwaniki D, Magnussen P, Friis H (2005) Earth-eating and reinfection with intestinal helminths among pregnant and lactating women in western Kenya. Tropical Med Int Health 10:220–227

Mahaney WC, Milner MW, Mulyono H, Hancock RGV, Aufreiter S, Reich M, Wink M (2000) Mineral and chemical analyses of soils eaten by humans in Indonesia. Int J Environ Health Res 10:93–109

Martini D, Fini M, Franchi M, Pasquale VD, Bacchelli B, Gamberini M, Tinti A, Taddei P, Giavaresi G, Ottani V, Raspanti M, Guizzardi S, Ruggeri A (2003) Detachment of titanium and fluorohydroxyapatite particles in unloaded endosseous implants. Biomaterials 24:1309–1316

Mcloughlin IJ (1987) The pica habit. Hosp Med 37:286–290

Meachim G, Williams D (1973) Changes in nonosseous tissue adjacent to titanium implants. J Biomed Mater Res 7:555–572

Mengel CE, Carter WA (1964) Geophagia diagnosed by roentgenograms. J Am Med Assoc 187:955–956

Miazzo R, Peralta MF, Magnoli C, Salvano M, Ferrero S, Chiacchiera SM, Carvalho EC, Rosa CA, Dalcero A (2005) Efficacy of sodium bentonite as a detoxifier of broiler feed contaminated with aflatoxin and fumonisin. Poult Sci 84(1):1–8

Mielke HW (1999) Lead in the inner cities. Am Sci 87:62–73

Miletic VJ, Santini A (2008) Remaining unreacted methacrylate groups in resin-based composite with respect to sample preparation and storing conditions using micro-Raman spectroscopy. J Biomed Mater Res B Appl Biomater 87:468–474

Minotti G, Menna P, Salvatorelli E, Cairo G, Gianni L (2004) Anthracyclines: molecular advances and pharmacologic developments in antitumor activity and cardiotoxicity. Pharmacol Rev 56:185–229

Mjor IA (1996) Glass-ionomer cement restorations and secondary caries: a preliminary report. Quintessence Int 27:171–174

Moellering RC (1998) Problems with antimicrobial resistance in Gram-positive cocci. Clin Infect Dis 26:1177–1178

Monier JM, Demaneche S, Delmont TO, Mathieu A, Vogel TM, Simonet P (2011) Metagenomic exploration of antibiotic resistance in soil. Curr Opin Microbiol 14:229–235
Mwangi J, Ombaka O (2010) Analyses of geophagic materials consumed by pregnant women in Embu, Meru and Chuka towns in Eastern Province, Kenya. J Environ Chem Ecotoxicol 3:340–344
Napierska D, Thomassen LC, Lison D, Martens JA, Hoet PH (2010) The nanosilica hazard: another variable entity. Part Fibre Toxicol 7(1):39
Nathwani D (2005) Tigecycline: clinical evidence and formulary positioning. Int J Antimicrob Agents 25:185–192
Newman DJ, Cragg GM (2007) Natural products as sources of new drugs over the last 25 years. J Nat Prod 70:461–477
Newman DJ, Cragg GM, Snader KM (2003) Natural products as sources of new drugs over the period 1981–2002. J Nat Prod 66:1022–1037
Ngole VM, Ekosse GE (2012) Physico-chemistry, mineralogy, geochemistry and nutrient bioaccessibility of geophagic soils from Eastern Cape, South Africa. Sci Res Essays 7:1319–1331
Ngole VM, Ekosse GE, De-Jager L, Songca SP (2010) Physicochemical characteristics of geophagic clayey soils from South Africa and Swaziland. Afr J Biotechnol 9:5929–5937
Nicholls SJ, Tuzcu EM, Sipahi I, Grasso AW, Schoenhagen P, Hu T, Wolski K, Crowe T, Desai MY, Hazen SL, Kapadia SR, Nissen SE (2007) Statins, high-density lipoprotein cholesterol, and regression of coronary atherosclerosis. J Am Med Assoc 297:499–508
Oberdorster G, Oberdorster E, Oberdorster J (2005) Nanotoxicology: an emerging discipline evolving from studies of ultrafine particles. Environ Health Perspect 113:823–839
Odewumi SC (2013) Mineralogy and geochemistry of geophagic clays from share area, Northern Bida Sedimentary Basin, Nigeria. J Geol Geosci 2:108
Oh WK, Kim S, Choi M, Kim C, Jeong YS, Cho BR, Hahn JS, Jang J (2010) Cellular uptake, cytotoxicity, and innate immune response of silica-titania hollow nanoparticles based on size and surface functionality. ACS Nano 4:5301–5313
Okabe T, Hero H (1995) The use of titanium in dentistry. Cell Mater (USA) 5:211–230
Okazaki Y, Gotoh E (2005) Comparison of metal release from various metallic biomaterials in vitro. Biomaterials 26:11–21
Okunlola OA, Owoyemi KA (2011) Compositional characteristics of geophagic clays in parts of southern Nigeria. Book of proceedings. 1st International conference on clays in Africa and 2nd conference on geophagic clays in South Africa, pp 290–305
Özcan M, Hämmerle C (2012) Titanium as a reconstruction and implant material in dentistry: advantages and Pitfalls. Materials 5:1528–1545
Patel N, Raiyani D, Kushwah N, Parmar P, Hetvi Hapani P, Jain H, Upadhyay U (2015) An introduction to mud therapy: a review. Int J Pharm Ther 6(4):227–231
Paterson RRM (2008) Fungal enzyme inhibitors as pharmaceuticals, toxins and scourge of PCR. Curr Enzym Inhib 4:46–59
Patrick L (2006) Lead toxicity, a review of the literature. Part 1: exposure, evaluation, and treatment. Altern Med Rev 11:2–22
Pawel A, Mucke D, Rehbein C, Fickenwirth W, Kluge K, Schulze W (2013) Mineralschlamm und Torf – Naturheilmittel in der Entwicklung des TeufelsbadesBergverein zu Hüttenrode e. V. Hüttenröder Edition Nr. 4, p 49
Pedersen K, Motamedi M, Karnland O, Sandén T (2000) Cultivability of microorganisms introduced into a compacted bentonite clay buffer under high-level radioactive waste repository conditions. Eng Geol 58(2):149–161
Peláez F (2006) The historical delivery of antibiotics from microbial natural products – can history repeat? Biochem Pharmacol 71:981–990
Peng Y, Demain AL (1998) A new hydroxylase system in Actinomadura sp. cells converting compactin to pravastatin. J Ind Microbiol Biotechnol 20:373–375
Pereverzev VN (2005) Peat soils of the Kola Peninsula. Euras Soil Sci 38:457–464

Peters DH, Fitton A, Plosker GL, Faulds D (1993) Tacrolimus. A review of its pharmacology, and therapeutic potential in hepatic and renal transplantation. Drugs 46(4):746–794
Piattelli A, Piattelli M, Mangano C, Scarano A (1998) A histologic evaluation of eight cases of failed dental implants: is bone overheating the most probable cause? Biomaterials 19:683–690
Pioletti DP, Takei H, Kwon SY, Wood D, Sung KL (1999) The cytotoxic effect of titanium particles phagocytosed by osteoblasts. J Biomed Mater Res 46:399–407
Price EW (1975) The management of endemic (non-filarial) elephantiasis of the lower legs. Trop Dr 5:70–75
Price EW (1976) The association of endemic elephantiasis of the lower legs in East Africa with soil derived from volcanic rocks. Trans R Soc Trop Med Hyg 70:288–295
Price EW (1977) The site of lymphatic blockage in endemic (nonfilarial) elephantiasis of the lower legs. J Trop Med Hyg 80:230–237
Price EW (1990) Podoconiosis: non-filarial elephantiasis. Oxford Medical Publications, Oxford
Price EW, Bailey D (1984) Environmental factors in the etiology of endemic elephantiasis of the lower legs in tropical Africa. Trop Geogr Med 36:1–5
Price EW, Henderson WJ (1978) The elemental content of lymphatic tissues in barefooted people in Ethiopia, with reference to endemic elephantiasis of the lower legs. Trans R. Soc Trop Med Hyg 72:132–136
Prince RJ, Luoba AI, Adhiambo P, Ng'uono J, Geissler PW (1999) Geophagy is common among Luo woman in western Kenya. Roy Soc Trop Med Hyg 93:515–516
Prusiner SB (1998) The prion diseases. Brain Pathol 8:499–513
Raaijmakers JM, Mazolli M (2012) Diversity and natural functions of antibiotics produced by beneficial and plant pathogenic bacteria. Annu Rev Phytopathol 50:403–424
Rastogi R (2012) Therapeutic uses of mud therapy in naturopathy. Indian J Tradit Knowl 11 (3):556–559
Reichl FX, Mohr K, Hein L, Hickel R (2014) Atlas der Pharmakologie und Toxikologie, vol 2., Nanopartikel. Thieme-Verlag, Stuttgart, pp 334–337
Reilly C, Henry J (2000) Geophagia: why do humans consume soil? Nutr Bull 25:141–144
Rejman J, Oberle V, Zuhorn IS, Hoekstra D (2004) Size-dependent internalization of particles via the pathways of clathrin- and caveolae-mediated endocytosis. Biochem J 377:159–169
Riyaz N, Arakkal FR (2011) Spa therapy in dermatology. Indian J Dermatol Venereol Leprol 7 (2):128–134
Romero D, Traxler MF, Lopez D, Kolter R (2011) Antibiotics as signal molecules. Chem Rev 111:5492–5505
Rosa CA, Miazzo R, Magnoli C, Salvano M, Chiacchiera SM, Ferrero S, Saenz M, Carvalho EC, Dalcero A (2001) Evaluation of the efficacy of bentonite from the south of Argentina to ameliorate the toxic effects of aflatoxin in broilers. Poult Sci 80(2):139–144
Sanders RKM (1996) The management of tetanus. Trop Dr 26:107–115
Schliephake H, Reiss G, Urban R, Neukam FW, Guckel S (1993) Metal release from titanium fixtures during placement in the mandible: an experimental study. Int J Oral Maxillofac Implants 8:502–511
Schloss PD, Handelsman J (2006) Toward a census of bacteria in soil. PLoS Comput Biol 2:786–793
Seeman E (2009) Bone modeling and remodeling. Crit Rev Eukaryot Gene Expr 19(3):219–233
Serizawa N, Matsuoka T (1991) A two-component-type cytochrome P-450 monooxygenase system in a prokaryote that catalyzes hydroxylation of ML-236B to pravastatin, a tissue-selective inhibitor of 3-hydroxy-3-methylglutaryl coenzyme A reductase. Biochim Biophys Acta 1084:35–40
Setcos JC, Babaei-Mahani A, Silvio LD, Mjor IA, Wilson NH (2006) The safety of nickel containing dental alloys. Dent Mater 22:1163–1168
Showalter HD, Bunge RH, French JC, Hurley TR, Leeds RL, Leja B, McDonnell PD, Edmunds CR (1992) Improved production of pentostatin and identification of fermentation cometabolites. J Antibiot 45:1914–1918

Sicilia A, Cuesta S, Coma G, Arregui I, Guisasola C, Ruiz E, Maestro A (2008) Titanium allergy in dental implant patients: a clinical study on 1500 consecutive patients. Clin Oral Implants Res 19:823–835

Soto-Alvaredo J, Blanco E, Bettmer J, Hevia D, Sainz RM, Lopez Chaves C, Sanchez C, Llopis J, Sanz-Medel A, Montes-Bayon M (2014) Evaluation of the biological effect of Ti generated debris from metal implants: ions and nanoparticles. Metallomics 6:1702–1708

Sukenik S, Buskila D, Neumann L, Kleiner-Baumgarten A (1992) Mud pack therapy in rheumatoid arthritis. Clin Rheumatol 11:243–247

Sum PE (2006) Case studies in current drug development: 'glycylcyclines'. Curr Opin Chem Biol 10:374–379

Sum PE, Sum FW, Projan SW (1998) Recent developments in tetracycline antibiotics. Curr Pharm Des 4:119–132

Terzano C, Di Stefano F, Conti V, Graziani E, Petroianni A (2010) Air pollution ultrafine particles; toxicity beyond the lung. Eur Rev Med Pharmacol Sci 14(10):809–821

Thompson SA (2000) An overview of nickel-titanium alloys used in dentistry. Int Endod J 33:297–310

Titzmann T, Balda BR (1996) Mineral water and spas in Germany. Clin Dermatol 14(6):611–613

Torsvik V, Ovreas L, Thingstad TF (2002) Prokaryotic diversity: magnitude, dynamics, and controlling factors. Science 296:1064–1066

Travlos GS (2006) Histopathology of bone marrow. Toxicol Pathol 34:566–598

Trckova M, Matlova L, Hudcova H, Faldyna M, Zraly Z, Dvorska L, Beran V, Pavlik I (2005) Peat as a feed supplement for animals: a review. Vet Med Czech 50(8):361–377

Truscheit E, Frommer W, Junge B, Müller L, Schmidt DD, Wingender W (1981) Chemistry and biochemistry of microbial a-glucosidase inhibitors. Angew Chem Int Ed Eng 20:744–761

Urban RM, Jacobs JJ, Sumner DR, Peters CL, Voss FR, Galante JO (1996) The bone-implant interface of femoral stems with non-circumferential porous coating. J Bone Joint Surg Am 78:1068–1081

Urban RM, Jacobs JJ, Tomlinson MJ, Gavrilovic J, Black J, Peoch M (2000) Dissemination of wear particles to the liver, spleen, and abdominal lymph nodes of patients with hip or knee replacement. J Bone Joint Surg Am 82:457–476

Urcan E, Scherthan H, Styllou M, Haertel U, Hickel R, Reichl FX (2010) Induction of DNA double-strand breaks in primary gingival fibroblasts by exposure to dental resin composites. Biomaterials 31:2010–2014

Valiev RZ, Semenova IP, Latysh VV, Rack H, Lowe TC, Petruzelka J, Dluhos L, Hrusak D, Sochova J (2008) Nanostructured titanium for biomedical applications. Adv Eng Mater 10: B15–BB7

Van Landuyt KL, Yoshihara K, Geebelen B, Peumans M, Godderis L, Hoet P, Van Meerbeek B (2012) Should we be concerned about composite (nano-)dust? Dent Mat 28:1162–1170

Vicente Rodríguez MA, López González JD, Bañares Muñoz MA (1994) Acid activation of a Spanish sepiolite: physicochemical characterization, free silica content and surface area of products obtained. Clay Miner 29:361–367

Voggenreiter G, Leiting S, Brauer H, Leiting P, Majetschak M, Bardenheuer M, Obertacke U (2003) Immuno-inflammatory tissue reaction to stainless-steel and titanium plates used for internal fixation of long bones. Biomaterials 24:247–254

Volk J, Ziemann C, Leyhausen G, Geurtsen W (2009) Non-irradiated campherquinone induces DNA damage in human gingival fibroblasts. Dent Mater 25:1556–1563

Waksman SA, Woodruff HB (1941) Actinomyces antibioticus, a new soil organism antagonistic to pathogenic and non-pathogenic bacteria. J Bacteriol 42:231–249

Walker S, Landovitz R, Ding WD, Ellestad GA, Kahne D (1992) Cleavage behavior of calicheamicin gamma 1 and calicheamicin T. Proc Natl Acad Sci U S A 89:4608–4612

Walker ARP, Walker BF, Sookaria FI, Canaan RJ (1997) Pica. J Roy Health 117:280–284

Wang ML, Nesti LJ, Tuli R, Lazatin J, Danielson KG, Sharkey PF, Tuan RS (2002) Titanium particles suppress expression of osteoblastic phenotype in human mesenchymal stem cells. J Orthop Res 20:1175–1184

Wang ML, Tuli R, Manner PA, Sharkey PF, Hall DJ, Tuan RS (2003) Direct and indirect induction of apoptosis in human mesenchymal stem cells in response to titanium particles. J Orthop Res 21:697–707

Wara-aswapati N, Krongnawakul D, Jiraviboon D, Adulyanon S, Karimbux N, Pitiphat W (2005) The effect of a new toothpaste containing potassium nitrate and triclosan on gingival health, plaque formation and dentine hypersensitivity. J Clin Periodontol 32(1):53–58

Watanabe K, Miyakawa O, Takada Y, Okuno O, Okabe T (2003) Casting behavior of titanium alloys in a centrifugal casting machine. Biomaterials 24:1737–1743

Weibel EK, Hadvary P, Hochuli E, Kupfer E, Lengsfeld H (1987) Lipstatin, an inhibitor of pancreatic lipase, produced by Streptomyces toxytricini. J Antibiot 40:1081–1085

Whittaker P (1998) Iron and zinc interactions in humans. Am J Clin Nutr 68:442–446

Whittaker CJ, Klier CM, Kolenbrander PE (1996) Mechanisms of adhesion by oral bacteria. Annu Rev Microbiol 50:513–552

WHO (World Health Organisation) (1996) Trace elements in human nutrition and health. World Health Organization, Geneva

Wiley AS, Katz SH (1998) Geophagy in pregnancy: a test of a hypothesis. Curr Anthropol 39:532–545

Williams LB, Haydel SE, Giese RF, Eberl DD (2008) Chemical and mineralogical characteristics of French green clays used for healing. Clay Clay Miner 56(4):437–452

Wilson MJ (2003) Clay mineralogical and related characteristics of geophagic materials. J Chem Ecol 29:1525–1547

Wilson NH, Burke FJ, Mjor IA (1997) Reasons for placement and replacement of restorations of direct restorative materials by a selected group of practitioners in the United Kingdom. Quintessence Int 28:245–248

Woywodt A, Kiss A (2002) Geophagia: the history of earth-eating. J Roy Soc Med 95:143–146

Yamamoto K, Ohashi S, Aono M, Kokubo T, Yamada I, Yamauchi J (1996) Antibacterial activity of silver ions implanted in SiO_2 filler on oral streptococci. Dent Mater 12:227–229

Yang J, Merritt K (1994) Detection of antibodies against corrosion products in patients after Co-Cr total joint replacements. J Biomed Mater Res 28:1249–1258

Young SL, Goodman D, Farag TH, Ali SM, Khatib MR, Khalfan SS, Tielsch JM, Stoltzfus RJ (2007) Geophagia is not associated with Trichuris or hookworm transmission in Zanzibar, Tanzania. Trans R Soc Trop Med Hyg 101(8):766–772

Young SL, Sherman PW, Lucks JB, Pelto GH (2011) Why on earth? Evaluating hypotheses about the physiological functions of human geophagy. Q Rev Biol 86:97–120

Zhao K, Zhou D, Ping W, Ge J (2004) Study on the preparation and regeneration of protoplast from taxol-producing fungus Nodulisporium sylviforme. Nat Sci 2:52–59

Zhu J, Liao L, Zhu L, Zhang P, Guo K, Kong J, Ji C, Liu B (2013) Size-dependent cellular uptake efficiency, mechanism, and cytotoxicity of silica nanoparticles toward HeLa cells. Talanta 107:408–415

Chapter 3
Soil-Borne Particles and Their Impact on Environment and Human Health

Abstract Dust particles can consist of either natural soil-borne particles or of particulate matter from human activities, or both of them. Particulate matter is a complex mixture of extremely small particles and liquid droplets consisting of soil or dust particles, metals, organic chemicals, and acids. Naturally generated particles consist of weathered rock materials, dryland soil and sediment materials, biogenic fibres and residues from forest fires, and ash developed during volcanic eruptions. World dust emissions from drylands amount to about 5 billion Mg per year. Dominant dust sources around the world are almost wholly in or adjacent to the great drylands of the northern hemisphere. The greatest of these includes a large belt from the western Sahara to the Yellow Sea, across North Africa, the Middle East, northwest India, and central and eastern Asia. Saharan dust, driven by the northeast trade winds, takes about a week to cross the Atlantic Ocean, reaching northeastern South America the Caribbean, Central America, and the southeastern USA. The mid latitude deserts of Asia are a source of substantial airborne dust, especially during spring and early summer. Mongolia and the Tarim Basin-Taklamakan Desert are the two major dust sources of China. They are also of worldwide importance, as fine dusts from these regions have been traced to North America, Greenland and Europe. Other notable sources of dusts include the Great Basin of the USA and, in the southern hemisphere, central and northern Argentina, parts of southern Africa and East-central Australia. Former lake basins are major sources of fine, readily wind-entrained mineral dusts, which may include salts and elevated levels of toxic elements. For example, the Bodélé depression in Chad (North Africa) and the numerous lake depressions in central Asia (e.g. Aral Sea region) and northern China are major dust sources of global significance. Sea spray produces aerosols containing particles that are commonly of salt, but can also contain radionuclides.

About 9% of the global population, more than 500 million people, lives within potential exposure range of a volcano that has been active within recorded history. There are at present an estimated 550 active volcanoes, many of which are in locations experiencing rapid population growth. Major urban centers are commonly found within close proximity to volcanoes, such as Naples in Italy and the capital cities of Mexico, Japan, and the Philippines. Population density generally decreases with increasing distance from the volcano, with the highest population densities in

R. Nieder et al., *Soil Components and Human Health*,
https://doi.org/10.1007/978-94-024-1222-2_3

close proximity to volcanoes in Southeast Asia and Central America. Of all eruptive hazards, ash-fall can affect most people because of the extent of areas that can be covered by fallout. Although eruptions are often short-lived, ash-fall deposits can remain in the local environment for years to decades, being remobilized by human activity or simply re-suspended by wind.

Potentially toxic natural particulate dusts include asbestos minerals and several species of crystalline silica and fibrous silicates, and dusts containing toxic trace elements such as volcanic ash particles, which hold transition metals and other toxic elements on their surfaces. The impact of high concentrations of naturally occurring silica-rich dust on human and animal health received little attention until recently, although the so-called desert lung syndrome (non-occupational silicosis with asthmatic symptoms) has been known for more than a century. Large quantities of silica and silicates, together with a range of chemicals including potentially toxic trace elements, are released during some volcanic eruptions. Inhaled ash can exacerbate symptoms in people who are susceptible to asthma and respiratory disease.

Anthropogenic sources of particulate matter include dusts from mining and quarrying, agricultural soils, and combustion of fossil fuel for energy generation and heating. Petrol and diesel-powered vehicles are an important source of particulate and gaseous atmospheric pollution. Residual ash from liquid fossil fuels has been categorized to be more harmful to human health than coal fly ash. Particulate matter released by biomass burning from forest clearance and agricultural practices continues to be important. The burning of biomass both natural and anthropogenic yields black carbon which adds to the opacity of the atmosphere. Smoke plumes from fires, are often carried thousands of kilometres from their sources. Potentially toxic particulate dust arising from anthropogenic activities includes quartz and other silicates from quarrying and mining, agricultural biomass burning and wild fires and higher-rank coal dust from coal extraction and processing. Workers employed in industries such as mining, quarrying, sand blasting, silica milling and stone masonry are particularly exposed to fine, crystalline quartz dust and can develop inflammation and fibrosis of the lung (silicosis), which is one of the most studied occupational lung diseases. Crystalline silica is also classed as a human carcinogen. Asbestosis is a progressive, incurable chronic lung disease which is attributable to prolonged exposure to asbestos. Unfortunately, the important insulation and fire proof properties of asbestos promoted its widespread use in construction, ship building and industrial refrigeration plants, despite the known link to serious lung disease. However, up to now it is not clear which components in coal cause pneumoconiosis. Coal fly ash can contain a component of unburnt organic matter and is widespread in industrial, urban and some natural environments. Human-health implications of fly ash in some regions of the world are still a subject of very high concern. For example, in northern China the domestic burning of local Permian coals has resulted in clusters of lung cancer. The relatively few studies of coal fly ash toxicity have yet to provide evidence of human lung inflammation and there is a continuing discussion about the importance of toxic trace elements being part of this material. There is evidence that oil fly ash (including diesel) is still more important to human health compared to coal fly ash because the former is smaller in diameter, chemically complex and rich in metals. In this chapter, main emphasis is given to the source, release, transportation, and deposition of mineral particulate aerosols derived from volcanoes, soils, sediments, and weathered rock surfaces and their impact on human health when in suspension in the atmosphere.

Keywords Health risk of mineral dusts • Volcanic ash • Asbestos • Crystalline silica • Coal ash • Dusts from coal and liquid fossil fuel burning • Dusts from landscape fires • Dust particles contaminated with toxic elements and pathogens • Clinical effects and therapy of diseases related to soil-borne particles exposure • Mitigation options

3.1 Background

Atmospheric dust (including fine mineral aggregates, fibrous minerals, and fibrous organic materials) reaches concentrations in many parts of the world sufficient to constitute a major influence upon both human and animal health. Following the "dust bowl years" of the 1930s, awareness of soil-derived atmospheric dusts increased considerably in the USA. Records of Saharan dust falls in Western Europe became increasingly common from the mid-nineteenth century. However, compared to work on artificially generated smoke and gases the impact of high concentrations of dust on human and animal health has received relatively little attention. Aerosols include particles injected into the atmosphere, such as mineral dust and sea salt, and those that form within the atmosphere, notably sulfates. Natural and man-made fires, including extensive burning of vegetation, generate smoke plumes that are often carried several thousands of kilometers from their sources, which adds to atmospheric health hazards. The finer components (<10 μm) of respirable atmospheric dusts include single particles, aggregates of very fine mineral grains (notably silica), fibrous minerals (e.g., the asbestos group), and fibrous organic materials. Particulate dusts may affect human health by way of direct and indirect pathways. Inhaled dusts derived from volcanos, soils or fine-grained sediment sources such as seasonally dry rivers and dry lakebeds make up a direct pathway, whereas an indirect pathway arises from generation of respirable mineral dusts by both natural and human induced erosion of soils (Fig. 3.1).

3.2 Definition of Dust Particles

Dust particles can consist of either natural soil-borne particles or of particulate matter from human activities, or both of them. Particles can be described by their "aerodynamic equivalent diameter" and are commonly subdivided into fractions based on how the particles are generated and where they deposit in human airways:

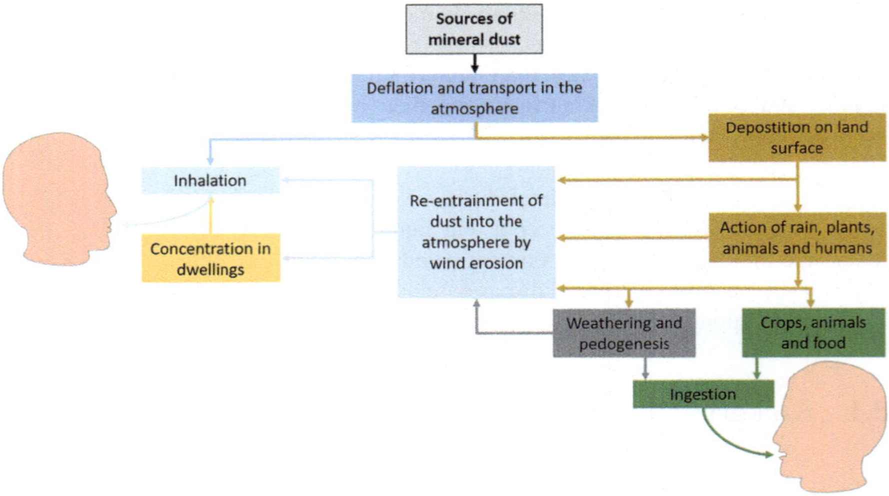

Fig. 3.1 Direct and indirect pathways from dust sources to human inhalation and ingestion (Adapted from Derbyshire 2005)

<10 μm (PM_{10}), <2.5 μm ($PM_{2.5}$), and <0.1 μm ($PM_{0.1}$) (Anderson et al. 2012). Since the industrial revolution, human activities have significantly increased the concentrations of $PM_{2.5}$ in both urban and rural regions (Parrish et al. 2012). These changes have been driven by direct changes in air pollutant emissions and by climate change that influences air quality through several mechanisms, including changes in photochemical reaction rates, biogenic emissions, deposition, and atmospheric circulation (Fiore et al. 2012). Diameters of particles in air including solids, liquids, gases, inorganic, and biologic earth materials as well as the techniques commonly employed in detection are summarized in Fig. 3.2.

3.3 Sources of Dust Particles

Sources of dusts are diverse, including volcanoes, natural dust, mining industry, mineral wool and asbestos production, combustion of fossil fuels, cement industry, and biomass burning. Estimates diverge greatly in terms of the total quantity of dust transported, although recent publications estimate annual volumes up to 1000 Tg from Sahara dust alone. Anthropogenic activities have contributed to dust formation by enhancing the process of desertification, with losses of millions of hectares of farmland per year.

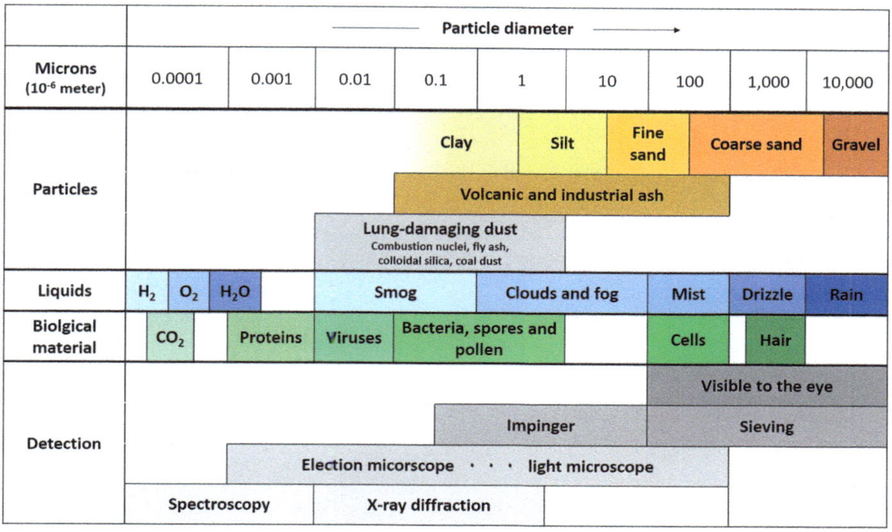

Fig. 3.2 Diameters of particles in air and techniques commonly employed in detection (Adapted from Skinner 2007)

3.3.1 Volcanic Ash

An estimated 800 million people live within 100 km of an active volcano in 86 countries worldwide (Auker et al. 2013) and more than 500 million people live within the potential exposure range of a volcano. Impacts of volcanic eruptions can be both short-term, e.g. physical damage, and long-term, e.g. sustained or permanent displacement of populations. The risk resulting from volcanic eruptions is often underestimated beyond areas within the immediate proximity of a volcano. Ash is the most frequent and often widespread volcanic hazard (Brown et al. 2015). Volcanic ash hazards can have effects hundreds of kilometers away from the vent and have an adverse impact on human and animal health, infrastructure, transport, agriculture and horticulture, the environment and economies. Most volcanoes are located close to the boundaries of tectonic plates and are the consequence of melting the Earth's interior at depths ranging mostly between 10 and 200 km (Brown et al. 2015). Numerous volcanoes form on the world's rifted plate boundaries, but mostly these are located deep below the ocean surface along submarine ocean ridges (Fig. 3.3).

Most active volcanoes that pose hazards are located at subduction zones, forming arc-shaped chains of volcanic islands like the Lesser Antilles in the Caribbean or lines of volcanoes parallel to the coasts of major continents as in the Andes (Brown et al. 2015). There are some dangerous volcanoes located where rifting plates form on land and in the shallow ocean, such as Iceland and the great East African rift valley. There are also active volcanoes within tectonic plates, the

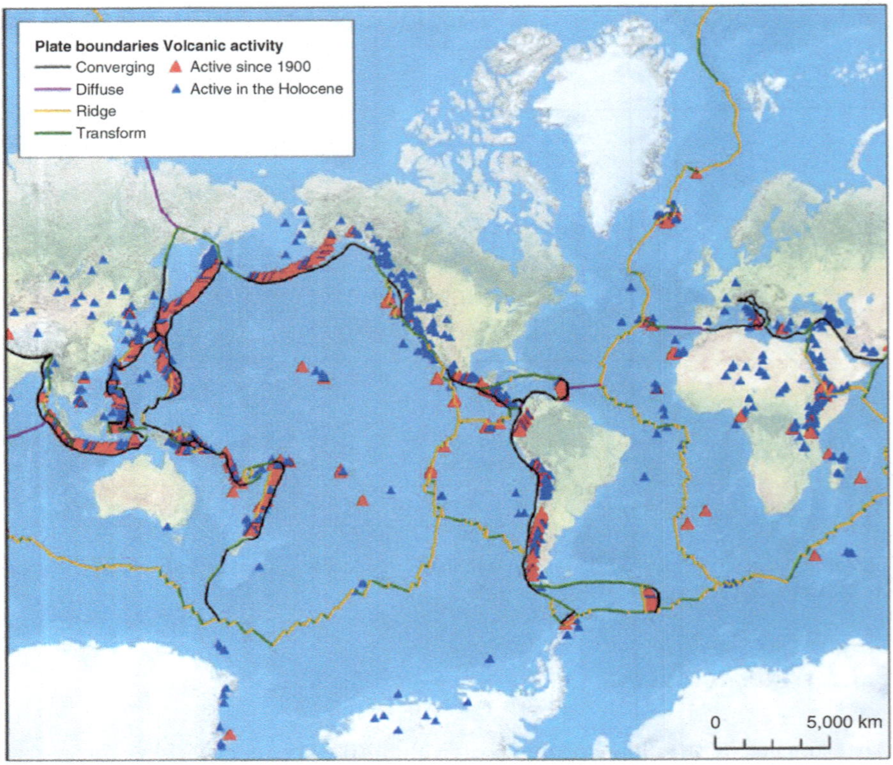

Fig. 3.3 Global map of the distribution and status of Holocene volcanoes (Brown et al. 2015, p. 85; with kind permission from Cambridge University Press)

Hawaiian islands being the best-known examples. Basaltic volcanoes are very common where the Earth crust is thin and above oceanic hot spots such as the Hawaiian islands. Volcanoes erupting more evolved magma, with more silica (such as andesitic and rhyolitic volcanoes) are generally found around Earth's subducting plate margins, e.g. the Pacific Ring of Fire which has given rise to the volcanic chains stretching from the western coast of the USA, Mexico, to Chile, and eastern Russia, Japan, the Philippines, Indonesia and south to New Zealand (Derbyshire et al. 2012).

Volcanoes produce multiple hazards including rocks ejected by volcanic explosions (also referred to as blocks or bombs), pyroclastic flows, lahars and floods, lava, volcanic gases, landslides, tsunamis, volcanic earthquakes and volcanic ash, the latter being the most widely-distributed product of explosive volcanic eruptions (Johnston et al. 2000). All explosive volcanic eruptions generate tephra, fragments of rock that are produced when magma or vent material is explosively disintegrated (Fig. 3.4).

Fig. 3.4 Eruption of the Pinatubo in 1991, one of the most spectacular volcano outbursts of the twentieth century (Available at http://www.topwelt.com/groessten-vulkanausbrueche/)

Table 3.1 Major types of magma

Type of magma	Melting point (°C)	SiO_2 content (wt%)
Basaltic	1000–1200	45–55
Andesitic	800–1000	55–65
Rhyolitic	650–1000	65–75

Adapted from Langmann (2013)

The chemical composition of the bulk volcanic ash is mainly determined by the magma from which it is generated. Three major types of magma can be distinguished from each other (Table 3.1) exhibiting different melting points and SiO_2 contents.

The mineral component of volcanic ash consists of about 45–75 wt% of silica (Heiken 1972). Basaltic (basic) magma eruptions produce silica-poor (<55 wt%) ash, which contains mafic minerals such as pyroxenes and olivine, calcium-rich feldspar, while acidic (e.g. rhyolitic) eruptions produce silica-rich (>65 wt%) ash with high concentrations of felsic minerals such as quartz, silica glass and potassium-rich feldspar (Derbyshire et al. 2012). These minerals are formed through successive crystallisation during cooling and decompression when the magma ascends from the Earth's mantle through the Earth's crust into the conduit and subsequently into the volcanic plume (Schmincke 2004). During the crystallisation process, the composition of the melt is changing due to depletion of crystallised components and enrichment of remaining components driving the successive generation of different minerals, including those without silicate, such as magnetite or ilmenite (Langmann 2013).

A major difference in the mineral composition of volcanic ash and mineral dust results from chemical weathering of mineral dust, generally on geological time scales. Mineral composition changes under the influence of water, oxygen, and acids. For example, feldspar will weather to clay and iron bearing minerals can form hematite and goethite. Therefore, primary iron containing minerals (e.g., amphiboles and pyroxenes) of volcanic rocks are not commonly found in mineral dust. Further multiphase chemical modifications at the surfaces of mineral dust and volcanic ash take place during atmospheric transport (Langmann 2013).

The quantity of respirable material (<4 μm aerodynamic diameter) varies greatly among different types of volcanic eruptions, related to magma composition, with increasing explosivity correlating with increasing silica content and viscosity (Derbyshire et al. 2012). Basaltic eruptions, which are generally effusive, produce coarse particles with <5 vol% respirable material. Andesitic or rhyolitic eruptions, which are usually more explosive, produce finer particles. The eruption of Vesuvius, Italy, in AD 79 produced ash with around 17 vol% respirable material (Derbyshire et al. 2012). Dome-forming eruptions such as the Soufrière Hills volcano, Montserrat, can generate ash with about 10–15 vol% respirable material during dome collapse (Horwell 2007). The crystalline silica content of volcanic ash is the main cause of concern in terms of health hazard but, in recent years, other potential hazards have been identified. For example, volcanic ash can generate substantial quantities of hydroxyl radicals through Fenton reaction. Iron-rich basaltic ash has greater reactivity than iron-poor ash (Horwell et al. 2007).

3.3.2 *Mineral Dust*

3.3.2.1 Natural Dust

Silt-sized particles, particularly those in the 10–50 μm range, are readily entrained by the wind from dry unvegetated surfaces. The clay-size (<2 μm) component of soils and sediments is not readily detached by the wind as individual particles because of the high interparticle cohesive forces typical of such colloidal materials (Derbyshire 2005). Entrainment of material finer than 2 μm usually occurs in association with the coarser (silt-sized) grains, and also in the form of coarse or medium silt-sized aggregates made up of variable mixtures of fine silt and clay-grade particles. Silicon (Si), which makes up more than one-quarter of the elements in the Earth's crust, is highly reactive and readily combines with oxygen to form free silica (SiO_2), the most common form of which is quartz. SiO_2 dominates the composition of dust from North Africa (60.95%) and China (60.26%). These values closely match the average content in the world's rocks (58.98%) (Derbyshire 2005). Si also combines with other elements in addition to oxygen to form the dominant mineral group known as the silicate family, which includes the group of fibrous

Fig. 3.5 Dust storm near Niamey, Niger (Photo: Andreas Bürkert, with kind permission)

amphibole minerals grouped together under the general term asbestos. In the finest (<2 mm) fractions of many dryland surface sediments and soils, quartz is an important, and sometimes a dominant mineral, ranging in type from lithic fragments to biogenic opal. Varying amounts of feldspars and clay minerals (notably kaolinite, illite, chlorite, vermiculite, smectite) are also common. Calcite, gypsum, and iron oxides result from weathering of primary minerals. The natural process by which substantial volumes of mineral dust are injected into the atmosphere is usually periodic, sometimes strongly seasonal (Fig. 3.5).

Dust storms have many impacts within the Earth System. Because of the large amount of mineral aerosol they place into the atmosphere, they have global implications in terms of climate change and biogeochemical cycling. Moreover, because of the large distances over which dust transport takes place, they have an impact on locations that are at great distances from the dust source regions which has implications for human health. In recent years the role of desert dust in the global system has become increasingly apparent (Goudie and Middleton 2006). For identification of major natural mineral dust sources, data from the Total Ozone Mapping Spectrometer (TOMS) have been of particular importance (Derbyshire et al. 2012). These data have demonstrated the great importance of the Sahara and some other drylands, including the Middle East, Taklamakan, southwest Asia, central Australia, the Etosha and Mkgadikgadi pans of southern Africa, the Salar de Uyuni of Bolivia and the Great Basin in the USA (Prospero et al. 2002).

Extension of deserts in semi-arid and arid lands occurs by both natural processes and human activity. Long-term natural processes include continental drift (semi-arid regions are transported into arid zones over a period of millions of years) and climate shifts due to changes in atmospheric chemistry or the planet's orbit. The Ice Ages of the Pleistocene Epoch in comparison to today's environment are good examples of climate variation. Low sea levels and arid environments are typical of ice ages. Ice core analyses have shown that during this period (more than 10,000 years ago), airborne dust was much more prevalent than at present. Most of the major dust source regions are large basins of internal drainage (Bodélé (former Lake Chad), Taoudenni, Tarim, Seistan, Eyre, Etosha, Mkgadikgadi, Uyuni and the Great Salt Lake) (Goudie 2009). Many of the world's major dust source regions are areas of hyper-aridity, with mean annual rainfalls of less than 100 mm (Goudie and Middleton 2001). The biggest dust source region in Asia is located in China (Tanaka and Chiba 2006), though estimates of its strength vary considerably. Concerning Chinese sources there has been considerable discussion as to the relative importance of the Taklamakan compared to the Gobi Desert (Goudie 2009). However, both sources are plainly very important (Laurent et al. 2006) and have been responsible for the development of the great loess deposits (more than 300 m thick) of wind-lain mineral dust known as loess in the Loess Plateau of Central China resulting from semi-continuous deposition of silt and sandy silt during the Pleistocene period. As these loess deposits are subject to natural erosion by rivers and streams since long, landslides and human activities such as deforestation, overgrazing, intensive agricultural actions and excessive irrigation, the Loess Plateau at present is very extreme in relief (Fig. 3.6). The main geomorphic landforms on the Loess Plateau include Yuan (flat surface with little erosion), ridge, hill and various gullies (Shi and Shao 2000). The surface of degraded loess soils is highly susceptible to wind erosion, making it a secondary but notable source of wind-blown dust.

Attempts to identify local dust sources within the various Chinese deserts were reported by Wang et al. (2006b), who stressed the importance of piedmont alluvial fans. More recently, Zhang et al. (2008) have found that the sandy lowlands of the eastern Mongolian Plateau are the principal contributor to long range dust transport to the North Pacific. Yang et al. (2007), using Rare Earth Element analysis, established that dust in the Beijing area had its main source in western China rather than closer sources. Surface erodibility is very important at a local and regional scale. Analysis of satellite images of areas like the Makran coast of Pakistan or the ephemeral rivers of Namibia indicate that dry river beds can be important point sources (Eckardt and Kuring 2005). Similarly, the nature of lake basin surfaces, including their texture, wetness and salt crust development may be highly significant (reviewed by Goudie 2009).

Future development of dust activity will depend mainly on anthropogenic modification of desert surfaces, natural climatic variability (e.g. in the North Atlantic Oscillation or the El Niño- Southern Oscillation or the) and changes in climate through global warming (Goudie 2009). Human activities will cause disturbance of desert surfaces by vehicular traffic, removal of vegetation cover for

Fig. 3.6 Extreme relief in the Loess Plateau of Central China, with altitude difference between ridges and valleys ranging from tens of meters to 200–300 m (Photo: Rolf Nieder)

wood supply, (over-)grazing and crop production, as well as desiccation of lakes and soil surfaces by inter-basin water transfers and ground water depletion. Natural climatic variability will continue as, but global warming has the potential to cause major changes in dust emissions. Under many scenarios dryland areas will suffer from lower rainfall levels and increased amounts of moisture deficits because of higher rates of evapotranspiration (IPCC 2007). If this is the case, dust storm activity could increase, though this is to a certain extent dependent on how wind energy levels may change (Goudie 2009).

3.3.2.2 Impact of Farming on Dust Generation

Human activities that have contributed to desertification and dust generation include agricultural land use practices. It has been estimated that the planet loses over 10 million hectares of farmland per year to unsustainable farming methods (Pimentel et al. 1995). One of the most prominent examples of the impact of farming on desertification is the Dust Bowl Days of the 1930s in the Midwest of the USA (Worster 1979). The promise of nutrient-rich soil and get-rich-quick farming caused mass migration to the region. The resulting combination of vast tracts of farmland, detrimental agricultural practices such as part-time farming, and the onset of what turned out to be a 10 year drought, resulted in an ecological disaster that drove over one-fourth of the population from the region (Griffin et al. 2001).

Africa has been more affected by desertification than any other continent. Over 60% of the continent is composed of deserts or drylands, and severe droughts are common (Griffin et al. 2001). In the Sahel (North Africa) dust emissions have increased significantly during the drought periods of the last decades (Ozer 2001). The population in the Sahel has increased by about 3% annually during the last decades, leading to an increased demand for food (Sterk 2003) which, in turn, has resulted in overcultivation, overgrazing, deforestisation and mismanagement of irrigated cropland which has been put forward as main reasons for desertification in the Sahel due to human activities (Thomas and Middleton 1994). The resulting increase of degraded soils has been attributed to be partly responsible for the increase in dust emissions (Mahowald et al. 2002). An area in Africa where desertification has been impacted by both nature and human activity, is Lake Chad (West Africa). The surface area of Lake Chad in 1963 was 25,000 km^2, and due to regional drought conditions and irrigation practices its surface area is now about 5% its 1963 size (~1350 km^2) (Coe and Foley 2001). Fifty percent of the lake's surface area decline has been attributed to water diversion/irrigation (Coe and Foley 2001).

The surface area of the Aral Sea (located within both Kazakhstan and Uzbekistan), which was the fourth largest lake in the world in 1960 (~68,000 km^2) (Micklin 1988), has decreased approximately by 50% until the 1990s (1992: ~33,800 km^2). This has been attributed to the irrigation schemes, which diverted river source waters for cotton production. Dust clouds originating from storm activity over approximately 27,000 km^2 of exposed seabed are common (Micklin 1988). Another country that has significant desertification problems is China. Approximately 27.3% of China (2,622,000 km^2) is affected by desertification. The average annual desertification rate was estimated at 2100 km^2 $year^{-1}$ between 1975 and 1987 (Zhenda and Tao 1993). Human factors such as population growth, deforestation (building, farming, firewood scavenging), and overgrazing have been identified as contributing to desertification in China (Fullen and Mitchell 1993).

The extent to which changes in dust emissions can be traced back to anthropogenic activities or natural climate variability (e.g. changes in precipitation or wind speed) is difficult to evaluate. There are only few studies estimating the contribution of anthropogenically disturbed soils on global dust emission by comparing scenarios of modeled dust emissions for different test cases allowing for higher emissions in regions with anthropogenically disrupted soils, e.g. croplands or pastures, with observations from either satellite aerosol retrievals or dust storm frequencies. The resulting contributions to global dust loads from anthropogenic soil dust sources range from 0% to 50% (reviewed by Engelstaedter et al. 2006). In a more recent study it was estimated that about 30% of the mineral dust load in the atmosphere could be ascribed to human activities through desertification and land misuse (Langmann 2013).

3.3.3 Asbestos and Asbestiform Minerals

Asbestos-rich dust occurs in the natural environment as well as in building materials. It originates from metamorphic rocks in orogenic belts around the world (Derbyshire et al. 2012). The term asbestos applies collectively to the six fibrous minerals viz. chrysotile, crocidolite, amosite, tremolite, actinolite and anthophyllite. Based on their chemistry and fiber morphology these minerals include the two groups, amphibole and serpentine asbestos (Frank and Joshi 2014). The serpentine group is comprised solely of chrysotile asbestos. Serpentine fibers appear wavy under low magnification (Fig. 3.7).

The amphibole group includes crocidolite, amosite, tremolite, actinolite, and anthophyllite asbestos. Amphibole fibers are all straight and needle-like in their microscopic appearance (Fig. 3.8).

Chrystolite carries the name "white asbestos", crocidolite is called "blue asbestos" and amosite is known as "brown asbestos". Chrysotile asbestos accounts for 90–95% of all the asbestos used worldwide. Crocidolite and amosite made up the bulk of the commercially used asbestos that was not chrysotile. In addition to the six asbestos minerals there are other fibrous minerals which are referred to as "asbestiform minerals" (Table 3.2). These minerals are similar structurally, but are not technically classified as asbestos (Frank and Joshi 2014).

All types of asbestos, including asbestiform minerals, cause asbestosis, a progressive, debilitating fibrotic disease of the lungs. Chrysotile represents nearly 100% of the asbestos produced and used worldwide today and 95% of all the asbestos used worldwide since 1900 (Virta 2005). Despite all that is known about

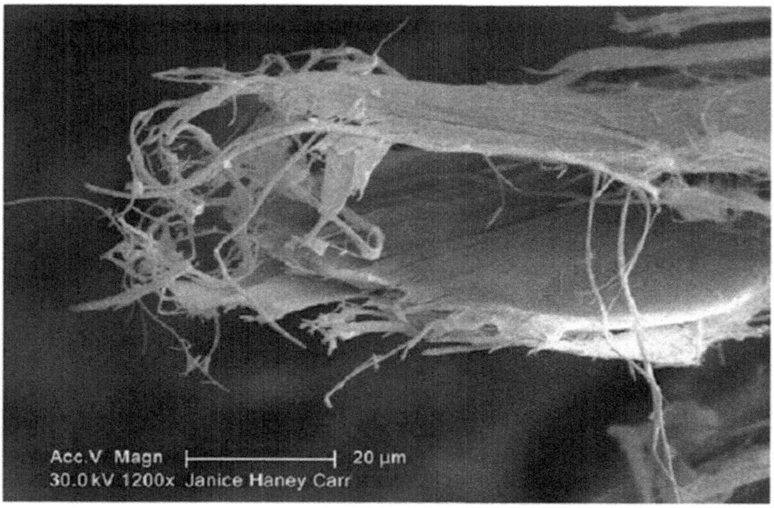

Fig. 3.7 Chrystolite asbestos (Frank and Joshi, 2014, p. 258; with kind permission from Elsevier)

Fig. 3.8 Amphibole asbestos (Frank and Joshi 2014, p. 258; with kind permission from Elsevier)

Table 3.2 Asbestos and asbestiform minerals

Name	Formula	Mineral type
Asbestos minerals		
Chrysotile (white asbestos)	$Mg_3(Si_2O_5)(OH)_4$	Serpentine
Amosite (brown asbestos)	$Fe_7Si_8O_{22}(OH)_2$	Amphibole
Crocidolite (blue asbestos)	$Na_2Fe^{2+}{}_3Fe^{3+}{}_2Si_8O_{22}(OH)_2$	Amphibole
Tremolite	$Ca_2Mg_5Si_8O_{22}(OH)_2$	Amphibole
Anthophyllite	$(Mg,Fe)_7Si_8O_{22}(OH)_2$	Amphibole
Actinolite	$Ca_2(Mg,Fe)_5(Si_8O_{22})(OH)_2$	Amphibole
Asbestiform minerals		
Erionite	$(Na_2,K_2,Ca)_2Al_4Si_{14}O_{36}{\bullet}15H_2O$	Zeolite
Fluoro-edenite	$NaCa_2Mg_5(Si_7Al)O_{22}F_2$	Endenite
Richterite	$Na(Ca,Na)(Mg,Fe^{2+})_5(Si_8O_{22})(OH)_2$	Amphibole
Balangeroite	$(Mg,Fe^{2+},Fe^{3+},Mn^{2+})_{42}Si_{16}O_{54}(OH)_{36}$	Serpentine
Nemalite	$Mg(OH)_2$	Magnesium hydroxide
Wollastonite	$CaSiO_3$	Pyroxenoid
Sepiolite	$Mg_4Si_6O_{15}(OH)_2{\bullet}6H_2O$	Clay

Adapted from Derbyshire et al. (2012)

the dangerous and adverse health effects of asbestos, annual world production remains at >2 million Mg (USGS 2009). Russia is now the leading producer of asbestos worldwide, followed by China, Kazakhstan, Brazil, Canada, Zimbabwe, and Colombia (LaDou et al. 2010). These six countries accounted for 96% of the world production of asbestos in 2007. Russia has mines rich enough in asbestos deposits to last for >100 years at current levels of production (Encyclopedia of the Nations 2010). Most of the 925,000 Mg of asbestos extracted annually in Russia is exported.

All forms of asbestos are now banned in 52 countries (International Ban Asbestos Secretariat 2010), including all EU member countries. However, these countries make up less than one-third of WHO member states. A much larger number of WHO member countries still use, import, and export asbestos and asbestos-containing products (LaDou et al. 2010). These are almost all countries in Asia, Eastern Europe, Latin America, and Africa. Most of the world's people still live in countries where asbestos use continues, usually with few safeguards. More than 85% of the world production of asbestos is used today to manufacture products in Asia and Eastern Europe (Virta 2005). In developing countries, where little or no protection of workers and communities exists, the asbestos cancer pandemic may be the most devastating. China is by far the largest consumer of asbestos in the world today, followed by Russia, India, Kazakhstan, Brazil, Indonesia, Thailand, Vietnam, and Ukraine (UNSD 2009).

3.3.4 Crystalline Silica

Silicon dioxide (SiO_2) or silica is the most abundant mineral on Earth and occurs in crystalline and amorphous forms (Leung et al. 2012). Because silicon and oxygen are so abundant in the earth's crust, many minerals contain SiO_2 usually combined with other elements, often cations. These minerals are known as silicates, rather than silica, and are ubiquitous. The most common free crystalline forms of silica in workplaces are quartz, tridymite, and cristobalite. Quartz can occur naturally and at varying concentrations in rocks such as granite (25–40%) and sandstone (67% silica) (Greenberg et al. 2007). Cristobalite and tridymite occur naturally in lava and are formed when quartz or amorphous silica is subjected to very high temperatures. They can also be formed in the manufacture of silica bricks (refractory bricks) used in industrial furnaces. Less common types include keatite, coesite, and stishovite. Opal, diatomaceous earth (tripolite), silica-rich fiberglass, fume silica, mineral wool, and silica glass (vitreous silica) which are common amorphous forms of silica (Greenberg et al. 2007). A number of operations or tasks involve exposure to free crystalline silica (Table 3.3).

Most common exposures occur in mining and mining-related occupations, such as milling ores, quarrying, tunnelling and excavation. Mining remains one of the most difficult, dirty and hazardous occupations, causing more fatalities than other occupations. Health and safety risks differ according to where the mines are, what

Table 3.3 Examples of industries and operations that involve exposure to crystalline silica

Industry	Operation	Source material
Agriculture	Plowing, harvesting, use of machinery	Soil
Arts, crafts, sculpture	Pottery firing, ceramics, clay mixing, abrasive blasting, sand blasting, engraving, cutting, grinding, polishing, buffing, etching, engraving, casting, etc.	Clays, glazes, bricks, stones, minerals,rocks, sand, silica flour
Cement	Raw materials processing	Clay, sand, limestone, diatomaceous earth
Construction	Abrasive blasting of buildings, highway and tunnel construction, excavation, earth moving, demolition	Sand, concrete, soil and rock, concrete, mortar
Dental material	Abrasive blasting, polishing	Sand, abrasives
Foundries	Abrasive blasting, fettling, furnace installation and repair	Sand, refractory material
Glass	Raw material processing Refractory installation and repair	Sand, crushed quartz, refractory materials
Iron and steel mills	Refractory preparation and furnace repair	Refractory material
Metal products	Abrasive blasting	Sand
Mining and milling	Underground and surface mining, milling	Ores and associated rock
Quarrying and milling	Crushing stone, sand and gravel processing, monumental stone cutting and abrasive blasting, slate work, diatomite calcination	Diatomaceous earth
Roofing and asphalt felt	Filling and granule application	Sand and aggregate, diatomaceous earth
Shipbuilding and repair	Abrasive blasting	Sand

Adapted from Madl et al. (2008)

products are being mined; who is involved and what processes are used. In many cases, the types of materials mined and their associated health effects are complex and interrelated. For example, studies focusing on the silicates attempt to distinguish fibrous silicates such as asbestos, asbestiform fibrous minerals (e.g. wollastonite), and non-fibrous silicates such as talc and kaolin (Short and Petsonk 1993). Mining activities providing minerals for the metallurgical, energy, aggregate, cement and brick-making industries, generate dusts during crushing and grinding (Derbyshire et al. 2012). Dust is also deflated from waste piles (tailings) if they are not kept moist. Such dust palls may represent a high risk to communities nearby and sometimes some distance away if toxic trace elements are present in the tailings dust.

Country rock, i.e., the rock in which the mined mineral is located, often determines the silica risk (Rees and Murray 2007). Coal mining, for example, is a silica risk in some regions. Industries with well-known silica risks include ceramics, construction and foundries. Occupations associated with furnace masonry, stone-working or cutting (e.g., monumental masonry and working tombstones), cutting and polishing gem stones and those in which fine silica materials are used (e.g., in fillers and abrasives) have a long history of producing silica-associated diseases. Abrasive blasting with sand, which has widespread applications in engineering, ship-building and in the metal and automotive repair industries, is particularly dangerous, and recommendations have been made to ban sand blasting in some countries (Rees and Murray 2007).

3.3.5 Coal Combustion

Coal has been an important source of energy for more than 150 years and will continue to be so for at least the first half of the twenty-first century to sustain the standard of living in developed countries. Coal dust, particularly in the process of extraction, has severe health impacts on coal miners. Despite the continued importance of occupational injuries in coal mining, it is the respiratory impacts due to inhaled coal dust particles that form the key health risk related to coal mining. Coal combustion is an important source of particulate matter, coal fly ash being widespread in industrial, many urban and some natural environments (Jones et al. 2009), and thus may affect non-occupationally exposed populations. The global energy demand is largely (>30%) met by coal fired power plants. World coal production is about 3.5×10^9 t year^{-1} (reviewed by Ram and Masto 2014). In many developing countries such as China, India and Mongolia, coal will be relied upon as the primary source of energy to fuel industrialization and to improve the standard of living of these large and growing populations (Fig. 3.9).

Burning coal for electricity is one of the biggest contributors to air pollution in China. However, primary sources of pollutants in Beijing include exhaust emission from more than five million motor vehicles, coal burning in neighbouring regions, dust storms from the north, and local construction dust (Fig. 3.10).

World coal consumption is expected to increase by 49% from 2006 to 2030 as per the projection of International Energy Outlook (IEO 2009). The coal consumption for electricity generation and the world total production in 2009 is given in Table 3.4.

The combustion of coal to generate electricity in thermal power plants produces solid wastes like fly ash, which can contain a component of unburnt organic matter. If this exceeds certain levels, the material is commonly reinjected into the furnace for secondary combustion (Derbyshire et al. 2012). The smaller-sized fly ash particles, typically around a few microns in diameter, are usually spherical and often display gas-expulsion holes. Larger particles tend to have a more blocky appearance and often have numerous smaller glass spheres adhering to the surface. In modern coal-burning factories or power stations, the vast majority of fly ash particles are collected post-furnace by electrostatic collectors or in bag rooms. The

Fig. 3.9 A coal-fired power plant spews smoke into the air in Ulan Bator, Mongolia Photo: Andreas Bürkert, with kind permission)

total amount of ashes produced worldwide is estimated to exceed 750 million Mg $year^{-1}$, but only less than 50% of fly ash production may be utilized (Table 3.5).

Fly ash consisting mainly of amorphous ferro-alumino silicates (Pandey and Singh 2010), can generally be used as an agricultural amendment, for example, to sewage sludge, biosolids, farmyard manure, press mud (a sugarcane industrial waste), biofertilizers, biochar, lime, gypsum and red mud (Ram and Masto 2014), or as amendment to building materials. However, fly ash may contain a range of toxic trace elements, including As, F, Hg, Pb, Cd, Cr, Cu, Ni, Zn and U, that can enter certain environments (Derbyshire et al. 2012). Both on a local and regional scale, human health has been adversely affected by coals containing these elements (Finkelmann et al. 2002). The overall chemistry of the fly ash is determined by the original chemistry of the coal (mineral impurities) and the operating conditions at the furnace. Fly ash that is not utilized is commonly dumped on nearby ash piles, which could be blown into the atmosphere (Derbyshire et al. 2012).

The human-health implications of fly ash produced by domestic coal burning, particularly in poorer countries or regions, are a subject of serious concern (Derbyshire et al. 2012). For example, in some parts of China, local Permian coals are burnt in poorly ventilated rooms in order to retain as much heat as possible and, consequently, the fumes are also retained.

Fig. 3.10 (**a**) Serious air pollution in Beijing with visibility limited to ~200 m. A severe smog engulfed the city for weeks in early 2013 (Photo: Lisa Heimann, with kind permission). (**b**) Photo taken from the same place without smog situation in summer with the Fragrant Hills in the background, located 25 km northwest of Beijing (Photo: Lisa Heimann, with kind permission)

Table 3.4 Coal consumption by electric plants in million Mg (country wise) and world total production in 2009

Country	Anthracite	Coking coal	Other bituminous coal	Sub-bituminous coal	Lignite	Peat
EU-27	4.21	5.58	115.27	2.02	289.07	4.17
Turkey	–	0.55	5.78	0.19	62.64	–
USA	0.61	–	332.09	438.74	58.04	–
Canada	–	–	3.58	28.75	10.08	–
Australia	–	–	27.40	28.38	644.60	–
China	–	0.17	1439.5	–	–	–
India	–	24.19	393.79	–	28.14	–
Japan	–	–	86.36	–	–	–
Indonesia	–	–	–	33.52	–	–
South Africa	–	–	121.39	–	–	–
World total	19.67	304.87	2722.9	558.2	575.2	4.17

Adapted from Ram and Masto (2014)

Table 3.5 Coal ash production (country wise)

Country	Year	Fly ash production (million Mg)	Fly ash utilization (million Mg)	Fly ash utilization (%)
EU-15	2010	48.0	43.87	91.4
Germany[a]	2010	15.26	15.26	100
Turkey	2012	24.0	–	10[b]
Israel	2012	1.45	1.36	94.05
USA	2011	59.9	22.98	38.36
Canada	2007–2008	6.09	1.88	31.0
Australia	2008	14.5	4.58	32.0
China	2010	480	321.6	67.0
India	2010–2011	131.09	73.13	55.79
Japan	2006	10.96	10.65	97.2

Adapted from Ram and Masto (2014)
[a]Limited to lignite fly ash
[b]According to data from 2003 to 2006

3.3.6 Liquid Fossil Fuel Combustion

Petrol and diesel-powered vehicles are an important source of particulate and gaseous atmospheric pollution. Residual ash from liquid fossil fuels has been categorized to be more harmful to human health than coal fly ash, the former being mostly less than 2.5 μm in diameter, chemically complex, largely inorganic and rich in metals. Liquid fossil fuels produce soot particles consisting of nanometer-scale carbon spheres that can exist individually but can also aggregate into larger chains or clusters corresponding to PM_{10} or more in diameter

(Derbyshire et al. 2012). The overall size of the particles is significant, as it has an impact on health issues such as the ability of the particles to penetrate deeply into the lung. However, it is well understood that these particles can chemically and biologically behave like the constituent nanoparticles and that the individual spheres have an onion-like structure of perturbed small graphitic structures (Derbyshire et al. 2012). The spaces between the graphitic molecules in the primary carbon spheres can trap metal sourced from the fuel, engine and exhaust systems. This includes platinum-group metals from the catalytic converters. The surfaces of the carbon spheres act as substrates for the condensation of a large variety of organic species, heavier-end hydrocarbons and inorganic species such as sulphates (Clague et al. 1999). In addition to the carbon sphere-based particles, other kinds of particles are generated by the direct condensation of organic and inorganic species due to cooling of the fumes in ambient air (Derbyshire et al. 2012).

Changing trends in both petrol and diesel engines have resulted in constantly changing composition of particles. For example, concerning petrol engines, the ban on the use of leaded petrol contributed to a significant decrease of atmospheric lead levels in most developed countries (Cook and Gale 2005). For diesel engines, changes include a significant increases in the number of private diesel cars, the introduction of biodiesel and the promotion of low-sulphur diesel, the development of diesel engines with sophisticated fuel-injection systems and tailpipe after-treatment devices (Derbyshire et al. 2012). In summary, the nature of the particles produced by diesel combustion depends on the type of fuel, the technology employed in the engine and exhaust systems, the running condition of the engine and ambient atmospheric conditions (Maricq 2007).

3.3.7 *Landscape Fires*

Landscape or vegetation fires represent an important source of atmospheric trace gases and aerosol particles. The term vegetation fire commonly used denotes open fires of various vegetation (forest fires, tropical deforestation fires, peat fires, agricultural burning, and grass fires) and peat that occur naturally or are set by humans, whereas the term biomass burning comprises prescribed and wild fires, as well as biofuel use, such as wood or peat for heating and cooking (Johnston et al. 2012; Langmann et al. 2009). The global extent of vegetation fire emissions was not fully recognised until the 1970s (e.g. Seiler and Crutzen 1980), but since the mid 1980s, remote sensing data became available for active fire and burned area detection (Langmann et al. 2009). However, fire emission estimates based on remote sensing are still affiliated with a high uncertainty (uncertainty range of at least $\pm$ 50%), mainly because of difficulties to accurately estimate the amounts of surface and subsurface biomass combusted per unit area (Andreae and Merlet 2001).

The most important fraction of landscape fire emissions is released as carbon with CO_2 and CO being responsible for about 90–95% of the total carbon emitted

(Andreae and Merlet 2001). A small fraction is released as CH_4 and other volatile organic carbon compounds. Less than 5% of the carbon is emitted as particulate matter (Reid et al. 2005). The major risk-related measure of smoke is particulate matter with an aerodynamic diameter $\leq$2.5 μm ($PM_{2.5}$). This PM particularly is composed of organic carbon and black carbon components, along with smaller contributions from inorganic species (Reid et al. 2005). Global fire-related carbon emissions have increased by around 50% (in the decadal mean) since the 1960s, mainly due to drastic intensification of forest fires in the tropics (Mouillot and Field 2005). Until recently, global and regional biomass burning emission inventories only included emissions from surface vegetation while omitting emissions from burning soil organic matter, in particular peat (van der Werf et al. 2003).

Peat soils (also referred to as Histosols) are characterized by a layer of partly decomposed organic material. Globally, peatlands cover an area of about 400 million ha (3% of the global land surface) and store approximately as much carbon as surface vegetation (around 460 Pg C) (IPCC 2001). In some ecosystems these emissions may equal or even exceed the emissions produced from burning of surface vegetation. The occurrence of peat fires and the resultant emissions have been increasing in several regions of the world over the last decades due to human perturbations, mainly drainage, and climatic change (Poulter et al. 2006). Kasischke et al. (2005) estimated that burning of soil organic matter, notably peat, contributed between 46% and 72% of all emissions from wildland fires in the boreal regions. Page et al. (2002) found that burning peat soil contributed between 79% and 84% of all carbon emitted (in total 0.5–2.6 Pg C) during the 1997 Indonesian wildland fire event. The most important disturbance controlling fire susceptibility and fire severity of peat layers is drainage. A permanent lowering of the water table by drainage renders the entire peat horizon above the drainage water level susceptible to fire. Drainage drastically increases fuel availability and fire severity because peat soils generally provide both, high fuel density and continuity. Both factors promote sustained, vertically and laterally spreading peat fires (Usup et al. 2000). Once the peat fire is ignited, it can keep burning for months and can even persist heavy rainfalls. Fires in deep peat deposits frequently spread below the surface, progressing around 30 cm per day (Usup et al. 2000). Observations show that fires may cause a decrease in surface elevation of the affected peat area by 1 m within a single burning season (Wösten et al. 2006).

Similar to vegetation and peat fires, coal fires are an environmental and economic problem of international magnitude. They occur worldwide in countries like China, Russia, the USA, Indonesia, Australia and South Africa (reviewed by Künzer et al. 2007). The term coal fire refers to a burning or smouldering coal seam, coal storage pile or coal waste pile. The adsorption of oxygen at the outer and inner surface of coal is an exothermic reaction leading to an increase in temperature within the coal accumulation. If the temperature of a coal accumulation exceeds approximately 80 °C the coal can ignite and starts to burn (Künzer 2005). This process is referred to as "spontaneous combustion". It is – after human influences – the second-most common cause for coal fires of large extent (Künzer et al. 2007). Coal fires can also be caused by through lightning, and forest- or peat fires.

However, mining activities, mining accidents and careless human interaction, such as improper mining techniques, thrown away cigarettes, small coal fire for heating in winter or for daily cooking are common courses for coal fire ignition. China is the leading country concerning coal production and faces the problem of numerous uncontrolled burning coal fires in more than 11 of its provinces (Walker 1999). The main coal fire areas stretch along the coal-mining belt in China. This belt extends for 5000 km from East to West along the North of the country. In this area more than 50 coal fires areas of larger extent are known. Area wise, China therefore faces the worldwide biggest problem of coal fires, including the negative impacts associated (Künzer et al. 2007).

Global landscape fire activity is highly variable in space and in time. For each geographic location on earth, a typical fire season exists (Carmona-Moreno et al. 2005). The timing of the fire season is linked to the typical seasonal variation of the prevailing climate conditions in a given region, namely precipitation and temperature, which directly influence fuel moisture and ignitability of the vegetation (Chuvieco et al. 2004). For example, in northern hemispheric winter, global fire activity concentrates across Africa's northern savanna belt, where particularly dry conditions prevail during that season (Cahoon et al. 1992). During northern hemispheric spring, in contrast, most fires occur between 3° and 17° N, namely in Central America and continental South Asia, as well as in southern Australia.

The fire seasonality is very stable with respect to climate variability in regions such as the African savannas, the high and medium latitudes in the Northern Hemisphere and the Far-East of China-Russia. In these areas interannual variations in climate have little or no effect on fire seasonality. In contrast, other regions such as Indonesia, Southern Europe, Southern and East Africa, California, Australia, Southern East Asia and areas in Latin America, are much more sensitive to interannual climate variability and show considerable interannual variations in regional fire activity and hence regional to global emission production (Carmona-Moreno et al. 2005). The latter authors demonstrated that the strong regional variability in the fire seasonality and overall fire activity is largely correlated with the El Niño Southern Oscillation, an interannual climate anomaly that significantly alters precipitation regimes in several regions of the world. Notably in the northern hemisphere and in the tropics between 5° N and 5° S, fire activity is strongest during the El Niño periods. This connection is particularly pronounced in Indonesia where unusually large fire events occur during El Niño years. Most emissions currently originate from fires in tropical rainforests and savannas where they cause severe pollution that affect some of the poorest regions of the world (van der Werf et al. 2010; Fig. 3.11). The annual global distribution of carbon released from vegetation fires (original data from the Global Fire Emission Database; van der Werf et al. 2006) is shown in Fig. 3.12. Langmann et al. (2009) used this data to describe the general global spatial and temporal distribution of vegetation fires (excluding coal fires).

In the tropics and subtropics, most vegetation fire emissions originate from savannah burning in Africa, with about 50% of the global total (Korontzki et al. 2003). Large-scale savannah burning also takes place in Australia (Hurst et al.

Fig. 3.11 Savanna fire in Zambia. The smoke forms a column that may rise hundreds of meters into the air above the origin of the fire (Photo: Friedrich Bailly)

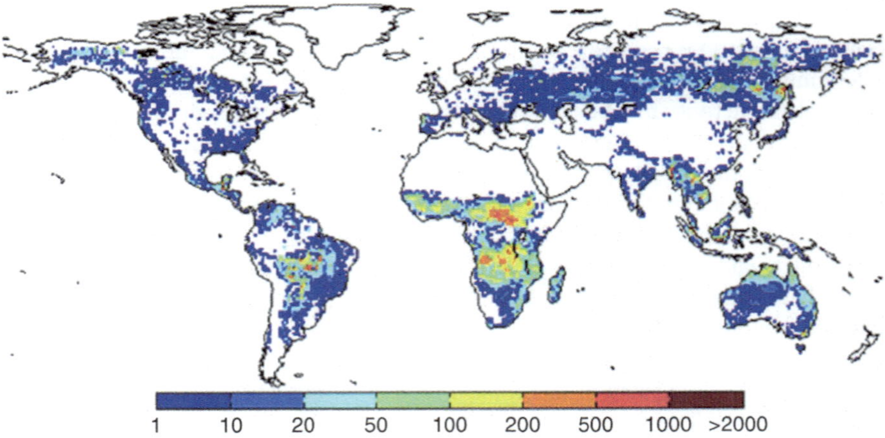

Fig. 3.12 Annual vegetation fire emissions averaged over 1997–2006 in g C m^{-2} $year^{-1}$ (Langmann et al. 2009, p. 110; with kind permission from Elsevier)

1994) and South America (Prins and Menzel 1994). Smoke from African and Brazilian savanna fires has been shown to contain substantial quantities of fine particles (Echalar et al. 1995). Mass concentrations ranged from 30 μg m^{-3} in areas not affected by biomass burning to 300 μg m^{-3} in large areas with intense burning. Additional studies of fine particle (<2 μm) composition associated with biomass burning in the Amazon Basin was reported by Artazo et al. (1994), who found 24-h average PM_{10} and $PM_{2.5}$ mass concentrations as high as 700 and 400 μg m^{-3}, respectively (Artazo et al. 1994).

Deforestation fires occur in Central and South America, Africa and Southeast Asia (Achard et al. 2002). These emissions are not compensated for by re-growth and provide a net source of CO_2 to the atmosphere. In the northern hemisphere midland high latitudes, vegetation fires in boreal forests, peat and grassland take place over much of North America and Eurasia (Kasischke et al. 2005). During one grass burning event in the state of Washington (USA), peak PM_{10} concentrations reached 100 μg m^{-3} (Norris 1998). Emissions from mid- and high latitude fires are about 10% of global carbon emissions, and somewhat higher for reduced gas emissions like CO and CH_4 due to more incomplete combustion compared to savannah regions.

There have been few studies of the impacts of agricultural burning, despite growing concern about its potential impact on human health (Tenenbaum 2000). In many countries, it is still a common practice to openly burn agricultural residues in fields after harvesting crops by mechanical harvesters. It is estimated that in India 22.3 million Mg of rice straw surplus is produced each year out of which 13.9 million Mg is estimated to be burnt in the field. The two states Punjab and Haryana alone contribute 48% of the total (Gadde et al. 2009). In China's rural areas, burning straw after harvest is common, which is a significant seasonal source of air pollution (Fig. 3.13).

The total quantity of straw (mainly wheat and rice straw) in China is about 600 million Mg each year, of which approximately one-third is burned in open air (Cao et al. 2006). Burning of straw is particularly important in Hunan, Hubei, Anhui, Jiangxi, and Jiangsu provinces (Qu et al. 2012). The air in numerous cities in China during the straw burning season was characterized by concentrations of fine particulate matter, including $PM_{2.5}$, that were much greater than during other seasons. Monitoring of the air quality in Nanjing showed that concentrations of $PM_{2.5}$ rose to 194 μg m^{-3}, which was threefold greater than normal. This concentration exceeded the nation's maximum allowable daily average concentration of 75 μg $PM_{2.5}$ m^{-3}. Data on quality of air collected in the city of Huaian showed maximum $PM_{2.5}$ concentration of 818 μg m^{-3}. In a Canadian study, 24-h average PM_{10} levels due to burning of wheat straw increased from 15–40 to 80–200 μg m^{-3} (Long et al. 1998).

Fig. 3.13 Burning of straw near Wuxi in Jiangsu Province, China (Photo: Marco Roelcke, with kind permission)

3.4 Global Pathways

3.4.1 Volcanic Ash

During volcanic eruption, ash (tephra <2 mm diameter) is convected upwards within the eruption column and carried downwind, falling out of suspension and potentially affecting communities across hundreds or even thousands of square kilometers. The products of volcanism and their impacts can extend beyond country borders, to be regional and even global in scale. Volcanic ash may be transported by prevailing winds hundreds or even thousands of kilometers away from a volcano and very large explosions can inject volcanic ash into the stratosphere (Fig. 3.4).

About 20 volcanoes on average erupt simultaneously at any given time worldwide, 50–70 volcanoes erupt throughout a year, and at least one large eruption occurs annually (reviewed by Langmann 2013). The dispersal of volcanic ash depends principally on meteorological conditions, including wind (speed and direction) and humidity, the grain size distribution of the ash, and the height of the volcanic plume, which depends on the intensity of the eruption. Total emissions by small volcanic eruptions with are usually removed from the atmosphere quickly and are therefore only of local interest in the vicinity of the volcanoes up to a distance of about 100 km. Stronger volcanic eruptions (e.g. Mount St. Helens, 1980; El Chichon, 1982; Pinatubo, 1991) are generally less frequent. However, depending on meteorological conditions and the injection height of the volcanic emissions, tephra from strong eruptions may be transported over thousands of kilometers in the atmosphere (Langmann et al. 2010), suggesting that volcanic ash is an abundant atmospheric species.

During the eruption of Eyjafjallajökull on Iceland in 2010, maximum ash concentrations up to 4000 μg m^{-3} were recorded from measurements of the volcanic ash cloud spreading over Europe, exceeding the threshold for safe aviation (2000 μg m^{-3}) (Langmann 2013). Daily mean near surface concentrations reached up to 400 μg m^{-3} in Scandinavia during the eruption. Close to Eyjafjallajökull, the maximum daily average near surface concentration exceeded 1230 μg m^{-3} during the ongoing eruption. After the eruption stopped, maximum daily average concentration still reached $>$1000 μg m^{-3} during resuspension events (Thorsteinsson et al. 2012). Volcanic ash concentrations within the volcanic eruption plume were expected to be even higher.

Volcanic ash is removed from the atmosphere by gravitational settling, turbulent dry deposition, and wet scavenging by rain called wet deposition (Langmann 2013). The ratio of dry-to-wet deposition differs considerably in mineral dust model estimates, with wet deposition over the ocean ranging from 30% to 95% of the total mineral dust deposition (Schulz et al. 2012). Olgun et al. (2011) estimated the millennial scale flux of volcanic ash into the Pacific Ocean based on marine sediment core data to be about 128–221 Tg year^{-1} which represents a conservative estimate, as already the volcanic eruption of the Kasatochi in 2008 alone can produce higher deposition fluxes. The estimated millennial volcanic ash flux is comparable to the mineral dust flux into the Pacific Ocean of around 100 Tg year^{-1} (reviewed by Langmann 2013).

Hazard footprints of ash-fall on the ground can be very large (up to millions of square kilometers in the largest eruptions) and can affect many different countries. The dispersal of volcanic ash is, therefore, of global concern (Brown et al. 2015). The impacts of ash fall are more complex and multi-faceted than for any of the other volcanic hazards. Near the volcano, thick accumulations of tephra and ash can cause roofs to collapse and lead to injuries and fatalities. Moderate ash-falls of several centimeters may damage infrastructure (e.g. power grids) and cause structural damage to buildings (Wilson et al. 2012). Even relatively thin falls ($<$1 mm) may damage crops and vegetation, disrupt critical infrastructure services, aviation, primary production and other socio-economic activities and may threaten public health over potentially large areas (Guffanti et al. 2010; Wilson et al. 2012).

3.4.2 *Mineral Dust*

Dust storms are commonly the result of turbulent winds which raise large quantities of dust from desert surfaces and may reduce visibility to less than 1 km. The dust can be transported over thousands of kilometers and is deposited downwind by wet and dry processes (O'Hara et al. 2006). Wet dust deposition is known regionally as 'loess rain' in eastern China and 'blood rain' in Mediterranean Europe (Derbyshire et al. 2012). An important precondition for dust formation is the availability of fine deflatable material which can be lifted from the ground when surface wind velocity exceeds a certain threshold wind speed (Engelstaedter et al. 2006; Marticorena and Bergametti 1995). This threshold depends on soil texture, moisture and surface roughness (e.g. rocks and vegetation). The European Center for Medium range Weather Forecasting calculated threshold wind speeds between 6.63 ± 0.67 and 9.08 ± 1.08 m s^{-1} for seven locations in North Africa (Chomette et al. 1999). Classification of dust concentrations in relation to wind velocities in north-east Asia showed that wind velocities of ≥ 10 m s^{-1} can generate total suspended solid loads in excess of 6000 mg m^{-3} with PM_{10} fractions as high as 5000 mg m^{-3} (Song et al. 2007). Silt-sized particles in the 10–50 mm range are readily picked up by the wind from dry, unvegetated surfaces, but, because of the high inter-particle cohesive forces typical of very fine (colloidal) materials, entrainment of clay particles (<2 mm) commonly occurs as cohering silt-size aggregates or as attachments to silt-size grains. The mean size of entrained particles diminishes with transport distance because of fallout of larger and denser particles.

The emissions of soil dust particles cannot be easily measured directly because the atmospheric lifetime of dust is relatively short and source regions, particularly in the vast expanses of the Sahara desert, often lie in remote desert areas where the installation and maintenance of equipment for continuous long-term measurements is not currently possible. In this setting, remote observations, either at the surface and downwind of sources, or from satellites, play a crucial role. Estimates of global dust emissions tend to vary between 1000 and 3000 Tg $year^{-1}$, and Saharan dust emissions between 500 and 1000 Tg $year^{-1}$, demonstrating that Sahara is the most important global source (Engelstaedter et al. 2006). By using different assumptions including erodibility factors, global mineral dust emissions were estimated to vary between 1500 and 1800 Tg $year^{-1}$ (summarized by Langmann 2013). Injection heights are usually restricted to the planetary boundary layer (2–4 km) but may reach up to 6 km dependent on meteorological conditions (Formenti et al. 2011). Within North Africa, the Bodélé depression may alone be responsible for 6–18% of global dust emissions, even though its surface area is relatively small (Todd et al. 2007). The lake deposits of the former lake Chad include abundant broken particles of diatomite (siliceous shells of eukaryotic algae) and fluxes from this source can reach 1.18 ± 0.45 Tg day^{-1} (Todd et al. 2007). Zhang et al. (2008) for China suggested an annual dust emission rate of 800 Tg, whereas Tanaka and Chiba (2006) indicated a value of only 214 Tg $year^{-1}$. Laurent et al. (2006) give values ranging between 100 and 460 Tg $year^{-1}$.

The natural processes by which mineral dust is injected into the atmosphere are commonly strongly seasonal and located mainly in arid and semi-arid regions (Choobari et al. 2014). Once in the atmosphere, dust particles can be transported over long distances by prevailing winds. Dust particles from North Africa are carried by the tropical easterly trade winds across the tropical Atlantic to the Caribbean, Central America and southern United States during summer, as well as spread westward as far away as South America during spring and winter (Griffin 2007; Griffin et al. 2002) (Fig. 3.14).

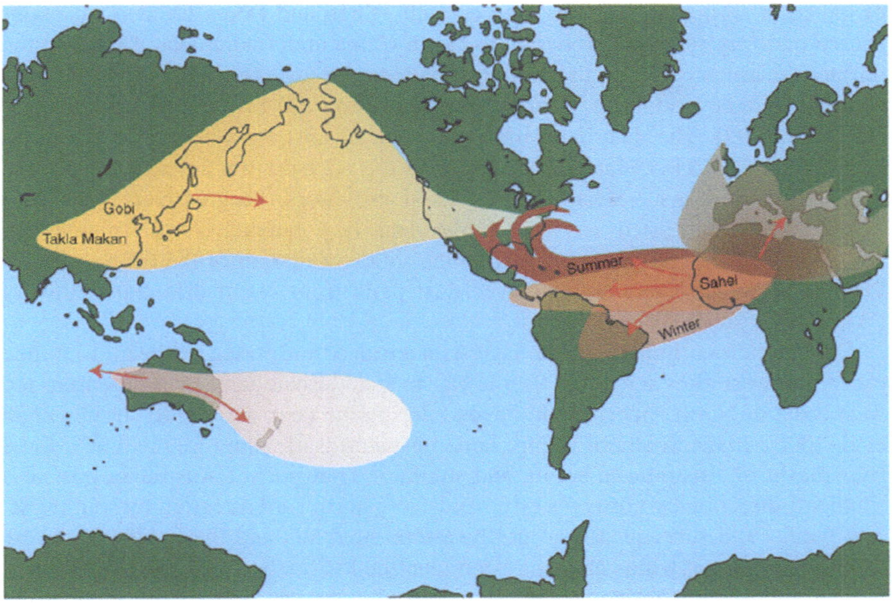

Fig. 3.14 Presentation of major global dust transport systems. Trade winds of the African dust system (red–orange) carry Saharan dust to the Caribbean and USA from May to November (summer). From December to April (winter) African dust flows to South America. Throughout the year, pulses of dust from northern Africa cross into the Mediterranean and Europe. The Asian dust system (yellow) exports dust primarily during March–May (spring). These dust events can incorporate emissions from industries in China, Korea and Japan, carrying a 'brown smog' across the Pacific to the west coast of North America. Large Asian dust events can also travel across the USA and then impact Europe. Australian deserts (pink) produce large dust storms that can reach New Zealand and the South Pacific (Kellog and Griffin 2006, p. 640; with kind permission from Elsevier)

Dust storm activity over North Africa is subject to a strong seasonal cycle, with maximum dust activity in spring and summer (Engelstaedter et al. 2006). Saharan dust may take about a week to cross the Atlantic Ocean, typically reaching north-eastern South America, including the lower Amazon basin, in the (northern) late winter and spring and the Caribbean, Central America and the south-eastern United States in summer and early autumn (Prospero and Nees 1986). Dust storms in Iran, north-eastern Iraq and Syria, the Persian Gulf and southern Arabian Peninsula are more frequent in summer (Furman 2003). In western Iraq and Syria, Jordan, Lebanon, northern Arabian Peninsula and southern Egypt, dust storms were found to be more active in spring, while in the Mediterranean parts of northern Egypt there are more frequent dust storms in winter and spring. Peak dust activity in East Asia is in spring (Choobari et al. 2014). For example, dust activity in the Gobi desert mainly occurs in spring, with the highest frequency in April, while dust storms are least frequent in summer. This seasonal variation of dust activity in the Gobi desert has a direct correlation with wind speed, with strong and weak winds observed in spring and summer, respectively (Shao and Dong 2006). Two major source-pathway systems (Mongolia to North China and Taklamakan Desert to the northern margins of Tibet) drawing dust palls in the lower troposphere to Beijing, Nanjing and beyond have been determined using Nd-Sr isotopic composition (Li et al. 2009). The dust deposition sequence in East Asia begins on the north and north-west China plateau regions, with the coarsest fractions falling on the loess lands of Gansu, Shaanxi and Shanxi provinces. Progressively finer dust fractions are transported eastward to the densely populated North China Plain, Korea and Japan. The city of Xi'An on the southern margin of the Chinese Loess Plateau in Shaanxi province receives dust palls from seven different pathways (Wang et al. 2006a).

Dust particles originating from East Asia are also transported by the mid-latitude prevailing westerlies across the Pacific Ocean and ultimately reach as far east as the west coast of North America and beyond during the peak activity in spring (Zhao et al. 2006). In the Southern Hemisphere, dust storms are more frequent over Lake Eyre Basin of Australia in spring and summer. Transport of Australian dust in a southeast direction by northwesterly winds, in a northward direction by southwesterly winds, and subsequently to northwestern Australia and the Indian Ocean by southeasterly trade winds has long been identified (Choobari et al. 2012).

The atmospheric residence time for aerosols depends on their location in the atmosphere, with a longer atmospheric lifetime for aerosol particles in the upper troposphere than those in the lower troposphere (Bolin et al. 1973). It is also size dependent, in which relatively large particles fall out rapidly near their sources due to the effect of gravity, while small particles can be suspended in the air for extended periods of time (Bolin et al. 1973). All aerosols, however, are finally extracted from the atmosphere by either dry deposition because of gravitational settling and/or turbulent fluxes or wet deposition through in-cloud (by cloud droplets) and below-cloud (by raindrops) scavenging (reviewed by Choobari et al. 2014).

Particulate dust transported westwards from the Sahara influences air quality in North America, the Caribbean, South America, Europe, the Middle East and Asia and affects the nutrient dynamics and biogeochemical cycles from northern Europe to South America (Derbyshire et al. 2012). It has been estimated that every year 13 Tg of African dust is deposited on the north Amazon Basin (Griffin et al. 2001, 2002). On the Caribbean island of Trinidad during influx of Saharan dust, PM_{10} values of 135–149 mg m^{-3} were registered which was four times the known values for non-Saharan days (Rajkumar and Chang 2000). At five sites in Spain, it has been shown that the daily mean concentration of $PM_{2.5}$ increased from 4 to 11 mg m^{-3} during influx of Saharan dust (Viana and Averol 2007). McKendry et al. (2007) reported detection of north-east African and Middle East dust over Japan, implying trans-Eurasian movement of natural mineral particles from Africa eastward as well as westward. In March 2005, for example, Saharan dust was traced across Asia and the Pacific, reaching the western coast of Canada in just 10 days (McKendry et al. 2007). In March 2003, rainwater samples in Japan contained dust and nanometre-scale particles that differed in both composition and shape from typical Chinese dust.

3.4.3 *Particles from Landscape Fires*

Particulate pollution generating smoke-haze is the most relevant impact of landscape fires on air quality and human health (Langmann et al. 2009). Biomass burning is presently the principle source of air pollution in the tropics. The smoke plumes emanating from the fires may be subject to long-range transport and can thereby affect air quality even more than thousand kilometres away from the fires (Bertschi and Jaffe 2005). During the burning seasons in Africa and South America, air pollution is substantially enhanced on regional scales (Bremer et al. 2004). In Southeast Asia, an extreme event of air pollution occurred during the strong El Niño of 1997/1998, when land-clearing fires became uncontrolled in Indonesia due to severe drought conditions. Total particulate matter concentrations of up to 4000 μg m^{-3} were reported close to the main fires on Borneo and Sumatra (Heil and Goldammer 2001). Cross-boundary transport of these polluted air masses led to dilution but still impacted air quality in neighbouring countries such as Singapore and Malaysia with total particulate matter concentrations up to 300–1000 μg m^{-3}. From September to November 1997, the smoke plume ranged from Papua New Guinea in the east into the Indian Ocean south of India (Nakajima et al. 1999). Smaller scale perturbations of air quality caused by vegetation fires occur worldwide, e.g. during summer 2003 in Portugal (Hodzic et al. 2007) and in California (Wu et al. 2006), and in summer 2002 in the Northeast US (DeBell et al. 2004). During dry weather condition, particularly for large injection heights, aerosol particles can be transported over thousands of kilometers before they are removed from the atmosphere by wet deposition. Intercontinental transport of vegetation fire aerosols in the atmosphere resulting from vegetation fires in Canada caused thin atmospheric layers with increased aerosol concentrations over Europe (Forster et al. 2001). Damoah et al. (2004) reported about hemispheric-scale transport of forest fire smoke from Russia in May 2003. Fires in Eastern Europe lead to record level pollution levels in the Arctic in spring 2006 (Stohl et al. 2007).

3.4.4 *Bacteria, Fungi and Viruses in Dust*

Dust events inject a large pulse of microorganisms and pollen into the atmosphere. Pathogenic microorganisms can be transferred by dust storms worldwide (Weir-Brush et al. 2004). Wind-borne bacteria are typically transported <1 km from their source (Bovallius et al. 1980), while dust-associated bacteria can be transported over 5000 km, e.g. from Africa to the Caribbean (Griffin et al. 2003). More than 40 fungal colony-forming units or spores have been found within dust palls (Griffin 2007); most of the genera known to contain pathogenic species are derived from the Sahara, Sahel and Middle East. According to a study in Mali, from 95 dust-borne bacteria identified, 25% were opportunistic pathogens (Kellogg et al. 2004). Griffin et al. (2003) reported that the number of airborne microorganisms (bacteria and fungi) in dusty days was nearly five times of normal days. In their study, *Bacillus* spp. and *Cladosporium* spp. were the most common species of bacteria and fungi, respectively. In South Korea, significant positive correlations were found between PM_{10} and culturable bacterial population levels during the days affected by Asian dust events (Jeon et al. 2011). Nourmoradi et al. (2015) in western Iran found that the mean number of bacteria and fungi of air in dusty days was 1.5 times and 3.83 times of normal days, respectively. The latter study also showed that *Bacillus* spp. (56.2–66.6%), *Microsporum* spp. (28.6%), and *Cladosporium* spp. (31.3%) were the predominant species of bacterial and fungal microorganisms, respectively, detected over normal and dusty days. Some genera of bacteria (e.g. *Bacillus* ssp.) and most fungi can form spores, a dormant state that is resistant to desiccation, heat, radiation and nutrient-poor conditions (Nicholson 2002). Many of the bacteria that are isolated from aerosol samples are highly pigmented, suggesting that pigmentation also helps to protect the microbes from UV radiation, in addition to the protection afforded by clouds, fog, smoke and desert dust particles (Griffin et al. 2003). A wide range of human diseases is caused by spores of pathogenic fungi. Like fungal spores, pollen is also adapted for aerosol dispersal and can be transported thousands of kilometers in the presence or absence of dust (Gregory 1973). While short-range transmission of infectious human viruses is well documented, little is known about long-range transmission within dust storms (Griffin 2007). Occurrence, clinical effects and therapy of diseases resulting from soil-borne pathogens are discussed comprehensively in Chap. 13.

3.4.5 Plant and Animal Pathogens in Dust

The effects of desertification on humans extend beyond the primary consequences of airborne particles and components attached to them. There are also effects of pathogens which infect plants and animals we depend upon for food or which hold important ecological roles. The estimated economic damage caused by invasions of non-native plants, animals, fungi and microbes was estimated to exceed $138 billion per year (Mack et al. 2000). Epidemics of infectious disease have been equated with the problem of invasive species (Kennedy 2001). Many of these 'invaders' are imported as a consequence of international trade and travel, however, there is an unknown percentage of microbes which are transported by the wind. There are a surprising number of reports of long-range transport of plant pathogens. The majority of these infectious invaders are fungi, whose dispersal spores provide protection from UV light and other harsh environmental conditions. For example, the potato blight (*Phytophthora infestans*) which caused widespread famine in the 1800s is thought to have originated in Mexico or South America, and was then transferred to Europe via infected potatoes (Bourke 1964). Once the fungus arrived in Europe, however, its rapid spread across the continent and to the British Isles is attributed to airborne transport (Griffin et al.

2001). This pathogen destroyed entire fields in a matter of days and rapidly spread downwind by releasing spores. 'Potato blackleg' is a potato disease caused by the bacteria *Erwinia carotovora* and *Erwinia chrysanthemi* which are dispersed by wind after being aerosolized by splashing raindrops, as well as being carried by insect vectors. These bacteria have been detected in air samples in Scotland several hundred meters from potato fields, as well as on insects trapped near vegetable refuse dumps (Perombelon 1992).

Seasonal atmospheric transport of spores is known to occur in the Americas. The "*Puccinia* Pathway" describes the 'migration' of the fungal wheat pathogen *Puccinia graminis* from southern Texas and northern Mexico to the northern USA and Canada in the spring, and then back again in late summer and autumn (Pedgley 1986). A series of blue-mold epidemics in the USA tobacco were attributed to aerosol transport of *Peronospora tabacina* spores (Davis 1987). The spread of sugarcane rust (*Puccinia melanocephala*) in the late 1970s was traced back to transoceanic transport from Africa to the Caribbean to the Americas (Purdy et al. 1985). Coffee rust (*Hemileia vastatrix*) is speculated to have traveled from Africa to Brazil within 5–7 days (Bowden et al. 1971). Within 1 year (1933–1934) the spread of banana leaf spot (*Mycospherella musicola*) was mapped as originating in Australia, then following a path to Africa and then to the Caribbean (Stover 1962).

Compared to plant pathogens there is less information available concerning the aerosol transmission of animal pathogens. Several studies of dust collected from surfaces on poultry, pig, and dairy farms have shown that the dust contained fungi such as *Aspergillus* and *Cladosporium* (Fiser et al. 1994) as well as the bacterium *Salmonella* (Letelier et al. 1999). Windblown desert dust carrying fungi caused an outbreak of aspergillosis in deserts (Venkatesh et al. 1975). Meteorological data and molecular techniques were employed to determine the source of the pseudorabies virus (cause of Aujeszky's disease in pigs) after outbreaks occurred in Denmark in December 1988. The evidence suggested the infections were a result of airborne transport of the viral pathogen from Germany (Christensen et al. 1993).

There were reports from Scandinavia that the virus responsible for foot-and-mouth disease (FMD) was being transmitted by air from Germany into Denmark and later into Norway and Sweden (>100 km) (Donaldson et al. 1982; Gloster 1982). Even considering that knowledge, it was thought unlikely that FMD virus would be able to cross the English Channel from France to Britain (distance: 250 km). However, just a year later a series of outbreaks on the coast of France proved that longer transport was possible. Beginning March 4, 1981, there were 14 outbreaks in France (13 in Brittany and 1 in Normandy), followed by outbreaks directly across the Channel in England (Jersey and the Isle of Wight) on March 19 and 22 (Donaldson et al. 1982). Favorable meteorological conditions combined with the earlier models predicted spread to those exact locations and biochemical analyses found no difference between the French and British viral isolates, which were all the same serotype (Donaldson et al. 1982).

3.5 Human Health-Affecting Properties of Dusts

The effect of inhaled particles on the human body varies widely, depending on certain characteristics of the particulate matter, the nature of any pathogens attached to them, the dust concentration, the exposure time and an individual's personal vulnerability (Derbyshire et al. 2012). The hazardous properties of airborne dusts vary with their size, shape, geochemistry, mineralogy, degree of crystallinity, density, solubility and reactivity with human fluids and tissue (Guthrie 1997). Fine particles may remain suspended in the atmosphere for hours to weeks depending on their size, form and density. Particles of the same diameter tend to have the same settling velocity. Those of the fraction PM_{10} with a relatively small suspension half-life and in the human body are largely filtered out by the nose and upper airway. Particles in the range of 2.5–10 μm diameter are commonly defined as "coarse", those with 2.5–0.1 μm diameter as "fine," and those with less than 0.1 μm as "ultrafine" (Anderson et al. 2012). In a mixed environmental sample, the total number and total surface area of these particles increases exponentially with decreasing diameter. In contrast, the total particulate mass of a substance generally decreases exponentially with decreasing particle diameter. For example, in a sample of PM_{10}, the numerical majority of particles would be ultrafine, but these particles would make up a negligible portion of the sample's total particulate mass. The composition of particles varies, as they can absorb and transfer a multitude of pollutants (Kampa and Castanas 2008). However, their major components are metals, organic compounds, material of biologic origin, ions, reactive gases, and the particle carbon core. Recent studies show an increase in morbidity and mortality related to particulate matter exposure. Epidemiological studies have shown that $PM_{2.5}$ has significant influences on human health, including premature mortality and $PM_{2.5}$ concentration contributes to approximately 800,000 premature deaths year^{-1} (WHO 2002), ranking it the 13th leading cause of mortality worldwide (Anderson et al. 2012).

3.5.1 *Crystalline Silica*

Inhalation of the crystalline form of silica has been associated with the development of severe respiratory diseases. Particles and fibers (e.g. asbestos fibers) can be cleared by several mechanisms, including the mucosal-ciliary escalator, engulfment, and removal by macrophages, or through splitting and chemical modification. Although the exact sequence of events from silica inhalation to disease is not known, it is generally accepted that the alveolar macrophage is the first cell of the body that will have significant contact with the inhaled silica particle (Davis 1986). It is the role of the macrophage to clear the lung of inhaled debris. Upon contact, the alveolar macrophage will bind to the silica and begin to engulf the particle. If the macrophage survives the silica encounter, it will likely migrate out of the lungs to either the proximal lymph nodes or through the mucosal-ciliary escalator and eventually out of the respiratory tract (Lapp and Castranova 1993). If the macrophage stays in the lung it will migrate to the interstitial space and become an activated interstitial macrophage that could contribute directly to pathogenesis (Migliaccio et al. 2005). Investigators studying the alveolar macrophage/silica particle interaction have developed several hypotheses to describe how silica is toxic to alveolar macrophage. Some explanations of toxicity focus on the surface qualities of the silica such as free radicals and surface charge (Hamilton et al. 2008). The formation of reactive oxygen species on the particle surface and free radical release are thought to be primary processes inducing inflammation. In the case of crystalline silica, breaking of the bond between Si and O by bond cleavage generates dangling bonds (reactive surface radicals) and surface charges, respectively (Fubini et al. 1995). Particle-derived reactive oxygen species from crystalline silica particles in contact with epithelial cells enhance both oxidative stress and inflammation (reviewed by Derbyshire et al. 2012). Crystalline silica particles, similar to asbestos fibres (Sect. 6.1), are thought to promote lung disease, in part, through iron-dependent generation of reactive metabolites, including superoxide, hydrogen peroxide, hydroxyl radical, and nitric oxide. Superoxide anion is formed by reduction of molecular oxygen (Eq. 3.1), which is then dismutated via the action of superoxide dismutases to hydrogen peroxide (Eq. 3.2). Highly reactive hydroxyl radicals may then be generated from superoxide anion and hydrogen peroxide via Haber-Weiss reaction (Eq. 3.3a), or via Fenton reaction (Eq. 3.3b).

$$Fe^{2+} + O_2 \rightarrow Fe^{3+} + O_2^{-\bullet} \tag{3.1}$$

$$O_2^{-\bullet} + 2H^+ + e^- \rightarrow H_2O_2 \tag{3.2}$$

$$O_2^{-\bullet} + H_2O_2 \rightarrow O_2 + OH^\bullet + OH^- \tag{3.3a}$$

$$Fe^{2+} + H_2O_2 \rightarrow Fe^{3+} + OH^- + OH^\bullet \tag{3.3b}$$

Highly reactive hydroxyl radicals lead to a cycle of increasing cell damage (Fubini and Otero Aréan 1999). In addition to reactive oxygen species, reactive nitrogen species, including nitric oxide (NO^-) and peroxynitrite ($ONOO^-$) are produced in the initial respiratory burst. An alternative explanation for silica toxicity includes lysosomal permeability, by which silica disrupts the normal internalization process leading to cytokine release and cell death (Hamilton et al. 2008). Besides the progression from chronic inflammation to fibrosis and silicosis long-term exposure to fine crystalline silica particles also has deleterious effects on the immune system (Derbyshire et al. 2012). For example, some rheumatic, as well as chronic renal diseases show higher than average incidence in individuals exposed to silica and it is likely that increased susceptibility to some mycobacterial diseases is due to impaired function of macrophages in silicotic lungs (Snider 1978). There is also a reduction in the ability of the macrophages to inhibit the growth of tubercle bacilli responsible for tuberculosis (Derbyshire et al. 2012), and the risk to silicotic patients of developing tuberculosis is up to 20 times the level found in the general population (Westerholm 1980).

3.5.2 Asbestos

Asbestos and asbestiform fibers constitute a severe health hazard. Human diseases associated with exposure to asbestos fibers include pleural fibrosis and plaques, pulmonary fibrosis (asbestosis), lung cancer, and diffuse malignant mesothelioma. The critical determinants of fiber bioactivity and toxicity include not only fiber dimensions, but also shape, surface reactivity, crystallinity, chemical composition, and presence of transition metals. Depending on their size and dimensions, inhaled fibers can penetrate the respiratory tract to the distal airways and into the alveolar spaces. Biopersistence of long asbestos fibers can lead to inflammation, granuloma formation, fibrosis, and cancer. Crocidolite is the most pathogenic form of asbestos. The increased toxicity of amphibole asbestos over that of serpentine asbestos (chrysotile) is due in part to the high aspect ratios (i.e. length to width ratio greater than 3) and durability of the amphibole asbestos, where the fibers are long, thin and needle-like and are biopersistent (Derbyshire et al. 2012). To be counted as a 'fiber', the World Health Organisation states that an asbestos particle must have an aspect ratio greater than 3:1. However, the most important single property for respirability is fibre diameter, which should be less than 3 μm (WHO 1986).

Asbestiform compounds have mineral aggregates with the distinctive features of amphibole asbestos, namely discrete, long thin, strong and flexible fibers in bunches or mats (Derbyshire et al. 2012). Fibrogenesis arising from inhalation of asbestos fibers spreads out along elements of the lung structure, rather than as discrete nodules as occurs with silicosis. Some fibers penetrate tissue and remain in the lungs, lung lining and abdominal cavity. On deposition within the conducting and alveolar regions, their surfaces are modified by adsorption of macromolecules which may enhance cell damage due to generation of free radicals (see Eqs. 3.1, 3.2 and 3.3b). Once in the lung, fibre length, surface chemistry, solubility and other physical and chemical properties control the course of disease.

3.5.3 Dust Particles Contaminated with Toxic Elements

Potentially toxic elements include mercury (Hg), lead (Pb), arsenic (As), cadmium (Cd), copper (Cu), zinc (Zn), iron (Fe) and radioactive elements (see also Chaps. 7 and 8). Particles contaminated with toxic elements can contribute to transport and accumulation of contaminants in soil, water and biota. Inhalation of contaminated particles may result in toxic element doses that could have deleterious consequences. Some metalloids and metals are known carcinogens (e.g. As, Cd, Cr) and can reduce mental and nervous system function, cause lower energy levels, DNA damage and damage vital organs (reviewed by Csavina et al. 2012), and long-term exposure to these contaminants may cause degenerative diseases such as Alzheimer's disease, Parkinson's disease, muscular dystrophy, and multiple sclerosis. High levels of Pb are known to cause effects ranging from comas, seizures and death (Woolf et al. 2007). Children are particularly at risk of neurotoxic effects following long-term Pb inhalation such as learning or behavioral problems including speech, intelligence, attention, behavior, and mental processing problems (Woolf et al. 2007). Childhood Pb poisoning at blood Pb levels >0.1 μg cm^3 is a common occurrence in mining and smelting towns, whose communities face an extraordinary exposure risk, with marked losses in intelligence quotient and also consequentially later childhood emotional and behavior problems (reviewed by Csavina et al. 2012). There is evidence to suggest that childhood disorders do not tend to remit with age but continue into adulthood. Atmospheric resuspension of Pb-enriched soil in urban environments represents a persistent source of Pb poisoning in children (Laidlaw and Filippelli 2008). Mercury and zinc and many other toxic elements can cause tracheitis, bronchitis and asthma, while inhalation of cadmium causes inflammation, oedema and fibrosis of the parenchyma (Derbyshire et al. 2012). Iron can act catalytically in free-radical generation, leading to greater oxidative stress and epithelial cell damage. Trace amounts of iron help to generate reactive oxygen species, leading to DNA damage, cell transformation and pulmonary reaction (Derbyshire et al. 2012). Released manganese-rich dust has been implicated in manganese-induced Parkinsonism (Koller et al. 2004) and excessive exposure to manganese can be associated with neurotoxicity, notably in mining, the ferroalloy and battery industries.

The potential toxicity of volcanic particles depends not only on mineral composition, grain size and surface area but also on the presence of toxic elements and complex interactions between toxic elements, particle surface, and lung tissue. Basaltic ash produces particularly high numbers of hydroxyl free radicals through interaction of reduced iron on the surface of ash particles with hydrogen peroxide (Horwell et al. 2007). Fluoride is released to the atmosphere by volcanic activity and other natural processes, as well as from anthropogenic inputs. Airborne fluoride is toxic if high concentrations are inhaled or ingested. It is rapidly absorbed following intake and has a high affinity for calcified tissue, including bones and teeth (Derbyshire et al. 2012).

Exposure to radioactive elements, such as radon, in dust has particularly been explored in occupational settings such as uranium ore mining, and mineral processing plants (Cook et al. 2005). Although the risk of radiation exposure from natural sources is not well characterized, volcanic dusts might have high uranium content and carry adhering particles of radon, an α-radioactive gas linked to the development of lung cancer. If these particles are inhaled and attach to bronchial epithelium, they potentially produce a high local radiation dose (Melloni et al. 2000). Exposure could also occur when the use of volcanic material is used for buildings (Cook et al. 2005).

The bioavailability of contaminants is significantly affected by the particle size. Coarse particles, such as those resulting from ore crushing and grinding, deposit in the upper respiratory system and are swallowed and go through the digestive system where contaminants may be absorbed, depending on their bioavailability. In contrast, fine particles, such as those originating from smelting operations, are respired deep into the lungs where they can be transported directly to the blood stream

(Krombach et al. 1997). Spear et al. (1998) in a study on the bioavailability of Pb according to size fraction from ambient aerosols around a Pb smelter found that the highest percentage of Pb concentration was in the fine size fraction, and moreover, it contained the largest percentage of bioavailable Pb. Therefore, finer size particles that deposit in the lungs and are transported to the blood stream via macrophages have a higher bioavailability (Krombach et al. 1997) and have been shown to contain the largest concentrations of metals and metalloids from smelter emissions (Spear et al. 1998; Csavina et al. 2011, 2012).

3.5.4 Bacteria, Fungi and Viruses in Dust

Dust carries a sizable inoculum of bacteria, fungi, and virus-like particles. As a rough approximation according to Griffin and Kellog (2004), a conservative estimate of 10^4 bacteria per gram of soil and 1 million Mg of airborne soil moving around the atmosphere each year may amount to about 10^{16} dustborne bacteria (fungi and viruses are not included in this estimate). Opportunistic pathogens (microbes that typically do not cause disease in healthy humans) can cause disease or colonize wounds in immunocompromised individuals. About 10–27% of the microorganisms cultured from African dust events have been identified as genetically similar to opportunistic human pathogens (Griffin and Kellog 2004). Around 10% from air samples collected in the Virgin Islands (Carribean) was composed mainly of a few fungi (*Aureobasidium* ssp., *Aspergillus* ssp.; *Cladosporium* ssp.) that are capable of causing skin or respiratory infections, and a few bacteria (*Kocuria* sp. and *Microbacterium arborescens*) that were originally identified from human noma lesions in Africa (Griffin et al. 2003). Examples of the bacteria isolated from Mali samples include many species of *Bacillus*, some of which are associated with gastrointestinal illness, several bacteria associated with septicemia, and two isolates (*Kocuria* ssp. and *Staphylococcus gallinarum*) identified from noma lesions in Nigerians (Griffin and Kellog 2004). Most of these potential pathogens do not cause respiratory diseases in healthy individuals, so inhalation of dust containing them is unlikely to trigger infection. As the major part of the drinking water in the Caribbean is collected from rooftop drainage and stored in cisterns, dust contamination of the water could result in numbers of microbes sufficient to cause disease by ingestion (Griffin and Kellog 2004). Several disease-causing microbes are typically transmitted on local scales (within continents) by dust. In the USA in the early 1990s, outbreaks of Valley Fever (caused by the fungus *Coccidioides immitis*) were associated with dust storms (Jinadu 1995). Hantavirus illnesses in the Midwest are also dust-associated (the inhalation of aerosolized rodent feces). In sub-Saharan Africa, the World Health Organization identified dust storm activity as a cause of regional outbreaks of bacterial meningitis (reviewed by Griffin and Kellog 2004). Analysis of a 28-year period of cerebrospinal meningitis outbreaks in Benin (West Africa) demonstrated a relation between the disease and Saharan dust storms (Besancenot et al. 1997). The impact of pathogens on human health is discussed in detail in Chap. 13.

3.5.5 *Dusts from Coal and Liquid Fossil Fuel Burning*

Fossil fuels are predominantly made up of organic matter but they also contain variable amounts of inorganic constituents retained in the ash left after the combustion process, called fly ash, which can contain many potentially toxic elements. Some of the inorganic components are emitted during combustion and can impact the environment in the vicinity of the power plant. The actual concentrations of individual elements depend on the source of the fuel. In Europe, oil and coal combustion has contributed significantly to atmospheric deposition of arsenic, cadmium, chromium, copper, nickel, and vanadium (Rühling 1994). Coal combustion is thought to have made a significant contribution to atmospheric lead deposition in the UK, with the coals containing up to 137 μg Pb g^{-1} (Farmer et al. 1999). Coal combustion is also the major anthropogenic source of selenium in the environment. In the USA coals contain up to 75 μg Se g^{-1} coal (Coleman et al. 1993). In China, the mean value of uranium in coal is 3 μg g^{-1}, which is near the published value for uranium in coal in the USA and the world coal average (Huang et al. 2012). There are a few examples of Chinese coal that have exceptionally high concentrations of uranium and may be considered as a uranium source. In Hubei province in southern China a special type of stone coal from the lower Silurian has a uranium concentration of 180–280 μg g^{-1} in the coal ash (Huang et al. 2012). Enhanced concentrations of uranium in many coals have resulted in enrichments of this element around coal-fired power stations. In the USA the mean concentration of mercury in coal is approximately 0.2 μg g^{-1}. Values for conterminous US coal areas range from 0.08 μg g^{-1} for coal in the San Juan and Uinta regions to 0.22 μg g^{-1} for the Gulf Coast lignites (Toole-O'Neil et al. 1999).

Domestic combustion of coal is very common in China. Both on a local and regional scale, human health in China has been adversely affected by coals containing arsenic, fluorine, selenium, and possibly, mercury (Finkelmann et al. 2002). Burning the coals in unvented stoves volatilizes the toxic elements and exposes the local population to the toxic elements in the emissions. The situation is exacerbated by the practice of drying crops directly over the coal fires. Chronic arsenic poisoning affected at least 3000 people in Guizhou Province (Zheng et al. 1996). Typical symptoms of As poisoning include hyperpigmentation, hyperkeratosis (scaly lesions on the skin, generally concentrated on hands and feet), Bowen's disease (dark, horny, precancerous lesions of the skin), and squamous cell carcinoma (Finkelmann et al. 2002). Chili peppers dried over open coal-burning stoves is a principal vehicle for As poisoning in China. Fresh chili peppers have less than 1 ppm As. In contrast, chili peppers dried over high-arsenic coal fires can have as much as 500 μg g^{-1} As. Significant amounts of As may also come from other tainted foods, ingestion of dust (samples of kitchen dust contained 3000 μg g^{-1} As), and from inhalation of indoor air polluted by As derived from coal combustion

(Finkelmann et al. 2002). Health problems caused by fluorine released during domestic coal use are much more extensive than those caused by As. More than ten million people in Guizhou Province and surrounding areas suffer from various forms of fluorosis (Zheng and Huang 1989; Zhang and Cao 1996) and this has also been reported from 13 other provinces, autonomous regions, and municipalities in China (Ando et al. 1998). Typical signs of fluorosis include mottling of tooth enamel (dental fluorosis) and various forms of skeletal fluorosis including osteosclerosis, limited movement of the joints, and outward manifestations such as knock-knees, bow legs, and spinal curvature. Fluorosis, combined with nutritional deficiencies in childhood, can result in severe bone deformation (Finkelmann et al. 2002). Dental fluorosis due to the use of high-fluorine coals was also reported to occur in cattle in the vicinity of large power stations in Yorkshire, England (Burns and Allcroft 1964). The etiology of fluorosis is similar to that of arsenism in that the disease is derived from foods dried over coal-burning stoves. Zheng and Huang (1989) have reported that adsorption of fluorine ($>200\ \mu g\ g^{-1}$) by corn dried over unvented ovens fluorine coal is the probable cause of the extensive dental and skeletal fluorosis in southwest China.

Residual oil fly ash may be more harmful to human health than coal fly ash (Derbyshire et al. 2012). Oil fly ash is mostly less than 2.5 μm in diameter, chemically complex, largely inorganic and rich in metals, especially vanadium. The soot produced by the burning of oil consists mainly of clusters of tiny carbon spheres. Elements enriched in oils such as vanadium have been found to be elevated in the environment in the vicinity of oil refineries (Rühling 1994). Lung inflammation, eye and throat irritation, cough, dyspnoea, rhinitis and bronchitis, have been found in workers (notably in power-generating plants) exposed to high oil fly ash concentrations (Derbyshire et al. 2012).

3.5.6 Dust from Terrestrial Biomass Burning

Although particles from biomass burning are usually within the size range thought to be most damaging to human health, their chemical composition is different from those derived from combustion of fossil fuel. Combustion of plant biomass releases solids, hydrocarbons, organic compounds and gases, but particulate matter and polycyclic aromatic hydrocarbons are of particular concern with respect to human health, especially as amorphous species which may be converted into crystalline minerals during combustion (Derbyshire et al. 2012). A majority of particles resulting from terrestrial biomass burning corresponds to the less than 2.5 μm ($PM_{2.5}$) range. These particles of the class $PM_{2.5}$ from landscape fires produce health effects different from those caused by similar sized particles in fossil fuel smoke, because their composition differs from those produced by fossil fuel combustion. Johnston et al. (2012) estimated on average 340,000 premature deaths every year attributed to $PM_{2.5}$ from landscape fire smoke around the world, the average exposure ranging from 0 to 180 days. The regions most affected by smoke as a chronic exposure (i.e. pollution lasts for whole seasons) were sub-Saharan Africa and Southeast Asia, with 160,000 and 110,000 deaths per year, respectively. In western and central Europe, which experienced sporadic exposure (i.e. pollution for a few days per year), the approximate number of deaths was 5000 per year.

3.6 Exposure

Inhalation is one of the major toxic pathways for dusts to enter the human body. The airways are highly vulnerable to injury from bioreactive compounds that could be contained within or adsorbed to the particles taken up by the lung. Human adults breathe tens of thousands of liters of air daily (Bates 1989). The gaseous mixture enters the lung and passes through a series of passages lined with cells and cilia to prevent inhalation or aid expulsion (Skinner 2007). The hairs in the nose, the bronchial tubes lined with mucous, and the ciliated cells that beat upward serve to expel any foreign materials through cough or sneeze. These mechanisms prevent particles from reaching the alveolar sacs where gas exchange (transfer of O_2 and CO_2) with the blood takes place (Skinner 2007). Breathing requires respiratory muscles, but expiration is largely passive, the lung being essentially a self-deflating balloon. A schematic scheme of the human respiratory system is given in Fig. 3.15.

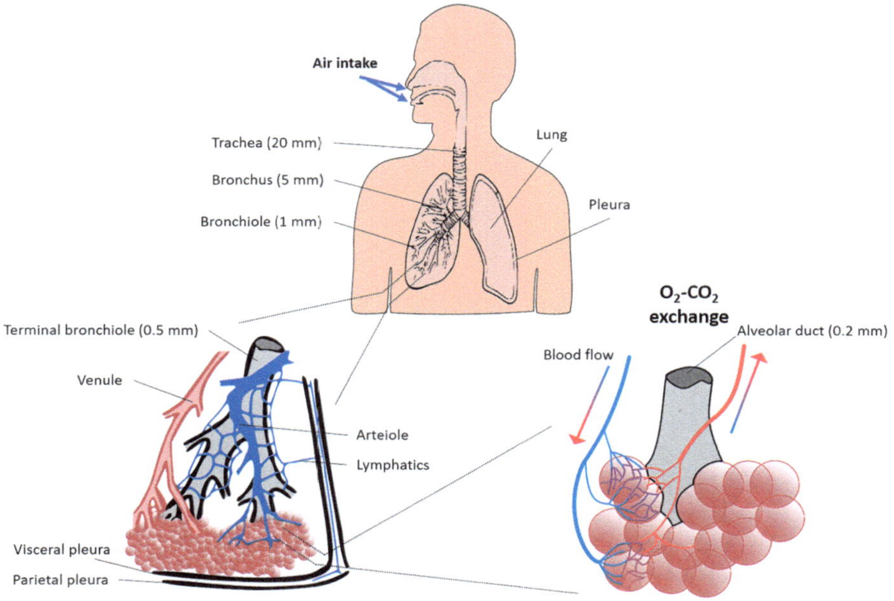

Fig. 3.15 Diagram of the human respiratory system (Adapted from Skinner 2007)

If small-sized particles (<10 μm) reach the alveolar sacs, there is a body defense mechanism, scarring or fibrosis, that isolates these foreign bodies. The cellular response through recruitment of other cells, macrophages and fibroblasts and seeks to encapsulate the offending material. The fibrosis effectively obliterates a portion of the sacs which results in reduced gas transfer to or from the blood and defective oxygenation of the red blood cells that carry oxygen to other tissues and organs (Skinner 2007). The consequence of increased particulate exposure may be cardiovascular disease. Size, shape, density, and reactivity of particles influence how far they will be transported and where they will be deposited (Wilson and Spengler 1996). Deposition of all foreign substances in the respiratory tract can initiate inflammatory responses and cause local proliferative fibrosis.

3.6.1 Exposure Levels

Increased natural dust emissions in connection with industrialization, urbanization, coupled with increased industrialization, emissions from vehicles as well as suspension from unpaved roads, and emissions from waste and biomass burning may all lead to substantial increase of particulate matter in ambient air. In Table 3.6 the mean annual exposure levels of PM_{10} in developing and developed countries are listed.

In most of the countries, PM concentrations exceeded the latest air quality guidelines set by the WHO for mean annual exposures to PM_{10} of 20 μg m^{-3}. As expected, the PM_{10} levels in developing countries are much higher than those in developed countries. It is very likely that increasing concentrations of particles should cause or contribute to premature mortality and morbidity-related health endpoints in countries with high PM_{10} levels.

3.7 Clinical Effects

3.7.1 Disease Associated with Inhalation of Dust from Silica and Coal Burning

For the most part dust from silica and coal burning is represented by particles finer than 5 μm, which readily access the deep lung during inhalation. Phagocytosis of particles and induction of macrophage cell death leads to release of the intracellular particles, which are then re-ingested by fresh macrophages. This cycle of particle phagocytosis and macrophage death contributes to inadequate particle clearance and retention in the lung, which, in turn, perpetuates the inflammatory process (Derbyshire et al. 2012). Figure 3.16 presents the appearance of dust particles in lung tissue from environmental dust pollution and coal combustion.

Table 3.6 Mean annual exposure to PM_{10} in developing and developed countries

Countries	Year	
	2009	2011
	PM_{10} ($\mu g\ m^{-3}$)	
Developing countries		
Argentina	39	36
Brazil	41	36
Egypt	129	120
Kenya	70	66
Nigeria	153	150
South Africa	42	40
China	86	82
Afghanistan	68	63
India	108	100
Pakistan	207	171
Bangladesh	118	121
Philippines	44	43
Thailand	45	45
Indonesia	50	47
Developed countries		
Ireland	18	18
UK	20	20
Finland	16	16
France	25	24
Germany	25	24
Italy	36	34
USA	20	18
Canada	15	14
Australia	15	14
New Zealand	18	16
Japan	20	19
South Korea	50	46
Russia	30	27
Saudi Arabia	113	108

Adapted from Kim et al. (2015)

Particles which escape removal by macrophage phagocytosis interact with and injure the epithelium, when the particles are then translocated into the lymph nodes. The variability in the bioreactivity of silica depends on its surface properties. For example, the clearance of cristobalite was far less than that found for quartz, while the inflammatory response was 30% greater, lasting over several months (Hemenway et al. 1990). The fibrotic processes result in the formation of silicotic nodules that are characterized by collagen arranged in a whorled fashion (Finkelman et al. 2002).

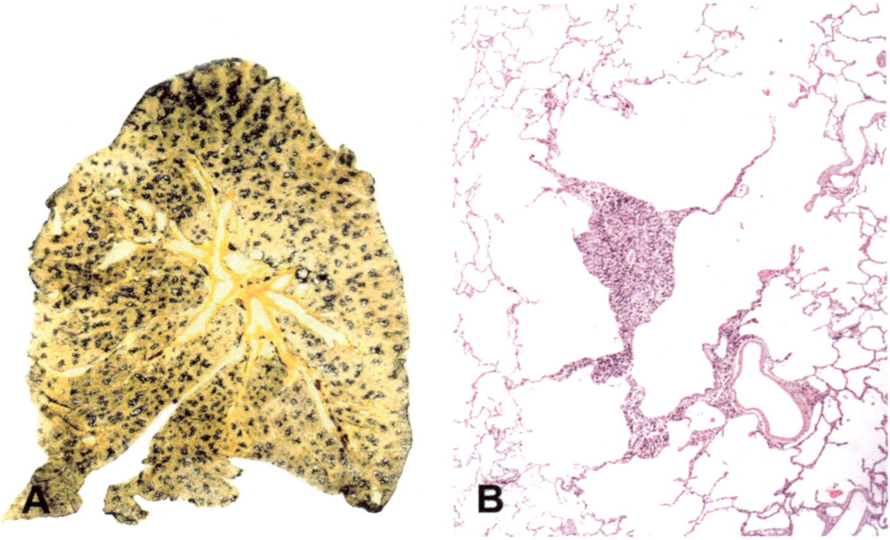

Fig. 3.16 Gross appearance of macules in whole lung section from a coal miner (**a**). Macules tend to predominate in the upper lung zones and occupy the central portions of the acinus. Microscopic section of typical macule in a coal miner showing black pigment enmeshed in a reticulin fibrous network within the walls of a respiratory bronchiole (**b**). The surrounding airspaces are dilated, indicative of focal emphysema (Green et al. 2007, p. 139; with kind permission from SAGE Publisher)

Human pathologic reactions to silica exposure include adverse pulmonary responses such as acute silicosis, accelerated silicosis, chronic silicosis, and conglomerate silicosis (Craighead et al. 1998). Acute silicosis (silicolipoproteinosis) results from exposure to relatively high levels of silica. The disease morphologically is characterized by pulmonary edema, interstitial inflammation, and the accumulation within the alveoli of proteinaceous fluid rich in surfactant (Craighead et al. 1998). Accelerated silicosis is similar in many respects to acute silicosis, exhibiting an exudative alveolar lipoproteinosis associated with chronic inflammation. In addition, accelerated silicosis is associated with fibrotic granulomas containing collagen, reticulin, and a large number of silica particles. As with acute silicosis, accelerated silicosis also is associated with an increased morbidity and mortality.

Exposure to crystalline silica over prolonged periods promotes the formation of the classic fibrotic nodules having a typical histologic appearance of concentric arrangements of collagen fibers with central hyalinized zones. Typical concentric silicotic lesions with the whorled fibers of collagen are characteristic of silicotic lesions produced by inhalation of crystalline silica and are morphologically distinct from lesions produced by other inorganic occupational exposures (Craighead et al. 1998). The nodules show variable degrees of calcification and necrosis. Dust-containing macrophages, fibroblasts, and lymphocytes are often restricted to the periphery of the nodules. Lesions of silicosis, which are sharply demarcated from the adjoining lung parenchyma, usually range in size from few millimeters to several centimeters in diameter. Nodules are often found predominantly in the upper zones of the lungs and in sub-pleural areas. Conglomerate silicosis results from the coalescence and agglomeration of several smaller nodules. In addition to the enlargement of nodules, profusion of nodular lesions increases and results in progressive massive fibrosis (Craighead et al. 1998). Cavitation and extensive destruction of the lung parenchyma, including bronchioles and blood vessels, are common with progressive massive fibrosis.

Health effects of volcanic ash depend upon a number of factors including particle size (particularly the proportion of respirable-sized material), mineralogical composition (including the crystalline silica content) and the physicochemical properties of the surfaces of the ash particles, all of which vary between volcanoes and even eruptions of the same volcano, but adequate information on these key characteristics is not reported for most eruptions (Horwell and Baxter 2006). The incidence of acute respiratory symptoms (e.g. asthma, bronchitis) varies greatly after ash falls, from very few reported cases to population outbreaks of asthma. Individuals with preexisting lung disease, including asthma, can be at increased risk of their symptoms being exacerbated after falls of fine ash (Horwell and Baxter 2006).

Silica exposure may be associated with systemic and autoimmune diseases such as scleroderma, rheumatoid arthritis, systemic lupus erythematosus, nephropathy, and proliferative glomerulonephritis (Craighead et al. 1998). Tuberculosis is a common complication of silicosis often seen in severe grades of the disease. The association between silicosis and intractable tuberculosis has been understood since long, and estimates place the risks of silicosis sufferers at up to 20 times that of the general population (Westerholm 1980). The depressant effect of crystalline silica on the ability of alveolar macrophages to kill *Mycobacterium tuberculosis* was reported by Policard et al. (1967). It is believed that silicosis leads to a reduction in cell-mediated immunity with alterations in lymphocyte subsets and serum immunoglobulin levels (Watanabe et al. 1987). A possible association between silicosis and lung cancer is being accepted on the basis of evidence for a role of silica exposure in increased lung tumor formation in experimental animals and exposed human populations (McDonald 1989).

3.7.2 Disease Associated with Inhalation of Dust Particles Contaminated with Toxic Elements

Dust particles contaminated with toxic elements such as certain metals and metalloids might produce a wide spectrum of respiratory effects, including upper airway injury or sensitivity (e.g. sinusitis from arsenic and mercury), lower airway inflammation (e.g. tracheitis, bronchitis, and asthma from Hg, Zn and other inhaled metals), and acute inflammation, edema, and fibrosis of the lung parenchyma with inhalation of elements such as Cd (Cook et al. 2005). At physiological pH, metals often tend to associate with organic molecules, generally through divalent bond formation. Genetic material such as DNA, functional molecules, and enzymes can be disturbed, whereas agents such as Fe are directly toxic through their capacity to catalyze oxidation and generate free radicals. In general, the solubility of the metal reflects the level at which it could potentially cause damage. High solubility equates with dissociation and dispersal well into lung tissues and relative insolubility is associated with airway deposition. Fibrosis, nodule formation, and granulomatous lesions are pathological processes shared by many forms of many trace element inhalation (Cook et al. 2005).

3.7.3 Disease Associated with Coal Mining

Inhalation of dust generated during mining is associated with coal workers' pneumoconiosis (Collis and Gilchrist 1928; Heppleston 1947) which is defined by coal dust-induced lesions in the gas exchange regions of the lung. A direct relationship could be confirmed between the mass of respirable coal mine dust inhaled and the incidence and severity of coal workers' pneumoconiosis (Hurley et al. 1982). Simple coal workers' pneumoconiosis is characterized by the formation of coal macules located at the bifurcations of respiratory bronchioles mainly in the upper lung lobes (Finkelmann et al. 2002). Coal macules are irregularly shaped lesions containing coal dust laden alveolar macrophages with a fine network of connective tissue (reticulin and collagen fibers) ranging in size from 1 to 6 mm in diameter (micro nodules). Coal nodules with continued exposure can develop to macro nodules (8 mm – 2 cm in diameter). Coal nodules have a centralized zone of concentric collagen fibers and are stellate in shape. Condensation of the lesion is associated with localized scar emphysema (Finkelmann et al. 2002). Complicated coal workers' pneumoconiosis is associated with progressive extensive fibrosis and emphysema. Fibrotic lesions become extensive forming coal dust laden lesions that are well demarcated, irregular, or rounded in shape, with collagen bundles. Dead cells are often found in the center of these lesions. Lesions associated with progressive massive fibrosis are 2 cm or greater in diameter and are often associated with vascular degeneration (Finkelmann et al. 2002). Silicosis is often found in conjunction with coal workers' pneumoconiosis. In miners' studies in the USA from 1972 to 1996, 23% of coal miners had pulmonary silicosis and 58% had lymph node silicosis (Green and Vallyathan 1998).

For early stage of simple coal workers' pneumoconiosis little decline in lung function was reported (Morgan and Lapp 1976). Miners at early stage of the disease showed only a small decline in gas exchange that did not compromise arterial oxygen content (Musk et al. 1981). However, increases in residual volume, suggesting collapse of small airways due to focal emphysema, have been reported in simple coal workers' pneumoconiosis (Legg et al. 1983). As the disease progresses to the complicated form, irreversible airway obstruction, noted as a fall in dynamic lung volumes measuring air flow and a decrease in arterial oxygen upon exertion, has been reported (reviewed by Finkelmann et al. 2002). These progressive developments can eventually lead to pulmonary hypertension and cor pulmonale (Lapp and Parker 1992).

The incidence of coal workers' pneumoconiosis is affected by coal rank. At similar dust exposures, the disease is five times more prevalent in anthracite miners than for miners of lower rank coal (Bennett et al. 1979). Little correlation existed between silica content of coal dust from various mines in Germany and the prevalence of coal workers' pneumoconiosis (Robock and Reisner 1982), while animal studies link development of pulmonary fibrosis with the amount of silica added to coal dust samples (Ross et al. 1962). Some clues to these different findings may lie in the observation that water-soluble extracts of coal dust and both low- and high-aluminium clays (kaolin and attapulgite) inhibit the cellular reactivity of quartz (Stone et al. 2004) which might explain why occupational exposure to coal dust containing less than 20% quartz does not induce classic silicosis (Derbyshire et al. 2012).

3.7.4 Disease Associated with Biomass Burning

A number of studies have found associations between wildfires and emergency room visits for both upper and lower respiratory tract illnesses (including asthma), respiratory symptoms, and decreased lung function (Naeher et al. 2007). For example, an analysis of emergency room visits for asthma in Singapore during a 1994 episode of regional pollution resulting from forest and plantation fires reported an association between PM_{10} and emergency room visits for childhood asthma (Chew et al. 1995). During the "haze" period, mean PM_{10} levels were 20% higher than the annual average. Although a time-series analysis was not conducted, Chew et al. (1995) suggested that the association remained significant for all concentrations above 158 $\mu g\ m^{-3}$. A study conducted in Indonesia during the 1997 haze episode also suggested acute impacts on respiratory and cardiovascular symptoms (Kunii et al. 2002). Of 539 interviewees, 91% reported respiratory symptoms (cough, sneezing, runny nose, sputum production, or sore throat), 44% reported shortness of breath on walking, 33% reported chest discomfort, and 23% reported palpitations. Sastry (2002) evaluated the population health effects in Malaysia of air pollution generated by a widespread series of fires that occurred mainly in Indonesia between April and November 1997. The results showed that the haze from these fires was associated with deleterious effects on population health in Malaysia and were in general agreement with the mortality impacts associated with particles in urban air (Sastry 2002). An increase in PM_{10} by 10 $\mu g\ m^{-3}$ measured in Kuala Lumpur was associated with 0.7% (all ages) and 1.8% (ages 65–74) increases in adjusted relative risks of nontraumatic mortality.

Agricultural burn smoke can be associated with serious exacerbations of asthma. For example, an association was found between asthma hospital admissions and the burning of rice field stubble and waste rice straw in Butte County, California, over a 10 year period (Jacobs et al. 1997). A relationship of rice stubble burning with asthma was also found in Niigata prefecture, Japan (Torigoe et al. 2000). A study in three rural villages in Iran also evaluated the relationship between rice stubble burning and respiratory morbidity, especially asthmatic symptoms (Golshan et al. 2002). During a burning period lasting several weeks, PM_{10} concentrations doubled. Based on responses to a physician-administered survey before and after this episode, the authors reported significant increases in the prevalence of asthma attacks, use of asthma medications, the occurrence of nocturnal sleep disturbances, and other respiratory symptoms among 994 residents of an agricultural region. Several measures of pulmonary function also decreased significantly. A few studies have specifically examined air pollution and health effects associated with the burning of sugar cane. In Brazil, daily indirect measurements (sedimentation of particle mass) of air pollution during the sugar cane burning season in 1995 were associated with the number of patients visiting hospitals for inhalation therapy for acute respiratory distress (Arbex et al. 2000).

Burning of sugarcane is not just a Brazilian problem and has been a common practice in other countries. In Brazil, however, the problem is more acute as the country is the biggest producer of sugar and alcohol in the world.

3.7.5 Asbestos-Related Lung Disease

Asbestos exposure has devastating consequences on the respiratory system, causing pulmonary fibrosis (asbestosis), bronchogenic lung cancer, malignant mesothelioma and pleural plaques. Due to the long latent period of asbestos-related diseases of several decades, the incidence of diffuse malignant mesothelioma, which has the longest latency, is predicted to reach its peak in Europe in 2020 (Peto et al. 1999). There are a number of other disorders presumed to be associated with or implicated to asbestos exposure, including benign pleural effusions, generalized pleural thickening, carcinoma of the larynx, carcinomas of the gastrointestinal tract, small airway disease, Caplan's syndrome, honeycombing, and bronchiectasis (Manning et al. 2002). However, these disorders are outside the scope of this chapter. The nature of asbestos-related lung disease depends, at least in part, on the type of asbestos to which individuals are exposed (Berman and Crump 2008). There are marked differences in lung retention between serpentine (chrysotile) and amphibole asbestos (Bernstein and Hoskins 2006). Amphibole asbestos is found in the lung in greater quantities than chrysotile asbestos. Inhalation studies in experimental animals show that, for the same exposure dose, chrysotile is cleared from the lung more effectively than amphibole asbestos, the former being cleared in a couple of days whereas the latter takes several years. Studies in people show that retention of amphibole asbestos, particularly fibres with a high aspect ratio and a longest diameter greater than 20 μm long, is significantly greater than chrysotile asbestos (Bernstein and Hoskins 2006). An important factor in the rapid clearance of chrysotile asbestos from the lung is its solubility. The crystal structure of serpentine consists of layers of octahedrally coordinated magnesium (brucite sheets) and layers of tetrahedrally coordinated silica. These sheets are mismatched so they curl and form scrolls, with the magnesium on the outside (Bernstein and Hoskins 2006). The structure is susceptible to environmental pH, in that the outer, magnesium, brucite-like sheets dissolve in mildly acid conditions. The silica matrix is susceptible to acid pH, a situation which exists in the phagolysosomal apparatus and in the immediate vicinity of the cell surface membrane of macrophages (Wypych et al. 2005). Consequently, regardless of the initial length of fibre, chrysotile deposited at the air-liquid interface of the lung disintegrates into fragments that can readily be internalised and cleared by macrophages and other clearance systems in the lung.

3.7.5.1 Asbestosis

Development of asbestosis (bilateral diffuse interstitial pulmonary fibrosis) is caused by fiber inhalation and deposition within the lung (Craighead et al. 1982; Becklake et al. 2007). The disease develops over many years or even several decades. Initially there are few symptoms, the risk and incidence relating to the cumulative dose of asbestos exposure and the time since the first exposure (Derbyshire et al. 2012). The most common clinical symptoms of asbestosis are dyspnea on exertion with progression over time leading to restrictive impairment and decreased diffusing capacity. The disease is usually confined to the lower zones of the lungs as reticulonodular infiltrates with the presence of calcified pleural plaques suggestive of exposure of asbestos. Gross pathology reveals bilateral interstitial fibrosis involving the lower zones of the lungs with severe disease closer to the pleura. In advanced asbestosis, honeycombing is common (Craighead et al. 1982).

3.7.5.2 Lung Cancer

Asbestos exposure can be the cause of all types of lung cancer, including small-cell and non-small-cell carcinomas (Derbyshire et al. 2012). Increased risk for developing lung cancer in workers with heavy occupational exposure to asbestos and a dose-response relationship between asbestos exposure and cancer incidence are well documented (McDonald 1989). However, the presence of asbestosis and the severity and extent of exposure required for an excess lung cancer risk are disputed especially when cancer risk is not observed for 10 or more years from initial exposure, whereas with longer exposures, increased risk is evident (Manning et al. 2002). On the other hand, it was shown that asbestosis is not a prerequisite for asbestos-associated increased lung cancer risk (Finkelstein 1997). This is further complicated by the difference in relative potency of chrysotile and amphibole asbestos. Indeed, both can induce lung cancer, but amphibole asbestos has at least six times greater carcinogenic potential (Berman and Crump 2008). Another important factor is fiber length. Asbestos fibers of 10 μm and above (but not shorter fibers) present in lung tissue have consistently been associated with development of lung cancer (Derbyshire et al. 2012).

3.7.5.3 Malignant Mesothelmioma

Malignant mesotheliomas of the pleura and peritoneum are well associated with asbestos exposure. The disease is represented by an aggressive tumor with a prognosis of only 6–18 months following diagnosis, depending on the cellular origin of the tumor (Churg 1998). Asbestos-induced mesothelioma has a long latency period, usually 30 or more years, and the latency increases with lower levels of exposure (Browne 1994). Unlike pulmonary fibrosis and cancer associated with asbestos exposure, mesothelioma is not associated with cigarette smoking (Harding et al. 2009). Exposure to very low levels of amphibole and exposure to low levels of amosite or crocidolite (but not chrysotile) induces mesothelioma (Derbyshire et al. 2012). Fibers above 10 μm are associated with mesothelioma, but short fibres are not (Berman and Crump 2008). Malignant mesothelioma rapidly spreads over the surfaces of the lung, thoracic, and abdominal cavities. Metastasis to other organs is very rare. The types of cellular features observed in mesotheliomas include epithelial, sarcomatous, and mixed types showing different cellular patterns (Manning et al. 2002). Patients suffering from malignant mesothelioma report breathlessness, cough and chest pain, possibly associated with weight loss. The tumour can be seen as dense white tissue encasing the lung (Derbyshire et al. 2012). Over time, it penetrates the respiratory tissue, destroying the pulmonary architecture and may impact on other structures such as the oesophagus and superior vena cava.

3.7.5.4 Pleural Plaques

Pleural plaques are often considered hallmarks of exposure to asbestos or other fibrous materials (Manning et al. 2002). They are discrete elevated areas of hyaline fibrosis almost invariably arising from the parietal pleura. Plaques can be elliptical, irregularly shaped, or round and they are usually limited to the parietal pleura but may also be located in the fissures (Norbet et al. 2015). Microscopically, they are cellular, nonvascular dense strands of hyalinized collagen showing a "basket-weave" pattern of mesh appearance (Manning et al. 2002). The collagen bundles may contain abundant numbers of asbestos fibers, almost exclusively chrysotile fibers, but asbestos bodies are absent (Peacock et al. 2000). The inner side is covered by normal mesothelial cells, and the costal side may demonstrate signs of low-grade inflammation (Peacock et al. 2000). Pleural plaques tend to lie adjacent to relatively rigid structures such as the ribs, vertebral column and tendinous portion of the diaphragm. They occur with lower inhaled fiber burdens, whereas asbestosis is associated with a higher fiber burden (Churg 1998). Pleural plaques can thus result from small temporally remote dust exposures. No correlation between fiber counts in the lung parenchyma and the parietal pleura associated with plaques has been found (Norbet et al. 2015). It was believed in the past that the fibers caused direct mechanical irritation of the parietal pleura (Hillerdal 1980). It is now thought that short asbestos fibers reach the parietal pleura by passage through lymphatic channels, where they excite an inflammatory reaction, whereas the largest fibers, amphiboles, are retained in the lung parenchyma (Norbet et al. 2015). Alternatively, fibers may reach the parietal pleura are via the blood supply or by direct migration of fibers through the visceral pleura (Norbet et al. 2015). Pleural plaques slowly progress in size and amount of calcification with time, independent of any further exposure. They do not usually cause any functional limitation and there is no evidence that pleural plaques undergo malignant degeneration into mesothelioma (Norbet et al. 2015).

3.8 Therapy

3.8.1 Silicosis

Silicosis is a major cause of morbidity and mortality in both developed and developing countries. However, up to now no proven curative treatment for silicosis exists A variety of treatment concepts aimed at decreasing the pulmonary inflammatory response to silica are available. Whole lung lavage techniques might remove large quantities of dust, cells, and soluble materials from the lungs and improve symptoms in some patients. However, sustained improvement in lung function parameters has not been shown in a clinical trial (Banks et al. 1993). Corticosteroids have been used in pharmacologic treatment protocols with varying success for silicosis. While one study demonstrated an improvement in both inflammatory broncho-alveolar lavage and pulmonary function tests when corticosteroids were administered (Sharma et al. 1991), another report indicated acute silicosis was reversed by corticosteroid therapy (Goodman et al. 1992). However, most authorities believe that corticosteroids have, at best, limited efficacy in the treatment of silicosis. Substrates coating silica particles such as aluminium citrate theoretically reduce the solubility of these particles. However, no sustained benefits in objective parameters of disease status have been reported for inhaled aluminum citrate powder (Kennedy 1956).

Polyvinylpyridine-*N*-oxide has been shown to concentrate silica particles inside of cells and consequently has improved the functional capabilities of some patients by slowing the course of disease (Chen and Lu 1970). The substance has been shown to coat the surface of silica particles and thereby decrease the potential for silica toxicity in both in vitro and in vivo models (Castranova 2004). Moreover, it decreases the generation of reactive oxygen species and possibly reduces silica-induced DNA damage by selectively blunting the active sites at the particle surface (Hoffmann et al. 1973). However, the efficacy of polyvinylpyridine-*N*-oxide may be limited by its potential for kidney and liver toxicity. Other therapies include the use of alveolar macrophage inhibitors and monoclonal antibodies (Castranova et al. 1991). Unfortunately, none of the proposed therapies have clearly reduced the mortality associated with silicosis. It is therefore most important that comorbid problems, such as mycobacterial infections, tuberculosis, and other pulmonary infections, be identified and treated promptly in all silicosis patients.

Several illnesses have been associated with silicosis. For example, silicosis and tuberculosis frequently coexisted. However, the risk of developing tuberculosis has been substantially reduced by improved respiratory dust protection as well as the development of antibiotics effective against mycobacteria. Nevertheless, mycobacterial infections continue as common complications associated with all forms of silicosis (Chang et al. 2001). Exposure to silica may also be associated with autoimmune diseases, such as rheumatoid arthritis, scleroderma, progressive systemic sclerosis and lupus erythematosis, as well as kidney disease (Derbyshire et al. 2012). However, it is difficult to know the exact risks since these diseases are also dependent on other environmental and genetic factors.

3.8.2 Coal Worker's Pneumoconiosis

Coal worker's pneumoconiosis results from inflammatory responses to dust in lung parenchyma and in large and small airways. As with other occupational lung diseases, the disease often occurs in patients with comorbidities, such as cardiovascular disease, smoking-related lung disease, and obesity (Petsonk et al. 2013). Heart failure or morbid abdominal obesity may cause or contribute to restrictive ventilatory defects and gas exchange abnormalities that must, based on a reasoned medical opinion, be considered in the overall clinical evaluation. No specific medical therapy has proven effective in reversing coal worker's pneumoconiosis. Whole-lung lavage can reduce lung dust burden (Wilt et al. 1996) and has recently been applied in China, but indications, risks, and benefits still require clarification. Current management practices mainly focus on prevention of disease progression, periodic medical monitoring, and recognition and treatment of complications and comorbid diseases. Moreover, awareness of relevant compensation and benefits programs is helpful in managing affected patients (Petsonk et al. 2013). For miners with impairment from coal worker's pneumoconiosis, a program of rehabilitation including medication, diet, exercise, and supplemental oxygen, targeted to the specific needs and capacities of the individual, is appropriate in addressing health-related quality of life (American Thoracic Society Documents 2004). Lung transplantation has been performed for coal miners with very advanced lung disease. However, opinions differ regarding the risks and benefits of this procedure (Enfield et al. 2012).

3.8.3 Asthma

Epidemiological studies on asthma suggest that its prevalence and severity increases constantly in many countries (Beasly 1998). Environmental exposure is one of the many proposed reasons for these increases. Chronic airway inflammation and airflow obstruction in individuals with asthma and increased airway hyperresponsiveness might cause lung remodelling from thickening and fibrosis of the airway walls (Vignola et al. 2000). This remodelling could result in irreversible and progressive airflow obstruction. Poorly treated chronic persistent asthma or severe asthma can cause changes in the lungs that are similar to those resulting from smoking (Silva et al. 2004). In patients with severe asthma, increased neutrophils, interleukin 8 (chemokine produced by macrophages), proteases and oxidative stress were observed (Barnes 2006). Patients with a new asthma diagnosis had lower initial values of lung function and greater rate of FEV_1 (Forced Expiratory Pressure in 1 Second) decline than did those without asthma (Ulrik and Lange 1994). Fifteen-year follow-up of these patients showed that those with self-reported asthma still had an increased rate of FEV_1 decline (Lange et al. 1998), indicating that asthma might be associated with reduced baseline FEV_1 and increased rate of decline in pulmonary function. However, many patients with asthma retain normal or near normal lung function throughout life, and only a subset of patients seem to show the pattern of progressive decline in lung function. Treatment of patients with new asthma diagnosis with oral bronchodilator drugs alone might not prevent progressive airflow obstruction, while corticosteroid treatment may do so (O'Byrne et al. 2006).

3.8.4 Asbestosis

In the case of asbestosis, a cure is rarely possible, leaving the physician with limited means for helping the patient. Management focuses on surveillance and prevention along with symptom abatement. Preventive therapy should include complete removal from exposure, smoking cessation, immunization against pneumococcal pneumonia, yearly influenza vaccination, and management of any coexisting respiratory disease, such as chronic obstructive pulmonary disease (Lazarus and Philip 2011). Unfortunately, the condition gets progressively worse, even after exposure has ceased. Palliative care is required in severe asbestosis with progressive respiratory failure and cor pulmonale (Becklake et al. 2007). If a patient is hypoxemic at less than 60 mm Hg oxygen, either at rest or with exertion, supplemental oxygen therapy should be initiated (Lazarus and Philip 2011).

3.8.5 Asbestos-Related Lung Cancer

Treatment for asbestos-related lung cancer is not different from any other lung cancer of the same histology. Surgery is the treatment of choice for early-stage cancers. For patients with chronic lung diseases, pulmonary function testing is important when assessing surgical risk (Suvatne and Browning 2011). When chemotherapy is applied, there is no particular agent unique to asbestos-related lung cancers. In terms of prognosis, there is no evidence that asbestos-induced lung cancer has a worse prognosis than lung cancer because of other causes (Suvatne and Browning 2011). It stands to reason that patients with asbestosis will have a worse prognosis given their underlying lung disease, but this has not been definitively proven in studies.

Prevention is probably the most important aspect of lung cancer. Smoking cessation is paramount. In many countries where asbestos still has not been banned, people are still being exposed to asbestos levels high enough to cause disease (Frost et al. 2008). An international ban on the mining and use of asbestos is, therefore, urgently needed. The risks of exposure to asbestos cannot be controlled by technology or by regulation of work practices. Even the best systems of workplace controls cannot prevent occupational and environmental exposures to products in use, or exposures to asbestos discarded as waste (La Dou et al. 2010).

3.8.6 Malignant Mesothelioma

Malignant mesothelioma is a rare form of cancer arising from the mesothelial cells of the pleura, peritoneum or pericardium with a strong causal link to asbestos exposure. The disease, due to its ubiquitous, aggressive nature and late diagnosis up to now does not have one widely accepted treatment modality since none reliably results in cure. Moreover, there is a striking lack of randomized, clinical trials comparing treatment regimens in this disease (Ismail-Khan et al. 2006). Based on the increasing incidence and the dismal prognosis, new therapeutic approaches on malignant mesothelmioma are long awaited. There are currently efforts to improve outcomes which are based on a better understanding of the stromal compartment and deregulated pathways leading to the design of clinical trial. These are based on immunotherapy or molecular-directed compounds (Remon et al. 2015).

Malignant mesothelioma has a strong negative impact on the quality of life of the people suffering from this disease. Although the disease can cause a large number of complaints, its symptom management is mainly aimed at pain relief and improving shortness of breath (van Thiel et al. 2011). Pain caused by chest wall involvement can be treated with opioids, as for the inflammatory part, non-steroidal anti-inflammatory drugs are useful. The cause of dyspnea is often multifactorial, including pleural fluid, a trapped lung or preexisting co-morbidity and a number of treatment modalities may be required to address this symptom (van Thiel et al. 2011).

3.8.7 Pleural Fibrosis

Pleural fibrosis can result from a vast array of inflammatory processes, including dust exposure (e.g. asbestos). Central to the pathogenesis of pleural fibrosis is inflammation of the pleural cavity. The causes of visceral pleural fibrosis include asbestos-associated diffuse pleural thickening, coronary bypass graft surgery, pleural infection (including tuberculous pleurisy), drug-induced pleuritis, rheumatoid pleurisy, uraemic pleurisy, and haemothorax (Huggins and Sahn 2004). However, a complete understanding of the pathogenesis of pleural fibrosis and why fibrosis occurs in some and not others remains unknown. Clinically significant pleural fibrosis requires involvement of the visceral pleura. Patients with isolated parietal pleural fibrosis, such as asbestos induced pleural plaques, do not develop respiratory symptoms. In contrast, visceral pleural fibrosis can lead to significant restriction. Decortication is an effective strategy for treating symptomatic patients regardless of cause, when significant underlying parenchymal disease has been excluded (Huggins and Sahn 2004). Corticosteroids should be considered in acute rheumatoid pleurisy, however, they have no effect in preventing pleural fibrosis in tuberculous pleurisy.

3.9 Measures to Minimize Exposure to Particulate Matter

Clean air is considered to be a basic requirement of human health and well-being. Particulate matter is a key indicator of air pollution brought into the air by a variety of natural and human activities. As it can be transported over long distances in the atmosphere, it can cause a wide range of diseases leading to a significant reduction of human life.

3.9.1 Regulation Guidelines for Particulate Matter

To protect public health, air quality standards have been set in many countries and as such have been an important component of national risk management and environmental policies. However, there is no evidence to support a safe or threshold level of exposure below which no adverse health effects occur or such effects are perceived. The exposure to particulate matter is ubiquitous and involuntary, which highlights its importance to human health. Regulation guidelines for particulate matter to reduce their potential harmful effect on public health and the environment and to offer guidance for protecting public health set by various governments around the world are given in Table 3.7.

In many areas in Europe and in the USA, air has become cleaner on average over the last quarter of the twentieth century (Kim et al. 2015). However, in urban areas

Table 3.7 Regulation guidelines for particulate matter

Region/country	Time scale	$PM_{2.5}$ (μg m^{-3})	PM_{10} (μg m^{-3})
European Union	Yearly mean	25	40
	Daily mean[a]	None	50
United States	Yearly mean	12	None
	Daily mean[a]	35	150
Australia	Yearly mean	8	None
	Daily mean[a]	25	50
Japan	Yearly mean	15	None
	Daily mean[a]	35	100
China	Yearly mean	35	70
	Daily mean[a]	75	150
Hong Kong	Yearly mean	35	50
	Daily mean[a]	75	100
South Korea	Yearly mean	25	50
	Daily mean[a]	25	100

Adapted from Kim et al. (2015)
[a]Mean of 24 h

of the EU and the USA, particulate matter levels are still found to frequently exceed the guideline levels set for the particulate standards. It is well known that the status of particulate matter pollution is highly significant in some Asian countries. As such, various tactics and techniques have been introduced in those areas to reduce or suppress its pollution in recent years. In China, the Ministry of Environmental Protection reported that the mean concentration of $PM_{2.5}$ in 74 cities was 76 μg m^{-3} which is far higher than the common guidelines (Cheng et al. 2013). Recently, 'The Airborne Pollution Prevention and Control Action Plan (2013–17)' has been established by the Chinese government to reduce particulate matter pollution by installing emissions-cutting exhaust filters, reducing coal use and tightening vehicle emission standards (Sheehan et al. 2014). The plan's goal is to reduce 25% of particulate matter emissions from 2012 levels by the year 2017. It has been proven that the worsening of air pollution in China affects the surrounding countries like Korea and Japan as well, with the fine particles moving with the westerly winds (Kim et al. 2015).

3.9.2 Measures to Reduce Occupational Exposure

3.9.2.1 Dust Exposure to Working Miners

Minimizing dust exposures is essential for working miners (Petsonk et al. 2013). Ventilation systems, water sprays, and other dust capture devices require continuous monitoring to assure they operate as intended. Respiratory protection using masks can offer a degree of health protection in certain industrial settings. However, the use of dust masks in coal mining can only be a short-term measure and is not a substitute for effective environmental controls. This means that reliance on protective masks to prevent chronic lung disease is neither practical nor legal. Personal electronic dust monitors have been approved for use in coal mines to continuously monitor an individual miner's dust exposure. These devices can provide real-time measurements of dust levels, allowing for immediate interventions to reduce harmful exposures (Page et al. 2008). Unfortunately, personal dust monitors are expensive, and their role in managing individual miners remains to be clarified. Underground coal miners with a diagnosis of pneumoconiosis may have a legal right to frequent exposure monitoring and transfer to a reduced dust job if one is available at the mine (Petsonk et al. 2013). As with other chronic lung diseases, patients with pneumoconiosis should adhere to recommended schedules for vaccinations against viral and bacterial pathogens, and for those who smoke, cessation efforts should be encouraged and supported.

3.9.2.2 Exposure to Respirable Crystalline Silica Dust Particles

Occupational exposure to respirable crystalline silica dust particles occurs in many industries and silica-associated diseases remain an important public health concern in the twenty-first century because crystalline silica is one of a handful of toxins that causes multiple serious diseases and increased mortality (Rees and Murray 2007). Primary prevention of silica-associated diseases, the control of dust to concentrations at which disease will not occur, is the optimum form of prevention and should be the primary goal of national and workplace occupational health programs. Regular monitoring of respirable silica should be done in all industries with silica exposure. For example, respirable dust can be collected by cyclone or impact dust sampler (Kromhout 2002). The content of free silica of respirable dust can be assessed by the phosphoric acid (Talvitie) method, infrared spectrophotometry, or x-ray diffraction method (reviewed by Leung et al. 2012). The limit of detection ranges between 5 and 10 μg per sample, but accuracy is poor at low filter loadings (<30 μg) that are typically collected when airborne concentrations of crystalline silica are similar to regulatory standards (Leung et al. 2012). Enforced or suggested permissible exposure limits for respirable silica were chosen according to the desired level of protection and available methods of dust control and monitoring technologies, and they vary between 0.025 and 0.35 mg m^{-3} in different countries (Leung et al. 2012). However, these standards have not been confirmed as fully protective by epidemiology studies.

The primary method of silicosis prevention is represented by avoidance or control of silica exposure by various measures directed at the source (WHO 2007). Source control can be represented by banning of sandblasting, isolation of the source or of workers, wet methods, enclosed processes, air curtain, water spray, local exhaust ventilation, general ventilation or air supply systems (Leung et al. 2012). If source control is not feasible or sufficient, other measures should be implemented to isolate or capture dust and introduce clean air to prevent workers being exposed to hazardous silica. Engineering controls such as substitution of materials, modification of processes and equipment, are also common methods (Akbar-Khanzadeh et al. 2010).

The best means to prevent exposure in the workplace are probably automating techniques, such as automated palletisers, bagging machines, and equipment monitored with programmable logic controllers and computer software systems (Leung et al. 2012). For workplaces with high dust levels, administrative measures should be used, such as short working hours or job rotation. Personal protective equipment is recommended for sandblasters and others working in similar environments. This equipment includes coveralls, boots, and properly fitted helmets supplied with filtered air, but it might not be fully effective in workplaces with high dust concentrations and should be the last resort for routine protection. Besides education about symptoms of silicosis, regular medical assessment might detect adverse health effects in exposed workers before disease reaches an advanced stage (Leung et al. 2012). Assessment commonly includes respiratory questionnaires, physical examination, chest radiography, and spirometry. No uniform standard exists for the frequency of such assessment because the decision may be affected by past and present respirable silica concentrations, dust particulate characteristics, and economic conditions. WHO recommends routine evaluation every 2–5 years, ideally for the rest of the lives of workers exposed to silica dust while the American College of Occupational and Environmental Medicine suggests tests at baseline and after 1 year, then every 3 years for the first 10 years, and every 2 years thereafter when silicosis is a concern and respirable silica concentrations are lower than 0.05 mg m^{-3} (Raymond and Wintermeyer 2006). The Institute for Occupational Safety and Health of the German Social Accident Insurance recommends examination every 3 years (Hessische Verwaltung für Bodenmanagement und Geoinformation 2004). New cases of silicosis should prompt a thorough assessment of silica exposure and control measures in workplaces (Leung et al. 2012). In addition to reports of new cases, occupational health physicians should regularly analyse health records of all exposed workers in an industry and assess the effects of prevention activities. Occupational hygiene and health records should also be properly maintained to enable calculation of disease rates and latency periods according to various exposure scenarios (Leung et al. 2012).

3.9.2.3 Exposure to Asbestos

Asbestos remains a human health hazard throughout the world. All forms of asbestos are now banned in more than 50 countries but the world's current production of asbestos continues at an alarming rate. However, even in countries where asbestos use has been discontinued, fibers are still present in homes and schools in the form of insulation, ceiling, and roofing tiles or in the environment as naturally-occurring deposits. For example, liberation of asbestos fibers during demolition of buildings or disturbance of previously unidentified asbestos deposits could lead to new exposure scenarios. Asbestos use continues particularly in developing countries. The risks of exposure to asbestos cannot be controlled by technology or by regulation of work practices. There is also no medical or scientific basis to exempt chrysotile (the dominant form of asbestos in use) from the worldwide ban of asbestos (LaDou et al. 2010). All countries of the world should have an obligation to their citizens to join in the international endeavor to ban the mining, manufacture, and use of all forms of asbestos. If global use of asbestos were to cease today, a decrease in the incidence of asbestos-related diseases would become evident in approximately 20 years (WHO 2006). The asbestos cancer pandemic may take as many as ten million lives before asbestos is banned worldwide and all exposure is brought to an end (LaDou 2004). Achieving a worldwide ban on the mining, sale and use of all forms of asbestos and the elimination of asbestos-related diseases will require that physicians and occupational health personnel responsibly and persistently express their concerns, raise awareness and take necessary action regarding the need to prevent asbestos-related diseases.

3.9.2.4 Protective Measures Due to Presence of Asbestos from Prior Use

Occupational exposures to asbestos will persist due to the continued presence of asbestos from prior use even after a total ban on production and use of asbestos is achieved. Workers who carry out maintenance, demolition and removal of asbestos-containing materials will thus continue to be at risk. Therefore a set of protective measures must be implemented to optimize effective prevention. Prevention also involves ensuring control of exposures to airborne asbestos fibers, monitoring concentrations according to established standards and reporting exposure levels to appropriate authorities. Exposed workers should be informed about their working conditions and associated hazards, and provided with appropriate respirators. While respirators should not be relied upon as the sole means of routinely limiting exposure to asbestos fibers, workers provided with them should be trained for their proper use, and encouraged to wear them when warranted. Adequate fitting, changing of filters, sanitary storage and maintenance of respirators are also required for optimal protection. Prevention also includes medical monitoring of exposed workers, early diagnosis and individual case management to prevent disease progression (American Thoracic Society Documents 2004).

3.9.2.5 Substitutes for Asbestos

As safer substitutes for asbestos exist, all illnesses and deaths related to asbestos are preventable. Asbestos cement products account for >85% of the world asbestos consumption (LaDou et al. 2010). In about 100 countries, asbestos-containing pipes and sheets are manufactured to be used as low-cost building materials (Tossavainen 1997). However, these asbestos cement water-pipe products could be replaced with ductile iron pipe, high-density polyethylene pipe, and metal-wire–reinforced concrete pipe (LaDou et al. 2010). Many substitutes exist for roofing as well as interior building walls and ceilings, including fiber-cement flat and corrugated sheet products that are made with polyvinyl alcohol fibers and cellulose fibers. Virtually all of the polymeric and cellulose fibers used instead of asbestos in fiber-cement sheets are >10 μm in diameter and therefore nonrespirable (WHO 2005). For roofing, lightweight concrete tiles can be made and used in the most remote locations using locally available plant fibers, such as jute, hemp, sisal, palm nut, coconut coir, and wood pulp. Galvanized iron roofing and clay tiles are among the other alternative materials (World Bank Group 2009).

3.9.3 Protective Measures to Minimize Exposure to Dust Storms

Dust storms sourced in North Africa occur throughout the year, with dust reaching North America and the northern Caribbean in the northern summer and South America and the southern Caribbean in the northern winter. In contrast, Asian dust storms are known to occur at any time of the year but the prime period is spring to early summer. Regional dust palls derived from smaller dust source regions may have a lower mean frequency. For example, the major dust storm that originated in Australia's Northern Territory in September 2009 and which extended beyond New Zealand into the Southern Ocean, was described as the largest dust storm event for over 40 years (Derbyshire et al. 2012). There is a clear link between frequent exposure to high-density airborne particles over several decades and pneumoconiosis and related conditions. The situation is still more complex through high human population density in relation to regional and global dust pathways. Almost half of the amount of dust from eastern Asia derived from the Taklamakan Desert in western China and the Mongolia/North China region is deposited in China, Korea and Japan, which have high urban population densities of between 100 and over 500 people km^{-2}. Parts of western China have a siliceous pneumoconiosis incidence of 7%, rising to three times that proportion in people over 40 years old (Xu et al. 1993). There is recognition that non-occupational pneumoconiosis is an increasingly serious problem in China. Regions in the Caribbean and North and South America affected by Saharan dust have moderately dense populations of >40 km^{-2} with some regional centers having higher densities of >100 km^{-2}, but dust palls are more frequent than the mean for eastern Asia. Saharan regional-scale dust also reaches moderately dense populated areas in Europe. However, the limited extent, low density and/or low frequency of many of these events limit their potential human health impact (Derbyshire et al. 2012).

Local reduction of risk can most likely be achieved by adopting mitigation measures such as relocation of populations, long-term use of dust masks and small-scale forestation projects. Direct action has been undertaken in China, designed to stem the frequency and magnitude of dust storms around dryland margins. Trees have been planted, including the Green Great Wall initiative (Derbyshire et al. 2012). However, the degree of success has been limited. At the urban scale, rapid growth and increasing dependence on motor traffic, as well as growth of manufacturing, power and other industries, has accelerated a rise in particle-rich air pollution in many cities in eastern China and in settlements in dryland areas of intense farming activity.

3.9.4 Measures to Reduce Dust Generation on Agricultural Land

Globally, 20% of emissions are from vegetated surfaces, primarily desert shrublands and agricultural lands. This implies that dust emission from these sources is particularly sensitive to land use practices (Ginoux et al. 2012). Projections of total agricultural land (crop plus pasture) prepared for the Intergovernmental Panel on Climate Change (Fifth Assessment Report) yield a range of estimates, from a projected increase by 13% to a decrease by 24% by 2100, depending on the scenarios used (Hurtt et al. 2011). Dust emission from agricultural land in specific regions is sensitive to climate change. Some regions that are currently dusty areas are likely to experience a decrease in precipitation. These include most of Mediterranean Europe and Africa, northern Sahara, central Asia, the southwestern USA, and southern Australia in spring. Conversely, in other currently dusty areas in East Africa and East Asia precipitation will likely increase (Ginoux et al. 2012). As a consequence of these precipitation changes a reduction in low-latitude dust sources and an intensification of sources in the tropics is likely to occur, unless implementation of better agricultural practices could mitigate the expected increase in dust emissions with reduced rainfall.

To mitigate the numerous negative effects of wind erosion, improved agricultural management practices have been developed (Ravi et al. 2011). Shelter belts and windbreaks have been used to decrease the erosive force of the wind in many local settings. Rows of trees and shrubs planted along the margins of fields are intended to protect the soil surface against soil erosion by wind. In several regions, earth berms, rock walls or fences are also used. Such barriers effectively decrease the wind speed for a distance of about 10–15 times their height downwind and about three times their height upwind (Oke 1978). Especially in semiarid regions, because of the limitations of tree growth, this distance rarely exceeds a few hundred meters downwind (Vigiak et al. 2003). Shelter belts and windbreaks are not as common as they were in the past. Although there were approximately 65,000 km of them planted in the Great Plains of North America by the 1960s, by the 1970s many were dying or were being removed (Ravi et al. 2011).

Conservation agriculture (CA) is a system of agronomic practices that include reduced tillage (RT) or no-till (NT), permanent organic soil cover by retaining crop residues, and crop rotations, including cover crops. Erosion control is a main objective of reduced tillage and residue retention (Palm et al. 2014). Reduced or

NT and surface applied residues directly reduce erosion by minimizing the time that the soil is bare and exposed to wind. Conservation agriculture and NT can reduce wind erosion due to the larger proportion of dry aggregates, less wind erodible fraction and greater crop residue cover of the soil surface (Singh et al. 2012). Crop residues protect the ground by offering elements that prevent saltating particles from cascading and by increasing the roughness height. Standing residues and growing crops provide greater protection than flat residues because they absorb much of the shear stress in the boundary layer (Skidmore 1994). The effects of vegetation on soil loss is estimated using the soil loss ratio, an index calculated by dividing the amount of soil loss from a residue-protected soil surface by the loss from a similar bare surface. The soil loss ratio decreases rapidly from 1.0, for a bare unprotected surface, to a value of approximately 0.2, with 40% soil cover (Fryrear 1985).

Increased soil organic matter contents and enhanced biological activity in the upper part of the surface soil horizon with surface residues and NT lead to increased stability of soil aggregates and greater macropore connectivity due to macrofaunal activity (Palm et al. 2014). Conservation agriculture practices result in more plant-available water than conventional practices which is a result of increased water infiltration and lower evaporation with reduced mixing of the surface soil, more residue cover and less exposure to drying compared to conventional tillage. The water holding capacity of the upper part of topsoil is also generally higher due to increased soil organic matter contents. Liu et al. (2013) found soil moisture content was most affected by residues compared to tillage practice; soil moisture remained highest throughout the growing season with NT plus residue, intermediate levels of soil moisture with tillage plus residues, and lowest levels with NT without residues. Similarly, Thierfelder et al. (2013) found higher soil moisture contents in Zambia for each of five years with NT and surface residue compared to conventional tillage (CT) with residues removed. This increased water content in the topsoil is especially important for crop growth during prolonged dry periods with less variable yields compared to conventional practices. In conservation agriculture, higher soil water content, residue retention and prolonged vegetation cover all contribute to reduced dust generation.

3.9.5 Protective Measures to Minimize Exposure to Volcanic Plumes

Although phenomena such as pyroclastic flow and surges, sector collapses, lahars and ballistic blocks are the most destructive and dangerous, volcanic ash is by far the most widely distributed eruption product. Volcanic ash can produce a wide range of health hazards. Volcanic risk management and risk reduction at a societal level is the official responsibility of civil authorities, but also relies on the engagement of communities, individuals, non-governmental organizations and the private sector. There are a variety of different disaster risk management options open to authorities. Short-term exposure to volcanic ash can be reduced directly through evacuation of people and long-term exposure can be reduced by transferring existing assets to geographical areas of lower risk.

Ash falling on local populations may remain in the environment for months, years or, occasionally, decades, if the eruption is sustained. For example, the Soufrière Hills volcano, Montserrat, West Indies, started erupting in July 1995 and is still active. The population has been subjected to frequent ash falls over the past 15 years. Risk analyses have been carried out to assess the likelihood of specific occupations developing chronic diseases (Hincks et al. 2006). Assuming continuing volcanic activity, outdoor workers, as the group most at risk, are thought to have a 2–4% risk of silicosis after 20 years of exposure to cristobalite-rich ash. Risk reduction measures for people that remain in locations affected by ash falls is achieved through safe ways to clear ash of roofs, to clean and seal houses and to maintain respiratory health through the use of dust masks.

3.9.6 Measures to Minimize Exposure to Particles Generated by Land Fires

Global fire emissions averaged over 1997–2009 corresponded to 2.0 Pg C year^{-1} with important contributions from Africa (52%), South America (15%), Equatorial Asia (10%), the boreal region (9%), and Australia (7%) (van der Werft et al. 2010). The largest contributor to global fire carbon emissions were fires in grasslands and savannas (44%), with another 16% emitted from woodland fires. Forest fires, mostly confined to temperate and boreal regions accounted for 15% (van der Werft et al. 2010). Landscape fire activity has been recognized as a global-scale environmental challenge because plumes transgress international boundaries and component gases and particles contribute to climate change (Johnston et al. 2012). Reducing population level exposure to air pollution from landscape fires is a challenge that is likely to have immediate and measurable health benefits. The burning of sugar cane at harvest has been common practice worldwide, but the environmental and human impact has led to new legislation in Brazil, banning pre- and post-harvesting burns from 2010 (Le Blond et al. 2008). In China's rural areas, the intense pollution of air due to burning of straw over a short period of time triggered public panic. Since 2000, the government of China has responded with increasing environmental regulations and put forward measures to control straw burning (Qu et al. 2012). As an example, in Jiangsu Province, the local government sent supervision teams to all cities to monitor straw burning. Unfortunately, some farmers choose to not burn straw during daytime but in the evening or on cloudy days. In summary, official temporary bans on straw burning have not achieved the desired results (Qu et al. 2012). The most important reason is that there is no visible profit in recycling straw and leaving it on farmland may negatively effect growth of the next season's crop. Thus, burning is still the easiest way to dispose the agricultural waste. There is thus an urgent need for the use of technologies and equipment for chopping straw and ploughing it into the soil. Recycling of straw into products is important as well. Local governments should push for the production of domestic fungus, clad plate, and cellulosic ethanol by using straw as raw materials. The collection and recycling of the straw is often a money-losing business without government support, resulting in poor economic performance that cannot be maintained. Therefore, financial incentives and subsidies are essential (Qu et al. 2012). Subsidies should be given to both farmers and recycling enterprises to counteract the negative effects caused by straw burning.

When biomass smoke reaches unhealthy levels, outdoor activities should be curtailed to decrease exposures to air pollutants. Public health officials recommend that people stay indoors during smoke episodes either in clean air shelters or in homes with clean air (Therriault 2001). In some cases, it is necessary to evacuate people who live in an area where biomass smoke has reached unhealthy levels. Evacuation reduces exposure to harmful air pollutants by moving people from sites with high levels of pollution to locations with better air quality. People who experience adverse health effects from air pollution seek care in hospital emergency rooms and are sometimes admitted to hospitals for respiratory (Patz et al. 2000) and other illnesses. Some people suffering from adverse consequences of biomass smoke seek care from private physicians.

One method for reducing human health risks is to extinguish wildfires. In reality it is not possible to extinguish all wildfires and completely eliminate health risks. Suppression of landscape fire by government authorities is increasingly driven by urban settlements in flammable landscapes (Hammer et al. 2009). Sophisticated technologies, such as aerial detections of ignitions and the use of aerial bombers to drop fire retardant, have been developed to fight fires on the 'wildland–urban interface'. Mechanized fuel treatments are also being carried out across landscapes and can be effective in reducing fire intensity in some dry forests that formerly sustained frequent, low-severity surface fires (Finney et al. 2005). However, mechanical treatments are more controversial in some forest types because there is debate about the ecological justification of this method and about its efficacy in reducing large-scale fires (Schoennagel et al. 2009). Recent catastrophic fires with tremendous losses in terms of property and lives have been experienced in the Mediterranean Basin, Israel, California, South Africa, southern Australia and Russia. These fires have demonstrated that fire management agencies cannot completely prevent fires from spreading into urban environments.

3.9.7 Measures to Minimize Exposure to Particles Generated from Fossil Fuel Combustion

3.9.7.1 Coal Burning

In Europe and North America, coal mining and release of particles from coal burning for energy generation have decreased significantly during the last half of the twentieth century. In contrast, coal production is increasing in developing countries, where it is an important source of energy. The predicted increase in mining and use of coal in Asia and Africa is causing concern that there will be a parallel increase in respiratory health problems (McCunney et al. 2009). In China, coal is used to generate approximately 70% of the electricity (Liu et al. 2009). In 2007, the volume of China's coal consumption amounted to 2580 Tg and accounted for 41% of the global coal consumption, 2.28 times that of America, and 6.30 times of that of India (Lin and Liu 2010). For reducing Chinese dependence on fossil fuels and slowing the growth of coal use and Chinese emissions of carbon and particles, various alternatives have been proposed, including boosting nuclear and wind power, increasing use of biofuels, rapidly developing domestic natural gas, and increasing energy efficiency in power generation and end uses through price and policy reform (Shealy and Dorian 2010). China has significant potential for increasing the energy efficiency of its economy, however, China's continuing drive to achieve robust economic growth and maintain jobs – particularly at the local and provincial levels – will keep the primary focus of industry on output and not on reduction in energy use. With alternative fuels and energy efficiency gains likely offering only marginal help in reducing Chinese coal reliance to 2025, slower economic growth may be the only viable solution to the recent rapid growth in Chinese energy demand (Shealy and Dorian 2010). If the Chinese government would not reign in economic growth in a controlled manner, various bottlenecks in environmental and health problems and in the energy supply system may show reduction of growth in a less-controlled manner.

3.9.7.2 Vehicular Traffic

Emissions due to road traffic are known to make a large contribution to total particulate matter concentrations in urban areas and exposure to particles from vehicular emissions has been demonstrated to have detrimental impacts on human health (Masiol et al. 2012; Rissler et al. 2012). Emissions from diesel and gasoline vehicles are different in terms of composition as diesel engines emit both a greater mass of particulate matter and a larger number of ultrafine particles compared to gasoline vehicles (Rose et al. 2006). The United Nations estimated that over 600 million people in urban areas worldwide were exposed to dangerous levels of traffic-generated air pollutants (Cacciola et al. 2002).

Legislation designed to encourage decreased use of cars and to create financial advantages by using smaller and less polluting vehicles have had mixed success. High petrol taxes have been justified on the grounds that they would decrease car use and, therefore, emission levels. However, a study in Southern California and Connecticut in the USA revealed that environmental taxes on fuel resulted in only minimal reductions in driving levels (Sipes and Mendelsohn 2001).

The development of new fuels offers the chance of reduced emission levels. The use of biofuels is rapidly increasing and there is generally a marked decrease in gaseous and particulate emissions. A study using a direct injection diesel engine fueled with canola-oil methyl ester and waste palm oil methyl ester, showed a decrease in engine power and increased fuel use, but smoke opacity decreased by up to 63% (Ozsezen and Canakci 2011). There are also significant developments in producing viable electrical and hybrid vehicles and promoting their use. For example, Germany's federal council, the Bundesrat, considers the target of 1 Million electric vehicles in 2020 (Schröder and Traber 2012) and recently passed a resolution to ban the internal-combustion engine by 2030 across the EU.

References

Achard F, Eva HD, Stibig HJ, Mayaux P, Gallego J, Richards T, Malingreau JP (2002) Determination of deforestation rates of the World's humid tropical forests. Science 297:999–1002

Akbar-Khanzadeh F, Milz SA, Wagner CD, Bisesi MS, Ames AL, Khuder S, Susi P, Akbar-Khanzadeh M (2010) Effectiveness of dust control methods for crystalline silica and respirable suspended particulate matter exposure during manual concrete surface grinding. J Occup Environ Hyg 7:700–711

American Thoracic Society Documents (2004) Diagnosis and initial management of nonmalignant diseases related to asbestos. Am J Respir Crit Care Med 170:691–715. http://ajrccm.atsjournals.org/content/170/6/691.full.pdf+html Accessed 17 Oct 2016

Anderson JO, Thundiyil JG, Stolbach A (2012) Clearing the air: a review of the effects of particulate matter air pollution on human health. J Med Toxicol 8:166–175

Ando M, Tadano M, Asanuma S, Matsushima S, Wanatabe T, Kondo T, Sakuai S, Ji R, Liang C, Cao S (1998) Health effects of indoor fluoride pollution from coal burning in China. Environ Health Perspect 106(5):239–244

Andreae MO, Merlet P (2001) Emission of trace gases and aerosols from biomass burning. Global Biogeochem Cycles 15:955–966

Arbex MA, Bohm GM, Saldiva PHN, Conceicao GMS, Pope CA, Braga ALF (2000) Assessment of the effects of sugar cane plantation burning on daily counts of inhalation therapy. J Air Waste Manage Assoc 50:1745–1749

Artazo P, Gerab F, Yamasoe MA, Martins J (1994) Fine mode aerosol composition at three long-term atmospheric monitoring sites in the Amazon Basin. J Geophys Res 99 (D11):22,857–22,868

Auker MR, Sparks RSJ, Siebert L, Crosweller HS, Ewert J (2013) A statistical analysis of the global historical volcanic fatalities record. J Appl Volcanol 2:1–24

Banks DE, Cheng YH, Weber SL, Ma JK (1993) Strategies for the treatment of pneumonconiosis. Occup Med 8(1):205–232

Barnes PJ (2006) Against the Dutch hypothesis: asthma and chronic obstructive pulmonary disease are distinct diseases. Am J Respir Crit Care Med 174:240–243

Bates DV (1989) Respiratory function in disease, 3rd edn. WB Saunders, Philadelphia

Beasly R (1998) Worldwide variation in prevalence of symptoms of asthma, allergic rhinoconjunctivitis, and atopic eczema: ISAAC. Lancet 351(9111):1220–1221

Becklake MR, Bagatin E, Neder JA (2007) Asbestos-related diseases of the lungs and pleura: uses, trends and management over the last century. Int J Tuberc Lung Dis 11(4):356–369

Bennett JG, Dick JA, Kaplan YS, Shand PA, Shennan DH, Thomas DJ, Washington JS (1979) The relationship between coal rank and the prevalence of pneumoconiosis. Br J Ind Med 36:206–210

Berman DW, Crump KS (2008) A meta-analysis of asbestosrelated cancer risk that addresses fiber size and mineral type. Crit Rev Toxicol 38(1):49–73

Bernstein DM, Hoskins JA (2006) The health effects of chrysotile: current perspective based upon recent data. Regul Toxicol Pharmacol 45(3):252–264

Bertschi IT, Jaffe DA (2005) Long-range transport of ozone, carbon monoxide, and aerosols to the NE Pacific troposphere during the summer of 2003: Observations of smoke plumes from Asian boreal fires. J Geophys Res 110:D05303. https://doi.org/10.1029/2004JD005135

Besancenot JP, Boko M, Oke PC (1997) Weather conditions and cerebrospinal meningitis in Benin (Gulf of Guinea, West Africa). Europ J Epidem 13:807–815

Bolin B, Aspling G, Persson C (1973) Residence time of atmospheric pollutants: as dependent on source characteristics, atmospheric diffusion processes and sink mechanisms. University of Stockholm, Stockholm Institute of Meteorology

Bourke PMA (1964) Emergence of potato blight, 1843–46. Nature 203:805

Bovallius A, Roffey R, Henningson E (1980) Long-range transmission of bacteria. Ann N Y Acad Sci 353(1):186–200

Bowden J, Gregory PH, Johnson CG (1971) Possible wind transport of coffee rust across the Atlantic Ocean. Nature 229:500–501

Bremer H, Kar J, Drummond JR, Nichitu F, Zou J, Liu J, Gille JC, Deeter MN, Francis G, Ziskin D, Warner J (2004) Spatial and temporal variation o MOPITT CO in Africa and South America: a comparison with SHADOZ ozone and MODIS aerosol. J Geophys Res 109:D12304

Brown SK, Loughlin SC, Sparks RSJ, Vye-Brown C, Barclay J, Calder E, Cottrell E, Jolly G, Komorowski J-C, Mandeville C, Newhall C, Palma J, Potter S, Valentine G (2015) Global volcanic hazard and risk. In: Loughlin SC, Sparks RSJ, Brown SK, Jenkins SF, Vye-Brown C (eds) Global volcanic hazards and risk. Cambridge University Press, Cambridge, pp 81–172

Browne K (1994) Asbestos-related disorders. In: Parkes WR (ed) Occupational lung disorders, 3rd edn. Butterworth-Heinemann, Oxford, pp 411–504

Burns KN, Allcroft R (1964) Fluorosis in cattle: occurrence and effects in industrial areas of England and Wales 1954–57, Industrial Disease Surveys, Reports 2, Part 1. Ministry of Agriculture Fisheries and Food, London

Cacciola RR, Sarva M, Polosa R (2002) Adverse respiratory effects and allergic susceptibility in relation to particulate air pollution: flirting with disaster. Allergy 57:281–286

Cahoon DR Jr, Stocks BJ, Levine JS, Cofer WR III, O'Neill KP (1992) Seasonal distribution of African savanna fires. Nature 359:812–815

Cao G, Zhang X, Zheng F, Wang Y (2006) Estimation the quantity of crop residues burnt in open field in China. Resour Sci 28:9–13

Carmona-Moreno C, Belward A, Malingreau JP, Hartley A, Garcia-Alegre M, Antonovskiy M, Buchshtaber V, Pivovarov V (2005) Characterizing interannual variations in global fire calendar using data from Earth observing satellites. Glob Change Biol 11(9):1537–1555

Castranova V (2004) Signalling pathways controlling the production of inflammatory mediators in response to crystalline silica exposure: role of reactive oxygen/ nitrogen species. Free Radic Biol Med 37(7):916–925

Castranova V, Kang JH, Ma JK, Mo CG, Malanga CJ, Moore MD, Schwegler-Berry D, Ma JY (1991) Effects of bisbenzylisoquinoloine alkaloids on alveolar macrophages. Correlation between binding affinity, inhibitory potency, and antifibrotic potential. Toxicol Appl Pharmacol 108:242–252

Chang KC, Leung CC, Tam CM (2001) Tuberculosis risk factors in a silicotic cohort in Hong Kong. Int J Tuberc Lung Dis 5(2):177–184

Chen SY, Lu XR (1970) Clinical studies of the therapeutic effect of kexiping on silicosis. In: Institute of Occupational Medicine: Proceedings of the Therapeutic Effect of Kexiping on Silicosis; CAPM Press, Beijing

Cheng Z, Jiang J, Fajardo O, Wang S, Hao J (2013) Characteristics and health impacts of particulate matter pollution in China 2001–2011. Atmos Environ 65:186–194

Chew FT, Ooi BC, Hui JKS, Saharom R, Goh DYT, Lee BW (1995) Singapore's haze and acute asthma in children. Lancet 346:1427

Chomette O, Legrand M, Marticorena B (1999) Determination of the wind speed threshold for the emission of desert dust using satellite remote sensing in the thermal infrared. J Geophys Res 104(31):207–231

Choobari OA, Zawar-Reza P, Sturman A (2012) Atmospheric forcing of the three-dimensional distribution of dust particles over Australia: a case study. J Geophys Res 117:D11206
Choobari OA, Zawar-Reza P, Sturman A (2014) The global distribution of mineral dust and its impacts on the climate system: a review. Atmosph Res 138:152–165
Christensen LS, Mortensen S, Botner A, Strandbygaard BS, Ronsholt L, Henricksen CA, Anderson JB (1993) Further evidence of long distance airborne transmission of Aujeszky's disease (pseudorabies) virus. Vet Rec 132:317–321
Churg A (1998) Neoplastic induced asbestos-related disease. In: Churg A, Green FHY (eds) Pathology of occupational lung disease, 2nd edn. Williams and Wilkins, Baltimore, pp 339–391
Chuvieco E, Aguado I, Dimitrakopoulos AP (2004) Conversion of fuel moisture content values to ignition potential for integrated fire danger assessment. Can J For Res 34(11):2284–2293
Clague ADH, Donnet JB, Wang TK, Peng JCM (1999) A comparison of diesel engine soot with carbon black. Carbon 37:1553–1565
Coe MT, Foley JA (2001) Human and natural impacts on the water resources of the Lake Chad basin. J Geophys Res 106:3349–3356
Coleman L, Bragg LJ, Finkelman RB (1993) Distribution and mode of occurrence of selenium in US coals. Environ Geochem Health 15:215–227
Collis EL, Gilchrist JC (1928) Effects of dust upon coal trimmers. J Ind Hyg Toxicol 10:101–109
Cook DE, Gale SJ (2005) The curious case of the date of introduction of leaded fuel to Australia: implications for the history of Southern Hemisphere atmospheric lead pollution. Atmosph Envir 39(14):2,553–2,557
Cook AG, Weinstein P, Centeno JA (2005) Health effects of natural dust. Biol Trace Elem Res 103:1–15
Craighead JE, Abraham JL, Churg A, Green FHY, Kleinerman J, Pratt PC, Seemayer TA, Vallyathan V, Weill H (1982) Asbestos-associated diseases. Arch Pathol Lab Med 106:541–596
Craighead JE, Kleinerman J, Abraham JL, Gibbs AR, Green FHY, Harley RA, Rüttner JR, Vallyathan NV, Juliano EB (1998) Diseases associated with exposure to silica and non-fibrous silicate minerals. Arch Pathol Lab Med 112:673–720
Csavina J, Landázuri A, Wonaschütz A, Rine K, Rheinheimer P, Barbaris B, Conant W, Sáez AE, Betterton EA (2011) Metal and metalloid contaminants in atmospheric aerosols from mining operations. Water Air Soil Pollut 221:145–157
Csavina J, Field J, Taylor MP, Gao S, Landázuri A, Betterton EA, Sáez AE (2012) A review on the importance of metals and metalloids in atmospheric dust and aerosol from mining operations. Sci Tot Environ 433:58–73
Damoah, Spichtinger N, Forster C, James P, Matthis I, Wandinger U, Beirle S, Wagner T, Stohl A (2004) Around the world in 17 days – hemisphere-scale transport of forest fire smoke from Russia in May 2003. Atmos Chem Phys 4:1311–1321
Davis GS (1986) The pathogenesis of silicosis: state of the art. Chest 89:166S–169S
Davis JM (1987) Modeling the long-range transport of plant pathogens in the atmosphere. Annu Rev Phytopathol 25:169–188
DeBell LJ, Talbot RW, Dibb JE, Munger JW, Fischer EV, Frolking SE (2004) A major regional air pollution event in the northeastern United States caused by extensive forest fires in Quebec, Canada. J Geophys Res 109:D19305
Derbyshire E (2005) Natural aerosolic mineral dust and human health. In: Essentials of medical geology. Selenius O, Alloway B, Centeno J, Finkelman R, Fuge R, Lindh U, Smedley P (Eds). Elsevier, pp 459–480
Derbyshire E, Horwell CL, Jones TP, Tetley TD (2012) Airborne particles. In: Pollutants, human health and the environment. In: Plant JA, Voulvoulis N, Ragnarsdottir KV (Eds). Wiley-Blackwell, pp 255–286

Donaldson AI, Gloster J, Harvey LDJ, Deans DH (1982) Use of prediction models to forecast and analyze airborne spread during the foot-and-mouthdisease outbreaks in Brittany, Jersey and the Isle of Wight in 1981. Vet Rec 110:53–57

Echalar F, Gaudichet A, Cachier H, Artaxo P (1995) Aerosol emission by tropical forest and savanna biomass burning: characteristic trace elements and fluxes. Geophys Res Lett 22 (22):3039–3042

Eckardt FD, Kuring N (2005) SeaWiFS identifies dust sources in the Namib Desert. Int J Remote Sens 26:4159–4167

Encyclopedia of the Nations (2010) Russia – mining. Available: http://www.nationsencyclopedia.com/Europe/Russia-MINING.html. Accessed 23 Sept 2016

Enfield KB, Floyd S, Barker B, Weder M, Kozower BD, Jones DR, Lau CL (2012) Survival after lung transplant for coal workers' pneumoconiosis. J Heart Lung Transplant 31:1315–1318

Engelstaedter S, Tegen I, Washington R (2006) North African dust emissions and transport. Earth Sci Rev 79:73–100

Farmer JG, Eades GLJ, Graham MC (1999) The lead content and isotopic composition of British coals and their implications for past and present releases of lead to the UK Environment. Environ Geochem Health 21:257–272

Finkelman RB, Orem W, Castranova V, Tatu CA, Belkin HE, Zheng B, Lerch HE, Maharaj SV, Bates AL (2002) Health impacts of coal and coal use: possible solutions. Int J Coal Geol 50:425–443

Finkelstein MM (1997) Radiographic asbestosis is not a prerequisite for asbestos-associated lung cancer in Ontario asbestos-cement workers. Am J Ind Med 32:341–348

Finney MA, McHugh CW, Grenfell IC (2005) Standard landscape-level effects of prescribed burning on two Arizona wildfires. Can J For Res 35:1714–1722

Fiore AM, Naik V, Spracklen DV, Steiner A, Unger N, Prather M, Bergmann D, Cameron-Smith PJ, Cionni I, Collins WJ, Dalsøren S, Eyring V, Folberth GA, Ginoux P, Horowitz LW, Josse B, Lamarque JF, MacKenzie IA, Nagashima T, O'Connor FM, Righi M, Rumbold ST, Shindell DT, Skeie RB, Sudo K, Szopa S, Takemura T, Guang Zeng G (2012) Global air quality and climate. Chem Soc Rev 41:6663–6683

Fiser A, Lanikova A, Novak P (1994) Mold and microbial contamination of dust deposition in cowsheds for heifers and dairy cows. Vet Med – Czech 39:245–253

Formenti P, Schuetz L, Balkanski Y, Desboeufs K, Ebert M, Kandler K, Petzold A, Scheuvens D, Weinbruch S, Zhang D (2011) Recent progress in understanding physical and chemical properties of African and Asian mineral dust. Atmos Chem Phys 11:8231–8256

Forster C, Wandinger U, Wotawa G, James P, Mattis I, Althausen D, Simmonds P, O'Doherty S, Jennings SG, Kleefeld C, Schneider J, Trickl T, Kreipl S, Jager H, Stohl A (2001) Transport of boreal forest fire emissions from Canada to Europe. J Geophys Res 106:22887–22906

Frank L, Joshi TK (2014) The global spread of asbestos. Ann Glob Health 80:257–262

Frost G, Harding AH, Darnton A, McElvenny D, Morgan D (2008) Occupational exposure to asbestos and mortality among asbestos removal workers: a poisson regression analysis. Br J Cancer 99(5):822–829

Fryrear DW (1985) Soil cover and wind erosion. Trans ASME 28:781–784

Fubini B, Otero Aréan C (1999) Chemical aspects of the toxicity of inhaled mineral dusts. Chem Soc Rev 28:373–381

Fubini B, Bolis V, Cavenago A, Volante M (1995) Physicochemical properties of crystalline silica dusts and their possible implication in various biological responses. Scand J Work Environ Health 21(2):9–14

Fullen M, Mitchell D (1993) Taming the Shamo dragon. Geogr Mag 63:26–29

Furman HKH (2003) Dust storms in the Middle East: sources of origin and their temporal characteristics. Indoor Build Environ 12(6):419–426

Gadde B, Bonnet S, Menke C, Garivait S (2009) Air pollutant emissions from rice straw open field burning in India, Thailand and the Philippines. Environ Pollut 157:1554–1558

Ginoux P, Prospero JM, Gill TE, Hsu NC, Zhao M (2012) Global scale attribution of anthropogenic and natural dust sources and their emission rates based on modis deep blue aerosol products. Rev Geophys 50:RG3005
Gloster J (1982) Risk of airborne spread of foot-and-mouth-disease from the continent to England. Vet Rec 111:290–295
Golshan M, Faghihi M, Roushan-Zamir T, Masood Marandi M, Esteki B, Dadvand P, Farahmand-Far H, Rahmati S, Islami F (2002) Early effects of burning rice farm residues on respiratory symptoms of villagers in suburbs of Isfahan, Iran. Int J Environ Health Res 12(2):125–131
Goodman GB, Kaplan PD, Stachura I, Castranova V, Pailes WH, Lapp NL (1992) Acute silicosis responding to corticosteroid therapy. Chest 101(2):366–370
Goudie AS (2009) Dust storms: recent developments. J Environ Manag 90:89–94
Goudie AS, Middleton NJ (2001) Saharan dust storms: nature and consequences. Earth Sci Rev 56:179–204
Goudie AS, Middleton NJ (2006) Desert dust in the Global system. Springer, Heidelberg
Green FH, Vallyathan V, Hahn FF (2007) Comparative pathology of environmental lung disease: an overview. Toxicol Pathol 35:136–147
Green FHY, Vallyathan V (1998) Coal workers' pneumoconiosis and pneumoconiosis due to other carbonaceous dusts. In: Churg A, Green FHY (eds) Pathology of occupational lung disease. Williams & Wilkins, Baltimore, pp 129–208
Greenberg MI, Waksman J, Curtis J (2007) Silicosis: a review. Dis Mon 53:394–416
Gregory PH (1973) The microbiology of the atmosphere (2nd edition). Leonard Hill Books
Griffin DW (2007) Atmospheric movement of microorganisms in clouds of desert dust and implications for human health. Clin Microbiol Rev 20(3):459–477
Griffin DW, Kellogg CA (2004) Dust storms and their impact on ocean and human health: dust in earth's atmosphere. EcoHealth 1:284–295
Griffin DW, Kellogg CA, Shinn EA (2001) Dust in the wind: long range transport of dust in the atmosphere and its implications for global public and ecosystem health. Global Chang Hum Health 2(1):20–33
Griffin DW, Kellogg CA, Garrison VH, Shinn EA (2002) The global transport of dust. Am Sci 90 (3):228
Griffin DW, Kellogg CA, Garrison VH, Lisle JT, Borden TC, Shinn EA (2003) Atmospheric microbiology in the northern Caribbean during African dust events. Aerobiologia 19:143–157
Guffanti M, Casadevall TJ, Budding K (2010) Encounters of aircraft with volcanic ash clouds: a compilation of known incidents, 1953–2009. US Geol Surv Data Ser 545:12
Guthrie GD (1997) Mineral properties and their contributions to particle toxicity. Environ Health Perspect 105(5):1003–1011
Hamilton RF Jr, Thakur SA, Holian A (2008) Silica binding and toxicity in alveolar macrophages. Free Radic Biol Med 44:1246–1258
Hammer RB, Stewart SI, Radeloff VC (2009) Demographic trends, the wildland–urban interface, and wildfire management. Soc Nat Res 22:777–782
Harding AH, Darnton A, Wegerdt J, McElvenny D (2009) Mortality among British asbestos workers undergoing regular medical examinations (1971–2005). Occup Envir Med 66 (7):487–495
Heiken G (1972) Morphology and Petrography of volcanic ashes. Geol Soc Am Bull 83:1961–1988
Heil A, Goldammer JG (2001) Smoke–haze pollution: a review of the 1997 episode in South-east Asia. Reg Environ Chang 2:24–37
Hemenway DR, Absher MP, Trombley L, Vacek PM (1990) Comparative clearance of quartz and cristobalite from the lung. Am Indust Hyg Assoc J 51(7):363–369
Heppleston AG (1947) The essential lesion of pneumoconiosis in Welsh coal workers. J Pathol Bacteriol 59:453–460
Hessische Verwaltung für Bodenmanagement und Geoinformation (2004) Arbeitsmedizinische Vorsorge. Druckerei Marquart GmbH, Sankt Augustin

Hillerdal G (1980) The pathogenesis of pleural plaques and pulmonary asbestosis: possibilities and impossibilities. Eur J Respir Dis 61:129–138

Hincks TK, Aspinall WP, Baxter PJ, Searl A, Sparks RSJ, Woo G (2006) Long-term exposure to respirable volcanic ash on Montserrat: a time series simulation. Bull Volcanol 68(3):266–284

Hodzic A, Madronich S, Bohn B, Massie S, Menut L, Wiedinmyer C (2007) Wildfire particulate matter in Europe during summer 2003: meso-scale modeling of smoke emissions, transport and radiative effects. Atmos Chem Phys 7:4043–4064

Hoffmann EO, Lamberty J, Pizzolato P, Coover J (1973) The ultrastructure of acute silicosis. Arch Pathol 96:104

Horwell CJ (2007) Grain size analysis of volcanic ash for the rapid assessment of respiratory health hazard. J Environ Monit 9(10):1107–1115

Horwell CJ, Baxter PJ (2006) The respiratory health hazards of volcanic ash: a review for volcanic risk mitigation. Bull Volcanol 69:1–24

Horwell CJ, Fenoglio I, Fubini B (2007) Iron-induced hydroxyl radical generation from basaltic volcanic ash. Earth Planet Sci Lett 261(3–4):662–669

Huang W, Wan H, Finkelman RB, Tang X, Zhao Z (2012) Distribution of uranium in the main coalfields of China. Energy Explor Exploit 30(5):819–836

Huggins JT, Sahn SA (2004) Causes and management of pleural fibrosis. Respirology 9 (4):441–447

Hurley JF, Burns J, Copland L, Dodgson J, Jacobsen M (1982) Coal workers' pneumoconiosis and exposure to dust at 10 British coal mines. Br J Ind Med 39:120–127

Hurst DF, Griffith DWT, Cook GD (1994) Trace gas emissions from biomass burning in tropical Australian savannas. J Geophys Res 99:16,441–16,456

Hurtt GC, Chini LP, Frolking S, Betts RA, Feddema J, Fischer G, Fisk JP, Hibbard K, Houghton RA, Janetos A, Jones CD, Kindermann G, Kinoshita T, Klein Goldewijk K, Riahi K, Shevliakova E, Smith S, Stehfest E, Thomson A, Thornton P, van Vuuren DP, Wang YP (2011), Harmonization of land-use scenarios for the period 1500–2100: 600 years of global gridded annual land use transitions, wood harvest, and resulting secondary lands Clim Change 109:117–161

IEO (International Energy Outlook) (2009) DOE/EIA-0484. Available: www.eia.doe.gov/oiaf/ieo/index.html Accessed 23 Sept 2016

International Ban Asbestos Secretariat (2010) Current asbestos bans and restrictions. Available: http://ibasecretariat.org/alpha_ban_list.php Accessed 23 Sept 2016

IPCC (Intergovernmental Panel on Climate Change) (2001) Climate change 2001: the scientific basis. Cambridge University Press, Cambridge

IPCC (Intergovernmental Panel on Climate Change) (2007) In: Solomon S (ed) Climate change: the physical science basis. Contribution of working group I to the fourth assessment report of the intergovernmental panel on climate change. Cambridge University Press, Cambridge

Ismail-Khan, Robinson LA, Williams CC, Garrett CR, Bepler G, Simon GR (2006) Malignant pleural mesothelioma: a comprehensive review. Cancer Control 13(4):255–263

Jacobs J, Kreutzer R, Smith D (1997) Rice burning and asthma hospitalizations, Butte County, California, 1983–1992. Environ Health Perspect 105(9):980–985

Jeon EM, Kim HJ, Jung K, Kim JH, Kim MY, Kim YP (2011) Impact of Asian dust events on airborne bacterial community assessed by molecular analyses. Atmos Environ 45 (25):4313–4321

Jinadu BA (1995) Valley fever task force report on the control of coccidioides immitis. Kern County Health Department, Bakersfield

Johnston DM, Houghton BF, Neall VE, Ronan KR, Paton D (2000) Impacts of the 1945 and 1995–1996 Ruapehu eruptions, New Zealand: an example of increasing societal vulnerability. Geol Soc Am Bull 112:720–726

Johnston FH, Henderson SB, Chen Y, Randerson JT, Marlier M, DeFries RS, Kinney P, Bowman DMJS, Brauer M (2012) Estimated global mortality attributable to smoke from landscape fires. Environ Health Perspect 120:695–701

Jones TP, Wlodarczyk A, Koshy L, Brown P, Longyi S, BéruBé KA (2009) The geochemistry and bioreactivity of fly-ash from coal-burning power stations. Biomarkers 14(1):45–48
Kampa M, Castanas E (2008) Human health effects of air pollution. Environ Pollut 151:362–367
Kasischke ES, Hyer EJ, Novelli PC, Bruhwiler LP, French NHF, Sukhinin AI, Hewson JH, Stocks BJ (2005) Influences of boreal fire emissions on Northern Hemisphere atmospheric carbon and carbon monoxide. Glob Biogeochem Cyc 19:GB1012
Kellogg CA, Griffin DW (2006) Aerobiology and the global transport of desert dust. Trends Ecol Evol 21(11):638–644
Kellogg CA, Griffin DW, Garrison VH, Peak KK, Royall N, Smith RR, Shinn EA (2004) Characterization of aerosolized bacteria and fungi from desert dust events in Mali, West Africa. Aerobiologia 20:99–110
Kennedy MC (1956) Aluminium powder inhalations in the treatment of silicosis of pottery workers and pneumonconiosis of coal-miners. Br J Ind Med 13(2):85–101
Kennedy D (2001) Black carp and sick cows. Science 292:169
Kim KH, Kabir E, Kabir S (2015) A review on the human health impact of airborne particulate matter. Environ Int 74:136–143
Koller WC, Lyons KE, Truly W (2004) Effect of levodopa treatment for parkinsonism in welders. Neurology 62:730–733
Korontzki S, Justice CO, Scholes RJ (2003) Influence of timing and spatial extent of savanna fires in southern Africa on atmospheric emissions. J Arid Environ 54:395–404
Krombach F, Munzing S, Allmeling AM, Gerlach JT, Behr J, Dorger M (1997) Cell size of alveolar macrophages: an interspecies comparison. Environ Health Perspect 105:1261–1263
Kromhout H (2002) Design of measurement strategies for workplace exposures. Occup Environ Med 59:349–354
Kunii O, Kanagawa S, Yajima I, Hisamatsu Y, Yamamura S, Amagai T, Ismail IT (2002) The 1997 haze disaster in Indonesia: its air quality and health effects. Arch Environ Health 57 (1):16–22
Künzer C (2005) Demarcating coal fire risk areas based on spectral test sequences and partial unmixing using multi sensor remote sensing data. PhD thesis. Austria: Technical University Vienna, 199 pp
Künzer C, Jianzhong Zhang J, Tetzlaff A, van Dijk P, Voigt S, Mehl H, Wagner W (2007) Uncontrolled coal fires and their environmental impacts: Investigating two arid mining regions in north-central China. Appl Geogr 27:42–62
LaDou J (2004) The asbestos cancer epidemic. Environ Health Perspect 112:285–290
LaDou J, Castleman B, Frank A, Gochfeld M, Greenberg M, Huff J, Joshi TK, Landrigan PJ, Lemen R, Myers J, Soffritti M, Soskolne CL, Takahashi K, Teitelbaum D, Terracini B, Watterson A (2010) The case for a Global Ban on asbestos. Environ Health Perspect 118 (7):897–901
Laidlaw MAS, Filippelli GM (2008) Resuspension of urban soils as a persistent source of lead poisoning in children: a review and new directions. Appl Geochem 23:2021–2039
Lange P, Parner J, Vestbo J, Schnohr P, Jensen G (1998) A 15-year followup study of ventilatory function in adults with asthma. N Engl J Med 339:1194–1200
Langmann B (2013) Volcanic ash versus mineral dust: atmospheric processing and environmental and climate impacts. ISRN Atmos Sci 2013:1–17
Langmann B, Duncan B, Textor C, Trentmann J, van der Werf GR (2009) Vegetation fire emissions and their impact on air pollution and climate. Atmos Environ 43:107–116
Langmann B, Zaksek K, Hort M (2010) Atmospheric distribution and removal of volcanic ash after the eruption of Kasatochi volcano: a regional model study. J Geophys Res 115:D2
Lapp NL, Castranova V (1993) How silicosis and coal workers' pneumoconiosis develop - a cellular assessment. Occup Med 8:35–56
Lapp NL, Parker JE (1992) Coal workers' pneumoconiosis. Occup Lung Dis 13:243–252
Laurent B, Marticorena B, Bergametti G, Mei F (2006) Modeling mineral dust emissions from Chinese and Mongolian deserts. Glob Planet Chang 52:121–141

Lazarus AA, Philip A (2011) Asbestosis. Dis Mon 57:14–26
Le Blond JS, Williamson BJ, Horwell CJ, Monro AK, Kirk CA, Oppenheimer C (2008) Production of potentially hazardous respirable silica airborne particulate from the burning of sugarcane. Atmos Environ 42(22):5558–5568
Legg SJ, Cotes JE, Bevan C (1983) Lung mechanics in relation to radiographic category of coal workers' simple pneumoconiosis. Br J Ind Med 40:20–33
Letellier A, Messier S, Pare J, Menard J, Quessy S (1999) Distribution of *Salmonella* in swine herds in Quebec. Vet Microbiol 67:299–306
Leung CC, Tak I, Yu S, Chen W (2012) Silicosis. Lancet 379:2008–2018
Li G, Chen J, Ji J, Yang J, Conway TM (2009) Natural and anthropogenic sources of East Asian dust. Geology 37:727–730
Lin BQ, Liu JH (2010) Estimating coal production peak and trends of coal imports in China. Energy Pol 38:512–519
Liu HB, Tang ZF, Yang YL, Weng D, Sun G, Duan ZW, Chen J (2009) Identification and classification of high risk groups for Coal Workers' Pneumoconiosis using an artificial neural network based on occupational histories: a retrospective cohort study. BMC Public Health 9:366
Liu Y, Gaoa M, Wua W, Tanveera SK, Wena X, Liaoa Y (2013) The effects of conservation tillage practices on the soil water-holding capacity of anon-irrigated apple orchard in the Loess Plateau, China. Soil Till Res 130:7–12
Long W, Tate RB, Neuman M, Manfreda J, Becker AB, Anthonisen NR (1998) Respiratory symptoms in a susceptible population due to burning of agricultural residue. Chest 113 (2):351–357
Mack RN, Simberloff D, Lonsdale WM, Evens H, Clout M, Bazzaz F (2000) Biotic invasions: causes, epidemiology, global consequences and control. Issues Ecol 5:1–25
Madl AK, Donovan EP, Gaffney SH, McKinley MA, Moody EC, Henshaw JL, Paustenbach DJ (2008) State of the science review of the occupational health hazards of crystalline silica in abrasive blasting operations and related requirements for respiratory protection. J Toxicol Environ Health 11:548–608
Mahowald N, Zender C, Luo C, Savoie D, Torres O, del Corral J (2002) Understanding the 30-year Barbados desert dust record. J Geophys Res 107(D21):4561
Manning CB, Vallyathan V, Mossman BT (2002) Diseases caused by asbestos: mechanisms of injury and disease development. Int Immunopharmacol 2:191–200
Maricq MM (2007) Chemical characterisation of particulate emissions from diesel engines: a review. Aerosol Sci 38:1079–1118
Marticorena B, Bergametti G (1995) Modeling the atmospheric dust cycle: 1. Design of a soil-derived dust emission scheme. J Geophys Res 100(8):16415–16430
Masiol M, Hofer A, Squizzato S, Piazza R, Rampazzo G, Pavoni B (2012) Carcinogenic and mutagenic risk associated to airborne particle-phase polycyclic aromatic hydrocarbons: a source apportionment. Atmos Environ 60:375–382
McCunney RJ, Morfeld P, Payne S (2009) What component of coal causes coal workers' pneumoconiosis? J Occup Environ Med 51(4):462–471
McDonald JC (1989) Silica, silicosis and lung cancer. Brit J Ind Med 46:289–291
McKendry I, Strawbridge KB, O'Neill NT, Macdonald AM, Liu PSK, Richard Leaitch W, Anlauf KG, Jaegle L, Fairlie D, Westphal DL (2007) Trans-pacific transport of Saharan dust to western North America: a case study. J Geophys Res 112:D01103
Melloni B, Vergnenegre A, Lagrange P, Bonnaud F (2000) Household radon exposure. Rev Malad Respir 17(6):1061–1071
Micklin PP (1988) Desiccation of the aral sea: a water management disaster in the Soviet Union. Science 241:1170–1176
Migliaccio CT, Hamilton RF Jr, Holian A (2005) Increase in a distinct pulmonary macrophage subset possessing an antigen-presenting cell phenotype and in vitro APC activity following silica exposure. Toxicol Appl Pharmacol 205:168–176

Morgan WKC, Lapp NLR (1976) Respiratory disease in coal miners. Am Rev Respir Dis 113:531–559
Mouillot F, Field CB (2005) Fire history and the global carbon budget: a 1°x1° fire history reconstruction for the 20th century. Glob Chang Biol 11:398–420
Musk AW, Cotes JE, Bevan C, Campbell MJ (1981) Relationship between type of simple coal workers' pneumoconiosis and lung function. A nine year follow-up study of subjects with small rounded opacities. Br J Ind Med 38:313–320
Naeher LP, Brauer M, Lipsett M, Zelikoff JT, Simpson CD, Koenig JK, Smith KR (2007) Woodsmoke health effects: a review. Inhal Toxicol 19(1):67–106
Nakajima T, Higurashi A, Takeuchi N, Herman JR (1999) Satellite and groundbased study of optical properties of 1997 Indonesien forest fire aerosol. Geophys Res Lett 26:2421–2424
Nicholson WL (2002) Roles of Bacillus endospores in the environment. Cell Mol Life Sci 59:410–416
Norbet C, Joseph A, Santiago SS, Bhalla S, Gutierrez FR (2015) Asbestos-related lung disease: a pictorial review. Curr Probl Diagn Radiol 44:371–382
Norris GA (1998) Air pollution and exacerbation of asthma in an arid, western US city. PhD thesis, University of Washington, Spokane
Nourmoradi H, Moradnejadi K, Moghadam FM, Khosravi B, Hemati L, Khoshniyat R, Kazembeigi F (2015) The effect of dust storm on the microbial quality of ambient air in Sanandaj: a city located in the west of Iran. Glob J Health Sci 7(7):114–119
O'Byrne PM, Pedersen S, Busse WW, Tan WC, Chen YZ, Ohlsson SV, Ullman A, Lamm CJ, Pauwels RA (2006) Effects of early intervention with inhaled budesonide on lung function in newly diagnosed asthma. Chest 129:1478–1485
O'Hara SL, Clarke ML, Elatrash MS (2006) Field measurements of desert dust deposition in Libya. Atmos Environ 40:3881–3891
Oke TR (1978) Boundary layer climates. Methuen, London, 359 pp
Olgun N, Duggen S, Croot PL, Delmelle P, Dietze H, Schacht U, Oskarsson N, Siebe C, Auer A, Garbe-Schönberg D (2011) Surface ocean iron fertilization: the role of airborne volcanic ash from subduction zone and hot spot volcanoes and related iron fluxes into the PacificOcean. Global Biogeochem Cycles 25:GB4001
Ozer P (2001) Les lithométéores en région sahélienne. Int J Trop Ecol Geog 24:1–317
Ozsezen AN, Canakci M (2011) Determination of performance and combustion characteristics of a diesel engine fueled with canola and waste palm oil methyl esters. Energy Conv Manag 52 (1):108–116
Page SE, Siegert F, Rieley JO, Boehm HDV, Jaya A, Limin S (2002) The amount of carbon release from peat and forest fires in Indonesia during 1997. Nature 420:61–65
Page SJ, Volkwein JC, Vinson RP, Joy GJ, Mischler SE, Tuchman DP, McWilliams LJ (2008) Equivalency of a personal dust monitor to the current United States coal mine respirable dust sampler. J Environ Monit 10:96–101
Palm C, Blanco-Canqui H, DeClerck F, Gatere L, Grace P (2014) Conservation agriculture and ecosystem services: an overview. Agric Ecosyst Environ 187:87–105
Pandey VC, Singh N (2010) Impact of fly ash incorporation in soil systems. Agric Ecosyst Environ 136:16–27
Parrish DD, Law KS, Staehelin J, Derwent R, Cooper OR, Tanimoto H, Volz-Thomas A, Gilge S, Scheel HE, Steinbacher M, Chan E (2012) Long-term changes in lower tropospheric baseline ozone concentrations at northern mid-latitudes. Atmos Chem Phys 12:11485–11504
Patz JA, Engelberg D, Last J (2000) The effects of changing weather on public health. Ann Rev Pub Health 21:271–307
Peacock C, Copley SJ, Hansell DM (2000) Asbestos-related benign pleural disease. Clin Radiol 55:422–432
Pedgley DE (1986) Long distance transport of spores. Macmillan Publishing Company, New York
Perombelon MCM (1992) Potato blackleg: epidemiology, host-pathogen interaction and control. Neth J Plant Pathol 98:135–146

Peto J, Decarli A, La VC, Levi F, Negri E (1999) The European mesothelioma epidemic. Brit J Cancer 79(3–4):666–672

Petsonk E, Rose C, Cohen R (2013) Coal mine dust lung disease. Am J Respir Crit Care Med 187 (11):1178–1185

Pimentel D, Harvey C, Resosudarmo P, Sinclair K, Kurz D, Mc Nair M, Crist S, Sphpritz L, Fitton L, Saffouri R, Blair R (1995) Environmental and economic costs of soil erosion and conservation benefits. Science 267:1117–1123

Policard A, Gernez-Rieux C, Tacquet A, Martin JC, Devulder B, LeBouffant L (1967) Influence of pulmonary dust load on the development of experimental infection by *Mycobacterium kansasii*. Nature 216:177–178

Poulter B, Christensen NL Jr, Halpin PN (2006) Carbon emissions from a temperate peat fire and its relevance to interannual variability of trace atmospheric greenhouse gases. J Geophys Res 111:D06301

Prins EM, Menzel WP (1994) Trends in South-American biomass burning detected with the GOES visible infrared spin scan radiometer atmospheric sounder from 1983 to 1991. J Geophys Res 99:16719–16735

Prospero JM, Nees RT (1986) Impact of the North African drought and El Niño on mineral dust in the Barbados trade winds. Nature 320:735–738

Prospero JM, Ginoux P, Torres O, Nicholson SE, Gill TE (2002) Environmental characterization of global sources of atmospheric soil dust identified with the NIMBUS-7 TOMS Absorbing Aerosol Product. Rev Geophys 40(1):1002

Purdy LH, Krupa SV, Dean JL (1985) Introduction of sugarcane rust into the Americas and its spread to Florida. Plant Dis 69:689–693

Qu CS, Li B, Wu H, Giesy JP (2012) Controlling air pollution from straw burning in China calls for efficient recycling. Environ Sci Technol 46:7934–7936

Rajkumar WS, Chang AS (2000) Suspended particulate concentrations along the East-West-Corridor. Trinidad, West Indies. Atmos Environ 34:1181–1187

Ram LC, Masto RE (2014) Fly ash for soil amelioration: a review on the influence of ash blending with inorganic and organic amendments. Earth-Sci Rev 128:52–74

Ravi S, D'Odorico P, Breshears DD, Field JP, Goudie AS, Huxman TS, Li J, Okin GS, Swap RJ, Thomas AD, Van Pelt S, Whicker JJ, Zobeck TM (2011) Aeolian processes and the biosphere. Rev Geophys 49:RG3001

Raymond LW, Wintermeyer S (2006) Medical surveillance of workers exposed to crystalline silica. J Occup Environ Med 48:95–101

Rees D, Murray J (2007) Silica, silicosis and tuberculosis. Int J Tuberc Lung Dis 11(5):474–484

Reid JS, Koppmann R, Eck TF, Eleuterio DP (2005) A review of biomass burning emissions part II: intensive physical properties of biomass burning particles. Atmos Chem Phys 5:799–825

Remon J, Reguart N, Corral Lianes JP (2015) Malignant pleural mesothelioma: new hope in the horizon with novel therapeutic strategies. Cancer Treat Rev 41:27–34

Rissler J, Swietlicki E, Bengtsson A, Boman C, Pagels J, Sandstrom T, Blomberg A, Londahl J (2012) Experimental determination of deposition of diesel exhaust particles in the human respiratory tract. J Aerosol Sci 48:18–33

Robock K, Reisner MTR (1982) Specific harmfulness of respirable dust from West Germany coal mines: I. Results of cell tests. Ann Occup Hyg 26:473–479

Rose D, Wehner B, Ketzel M, Engler C, Voigtlander J, Tuch T, Wiedensohler A (2006) Atmospheric number size distributions of soot particles and estimation of emission factors. Atmos Chem Phys 6(4):1021–1031

Ross HF, King EJ, Yogunathan M, Naggelschmidt G (1962) Inhalation experiments with coal dust containing 5 percent, 10 percent, 20 percent, and 40 percent quartz. Tissue reactions in the lungs of rats. Ann Occup Hyg 5:149–161

Rühling A (ed) (1994) Atmospheric heavy metal deposition in Europe – estimations based on moss analysis. Nordic Council of Ministers, Copenhagen

Sastry N (2002) Forest fires, air pollution, and mortality in southeast Asia. Demography 39 (1):1–23
Schmincke HU (2004) Volcanism. Springer, Berlin
Schoennagel T, Nelson CR, Theobald DM, Carnwath GC, Chapman TB (2009) Implementation of National Fire Plan treatments near the wildland–urban interface in the western United States. Proc Nat Acad Sci USA 106:10706–10711
Schröder A, Traber T (2012) The economics of fast charging infrastructure for electric vehicles. Energy Policy 43:136–144
Schulz M, Prospero JM, Baker AR, Dentener F, Ickes L, Liss PS, Mahowald NM, Nickovic S, García-Pando CP, Rodríguez S, Sarin M, Tegen I, Duce RA (2012) Atmospheric transport and deposition of mineral dust to the ocean: implications for research need. Environ Sci Technol 46:10390–10404
Seiler W, Crutzen PJ (1980) Estimates of gross and net fluxes of carbon between the biosphere and the atmosphere from biomass burning. Clim Chang 2:207–247
Shao Y, Dong CH (2006) A review on East Asian dust storm climate, modelling and monitoring. Glob Planet Chang 52(1):1–22
Sharma SK, Pane JN, Verma K (1991) Effect of prednisolone treatment in chronic silicosis. Am Rev Respir Dis 143:814–821
Shealy M, Dorian JP (2010) Growing Chinese coal use: dramatic resource and environmental implications. Energy Policy 38:2116–2122
Sheehan P, Cheng E, English A, Sun F (2014) China's response to the air pollution shock. Nat Clim Chang 4:306–309
Shi H, Shao M (2000) Soil and water loss from the Loess Plateau in China. J Arid Environ 45:9–20
Short SR, Petsonk EL (1993) Respiratory health risks among nonmetal miners. Occup Med 8 (1):57–70
Silva GE, Sherrill DL, Guerra S, Barbee RA (2004) Asthma as a risk factor for COPD in a longitudinal study. Chest 126:59–65
Singh P, Sharratt B, Schillinger WF (2012) Wind erosion and PM10 emissionaffected by tillage systems in the world's driest rainfed wheat region. Soil Till Res 124:219–225
Sipes KL, Mendelsohn R (2001) The effectiveness of gasoline taxation to manage air pollution. Ecol Econ 36:299–309
Skidmore EL (1994) Wind erosion. In: Lal R (ed) Soil erosion research methods, 2nd edn. Soil and Water Conserv Soc Ankeny, Iowa, pp 265–293
Skinner HCW (2007) The earth, source of health and hazards: an introduction to medical geology. Annu Rev Earth Planet Sci 35:177–213
Snider DE (1978) The relationship between tuberculosis and silicosis. Am Rev Resp Dis 118:455–460
Song Z, Wang J, Wang S (2007) Quantitative classification of northeast Asian dust events. J Geophys Res 112:D04211
Spear TM, Svee W, Vincent JH, Stanisich N (1998) Chemical speciation of lead dust associated with primary lead smelting. Environ Health Persp 106:565–571
Sterk G (2003) Causes, consequences and control of wind erosion in Sahelian Africa: a review. Land Degrad Dev 14(1):95–108
Stohl A, Berg T, Burkhart JF, Fjaeraa AM, Forster C, Herber A, Hov O, Lunder C, McMillan WW, Oltmans S, Shiobara M, Simpson D, Solberg S, Stebel K, Strom J, Torseth K, Treffeisen R, Virkkunen K, Yttri KE (2007) Arctic smoke – record high air pollution levels in the European Arctic due to agricultural fires in Eastern Europe in spring 2006. Atmos Chem Phys 7:511–534
Stone V, Jones R, Rollo K, Duffin R, Donaldson K, Brown DM (2004) Effect of coal mine dust and clay extracts on the biological activity of the quartz surface. Toxicol Lett 149(1–3):255–259
Stover RH (1962) Intercontinental spread of banana leaf spot (*Mycospherella musicola*). Trop Agric – Trinidad 39:327–338
Suvatne J, Browning RF (2011) Asbestos and lung cancer. Dis Mon 57:55–68

Tanaka TY, Chiba M (2006) A numerical study of the contributions of dust source regions to the global dust budget. Glob Planet Chang 52:88–104

Tenenbaum DJ (2000) A burning question: do farmer-set fires endanger health? Environ Health Perspect 108(3):A117–A118

Therriault S (2001) Wildfire smoke: a guide for public health officials. Missoula City-County Health Department, Missoula

Thierfelder C, Mwila M, Rusinamhodzi L (2013) Conservation agriculture in eastern and southern provinces of Zambia: long-term effects on soil quality and maize productivity. Soil Till Res 126:246–258

Thomas DSG, Middleton NJ (1994) Desertification: exploding the myth. Wiley, Chichester

Thorsteinsson T, Johannsson T, Stohl A, Kristiansen NI (2012) High levels of particulate matter in Iceland due to direct ash emissions by the Eyjafjallajökull eruption and resuspension of deposited ash. J Geophys Res 117:B9

Todd MC, Washington R, Martins JV, Dubovik O, Lizcano G, M'Bainayel S, Engelstaedter S (2007) Mineral dust emission from the Bodélé Depression, northern Chad, during BoDEx 2005. J Geophys Res 112:D06207

Toole-O'Neil B, Tewalt SJ, Finkelman RB, Akers DJ (1999) Mercury concentration in coal – unraveling the puzzle. Fuel 78(1):47–54

Torigoe K, Hasegawa S, Numata O, Yazaki S, Matsunaga M, Boku N, Hiura M, Ino H (2000) Influence of emission from rice straw burning on bronchial asthma in children. Pediatr Int 42 (2):143–150

Tossavainen A (1997) Asbestos, asbestosis and cancer: the Helsinki criteria for diagnosis and attribution. Consensus report. Scand J Work Environ Health 23:311–316

Ulrik CS, Lange P (1994) Decline of lung function in adults with bronchial asthma. Am J Respir Crit Care Med 150:629–634

UNSD (United Nations Statistics Division) (2009) United Nations Statistics Division Homepage. Available: http://unstats.un.org/unsd/default.htm. Accessed 23 Sept 2016

USGS (U.S. Geological Survey) (2009) Asbestos. In: 2008 Minerals Yearbook. Reston, VA: U.S. Geological Survey, 8.1–8.6. Available: http://minerals.usgs.gov/minerals/pubs/commodity/asbestos/myb1-2008-asbes.pdf Accessed 23 Sept 2016

Usup A, Takahashi H, Limin SH (2000) Aspect and mechanism of peat fire in tropical peat land: a case study in Central Kalimantan 1997, Proceedings of the International Symposium on Tropical Peatlands. Bogor, Indonesia, Hokkaido University and Indonesian Institute of Science, pp 79–88

van der Werf GR, Randerson JT, Collattz GJ, Giglio L (2003) Carbon emissions from fires in tropical and subtropical ecosystems. Glob Change Biol 9:547–562

van der Werf GR, Randerson JT, Giglio L, Collatz GJ, Kasibhatla PS, Arellano AF Jr (2006) Interannual variability in global biomass burning emissions from 1997 to 2004. Atmos Chem Phys 6:3423–3441

van der Werf GR, Randerson JT, Giglio L, Collatz GJ, Mu M, Kasibhatla PS, Morton DC, DeFries RS, Jin Y, van Leeuwen TT (2010) Global fire emissions and the contribution of deforestation, savanna, forest, agricultural, and peat fires (1997–2009). Atmos Chem Phys 10:11707–11735

van Thiel E, Gaafar R, van Meerbeeck JP (2011) European guidelines for the management of malignant pleural mesothelioma. J Adv Res 2:281–288

Venkatesh MV, Joshi KR, Harjai SC, Ramdeo IN (1975) Aspergillosis in desert locust. Mycopathologia 57:135–138

Viana M, Averol X (2007) Source apportionment of ambient PM2.5 at 5 Spanish centers of the European community respiratory health survey (ECRHS II). Atmos Environ 41(7):1395–1406

Vigiak O, Sterk G, Warren A, Hagen LJ (2003) Spatial modeling of wind speed around windbreaks. Catena 52:273–288

Vignola AM, Kips J, Bousquet J (2000) Tissue remodeling as a feature of persistent asthma. J Allergy Clin Immunol 105:1041–1053

Virta RL (2005) Mineral commodity profiles – asbestos. U.S. Geological Survey Circular 1255-KK. Available: http://pubs.usgs.gov/circ/2005/1255/kk/ Accessed 23 Sept 2016
Walker S (1999) Uncontrolled fires in coal and coal wastes. International Energy Agency, IEA on Coal Research, London. 73 pp
Wang YQ, Zhang XY, Arimoto R (2006a) The contribution from distant dust sources to the atmospheric particulate matter loadings at Xi'An, China during spring. Sci Tot Envir 368:875–883
Wang X, Zhou Z, Dong Z (2006b) Control of dust emissions by geomorphic conditions, wind environments and land use in northern China: an examination based on dust storm frequency from 1960–2003. Geomorphology 81:292–308
Watanabe S, Shirakami A, Takeichi T, Ohara T, Saito S (1987) Alterations in lymphocyte subsets and serum immunoglobulin levels in patients with silicosis. J Clin Lab Immunol 23:45–51
Weir-Brush J, Garrison V, Smith G, Shinn E (2004) The relationship between gorgonian coral (*Cnidaria Gorgonacea*) diseases and African dust storms. Aerobiologia 20(2):119–126
Westerholm P (1980) Silicosis. Observations on a case register. Scand J Work Environ Med 9:523–531
WHO (World Health Organisation) (1986) Asbestos and other natural mineral fibers: environmental health criteria 53. World Health Organisation, Geneva
WHO (World Health Organization) (2002) World health report 2002. World Health Organization, Geneva
WHO (World Health Organization) (2005) WHO Workshop on Mechanisms of Fibre Carcinogenesis and Assessment of Chrysotile Asbestos Substitutes, November 8–12, 2005, Lyon, France
WHO (World Health Organization) (2006) Elimination of asbestos-related diseases. Geneva
WHO (World Health Organization) (2007) The Global Occupational Health Network newsletter: elimination of silicosis. 2007. http://www.who.int/occupational_health/publications/newsletter/gohnet12e.pdf. Accessed 19 Oct 2016
Wilson R, Spengler JD (1996) Particles in our air. Harvard Sch. Public Health, Cambridge
Wilson TM, Stewart C, Sword-Daniels V, Leonard GS, Johnston DM, Cole JW, Wardman J, Wilson G, Barnard ST (2012) Volcanic ash impacts on critical infrastructure. Phys Chem Earth 45–46:5–23
Wilt JL, Banks DE, Weissman DN, Parker JE, Vallyathan V, Castranova V, Dedhia HV, Stulken E, Ma JKH, Ma JYC, Cruzzavala J, Shumaker J, Childress CP, Lapp NL (1996) Reduction of lung dust burden in pneumoconiosis by whole-lung lavage. J Occup Environ Med 38:619–624
Woolf AD, Goldman R, Bellinger DC (2007) Update on the clinical management of childhood lead poisoning. Pediatr Clin N Am 54:271–294
World Bank Group (2009) Good practice note: asbestos: occupational and community health issues. Available: http://siteresources.worldbank.org/EXTPOPS/Resources/AsbestosGuidanceNoteFinal.pdf. Accessed 17 Oct 2016
Worster D (1979) Dust bowl: the Southern Plains in the 1930s. Oxford University Press, New York
Wösten JHM, Van Den Berg J, Van Eijk P, Gevers GJM, Giesen WBJT, Hooijer A, Idris A, Leenman PH, Rais DS, Siderius C, Silvius MJ, Suryadiputra N, Wibisono IT (2006) Interrelationships between hydrology and ecology in fire degraded tropical peat swamp forests. Int J Water Res Dev 22(1):157–174
Wu J, Winer AM, Delfino RJ (2006) Exposure assessment of particulate matter air pollution before, during, and after the 2003 Southern California wildfires. Atmos Environ 40:3,333–3,348
Wypych F, Adad LB, Mattoso N, Marangon AA, Schreiner WH (2005) Synthesis and characterization of disordered layered silica obtained by selective leaching of octahedral sheets from chrysotile and phlogopite structures. J Colloid Interf Sci 283(1):107–112
Xu XZ, Cai XG, Men XS (1993) A study of siliceous pneumoconiosis in a desert area of Sunan County, Gansu province, China. Biomed Environ Sci 6:217–222

Yang B, Bräuning A, Zhang Z, Dong Z, Epser J (2007) Dust storm frequency and its relation to climate changes in Northern China during the past 1000 years. Atmos Environ 41:9288–9299

Zhang Y, Cao SR (1996) Coal burning induced endemic fluorosis in China. Fluoride 29 (4):207–211

Zhang B, Tsunekawa A, Tsubo M (2008) Contributions of sandy lands and stony deserts to long-distance dust emission in China and Mongolia during 2000–2006. Glob Planet Chang 60:487–504

Zhao TL, Gong SL, Zhang XY, Blanchet JP, McKendry IG, Zhou ZJ (2006) A simulated climatology of Asian dust aerosol and its trans-Pacific transport. Part I: Mean climate and validation. J Clim 19(1):88–103

Zhenda Z, Tao W (1993) The trends of desertification and its rehabilitation in China. Desertification Control Bull 22:27–29

Zheng B, Huang R (1989) Human fluorosis and environmental geochemistry in southwest China. Developments in Geoscience, Contributions to 28th International Geologic Congress. Washington, DC Science Press, Beijing, China, pp 171–176

Zheng B, Yu X, Zhand J, Zhou D (1996) Environmental geochemistry of coal and endemic arsenism in southwest Guizhou, PR China. 30th Int Geol Congr Abstr 3:410

Chapter 4
Soil-Borne Gases and Their Influence on Environment and Human Health

Abstract The major soil-borne gases produced as a consequence of chemical and biological processes in soil include carbon dioxide (CO_2), methane (CH_4), nitrous oxide (N_2O), and ammonia (NH_3). The first three gases are referred to as greenhouse gases (GHGs) as these are considered to force global climate change. These GHGs respectively account for 76%, 16% and 6% of total anthropogenic emissions. Most of the anthropogenic emissions of GHGs come from the combustion and production of fossil fuel and from industrial processes, but there is considerable contribution from agriculture and land use changes. Emission of CO_2 from soil to the atmosphere is governed by the mineralization of soil organic carbon and the rates of soil CO_2 efflux are significantly related to climatic factors such as temperature and precipitation. Most dominant soil-borne source of global CH_4 flux is emissions from wetlands and rice cultivation. Denitrification process in soil is the major source of nitrogen oxide emission and the soil-borne flux accounts for 60% of the total emissions. Emissions of GHGs besides causing global warming, deplete the concentration of ozone in the stratosphere and contribute to acid deposition, which adversely impacts human health. Climate change can impact human health directly by affecting body's physiological functions because of high temperature and extreme weather events; and indirectly by affecting the spread of vector-borne pathogens and increased risk of water-, food-, and rodent-borne diseases. Shifts in temperature and precipitation patterns can also impact agriculture productivity and thus affect food security in many parts of the world. Besides agricultural production and food availability, elevated CO_2 concentration in the atmosphere may affect human health by altering nutrient content of food crops. Ammonia influences the environment and human health through its role in the formation of aerosols and particulate matter (PM). Both short-term and long-term exposure to inhalable PM cause adverse health effects including aggravation of asthma, respiratory symptoms and mortality from cardiovascular and respiratory diseases. Adoption of improved plant-, nutrient-, water- and soil management practices and mitigation technologies could help in reducing emissions and mitigating adverse effect of climate change on human health. In this chapter, we present information on sources of soil-borne gases, their impacts on climate and ecosystems. Further, we discuss the probable impacts of climate change and climate variability as well as atmospheric pollutants, NH_3 and aerosols, on human health and delve over opportunities for mitigation.

R. Nieder et al., *Soil Components and Human Health*,
https://doi.org/10.1007/978-94-024-1222-2_4

Keywords Greenhouse gas emissions • Climate change • Ammonia emissions • Aerosols • Particulate matter • Vector-borne diseases • Mitigation options • Mitigating climate change • Methane • Carbon dioxide • Nitrous oxide

4.1 Overview of Soil-Borne Gases

Chemical and biological processes occurring in the soil result in emission of gases that can contribute to global warming and impact animal and human health. The major soil-borne gases include carbon dioxide (CO_2), methane (CH_4), nitrous oxide (N_2O), and ammonia (NH_3). Mercury can also be emitted in the form of mercury vapors (see Chap. 8). The first three gases are referred to as greenhouse gases (GHGs) as these are considered to force global climate change. Ammonia influences the environment and the human health through its role in the formation of aerosols and particulate matter (PM). Oxidation of CH_4 is an important source of atmospheric carbon monoxide (CO). The CO concentration in the atmosphere ranges from 0.05 to 0.2 ppm (parts per million), with considerable differences between the northern and southern hemispheres. The global storage of CO is estimated at 0.2 Pg (Holmen 2000). Besides CO_2, CH_4 and CO, other C-containing gases present in the atmosphere include terpenes, isoprenes and compounds of petrochemical origin and these are estimated at 0.05 Pg (Holmen 2000).

4.2 Sources of Soil-Borne Gases

The GHGs originate from several natural and anthropogenic sources (Table 4.1). Because of human interventions in the global carbon (C) and nitrogen (N) cycles driven largely by economic and population growth, the atmospheric concentrations of CO_2, CH_4 and N_2O have increased to levels unprecedented in at least the last 800,000 years (IPCC 2013). CO_2 remains the major anthropogenic GHG accounting for 76% (38 ± 3.8 Pg CO_2eq year^{-1}) of total anthropogenic GHG emissions in 2010. Methane and N_2O accounted for 16% (7.8 ± 1.6 Pg CO_2eq year^{-1}) and 6% (3.1 ± 1.9 Pg CO_2 eq year^{-1}) of the anthropogenic emissions, respectively. Remaining 2% (1.0 ± 0.2 Pg CO_2eq year^{-1}) of the emissions came from fluorinated gases. Though CO_2 has the highest atmospheric concentration among the GHGs, N_2O and CH_4 have several times higher global warming potentials than CO_2 (Table 4.2). Most of the anthropogenic emissions come from the combustion and production of fossil fuel and from industrial processes, but there is considerable contribution from agriculture and land use changes. During 1970 to 2010, emission of GHGs in the agriculture sector increased by 35%, from 4.2 to 5.7 Pg CO_2 eq year^{-1} (IPCC 2014a). Though total global emission increased, yet per capita emissions decreased from 2.5 Mg in 1970 to

Table 4.1 Sources of principal greenhouse gases

Gas	Natural source	Anthropogenic source
Carbon dioxide (CO_2)	Terrestrial biosphere	Fossil fuel combustion (coal, oil and gas)
	Oceans	Cement production
		Land use changes (mainly deforestation)
Methane (CH_4)	Natural wetlands	Fossil fuels (natural gas production, coal mines, petroleum industry, coal combustion)
	Geological sources	
	Oceans and freshwater lakes	Ruminants (enteric fermentation)
	Termites	Landfills and waste
	Wild animals	Rice paddies
	Hydrates	Biomass burning
	Wild fires	
	Permafrost	
Nitrous oxide (N_2O)	Oceans	Nitrogen fertilizers
	Atmosphere (NH_3 oxidation)	Industrial sources (adipic acid/nylon, nitric acid)
	Tropical soils (wet forests, dry savannas)	Land use changes (biomass burning, forest clearing)
	Temperate soils (forests, grasslands)	Cattle and feedlots
		Human excreta
		Rivers, estuaries, coastal zones

Table 4.2 Characteristics of three principal greenhouse gases

Characteristic	CO_2	CH_4	N_2O
Pre-industrial concentration (1750)	278 ppm	722 ppb	271 ppb
Concentration in 2011	391 ppm	1803 ppb	324.2 ppb
Rate of concentration change per year (2002–2011)	2 ± 0.1 ppm	6 ppb[a]	0.73 ± 0.03 ppb
Increase in concentration since 1750 (%)	40	150	20
Atmospheric life time (years)	~100–300[b]	12.4 ± 14[c]	121 ± 10[c]
100-year global warming potential	1	28	265
Radiative forcing ($W\ m^{-2}$) in 2011	1.82 ± 0.19	0.48 ± 0.05	0.17 ± 0.03

Compiled from IPCC (2013)
[a]Time period 2007–2011
[b]No single lifetime can be defined because of different rates of uptake by different sink processes
[c]This life time has been defined as an adjustment time that takes into account the indirect effect of the gas on its own residence time

1.6 Mg in 2010 because of growth in population. Per capita emission decreased in Latin America, Middle East and Africa and Economies in Transition, whereas in Asia and Organization for Economic Co-operation and Development (OECD-1990) countries, it remained almost unchanged.

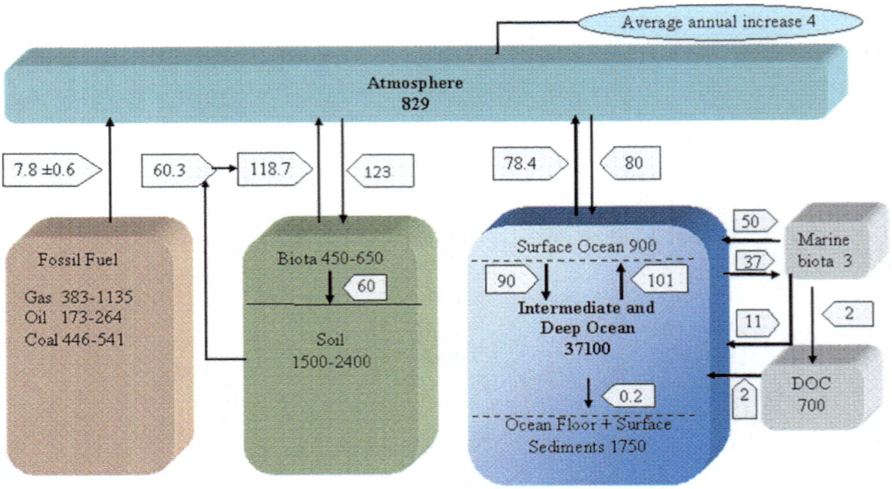

Fig. 4.1 The major reservoirs (Pg) and flows of carbon (Pg C $year^{-1}$) for the globe, during the 2000s (Adapted from Ciais et al. 2013)

4.2.1 Carbon Dioxide Emission

Large amounts of carbon are found in the terrestrial ecosystems and it cycles globally between the atmosphere, the ocean and the land biosphere (Fig. 4.1). Ocean is the biggest reservoir of C and stocks about 38,000 Pg C. Soils and vegetation stock about 3000 Pg C. Atmosphere contains 828 Pg C as CO_2 and the amount of CO_2-C exchanged annually between the land and the atmosphere as gross primary productivity is estimated at ~120 Pg C and about half of it is released by plant respiration giving a net primary productivity of ~60 Pg C $year^{-1}$. About 60 Pg C is returned to the atmosphere annually through heterotrophic soil respiration and fire (Denman et al. 2007).

An imbalance between emissions and uptake, caused by anthropogenic activities leads to increased concentration of CO_2 in the atmosphere. During the last over 250 years the atmospheric concentration of CO_2 has increased globally by more than 100 ppm (40%) from about 275 ppm in the pre-industrial era (AD 1000–1750) to 391 ppm in 2011 (Fig. 4.2) and 409 ppm in April 2017 (NOAA 2017).

Global C budget shows that during the years 2002–2011 the atmospheric load increased at 4.3 ± 0.2 Pg C $year^{-1}$ compared to 3.1 ± 0.2 Pg C $year^{-1}$ in the 1990s (Table 4.3). Major anthropogenic sources of CO_2 to the atmosphere include fossil fuel combustion, cement production and land use changes. Until the beginning of the twentieth century, the emissions from changes in land use and management were greater than those from fossil-fuel burning, but the latter now dominates by a factor of about 3 (Nieder and Benbi 2008). CO_2 is emitted to the atmosphere by land use and land use changes, particularly deforestation, and taken up from the atmosphere by other land uses such as afforestation and vegetation regrowth on

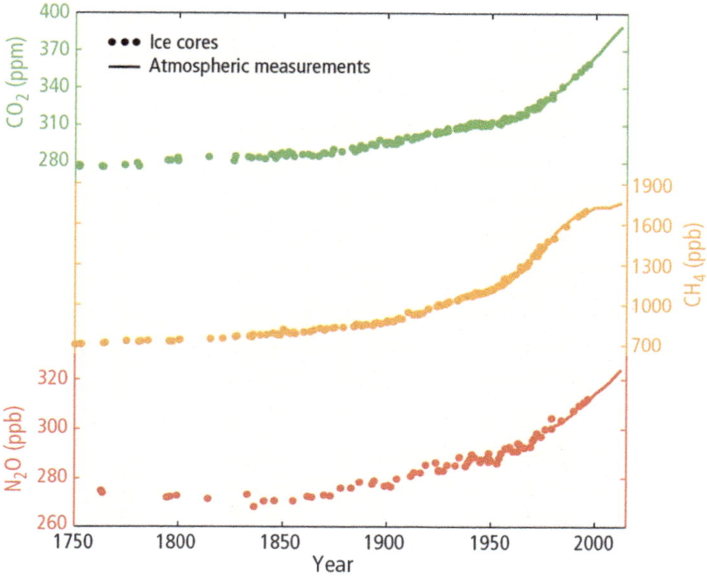

Fig. 4.2 Atmospheric concentrations of carbon dioxide (CO_2), methane (CH_4) and nitrous oxide (N_2O) since 1750 (IPCC 2014b, p. 44; with kind permission from IPCC)

abandoned lands. Global net CO_2 emissions from land use change are estimated at 0.9 Pg C year^{-1}for the period 2002–2011 (Houghton et al. 2012). During the years 1750–2011 cumulative emissions of CO_2 from land use changes amounted to 180 Pg C (Houghton 2003).

Cumulative anthropogenic emissions of CO_2 between 1750 and 2011 amount to 555 ± 85 Pg C with 375 ± 30 Pg C coming from fossil fuel combustion and cement production and 180 ± 80 Pg C from land use changes (IPCC 2013). Less than half of these anthropogenic emissions have accumulated in the atmosphere (240 ± 10 Pg C) and the remaining has been absorbed by the ocean (155 ± 30 Pg C) and in terrestrial ecosystems (160 ± 90 Pg C): the residual 'sinks'. The increased terrestrial carbon storage in ecosystems not affected by land use change is called the Residual land sink (Table 4.3). This increased storage in terrestrial ecosystems is probably caused by enhanced photosynthesis at higher CO_2 levels and nitrogen deposition, and changes in climate favouring carbon sinks such as longer growing seasons in mid-to-high latitudes. Forest area expansion and increased biomass density of forests that result from changes in land use are also carbon sinks.

4.2.1.1 Soil-Borne CO_2 Emission

Emission of CO_2 from soil to the atmosphere is governed by the mineralization of SOC through microbial processes that use C as a source of energy and combine it

Table 4.3 Global anthropogenic carbon budget (Pg C $year^{-1}$) during 1990s and 2002 to 2011 and accumulated since 1750 (onset of Industrial Revolution)

	1750–2011 cumulative Pg C	1990s	2002–2011
Fossil fuel combustion and cement production	375 ± 30	+6.4 ± 0.5	+8.3 ± 0.7
Net ocean to atmosphere flux	−155 ± 30	−2.2 ± 0.7	−2.4 ± 0.7
Net land to atmosphere flux	30 ± 45	−1.1 ± 0.9	−1.6 ± 1.0
Land use change	180 ± 80	+1.5 ± 0.8	+0.9 ± 0.8
Residual terrestrial sink	−160 ± 90	−2.6 ± 1.2	−2.5 ± 1.3
Atmospheric increase	240 ± 10	+3.1 ± 0.2	+4.3 ± 0.2

Errors represent ± standard deviation. Positive fluxes indicate emissions to the atmosphere and negative fluxes are losses from the atmosphere (sinks) (Adapted from Ciais et al. 2013)

with O_2 leading to release of CO_2 and H_2O. The global annual flux of CO_2 from soils to the atmosphere due to respiration of soil organisms and plant roots has been estimated at 76.5 Pg C (Raich and Potter 1995). The CO_2 emission from soil is enlarged because of ploughing and mixing of crop residues and other biomass in the soil surface. Soil management practices such as tillage enhance emission of CO_2 through microbial decomposition of soil organic matter. Tillage breaks the soil aggregates, increases the oxygen supply and exposes the surface area of organic material, thereby promoting the decomposition of organic matter. Despite large annual exchange of CO_2 between soil and atmosphere, the net flux of CO_2 from agricultural lands is small and has been estimated to be around 40 Tg CO_2 $year^{-1}$ (Smith et al. 2007). At the global scale, rates of soil CO_2 efflux are significantly related to temperature and precipitation. Soil erosion by water could be another important source of atmospheric CO_2. Tropical soils are particularly prone to water erosion due to high rainfall intensity. It has been estimated that globally about 1.14 Pg C $year^{-1}$ may be emitted to the atmosphere through soil erosion (Lal 2004).

Agriculture also contributes to C emissions through direct and indirect consumption of fossil fuel. The amount of C emitted per unit of energy use varies from 14 kg C GJ^{-1} for natural gas to 24 kg C GJ^{-1} for coal (DTI 2001). Direct and indirect emissions of C through fertilizer N, P and K application vary between 0.1 and 1.8 kg C per unit nutrient used. Emissions due to application of pesticides and different farm operations are still higher. Apparently, increased use of fertilizers, pumped irrigation and mechanical power, account for 90% or more of the total direct and indirect energy inputs to most farming systems (Pretty et al. 2002). Thus C emissions are greater from high-input systems than the low-input systems; for each Mg of cereal or vegetable production 3–10 GJ of energy are consumed in high-input systems compared to 0.5–1.0 GJ of energy in low-input systems (Pretty 1995). Globally, energy use in agriculture is estimated to account for 1.5% of the total GHG emissions (Timothy 2009).

4.2.2 Methane Emission

The atmospheric concentration of CH_4 has increased from 722 parts per billion (ppb) in pre-industrial times to 1803 ppb in 2011 (Ciais et al. 2013). However, the atmospheric growth rate of CH_4 during mid 1980s and mid 2000s was almost zero, which has started to increase again since 2006 (Fig. 4.2). Atmospheric CH_4 originates from both natural and anthropogenic sources (Table 4.1). The natural sources of CH_4 include wetlands, oceans, forests, wildfires, and termites. The main anthropogenic sources of CH_4 include agriculture, energy production and transmission, leakages from gas fields, coal mines, waste and landfills. While natural sources accounted for the major share of the pre-industrial global CH_4 budget, the anthropogenic emissions dominate the contemporary budget. Total global pre-industrial emissions of CH_4 are estimated to be 200–250 Tg CH_4 $year^{-1}$, of which the natural sources emitted 190–220 Tg CH_4 $year^{-1}$ and the anthropogenic sources accounted for the rest. In contrast, anthropogenic emissions account for 50–65% of the current global CH_4 budget (Table 4.4). During the period 2000–2011, agriculture and waste accounted for 63% of the anthropogenic emissions, fossil fuels for 29% and rest of the emissions originated from biomass burning, landfills, wastewater and others. Enteric fermentation was the largest contributor to methane emission followed by rice cultivation and manure management (IPCC 2014a). Ruminants and rice agriculture together contribute ~125 Tg CH_4 $year^{-1}$. Ruminants emit between 87 and 94 Tg CH_4 $year^{-1}$ with majority of contributions coming from India, China, Brazil and the USA (Olivier and Janssens-Maenhout 2012). India, with the world's largest livestock population emitted 11.8 Tg CH_4 $year^{-1}$ in 2003, including emission from enteric fermentation (10.7 Tg CH_4 $year^{-1}$) and manure management (1.1 Tg CH_4 $year^{-1}$; Chhabra et al. 2013). Because of anoxic conditions and a high availability of acetate, CO_2 and H_2, methanogenesis in landfills, livestock manure and waste waters produces between 67 and 90 Tg CH_4 $year^{-1}$.

Natural and anthropogenic sources of CH_4 could be biogenic, thermogenic or pyrogenic in origin (Neef et al. 2010). Biogenic sources are due to degradation of organic matter under anaerobic conditions and include natural wetlands, ruminants, waste, landfills, rice paddies, fresh waters and termites. Thermogenic sources arise from the slow transformation of organic matter into fossil fuels (natural gas, coal, oil) on geological time scales and pyrogenic sources result from incomplete combustion of organic matter (biomass and biofuel burning). Pyrogenic sources of CH_4 (biomass burning) had a small contribution in the global flux for the 2000s (32–39 Tg CH_4 $year^{-1}$). Some sources can eventually combine a biogenic and a thermogenic origin (e.g., natural geological sources such as oceanic seeps, mud volcanoes or hydrates). Each of these three types of emissions is characterized by its isotopic composition in ^{13}C-CH_4: typically −55 to −70% for biogenic, −25 to −45% for thermogenic, and −13 to −25% for pyrogenic. The isotopic composition provides a basis for separating relative contribution of different methane sources using the top-down approach (Bousquet et al. 2006; Monteil et al. 2011). The CH_4

Table 4.4 Global annual CH_4 budget (Tg CH_4 year^{-1}) for the decade 2000–2009 using top-down and bottom-up approach

Source	Top-down	Bottom-up
Natural sources	**218**	**347**
Wetlands	175	217
Other sources	*43*	*130*
Geological sources (including oceans)		54
Freshwater (lakes and rivers)		40
Wild animal		15
Termites		11
Hydrates		6
Wildfires		3
Permafrost (Excl lakes and wetlands)		1
Anthropogenic sources	**335**	**331**
Fossil fuels	96	96
Agriculture and waste	*209*	*200*
Ruminants		89
Rice agriculture		36
Landfills and waste		75
Biomass burning	30	35
Global total	**553**	**678**
Sinks		
Soils	32	28
Total chemical loss	*518*	*604*
Tropospheric OH		528
Stratospheric OH		51
Troposheric Cl		25
Total sinks	**550**	**632**
Imbalance (sources minus sinks)	**3**	
Atmospheric growth rate	**6**	

Adapted from Ciais et al. (2013)

growth rate results from the balance between emissions and sinks. The main sink for methane in the atmosphere is the hydroxyl radical (OH).

4.2.2.1 Soil-Borne Methane Emission

Most dominant soil-borne source of global CH_4 flux is emissions from wetlands (177–284 Tg CH_4 year^{-1}), which include wet soils, swamps, bogs and peatlands. Rates of CH_4 emission from wetlands are influenced by a number of factors including soil water status and temperature, soil type, pH, soil redox potential (Eh), nutrient inputs, and the presence of adapted vascular plants having a well-developed system of intracellular air spaces (aerenchyma). The aerenchyma serves as a pathway for the movement of methane from the soil into the atmosphere and also permits the transport of oxygen from the atmosphere to the root meristem (Lloyd et al. 1998).

Rice cultivation is a major source of CH_4 emission from agricultural soils and it contributes between 33 and 40 Tg CH_4 year^{-1} and 90% of these emissions come from tropical Asia, with more than 50% from China and India (Yan et al. 2009). There is considerable uncertainty in the CH_4 emission estimates probably because of different hydrological environments under which rice is grown, wide range of agricultural and water management practices, climatic conditions, and complexity of the role of rice plants in CH_4 transport (Nieder and Benbi 2008). On a global scale, irrigated rice accounts for 70–80%, rainfed rice ~15% and deep-water rice ~10% of CH_4 emission from rice agriculture (Wassmann et al. 2000). Methane is produced in rice fields under anaerobic conditions after the sequential reduction of O_2, nitrate, manganese, iron and sulphate, which serve as electron acceptors for oxidation of organic matter to CO_2. The decomposition of organic matter occurs through methanogenic fermentation, which produces CH_4 and CO_2 according to the following reaction:

$$C_6H_{12}O_6 \rightarrow 3CO_2 + 3CH_4$$

Methanogenesis, which requires low redox potentials (Eh < -200 mV) is carried out by a specialised, strictly anaerobic microorganisms, called methanogenic archaea that can develop in synergy or in syntrophy with other anaerobic bacteria. In paddy soil, methanogens produce CH_4 from either the reduction of CO_2 with H_2 (hydrogenotrophic) or from the fermentation of acetate to CH_4 and CO_2 (acetoclastic) (Deppenmeir et al. 1996). Methane escapes to the atmosphere via molecular diffusion, ebullition (gas transport via gas bubbles) and plant transport (Fig. 4.3). Plant mediated transport is the primary mechanism for CH_4 emission from paddy fields and contributes 50–90% of the total CH_4 flux (Wassmann and Aulakh 2000).

The rate and amount of CH_4 emitted from a rice field depends upon several soil, plant, management and climatic factors including soil redox potential (Eh), pH and organic matter content. Soil redox potential (Eh) is the most important factor that controls the production and emission of CH_4 in soils. The production of CH_4 in paddy soils is initiated at Eh values between -100 and -200 and it is emitted to the atmosphere as Eh falls below -200 mV (Yamane and Sato 1964; Yagi and Minami 1990). There exists a negative relationship between Eh and methane emission; a decrease in Eh from -200 to -300 mV has been reported to cause a tenfold increase in CH_4 production and a 17-fold increase in its emission (Kludze et al. 1993). The development of soil redox conditions are controlled by degree of submergence, the amount and nature of electron donors and acceptors, and temperature (Ponnamperuma 1972). Soil submergence reduces the size of the oxidised zones and allows the development of methanogenic activity. Soils containing large amounts of readily decomposable organic substrates (e.g. acetate, formate, methanol, methylated amines etc.) and small amounts of electron acceptors (NO_3^-, Mn^{4+}, SO_4^{2-}) exhibit high production of CH_4 (Parashar et al. 1991). Addition of organic

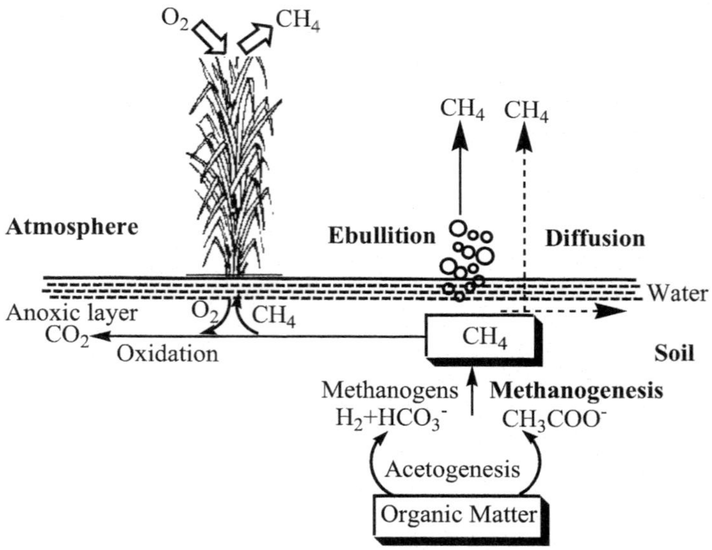

Fig. 4.3 Schematic representation of methane emission from rice fields

matter increases CH_4 emission and the magnitude of increase depends on C:N ratio, biochemical composition and amount of the organic material added (Merr and Roger 2001). Temperature influences CH_4 emission through its effect on the activity of soil microorganisms and decomposition of organic materials. Methanogenesis is considered to be optimum between 30 and 40 °C. Temperature also affects CH_4 transport through the rice plant (Nouchi et al. 1994). The other soil properties that influence CH_4 emission include soil pH and texture. Soil pH influences the activity of methanogens and the optimum pH for methane production is considered to be between 6.7 and 7.1 (Wang et al. 1993). The flooded paddy soils provide favourable pH environment for CH_4 production as the soil pH on flooding tends to be towards neutrality. Soil texture influences CH_4 production indirectly through its effect on several physico-chemical properties of soil. Besides soil properties, plants exert a major influence on the magnitude and seasonality of emissions. The presence of rice plants increases CH_4 emission by providing C source and by facilitating CH_4 transfer to the atmosphere (Dannenberg and Conrad 1999). The plant variety/cultivar and growth stage also influence methane flux from rice fields. Lower CH_4 fluxes are recorded in the early growth period of rice plant, which increases gradually until flowering and drops to very low level before harvest. High emission rates at flowering have been attributed to root exudates and decaying tissue (Watanabe et al. 1997). Application of inorganic fertilizers, depending on rate, type and mode of application, influences CH_4 emission from soils. While urea application enhances CH_4 fluxes by increasing soil pH; application of sulphate or nitrate containing fertilizers can suppress the production of CH_4 by affecting the activity of methanogens.

4.2.3 *Nitrous Oxide Emission*

Nitrous oxide accounts for 6% of the anthropogenic greenhouse effect and its concentration in the atmosphere has increased by about 20%, from about 271 ppb in pre-industrial times to 324 ppb in 2011 (Fig. 4.2). The concentration of N_2O in the atmosphere increased at a rate of 0.73 ± 0.03 ppb year^{-1}. Nitrous oxide is emitted into the atmosphere both from natural and anthropogenic sources but there is considerable uncertainty about the contribution of different sources to the global N_2O emissions. Natural sources include soils, ocean and atmospheric NH_3 oxidation. Anthropogenic sources of N_2O can be both biogenic (biological nitrification and denitrification) and abiogenic (e.g. during biomass burning). During biomass burning, the nitrogen in fuel in end groups, open chains and heterocyclic rings can be converted into gaseous forms such as ammonia, nitric oxide, nitrous oxide, dinitrogen and hydrogen cyanide (Galbally and Gillett 1988) and emitted to the atmosphere. Estimates of Global N_2O emissions from natural and anthropogenic sources are estimated at 17.9 Tg N year^{-1} (Table 4.5; IPCC 2013). About 70% of the anthropogenic N_2O emissions stem from agricultural soils and biomass burning. Soil, application of fertilizer and manure were the major contributors to N_2O emissions. Most of the biomass burning (about 90%) occurs in the tropics as a result of forest clearing, savanna and sugarcane fires, and burning of agricultural wastes and it contributes about 0.7 Tg N year^{-1} to the global atmospheric N_2O budget. Crop production is responsible for about 50% of N_2O emissions from the agricultural sector. Global emission from fertilized arable land has been estimated at 3.3 Tg N_2O-N year^{-1} and 1.4 Tg NO-N year^{-1} (Stehfest 2006). The only significant process that removes N_2O is its reaction in the stratosphere with excited

Table 4.5 Estimates of the global nitrous oxide budget in the 2000s

Sources	Tg N (N_2O) year^{-1}
Natural sources	**11.0**
Ocean	3.8
Atmosphere (NH_3 oxidation)	0.6
All soils	6.6
Anthropogenic sources	**6.9**
Agricultural soils	4.1
Biomass burning	0.7
Fossil fuel combustion and industrial sources	0.7
Cattle and feedlots	
Human excreta	0.2
Rivers, estuaries, coastal zones	0.6
Atmospheric deposition	0.6
Total sources	**17.9**
Stratospheric sink	**14.3**
Observed growth rate	3.6

Adapted from Ciais et al. (2013)

oxygen atoms formed by photolysis of ozone (Crutzen 1981). Microorganisms in soils can reduce N_2O into N_2 under anaerobic conditions (Ryden 1981) but the significance of soil as a sink for N_2O remains uncertain and probably very small (Freney et al. 1978).

4.2.3.1 Soil-Borne Nitrogen Oxide Emission

Soils are the major source of nitrogen oxide emission and account for 60% of the total emissions. The emissions from soils result from nitrification- denitrification processes. Nitrification is the biological oxidation of NH_4^+ to NO_3^- through NO_2^- under aerobic conditions. Nitrification is mediated by the aerobic and chemolithoautotrophic microorganisms *Nitrosomonas* and *Nitrobacter* that oxidize NH_4^+ to NO_2^-, and NO_2^- to NO_3^-, respectively. During nitrification, some conversion of NO_2^- to NO_x can occur that may contribute to greenhouse gas production (Xu et al. 1998). Nitrification rate depends strongly on NH_4^+ concentration, temperature, soil water content and pH (Benbi and Richter 2003). Under oxygen limited conditions nitrifiers can use NO_2^- as a terminal electron acceptor and result in the production of N_2O and NO. Denitrification is an anaerobic bacterial process by which nitrate is reduced to nitrite (NO_2^-) and further reduced to nitrous oxide (N_2O) or dinitrogen (N_2), which is lost to atmosphere as a gas:

$$NO_{3^-} \rightarrow NO_{2^-} \rightarrow NO \rightarrow N_2O \rightarrow N_2$$

Non-biological denitrification, called chemodenitrification occurs in subsoil and is relatively small source of emissions. N_2O is also emitted from manure, soil-borne N, legumes, plant residues and compost. Emissions of N_2O and NO_x from agricultural soils are significantly affected by rate and type of nitrogen fertilizer application, presence of NO_3^-, soil organic matter content, soil pH, texture, soil moisture, soil aeration status, soil temperature and land use. The mineral N concentration in agricultural soils is increased by fertilizer application and mineralization of organic matter, which leads to increased emissions of nitrogen oxides. It is estimated that on an average about 1.25% of the added fertilizer is lost through denitrification (Bouwman 1996). However, the amount of fertilizer N emitted as nitrogen oxide vary considerably depending on soil, crop, climatic conditions and the type of fertilizer used. The mean global fertilizer induced emissions of N_2O and NO, respectively are 0.91% and 0.55% of the N applied in cropland (excluding legumes) and grassland. The denitrification rate increases with increasing soil NO_3^- concentrations up to a certain level and then becomes constant. Production of N_2O by nitrification is also enhanced as the soil concentration of NH_4^+ increases. However, the magnitude of NH_3 volatilization determines the availability of N for nitrification and denitrification; greater the NH_3 volatilization lesser is the potential for denitrification. Generally, N_2O emissions increase with increasing water filled pore space (WFPS) but the effect depends on the mechanism of N_2O and NO production and the diffusion properties of the soil. The rate of N_2O production from nitrification increases with increasing water content up to 55–65% of WFPS beyond which denitrification increases because of impeded aeration. Factors such as temperature, NO_3^- concentration, soil texture, and compaction modify the effect of soil water on denitrification.

Soil denitrification capacity is reported to be positively related to organic C, water soluble C, and total C concentration in soil provided other factors are favorable. The N_2O emissions may be enhanced by the addition of organic materials such as crop residues. The N_2O emission from crop residue incorporation has been reported to range from <0.1 to 8 kg N ha^{-1} depending on the type, C:N ratio and management of the residue and the measurement period. In general, addition of degradable organic materials such as animal and green manures increases N_2O emissions from soils containing NO_3^-. Soil pH is another important factor that affects gaseous emissions of N_2O, NO and N_2. The emissions are small in acidic than in neutral or slightly alkaline soils; and the ratio N_2O:N_2 is increased when the pH of soil is reduced (Šimek and Cooper 2002). Soil pH around 6.5 is considered optimum for reducing denitrification. Greater rates of denitrification are usually observed with zero tillage compared to ploughed soils (Nieder et al. 1989). The increase is related to increased soil organic matter and higher levels of available C in the topsoil, as well as to greater soil densities and decreased soil aeration (Myrold 1988).

Estimates of N_2O emissions for a range of land uses across Europe (Machefert et al. 2002) showed that the emission rates were higher for agricultural lands (2.1–38.3 kg N_2O-N ha^{-1} $year^{-1}$) compared to forests and grasslands (0.122–7.3 kg N_2O-N ha^{-1} $year^{-1}$). Conversion of tropical forests to crop production and pasture significantly affects nitrogen oxides emissions because of changes in nitrogen availability and bacterial population. Conversion of natural grassland to cropland also results in increased NO emissions. Grazing stimulates NO emissions from grasslands, by increasing the availability of inorganic N and N input to soils. The N_2O emission factors reported for grazed pastures range between 0.2% and 10% of N excreted with highest values from intensively managed dairy pastures in the UK and Netherlands (Velthof et al. 1996). N_2O emissions from soils are likely to increase in future due to greater demand for feed/food and the dependence of agriculture on nitrogen fertilizers. In addition to N_2O, two other nitrogen compounds emitted to the atmosphere which are of importance to climate change and human health include NH_3 and NO_x. The NO_x emissions, which are estimated at ~49 Tg N $year^{-1}$ are about four times greater from anthropogenic sources than natural emissions (Table 4.6). The anthropogenic fluxes mainly arise from fossil fuel combustion and agriculture.

4.2.4 Ammonia Emission

Ammonia is the major component of total reactive nitrogen and plays an important role in the global N cycle. Ammonia is emitted to the atmosphere from both natural

Table 4.6 Emission of NO_x to atmosphere in the 2000s

Source	NO_x (Tg N year^{-1})
Emissions to atmosphere	
Anthropogenic total	**37.5**
Fossil fuel combustion industrial processes	28.3
Agriculture	3.7
Biomass and biofuel burning	5.5
Natural sources	**11.3**
Soils under natural vegetation	7.3
Oceans	–
Lightning	4
Total sources	**48.8**

Adapted from Ciais et al. (2013)

and anthropogenic sources but the latter account for ~80% of the total NH_3 emissions. Agriculture accounts for about two-thirds of the total global NH_3 emissions. The anthropogenic sources of NH_3 include excreta from domestic animals, use of synthetic fertilizers, biomass burning and crop residue decomposition. Animal excreta contain nitrogen in the form of NH_4^+, urea and organic nitrogen. Urea and most of the organic nitrogen in animal excreta is rapidly converted to NH_3, which can be directly volatilized from the animal production system or after application to the soil. About 30% of the nitrogen excreted by farm animals is released to the atmosphere from animal houses, during storage, grazing and after application of animal waste to soil. When urea and ammonium forms of fertilizers such as ammonium chloride, ammonium bicarbonate, ammonium sulfate are applied to moist soil surfaces, they undergo a series of chemical conversions to NH_3, which escapes to the atmosphere. During burning of biomass and fossil fuel, a part of the N contained in these fuels is converted to NH_3 and emitted to the atmosphere. The natural sources of NH_3 include oceans and soils under natural vegetation.

Globally, NH_3 emissions are estimated to be ~58 Tg N year^{-1}, though there is considerable uncertainty in these estimates (Table 4.7). NH_3 emissions have been increasing over the last few decades on a global scale and the current emissions are three times higher than the pre-industrial emissions estimated at ~20 Tg N year^{-1}. Calculations based on the energy balance statistics of IEA (2010) showed that during the period 1970–2008 global NH_3 emission increased from 28 Tg NH_3-N in 1970 to 49 Tg NH_3-N in 2008, and the emission from agricultural soils in the corresponding period increased from 13 to 30 Tg NH_3-N representing an increase of ~450 Gg NH_3-N year^{-1} (Fig. 4.4).

Of the current global emissions, ~43–47 Tg N year^{-1} originate from anthropogenic sources with excreta from domestic animals being the highest emitter, followed by synthetic N fertilizers, and biomass burning. The emissions from

Table 4.7 Estimates of global NH_3 emissions (Tg N $year^{-1}$) to the atmosphere

Source	Bouwman et al. (1997)	Holland et al. (1999)	van Aardenne et al. (2001)	Denman et al. (2007)	Ciais et al. (2013)
Natural sources					
Soils under natural vegetation	2.4	2.4–10	4.6	2.4	2.4
Oceans	8.2	8.2–13	5.6	8.2	8.2
Wild animal excreta	0.1	0.1–6	–	–	–
Natural burning at high altitudes	–	–	0.8	–	–
Anthropogenic sources					
Fossil fuel combustion and industrial process	0.3	0.3–2.2	0.3	2.5	0.5
Agriculture	–	–	–	35.0	30.4
Domestic animal excreta	21.6	20–43	22.9	–	–
Synthetic fertilizer use	9.0	1.2–9.0	9.7	–	–
Biomass and biofuel burning	5.9	2.0–8.0	7.2	5.4	9.2
Crops	3.6	3.6	4.0	–	–
Humans and pets	2.6	2.6–4	3.1	2.6	–
Total	54	45–83	58.2	56.1	50.7

Adapted from Nieder and Benbi (2008) and Ciais et al. (2013)

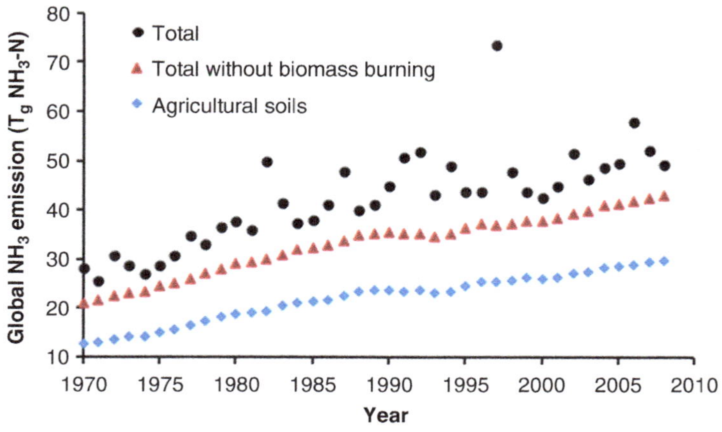

Fig. 4.4 Temporal trends of global NH_3 emission from all sources (Total) and agricultural soils during the period 1970–2008 (Compiled from EDGARv4.2; http://edgar.jrc.ec.europa.eu/, 2011)

industrial processes such as chemical and fertilizer manufacture are rather small. Among the natural sources, the oceans contribute the maximum followed by soils under natural vegetation. The regions with highest emission rates are located in Europe, the Indian sub-continent and China, reflecting the patterns of animal densities, type and intensity of synthetic fertilizer use (Bouwman et al. 1997). In Western Europe, 74% of total NH_3 emissions (4.02 Gg N $year^{-1}$) originate from animal excreta, 12% from synthetic fertilizers and ~6% from crops (Ferm 1998). In contrast, in east, southeast and south Asia, ammonia emissions from croplands are estimated to be 11.8 Tg N $year^{-1}$ with 6.5 Tg N originating from the use of chemical N fertilizer and 4.7 Tg N from the use of animal manures. The average NH_3 loss rate from chemical N fertilizer in the Asian region is 16.8%, which is higher than the global average of 10%; mainly because of widespread use of urea and ammonium form of fertilizers, occurrence of alkaline soils in some parts of the region (such as semi-arid India and north China), and cultivation of rice on large area. It is estimated that 22% of the urea fertilizer applied to rice fields is lost through ammonia volatilization as compared to 13.7% from upland crops. Besides fertilizer type, a number of soil (buffering capacity, texture, calcium carbonate and organic matter contents), environmental and management factors including wind speed, temperature, soil moisture, method of N application influence the emission of NH_3 from soil to atmosphere. In submerged rice soils, losses of NH_3 are directly related to the concentration of aqueous NH_3 in the floodwater, which in turn depends on total ammoniacal N, the pH, and the temperature of the floodwater. The process of NH_3 volatilization and the variables that influence NH_3 emissions from agricultural soils have been discussed in detail elsewhere (Benbi and Richter 2003).

4.2.5 *Aerosols or Particulate Matter*

Aerosols or particulate matter (PM) represent suspended airborne solid or liquid particles, with a size between a few nanometers and 10 μm that reside in the atmosphere for at least several hours. Mass concentration of particles with aerodynamic diameter of less than 10 μm (PM_{10}) and less than 2.5 μm ($PM_{2.5}$) is considered relevant to human health. As a comparison, the size of PM_{10} is roughly equivalent to 1/10th the diameter of a human hair. $PM_{2.5}$, usually called fine PM, also includes ultrafine particles with diameter less than 0.1 μm. In most locations in Europe, $PM_{2.5}$ constitutes 50–70% of PM_{10} (WHO 2013). Particulate matter with diameter between 0.1 and 1 μm can remain in the atmosphere for days or weeks and is subject to long-range transboundary transport in the air (WHO 2013). Particulate matter is a mixture with location-specific physical and chemical characteristics. Common chemical constituents of PM include sulfates, nitrates, ammonium, other inorganic ions such as ions of sodium, potassium, calcium, magnesium and chloride, organic and elemental carbon, crustal material, particle-bound water, metals

Table 4.8 Global emission of $PM_{2.5}$

Source	Emission (Tg $year^{-1}$)
Natural	
Sulfate from biological gases	130
Volcanic sulphates	20
Biogenics (terpenes)	13–60
Nitrates	60
Sub-total	223–270
Anthropogenic	
Black carbon	13
Sulfate from SO_2	190
Organic carbon, biomass, fossil fuel burning	70
Volatile organic compounds	10
Sub-total	283

Adapted from Scheffe (2003)

(including cadmium, copper, nickel, vanadium and zinc) and polycyclic aromatic hydrocarbons (PAHs). In addition, biological components such as allergens and microbial compounds are found in PM. Particles can either be directly emitted into the air (primary PM) or be formed in the atmosphere from gaseous precursors such as sulfur dioxide, oxides of nitrogen, ammonia and non-methane volatile organic compounds (secondary particles). Primary PM and the precursor gases can be of natural and anthropogenic origin. About half of the $PM_{2.5}$ global emissions (~500 Tg $year^{-1}$) originate from anthropogenic sources and the other half from natural sources (Table 4.8; Scheffe 2003). Anthropogenic sources include coal and oil acids, elemental carbon, heavy metals, and organic species emitted from coal and oil combustion (Davidson et al. 2005). The distribution of these emissions in air and the resultant PM varies greatly over space and time. Suspended particles are present in the atmosphere throughout the year in the arid and semi-arid ecosystems and beyond in a radius of 500 km or more (Jackson et al. 1973). Soil and dust re-suspension is also a contributing source of PM, particularly in arid areas or during episodes of long-range transport of dust, for example from the Sahara to southern Europe. The natural emissions include gaseous sulfur from volcanoes and decaying vegetation and nitrates. Among the natural sources, soil dust represents 9% of the total global particulate emissions, and it is generated at an estimated rate of 182 Tg $year^{-1}$ (Su 1996). The factors determining generation of dust include land use, climate and soil properties. Agricultural land contributes more dust than forest or wetland. Dust generation from cropland is more than orchards. Conventional tillage (Garrity 1998), grazing, and mining contribute dust. Although air-borne dusts are similar to source soils in terms of their mineralogical constituents; but the composition of source soils and respirable dusts are different (Clausnitzer and Singer 1999). Dust contributes to the loess parent materials; only finer particles move further from source than their coarser associates.

4.3 Impact on Atmosphere and Climate

Emissions of GHGs are bringing about major changes to the global environment. The increased concentration of GHGs cause global warming, deplete the concentration of ozone in the stratosphere and contribute to acid deposition. During the period 1880–2012 the globally averaged combined land and ocean surface temperature increased by 0.65–1.06 °C (Fig. 4.5). The total increase between the average of the 1850–1900 period and the 2003–2012 period is 0.72–0.85 °C (IPCC 2013). Each of the last three decades has been successively warmer at the Earth's surface than any preceding decade since 1850 (IPCC 2013). The period from 1983 to 2012 was the warmest 30-year period of the last 1400 years in the Northern Hemisphere. Increase in global surface temperature is projected to exceed 1.5 °C by the end of the twenty-first century compared to average for the years 1850–1900. Available evidence indicates that since about 1950 there has been change in precipitation and a number of extreme weather and climate events. The frequency, intensity and/or

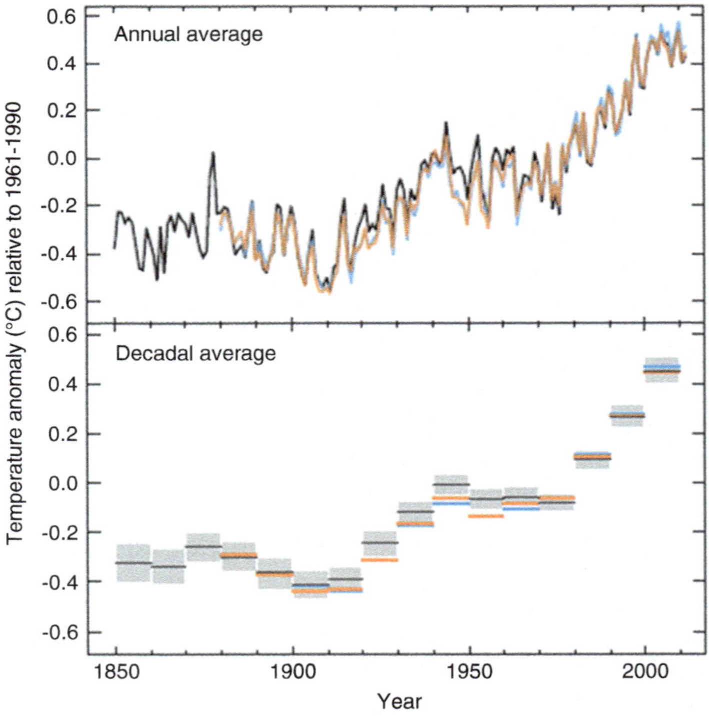

Fig. 4.5 Observed global mean combined land and ocean surface temperature anomalies, from 1850 to 2012 relative to the mean of 1961–1990. Top panel: annual mean values. Bottom panel: decadal mean values including the estimate of uncertainty for one dataset (black) (IPCC 2014b, p. 41; with kind permission from IPCC)

amount of heavy rainfall has increased. Apparently, the number of cold days and nights has decreased and the number of warm days and nights has increased on the global scale. Frequency and duration of warm spells or heat waves has probably increased in large parts of Europe, Asia and Australia. Projections indicate that there will be more frequent hot and fewer cold temperature extremes over most land areas on daily and seasonal time scales as global mean temperatures increase.

Though CO_2 concentration is the greatest in the atmosphere but CH_4 and N_2O are more potent than CO_2 in warming the atmosphere (25 and 310 times, respectively) over a 100-year period. Emissions of N_2O cause the depletion of stratospheric ozone, which permits increased transmission of the Sun's ultraviolet radiation to the Earth's surface. However, N oxides and NH_3 in air may also counteract warming by combining with other air-borne components to form aerosols that reflect incoming radiation, and by stimulating plant growth in N-limited forests, thereby taking more CO_2 out of the atmosphere. The balance between heating and cooling effects of N_2O, N-oxides and NH_3 on the atmosphere is not well understood (Mosier et al. 2004). Besides the above mentioned direct effects, creation of reactive nitrogen (Nr) indirectly influences climate change through (i) changes in soil organic matter decomposition and the resulting CO_2 emissions; (ii) altering the biospheric CO_2 sink due to increased supply of Nr; (iii) increasing marine primary productivity due to Nr deposition; and (iv) formation of O_3 in the troposphere because of NO_x and volatile organic compound emissions. Reactive Nr also contributes to smog and increases the haziness of the troposphere. Globally, the net influence of the direct and indirect contributions of Nr on the radiative balance was estimated to be -0.24 W m^{-2} (Erisman et al. 2011).

4.4 Impact on Ecosystems

4.4.1 Impact of Greenhouse Gases

The biogeochemical cycles of nitrogen and carbon are coupled because of the metabolic needs of organisms for the two elements. Change in the availability of one element influences not only biological productivity but also availability and requirement for the other element (Gruber and Galloway 2008) and in the longer term, the structure and functioning of ecosystems as well. Microbial growth can be limited by the availability of nitrogen, particularly in cold, wet environments. The increasing concentration of atmospheric CO_2 is known to increase plant photosynthesis and plant growth, which leads to increased carbon storage in terrestrial ecosystems (IPCC 2014a). However, plant growth is limited by nitrogen availability in soils, suggesting that in some nitrogen-deficient ecosystems, inadequate availability of reactive nitrogen will limit carbon sinks. On the other hand, nitrogen deposition may alleviate its deficiency and enable larger carbon sinks. A portion of the emitted NH_3 and NO_x is deposited over the continents, while the rest gets transported by winds and deposited over the oceans (Table 4.9). Deposition of

Table 4.9 Deposition of NO_x and NH_3 (Tg N $year^{-1}$) from the atmosphere in the 2000s to continents and oceans

	NO_x	NH_3
Continents	27.1	36.1
Oceans	19.8	17.0
Total	46.9	53.1

Adapted from Ciais et al. (2013)

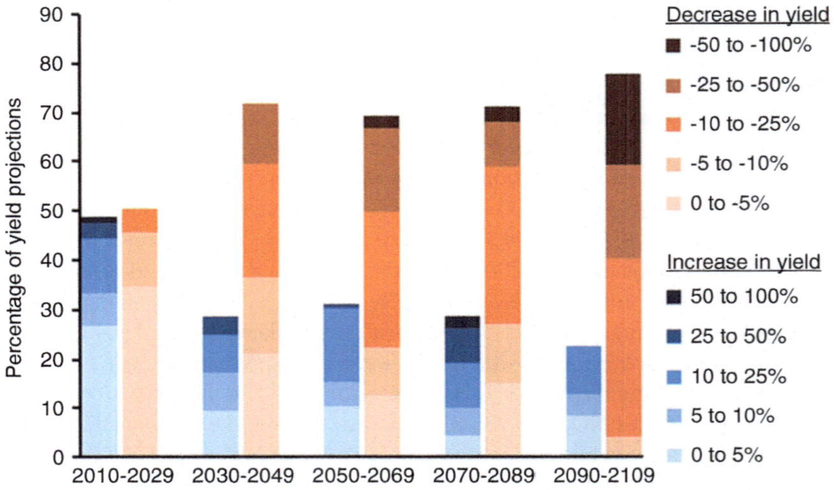

Fig. 4.6 Summary of studies projecting increase or decrease in crop yields (mainly wheat, maize, rice and soybeans) due to climate change during the twenty-first century. Data for each time frame sum to 100% indicating the percentage of projections showing yield increases versus decreases. Changes in crop yields are relative to late twentieth century levels (Adapted from IPCC 2014a)

ammonia and oxidized N from air fertilizes crops and native vegetation in N-deficit areas, but in already N-rich soils, it causes ecosystem acidification which destroys forests, reduces crop yields, and leads to unsustainable losses of biodiversity. The availability of nitrogen also changes in response to climate change, generally increasing with warmer temperatures and increased precipitation, but with complex interactions in the case of seasonally inundated environments.

Both positive and negative effects of climate change on crop productivity have been predicted though majority of the studies have projected negative impacts than positive impacts (Fig. 4.6). Elevated atmospheric concentration of CO_2 besides directly affecting the process of photosynthesis influences crop growth and productivity through its effect on temperature, precipitation and incidence of extreme weather events. The availability of additional photosynthates under elevated CO_2 enables most plants to grow faster and yield more. However, the effect may be modified when combined with other climatic variables. High temperature and altered precipitation can shorten total crop duration especially grain-fill period, create water or aeration stress and increase the incidence of disease and insect attacks. Besides the effect of climatic variables on plant physiological processes, ground-level ozone impairs the ability of plants to absorb CO_2, thereby inhibiting plant growth and damages plant tissues as well. It injures crops, forests and natural ecosystems and predisposes plants to attack by insects and pathogens (Peoples et al. 2004). The net result is loss in crop production and the increased inability of plants to absorb CO_2 and thereby offset global warming.

Nitrogen oxides emissions to the atmosphere contribute to acid rain and cause acidification of ecosystems. Acidification of soils, lakes, and streams in Scandinavia, Western Europe, and eastern North America, has caused extensive defoliation and dieback of trees and mortality of fish (Mathews and Hammond 1999; IFA 2007). More than 25% of the plant species in grasslands in Europe has been lost due to atmospheric N deposition and acid rain. Ecosystem acidification is increasing because of rapidly growing vehicular traffic and intensification of agriculture with large application of fertilizer N in parts of Asia and the Pacific Region, and due to biomass burning in Africa and Latin America (Mathews and Hammond 1999). Not only terrestrial but also aquatic systems are being acidified. The absorption of CO_2 by ocean has resulted in ocean acidification (IPCC 2013). The pH of the ocean surface water has decreased by 0.1 unit (= 26% increase in hydrogen ion concentration) since the beginning of the industrial age.

In addition to the direct effects of increased concentration of GHGs in atmosphere, there are indirect effects caused by global climate change on soil and plant processes. Changes in climate are likely to influence the rates of accumulation and decomposition of SOM, both directly through changes in temperature and water balance, and indirectly through changes in primary productivity and rhizodepositions. Increase in temperature is expected to accelerate the decomposition of SOM and consequently increase the release of CO_2. The quality of soil organic matter may decline with increasing proportion of inert components in the soil carbon pool. Increased temperature can enhance N availability to plants through higher turnover of soil organic N but it may also lead to greater emissions of NH_3 and nitrogen oxides through processes of volatilization and denitrification. Increased precipitation will increase the surface runoff in hilly lands; increase infiltration and water storage within the soil in flat lands; enhance groundwater recharge in the highly permeable and well-drained soils; increase evaporation in soils having low infiltration (Varallyay 2007). Rise in temperature will increase the potential evapotranspiration and decrease surface runoff, infiltration, water storage and groundwater recharge, especially if accompanied by low precipitation. Change in rainfall amount and frequency, and wind intensity may alter the severity and extent of soil erosion. Climate change can impact the overall soil quality through its influence on a number of physical, chemical and biological indicators.

Because of the impact of increasing CO_2 concentration on soil temperature, moisture and microbial activity, production and consumption of CH_4 and N_2O may be altered (Smith et al. 2003; Pendall et al. 2004). Meta analysis of several studies showed that increased CO_2 (ranging from 463 to 780 ppm) stimulates N_2O emissions from upland soils and CH_4 emissions from rice paddies and natural wetlands (van Groenigen et al. 2011). The estimated stimulation of soil N_2O emissions may correspond to an additional source of 0.33 and 0.24 Pg CO_2-eq year^{-1} from agricultural and all other upland ecosystems, respectively. The CO_2-stimulation of CH_4 emissions may correspond to an additional source of 0.25 and 0.31 Pg CO_2-eq $year^{-1}$ from rice paddies and natural wetlands, respectively.

4.4.2 Impact of Ammonia and Aerosols

Ammonia and ammonium (collectively termed NH_x) undergo dry and wet deposition in the areas downwind of their major sources (Sutton and Fowler 2002). The deposited NH_x species act as plant nutrients as they fertilize plants. However, high anthropogenic input of N to the environment may lead to eutrophication of terrestrial and aquatic ecosystems and thus threaten the biodiversity (Asman et al. 1998; Galloway et al. 2003; Erisman et al. 2005). Eutrophication can also result in severe reductions in water quality, changes in species composition and dominance, and toxicity effects. Ammonia is also an important atmospheric pollutant with several effects. It is involved in aerosol formation and is the most abundant atmospheric base with the ability to neutralize harmful acids. It plays an important role in determining the overall acidity of precipitation, cloud water and airborne particulate matter (PM) or aerosols (Shukla and Sharma 2010; Xue et al. 2011; Behera et al. 2013). Aerosols may influence climate in several ways: directly through scattering and absorbing radiation and indirectly by acting as cloud condensation nuclei or ice nuclei, modifying the optical properties and lifetime of clouds. The aerosols play a role in human health, in acid rain that threatens land and aquatic eco-systems, and soil fertility.

4.5 Exposure and Human Health Effects

Elevated atmospheric concentration of greenhouse gases may not directly impact human health but these can indirectly influence it through their effects on atmosphere, climate and ecosystems. As discussed in the preceding sections, GHG emissions are responsible for global warming, change in precipitation and a number of extreme weather and climate events, destruction of stratospheric ozone, and formation of ground-level ozone (O_3) that can adversely impact human health. Ground-level O_3 or smog is formed when N oxides react with volatile organic compounds in the presence of sunlight. It has severe impacts on human health, crops and natural vegetation, and global warming. Ground-level ozone damages lung tissues and reduces lung function, especially in children and adults who work outdoors. Ozone has also been implicated in increased bronchial reactivity, and airway inflammation in both healthy and asthmatic populations (Koren and Utell 1997). People with respiratory diseases like asthma suffer more from smog (Peoples et al. 2004). Excess N (and P) in the ecosystem may increase the release of air-borne allergens that affect many people. Because of longer pollen season and production of aero-allergens there could be increase in allergic disorder such as hay fever and asthma. Nitrogen oxides react with ammonia, other compounds and moisture in the air to form particles such as ammonium nitrate (NH_4NO_3). When inhaled into the lungs, these particles damage lung tissues, affect breathing and respiratory systems, and can cause premature death. Breathing ammonium nitrate particles can also worsen respiratory diseases such as bronchitis and aggravate existing heart diseases.

4.5.1 Climate Change Effects

The impacts of climatic change on human health are likely to be threefold: (i) direct effects of heat and cold and other extreme weather events on physiological functions, (ii) indirect effects such as spread of vector-borne pathogens into areas where disease currently does not exist and increased risk of water-, food-, and rodent-borne diseases, and (iii) indirect impacts mediated through societal systems, such as under-nutrition and mental illness from food insecurity, stress, and violent conflict caused by population displacement; economic losses due to widespread "heat exhaustion" of the workforce; or other environmental stresses, and damage to health care systems by extreme weather events (Smith et al. 2014). Higher temperatures and greater frequency of extreme weather events increase the risk of deaths from dehydration and heat stroke, and of injuries from intense local weather changes. There may be a higher risk of respiratory and cardiovascular illnesses and certain types of cancers. Vulnerable populations with diminished capacity for thermoregulation such as children and the elderly people especially women, mentally ill people and people with pre-existing illnesses such as cardiovascular disease or chronic respiratory diseases and others in thermally stressful occupations are expected to be most affected by climatic changes (McMichael et al. 2006). In large urban areas, heat waves are associated with pollution linked to the formation of tropospheric ozone which has adverse human health effects (Beniston 2002).

Transmission of infectious disease is determined by intrinsic and extrinsic factors, which *inter alia* include climatic and ecological conditions (Weiss and McMichael 2004). Many infectious agents, vectors, pathogens and hosts survive and reproduce within a range of optimal climatic conditions, particularly temperature and precipitation (WHO 2003). Climate change will influence the prevalence of several vector-borne diseases in humans (Table 4.10). For example, the infection rate for malaria has an exponential relationship with temperature (WHO 1990); small increases in temperature can lead to a sharp reduction in the number of days of incubation. Regions at higher altitudes or latitudes may thus become conducive to the vectors because of an increase in the annual temperature. Similarly, *Salmonella* and cholera bacteria proliferate more rapidly at higher temperatures. Strong associations have been observed between temperature and occurrence of salmonellosis in European countries (Kovats et al. 2004) and Australia (D'Souza et al. 2004). Climate change can shift the spread of vector-borne diseases to regions where currently low temperature,

Table 4.10 Vector-borne diseases considered sensitive to climate change

Vector	Diseases	People at risk (million)
Mosquito	Malaria	2020
Mosquito	Yellow fever, West Nile fever	–
Mosquito	Lymphatic filariasis	1100
Mosquito	Dengue fever	2500–3000
Hanta virus	Hemorrhagic fever	–
Ixodes (hard ticks)	Lyme disease and tick borne encephalitis	–
Triatomines bug	American trypanosomasis ("changas disease")	100
Tsetse fly	African trypanosomasis ("sleeping sickness")	55
Phlebotomine sandfly	Leishmaniasis	350
Blackfly (Simulium species)	Onchocerciasis (river blindness)	120
Snail (intermediate host)	Schistosomiasis	600

Compiled from WHO (1997) and Haines et al. (2006)

low rainfall, or absence of vector habitat restricts transmission. Change in precipitation pattern and warming can influence the spread of water-borne diseases. Human exposure to water-borne infections occurs by contact with contaminated drinking water or food. Rainfall can impact the transport and spread of infectious agents, while temperature influences their growth and survival (WHO 2003).

Shifts in temperature and precipitation patterns can also impact agriculture productivity and thus affect food security in many parts of the world. Projections suggest that, while global food supply may be maintained upto the middle of the twenty-first century, many regions of the world will experience adverse effects of heat waves, droughts, and excessive moisture on crops. The developing countries, in particular are likely to experience shortfalls of up to 30% of current food production (IPCC 2001). This will adversely impact human health and capacity to work. Besides agricultural production and food availability, climate change may affect human health by altering nutrient content of food crops. A number of studies have suggested lower elemental concentration in soybeans, sorghum, potatoes, wheat or barley grown at elevated CO_2 concentration (Prior et al. 2008; Högy and Fangmeier 2009; Högy et al. 2009; Erbs et al. 2010). Several studies during the last over one decade observed decreases in zinc, sulfur, phosphorus, magnesium, and iron in wheat and barley grain; increase in copper, molybdenum, and mixed results for calcium and potassium (Högy et al. 2009; Erbs et al. 2010; Fernando et al. 2012). Nitrogen concentration, a proxy for protein concentration, has been found to be lower by 10–14% in edible portions of wheat, rice, barley, and potato, and by 1.5% in soybeans when grown under elevated CO_2 (Taub et al. 2008). Meta analysis of results from 143 comparisons of edible portions of crops grown at ambient (~380 ppm) and elevated CO_2 concentration (546–584 ppm) from seven different free-air CO_2 enrichment (FACE) experiments in Japan, Australia and

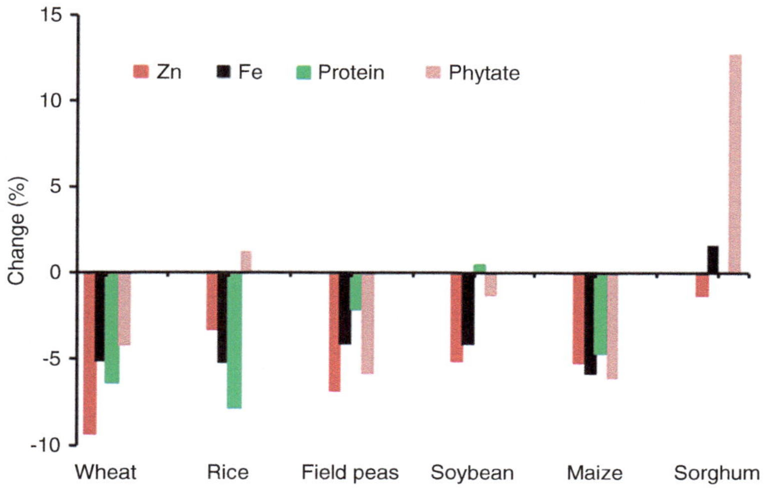

Fig. 4.7 Percentage change in nutrients at elevated concentration of CO_2 relative to ambient concentration (Adapted from Myers et al. 2014)

USA involving six food crops showed that C_3 grains (wheat and rice) and legumes (field peas and soybean) had lower concentrations of zinc and iron, when grown under elevated atmospheric CO_2 concentrations; C_3 crops other than legumes also had significantly lower concentrations of protein, whereas C_4 crops were less affected (Fig. 4.7). Besides nutrient concentration, elevated concentration of CO_2 also influenced phytate content, an antinutrient that inhibits the bioavailability of zinc and iron in the human gut (see Chap. 7). Elevated CO_2 concentration decreased significantly phytate content in wheat which may offset some of the decline in zinc concentration, although the decrease was slightly less than half of the decrease in zinc. For other crops examined, there was no decrease in phytate content that may further aggravate problems of zinc deficiency in food crops. Reduction in the nutrient content of staple crops with elevated CO_2 concentration could have considerable implications for public health, particularly in regions where people depend on those crops for meeting their nutrient requirements. Mechanism responsible for the decline in nutrient concentrations with elevated CO_2 concentration is not precisely known but it is usually ascribed to "carbohydrate dilution" whereby CO_2-stimulated carbohydrate production by plants dilutes the rest of the grain components (Gifford et al. 2000) and improved nutrient use efficiency. However, the carbohydrate dilution mechanism cannot be generalized as the effect of elevated CO_2 on plant nutrient status differs among the nutrient elements, plant functional groups, plant organs, CO_2 concentration, and N availability (Duval et al. 2011). Besides CO_2 concentration, increasing temperature has been reported to impact grain quality. High temperature regime after anthesis in wheat affected the accumulation of protein during grain-fill (Hurkuman et al. 2009) and gluten quality (Moldestad et al. 2011).

Though most of the health impacts of climate change are considered to be adverse but there could be some beneficial health effects in certain regions. For example, in temperate countries, milder winters would reduce the seasonal wintertime deaths and in currently hot regions an increase in temperature might reduce the population of disease-transmitting mosquitoes (WHO 2003; McMichael et al. 2006). Populations in some temperate areas with extreme cold may benefit from increased agricultural productivity due to moderate warming (IPCC 2013). Some flood-prone areas may experience fewer floods. However, for the world as a whole and for nearly all populations the overall impact is expected to be more negative than positive, increasingly so as climate change progresses (IPCC 2013). Until mid-century, projected climate change will impact human health mainly by aggravating health problems that already exist, particularly in developing countries with low income. By 2100, high temperature coupled with high humidity in some areas for parts of the year is expected to influence common human activities, including growing food and working outdoors (IPCC 2014a). Climatic changes have probably already affected some health outcomes. The World Health Organization in its "World Health Report 2002" estimated that climate change was responsible in 2000 for approximately 2.4% of worldwide diarrhoea, and 6% of malaria in some middle-income countries (WHO 2002).

4.5.2 Exposure to Ammonia

Ammonia is rapidly absorbed in the upper airways and thus damages upper airway epithelia. Moderate concentrations of NH_3 (50–150 ppm) can lead to severe cough and mucous production; and concentrations greater than 150 ppm may cause scarring of the upper and lower airways (Close et al. 1980; Leduc et al.1992). These inflammatory responses can lead to reactive airways dysfunction syndrome (RADS) and persistent airway hyper responsiveness (Flury et al. 1983; Bernstein and Bernstein 1989). At higher concentrations, NH_3 may bypass the upper airways to cause lower lung inflammation and pulmonary oedema (Sobonya 1977; Close et al. 1980). Huge exposure to ammonia such as from the disruption of tanks of anhydrous NH_3 in agriculture can be fatal (Sobonya 1977). Occurrence of chronic lung disease following exposure to high concentrations of NH_3, even for 2 min, may lead to the development of bronchiolitis obliterans (Kass et al. 1972; Walton 1973; Sobonya 1977; de la Hoz et al. 1996), restrictive lung disease (de la Hoz et al. 1996), and bronchiectasis (Leduc et al. 1992). Exposure to 100 ppm NH_3 even for 30 s results in nasal irritation and increase in nasal airway resistance (McLean et al. 1979). In addition to pulmonary disease, exposure to NH_3 also leads to irritation of the eyes, sinuses, and skin. Chemical burns to the skin and eyes are commonly seen following exposure to high-concentration of ammonia (Latenser and Loucktong 2000). Although the most serious adverse effects of NH_3 inhalation are usually observed at concentrations around 500 ppm, but NH_3 can reach the alveoli and may be adsorbed to respirable particulates even at lower concentrations. Therefore, an occupational exposure limit of 7 ppm in agriculture settings has been recommended. US EPA has recommended as reference concentration for chronic inhalation of ammonia of 1.4 ppm.

4.5.3 Exposure to Aerosols and Particulate Matter

Air pollution (both ambient and household) poses the biggest environmental risk to health, accounting for about one in every nine deaths annually. Ambient air pollution alone kills around three million people each year, mainly from non-communicable diseases (WHO 2016). Since sources of air pollution also produce climate-modifying pollutants (e.g. CO_2 or black carbon), it is used as a marker of sustainable development. The aerosols comprising organic and inorganic materials are responsible for carrying pollutant gases, pollens, pesticides, a large numbers of bacteria and fungi (Giese and van Oss 1993), endotoxins (Dutkiewicz 1978), and biogenic silica (Epstein 1994). Categories of chemical constituents that could be responsible for adverse health effects include emissions from combustion of fossil and biomass fuel, particles generated by high temperature industrial processes such as smelting, products of chemical reactions in the atmosphere (such as SO_4^{2-} and NO_3^-), and fine particles from soil and other sources (Pope 2000).

People are exposed to PM from many sources during their daily activities, in their residences and at workplaces, commuting, during recreation and leisure activities. People exposed to a high ambient PM concentration appear to have a range of responses: most people show no clinical effects, others may become mildly or even seriously ill, and a few may die. Individuals or population groups vary in their susceptibility to PM due to factors such as respiratory habits (e.g. mouth breathing versus nose breathing), pre-existing diseases (lungs, cardiovascular, diabetes), and nutritional deficiencies. Susceptibility also varies with race, genetic factors, gender and age (Schwartz 2003). Elderly people and children are particularly vulnerable.

The air pollutants enter the human body primarily via inhalation and ingestion and thus may be absorbed through the gastrointestinal and respiratory tract. PM_{10} and $PM_{2.5}$ particles are small enough to penetrate the thoracic region of the respiratory system. Both short-term (hours, days) and long-term (months, years) exposure to inhalable PM cause adverse health effects including aggravation of asthma, respiratory symptoms and mortality from cardiovascular and respiratory diseases and, lung cancer. The other adverse effects of long-term exposure include neurological disorder in adults, adverse birth outcomes and exacerbation of allergies (Table 4.11). Exposure to PM affects lung development in children, including reversible deficits in lung function as well as chronically reduced lung growth rate and a deficiency in long-term lung function (WHO 2011). The possible mechanisms by which PM could cause adverse human health effects include pulmonary inflammation, systemic inflammation, oxidative stress response, protein modification,

Table 4.11 Health effects associated with long-term and short-term exposure to ambient particulate matter (PM)

Disease	Health effect
Long-term PM exposure (months, years)	
Cardiovascular	Cardiovascular-related mortality; atherosclerosis; ischaemic heart disease; complications of diabetes
Respiratory	Respiratory- related mortality; asthma symptoms; reduced lung function in children; reduced lung function in susceptible adults (elderly, people with chronic obstructive pulmonary disease (COPD) or asthma); respiratory infections in children
Cancer	Lung cancer mortality
Neurological	Neurological disorders; impaired cognitive function
Development	Lung development; neurological development in children
Reproduction	Adverse birth outcomes; sperm quality
Allergies	Exacerbation of allergies; allergic sensitization
Short-term PM exposure (daily)	
Cardiovascular	Cardiovascular-related mortality; ischaemic heart disease; ischaemic stroke; myocardial infarction; cognitive heart failure
Respiratory	Respiratory- related mortality; asthma symptoms; respiratory infections; bronchitis in children; COPD symptoms
Allergies	Exacerbation of allergies

Adapted from Hime et al. (2015)

stimulation of autonomic nervous system, pro-coagulant activity and suppression of immune defense in the lung (Nel 2005; Poschl 2005; US EPA 2009).

The adverse effects of PM on health are considered to be associated with particle size, composition, and concentration. A range of chemical and physical properties of aerosols, such as particle number, surface area, and mass influence human health (Harrison and Yin 2000). Greater numbers of people in cities become ill when airborne concentrations of $PM_{2.5}$ mass and $PM_{2.5}$ SO_4^{2-} increase (Pope 2000). The size of the particles determines the site in the respiratory tract where these will be deposited. While PM_{10} particles are deposited mainly in the upper respiratory tract, the fine and ultra fine particles reach lung alveoli (Kampa and Castanas 2008). Long-term exposure to $PM_{2.5}$ is a greater risk factor for mortality than PM_{10}. Exposure to $PM_{2.5}$ is associated with an increase in the long-term risk of cardiopulmonary mortality by 6–13% per 10 μg m^{-3} of $PM_{2.5}$ (Pope et al. 2002; Beelen et al. 2008). All-cause daily mortality is estimated to increase by 0.2–0.6% per 10 μg m^{-3} of PM_{10} (WHO 2006; Samoli et al. 2008). It is estimated that $PM_{2.5}$ exposure causes 3.1 million deaths a year, globally, and any level above zero is considered unsafe implying that there is no threshold above zero below which negative health effects do not occur (WHO 2013). However, some guidelines for air-borne PM_{10} and $PM_{2.5}$ have been suggested by different agencies (Table 4.12). The air quality guidelines recommended by WHO are lower than those proposed by other agencies. The PM pollution problem is further complicated by their effects on climate change and visibility degradation.

Table 4.12 Air quality guidelines for PM_{10} and $PM_{2.5}$

	PM_{10} ($\mu g\ m^{-3}$)		$PM_{2.5}$ ($\mu g\ m^{-3}$)	
Authority	Annual average	24 h average	Annual average	24 h average
WHO (2006)	20	50	10	25
US EPA (2006)	–	150	15	35
European commission (2008)	40	50	25	–

4.6 Mitigation Options

Adoption of improved management practices and mitigation technologies could help in reducing emissions. A number of technologies that offer opportunities for mitigation from both supply and demand- side, in agricultural ecosystems, have been advocated. The supply-side opportunities include sustainable intensification, improving N fertilizer production and management, reducing emissions from enteric fermentation, reducing methane emissions from rice fields and improving manure management. The demand-side opportunities include sequestering carbon in agricultural systems, reducing food waste and shifting dietary trends. The IPCC (2014a) identified the following mitigation options in agriculture: (i) plant management: improved variety, rotation, cropping system; (ii) nutrient management: type of fertilizer, urease/nitrification inhibitor; (iii) Tillage/residue management: reduced tillage, residue retention; (iv) Water management: improved water application, drainage; (v) Land use change: agroforestry, bio-energy crops, and (vi) Biochar application. The net benefit from these technologies in mitigating GHGs will depend on several factors including trade-off between different GHGs.

4.6.1 Carbon Dioxide

Several strategies have been advocated for stabilizing atmospheric abundance of CO_2. The main strategies to lower CO_2 emissions include (i) reducing the global energy use by increasing the energy conversion and utilization efficiency, (ii) developing low or no-carbon fuel, (iii) replacing fossil fuel with biofuels that recycle recently photosynthesized atmospheric CO_2 rather than introducing previously dormant C into active cycling, (iv) reducing emissions from agriculture related energy-dependent operations, and (v) sequestering CO_2 through natural and engineering techniques. Since engineering techniques are still in development stage and are yet not commercialized, C sequestration in soil and vegetation is considered a practical option. Practices such as intensification of agriculture, which result in increased crop productivity and input use efficiency invariably lead to

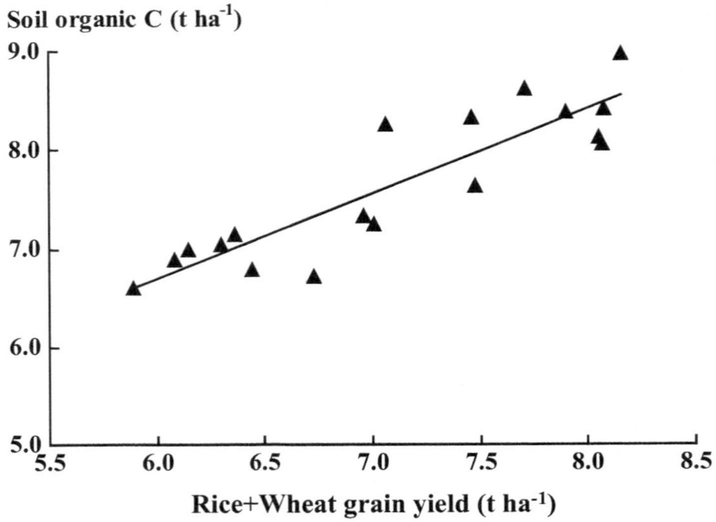

Fig. 4.8 Relationship between soil organic carbon and total rice and wheat grain yield in Punjab (Adapted from Benbi and Brar 2009)

emission reductions. According to estimates presented by Burney et al. (2010), global yield gains in agriculture during 1961–2005 have avoided emissions of 161 Pg C; an average of 3.6 Pg C year^{-1}. This corresponds to 34% of the total anthropogenic emissions (478 Pg C) between 1850 and 2005 (Canadell et al. 2007). If the yield and fertilizer intensities were held constant as in 1961, an additional 1514 million ha cropland area would have been required to meet the food requirement in 2005. Intensive agriculture, in addition to improving crop productivity results in enhanced C sequestration due to greater return of crop residues, root biomass and root exudates to soil. Results of a 25-year study from India showed (Fig. 4.8) that intensive agriculture improved SOC concentration by 38% due to progressive increase in crop yields (Benbi and Brar 2009). The increased SOC not only mitigates CO_2 emission but also enhances soil productivity and nutrient use efficiency. Increased SOC content reduces the amount of fertilizer N required to attain a yield goal; one Mg of organic C sequestered in the plough layer could compensate for 4.7 kg N ha^{-1} (Benbi and Chand 2007).

Carbon emissions from agricultural soils could be reduced by adopting conservation agriculture and minimizing soil disturbance, checking erosion through reduced tillage intensity and using low quality organic inputs. No- till systems not only improve soil C status but also require less fossil fuel for machinery use thus reducing C emissions (Pretty et al. 2002). For example, fuel use in conventional tillage systems in the UK and Germany range between 0.046 and 0.053 t C ha^{-1} year^{-1}, compared to 0.007–0.029 t C ha^{-1} year^{-1} for no-till systems (Smith et al. 1998). Minimizing conversion of new land to agriculture in the tropics and using less energy in agricultural operations (such as through reduced tillage, optimal fertilizer use efficiency, improved irrigation techniques and solar drying) could reduce agriculture related emissions of CO_2.

4.6.2 Methane

Appropriate water and nutrient management, cultural practices and choice of crop cultivar can help reduce CH_4 emission from rice fields. Mid-season drainage and intermittent irrigation decrease CH_4 emission as it prevents the development of soil reductive conditions. A number of studies have documented a significant decrease (50–90%) in CH_4 emission from rice fields that are drained once or several times during the crop cycle compared to continuous flooding (Nieder and Benbi 2008; Khosa et al. 2011). In major rice growing areas in Asia, the average CH_4 flux with single and multiple drainages has been estimated to be 60% and 52%, respectively of the flux from continuously flooded rice fields (Yan et al. 2005). Studies from India show that compared to seasonal CH_4 flux of 15.3 g m^{-2} with continuous flooding, the introduction of single and multiple mid-season drainage reduced methane flux to 6.9 and 2.2 g m^{-2}, respectively (Gupta et al. 2002). Water management between crops also influences CH_4 emission during rice season. A dry fallow emitted less CH_4 during the next crop cycle than a wet fallow (Trolldenier 1995). Water management as a mitigation practice is only feasible in areas with lowland and irrigated rice fields that have assured and manageable water supplies. In rainfed rice, drainage may be less feasible because farmers depend on the water stored in the bunded field. Draining the fields may lead to escalated emission of N_2O. In addition to water management, the other practices that are effective in mitigating CH_4 emission from rice fields include (i) fertilizing with ammonium sulphate, (ii) application of gypsum in saline and alkaline soils, (iii) use of nitrification inhibitors, (iv) integrated use of organic and mineral fertilizers, and (v) addition of redox species. Application of ammonium sulphate fertilizer has been reported to reduce CH_4 emission by 50–60% as compared to urea. Ammonium sulphate supplies N and sulphate, which maintains soil redox potential above that required to produce CH_4. Application of gypsum (at 6.7 t ha^{-1}) in saline and alkaline soils can reduce CH_4 emission by 50% and 70% in rice fields fertilized with urea or green manure, respectively (Denier van der Gon and Neue 1994). However, sulphate addition might be detrimental to rice by favouring rhizosphere sulphate-reduction. Use of nitrification inhibitor, coated calcium carbide can reduce CH_4 production by producing small quantities of acetylene slowly over time (Banerjee and Mosier 1989). Integrated use of organic and mineral fertilizers can mitigate the increased CH_4 emission due to sole application of organic manure. Another mitigation option is fertilization with iron. Increased content of iron in the rhizosphere helps suppress CH_4 formation (Jäckel and Schnell 2000). Addition of iron containing furnace slag has been reported to suppress CH_4 emission from paddy soils (Furukawa and

Inubushi 2002). Addition of redox species such as ferric oxide, potassium nitrate and potassium sulphate has been reported to reduce CH_4 emission from paddy fields by 65–75% (Khosa et al. 2010). An evaluation of different mitigation strategies showed that while mid-season drainage is an effective strategy, it may promote N_2O production. Similarly, high percolation rates may promote NO_3^- leaching, and addition of sulphate may cause H_2S injury (Yagi 2002). Methane emissions due to burning of biomass may be reduced through sustained land management programs that aim at (i) increasing the productivity of existing agricultural lands and restoring the degraded lands, (ii) lengthening the rotation times and improving the productivity of shifting agriculture, (iii) improving grassland management to reduce frequency of fires, and (iv) returning crop residues to the field instead of burning. Methane emissions from domesticated ruminants can be reduced through dietary supplementation to improve the nutrition and animal productivity for milk and growth (Mosier et al. 1998). Manure management and treatment practices can be adapted to collect CH_4 produced during anaerobic digestion and utilize it as an energy source (Hogan 1993). It is estimated that with existing technology, CH_4 emissions from manures can be reduced by 25–80%. The total potential for reducing methane emissions in agriculture is estimated to be 24–92 Tg year^{-1} (Cole et al. 1997).

4.6.3 Nitrous Oxide

Options for reducing N_2O emission from soil include enhancing N-use efficiency and reducing loss of applied N in soil, plant breeding and genetic modifications to increase the N uptake, use of manure or integrated use of manure and fertilizer to reduce reliance on chemical fertilizers, and use of technologically advanced fertilizers such as slow-release fertilizers or nitrification inhibitors. Improving N use efficiency can reduce N_2O emissions directly from the soil and indirectly from N fertilizer manufacture. Improved N use efficiency can also decrease off-site N_2O emissions by reducing leaching and volatilization losses. Fertilizer N use efficiency can be improved by achieving greater synchrony between crop N demand and the N supply from all sources throughout the growing season (Cassman et al. 2002). Strategies to improve N use efficiency include site-specific prescription algorithms, improved timing of N applications to achieve better synchrony between crop N demand and supply, more efficient fertilizer N application methods, more efficient fertilizers and appropriate crop residue and organic manure management for sustaining high levels of indigenous soil N supply. Besides time of N application and distribution of N above or in the soil, the time and intensity of soil tillage affect the efficiency by which N is utilized in the cropping system. Modern concepts for tactical N management should involve a combination of anticipatory (before planting) and responsive (during the growing season) decisions. Crop-based approaches for in-season N management such as optical sensors are now becoming widely available. Other practices that can mitigate emissions include growing of cover crops, replacement of the traditional slash and burn type agriculture with a more intensive agriculture and proper residue management. Higher rates of denitrification and N_2O fluxes are observed after application of organic residues on the soil surface as mulch compared to sites where the residues are removed (IFA/FAO 2001). Incorporation of residues may stimulate the mineralization of soil organic matter and, therefore, accelerate emissions of N_2O. It has been estimated that if all of the suggested strategies are implemented, then the likely reduction in soil NO emissions from fertilizer application would be of the order of 0.4 Tg N year^{-1}, or 4% of the total biogenic NO emissions (Skiba et al. 1997). Since adoption of the suggested strategies will require substantial and expensive changes in current agricultural practices, these are unlikely to occur in the foreseeable future. Therefore, reducing emissions from vehicles and fossil fuels burning could be a more realistic and straightforward option.

The global technical GHG mitigation potential from agriculture by 2030, considering all gases, is estimated to be 5.5–6.0 Pg CO_2-eq year^{-1}. Realization of this potential will require adoption of best available management practices with reference to soil type and land use system, social acceptability and economic feasibility. Most mitigation practices in agriculture have synergies with sustainable development. If full biophysical potential is achieved, agriculture could mitigate about 20% of the total annual CO_2 emissions (29 Pg CO_2 eq year^{-1}) in 1990 (Smith et al. 2007). The major reduction will come from reduced soil emissions (89%) followed by mitigation of methane (9%). Mitigation of soil N_2O emissions will compensate for only 2% of emissions. Agriculture may also contribute to mitigation in energy sector through production of biomass as feed stocks and energy efficiency measures.

4.6.4 Ammonia

Since the animal excreta and croplands constitute the main sources of anthropogenic emissions of NH_3, the mitigation options need to focus on improved management of these sources. The measures to reduce NH_3 emissions from animal wastes include optimizing livestock densities, reducing the excretion of urea and urea like products by optimizing N intake and retention, dilution or acidification of slurry, covering of slurry storage tanks, keeping the temperature of animal waste as low as possible, and storing solid and liquid wastes anaerobically, injecting or band placing slurry in the soil instead of surface application, irrigating the field after slurry application. NH_3 emissions from croplands can be reduced by efficient fertilizer nitrogen management. Currently, there are no regulations in most countries around the world for reductions in NH_3 emissions. Compared with regulations for other primary gaseous pollutants such as SO_2, NO_x and volatile organic compounds (VOCs), extensive control measures have not been taken to mitigate emissions of NH_3 though all these pollutants make similar contributions to PM mass loading, visibility degradation and/or acidification/eutrophication.

4.6.5 Particulate Matter and Aerosols

Exposure to total ambient PM can be minimized by reducing emissions from sources that are in close proximity to people. As per European Union, pre-existing PM levels are considered to be responsible for at least a fraction of the illness observed in the polluted areas and it is assumed that the incidence of illness can de decreased by reducing $PM_{2.5}$ concentration. Since $PM_{2.5}$ is correlated with PM_{10} in Europe, the EU is considering whether a single standard for PM is sufficient, unlike the USA which has separate standards for $PM_{2.5}$ and PM_{10} (US EPA 2004). A number of factors make it difficult to frame policies for reducing levels of PM in the atmosphere (Davidson et al. 2005). These factors include (i) inability to identify sources producing the primary particulate matter and precursor gases as these can be transported over long-distances resulting in their mixing over space and time, (ii) difficulty in estimating the emissions under all possible conditions, particularly for NH_3 as its emissions from livestock manure, fertilizer and soil can vary by orders of magnitude depending on the surface conditions (Anderson et al. 2003), (iii) non-linear atmospheric chemistry leading to counterintuitive chemical interaction effects, (iv) events leading to excessive $PM_{2.5}$ concentrations that cannot be readily lessened by human interventions, and (v) economic and other tradeoffs associated with control measures. Policies to address air pollution also generate a range of benefits to human health, not only through air quality improvements but also other health benefits, such as injury prevention or enabling physical activity.

4.6.6 Mitigating Climate Change Effects

Physiological and behavioural adaptations and changes in public health preparedness can reduce heat wave morbidity and mortality. Adaptations measures such as heat wave and extreme weather events warning systems, disease surveillance and monitoring, improved health care, food security and implementation of nutritional action plans, efficient water management and public awareness can reduce health risks associated with climate change (Table 4.13).

WHO (2001) has made certain recommendations for alleviating some of the negative effects of climatic change on human health. The recommendations include: (i) Increasing the flexibility of managed systems, by making incremental adjustments, avoiding practices that encourage deforestation, desertification,

Table 4.13 Adaptive measures to reduce climate change health impacts

Impact	Adaptive measures
Heat stress	Heat wave warning systems and treatment of heat stress
	Urban planning
Extreme weather events	Disaster preparedness, mitigation and response
	Early warning systems
	Disaster protection measures
Infectious diseases	Integrated environmental management
	Disease surveillance and monitoring
	Control of vector-, food- and water- borne diseases
Food security	International mechanisms of agriculture, trade and finance
	Seasonal climate forecasting
	Famine early warning systems
	National and local improved agriculture and land use measures, upgraded food storage and distribution systems
	Monitoring and surveillance
	Implementation of nutrition action plans
Water	Pollution reduction and pollution control policies
	Demand management and water allocation policies
	Waste water treatment
	Economic and regulatory measures to increase irrigation efficiency
	Capacity building

Adapted from Menne (2000)

loss of agricultural soils, and enhancing the adaptability of natural systems. (ii) Repealing trends that increase vulnerability, for example avoiding settlement and economic activity in risk-prone areas such as floodplains, coastal zones or landslide zones. (iii) Improving societal awareness and preparedness, particularly with respect to risks associated with climate change and health, early-warning systems and public education programs. Carefully selected indicators are needed to monitor climate related environmental and human health impacts at regional and national levels. Strategies to mitigate the impacts of climate change on human health require adaptation measures across different sectors; health sector alone cannot provide sufficient adaptation measures to cope with climate change.

References

Anderson N, Strader R, Davidson C (2003) Airborne reduced nitrogen: ammonia emissions from agriculture and other sources. Environ Int 29:277–286

Asman WAH, Sutton MA, Schjørring JK (1998) Ammonia: emission, atmospheric transport and deposition. New Phytol 139:27–48

Banerjee NK, Mosier AR (1989) Coated calcium carbide as a nitrification inhibitor in upland and flooded soils. J Indian Soc Soil Sci 37:306–313

Beelen R, Hoek G, van Den Brandt PA, Goldbohm RA, Fischer P, Schouten LJ, Jerrett M, Hughes E, Armstrong B, Brunekreef B (2008) Long-term effects of traffic-related air pollution on mortality in a Dutch cohort (NLCS-AIR study). Environ Health Perspect 116:196–202

Behera SN, Sharma M, Aneja VP, Balasubramanian R (2013) Ammonia in the atmosphere: a review on emission sources, atmospheric chemistry and deposition on terrestrial bodies. Environ Sci Pollut Res 20:8092–8131

Benbi DK, Brar JS (2009) A 25-year record of carbon sequestration and soil properties in intensive agriculture. Agron Sustain Dev 29:257–265

Benbi DK, Chand M (2007) Quantifying the effect of soil organic matter on indigenous soil N supply and wheat productivity in semiarid sub-tropical India. Nutr Cycl Agroecosyst 79:103–111

Benbi DK, Richter J (2003) Nitrogen dynamics. In: Benbi DK, Nieder R (eds) Handbook of processes and modeling in the soil-plant system. The Haworth Press, New York, pp 409–481

Beniston M (2002) Climatic change: possible impacts on human health. Swiss Med Wkly 132:332–337

Bernstein IL, Bernstein DI (1989) Reactive airways disease syndrome (RADS) after exposure to toxic ammonia fumes. J Allergy Clin Immunol 83:173–179

Bousquet P, Ciais P, Miller JB, Dlugokencky EJ, Hauglustaine DA, Prigent C, van der Werf GR, Peylin P, Brunke EG, Carouge C, Langenfelds RL, Lathière J, Papa F, Ramonet M, Schmidt M, Steele LP, Tyler SC, White J (2006) Contribution of anthropogenic and natural sources to atmospheric methane variability. Nature 443:439–443

Bouwman AF (1996) Direct emissions of nitrous oxide from agricultural soils. Nutr Cycl Agroecosyst 46:53–70

Bouwman AF, Lee DS, Asman WAH, Dentener FJ, van der Hoek KW, Olivier JGJ (1997) A global high-resolution emission inventory for ammonia. Glob Biogeochem Cycles 11:561–587

Burney JA, Davis SJ, Lobell DB (2010) Greenhouse gas mitigation by agricultural intensification. Proc Nat Acad Sci USA 107:12052–12057

Canadell JG, Corinne LQ, Raupach MR, Field CB, Buitenhuis ET, Ciais P, Conway TJ, Gillett NP, Houghton RA, Marland G (2007) Contributions to accelerating atmospheric CO_2 growth from economic activity, carbon intensity, and efficiency of natural sinks. Proc Nat Acad Sci USA 104:18866–18870

Cassman KG, Dobermann A, Walters D (2002) Agroecosystems, nitrogen-use efficiency and nitrogen management. Ambio 31:132–140

Chhabra A, Manjunath KR, Panigrahy S, Parihar JS (2013) Greenhouse gas emissions from Indian livestock. Clim Chang 117:329–344

Ciais PC, Sabine G, Bala L, Bopp V, Brovkin J, Canadell A, Chhabra R, De Fries J, Galloway M, Heimann C, Jones C, Le Quéré RB, Piao MS, Thornton P (2013) Carbon and other biogeochemical cycles. In: Stocker TF, Qin D, Plattner GK, Tignor M, Allen SK, Boschung J, Nauels A, Xia Y, Bex V, Midgley PM (eds) Climate change 2013: the physical science basis, Contribution of Working Group I to the Fifth Assessment Report of the Intergovernmental Panel on Climate Change. Cambridge University Press, Cambridge, UK

Clausnitzer H, Singer MJ (1999) Mineralogy of agricultural source soils and respirable dust in California. J Environ Qual V28:1619–1629

Close LG, Catlin FI, Cohn AM (1980) Acute and chronic effects of ammonia burns of the respiratory tract. Arch Otolaryngol 106:151–158

Cole CV, Duxbury J, Freney J, Heinemeyer O, Minami K, Mosier A, Paustian K, Rosenberg N, Sampson N, Sauerbeck D, Zhao Q (1997) Global estimates of potential mitigation of greenhouse gas emissions by agriculture. Nutr Cycl Agroecosyst 49:221–228

Crutzen PJ (1981) Atmospheric chemical processes of the oxides of nitrogen, including nitrous oxide. In: Delwiche CC (ed) Denitrification, nitrification and atmospheric N_2O. Wiley, New York, pp 17–44

D'Souza RM, Becker NG, Hall G, Moodie K (2004) Does ambient temperature affect foodborne disease? Epidemiology 15:86–92

Dannenberg S, Conrad R (1999) Effect of rice plants on methane production and rhizospheric metabolism in paddy soils. Biogeochemistry 45:53–71

Davidson CI, Phalen RF, Solomon PA (2005) Airborne particulate matter and human health: a review. Aerosol Sci Technol, ibid 39:737–749

de la Hoz RE, Schlueter DP, Rom WN (1996) Chronic lung disease secondary to ammonia inhalation injury: a report on three cases. Am J Ind Med 29:209–214

Denier van der Gon HAC, Neue HU (1994) Impact of gypsum application on the methane emission from a wetland ricefield. Glob Biogeochem Cycles 8:127–134

Denman KL, Brasseur GP, Chidthaisong A, Ciais P, Cox PM, Dickinson RE, Hauglustaine DA, Heinze C, Holland EA, Jacob DJ, Lohmann U (2007) Couplings between changes in the climate system and biogeochemistry. In: Solomon S, Qin D, Manning M, Chen Z, Marquis M, Averyt KB, Tignor M, Miller HL (eds) Climate change 2007: the physical science basis, Contribution of Working Group I to the Fourth Assessment Report of the Intergovernmental Panel on Climate Change. Cambridge University Press, Cambridge, UK, pp 499–587

Deppenmeir U, Müller V, Gottschalk G (1996) Pathways of energy conservation in methanogenic archae. Arch Microbiol 165:149–163

DTI (Department of Trade and Industry) (2001) Digest of United Kingdom energy statistics. London

Dutkiewicz J (1978) Exposure to dust borne bacteria in agriculture. I. Environmental studies. Arch Environ Health 33:250–259

Duval BD, Blankinship JC, Dijkstra P, Hungate BA (2011) CO_2 effects on plant nutrient concentration depend on plant functional group and available nitrogen: a meta-analysis. Plant Ecol 213:505–521

EDGAR (Emissions Database for Global Atmospheric Research) (2011) http://edgar.jrc.ec.europa.eu/, Emission, release version 4.2, 2011. Washington, DC

Epstein E (1994) The anomaly of silicon in plant biology. Proc Natl Acad Sci U S A 91:11–17

Erbs M, Manderscheid R, Jansen G, Seddig S, Pacholski A, Weigel HJ (2010) Effects of free-air CO_2 enrichment and nitrogen supply on grain quality parameters and elemental composition of wheat and barley grown in a crop rotation. Agric Ecosyst Environ 136:59–68

Erisman JW, Domburg N, de Vries W, Kros H, de Haan B, Sanders K (2005) The Dutch N-cascade in the European perspective. Sci China 48:827–842

Erisman JW, Galloway J, Seitzinger S, Bleeker A, Butterbach-Bahl K (2011) Reactive nitrogen in the environment and its effect on climate change. Curr Opin Environ Sustain 3:281–290

European Commission (2008) Directive 2008/50/EC of the European Parliament and of the council: ambient air quality and cleaner air for Europe. Brussels. Official Journal of the European Union 11.6.2008: L152/1-L152/44

Ferm M (1998) Atmospheric ammonia and ammonium transport in Europe and critical loads: a review. Nutr Cycl Agroecosyst 51:5–17

Fernando N, Panozzo J, Tausz M, Norton R, Fitzgerald G, Seneweera S (2012) Rising atmospheric CO_2 concentration affects mineral content and protein concentration of wheat grain. Food Chem 133:1307–1311

Flury KE, Dines DE, Rodarte JR, Rodgers R (1983) Airway obstruction due to inhalation of ammonia. Mayo Clin Proc 58:389–393

Freney JR, Denmead OT, Watanabe I, Simpson JR (1978) Soil as a source or sink for atmospheric nitrous oxide. Nature 273:530–532

Furukawa Y, Inubushi K (2002) Feasible suppression technique of methane emission from paddy soil by iron amendment. Nutr Cycl Agroecosyst 64:193–201

Galbally IE, Gillett RW (1988) Processes regulating nitrogen compounds in tropical atmosphere. In: Rodhe H, Herrera R (eds) Acidification in tropical countries. Wiley, Chichester, pp 73–115

Galloway JN, Aber JD, Erisman JW, Seitzinger SP, Howarth RW, Cowling EB, Cosby BJ (2003) The nitrogen cascade. Bioscience 53:341–356

Garrity DP (1998) Addressing key natural resources management challenges in humid tropics through agroforestry research. In: Lal R (ed) Soil quality and agricultural sustainability. CRC Press, Ann Arbor, pp 86–111

Giese RF Jr, van Oss CJ (1993) The surface thermodynamic properties of silicates and their interactions with biological materials. In: Mossman BT (ed) Health effects of mineral dusts, vol 28. Mineral Soc Am, Washington DC, pp 327–346

Gifford R, Barrett D, Lutze J (2000) The effects of elevated [CO_2] on the C:N and C:P mass ratios of plant tissues. Plant Soil 224:1–14

Gruber N, Galloway JN (2008) An Earth-system perspective of the global nitrogen cycle. Nature 451:293–296

Gupta PK, Sharma C, Bhattacharya S, Mitra AP (2002) Scientific basis for establishing country greenhouse gas estimates for rice-based agriculture: an Indian case study. Nutr Cycl Agroecosyst 64:19–31

Haines A, Kovates RS, Campbell-Lendrum D, Corvalan C (2006) Climate change and human health: impacts, vulnerability and public health. Public Health 120:585–596

Harrison RM, Yin J (2000) Particulate matter in the atmosphere: which particle properties are important for its effects on health? Sci Total Environ 249:85–101

Hime N, Cowie C, Marks G (2015) Review of the health impacts of emission sources, types and levels of particulate matter air pollution in ambient air in NSW. Centre for Air Quality and Health Research and Evaluation, Woolcock Institute of Medical Research, NSW

Hogan KB (1993) Methane reductions are a cost-effective approach for reducing emissions of greenhouse gases. In: van Amstel AR (ed) Methane and nitrous oxide: methods in National Emissions Inventories and Options for Control, RIVM report no. 481507003. RIVM, Bilthoven, p 187–201

Högy P, Fangmeier A (2009) Atmospheric CO_2 enrichment affects potatoes: 2. Tuber quality traits. Eur J Agron 30:85–94

Högy P, Wieser H, Köhler P, Schwadorf K, Breuer J, Franzaring J, Muntifering R, Fangmeier A (2009) Effects of elevated CO_2 on grain yield and quality of wheat: results from a 3-year free-air CO_2 enrichment experiment. Plant Biol 1:60–69

Holland EA, Dentener FJ, Braswell BH, Sulzman JM (1999) Contemporary and pre-industrial global reactive nitrogen budgets. Biogeochemistry 46:7–43

Holmen K (2000) The global carbon cycle. In: Jacobson MC, Charlston RJ, Rodhe H, Orians GH (eds) Earth system science. Academic, Amsterdam, pp 282–321

Houghton RA (2003) Revised estimates of the annual net flux of carbon to the atmosphere from changes in land use and land management 1850–2000. Tellus B 55:378–390

Houghton RA, House JI, Pongratz J, van der Werf GR, DeFries RS, Hansen MC, Le Quéré C, Ramankutty N (2012) Carbon emissions from land use and land-cover change. Biogeosciences 9:5125–5142

Hurkuman WJ, Vensel WH, Tanaka CK, Whitehand L, Altenbach SB (2009) Effects of high temperature on albumin and globulin accumulation in endosperm proteome of the developing wheat grain. J Cereal Sci 49:12–23

IEA (International Energy Agency) (2010) CO_2 emissions from fuel combustion. International Energy Agency, Paris, p 540

IFA (International Fertilizer Industry Association) (2007) Sustainable management of the nitrogen cycle in agriculture and mitigation of reactive nitrogen side effects. IFA task force on reactive nitrogen. IFA, Paris, p 53

IFA/FAO (International Fertilizer Industry Association/Food and Agriculture Organization of the United Nations) (2001) Global estimates of gaseous emissions of NH_3, NO and N_2O from agricultural land. IFA and FAO, Rome

IPCC (Intergovernmental Panel on Climate Change) (2001) Climate change 2001: the scientific basis. Contribution of working group I to the third assessment report of the IPCC. Cambridge University Press, Cambridge, UK

IPCC (Intergovernmental Panel on Climate Change) (2013) Climate change 2013: the physical science basis. In: Stocker TF, Qin D, Plattner GK, Tignor M, Allen SK, Boschung J, Nauels A, Xia Y, Bex V, Midgley PM (eds) Contribution of working group I to the fifth assessment report of the IPCC. Cambridge University Press, Cambridge, UK, p 1535

IPCC (Intergovernmental Panel on Climate Change) (2014a) In: Field CB, Barros VR, Dokken DJ, Mach KJ, Mastrandrea MD, Bilir TE, Chatterjee M, Ebi KL, Estrada YO, Genova RC, Girma B, Kissel ES, Levy AN, Mac Cracken S, Mastrandrea PR, White LL (eds) Climate change 2014: impacts, adaptation, and vulnerability. Part A: global and sectoral aspects, Contribution of Working Group II to the Fifth Assessment Report of the IPCC. Cambridge University Press, Cambridge, UK, p 1132

IPCC (Intergovernmental Panel on Climate Change) (2014b) In: Pachauri RK, Meyer LA (eds) Climate change 2014: synthesis report, Contribution of Working Groups I, II and III to the Fifth Assessment Report of the Intergovernmental Panel on Climate Change. IPCC, Geneva, p 151

Jäckel U, Schnell S (2000) Suppression of methane emission from rice paddies by ferric iron fertilization. Soil Biol Biochem 32:1811–1814

Jackson ML, Gillette DA, Danielson EF, Blifford IH, Bryson RA, Syers JK (1973) Global dustfall during the quaternary as related to environments. Soil Sci 116:135–145

Kampa M, Castanas E (2008) Human health effects of air pollution. Environ Pollut 151:362–367

Kass I, Zamel N, Dobry CA, Holzer M (1972) Bronchiectasis following ammonia burns of the respiratory tract. A review of two cases. Chest 62:282–285

Khosa MK, Sidhu BS, Benbi DK (2010) Effect of redox species on methane emission from submerged rice soils. Ind J Ecol 37:51–55

Khosa MK, Sidhu BS, Benbi DK (2011) Methane emission from rice fields in relation to management of irrigation water. J Environ Biol 32:169–172

Kludze HK, Delaune RD, Patric WH (1993) Aerenchyma formation and methane and oxygen exchange in rice. Soil Sci Soc Am J 57:386–391

Koren HS, Utell (1997) Asthma and the environment. Environ Health Perspect 105:534–537

Kovats RS, Edwards SJ, Hajat S, Armstrong BG, Ebi KL, Menne B (2004) The effect of temperature on food poisoning: a time-series analysis of salmonellosis in ten European countries. Epidemiol Infect 132:443–453

Lal R (2004) Agricultural activities and the global carbon cycle. Nutr Cycl Agroecosyst 70:103–116

Latenser BA, Loucktong TA (2000) Anhydrous ammonia burns: case presentation and literature review. J Burn Care Rehabil 21:40–42

Leduc D, Gris P, L'heureaux P, Gevenois PA, De Vuyst P, Yernault JC (1992) Acute and long-term respiratory damage following inhalation of ammonia. Thorax 47:755–757

Lloyd D, Thomas KL, Benstead J, Davies KL, Lloyd SH, Arah JRM, Stephen KD (1998) Methanogenesis and CO_2 exchange in an ombrotrophic peat bog. Atmos Environ 2:3229–3238

Machefert SE, Dise NB, Goulding KWT, Whitehead PG (2002) Nitrous oxide emission from a range of land uses across Europe. Hydrol Earth Syst Sci 63:325–337

Mathews E, Hammond A (1999) Critical consumption trends and implications: degrading Earth's ecosystems. World Resources Institute, Washington, DC

McLean JA, Mathews KP, Brayton PR, Solomon ER, Bayne NK (1979) Effects of ammonia on nasal resistance in atopic and non-atopic subjects. Ann Otol Rhinol Laryngol 88:228–234

McMichael AJ, Woodruff RE, Hales S (2006) Climate change and human health: present and future risks. Lancet 367:859–869

Menne B (2000) Can the health sector adapt to climate variability/change? In: European bulletin on environment and health. WHO Regional Office for Europe, Copenhagen

Merr JL, Roger P (2001) Production, oxidation, emission and consumption of methane by soils: a review. Eur J Soil Biol 37:25–50

Moldestad A, Fergestad EM, Hoel B, Skjelvag AO (2011) Effect of temperature variation during grain filling on wheat gluten resistance. J Cereal Sci 53:1–8

Monteil G, Houweling S, Dlugockenky EJ, Maenhout G, Vaughn BH, White JCW, Rockmann T (2011) Interpreting methane variations in the past two decades using measurements of CH_4 mixing ratio and isotopic composition. Atmos Chem Phys 11:9141–9153

Mosier AR, Duxbury JM, Freney JR, Heinemeyer O, Minami K, Johnson DE (1998) Mitigating agricultural emissions of methane. Clim Chang 40:39–80

Mosier AR, Syers JK, Freney JR (eds) (2004) Agriculture and the nitrogen cycle. Island Press, Washington, DC, p 296

Myers SS, Zanobetti A, Kloog I, Huybers P, Leakey ADB, Bloom AJ, Carlisle E, Dietterich LH, Fitzgerald G, Hasegawa T, Holbrook NM, Nelson RL, Ottman MJ, Raboy V, Sakai H, Sartor KA, Schwartz J, Seneweera S, Tausz M, Usui Y (2014) Increasing CO_2 threatens human nutrition. Nature 510:139–149

Myrold DD (1988) Denitrification in ryegrass and winter wheat cropping systems of western Oregon. Soil Sci Soc Am J 52:412–416

Neef LM, van Weele, van Velthoven P (2010) Optimal estimation of the present-day global methane budget. Glob Biogeochem Cycles 24:GB4024

Nel A (2005) Air pollution-related illness: effects of particles. Science 308:804–806

Nieder R, Benbi DK (2008) Carbon and nitrogen in the terrestrial environment. Springer, Heidelberg, p 432

Nieder R, Schollmayer G, Richter J (1989) Denitrification in the rooting zone of cropped soils with regard to methodology and climate: a review. Biol Fertil Soils 8:219–226

NOAA (2017) NOAA-ESRL Global Monitoring Mauna Loa CO_2: April 2017. Available online https://www.co2.earth/monthly-co2. Accessed 20 June 2017

Nouchi I, Hosono T, Aoki K, Minami K (1994) Seasonal variation in methane flux from rice paddies associated with methane concentration in soil water, rice biomass and temperature, and its modeling. Plant Soil 161:195–208

Olivier JGJ, Janssens-Maenhout G (2012) Part III: greenhouse gas emissions: 1. Shares and trends in greenhouse gas emissions; 2. Sources and methods: total greenhouse gas emissions. In: CO2 emissions from fuel combustion, 2012. International Energy Agency (IEA), Paris, p III1–III51

Parashar DC, Rai J, Gupta PK, Singh N (1991) Parameters affecting methane emission from paddy fields. Indian J Radio Space 20:12–17

Pendall E, Bridgham S, Hanson PJ, Hungate B, Kicklighter DW, Johnson DW, Law BE, Luo Y, Megonigal JP, Olsrud M, Ryan MG (2004) Below-ground process responses to elevated CO_2 and temperature: a discussion of observations, measurement methods, and models. New Phytol 162:311–322

Peoples MB, Boyer EW, Goulding KWT, Heffer P, Ochwoh VA, Vanlauwe B, Wood S, Yagi K, van Cleemput O (2004) Pathways of nitrogen loss and their impacts on human health and the environment. In: Mosier AR, Syers JK, Freney JR (eds) Agriculture and the nitrogen cycle: assessing the impacts of fertilizer use on food production and the environment, SCOPE 65. Scientific Committee on Problems of the Environment (SCOPE), Paris, p 53–69

Ponnamperuma FN (1972) The chemistry of submerged soils. Adv Agron 24:29–96

Pope CA III (2000) What do epidemiologic findings tell us about health effects of environmental aerosols? J Aeros Med 13:335–354

Pope CA III, Burnett RT, Thun MJ, Calle EE, Krewski D, Ito K, Thurston GD (2002) Lung cancer, cardiopulmonary mortality, and long-term exposure to fine particulate air pollution. J Am Med Assoc 287:1132–1141

Poschl U (2005) Atmospheric aerosols: composition, transformation, climate and health effects. Angew Chem Int Ed Eng 44:7520–7540

Pretty JN (1995) Regenerating agriculture. Earthscan, London

Pretty JN, Ball AS, Xiaoyun L, Ravindranathan NH (2002) The role of sustainable agriculture and renewable-resource management in reducing greenhouse-gas emissions and increasing sinks in China and India. Phil Trans R Soc A 360:1741–1761

Prior SA, Runion GB, Rogers HH, Torbert HA (2008) Effects of atmospheric CO_2 enrichment on crop nutrient dynamics under no-till conditions. J Plant Nutr 31:758–773
Raich JW, Potter CS (1995) Global patterns of carbon dioxide emissions from soils. Glob Biogeochem Cycles 9:23–36
Ryden JC (1981) Nitrous oxide exchange between a grassland soil and the atmosphere. Nature 292:235–237
Samoli E, Peng R, Ramsay T, Pipikou M, Touloumi G, Dominici F, Burnett R, Cohen A, Krewski D, Samet J, Katsouyanni K (2008) Acute effects of ambient particulate matter on mortality in Europe and North America: results from the APHENA study. Environ Health Perspect 116:1480–1486
Scheffe R (2003) Presentation at the conference "Particulate matter: atmospheric sciences, exposure, and the fourth colloquium on PM and human health," Pittsburgh. March 31–April 4 (p 219). Cited in Davidson et al. (2005): *ibid.*
Schwartz J (2003) Presentation at the conference "Particulate matter: atmospheric sciences, exposure, and the fourth colloquium on PM and human health," Pittsburgh. March 31–April 4 (p 219). Cited in Davidson et al. (2005): *ibid.*
Shukla SP, Sharma M (2010) Neutralization of rainwater acidity at Kanpur, India. Tellus B 62:172–180
Šimek M, Cooper JE (2002) The influence of soil pH on denitrification: progress towards the understanding of this interaction over the last 50 years. Eur J Soil Sci 53:345–354
Skiba U, Fowler D, Smith KA (1997) Nitric oxide emissions from agricultural soils in temperate and tropical climates: sources, controls and mitigation options. Nutr Cycl Agroecosyst 48:139–153
Smith P, Powlson DS, Glending MJ, Smith JOU (1998) Preliminary estimates of the potential for carbon mitigation in European soils through no-till farming. Glob Chang Biol 4:679–685
Smith KA, Ball T, Conen F, Dobbie KE, Massheder J, Rey A (2003) Exchange of greenhouse gases between soil and atmosphere: interactions of soil physical factors and biological processes. Eur J Soil Sci 54:779–791
Smith J, Smith P, Wattenbach M, Gottschalk P, Romanenkov VA, Shevtsova LK, Sirotenko OD, Rukhovich DI, Koroleva PV, Romaneko IA, Lisovo NV (2007) Projected changes in the organic carbon stocks of cropland mineral soils of European Russia and Ukraine, 1990–2070. Glob Chang Biol 13:342–356
Smith KR, Woodward A, Campbell-Lendrum D, Chadee DD, Honda Y, Liu Q, Olwoch JM, Revich B, Sauerborn R (2014) Human health: impacts, adaptation, and co-benefits. In: Field CB, Barros VR, Dokken DJ, Mach KJ, Mastrandrea MD, Bilir TE, Chatterjee M, Ebi KL, Estrada YO, Genova RC, Girma B, Kissel ES, Levy AN, Mac Cracken S, Mastrandrea PR, White LL (eds) Climate change 2014: impacts, adaptation, and vulnerability. Part A: global and sectoral aspects, Contribution of Working Group II to the Fifth Assessment Report of the Intergovernmental Panel on Climate Change. Cambridge University Press, Cambridge, UK, pp 709–754
Sobonya R (1977) Fatal anhydrous ammonia inhalation. Hum Pathol 8:293–299
Stehfest E (2006) Modelling of global crop production and resulting N_2O emissions. Ph.D. Thesis. University of Kassel, p 150
Su WH (1996) Dust and atmospheric aerosols. Resour Conserv Recycl 16:1–14
Sutton MA, Fowler D (2002) Introduction: fluxes and impacts of atmospheric ammonia on national, landscape and farm scales. Environ Pollut 119:7–8
Taub DR, Miller B, Allen H (2008) Effect of elevated CO_2 on the protein concentration of food crops: a meta-analysis. Glob Chang Biol 14:565–575
Timothy H (2009) World greenhouse gas emissions in 2005.WRI Working Paper. World Resources Institute
Trolldenier G (1995) Methanogenesis during rice growth as related to the water regime between crop seasons. Biol Fertil Soils 19:84–86

US EPA (2004) Environmental Protection Agency, Part II, 40 CFR Part 50, National Ambient Air Quality Standards for Particulate Matter; Final Rule. In: Federal Register, 62 (138), July 18, 1997. Available at https://archive.epa.gov/ttn/pm/web/pdf/pmnaaqs.pdf

US EPA (United States Environmental Protection Agency) (2006) National ambient air quality standards for particulate matter: final rule. Fed Regist 71:61144–61233

US EPA (United States Environmental Protection Agency) (2009) Integrated science assessment for particulate matter. Research Triangle Park, North Carolina

van Aardenne JA, Dentener FJ, Olivier JGJ, Goldewijk K, Klein CGM, Lelieveld J (2001) A $1° \times 1°$ resolution dataset of historical anthropogenic trace gas emissions for the period 1890–1990. Glob Biogeochem Cycles 15:909–928

van Groenigen KJ, Osenberg CW, Hungate BA (2011) Increased soil emissions of potent greenhouse gases under increased atmospheric CO_2. Nature 475:214–216

Várallyay G (2007) Potential impacts of climate change on agro-ecosystems. Rev Agric Consp Sci 72:1–8

Velthof GL, Brader AB, Oenema O (1996) Seasonal variations in nitrous oxide losses from managed grasslands in the Netherlands. Plant Soil 181:263–274

Walton M (1973) Industrial ammonia gassing. Br J Ind Med 30:78–86

Wang ZP, Delaune RD, Masscheleyn PB, Patric WHJ (1993) Soil redox and pH effects on methane production in a flooded rice soil. Soil Sci Soc Am J 57:382–385

Wassmann R, Aulakh MS (2000) The role of rice plants in regulating mechanisms of methane emissions. Biol Fertil Soils 31:20–29

Wassmann R, Neue HU, Lantin R, Makarim K, Chareonsilp N, Buendia LV, Rennenberg H (2000) Characterization of methane emissions from rice fields in Asia. II. Differences among irrigated, rainfed, and deepwater rice. Nutr Cycl Agroecosyst 58:13–22

Watanabe I, Hashimoto T, Shimoyama A (1997) Methane oxidizing activities and methanotrophic populations associated with wetland rice soils. Biol Fertil Soils 24:261–265

Weiss R, McMichael AJ (2004) Social and environmental risk factors in the emergence of infectious diseases. Nat Med 10:S70–S76

WHO (World Health Organization) (1990) Potential health effects of climatic change. Report of a WHO Task Group. WHO, Geneva

WHO (World Health Organization) (1997) Health and environment in sustainable development: five years after the Earth Summit. WHO, Geneva

WHO (World Health Organization) (2001) World health report 2001. WHO, Geneva

WHO (World Health Organization) (2002) World health report 2002. Reducing risks, promoting healthy life. WHO, Geneva

WHO (World Health Organization) (2003) Climate change and human health: risks and responses: summary. WHO, Geneva

WHO (World Health Organization) (2006) Air quality guidelines: global update 2005. Particulate matter, ozone, nitrogen dioxide and sulfur dioxide. WHO Regional Office for Europe, Copenhagen

WHO (World Health Organization) (2011) Exposure to air pollution (particulate matter) in outdoor air. (ENHIS Factsheet 3.3). WHO Regional Office for Europe, Copenhagen

WHO (World Health Organization) (2013) Health effects of particulate matter: policy implications for countries in Eastern Europe, Causcas and Central Asia. WHO, Regional Office for Europe, Copenhagen

WHO (World Health Organization) (2016) Ambient air pollution: a global assessment of exposure and burden of disease. WHO, Geneva

Xu C, Shaffer MJ, Al-Kaisi M (1998) Simulating the impact of management practices on nitrous oxide emissions. Soil Sci Soc Am J 62:736–742

Xue J, Lau AK, Yu JZ (2011) A study of acidity on PM2.5 in Hong Kong using online ionic chemical composition measurements. Atmos Environ 45:7081–7088

Yagi K (2002) Methane emission in rice, mitigation options. In: Lal M (ed) Encyclopedia of soil science. Marcel Dekker, New York, p 814–818

Yagi K, Minami K (1990) Effects of organic matter application on methane emission from some Japanese paddy fields. Soil Sci Plant Nutr 36:599–610

Yamane I, Sato K (1964) Decomposition of glucose and gas formation in flooded soil. Soil Sci Plant Nutr 10:127–133

Yan X, Yagi K, Akiyama H, Akimoto H (2005) Statistical analysis of the major variables controlling methane emission from rice fields. Glob Chang Biol 11:1131–1141

Yan X, Akiyama H, Yagi K, Akimoto H (2009) Global estimations of the inventory and mitigation potential of methane emissions from rice cultivation conducted using the 2006 Intergovernmental Panel on Climate Change Guidelines. Glob Biogeochem Cycles 23:GB2002

Chapter 5
Reactive Water-Soluble Forms of Nitrogen and Phosphorus and Their Impacts on Environment and Human Health

Abstract Water is a basic necessity of life and access to clean water is vital for sanitation, hygiene, agriculture, and industry. Yet the world faces threats of scarcity of clean drinking water and the pollution of the water resources. Groundwater is an important source of drinking-water in many regions of the world, particularly in areas with limited or polluted surface water sources. It could be contaminated with chemicals that may affect human health. Nitrate is the most common chemical contaminant in the world's groundwater aquifers, and surface water are particularly affected by the presence of phosphorus. Occurrence of reactive N and P species in water can have serious environmental and human health impacts. Nitrate by itself is usually non-toxic. Its adverse effects on human health are due to conversion of nitrate into nitrite, nitric oxide and N-nitroso compounds. In the human body, NO_3^- is converted to NO_2^-, which can cause methaemoglobinemia by interfering with the ability of haemoglobin to take up O_2. Infants younger than 3 months of age are particularly prone to adverse effects of nitrate exposure. Opinions differ on the effects of high dietary intake of nitrate on human health and the evidence linking high nitrate level in drinking water with methaemoglobinemia is still controversial. High nitrate level in drinking water has also been implicated, but not incontrovertibly, in the incidence of cancers of the digestive track. Besides the possible adverse human health effects associated with high nitrate intake, evidence is emerging of its potential benefits in cardiovascular health and providing protection against infections. Enrichment of surface waters with nitrogen and phosphorus contributes to the phenomenon of eutrophication leading to harmful algal blooms, which can impact several ecosystem services. Therefore, it is important to protect the quality of groundwater and surface water by proper management of sources of pollutants and reducing input of nitrogen and phosphorus in agricultural systems. In this chapter, we provide information on sources of reactive nitrogen and phosphorus species in water, their effects on ecosystems and human health as well as mitigation strategies.

Keywords Water-soluble nitrogen compounds • Nitrate in drinking water • Leaching • Runoff • Nitrate metabolism • Methaemoglobinemia • Blue-baby syndrome • Eutrophication • Cyanotoxins • Alagal blooms • Preventive health care

R. Nieder et al., *Soil Components and Human Health*,
https://doi.org/10.1007/978-94-024-1222-2_5

5.1 Human Dependence on Water

Water is a basic necessity of life. Besides its requirement for drinking, sanitation, hygiene, and other day-to-day activities, access to clean and reliable water supplies is vital for agriculture, industry and energy production. Yet the world faces threats of scarcity of clean drinking water and pollution of the water resources. Approximately 884 million people, almost all in the developing countries lack access to improved drinking-water supply (WHO/UNICEF 2010). About 2.6 billion people do not use improved sanitation facilities and most of these live in Southern Asia, Eastern Asia and Sub-Saharan Africa. Liquid freshwater exists on Earth either as groundwater or as surface water (lakes, rivers, streams, ponds, etc.). Groundwater, which constitutes 97% of global fresh water, is an important source of drinking-water in many regions of the world, particularly in areas with limited or polluted surface water sources. In India, China, Bangladesh, Thailand, Indonesia and Vietnam more than 50% of potable supplies are provided from groundwater. In Africa and Asia, most of the large cities use surface water, but many millions of people in rural areas and in low-income peri-urban areas are dependent on groundwater. An estimated 80% of the drinking-water used by these communities is drawn from groundwater sources (Pedley and Howard 1997).

Groundwater could be contaminated with chemicals such as fluoride, arsenic, nitrate, selenium, uranium, metals and radionuclides (e.g. radon) that may be hazardous to human health. The chemical substances in water may arise from natural processes as well as by anthropogenic activities. Natural processes include decomposition of organic matter in soils and leaching of mineral deposits. Anthropogenic activities that can impact quality of water include land application of animal wastes and agrochemicals in agriculture, disposal practices of human excreta and wastes, landfills, mining and traffic. Nitrate is the most common chemical contaminant in the world's groundwater aquifers (Spalding and Exner 1993) and surface waters are particularly affected by the presence of phosphorus. It is important to adopt measures for mitigating contamination of water resources and for protecting human health.

5.2 Sources of Water-Soluble Nitrogen and Phosphorus Compounds

5.2.1 *Nitrogen*

Nitrogen enters water in organic and inorganic forms. The organic forms arise from living or dead organisms such as algae and bacteria and decaying plant material. The inorganic forms of N *viz.* ammonia (NH_3), ammonium (NH_4^+), nitrate (NO_3^-), and nitrite (NO_2^-) in surface and groundwater arise from weathering, natural soil nitrate, soil organic matter (SOM), nitrogenous fertilizers, atmospheric N deposition, livestock waste, sewage effluent, septic systems and wastewater treatment plants. These organic and inorganic sources of nitrogen are transformed to nitrate by mineralization, hydrolysis, and bacterial nitrification. Therefore, nitrate is the most abundant form of reactive N (N_r) in surface and groundwater. Ammonium being positively charged is adsorbed to negatively charged soil particles and is thus not easily leached from the soil. However, under aerobic conditions NH_4^+ is nitrified to NO_3^- through NO_2^-. Nitrite is chemically unstable and is rapidly oxidized to nitrate, but it can accumulate to a limited extent in surface and groundwater in a reducing environment. Nitrite can be formed by the microbial reduction of nitrate and can also be formed chemically in distribution pipes by *Nitrosomonas* bacteria during stagnation of nitrate-containing and oxygen-poor drinking-water in galvanized steel pipes or if chloramination is used to provide a residual disinfectant (WHO 2011). But the occurrence is invariably random. Nitrate being negatively charged moves freely with the soil-water unless the soils have significant anion exchange capacity. Therefore, movement of water through soil can result in the transport of N down the profile (leaching) and contaminate the groundwater. Nitrates in soil originate from mineralization of SOM and crop and animal residue, applied fertilizer N and to a smaller extent from atmospheric deposition. Domestic sewage effluents are important source of nitrate in groundwater, particularly in suburban and peri-urban areas with shallow water table. High NO_3^-concentration in ground and surface water near big cities could be due to discharge of inadequately treated sewage effluent into pits or depressions or irrigation channels or rivers. Nitrate can reach both surface water and groundwater as a consequence of agricultural activities (including excess application of inorganic nitrogenous fertilizers and manures), from wastewater disposal and from oxidation of nitrogenous waste products in human and animal excreta, including septic tanks. Globally, fresh waters receive about 39–95 Tg N $year^{-1}$ from agricultural soils (Bouwman et al. 2011; Billen et al. 2013); a part of which remains in superficial aquifers, part is lost to the atmosphere by denitrification and part (40–66 Tg N $year^{-1}$) is discharged to coastal waters (Seitzinger et al. 2005; Voss et al. 2011). However, the rate of denitrification is slow under aerobic conditions and low organic carbon content. It takes several decades to decrease the high groundwater nitrate loads (Green et al. 2008). Besides NO_3-N, transport of dissolved organic nitrogen (DON) from surface soils to groundwater and streams is important from a nitrogen balance and ecological point of view. Though the existence of soluble organic forms of N in rain and drainage waters has been known for several years, but these were not regarded as significant pools of N in agricultural soils probably because of the difficulties with its measurement. In forested watersheds, DON has been recognised as a major contributor of nitrogen to surface water. The annual transport of DON from the forest floor into the mineral soil amounts to an average of 26% of the litter N input (Michalzik et al. 2001). The DON flux in leachates from forest floor has been shown to increase with increasing precipitation.

5.2.2 *Phosphorus*

Phosphorus is one of the key elements necessary for plants and animals; no life is possible without phosphorus. After nitrogen, it is the second most widely used fertilizer nutrient in agriculture. Phosphorus is extremely reactive and does not occur in elemental form in nature and combines with oxygen when exposed to air. In natural systems like soil and water, phosphorus exists as orthophosphate. In water, orthophosphate mostly exists as $H_2PO_4^-$ in acidic conditions and as HPO_4^{2-} in alkaline conditions. Phosphate compounds are not very soluble in water; therefore, most of the phosphate in natural systems exists in solid form. Small amount of P are naturally present in soil and water. Natural process, such as weathering of rocks and minerals releases 15–20 Tg P year^{-1}, of which 2–7 Tg P year^{-1} reaches fresh water by runoff and leaching (Bennett et al. 2001; Bouwman et al. 2009; van Vuuren et al. 2010). The most important anthropogenic sources of P transfer to water are untreated or inadequately treated wastewater and manure stores (point sources), soil erosion and runoff from crop fields and pastures (diffuse source). Globally, about 1.5 Tg P year^{-1} is released to surface waters from waste water treatment (Smil 2000). Excess fertilization in agriculture and manure production causes phosphorus surplus in soil, some of which is transported to aquatic ecosystems. Worldwide about 12 Tg P year^{-1} accumulates in agricultural soils (Bouwman et al. 2009). Accumulation of phosphorus in soil and the resulting increase in soil solution P concentration leads to small but important increases in the amounts of phosphate in water that passes over or through soils (Fig. 5.1). Rainfall can cause varying amounts of phosphates to wash from agricultural soils into nearby

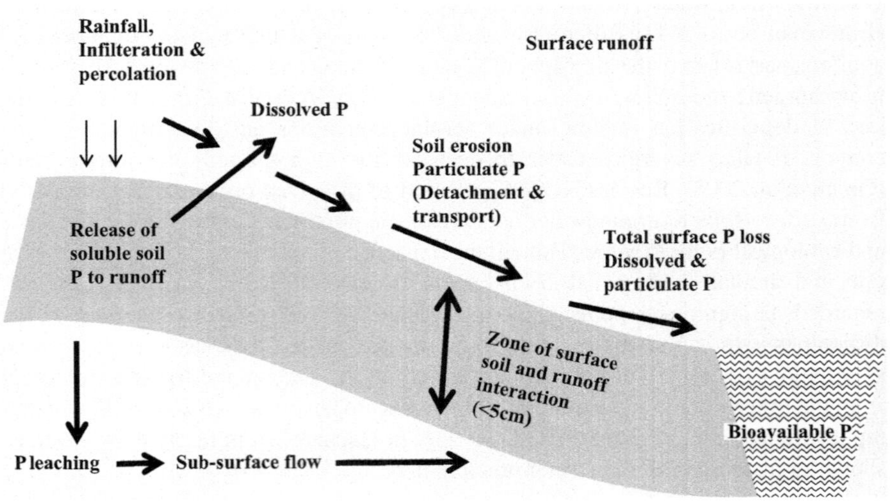

Fig. 5.1 Conceptual representation of the transport of phosphorus from agricultural fields to surface water (Adapted from Mullins 2009)

waterways. Overland flow of water or through soil can remove both dissolved and particulate (associated with soil particles) forms of soil phosphorus Phosphate in soils is associated more with fine than coarse particles. When soil erosion occurs, more fine particles are removed than coarse particles, causing sediment leaving a soil through erosion to be enriched in P. Detergents are another source of P in water. Sodium tripolyphosphates ($Na_5P_3O_{10}$) and potassium pyrophosphate ($K_4P_2O_7$) have been widely used in production of detergents. Though the use of phosphate-based detergents has been banned in some countries, still they are being used in many countries. Phosphates may also arise from the breakdown of organic pesticides which contain phosphates. They may exist in solution, as particles, loose fragments, or in the bodies of aquatic organisms. Water bodies may also contain organic P and phosphate attached to small particles of sediment.

5.3 Impacts on Water Quality

5.3.1 Nitrogen

Nitrate level in relatively pollution-free regions of world such as high altitude lakes and rivers and snow-clad mountains is about 0.50 mg nitrate L^{-1} of water. In oceans, where all rivers end up, the average nitrate level is 0.68 mg nitrate L^{-1}. Nitrate concentrations higher than this reflect anthropogenic or geogenic contributions. Since surface water originates either as runoff which drains the land surface, or as groundwater that has come back to the surface, the contamination of surface water depends on the activities that occur on land adjacent to lakes and rivers and the N content of the groundwater that feeds surface water bodies. The groundwater that discharges into surface water bodies contains N from both natural and anthropogenic sources. Land use and management can influence the amount of N transported to surface waters. Rates of N transport to surface water bodies range from less than 1 kg N ha^{-1} $year^{-1}$ on some natural lands to 20 kg N ha^{-1} $year^{-1}$ on agricultural lands (Beaulac and Reckhow 1982; Jordan et al. 1997). Nitrate concentrations in surface water can change rapidly due to surface runoff of fertilizer, uptake by phytoplankton and denitrification by bacteria, but groundwater concentrations generally show relatively slow changes (WHO 2011).

Nitrate levels in drinking water are high and still increasing in areas with N fertilizer-intensive irrigated farming and concentrated livestock industries. Globally, 170–180 Tg of N_r is added to agricultural systems every year and more than half of it is leaked into groundwater, lakes and rivers and carried to oceans. This results in nitrate pollution of drinking water. The N_r from agroecosystems leaches into the groundwater where it may accumulate and distributed to other systems through hydrologic (e.g., NO_3^-) or atmospheric pathways (e.g., N_2O or NO) (Puckett et al. 1999; Refsgaard et al. 1999). Though the accumulation of N_r in groundwater is a minor sink relative to the amount of N_r created (Schlesinger 2009), yet it poses a significant human health risk, because of contamination of drinking water. Nitrogen concentrations in freshwater have mostly been studied in developed countries. A national survey in the United States, in 1994, revealed that nearly 40% of the country's lakes and rivers were too polluted with nitrate for basic uses such as fishing or swimming (Mathews and Hammond 1999). However, nitrogen is fast accumulating in developing countries and emerging nations too. In China, between 1963 and 1980, there was a fourfold increase in nitrate and twofold increase in ammonium concentrations in Yangtze River (Galloway et al. 1998). Majority of this increase is because of overuse of nitrogen fertilizers, which increased 12-fold over the period. Nitrate pollution of drinking water is common in Australia, Europe, and parts of Africa, Asia and the Middle East.

5.3.2 *Phosphorus*

Groundwater and surface water usually contain relatively low concentrations of dissolved or soluble phosphorus. Depending on the types of minerals in a given area, water bodies usually contain about 10 ppb (parts per billion) or more of dissolved P as orthophosphate. Dissolved P constitutes 0.1–5% of the nutrient applied in inorganic fertilizers, and the annual loss rates are usually between 1% and 2% (Lennox et al. 1997; Rekolainen et al. 1997). Therefore, watersheds heavily fertilized by manures and phosphates may be discharging several kg P ha^{-1} every year. Loss of dissolved P from land is about 1 Tg P $year^{-1}$. Global flux of particulate and dissolved P respectively, to the ocean is estimated to be 22 and 3 Tg P $year^{-1}$ (Smil 2000). Anthropogenic activities, which lead to greater flux of P to water bodies include, (i) accelerated erosion and runoff due to land use changes, (ii) limited recycling of organic wastes in agricultural systems, (iii) disposal of untreated human waste and urban sewage, and (iv) excessive use of inorganic fertilizers (Smil 2000). The inadvertent fertilization of inland and coastal waters can change them from nutrient poor oligotrophic to nutrient rich hypertrophic (= hypereutrophic) state (Table 5.1). There is considerable concern about P being lost from soils and transported to nearby streams and lakes. Several chemical properties

Table 5.1 Relationship between trophic levels and lake characteristics

Trophic status	Organic matter	Mean phosphorus (ppb)	Secchi Depth[a] (m)	Chlorophyll maximum (ppb)
Oligotrophic	Low	8.0	9.9	4.2
Mesotrophic	Medium	26.7	4.2	16.1
Eutrophic	High	84.4	2.45	42.6
Hypertrophic	Very high	750–1200	0.4–0.5	

Adapted from Janus and Vollenweider (1981) and FAO (1996)

[a]Secchi depth is a measure of transparency of water in a lake; lower the Secchi depth higher the algal concentration

of soil P have important implications for the potential loss of P to surface water. Phosphorus in soils is almost entirely associated with soil particles. When soil particles are carried to a river or lake, P will be contained in this sediment. When the sediment reaches a body of water it may act as a sink or a source of P in solution. In either case, it is a potential source of P that may eventually be released. Ocean is the major biospheric reservoir of phosphorus and contains about 93 Pg P out of which only 8 Pg is in the surface ocean (Mackenzie et al. 1998). Less than 0.2% of the oceanic P is in coastal waters where P concentration can be up to 0.3 mg L^{-1} (Smil 2000). In open ocean surface waters, the dissolved P is usually undetectable.

5.4 Effects on Ecosystems

Excessive application of nitrogenous fertilizers and their accumulation in soil and water can have threefold environmental impacts (Mensinga et al. 2003). Firstly, there could be greater accumulation of nitrate in plants, particularly in vegetables grown under sunlight-deprived conditions such as greenhouses in autumn and winter, cloudy skies and shady locations. This can result in higher dietary intake of nitrate by humans (see Chap. 6). About 80–85% of dietary intake of nitrate originates from consumption of vegetables and the remaining from drinking water (Walker 1996). The other dietary sources could be cured meats and certain types of cheese wherein nitrates are added as preservative (EC 1995). Secondly, there could be excessive leaching of nitrates into groundwater and runoff into surface water, which not only affects the quality of drinking water but also stimulates the growth of algae (Fig. 5.2). Moreover, nitrate leaching in croplands can result in low N use efficiency necessitating higher input of fertilizer nitrogen, which can further increase nitrate availability in soil. Thirdly, higher availability of nitrate exacerbates denitrification particularly under oxygen-limited conditions leading to greater emissions of nitrous oxide and nitric oxide to the atmosphere. The process of denitrification is mediated by a wide range of facultative anaerobic and heterotrophic bacteria. They oxidize organic carbon and reduce NO_3^- and NO_2^- in absence of O_2 resulting in the production of N_2O, NO_x and finally N_2. The process of denitrification and the underlying factors are described in detail in Chap. 6. Some of the nitrogen oxides (NO_x) in the air are converted to nitrate, which precipitates in the form of rain and further contributes to the excess burden of nitrogenous compounds in soil. Atmospheric addition of N compounds to surface waters can cause acidification of freshwater aquatic ecosystems. Nitrous oxide (N_2O), an intermediate product of nitrification and denitrification processes is a potent greenhouse gas and an important driver of global climate change. More information on the effect of reactive N compounds on atmosphere and ecosystems is given in Chap. 4.

Fig. 5.2 Stimulation of algal blooms through high nutrient concentration in Tsunkhul Lake, Mongolia. The nutrient (particularly nitrogen) input results from high livestock densities on the pasture land. The algal blooms cloud the water and block sunlight, causing underwater plants to die. When the algae and underwater plants die and decompose, oxygen is used up. Dissolved oxygen in the water is essential to most organisms living in the water (Photo: Andreas Bürkert, with kind permission)

5.4.1 Leaching and Runoff

During the last few decades, loss of nitrate to groundwater has become a significant fate of nitrogen in agroecosystems. Leaching of NO_3^- to the groundwater is controlled by the concentration of NO_3^- in the soil solution and the amount of water percolating through the soil profile. A number of soil, environmental and management factors control nitrate leaching from agricultural soils. The most important soil factors that influence NO_3^-leaching include texture and structure. Leaching losses of N, under conditions of heavy rainfall or excessive irrigation, could be higher in rapidly percolating sandy soils than in silt and clay soils (Nieder and Benbi 2008). The effect of texture is greatly modified by soil structure and the micro-scale distribution of NO_3^- in the soil (White 1985). If NO_3^- is held within soil aggregates, it will be protected from leaching when bypass or preferential flow occurs (Thomas and Phillips 1979). On the other hand, if NO_3^- is held on the outside of aggregates bypass flow causes it to leach faster (Addiscott and Cox

1976). The environmental factors that influence NO_3^- leaching include rainfall, evaporation and temperature. These factors influence NO_3^- leaching directly through their effect on water drainage and indirectly by affecting the process of nitrification. Fertilizer N application greatly influences NO_3^- leaching in soil; losses increase with increasing rate of fertilizer N (Benbi 1990; Benbi et al. 1991). In addition to overuse of fertilizers, nitrate leaching into groundwater can arise from manure, as in the Netherlands. Nitrate leaching is likely to be higher in agricultural systems based largely on organic inputs mainly because of lack of synchrony between N release from organic sources and crop demand. Nitrate leaching losses vary considerably between different land uses and these are generally higher under arable cropping compared to forest ecosystems. The potential for NO_3^- leaching in different land use systems typically follow the order forest $<$ cut grassland $<$ grazed pastures, arable cropping $<$ ploughing of pasture $<$ vegetables (Di and Cameron 2002). On the basis of the global annual flux of groundwater of 11,000 km^3 $year^{-1}$, global sink of nitrogen in ground water are estimated at ~11 Tg N $year^{-1}$ (Schlesinger 1997). The amount of N leached from agricultural soils, on a global scale, is estimated at 55 Tg $year^{-1}$, contributing about 22 Tg $year^{-1}$ to the river export at the river estuary (van Drecht et al. 2003). The current contribution of deep groundwater flow to NO_3^- leaching, which is influenced largely by historical fertilizer use is in the order of 10%. Therefore, most of the current river export is due to recent development in fertilizer use such as Europe and Asia. Global estimates of nitrate leaching are complicated by numerous factors.

During the last few decades, N concentrations and loads have increased greatly in several rivers of the world (Seitzinger and Kroeze 1998), particularly of Europe (Tsirkunov et al. 1992; Isermann and Isermann 1997), North America (Gilliom et al. 1995) and Asia (Duan et al. 2000; Galloway 2000). The nitrogen transport to coastal systems by rivers increased from ~27 Tg N $year^{-1}$ in 1860 to ~50 Tg N $year^{-1}$ in 1990s (Galloway et al. 2004). The transport of inorganic nitrogen was estimated at ~35 Tg N $year^{-1}$, that of organic N at ~15 Tg N $year^{-1}$. Though this flux in rivers (~50 Tg N $year^{-1}$) is a small component of the terrestrial N cycle, yet it accounts for 60% of the total nitrogen delivered annually to oceans. On a global basis, the river-ocean continuum is a permanent nitrogen sink. However, the role of denitrification in stream systems is not completely understood.

Phosphorus being poorly soluble, large amounts of P cannot be lost through leaching except for soils with high water tables and those saturated with phosphorus. Annual losses of P through leaching and runoff are estimated to be only 0.01–0.6 kg P ha^{-1} in forests and grasslands (Cole and Rapp 1981; Taylor and Kilmer 1980). The natural river-borne export to ocean is about 1 Tg P $year^{-1}$ (Woodwell and Mackenzie 1995). Phosphorus is less abundant than N in fresh waters relative to plant needs, and its concentrations are reduced to very low levels by uptake during the growing season. It is, therefore, P that often regulates the extent of algal and other plant growth in the aquatic environment.

5.4.2 Eutrophication

Nutrient enrichment, particularly of nitrogen and phosphorus, is contributing to the phenomenon of eutrophication of headwater streams, freshwater lakes, rivers, and coastal and marine eco-systems. Submarine groundwater discharge or direct atmospheric deposition is the key source of nutrient loading in coastal systems lacking direct riverine discharge. Loss or degradation of riverine and coastal wetlands that trap and retain nutrients, can also contribute to nutrient loading in coastal waters. In freshwater systems, phosphorus is often the main cause of impairment, while nitrogen is generally associated with the impairment of coastal systems. Concentrations of 1–2 mg N L^{-1} and 0.1 mg P L^{-1} or lower in freshwater ecosystems, are considered to cause eutrophication, though specific concentrations depend on the local hydrological and climatic conditions (Camargo and Alonso 2006; Grizzetti et al. 2011; van de Bund 2009; Phillips et al. 2008). Pollution of marine waters by human activities besides threatening marine life is seriously damaging oceanic ecological services (Table 5.2). Elevated nutrient loads stimulate abundant growth of algae, phytoplankton, proliferation of gelatinous organisms and other aquatic plants leading to the development of harmful algal blooms, hypoxia and dead zones in inland lakes and coastal waters. Eutrophication of freshwaters depletes the dissolved oxygen to hypoxic levels with a dissolved oxygen concentration of 2–3 ml L^{-1}, which can suffocate all life in affected areas. Further depletion of dissolved oxygen concentration to lethal levels leads to development of dead zones, killing all marine organisms except bacteria. Overall, about half of the 650 waterways studied in the USA are affected by varying degrees of hypoxia. The occurrence of hypoxia has increased tenfold globally and 30-fold in the USA (Committee on Environment and Natural Resources 2010). Globally, a large number of mostly coastal and a few inland lake areas (Fig. 5.3) have been experiencing severe hypoxic conditions. Worldwide 415 coastal areas have been identified as eutrophic and hypoxic by the World Resources Institute's (WRI) Water Quality team. Of these, 169 are documented hypoxic areas, 233 are areas of concern and 13 are systems in recovery. The hypoxic

Table 5.2 Effects of eutrophication on coastal and inland waters

Increased biomass of phytoplankton
Shifts in phytoplankton to bloom-forming species that may be toxic or inedible
Increases in blooms of gelatinous zooplankton in marine environments
Increased biomass of benthic and epiphytic algae and bacteria
Development of rooted macrophytes and macroalgae along the shores
Oxygen depletion
Increased incidence of fish and shellfish mortality and reduced harvest
Loss of desirable fish species
Death of coral reefs and loss of coral reef communities
Decreases in water transparency
Taste, odor, and water treatment problems
Decrease in perceived aesthetic value of the water body

Adapted from Smith (2003) and Committee on Environment and Natural Resources (2010)

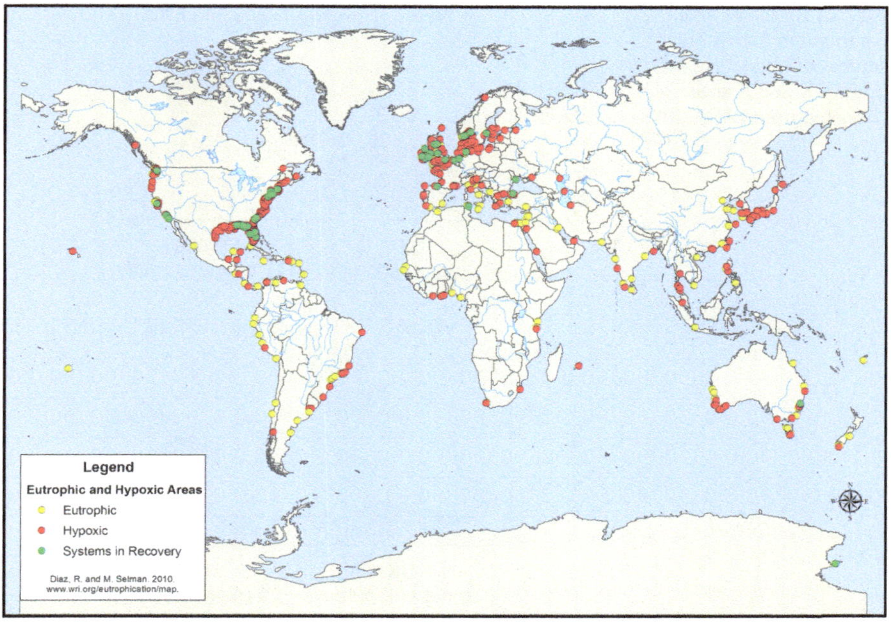

Fig. 5.3 Global hypoxic and eutrophic coastal areas (Source: WRI 2008, p. 3; with kind permission from World Resources Institute)

sites are located mostly adjacent to large cities with high population density and large discharge of treated or partially treated sewage and industrial effluents into water bodies, and sites receiving riverine nutrient loads from near or distant watersheds with N-intensive farming (Mathews and Hammond 1999). Both types of dead zones reduce the ecological health of coastal waters, thereby decreasing their ability to provide ecological services; they are also a major threat to all marine life including fish (Anderson et al. 2008). Fish kills and large migration of fishes out of hypoxic or dead zones have serious ecological consequences. Though it is difficult to determine the consequences of hypoxia on ecosystem services because of interaction among several stresses and impacts, yet the main ecosystems services influenced include loss of biomass, loss of biodiversity, loss of habitat, and alteration of energy and biochemical cycling (Committee on Environment and Natural Resources 2010).

5.5 Exposure and Health Risks

5.5.1 Nitrogen

The average daily intake of nitrate ions ranges between 50–140 mg in Europe and 40–100 mg in the US (MAFF 1995; Gregory et al. 1990; Gangolli et al. 1994). The median chronic nitrate exposure in children aged 1–3 years and 1–18 years from all

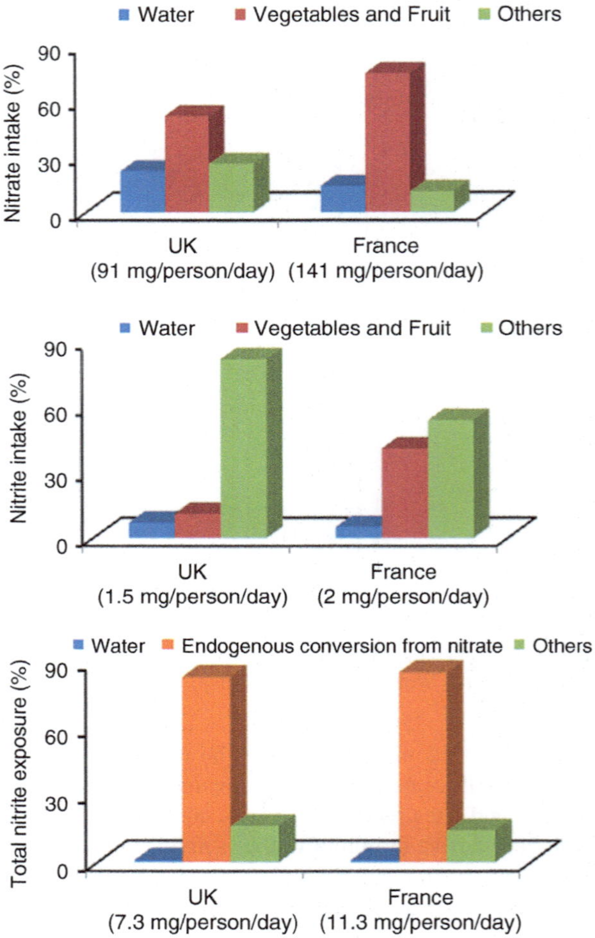

Fig. 5.4 Relative intake contribution for different sources of nitrate and nitrite in the UK and France (Adapted from EFSA 2008)

foods is estimated to be 1.00 and 1.39 mg kg^{-1} body weight day^{-1}, respectively. Vegetarians consume two to four times more nitrate than non-vegetarians (Walker 1996). Though vegetables and meat in the diet (nitrite is used as a preservative in many cured meats) are the most important sources of human exposure to nitrate and nitrite (see Chap. 6), yet drinking-water, in some situations such as bottle-fed infants, can make a significant contribution to nitrate intake. In addition to drinking water, nitrates may come from infant food prepared with nitrate contaminated water. The NO_3^- intake from water in the UK and France ranges between 14% and 22%, of the daily intake compared to 52% and 75% from fruits and vegetables (Fig. 5.4). This proportion is relatively higher (30–40%) in Asia and Africa because of lower intake from food sources (Santamaria 2006). Nitrite intake from water makes a small contribution; the majority of human exposure to nitrite comes from

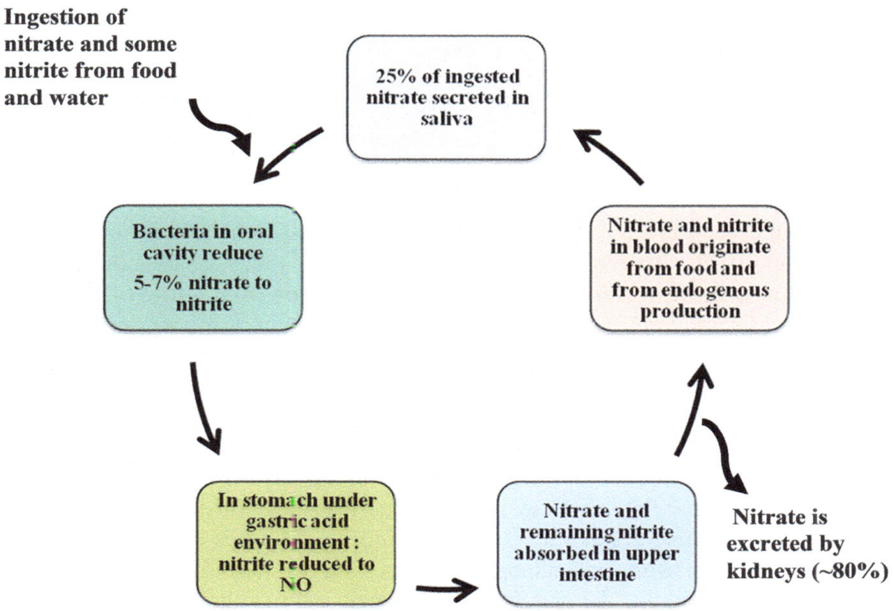

Fig. 5.5 Nitrate metabolism in human beings (Adapted from Graves 2012)

endogenous conversion of nitrate to nitrite (EFSA 2010). Excessive intake of nitrate may affect human health in several ways.

5.5.1.1 Nitrate Metabolism

Following ingestion in food or water, nitrate is absorbed rapidly and almost completely in the duodenum and jejunum (Spiegelhalder et al. 1976; Turek et al. 1980; Bartholomew and Hill 1984) and it does not reach the large intestine (Ellen and Schuller 1983). Peak nitrate concentrations in human serum, saliva and urine are attained within 1–3 h (Zetterquist et al. 1999). About 60–70% of the nitrate absorbed is excreted in urine. The half life of nitrate in plasma is 5–7 h (Wagner et al. 1983). In human beings, approximately 25% of the ingested nitrate is secreted in saliva by an active transport system in salivary glands (Fig. 5.5). About 20% of the secreted salivary nitrate (5–7% of the total ingested) is reduced to nitrite in the oral cavity by the action of nitrate-reducing bacteria at the base of the tongue (McKnight et al. 1999; Duncan et al. 1995). The oral microbial flora are influenced by nutritional status, infection, environmental temperature and age. Factors such as antibacterial mouth wash may lower the transformation of nitrate to nitrite (van Maanen et al. 1998). Under normal physiological conditions, the reduction of nitrate in the mouth is the most important source of nitrite, which accounts for approximately 70–80% of the human total nitrite exposure (Stephany and Schuller

1980). Salivary nitrite levels are usually higher in older age persons, though there could be considerable variations among individuals (Eisenbrand et al. 1980; Forman et al. 1985a, b). Significant bacterial reduction of nitrate to nitrite does not normally occur in the stomach, except in individuals with low gastric acidity or with gastrointestinal infections and those using antacids, particularly that block acid secretion. Infants younger than 3 months are highly susceptible to bacterial reduction of nitrate to nitrite in the stomach because of very little production of gastric acid and the occurrence of gastrointestinal infections (Ellen and Schuller 1983; Kross et al. 1992). Plasma nitrite levels are normally much smaller than nitrate levels because of the lower exposure and rapid oxidation from nitrite to nitrate by oxygenated haemoglobin in the blood. Therefore, the sum of nitrate and nitrite in blood is almost equal to the nitrate levels (Lundberg and Weitzberg 2005). In the stomach, the acidic conditions rapidly transform nitrite to nitrous acid, which in turn spontaneously decomposes to nitrogen oxides including nitric oxide.

In humans, nitrate is also synthesized endogenously by a biochemical pathway that involves the formation of nitric oxide from amino acid L-arginine breakdown catalysed by nitric oxide synthase, as part of normal metabolism (Eq. 5.1). Nitric oxide, being unstable reacts quickly to form nitrite, which is oxidized to nitrate through a reaction with oxyhaemoglobin (Eqs. 5.2, 5.3, and 5.4; Wishnok et al. 1995). In healthy adults, the endogenous synthesis leads to the excretion of about 62 mg of NO_3^- per day in the urine. Endogenous synthesis of nitrate or nitrite can be significantly increased in the presence of gastrointestinal infections (WHO 2011). Conditions such as low oxygen pressure, low pH and high nitrite concentration favour reduction of nitrite to nitric oxide by deoxyhaemoglobin (Cosby et al. 2003; Gladwin et al. 2005). Contrarily, oxidized haemoglobin will react with nitrite to form nitrate and methaemoglobin (Kosaka and Tyuma 1987). Normal levels of methaemoglobin in human blood are 1–3%, and reduced oxygen transport is observed at methaemoglobin concentration of 10% or more (Walker 1990; FAO/WHO 2003a, b). The balance between the two haemoglobin reactions produces nitric oxide at low oxygen pressure and the vasodilation induced by nitric oxide will increase blood flow to reverse the process (Jensen 2005).

$$\text{Arginine} + O_2 \xrightarrow[\text{Nitric oxide synthase}]{} \text{NO} + \text{Citrulline} \tag{5.1}$$

$$2\text{NO} + O_2 \rightarrow N_2O_2 \tag{5.2}$$

$$N_2O_4 + H_2O \rightarrow NO_3^- + NO_2^- + 2H^+ \tag{5.3}$$

$$NO_2^- + \text{OxyHb}\left(Fe^{2+}\right) \rightarrow \text{MetHb}\left(Fe^{3+}\right) + NO_3^- \tag{5.4}$$

5.5.1.2 Nitrate Toxicity

Nitrate *per se* is generally considered non-toxic. Its adverse effects on human health are due to conversion of nitrate into nitrite, nitric oxide and N-nitroso compounds. In the human body, NO_3^- is converted to NO_2^-, which can cause

Table 5.3 Methaemoglobin concentrations and clinical symptoms in humans

Methaemoglobin concentration (%)	Clinical symptoms
10–20	Central cyanosis of limbs/trunk; often asymptomatic, may have weakness, tachycardia
20–35	Central nervous system depression, dyspnoea, nausea, dizziness, headache, fatigue
35–55	Lethargy, syncope, coma, arrhythmias, shock, convulsions
>70	High risk of mortality

Source: Kross et al. 1992; adapted from Dabney et al. 1990; EFSA 2010

methaemoglobinemia (MHG), by interfering with the ability of haemoglobin to take up O_2. Nitrite ions react with haemoglobin in the red blood cells to form methaemoglobin (MetHb) by oxidizing the iron moiety in haemoglobin from Fe^{2+} to Fe^{3+}, which binds oxygen and blocks its transport. Normal levels of MetHb in human blood are 1–3%, and reduced oxygen transport is observed at MetHb concentration of 10% or more (Walker 1990; FAO/WHO 2003a, b). Methaemoglobin concentration greater than 10% in infants can give rise to cyanosis, commonly referred to as "blue-baby syndrome". The clinical symptoms of MHG vary depending on MetHb level and there is a high risk of mortality at concentration greater than 70% (Table 5.3). Though MHG can occur in all age groups, but bottle-fed infants and elderly (>45 years) are more susceptible (Gupta et al. 2000). Infants younger than 3 months of age are particularly prone to adverse effects of nitrate exposure, usually due to formula baby-food diluted with water from rural domestic wells (Dusdieker and Dungy 1996) and interplay of a number of physiological factors including, (i) presence of a proportion of infant haemoglobin in the form of foetal haemoglobin that is readily oxidised to MetHb by nitrite, (ii) low activity of the enzyme NADH-dependent methaemoglobin reductase responsible for reduction of MetHb to normal haemoglobin, and (iii) relatively high pH of the infant stomach favouring the growth of nitrate-reducing bacteria that can lead to gastroenteritis with concomitant increase in nitrite formation (Johnson and Kross 1990; Smith 1991; Zeman et al. 2002; ATSDR 2004). These factors rapidly diminish in importance after 3 months of age.

Opinions differ on the effects of high dietary intake of nitrate on human health and the evidence linking high nitrate level in drinking water with MHG is still controversial (Powlson et al. 2008). The occurrence of MHG was first reported in bottle-fed infants from Iowa, USA (Comly 1945). Later, it was found that the wells that provided water for bottle-feeding infants were contaminated with human and animal excreta and the water contained considerable number of bacteria as well as high concentration of nitrate (Avery 1999). This indicates that MHG in bottle-fed infants resulted from the presence of bacteria rather than nitrate *per se*. It has been argued that gastroenteritis resulting from bacteria in the well-water, stimulated nitric oxide production in the gut, which reacted with oxyhaemoglobin in blood and converted it into methaemoglobin (Addiscott 2005). Therefore, the available evidence indicates that though drinking-water nitrate may be an important risk factor for MHG in bottle-fed infants, the risk is increased in the presence of simultaneous gastrointestinal infections, because of increased endogenous nitrite formation and reduction of nitrate to nitrite and higher intake of water for combating dehydration.

Besides gastro-intestinal infections, MHG has also been associated with a number of other conditions including inflammation, acidosis, exposure to a number of drugs including topical anaesthetic agents, silver nitrate, chloroquin, sulfonamides, dapsone, phenacetin, sodium valproate, phenazopyridine, inhaled nitrous oxide, amyl nitrite, as well as acute nitrite toxicity resulting from accidental exposure to aniline dyes, colouring compounds or cleaning solutions (Gupta et al. 1998; Levine et al. 1998; Sanchez-Echaniz et al. 2001). Diarrhoea produces acidosis that increases MetHb production and impairs the MetHb reductase systems (Yano et al. 1982).

Nitrites formed on reduction of ingested nitrates due to the saliva at the back of the tongue (Eisenbrand et al. 1980) and in other parts of the alimentary canal react with amines and amides and form *N*-nitroso compounds, which have been linked with the risk of cancer of stomach, oesophagus and urinary bladder in humans (IARC 1978). Several studies have been carried out on the formation of *N*-nitroso compounds in relation to nitrate intake in humans, but there is large variation in the intake of nitrosatable compounds and in gastric physiology. Higher levels of *N*-nitroso compounds, along with high nitrate levels, have been found in the gastric juice of individuals who are achlorhydric (very low levels of hydrochloric acid in the stomach). However, other studies have been largely inconclusive, and there appears to be no clear relationship with drinking-water nitrate compared with overall nitrate intake. Recent studies suggest that concerns about the formation of potentially carcinogenic N-nitrosamines from dietary NO_2^- and NO_3^- are exaggerated, especially considering the relatively high concentration of these anions in the body from diets associated with health, not disease (Lundberg et al. 2004). High nitrate level in drinking water has also been implicated, but not proved beyond doubt, in the incidence of cancers of the digestive track. About 50 epidemiological studies conducted since 1975 examining the association between nitrate and stomach cancer did not find a convincing link between nitrate and incidence of stomach cancer (Powlson et al. 2008).

Besides MHG, high nitrate ingestion has also been associated with acute respiratory tract infection (Gupta et al. 2000), increased risk of spontaneous abortion or certain birth defects in infants (Fewtrell 2004), reproductive problems and cancer (Kramer et al. 1996; Nolan 1999). Nitrate has been associated with ovarian and liver cancers (Tsezou et al. 1996; Mueller et al. 2001; Weyer et al. 2001). In England, an ecological study found an increased incidence of adult brain and central nervous system tumours in areas with high drinking water nitrate levels, but a number of case-control studies did not find any relationship (Johnson and Kross 1990).

Though there is no clear association between nitrate in drinking water and the adverse human health effects, evidence is emerging of possible benefits in cardiovascular health and providing protection against infections. Nitrate is considered to have a role in protecting the gastrointestinal tract against a variety of pathogens, as nitrous oxide and acidified nitrite have antibacterial properties. It is particularly effective in the stomach against *Salmonella*, *Escherichia coli* and other organisms that cause gastroenteritis. It also acts in our mouth against dental caries and even on our skin against fungal pathogens such as *Tinea pedis* (athlete's foot). It may have other beneficial physiological roles. Recent research indicates that nitrite participates in host defence because of antimicrobial activity and it protects against a number of problems including ischemia-reperfusion injury (in heart attack and stroke, for example), kidney injuries and hypertension (Carlström et al. 2011). Other nitrate metabolites e.g. nitric oxide, have important physiological roles such as vasoregulation (EFSA 2008). It is generally suspected that because of antimicrobial salivary NO and other compounds, animals lick their wounds. The presence of NO has been shown to increase both mucosal blood flow and mucus generation in the gastrointestinal tract (Petersson et al. 2007). Nitrite and its conversion to NO in the stomach also have an important role in promoting neonatal health (Berens and Bryan 2011). Human breast milk is known to contain nitrate and nitrite, but the breast milk during the first 3 days (colostrum) contains more nitrite than nitrate, which presumably provides more NO in the baby's stomach because of lack of nitrate reducing bacteria in the mouth and stomach of the baby. This ensures protection against infection to the newborn. Finally, it is important to note that increased consumption of vegetables is widely recommended because of their beneficial effects for health notwithstanding that these are a major source of nitrate (EFSA 2008). There may, therefore, be a benefit from exogenous nitrate uptake, but there is a need to balance the potential risks with the potential benefits. Some scientists have suggested that the regulatory limit for nitrate is very conservative and needs to be increased without enhancing the risk to human health. Rather it would relieve many small rural communities of a significant economic burden without adding appreciably to any known health risks (L'hirondel et al. 2006). On the other hand, it has been argued that sub-groups within a population may be more susceptible than others to the adverse health effects of nitrate and individuals with high rates of endogenous formation of N-nitroso compounds are likely to be susceptible to the development of cancers in the digestive track. Adverse health effects may be the result of a complex interaction of the amount of nitrate ingested, the concomitant ingestion of nitrosation cofactors and precursors, and specific medical conditions that increase nitrosation (Ward et al. 2005).

5.5.1.3 Indirect Effects on Human Health

Greater availability of nitrogen leads to a cascade of ecological impacts, which may induce varied changes in the epidemiology of vector-borne human diseases. High concentrations of N and P in surface waters may help the growth of some disease-causing agents such as mosquito, snail, etc. that promote the spread of certain infectious diseases such as malaria, cholera, schistosomiasis and West Nile virus. Positive relationships between concentration of inorganic N in surface water and larval abundance of *Anopheles* sp. mosquitoes carriers of malaria, as well as for *Culex* sp. and *Aedes* sp., carriers of La Crosse encephalitis, Japanese encephalitis, and West Nile virus have been reported (Rejmankova et al. 1991; Walker et al. 1991; Teng et al. 1998).

5.5.2 Phosphorus

Phosphates are not toxic to people or animals unless they are present in very high levels. Because of this, no drinking water standards have been established for phosphorus. Digestive problem could occur from extremely high level of phosphate. Occurrence of phosphorus in water can influence human health indirectly through its role in eutrophication. The eutrophication of coastal and marine eco-systems leading to harmful algal blooms (HABs) has been linked to neurological, amnesic, paralytic, and/or diarrheic shellfish poisoning, as well as toxins produced by various cyanobacteria (known as blue green algae), and by the estuarine dinoflagellates *Pfiesteria piscicida* and *P. shumwayii* (Burkholder 1998). Under favorable light and nutrient conditions, some species of cyanobacteria produce secondary toxic metabolites, known as cyanotoxins. There are about 40 genera that produce toxins but the main ones are *Anabaena*, *Aphanizomenon*, *Cylindrospermopsis*, *Lyngbya*, *Microcystis*, *Planktothrix* (*Oscillatoria*) and *Nostoc* (Carmichael 2001). Both non-toxic and toxic species of the common toxin-producing cyanobacteria exist, but it is not possible to visually differentiate a toxic species from a non- toxic one. Moreover even if toxin- producing cyanobacteria are present, they may not actually produce toxins (US EPA 2014).

Cyanotoxins fall into three broad groups of chemical structure *viz.* cyclic peptides, alkaloids, and lipopolysaccharides. Exposure to cyanobacteria and their toxins could occur by ingestion of cyanotoxins-contaminated drinking water and through direct contact, inhalation and/or ingestion during recreational activities (US EPA 2014). These toxic substances can cause diverse human health effects ranging from hepatotoxic, neurotoxic and dermatotoxic to general inhibition of protein synthesis (Table 5.4). Recently, there have been studies of their effects in other systems, including hematological, kidney, cardiac, reproductive, and gastrointestinal effects. Hepatotoxins, the commonly occurring cyanotoxins that affect liver, can cause death by liver haemorrhage or cardiac failure within hours of acute exposure (Stillman 2010). Neurotoxins, which are relatively less common in occurrence, affect the nervous systems and can cause death in mice and aquatic birds due to respiratory arrest. Dermatotoxins that affect the skin induce irritant and allergenic responses in any tissue they come in contact and they may also cause damage to nervous, digestive, respiratory, and cutaneous systems (Stillman 2010). Available evidence suggests that long-term exposure to low levels of microcystins and cylindrospermopsin may promote cell proliferation and the growth of tumors.

Table 5.4 Cyanotoxins, cyanobacterial genera and the primary organ affected in mammals

Chemical structure group	Cyanotoxin	Primary organ affected	Cyanobacterial genera
Cyclic peptides	Microcystins	Liver	*Microcystis, Anabaena, Planktothrix (Oscillatoria), Nostoc, Hapalosiphon, Anabaenopsis, Aphanizomenon*
	Nodularin	Liver	*Nodularia*
Alkaloids	Anatoxin-a	Nerve synapse	*Anabaena, Planktothrix (Oscillatoria), Aphanizomenon*
	Anatoxin-a (S)	Nerve synapse	*Anabaena*
	Aplysiatoxins	Skin	*Lyngbya, Schizothrix, Planktothrix (Oscillatoria)*
	Cylindrospermopsins	Liver	*Cylindrospermopsis, Aphanizomenon, Umezakia, Anabaena, Rhaphidiopsis*
	Lyngbyatoxin-a	Skin, gastro-intestinal tract	*Lyngbya*
	Saxitoxins	Nerve axons	*Anabaena, Aphanizomenon, Lyngbya, Cylindrospermopsis*
Lipopolysaccharides	–	Irritant, affects any exposed tissue	All

Adapted from Chorus and Bartram (1999)

However, more information is required to determine the carcinogenicity of microcystins and cylindrospermopsin (US EPA 2014).

Symptoms of cyanotoxity include fatigue, headaches, fever, muscle and joint pain, blisters, stomach cramps, diarrhoea, vomiting, mouth ulcers, sore throat, skin irritations and allergic reactions (Volterra et al. 2002). Such effects can occur within minutes to days after exposure. In severe cases, seizures, liver failure, respiratory arrest, and in rare cases death may occur (US EPA 2014). The HABs can also indirectly affect humans by disrupting freshwater and marine ecosystems and sources of nutrition derived from them. For example, eating of shellfish and other seafood containing bioaccumulated toxins can cause acute and potentially fatal effects. The symptoms of diarrheal shellfish poisoning include diarrhea, vomiting, and abdominal pain and that of paralytic shellfish poisoning include muscular paralysis, difficulty in breathing, shock, and, in extreme cases, death by respiratory arrest.

Harmful algal blooms are often referred to as "red tides" or "brown tides" because of the appearance of the water when these blooms occur. Decaying algal biomass produces surface scum, odors, and increased populations of insect pests. Increased N and P can also increase the availability of other nutrients that can facilitate algal blooms of many harmful species (NRC 2000) and enhanced survival of pathogenic bacteria such as *Escherichia coli*, *Salmonella* spp., and *Vibrio cholerae*. Under normal conditions, these pathogenic species do not survive because of lack of nutrients, damage from UV light, and differences in osmolarity between sea water and the bacteria themselves, which cause the bacteria cells to rupture (Stillman 2010). During an algal bloom, these conditions are reversed; food is abundant, light is diminished, and osmo-protective chemicals are produced by algae. The bacterium *Vibrio cholerae* is associated with a wide range of marine life, and outbreaks of cholera have been associated with coastal algal blooms (Colwell and Huq 2001; Cottingham et al. 2003).

5.6 Mitigation Options

Many countries have set limits on permissible concentration of nitrate in drinking water and in surface waters. The safe limit of nitrates in drinking water is set at 50 mg of nitrate per litre (11.3 mg L^{-1} as nitrogen) in the EU and 44 mg L^{-1} (10 mg L^{-1} as nitrogen) in the USA. These limits are in accordance with the recommendations of World Health Organization (WHO 2004). The guideline values for nitrate are derived from epidemiological studies wherein MHG was not reported in bottle-fed infants in areas where drinking-water consistently contained less than 50 mg of nitrate per litre. The available data related to nitrate in drinking water and in vegetables indicated that MetHb is not elevated in children or infants above 3 months old when exposure to nitrate is below at least 15 mg kg^{-1} body weight per day. The value set for nitrite is 3 mg nitrite or 1 mg as nitrogen per litre of water (EFSA 2008). This is supported by accepting a relative potency for nitrite and nitrate with respect to MetHb formation of 10:1 (on a molar basis). The guideline values for nitrite in bottle-fed infants have been derived based on human data showing that doses of nitrite that cause MHG in infants range from 0.4 to more than 200 mg kg^{-1} body weight. Using the lowest level of the range (0.4 mg kg^{-1} body weight), a body weight of 5 kg for an infant and a drinking-water consumption of 0.75 l, a guideline value of 3 mg L^{-1} is derived. Since nitrate and nitrite can occur simultaneously in drinking-water, the sum of the ratios of the concentration (*C*) of each to its guideline value (*GV*) should not exceed 1 (Eq. 5.5; WHO 2011):

$$\frac{C_{nitrate}}{GV_{nitrate}} + \frac{C_{nitrite}}{GV_{nitrite}} \leq 1 \tag{5.5}$$

Joint FAO/WHO Expert Committee on Food Additives proposed acceptable daily intakes of 0–3.7 mg kg^{-1} body weight for nitrate and 0–0.07 mg kg^{-1} body weight for nitrite (EFSA 2008). For a 60 kg adult, this translates into an exposure of 222 mg nitrate and 3.6 mg nitrite. Since these values were derived from laboratory animals, there is considerable uncertainty with regard to their relevance to humans. Therefore, these values are considered provisional and have now been suspended. Similar to these guideline values, US EPA has set a reference dose of 7 mg nitrate ion per kg body weight per day. The regulatory level is usually met for public water supplies, which are routinely monitored. Currently, 2–3% of the population in the USA and the EU are potentially exposed to drinking water exceeding the standards for nitrate in drinking water. The proportion of population exposed to high nitrate concentration in drinking water is probably higher in developing countries (van Grinsven et al. 2006).

The US EPA has also issued standards for both acute and chronic toxicity of certain juvenile fish species to elevated levels of dissolved ammonia in surface water (US EPA 1999). Dissolved ammonia is not considered a direct human health concern in drinking water because it rarely occurs at very high concentrations (WHO 1996). The presence of ammonia in excess of 0.165 mg ammonia-N L^{-1} (0.2 mg NH_3 L^{-1}) can significantly reduce the efficiency of chlorination of drinking water supplies (WHO 1996). Elevated levels of dissolved ammonia are sometimes found downstream of wastewater treatment sites, septic systems, and agricultural sites receiving fertilizer (Parker et al. 2012; Lehman et al. 2004).

Though there are no guidelines for phosphorus in drinking water, but the USEPA has recommended a limit of 0.05 mg L^{-1} for total phosphorus in streams that enter lakes and 0.1 mg L^{-1} for total phosphorus in flowing streams for controlling eutrophication (US EPA 1986). Acceptable levels of phosphorus in surface runoff from agricultural fields have not been established.

5.6.1 *Environmental Protection Measures*

5.6.1.1 Nitrogen

The principal means of controlling nitrate contamination in groundwater is by proper management of sources of nitrate. It is easier to address nutrient load problems from point sources such as municipal and industrial facilities than from the diffuse, nonpoint sources like agricultural and urban land runoff (Committee on Environment and Natural Resources 2010). The strategies to reduce leakage of nitrate to surface and groundwater include adoption of appropriate agricultural management practices, improved fertilizer and manure management as well as storage of animal manures careful location of pit latrines and septic tanks, and control of sewer leakage (WHO 2011). Septic tanks and pit latrines should not be located near a well or where a well is to be dug. Animal manure must be kept at a distance from the well so that runoff does not enter the well or the areas near the well. The well should be suitably protected to prevent runoff from entering the well. Manures and fertilizers on land area near wells should be carefully managed to avoid potential contamination. Approaches to mitigate nitrate contamination in piped water supplies include dilution of contaminated water with a low-nitrate source water, disinfection of water to oxidize nitrite to the less toxic nitrate and to minimize the pathogenic and non-pathogenic reducing bacterial population in the water, and removal of nitrate through ion exchange (usually for groundwater) and biological denitrification (usually for surface waters). Ion exchange method has problems with regeneration and disposal of spent generated. On the other hand, operation of biological denitrification is complex and there is a possibility for microbial and carbon feed contamination of the final water. As per WHO (2011) a concentration of 5 mg NO_3^- L^{-1} or lower should be achievable using biological denitrification or ion exchange and a concentration of 0.1 mg NO_2^- L^{-1} should be achievable using chlorination (to form nitrate).

Another approach to control nitrate contamination of groundwater and surface water is to minimize agriculture induced leaching and run-off losses of nitrogen. Practices that maximize N and water use efficiency help reduce the leaching and run-off losses. Measures to reduce NO_3^- leaching include optimal and balanced fertilization, synchronizing N supply with plant demand, manipulation of water applications and rooting depth, appropriate cropping sequence, use of cover crops, and the use of slow release fertilizers and nitrification inhibitors (Benbi et al. 1991; Prihar et al. 2000). Balanced application of N, P and K under intensive agriculture has been shown to significantly reduce the amount of unutilized nitrates in the soil profile by enhancing N recovery in crop plants (Benbi et al. 1991). Development of suitable irrigation schedules with respect to timing and amount so as to synchronize the NO_3^- rich zone with the moist zone of high root activity can help efficient N uptake by plant. It has been reported (Pratt 1984) that when roots have access to the entire soil solution, nitrate is not leached unless excess fertilizer N is added or the soil is over irrigated. Other options to reduce leaching losses of nitrate from agricultural soil include use of slow release fertilizers and nitrification inhibitors. Slow release fertilizers such as urea-formaldehyde, isobutylidene diurea, sulphur coated urea control nitrification by slowing down the rate at which NH_4^+ is made available for nitrification. However, the use of slow release fertilizers has been limited because of their high cost and possible mismatch between nitrogen availability and crop demand. Nitrification inhibitors such as Dicyandiamide (DCD), N-serve and calcium carbide, which retard the formation of NO_3^- by nitrifying bacteria, are known to increase the fertilizer use efficiency provided that maintaining the added fertilizer N as NH_4^+ does not lead to increased ammonia volatilization. The beneficial effect of nitrification inhibitors in reducing N leaching losses depend on soil type and time and rate of N application. In addition to fertilizer management, the choice of appropriate cropping sequence wherein heavily fertilized shallow-rooted crops are followed by low-nutrient requiring deep-rooted crops can minimize residual NO_3-N accumulation in the soil profile and thus reduce the risk of leaching. Measures to reduce NO_3^- leaching from grasslands and pastures include avoiding ploughing or better timing of ploughing pasture leys, removing stock from the fields earlier in the grazing seasons, improved stock management, precision farming and split application of fertilizer nitrogen. Leakage of nitrate from stored animal manure and livestock waste into local water bodies can be minimized by integrating livestock and crop farming. Animal manure can be treated in mesophilic or thermophilic digesters to recover plant nutrients and energy contained in manure. Though a number of management options are available for reducing nitrate leaching but the choice of a strategy will depend on soil, environmental and cultural variables.

Nutrient transport to surface waters from agricultural lands can be reduced by decreasing runoff and increasing the rate and duration of infiltration of water into the soil. On sloping lands, the runoff can be reduced by decreasing the velocity of runoff water and/or increasing the surface roughness. Velocity of runoff can be decreased by mechanical and biological interventions. While mechanical interventions include field and contour bunding, land levelling, terracing and contour cultivation of fields; biological measures include growing of narrow-row crops, strip cropping, mulching and vegetative bunds (Prihar et al. 2000). Conservation and minimum tillage, mulches, and cover crops prevent runoff initiation by intercepting raindrops. High infiltration rates are maintained and, therefore, runoff and erosion are minimized. Infiltration of water into soil can be increased by improving and protecting the structure of surface soil and increasing the proportion of large continuous pores in the profile by breaking impervious soil layers. Practices for retaining runoff on cropland include contour tillage, furrow dikes, level terraces, and land leveling.

5.6.1.2 Phosphorus

Since agricultural P involving application of fertilizers and manures is the principal source of P in water, the mitigation strategies should aim at minimizing losses of phosphorus and maximizing its use efficiency in arable land and the food chain. In the short term, recovery of fertilizer P in crops is often only 10–20%. Best management practices should be followed to maximize recovery and reuse of phosphorus, mostly of animal and human excreta. Strategies to improve P use efficiency and minimizing accumulation and losses in soil include (i) judicious application of fertilizer P based on soil test, (ii) placement or incorporation rather than broadcasting of fertilizer and manure, (iii) proper time of fertilizer application, and (iv) precision farming to account for spatial variability of P distribution in a field. Indirect measures to reduce P input and losses include proper choice of land use, balanced fertilization, and adoption of conservation agriculture to control soil erosion. Wherever found effective, use of P solubilising bacteria and suitable mycorrhizal fungi can help reduce fertilizer input. In livestock systems, reducing P losses will require maximizing use of P in manure for soil fertility in croplands and pastures, and manipulating livestock diets such as by adding phytase enzyme to enhance utilization of phytate that could subsantially reduce P excretion by non-ruminants (pigs). Another important strategy is to reduce the intake of animal foods whose production requires first high P inputs in growing the requisite feed and then entails unavoidably large P losses in animal wastes. Efforts should be made to reduce or ban P in detergents so as to reduce eutrophication of lakes. However, it has been argued that since detergents contribute only small fraction of P load compared with other urban or agricultural sources this may not significantly influence eutrophication (Lee and Jones 1986). While phosphorus use in detergents has been totally banned in some countries (Switzerland and some states of the

United States), and partial restrictions occur in others, the use of high-P detergents is unrestricted in a number of other countries.

Nutrient inputs to rivers and streams and pollution of aquatic systems may be reduced by establishing nutrient criteria, developing tools and management strategies for reducing nutrient loadings, and restoring and enhancing natural nutrient retention processes. For example, establishing buffer zones with trees around heavily fertilized fields or along streams and water bodies can reduce the runoff into streams. Similarly, checking uncontrolled runoff of livestock waste can substantially reduce the P loading of water bodies. Nutrient fluxes to coastal ecosystems can be reduced when nutrients present in streams are attenuated by natural processes occurring in riverine wetlands. Eutrophication can be reversed by decreasing input rates of P and N to aquatic ecosystems, but rates of recovery are slow and highly variable among water bodies (Carpenter et al. 1998). In fact, diverse actions are required to mitigate nutrient losses and reduce eutrophication. Based on review of literature and interviews with eutrophication experts, the World Resources Institute (WRI) proposed several policies and action plans to reduce eutrophication (Selman and Greenhalgh 2009). These include (i) implementation of research and monitoring programs to characterize the effects of eutrophication and enable the adoption of management strategies; (ii) creating awareness among public and policymakers about eutrophication and its harmful effects; (iii) enforcing regulation to mitigate nutrient losses; (iv) incentivising nutrient-reducing practices; (v) preserving and restoring natural ecosystems such as wetlands and forests; (vi) creating institutional support to implement and enforce policies; (vii) drawing advantage from environmental synergies while designing eutrophication policies. While developing strategies and policies for addressing eutrophication, it is important to reckon other environmental aspects that have synergies with water quality.

5.6.2 *Preventive Health Care*

A number of preventive actions can be taken to minimize the adverse effects associated with elevated concentrations of nitrate and microbial contamination in drinking water (WHO 2011). Since MHG in infants appears to be associated with simultaneous diarrhoeal disease, the authorities should make it sure that water to be used for bottle-fed infants is microbiologically safe when nitrate is present at concentrations near the guideline value or in the presence of endemic infantile diarrhoea. As a preventive health care strategy, water should be boiled or disinfected before consumption. Where alternative microbiologically-safe supplies are available, these can be used for bottle-fed infants. Water with nitrate concentration above 100 mg L^{-1} should not be used for bottle-fed infants. However, the water can be used, under increased vigilance by medical authorities, if the concentration is between 50 and 100 mg L^{-1} and the water is microbiologically safe. Steps should then be taken to prevent pollution of well-water by removing sources of nitrate and microbial contamination from the vicinity of the well. In areas where household wells are common, health authorities should provide appropriate information about water safety, assist with visual inspection of wells, provide testing facilities where a problem is suspected, provide guidance on disinfecting water or where nitrate levels are high, provide bottled water from safe sources or disseminate information as to where such water can be obtained. Since contamination of drinking water is also associated with leaky sewage systems, there is a need to improve these systems for general hygienic conditions. This need is particularly important for developing countries which do not have well- established sewage and waste disposal infrastructure (Powlson et al. 2008).

A number of measures can be taken to remove or inactivate cyanotoxins in the water supply system (Table 5.5). However, effectiveness of different treatment

Table 5.5 Effectiveness of different treatments for removal of cyanotoxins

Treatment process	Effectiveness
Removal of intracellular cyanotoxins (intact cells)	
Pre-treatment oxidation	Oxidation often lyses cyanobacteria cells releasing the cyanotoxin in water. If oxidation is necessary to meet other treatment objectives, use low dose of an oxidant less likely to lyse cells (e.g. potassium permanganate); otherwise use sufficiently high dose to lyse cells as well as destroy total toxins
Coagulation/sedimentation/filtration	Effective when cells accumulated in sludge are isolated from the plant and the sludge is not returned to the supply after sludge separation
Membranes	Microfiltration and ultrafiltration effective when cells are not allowed to accumulate on membranes for long periods of time
Flotation	Effective for removal of intracellular cyanotoxins
Removal of extracellular cyanotoxins (dissolved)	
Membranes	Depends on the material, membrane pore size distribution, and water quality. Nano-filtration effective in removing extracellular microcystin. Reverse osmosis filtration effective for removal of extracellular microcystin and cylindrospermopsin. Cell lysis highly likely
Potassium permanganate	Effective for oxidizing microcystins and anatoxins
Ozone	Very effective for oxidizing extracellular microcystin, anatoxin-a, and cylindrospermopsin
Chlorination	Effective for oxidizing extracellular cyanotoxins if pH less than 8; ineffective for anatoxin-a
UV Radiation	Effective at degrading microcystin and cylindrospermopsin but at impractically high doses
Powdered activated carbon	Effectiveness depends on type of carbon and pore size. Wood-based activated carbon most effective at microcystin adsorption. Not as effective at adsorbing saxitoxin or taste and odor compounds. Doses in excess of 20 mg l^{-1} required for complete toxin removal
Granular activated carbon	Effective for microcystin but less effective for anatoxin-a and cylindrospermopsins
Chloramines	Not effective
Chlorine dioxide	Not effective at doses used in drinking water treatment

Adapted from US EPA (2014)

strategies varies depending on the growth pattern and species of cyanobacteria that dominates the bloom, as well as the properties of the cyanotoxins (i.e., intracellular or extracellular). Application of an inappropriate treatment process can damage cells and lead to release rather than removal of cyanotoxins (US EPA 2014). The strategies that are effective in removing intracellular cyanotoxins include coagulation, flocculation, sedimentation and filtration (micro- and ultra-filtration). Coagulation, flocculation and dissolved air flotation are more effective than sedimentation. Effective techniques for the removal of extracellular toxins include activated carbon, membrane filtration (nano-filtration and reverse osmosis) and chemical inactivation (Ultraviolet (UV), disinfectants and oxidants). Besides treatment for removal of cyanotoxins, drinking water utilities having access to more than one source can switch to an alternate one that is not as severely impacted by the bloom or adjust intake depth to avoid drawing contaminated water and cells into the treatment plant.

References

Addiscott TM (2005) Nitrate, agriculture, and environment. CABI Publishers, Wallingford

Addiscott TM, Cox D (1976) Winter leaching of nitrate from autumn-applied calcium nitrate, ammonium sulphate, urea and sulphur-coated urea in bare soil. J Agric Sci (Camb) 87:381–389

Anderson DM, Burkholder JM, Cochlan WP, Glibert PM, Gobler CJ, Heil CA, Kudela R, Parsons ML, Jack Rensel JE, Townsend DW, Trainer VL, Vargo GA (2008) Harmful algae blooms and eutrophication: examining linkages from selected coastal regions of the United States. Harmful Algae 8:39–53

ATSDR (Agency for Toxic Substances and Disease Registry) (2004) Interaction profile for cyanide, fluoride, nitrate, and uranium. US Department of Health and Human Services, Atlanta

Avery AA (1999) Infantile methemoglobinemia: reexamining the role of drinking water nitrates. Child Health 107:583–586

Bartholomew B, Hill MJ (1984) The pharmacology of dietary nitrate and the origin of urinary nitrate. Food Chem Toxicol 22:789–795

Beaulac MN, Reckhow KH (1982) An examination of land use-nutrient export relationships. Water Resour Bull 18:1013–1022

Benbi DK (1990) Efficiency of nitrogen use by dryland wheat in a subhumid region in relation to optimizing the amount of available water. J Agric Sci (Camb) 115:7–10

Benbi DK, Biswas CR, Kalkat JS (1991) Nitrate distribution and accumulation in an Ustochrept soil profile in a long-term fertilizer experiment. Fertil Res 28:173–178

Bennett EM, Carpenter SR, Caraco NF (2001) Human impact on erodable phosphorus and eutrophication: a global perspective. Bioscience 51:227–234

Berens PD, Bryan NS (2011) Nitrite and nitrate in human breast milk: implications for development. In: Bryan NS, Loscalzo J (eds) Nitrite and nitrate in human health and disease. Humana Press, New York, pp 139–153

Billen G, Garnier J, Lassaletta L (2013) The nitrogen cascade from agricultural soils to the sea: modelling N transfers at regional watershed and global scales. Philos Trans R Soc B 368:20130123. https://doi.org/10.1098/rstb.2013.0123

Bouwman AF, Beusen AHW, Billen G (2009) Human alteration of the global nitrogen and phosphorus soil balances for the period 1970–2050. Glob Biogeochem Cycles 23:GB0A04

Bouwman L, Goldewijk KK, van der Hoek KW, Beusen AHW, van Vuuren DP, Willems J, Rufino MC, Stehfest E (2011) Exploring global changes in nitrogen and phosphorus cycles in agriculture induced by livestock production over the 1900–2050 period. Proc Natl Acad Sci US 109:6348–6353

Burkholder JM (1998) Implications of harmful microalgae and heterotrophic dinoflagellates in management of sustainable marine fisheries. Ecol Appl 8:S37–S62

Camargo JA, Alonso A (2006) Ecological and toxicological effects of inorganic nitrogen pollution in aquatic ecosystems: a global assessment. Environ Int 32:831–849

Carlström M, Persson AE, Larsson E, Hezel M, Scheffer PG, Teerlink T, Weitzberg E, Lundberg JO (2011) Dietary nitrate attenuates oxidative stress, prevents cardiac and renal injuries, and reduces blood pressure in salt-induced hypertension. Cardiovasc Res 89:574–585

Carmichael WW (2001) Health effects of toxin-producing cyanobacteria: "The CyanoHABs". Hum Ecol Risk Assess 7:1393–1407

Carpenter SR, Caraco NF, Correll DL, Howarth RW, Sharpley AN, Smith VH (1998) Nonpoint pollution of surface waters with phosphorus and nitrogen. Ecol Appl 8:559–569

Chorus I, Bartram J (eds) (1999) Toxic cyanobacteria in water: a guide to their public health consequences, monitoring and management. World Health Organization, Geneva

Cole DW, Rapp M (1981) Elemental cycling in forest ecosystems. In: Reichle DE (ed) Dynamic properties of forest ecosystems. Cambridge University Press, New York, pp 341–403

Colwell R, Huq A (2001) Marine ecosystems and cholera. Hydrobiologia 460:141–145

Comly HH (1945) Cyanosis in infants caused by nitrates in well water. J Am Med Assoc 129:112–116

Committee on Environment and Natural Resources (2010) Scientific assessment of hypoxia in U.S. coastal waters. Interagency Working Group on Harmful Algal Blooms, Hypoxia, and Human Health of the Joint Subcommittee on Ocean Science and Technology, Washington, DC

Cosby K, Partovi KS, Crawford JH, Patel RP, Reiter CD, Martyr S, Yang BK, Waclawiw MA, Zalos G, Xu X, Huang KT, Shields H, Kim-Shapiro DB, Schechter AN, Cannon RO III, Gladwin MT (2003) Nitrite reduction to nitric oxide by deoxyhemoglobin vasodilates the human circulation. Nat Med 9:1498–1505

Cottingham KL, Chiavelli DA, Taylor RA (2003) Environmental microbe and human pathogen: the ecology and microbiology of *Vibrio cholerae*. Front Ecol Environ 2:80–86

Dabney BJ, Zelarney PT, Hall AH (1990) Evaluation and treatment of patients exposed to systemic asphyxiants. Emerg Care Q 6:65–80

Di HJ, Cameron KC (2002) Nitrate leaching in temperate agroecosystems: sources, factors and mitigating strategies. Nutr Cycl Agroecosyst 46:237–256

Duan S, Zhang S, Huang H (2000) Transport of dissolved inorganic nitrogen from the major rivers to estuaries in China. Nutr Cycl Agroecosyst 57:13–22

Duncan C, Dougall H, Johnston P, Greens S, Brogan R, Leifert C, Smith L, Golden M, Benjamin N (1995) Chemical generation of nitric oxide in the mouth from the enterosalivary circulation of the dietary nitrate. Nat Med 1:546–551

Dusdieker LB, Dungy CI (1996) Nitrates and babies: a dangerous combination. Contemp Pediatr 13:91–102

EC (1995) European Parliament and Council Directive No. 95/2/EC of 20 February 1995 on food additives other than colours and sweeteners. Official J Eur Commun L61:1–53

EFSA (European Food Safety Authority) (2008) Nitrate in vegetables. Scientific opinion of the panel on contaminants in the food chain. EFSA J 689:1–79

EFSA (European Food Safety Authority) (2010) EFSA panel on contaminants in the food chain (CONTAM); scientific opinion on possible health risks for infants and young children from the presence of nitrates in leafy vegetables. EFSA J 8(12):42

Eisenbrand G, Schmähl D, Preussmann R (1980) Carcinogenicity of N-nitroso-3-hydroxypyrrolidine and dose-response study with N-nitrosopiperidine in rats. IARC Sci Publ 31:657–666

Ellen G, Schuller PL (1983) Nitrate, origin of continuous anxiety. In: Preusmann R (ed) Das nitrosamin problem. Deutsche Forschungsgemeinschaft, Verlag Chemie GmbH, Weinheim, pp 97–134

FAO (Food and Agriculture Organisation) (1996) Control of water pollution from agriculture – FAO irrigation and drainage paper 55. Food and Agriculture Organization of the United Nations, Rome

FAO/WHO (Food and Agriculture Organisation of the United Nations/World Health Organization) (2003a) Nitrate (and potential endogenous formation of N-nitroso compounds), WHO Food Additive series 50. World Health Organisation, Geneva

FAO/WHO (Food and Agriculture Organisation of the United Nations/World Health Organization) (2003b) Nitrite (and potential endogenous formation of N-nitroso compounds), WHO Food Additive series 50. World Health Organisation, Geneva

Fewtrell L (2004) Drinking-water nitrate, methemoglobinemia, and global burden of disease: a discussion. Environ Health Perspect 112:1371–1374

Forman D, Al-Dabbagh S, Doll R (1985a) Nitrates, nitrites and gastric cancer in Great Britain. Nature 313:620–625

Forman D, Al-Dabbagh S, Doll R (1985b) Nitrate and gastric cancer risks. Nature 317:675–676

Galloway JN (2000) Nitrogen mobilization in Asia. Nutr Cycl Agroecosyst 57:1–12

Galloway JN, Ojima DS, Melillo JM (1998) Asian change in the context of global change: an overview. In: Galloway J, Melillo J (eds) Asian change in the context of global climate change: impact of natural and anthropogenic changes in Asia on global biogeochemistry. Cambridge University Press, Cambridge, pp 1–18

Galloway JN, Dentener FJ, Capone DG, Boyer EW, Howarth RW, Seitzinger SP, Asier GP, Cleveland C, Green P, Holland E, Karl DM, Michaels AF, Porter JH, Townsend A, Vorosmary C (2004) Nitrogen cycles: past, present, and future. Biogeochemistry 70:153–226

Gangolli SD, van den Brandt P, Feron V, Janzowsky C, Koeman J, Speijers G, Speigelhalder B, Walker R, Winshnok J (1994) Assessment of nitrate, nitrite, and N-nitroso compounds. Eur J Pharmacol Environ Toxicol Pharmacol 292:1–38

Gilliom RJ, Alley WM, Gurtz ME (1995) Design of the national water quality assessment program: occurrence and distribution of water quality condition. US Geol Sur Circ 1112:33

Gladwin MT, Schechter AN, Kim-Shapiro DB, Patel RP, Hogg N, Shiva S, Cannon RO III, Kelm M, Wink DA, Espey MG, Oldfield EH, Pluta RM, Freeman BA, Lancaster JR Jr, Feelisch M, Lundberg J (2005) Nat Chem Biol 1:308–314

Graves DV (2012) The emerging role of reactive oxygen and nitrogen species in redox biology and some implications for plasma applications to medicine and biology. J Phys D Appl Phys 45 (26):42

Green CT, Puckett LJ, Böhlke JK, Bekins BA, Phillips SP, Kauffman LJ, Denver JM, Johnson HM (2008) Limited occurrence of denitrification in four shallow aquifers in agricultural areas of the United States. J Environ Qual 37:994–1009

Gregory J, Foster K, Tyler H, Wiseman M (1990) The dietary and nutritional survey of British adults. Her Majesty's Stationary Office, London

Grizzetti B, Bouraoui F, Billen G, van Grinsven H, Cardoso AC, Thieu V, Garnier J (2011) Nitrogen processes in aquatic ecosystems. In: Sutton MA, Howard CM, Erisman JW, Billen G, Bleeker A, Grennfelt P, van Grinsven H, Grizzetti B (eds) The European nitrogen assessment. Cambridge University Press, Cambridge, pp 379–404

Gupta SK, Fitzgerald JF, Chong SK, Croffie JM, Garcia JG (1998) Expression of inducible nitric oxide synthase (iNOS) mRNA in inflamed esophageal and colonic mucosa in a pediatric population. Am J Gastroenterol 93:795–798

Gupta SK, Gupta RC, Gupta AB, Seth AK, Bassin JK, Gupta A (2000) Recurrent acute respiratory infections in areas with high nitrate concentrations in drinking water. Environ Health Perspect 108:363–366

IARC (International Agency for Research on Cancer) (1978) Monograph on the evaluation of the carcinogenic risk of chemicals to humans: some N-nitroso compounds, Monograph no. 17. International Agency for Research on Cancer (IARC), Lyon

Isermann K, Isermann R (1997) Ausgangslage, Lösungsansätze und Lösungsaussichten zur nachhaltigen Landnutzung des deutschen Donaueinzugsgebietes auf der Grundlage seiner Stickstoff- und Phosphorbilanz. VDLUFA-Schriftenreihe 45:623–626

Janus LL, Vollenweider RA (1981) The OECD Cooperative programme on eutrophication: summary report – Canadian contribution, Inland Waters Directorate Scientific Series No. 131, Environment Canada, Burlington

Jensen FB (2005) Nitrite transport into pig erythrocytes and its potential biological role. Acta Physiol Scand 184:243–251

Johnson CJ, Kross BC (1990) Continuing importance of nitrate contamination of groundwater and wells in rural areas. Am J Ind Med 18(4):449–456

Jordan TE, Correll DL, Weller DE (1997) Effects of agriculture on discharges of nutrients from coastal plain watersheds of Chesapeake Bay. J Environ Qual 26:836–848

Kosaka H, Tyuma I (1987) Mechanism of autocatalytic oxidation of oxyhemoglobin by nitrite. Environ Health Perspect 73:147–151

Kramer MH, Herwaldt BL, Craun GF (1996) Surveillance of waterborne-disease outbreaks-United States, 1993–1994. CDCP Surveill Summ MMWR 45:1–33

Kross BC, Ayebo AD, Fuortes LJ (1992) Methemoglobinemia: nitrate toxicity in rural America. Am Fam Physician 46(1):183–188

L'hirondel J-L, Avery A, Addiscott T (2006) Dietary nitrate: where is the risk? Environ Health Perspect 114:A458–A459

Lee GF, Jones RA (1986) Detergent phosphate bans and eutrophication. Environ Sci Technol 20 (4):330–331

Lehman PW, Sevier J, Giulianotti J, Johnson M (2004) Sources of oxygen demand in the lower 1176 San Joaquin River, California. Estuaries 27(3):405–418

Lennox SD, Foy RH, Smith RV, Jordan C (1997) Estimating the contribution from agriculture to the phosphorus load in surface water. In: Tunney H, Carton OT, Brookes PC, Johnson AE (eds) Phosphorus loss from soil to water. CAB Int, Wallingford, pp 55–75

Levine JJ, Pettei MJ, Valderrama E, Gold DM, Kessler BH, Trachtman H (1998) Nitric oxide and inflammatory bowel disease: evidence for local intestinal production in children with active colonic disease. J Pediatr Gastroenterol Nutr 26(1):34–38

Lundberg JO, Weitzberg E (2005) NO generation from nitrite and its role in vascular control. Arterioscler Thromb Vasc Biol 25:915–922

Lundberg JO, Weitzberg E, Cole JA, Benjamin N (2004) Nitrate, bacteria and human health. Nat Rev Microbiol 2:593–602

Mackenzie FT, Ver LM, Lerman A (1998) Coupled biogeochemical cycles of carbon, nitrogen, phosphorus, and sulfur in the land-ocean-atmosphere system. In: Galloway JN, Melillo JM (eds) Asian change in the context of global change. Cambridge University Press, New York, pp 42–100

MAFF (Ministry of Agriculture, Fisheries and Food) (1995) National food survey 1994: annual report on household food consumption and expenditure. Her Majesty's Stationary Office, London

Mathews E, Hammond A (1999) Critical consumption trends and implications-degrading earth's ecosystems: 1. Food consumption and disruption of the nitrogen cycle. World Resource Institute (WRI), Washington, DC, pp 11–24

McKnight GM, Duncan CW, Leifert C, Golden MH (1999) Dietary nitrate in man: friend or foe? Br J Nutr 81:349–335

Mensinga TT, Speijers GJA, Meulenbelt J (2003) Health implications of exposure to environmental nitrogen compounds. Toxicol Rev 22:41–51

Michalzik B, Kalbitz K, Park JH, Solinger S, Matzner E (2001) Fluxes and concentrations of dissolved organic carbon and nitrogen- a synthesis for temperate forests. Biogeochemistry 52:173–205

Mueller BA, Newton K, Holly EA, Preston-Martin S (2001) Residential water source and the risk of childhood brain tumors. Environ Health Perspect 109:551–556

Mullins G (2009) Phosphorus, agriculture and the environment. Virginia Cooperative Extension, Virginia State University, Blacksburg

Nieder R, Benbi DK (2008) Carbon and nitrogen in the terrestrial environment. Springer, Heidelberg, p 432

Nolan BT (1999) Nitrate behavior in ground waters of the Southeastern USA. J Environ Qual 28:1518–1527

NRC (National Research Council) (2000) Clean coastal waters. National Academy Press, Washington, DC

Parker AE, Dugdale RC, Wilkerson FP (2012) Elevated ammonium concentrations from 1240 wastewater discharge depress primary productivity in the Sacramento River and the Northern 1241 San Francisco Estuary. Mar Pollut Bull 64(3):574–586

Pedley S, Howard G (1997) The public health implications of microbiological contamination of groundwater. Q J Eng Geol 30:179–188

Petersson J, Phillipson M, Jansson E, Patzak A, Lundberg JO, Holm L (2007) Dietary nitrate increases gastric mucosal blood flow and mucosal defense. Am J Physiol Gastrointest Liver Physiol 292:G718

Phillips G, Pietilainen OP, Carvalho L, Solimini A, Solheim AL, Cardoso AC (2008) Chlorophyll-nutrient relationships of different lake types using a large European dataset. Aquat Ecol 42:213–226

Powlson DS, Addiscott TM, Benjamin N, de Kok TM, van Grinsven H, L'hirondel J-L, Avery AA, van Kassel C (2008) When does nitrate become a risk for humans. J Environ Qual 37:291–295

Pratt PF (1984) Nitrogen use and nitrate leaching in irrigated agriculture. In: Hauck RD (ed) Nitrogen in crop production. ASA, CSSA, SSSA, Madison, pp 319–333

Prihar SS, Gajri PR, Benbi DK, Arora VK (2000) Intensive cropping: efficient use of water, nutrients and tillage. Food Products Press, New York, p 264

Puckett LJ, Cowdery TK, Lorenz DL, Stoner JD (1999) Estimation of nitrate contamination of an agro-ecosystem outwash aquifer using a nitrogen mass-balance budget. J Environ Qual 28:2015–2025

Refsgaard JC, Thorsen M, Jensen JB, Kleeschulte S, Hansen S (1999) Large scale modelling of groundwater contamination from nitrate leaching. J Hydrol 221:117–140

Rejmankova E, Savage HM, Rejmanek M, Arredondo-Jimenez JI, Roberts DR (1991) Multivariate analysis of relationships between habitats, environmental factors and occurrence of Anopheline mosquito larvae *Anopheles albimanus* and *A. pseuodopunctipennis* in southern Chiapas, Mexico. J Appl Ecol 28:827–841

Rekolainen S, Ekholm P, Ulen B, Gustafson A (1997) Phosphorus losses from agriculture to surface waters in the Nordic countries. In: Tunney H, Carton OT, Brookes PC, Johnson AE (eds) Phosphorus loss from soil to water. CAB Int, Wallingford, pp 77–93

Sanchez-Echaniz J, Benito-Fernandez J, Mintegui-Raso S (2001) Methemoglobinemia and the consumption of vegetables in infants. Pediatrics 107:1024–1028

Santamaria P (2006) Nitrate in vegetables: toxicity, content, intake and EC regulation. J Sci Food Agric 86:10–17

Schlesinger WH (1997) Biogeochemistry, an analysis of global change, 2nd edn. Academic, San Diego

Schlesinger WH (2009) On the fate of anthropogenic nitrogen. Proc Natl Acad Sci US 106:203–208

Seitzinger SP, Kroeze C (1998) Global distribution in nitrous oxide production and N inputs in freshwater and coastal marine ecosystems. Glob Biogeochem Cycles 12:93–113

Seitzinger SP, Harrison JA, Dumont E, Beusen AHW, Bouwman AF (2005) Sources and delivery of carbon, nitrogen, and phosphorus to the coastal zone: an overview of global nutrient export from watersheds (NEWS) models and their application. Glob Biogeochem Cycles 19(4):GB4S01

Selman M, Greenhalgh S (2009) Eutrophication: policies, actions, and strategies to address nutrient pollution. WRI policy note no. 3 water quality: eutrophication and hypoxia. World Resources Institute, Washington, DC

Smil V (2000) Phosphorus in the environment: natural flows and human interferences. Annu Rev Energy Environ 25:53–88

Smith RP (1991) Toxic responses of the blood. In: Amdur MO, Doull J, Klaassen CD (eds) Casarett and Doulls toxicology: the basic science of poisons. Pergamon Press, New York, pp 257–281

Smith VH (2003) Eutrophication of freshwater and marine ecosystems: a global problem. Environ Sci Pollut Res Int 10:126–139

Spalding RF, Exner ME (1993) Occurrence of nitrate in groundwater-a review. J Environ Qual 22:392–402

Spiegelhalder B, Eisenbrand G, Preussmann R (1976) Influence of dietary nitrate on nitrite content of human saliva: possible relevance to in vivo formation of N-nitroso compounds. Food Cosmet Toxicol 14:545–548

Stephany RW, Schuller PL (1980) Daily dietary intakes of nitrate, nitrite and volative N-nitrosamines in the Netherlands using the duplicate portion sampling technique. Oncology 37:203–210

Stillman L (2010) Pollution and public health in a shrinking world: concentrated animal feeding operations as a paradigm for emergent needs in environmental and public health policy. Self-designed majors honors papers, Paper 2. http://digitalcommons.conncoll.edu/selfdesignedhp/2

Taylor AW, Kilmer VJ (1980) Agricultural phosphorus in the environment. In: Khasawneh FE, Sample EC, Kamprath EJ (eds) The role of phosphorus in agriculture. Am Soc Agron, Madison, pp 545–557

Teng HJ, Wu YL, Wang SJ, Lin C (1998) Effects of environmental factors on abundance of Anopheles minimus larvae and their seasonal fluctuations in Taiwan. Environ Entomol 27:324–328

Thomas GW, Phillips RE (1979) Consequences of water movement in macropores. J Environ Qual 8:149–152

Tsezou A, Kitsiou-Tzeli Galla A, Gourgiotis D, Mitrou S, Molybdas P, Sinaniotis C (1996) High nitrate content in drinking water: cytogenetic effects in exposed children. Arch Environ Health 51:458–461

Tsirkunov VV, Nikrangrov AM, Laznik MM, Zhu DW (1992) Analysis of long-term and seasonal river quality changes in Latvia. Water Res 26:1203–1216

Turek B, Hlavsova D, Tucek J, Waldman J, Cerna J (1980) The fate of nitrates and nitrites in the organism. IARC Sci Publ:625–632

US EPA (US Environmental Protection Agency) (1986) Quality criteria for water 1986, Report 440/5-86-001. US, Environmental Protection Agency , Washington, DC

US EPA (US Environmental Protection Agency) (1999) Update of ambient water quality criteria for ammonia. 1300 EPA-822-R-99-014. Office of Water, Washington DC

US EPA (US Environmental Protection Agency) (2014) Cyanobacteria and cyanotoxins: information for drinking water systems. EPA-810F11001, Office of Water, Washington DC

van de Bund W (2009) Water framework directive intercalibration technical report part 1: rivers. JRC Sci Tech Rep, EUR 23838 EN/1

van Drecht G, Bouwman AF, Knoop JM, Beusen AHW, Meinardi CR (2003) Global modeling of the fate of nitrogen from point and nonpoint sources in soils, groundwater, and surface water. Glob Biogeochem Cycles 17: doi: https://doi.org/10.1029/2003GB002060

van Grinsven HJ, Ward MH, Benjamin N, de Kok TM (2006) Does the evidence about health risks associated with nitrate ingestion warrant an increase of the nitrate standard for drinking water? Environ Health 5:26

van Maanen JM, Pachen DM, Dallinga JW, Kleinjans JC (1998) Formation of nitrosamines during consumption of nitrate and amine-rich foods, and the influence of mouthwashes. Cancer Detect Prev 22:204–212

van Vuuren DP, Bouwman AF, Beusen AHW (2010) Phosphorus demand for the 1970–2100 period: a scenario analysis of resource depletion. Glob Environ Chang Hum Policy Dimens 20:428–439

Volterra L, Boualam M, Menesguen A, Duguet J, Duchemin J, Bonnefoy, X (2002) Eutrophication and health. World Health Organization Regional Office for Europe. European Communities. Available from http://europa.eu.int

Voss M, Baker A, Bange HW, Conley D, Cornell S, Deutsch B, Engel A, Ganeshram R, Garnier J, Heiskanen A-S, Jickells T, Lancelot C, McQuatters-Gollop A, Middelburg J, Schiedek D, Slomp CP (2011) Nitrogen processes in coastal and marine systems. In: Sutton MA, Howard CM, Erisman JW, Billen G, Bleeker A, Greenfelt P, Van Grinsven H, Grizzetti B (eds) The European nitrogen assessment. Cambridge University Press, Cambridge

Wagner DA, Schultz DS, Deen WM, Young VR, Tannenbaum SR (1983) Metabolic fate of an oral dose of 15N-labeled nitrate in humans: effect of diet supplementation with ascorbic acid. Cancer Res 43:1921–1925

Walker R (1990) Nitrates, nitrites and N-nitrosocompounds: a review of the occurrence in food and diet and the toxicological implications. Food Addit Contam 7:717–768

Walker R (1996) The metabolism of dietary nitrites and nitrates. Biochem Soc Trans 24 (3):780–785

Walker ED, Lawson DL, Merritt RW, Morgan WT, Klug MJ (1991) Nutrient dynamics, bacterial populations and mosquito productivity in treehole ecosystems and microcosms. Ecology 72:1529–1546

Ward MH, deKok TM, Levallois P, Brender J, Gulis G, Nolan BT, Van Derslice J (2005) Workgroup report: drinking-water nitrate and health-recent findings and research needs. Environ Health Perspect 113:1608–1614

Weyer PJ, Cerhan JR, Kross BC, Hallberg GR, Kantamneni J, Breuer G, Jones MP, Zheng W, Lynch CF (2001) Municipal drinking water nitrate level and cancer risk in older women: the Iowa women's health study. Epidemiology 11:327–338

White RE (1985) A model for nitrate leaching in undisturbed structured clay soil during unsteady flow. J Hydrol 79:37–51

WHO (World Health Organization) (1996) Toxicological evaluation of certain food additives and contaminants. Prepared by the Forty-Fourth Meeting of the Joint FAO/WHO Expert Committee on Food Additives (JECFA), International Programme on Chemical Safety (WHO Food Additives Series 35). World Health Organization, Geneva

WHO (World Health Organization) (2004) Guidelines for drinking water quality, vol 1., Recommendations, 3rd edn. World Health Organization, Geneva

WHO (World Health Organization) (2011) Guidelines for drinking-water quality, 4th edn. World Health Organization, Geneva

WHO/UNICEF (2010) Progress on sanitation and drinking-water: 2010 Update. WHO/UNICEF joint monitoring programme for water supply and sanitation. World Health Organization, Geneva

Wishnok JS, Tannenbaum SR, Tamir S, De Rojas-Walker T (1995) Endogenous formation of nitrate. In: Health aspects of nitrate and its metabolites (particularly nitrite). Proc International Workshop, Bilthoven, Netherlands, 8–10 November 1994, Council of Europe Press, Strasbourg, pp 151–179

Woodwell GM, Mackenzie FT (eds) (1995) Biotic feedbacks in the global climatic system. Oxford University Press, New York

WRI (World Resources Institute) (2008) WRI policy note. Water quality: eutrophication and hypoxia 1:1–6. Available online: http://www.wri.org/publication/eutrophication-and-hypoxia-coastal-areas; Accessed 11 Jul 2017

Yano SS, Danish EH, Hsia YE (1982) Transient methemoglobinemia with acidosis in infants. J Pediatr 100:415–418

Zeman CL, Kross B, Vlad M (2002) A nested case-control study of methemoglobinemia risk factors in children of Transylvania, Romania. Environ Health Perspect 110:817–822

Zetterquist W, Pedroletti C, Lundberg JON, Alving K (1999) Salivary contribution to exhaled nitric oxide. Eur Respir J 13:327–333

Chapter 6
Macro- and Secondary Elements and Their Role in Human Health

Abstract Sixteen elements are known to be essential for plants and animals. Based on their concentration in various plant tissues required for adequate growth, nine elements are categorized as macronutrients (>0.1% of dry plant tissue) and the rest as micronutrients. The human body cannot biosynthesize these essential elements and acquire them from food. Except for carbon (C), hydrogen (H) and oxygen (O), which are obtained from air and water, plants take up other essential nutrients from soil. Nutrient input to soil occurs from minerals in the parent material, soil organic matter, fertilization and in some cases (e.g. nitrogen and sulphur) through atmospheric deposition. Soils are most often deficient in nitrogen (N), phosphorus (P) and potassium (K) and, therefore, require their regular input for optimum plant production. In recent times, deficiencies of sulphur (S) are increasing worldwide. The nutrients from soil are taken up in ionic form by the plant and abundance of ionic forms determines the nutrient availability, which besides depending on total quantity of a nutrient is governed by a number of physical, chemical and biological processes and transformations in soil. Nutrient availability and utilization by plant influences human health by providing food and nutritional security in terms of quantity and quality of food consumed to meet dietary requirements and food preferences of people. Essential nutrients play varied roles in the human body ranging from being constituent of structural components (e.g. bones, teeth, cell wall) and biomolecules (amino acids, proteins, enzymes, vitamins, hormones etc.) to performing a variety of physiological functions such as enzyme activation, protein synthesis, energy transfer, transport of sugars, secretion of insulin, creatinine phosphorylation, carbohydrate metabolism, electrical activity of heart, maintenance of acid-base balance etc. Excessive and inadequate nutrient intake and accumulation can disturb key body functions which may lead to severe human health problems. In this chapter, a brief description of sources and transformation of macronutrients in soil as a background to subsequent discussion on their functions in plants and humans and the impact of excessive or inadequate intake on human health are presented.

R. Nieder et al., *Soil Components and Human Health*,
https://doi.org/10.1007/978-94-024-1222-2_6

Keywords Macronutrients • Secondary nutrients • Sources of macroelements • Biological nitrogen fixation • Fertilizers • Macroelement transformations • Denitrification • Nitrogen cycle • Phosphorus cycle • Macroelement functions • Micronutrient deficiency • Optimizing macronutrients • Food sources of macronutrients

6.1 Overview of Macro- and Secondary Elements

Sixteen elements are known to be essential for crop plants. Based on their concentration in various plant tissues required for adequate growth, these elements are, generally categorized as macronutrients and micronutrients. The macronutrients include carbon (C), hydrogen (H), oxygen (O), nitrogen (N), phosphorus (P), potassium (K), calcium (Ca), magnesium (Mg) and sulphur (S). These are used in relatively large amounts to form constituents of organic compounds that act as building blocks for cells or act as osmotica and constitute greater than 0.1% of dry plant tissue. The micronutrients, on the other hand, used in relatively small amounts and constituting less than 0.1% of dry plant tissue include iron (Fe), zinc (Zn), manganese (Mn), boron (B), copper (Cu), molybdenum (Mo) and chlorine (Cl). In addition, there are some quasi-essential elements viz. silicon (Si), sodium (Na), cobalt (Co) and nickel (Ni) that are required by a particular group of plants for optimum growth and production. The essentiality of Si as a macronutrient has been established only for diatoms and plants in the *Equisetaceae* family. While Co has been found essential in small amounts for symbiotic nitrogen fixation in legumes; Na in small amounts has been proved essential for tropical grasses employing C_4 photosynthetic pathway (Brady and Weil 2015). Some of the micronutrients viz. chromium (Cr), iodine (I) and selenium (Se) are required by animals only. Except for Mo and I, all the other essential elements belong to first four periods of the Periodic Table of Elements (Fig. 6.1). Essentiality of arsenic (As), fluorine (F), tin (Sn) and vanadium (V) for animals has been observed under highly specialized conditions. Some of these elements may be toxic when consumed at high amounts.

Among the macronutrients, N, P and K are termed major or fertilizer nutrients; Ca, Mg and S are called secondary nutrients. Though more than 16 nutrients, as listed above, are essential for plants but only three nutrients (C, H and O) make up 94–99.5% of the plant tissue and in normal agricultural conditions these seldom limit plant growth. It is the other essential elements, also called mineral nutrients that most often limit plant growth and development and quality or nutrient composition of food products. The human body cannot biosynthesize these essential mineral elements and obtain these from food, which in turn gets it from the soil and the parent material from which the soil is derived. Macronutrient composition of soil, plant and human body shows that both the plants and the human body have the highest content of C followed by O and H (Table 6.1). The general values shown in Table 6.1 are only indicative of the relative abundance of mineral nutrients in plants and soils; these differ greatly among soils, plants and plant products.

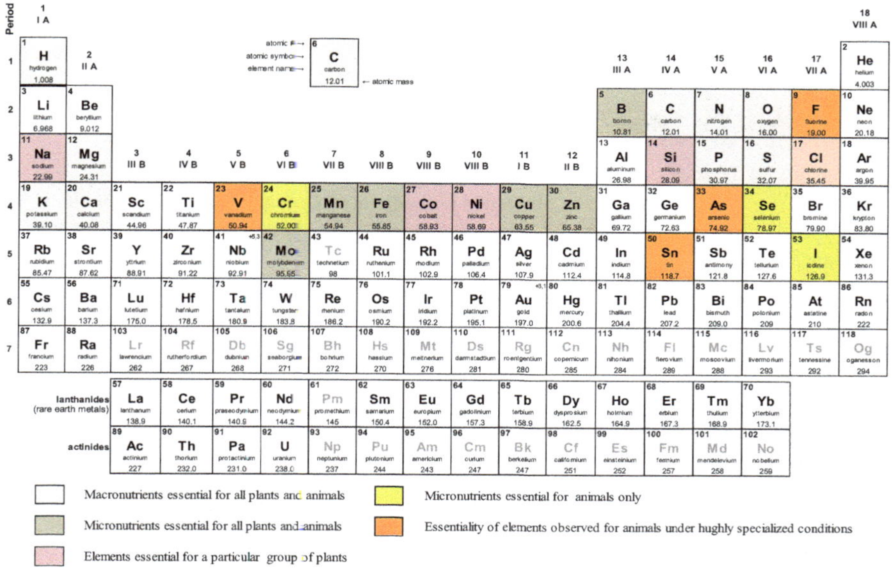

Fig. 6.1 Periodic table of elements highlighting macro- and micronutrients essential for plants, animals and humans

Table 6.1 Macronutrients concentrations in plants, soil (1 m depth) and human body

	Content in elemental form (%)		
Element	Plant	Soil	Human body
C	45.4	2.0	65
O	41.0	49	18
H	5.5	0.065	10
N	3.0	0.1	3
K	1.4	1.0	0.34
P	0.23	0.08	1.0
Ca	1.8	1.0	1.4
Mg	0.32	0.6	0.50
S	0.34	0.05	0.26
Cl	0.2	0.01	0.14

Compiled from Bohn et al. (1979) and Lindh (2005)

6.2 Sources of Macroelements

Macro- and secondary nutrients in soil exist in organic and inorganic forms and these originate from varied sources. The elements C, H and O, which are involved in the synthesis of biomolecules are taken up from air and water. Plants capture C and O from air and H from water through the process of photosynthesis. The other mineral nutrients are taken up from soil and/or supplemented through fertilizers

Table 6.2 Sources and forms of macronutrients absorbed by plants

Element	Source	Form available to plants
Carbon (C)	Air	CO_2
Oxygen (O)	Air	CO_2
Hydrogen	Water	H_2O
Nitrogen (N)	Soil organic matter, biological nitrogen fixation, fertilizer and manure application, atmospheric deposition	NH_4^+ (ammonium) and NO_3^- (nitrate)
Phosphorus (P)	Soil (minerals, fertilizers and manure)	$H_2PO_4^-$ and HPO_4^{2-} (orthophosphate)
Potassium (K)	Soil (minerals, fertilizers and manure)	K^+
Calcium (Ca)	Soil (minerals and fertilizers)	Ca^{2+}
Magnesium (Mg)	Soil (minerals and fertilizers)	Mg^{2+}
Sulfur (S)	Soil organic matter, fertilizers and atmospheric deposition	SO_4^{2-}(sulfate)
Silicon (Si)	Soil minerals, chemical products or slag-based silicate fertilizers	H_4SiO_4, $H_3SiO_4^{2-}$

and manures. Soils are most often deficient in N, P and K and, therefore, require their regular input for optimum plant production. The amount of fertilizer or manure input required depends on soil's capacity to supply a particular nutrient and its availability from other sources. The nutrients from soil are taken up in ionic form by the plant (Table 6.2) and abundance of ionic forms determines the nutrient availability, which besides depending on total quantity of a nutrient is governed by a number of physical, chemical and biological processes and transformations in soil.

6.2.1 *Nitrogen*

Of all the essential nutrients, N is required in the largest amount and is most often the limiting nutrient. Nitrogen was isolated by the British physician Daniel Rutherford (in 1772) and recognized as an elemental gas by Antoine Laurent Lavoisier (in 1776). Gaseous nitrogen makes up the greatest portion of the earth's atmosphere, which is estimated to contain 3.9×10^{12} Tg N (Table 6.3). Total N content in soil ranges between 200 and 4000 ppm (parts per million) with a mean value of 1400 ppm. Soil organic matter represents the second largest pool of N estimated at 1.33×10^5 Tg N. Nitrogen in soil is present mainly as organic and only a small amount occurs as inorganic. Of the total N in soil, around 90% is part of organic matrix, 6–12% is mineral fixed NH_4^+, and 1–3% is plant-available mineral N (NH_4^+ and NO_3^-) (Benbi and Richter 2003). In coarse-textured soils with little capacity to

Table 6.3 Global pool sizes of nitrogen

Reservoir	Tg N	Source
Atmosphere	3.9×10^{12}	Schlesinger (1997)
Terrestrial biomass	3.5×10^{3}	"
Plant biomass	1.0×10^{3}	Davidson (1994)
Microbial biomass	2.0×10^{3}	"
Soil organic matter	1.33×10^{5}	Batjes (1997)

fix NH_4^+ in clay minerals, organic N constitutes more than 97% of total soil N (Baldock and Nelson 2000). On a global scale, the organic N may account for 95% of the total soil N pool. The soil organic N (SON) pool is smaller than atmospheric N_2, but larger than the amounts of N in biomass and the surface oceans. Estimates of global SON in 100-cm soil profile range between 92 and 133 Pg N (Zinke et al. 1984; Davidson 1994; Batjes 1997). In mineral soils, the major part of SON is stored in the uppermost A horizon and up to 50% of the total SON stocks in the 0–100 cm profile could be present in the upper 30 cm soil (Bird et al. 2001). Mean SON content in upper 100 cm of different soils ranges between 0.52 (sandy Arenosols) and 4.01 kg N m^{-2} (Histosols). The higher SON content in Histosols is because of slow decomposition of organic matter under water saturated conditions, particularly when mean soil temperature is low. Very small amounts of N are present in soils of half-deserts (Xerosols: 0.58 kg N m^{-2}) and deserts (Yermosols: 0.37 kg N m^{-2}) because of limited plant growth and residue return to the soil. Nitrogen input to soil occurs through atmospheric deposition, biological nitrogen fixation (BNF), industrial fixation (fertilizer application), and recycling of organic sources.

6.2.1.1 Atmospheric Deposition

Nitrogen can be added from the atmosphere through wet (rain, snow, and hail) and dry (dust and aerosols) deposition in the form of NH_3 and NH_4^+ (collectively termed NH_x) and NO_x and its reaction products: gaseous nitric acid (HNO_3) and particulate NO_3^-. The atmospheric deposition originates mainly from previously emitted NH_3 and NO_x. Dry deposition of NH_3 is important close to a source and wet deposition of NH_4^+ is most important some distance downwind from the source. In parts of Europe with high NH_3 emissions, such as the Netherlands, Belgium, and Denmark, dry deposition of NH_3 represents the largest contribution to total NH_x deposition. In countries with low NH_3 emission densities, only wet deposition of NH_4^+ dominates. The quantities of NO_3^- and NH_4^+ in precipitation over land have increased during the last century. Annually 46.9 Tg N as NO_x and 53.1 Tg N as NH_x are deposited to continents and oceans (Lamarque et al. 2010). Depending on the magnitude of deposition and the type of ecosystem, the atmospheric N deposition can lead to negative effects such as soil acidification, saturation of woodland ecosystems, increased nitrous oxide emissions, reduced methane oxidation rates, altered balance of nitrification and mineralization/ immobilization and change in biodiversity (Nieder and Benbi 2008). Nitrogen deposition can increase the risk of damage from abiotic factors (such as drought and frost) and pest and pathogens. The adverse effects can essentially be reduced by controlling the emissions.

6.2.1.2 Biological Nitrogen Fixation

Biological nitrogen fixation (BNF) is the process of reducing atmospheric N_2 to NH_3 in the presence of nitrogenase enzyme. It is brought-about by prokaryotic organisms living either freely (non-symbiotic fixation) or in association with higher plants (symbiotic fixation). Almost one-fourth of the estimated BNF is accomplished by the root nodule bacterium, *Rhizobium,* in symbiotic association with forages and grain legumes. The rest is fixed by blue-green algae, bacteria, and actinomycetes living either freely in soil or in association with aquatic ferns, grasses, or shrubs. Recent research shows that a range of microbes including archaea and a number of previously undiscovered bacteria are capable of BNF (Reed et al. 2011). In addition to *Rhizobia*, other microbial partners such as *Burkholderia* have been found to nodulate legumes and fix atmospheric N_2 (Gyaneshwar et al. 2011). The amount of atmospheric N_2 fixed varies greatly with the plant species grown and is influenced by a number of soil and environmental conditions. The soil conditions which could limit BNF include excessive moisture, drought, acidity, excess mineral N and deficiency of P, Ca, Mo, Co and B. Presence of metals such as Co and Cu is essential for legume-bacteria association. Absence of these trace elements may restrict development of free-living *Rhizobia* in rhizosphere and thus impair nodule functions. The environmental factors include appropriate temperature for optimum enzyme activity and sunlight for photosynthesis. The magnitude of BNF (kg N ha^{-1}) ranges between 0 and 320 for soybean, 0–125 for common bean, 47–120 for cowpea, 94–222 for groundnut, and 13–69 for pigeon pea (Herridge and Bergersen 1988). Average annual fixation is estimated to range between 75 and 140 kg N ha^{-1} (LaRue and Patterson 1981; Burns and Hardy 1975). On a global scale, estimates of BNF range between 44 and 200 Tg N $year^{-1}$ (Søderlund and Rosswall 1982; Cleveland et al. 1999); 128–140 Tg N $year^{-1}$ being the most commonly used estimate (Galloway et al. 2004) (Table 6.4). These values are substantially higher than 44 Tg N $year^{-1}$ (range 40–100) estimated for pre-industrial world (Vitousek et al. 2013).

Biological nitrogen fixation not only serves as plant N source but also provides means for improving N fertility of the soil if the amount of N fixed exceeds crop removal. Improved soil fertility can result in lowering of fertilizer N use ranging between 50 and 100 kg N ha^{-1} for the succeeding crop (Herridge and Bergersen 1988). Fixed N enters the soil-N pool as stable organic N and is resistant to short-term losses via leaching and denitrification. Therefore, accurate assessment of BNF can help in efficient N management. Though it is difficult to precisely estimate the amount of N_2 fixed by a legume crop, advances in methodology in recent years have allowed estimates of seasonal fixation to be made for most of the common species.

Table 6.4 Estimates of biological nitrogen fixation

References	Tg N $year^{-1}$	Remarks
Delwiche (1970)	100	–
Burns and Hardy (1975)	175	–
Burris (1980)	122	–
Galloway et al. (1995)	32–53 (mid value 43)	Cultivated agricultural systems excluding the tropical savannas
Cleveland et al. (1999)	100–290 (195, intermediate and preferred)	23 biome types covering the whole planet; agricultural activity or cultivated legumes not considered
Smil (1999)	25–41 (33)	Cultivated agricultural systems
Galloway et al. (2004)	128	
Galloway et al. (2004)	32	
Vitousek et al. (2013)	40–100; preferred estimate of 58 (44 after accounting for geological N)	Pre-industrial terrestrial fixation
Herridge et al. (2008)	50–70	BNF in agricultural systems
Herridge et al. (2008)	21.45	Total crop legumes (Pulses 2.95 and oilseed legumes 18.5 Tg N $year^{-1}$)
Herridge et al. (2008)	12–25	Pasture and fodder legumes
Herridge et al. (2008)	5	Rice
Herridge et al. (2008)	0.5	Sugarcane
Herridge et al. (2008)	<4	Non-legume croplands
Herridge et al. (2008)	<14	Extensive savannas

6.2.1.3 Industrial Nitrogen Fixation

Invention of Haber-Bosch process, in 1913, provided much-needed breakthrough to convert atmospheric N_2 to NH_3, which could be used to manufacture N fertilizers for food production. Currently, mineral fertilizers are the main source of N supply to plants and to maintain optimum soil fertility. Plants take up nitrogen either as NH_4^+ or NO_3^-. The inorganic nitrogenous fertilizers may contain either of the ion or both and are thus categorized as ammonical, nitrate, and ammonium- and nitrate-containing fertilizers. The other forms of inorganic N fertilizers include amide

Table 6.5 Major nitrogen fertilizers

Compound	Formula	N-form	N content (%)
Ammonium chloride	NH_4Cl	NH_4^+	25
Ammonium sulphate	$(NH_4)_2SO_4$	NH_4^+	20 (23% S)
Anhydrous ammonia	NH_3	NH_3	82
Ammonium solution	NH_4OH	NH_4^+	20–24
Calcium nitrate	$Ca(NO_3)_2$	NO_3^-	16 (20% Ca)
Sodium nitrate	$NaNO_3$	NO_3^-	16
Ammonium nitrate	NH_4NO_3	NH_4^+ and NO_3^-	33–34
Ammonium sulphate nitrate	$H_{12}N_4O_7S$	NH_4^+ and NO_3^-	26 (13% S)
Calcium ammonium nitrate	$NH_4NO_3 + CaCO_3 * MgCO_3$	NH_4^+ and NO_3^-	27 (8% Ca and/or Mg)
Urea	$CO(NH_2)_2$	NH_2 (amide)	46
Calcium cyanamide	CaNCN	NH_2	21

fertilizers, nitrogen solutions or liquid fertilizers, and slowly available or slow-release fertilizers (Table 6.5). These could be straight nitrogenous fertilizers containing only N or complex/compound fertilizers containing another nutrient in addition to N. Ammonium forms of fertilizers are frequently used in flooded rice soils. The long-term use of ammonical fertilizers leads to soil acidification resulting in yield loss due to aluminum (Al) and Mn toxicities. Urea is the most common form of chemical fertilizer used in agriculture. Urea contains N as amide, which in the soil is rapidly converted into NH_4^+. The NH_4^+ ions released on urea hydrolysis are nitrified to NO_3^- in most soils. Nitrification inhibitors are used to slow down the process of conversion of NH_4^+ to NO_3^-. There are at least eight compounds recognized commercially as nitrification inhibitors; the most commonly used are 2-chloro-6-(trichloromethyl)- pyridine (Nitrapyrin), dicyandiamide (DCD) and 3,4-dimethylpyrazole phosphate (DMPP). These compounds suppress nitrification activity for a few days to weeks depending on soil moisture and soil type. In general, nitrification inhibitors are more effective in sandy or low organic matter soils under high temperature regime.

Because of their role in cereal production, consumption of nitrogen fertilizers has helped sustain global population. Analysis of global cereal production and fertilizer N consumption trends since 1960 show a strong relationship between the two variables (Fig. 6.2). With increase in fertilizer N use from 18 kg ha^{-1} in 1961 to 157 kg ha^{-1} in 2014, the average cereal yields increased from 1.35 to 3.89 t ha^{-1}. Apparently, while cereal yields increased threefold the fertilizer N use increased almost ninefold suggesting that much of the industrially fixed N_2 has been added to the environment as reactive N (Nr) species (nitrogen forms other than N_2).

The reactive N produced disturbs the natural N cycle, which directly and indirectly impacts the environment and human health. Because of greater demand for food for the increasing population, the fertilizer N consumption is expected to

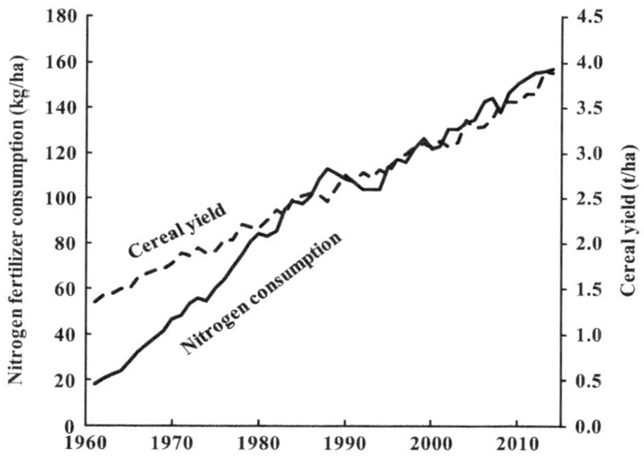

Fig. 6.2 Nitrogen fertilizer use and cereal yield in the world during 1960–2014 (Data source: FAOSTAT n.d.)

increase further, more so in the developing world, where fertilizer use efficiency is generally low. The addition of N fertilizers can play both a constructive and a destructive role in the maintenance of soil health. The constructive role includes production of increased biomass and C sequestration in soil and the destructive role involves mineral N induced mineralization of soil organic matter (SOM) and release of CO_2 to the atmosphere. The long-term use of N fertilizers besides causing soil acidification can lead to leaching of soil cations (Ca, Mg, K) with NO_3^- from applied fertilizers or removal of soil cations in increased harvests of agricultural produce (Kennedy 1992; Peoples et al. 2004). If soil acidity is not controlled this will lead to deterioration of soil health and decline in crop yields, which can negatively impact human health.

6.2.2 *Phosphorus*

Phosphorus was identified in the late seventeenth century by the German alchemist Henning Brandt. It is a non-metal, solid chemical element and has an atomic number of 15 and an atomic mass of 30.97. Phosphorus was recognised as a plant growth limiting nutrient by German chemist Liebig in 1840. In the twentieth century, phosphorus became one of three (N, P and K) essential fertilizer nutrient worldwide (Brink 1977). Ocean comprises the largest biospheric pool estimated at 93,000 Tg of P whereas soils (40–50 Tg P), phytomass (570–625 Tg P) and zoomass (30–50 Tg P) collectively account for only about 640–725 Tg of P (Smil 2000a). In the lithosphere, phosphorus occurs in small quantity and is the 11th most abundant element in the earth's crust. Its concentration in the upper crust averages ~0.1% P (Cordell and White 2011). All except a very small fraction of this huge reservoir, containing approximately 4×10^{15} Mg of total P, is not accessible to plants (Smil 2000a). The total P content of surface 15-cm soil ranges between

100 and 1000 ppm with a mean value of 500 ppm (Brady and Weil 2015). Total P content of soils is generally highest in soils developed on granite gneiss, followed by shales with basic intrusion, limestones with intrusion of micaceous schist and quartzite. In the hydrosphere, typical P concentrations (mg L^{-1}) are 3–15 in domestic waste water, 0.05–1 in agricultural drainage and 0.01–0.04 in lake water (Arai and Sparks 2007). Total P in soil exists in organic and inorganic forms. Organic P in soil constitutes 20–80% of total P depending on age of soil and texture, organic matter content, climate, vegetation, land use, etc. (Arai and Sparks 2007). The principal organic forms of P in soils include monoester inositol hexaphosphate ($C_6H_6(H_2PO_4)_6$), diesters of orthophosphate, nucleic acids, phospholipids and phosphoglycerides and other unidentified esters and phosphoproteins. Inositol phosphate constitutes more than 50% of the total organic P whereas the phospholipids comprise 0.5–7% of total organic P. Nucleic acids, which originate from decomposition of SOM constitute the smallest (<3%) fraction of the total organic P (Dalal 1977). The remainder organic P occurs as solid organic P. The rate of mineralization of these organic compounds is not known. Inorganic P includes iron and aluminium phosphates, P adsorbed onto clay particles and P-bearing minerals.

Phosphorus does not occur in nature as a free element because of its high reactivity, but is found in the form of phosphate minerals. It forms complex minerals with a variety of elements and about 150 P-containing minerals are known (Cathcart 1980). Phosphorus in most soils originates from the weathering of apatite, which contains 95% of all P in the Earth's crust. Apatite is a group of minerals, which may exist as hydroxyapatite ($Ca_5(PO_4)_3OH$), fluorapatite ($Ca_5(PO_4)_3F$), chlorapatite ($Ca_5(PO_4)_3Cl$) and francolite ($Ca_5(PO_4)_3CO_3$). These minerals respectively contain high concentrations of hydroxyl (OH^-), fluoride (F^-), chloride (Cl^-) and carbonate (CO_3^{2-}) ions in the crystal. Some other P bearing minerals are wavellite ($Al_3(PO_4)_2(OH)_3 \cdot 5H_2O$), vivianite ($Fe_3(PO_4)_2 \cdot 8H_2O$), dufrenite ($FePO_4 \cdot Fe(OH)_3$), strengite ($Fe(PO_4) \cdot H_2O$) and variscite ($Al(PO_4) \cdot 2H_2O$).

Phosphorus in soil minerals per se is not available for plant uptake and it must be released from the mineral particles to be available to plants. In soils, soluble P is released by weathering of apatite. As soil weathers, Al and Fe phosphates and organic forms of P become prevalent. The factors that control P solubility and availability in soil include amount of P in parent minerals, pH and content of Ca, Fe and Al in soil, content and type of clay minerals, and amount of SOM. Because of complex chemistry, the plant availability of P in soils is usually low and requires regular input of P through fertilizers. The world's main source of phosphatic fertilizer is rock phosphate, a naturally-occurring P-rich sedimentary or igneous rock containing about 5–13% P (Gilbert 2009). All rock phosphates contain hazardous elements including heavy metals, such as cadmium (Cd), chromium (Cr), mercury (Hg) and lead (Pb), and radioactive elements, e.g. uranium (U), that are considered to be toxic to human and animal health (Mortvedt and Sikora 1992). The amounts of these toxic elements vary widely among rock phosphate sources and even in the same deposit. Their effects on human health are discussed in Chap. 8. Naturally occurring phosphate rock is often upgraded to a rock production containing 11–15% P. About 7 kg of rock phosphate is required to produce 1 kg

Table 6.6 Major phosphatic fertilizers

Compound	Formula	Nutrient content (% P_2O_5)
Monocalcium phosphate (MCP) or Single super phosphate (SSP)	$Ca(H_2PO_4)_2$	16–18 (11% S, 21% Ca)
Dicalcium phosphate (DCP)	$CaHPO_4.H_2O$	34–39
Triple superphosphate (TSP)	$Ca(H_2PO_4)_2.H_2O$	46–48
Double superphosphate	$Ca(H_20P_4)_2.H_2O$	38
Monoammonoium phosphate (MAP)	$NH_4H_2PO_4$	48–61 (10–12% N)
Diammonoium phosphate (DAP)	$(NH_4)_2HPO_4$	46 (18% N)
Monopotassium phosphate (MKP)	KH_2PO_4	39
Rock phosphate	$(Ca_5(PO_4)_3OH)$	20–40
Basic slag	$(CaO)_5P_2O_5SiO_2$	10–18
Bone-meal		20–25

of fertilizer P. The world reserves of rock phosphate are about 67 Pg, located mainly in Morocco and Western Sahara (75%), China (6%), Syria (3%), South Africa (2%), and the remaining ~19% exist in several other countries. Large deposits have also been identified on the continental shelves and on seamounts in the Atlantic Ocean and Pacific Ocean. The world production of rock phosphate increased to 215 Tg in 2012 from 202 Tg in 2011. China (44%), Morocco (13%), USA (14%), Russia (5%) and Jordan (4%) were the major producers. Eighty-two percent of rock phosphate produced is used as fertilizer, either directly or in the manufacture of mineral fertilizers. The remaining 18% is used for other industrial purposes, such as detergents, insecticides, matches, fireworks, military smoke screens, incendiary bombs. There are a number of commercially available phosphatic fertilizers, which contain P in the form of Ca, ammonium or potassium phosphate (Table 6.6). The phosphate in fertilizers is either fully water soluble or partly water soluble and partly citrate soluble, both are considered plant available. Monoammonium phosphate (MAP) has the highest P content of any common solid fertilizer and is water soluble. When it dissolves in soil, the pH of the solution surrounding the granule is moderately acidic, making it an effective fertilizer in neutral and alkaline soils. Diammonium phosphate (DAP) having higher N content than MAP is a commonly used source of P and N in agricultural soils. It is highly water soluble and dissolves quickly in soil to release plant-available phosphate and ammonium. Contrary to MAP, alkaline pH develops around the dissolving granule of DAP, which can influence microsite reactions of phosphate and SOM. Other commonly used phosphate fertilizers include basic slag, single superphosphate, dicalcium phosphate and triple superphosphate. Basic slag, a by-product of smelting phosphatic iron ores, was commercially introduced in 1870s but has low P concentration. Basic slag may be used both as a fertilizer and a soil conditioner because it contains lime and citric-acid-soluble P. High analysis phosphate fertilizers are usually produced by acidulation of phosphate rock. Single superphosphate (SSP) is produced by the use of sulphuric acid and triple superphosphate uses phosphoric acid for acidulation.

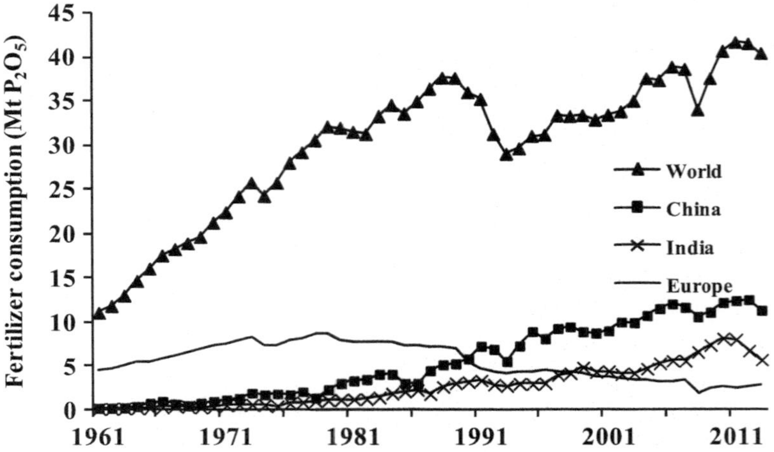

Fig. 6.3 Global phosphorus fertilizer consumption between 1961 and 2013 (Tg P_2O_5) (Data source: FAOSTAT n.d.)

Global P-fertilizer consumption increased almost fourfold between 1960 and 2013 (IFA, Fig. 6.3). China, India and Europe together consume about 50% of the global use of P-fertilizers. China is the largest consumer of P fertilizers with 28% of world total and India is second with 14% of global consumption (IFA). Between 1990 and 2013, global use of phosphate fertilizers increased by 12%. While fertilizer use in Europe showed a declining trend since 1990s, India and China showed an increase of 80–94% in fertilizer P_2O_5 use in 2013.

Population growth, changes towards meat-rich diets, and growing demands for bio-energy crops will further push demand for phosphate fertilizers (Cordell and White 2011). By the year 2050, the demand for phosphorus is predicted to increase by 50–100% (EFMA 2000; Steen 1998) and a peak in phosphorus production is predicted to occur around 2030 (Cordell et al. 2009). Phosphate rock is a finite non-renewable resource and at the current rate of extraction the global commercial phosphate reserves are likely to be depleted in 50–100 years (Runge-Metzger 1995; Steen 1998; EcoSanRes 2003). Global food production will need to increase by about 70% (Fraiture 2007) by 2050 to feed the growing population. Under these circumstances acquiring enough phosphorus to grow food will be a real challenge.

6.2.3 *Potassium*

Potassium is one of the three major essential nutrient elements required by plants. The symbol K for potassium is derived from the Latin word *'Kalium'* and German word *'Kali'*. Elemental K is a soft silvery-white metal that oxidizes rapidly when exposed to air. It has an atomic mass of 39.1 and an atomic number of 19 and is a member of alkali group in the periodic table. Potassium has a naturally occurring

Table 6.7 Potash bearing minerals in soil

Mineral	Chemical formula	Nutrient content (% K_2O)
Sylvite	KCl	63.1
Carnallite	$KCl.MgCl_2.6H_2O$	17
Kainite	$KCl.MgSO_4.3H_2O$	18.9
Langbeinite	$K_2SO_4.2MgSO_4$	22.6
Muscovite	$KAl_3Si_3O_{10}(OH)_2$	11.8
Glauconite	$(K,Na)(Mg,Fe^{2+},Fe^{3+})(Fe^{3+},Al)(Si,Al)_4O_{10}(OH)_2$	6.6
Illite	$(K,H_3O)(Al,Mg,Fe)_2(Si,Al)_4O_{10}[(OH)_2,(H_2O)]$	7.3
Biotite	$KAl(Mg,Fe)_3Si_3O_{10}(OH)_2$	10.9
Phlogopite	$KMg_3Al_2Si_3O_{10}(OH, F)_2$	11.2
Sanidine	$KAl_3Si_3O_8$	12.9
Orthoclase	$KAlSi_3O_8$	16.9
Microcline	$KAl_3Si_3O_8$	16.9
Niter	KNO_3	46.4

stable isotope ^{41}K (natural abundance 6.7%) and a radioactive isotope ^{40}K (abundance 0.01%) with a very long half-life (1.251 billion years). It is a powerful reducing agent, which is easily oxidised. Pure elemental K is never found in nature, but is always combined with other elements. Generally, in plant nutrition and for composition of fertilizers, potassium is expressed in terms of K_2O, though the nutrient is absorbed by plants in the ionic form K^+. Potassium comprises 2.4% of the earth's crust and ranks seventh in order of abundance in the earth's crust. Potassium is the fourth most abundant mineral plant nutrient after iron, calcium and sodium. More than 98% of total K is bound in minerals while less than 2% is organically bound or adsorbed or in soil solution. In minerals, K is found predominantly in primary and secondary crystalline silicates and to a small extent in non-crystalline (amorphous) or para-crystalline compounds (Table 6.7). Most of the K-bearing minerals in soils are dioctahedral micas such as muscovite, gluconite, illite (hydrous mica) and trioctahedral micas viz. biotite and phlogopite, and the feldspars viz. saidine, orthoclase and microcline. Transitional clay minerals (edge expanded illite, illite + monmorillonite or illite + vermiculite) and allophones also contain little amount of K (approx 1%). The K-feldspars, at about 16% compared with the K micas at about 5.2% predominate. Among the micas, biotite (3.8%) is more abundant than muscovite (1.4%). However, in sediments formed by weathering, transport and deposition, this situation is usually reversed because biotite is less stable to weathering. Secondary K minerals, exemplified by illite and the transitional clay minerals, are found in variable proportions. Their contents vary greatly depending on parent rock and the conditions under which the soils have been formed. Potassium is released in the soil through weathering of minerals. The K content in soils ranges between 0.1% and 3% K, most frequently about 1%. Soils with diverse climatic regimes, mineralogy, and texture may contain different amounts of total K. Neutral soils in temperate regions may contain up to about 3.6% K in the surface 30 cm soil (Lawton 1955), whereas K content of soils in wet, humid tropics can be as low as 0.05% (Phetchawee et al. 1985). Only a very small

Table 6.8 Major potassium fertilizers

Fertilizer	Chemical formula	Water soluble K_2O (%)	Other nutrients
Muriate of potash	KCl	60	48% Cl
Sulphate of potash	K_2SO_4	48	18% S
Potassium nitrate	KNO_3	44	13% N
Potassium schoenite	$K_2SO_4.MgSO_4.6H_2O$	23	16% S
Langbeinite	$K_2Mg_2(SO_4)_3$	22.7	11% Mg; 23% S

proportion of the total amount of K in soils is exchangeable or in solution and is thus readily available to crops. Therefore, in many soils, the readily available K is inadequate to meet plant K requirement. But large reserve of mineral K in soil becomes available over a period of time, in response to plant removal. However, the rate of mineral K mobilization varies according to the structure and morphology of soil minerals. Micas especially biotite weather faster than feldspars and release K more readily. However, the rate of release is too slow to provide the large amounts of K required by crops and has to be supplemented with mineral fertilizers.

The commonly used potassium fertilizers, listed in Table 6.8, differ with respect to the associated anion, Cl^-, SO_4^{2-} or NO_3^-. Potassium chloride (KCl) or muriate of potash, with highest K content of all mineral fertilizers, is the most commonly used fertilizer accounting for 90% of mineral K applied to crops throughout the world (IFA). The fertilizer provides two essential elements, the macronutrient potassium, as a cation (K^+) and the micronutrient chlorine, as an anion (Cl^-). Potassium chloride is the only suitable source of K that can be used in irrigation and fertigation systems. However, some crops such as tobacco, grapes, fruit trees, cotton, sugarcane, potatoes, tomatoes, straw berries are sensitive to high amount of KCl. Potassium nitrate (KNO_3) is a preferred fertilizer for spraying on fruit trees and horticultural crops. Global consumption of potassium fertilizers increased more than threefold between 1961 and 2013 (from ~9 Mt in 1961 to 30 Mt K_2O in 2013) (Fig. 6.4).

During the period 1990–1996, the global consumption of K-fertilizers decreased but steadily increased thereafter. However, in Europe the fertilizer K consumption has been steadily declining. On the contrary, fertilizer K use in China is increasing, which alone accounted for 22% of the world-wide fertilizer K consumption in 2013. Fertilizer K consumption in India is relatively low and accounts for about 7% of global consumption.

6.2.4 Secondary Nutrients: Sulphur, Calcium and Magnesium

Sulphur was identified by Antoine Lavoisier in 1777 as a non-metallic element. Solid sulphur is yellow, brittle, odorless, tasteless, and insoluble in water. It belongs to group VIA of the periodic table of elements and has an atomic weight of 32.06

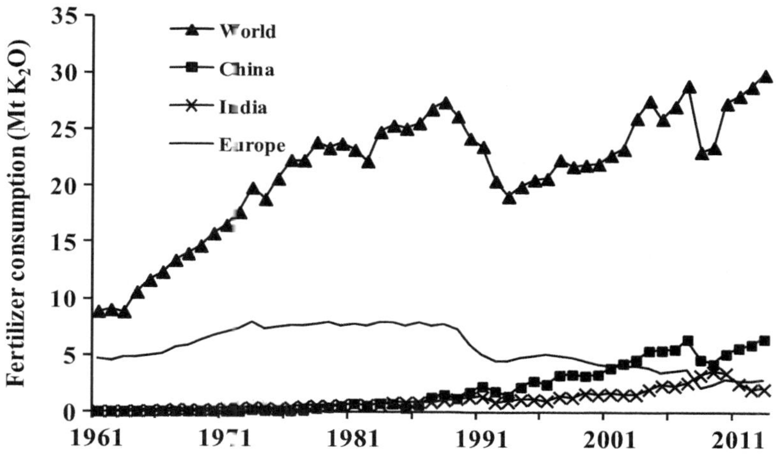

Fig. 6.4 Global consumption of potassium fertilizers between 1961 and 2013 (Tg K_2O) (Data source: FAOSTAT n.d.)

and an atomic number of 16. It is the 13th most abundant element in the earth's crust. In the parent material of soils, S may occur in some primary minerals as sulphide (S^-) or sulphate (SO_4^{2-}). More than 2000 S-bearing minerals with S contents ranging between 7% and 53% are known to occur in soils. Silicate minerals in soil generally contain <100 mg S kg^{-1}. Some of the S-containing minerals in soils include pyrite (FeS_2), chalcopyrite ($CuFeS_2$), gypsum ($CaSO_4 \cdot 2H_2O$), cobaltite (CoAsS) and epsomite ($MgSO_4 \cdot 6H_2O$). While organic S is the dominant form in humid regions, the sulphates of Ca, Mg, K and Na dominate in arid regions. Total S in cultivated soils ranges between 30 and 50 mg kg^{-1}, which mainly occurs in organic combinations. Soils rich in organic matter generally have relative high levels of S. Salt-affected and acid-sulphate soils can have very high levels of S. Input of S to soil occurs through atmospheric deposition (rainfall and dry deposition) and application of S-containing fertilizers. Atmospheric accretions, ranging between 1 and 30 kg ha^{-1} $year^{-1}$ may come through rainfall and dry deposition (Syers et al. 1987). The magnitude of deposition varies with distance from coast and proximity to industrialized areas. Industrial and other anthropogenic activities such as power generation, fuel production and refining, metallurgical processes, and paper manufacturing result in the emission of SO_2 to the atmosphere and may be deposited as SO_2 and SO_4^{2-} particulates (collectively called SO_x). In the atmosphere, SO_2 can react with atmospheric moisture and form sulphurous acid or acid rain. It can also react with oxides of nitrogen (NO_x species) to form ammonium sulphate. Fertilizers are the main source of S input to the soil. Earlier, large amounts of S were applied as incidental constituents of N, P, and K fertilizers. Currently, a wide range of S-containing fertilizers suited to different environmental and farming conditions are available. Materials such as gypsum, pyrite, elemental S, and sulphuric acid have been used as S fertilizers. Fertilizers containing primary nutrients, such as single superphosphate, ammonium sulphate, potassium sulphate and sodium sulphate have also been used to supply S. Gypsum, a less soluble material, has been found comparable to ammonium sulphate and sodium sulphate or slightly less as compared to single superphosphate. The latter is a preferred source for those crops, which need to be fertilized with S and P, instead of other P sources such as diammonium phosphate (DAP) and triple superphosphate. The use of elemental sulphur (S^0) is increasing because of several advantages such as increase in soil P availability owing to lowering of pH and release of phosphate fixed by Ca^{2+} in neutral and alkaline soils and by Fe and Al ions in acidic soils; reduction in leaching and runoff losses of SO_4^{2-} as it is a slow release SO_4^{2-} source; and minimum handling and transportation costs.

Calcium is a divalent cation with an atomic mass of 40.08 and an atomic number of 20. It belongs to alkaline earth metals in group IIA of the Periodic Table of Elements and is the fifth most abundant element in the earth's crust. Soils usually contain about 0.1–5.0% Ca but some calcareous soils may contain more than 20% Ca (Bruce 1999). Calcium content of soil depends on the type of parent materials and the extent of weathering. Calcium in soils is found mainly in minerals such as feldspar, calcite, dolomite, apatite, gypsum and augite. Calcium sulphate (gypsum), which occurs in arid soils, and calcium carbonate (calcite), which occurs in calcareous soils, are the two important calcium minerals controlling Ca concentration in these soils. Soils developed from calcite and dolomite, are alkaline in reaction. High pH and the presence of Ca favor the formation of Ca-humate complexes, which impart dark color to soils. In acid and non-calcareous soils, where Ca content is low, soils may not be able to meet plant Ca requirement and thus necessitate its application through Ca-containing compounds. Superphosphate used as a source of P, also supplies Ca to soils as it contains gypsum ($CaSO_4 \cdot 2H_2O$). Besides superphosphate, the other two commonly used Ca compounds are lime and gypsum. Lime is added mainly to overcome the problems associated with soil acidification; gypsum is used both as a source of S and as an amendment to improve soil physical conditions.

Magnesium with an atomic weight of 24.3 and an atomic number of 12 is also an alkaline earth metal. It is a normal component of both igneous and sedimentary rocks and of the soils developed from such rocks and is the eighth most abundant element in the earth's crust, which contains approximately 2% Mg. Soils developed from basic rocks such as diabase, basalts, and limestone generally contain high levels of Mg (0.27–2.86%) and those developed on coastal sand and granite and sandstones contain low levels of Mg (0.01–0.34%) (Aitken and Scott 1999). Magnesium exists in primary minerals such as biotite, serpentine, olivine, and hornblende, and in the secondary silicate clay minerals chlorite, vermicullite, illite, and montmorillonite and its content ranges between 1% and 35% in different minerals (Table 6.9). In nature, the composition varies depending on the extent of substitution in the crystalline structure of these minerals. Most of the Mg in soil is not readily available to plant and in areas where Ca and Mg content in soils is low such as acid soils or where high Mg-demanding crops (e.g. forages, coffee, vegetables, tree fruits, oilpalm, rubber plantations) are grown, input of Mg may be

Table 6.9 Magnesium minerals in soils

Magnesium mineral	Chemical formula	Magnesium content (%)
Fosterite	Mg_2SiO_4	32–35
Enstatite	$Mg(SiO_3)_2$	18–22
Talc	$H_2Mg_3Si_4O_{12}$ or $Mg_3Si_4O_{10}(OH)_2$	16–20
Phlogopite	$KMg_3(AlSi_3O_{10})(OH,F)_2$	13–18
Actinolite	$Ca(Mg,Fe)_3Si_4O_{12}$	10–16
Clinochlore	$H_8(Mg,Fe)_5Al_2Si_3O_{18}$	10–12
Pyrope	$3MgO.Al_2O_3.3SiO_2$	6–13
Augite	$CaMg(SiO_3)_2$	4.5–10
Iolite	$H_2(Mg,Fe)_4Al_8Si_{10}O_{37}$	5–8
Diopside	$CaMg(SiO_3)_2$	2–14
Hornblende	CaMg metasilicate	1–9
Serpentine	$H_4Mg_3Si_2O_9$	1.9–2.6
Biotite	$(H,K)_2(Mg,Fe)_2Al_2Si_3O_{12}$	1–16

Table 6.10 Magnesium containing fertilizers

Magnesium fertilizer	Chemical formula	Magnesium content (%)
Dolomite	$MgCO_3.CaCO_3$	10
Calcined dolomite	$MgO.CaCO_3$	16
Hydrated dolomite	$MgO.Ca(OH)_2$	17
Magnesite	$MgCO_3$	26
Brucite	$Mg(OH)_2$	36
Magnesia	MgO	56
Kieserite	$MgSO_4.H_2O$	16
Epsom salt	$MgSO_4.7H_2O$	9
Kainite	$MgSO_4.KCl.3H_2O$	7
Langbeinite	$2MgSO_4.K_2SO_4$	11

required. Some of the Mg minerals or other compounds, which may be used as fertilizers, include serpentine superphosphate, epsom salt, dolomite, and calcined magnesite (Table 6.10) (Augustin et al. 1997). Epsom salt is used as a fast-release Mg source, and is the most satisfactory Mg source used for foliar application to plants as it is soluble in water. The other fertilizers being insoluble in water are used as a slow-release source. Dolomite, which contains both Ca and Mg, is more effective in acid soils because Mg is brought into solution by the acid soil. It is the most widely used source of Mg for application to soils, both as an ingredient of mixed fertilizers and as a separate amendment for liming. All the slightly soluble Mg fertilizers listed in Table 6.10 are acid neutralizing materials and would thus reduce the acid-forming potential of the fertilizer to which they might be added. Magnesium silicates, including magnesite, are not effective sources of Mg for plants. Selectively, calcined dolomite in which the Mg component is oxidized to

MgO is more reactive than dolomite. Calcined magnesite, currently available as a granular product (Granmag), can be added to soils or dusted onto pastures as a source of Mg. Both these materials are not suitable for mixing with ammonium containing fertilizers.

6.3 Macroelement Transformations in Soil

6.3.1 Nitrogen

Nitrogen exists in many different forms in soils, plants, animals, and the atmosphere with an oxidation state between +5 and −3 (Table 6.11).

A number of compounds have N bonded to C, H or O. When N is bonded to C or H its oxidation state is negative because it is more electronegative than C or H. In contrast, N bound to O, has a positive oxidation state. Three major forms of N in soil are organic, ammonium (NH_4^+), and nitrate (NO_3^-). Most organic N is covalently bonded to C and is bound up by protein-precipitating compounds. Of the total N in soils, proteins, peptides, and amino acids constitute about 40%; amino sugars 5–6%; heterocyclic N compounds (including purines and pyrimidines) about 35%; and NH_3 about 19% (Schulten and Schnitzer 1998). Ammonical N occurs in two forms: ammonium ions (NH_4^+) and ammonia gas (NH_3). At high pH the equilibrium between NH_4^+ and NH_3 is in favor of NH_3, especially at high temperatures. NH_4^+ can be held on negatively charged sites of clay minerals and organic compounds. This reduces its mobility in soil. NO_3^- is not held by negatively charged soil and organic matter particles. Therefore, NO_3^- in excess of plant uptake is at risk of loss through leaching. The inorganic and organic forms of N undergo several transformations in soil and environment as part of what is called the N cycle (Fig. 6.5). In soil, N transforms continuously from one form to another through processes involving mineralization-immobilization, nitrification-denitrification and fixation-defixation of NH_4.

Table 6.11 Oxidation states of nitrogen

Ion	Oxidation state
Nitrate (NO_3^-)	+5
Nitrogen dioxide (NO_2)	+4
Nitrite (NO_2^-)	+3
Nitric oxide (NO)	+2
Nitrous oxide (N_2O)	+1
Dinitrogen (N_2)	0
Azide (N_3^-)	−1/3
Imidogen (NH)	−1
Amidogen (NH_2)	−2
Ammonia (NH_3)	−3

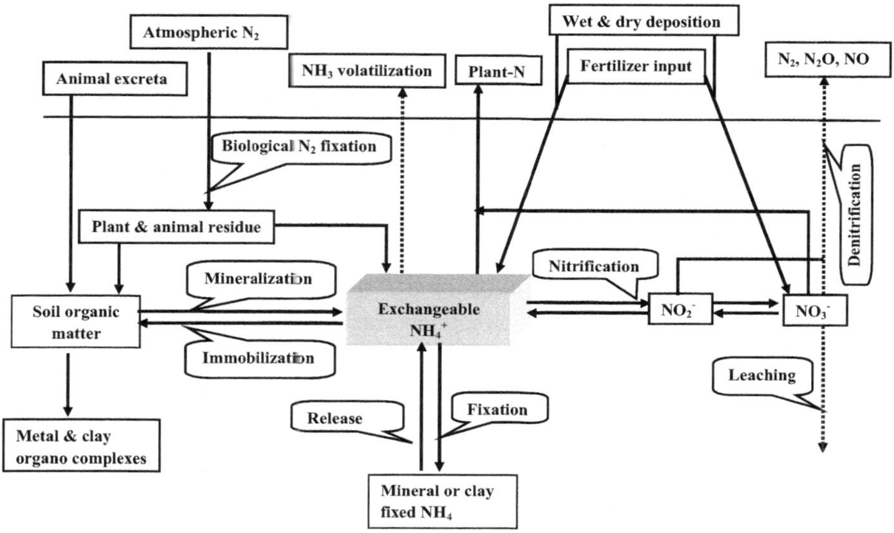

Fig. 6.5 Nitrogen cycle in soil-plant atmosphere system (Adapted from Benbi and Richter 2003)

6.3.1.1 Nitrogen Mineralization and Immobilization

There is an internal N cycle in soil, which involves the conversion of organic forms of N to NH_3 or NH_4^+ and NO_3^- by a process called mineralization. The first step in the process, called ammonification, is an enzymatic process, and involves conversion of organic N to NH_3. It is carried out exclusively by heterotrophic microorganisms. The second step involving conversion of NH_3 to NO_3^-, termed nitrification, is mediated primarily through two groups of autotrophic bacteria viz. *Nitrosomonas* and *Nitrobacter*. The process of nitrification is generally fast and NH_3 produced through ammonification is rapidly nitrified to NO_3^-. Besides the availability of substrates (NH_4^+, O_2, and CO_2), nitrification in soil is influenced by temperature, moisture and pH. Nitrification is optimal at mesophilic temperatures and slightly alkaline pH. The process may be retarded at low and high pH or at water potentials below 1.5 mega pascals, MPa (Justine and Smith 1962) or in the presence of high concentrations of heavy metals (Benbi and Richter 1996). Mineralization is always coupled with immobilization, which operates in the reverse direction with the soil microbial biomass (SMB) assimilating inorganic N forms and transforming them into organic N constituents in their cells and tissues during the oxidation of suitable C substrates. However, immobilized N is subsequently available for mineralization as the microbial population turns over. The continuous transfer of mineralized N into synthesized organic matter and the release of immobilized N back into inorganic forms is known as mineralization-immobilization turnover or MIT (Jansson and Persson 1982). Immobilization primarily occurs from NH_4^+ pool. Total release of NH_4^+ through microbial activity prior to any immobilization back into the organic forms is termed gross mineralization. The difference between gross mineralization and immobilization constitutes net mineralization (or net immobilization). Generally, during an annual cycle there is net mineralization, although over the shorter-term immobilization may occur when available C concentration in the soil is high, for example after addition of organic materials such as crop residues with wider C/N ratios and/or release of C through root exudates. Organic materials having C/N ratios less than 20 lead to net mineralization and those with C/N greater than 30 result in net immobilization by microorganisms. Under conditions favorable for microbial activity, rapid decomposition occurs with concurrent release of CO_2 and consumption of mineral N by microorganisms resulting in net immobilization of N. As the C/N ratio of the decomposing material is lowered to about 20, net mineralization occurs. Besides substrate availability, microbial biomass, soil texture, temperature and moisture are the most important factors that govern N mineralization rates in soils. Water potentials between −0.01 and −0.03 MPa and temperature around 35 °C are considered optimum for N mineralization in soil. However, the two are known to interact and the effect of moisture on microbial activity is enhanced at higher temperatures.

6.3.1.2 Denitrification

The NO_3^- produced as a result of mineralization and nitrification processes in soil may be reduced to dinitrogen gas via denitrification. Denitrification is a microbial respiratory process carried out by various eubacterial genera (Sprent 1999). Reactions proceed via N_2O and other intermediates (NO_2, NO), which are used as alternative terminal electron acceptors to O_2. Therefore, the process is generally facilitated under anaerobic conditions, high levels of soil NO_3^-, and ready availability of carbon source as electron donor. Though anaerobic conditions promote the process yet this is not essential as different species show a wide variation of oxygen tolerance. In addition aerobic soils often have anaerobic microsites (Sprent 1999). Denitrification leads to substantial nitrogen losses in agriculture, which may range between 0 and 200 kg N ha^{-1} $year^{-1}$ (Hofstra and Bouwman 2005) depending on several soil, plant, environmental, and management factors. Among the management and environmental factors fertilizer N application, soil moisture and aeration status and soil temperature are the most important that influence nitrogen oxide emissions. Application of N fertilizers or manures is usually followed by an increase in N_2O emission, which is likely to be greater with urea than nitrate as the fertilizer source (Nieder and Benbi 2008). The mean global fertilizer induced emissions as N_2O and NO are respectively estimated to be 0.91% and 0.55% of the N applied in cropland (excluding legumes) and grassland (Yan et al. 2003). Nitrogen oxide emissions increase with increasing water filled pore space (Skiba and Ball 2002; Bateman and Baggs 2005) though the influence depends on the pathway responsible for the production of N_2O and NO and the diffusion properties of the soil. The rate of N_2O production increases rapidly with increasing water content up to 55–65% water-filled pore space and an increase in water content beyond this hinders aeration and promotes denitrification. Such situations could be encountered under waterlogged conditions, such as in rice paddies and in pasture systems with compacted soil. Total annual denitrification losses for the global agricultural area (excluding leguminous crops) are estimated to be 22–87 Tg N (Drecht et al. 2003; Hofstra and Bouwman 2005).

6.3.1.3 Fixation and Release of NH_4^+

NH_4^+-N in soil may undergo fixation reactions, which involve entrapment of NH_4^+ ions in interlayer spaces of phyllosilicates, in sites similar to K^+ in micas. The fixed NH_4^+, also called non-exchangeable or interlayer NH_4^+, cannot be extracted with neutral normal potassium salt solutions. Soils vary in their capacity to fix NH_4^+ ranging from a few kilograms to several thousand kilograms ha^{-1} in the plow layer depending on the parent material, texture, clay content, clay mineral composition, soil K content and K saturation of the interlayers of 2:1 clay minerals, and moisture conditions (Nieder et al. 2011). The amount of non-exchangeable NH_4^+-N as proportion of the total N, generally increases with soil depth. The non-exchangeable NH_4^+ can either be indigenous/native-fixed during the genesis of silicate minerals or recently fixed as a result of organic and inorganic fertilizer application and organic N mineralization. While the native fixed NH_4^+ is generally unavailable for plant uptake, the recently fixed NH_4^+ may be released during crop growth period for plant uptake (Nieder et al. 2011). The phenomenon of temporary fixation of added fertilizer NH_4^+ helps in reducing N losses through nitrate leaching and NH_3 volatilization.

6.3.2 Phosphorus

Phosphorus is rarely found in its elemental form (P). It has several oxidation states, the most important being +3 and +5 and is generally expressed as phosphorus pentoxide (P_2O_5), which contains 44% P. Phosphorus transformation and cycling processes in soil include (i) release through weathering of soil minerals, mineralization of organic matter and fertilizer P input, (ii) fixation, dissolution and precipitation reactions with Fe, Al and Ca, and (iii) output or losses through crop harvest, soil erosion and leaching (Fig. 6.6). In soils, soluble P released by weathering of apatite or applied through water-soluble P fertilizers is rapidly immobilized into insoluble forms by fixation and precipitation reactions (Brady 1990). Soils rich in soluble iron (Fe) or aluminium (Al), clay minerals like kaolinite, or with high calcium (Ca) activity, react with P to form insoluble compounds inaccessible to plant roots. This is often referred to as P fixation, which is particularly important in many weathered tropical soils, such as Ultisols and Oxisols, and volcanic ash soils (Andosols) (Fairhurst et al. 1999; Sanchez 1976).

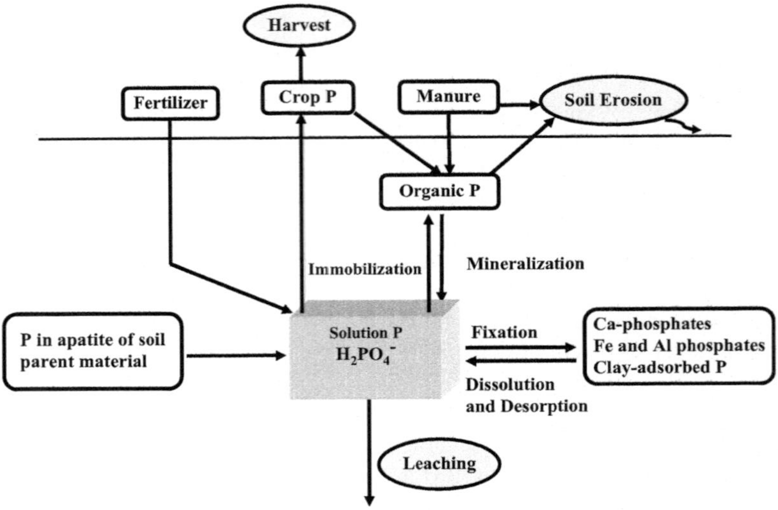

Fig. 6.6 Phosphorus cycle in the soil-plant system

The fixation reactions are pH-dependent; while precipitation reactions with Fe and Al dominate in acidic soils; calcium phosphates are formed in alkaline soils. As a result, only a very small fraction of P exists in soil solution to be directly available to plants. The highest P-availability to crops is in the pH range 6.5–7.5 when monovalent ($H_2PO_4^-$) and divalent (HPO_4^{2-}) orthophosphate anions, the forms absorbed by plants, co-exist each representing 50% of total P in soil solution. At pH 4–6, $H_2PO_4^-$ is about 100% of total P in solution and at pH 8, $H_2PO_4^-$ and HPO_4^{2-} represent 20% and 80% of total P, respectively (Black 1968). Soil solution P concentration typically ranges from <0.01 to 1 mg P L^{-1}. A soil solution P concentration of 0.2 mg L^{-1} is considered optimum for plant growth (Fox and Kamprath 1970; Barber 1995). However, for plants to absorb the total amounts of P required to produce good yields, the P concentration of the soil solution in contact with the roots requires continuous renewal during the growth cycle. Phosphorus in soil moves through diffusion at a very slow rate (10^{-1} to 10^{-15} m^2 s^{-1}) and high plant uptake creates a zone around the root that is depleted of P (Lynch 1995). In soils of low fertility, the soil solution P concentration is usually inadequate to meet crop P requirement. Such soils need regular application of water-soluble P fertilizers.

Besides concentration of phosphate ions in soil solution, the capacity of a soil to replenish or maintain P concentration in soil solution (P-buffering capacity) determines P availability to crops. Organic matter input to the soil increases the soil P availability by competing for binding sites on clay with phosphate. In soils, inorganic P can be absorbed by diffusive penetration into soil components. This may result in a reversible transfer of P between plant-available and non-available forms. The total amount of crop P taken up from soil ranges between 10 and 35 kg P ha^{-1} per crop, though it varies widely with species, cultivars, crop yields and fertilizer application. Typical harvests of cereals, legumes and root crops and vegetables and fruits respectively take up 15–35, 15–25 and 5–15 kg P ha^{-1} (Pierzynski and Loan 1993). Increase in crop yields over the years has resulted in two to three times greater P uptake. The recovery of fertilizer P in crops is usually 15–20% in the short-term. The unused P accumulates in the soil as residual P. The residual P can contribute to soil solution in the subsequent years and be taken up by the succeeding crops. In paddy fields the availability of residual P is higher because of reducing conditions that solubilise iron phosphates. In situations where the amount of readily available P is below the critical level, the rate of release from residual P may not be rapid enough to sustain optimal crop yields.

Phosphorus is mainly lost by surface runoff as P dissolved in runoff water and adsorbed to the eroded particles. Erosion and runoff transfer soluble and particulate P to the ocean where it is eventually buried in sediments (Mackenzie et al. 2002). Heavy rainfall immediately after surface application of fertilizer or manure and intensive tillage can increase erosion losses of phosphorus. Incorporation of crop residue and placement of fertilizers below the soil surface can reduce the risk of P losses through erosion. Because of low mobility the leaching losses of P are much smaller than the losses by erosion and surface runoff. Leaching of P may only occur in P-saturated soils in some industrialized countries (Smil 2000a).

6.3.3 *Potassium*

Potassium exists in soil in four different forms *viz.* soil solution or water soluble K, exchangeable K held on cation exchange sites of surface of clay minerals and organic matter, fixed K held on adsorption sites with high specificity for K ions that are mainly in inter-lattice positions of expanded illites, montmorillonites, and vermiculites, and lattice or matrix K bonded in the crystal structure of clay minerals (Fig. 6.7). The various K forms, which are in dynamic equilibrium differ in amount, mobility and availability to plants. The soil solution K and the exchangeable K are readily available to plants, the fixed K is slowly available and the lattice K is released only through weathering of the minerals. The relative mobility, availability and amount of the different forms of K to plants are in the order: soil solution K > exchangeable K > fixed K > lattice K. While the release of fixed K is reversible that of matrix K is irreversible (Kirkman et al. 1994).

Soils with different climatic regimes, mineralogy, and texture may contain very different amounts of total and other forms of K. Neutral soils in temperate regions may contain up to about 3.6% or about 140 Mg K ha^{-1} in the top 30 cm soil (Lawton 1955), whereas in wet, humid tropics K levels can be as low as 0.05% or about 0.6 Mg ha^{-1} (Phetchawee et al. 1985). Only a very small proportion of the total amount of K in soils is exchangeable or in solution and is thus readily available to crops. Levels of exchangeable K in soil vary enormously as a result of plant

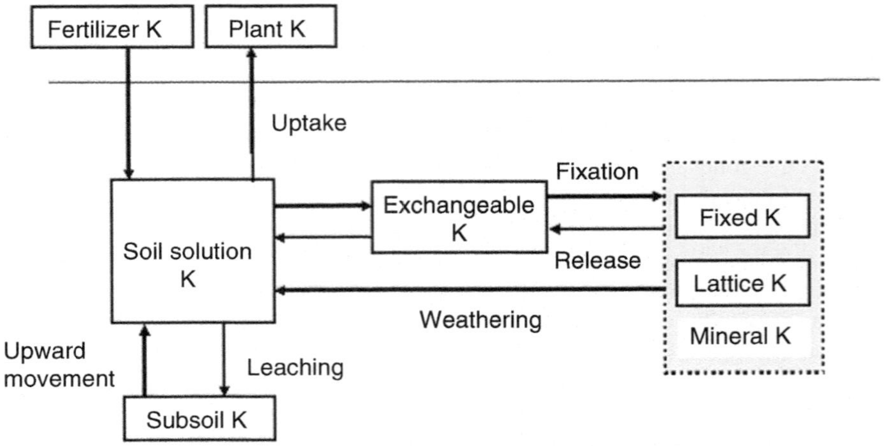

Fig. 6.7 Main components of K regime in soil-plant systems and the crucial transformations between them

growth and nutrient uptake, and fertilizer and manurial practices; levels vary from 20 to 1500 mg K kg^{-1} soil. Soils containing substantial quantities of vermiculites, montmorillonites or smectites have much fixed-K and have a considerable K-buffering capacity. The amount of K held on the cation exchange sites on the surface of particles and the K held between the layers of 2:1 minerals is especially important for crop growth. The larger the cation exchange capacity the greater is the exchangeable K holding capacity of a soil. Typical cation exchange capacities (meq kg^{-1}) of clay minerals are 800–1500 for montmorillonite, 100–400 for illite and 30–150 for kaolinite (Toth 1955). Soil organic matter can have exchange capacities as high as 3000 meq kg^{-1} of dry organic matter (Russell 1973) which underlines the importance of soil organic matter in K nutrition of crops.

6.3.4 Secondary Nutrients: Sulphur, Calcium and Magnesium

Sulphur in soil is present both as organic and inorganic, the amounts and proportion of which depend on soil type, soil depth, climate, and cultural conditions. Inorganic S in soil usually represents less than 25% of total S, comprising mainly soluble, adsorbed, insoluble, and co-precipitated sulphate (SO_4^{2-}) and sulphide (S^-). Elemental S (S°) and other compounds of lower oxidation state than SO_4^{2-}, mostly present as transitory reaction intermediates, are less prevalent in the soil environment. Organic S mainly originates as plant and animal residues that are decomposed by soil organisms (Fig. 6.8). The main forms of organic S in soils include S-containing amino acids and sulfonates, in which S is directly bonded to

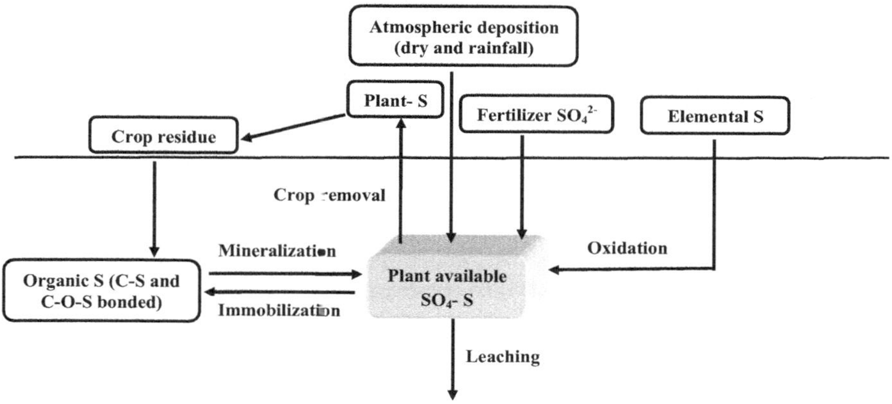

Fig. 6.8 Sulphur cycle in the soil-plant systems

carbon (C-S), and also the true organic esters of sulphuric acid (C-O-S), in which S is bonded to oxygen in the form of C-O-SO_3 – linkages. Sulphamates may also be found, in which S occurs in the form of N-O-SO_3^- and N-SO_3^- groups. Although most soil S may be present in the organic form, only a small proportion (0.5–3%) of this fraction may be mineralized in a year (Eriksen 2008).

The process of mineralization may occur through biological and biochemical pathways (McGill and Cole 1981). In the biological pathway, SO_4^{2-} is released as a byproduct from the microbial decomposition of organic carbon. Biochemical pathway involves hydrolysis of sulphate-esters by sulphatase enzymes leading to release of sulphate-sulphur. Available sulphate in soil can also be assimilated/ immobilized by microorganisms rendering it unavailable for plant uptake (Wu et al. 1995; TSI Bulletin No. 23). Plants take up S in the form of SO_4^{2-} ions but a small amount may be absorbed through the leaves in the form of H_2S or SO_2, the high levels of which could be toxic to the plant. Sulphur supply to plants in neutral to slightly alkaline soils can be augmented by adding easily soluble fertilizers containing SO_4^{2-}S such as ammonium sulphate, potassium sulphate and sodium sulphate. On the contrary, availability of S to plants from S^0 is usually less than sulphate sources as it has to be oxidized by soil microbes to SO_4^{-2} in the soil before plants can utilize it. Oxidation of S^0 is mediated by autotrophic bacteria such as *Thiobacillus*, heterotropic bacteria, fungi and actinomycetes. The activity of these organisms is regulated by soil pH, oxygen and moisture content, and temperature. Rate of oxidation is higher in alkaline soils than acidic soils suggesting greater efficiency of elemental sulphur in the former soils. However, SO_4^{2-} ions in soil solution are prone to leaching as these are not adsorbed on clay and organic matter surfaces. In soils devoid of oxygen such as marshylands and lowland paddy fields, SO_4^{2-} ions are reduced and utilized by organisms like *Desulphovibrio* and *Desulphotomaculum*. Sulphur deficiency observed in submerged rice-fields may sometimes be because of reduction of SO_4^{2-} to sulphide.

Calcium is present in soil as $CaCO_3$, Ca-salts, soluble (in soil solution) and exchangeable Ca. The soluble and exchangeable forms of Ca are in dynamic equilibrium with each other. Calcium and Mg being strongly electrovalent are the most abundant cations occupying the exchange sites of the soil colloid. In sandy soils with low cation exchange capacity (CEC) the amount of exchangeable Ca is low. Calcium availability in soil depends on nature of clay minerals; 2:1 clay minerals requiring relatively high degree of Ca-saturation compared to 1:1 clays to provide similar levels of Ca for plant uptake. Soils usually contain less Mg than Ca because Mg^{2+} ions are not adsorbed as strongly by clay and organic matter as Ca^{2+} ions. Similar to Ca, Mg is also present in soil as water soluble, exchangeable and non-exchangeable. The water soluble and exchangeable forms of Mg are in equilibrium with non-exchangeable Mg. The availability of Mg in soils is influenced by pH and clay content. Leaching of Mg in coarse-textured soils with low pH may result in its deficiency. Mg availability in soil is diminished because of its fixation in clay minerals such as chlorite and vermiculite.

6.4 Cycling of Nitrogen and Phosphorus

Although most of the N transformation processes discussed in Sect. 6.3.1 occur in soil ecosystem (pedosphere), but different N compounds circulate through the earth's air, water, and biosphere, which are closely connected to the pedosphere. Mineralization of SOM and release of N from organic and mineral fertilizers and subsequent nitrification provide reactive inorganic N (Nr) forms such as NH_4^+, nitrite (NO_2^-) and NO_3^-. Nitrate and exchangeable NH_4^+ are readily available for plant and microbial use but most of these are leaked into the environment, groundwater, lakes and rivers and carried to oceans. Various anthropogenic activities such as cultivation, mineral fertilizers, industrial processes and energy generation lead to Nr creation (Galloway and Cowling 2002). In 2005, anthropogenic sources resulted in creation of 210 Tg N $year^{-1}$ and food production for a population of 6.5 billion accounted for most of it (Table 6.12). Human body requires about 2 kg N $year^{-1}$ of protein to survive (Smil 2000b), therefore a surplus of 147 Tg N $year^{-1}$ is produced. Most of the surplus Nr is distributed to the environment at different steps in the N cycle and only about 14% of N fertilizer produced enters the human mouth in a

Table 6.12 Anthropogenic activities and Nr creation in 2005

Activity	Nr created (Tg N $year^{-1}$)
Cultivation	60 (50–70)
Haber-Bosch process	
Fertilizer	100
Industrial activities	24
Energy production	30 (27–32)
Total	210

Adapted from Ciais et al. (2013)

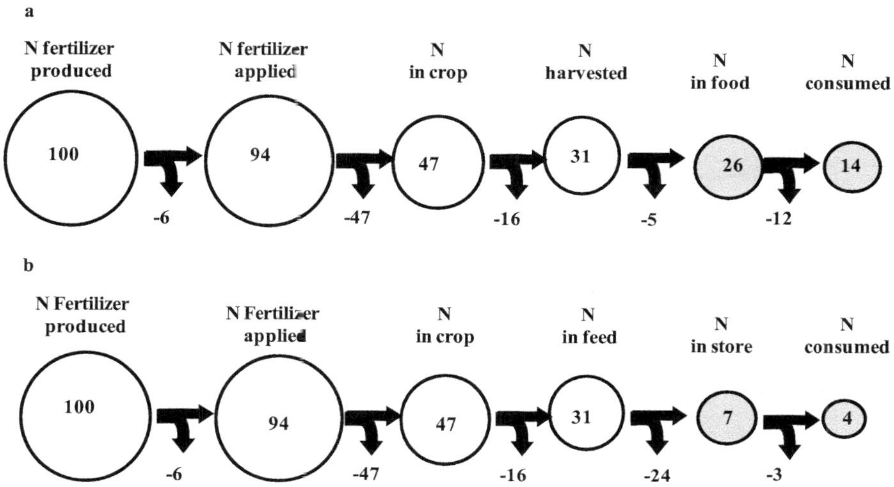

Fig. 6.9 The fate of fertilizer N produced from the factory to the mouth for (**a**) vegetarian diet, and (**b**) non-vegetarian diet (Adapted from Galloway and Cowling 2002)

vegetarian diet (Fig. 6.9; Galloway and Cowling 2002). The remaining 86% of the N is either recycled to agroecosystems (as crop residues and manure) or lost to the environment at different stages of food production. The losses may take place during fertilizer storage, transport, emission to the atmosphere as NH_3, NO, or N_2O, or transport to groundwater or surface water primarily as nitrate, food spoilage, product preparation, and animal metabolism. The amount of N entering human mouth in a non-vegetarian diet is still smaller (4%) because of additional losses during animal metabolism and product preparation for animal feed.

Unlike nitrogen cycle, the P cycle does not have a gaseous phase and there is only a small atmospheric reservoir of P (Mackenzie et al. 1998). Therefore, movement of P to and from the atmosphere is of minor importance. On a time scale of thousands of years, the natural P cycle appears to be a one-way flow (Bouwman et al. 2009). Anthropogenic changes to global P cycle, largely due to mining of P for use as fertilizer, have augmented the rate of P movement from mineral deposits to the ocean fourfold (Smil 2000a; Falkowski et al. 2000). Such increases are mainly due to increase in food production for growing population (Cordell et al. 2009), change in food habits to more P intensive products (Keyzer et al. 2005) and greater fertilizer input to enhance yields (Tilman et al. 2002; Godfray et al. 2010). The overuse of P resources is both a threat to food security and to downstream ecosystems. Figure 6.10 shows phosphorus flows in terms of Mt of phosphorus per year through the global food production and consumption system and enumerates estimated losses at different stages.

Actual losses from agricultural fields attributed to applied phosphate fertilizer are difficult to estimate because of complex phosphate chemistry in soil. Only a

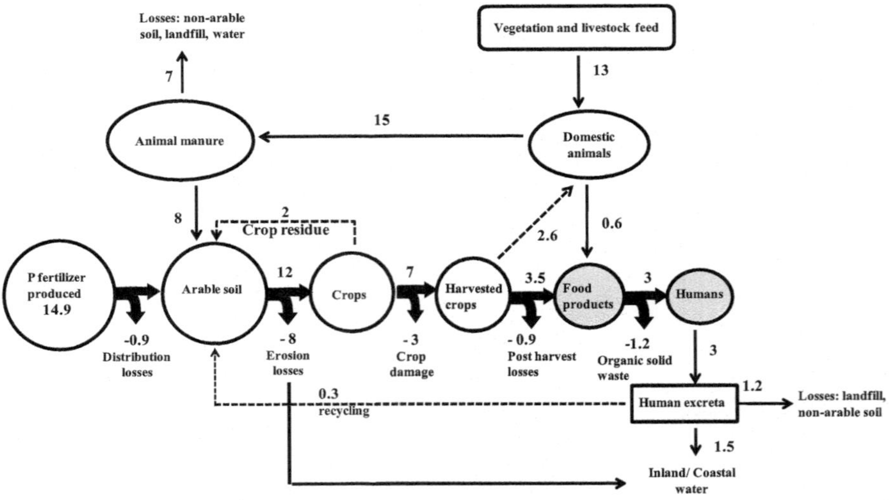

Fig. 6.10 Significant phosphorus flows (Tg P year^{-1}) through the global food production and consumption system (Adapted from Cordell et al. 2009)

fraction of the added P is taken by crops in a year and the balance comes from soil. Total phosphorus content in annual agricultural harvest is ~12 Tg P, of which 7 Tg P is processed for food, feed and fibre and the remaining 5 Tg P is returned to the land as crop residues (Smil 2002a). Cereals and legumes containing 0.25–0.45% in their grains and 0.05–0.1% P in their straw, account for major proportion of crop P (Smil 1999). Studies on post-harvest losses of food from the global food production and consumption chain (Smil 2000a) show that approximately 55% of phosphorus in food is lost between "farm and fork". Only about 20% of the P added as fertilizer is consumed by humans. An estimated 50% of the phosphorus consumed and hence excreted by livestock is returned to agriculture (Smil 2000a). However, there are significant regional imbalances because of differences in manure supply and soil phosphorus status (Cordell et al. 2009). Estimates of phosphorus flows in the food production and consumption system indicate that significantly less mineral phosphate fertilizer per person is required for a vegetarian diet (0.6 kg P year^{-1}) than a non-vegetarian diet (1.6 kg P year^{-1}).

6.5 Beneficial Effects of Macronutrients

6.5.1 Carbon, Oxygen and Hydrogen

These elements are major constituents of all organic chemical compounds of which the plant is made and they are involved in a number of metabolic reactions vital for growth and development. The elements C, H and O play a key role in providing energy for growth and metabolism. Majority of the energy required for these processes is derived from the oxidative breakdown of carbohydrates, fats and proteins during cellular respiration.

6.5.2 *Nitrogen*

Plants, animals and humans require N to survive and it directly affects human health by providing food and nutritional security in terms of quantity and quality of food consumed to meet dietary requirements and food preferences of growing population. While production of adequate food was a major concern during twentieth century, it is the nutritional quality of food that is of contemporary importance (Graham and Welch 2000). Nitrogen is a component of substances such as proteins and enzymes that catalyze various biochemical processes, the nucleic acids (DNA and RNA) that encode the genetic character of all living beings and the essential cell constituents such as cell wall. In plants, nitrogen is a component of chlorophyll and growth hormones and stimulates plant growth, branching, tillering, leaf number and leaf area expansion (Table 6.13).

Nitrogen delays leaf senescence and promotes the process of grain setting and grain filling. Nitrogen content in plants ranges between 1% and 6%. Application of N at grain filling stage results in improved protein content of grain. Proteins in different plant products contain about 15.8% (wheat bran and millets) to 18.9% (in nuts) N with a mean value of 16% (FAO/WHO 1973). Soil per se cannot supply the requisite amount of N for food production and it has to be supplemented through N-fertilizers. More than half of the food eaten by the world's population is produced using N-fertilizers (Smil 2001). Without consumption of fertilizer N, it would not have been possible to sustain growing population (Fig. 6.11).

World population during 1960–2014 increased from three to seven billion and N consumption and cereal production during the same period increased from 11.8 to 113 Mt N and 877 to 2801 Mt grains, respectively. Besides increasing productivity, nitrogen also provides nutritional security through its role in building plant proteins, which are directly consumed by humans or fed to animals to produce quality milk, egg and meat products for human consumption The human body requires ~2 kg N year^{-1} of protein to survive (Smil 2000b) and globally humans ingest ~20 Tg N year^{-1} in their food (Galloway et al. 2002). Protein is the second most abundant substance, after water, in human body comprising one-fifth of the total body weight (Garrison and Somer 1995). Proteins are important components of enzymes, muscles, body organs, skin, hair, and nails and immune system compounds (Garrison and Somer 1995). Proteins are involved in regulation of fluid balance and blood pH; and can serve as an energy source. The integrity of the gastrointestinal tract is maintained through net secretion of nitrogen-containing compounds such as mucins and antibodies, and the shedding of enterocytes (WHO 2007).

Table 6.13 Major role of essential macronutrient elements in plants and humans

Element	Role in plants	Role in humans
Nitrogen (N)	Component of chlorophyll, proteins, nucleic acids, enzymes and growth hormones; stimulates plant growth, delays leaf senescence and promotes the process of grain setting and grain filling	Amino acid and protein synthesis, creation of compounds that influence growth hormones, brain functions and immune system
Phosphorus (P)	Constituent of cell membrane, nucleic acids phospholipids, co-enzymes, and metabolic substrates and; important in energy transfer	Bone structure, membrane structure, metabolic regulation and signalling, component of DNA, RNA and ATP
Potassium (K)	Activation of enzymes of photosynthesis and protein synthesis; involved in ATP production and transport of sugars to storage organs, control of stomatal aperture	Electrochemical regulation, acid-base balance, osmotoic control of water distribution
Calcium (Ca)	Constituent of cell walls and membranes; helps to maintain membrane stability and permeability; promotes cell division, cell elongation; retards senescence and abscission of leaves	Bone structure, nervous transduction
Magnesium (Mg)	Component of chlorophyll, activator of enzymes involved in photosynthesis, starch and sugar production; involved in energy (ATP) production and transfer processes	Bone structure, electrochemical regulation, enzyme catalysis
Sulphur (S)	Constituent of amino acids (cysteine and methionine), proteins and coenzymes; imparts stability to chlorophyll by forming complex with protein	Constituent of amino acids (methionine and cysteine), vitamins (thiamine and biotin), insulin, heparin, fibrinogen; involved in detoxification of aromatic compounds

6.5.3 Phosphorus

Phosphorus being one of the fundamental building blocks of life is essential for all types of life, including plants, animals and microorganisms (Cordell and White 2011). It is essential for food production and there is no substitute for phosphorus in crop growth. Phosphate regulates photosynthesis, sucrose translocation to fruit/seeds, starch synthesis in storage tissues and fruit ripening. These functions of P help in the processes leading to higher crop yields viz. plant growth, higher leaf number and expansion, bud development and opening, fruit and seed formation, improvement in quality, straw strength to prevent lodging in cereals, and crop maturation. Phosphorus concentration in plants and plant parts commonly varies between ~2 and 20 g kg^{-1} dry matter, of which 29–88% exists as inorganic, 5–6% as phospholipids, 5–50% as nucleic acid, and 1–12% as inositol phosphate-P. Intermediate metabolites such as sugar and adenosine phosphates and

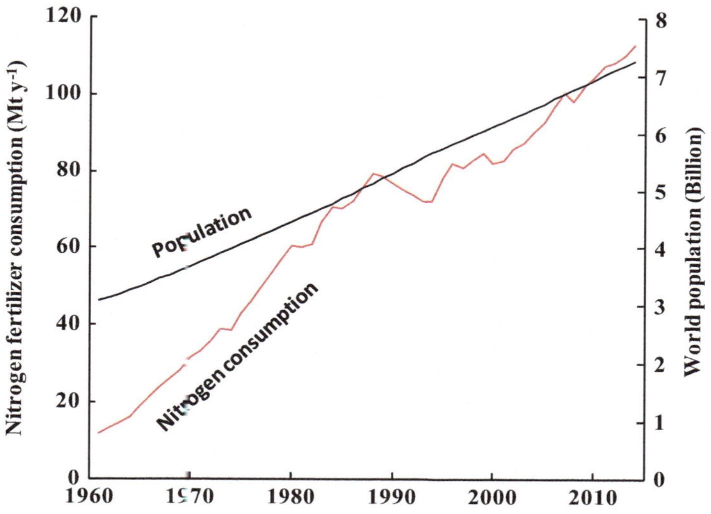

Fig. 6.11 Parallelism between global population and fertilizer N consumption (Data source: FAOSTAT n.d.)

phosphoproteins comprise only trace concentrations. Phosphorus is absent in cellulose, hemicelluloses, lignin and N-rich amino acids (Smil 2000a). In animals and humans, phosphorus is an essential constituent of phospholipids that form membrane bilayers, many sugars, proteins, enzymes, metabolic substrates and nucleic acids that carry genetic information. As part of Adenosine Triphosphate (ATP), it is involved in energy transfer and utilization. Phosphorus is also involved in many physiological processes such as pH regulation of the body, intracellular signalling, brain cell communication and dissociation of oxygen from haemoglobin. Approximately 85% of the body's phosphorus is in bones and teeth in the form of hydroxyapetite containing 18.5% P and making up almost 60% of bone and 70% of teeth and fibrous collagen, a biopolymer (Marieb 1998). The remaining 15% of phosphorus present in the body is integral to diverse functions ranging from transfer of genetic information to energy utilisation. Total body phosphorus in adults has been reported to be in the order of 400–800 g (Moe 2008) yielding an estimated global anthropomass pool of approximately 2.5 Tg P (Smil 2000a). At birth, a neonate contains roughly 20 g phosphorus (0.5 g 100 g^{-1} fat free tissue) (Widdowson and Spray 1951). Assuming continuous growth and maturity at 18 years, the estimated phosphorus accretion rates are 107 mg day^{-1} in boys and 80 mg day^{-1} in girls, with a peak rate of 214 mg day^{-1} in adolescence (Prentice and Bates 1994). Total phosphorus concentration in blood is ~13 mmol L^{-1} (40 mg dL^{-1}); 70% being organic phosphates in the phospholipids of red blood cells and in the plasma lipoproteins and the remaining 30% as inorganic phosphate, of which 15% is protein bound. About 50% of the inorganic phosphate is present as HPO_4^{2-}, the remaining as $H_2PO_4^-$ and PO_4^{3-}, or as HPO_4^{2-} complexed with Na, Ca and Mg salts. These anion forms being inter-convertible are effective buffers of blood pH and are involved in regulation of acid-base balance of the whole body.

6.5.4 *Potassium*

Potassium in plants is known to activate about 80 enzymes involved in various physiological processes such as photosynthesis, energy metabolism, starch synthesis, protein synthesis, transport of sugars to storage organs, nitrate reduction etc. Potassium is associated with increased growth rate of cambium, formation and lignification of vascular bundles, lodging resistance, mitigation of abiotic (e.g. frost, waterlogging and droughts) and biotic stresses (e.g. insects, pest and fungal infections). Potassium ions protect cells against reactive oxygen species induced under stress conditions (Cakmak 2005). Potassium is also known to improve nutritional and processing quality of fleshy fruits and tubers. In the human body, potassium is the third most abundant mineral and more than 90% of it is intracellular. The total body content of potassium is about 1.6–2.1 g (40–55 mmol) kg^{-1} body weight (Rastegar 1990; Agarwal et al. 1994; Crook 2012; Bailey et al. 2014) that corresponds to 110–150 g (3–4 moles) K for a 70 kg adult. Similar potassium contents per kg body weight have been reported for infants and children (Fomon et al. 1982; Butte et al. 2000). Most of the body potassium is located in muscle, with lower amounts in bone, liver, skin and red blood cells. In humans, potassium is a cofactor for a number of enzymes and is required for the secretion of insulin, creatinine phosphorylation, carbohydrate metabolism, electrical activity of heart, protein synthesis and muscular strength (WHO 2009a). Potassium is important for maintaining acid-base balance and functions as an antioxidant. Potassium together with sodium is responsible for maintaining normal osmotic pressure in cells. Potassium is the principal cation in the fluid inside the cells and its concentration is about 30 times higher than that of plasma and interstitial fluid (EFSA NDA Panel 2016). It contributes to establishing a trans-membrane gradient, which is important for the transmission of electrical activity in nerve fibres and muscle cells. Maintenance of the trans-membrane gradient is important for electrolytes and fluid homeostasis, a critical factor in blood pressure regulation (Bailey et al. 2014; Gumz et al. 2015). Potassium ensures proper growth of muscle tissue and utilization of energy released during metabolism (Singh et al. 1996). Potassium has a role in cell metabolism, participating in energy transduction, hormone secretion and the regulation of protein and glycogen synthesis. Potassium in the body is regulated by the balance between dietary intake and renal excretion. Maintaining consistent levels of potassium in the blood and cells is vital to body functions. Though there is no recommended daily allowance (RDA) for potassium but the adequate intake of potassium is 4700 mg day^{-1} for youth and adults (14–70 years of age), which is equivalent to 78 mg kg^{-1} body weight per day for a person weighing 60 kg (IOM 2005). For children in the age group of 1–13 years, the adequate intake is 3000–4500 mg day^{-1}. Dietary surveys in European Union showed that the average potassium intakes per day ranged between 821 and 1535 mg in infants (<1 year), between 1516 and 2750 mg in children aged 1 to <10 years, between 2093 and 3712 mg in children aged 10 to <18 years, and between 2463 and 3991 mg in adults (≥18 years). Based on the relationships between potassium intake and blood pressure and stroke, European Food Safety Authority (EFSA NDA Panel 2016) recommended an adequate intake (AI) of 3500 mg day^{-1} for youths and adults, and 800 mg for children aged 1 to 3 years.

6.5.5 Sulphur, Calcium and Magnesium

It is now well-recognized that N, P and K fertilizers alone are not always sufficient to provide balanced plant nutrition for optimal crop yields and quality. Deficiencies of S are frequently observed in many parts of the world and S is now considered as the fourth major limiting nutrient after N, P and K. The rise in S deficiencies is possibly because of large removal of S in crop harvests, use of high analysis N, P and K fertilizers, decrease in S-based pesticide use, and reduction in the industrial emissions and subsequent depositions. Sulphur is a constituent of amino acids (cysteine and methionine) and is involved in protein synthesis. It improves nutritional quality of seeds due to higher content of methionine. Higher cysteine content of cereal grains improves baking quality of flour. Its application increases S-rich proteins (Glutelins) in maize. Sulphur imparts stability to chlorophyll molecule by forming complex with protein and increases leaf area. It acts as a functional group (SH) of enzymes of photosynthesis, respiration and fatty acid synthesis and N_2 fixation in legumes. Being a component of biomembranes, S regulates ion transport across membrane and induces salt tolerance. It decreases heavy metal toxicity (Cu, Cd, Zn) by complexing with metallothioneins.

In humans, S-containing compounds are present in all the body cells and constitute essential components of amino acids (methionine, cysteine), vitamins (thiamin, biotin), hormone (insulin), heparin, fibrinogen, taurine, glutathione (GSH), S-adenosylmethionine, α-keto-γ-methyl-thiobutyrate, methanethiol, alpha-lipoic acid, coenzyme A, chondroitin sulphate, glucosamine sulphate, metallothionein, and inorganic sulphate. Except for the S-containing vitamins, all the S compounds are synthesized from methionine (Baker 1986). Sulphur is also needed for a number of chemical reactions involved in the detoxification of aromatic compounds in the body and for metabolism of drugs, steroids, and xenobiotics (Parcell 2002). The recommended daily combined intake of S-containing amino acids varies from 13 to 25 mg kg^{-1} body weight, which is equivalent to approximately 910–1750 mg day^{-1} for a 70 kg adult (Storch et al. 1988). Presence of organosulphur compounds (isothiocyanates, diallyl sulfide, allicin) in garlic, onions, and other vegetables are perceived to have beneficial human health effects. Sulphur-containing compounds found in onion including alk(en)yl cysteine sulphoxides, thiosulphonates, mono-, di- and tri-sulphides besides providing characteristic flavour supposedly have anticarcinogenic, antiplatelet and antithrombotic properties, and antiasthmatic and antibiotic effects (Griffiths et al. 2002).

Calcium is a constituent of cell wall and membranes and it helps to maintain membrane permeability and stability. It promotes cell division, cell elongation and hence growth. It also retards senescence and abscission of leaves. It is helpful in pollen germination and pollen tube growth. In the human body, Ca is fifth most abundant element after C, H, O and N. It makes 1.9% of the body by weight (Nordin 1976). Calcium accumulates in the body at an average rate of 180 mg per day during the first 20 years of growth (FAO/WHO 2001). Calcium is the largest constituent of bone, comprising 32% by weight. Over 99% of total body Ca is found in bones and teeth, where it functions as a key structural element. The remaining body Ca functions in metabolism, serving as a signal for vital physiological processes, including vascular contraction, blood clotting, muscle contraction and nerve transmission (WHO 2009b). Calcium intake has been associated with both the prevention and formation of kidney stones depending on whether Ca is consumed with food or separately. Dietary Ca reduces the incidence of kidney stones by binding oxalic acid (a precursor to kidney stones) in the lower small intestine and decreasing its absorption. On the other hand Ca supplements not ingested with food or taken at high levels (>2500 mg day^{-1}) could increase the risk of kidney stone formation (WHO 2009b). There is no recommended dietary allowance (RDA) for Ca but the suggested adequate intake (IOM 1997) ranges between 500 and 1300 mg day^{-1} depending on the age of the individual (Table 6.14).

Magnesium is a component of chlorophyll molecule in plants. It is involved in the activation of enzymes of photosynthesis, starch and sugar production. It regulates protein synthesis and is involved in energy (ATP) production and transfer processes. Magnesium helps to increase plant growth and yield by enhancing leaf area duration, and increasing root growth. It has a role in translocation of carbohydrates from source to sink resulting in increased size of potato tubers and cereal grains. Magnesium is the fourth most abundant cation in human body and the second most abundant cation in intracellular fluid. It comprises about 0.034% of body weight and an adult weighing 70 kg contains 21–38 g Mg. About 50–60% of

Table 6.14 Dietary reference values for adequate intake (AI) of Ca and estimated average requirement (EAR) and recommended dietary allowance (RDA) of Mg

Age group (years)	Ca (mg day^{-1})	Mg (mg day^{-1})	
	AI	EAR	RDA
1–3	500	65	80
4–8	800	110	130
9–13	1300	200	240
14–18	1300	340 (M)/300(F)	410 (M)/360 (F)
19–50	1000	330 (M)/255 (F)	400 (M)/310 (F)
>50	1200	350 (M)/265 (F)	400 (M)/310 (F)

Compiled from IOM (1997)
M Male, *F* Female

the body's Mg is found in bone; 30–40% in muscles and soft tissues, and 1% in extracellular fluid (Webster 1987). It is a cofactor for about 350 enzymes, many of which are involved in energy metabolism. It is also involved in protein and nucleic acid synthesis, cardiac and vascular functions, insulin sensitivity, and in stabilizing and protecting the membranes. Both dietary Ca and Mg probably play a role in the etiology of osteoporosis and cardiovascular disease (IOM 1997). Extracellular and intracellular Mg^{2+} concentrations significantly affect cardiac excitability and vascular tone, contractility, reactivity and growth (Laurant and Touyz 2000; Touyz and Yao 2003). Several studies have shown relationship between dietary Mg intake and blood pressure in humans (Joffres et al. 1987) leading to the proposition that increased Mg intake can help in the prevention of hypertension and cardiovascular disease (Whelton and Klag 1989; Van Leer et al. 1995; Simons-Morton et al. 1997). On the contrary, some studies did not find any association between Mg intake and blood pressure or cardiovascular disease (Whelton and Klag 1989; Rosenlund et al. 2005). It has been suggested that Mg^{2+} may not be universally effective against hypertension and it may benefit a specific group of patients. Current international guidelines recommend maintenance of adequate dietary Mg intake for prevention management of hypertension (Stergiou and Salgami 2004; Khan et al. 2005). Mg is also considered essential for maintaining Ca, K and Na homeostasis (Seelig 1989; Wacker 1980). Magnesia is commonly used as antacid and as an antidote against poisons such as acids and arsenic. Epsom salt (magnesium sulphate) is used as a laxative (EFSA 2006). Depending on the age and gender, the Estimated Average Requirement (EAR) and Recommended Dietary Allowance (RDA) for Mg range between 65–350 and 80–400 mg day^{-1}, respectively (Table 6.13). EFSA's Panel on Dietetic Products, Nutrition and Allergies (NDA) set an AI for magnesium of 350 mg day^{-1} for men and 300 mg day^{-1} for women (EFSA 2015a). For children the AI ranges between 170 and 300 mg day^{-1}, according to age.

6.6 Effects of Excessive Macronutrient Uptake

6.6.1 Nitrogen

Excessive nutrient uptake and accumulation in plants can have adverse effects both for food production and human health. Enhanced stem elongation of plants by N application is a negative side effect that may lead to yield reduction by lodging. High nitrogen produces succulence in plants and enhances their sensitivity to water and temperature stress. Plants with high nitrogen accumulation are susceptible to lodging, pathogens and pests. Excessive N fertilizer application could increase the concentration of nitrate in crops, especially vegetables. Vegetables accumulate different amounts of nitrate ranging between less than 20 and more than 250 mg nitrate per 100 g fresh weight (Table 6.15). Vegetables that accumulate nitrate include spinach, parsley, lettuce (1000–6000 mg kg^{-1}); cabbage, cauliflower, celery, artichoke, kale, leek (a few hundred to 1000 mg kg^{-1}); and beetroot,

Table 6.15 Classification of vegetables according to nitrate content

Nitrate content (mg 100 g^{-1} fresh weight)	Vegetables
Very low (<20)	Artichoke, asparagus, broad bean, Brussels sprouts, eggplant, garlic, green bean, melon, mushroom, onion, pea, pepper, potato, summer squash, sweet potato, tomato, watermelon
Low (20- <50)	Broccoli, carrot, cauliflower, cucumber, pumpkin, chicory
Medium (50- <100)	Cabbage, dill, savoy cabbage, turnip
High (100- <250)	Celeriac, Chinese cabbage, endive, fennel, kohlrabi, leek, parsley
Very high (> 250)	Celery, cress, chervil, lettuce, red beetroot, spinach, rocket (rucola)

Adapted from Gorenjak and Cencič (2013) and Santamaria (2006)

broccoli, and carrot (200–300 mg kg^{-1}) (Ayaz et al. 2007). Nitrogen fertilization and light intensity are the major factors that influence nitrate content in vegetables (Santamaria 2006). Presence of free nitrate and nitrite in leaf, stem and flower-vegetables can be a health risk to humans. Besides the nitrate content per se, the amount of nitrate consumed through vegetables, depending on dietary habits and preparation of the vegetable (Pennington 1998; Thomson et al. 2007), can reach up to 85% of the total intake. The intake of nitrate through vegetables accounts for more than 80% of the total nitrate ingested by humans in the USA and 60% in the UK (Peoples et al. 2004). Nitrates in vegetables are significantly reduced during handling, storage, processing including washing, peeling and cooking. This holds true for vegetables eaten cooked, such as potato, spinach, and cabbage. For vegetables eaten raw only handling and storage impact nitrate levels (EFSA 2008a). Mozolewski and Smoczynski (2004) showed that levels of nitrate and nitrite in potatoes can also be decreased by 18–40% and 25–75%, respectively, after washing, peeling and rinsing. Cooking of potato tubers by different methods such as boiling, microwave, steaming, and deep frying can lessen nitrate (16–62%) and nitrite content (61–98%). Generally, about 45% of the total nitrate in the vegetables can be reduced by pickling, and 75% reduced by cooking (Du et al. 2007).

The excessive intake of nitrate in vegetables is often associated with harmful effects on human health such as methaemoglobinaemia and the possibility of endogenous formation of carcinogenic N-nitroso compounds (Gorenjak and Cencič 2013). However, views differ on the effects of high dietary intake of nitrate on human health. As discussed in Chap. 5, though high nitrate level in food has been implicated in causing methemoglobinemia yet its role has not been established beyond doubt. Earlier, concerns were raised for restrictions on nitrate levels in food due to the formation of carcinogenic nitrosamines by nitrosation of amines in the gastrointestinal tract of humans; but this view is no more pursued because of decline in the incidence of gastric cancers despite increased consumption of nitrate-rich green vegetables. Some studies indicate a positive protective effect of nitrate against gastrointestinal pathogens (Mosier et al. 2004). Nitrates per se are relatively non-toxic. They are discussed in toxicology due to their metabolic products, which are converted via nitrite to N-nitro compounds (Gangolli et al. 1994). Several authors (McKnight et al. 1999; Ying and Hofseth 2007; Lundberg et al. 2008; Hord et al. 2009; Ralt 2009) have emphasized the positive physiological effects of metabolic products of nitrate and therefore advocate a high concentration of nitrate in vegetable. Nitrate offers an alternative pathway to bioactive NO and has important physiological role in vascular and immune functions. Excessive nitrogen intake could also be in the form of proteins and amino acids particularly in the developed countries where not only the diets are rich in protein but also these are taken as dietary supplements. Though the current knowledge about the relationship between protein intake and health is insufficient to define safe upper limit yet it has been suggested to avoid protein intakes of more than twice the reference dietary amount. High intakes of protein by patients with renal disease have been found to contribute towards deterioration of kidney function (Millward 1999; WHO 2007).

6.6.2 *Phosphorus*

There is no suitable biomarker of phosphorus intake or status that can be used for setting dietary reference values (DRVs); the adequate intake (AI) for phosphorus is set at 550 mg day^{-1} for adults and 250–640 mg day^{-1} for children (EFSA 2015b). The average intake from foods and supplements in adults is usually between 1000 and 2000 mg day^{-1} and there is no evidence of adverse effects associated with the current intakes of phosphorus (EFSA 2006). In healthy individuals, excess phosphorus does not accumulate in the body as excess phosphate is excreted. Data for USA show that for all age groups, the usual mean phosphorus intake in both men and women exceeded the EAR and RDA (Table 6.16; Calvo et al. 2014). The phosphorus requirement has often been linked to the calcium requirement, allowing a Ca: P molar intake ratio of about 1 or mass intake ratio of 1.5. Consuming too much P causes transfer of Ca from the bones to the blood for maintaining balance. This leads to weakening of bones, calcification of internal organs,

Table 6.16 Mean intake, estimated average requirement (EAR), tolerable upper intake level (UL) and recommended dietary allowance (RDA) for phosphorus (mg day^{-1}) by age in the USA

	Mean P intake	EAR			
Age (years)	Men	Women	UL	RDA	
1–8	1030–1145	1030–1145	380–405	3000	460–500
9–18	1321–1681	1067–1176	1055	4000	1250
19–30	1656	1120	580	4000	700
31–50	1727	1197	580	4000	700
51–70	1492	1106	580	4000	700
71 and above	1270	985	580	3000	700

Source: IOM 1997, adapted from Calvo et al. 2014

hyperparathyroidism, and increased risk of heart attack and other vascular diseases. Widespread use of P-containing food additives by the food industry can lead to excessive consumption of P with food. Most P-containing additives are inorganic salts of phosphorus that are widely used in the processing of different foods, ranging from baked goods and restructured meats to cola beverages. Cola soft drinks contain between 120 and 200 mg L^{-1} phosphorus. In the US the contribution from phosphorus-containing food additives is estimated at 320 mg day^{-1}, constituting twenty to thirty percent of the adult phosphorus intake (Calvo and Park 1996). However, the amount of P contributed by the use of phosphorus-containing food additives in processed and prepared foods is difficult to quantify as these are never mentioned on the label/package (Calvo and Uribarri 2013). Other sources of dietary phosphorus such as dietary supplements, vitamins and minerals may contribute on average 108 mg phosphorus day^{-1} (Calvo et al. 2014).

6.6.3 Potassium

Excess of potassium supply to plants can lead to reduced uptake of Ca and Mg, resulting in deficiencies of these nutrients in the plant. In humans, occurrence of high potassium level in serum following excessive dietary intake of potassium is rare because of effective homeostatis mediated by increased cellular uptake of potassium from the bloodstream by various organs and increased urinary excretion (Lehnhardt and Kemper 2011). There is no biomarker of potassium status that can be used for setting DRVs for general population and as such there is no fixed Tolerable Upper Intake Level (UL) (EFSA NDA Panel 2005). There are contradictory reports on the adverse effects of high dietary potassium intake on human health. Some studies suggest little or no adverse effects of dietary intake of potassium up to 5000–6000 mg (129–154 mmol) day^{-1} in healthy adults. In addition to dietary intake, long-term supplemental intakes of about 3000 mg K (77 mmol) day^{-1} as KCl are reported to have no adverse effects in healthy adults. On the other hand, some studies found that supplemental potassium in doses ranging between 1000 and 5000 mg (26–128 mmol) day^{-1} can cause gastrointestinal problems and doses ranging between 5000 and 7000 mg (128–179 mmol) day^{-1} can adversely affect heart function (Perazella and Mahnensmith 1997; EFSA NDA Panel 2005; Cicero and Borghi 2013). However, hyperkalaemia, the condition when the potassium concentration in serum is greater than 5.5 mmol L^{-1} usually does not occur in healthy adults (Pepin and Shields 2012; Michel et al. 2015) It may occur in individuals with impaired renal function (Lehnhardt and Kemper 2011; Crook 2012), unusually high intakes of oral potassium supplements, parenteral potassium administration and a potassium shift from cells (e.g. metabolic acidosis, hypoxia, severe tissue damage). Symptoms of mild to moderate hyperkalaemia are usually non-specific and may include generalised weakness, paralysis, nausea, vomiting, and diarrhoea (Pepin and Shields 2012). Severe hyperkalaemia may lead to cardiac arrhythmias (Paice et al. 1983; Lehnhardt and Kemper 2011).

6.6.4 Sulphur, Calcium and Magnesium

Crops differ greatly in S content and crucifers have higher content (1.1–1.7% of seed dry weight) than legumes (0.25–0.3%) and cereals (0.18–0.19%). Similarly, crops differ in S requirement; for instance the members of the *Brassicaceae* family, protein-rich (lucerne, clover) and fast growing crops (e.g. maize, sugarcane) require higher amounts. Typically oilseed rape crop removes between 20 and 30 kg S ha^{-1}, while cereals remove about 10–15 kg S ha^{-1} (Walker and Booth 1992). Members of *Brassicaceae* family including crops grown as vegetables, spices and sources of oil, require high amounts of S to form special S-containing compounds, Glucosinolates, considered to provide characteristic flavor to *Brassica* vegetables, the intensity of which depends on the concentration and the degree of hydrolysis of Glucosinolates. Commonly grown *Brassica* vegetables include broccoli, Brussels sprouts, cabbage, cauliflower, collards, kale, turnip greens or leaf rape. Glucosinolates are secondary metabolites representing a diverse class of alkyl aldoxime-O-sulphate esters with a β-D-thioglucopyranoside group attached to the hydroximine carbon in Z-configuration to the sulphate group (EFSA 2008b). They can be synthesized from seven amino acids viz. alanine, (iso) leucine, tyrosine, tryptophan, valine, phenylalanine, methionine and chain elongated homologues of methionine. More than 120 individual compounds have been characterized and depending on their structure, these have been classified as aliphatic, aromatic, ω-methylthioalkyl and heterocyclic (indole-) glucosinolates (Fahey et al. 2001). While all types of glucosinolates have been found in *Brassica* crops, the methionine-derived aliphatic compounds are the most important in vegetables (Mithen et al. 2003). Glucosinolates and their breakdown products commonly known as mustard oil glucosides or thioglucosides have been widely researched for their beneficial and harmful effects on plants, humans and animals. Being source of bioactive compounds (isothiocyanates, thiocyanates, nitriles and epithionitriles) glusosinolates provide distinctive benefits to human nutrition and plant defence. In plants, glucosinolates are part of an innate defense system against insects, fungi and bacteria (Rask et al. 2000; Kliebenstein et al. 2005). In humans, glucosinolates have potential anti-carcinogenic properties. Isothyocyanate and indole products formed from glucosinolates may regulate cancer cell development by regulating target enzymes, controlling apoptosis and blocking the cell cycle (Cartea and Velasco 2008). Human obtain these compounds through consumption of glucosinolates containing vegetables. Daily dietary intake of these compounds could be up to several milligrams depending on the vegetable taken. Glucosinolates are the highest in garden cress (*Lepidium sativum*) (3.89 mg g^{-1}) and the lowest in Pe-tsai Chinese cabbage (*Brassica rapa*) (0.20 mg g^{-1}) though the values vary considerably for the same vegetable (EFSA 2008b). Animal-derived products such as milk contain much less thiocyanate (0.18 mmol L^{-1}) than the vegetables (0.8–20 mmol kg^{-1} dry weight) suggesting that animal products contribute little to dietary human intake of these compounds. Glucosiolates could be toxic to humans and animals because of the formation of isothiocyanates, thiocyanates, oxazolidinethiones and nitriles (Burel et al. 2001), which interfere with iodine uptake (thiocyanates) and the synthesis of the thyroid hormones (oxazolidinethiones) leading to hypothyroidism and enlargement of the thyroid gland or goitre (Griffiths et al. 1998; Halkier and Gershenzon 2006). Prominent signs of glucosinolates toxicity in farm animals include growth retardation, reduction in milk and egg production, weakened reproductive activity, and impaired liver and kidney functions, (Mawson et al. 1994). However, there is no evidence of any goitrogenic effect on humans because of *Brassica* consumption (Mithen 2001) and there is no relationship between glucosinolate intake and the incidence of thyroid cancer.

Calcium is a non-toxic mineral nutrient even in high concentrations and is very effective in detoxifying elevated concentrations of other mineral elements in plants. Excessive Ca is precipitated in the cell walls and the vacuoles as Ca-oxalate. Most plant species can accumulate high Ca contents in leaf blades (100 g kg^{-1} dry weight) without any symptoms of toxicity (calcicole plant species). In these species, excessive Ca is sequestered as insoluble Ca oxalate and deposited either in the cell wall or in the vacuole. However, some species may have insufficient capacity for this mechanism of detoxification (calcifuge species) and their growth is severely depressed at high tissue contents of Ca. Humans are protected from excess intake of Ca by a closely regulated intestinal absorption mechanism through the action of vitamin D. In healthy individuals, Ca absorbed in excess of need is excreted by the kidney. However, excess Ca intake could be a cause of concern for individuals susceptible to milk alkali syndrome (the simultaneous presence of hypercalcaemia, metabolic alkalosis and renal insufficiency). High amounts of Ca can reduce the absorption of Fe, Zn, Mg and P in the intestine, but evidence confirming the depletion of these minerals with consumption of Ca-rich diets is lacking (WHO 2009b). Sufficiency and deficiency of dietary Ca probably could influence insulin resistance because of fluctuations in Ca-regulating hormones. High Ca intake can reduce intestinal magnesium absorption and decrease renal magnesium excretion.

High dietary intake of Mg seldom causes hypermagnesaemia or high levels of serum Mg in individuals with normal kidney function. Hypermagnesaemia may occur in persons with impaired renal function associated with decreased ability to excrete Mg. However, increased intake of Mg as supplements and drinking water rich in Mg and sulphate can have a laxative effect. Excessive intake of Mg salts may cause diarrhoea.

6.7 Effects of Deficient Macronutrient Uptake

Nitrogen being mobile, its deficiency first appears on the older leaves of the plant which show pale yellow color initiating at the tip and extending towards the base of the leaf (Fig. 6.12). With time, the yellowing extends to upper leaves and ultimately the whole plant looks yellow. The affected leaves are narrow and look erect due to their reduced angle with the stem, particularly in wheat and rice crops. Nitrogen-

Fig. 6.12 Symptoms of nitrogen deficiency in wheat, maize and rice (Source: Nayyar and Chhibba 2000; with kind permission)

Fig. 6.13 Symptoms of phosphorus deficiency in maize and sorghum (Source: Nayyar and Chhibba 2000; with kind permission)

deficient plants have poor tillering and shorter internodal length leading to poor crop growth and yield.

Nitrogen deficiency depresses the rate and the extent of protein synthesis in plants. During early stages of nitrogen deficiency, usually there is accumulation of carbohydrates because of build-up of total sugars. Nitrogen deficiency also results in the accumulation of anthocyanin pigments in the epidermal and sub-epidermal cells and the stem. Nitrogen deficiency adversely affects flowering and fruit setting resulting in small and poor quality fruits. Nitrogen deficiency, in addition to influencing crop productivity, affects the nutritional quality of the produce. For example, N in plant is a source of amino acids and proteins to the human body; lower the N content in the produce lesser will be the amount of amino acids supplied through consumption of nutrient deficient food. When the dietary intake of nitrogen is zero, and energy and all other nutrients are consumed in adequate amounts, there are obligatory nitrogen losses of about 48 mg kg^{-1} day^{-1} from the body (WHO 2007; Rand et al. 2003).

Similar to nitrogen, phosphorus deficiency also appears first on older leaves. Initially, the P-deficient plants have bright dark green color, which subsequently become purplish from tip towards base (Fig. 6.13). In case of severe deficiency, the stem also turns purplish. Phosphorus deficiency retards plant growth and tillering and reduces produce quality and storage properties. Since plants require phosphorus for cell growth and formation and ripening of fruits and seeds, its deficiency can severely hinder crop yields and fruit/seed development. Plants deficient in phosphorus cannot utilize other nutrients efficiently.

Phosphorus deficiency disturbs nitrogen metabolism resulting in accumulation of soluble nitrogen compounds (free amino acids and amides) and a decrease in protein content. Very often phosphorus deficiency results in increased accumulation of free reducing sugars. Phosphorus is found in almost every food and as such its deficiency is rare in humans. Typical daily consumption of phosphorus is about 1.5 g per capita for adults, well above the recommended daily allowances, which are 0.8 g per capita for adults over 24 years of age and children, and 1.2 g for young adults (NRC 1989). Phosphorus deficiency (hypophosphatemia) can occur in patients suffering from liver disease, sepsis, antacid therapy with aluminium containing drugs, and in diabetic ketoacidosis. A deficiency in phosphorus can lead to lowered appetite, anaemia, muscle weakness, improper bone formation (rickets) in children and osteomalacia in adults, numbness, hyperthyroidism, De Toni-Fanconi syndrome and a weakened immune system (Lotz et al. 1968). Inadequate intake of calcium and phosphorus has been associated with pathogenesis of bone disease in newborn infants (Bishop 1989).

Potassium being highly mobile in the plant, its deficiency appears first in the lower, older leaves. Most common symptoms of potassium deficiency are scorching or firing along leaf margins. Later, these necrotic areas undergo scorching and they look brownish and become brittle. Potassium deficient plants have poorly developed root system, grow slowly and are susceptible to lodging because of weak stalks and stems and accumulate unused nitrogen. Potassium deficiency results in a decrease in total and reducing sugars. Seeds and fruits are small and shrivelled and plant possesses low resistance to diseases. In humans, symptoms of potassium deficiency in the body include water retention, heart arrhythmias, continual thirst, high blood pressure, nerve and muscle dysfunction and vomiting. The vomiting will lead to a further deficiency in the mineral. Potassium deficiency or hypokalaemia, when serum potassium concentration is less than 3.5 mmol L^{-1} (Pepin and Shields 2012), may be caused by increased potassium losses (e.g. via diarrhoea, vomiting, intense and prolonged sweating or excessive renal losses) or intracellular shift of potassium (e.g. during alkalosis) (Rastegar 1990). Use of diuretics without potassium compensation can lead to its deficiency. Hypokalaemia resulting from insufficient dietary intake is rare; it may only occur due to extremely low calorie diets or during recovery from malnutrition. Potassium deficiency, which leads to electrolyte imbalance in the body can cause a range of effects, including cardiac arrhythmia, atrial fibrillation, muscle weakness, nausea and vomiting, low muscle tone in the gut, polyuria, decreased peristalsis possibly leading to intestinal ileus, mental depression and respiratory paralysis (Krijthe et al. 2013; Rodenburg et al. 2014). Longer-term hypokalaemia is believed to cause a predisposition to hypertension (UKEVM 2003). Low potassium intake can induce Na retention and an increase in blood pressure. Increased potassium intake as a supplement can lower blood pressure and increase urinary sodium excretion (Whelton et al. 1997; Naismith and Braschi 2003).

As discussed in Sect. 6.5.5, deficiency of S in plants is increasing worldwide and this could significantly impact human nutrition and health. Legumes are

Fig. 6.14 Sulphur deficiency in field peas (Source: Nayyar and Chhibba 2000 with kind permission)

particularly sensitive, but other crops, including cereals, brassica and tea are affected by S deficiency. Symptoms of S deficiency in plants are similar to those caused by insufficient N: pale green plants and reduced or stunted growth (Fig. 6.14). However, with N deficiency pale leaves are seen first on older leaves, while with S deficiency they occur first on younger leaves. Sulphur deficiency adversely affects the formation of proteins, which reduces nitrogen utilization in the plant.

The unused NO_3^- may accumulate in the crop tissue and consumption of plant foods with excessive NO_3^- may be hazardous to human health (see Sect. 6.6.1). Wheat flour from S-deficient plants has been found to contain substantially higher levels of free amino acids, particularly asparagine, a precursor of the potentially carcinogen acrylamide, than flour from wheat grown under sufficient S nutrition. Elevated levels of asparagine resulted in acrylamide levels up to six times higher in S-deprived wheat flour, compared with S-sufficient wheat flour (Muttucumaru et al. 2006). Sulphur deficiency in wheat, however, not only reduces the nutritional quality, but also adversely affects the baking quality of flour. In contrast, it has been shown that S-deficient potatoes had lower acrylamide production and fertilization of potato with sulphur leads to an increase in acrylamide formation during processing (Elmore et al. 2007). Sulphur application has been reported to reduce glucose concentrations and mitigate the effect of high N application on the acrylamide-forming potential of some of the French fry-type potatoes (Muttucumaru et al. 2013).

Most soils, except highly weathered and leached acid soils, contain adequate amounts of Ca and Mg for crop growth. Ca deficiency in plants can occur in saline and sodic soils. These soils cover 260 Mha world-wide and are mostly found in the arid subtropics. In these soils, excessive Na concentrations impair plant Ca uptake. Temporary unavailability of Ca to developing tissue can cause Ca-deficiency disorders in horticulture crops. Calcium deficiency is characterized by a reduction in the growth of meristematic tissue, in which the affected tissue becomes soft because of dissolution of cell walls. Calcium deficiency is known to result in chromosome abnormality. In the roots of Ca-deficient plants, mitosis is abnormal and is often not followed by cell wall formation resulting in two nuclei within a cell. Although Ca requirements of plants are low, yet it has great significance in balancing other nutrients including N. Application of NO_3^- containing fertilizers generally enhances Ca concentration in plants for maintaining cation-anion balance. Calcium deficiency related disorders occur in plant organs having low transpiration and high growth rate (e.g. fleshy fruit and tubers) that become prone to fungal infections and early tissue senescence. Calcium deficiency disease is called bitter pit in apples and blossom-end rot in tomatoes. Even a small increase in Ca level can maintain quality during storage. Both Ca and Mg are essential to human health and inadequate intake of either nutrient can impair health. Inadequate intake of Ca has been associated with increased risks of osteoporosis, nephrolithiasis (kidney stones), colorectal cancer, hypertension and stroke, coronary artery disease, insulin resistance and obesity (WHO 2009b). Individuals who avoid dairy products or lack access to them may be at increased risk of calcium deficiency.

Magnesium deficiency in plants occurs on strongly acidic soils. High concentrations of competing cations such as Al^{3+} and Mn^{2+} in the soil solution can aggravate Mg deficiency. In alkaline soils, carbonate formation and excess Ca, K and Na reduce Mg availability to plants. Plants differ greatly in their response to Mg deficiency in soil; buck wheat being sensitive and small grains and grasses being slightly responsive to Mg fertilization. Magnesium deficiency is characterized by interveinal chlorosis and necrosis of the affected leaves. In plants such as cotton, Mg deficiency may induce formation of anthocyanins. In tobacco, Mg deficiency diseases, known as sand drown occurs when leaf Mg content is less than 0.25% (dry weight basis). Deficiency of Mg in pastures leads to grass tetany disease in grazing animals (McNaught and Dorofaeff 1965). Magnesium deficiency results in decreased rate of photosynthesis. In humans, severe Mg deficiency is rarely seen in healthy people because it is present and consumed through a variety of foods including fruits, vegetables, nuts, dairy products and fish (see Sect. 6.4). Low Mg levels in humans are associated with endothelial dysfunction, increased vascular reactions, hypertension, coronary heart disease, type 2 diabetes mellitus, elevated circulating levels of C-reactive protein and decreased insulin sensitivity and metabolic syndrome (WHO 2009b). Elderly people are particularly at risk of Ca and Mg deficiencies because of inadequate nutrient intake from food and water, multiple drug use, and altered gastrointestinal function. Inadequate intake of Ca and Mg may affect the aging process; however, Mg-deficient conditions have been associated with neuromuscular and cardiovascular disorders, endocrine disturbances and insulin resistance. It has been suggested that drinking of Mg-containing water could play an important role in protecting against cardiovascular trauma and other ailments (Marier and Neri 1985).

6.8 Optimizing Macro- and Secondary Element Status

6.8.1 Soils

As a result of nutrients' critical roles and low supply, the management of nutrient resources is extremely important for crop production and human nutrition. The productivity and the quality of food grown and soil health is improved by balanced application of fertilizer nutrients. Nutrients in plants can be optimized through the application of inorganic and organic fertilizers and on-farm recycling of crop residues and agricultural waste. The choice of fertilizer material and amount, time, and method of its application varies according to the differing need and perceptions of the soils and crops, and to factors controlling transformations and utilization of fertilizer nutrients. Innovative tools and technologies are now available to effectively use the reactive N from all sources (soil, organic materials and urban wastes, BNF from legumes, and fertilizers) for crop production, and to meet our future food and nutrition needs with the same or less amount of fertilizer N and other resources. Adequate supply of nitrogen improves not only proteins but also other essential nutrients that are obtained from food grown with balanced application of N. However, excess of reactive N may have detrimental effects on human health (see Chap. 5). Synchronization of external N applications with crop needs and indigenous/soil N supply is the best way to enhance N use efficiency and reduce N losses from farms to the environment. Development of perennial agricultural systems may help reduce future N inputs into the environment. Phosphorus deficiency in soils can be overcome by applying P-containing fertilizers (see Table 6.6) and balancing it with nitrogen and potassium. Water soluble P fertilizers provide readily available source of P and it is important to apply these fertilizers at the time of seeding as majority of the phosphorus is taken up during initial stages of crop growth. Fertilizers enriched with S are now commonly used to correct S deficiencies, but gypsum can also be applied. A wide range of S-fertilizers, suitable for different environmental and farming conditions, are now available. Rates of S application to crops and pastures generally range between 10 and 30 kg ha^{-1} $year^{-1}$. Application of S containing fertilizers to certain crops can reduce the undesirable NO_3^- contained in their edible parts. An increased soil S level can significantly reduce NO_3^- concentration in tubers and leaves of kohlrabi (Losak et al. 2008), and turnip tops (De Pascale et al. 2007). N and S supplies exert a strong interactive impact not only on plant growth and mineral composition but also on the concentration of isothiocyanate in Kohlrabi (*Brassica oleracea* L. Var. *gongylodes*) (Gerendás et al. 2008). Calcium deficiency in soils can be overcome by adding Ca-containing compounds such as lime and gypsum. While lime is added mainly to overcome soil acidity, gypsum is used both as an S source and as an amendment to improve soil physical conditions. Superphosphate used as a source of phosphorus also supplies Ca as it contains gypsum ($CaSO_4$ $2H_2O$). Magnesium deficiency in soils can be managed by adding Mg compounds such as dolomite, Epsom salt, magnesite, serpentine etc. (see Table 6.10). Epsom salt being soluble in water is used as a fast-release Mg source both as soil and foliar application.

The other fertilizers being insoluble in water are used as slow-release fertilizers. Dolomite is the most widely used source of Mg, particularly in acid soils. It neutralizes the free acid present in certain phosphate materials and potential acidity of fertilizer N.

6.8.2 Humans

6.8.2.1 Nitrogen

Healthy adults on an average require 105 mg nitrogen kg^{-1} day^{-1} (0.66 g kg^{-1} day^{-1} of protein); and a safe intake level of 133 mg nitrogen kg^{-1} day^{-1} (0.83 g kg^{-1} day^{-1} protein) is recommended (Rand et al. 2003). An additional intake of 1, 9 and 31 g protein day^{-1} is recommended for pregnant women during the first, second and third trimesters, respectively. For lactating women, an average of 19 g protein day^{-1} is required, which may be reduced to 12.5 g protein day^{-1} after 6 months. Diet is the primary source of nitrogen containing compounds and it supplies protein, free amino acids, nucleotides and creatine to the human body. As humans are not capable of utilizing simple forms of nitrogen to synthesize protein, adequate amounts of amino acids of a suitable pattern must be present in the diet, either in a preformed state or as precursors that can be endogenously transformed to a suitable mix of amino acids for producing required amount of protein (WHO 2007). Of the 20 amino acids required by the human body, 9 of these called "indispensable amino acids" viz. leucine, isoleucine, lysine, methionine, phenylalanine, threonine, tryptophan, valine, histidine have to be obtained from dietary sources (Wardlaw and Insel 1996; Wildman and Medeiros 2000). The estimated daily requirement of different amino acids ranges between 4 (Cysteine and Tryptophan) and 39 (Leucine) mg kg^{-1} (Table 6.17). In addition to indispensable amino acids, the diet must also provide "dispensable"

Table 6.17 Indispensable amino acid requirements of adults based on mean nitrogen requirement (mg^{-1} day^{-1}) of 105 mg nitrogen or 0.66 g protein

Amino acid	mg kg^{-1} day^{-1}	mg g^{-1} protein
Histidine	10	15
Isoleucine	20	30
Leucine	39	59
Lysine	30	45
Methionine	10	16
Cysteine	4	6
Phenylalanine + tyrosine	25	38
Threonine	16	323
Tryptophan	4	6
Valine	25	39
Total	184	277

Adapted from WHO (2007)

Table 6.18 On-farm conversion efficiency for production of human-digestible protein from feed grains and forages

Source	Conversion efficiency (%)[a]	Conversion efficiency (%)[b]	Protein content (% of edible weight)[b]
Fish	50–60	30	18
Poultry and eggs	40–50	30	13
Chicken	–	25	20
Dairy/Milk	35–40	40	3.5
Swine	30–40	13	14
Beef	15–30	5	15

Sources: [a]Galloway et al. (2002), [b]Smil (2002b)

amino acids or some utilizable source of "non-specific" nitrogen to enable their synthesis and that of other physiologically important nitrogen-containing compounds, such as purines and pyrimidines, glutathione and creatine.

Amino acids and proteins in plant foods can be ingested either directly by humans or indirectly through consumption of animal products such as milk, eggs, and meat. Rates of on-farm conversion efficiencies for production of human-digestible protein from feed grains and forages by animals range between 5% and 60%, the lowest being for beef and the highest for fish (Table 6.18). Effective utilization of human-digestible dietary protein requires proper balance among the indispensable and dispensable amino acids and other nitrogen-containing compounds. An intake (kg day^{-1}) of 0.18 g indispensable amino acids and 0.48 g of dispensable amino acids is considered adequate to maintain body nitrogen homeostasis in healthy adults (WHO 2007). Excessive intake of indispensable amino acids leads to consumption of dispensable amino acids in detoxifying them, resulting in increased dietary nitrogen requirement. The rate of formation of dispensable amino acids in the body is determined by the total intake of nitrogen and recovery of adequate amounts of urea. At lower levels of nitrogen intake, the formation of dispensable amino acids is impaired (Jackson 1983). Unlike animal protein sources, plant protein sources are generally deficient in lysine, tryptophan, or methionine (Garrison and Somer 1995). However, for a vegetarian diet it is not essential to take an animal protein, rather a balance of essential amino acids could be achieved by combining different plant protein sources, for instance beans (low in methionine) with rice (low in lysine).

6.8.2.2 Phosphorus

Phosphorus is widely found in foods as phosphates and the foods rich in proteins are generally high in phosphorus. The major dietary sources of phosphorus intake include milk and dairy products (100–900 mg 100 g^{-1}), nuts, beans, peas, lentils, fruits especially banana, grains and grain-based products (100–300 mg 100 g^{-1}), and fish, meat and meat products (200 mg 100 g^{-1}). In children and adults, milk and

Table 6.19 Phytic acid content of different foods

Product	Phytic acid (g kg^{-1} dry weight)
Lettuce leaves	0.1
Roots/tubers	0.1–2.2
Tomato	0.4
Plantain	1.8
Seeds	
Wheat	3.9–13.5
Corn	8.3–22.2
Oat and barley	7.0–11.6
Rye	5.4–14.6
Sorghum	9.1–13.5
Rice (brown)	8.4–8.9
Rice (white)	3.4–5.0
Cowpea	9.4
Soybean	10.0–22.2
Pigeon pea	7.1–70.0

Adapted from Frossard et al. (2000)

dairy products contribute up to 30–53% and grains and grain-based products up to 27–38% towards phosphorus intake. The meat and meat products contribute 10–25% of phosphorus intake in the age groups from 10 years and above. Phosphorus is absorbed easily from meat products, and only about half of the phosphorus contained in plant foods is absorbed. Inorganic salts of phosphorus added to foods during preparation or processing are rapidly and efficiently absorbed (80–100%) compared to organic sources which are slowly and less efficiently (40–60%) absorbed (Calvo and Uribarri 2013).

Some forms of dietary phosphorus such as phytic acid or phytin are less bioavailable because the humans lack phytase enzyme required to hydrolyse phytin and release phosphorus. In cereal grains, legume seeds, and nuts about 50–80% of total P accumulates as phytic acid (myo-inositol hexakisphosphoric acid). In potato tubers, phytic acid P may be higher than 20% of total P during early tuber development (Samotus and Schwimmer 1962). In seeds, where it is primarily present as phytin, a mixed K, Ca, Mg, Zn and Fe salt of inositol may account for up to 7% of their dry weight (Table 6.19; Frossard et al. 2000). It is estimated that worldwide 35 Tg of phytic acid containing 9.9 Tg P is combined with about 12.5 Tg K and 3.9 Tg Mg to form 51 Tg of phytate in 4.1 Pg of crop seeds and fruits produced in a year (Lott et al. 2000). Cereal grains dominate, accounting for 77% of the total phytic acid stored in crop seeds and fruits. Phytic acid is considered anti-nutrient because of its inhibitory effect on the bioavailability of mineral nutrients. Phytic acid has the ability to chelate with multivalent metal ions, particularly Zn, Ca and Fe and form insoluble salts that are poorly absorbed from the gastrointestinal tract leading to their deficiency (Reinhold et al. 1973; Brune et al. 1992; Lei and Liu 1989). The phytate-mineral complexes being insoluble at physiological pH render the minerals biologically unavailable to humans. There are several reports of

Table 6.20 Approximate potassium content of some food groups

Food group	Potassium (mg 100 g^{-1} fresh weight)
Beans and peas (e.g. cowpeas, pigeon peas, lima beans, African yam beans)	1300
Nuts (e.g. hazelnuts, walnuts, cashew nuts, brazil nuts)	600
Green vegetables (e.g. spinach, cabbage, parsley)	550
Root vegetables (carrots, onions beetroot)	200
Other vegetables (e.g. tomatoes, cucumbers, pumpkins)	300
Fruits (e.g. banana, papaya, dates)	300

Adapted from WHO (2012)

phytic acid-induced Zn deficiency in humans (Maga 1982; Reddy et al. 1982). Therefore, regulating dietary phytic acid intake could help in optimizing bioavailability of essential nutrients. For instance, people with higher Ca requirement or those with low Ca intake or having higher incidence of Zn or Fe deficiency should lessen dietary phytic acid consumption so as to improve bioavailability of these nutrients. Availability of low phytate mutants for crops, such as maize and barley, offers option to increase bioavailability of metal ions (Lott et al. 2000). Contrarily, some in vitro animal and epidemiological studies indicate beneficial effects of chelation properties of phytic acid, such as lowering of serum cholesterol and triglycerides and preventing heart disease, renal stone formation and colon cancer (Sharma 1986). Some studies show that metabolites of phytic acid may function as second messengers, whereas higher phophorylated inositols may act as neuromodulators (Zhou and Erdman 1995). However, studies are inconclusive with respect to the adverse and potentially beneficial effects of dietary phytic acid on human health.

6.8.2.3 Potassium

Diet is the primary source of potassium and it is found in all foods with higher content in beans, nuts and leafy vegetables (Table 6.20). The main food groups contributing to potassium intake include starchy roots or tubers and vegetables, fruits and berries, whole grains and grain-based products, milk and dairy products, nuts and coffee (EFSA NDA Panel 2016). Potassium occurs in foods mainly associated with weak organic acids. WHO (2012) recommended that potassium be consumed through food and the best dietary sources of potassium are fresh unprocessed foods as some food additives also contain potassium salts (e.g. potassium iodide). Several potassium supplements are available, including potassium acetate, potassium bicarbonate, potassium citrate, potassium chloride, potassium gluconate, potassium-L-ascorbate, potassium glycerophosphate, potassium lactate, potassium hydroxide, which may be added to both foods and food supplements. Salts such as potassium sulphate, potassium L-pidolate, potassium malate and potassium molybdate may only be used in the manufacture of food supplements. Potassium is also found in combination with multivitamins.

6.8.2.4 Sulphur, Calcium and Magnesium

Amino acids, methionine and cysteine, are the primary constituents supplied by dietary sources of S and these are more abundant in animal and cereal proteins than in legume proteins. Glutathione is another dietary source of S contributed mainly by fruits and vegetables (~50% of dietary glutathione) and to a lesser extent (<25%) by meat (Flagg et al. 1994). Methylsulfonylmethane (MSM), a volatile compound, is also source of dietary S found in fruits, alfalfa, corn, tomatoes, tea, and coffee (Richmond 1986) and in human and bovine milk. The compound (MSM) is being investigated for its role in treating arthritis.

Food is the principal source of Ca; the dairy products being the most concentrated source contributing more than 50% of the total Ca in many diets. Typically, a litre of milk contains 1276 mg Ca, 118 mg Mg, 1044 mg P, 1715 mg K and 35.5 g protein (Weaver and Nieves 2009). Other plant foods contributing to dietary Ca include legumes, green leafy vegetables, broccoli, almonds and dried apricots. In plant foods containing high concentrations of oxalate (rhubarb, spinach) or phytate (legumes, cereal grains) the bioavailability of Ca can be low. The drinking water is another source of Ca e.g. bottled water in Europe generally contains 1.5–600 mg Ca^{2+} L^{-1} (Ong et al. 2009). In the European Union, the calcium compounds permitted as source of calcium in foods and in food supplements include carbonate, chloride, citrates, gluconate, glycerophosphate, lactate, orthophosphates, hydroxide and oxide (EFSA 2006).

Magnesium is primarily supplied through food and the important dietary sources include milk and dairy products, seeds, dark green vegetables, beans, peas, whole grains, fruits and nuts all of which differ substantially in Mg content. Cocoa and bitter chocolate, conches, shrimps, soybeans, butter beans, and beet greens contain more than 100 mg Mg 100 g^{-1}. Whole grains barley, rye, wheat or brown rice contain 110–180 mg Mg 100 g^{-1} (Seelig 1980) but high amounts of phytic acid as well as high levels of dietary fibre probably decrease its bioavailability (Schümann et al. 1997). Meat, fish, fruits, vegetables and dairy products contain less than 25 mg Mg 100 g^{-1} fresh weight (Seelig 1980). The plant and animal derived foods contain Mg mostly in bound or chelated forms, e.g. to phytic acid, phosphates, chlorophylls or in biological apatites (skeleton). The drinking water is another source of Mg and the bottled water in Europe may contain 0.5–90 mg Mg^{2+} L^{-1} (Ong et al. 2009). The dietary components along with phosphorus are strongly associated with the development of strong bones and teeth and are essential to cardiovascular function. Finally a word of caution, because of potential side effects and interactions with other medicines, one should take dietary supplements only under the supervision of a medical professional.

References

Agarwal R, Afzalpurkar R, Fordtran JS (1994) Pathophysiology of potassium absorption and secretion by the human intestine. Gastroenterology 107:548–571

Aitken RL, Scott BJ (1999) Magnesium. In: Peverill KI, Sparrow LA, Reuter DI (eds) Soil analysis: an interpretation manual. CSIRO Publishing, Collingwood, pp 255–262

Arai Y, Sparks DL (2007) Phosphate reaction dynamics in soils and soil components: a multiscale approach. Adv Agron 94:135–179

Augustin S, Mindrup M, Meiwes KJ (1997) Soil chemistry. In: Huttl RF, Schaaf W (eds) Magnesium deficiency in forest ecosystem. Kluwer Academic Publishers, London, pp 255–273

Ayaz A, Topcu A, Yurttagul M (2007) Survey of nitrate and nitrite level of fresh vegetables in Turkey. J Food Tech 5:177–179

Bailey J, Sands J, Franch H (2014) Water, electrolytes, and acid-base metabolism. In: Ross AC, Caballero B, Cousins RJ, Tucker KL, Ziegler TR (eds) Modern nutrition in health and disease. Lippincott Williams & Wilkins, Philadelphia, pp 102–132

Baker DH (1986) Utilization of isomers and analogs of amino acids and other sulfur- containing compounds. Prog Food Nutr Sc 10:133–178

Baldock JA, Nelson PN (2000) Soil organic matter. In: Sumner ME (ed) Handbook of soil science. CRC, Boca Raton, pp B-25–B-84

Barber SA (1995) Soil nutrient bioavailability. A mechanistic approach. Wiley, New York

Bateman EJ, Baggs EM (2005) Contribution of nitrification and denitrification to N_2O emissions from soils at different water-filled pore space. Biol Fertil Soils 41:379–388

Batjes NH (1997) World soil carbon stocks and global change. In: Squires VR, Glenn EP, Ayoub AT (eds) Proceedings of the workshop combating global climate change by combating land degradation. International Soil Reference and Information Centre (ISRIC), Wageningen, pp 51–78

Benbi DK, Richter J (1996) Nitrogen mineralization kinetics in sewage water irrigated and heavy metal treated sandy soils. In: Van Cleemput O, Hofmann G, Vermoesen A (eds) Progress in nitrogen cycling studies. Kluwer Academic Publishers, Dordrecht, pp 17–22

Benbi DK, Richter J (2003) Nitrogen dynamics. In: Benbi DK, Nieder R (eds) Handbook of processes and modeling in the soil-plant system. Haworth, New York, pp 409–481

Bird M, Santruckova H, Lloyd J, Veenendahl E (2001) The soil carbon pool and global change. In: Schulze ED, Heimann M, Harrison S, Holland E, Lloyd J, Prentice I, Schimel D (eds) Global biogeochemical cycles in the climate system. Academic, San Diego

Bishop N (1989) Bone disease in preterm infants. Arch Dis Child 64:1403–1409

Black CA (1968) Soil-plant relationships. Wiley, New York

Bohn HL, Mc Neal BL, O'Connor GA (1979) Soil chemistry. Wiley

Bouwman AF, Beusen AHW, Billen G (2009) Human alteration of the global nitrogen and phosphorus soil balances for the period 1970–2050. Global Biogeochem Cycles 23:GB0A04. https://doi.org/10.1029/2009GB003576

Brady NC (1990) The nature and properties of soils. Macmillan Publishers, New York

Brady NC, Weil RR (2015) The Nature and properties of soils, 14th edn. Pearson Education Inc., Pearson India Education services Pvt. Ltd, Noida, p 1046

Brink JW (1977) World resources of phosphorus. Ciba Found Symp 57:23–48

Bruce RC (1999) Calcium. In: Peverill KI, Sparrow LA, Reuter DJ (eds) Soil analysis: an interpretation manual. CSIRO Publishing, Collingwood, pp 247–254

Brune M, Rossander L, Hallberg L, Gleerup A, Sandberg AS (1992) Iron absorption from bread in humans: inhibiting effects of cereal fibre, phytate and inositol phosphates with different numbers of phosphate groups. J Nutr 122:442–449

Burel C, Boujard T, Kaushik SJ, Boeuf G, Mol KA, Geyten SV, Darras VM, Kuhn ER, Pradet-Balade B, Querat B, Quinsac A, Krouti M, Ribaillier D (2001) Effects of rapeseed meal-glucosinolates on thyroid metabolism and feed utilization in rainbow trout. Gen Comp Endocrinol 124:343–358

Burns RC, Hardy RWF (1975) Nitrogen fixation in bacteria and higher plants. Springer-Verlag, Berlin

Burris RH (1980) The global nitrogen budget – science or séance? In: Newton WE, Orme-Johnson WH (eds) Nitrogen fixation, vol I. University Park Press, Baltimore, pp 7–16

Butte NF, Hopkinson JM, Wong WW, Smith EO, Ellis KJ (2000) Body composition during the 1682 first 2 years of life: an updated reference. Pediatr Res 47:578–585

Cakmak I (2005) The role of potassium in alleviating detrimental effects of abiotic stresses in plants. J Plant Nutr Soil Sci 168:521–530

Calvo MS, Park YK (1996) Changing phosphorus content of the U.S. diet: potential for adverse effects on bone. J Nutr 126:1168S–1180S

Calvo MS, Uribarri J (2013) Contributions to total phosphorus intake: all sources considered. Semin Dial 26:54–61

Calvo MS, Moshfegh AJ, Tucker KL (2014) Assessing the health impact of phosphorus in the food supply: issues and considerations. Adv Nutr 5:104–113

Cartea ME, Velasco P (2008) Glucosinolates in Brassica foods: bioavailability in food and significance for human health. Phytochem Rev 7:213–229

Cathcart JB (1980) World phosphate reserve and resources. In: Khasawneh FE, Sample EC, Kamprat EJ (eds) The role of phosphorus in agriculture. Soil Science Society of America, Madison, pp 1–18

Ciais P, Sabine C, Bala G, Bopp L, Brovkin V, Canadell J, Chhabra A, Defries R, Galloway J, Heimann M, Jones C, Le Quere C, Myneni RB, Piao S, Thornton P (2013) Carbon and other biogeochemical cycles. In: Stocker TF, Qin D, Plattner GK, Tignor M, Allen SK, Boschung J, Nauels A, Xia Y, Bex V, Midgley PM (eds) Climate change 2013: the physical science basis. Contribution of working group I to the fifth assessment report of the intergovernmental panel on climate change. Cambridge University Press, Cambridge/New York

Cicero AF, Borghi C (2013) Evidence of clinically relevant efficacy for dietary supplements and nutraceuticals. Curr Hypertens Rep 15:260–267

Cleveland CC, Townsend AR, Schimel DS, Fisher H, Howarth RW, Hedin LO, Perakis SS, Latty EF, Von Fischer JC, Elseroad A, Wasson MF (1999) Global patterns of terrestrial biological nitrogen (N_2) fixation in natural ecosystems. Global Biogeochem Cycles 13:623–645

Cordell D, White S (2011) Peak phosphorus: clarifying the key issues of a vigorous debate about long-term phosphorus security. Sustainability 3:2027–2049

Cordell D, Jan-Olof D, White S (2009) The story of phosphorus: global food security and food for thought. Glob Environ Chang 19:292–305

Crook MA (2012) Potassium. In: Koster J, Arnold WJH (eds) Clinical biochemistry and metabolic medicine. University of Greenwich, London, pp 86–94

Dalal RC (1977) Soil organic phosphorus. In: Brady NC (ed) Advances in agronomy 29:83–117

Davidson EA (1994) Climate change and soil microbial processes: secondary effects are hypothesised from better known interacting effects. In: MDA R, Lovelend PJ (eds) Soil responses to climate change, NATO ASI Series, vol 23. Springer, Heidelberg

De Pascale S, Maggio A, Pernice R, Fogliano V, Barbieri G (2007) Sulphur fertilization may improve the nutritional value of *Brassica rapa* L. subsp *sylvestris*. Eur J Agron 26:418–424

Delwiche CC (1970) The nitrogen cycle. Sci Am 223:136–146

Drecht GV, Bouwman AF, Knoop JM, Beusen AHW, Meinardi CR (2003) Global modeling of the fate of nitrogen from point and nonpoint sources in soils, groundwater and surface water. Global Biochem Cycles 17:1115

Du ST, Zhang YS, Lin XY (2007) Accumulation of nitrate in vegetables and its possible implications to human health. Agric Sci China 6:1246–1255

EcoSanRes (2003) Closing the loop on phosphorus. Stockholm Environment Institute (SEI) funded by SIDA Stockholm

EFMA (European Fertilizer Manufacturers Association) (2000) Phosphorus: essential element for food production. EFMA, Brussels

EFSA (European Food Safety Authority) (2006) Tolerable upper intake levels for vitamins and minerals: scientific committee on food and scientific panel on dietetic products, nutrition and allergies. European Food Safety Authority, p 480
EFSA (European Food Safety Authority) (2008a) Nitrate in vegetables. Scientific opinion of the panel on contaminants in the food chain. EFSA J 689:1–79
EFSA (European Food Safety Authority) (2008b) Glucosinolates as undesirable substances in animal feed: scientific panel on contaminants in the food chain. EFSA J 590:1–76
EFSA (European Food Safety Authority) (2015a) Scientific opinion on dietary reference values for phosphorus. EFSA J 13:4185
EFSA (European Food Safety Authority) (2015b) EFSA panel on dietetic products, nutrition and allergies (NDA). Scientific opinion on dietary reference values for magnesium. EFSA J 13:4186
EFSA NDA Panel (EFSA Panel on Dietetic Products, Nutrition and Allergies) (2016) Scientific opinion on dietary reference values for potassium. EFSA J 14(10):4592
EFSA NDA Panel (EFSA Panel on Dietetic Products, Nutrition and Allergy) (2005) Opinion of the scientific panel on dietetic products, nutrition and allergies on a request from the commission related to the tolerable upper intake level of potassium. EFSA J 193:1–19
Elmore JS, Mottram DS, Muttucumaru N, Dodson AT, Parry MAJ, Halford NG (2007) Changes in free amino acids and sugars in potatoes due to sulfate fertilization and the effect on acrylamide formation. J Agric Food Chem 55:5363–5366
Eriksen J (2008) Soil sulphur cycling in temperate agricultural systems. In: Sulfur: a missing link between soils, crops and nutrition. Agron Monograph 50
Fahey JW, Zalcman AT, Talalay P (2001) The chemical diversity and distribution of gluocosinolates and isothiocyanates among plants. Phytochemistry 56:5–51
Fairhurst T, Lefroy R, Mutert E, Batjes N (1999) The importance, distribution and causes of phosphorus deficiency as a constraint to crop production in the tropics. Agrofor Forum 9:2–8
Falkowski P, Scholes RJ, Boyle E, Canadell J, Canadell D, Canfield D (2000) The global C cycle: a test of our knowledge of earth as a system. Science 290:291–296
FAO/WHO (2001) Expert consultation on human vitamin and mineral requirements. Food and Agriculture Organization of the United Nations; World Health Organization; Food and Nutrition Division, FAO, Rome
FAO/WHO (Food and Agriculture Organization/World Health Organization) (1973) Energy and protein requirements. Report of a joint FAO/WHO ad hoc Expert committee. FAO nutrition meeting report series no. 52. FAO, Rome
FAOSTAT Database (n.d.) Food and Agriculture organization (FAO), Rome
Flagg EW, Coates RJ, Eley JW (1994) Dietary glutathione intake in humans and the relationship between intake and plasma total glutathione level. Nutr Cancer 21:33–46
Fomon SJ, Haschke F, Ziegler EE, Nelson SE (1982) Body composition of reference children from 1795 birth to age 10 years. Am J Clin Nutr 35:1169–1175
Fox RL, Kamprath EJ (1970) Phosphorus sorption isotherms for evaluating the phosphate requirements of soils. Soil Sci Soc Am Proc 34:902–907
Fraiture CD (2007) Future water requirements for food-three scenarios, International Water Management Institute (IWMI), SIWI Seminar: water for food, bio-fuels or ecosystems? World water week 2007, August 12th–18th 2007, Stockholm
Frossard E, Bucher M, Mächler F, Mozafar A, Hurrell R (2000) Potential for increasing the content and bioavailability of Fe, Zn and Ca in plants for human nutrition. J Sci Food Agric 80:861–879
Galloway JN, Cowling EB (2002) Reactive nitrogen and the world: 200 years of change. Ambio 31:64–71
Galloway JN, Schlesinger WH, Levy H II, Michaels A, Schnoor JL (1995) Nitrogen fixation: atmospheric enhancement- environmental response. Global Biogeochem Cycles 9:235–252
Galloway JN, Cowling EB, Seitzinger SP, Socolow RH (2002) Reactive nitrogen: too much of a good thing? Ambio 31:60–63

Galloway JN, Dentener FJ, Capone DG, Boyer EW, Howarth RW, Seitzinger SP (2004) Nitrogen cycles: past, present and future. Biogeochemical 70:153–226
Gangolli SD, Van den Brandt PA, Feron VJ, Janzowsky C, Koeman JH, Speijers GJA, Spiegelhalder B, Walker R, Wishnok JS (1994) Assessment of nitrate, nitrite and N-nitroso compounds. Eur J Pharmacol Environ Toxicol Pharmacol 292:1–38
Garrison RH, Somer E (1995) The nutrition desk reference. Keats, New Canaan
Gerendás J, Breuning S, Stahl T, Mersch-Sundermann V, Mühling KH (2008) Isothiocyanate concentration in Kohlrabi (*Brassica oleracea* L., Var. *gongylodes*) plants as influnenced by sulphur and nitrogen supply. J Agric Food Chem 56:8334–8342
Gilbert N (2009) The disappearing nutrient. Nature 461:716–718
Godfray HCJ, Beddington JR, Crute IR, Haddad L, Lawrence D, Muir JF, Pretty J, Robinson S, Thomas SM, Toulmin C (2010) Food security: the challenge of feeding 9 billion people. Science 327:812–818
Gorenjak AH, Cencič A (2013) Nitrate in vegetables and their impact on human health. Acta Aliment 42:158–172
Graham RD, Welch RM (2000) Plant food micronutrient composition and human nutrition. Commun Soil Sci Plant Anal 31:1627–1640
Griffiths DW, Birch ANE, Hillman JR (1998) Antinutritional compounds in the Brassicaceae. Analysis, biosynthesis, chemistry and dietary effects. J Hort Sci Biotech 73:1–18
Griffiths G, Trueman I, Crowther T, Thomas B, Smith B (2002) Onions-a global benefit to health. Phytother Res 16:603–615
Gumz ML, Rabinowitz L, Wingo CS (2015) An integrated view of potassium homeostasis. N Engl J Med 373:1787–1788
Gyaneshwar P, Hirsch AM, Moulin L, Chen WM, Elliott GN, Bontemps C, Estrada-de Los Santos P, Gross E, Dos Reis FB, Sprent JI, Young JP, James EK (2011) Legume-nodulating betaproteobacteria: diversity, host range, and future prospects. Mol Plant-Microbe Interact 24:1276–1288
Halkier BA, Gershenzon J (2006) Biology and biochemistry of glucosinolates. Annu Rev Plant Biol 57:303–333
Herridge DF, Bergersen FJ (1988) Symbiotic nitrogen fixation. In: Wilson JR (ed) Advances in nitrogen cycling. CAB International, Wallingford, pp 46–65
Herridge DF, Peoples MB, Boddey RM (2008) Global inputs of biological nitrogen fixation in agricultural systems. Plant Soil 311:1–18
Hofstra N, Bouwman AF (2005) Denitrification in agricultural soils: summarizing published data and estimating global annual rates. Nutr Cycl Agroecosys 72:267–278
Hord NG, Tang Y, Bryan NS (2009) Food sources of nitrates and nitrites: the physiologic context for potential health benefits. Am J Clin Nutr 90:1–10
IFA (International Fertilizer Association) (n.d.) Statistics. Available online http://www.fertilizer.org/Statistics?
IOM (Institute of Medicine) (1997) Dietary reference intakes for calcium, phosphorus, magnesium, vitamin D, and fluoride. Prepared by the standing committee on the scientific evaluation of dietary reference intakes, food and nutrition board. Institute of Medicine. National Academy Press, Washington, DC, p 454
IOM (Institute of Medicine) (2005) Dietary reference intakes for water, potassium, sodium, chloride, and sulfate. National Academies Press, Washington, DC, p 617
Jackson AA (1983) Amino acids: essential and non-essential. Lancet:1034–1037
Jansson SL, Persson J (1982) Mineralization and immobilization of soil nitrogen. In: Stevenson FJ (ed) Nitrogen in agricultural soils. American Society of Agronomy, Madison, pp 229–252
Joffres MR, Reed DM, Yano K (1987) Relation of magnesium intake and other dietary factors to blood pressure: the Honolulu Heart Study. Am J Clin Nutr 45:469–475
Justine JK, Smith RL (1962) Nitrification of ammonium sulphate in a calcareous soil as influenced by combination of moisture, temperature and levels of added N. Soil Sci Soc Am Proc 26:246–250

Kennedy IR (1992) Acid soil and acid rain. Wiley, New York
Keyzer MA, Merbis M, Pavel I, Van Wesenbeeck C (2005) Diet shifts towards meat and the effects on cereal use: can we feed the animals in 2030? Ecol Econ 55:187–202
Khan NA, McAlister FA, Lewanczuk RZ, Touyz RM, Padwal R, Rabkin SW, Leiter LA, Lebel M, Herbert C, Schiffrin EL, Herman RJ, Hamet P, Fodor G, Carruthers G, Culleton B, DeChamplain J, Pylypchuk G, Logan AG, Gledhill N, Petrella R, Campbell NR, Arnold M, Moe G, Hill MD, Jones C, Larochelle P, Ogilvie RI, Tobe S, Houlden R, Burgess E, Feldman RD, Canadian Hypertension Education Program (2005) The 2005 Canadian Hypertension Education Program recommendations for the management of hypertension: part II – therapy. Can J Cardiol 21(8):657–672
Kirkman JH, Basker A, Surapaneni A, MacGregor AN (1994) Potassium in the soils of New Zealand-a review. New Zeal J Agric Res 37:207–227
Kliebenstein DJ, Kroymann J, Mitchell-Olds T (2005) The glucosinolate-myrosinase system in an ecological and evolutionary context. Curr Opin Plant Biol 8:264–271
Krijthe BP, Heeringa J, Kors JA, Hofman A, Franco OH, Witteman JC, Stricker BH (2013) Serum potassium levels and the risk of atrial fibrillation: the Rotterdam Study. Int J Cardiol 168:5411–5415
Lamarque JF, Bond TC, Eyring V, Granier C, Heil A, Klimont Z, Lee D, Liousse C, Mieville A, Owen B, Schultz MG, Shindell D, Smith SJ, Stehfest E, Van Aardenne J, Cooper OR, Kainuma M, Mahowald N, McConnell JR, Naik V, Riahi K, van Vuuren DP (2010) Historical (1850–2000) gridded anthropogenic and biomass burning emissions of reactive gases and aerosols: methodology and application. Atmos Chem Phys 10:7017–7039
LaRue TA, Patterson TG (1981) How much nitrogen do legumes fix? Adv Agron 34:15–38
Laurant P, Touyz RM (2000) Physiological and pathophysiological role of magnesium in the cardiovascular system: implications in hypertension. J Hypertens 18(9):1177–1191
Lawton K (1955) Chemical composition of soils. In: Bear FE (ed) Chemistry of the soil. Reinhold Publishing, New York, pp 53–84
Lehnhardt A, Kemper MJ (2011) Pathogenesis, diagnosis and management of hyperkalemia. Pediatr Nephrol 26:377–384
Lei S, Liu S (1989) Phytic acid intake and its effect on the bioavailability of zinc in preschool children. Acta Nutr Sin 11:211
Lindh U (2005) Biological functions of the elements. In: Selinus O (ed) Essentials of medical geology: impacts of the natural environment on public health. Elsevier Academic Press, Burlington, pp 115–160
Losak T, Hlusek J, Kra S, Varga L (2008) The effect of nitrogen and sulphur fertilization on yield and quality of kohlrabi (*Brassica oleracea*). R Bras Ci Solo 32:697–703
Lott JNA, Ockenden I, Raboy V, Batten GD (2000) Phytic acid and phosphorus in crop seeds and fruits: a global estimate. Seed Sci Res 10:11–33
Lotz M, Zisman E, Bartter FC (1968) Evidence for a phosphorus-depletion syndrome in man. N Engl J Med 278:409–415
Lundberg JO, Weitzberg E, Gladwin MT (2008) The nitrate-nitrite-nitric oxide pathway in physiology and therapeutics. Nat Rev Drug Discov 7:156–167
Lynch J (1995) Root architecture and plant productivity. Plant Physiol 109:7–13
Mackenzie FT, Ver LM, Lerman A (1998) Coupled biogeochemical cycles of carbon, nitrogen, phosphorus, and sulfur in the land-ocean-atmosphere system. In: Galloway JN, Melillo JM (eds) Asian change in the context of global change. Cambridge University Press, New York, pp 42–100
Mackenzie IJ, Ver LM, Lennan A (2002) Century scale nitrogen and phosphorus controls of the C cycle. Chem Geol 190:13–32
Maga JA (1982) Phytate: its chemistry, occurrence, food interactions, nutritional significance and methods of analysis. J Agric Food Chem 30:1
Marieb EN (1998) Human anatomy and physiology. Benjamin/Cummings, Menlo Park

Marier JR, Neri LC (1985) Quantifying the role of magnesium in the interrelationship between human mortality. Magnesium 4:53–59
Mawson R, Heaney RK, Zdunczyk Z, Kozlowska H (1994) Rapeseed mealglucosinolates and their antinutritional effects Part 3. Anim Growth Perform Nahrung 38:167–177
McGill WB, Cole CV (1981) Comparative aspects of cycling of organic C, N, sulphur and P through soil organic matter. Geoderma 26:267–286
McKnight GM, Duncan CW, Leifert C, Golden MH (1999) Dietary nitrate in man: friend or foe? Br J Nutr 81:349–358
McNaught KJ, Dorofaeff FD (1965) Magnesium deficiency in pastures. New Zeal J Agric Res 8:555–572
Michel A, Martin-Perez M, Ruigomez A, Garcia Rodriguez LA (2015) Risk factors for hyperkalaemia in a cohort of patients with newly diagnosed heart failure: a nested case-control study in UK general practice. Eur J Heart Fail 17:205–213
Millward DJ (1999) Optimal intakes of protein in the human diet. Proc Nutr Soc 58:403–413
Mithen R (2001) Glucosinolates and their degradation products. Adv Bot Res 35:213–262
Mithen R, Faulkner K, Magrath R, Rose P, Williamson G, Marquez J (2003) Development of isothiocyanate-enriched broccoli and its enhanced ability to induce phase 2 detoxification enzymes in mammalian cells. Theor Appl Genet 106:727–734
Moe SM (2008) Disorders involving calcium, phosphorus, and magnesium. Prim Care 35:215–237
Mortvedt JJ, Sikora FJ (1992) Heavy metal, radionuclides, and fluorides in phosphorus fertilizers. In: Sikora FJ (ed) Future directions for agricultural phosphorus research. TVA Bulletin Y-224. Muscle Shoals, pp 69–73
Mosier AR, Syers JK, Freney JR (2004) Nitrogen fertilizer: An essential component of increased fodder, feed and fiber production. In: Mosier AR, Syers JK, Freney JR (eds) Agriculture and the nitrogen cycle: assessing the impacts of fertiliser use on food production and the environment, vol 65. SCOPE, Paris, pp 3–15
Mozolewski W, Smoczynski S (2004) Effect of culinary processes on the content of nitrates and nitrites in potatoes. Pak J Nutr 3:357–361
Muttucumaru N, Halford NG, Elmore JS, Dodson AT, Parry MAJ, Shewry PR, Mottram DS (2006) The formation of high levels of acrylamide during the processing of flour derived from sulfate-deprived wheat. J Agric Food Chem 54:8951–8955
Muttucumaru N, Powers SJ, Elmore JS, Mottram DS (2013) Effects of nitrogen and sulphur fertilization on free amino acids, sugars and acrylamide-forming potential in potato. J Agric Food Chem 61:6734–6742
Naismith DJ, Braschi A (2003) The effect of low-dose potassium supplementation on blood pressure in apparently healthy volunteers. Br J Nutr 90:53–60
Nayyar VK, Chhibba IM (2000) Nutritional disorders in field crops – visual symptoms and remedial measures. Department of Soils, Punjab Agricultural University, Ludhiana
Nieder R, Benbi DK (2008) Carbon and nitrogen in the terrestrial environment. Springer, Heidelberg/New York, p 432
Nieder R, Benbi DK, Scherer W (2011) Fixation and defixation of ammonium in soils: a review. Biol Fertil Soils 47:1–14
Nordin BEC (1976) Nutritional considerations. In: Nordin BEC (ed) Calcium, phosphate and magnesium metabolism. Churchill Livingstone, Edinburgh, pp 1–35
NRC (National Research Council) (1989) Recommended daily allowances. National Research Council, Washington, DC
Ong CN, Grandjean AC, Heaney RP (2009) The mineral composition of water and its contribution to calcium and magnesium intake. In: Calcium and magnesium in drinking-water: public health significance. World Health Organization, Geneva, pp 37–58
Paice B, Gray JM, McBride D, Donnelly T, Lawson DH (1983) Hyperkalaemia in patients in hospital. BMJ (Clin Res ed) 286:1189–1192
Parcell SW (2002) Sulfur in human nutrition and applications in medicine. Altern Med Rev 7:22–44

Pennington JAT (1998) Dietary exposure models for nitrates and nitrites. Food Control 9:385–395

Peoples MB, Boyer EW, Goulding KWT, Heffer P, Ochwoh VA, Vanlauwe B, Wood S, Yagi K, Cleemput OV (2004) Pathways of nitrogen loss and their impacts on human health and the environment. In: Mosier AR, Syers JK, Freney JR (eds) Agriculture and the nitrogen cycle: assessing the impacts of fertiliser use on food production and the environment, SCOPE 65. Permaculture, Paris, pp 53–69

Pepin J, Shields C (2012) Advances in diagnosis and management of hypokalemic and hyperkalemic emergencies. Emerg Med Pract 14:1–17

Perazella M, Mahnensmith R (1997) Hyperkalemia in the elderly. J Gen Intern Med 12:646–656

Phetchawee S, Kanareugsa C, Sittibusaya C, Khunathai H (1985) Potassium availability in the soils of Thailand. In: Proceedings of the 19th colloquium of the International Potash Institute. International Potash Institute, Worblaufen, pp 167–196

Pierzynski GM, Loan TJ (1993) Crop, soil, and management effects of phosphorus test levels. J Prod Agric 6:513–520

Prentice A, Bates CJ (1994) Adequacy of dietary mineral supply for human bone growth and mineralisation. Eur J Clin Nutr 48(Suppl 1):S161–S176

Ralt D (2009) Does NO metabolism play a role in the effects of vegetables in health? Nitric oxide formation via the reduction of nitrites and nitrates. Med Hypotheses 73:794–796

Rand WM, Pellett PL, Young VR (2003) Meta-analysis of nitrogen balance studies for estimating protein requirements in healthy adults. Am J Clin Nutr 77:109–127

Rask L, Andreasson E, Ekbom B, Eriksson S, Pontoppidan B, Meijer J (2000) Myrosinase: gene family evolution and herbivore defense in Brassicaceae. Plant Mol Biol 42:93–113

Rastegar A (1990) Serum potassium. In: Walker HK, Hall WD, Hurst JW (eds) Clinical methods: the history, physical, and laboratory examinations. Butterworths, Boston, pp 884–887

Reddy NR, Sathe SK, Salunkhe DK (1982) Phytates in legumes and cereals. Adv Food Res 28:1

Reed SC, Cleveland CC, Townsend AR (2011) Functional ecology of free-living nitrogen fixation: a contemporary perspective. Annu Rev Ecol Evol Syst 42:489–512

Reinhold JG, Nasr K, Lanhimagarzedeh A, Hedayati H (1973) Effect of purified phytate and phytate-rich breads upon metabolism of zinc, calcium, phosphorus, and nitrogen in man. Lancet 1:283

Richmond VL (1986) Incorporation of methylsulfonylmethane sulfur into guinea pig serum proteins. Life Sci 39:263–268

Rodenburg EM, Visser LE, Hoorn EJ, Ruiter R, Lous JJ, Hofman A, Uitterlinden AG, Stricker BH (2014) Thiazides and the risk of hypokalemia in the general population. J Hypertens 32:2092–2097

Rosenlund M, Berglind N, Hallqvist J, Bellander T, Bluhm G (2005) Daily intake of magnesium and calcium from drinking water in relation to myocardial infarction. Epidemiology 16 (4):570–576

Runge-Metzger A (1995) Closing the cycle: obstacles to efficient P management for improved global food security. SCOPE 54-phosphorus in the global environment-transfers, cycles and management

Russell EW (1973) Soil conditions and plant growth. Longman, London

Samotus B, Schwimmer S (1962) Phytic acid as a phosphorus reservoir in the developing potato tuber. Nature 194:578–579

Sanchez PA (1976) Properties and management of soils in the tropics. Wiley Interscience Publishers, New York

Santamaria P (2006) Nitrate in vegetables: toxicity, content, intake and EC regulation. J Sci Food Agric 86:10–17

Schlesinger HW (1997) Biogeochemistry, an analysis of global change, 2nd edn. Academic, San Diego

Schulten HR, Schnitzer M (1998) The chemistry of soil organic nitrogen: a review. Biol Fertil Soils 26:1–15

Schümann K, Classen HG, Hages M, Prinz-Langenohl R, Pietrzik K, Biesalski HK (1997) Bioavailability of oral vitamins, minerals, and trace elements in perspective. Arzneim-Forsch/Drug Res 47:369–380

Seelig MS (1980) Magnesium deficiency in the pathogenesis of disease. Plenum Med Book Comp, New York

Seelig MS (1989) Cardiovascular consequences of magnesium deficiency and loss: pathogenesis, prevalence and manifestations – magnesium and chloride loss in refractory potassium repletion. Am J Cardiol 63:46–216

Sharma RD (1986) Phytate and the epidemiology of heart disease, renal calculi and colon cancer. In: Graf E (ed) Phytic acid chemistry and applications. Pilatus Press, Minneapolis, p 161

Simons-Morton DG, Hunsberger SA, Van Horn L, Barton BA, Robson AM, McMahon RP, Muhonen LE, Kwiterovich PO, Lasser NL, Kimm SYS, Greenlick MR (1997) Nutrient intake and blood pressure in the Dietary Intervention Study in children. Hypertension 29:930–936

Singh RB, Singh NK, Niaz MA, Sharma JP (1996) Effect of treatment with magnesium and potassium on mortality and reinfarction rate of patients with suspected acute myocardial infarction. Int J Clin Pharmacol Ther 34:219–225

Skiba U, Ball B (2002) The effect of soil texture and soil drainage on emissions of nitric oxide and nitrous oxide. Soil Use Manag 18:56–60

Smil V (1999) Crop residues: agriculture's largest harvest. Bioscience 49:299–308

Smil V (2000a) Phosphorus in the environment: natural flows and human interferences. Annu Rev Energy Environ 25:53–88

Smil V (2000b) Feeding the world: a challenge for the 21st century. The MIT Press, Cambridge, MA

Smil V (2001) Enriching the earth. The MIT Press, Cambridge, p 338

Smil V (2002a) Phosphorus: global transfers. In: Douglas PI (ed) Encyclopedia of global environmental change. Wiley, Chichester

Smil V (2002b) Nitrogen and food production: proteins and human diet. Ambio 31:126–131

Søderlund R, Rosswall T (1982) The nitrogen cycle. In: Hutzinger O (ed) The handbook of environmental chemistry, The natural environment and the biogeochemical cycles, vol 1B. Springer Verlag, Heidelberg, pp 60–81

Sprent JI (1999) The biology of nitrogen transformations. Soil Use Manag 6:74–77

Steen I (1998) Phosphorus availability in the 21st century: management of a non renewable resource. Phosphor Potassium 217:25–31

Stergiou GS, Salgami EV (2004) World Health Organization-International Society of Hypertension (WHO-ISH), USA Joint National Committee on Prevention, Detection, Evaluation, and Treatment of High Blood Pressure (JNC-7) and European Society of Hypertension-European Society of Cardiology (ESH-ESC) New European, American and international guidelines for hypertension management: agreement and disagreement. Expert Rev Cardiovasc Ther 2 (3):359–368

Storch KJ, Wagner DA, Burke JF, Young VR (1988) Quantitative study in vivo of methionine cycle in humans using [methyl-2H3]- and [1-13C]methionine. Am J Phys 255:E322–E331

Syers JK, Skinner RJ, Curtin D (1987) Soil and fertilizer sulphur in UK agriculture. Proc Fertil Soc Lond 264:43

Thomson BM, Nokes CJ, Cressey PJ (2007) Intake and risk assessment of nitrate and nitrite from New Zealand foods and drinking water. Food Addit Contam 24:113–121

Tilman GD, Cassman KG, Matson PA, Naylor RL, Polasky S (2002) Agricultural sustainability and intensive production practices. Nature 418:671–677

Toth SJ (1955) Colloid chemistry of soils. In: Bear FE (ed) Chemistry of the soil. Reinhold Publishing, New York, pp 85–106

Touyz RM, Yao G (2003) Modulation of vascular smooth muscle cell growth by magnesium -role of mitogen-activated protein kinases. J Cell Physiol 197(3):326–335

TSI (The Sulphur Institute) (n.d.) Atmospheric sulphur – the agronomic aspects. TSI Bull. No. 23

UKEVM (United Kingdom Food Standards Agency, Expert Group on Vitamins and Minerals) (2003) Risk assessments: potassium. In: Safe upper levels for vitamins and minerals. London, p 299

Van Leer EM, Seidell JC, Kromhout D (1995) Dietary calcium, potassium, magnesium and blood pressure in the Netherlands. Int J Epidemiol 24:1117–1123

Vitousek PM, Menge DNL, Reed SC, Cleveland CC (2013) Biological nitrogen fixation: rates, patterns and ecological controls in terrestrial ecosystems. Philos Trans R Soc B: Biol Sci 368 (1621):20130119

Wacker WEC (1980) Magnesium and man. Harvard University Press, Cambridge, UK

Walker KC, Booth EJ (1992) Sulphur research on oilseed rape in Scotland. Sulphur Agric 16:15–19

Wardlaw GM, Insel PM (1996) Perspectives in nutrition. Mosby, St. Louis

Weaver CM, Nieves JW (2009) Calcium and magnesium: role of drinking-water in relation to bone metabolism. In: Calcium and magnesium in drinking-water: public health significance. World Health Organization, Geneva, pp 96–109

Webster PO (1987) Magnesium. Am J Clin Nutr 45:1305–1312

Whelton PK, Klag MJ (1989) Magnesium and blood pressure: review of the epidemiologic and clinical trial experience. Am J Cardiol 63:26G–30G

Whelton PK, He J, Cutler JA, Brancati FL, Appel LJ, Follmann D, Klag MJ (1997) Effects of oral potassium on blood pressure. Meta-analysis of randomized controlled clinical trials. J Am Med Assoc 277:1624–1632

WHO (World Health Organization) (2007) Protein and amino acid requirements in human nutrition: report of a joint FAO/WHO/UNU expert consultation. WHO technical report series No. 935, p 265

WHO (World Health Organization) (2009a) Potassium in drinking-water: background document for development of WHO. Guidelines for drinking-water quality. WHO, Geneva, p 12

WHO (World Health Organization) (2009b) Calcium and magnesium in drinking-water: public health significance. WHO, Geneva, p 194

WHO (World Health Organization) (2012) Guidelines: potassium intake for adults and children. WHO, Geneva

Widdowson EM, Spray CM (1951) Chemical development in utero. Arch Dis Child 26:205–214

Wildman REC, Medeiros DM (2000) Protein. In: Advanced human nutrition. CRC Press, New York, pp 123–150

Wu J, O'Donnell GO, Syers JK (1995) Influence of glucose, nitrogen and plant residues on the immobilisation of sulphate S in soil. Soil Biol Biochem 25:1567–1573

Yan X, Akimoto H, Ohara T (2003) Estimation of nitrous oxide, nitric oxide and ammonia emissions from croplands in East, Southeast and South Asia. Glob Chang Biol 9:1080–1096

Ying L, Hofseth LJ (2007) An emerging role for endothelial nitric oxide synthase in chronic inflammation and cancer. Cancer Res 67:1407–1410

Zhou JR, Erdman JW Jr (1995) Phytic acid in health and disease. Crit Rev Food Sci Nutr 35:495–508

Zinke PJ, Stangenberger AG, Post WM, Emmanuel WR, Olson JS (1984) Worldwide organic soil carbon and nitrogen data. Oak Ridge National Laboratory, Oak Ridge. ORNL/TM-8857

Chapter 7
Microelements and Their Role in Human Health

Abstract Microelements viz. zinc (Zn), iron (Fe), manganese (Mn), copper (Cu), molybdenum (Mo), boron (B), chlorine (Cl) and nickel (Ni) have been recognized as essential and silicon (Si), sodium (Na), cobalt (Co), and strontium (Sr) as beneficial or quasi-essential for plants. In addition to these plant essential microelements, human beings and animals require chromium (Cr), iodine (I) and selenium (Se). Microelements are used in relatively small amounts and constitute less than 0.1% of dry plant tissue. Some of the microelements may be toxic when consumed at high amounts. Soil is the main source of microelements for plants, except in situations of large atmospheric deposition or from flooding by contaminated waters. The microelements in soils undergo several transformations and their availability to plants depends on the chemical form and distribution between soil's solid and liquid phases, which is influenced by soil conditions, particularly pH, texture and soil aeration status. Microelements perform a variety of functions in plants. Besides being component of enzymes, certain microelements are involved in activation of enzymes and play a role in oxidation-reduction reactions of plant metabolism. Micronutrient deficiencies in plants not only limit agricultural production but also affect human nutrition as plant food is the main source of dietary intake. Microelements in humans play several physiological functions including synthesis of enzymes, hormones and other substances, helping to regulate growth, development and functioning of the immune and the reproductive systems. Deficiencies of microelements in soil and plants can be corrected by applying chemical fertilizers either alone or in combination with organic manures. Micronutrient level in humans can be optimized through dietary diversification, mineral supplementation, food fortification, or increasing their concentrations and/or bioavailability in food products. Correction of deficiencies and optimization of micronutrient levels in humans depends on several factors including current dietary intake, food habits and the nutrient content of the food items consumed, metabolic pathway of a nutrient, current body stocks, age, gender, and body weight. In this chapter, we discuss sources of microelements, their transformations in soil, functions in plants and humans, effects of their deficient and excessive uptake on plants and humans, and the approaches to optimize their levels in plants and humans.

R. Nieder et al., *Soil Components and Human Health*,
https://doi.org/10.1007/978-94-024-1222-2_7

Keywords Micronutrients • Quasi-essential elements • Micronutrient sources • Microelement transformations • Micronutrient functions • Excessive micronutrient uptake • Deficient micronutrient uptake • Micronutrient deficiency symptoms • Optimizing microelement status • Food fortification • Biofortification • Biovailability of iron • Bioavailability of zinc

7.1 Overview of Microelements

Eight microelements viz. zinc (Zn), iron (Fe), manganese (Mn), copper (Cu), molybdenum (Mo), boron (B), chlorine (Cl), and nickel (Ni) have been recognized as essential for plants. In addition, there are some quasi-essential or beneficial elements including silicon (Si), sodium (Na), cobalt (Co), and strontium (Sr) that are required by a particular group of plants for optimum growth and production. Besides these plant essential microelements, human beings and animals require chromium (Cr), iodine (I) and selenium (Se) in small amounts. Essentiality of arsenic (As), fluorine (F), tin (Sn) and vanadium (V) for animals has been observed under highly specialized conditions and these could be considered beneficial. While plants need mineral elements, air and water and can synthesize the necessary biomolecules such as amino acids and vitamins on their own, the animals and humans cannot synthesize organic molecules and thus require micronutrients in the form of vitamins (A, B, C, D, E and K) as well as minerals. In this chapter, discussion will be primarily confined to microelements B, Cu, Co, I, Mn, Mo, Fe and Zn, which are of most practical significance in agriculture and human health. We have added iodine in the list even though it has not been found essential for plants, but it is very important for humans and its deficiency has been reported worldwide causing several illnesses or disorders. Besides being essential, some of the microelements such as Cu, Zn, Se and Cr may be toxic when consumed at high amounts and these are discussed in Chap. 8.

Microelements are used in relatively small amounts and constitute less than 0.1% of dry plant tissue (Table 7.1). Their uptake is usually expressed in parts per million (ppm) or mg kg^{-1} but this does not mean that microelements are of lesser importance. They have enormous impact on human health. Insufficient dietary intakes of microelements impair several functions of the central nervous system, reproductive system, enzyme activities and energy metabolism and thus lead to serious illnesses. The concentrations of microelements differ greatly among soils, plants and human body and the average concentrations shown in Table 7.1 are only indicative of their relative abundance.

Table 7.1 Mean concentrations (mg kg^{-1}) of microelements in Earth's crust, soil (the upper meter), plants and human body

Element	Crust	Soil	Plant	Human body
Boron (B)	17	10	50	0.30
Chromium (Cr)	35	70	–	0.094
Cobalt (Co)	12	8	0.5	0.021
Copper (Cu)	14	20	14	1.0
Fluorine (F)	610	200	–	37
Iodine (I)	1.4	5	0.005	0.19
Iron (Fe)	31,000	40,000	140	60
Lead (Pb)	17	35	–	–
Manganese (Mn)	530	800	630	0.17
Molybdenum (Mo)	1.4	3	0.05	0.08
Nickel (Ni)	–	–	–	0.14
Selenium (Se)	0.083	0.01	0.05	0.11
Silicon (Si)	–	330,000	1000	260
Sodium (Na)	–	7000	1200	1400
Vanadium (V)	53	90	–	0.11
Zinc (Zn)	52	50	100	33.0

Compiled from Bohn et al. (1979), Fortescue (1980) and Lindh (2005)

7.2 Sources of Microelements

Soil is the main source of microelements for plants, except in situations of heavy atmospheric deposition or from flooding by contaminated waters. In soils, microelements are generally derived from the minerals present in the parent material, and these become plant available by weathering of soil minerals. Based on the affinity of elements for different minerals, these may be classified as Lithophile, Chalcophile and Siderophile elements. Lithophiles (B, Zn, Mn and Fe) show affinity for silicate minerals, Chalcophiles (Zn, Cu and Ni) for sulphide minerals and Siderophiles (Fe, Cu, Mo and Ni) for iron minerals. Some elements also show secondary affinity for silicate (Fe) and sulphide (Fe and Mo) minerals. Of the essential microelements, Co, Cu and Zn (also called heavy metals) are in lower concentration in soils than in the material from which they are derived. All the essential elements are derived not only from the soil minerals but some trace elements such as boron, iodine, and selenium are supplied in significant amounts to soils by atmospheric transport from the marine environment. The microelements are usually taken up from soil in ionic form by the plant (Table 7.2) and their availability to plants is controlled by a number of physical, chemical and biological processes and transformations in soil. Boron is usually absorbed in non-ionic form as boric acid, which has a strong ability to form complexes with diols and polyols inside the plant (Loomis and Durst 1992). However, in certain situations, B is also absorbed in ionic form as borate.

Table 7.2 Sources and forms of micronutrients absorbed by plants

Element	Symbol	Form absorbed
Boron	B	H_3BO_3 (boric acid) and $H_2BO_3^-$ (borate)
Copper	Cu	Cu^{+2}
Cobalt	Co	Co^{2+}
Iron	Fe	Fe^{+2} (ferrous) and Fe^{+3} (ferric)
Manganese	Mn	Mn^{+2}
Molybdenum	Mo	MoO_4^{-2} (molybdate)
Silicon	Si	$Si(OH)_4$ (silicate)
Zinc	Zn	Zn^{+2}

Boron

Boron is a metalloid belonging to Group IIIA and period 2 of the Periodic Table of Elements. It has an atomic mass of 10.81 and an atomic number 5. It ranks 51st in the order of abundance in the Earth's crust with an average concentration of 17 mg kg^{-1} (Steinnes 2009). Its concentration, in rocks, averages about 10–20 mg B kg^{-1}. In nature, B is always bound to oxygen as borate, less frequently as boric acid and rarely to fluorine as in BF_4^- (Power and Woods 1997). Boron, in the environment, originates mainly from marine salts, volcanic activity, and industrial pollution. Most soils have a low B content. Soils high in B content are associated with recent volcanic activity (Power and Woods 1997). Based on B content, soils may be divided into low (<10 mg B kg^{-1}) and high B containing (10–100 mg B kg^{-1}). The sources of B include concentrated mineral deposits that are associated with former geologic volcanic activity and were formed because of steam volatility of boric acid. The primary B minerals include ulexite $NaCa[B_5O_6(OH)_6]{\cdot}5H_2O$, borax (tincal) $Na_2[B_4O_5(OH)_4]{\cdot}8H_2O$, natural boric acid (sassolite), colemanite $Ca[B_3O_4(OH)_2]{\cdot}2H_2O$, kernite $Na_2[B_4O_5(OH)_4]{\cdot}2H_2O$; datolite $2CaO{\cdot}B_2O_3{\cdot}2SiO_2{\cdot}H_2O$, hydroboracite $CaO{\cdot}MgO{\cdot}3B_2O_3{\cdot}6H_2O$; tourmaline $Na(FeMg)_3Al_6(OH)_4(BO_3)_3Si_6O_8$ and ascharite $2MgO{\cdot}B_2O_3{\cdot}H_2O$ (Erd 1980). Besides these natural minerals, some soluble borates are used as sources of B in agriculture.

Cobalt

Cobalt is a lustrous silver-white, hard and brittle transition metal belonging to Group VIII and period 4 of the Periodic Table of Elements. It has an atomic mass of 58.93 and an atomic number 27. Cobalt is of relatively low abundance in the Earth's crust and most of it occurs in the Earth's core. Its concentration averages 12 mg kg^{-1} in Earth's crust and 5–45 mg kg^{-1} in rocks. The concentration of Co in soils ranges between 4.5 and 12 mg kg^{-1} and averages 8 mg kg^{-1}. Cobalt and iron occur together in many geological materials and have several similar properties such as ionic radii and charge and can replace each other. Similar to iron, cobalt can also be magnetized. Cobalt has oxidation states varying from +1 to +5; the most common being +2 and +3. Cobalt is stable in air and unaffected by water, but is slowly attacked by dilute acids. Cobalt is generally found in the form of ores and the main ores include cobaltite (CoAsS), erythrite ($Co_2(AsO_4)_2{\cdot}8\ H_2O$), skutterudite also called smaltite ($CoAs_{2-3}$), linneite (Co_3S_4), and glaucodot (Co,Fe)AsS.

Table 7.3 Copper containing minerals

Mineral	Compound
Chalcocite	Cu_2S
Covellite	CuS
Chalcopyrite	$CuFeS_2$
Chrysocolla	$CuSiO_3.2H_2O$
Cuprite	Cu_2O
Malanchite	$Cu_2(CH_2)CO_3$
Azurite	$Cu_3(OH)_2(CO_3)_2$
Bornite	$CuFeS_4$
Cupric ferrite	$CuFe_2O_4$

Copper

Copper is a reddish transition metal belonging to Group IB and period 4 of the Periodic Table of Elements. It has an atomic mass of 63.546 and an atomic number 29. It forms monovalent (cuprous) and divalent (cupric) cations. It is mined as a primary ore product from CuS and oxide ores (Table 7.3). The mean concentration of Cu in earth's crust is 14 mg kg^{-1}, which is lower than the median value of 30 mg kg^{-1} reported for soils. Its concentration in rocks averages between 30 and 140 mg kg^{-1}.

Iodine

Iodine is a dark-grey/purple-black non-metal and belongs to Group VIIA and period 5 of the Periodic Table of Elements. It is the most electropositive halogen and is the least reactive of the elements in this group. It has an atomic mass of 126.9 and an atomic number of 53. Iodine has two radioactive isotopes, ^{129}I (half-life 16 million years) and ^{131}I (8 days), which as by-products of atomic reactors are released into the environment. Iodine forms simple anion with oxidation state 1 and exhibits lithophilic properties. It is highly mobile in the Earth's crust and its content in upper continent crust is estimated to vary between 0.15 and 1.4 mg kg^{-1} (Kabata-Pendias and Mukherjee 2007). Iodine content is highest in shales (~1.5 mg kg^{-1}), intermediate in lime stones (~1 mg kg^{-1}), and the lowest in acid igneous rocks (~ 0.17 mg kg^{-1}). Deep-sea carbonates contain high amounts (~30 mg kg^{-1}) of iodine (Muramatsu and Wedepohl 1998). Iodine content in soil varies over a wide range (average 2.8 mg kg^{-1}) from less than 1 mg kg^{-1} in areas far inland to about 15 mg kg^{-1} in organic rich soils near the coast (Låg and Steinnes 1976). Volcanic ash soils contain high amounts of iodine, up to 104 mg kg^{-1}, and solonchak soils of arid and semiarid regions contain up to 340 mg kg^{-1} (Kabata-Pendias and Pendias 2001). Iodine content is usually higher in fine-textured soils rich in humus compared to light soils of humid climates. But there could be large variations in soil iodine content because of difference in atmospheric deposition depending on the distance from the coast. Soils close to the coast are rich in iodine, whereas those derived from recent glacial deposits are generally poor in this element (Kabata-Pendias and Mukherjee 2007). The major source of iodine appears to be volatile methyl iodide released by marine organisms (Yoshida and Muramatsu 1995) and transferred from ocean to land through volatilization of seawater into the atmosphere (Fuge 2005).

Table 7.4 Iron containing minerals

Mineral	Compound
Bornite	$CuFeS_4$
Chalcopyrite	$CuFeS_2$
Goethite	FeOOH
Hematite	Fe_2O_3
Ilmenite	$FeTiO_3$
Limonite	$FeO(OH).\ nH_2O + Fe_2O_3.\ nH_2O$
Magnetite	Fe_3O_4
Olivine	$(Mg,Fe)_2SiO_4$
Pyrite	FeS
Siderite	$FeCO_3$

Iron

Iron is a lustrous, ductile, malleable, silver-grey transition metal belonging to Group VIII and period 4 of the Periodic Table of Elements. It has an atomic mass of 55.85 and an atomic number 26. It is known to exist in four distinct crystalline forms. Iron is the tenth most abundant element in the universe and is the most abundant (by mass 34.6%) element making up the Earth. The concentration of iron in the various layers of the Earth ranges from high at the inner core to about 5% in the outer crust. Most of this iron is found in various oxides and hydroxides and occurs in the crystal lattices of a number of primary and secondary minerals (Table 7.4). Iron dissolves readily in dilute acids. It is chemically active and forms the divalent iron (II), or ferrous compounds and the trivalent iron (III), or ferric compounds.

Manganese

Manganese is a pinkish-grey, hard and very brittle transition metal belonging to Group VIIB and period 4 of the Periodic Table of Elements. It has an atomic mass of 54.94 and an atomic number 25. It is easily oxidized but is difficult to melt. Manganese is reactive when pure, and as a powder it burns in oxygen. Similar to iron, it reacts with water and rusts and dissolves in dilute acids. The average content of Mn in lithosphere is 1000 mg kg^{-1}. Its total quantity in soils vary from less than 100 to several thousand mg kg^{-1}, where it occurs in the form of oxides, carbonates and silicates such as pyrolusite (MnO_2), rhodochrosite ($MnCO_3$), hausmannite (Mn_3O_4), manganite (MnOOH) and rhodonite ($MnSiO_3$). It cycles through various oxidation states Mn(II), Mn (III) and Mn (IB). Besides these, Mn also forms hydrated oxides with mixed valence states.

Molybdenum

Molybdenum is silvery white, very hard transition metal belonging to Group VIB and period 5 of the Periodic Table of Elements. It has an atomic weight of 95.94 and an atomic number 42. The mean crustal concentration of Mo is 2.3 mg kg^{-1} and averages 1.4–2.0 mg kg^{-1} in rocks. In soils, Mo content averages 1.4 mg kg^{-1} and ranges between 0.1 and 7.4 mg kg^{-1} in different soils. Molybdenum is attacked slowly by acids. Mineral form of molybdenum in rocks include molybdenite

Table 7.5 Zinc containing minerals

Mineral	Compound
Sphalerite	ZnS
Smithsonite	$ZnCO_3$
Hemimorphite	$Zn_4(OH_2)Si_2O_7.H_2O$
Willemite	Zn_2SiO_4
Franklenite	$ZnFe_2O_4$
Zinc bloom	$Zn_5(OH)_5(CO_3)_2$
Zincite	$ZnSiO_4$

(MoS_2), ilsemannite ($Mo_3O_8{\cdot}8H_2O$), wulfenite ($PbMoO_4$), ferrimolybdite ($Fe_2(MoO_4)_3{\cdot}8H_2O$) and powellite ($CaMoO_4$). Molybdenum is released from mineral forms through weathering. In soils, Mo exists in several oxidation states ranging from 0 to VI, the latter being the most common form in agricultural soils.

Zinc

Zinc is a bluish-white transition metal belonging to Group IIB and period 4 of the Periodic Table. It has an atomic mass of 65.4 and an atomic number 30. Zinc makes up about 0.02% of the Earth's crust and is the 23rd most abundant element. The mean concentration of Zn in Earth's crust is 52 mg kg^{-1}, which is lower than the median value of 90 mg kg^{-1} for soils (Steinnes 2009). Total Zn concentration in soils can range between 10 and 100 mg kg^{-1} (Mertens and Smolders 2013). In nature, Zn does not occur in elemental form and is usually extracted from mineral ores to form ZnO (Table 7.5). Zinc usually exists in +II oxidation state and forms complexes with a number of anions, amino acids and organic acids. Zinc may precipitate as $Zn(OH)_2$, $ZnCO_3$, or ZnS. At higher pH values, Zn can form carbonate and hydroxide complexes which control its solubility. Zinc readily precipitates under reducing conditions and in highly polluted systems when it is present at very high concentrations, and may co-precipitate with hydrous oxides of iron or manganese (Cynthia and David 1997).

7.3 Microelement Transformations in Soil

The microelements occur in soils in several forms and their transformations and availability to plants depends on their distribution between solid and liquid phases of soil. Based on exchange, complexation and chelation reactions between soil's liquid and solid phases, Viets (1962) proposed the existence of five distinct pools viz. (i) soil solution or water soluble, (ii) exchangeable, (iii) adsorbed, complexed and chelated, (iv) associated with secondary minerals and as insoluble metal oxides and hydroxides, and (v) associated with primary minerals (Fig. 7.1). The water-soluble pool is almost negligible for Cu and Zn and very small for Fe and Mn in well aerated neutral soils. However, poor aeration, low redox potential and low pH can increase the amount of Mn and Fe in this pool, but may not affect Cu and Zn. The exchangeable pool, which also includes water soluble pool, is

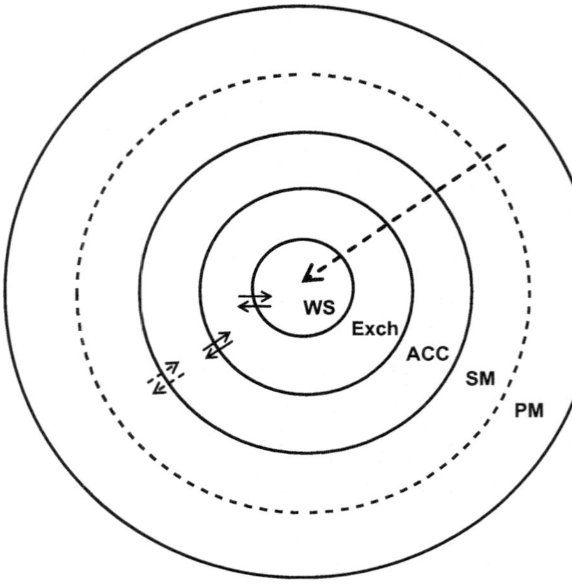

Fig. 7.1 Schematic representation of five pools of micronutrient cations in soil (Adapted from Viets 1962). *WS* water soluble, *Exch* Cations exchangeable by a weak exchanger, *ACC* Adsorbed, chelated or complexed ions exchangeable by cations having high affinity for exchange sites, *SM* Cations in secondary clay minerals and insoluble metal oxides, *PM* cations held in primary minerals

relatively large in size except for Cu and Mn. The third pool comprises cations that can be replaced by mass action of cations with similar affinities or by extraction with chelating agents such as ethylenediamine tetraacetic acid (EDTA). The first three pools exist in dynamic equilibrium and constitute the labile pool from which the plants draw micronutrients. The chemical composition of liquid phase keeps on changing because of contact with highly diverse soil solid phase and the uptake of water and microelement ions by plants (Kabata-Pendias 2004).

The available forms of micronutrient cations (Fe, Zn, Mn, Cu and Co) are usually present in their divalent form and are either adsorbed on the soil colloids or become part of the secondary silicate minerals through isomorphic substitution. The micronutrient anions (Mo and B) combine with soil constituents and form new reaction products and may also precipitate in some soils. The micronutrients, held strongly in organic and inorganic forms, become slowly available by weathering and microbial decomposition. The availability of micronutrients is greatly influenced by soil conditions, particularly pH and soil aeration status. The solubility of divalent and trivalent cations changes 100 and 1000-fold with a unit change in pH (Lindsay 1979). Generally, in well-aerated acid soils, several essential trace elements, including Fe, Al, Mn, and Zn are mobile and readily available to plants. In the poorly aerated neutral or alkaline soils, metals are considerably less available. On the contrary, elements such as Mo and Se are strongly bound in the pH range 5.5–7.5 and are readily available in alkaline soils. In addition to soil conditions, competition between different metals may inhibit uptake of a particular nutrient, such as high concentrations of Cu, Fe and Ca may depress the availability of Zn (Kiekens 1995).

Boron

Boron in soil exists in organic and inorganic forms, both of which are in equilibrium with each other. In acid soils, much of the B is in inorganic form and is lost by leaching rendering these soils low in B. In arid and semi-arid regions, inorganic B may occur as Na and Ca salts. Usually less than 10% of the total B in soil is available to plants. Soil factors affecting availability of B to plants include pH, texture, moisture, temperature, organic matter and clay mineralogy. Boron adsorbing surfaces in soils are: aluminium and iron oxides, magnesium hydroxide, clay minerals, calcium carbonate, and organic matter (Goldberg 1997). With increase in soil pH, B availability to plants decreases. Therefore, liming of acid soils can result in reduced B availability and its deficiency in plants. Native B is positively correlated with soil clay content (Elrashidi and O'Connor 1982) and coarse-textured soils contain less available B than fine textured soils. Therefore, B deficiency often occurs in plants growing in sandy soils (Fleming 1980; Gupta 1968). Total and plant available B content is reported to be significantly related to soil organic matter (Berger and Truog 1945; Elrashidi and O'Connor 1982; Miljkovic et al. 1966). Boron availability generally decreases as soils become dry. Soils may also have high concentrations of B that could cause B toxicity and decrease crop yields. The highest naturally occurring concentrations of B are in soils derived from marine evaporites and marine argillaceous sediment (Nable et al. 1997). In addition, anthropogenic sources of B including irrigation water, waste from surface mining, fly ash and industrial chemicals may increase soil B to toxic levels for plants.

Cobalt

A number of factors influence Co distribution in soils, the most important ones being hydrous oxides of Fe and Mn. Oxides of Fe possess strong affinity for the selective adsorption of Co. The oxidation of Co^{2+} to Co^{3+}, particularly by Mn oxides, commonly occurs in soils. Interchange reactions are involved in the sorption of Co by Mn oxides, resulting in the formation of hydroxyl species, $Co(OH)_2$ that precipitate at the oxide surface (Kabata-Pendias and Mukherjee 2007). In most soils, Co is slowly mobilized and its concentration in soil solution ranges between 0.3 and 87 μg L^{-1} (Kabata-Pendias and Pendias 2001). Common ionic species in soil solution include Co^{2+}, Co^{3+}, $CoOH^{+}$, and $Co(OH)_3^{-}$. Soil redox potential (Eh) and pH influence speciation and availability of Co to plants. The Co availability is higher at low pH than at high pH values. Liming decreases Co availability. Soil texture and organic matter also influence behaviour of cobalt in soils. Generally, soils high in organic matter have low total and plant available Co. Cobalt occurs in both inorganic and organic forms. The inorganic form is essential for human body and the organic form is present in green parts of plants, fish, cereals, and water.

Copper

The total amount of Cu in soil depends on the parent material, the degree of weathering, soil texture, extent of leaching and soil acidity. Solution and soil chemistry have a strong influence on the speciation of Cu. In alkaline aerobic systems, $CuCO_3$ is the dominant Cu species. The cupric ion, Cu^{2+}, and Cu hydroxides, $CuOH^+$ and $Cu(OH)_2$, are also commonly present. The factors affecting Cu availability to plants include organic matter content, soil pH, $CaCO_3$ and clay content. Copper in soil is strongly bound to soil organic matter and clay minerals. As a result, it is not leached to the groundwater (Cynthia and David 1997). The affinity of Cu to SOM increases as pH increases. If sulfur is present in anaerobic environments, CuS is formed.

Iodine

In soil, iodine is oxidized to iodate and the most common species in aqueous phase include I^-, I_3^-, IO_3^-, and $H_4IO_6^-$ (Kabata-Pendias and Sadurski 2004). Majority of iodine in soil occurs either sorbed with organic matter and clay or fixed in minerals and only a small fraction (<1–25%) is available to plants; the species I^- being absorbed at a higher rate than iodate (IO_3^-) (Yuita 1983). Iodate (IO_3^-) adsorption is negatively related to soil organic matter and positively to free Fe oxides (Dai et al. 2004). Volatile iodine compounds are commonly exchanged between soils and the atmosphere.

Iron

Iron occurs in soils in various forms including water soluble and exchangeable, insoluble Fe^{2+} compounds and Fe-organic complexes and active iron oxides. Iron is absorbed by plants as Fe^{2+} and the low solubility of Fe^{3+} containing compounds limits its availability for plant uptake. Compared to total Fe content in soil, the amount of soluble Fe is very low and in well-aerated soils iron solubility is mainly controlled by the solubility of Fe^{3+} hydroxides. The reaction is pH dependent and the activity of Fe^{3+} decreases with increasing pH and the solubility is minimum at pH 7.4–8.5 (Lindsay and Schwab 1982). Therefore, iron availability is higher in acid soils compared to calcareous and alkali soils. Under anaerobic/waterlogged conditions, Fe^{3+} is reduced to Fe^{2+} by anaerobic bacteria which use iron oxides as electron acceptors during respiration. The extent of reduction depends on period of submergence, soil organic matter content and active Fe, Mn and NO_3-N contents. Higher the organic matter content, greater is the reduction of Fe. Iron availability in soils, besides depending on pH, is influenced by presence of CO_3^{2-} and HCO_3^- ions, organic matter content, soil aeration status, soil moisture regime, soil texture and nutrient interactions. Plant genotypes also differ in their ability to absorb iron from soil. Some crop varieties are genetically more efficient in taking up iron from the soil than the other varieties.

Manganese

Manganese exists in soil as water soluble, organically bound, clay minerals bound crystalline and amorphous oxides. Different forms of Mn are in dynamic equilibrium with each other and the form in which Mn occurs in soil is influenced by pH and the oxidation reduction conditions. The tetravalent form is likely to be more prevalent in alkaline soils, the divalent form in acid soils and the trivalent form mainly occurs in neutral or near neutral soils. Available or active Mn in soils comprises water soluble, exchangeable and reducible forms. The water soluble and exchangeable forms usually exist in divalent form and are considered readily available to plants. Reducible Mn represents higher oxides that can be reduced by easily oxidizable organic substances such as quinol or hydroquinol. Factors affecting Mn availability in soil include pH, organic matter content, presence of $CaCO_3$, soil texture, and soil submergence. In soils with pH less than 5.5, large amount of Mn may be present as water soluble and exchangeable forms and their amount decreases with increase in pH. High organic matter content in neutral and alkaline soils decreases the availability of Mn by forming insoluble complexes with Mn^{3+}. The organic matter can influence Mn transformations in soil through (i) the production of complexing agents that effectively reduce the activity of free ions in solution, (ii) decrease in the oxidation potential of the soil, and (iii) the stimulation of microbial activity that results in the incorporation of Mn in biological tissue (Hodgson 1963). Presence of $CaCO_3$ reduces the availability of Mn by surface adsorption. Soil submergence, as in paddy fields, leads to reduction of Mn oxides and thus increases their availability to plants.

Molybdenum

Compared to other microelements, Mo behaves differently in soils. Unlike other microelements, it is highly mobile in alkaline soils (pH >6.5) and only slightly mobile in acid soils (pH <4–5). The MoO_4^{2-} anion dominates in the neutral and moderate alkaline pH range, whereas $HMoO_4^{-}$ occurs at lower pH values. Both anions occur in soil solution (Kabata-Pendias and Sadurski 2004). However, soluble Mo anions are readily precipitated by organic matter, $CaCO_3$, and a number of other cations (Fe, Mn, Cu, Zn and Pb). Under reducing conditions, Mo can form soluble thiomolybdates (e.g. MoS_4^{2-}, $MoO_2S_2^{2-}$). All these reactions being soil Eh and pH dependent may vary considerably. Every unit increase in pH above 3.0, increases MoO_4^{2-} solubility 100-fold mainly because of decreased adsorption of metal oxides (Lindsay 1979).

Zinc

Zinc in the solid phase exists as adsorbed to organic and inorganic particles and as Zn precipitates. Clay and soil organic matter are capable of holding Zn quite strongly, especially at neutral and alkaline pH regimes (Kabata-Pendias and Pendias 2001). Two different mechanisms of Zn sorption exist, one in acid soils related to cation exchange sites, and the other in alkaline soils (chemisorptions) that is greatly influenced by organic ligands. Hydroxides of Al, Fe, and Mn may also be involved in binding Zn in some soils. In the soil solution, Zn occurs in the form of free hydrated $[Zn(6H_2O)^{2+}]$ and complexed species as cations (Zn^{2+}, $ZnCl^+$, $ZnOH^+$, $ZnHCO_3^+$) and anions (ZnO_2^{2-}, $Zn(OH)_3^-$, and $ZnCl_3^-$) (Kabata-Pendias and Sadurski 2004). The concentration of Zn in soil solution is very low, and may range between 4×10^{-10} and 4×10^{-6} M (Barber 1995). Zinc concentration in soil solution is negatively correlated to soil pH because of its strong sorption on the solid phase at high pH values (Jeffery and Uren 1983). Availability of Zn to plants from soil and fertilizers depends on soil pH, organic matter and $CaCO_3$ contents, soil P status and adsorption by clay. Generally, Zn availability is higher in acid soils than in alkaline soils. In alkaline soils having pH greater than 7.85, Zn forms negatively charged zincate ions, $Zn(OH)_4^{2-}$. Because of interaction between zinc and calcium, the availability of zinc decreases in alkaline soils. Organic matter increases the availability of Zn by complexing with other substances that may fix Zn.

7.4 Beneficial Effects of Microelements

Microelements perform a number of functions in plants and humans; the major ones are summarized in Table 7.6. Besides being component of enzymes, certain micro-elements are involved in cellular functions, activation of enzymes and function in oxidation-reduction reactions of plant metabolism.

7.4.1 Plants

Boron

The essentiality of B for higher plants was established in 1923 (Warington 1923). The possible roles of B in plants include sugar transport, protein synthesis, cell wall synthesis, lignifications, cell wall structure integrity, carbohydrate metabolism, ribose nucleic acid (RNA) metabolism, respiration, indole acetic acid (IAA) metabolism, phenol metabolism, and as part of the cell membranes (Parr and Loughman 1983; Welch 1995; Ahmad et al. 2009). Boron increases the transport of chlorine and phosphorus becuase of plasmalemma ATPase induction.

Cobalt

Cobalt is involved in N_2-fixation in legumes. Three cobalamin dependent enzyme systems in *Rhizobium* may account for the influence of cobalt on nodulation and N_2-fixation in legumes. Cobalt increases the weight and cobalt content of nodules, number of bacteroides per nodule, cobalamin and leghemoglobin concentration.

Copper

Presence of Cu in chloroplasts is related to binding of solar energy during photosynthesis. The Cu-Zn- superoxide dismutase (SOD) enzyme in green leaves is involved in the detoxification of O^{2-} produced in photorespiration. Cytochrome oxidase, an enzyme involved in respiration, contains Cu. Both phenolase and laccase contain Cu and are involved in lignin synthesis. Cell wall loosening involving Cu plays a role in cell expansion and fruit softening. Copper is an essential element of various enzymes, such as polyphenol oxidases, plastocyanin of the photosynthetic transport chain, and cytochrome-c oxidase, the terminal

Table 7.6 Functions of essential microelements in plants and humans

	Plants	Humans
B	Involved in sugar transport; synthesis of plant hormones and lignin; formation of cell walls	Well defined biochemical function not identified; probably affects the metabolism of macrominerals, energy, nitrogen, and reactive oxygen
Cu	Involved in binding of solar energy during photosynthesis; part of cytochrome oxidase, phenolase and laccase; component of enzymes important for lignin formation; component of electron carriers in chloroplast	Essential for the development of connective tissue, nerve coverings, and bone;. participates in both Fe and energy metabolism;. acts as a reductant in several oxidases that reduce molecular oxygen; involved in enzyme catalysis
Co	Required by N_2-fixing plants for activity of cobalamin dependent enzymes in *Rhizobium*	Component of cobalamin, a functional unit of vitamin B_{12} thus required for red blood cell formation
Fe	Required for chlorophyll synthesis and metabolism of N and S; component of enzymes involved in detoxification of H_2O_2 and superoxide; component of enzymes involved in the synthesis of suberin and lignin for cell wall formation	Functions in oxygen transport and electron transport; occurs in Fe-heme proteins; Fe–sulphur enzymes; proteins for Fe storage and transport, and other Fe-containing or Fe-activated enzymes
I	Beneficial to higher plants	Metabolic regulation
Mn	Involved in conversion of light energy into chemical energy, reduction of CO_2, sulphite and nitrite; involved in the synthesis of fatty acids as a component of biotin enzyme	Associated with bone development; involved in amino acid, lipid, and carbohydrate metabolism; part of enzymes, e.g. mitochondrial Mn superoxide dismutase, glutamine synthetase, arginase, and activates several hydrolases, transferases and carboxylases
Mo	Nitrogen metabolism, required for nitrogen fixation	Potentiation of insulin action in the maintenance of glucose tolerance, cofactor for some enzymes
Se	Beneficial to higher plants	Se is incorporated into proteins to make selenoproteins, which are important antioxidant enzymes. Se is found in glutathione peroxidase, thioredoxins, and selenoprotein P, involved in antioxidant protection, redox regulation, essential for testosterone biosynthesis and formation and development of spermatozoa
Zn	Component of enzyme superoxide dismutase, carbonic anhydrase; component of many organic complexes and DNA; involved in growth hormone production and seed development	Involved in the activity of about 100 enzymes, e.g. RNA polymerase, carbonic anhydrase, Cu–Zn superoxide dismutase, angiotensin I converting enzyme; present in Zn-fingers associated with DNA; supports normal growth and development in pregnancy, childhood, and adolescence

Compiled from Fraga (2005) and Combs (2005)

oxidase in the mitochondrial electron transport chain. Deficiency of Cu leads to pollen sterility which affects fruiting of plants. Copper compounds are used as/or in fungicides and insecticides.

Iron

Iron is a component of chloroplast and 63% of the total leaf iron is found in chloroplast. It participates in the process of photosynthesis and the metabolism of nitrogen and sulphur. Some steps in chlorophyll biosynthesis are Fe-dependent. It is also a component of enzymes involved in detoxification of H_2O_2 and superoxide and thus protects the photosynthetic apparatus. Iron is a necessary component of many enzymes involved in the synthesis of suberin and lignin for cell wall formation. It is an essential element for heme and ferredoxin groups.

Manganese

The essentiality of Mn for plants was established in 1922. Manganese is involved in conversion of light energy into chemical energy, reduction of CO_2, sulphite and nitrite. It plays an important role in water splitting and O_2 evolving system in photosynthesis. It is an integral part of the superoxide dismutase and of the electron donor complex of photosystem II. It may activate enzymes by bridging the phosphate group with the enzyme or the substrate. Manganese containing enzyme protects the photosynthetic apparatus from photo destruction. Manganese content of leaves is positively correlated with seed yield and oil content, may be due to supply of carbon skeleton for fatty acid synthesis. Another factor may be direct involvement of Mn in the synthesis of fatty acids as a component of biotin enzyme.

Molybdenum

The essentiality of Mo for plants was first demonstrated by Arnon and Stout (1939). Molybdenum is involved in reduction and oxidation-reduction reactions of plant metabolism. It is mainly present as an integral part of the organic pterin complex, termed the molybdenum co-factor (Moco), which binds to Mo requiring enzymes (molybdoenzymes) (Williams and Frausto da Silva 2002). Molybdenum is an essential component of two major enzymes of nitrogen metabolism: *nitrogenase* involved in N_2 fixation in legumes and *Nitrate reductase* that catalyzes the reduction of nitrate to nitrite. Response to Mo varies with the source of N supply. When nitrate is supplied in the absence of Mo, plants show typical deficiency symptoms (Whiptail). When ammonium is supplied, the response to Mo is less marked. Other molybdoenzymes that have been identified include xanthine dehydrogenase, aldehyde oxidase and sulfite oxidase (Williams and Frausto da Silva 2002).

Zinc

Zinc is a component of enzyme superoxide dismutase that protects the photosynthetic apparatus and carbonic anhydrase that increases the pool size of CO_2 in the chloroplast in C_3-plants, thus increasing the rate of photosynthesis. Zinc is a part of RNA polymerase, and various dehydrogenases and is required for the activity of enzymes involved in carbohydrate metabolism, protein synthesis, maintenance of the integrity of cellular membranes, auxin synthesis and pollen formation (Marschner 1995). Zinc is intimately involved in nitrogen metabolism of plants.

It is considered to be involved in the synthesis of tryptophan, which is a precursor of indole acetic acid, an important phytohormone. Zinc is required for tolerance of environmental stresses such as high light intensity and high temperature through its role in regulation of gene expression (Cakmak 2000).

7.4.2 *Humans*

Boron

Well defined biochemical function for B in humans has not been identified; it probably affects the metabolism of macrominerals, energy, nitrogen, and reactive oxygen. Data from animal studies indicate a functional role for B. Boron forms complexes with some biologically important substances such as diadenosine polyphosphates, S-adenosylmethionine, pyridoxine, riboflavin, dehydroascorbic acid and pyridine nucleotides which may regulate certain functions (Nielsen 2006). A number of organoboron compounds (boroesters) including antibiotics produced by microorganism, plant cell wall component (rhamnogalacturonan-II) and a bacterial extracellular signaling molecule have been identified (Sato et al. 1978; Schummer et al. 1994; O'Neill et al. 1996; Matoh 1997; Chen et al. 2002).

Cobalt

Cobalt is essential for humans as a component of cobalamin, the functional unit of Vitamin B_{12}, which is the only vitamin containing a metal ion (Carmel 2007). Vitamin B_{12} is synthesized by microorganisms such as algae and yeast which are abundantly present in soil. Two forms of vitamin B_{12} viz. methylcobalamin and 5-deoxyadenosyl cobalamin are utilized in the human body. Cobalamin acts as a cofactor for two enzymes L-methylmalonyl-CoA mutase involved in the synthesis of haemoglobin, and methionine synthase important in cancer prevention. Cobalt binds to Fe-transport proteins and is thus involved in the synthesis of haemoglobin. It seems that some organic Co compounds are involved in processes of stabilizing the DNA structure, which has a role in cancer prevention (Munno De et al. 1996). Vitamin B_{12} is also required for the development of nerve cells, growth and repair of cells and for the movement of carbohydrates and fats in the body (Seatharam and Alpers 1982). Cobalt is important for forming amino acids and some proteins to create myelin sheath in nerve cells (Ortega et al. 2009). The salts of cobalt stimulate the synthesis of erythropoietin, which is involved in the formation of erythrocytes in bone marrow (Simonsen et al. 2012). Cobalt may be bound by some proteins to replace other divalent cations such as Zn and Mn in different enzymes.

Copper

The essentiality of copper for humans was established during the 1960s in malnourished children from Peru (Cordano et al. 1964). Copper is essential for the development of connective tissue, nerve coverings, and bone; participates in both Fe and energy metabolism; acts as a reductant in several oxidases that reduce molecular oxygen; and is involved in enzyme catalysis. Copper-containing enzymes include ceruloplasmin, superoxide dismutase, cytochrome-c oxidase, tyrosinase, monoamine oxidase, lysyl oxidase and phenylalanine hydroxylase (Linder and Hazegh-Azam 1996). Copper is required for infant growth, host defence mechanisms, bone strength, red and white cell maturation, iron transport, cholesterol and glucose metabolism, myocardial contractility, and brain development (Danks 1988; Olivares and Uauy 1996). Copper functions in angiogenesis, neuro- hormone release, oxygen transport, and the regulation of genetic expression. Copper compounds can be used as food and animal feed additives as a nutrient or a colouring agent.

Iodine
Iodine is required for growth and survival. It is a component of the thyroid hormone thyroxene. The hormones thyroxine (T4) and triiodothyronine (T3) are iodinated molecules of the amino acid tyrosine. The thyroid hormones regulate a variety of important physiological processes including the cellular oxidation. Iodine is also essential to the normal development of the foetal nervous system and regulates the effect of oestrogen on breast tissue (Knez and Graham 2013). Iodine removes toxic chemicals and biological toxins, suppresses autoimmunity, strengthens the T-cell adaptive immune system and protects against abnormal growth of bacteria in the stomach, particularly *Helicobacter pylori* (Miller 2006). Iodine is also used in a number of chemicals and pharmaceuticals. Radioactive ^{131}I is used in medical diagnosis. KI-water solution and I-alcohol solution are strong disinfectants and used for external wounds. In Chemistry, iodine compounds are used as catalysts. Iodine is added to colorants and inks, and AgI, a photosensitive compound is used in photography (Kabata-Pendias and Mukherjee 2007).

Iron
Iron (Fe) plays an important role in the transport of oxygen, several oxidation-reduction pathways and cellular growth. It is required for making haemoglobin (Hb) and it is a prooxidant which is also needed by microorganisms for proliferation (Galan et al. 2005). Biologically important compounds of iron are haemoglobin, myoglobin, cytochromes, catalases and peroxidase (Malhotra 1998). Iron is an integral component of several heme (-1.0%) and non-heme metalloenzymes that are involved in various respiratory, oxidative, and phosphorylating functions (Melki et al. 1987).

Manganese
Manganese is involved in glycoprotein and protoglycan synthesis and is a component of mitochondrial superoxide dismutase. Manganese is a cofactor of phosphohydrolases, phosphotransferases, and decarboxylase enzymes (Murray et al. 2000) and it is a part of the enzymes involved in urea formation, pyruvate metabolism and the galactotransferases of connective tissue biosynthesis (Chandra 1990). Manganese activates several enzymes involved in amino acid, cholesterol and carbohydrate metabolism. It is required for the synthesis of acid mucopolysaccharides, such as chondroitin sulphate, to form bone matrix. Manganese functions with vitamin K for the formation of prothrombin required for blood clotting.

Molybdenum

Molybdenum is a component of several metalloenzymes including xanthine oxidase, aldehyde oxidase, nitrate reductase, and hydrogenase (Soetan et al. 2010). Xanthine oxidase and aldehyde oxidase play a role in iron utilization as well as in cellular metabolism in electron transport. Mo is a co-factor for enzymes necessary for the metabolism of S-containing amino acids and nitrogen-containing compounds present in DNA and RNA, the production of uric acid, and the oxidation and detoxification of various other compounds.

Zinc

The essentiality of Zn was established in 1869 for plants, in 1934 for animals and in 1961 for humans (King and Cousins 2006). Zinc occurs in all living cells and is an important constituent of plasma (Murray et al. 2000). It is involved in the functioning of a number of enzymes and serves as a structural, catalytic, and regulatory ion. Zinc is the only metal involved in all six classes of enzymes viz. oxido-reductases, transferases, hydrolases, lyases, isomerases and ligases (Barak and Helmke 1993). Some of the Zn-dependent enzymes include lactate dehydrogenase, alcohol dehydrogenase, glutamic dehydrogenase, alkaline phosphatase, carbonic anhydrase, carboxypeptidase, superoxide dismutase, retinene reductase, DNA and RNA polymerase (Soetan et al. 2010). As a structural component, Zn allows the binding of amino acids (cysteine and histidine residues) in the protein chain to form finger like structure (Knez and Graham 2013). Zinc is involved in a number of biochemical, immunological and clinical functions and is considered the most important metabolic promoter (Hotz and Brown 2004). It is involved in the metabolism of nucleic acids, amino acids, vitamin A, vitamin E and glucose. Zinc is needed for cellular differentiation and replication, gene expression, protein synthesis, testicular development, and functioning of taste buds (Merck 1986; Vallee and Falchuk 1993; Szabo et al. 1999). Zinc is required for regulating immune functions including tissue repair and wound healing. It plays a role in preventing free radical formation and protecting biological structures from damage, and promotes the synthesis of metallothionein, a protein present in intestinal mucosa, liver and kidneys that is essential for heavy metal metabolism (Hotz and Brown 2004). Zinc, being an integral constituent of insulin, is essential for its optimum activity. Zinc is essential for thymic functions as it influences thymic hormone, thymulin, needed for T-cell maturation and differentiation (Mocchegiani et al. 2000).

7.5 Effects of Excessive Micronutrient Uptake

7.5.1 Plants

Under similar soil conditions, different plant species may take up widely different amount of the same microelement from the soil (Alloway 2005). There exists a certain concentration range representing a safe and adequate intake of a

Table 7.7 Critical leaf concentrations for sufficiency and toxicity of microelements in non-tolerant crop plants

Element	Critical leaf concentrations ($mg\ kg^{-1}$dry matter)	
	Sufficiency	Toxicity
Boron (B)	5–100	100–1000
Cobalt (Co)	–	10–20
Copper (Cu)	1–5	15–30
Iodine (I)	–	1–20
Iron (Fe)	50–150	>500
Manganese (Mn)	10–20	200–5300
Molybdenum (Mo)	0.1–1	1000
Zinc (Zn)	15–30	100–300

Adapted from White and Brown (2010)

micronutrient. Nutrient concentration above or below a critical range may lead to toxicity or deficiency of an element. This may adversely impact plant physiological functions and productivity. Critical leaf concentrations for sufficiency and toxicity of microelements in non-tolerant crop plants are presented in Table 7.7. The tendency of plants to accumulate and translocate essential nutrients to edible parts mainly depends on plant genotype and soil, climate and management factors. The nutrient concentration in edible plant parts such as cereal grains is generally termed as "micronutrient density", which refers to amount of a micronutrient per unit of seed weight.

Among the essential microelements toxic concentrations of B, Fe and Mn often occur in agricultural soils under specific conditions. While toxic concentration of B could occur in sodic soils, that of Mn occurs in acid soils. Toxicities of Fe and Mn can occur in waterlogged or flooded soils such as rice fields. In addition to specific soil conditions, toxic concentrations of some essential microelements such as Zn and Cu may occur because of anthropogenic activities (see Chap. 8). High bioavailability of essential microelements in soil can hamper plant growth and reduce crop yields. For example, high tissue concentrations of B are toxic to plants but plants maintain tissue B concentration within an optimum range by modulating B transport processes (Miwa and Fujiwara 2010). Excessive supply of Co to plants can lead to depressed growth, chlorosis, necrosis and even death of plants. Copper is known to interact with other nutrients. Copper toxicity may occur in acid soils or in soils receiving high amounts of Bordeaux mixture for a long time. However, because of high Cu fixation capacity of soils, its toxicity rarely occurs in plants. But foliar spray may be injurious to plants and the Cu solution needs to be neutralized with lime before application.

Excessive uptake of iron by aerobic organisms can be toxic because it can undergo a number of chemical reactions with oxygen and produce highly reactive and damaging oxygen free radicals. Excessive uptake of Mn (Mn^{2+}) inhibits crop production in acid soils as it displaces Ca^{2+}, Mg^{2+}, Fe^{2+} and Zn^{2+} in their essential cellular functions (White and Brown 2010). Therefore, symptoms of Mn-induced Ca, Mg and Fe deficiency resemble toxicity symptoms of Mn. Molybdenum toxicity in plants is rarely seen, but the effect of excessive Mo uptake in forages and pasture grasses is manifested in animals. Ruminants that consume plant tissue high in Mo can develop molybdenosis, a disorder that induces Cu deficiency (Scott 1972). High concentration of Mo in tomato and cauliflower plants can lead to accumulation of anthocyanins in leaves turning them purple. Molybdenum toxicity commonly occurs in calcareous soils and peaty soils or as a result of over liming or excessive application of Mo fertilizers.

Zn toxicity rarely occurs in plants under field conditions. However, continued application of Zn fertilizers to alkaline sandy soils low in organic matter content may lead to Zn toxicity in plants. The general symptoms of Zn toxicity in plants include stunting of shoot, curling and rolling of young leaves, chlorosis and death of leaf tips (Rout and Das 2003). Excessive uptake of Zn may cause growth inhibition and oxidative damage in plants (Vaillant et al. 2005; Panda et al. 2003). Excessive Zn can affect cell mitotic activity, membrane integrity and permeability and can even lead to death of cell (Stoyanova and Doncheva 2002; Chang et al. 2005). Zinc toxicity affects metabolic processes through competition for uptake, inactivation of enzymes, displacement of essential elements from functional sites (Rout and Das 2003).

7.5.2 Humans

Studies on animals have shown that excessive consumption of B can have adverse reproductive and developmental effects. The excessive intake of Co may cause polycythemia (increased red blood cells), cardiomyopathy, hypothyroidism, failure of pancreas, bone marrow hyperplasia, and some types of cancer (Plumlee and Ziegler 2003). In occupational settings, inhalation of dust containing Co could affect respiratory system leading to asthma, fibrosing alveolitis and lung cancer in humans. Cobalt inhalation could also affect the hematopoietic system, the myocardium, the thyroid gland, and the nervous system (Yamada 2013). The recommended safety threshold for cobalt in workplace air ranges between 0.05 and 0.1 mg m^{-3} for an 8–10-h work day, 40-h work week.

Excessive ingestion of Cu can cause gastrointestinal distress and liver damage characterized by diarrhea, vomiting, abdominal pain and nausea. Chronic high Cu intake can lead to the hepatic accumulation of Cu, which has been suspected in juvenile cases of hepatic cirrhosis in India. Individuals with Wilson's disease may be at increased risk of adverse effects of excessive Cu intake.

High intake of iodine may occur through consumption of iodine-rich diet such as seafood or iodine containing supplements. Excessive intake of iodine causes a decrease of thyroid hormone production resulting in formation of high iodine goiter. Individuals with autoimmune thyroid disease or with history of iodine deficiency are particularly susceptible to adverse effects of excessive iodine intake.

Iron toxicity is mostly associated with undue oral intake, usually in combination with high consumption of alcohol and vitamin C. Other causes could be multiple transfusions, as in thalassemic and hemophiliac patients, or excessive parenteral administration of unbound iron preparations. Susceptibility to increased iron absorption from the intestinal tract occurs in hereditary hemochromatosis. Excessive iron intake can cause gastrointestinal distress and is considered to increase risk of colorectal cancer and reduced insulin sensitivity; though these have not been confirmed yet. Therefore, iron uptake is highly regulated to prevent excessive accumulation of iron in cells (Graham et al. 2001). Iron toxicity leads to deposition of excess iron in parenchyma tissues of the liver, heart, pancreas, spleen, and other organs leading to multiple derangements.

Excessive exposure to Mn could occur due to inhalation. This may have an adverse effect on functions of central nervous system and produce psychotic symptoms and Parkinsonism (Tan et al. 2006). High intakes of Mn could be of concern for individuals not consuming adequate amounts of magnesium. Molybdenum has fairly low toxic effects. High intake of Mo can have adverse reproductive effects in animals and can precipitate Cu deficiency in cattle and sheep (Underwood 1971). Dietary Mo affects Cu metabolism in man (Deosthale and Gopalan 1974). Individuals with impaired Cu metabolism or deficient in Cu are at increased risk of Mo toxicity. Excessive intake of Zn may occur because of inappropriate use (more than 100 mg Zn day^{-1}) of zinc supplements. This can lead to gastrointestinal irritation, vomiting, decreased immune function, reduction in high density lipoprotein (HDL) cholesterol, diminution of Cu status in the body and altered Fe function (Combs 2005; Hamilton et al. 2000). Toxic effects of elements in humans are further discussed in Chap. 8.

7.6 Effects of Deficient Micronutrient Uptake

7.6.1 Plants

Deficiencies of microelements in soils and plants are of global occurrence and major staple foods are highly susceptible to such deficiencies. Micronutrient deficiency in plants not only limit agricultural production but also impact, directly or indirectly, human nutrition (Alloway 2005; Andersen 2007). Micronutrient deficiencies in plants have been reported to be related to soil properties, being more pronounced in coarse-textured compared to fine textured ones (Benbi and Brar 1992).

Boron and Cobalt

Several soil properties could lead to B deficiency including low soil organic matter, coarse/sandy texture, high pH, liming, drought, intensive cultivation and more nutrient uptake than application (Ahmad et al. 2012). Boron deficiency has adverse effects on cellular functions and physiological processes (Cakmak and Römheld 1997; Dugger 1983; Marschner 1995). It inhibits the growth of vegetative and reproductive plant parts. During vegetative growth, B deficiency causes stunted root and shoot tips. In root crops, B deficiency has been associated with internal tissue breakdown. Soils showing Co deficiency include highly leached acid soils low in total Co, soils derived from granite, highly calcareous soils, coastal sandy soils and peaty soils, and soils derived from rocks rich in Mg.

Iron

Despite being present in abundant amount in Earth's crust and soil, the iron deficiency is generally prevalent because of the insoluble nature of iron in an oxidizing environment. Plants have their own mechanism to acquire iron from soil. Higher plants, except cereals, depend on root-cell membrane bound ferric reductases to take-up iron in ferrous form from soil solution (Frossard et al. 2000). In cereals the chelates excreted by their roots are absorbed intact by root cells via a specific ferric-phytometallophore transport system in the cell plasma membrane. Deficient uptake of iron leads to chloroplast disorders; the synthesis of thylakoid membranes is disturbed and the photochemical activity affected (Spiller and Terry 1980).

Manganese

Manganese deficiency is more prevalent in near neutral to alkaline (pH 6.5–8.0) soils. Frequent fluctuation in water table or alternate submergence and drying may lead increased incidence of Mn deficiency. This has commonly been observed in wheat grown after lowland rice in northern India. Deficiency of Mn^{2+} leads to the breakdown of chloroplasts.

Molybdenum

Growing of plants under Mo deficient conditions leads to the development of several different phenotypes that are associated with reduced molybdoenzymes and their presence hampers plant growth. Deficient uptake of Mo impacts plant metabolism and the magnitude of effect depends on the requirement for various types of molybdoenzymes present in plants. It is not simple to assign a well-defined plant response to Mo deficiency because of its involvement in a number of enzymatic processes. However, Mo deficiencies are primarily associated with impaired nitrogen metabolism particularly when nitrate is the major form available for plant growth. Moco decreases the activity of nitrogen reducing and assimilatory enzymes including nitrate reductase and xanthine deyhydrogenase/oxidase (Kaiser et al. 2005). In most plant species, the loss of nitrate reductase activity is usually associated with increased tissue nitrate concentration and decreased plant growth and yield (Spencer and Wood 1954).

Zinc

Zinc deficiency occurs worldwide in soils and crops. Soils with high pH, high $CaCO_3$ and low organic matter contents are more prone to Zn deficiency. About 30% of the agricultural soils in the world are estimated to be deficient in Zn (Sillanpää 1982). Countries with large Zn-deficient areas include China, India, Iran, Pakistan, and Turkey where 50–70% of the cultivated soils are Zn deficient (Alloway 2009). Zinc deficiency in soils has also been observed in Australia and Brazil. Besides low concentration of total Zn in soil, the soil properties that lead to

Zn deficiency in crops include high $CaCO_3$ content, alkaline pH, high salt concentration, high levels of available P, and high concentration of Mg. Other conditions that may induce Zn deficiency include prolonged flooding such as in paddy soils or waterlogging and high concentration of bicarbonate in irrigation water (Alloway 2008). Zinc deficiency in agricultural soils results in poor yields and nutritional quality. Cereals are particularly prone to Zn deficiency because of their inherently low concentrations. It is estimated that 50% of world soils growing cereals are Zn deficient.

7.6.2 *Humans*

It is estimated that more than 60% of the world population is Fe deficient, over 30% Zn deficient, 30% I deficient and 15% are selenium (Se) deficient (White and Broadley 2009). This situation is attributed to shift in cultivation towards cereals, crop production in areas with low nutrient availability and consumption of food crops with inherently low nutrient concentrations. Cereals are inherently low in micronutrients, compared to many other food crops and their consumption as staple food leads to micronutrient deficiencies. Because of increasing population the dependence on cereals is increasing as cereals are often the most productive and remunerative food crops. This is causing decline in diet diversity, particularly in the developing countries leading to micronutrient deficiencies. The deficiencies are exacerbated by a lack of animal products or fish in the diet (Gibson 2006). The deficiency of a nutrient can cause impairment of some specific biochemical function and illnesses in human beings.

Boron

Though no specific biochemical function has been associated with B in humans but it appears to have beneficial effects on bone. Therefore, B deficiency without modification by other nutrients affects bone strength. Its deficiency together with low dietary intake of Cu and Mg can aggravate the increase in serum calcitonin (Nielsen 1996).

Cobalt

Endemic problems of Co deficiency in humans are not known but the deficiency symptoms are evident as vitamin B_{12} deficiency. Cobalt deficiency can cause anaemia, structural alteration of buccal cavity tissue, hypofunction of thyroid, degeneration of the peripheral nervous system, dermal hypersensitivity and increased risk of developmental abnormalities in infants. Its deficiency has been observed in ruminant mammals including sheep, goat and cattle. Cobalt deficiency is widespread in New Zealand, Australia, Great Britain, and parts of Scandinavia (Frøslie 1990). The deficiency symptoms include emaciation, loss of appetite and thriftiness. This condition is often called pining, salt sickness, bush sickness or coast disease.

Copper

Copper deficiency in humans is associated with malnutrition in children. Copper deficiency can result in the expression of a genetic defect such as Menkes syndrome that results in growth retardation, hypothermia, skin and hair depigmentation and abnormal spiral twisting of the hair, lax skin and articulations, tortuosity and dilatation of major arteries, varicosities of veins, osteoporosis, flaring of metaphyses, bone fractures, retinal dystrophy, and damage to central nervous system (Danks 1995). The syndrome appears before 3 months of age and usually causes death of the child before 5 or 6 years of age. Acquired Cu deficiency results in hypochromic, normocytic or macrocytic anaemia, neutropenia, bone abnormalities, hypopigmentation of the hair, hypotonia, poor growth, increased susceptibility to infections, abnormalities of cholesterol and glucose metabolism, and cardiovascular alterations.

Iodine

Iodine deficiency in humans may cause a number of illnesses or disorders including thyroid hypertrophy or goitre and mental retardation. Under iodine deficiency condition, the thyroid may not be able to synthesize sufficient amounts of thyroid hormone. The resulting low level of thyroid hormones in the blood (hypothyroidism) is mainly responsible for damage to the developing brain and other harmful effects known collectively as "iodine deficiency disorders" (Table 7.8; Hetzel 1983). Iodine deficiency during pre-natal development and the first year of life can result in endemic creatinism, a disease which causes stunted growth and general development along with brain damage. This mental deficiency has an immediate effect on child's learning capacity and women's health. An estimated two billion people worldwide have insufficient iodine intake, which is the single most important preventable cause of brain damage (WHO 2007; Zimmermann et al. 2008). The regions of the world mostly affected by iodine deficiency include developing countries of Africa, Asia, and Latin America particularly located far from the ocean (Fuge 2005). Even in some developed countries of Western Europe it has been suggested that 50–100 million people may be at risk of iodine deficiency (Delange 1994). Population mainly dependent on vegetarian diet have low iodine intake, which could lead to iodine deficiency (Davidsson 1999; Remer et al. 1999).

Table 7.8 Iodine deficiency disorders in humans

Physiological group	Health consequences
Fetus	Spontaneous abortion, stillbirth, congenital anomalies, increased perinatal mortality, deaf mutism
Neonate	Endemic cretinism, neonatal mental retardation, spastic diplegia, squint, neonatal hypothyroidism and short stature, infant mortality
Children, adolescent and adults	Retarded physical development, impaired mental function
	Iodine-induced hyperthyroidism (IIH)
All ages	Goitre, hypothyroidism, increased susceptibility to nuclear radiation

Adapted from Hetzel (1983)

Iron

Iron deficiency mainly occurs in infants, children, pregnant women that are dependent on plant foods as their main source of iron. This deficiency is caused by malnutrition and is associated with inadequate intake, poor absorption from diets high in phytate or phenolic compounds or excessive losses such as through menstruation, gastrointestinal haemorrhage, etc. Iron deficiency results in decreased blood haemoglobin and the appearance of microcytic hypochromic anemia. Red blood cells (RBC) have a life span of about 4 months and a continuous supply of iron from the diet (together with folate and proteins) is required to make haemoglobin and replace dead RBCs. Iron deficiency is the most significant cause of anaemia worldwide so that iron deficiency anaemia (IDA) and anaemia are often used synonymously. Approximately, 50% of the cases of anaemia are assumed to be due to iron deficiency (WHO 2001). However, the proportion may vary among population groups and in different regions depending on the local conditions (Fig. 7.2). According to a recent family health survey (WHO 2015) more than half of the children (58%) below the age of 5 in India are anaemic because of insufficient intake of iron. Other causes of anaemia include parasite infections (such as hookworms, ascaris, and schistosomiasis); acute and chronic infections (malaria, cancer, tuberculosis, and HIV) and the occurrence of other micronutrient deficiencies such as vitamins A and B_{12}, folate, riboflavin, and copper (WHO 2008).

Besides anaemia, the other effects of iron deficiency include muscle dysfunction due to decrease in the concentration of α-glycerol-phosphate oxidase in muscle and increased intracellular formation of lactic acid (Ohira et al. 1979); decreased

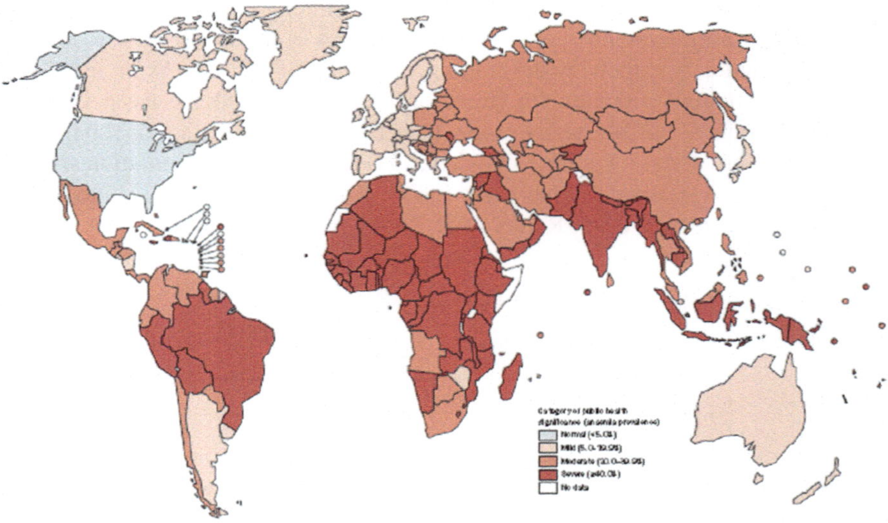

Fig. 7.2 Global distribution of anaemia in preschool-age children (Source: WHO 2008, p. 9; reprinted with kind permission from WHO)

immunity and enhanced susceptibility to infection because of abnormalities in cellular morphology and functions (Dallman 1976); angular stomatitis, webbing of the oesophagus, and atrophy of the gastric mucosa due to disruption of epithelial structures (Melki et al. 1987). Severe deficiency of iron can lead to defects in attention and cognition because of alterations in the central nervous system dopaminergic pathways (Tucker et al. 1984). Iron deficiency has been reported to play a role in brain development and functions involving neurotransmitters metabolism, protein synthesis, organogenesis etc. and in the pathophysiology of restless legs syndrome (Beard 1999; Tan et al. 2006).

Manganese
Manganese deficiency in humans is unknown, though its deficiency has been demonstrated in several animal species. Low dietary Mn or low blood and tissue Mn in humans has been associated with osteoporosis, diabetes, epilepsy, artherosclerosis, impaired wound healing and cataracts (Klimis-Tavantzis 1994).

Molybdenum
Molybdenum deficiency has not been unambiguously recognized in humans and is considered to be of little nutritional concern. However, low Mo intake is considered a predisposing cause of renal xanthine caliculi (Sardesai 1993). It also causes gout. Deficient Mo supply in the diet of pregnant and lactating women is considered to contribute to impaired growth and brain development of the foetus and infant.

Zinc
About one-third of the world population is estimated to be suffering from Zn deficiency and there is a close relationship between Zn deficiency in soils and crops and in humans (Fig.7.3).

The Zn deficiency is more prevalent in underdeveloped countries and is primarily associated with malnutrition caused by low dietary intake or too much dependence on foods with little or low bioavailable Zn (Cakmak et al. 2017). Generally, regions having Zn-deficient soils (Alloway 2008; Fig 7.3a) are also regions where Zn deficiency in humans is prevalent (Wessells and Brown 2012; Fig. 7.3b). Other causes of Zn deficiency include increased requirement, poor absorption, higher losses and low utilization (King and Cousins 2006). Susceptibility to Zn deficiency is high during periods of greatest protein synthesis. Therefore, clinical manifestation of Zn deficiency generally varies with age. Zinc deficiency affects epidermal, gastrointestinal, central nervous, immune, skeletal, and reproductive systems (Hambidge and Walravens 1982). Severe Zn deficiency is characterized by growth retardation and dwarfism, diarrhoea, alopecia anorexia, parakeratotic skin lesions, pustular dermatitis, abnormal dark adaptation, impaired testicular development and male hypo-gonadism, impaired immune functions including delayed wound healing, impaired cognitive functions, mental disturbance/lethargy, behavioral problems, impaired memory, learning disability, neuronal atrophy, impaired DNA synthesis, and recurrent infections (Barceloux 1999; Combs 2005; Knez and Graham 2013). Zinc deficiency is considered to increase risk to osteoporosis and susceptibility to oxidative stress. Zinc deficiency could lead to losses in activities

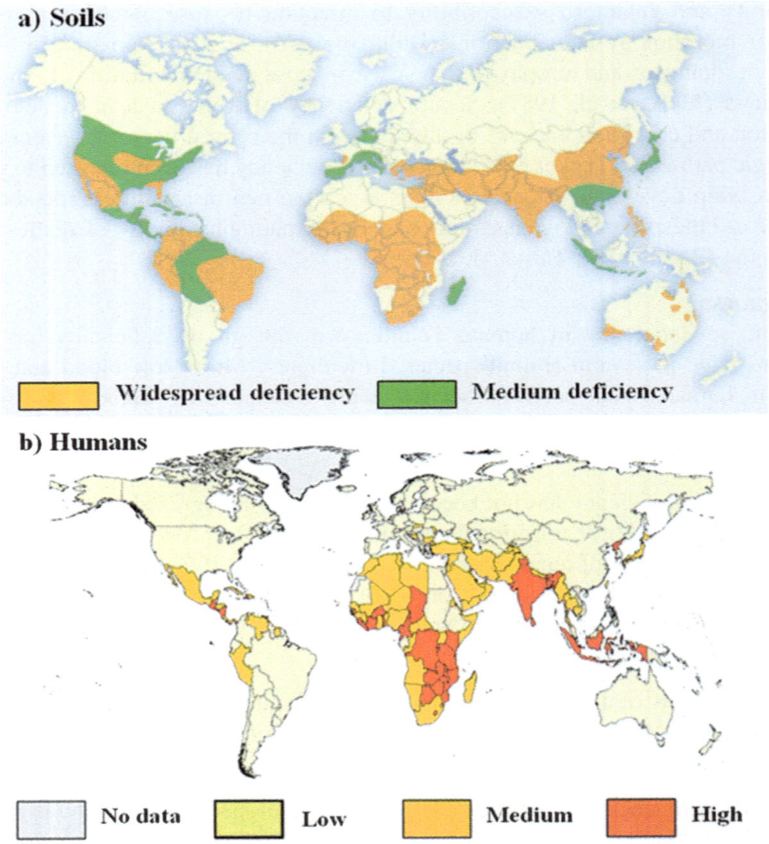

Fig. 7.3 Global distribution of Zn deficiency in (**a**) soils (Alloway 2008, p. 109; with kind permission from International Zinc Association) and (**b**) humans (Wessells and Brown 2012, p. 2; unrestricted reproduction permitted by PLOS ONE)

of some Zn-dependent enzymes. A mild deficiency of Zn is characterized by neurosensory changes, oligospermia in males, hyperammonemia, decreased serum thymulin activity, decreased natural killer cell activity, alterations in T cell subpopulation, impaired neurophysiological functions, and decreased ethanol clearance (Beck et al. 1997; Prasad 2002). Zinc deficiency is associated with several diseases including malabsorption syndrome, chronic liver disease, chronic renal disease, diabetes, and malignancy (Prasad 2003). Systemic intestinal inflammation associated with Zn deficiency can lead to Fe deficiency anemia (Roy 2010). Degenerative changes associated with aging such as decline in immunocompetence and some neurological and psychological shifts may partly be due to Zn deficiency (Whittaker 1998). In cattle, the signs of Zn deficiency occur when the pasture or fodder contains less than 18–42 mg Zn kg^{-1} dry matter (Prasad 2010).

7.7 Symptoms of Micronutrient Deficiencies

7.7.1 *Plants*

Boron

Boron deficiency first appears on young leaves which are thickened and may curl and become brittle. The deficiency symptoms of B differ in different crop plants. Boron deficiency results in hollow hearts in peanuts, black hearts in beets, distorted and lumpy fruit in papaya, and pith in hollow stem in cabbage and cauliflower. Boron deficiency commonly occurs in soil which are highly leached or developed from calcareous, alluvial and loess deposits.

Copper

Copper deficiency mostly occurs in soils with high organic matter content. Copper deficiency symptoms appear as interveinal chlorosis of new leaves followed by chlorosis of the veins and rapid and extensive necrosis of leaf blades. Copper deficiency leads to reduced and stunted growth of plants and give a bushy appearance. Ears may or may not form depending upon the severity of the deficiency.

Iron

Iron deficiency in plants appears as interveinal chlorosis of the young or new leaves (Fig. 7.4). Soon after the veins also loose green colour and the whole leaf turns yellow. Under acute deficiency, there is bleaching of the affected leaves and the newly emerging leaves also look bleached. In sugarcane, red pigmentation/lesions may also develop along the bleached area. With time the affected leaves turn papery, necrotic and ultimately die. The symptoms of Fe deficiency are similar in most of the crops. Iron deficiency in plants growing on calcareous soils is not due to insufficient Fe uptake from the soil but due to a physiological disorder in leaves, affecting the Fe distribution in the leaf tissue (Mengel and Geurtzen 1986).

Fig. 7.4 Symptoms of iron deficiency in rice, chickpea (*Cicer aeriertinum* L.), groundnut (*Arachis hypogaea* L.), sugarcane (*Sachharum officinarum*) and kinnow (*Citrus reticulate* x *C. deliciosa*) (Source: Nayyar and Chhibba 2000; with kind permission)

Fig. 7.5 Manganese deficiency symptoms in wheat (Source: Nayyar and Chhibba 2000; with kind permission)

Manganese

Manganese deficient plants are stunted with restricted foliage and root system. The symptoms of Mn deficiency first appear on the lower and middle leaves as small chlorotic spots in the interveinal area and extend from base of the leaf towards the tip (Fig. 7.5). Under acute deficiency the spots join together to form streaks of pinkish brown colour in the interveinal areas. At ear formation, these symptoms are conspicuous on the flag leaf of wheat. In Mn deficient plants, ears emerge with great difficulty and are deformed/sickle shaped. Similar symptoms of Mn deficiency are observed in most of the crops.

Molybdenum

Molybdenum deficiency usually appears in legumes and the symptoms resemble N deficiency because Mo is involved in nitrogen assimilation in plants. Mo deficiency appears on old and middle leaves, which become chlorotic with inward rolling of leaf margins. The most well-known Mo deficiency is the "whiptail" of cauliflower. The characteristic symptoms include twisted and elongated leaves with cupped lamina showing various degrees of narrowness and irregularities. Citrus plants suffering from Mo deficiency develop yellow spots.

Zinc

The general symptoms of Zn deficiency in plants include stunted growth, leaf chlorosis, small leaves, spikelet sterility, poor quality of crop products, and susceptibility to biotic and abiotic stresses. However, specific symptoms of Zn deficiency are different in different crops (Fig. 7.6). In rice, Zn deficiency first appears on old leaves as light yellow spots in the interveinal area about 15–20 days after transplanting. These spots later turn yellowish-brown and join together giving reddish brown or rusty look. With intensification of pigmentation, the midrib and the leaf sheaths also look like the rest of the leaf. Under acute deficiency, the plants give rusty look. The affected leaves wither with time and further growth of plant is restricted giving bushy appearance. Most of the tillers fail to develop panicles resulting in very poor grain yield. In maize, Zn deficiency first appears on 3 weeks old plants as a white patch between the mid-rib and the margin at the basal part of the lamina of the second or third leaf from top. The white patch extends along the mid-rib towards tip and the colour changes to reddish/bluish red. The

Fig. 7.6 Symptoms of zinc deficiency in rice, maize, chickpea (*Cicer aeriertinum* L.), black gram (*Phaseolus mungo* L.) and kinnow (*Citrus reticulate x C. deliciosa*) (Source: Nayyar and Chhibba 2000; with kind permission)

mid-rib and the margins, however, remain green for a longer time. Sometimes even the leaf sheaths and the internodes also develop reddish color. The internodal length decreases and the leaves appear to arise from points close to each other. Under moderate Zn deficiency, tasselling and silking are delayed and in severe deficiency tasselling may not take place and the crop virtually fails. In wheat, the symptoms of Zn deficiency appear first on the second or third leaf from top of the plant at tillering stage. They show up as light yellow white tissue between the mid-rib and margins in the middle or lower half of the affected leaf. The effects of Zn deficiency in plant include short internodal length, less number of productive tillers, poor foliage expansion, bushy growth, small ears, poor grain formation and delayed maturity. In Zn-deficient plants, protein synthesis is hampered and free amino acids accumulate.

7.7.2 Humans

The visual symptoms of microelement deficiencies in humans are not as distinct as in plants because a nutrient may be involved in several biochemical functions and interfere with metabolic processes and interact with other nutrients. The diagnosis is complicated due to the occurrence of multi-nutrient deficiencies and the appearance of specific symptoms also depends on the severity of deficiency. The acute symptoms are usually seen in profoundly ill patients, whereas subclinical problems may not be so distinct. The following symptoms of deficient state of different essential microelements in humans are usually observed (Wada 2004).

Although a definite biochemical role for B in humans has not been identified, yet it is considered a bioactive food component essential for humans. Boron deficiency has been reported to increase serum glucose and decrease serum triglyceride concentration (Nielsen 1996). Cobalt deficiency leads to perinicious anemia and methylmalonic academia. The deficiency symptoms of copper include anemia, leucopenia, neutropenia, disturbed maturation of myeloleukocytes. The deficiency of copper also leads to bone changes in children i.e. reduced osseous age, irregular/ spurring metaphysic, bone radiolucency and bone cortex thinning. Associated changes due to Cu deficiency may include subperiosteal hemorrhages, hair and skin depigmentation and defective elastin formation (Lonnerdal and Uauy 1998). Iodine deficiency causes goiter (enlarged thyroid gland) and hypothyroidism. Iron deficiency leads to development of anemia when iron supply is inadequate to synthesize normal level of hemoglobin. Common symptoms of Fe deficiency include tiredness, lethargy, breathlessness (dyspnoea) and palpitation. The deficiency of manganese leads to reduced serum cholesterol, reduced coagulation, hair reddening, dermatitis, growth retardation and increased radiolucency at the epiphyses of long bones. Molybdenum deficiency symptoms are tachycardia, polypnea, night blindness, scotoma, irritability, somnolence, disorientation and coma. Several physiological signs of Zn deficiency, depending on its severity and age of the individual, are discernible. Signs of severe Zn deficiency were identified in persons suffering from acrodermatitis enteropathica, a genetic disorder that affects Zn absorption (Van Wouwe 1989). The major symptoms of zinc deficiency appear as gradually worsening eruptions (acne), which first affect the face and perineum. The associated symptoms of zinc deficiency include stomatitis, glossitis, alopecia, white spots on nails, diarrhea, vomiting, fever, delayed wound healing, dwarfism, growth retardation, negative nitrogen balance, immune-suppression, depression, taste disorder and anorexia. Adverse effects of Zn deficiency generally vary with age. While diarrhea, anorexia, low weight gain and neurobehavioral changes are more frequent during infancy, the skin changes, blepharoconjuctivities and growth retardation are common among school children (Hambidge 1997). The other symptoms are more frequent in elderly people.

7.8 Optimizing Microelement Status

7.8.1 Soils and Plants

Deficiencies of microelements in soil and plants can be corrected by applying chemical fertilizers either alone or in combination with organic manures. Fertilizes may be applied to the soil or in aqueous solutions to the leaves of the plants as foliar spray. The use of fertilizers containing some combination of macro- and micronutrients is a standard practice in highly productive agriculture. Besides correcting nutrient deficiencies, fertilizers are also applied to increase the concentration of essential element in the edible part of the plant to augment the dietary supply of nutrients to humans and animals. In order to maximize use efficiency of the applied fertilizer, several management practices involving source, method and amount of application of a nutrient have been standardized for different soils, crops and environmental conditions.

Boron

Plant species differ greatly in their B requirement and the differences are reflected in optimum concentration of B in leaf tissue (Marschner 1995). Boron concentration in plants generally ranges between 3 and 100 μg g^{-1} dry matter. Based on B requirement and optimum leaf concentration, the plants could be grouped (Goldbach 1997) into (i) very low B requiring, which rarely show B deficiency (e.g. Graminaceous species) and have optimum leaf concentrations between 2 and 5 μg g^{-1} dry matter, (ii) severely affected by high B concentration in soil (e.g. non-grass monocotyledons and dicotyledons) and have optimum leaf concentrations between 20 and 80 μg g^{-1} dry matter and (iii) high B requiring (e.g. latex containing plants) with optimum leaf concentration greater than 80 μg g^{-1} dry matter. Boron deficiency can be overcome by adding soluble borates to soils or as foliar application (Table 7.9). Boron to soils can also be added in the form of crushed ores but these are mostly insoluble and have variable B content. Reclaiming high-B soils is really difficult because of narrow range between deficiency and toxicity of B in plants (Marschner 1995). Commonly used method of reclaiming high B soils include extensive leaching with low B water, the use of soil amendments such as lime and gypsum and the planting of B-tolerant plant genotypes (Nable et al. 1997).

Cobalt

Cobalt availability to plants increases with increasing acidity and waterlogging. Cobalt deficiency in soil can be corrected by adding $CoCl_2$, $CoSO_4$, cobaltized superphosphate, and pulverized serpentine rock. The Co containing fertilizers can be applied as soil, foliar and seed treatment.

Copper

Copper deficiency has rarely been encountered in soils, which are sandy or have high organic matter content. The deficiency of Cu is generally corrected by application of copper sulphate ($CuSO_4{\cdot}5H_2O$) either to soil or foliage depending on the plant species. In wheat, soil application of Cu has been found to be more beneficial. In vegetables and food crops, the Cu deficiency is controlled by foliar application of boardeux mixture, which is generally used for control of diseases.

Table 7.9 Commonly used soluble borates in agriculture

Compound	Formula	Nutrient content (% B)
Borax	$Na_2B_4O_7{\cdot}10H_2O$	10.5
Solubor	$Na_2B_4O_7{\cdot}5H_2O$ $Na_2B_{10}O_{16}{\cdot}10H_2O$	19
Boric acid	H_3BO_3	17
Sodium tetraborate (Borate-45)	$Na_2B_4O_7{\cdot}5H_2O$	14
Sodium tetraborate (Borate-65)	$Na_2B_4O_7$	20
Sodium pentaborate	$Na_2B_{10}O_{16}{\cdot}10H_2O$	18
Colemanite	$Ca_2B_6O_{11}{\cdot}$ $5H_2O$	10–16
Boron frits	Fritted glass	3–6

Compiled from Shorrocks (1997)

Table 7.10 Copper fertilizers

Compound	Formula	Nutrient content (% Cu)
Copper sulphate pentahydrate	$CuSO_4{\cdot}5H_2O$	25
Copper sulphate monohydrate	$CuSO_4{\cdot}H_2O$	35
Cuprous oxide	Cu_2O	89
Cupric oxide	CuO	75
Copper ammonium phosphate	$Cu(NH_4)PO_4{\cdot}H_2O$	32
Cupric chloride	$CuCl_2$	47
Copper nitrate	$Cu(NO_3)_2{\cdot}3H_2O$	
Copper acetate	$Cu(C_2H_3O_2)_2{\cdot}H_2O$	32
Copper chelate	$Na_2CuEDTA$	13
	NaCuHEDTA	9
Basic copper sulphate	$CuSO_4{\cdot}3Cu(OH)_2$	13–53
Copper lignosulphate		5–8
Copper polyfavonoid		5–7

EDTA Ethylene diamine tetraacetic acid, *HEDTA* Hydroxyethylethylene diamine triacetic acid

Copper being immobile in plant, multiple sprays are required to ameliorate its deficiency and maximise productivity. Other carriers of Cu include Cuprous/Cupric oxide, Cupric chloride and synthetic chelates (Table 7.10).

Iron

Out of a number of available Fe containing fertilizers (Table 7.11), ferrous sulphate ($FeSO_4{\cdot}7H_2O$) has been rated as the most efficient source and is commonly used for correcting its deficiency in different crops. Use of organic manures alone or in combination with $FeSO_4{\cdot}7H_2O$ has also been found to be effective in correcting Fe deficiency. Since inorganic Fe-carriers are less effective when applied to soil because of their quick reversion, the amelioration of Fe deficiency can be best accomplished by its foliar application. Ineffectiveness of the soil applied Fe is because of inadequate degree of reduction of Fe^{3+} to Fe^{2+} under aerobic conditions. However, in lowland rice prevalence of reduced conditions due to standing water help reduce Fe deficiency to some extent.

Manganese

Soil management practices, which transform insoluble form of manganese to mobile form (Mn^{2+}) increase the availability of Mn to plants. Manganese deficiency generally occurs in highly permeable coarse textured soils. Ameliorative measures include soil or foliar application of $MnSO_4$, but foliar application is generally superior. The results of the studies relating to the time of foliar application of Mn have revealed that an equal number of sprays of a particular concentration of $MnSO_4$ initiated prior to the first irrigation to wheat crop helped increase the yield significantly over that obtained when the sprays were initiated after the first irrigation. Besides $MnSO_4$, the other sources include oxide, carbonate, chloride and phosphate of manganese (Table 7.12).

Table 7.11 Iron fertilizers

Compound	Formula	Nutrient content (% Fe)
Ferrous sulphate heptahydrate	$FeSO_4{\cdot}7H_2O$	19/20.5
Ferric sulphate	$Fe_2(SO_4)_3{\cdot}4H_2O$	23/20
Ferrous carbonate	$FeCO_3$	42
Ferrous oxide	FeO	77/75
Ferric oxide	Fe_2O_3	69
Ferrous ammonium phosphate	$Fe(NH_4)PO_4{\cdot}H_2O$	29
Ferrous ammonium sulphate	$FeSO_4(NH_4)_2SO_4{\cdot}6H_2O$	14
Iron ammonium polyphosphate	$Fe(NH_4)HP_2O_7$	22
Iron chelate	NaFeDTPA	10
	NaFeEDTA	5–9
	NaFeEDDHA	6
	NaFeHEDTA	5–9
Iron methoxy phenylpropane complex	FeMPP	5
Iron frits	Fritted glass	20–40
Iron polyflavonoid	–	6–10
Iron lignosulphonate	–	5–8

DTPA Diethylene triamine pentaacetic acid, *EDTA* Ethylenediamine tetra-acetic acid, *EDDHA* Ethylene diamine di(o-hydroxy phenyl acetic acid), *HEDTA* Hydroxyethyl ethylene diamine triacetic acid

Table 7.12 Manganese fertilizers

Compound	Formula	Nutrient content (% Mn)
Manganese sulphate	$MnSO_4.4H_2O$	26–28
Manganese sulphate	$MnSO_4.H_2O$	32
Manganous oxide	MnO	41–68
Manganese carbonate	$MnCO_3$	31
Manganese chloride	$MnCl_2$	17
Manganese phosphate	$Mn_3(PO_4)_2$	20
Manganese oxide	MnO_2	63
Manganese chelate	$Na_2MnEDTA$	5–12
Manganese methoxyphenyl propane	–	10–12
Manganese frits	Fritted glass	10–35
Manganese ligniosulphate	–	5
Manganese polyflavonoid	–	5–7

EDTA Ethylene diamine tetra-acetic acid

Molybdenum

Molybdenum deficiency can be corrected by soil application of sodium molybdate along with superphosphate or by foliar application of molybdate. A number of other sources can be used as Mo fertilizers (Table 7.13). Frequently, phosphate, sulphate and Mo deficiencies occur together, which can be corrected by applying

Table 7.13 Molybdenum fertilizers

Compound	Formula	Nutrient content (% Mo)
Ammonium molybdate	$(NH_4)_6Mo_7O_{24}\cdot 4H_2O$	52–54
Sodium molybdate	$Na_2MoO_4\cdot 2H_2O$	39
Molybdenum trioxide	MoO_3	66
Molybdenite	MoS_2	60
Molybdic acid	$H_2MoO_4\cdot H_2O$	53
Calcium molybdate	$CaMoO_4$	48
Molybdenum frits	Fritted glass	2–3

molybdenized superphosphate. In Australia, molybdenum trioxide along with superphosphate is commonly used. Liming of acid soils also helps in controlling Mo deficiency. It has been found that applying a few kilogram of Mo to clover in New Zealand and Australia produced the similar effect as several Mg of limestone. Treatment (soaking) of seeds with sodium molybdate has been found to be as effective as application of Mo fertilizers to soil in controlling Mo deficiency. In the USA, seed treatment with sodium molybdate together with a sticking agent is recommended practice for correcting Mo deficiency.

Zinc

Zinc deficiency in soils and crops can be corrected by application of Zn fertilizers. A number of Zn carriers are available that can be used as fertilizers (Table 7.14). Evaluation of different inorganic Zn fertilizers has revealed that $ZnSO_4\_7H_2O$, ZnO, $ZnCO_3$, $Zn_3(PO_4)_2$ and Zn-frits are efficient sources for correcting Zn deficiency. Other forms of Zn include organic, such as lignosulfonate and synthetic chelates. Multi-micronutrient mixtures, used mainly for foliar application or seed treatment are considered inferior sources of Zn because of their relatively low Zn content. A number of studies have shown the superiority of soil compared to the foliar application of Zn in field crops. In the soil mode of Zn application, the broadcast and mix method has been found to be the more efficient compared to band placement and top-dressing. In addition to synthetic fertilizers other sources of Zn include animal manures and sewage sludge. Organic manures have been found to enhance the availability of the native soil Zn through chelation besides being a direct source of micronutrients. Cultivation of Zn-efficient cultivars under Zn stress conditions is another alternative to combat Zn deficiency.

7.8.2 Humans

Alleviation of deficiency and optimization of micronutrient levels in humans depends on several factors including current dietary intake, food habits and the nutrient content of the food items consumed, metabolic pathway of a nutrient, current body stocks, age, gender, and body weight. A number of Dietary Reference Intake (DRI) criteria viz. Recommended Dietary Allowance (RDA), Estimated

Table 7.14 Zinc fertilizers

Compound	Chemical formula	Nutrient content (% Zn)
Zinc sulphate monohydrate	$ZnSO_4 \cdot H_2O$	35
Zinc sulphate heptahydrate	$ZnSO_4 \cdot 7H_2O$	23
Basic zinc suplhate	$ZnSO_4 \cdot 4Zn(OH)_2$	55
Zinx oxysulphate	$ZnO + ZnSO_4$	40–55
Zinc oxide	ZnO	50–80
Zinc carbonate	$ZnCO_3$	52–56
Zinc phosphate	$Zn_3(PO_4)_2$	51
Zinc nitrate	$Zn_3(NO_3)_2 \cdot 6H_2O$	22
Zinc chloride	$ZnCl_2$	48–50
Zinc ammonia complex	$Zn\text{-}NH_3$	10
Zinc chelate	$Na_2ZnEDTA$	14
	NaZnHEDTA	9
	NaZnNTA	9
Zinc frits	Fritted glass	10–30
Zinc lignosulphonate	–	10–30
Zinc polyflavonoid	–	40–55

EDTA Ethylene diamine tetraacetic acid, *HEDTA* Hydroxyethylethylene diamine triacetic acid, *NTA* Nitrilotriacetate

Average Requirement (EAR), Adequate intake (AI) and Tolerable upper limit (TUL) for different life stage groups and gender are advocated for optimizing level of nutrients in humans. Recommended Dietary Allowance (RDA) is the average daily dietary intake sufficient to meet the nutrient requirements of nearly all (97–98%) healthy individuals in a life stage and gender group (Table 7.15). It is calculated from EAR, which represents average daily nutrient intake level estimated to meet the requirement of half of the healthy individuals in an age group (Table 7.16). If information on EAR is not available, then AI for different life stage groups and gender is established, which is considered to cover the needs of all healthy individuals in the group. In order to avoid excessive consumption, a TUL is fixed, which represents the maximum level of daily nutrient intake that is unlikely to pose any risk of adverse effects to almost all individual in the population (Table 7.17). It represents total intake from all sources including food, water, and supplements.

Boron

In the human body, B is distributed throughout soft tissue and fluids at concentrations between 0.015 and 0.6 μg g^{-1} fresh tissue (Bai and Hunt 1996; Ward 1993; Shuler et al. 1990). Bone, fingernails, hair, teeth and spleen usually contain high amounts of B. World Health Organization (1996) suggested an acceptable safe intake level of 1–13 mg day^{-1} for adults. In the US, the TUL of B is 20 mg day^{-1} for adults above 18 years of age and 11–17 mg day^{-1} for 9–18 years of age (Table 7.17; IOM 2001). Diet is the main source of B and typical daily intake of

Table 7.15 Recommended Dietary Allowances (RDA) of microelements for different life stage groups

Life stage group	Copper ($\mu g\ day^{-1}$)	Iodine ($\mu g\ day^{-1}$)	Iron ($mg\ day^{-1}$)	Molybdenum ($\mu g\ day^{-1}$)	Zinc ($mg\ day^{-1}$)
Children					
1–3 year	340	90	7	17	3
4–8 year	440	90	10	22	5
Males					
9–13 year	700	120	8	34	8
14–18 year	890	150	11	43	11
19–50 year	900	150	8	45	11
>51 year	900	150	8	45	11
Females					
9–13 year	700	120	8	34	8
14–18 year	890	150	15	43	9
19–50 year	900	150	18	45	8
>51 year	900	150	8	45	8
Pregnancy	1000	220	27	50	12
Lactation	1300	290	9–10	50	13

Sources: Dietary reference intakes for vitamin A, vitamin K, arsenic, boron, chromium, copper, iodine, iron, manganese, molybdenum, nickel, silicon, vanadium, and zinc (2001); Available at www.nap.edu

Compiled from Food and Nutrition Board, Institute of Medicine, National Academies

B ranges between 0.75 and 1.35 mg depending on the food products consumed (Nielsen 2006). Fruits, vegetables, pulses, legumes, nuts and chocolates are rich sources of B compared to dairy products, fish, meat and grains (Table 7.18). Wine, cider and beer also have high B content.

Cobalt

The adult human body (70 kg) contains about 1.1 mg Co mainly in the heart, liver, kidney, and spleen, with relatively small amounts in the pancreas, brain, and serum. The daily requirement of cobalt is 0.1 $\mu g\ day^{-1}$. The requirement for Co is not for the element but for a preformed compound, Cobalamin (Vitamin B_{12}) produced by microorganisms, including bacteria and fungi. Higher plants and animals are unable to produce cobalamin. The cobalamin synthesized in microorganisms enters the human food chain through incorporation into food of animal origin. In many animals, gastrointestinal fermentation supports the growth of vitamin B_{12} synthesizing microorganisms, and subsequently the vitamin is absorbed and incorporated into the animal tissues. The RDA for cobalamin for adults is 2.4 $\mu g\ day^{-1}$, the lowest for all essential nutrients (Yamada 2013). In humans, most of the dietary intake of Co is inorganic with vitamin B_{12} representing only a small fraction. The bioavailability of inorganic Co compounds varies between 5 and 45% depending on their water solubility. Dietary sources of Co are the same as for vitamin B_{12}. The vitamin B_{12} content of food is very low and it is generally absent in fruits and

Table 7.16 Estimated Average Requirements (EAR) of microelements for different life stage groups

Life stage group	Copper (μg day^{-1})	Iodine (μg day^{-1})	Iron (mg day^{-1})	Molybdenum (μg day^{-1})	Zinc (mg day^{-1})
Children					
1–3 year	260	65	3.0	13	2.5
4–8 year	340	65	4.1	17	4.0
Males					
9–13 year	540	73	5.9	26	7.0
14–18 year	685	95	7.7	33	8.5
19–50 year	700	95	6	34	9.4
>51 year	700	95	6	34	9.4
Females					
9–13 year	540	73	5.7	26	7.0
14–18 year	685	95	7.9	33	7.3
19–50 year	700	95	8.1	34	6.8
>51 year	700	95	5	34	6.8
Pregnancy					
14–18 year	785	160	23	40	10.5
19–50 year	800	160	22	40	9.5
Lactation					
14–18 year	985	209	7	35	10.9
19–50 year	1000	209	6.5	36	10.4

Source: Dietary reference intakes for vitamin A, vitamin K, arsenic, boron, chromium, copper, iodine, iron, manganese, molybdenum, nickel, silicon, vanadium, and zinc (2001); Available at www.nap.edu

Compiled from Food and Nutrition Board, Institute of Medicine, National Academies

vegetables. Organ meats (liver, kidney, heart and pancreas), clams, oysters, fish, seafood, chicken, eggs, milk products and an edible green and purple seaweed- *nori* contain significant amounts of vitamin B_{12}. Apparently, humans get dietary vitamin B_{12} almost exclusively from animal tissues or products such as milk, butter, cheese, eggs, meat, poultry. Dietary Co absorption is reduced in the presence of high dietary intakes of iodine. Ionic forms of iinorganic cobalt are toxic for the humans.

Copper

The human body weighing 70 kg contains about 120 mg Cu, which is widely distributed in tissues and fluids bound to proteins or to organic compounds. About two-thirds of body copper is found in the skeleton and muscles. Copper homeostasis is accomplished at enteric absorption level. It is excreted in the bile and only a small amount is lost through urine. Copper absorption in the body is inversely related to its intake and about 50–60% of Cu intake is retained. The recommended dietary allowance (RDA) of Cu ranges between 340 and 440 μg day^{-1} for children up to 8 years of age; 700–890 μg day^{-1} for ages 14 through 18, and 900 μg day^{-1} for adults (IOM 2001). The RDA of Cu is higher during pregnancy and lactation. The average daily Cu requirements are 12.5 μg kg^{-1} of body weight

Table 7.17 Tolerable Upper Intake Levels (TUL) of microelements for different life stage groups

Life stage group	Boron (mg day^{-1})	Copper (μg day^{-1})	Iodine (μg day^{-1})	Iron (mg day^{-1})	Manganese (mg day^{-1})	Molybdenum (μg day^{-1})
Children						
1–3 year	3	1000	200	40	2	300
4–8 year	6	3000	300	40	3	600
Males and females						
9–13 year	11	5000	600	40	6	1100
14–18 year	17	8000	900	45	9	1700
>19 year	20	10,000	1100	45	11	2000
Pregnancy						
14–18 year	17	8000	900	45	9	1700
19–50 year	20	10,000	1100	45	11	2000
Lactation						
14–18 year	17	8000	900	45	9	1700
19–50 year	20	10,000	1100	45	11	2000

Source: Dietary reference intakes for vitamin A, vitamin K, arsenic, boron, chromium, copper, iodine, iron, manganese, molybdenum, nickel, silicon, vanadium, and zinc (2001); Available at www.nap.edu

Compiled from Food and Nutrition Board, Institute of Medicine, National Academies

for adults and about 50 μg kg^{-1} of body weight for infants (WHO 1996). The recommended tolerable upper intake level from foods and supplements varies with age group and ranges between 1–3 mg day^{-1} for children below 8 years of age and 10 mg day^{-1} for adults (Table 7.17; IOM 2001). Generally, dietary Cu intakes for adults range between 1 and 3 mg day^{-1} (IPCS 1998; IOM 2001) and together with uptake from water it ranges between 1 and 5 mg day^{-1}. Food is the main source of Cu for humans and about 40% of dietary copper is derived from yeast breads, potatoes, tomatoes, cereals, beef and dried beans and lentils (Subar et al. 1998). Major sources of dietary Cu include liver and other organ meats, seafood, nuts, dark chocolate and whole grains (Table 7.19). In most foods, Cu is bound to macromolecules and does not exist as a free ion (IOM 2001).

Iodine

The human body contains approximately 5 mg of iodine, which functions only in the iodine containing thyroid hormones. Iodine is absorbed through the gut as iodide, the chemically bound form of iodine. Iodine homeostasis is realized by the kidney through urinary excretion comprising 90% of iodine absorbed by iodine-adequate individuals (Combs 2005). Therefore, analysis of urinary iodine concentration is the most commonly used biochemical indicator of I status of an individual. Median urinary iodine concentration is used to assess iodine status of a population. Other assessment methods include goitre, measurement of thyroid-stimulating hormone (TSH) levels in neonates and blood thyroglobulin in school-age children (WHO 2007; Zimmermann et al. 2008). The daily recommended

Table 7.18 Boron content of selected foods

Food	B ($\mu g\ g^{-1}$) fresh weight	Food	B ($\mu g\ g^{-1}$) fresh weight
Fruits		*Nuts*	
Apple	2.73	Peanuts	13.8
Banana	1.04	Pecans	6.6
Cherries	7.0	*Animal products*	
Grapes	4.6	Meats	$<$0.05–0.34
Orange	2.17	Milk	0.23
Avocado	11.1	Cheese	0.19
Fruits, dried		Eggs	0.12
Prunes	21.5	*Cereal grain products*	
Raisins	19.0	Bread	0.48
Vegetables		Corn flakes	0.92
Beans, green	1.56	Oatmeal	0.10
Broccoli, flowers	2.47	Rice	0.32
Carrots	2.59	*Others*	
Peas	1.28	Chocolate powder	4.25
Potato	1.25	Honey	6.07
Squash, winter	2.65	Sugar	0.29
Sweet potato	1.08	Vegetable oil	$<$0.04
Tomato	0.75	Wine	3.52
Pulses (cowpeas, lima beans, red beans)	3.14–4.76		

Adapted from Anderson et al. (1994)

dietary intake of I is 110–130 μg day^{-1} for children under the age of 1, 90–120 μg day^{-1} for children aged 1–10, and 150 μg day^{-1} for adults and adolescents, with higher concentrations (250 μg day^{-1}) required during pregnancy and lactation (WHO 1996). The humans can obtain iodine through direct inhalation, drinking water and beverages' and food consumption. Inhalation cannot be a significant pathway of iodine intake, except in near-coastal environments, where humans may obtain some iodine by inhalation. Similarly, drinking water is unlikely to supply more than 10% of the daily adult iodine requirement. Apparently, food is the main source of dietary iodine. Among the food items, the major sources of dietary iodine include iodized salt, sea fish, and kelp. The food sources having high concentrations of iodine include marine fish, egg, milk and dairy products (Table 7.20; Haldimann et al. 2005). Dairy products are rich sources due to the addition of iodine to cattle feed, the use of iodine containing disinfectants in the dairy industry, and the inadvertent ingestion of soil by the animals that could add iodine. Baked products have high iodine concentration mainly because of addition of iodized salt during manufacturing process. Plant foods, with the exception of leafy vegetables, are low in iodine and the content varies considerably from species to species (Table 7.20).

Table 7.19 Concentrations of Cu in selected food products (Compiled from Onianwa et al. 2001)

Food product	Cu concentrations (μg g^{-1})
Food crops	
Lettuce[a]	0.72
Cabbage[a]	0.41
Carrot[a]	0.40
Tomatoes[a]	0.36
Apple[a]	0.25
Orange[a]	2.13
Beans[b]	1.3–8.8
Potato[b]	0.60–0.72
Corn[b]	0.38–2.33
Rice[b]	0.73–2.3
Animal products	
Beef[b]	0.58–7.24
Pork[b]	0.80–2.87
Goat[a]	3.8
Chicken[a]	1.0
Cow liver[a]	3.8
Cow kidney[a]	9.67
Eggs[a]	1.13
Fish and seafood	
Stock fish[a]	2.00
Dried fish[a]	3.33
Crayfish[a]	2.67

[a]Data from Nigeria (means)
[b]Data from Nigeria, USA, Sweden, East Asia and Egypt (minima and maxima)

Even the same food product has widely different iodine content depending on the location where it is produced as the iodine uptake is in proportion to the iodine present in the environment. Plants grown in iodine-rich soil contain substantial amounts of iodine. Iodine in soil is often leached and carried to the sea by repeated flooding and glacial activity, thereby making seawater, seaweed and marine fish rich sources of iodine. The most practical and cost-effective strategy to control iodine deficiency is iodisation of salt. Salt is iodized by adding potassium iodate (KIO_3) or potassium iodide (KI), either in powder form or as aqueous solution. Iodate is recommended as fortificant in preference to iodide because it is more stable, particularly in warm, damp, or tropical climates (WHO 1991; Mannar and Dunn 1995).

Iron

Iron is the most abundant trace element in the human body that contains ~3–4 g of elemental iron, with almost 50% present in haemoglobin and the rest in the myoglobin of muscle, in ferritin, and in hemosiderin (a breakdown product of ferritin). Serum Fe is about 1.3 mg L^{-1}, mostly bound to transferrin. The average

Table 7.20 Mean iodine content in food products (Compiled from Haldimann et al. 2005)

Food product	Iodine content ($\mu g\ kg^{-1}$)
Cereals and grains	
Wheat	245
Rice	333
Breakfast cereals	42
Animal products	
Red meat	59
Poultry	66
Game	34
Processed meat	335
Egg	1625
Fish, marine	2112
Fish, freshwater	375
Milk	690
Yoghurt	670
Cheese	473
Fruits and vegetables	
Fresh fruits	18
Fresh vegetables	47
Potatoes	16
Leafy vegetables	236
Frozen or canned vegetables	1203
Mushrooms	211
Nuts	218

daily excretion of endogenous iron is ~1 mg (range 0.4–2.0 mg). Majority of the losses occur through faeces (0.6 mg) followed by skin shedding/peeling (0.2–0.3 mg) and urine (0.1 mg). The RDA of iron varies with age and sex of the subjects (Table 7.15). The RDA for adults is 8 mg day^{-1} for males and 18 mg day^{-1} for females. During pregnancy the RDA increases to 27 mg day^{-1}. However, the estimated intake is much lower than the recommended allowance. Total iron intake ranges from 14.4 to 20.2 mg day^{-1} (Chanarin 1999). The tolerable upper limit for different age groups ranges between 40 and 45 mg day^{-1} (Table 7.17). Iron deficiency anaemia is considered to be among the most important contributing factors to the global burden of disease (WHO 2002). Globally, anaemia affects 1.62 billion people, which corresponds to 24.8% of the population (Fig. 7.2; WHO 2008); the greatest prevalence being in preschool-age children (47.4%) and the lowest in men (12.7%). However, the population group with the greatest number of individuals affected is non-pregnant women (468.4 million). WHO regional estimates for preschool-age children and pregnant and non-pregnant women indicate that the highest proportion of individuals affected are in Africa (47.5–67.6%), while the greatest number affected are in South-East Asia (315 million individuals).

Food is the main source of both nonheme and heme iron. Nonheme sources of dietary iron include fruits, vegetables, cereals, legume seeds, fortified bread and dairy products. Heme sources of iron are haemoglobin and myoglobin from meat and poultry. The average diet in the United States contains-15 g of elemental iron, most of which is chelated to various phytates, phosphates, and other molecules such as oxalate and tannin and is thus unavailable for absorption. Only 1.5–2.0 g of dietary iron is absorbed in the gastrointestinal tract. Therefore, low bioavailability of dietary iron is one of the reasons for low iron level in the human body, particularly for individual dependent on vegetarian plant-based foods. To compensate for lower iron bioavailability, about 10%, from vegetarian diets compared with 18% from a mixed Western diet, vegetarians need to increase dietary iron by 80% (FNB 2001). In vegetarian diets, iron intake can be augmented by increased consumption of dried beans and legumes, fruits and vegetables, and whole-grain rather than refined-grain products. Such food choices can considerably alter the dietary components that enhance or inhibit the bioavailbility and absorption of nonheme iron. Iron intake can also be optimized through supplementation and food fortification. Iron compounds recommended for food fortification include ferrous sulphate, ferrous fumarate, ferric pyrophosphate and electrolytic iron powder (WHO 2006). Low cost elemental iron powders, which are sometimes used to fortify cereal foods, are not recommended. Analysis of the previously published (1995–2011) estimates on anaemia showed that iron supplementation could increase the mean blood haemoglobin concentration by 8.0 g L^{-1} in children, 10.2 g L^{-1} in pregnant women and 8.6 g L^{-1} in non-pregnant women indicating that about 42% of anaemia in children and about 50% in women could be eliminated by iron supplementation (WHO 2015).

Manganese

The human body contains approximately 20 mg of Mn distributed mainly in the liver, bones and kidneys. Because of lack of information on Mn deficiency induced illnesses, not many studies have characterized the effect of Mn supplementation in humans. Adequate intakes of Mn (mg day^{-1}) are 1.2–1.5 for children below 9 years of age, 1.6–2.2 for age group between 14 and 18 years. Adequate intakes for adults above 18 years of age are 2.3 and 1.8 mg day^{-1} for men and women, respectively (FNB 2001). The recommended intakes for pregnant and lactating women are 2 and 2.6 mg day^{-1}, respectively. The UL for different life stage groups ranges between 2 mg day^{-1} for ages 1–3 years and 11 mg day^{-1} for ages above 19 years (Table 7.17). Mean daily intakes of Mn range between 0.52 and 10.8 mg (Freeland-Graves and Lianis 1994). Food sources of Mn include nuts, legumes, leafy vegetables, tea and whole grains. Meats, refined grains and dairy products are low in Mn.

Molybdenum

Food is the major source of Mo intake in humans. Plant foods such as legumes, grain products and nuts and milk and milk products and organ meats (especially liver and kidney) are rich sources of Mo in the diet. Non-leguminous vegetables, fruits, oils, fats and fish are poor sources of Mo (Nielsen 1996). Dietary Mo and supplements in the form of soluble complexes such as ammonium molybdate are easily absorbed. The RDA for Mo is 17–22 μg day^{-1} for 1–8 years of ages, 34 μg day^{-1} for 4–8 years, 43–45 μg day^{-1} for ages 14 years or more (Table 7.15). For pregnant and lactating women the RDA is set at 50 μg day^{-1}. The estimated intake (EAR) for different life stage groups ranges between 13 μg day^{-1} for age 1–3 years and 34 μg day^{-1} for age over 18 years (Table 7.16). Though Mo has low toxicity in humans, yet the ULs are set because of its harmful effects on reproduction and fetal development in animals. The ULs for different life stage groups range between 300 μg day^{-1} for age 1–3 years and 2000 μg day^{-1} for age over 18 years (Table 7.17).

Zinc

The amount of Zn in the adult human body ranges between 1.4 and 2.3 g Zn (Calesnick and Dinam 1988). Zinc is present in all body tissues and fluids in relatively high concentrations, with 85% of the total Zn in muscle and bone, 11% in the skin and the liver and the remaining in all the other tissues (Kawashima et al. 2001). Less than 0.1% of the total body Zn is in the plasma and it has a rapid turnovers rate and thus cannot be used as an indicator of overall Zn status except under conditions of apparent deficiency. Blood Zinc concentration does not decrease in proportion to the magnitude of deficiency rather the physical growth is slowed down and excretion is reduced. Therefore, Zn homeostasis is primarily maintained by adjustments in total Zn absorption and endogenous intestinal excretion (Hambidge and Krebs 2001). The RDA of Zn is 3–5 mg day^{-1} for children below 9 years of age and 8 mg day^{-1} for children up to 13 years of age. For adults, the dietary allowance is 11 mg day^{-1} for men, 8 mg day^{-1} for women and 12–13 mg day^{-1} for pregnant and breast feeding women (Table 7.15). The estimated intake for different life stage groups varies between 7–9.4 mg day^{-1} for men and 6.8–10.9 mg day^{-1} for women (Table 7.16). About one-third of the world population is estimated to be suffering from Zn deficiency and there is a close relationship between Zn deficiency in soils and crops and in humans (Fig. 7.3). Generally, regions such as south and middle-east Asia, Africa and Latin American countries having high incidence of Zn deficiency in humans are also deficient in plant available Zn in soils. Food is the main source of Zn for humans and is present in a number of foods including red meat and poultry, beans, nuts, legumes, seafood (oysters), whole grains, fortified breakfast cereals, and dairy products. Zinc from animal foods is more bioavailable than that from plant foods which contain certain antinutrients such as fiber and phytate that inhibit Zn uptake by the intestine. Zinc is generally associated with the protein fraction and nucleic fraction of food. Strategies to manage Zn deficiency in humans include dietary diversification/modification, supplementation (giving zinc tablets to certain target groups), food fortification and bio-fortification. The choice of a strategy depends on the available resources and technical feasibility.

7.8.2.1 Dietary Diversification

Individuals in developing countries, whose diet is primarily cereal-based, are susceptible to mineral deficiencies because cereals not only have low Zn density, but also high levels of phytate, which chelates Zn ions and inhibits its absorption. Though legumes have high Zn content (~1 mg 100 g^{-1} cooked), yet only about 15% can be absorbed because of their high phytate content. Animal foods such as, meat, fish and poultry contain more Zn than cereals (Table 7.21) and do not have antinutrients such as phytates. Therefore, the easiest approach to provide enough Zn is to formulate balanced diets that combine food products together to deliver known dietary requirements in an appropriate proportion, for example, increasing the intake of proteinaceous foods such as meat, fish, poultry, cheese, milk, etc. along with fruits and vegetables. Consumption of animal proteins improve the bioavailability of Zn from plant food sources probably because amino acids released from the animal protein keep zinc in solution (Lonnerdal 2000) or the protein binds the phytate. Increasing the proportion of fruits and vegetables in diet can also help in meeting Zn requirements. A varied diet comprising fresh fruit, vegetables, fish and meat provides sufficient nutrients and enhancers to promote adequate mineral absorption in the gut. Concurrent consumption of foods with low levels of phytate and other antinutrients and those rich in nutrient enhancers can increase Zn bioavailability.

Table 7.21 Zinc and phytate contents of selected foods (Compiled from IZiNCG 2004)

Food	Zinc content mg 100 g^{-1}	Phytate content mg 100 g^{-1}	Phytate: zinc Molar ratio
Liver, kidney (beef poultry)	4.2–6.1	0	0
Meat (beef, pork)	2.9–4.7	0	0
Poultry	1.8–3.0	0	0
Sea food	0.5–5.2	0	0
Eggs (chicken, duck)	1.1–1.4	0	0
Dairy (milk, cheese)	0.4–3.1	0	0
Seeds, nuts	2.9–7.8	1760–4710	22–88
Beans, lentils	1.0–2.0	110–617	19–56
Whole-grain cereals	0.5–3.2	211–618	22–53
Refined cereal grains	0.4–0.8	40–349	16–54
Bread (white flour, yeast)	0.9	30	3
Fermented cassava root	0.7	70	10
Tubers	0.3–0.5	93–131	26–31
Vegetables	0.1–0.8	0–116	0–42
Fruits	0–0.2	0–63	0–31

7.8.2.2 Supplementation and Food Fortification

Zinc supplementation and food fortification programs are used to correct Zn deficiency in vulnerable population subgroups. Zinc compounds that are available for supplementation and fortification include the oxide, carbonate, sulphate, gluconate and acetate forms of Zn as these are efficiently absorbed; the sulfide form is poorly utilized. The recommended dosages of Zn supplements are 5 mg day^{-1} for children between 7 months and 3 years and 10 mg day^{-1} for older children (Müller et al. 2001; Brown et al. 2002). Proposed levels of fortification of flour are 30–70 mg Zn kg^{-1} (WHO 2009). Zinc, either ingested or injected, is primarily excreted in the faeces. Urinary excretion of Zn and several other metals is increased if chelating agents such as EDTA are administered in combination with Zn (Hays and Swenson 1985). Use of Zn supplements among children below 2 years of age provides protective effect against pneumonia, suppurative otitis media and pneumonia-related mortality. Zinc supplementation has a mild effect on reducing the frequency of diarrhoea and improving growth (WHO 2013). While supplementation and food fortification approaches are widely applied in some countries, the success of these approaches requires infrastructure, health care systems, purchasing power, which are often not available to susceptible population sub-groups in developing countries (Bouis 2003; Stein et al. 2007). Food fortification programs depend on industrially processed food items, which are expensive and hardly affordable by the poor people predominantly susceptible to micronutrient deficiencies (Mayer et al. 2008).

7.8.2.3 Biofortification

Biofortification refers to an agricultural strategy intended to increase the content of Zn or any other microelement in food crops such as rice, wheat, maize, pearl millet etc. (Hotz 2009). Unlike ordinary food fortification, biofortification aims at inherent enrichment of microelements in edible plant parts while the plants are still growing, rather than exogenously adding nutrients to the foods during processing. Ordinary food fortification may not be suitable for combating microelement deficiencies if a suitable food vehicle to reach the vulnerable population is not available. On the other hand biofortification of staple foods is a promising strategy for addressing microelement deficiencies in the developing countries with co-benefits of higher yields and increased vitality of seedlings emerging from Zn-enriched seeds.

Biofortification strategies could be agronomic or genetic. Agronomic biofortification of Zn is achieved by increasing plant availability of Zn present in soil or by the application of Zn fertilizers to soil and crop. Studies have shown that increasing the Zn content of soils through fertilizer application or augmenting the soil Zn supply to plants by managing the constraining factors not only results in improved productivity but also increases grain Zn density. Genetic biofortification involves the development of crop genotypes that acquire more Zn from the soil and accumulate in edible portions (White and Broadley 2011). This can be achieved by conventional plant breeding or genetic engineering techniques. The breeding steps include (i) identification of meaningful genetic variation, (ii) long-term crossing and back-crossing, and (iii) stability and adaptation of the high grain Zn density target traits across a range of soil, crop, climate and management environments (Cakmak 2008). Recent achievements in genetic engineering of new crop plants hold promise of improving nutritional balance from smaller number of dietary components.

Besides enriching the edible portion of food crops with a microelement, the biofortification strategies should aim to improve its bioavailability by increasing the concentrations of promoter substances, which enhance the absorption of the essential mineral elements; and by reducing the concentrations of antinutrients that inhibit their absorption in the gut. The HarvestPlus project of CGIAR (Consultative Group on International Agricultural Research) is coordinating research efforts in breeding for increasing concentration and bioavailable levels of Zn and Fe in seeds of major staple food crops (Bouis 2003; Pfeiffer and McClafferty 2007). Considerable progress has been made in developing transgenic plant genotypes with high concentrations of Zn and Fe. In spite of being promising strategy, genetic fortification has some limitations including long-time required for breeding suitable genotypes, uncertainty about adaptation or success across environments and target regions, and the need for enormous resources (Cakmak 2008). Moreover the genetic capacity of the newly developed high-Zn genotypes to accumulate sufficient amount of Zn may depend on the soil's capacity to supply Zn to plants. Mostly the soils in cereal growing regions have unfavourable soil properties, such as high pH and low organic matter content, which can constrain the expression of high Zn trait in biofortified genotype. Further, the acceptance of genetically biofortified crops by producers and consumers is another important consideration for successful implementation of this strategy. Though agronomic biofortification, apart from being simple offers a promising solution in the short term but its success depends on regular addition of fertilizers and may have adverse environmental impacts. Apparently, both the agronomic and the breeding strategies have strengths and weaknesses (Table 7.22), the agronomic approach may be used complementary to the breeding approach to derive maximum benefits. In resource poor settings, dietary diversification together with biofortification may be the preferred strategy to overcome micronutrient deficiencies.

Table 7.22 Strengths and weaknesses of different biofortification strategies to increase nutrient density of plant based foods (Adapted from Carvalho and Vasconcelos 2013)

Biofortification strategy	Strengths	Weaknesses
Agronomic	Simple method with immediate results; can be complementary to other strategies	Short-term strategy, needs regular application, successful only for minerals, difficult distribution, success dependent on several factors, expensive, could have adverse environmental effects
Conventional plant breeding	Successful for minerals and vitamins, one-off cost, easier distribution, long-term strategy, wide public acceptance	Long development time, success limited to minerals available in the soil, requires genetic variation
Genetic engineering	Successful for minerals and vitamins, one-off cost, easier distribution, long-term strategy, speeds up process of conventional breeding	Long development time, success limited to minerals available in the soil, interaction among transgenes may limit the process, low public acceptance, complex regulatory approval, environmental impact

7.8.3 Factors Affecting Bioavailability of Iron and Zinc

7.8.3.1 Iron

The total Fe content of a diet does not provide much information about its bioavailability, which can vary by an order of a magnitude for different diets with similar iron content (Hallberg and Hulthen 2000). The differences in bioavailability of iron among different foods are attributed to the chemical form of Fe present in a food. Dietary iron is of two types: nonheme iron present in both plant foods and animal tissue, and heme iron present only in animal foods and is derived from hemoglobin and myoglobin. The heme iron is better absorbed (15–35%) than the nonheme iron. Heme iron is estimated to contribute 10–15% of total iron intake in meat-eating population but because of higher absorption it constitutes more than 40% of the total iron absorbed (Carpenter and Mahoney 1992; Hunt 2002). Based on intake data, iron bioavailability has been estimated between 14–18% for mixed diets and 5–12% for vegetarian diets (Hurell and Egli 2010). In addition to the chemical form, the amount of iron absorbed from a food by an individual is determined by the body iron store and the presence of dietary inhibitors or antinutrients and enhancers or promoters (Table 7.23). Both heme and non-heme forms of iron are absorbed in inverse logarithmic proportion to body iron stores (Hunt 2003). The main dietary antinutrients that inhibit absorption of nonheme iron include phytic acid or phytate (inositol phosphate) present in whole grains, legumes, lentils, and nuts; polyphenols and tannins found in tea, coffee, red wines, a variety of cereals, vegetables, and spices; oxalates found in certain vegetables and herbs, nuts, chocolate; and other microelements, e.g., zinc and copper (Hallberg and Hulthen 2000; Table 7.23). The extent of inhibitory effect

Table 7.23 Dietary components that can inhibit (antinutrients) and enhance (promoters) the bioavailability of Zn and Fe in humans, if consumed concurrently (Compiled from Graham et al. 2001; Gibson 2007 and several other sources)

Dietary component	Major dietary sources	Main technical influence
Inhibitors/antinutrients		
Phytic acid or phytin (*myo*-inositol hexaphosphate)	Unrefined cereals, legumes, nuts, and oil seeds	Binds positively charged Fe and Zn ions to form insoluble complexes in gut
Dietary fibre (e.g. cellulose, hemicelluloses, lignin, cutin, suberin etc.)	Whole cereal grain products (e.g. wheat, rice, maize, oat, barley, rye), nuts, oilseeds, fruits and vegetables	Effect compounded by the quantity of minerals and protein, presence of phytate or oxalic acid and type of fibre in the diet
Polyphenols and tannins	Certain cereals (red sorghum), legumes (red kidney beans, black beans, black gram, lentils), spinach, betel leaves, oregano, beverages (tea, coffee, cocoa, red wine)	Form insoluble complexes with iron (e.g. iron-tanin);
Oxalic acid/oxalates	Amaranth, spinach, rhubarb, yam, beets, sweet potato, kale, sesame seeds, nuts, chocolate, tea, wheat bran, strawberries and herbs such as oregano, basil, parsley, taro, sorrel	Probably oxalate forms insoluble complexes with Fe
Egg factor	Eggs	Phosvitin, a Phosphoprotein binds iron and decreases its bioavailability
Soy protein	Tofu, soybeans, tempeh, bean curd, soy milk	Presence of phytic acid and a functional group that makes up part of the protein conglycinin.
Hemagglutinins (e.g. lectins)	Most legumes and wheat grain	Lectins bind to glycoprotein receptors on the epithelial cells lining the intestinal mucosa and inhibit the absorption of nutrients
Enhancers/promoters		
Organic acids (citric, lactic, acetic, butyric, propionic, formic acids)	Fermented milk products (e.g., yogurt), vegetables (e.g., sauerkraut), soy sauce, fresh fruits	May form soluble ligands with some trace minerals in the gut
Ascorbic acid	Citrus fruits and juices, other fruits (e.g., guava, mango, papaya kiwi, strawberry), vegetables (e.g., tomato, asparagus, Brussels sprouts)	Reduces ferric iron to more soluble ferrous iron; forms iron-ascorbate chelate; may counteract inhibitory effect of phytate and dietary fibre
Amino acids (e.g. methionine, cysteine, histidine and lysine); meat factors	Animal meats (Beef, pork, chicken, fish, etc)	Possibly peptides form soluble ligands; reduce or chelate iron;

(continued)

Table 7.23 (continued)

Dietary component	Major dietary sources	Main technical influence
β-carotene	Green and orange vegetables: (e.g. carrots, collard greens, beets and beet greens, spinach, sweet potatoes, tomatoes, turnip greens and yellow squash), fruits (e.g. red grapes, oranges, peaches, prunes, apricots), red peppers, red palm oil, corn	Decreases the inhibitory effects of phytates and tannins depending on their concentrations
Riboflavin (e.g. falvin mononucleotide, flavin adenine dinucleotide)	Milk and dairy products, meat and fish, and certain fruit and vegetables especially dark-green vegetables	Influences iron absorption or mobilization from existing stores

of phytate on iron absorption depends on phytate to iron molar ratio. For cereals or legume based diets that do not contain any enhancers, the molar ratio should preferably be less than 0.4:1 so as to significantly improve iron absorption. For composite diets comprising vegetables containing ascorbic acid and meat as enhancers the phytate to iron molar ratio should be less than 6:1 for better iron absorption (Hurrell 2004; Tuntawiroon et al. 1990). In cereals and legumes, polyphenols add to the inhibitory effect of phytate. The inhibitory effect of polyphenols depends not only on its quantity but also the type. For example, while polyphenols in chilli are inhibitory to iron absorption that in turmeric are not (Tuntipopipat et al. 2006). Dietary fibre per se does not clearly inhibit iron absorption and the effect is compounded by the type of fibre, the presence of phytate or oxalic acid, and the quantity of minerals and protein in the diet (Anderson 1990; Freeland-Graves 1988).

Food processing and preparation methods including milling, thermal processing, soaking, malting, germination, and fermentation can be used to remove or degrade phytate (Table 7.24; Egli et al. 2002). Food processing methods involving soaking, malting, and fermentation hydrolyse the phytate through the activation of plant and microbial phytases. Exogenous addition of phytase has also been shown to improve iron absorption (Sandberg and Andersson 1988). Besides these food processing methods, iron absorption from vegetarian diets can probably be improved by modifying food preparation method, food selection, and food combination (Hunt 2003). Such modifications can include the use of iron cookware particularly for cooking acidic foods that solubilise iron from the pan; the concurrent consumption of iron and ascorbic acid containing foods while limiting inhibitory foods such as coffee and tea in between meals; and the selection of low phytate foods (Martinez and Vannucchi 1986; Gibson et al. 1997).

The main dietary components that enhance iron absorption include organic acids, vitamin C or ascorbic acid, amino acids and meat, poultry, and fish (Table 7.23). Ascorbic acid reduces ferric iron to more soluble ferrous iron and also forms iron-ascorbate chelate. Ascorbic acid is considered to counter the

Table 7.24 Categorization of diets according to their potential bioavailability of zinc (Source: FAO/WHO 2002)

Zinc absorption	Phytate: Zn molar ratio	Diet types
High (50%)	<5	Highly refined low in cereal fibre and animal foods; includes semi-purified formula diets
Moderate (30%)	5–15	Mixed diets with animal or fish protein and lacto-ovo-vegetarian diets not based on unrefined cereals grains or high extraction rate (>90%) flours
Low (15%)	>15	Cereal based diets with >50% energy intake from unrefined cereal grains or legumes and negligible animal protein intake

inhibitory effects of phytate and polyphenols and enhance the absorption of iron in vegetarian and vegan diets. However, cooking, processing and storage of food degrade ascorbic acid nullifying its promontory effect on iron absorption. Moderate consumption of alcohol can enhance the absorption of iron, but heavy or excessive drinking has adverse health effects. Iron absorption from cereals and legumes can be improved by degrading inositol phosphates to less phosphorylated inositol phosphates because the phytates with less than three phosphate groups do not inhibit nonheme iron absorption (Sandberg et al. 1999). Presence of β-carotene along with phytates or tannic acid usually counters the inhibitory effect depending on their concentrations.

7.8.3.2 Zinc

Similar to iron, the bioavailability of dietary Zn can be reduced by the phytic acid and some other constituents of plant food. Phytic acid binds minerals, including Zn, in the human gastrointestinal tract, forming insoluble complexes that prevent Zn absorption (Cheryan 1980). Usually, phytic acid to zinc molar ratio of a meal is used to estimate the fractional absorption of the ingested zinc. Based on the Zn concentration and contents of promoters and inhibitors of Zn absorption, three categories of diet viz. high, moderate and low in bioavailable Zn with respective fractional absorption rates of 50%, 30% and 15% were identified (Table 7.24; FAO/WHO 2002). Molar ratios greater than 15:1 according to World Health Organization (2006) or 18:1 according to International Zinc Nutrition Consultative Group (2004) severely inhibit Zn absorption and have been associated with low Zn status in humans. Phytate with less than five phosphate groups do not inhibit Zn absorption (Lonnerdal et al. 1989). The inhibitory effect of phytate can be countered by the addition of ascorbic acid (Siegenberg et al. 1991). However, it has been suggested that the chelation properties of phytic acid may have some potential beneficial effect such as lowering serum cholesterol and triglycerides, and preventing heart disease, renal stone formation, and certain types of cancer, such as colon cancer (Zhou and Erdman 1995).

Proteins generally have positive influence on Zn absorption and its absorption tends to increase with protein intake (Lonnerdal 2000; Sandström et al. 1989). However, the amount and sources of protein influence dietary Zn absorption. Zinc absorption is high from a diet rich in animal proteins than from a diet high in plant proteins such as soy and legumes. The amount of Zn in a meal will affect zinc absorption in that as the Zn amount in a meal increases the fractional Zn absorption will decrease. High Fe concentrations that are present in some food supplements can reduce Zn absorption. The effect of Fe on Zn only occur when iron to Zn ratio is very high (e.g. 25:1) and both are administered in solution. However, Zn-Fe interaction may not likely have a major influence on Zn requirement under most dietary conditions.

References

Ahmad W, Niaz A, Kanwal S, Rahmatullah, Rasheed MK (2009) Role of boron in plant growth: a review. J Agric Res 47:329–338

Ahmad W, Zia MH, Malhi SS, Niaz A, Saifullah (2012) Boron deficiency in soils and crops: a review. In: Goyal A (ed) Crop plant, InTech, Available from: http://www.intechopen.com/books/crop-plant/boron-deficiency-in-soils-and-crops-a-review, pp 77–114

Alloway BJ (2005) Bioavailability of elements in soil. In: Selinus O, Alloway B, Centeno JA, Finkelman RB, Fuge R, Lindh U, Smedley P (eds) Essentials of medical geology-impacts of the natural environment on public health. Elsevier Academic Press, London, pp 347–372

Alloway BJ (2008) Zinc in soils and crop nutrition. IFA/IZA, Paris/Brussels

Alloway BJ (2009) Soil factors associated with zinc deficiency in crops and humans. Environ Geochem Health 31:537–548

Andersen P (2007) A review of micronutrient problems in the cultivated soil of Nepal. Mt Res Dev 27:331–335

Anderson JW (1990) Dietary fibre and human health. Hortic Sci 25:1488–1495

Anderson DL, Cunningham WC, Lindstrom TR (1994) Concentrations and intakes of H, B, S, K, Na, Cl and NaCl in foods. J Food Compos Anal 7:59–82

Arnon DL, Stout PR (1939) Molybdenum as an essential element for higher plants. Plant Physiol 14:599–602

Bai Y, Hunt CD (1996) Dietary boron enhances efficacy of cholecalciferol in broiler chicks. J Trace Elem Exp Med 9:117–132

Barak P, Helmke PA (1993) The chemistry of zinc. In: Robson AD (ed) Zinc in soils and plants. Kluwer Academic Publishers, Dordrecht, pp 90–106

Barber S (1995) Soil nutrient bioavailability, 2nd edn. Wiley, New York

Barceloux DG (1999) Zinc. Clin Toxicol 37:279–292

Beard JL (1999) Iron deficiency and neural development: an update. Arch Latinoam Nutr 49 ((3) Suppl (2)):34–39

Beck FW, Kaplan J, Fitzgerald JT, Brewer GJ (1997) Changes in cytokine production and T cell subpopulations in experimentally induced zinc-deficient humans. Am J Phys 272:E1002–E1007

Benbi DK, Brar SPS (1992) Dependence of DTPA extractable Zn Fe Mn and Cu availability on organic carbon presence in arid and semiarid soil of Punjab. Arid Soil Res Rehabil 6:207–216

Berger KC, Truog E (1945) Boron availability in relation to soil reaction and organic matter content. Soil Sci Soc Am Proc 10:113–116

Bohn HL, McNeal BL, O'Connor GA (1979) Soil chemistry. Wiley, New York

Bouis HE (2003) Micronutrient fortification of plants through plant breeding: can it improve nutrition in man at low cost? Proc Nutr Soc 62:403–411

Brown KH, Peerson JM, Rivera J, Allen LH (2002) Effect of supplemental zinc on the growth and serum zinc concentrations of prepubertal children: a meta-analysis of randomized controlled trails. Am J Clin Nutr 75:1062–1071

Cakmak I (2000) Role of zinc in protecting plant cells from reactive oxygen species. New Phytol 146:185–205

Cakmak I (2008) Enrichment of cereal grains with zinc: agronomic or genetic biofortification? Plant Soil 302:1–17

Cakmak I, Römheld V (1997) Boron deficiency-induced impairments of cellular functions in plants. Plant Soil 193:71–83

Cakmak I, McLaughlin MJ, White P (2017) Zinc for better crop production and human health. Plant Soil 411:1–4

Calesnick B, Dinam AM (1988) Zinc deficiency and zinc toxicity. Am Fam Physician 37:267–270

Carmel R (2007) Haptocorrin (transcobalamin I) and cobalamin deficiencies. Clin Chem 53:367–368

Carpenter CE, Mahoney AE (1992) Contributions of heme and nonheme iron to human nutrition. Crit Rev Food Sci Nutr 31:333–367

Carvalho SMP, Vasconcelos MW (2013) Producing more with less: strategies and novel technologies for plant-based food biofortificaion. Food Res Int 54:961–971

Chanarin I (1999) Nutritional aspects of hematological disorders. In: Shils ME, Olson JA, Shike M, Ross AC (eds) Modern nutrition in health and disease. Lippincot, Williams & Wilkins, Baltimore, pp 1419–1436

Chandra RK (1990) Micro-nutrients and immune functions: an overview. Ann N Y Acad Sci 587:9–16

Chang HB, Lin CW, Huang HJ (2005) Zinc-induced cell death in rice (*Oryza sativa* L) roots. Plant Growth Regul 46:261–266

Chen X, Schauder S, Potier N, Van Dorsselaer A, Pelczer I, Bassier BL, Hughson FM (2002) Structural identification of a bacterial quorum-sensing signal containing boron. Nature (London) 415:545

Cheryan M (1980) Phytic acid interactions in food systems. Crit Rev Food Sci Nutr 13:297–335

Combs GF Jr (2005) Geological impacts on nutrition. In: Selinus O, Alloway B, Centeno JA, Finkelman RB, Fuge R, Lindh U, Smedley P (eds) Essentials of medical geology – impacts of the natural environment on public health. Elsevier Academic Press, London, pp 161–177

Cordano A, Baert JM, Graham G (1964) Copper deficiency in infancy. Pediatrics 34:324–326

Cynthia E, David D (1997) Remediation of metals-contaminated soil and groundwater. Technology evaluation report (TE-97-01) in groundwater remediation technologies. Analysis Centre (GWRTAC) E series

Dai JL, Zhang M, Zhu YG (2004) Adsorption and desorption of iodine by various Chinese soil. I Iodate Environ Intern 30:525–530

Dallman PR (1976) Tissue effects on iron deficiency. In: Jacobs A, Worwood M (eds) Iron in biochemistry and medicine. Academic, New York, pp 437–475

Danks DM (1988) Copper deficiency in humans. Annu Rev Nutr 8:235–257

Danks DM (1995) Disorders of copper transport. In: Scriver CL, Beaudet AL, Sly WS, Valle D (eds) The metabolic and molecular bases of inherited disease. McGraw-Hill, New York, pp 2211–2235

Davidsson L (1999) Are vegetarians an "At Risk Group" to iodine deficiency? Brit J Nutr 81:3–4

Delange F (1994) The disorders induced by iodine deficiency. Thyroid 4:107–128

Deosthale YG, Gopalan C (1974) The effect of molybdenum levels in sorghum (*Sorghum vulgare* Pers.) on uric acid and copper excretion in man. Brit J Nutr 31:351–355

Dugger WM (1983) Boron in plant metabolism. In: Lauchli A, Bieleski RL (eds) Encyclopedia of plant physiology, New Series, vol 15. Springer, Berlin, pp 626–650

Egli I, Davidsson L, Juillerat MA, Barclay D, Hurrell RF (2002) The influence of soaking and germination on the phytase activity and phytic acid content of grains and seeds potentially useful for complementary feeding. J Food Sci 67:3484–3488

Elrashidi MA, O'Connor GA (1982) Boron sorption and desorption in soils. Soil Sci Soc Am J 46:27–31
Erd RC (1980) The minerals of Boron. In: Mellor S (ed) Boron Oxygen Compounds. Longman, New York
FAO/WHO (2002) Zinc. In: Human vitamin and mineral requirements. Report of a joint FAO/WHO expert consultation. FAO, Rome, pp 257–270
Fleming GA (1980) Essential micronutrients. I. Boron and molybdenum. In: Davies BE (ed) Applied soil trace elements. Wiley, New York, pp 155–197
FNB (Food and Nutrition Board), Institute of Medicine (2001) Dietary reference intakes for vitamin A, vitamin K, arsenic, boron, chromium, copper, iodine, iron, manganese, molybdenum, nickel, silicon, vanadium, and zinc. National Academy Press, Washington, DC
Fortescue JAC (1980) Environmental geochemistry: a holistic approach. Springer, New York
Fraga CG (2005) Relevance, essentiality and toxicity of trace elements in human health. Mol Asp Med 26:235–244
Freeland-Graves J (1988) Mineral adequacy of vegetarian diets. Am J Clin Nutr 48:859–862
Freeland-Graves J, Lianis C (1994) Models to study manganese deficiency. In: Klimis-Tavantzis DJ (ed) Manganese in health and disease. CRC Press, Boca Raton, pp 59–86
Frøslie A (1990) Problems on deficiency and excess of minerals in animal nutrition. In: Låg J (ed) Geomed. CRC Press, Boca Raton, pp 37–60
Frossard E, Bucher M, Mächler F, Mozafar A, Hurrell R (2000) Potential for increasing the content and bioavailability of Fe, Zn and Ca in plants for human nutrition. J Sci Food Agric 80:861–879
Fuge R (2005) Soils and iodine deficiency. In: Selinus O, Alloway B, Centeno JA, Finkelman RB, Fuge R, Lindh U, Smedley P (eds) Essentials of medical geology – impacts of the natural environment on public health. Elsevier Academic Press, London, pp 417–433
Galan P, Viteri F, Bertrais S, Czernichow S, Fature H (2005) Serum concentrations of beta carotene, vitamins C and E, zinc and selenium are influenced by sex, age, diet, smoking status, alcohol consumption and corpulence in a general French adult population. Er J Clin Nutr 59:1181–1190
Gibson RS (2006) Zinc: the missing link in combating micronutrient malnutrition in developing countries. Proc Nutr Soc 65:51–60
Gibson RS (2007) The role of diet- and host-related factors in nutrient bioavailability and thus in nutrient-based dietary requirement estimates. Food Nutr Bull 28(1):S77–S100
Gibson RS, Donovan UM, Heath AL (1997) Dietary strategies to improve the iron and zinc nutriture of young women following a vegetarian diet. Plant Foods Hum Nutr 51:1–16
Goldbach HE (1997) A critical review on current hypotheses concerning the role of boron in higher plants: suggestions for further research and methodological requirements. J Trace Microprobe Tech 15:51–91
Goldberg S (1997) Reactions of boron with soils. Plant Soil 193:35–48
Graham RD, Welch RM, Bouis HE (2001) Addressing micronutrient malnutrition through enhancing the nutritional quality of staple foods: principles, perspectives and knowledge gaps. Adv Agron 70:77–142
Gupta UC (1968) Relationship of total and hot-water soluble boron and fixation of added boron, to properties of podzol soils. Soil Sci Soc Am Proc 36:332–334
Haldimann M, Alt A, Blanc A. Blondeau K (2005) Iodine content of food groups. J Food Compos Anal 18:461–471
Hallberg L, Hulthen L (2000) Prediction of dietary iron absorption: an algorithm for calculating absorption and bioavailability of dietary iron. Am J Clin Nutr 71:1147–1160
Hambidge KM (1997) Zinc deficiency in young children. Am J Clin Nutr 65:160–161
Hambidge KM, Krebs NF (2001) Interrelationships of key variables of human zinc homeostasis: relevance to dietary zinc requirements. Annu Rev Nutr 21:429–452
Hambidge KM, Walravens PA (1982) Disorders of mineral metabolism. Clin Gastroenterol 11:87–117
Hamilton IM, Gilmore WS, Strain JJ (2000) Marginal copper deficiency and atherosclerosis. Biol Trace Elem Res 78:179–189

Hays VW, Swenson MJ (1985) Minerals and bones. In: Dukes' physiology of domestic animals, 10th edn. Cornell University Press, pp 449–466

Hetzel BS (1983) Iodine deficiency disorders (IDD) and their eradication. Lancet 2:1126–1129

Hodgson SF (1963) Chemistry of the micronutrient elements in soils. Adv Agron 15:119–159

Hotz C (2009) The potential to improve zinc status through biofortification of staple food crops with zinc. Food Nutr Bull 30:S172–S178

Hotz C, Brown KM (2004) Assessment of the risk of zinc deficiency in populations and options for its control. Food Nutr Bull 25:S99–S199

Hunt JR (2002) Moving toward a plant-based diet: are iron and zinc at risk? Nutr Rev 60:127–134

Hunt JR (2003) Bioavailability of iron, zinc, and other trace minerals from vegetarian diets. Am J Clin Nutr 78:633S–639S

Hurell R, Egli I (2010) Iron bioavailability and dietary reference values. Am J Clin Nutr 91 (5):1461S–1467S

Hurrell RF (2004) Phytic acid degradation as a means of improving iron absorption. Int J Vitam Nutr Res 74:445–452

IOM (2001) Zinc. In: Dietary reference intakes for vitamin A, vitamin K, arsenic, boron, chromium, copper, iodine, iron, manganese, molybdenum, nickel, silicon, vanadium and zinc. Food and Nutrition Board, Institute of Medicine. National Academy Press, Washington, DC, pp 442–501

IPCS (International Programme on Chemical Safety) (1998) Copper. Environmental health criteria 200. World Health Organization, Geneva

IZiNCG (International Zinc Nutrition Consultative Group), Brown KH, Rivera JA, Bhutta Z, Gibson RS, King JC, Lönnerdal B, Ruel MT, Sandtröm B, Wasantwisut E, Hotz C, de Romaña DL, Peerson JM (2004) Technical document # 1. Assessment of the risk of zinc deficiency in populations and options for its control. Food Nutr Bull 25:94S–203S

Jeffery J, Uren N (1983) Copper and zinc species in the soil solution and the effects of soil pH. Soil Res 21:479–488

Kabata-Pendias A (2004) Soil–plant transfer of trace elements-an environmental issue. Geoderma 122:143–149

Kabata-Pendias A, Mukherjee AB (2007) Trace elements from soil to human. Spinger, Berlin

Kabata-Pendias A, Pendias H (2001) Trace metals in soils and plants, 2nd edn. CRC Press, Boca Raton

Kabata-Pendias A, Sadurski W (2004) Trace elements and compounds in soil. In: Merian E, Anke M, Ihnat M, Stoepppler M (eds) Elements and their compounds in the environment, 2nd edn. Wiley-VCH, Weinheim, pp 79–99

Kaiser BN, Bridley KL, Brady JN, Phillips T, Tyerman SD (2005) The role of molybdenum in agricultural production. Ann Bot 96:745–754

Kawashima Y, Someya Y, Sato S, Shirato K, Jinde M, Ishida S, Akimoto S, Kobayashi K, Sakakibara Y, Suzuki Y, Tachiyashiki K, Imaizumi K (2001) Dietary zinc-deficiency and its recovery responses in rat liver cytosolic alcohol dehydrogenase activities. J Toxicol Sci 36:101–108

Kiekens L (1995) Zinc. In: Alloway BJ (ed) Heavy metals in soils, 2nd edn. Blackie Academic and Professional, Glasgow, pp 284–305

King JC, Cousins RJ (2006) Zinc. In: Shils ME, Shike M, Ross AC, Caballero B, Cousins RJ (eds) Modern nutrition in health and disease, 10th edn. Lippincott Williams and Wilkins, Baltimore, pp 271–285

Klimis-Tavantzis DJ (ed) (1994) Manganese in health and disease. CRC Press, Boca Raton

Knez M, Graham RD (2013) The impact of micronutrient deficiencies in agricultural soils and crops on the nutritional health of humans. In: Selinus O, Alloway B, Centeno JA, Finkelman RB, Fuge R, Lindh U, Smedley P (eds) Essentials of medical geology, revised edn. Springer, Dordrecht, pp 517–533

Låg J, Steinnes E (1976) Regional distribution of halogens in Norwegian forest soils. Geoderma 16:317–325

Linder MC, Hazegh-Azam M (1996) Copper biochemistry and molecular biology. Am J Clin Nutr 63:797S–811S
Lindh U (2005) Biological functions of the elements. In: Selinus O (ed) Essentials of medical geology: impacts of the natural environment on public health. Elsevier Academic Press, Burlington, pp 115–160
Lindsay WL (1979) Chemical equilibria in soils. Wiley Interscience, New York
Lindsay WL, Schwab AP (1982) The chemistry of iron in soils and its availability to plants. J Plant Nutr 5:821–840
Lonnerdal B (2000) Dietary factors influencing zinc absorption. J Nutr 130:S1378–S1383
Lonnerdal B, Uauy R (1998) Genetic and environmental determinants of copper metabolism. Am J Clin Nutr 67:951S–1102S
Lonnerdal B, Sandberg A-S, Sandstrom B, Kunz C (1989) Inhibitory effects of phytic acid and other inositol phosphates on zinc and calcium absorption in suckling rats. J Nutr 119:211–214
Loomis WD, Durst RW (1992) Chemistry and biology of boron. Bio Factors 4:229–239
Malhotra VK (1998) Biochemistry for students, 10th edn. Jaypee Brothers Medical Publishers (P) Ltd, New Delhi
Mannar V, Dunn JT (1995) Salt iodization for the elimination of iodine deficiency. International Council for Control of Iodine Deficiency Disorders, Dordrecht
Marschner H (1995) Mineral nutrition of higher plants. Academic, San Diego, pp 379–396
Martinez FE, Vannucchi H (1986) Bioavailability of iron added to the diet by cooking food in an iron pot. Nutr Res 6:421–428
Matoh T (1997) Boron in plant cell walls. Plant Soil 193:59–70
Mayer JE, Pfeiffer WH, Beyer P (2008) Biofortified crops to alleviate micronutrient malnutrition. Curr Opin Plant Biol 11:166–170
Melki IA, Bulus NM, Abumrad NN (1987) Invited review: trace elements in nutrition. Nutr Clin Pract 2(6):230–240
Mengel K, Geurtzen G (1986) Iron chlorosis on calcareous soils. Alkaline nutritional conditions as the cause of chlorosis. J Plant Nutr 9:161–173
Merck VM (1986) The Merck veterinary manual. 6th edn. A handbook of diagnosis, therapy and disease prevention and control for the veterinarian. Merck and Co, Rahway
Mertens J, Smolders E (2013) Zinc. In: Alloway BJ (ed) Heavy metals in soils, vol 22. Springer, Dordrecht, pp 465–493
Miljkovic NS, Mathews BC, Miller MH (1966) The available B content of the genetic horizons of some Ontario soils. I. The relationship between water-soluble B and other soil properties. Can J Soil Sci 46:133–138
Miller W (2006) Extrathyroidal benefits of iodine. J Am Physician Surg 11(4):106–110
Miwa K, Fujiwara T (2010) Boron transport in plants: coordinated regulation of transporters. Ann Bot 105:1103–1108
Mocchegiani E, Giacconi R, Muzzioli M, Cipriano C (2000) Zinc infections and immunoscence. Mech Ageing Dev 121:21–35
Müller O, Becher H, van Zweeden AB, Ye Y, Diallo DA, Konate AT, Gbangou A, Kouyate B, Garenne M (2001) Effect of zinc supplementation on malaria and other causes of morbidity in West African children: randomised double blind placebo controlled trial. BMJ 322:1567
Munno De G, Geday MA, Medaglia M, Anastassopoulou J, Theophanides T (1996) Manganese and cobalt cytosine (Cyt) and 1-methycytosine (1-Mecyt) complexes. In: Collery P, Corbella J, Dominnngo JL, Etienne JC, Llobet JM (eds) Metal ions in biology and medicine 4. Libbey Eurotext, Paris, pp 3–5
Muramatsu Y, Wedepohl KH (1998) The distribution of iodine in the Earth's crust. Chem Geol 147:201–216
Murray RK, Granner DK, Mayes PA, Rodwell VW (2000) Harper's biochemistry, 25th edn. McGraw-Hill, Health Profession Division, New York
Nable RO, Bañuelos GS, Paull JG (1997) Boron toxicity. Plant Soil 193:181–198
Nayyar VK, Chhibba IM (2000) Nutritional disorders in field crops – visual symptoms and remedial measures. Department of Soils, Punjab Agricultural University, Ludhiana

Nielsen FH (1996) Evidence for the nutritional essentiality of boron. J Trace Elem Exp Med 9:215–229
Nielsen FH (2006) Boron, manganese, nickel, silicon and vanadium. In: Driskell JA, Wolinsky I (eds) Sports nutrition: vitamins and trace elements. CRC Press/Taylor and Francis Group, Boca Raton, pp 287–320
O'Neill MA, Warrenfeltz D, Kates K, Pellerin P, Doco T, Darvill AG, Albersheim P (1996) Rhamnogalacturonan II, a pectic polysaccharide in the wall of growing plant cells, forms a dimer that is covalently cross-linked by a borate ester. In vitro conditions for the formation and hydrolysis of the dimer. J Biol Chem 271:22923–22930
Ohira Y, Edgerton VR, Gardner GW, Senewiratne B, Barnard RJ, Simpson DR (1979) Work capacity, heart rate and blood lactate responses to iron treatment. Br J Haematol 41:365–372
Olivares M, Uauy R (1996) Copper as an essential nutrient. Am J Clin Nutr 63:791S–796S
Onianwa PC, Adeyemo AO, Idowu OE, Ogabiela EE (2001) Copper and zinc contents of Nigerian foods and estimates of the adult dietary intakes. Food Chem 72:89–95
Ortega R, Bresson C, Fraysse A, Sandre C, Devès G, Gombert C, Tabarant M, Bleuet P, Seznec H, Simionovici A, Moretto P, Moulin C (2009) Cobalt distribution in keratinocyte cells indicates nuclear and perinuclear accumulation and interaction with magnesium and zinc homeostasis. Toxicol Lett 188:26–32
Panda SK, Chaudhury I, Khan MH (2003) Heavy metals induced lipid peroxidation and effect antioxidants in wheat leaves. Biol Plant 46:289–294
Parr AJ, Loughman BC (1983) Boron and membrane function in plants. In: Robb DA, Pierpoint WS (eds) Metals and micronutrients, uptake and utilisation by plants. Academic, New York, pp 87–107
Pfeiffer WH, McClafferty B (2007) Biofortification: breeding micronutrient-dense crops. In: Kang MS (ed) Breeding major food staples. Blackwell Science Ltd, Oxford
Plumlee GS, Ziegler TL (2003) The medical geochemistry of dusts, soils, and other earth materials. In: Lollar BS (ed) Treatise on geochemistry, vol 9. Elsevier, Oxford, pp 263–310
Power PP, Woods WG (1997) The chemistry of boron and its speciation in plants. Plant Soil 193:1–13
Prasad AS (2002) Zinc deficiency in patients with sickle cell disease. Am J Clin Nutr 75:181–182
Prasad AS (2003) Zinc deficiency. BMJ 326:409–410
Prasad R (2010) Zinc biofortification of food grains in relation to food security and alleviation of zinc malnutrition. Curr Sci 98:10–25
Remer T, Neubert A, Manz F (1999) Increased risk of iodine deficiency with vegetarian nutrition. Br J Nutr 81:45–49
Rout GR, Das P (2003) Effect of metal toxicity on plant growth and metabolism: I. Zinc Agronomie 23:3–11
Roy C (2010) Anaemia of inflammation. Hematology 30:276–280
Sandberg AS, Andersson H (1988) Effect of dietary phytase on the digestion of phytate in the stomach and small intestine of humans. J Nutr 118:469–473
Sandberg A, Brune M, Carlsson N, Hallberg L, Skoglund E, Rossander-Hulthen L (1999) Inositol phosphates with different numbers of phosphate groups influence iron absorption in humans. Am J Clin Nutr 70:240–246
Sandström B, Almgren A, Kivistö B, Cederblad Å (1989) Effect of protein level and protein source on zinc absorption in humans. J Nutr 119:48–53
Sardesai OK (1993) Molybdenum: an essential trace element. Nutr Clin Pract 8:277–281
Sato K, Okazaki T, Maeda K, Okami Y (1978) New antibiotics, aplasmomycins B and C. J Antibiot (Tokyo) 31:632–635
Schummer D, Irschik H, Reichenbach H, Hofle G (1994) Antibiotics from gliding bacteria. LVII. Tartrolons: new boron-containing macrodiolides from *Sorangium cellulosum*. Liebigs Ann Chem 1994:283–289
Scott M (1972) Trace elements in animal nutrition. In: Mortvedt JJ, Giordano OM, Lindsay WL (eds) Micronutrients in agriculture. Soil Science Society of America, Madison

Seatharam B, Alpers DH (1982) Absorption and transport of cobalamin (vitamin B_{12}). Annu Rev Nutr 2:343–349
Shorrocks VM (1997) The occurrence and correction of boron deficiency. Plant Soil 193:121–148
Shuler TR, Pootrakul P, Yarnsukon P, Neilson FH (1990) Effect of thalassemia/haemoglobin E disease on macro, trace, and ultra-trace element concentrations in human tissues. J Trace Elem Exp Med 3:31–43
Siegenberg D, Baynes RD, Bothwell TH, Macfarlane BJ, Lamparelli RD, Car NG, Macphail P, Schmidt U, Tal A, Mayet F (1991) Ascorbic acid prevents the dose-dependent inhibitory effects of polyphenols and phytates on nonheme-iron absorption. Am J Clin Nutr 53:537–541
Sillanpää M (1982) Micronutrients and the nutrient status of soils: a global study, Soil Bulletin, vol 48. Food and Agriculture Organization (FAO), Rome
Simonsen LO, Harbak H, Bennekou P (2012) Cobalt metabolism and toxicology – a brief update. Sci Total Environ 432:210–215
Soetan KO, Olaiya CO, Oyewole OE (2010) The importance of mineral elements for humans, domestic animals and plants: a review. Afr J Food Sci 4:200–222
Spencer D, Wood JG (1954) The role of molybdenum in nitrate reduction in higher plants. Aust J Biol Sci 7:425–434
Spiller S, Terry N (1980) Limiting factors in photosynthesis II. Iron stress diminishes photochemical capacity by reducing the number of photosynthetic units. Plant Physiol 65:121–125
Stein AJ, Nestel P, Meenakshi JV, Qaim M, Sachdev HPS, Bhutta ZA (2007) Plant breeding to control zinc deficiency in India: how cost-effective is biofortification? Public Health Nutr 10:492–501
Steinnes E (2009) Soils and geomedicine. Environ Geochem Health 31:523–535
Stoyanova Z, Doncheva S (2002) The effect of zinc supply and succinate treatment on plant growth and mineral uptake in pea plant. Braz J Plant Physiol 14:111–116
Subar AF, Krebs-Smith SM, Cook A, Kahle LL (1998) Dietary sources of nutrients among US adults, 1989 to 1991. J Am Diet Assoc 98:537–547
Szabo G, Chavan S, Mandrekar P, Catalano D (1999) Acute alcoholic consumption attenuates IL-8 and MCP-1 introduction in response to ex vivo simulation. J Clin Immunol 19:67–76
Tan JC, Burns DL, Jones HR (2006) Severe ataxia, myelopathy and peripheral neuropathy due to acquired copper deficiency in a patient with history of gastrectomy. J Paenteral Nutr 30:446–450
Tucker DM, Sandstead HH, Penland JG, Dawson SL, Milne DB (1984) Iron status and brain function: serum ferritin levels associated with asymmetries of cortical electrophysiology and cognitive performance. Am J Clin Nutr 39:105–113
Tuntawiroon M, Sritongkul N, Rossander HL, Pleehachinda R, Suwanik R, Brune M, Hallberg L (1990) Rice and iron absorption in man. Eur J Clin Nutr 44:489–497
Tuntipopipat S, Judprasong K, Zeder C, Wasantwisut E, Winichagoon P, Charoenkiatkul S, Hurrell R, Walczyk T (2006) Chilli, but not turmeric, inhibits iron absorption in young women from an iron-fortified composite meal. J Nutr 136:2970–2974
Underwood EJ (1971) Trace elements in human and animal nutrition, 3rd edn. Academic, New York, p 116
Vaillant N, Monnet F, Hitmi A, Sallanon H, Coudret A (2005) Comparative study of responses in four *Datura* species to a zinc stress. Chemosphere 59:1005–1013
Vallee B, Falchuk KH (1993) The biochemical basis of zinc physiology. Physiol Rev 73:79–118
Van Wouwe JP (1989) Clinical and laboratory diagnosis of acrodermatitis enteropathica. Eur J Pediatr 149:2–8
Viets FG Jr (1962) Chemistry and availability of micronutrients in soils. J Agric Food Chem 10:174–178
Wada O (2004) What are trace elements- their deficiency and excess states. JMAJ 47:351–358
Ward NI (1993) Boron levels in human tissues and fluids. In: Anke M, Meissner D, Mills CF (eds) Trace elents in man and animals, TEMA-8. Verlag Media Touristik, Gersdorf, pp 724–728
Warington K (1923) The effect of boric acid and borax on broad bean and certain other plants. Ann Bot 37:629–672
Welch RM (1995) Micronutrient nutrition of plants. Crit Rev Plant Sci 14:49–82

Wessells KR, Brown KH (2012) Estimating the global prevalence of zinc deficiency: results based on zinc availability in national food supplies and the prevalence of stunting. PLoS One 7: e50568. https://doi.org/10.1371/journal.pone.0050568

White PJ, Broadley MR (2009) Biofortification of crops with seven mineral elements often lacking in human diets-iron, zinc, copper, calcium, magnesium, selenium and iodine. New Phytol 182:49–84

White PJ, Broadley MR (2011) Physiological limits to zinc biofortification of edible crops. Front Plant Sci 30:S172–S178

White PJ, Brown PH (2010) Plant nutrition for sustainable development and global health. Ann Bot 105:1073–1080

Whittaker P (1998) Iron and zinc interactions in humans. Am J Clin Nutr 68:S442–S446

WHO (World Health Organisation) (1991) Evaluation of certain food additives and contaminants. 37th report of the Joint FAO/WHO Expert Committee on Food Additives. WHO Technical Series No. 806. World Health Organization, Geneva

WHO (World Health Organisation) (2001) Iron deficiency anaemia: assessment, prevention, and control. A guide for program managers. World Health Organisation, Geneva

WHO (World Health Organisation) (2002) The world health report. WHO, Geneva

WHO (World Health Organisation) (2006) Guidelines on food fortification with micro-nutrients. World Health Organization, Geneva

WHO (World Health Organization) (1996) Trace elements in human nutrition and health. World Health Organization, Geneva, pp 49–71

WHO (World Health Organization) (2007) Assessment of iodine deficiency disorders and monitoring their elimination: a guide for programme managers, 3rd edn. WHO, Geneva

WHO (World Health Organization) (2008) In: de Benoist B, McLean E, Egli I, Cogswell M (eds) Worldwide prevalence of anaemia 1993–2005: WHO global database on anaemia. WHO, Geneva

WHO (World Health Organization) (2009) Trace elements in human nutrition and health. WHO, Geneva

WHO (World Health Organization) (2013) The world health report 2013: research for universal health coverage. WHO, Geneva

WHO (World Health Organization) (2015) The global prevalence of anaemia in 2011. WHO, Geneva

Williams RJP, Frausto da Silva JJR (2002) The involvement of molybdenum in life. Biochem Biophys Res Commun 292:293–299

Yamada K (2013) Cobalt: its role in health and disease. In: Sigel A, Sigel H, Sigel RKO (eds) Interrelations between essential metal ions and human diseases, metal ions in life sciences. Springer, Dordrecht

Yoshida S, Muramatsu Y (1995) Determination of organic, inorganic, and particulate iodine in the coastal atmosphere of Japan. J Radioanal Nucl Chem 196:295–302

Yuita K (1983) Iodine, bromine and chlorine in soils and plants of Japan. Soil Sci Plant Nutr 9:403–407

Zhou JR, Erdman JW (1995) Phytic acid in health and disease. Crit Rev Food Sci Nutr 35 (6):495–508

Zimmermann MB, Jooste PL, Panday CS (2008) Iodine-deficiency disorders. Lancet 372:1251–1262

Chapter 8
Role of Potentially Toxic Elements in Soils

Abstract There are numerous sources of toxic elements in soils. In naturally occurring soils they can accumulate during weathering of rocks. Background concentrations in soils are thus determined by the concentrations in the underlying parent materials. Most toxic elements in soils exhibit strong adsorption by clays. Under certain conditions, small portions become soluble. Humans cause accumulation of toxic elements in soils from different sources. They are common in industrial goods, as components of paints and pesticides, and as a constituent of land influenced by municipal and industrial waste or by mining activities. Trace elements have been added to soils by applying fertilizers and from atmospheric deposition. A significant relationship exists between the presence of toxic elements and the incidence of some serious human diseases. Toxic elements may enter the human body through inhalation of dust, direct ingestion of soil and water, dermal contact of contaminated soil and water, and consumption of vegetables grown in contaminated fields. They are known to be persistent in the human body and can lead to a wide range of toxic effects. Specific antidotes exist for treatment of intoxication. For remediation of contaminated soils, a number of technologies have been developed mainly using mechanically, physico-chemically or biologically based methods. In this chapter, sources of toxic element contamination, toxic element speciation, transformation and transport behavior in soils, bioavailability, and the associated environmental and health risks, medical treatment and mitigation options are discussed.

Keywords Geochemical background • Anthropogenic sources of toxic elements • Chemistry of toxic elements • Human exposure • Clinical effects and therapy • Measures to reduce human toxic element exposure • Remediation of heavy metal contaminated soils

R. Nieder et al., *Soil Components and Human Health*,
https://doi.org/10.1007/978-94-024-1222-2_8

8.1 Overview of Potentially Toxic Elements

Potentially toxic elements include metals and non-metals such as arsenic (As) and selenium (Se) (Table 8.1). Some are of greatest and others are of relatively low environmental significance. Metals such as copper (Cu), nickel (Ni) and zinc (Zn) are also known as micronutrients (Chap. 7). They may be toxic to living organisms if present in excessive concentrations. Toxic elements and micronutrients are categorized as trace elements that in non-contaminated soils are present in concentrations <1000 mg kg^{-1}, being infrequently poisonous (Kabata-Pendias and Pendias 2001).

Toxic elements in naturally occurring soils accumulate during weathering of rocks and ores that compose geologic parent materials. Background concentrations of toxic elements in young or immature soils are thus mainly determined by the concentrations in the underlying parent materials (Mitchell 1964). In more developed soils, the degree of weathering undergone during soil development may make the correlation less obvious.

Humans cause additional accumulation of toxic elements in soils through emissions from industrial areas, mine tailings, disposal of wastes, leaded gasoline and paints, application of mineral fertilizers, animal manures, sewage sludge, pesticides, wastewater irrigation, spillage of petrochemicals, coal combustion residues and atmospheric deposition (Khan et al. 2008). Among toxic elements commonly found at contaminated sites are As, Cd, Cr, Cu, Hg, Ni, Pb and Zn (GWRTAC 1997).

Soils are the major sink for toxic elements released into the environment (Wuana and Okieimen 2011). However, most toxic elements in soils exhibit strong

Table 8.1 Toxic elements of greatest as well as of relatively low environmental significance

Toxic elements of greatest environmental significance	Toxic elements of relatively low environmental significance[a]
Arsenic (As)	Antimony (Sb)
Cadmium (Cd)	Beryllium (Be)
Chromium (Cr)	Gold (Au)
Copper (Cu)	Rubidium (Ru)
Fluorine (F)	Silver (Ag)
Lead (Pb)	Thallium (TI)
Mercury (Hg)	Tin (Sn)
Molybdenum (Mo)	Uranium (U)
Nickel (Ni)	Vanadium (V)
Selenium (Se)	
Zinc (Zn)	

Compiled from Mengel and Kirkby (1978), Underwood (1981), Kabata-Pendias and Pendias (1992) and McLaren (2003)

[a]May be of importance in specific situations

adsorption by clays and do not undergo microbial or chemical degradation. Their total concentration in soils persists for a long time (Adriano 2003), with residence times up to thousands of years. However, changes in their valence and chemical form are possible, and under certain conditions, small portions become soluble. Among the factors that determine trace element solubility and bioavailability are pH, cation exchange capacity (CEC), anion exchange capacity (AEC), soil organic matter (SOM) content, clay content and quality, oxide content and type, and redox potential (Gregor 2004).

Toxic elements in soils pose numerous health hazards to higher organisms, causing various diseases and disorders even in low concentrations. Bioavailability of toxic elements is important because bioavailable fractions are most likely to harm plants, animals and humans. Due to their toxicity and long-term persistence, toxic elements are a serious threat to the functioning of ecosystems. They are known to affect plant growth and have a negative impact on soil microflora (Giller et al. 1998). Different sources of toxic elements may have different effects regarding their toxicity to soil microorganisms, depending on variation in the mineralogical composition of the source, chemical forms of elements added to soil, and duration of exposure (Castaldi et al. 2004). Exposure to toxic elements causes reduction of microbial diversity and activity (Lasat 2002; Mc Grath et al. 2001). Toxic metals in soil can severely inhibit the biodegradation of organic contaminants (Maslin and Maier 2000).

Plants are known to take up trace elements from contaminated soils. Some 450 species are even known as hyperaccumulators. Although accounting for less than 0.2% of all known species, they have a wide taxonomic range (Visioli and Marmiroli 2013). For animals, increased toxic element concentrations result in reduced vitality, reproduction problems, and the occurrence of cancareous, mutagenic and teratogenic diseases (Bires et al. 1995).

Toxic contamination of soil may also pose risks and hazards to humans (Magbagbeola and Oyeleke 2003; Lottermoser 2007). Toxic elements enter the human body through the food chain (soil-plant-human or soil-plant-animal human), inhalation of dust, ingestion of soil and water and dermal contact of contaminated soil and water. They may also affect humans through reduction in food quality (safety and marketability) via phytotoxicity (Wuana and Okieimen 2011). Toxic elements are known to be persistent in the human body, with excretion half-lives that last for decades and can lead to a wide range of toxic effects (e.g. Tong et al. 2000; Jaerup et al. 2000; Thomas et al. 2009; Putila and Guo 2011).

Adequate protection and restoration of soils contaminated by toxic elements require environmental risk assessment including metal toxicity monitoring, especially for land reclamation and/or recultivation for agriculture (Nieder et al. 2014). This would require knowledge of the source of contamination, basic chemistry, and environmental and associated health risks of toxic elements. Risk assessment enables decision makers to manage contaminated sites in a cost-effective manner while preserving public and ecosystem health. Among the technologies for remediation of heavy metal-contaminated sites most frequently listed are immobilization, soil washing and phytoremediation (Wuana and Okieimen 2011).

8.2 Sources of Potentially Harmful Trace Elements

8.2.1 Geochemical Background

The term geochemical background became an important reference with increasing environmental awareness of toxic elements (Matschullat et al. 2000). Heavy metal toxicity is exacerbated by the fact that plants may take up and accumulate toxic elements in concentrations that might affect human health (Harris et al. 1996). Contents of potentially toxic elements in soils are to a major part dependent on the nature of parent material from which the soils have developed (Mitchell 1964). A significant correlation exists between total concentrations of elements between parent material and the derived soils (Prabhakaran and Cottenie 1971). Trace element concentrations in rocks are dependent on the trace element composition of the rock forming minerals. Table 8.2 shows ranges of toxic element concentrations in rock-forming minerals.

As to the varying mineral composition of different rock types, soil parent materials vary widely in trace element concentrations. Table 8.3 shows commonly reported concentrations of trace elements in igneous and metamorphic rocks. Soil

Table 8.2 Concentration ranges of trace elements present in rock-forming minerals

Mineral	0.X%	0.0X%	0.00X% or less
Quartz			Zn
Muscovite		Cr, V	Cu
Potassium feldspar			Cu, Ni, Pb, V, Zn
Biotite		Cr, Ni, V, Zn	Co, Cu, Pb
Amphibole		Cr, Ni, V, Zn	Co, Cu, Pb
Plagioclase feldspar			Cu, Ni, Pb, Zn
Pyroxene		Cr, Ni	Co, Cu, Zn
Olivine	Ni	Co, Cr	Cu, Zn

Adapted from Smith and Huyck (1999)

Table 8.3 Concentrations (mg kg^{-1}) of trace elements (ranges) in igneous and sedimentary rocks

Element	Basaltic igneous	Granitic igneous	Shales and clays	Black shales	Sandstone
As	0.2–10	0.2–13.8	–	1–900	0.6–9.7
Cd	0.006–0.6	0.003–0.18	<11	0.3–8.4	–
Co	24–90	1–15	5–25	7–100	–
Cr	40–600	2–90	30–590	2–1000	–
Cu	30–160	4–30	18–120	20–200	–
Mo	0.9–7	1–6	–	1–300	–
Ni	45–410	2–20	20–250	10–500	–
Pb	2–18	6–30	16–50	7–150	<1–31
Zn	48–240	5–140	18–180	34–1500	2–41

Compiled from Alloway (1995) and Cannon et al. (1978)

Table 8.4 Concentrations ($mg\ kg^{-1}$) of trace elements (ranges) in typical uncontaminated soils

Element	Range in soil ($mg\ kg^{-1}$)
As	5–10
Cd	0.01–0.7
Co	1–40
Cr	5–3000
Cu	2–100
Fe	7000–55,000
Mo	0.2–5
Ni	10–100
Pb	2–200
Zn	10–300

Compiled from Allaway (1968) and Feng et al. (2009)

forming processes modify the regolith and thus redistribute the content of trace elements in soils. Stronger the soils are developed, lesser may be the influence of the parent materials on toxic elements (Zhang et al. 2002). Soils may be influenced by intrasolum displacement (e.g. SOM, clay metals), leaching, erosion, deposition and turbation processes (Bronger and Bruhn-Lobin 1993), that each may have a significant influence on the distribution of toxic elements in the soil profile (Bech et al. 1998).

Part of the elements reprecipitated as, or incorporated into, new secondary minerals such as phyllosilicate clays and Fe and Mn hydrous oxides and oxides. Another part may be sorbed onto the surfaces of such materials, or be bound by accumulating organic materials (McLaren 2003). To summarize, the concentration and forms of trace elements in mature soils may differ substantially from those in the soil parent material. Table 8.4 gives ranges of heavy metal contents in uncontaminated naturally occurring soils.

8.2.2 Anthropogenic Sources

Most soils of rural and urban environments may accumulate toxic elements above defined background values due to the disturbance and acceleration of nature's geochemical cycle of metals by man. This may cause risks to human health, plants, animals, ecosystems, other environmental media and humans (D'Amore et al. 2005). Toxic elements become contaminants in soils if (i) their rates of generation through anthropogenic inputs are more rapid compared to natural cycles, (ii) they are transferred from mines to random ecosystems or environmental compartments where direct exposure may occur, (iii) the concentrations of toxic elements in deposited products are higher compared to those in the environment, and (iv) the chemical form in which an element is present in the environmental system may render it more mobile and bioavailable. Toxic elements in the soil originating from anthropogenic sources compared to pedogenic ones tend to be more mobile

(Kaasalainen and Yli-Halla 2003). It is projected that the anthropogenic emission into the atmosphere, for several heavy metals, is one-to-three orders of magnitude higher than natural fluxes (Sposito and Page 1984). A simple mass balance of toxic elements (TE) in the soil can be expressed as follows:

$$TE\text{total} = TEpa + TEat + TEfert + TE\text{ag} + TE\text{ow} + TE\text{ip} - (TE\text{cr} + TE\text{loss}),$$

where "*TE*" is the toxic element, "*pa*" is the soil parent material, "*at*" is the atmospheric deposition, "*fert*" is the fertilizer source, "ag" is the agrochemical source, "ow" is the organic waste source, "ip" are other inorganic pollutants, "cr" is crop removal, and "*loss*" is the loss by leaching and volatilization (Wuana and Okieimen 2011).

Toxic elements can originate from a wide spectrum of anthropogenic sources including airborne sources, metal mine tailings, land application of (in)organic fertilizers and pesticides, disposal of wastes in landfills, coal ashes, lead-based paints, leaded gasoline and petrochemicals (Khan et al. 2008; Zhang et al. 2010; Basta et al. 2005; Wuana and Okieimen 2011). The major sources are discussed as follows.

8.2.2.1 Airborne Sources

Airborne sources of toxic elements include natural and anthropogenic. Volcanic eruptions and wind-blown dusts are the most important natural sources. Volcanoes have been reported to emit high levels of Al, Zn, Pb, Ni, Cu and Hg along with toxic and harmful gases (Seaward and Richardson 1990).Wind dusts arising from desert regions such as Sahara, have high levels of Fe and lesser amounts of Zn, Cr, Ni and Pb (Ross 1994). Marine aerosols and forest and steppe fires also exert a major influence in the transport of some toxic elements in many ecosystems. Table 8.5 gives an overview of the worldwide emissions of heavy metals from natural sources.

Table 8.5 Global emissions (in 10^6 kg year^{-1}) of heavy metals from natural sources

Source	Global production (10^6 kg year^{-1})	Emission (10^6 kg year^{-1})							
		Cd	Co	Cu	Cr	Hg	Ni	Pb	Zn
Volcanic particles	6.5–150	0.5	1.4	4	3.9	0.03	3.8	6.4	10
Windblown dust	6–1100	0.25	4	12	5	0.03	20	10	25
Forest fires	2–200	0.01	–	0.3	–	0.1	0.6	0.5	0.5
Vegetation	75–1000	0.2	–	2.5	–	–	1.6	1.6	10
Sea salt	300–2000	0.002	–	0.1	–	0.003	0.04	0.1	0.02
Total		0.96	5.4	18.9	8.9	0.16	26	18.6	45.5

Adapted from Pacyna (1986)

Coal and refuse burning, metal smelting, and automobile emissions are the most important anthropogenic airborne sources of toxic elements (Fergusson 1990; McLaren 2003). Metals from airborne sources are usually released as particulates contained in the gas stream. Metals like Cd, As and Pb can also volatilize by high temperature activities, may covert to oxides and subsequently condense as fine particulates unless a reducing atmosphere is maintained (Smith et al. 1995). Emissions from chimney (stack emissions) may be circulated over extensive areas by natural air flow until dry or wet precipitation remove them from the gas stream. Fugitive emissions are often distributed over a much smaller area. Contaminant concentrations are generally lower in fugitive compared to stack emissions (Wuana and Okieimen 2011).

All solid particles in smoke from fires and in other emissions from factory chimneys are eventually deposited on land or sea. Most fossil fuels contain some heavy metals. This form of contamination has been continuing on a large scale since the beginning of industrialization. For example, very high concentration of Cd, Pb, and Zn has been found in plants and soils adjacent to smelting works (Wuana and Okieimen 2011). Another source of soil pollution is the aerial emission of Pb from the burning of petrol containing tetraethyl lead which increases Pb concentrations in soils of urban areas and next to main roads. Zinc and Cadmium contained in tyres and lubricant oils may also be deposited on soils next to roads (US EPA 1996). Figure 8.1 shows the distribution of Cd, Cu and Pb at various distances along a roadside.

Atmospheric As usually arises in particulate form from both natural sources, such as volcanic activity or forest fires, and man-made (anthropogenic) sources, such as the burning of fossil fuels, automobile exhaust, and tobacco smoke. These particles (mostly in diameters of <2 μm) have a typical atmospheric residence time of approximately 9 day, during which time they are transported by wind currents until they fall to earth. It has been estimated that on an annual basis, volcanoes, microbial activity, and the burning of fossil fuels release 3000, 20,000, and 80,000 mt of atmospheric As, respectively (Van den Enden 1999). It is estimated that the median values of worldwide emissions of Cd, Cu, Pb, and Zn into soils were 22, 954, 796 and $1372 \cdot 10^6$ kg year^{-1}, respectively; more than half of those metals were associated with base metal mining and smelting activities (Nriagu and Pacyna 1988; Lone et al. 2008).

8.2.2.2 Metal Mining and Industrial Wastes

Mining activities, including crushing, grinding, washing, smelting and all other metal extraction and concentration processes generate high amounts of wastes (Fig. 8.2). Mining wastes often contain high concentrations of potentially toxic elements such as Al, As, Cd, Cu, Hg, Mo, U, V, Zn and Cd (Jung 2001; Naicker et al. 2003; Nagajyoti et al. 2010). Deposits outside the mine area can cause deleterious effects on the soils (Dudka and Adriano 1997). Negative impact of the mining activities on the surroundings is mainly due to the presence of high

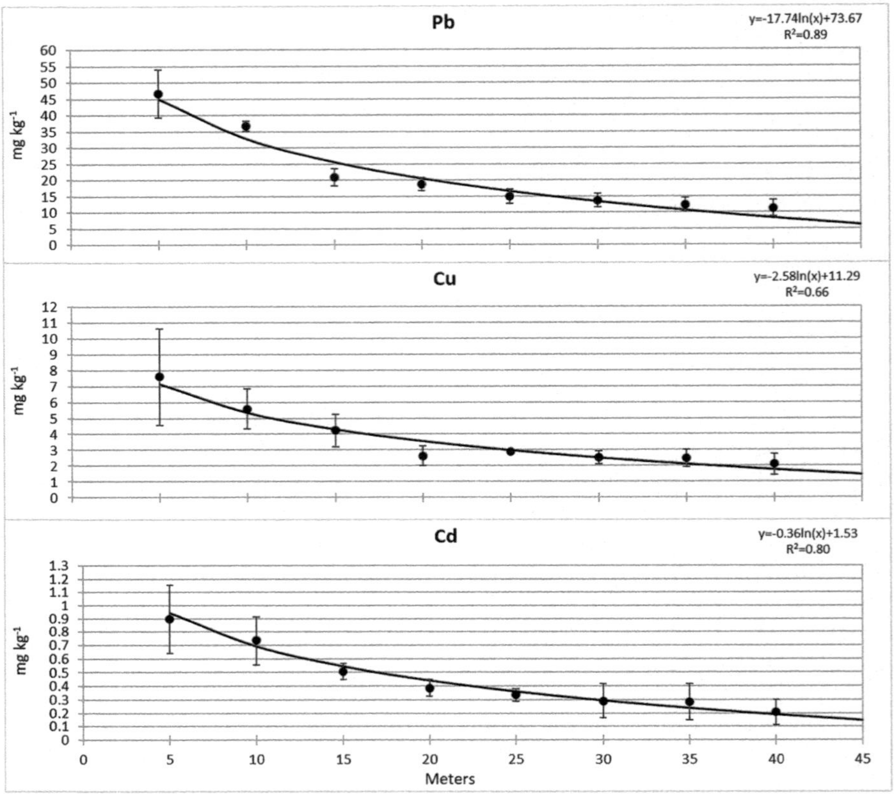

Fig. 8.1 Heavy metal (Pb, Cu, Cd) concentrations at various distance from a highway in Lithuania (Adapted from Grigalavičienė et al. 2005)

volumes of tailings which usually have unfavorable conditions to most plant species growing on them, such as low pH (Obrador et al. 2007), toxic metal concentrations (Norland and Veith 1995), low water retention capacity (Henriques and Fernandez 1991), and low levels of plant nutrients (Nieder et al. 2014). Wells located near mining sites have been reported to contain toxic elements at levels that exceed drinking water criteria (Garbarino et al. 1995; Peplow 1999). The management of waste materials is thus an important issue for mining industry worldwide.

Release of metals from mine sites takes place mainly through drainage and erosion of waste dumps. Heavy metal-bearing sediment particles enter river systems by discharge of mine or processing waste, tailings dam failures, remobilization of mining-contaminated alluvium and mine drainage. The mineralogy and geochemistry of these particles is dependent upon the original ore mineralogy, and on processes that have occurred in the source areas, during transport and deposition and during post depositional early diagenesis (Wuana and Okieimen

Fig. 8.2 Mining waste material eroding from a mining dump in the Gatumba Coltan Mining District of Rwanda. The mining waste is partly exported from the area via streams; another part is deposited on farmers' fields (Photo: Rolf Nieder)

2011). Soils are also prone to contamination from smelter emissions as well as wind-blown dust from mine tailings and smelter slag dumps which are important point sources of toxic elements like such as As, Cd and Cr (e.g. Ettler et al. 2009; Kribek et al. 2010). Human exposure to these hazardous elements may have different pathways, through ingestion of vegetables grown on contaminated soils (food chain) or through dust inhalation and dust adhering to plants.

Artisanal small-scale gold mining has expanded during the last decades and is often carried out under hazardous conditions. Extraction of gold using mercury has been a way out of poverty for millions of people in developing countries. An estimated 640–1350 Mg of mercury (Hg) is released per annum into the environment because of gold mining averaging 1000 Mg per annum from at least 70 different countries. This equals approximately a third of the total global anthropogenic release of Hg into the environment (Telmer and Veiga 2009). Thus, workers in this industry may be exposed to high levels of mercury and suffer from toxic effects from mercury exposure (Kristensen et al. 2013). Other materials containing toxic elements are generated by a variety of industrial uses (Table 8.6).

8.2.2.3 Inorganic Fertilizers

Fertilizers (inorganic and organic) are the most important sources of toxic elements to agricultural soils. Large quantities of fertilizers are regularly added to soils in

Table 8.6 Industrial uses of potentially toxic elements

Element	Industrial uses
As	Glass and ceramic works, medicines, pesticides, feed additives
Cd	Storage batteries, electroplating, stabilizer, solar cells, pigments, alloys for telephone and telegraph wires, electron optical and photoelectrical devices, nuclear reactors
Cr	Chrome plating, alloying additives to stainless steel, dyes, tanning for leathers, textiles, cassette tapes, magnetic storage media for computers, matches, pyrotechnics, photography, seed disinfection, pigments
Cu	Kitchenware, electrical goods, alloys, fungicides, algaecides, pigments, intra-uterine conceptive devices
Ni	Cooking utensils, alloys for corrosion resistant equipment, coinage, heating elements, gas turbines, jet engines, batteries, electroplating, paints, pigments
Pb	Antiknock agent in petroleum/gasoline, storage batteries, paints, ammunition, glass-ware, ceramics, protection from radiation, bearing alloys, printing press, rubber industry
Zn	Alloys, galvanizing of iron and steel for corrosion protection, vulcanization of rubber, photocopying paper, paints, TV tubes, rayon glass, enamel and plastic industries, fertilizers, feed additives, medicine, cosmetics

Adapted from Alloway (1995)

Table 8.7 Concentration ranges (μg g^{-1}) of heavy metals present in inorganic fertilizers

	Inorganic fertilizers		
Metal	Phosphate fertilizer	Nitrogen fertilizers	Lime
Cr	66–245	3.2–19	10–15
Ni	7–38	7–34	10–20
Cu	1–300	–	2–125
Zn	1–1450	1–42	10–450
Cd	0.1–190	0.05–8.5	0.04–0.1
Pb	4–1000	2–120	20–1250

Adapted from Ross (1994)

intensive farming systems to provide adequate N, P, and K for crop growth. The compounds used to supply these elements contain trace amounts of heavy metals (e.g., Cd and Pb) as impurities, which, after continued fertilizer application may significantly increase their content in soils (Verkleji 1993). Application of certain phosphatic fertilizers inadvertently adds Cd and other potentially toxic elements to the soil, including F, Hg, and Pb. Liming increases heavy metal levels of soils more than nitrate fertilizers. Heavy metal concentrations of some inorganic fertilizers are given in Table 8.7.

8.2.2.4 Biosolids and Manures

The application of numerous biosolids (e.g., livestock manures, composts, and municipal sewage sludge) to land leads to the accumulation of heavy metals in

Fig. 8.3 Liquid manure leaving a pig fattening facility near Beijing, China. The manures pollute canals, streams, rivers and groundwater with nutrients and toxic elements (Photo: Marco Roelcke, with kind permission)

soils (Wuana and Okieimen 2011). Animal wastes such as poultry, cattle, and pig manures produced in agriculture are commonly applied to crops and pastures either as solids or slurries. In some cases liquid manures are dumped into surface waters and directly affect the quality of potential drinking water (Fig. 8.3). Although most manures are seen as valuable fertilizers, in pig and poultry industry, Cu and Zn added to diets may have the potential to cause metal accumulation in soils (Sumner 2000). There is also evidence for the use of Cr and As as feed additives (Zhang et al. 2005). Hence, regions with intensive livestock farming and excessive disposal of animal wastes on arable land are particularly concerned by toxic element contamination (Ostermann et al. 2014).

Sewage sludge, produced by wastewater treatment processes, is one of the most important sources of heavy metal contamination of soils. The term biosolids has become more common as a replacement for sewage sludge because it is thought to reflect more accurately the beneficial characteristics inherent to sewage sludge (Silveira et al. 2003). In the EU, more than 30% of the sewage sludge is used as fertilizer in agriculture (Silveira et al. 2003). In the USA, more than half of approximately 5.6 million Mg of sewage sludge (dry matter) produced annually is land applied (Wuana and Okieimen 2011). In Australia, over 175,000 Mg of dry biosolids are produced annually, most of which are applied agricultural land (McLaughlin et al. 2000). There is also considerable interest in the potential for composting biosolids together with other organic materials such as sawdust, straw, or garden waste. Concentrations of common organic fertilizers are given in Table 8.8.

Table 8.8 Concentration ranges ($\mu g\ g^{-1}$) of heavy metals present in organic fertilizers

	Organic fertilizers		
Metal	Farmyard manure	Compost refuse	Sewage sludge
Cr	1.1–55	1.8–400	8.4–600
Ni	2.1–30	0.9–279	6–5300
Cu	2–172	13–3500	50–8000
Zn	15–556	82–5894	91–49,000
Cd	0.1–0.8	0.01–100	<1–3400
Pb	0.4–27	1.3–2240	2–7000

Adapted from Ross (1994)

8.2.2.5 Pesticides

Several heavy metal-containing pesticides are used to control diseases of grain and fruit crops and vegetables. Examples of such pesticides are copper-containing fungicidal sprays such as copper sulphate (*Bordeaux mixture*) and copper oxychloride (Jones and Jarvis 1981). Particularly in orchard plantations these compounds have been used frequently. Many orchard soils may be thus highly contaminated particularly with Cu, As, Pb, Zn, Fe, and Hg (Ross 1994). In the UK about 10% of the chemicals that have approved for use as insecticides and fungicides were based on compounds which contain Cu, Hg, Mn, Pb, or Zn. In Canadian orchards, lead arsenate was used for many decades for the control of parasitic insects and soils were found to be highly enriched with Pb, As and Zn (Nagajyoti et al. 2010). Arsenic containing compounds were also used extensively to control cattle ticks and to control pests in banana (Wuana and Okieimen 2011). In New Zealand and Australia, timbers have been preserved with formulations of Cu, Cr, and As, and there are now many sites where soil concentrations of these elements greatly exceed background concentrations. Part of the soil contamination with heavy metals may also be from irrigation water sources such as lakes, rivers, deep wells or canals (Ross 1994).

8.2.2.6 Wastewater

Waste waters probably constitute the largest single source of increased metal concentrations in surface waters (Nagajyoti et al. 2010). Domestic effluents may consists of (i) untreated or solely mechanically treated waste waters, (ii) substances which have passed through filters of biological treatment plants and (iii) waste substances passed over sewage outfalls and discharged to receiving water bodies. The use of detergents creates a possible pollution hazard, since common household detergent products can affect the water quality. For example, Angino et al. (1970) found that most enzyme detergents contained trace amounts of Fe, Cr, Co, Zn, Sr and B.

Globally, it is estimated that 20 million hectares of arable land are irrigated with waste water. In several Asian and African cities, studies suggest that agriculture based on wastewater irrigation accounts for 50% of the vegetable supply to urban areas (Bjuhr 2007). Although the metal concentrations in wastewater effluents are commonly relatively low, long-term irrigation of land with such can possibly result in heavy metal accumulation in the soil. With regard to pollution resulting from urbanized areas, there is an increasing awareness that urban runoff presents a serious problem of heavy metal contamination and is recognized as a major source of pollutants to surface waters (Nagajyoti et al. 2010).

8.3 Chemistry of Toxic Elements in Soils

The knowledge of the basic chemistry, environmental and associated health effects of toxic elements is necessary for understanding their speciation, bioavailability, and remedial options. The fate in soil depends significantly on the chemical form and speciation of the element. Toxic elements in soil may be adsorbed by different reactions and later may be redistributed into different chemical forms with varying mobility, bioavailability and toxicity (Wuana and Okieimen 2011). This distribution is controlled by reactions such as (i) ion exchange, (ii) mineral precipitation and dissolution, (iii) aqueous complexation, (iv) biological immobilization and mobilization, and (v) plant uptake (Levy et al. 1992). The general relationships and equilibria between different forms of toxic elements in soils are given in Fig. 8.4.

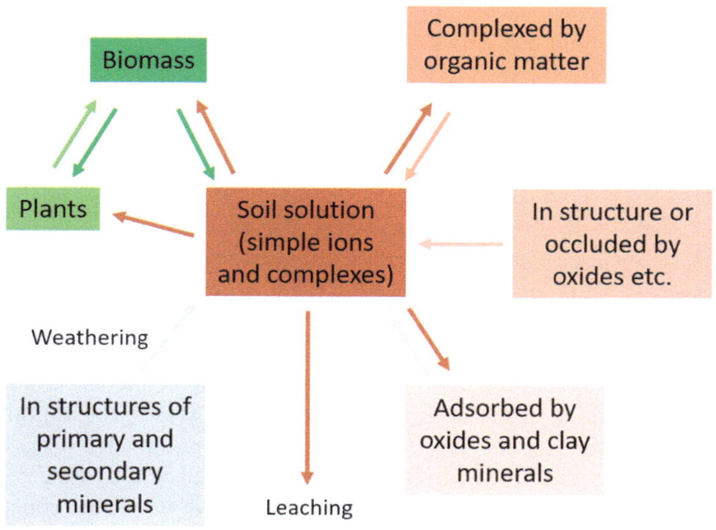

Fig. 8.4 Relationships between different forms of trace elements in soils (Adapted from McLaren 2003)

Toxic element concentrations in soil solution are mainly controlled by (i) solubility of solid phase compounds containing the element of interest and (ii) by sorption/desorption reactions at the surface of soil colloids, such as clay minerals and soil organic matter. The hazard imposed by toxic elements in soils depends on their ability to migrate into the water system and their availability for biological uptake. The degree to which toxic elements may dissociate from soil solids and become bioavailable is therefore an important risk factor (Scott et al. 2004).

8.3.1 Factors Influencing Toxic Element Bioavailability in Soils

The presence of soil colloids has a major influence on the bioavailability of toxic elements. Soil organic matter (SOM), some clay minerals and the oxides and hydroxides of Al, Fe and Mn have a large capacity to sorb or complex toxic elements. The main soil properties responsible for sorption and desorption of toxic elements include pH, redox potential (Eh), cation exchange capacity (CEC), clay content and quality of clay minerals, soil organic matter content, and Fe, Mn, and Al oxides and hydroxides (Du Laing et al. 2009). Depending on the total amounts and the relative proportions of organic and inorganic constituents, a soil will have higher or lower capacity to bind a particular toxic element species. Many factors influence the bioavailability of toxic elements in soils. The most important are given in the following sections.

pH

The pH in soil solution influences almost all chemical processes, and particularly the behavior of toxic elements (McLaren 2003). The solubility of toxic elements that can occur as free hydrated cations generally increases with decreasing pH (e.g. Cd, Cu, Pb and Zn). Fig. 8.5 shows the influence of pH on the desorption of native soil Cd through repeated equilibration of soil in a weak electrolyte (0.01 M $Ca[NO_3]_2$).

Factors such as competition for sorption, decreasing pH dependent negative charge of the sorption complex and dissolution of soil components explain this behaviour (Shaheen et al. 2013). However, there is evidence that with increasing contact time the ability of toxic elements to desorb from soil may decrease (McLaren et al. 1998). Soil pH directly controls the solubilities of element hydroxides and thus plays a major role in the sorption of toxic elements. When soil pH increases, the retention of the toxic element cations increases via sorption, inner-sphere surface complexation, and/or precipitation and multinuclear type reactions (Sposito 1989). For example, sorption experiments conducted by Shaheen (2009) have shown that the total amounts of Pb and Cd sorbed within the concentration range used were larger in alkaline than in acidic soils. The charge of variable charge soils is highly pH-dependent. Many researchers (cited by Shaheen et al. 2013) have

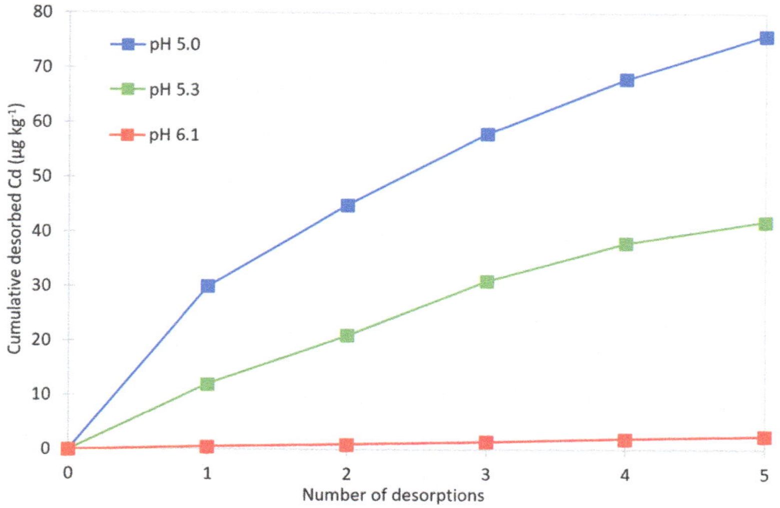

Fig. 8.5 Influence of pH on Cd desorption from a silt loam soil (Adapted from Gray et al. 1998)

shown increased Cd and/or Zn sorption in variable charge soils with increasing pH, mainly due to an increased negative surface charge.

Soil Texture and Soil Mineral Composition
The ability of soils to exchange cations, expressed as CEC, is one of the most important soil properties governing the sorption of toxic elements. Coarse-textured soils exhibit a lower CEC compared to fine-textured ones. The fine-textured soil fraction which mainly consists of clay minerals, iron and manganese oxides and hydroxides, is characterized by large surface reactivities and surface areas. Clay particles are negatively charged silicate minerals and are thus a major contributor to CEC. Therefore, they preferentially adsorb positively charged ions. Besides the amount, the type of clay mineral is of great importance. Some clay minerals such as vermiculite, montmorillonite, imogolite, and amorphous allophanes reveal the highest CEC (Du Laing et al. 2009). However, it has been reported that sorption of As occurs by chemisorption or ligand exchange on clay surfaces, mainly by replacing or competing with phosphate (Sadiq 1997).

Soil Organic Matter
Soil organic matter plays a key role in the various physical, biological, and chemical processes in soil, including the sorption of toxic elements. The retention mechanisms of toxic elements by SOM involve the formation of inner-sphere complexes as well as ion exchange and precipitation reactions (Stevenson 1982). Figure 8.6 shows an example of the complexing of Cu by organic colloids involving carboxyl and phenolic groups.

The stability constants (log K) of element-fulvic acid complexes at pH 3.5 decrease in the order (with log K in parentheses) Cu (5.8) > Fe (5.1) > Ni (3.5) > Pb (3.1) > Co (2.2) > Ca (2.0) > Zn (1.7) > Mn (1.5) > Mg (1.2). At pH 5, the

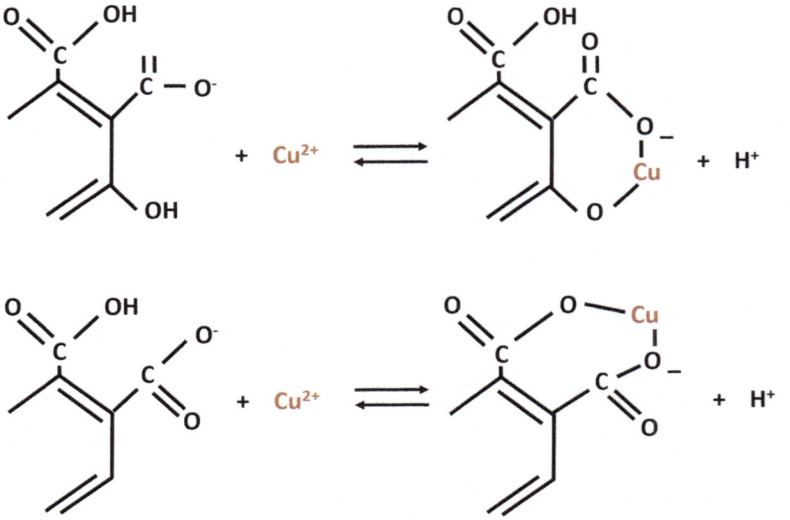

Fig. 8.6 Complexation of Cu by organic soil colloids

sequence is Cu (8.7) > Pb (6.1) > Fe (5.8) > Ni (4.1) > Mn (3.8) > Co (3.7) > Ca (2.9) > Zn (2.3) > Mg (2.1) (Stevenson 1982). Thus, the stability of element-organic complexes increases with pH. The stability constants reflect the great affinity for organic complex formation particularly of Cu, Pb, and Fe. Above pH 6–7, most elements in solution exist as organic complexes. Organic complexes of Cu and Pb remain stable until pH 4, while complexes of Cd and Zn are less stable and dissociate when the pH is below 6. Moreover, the solubility of humic acids decreases with decreasing pH. Both factors explain why organic complexes of elements might still be present in acidic soils (pH <4).

Soil Carbonates

The presence of free $CaCO_3$ generally reduces the solubility of toxic elements, as it increases the soil pH. Moreover, the accompanying carbonate/bicarbonate ions will form element carbonates that may be precipitated. Free $CaCO_3$ in soils therefore controls the solubility and Kd of toxic elements via its influence on pH as well as the formation of element carbonates (Alloway 1995). The greatest affinity for reaction with carbonates has been observed for Co, Cd, Cu, Fe, Mn, Ni, Pb, Sr, U, and Zn. Boron and arsenate are significantly sorbed to carbonate mineral surfaces by ligand exchange or chemisorption at pH below 10 (Sadiq 1997). Sorption decreases with higher pH, owing to carbonates acquiring a negative charge above pH 10 (Sadiq 1997). Trace elements may coprecipitate with carbonates, be incorporated into their structure, or may be sorbed by oxides (mainly Fe and Mn) that precipitate onto carbonates or other soil particles. Carbonates can be the dominant toxic element sink in a particular soil, but the most important mechanisms for regulating toxic elements' stability constant by carbonates are related to the variation of soil pH.

Soil Redox Conditions

When soils become anaerobic due to waterlogging, the redox potential (Eh) decreases. Several trace elements such as Fe and Mn occur in more than one oxidation state. In aerobic soils they are present particularly in their higher oxidation states (McLaren 2003). Under anaerobic conditions, Fe and Mn are reduced to the more soluble forms Fe^{2+} and Mn^{2+}. This effectively results in substantial increases in solution concentrations of Fe and Mn. In addition, since substantial concentrations of many other trace elements, including toxic elements, are often sorbed or occluded by Fe and Mn oxides in soils (Taylor and McKenzie 1966), solubilisation of the oxides also releases the other elements into solution.

8.3.2 Toxic Elements Speciation

Toxic elements most frequently found at contaminated sites, in order of abundance are As, Cd, Cr, Cu, Hg, Ni, Pb and Zn (US EPA 1996). The hazard imposed by toxic elements in soils is dependent on their ability to migrate into the water system and their availability for biological uptake. The degree to which a contaminant may dissociate from soil solids and become available to a target organism is therefore a determining risk factor (Scott et al. 2004).

Arsenic

Arsenic is a metalloid in group VA and period 4. It has the following properties: atomic number 33, atomic mass 75, density 5.72 g cm^{-3}, melting point 817 °C, and boiling point 613 °C (Wuana and Okieimen 2011). It is present in a wide variety of minerals, mainly as As_2O_3, and may originate from processing of ores containing mostly Cu, Pb, Zn, Ag and Au. It occurs also in ashes from coal combustion. In nature, As occurs as arsenate (As^{5+}), arsenite (As^{3+}), arsenic metal (As^0), and arsine (As^{3-}). Most investigations on As have focused on As^{5+} and As^{3+} because As^0 rarely occurs in nature and As^{3-} exists only under highly reducing conditions (Feng et al. 2009). Agricultural practices like applications of As-containing phosphorus fertilizers, pig manure, pesticides and herbicides may contribute to As accumulation in soils. Arsenic in soils can interact with different components, such as SOM, Fe and Mn oxides, carbonates and sulphides which influences its mobility, bioavailability and toxicity (Cumming et al. 1999).

The most toxic forms of As are aqueous As^{5+} and As^{3+}. However, As^{3+} at the same concentration is 200 more toxic compared to As^{5+} (Jung et al. 2009). Moreover, As^{3+} is highly mobile in aqueous systems and the mobility increases with increasing pH (Smith et al. 1995). Under anaerobic conditions As^{3+} dominates, existing as arsenite (AsO_3^{3-}) and its protonated forms H_3AsO_3, $H_2AsO_3^-$ and $HAsO_3^{2-}$. Arsenite can adsorb or co-precipitate with metal sulphides and has a high affinity for other sulphur compounds. Under aerobic conditions, As^{5+} dominates, usually as arsenate (AsO_4^{3-}) and it's protonated forms H_3AsO_4, $H_2AsO_4^-$

and $HAsO_4^{2-}$. Arsenate and other anionic forms of As can precipitate in the presence of metal cations (Bodek et al. 1988).

Arsenate (As^{5+}) is readily immobilized on soil and sediment particles by adsorption on and co-precipitation with oxides of Mn, Al and Fe (Jung et al. 2009). However, it can be mobilized under reduced conditions that encourage the formation of As^{3+}. The presence of other ions competing for sorption sites and organic compounds that form complexes with As further enhance mobilization (Smith et al. 1995). Arsenate can be leached easily if the amount of reactive metals in the soil is low. Microorganisms play a significant role in electrochemical speciation and cycling of As by mediating the transformation of As and soil compounds that adsorb As (Newman et al. 1998). Many common microbial genera (*Bacillus, Clostridium, Alcaligenes,* and *Citrobacter*) have been shown to convert As compounds to arsine or methylarsine gases, reducing the amount of arsine present in soils (Oremland and Stolz 2003). Arsine (AsH_3) and elemental arsenic may be only present under extreme reducing conditions. The most important removal mechanisms are sorption and coprecipitation with hydrous iron oxides under most environmental conditions (Pierce and Moore 1982). Figure 8.7 outlines the effect of pH and redox conditions (Eh) on the behaviour of As.

Cadmium

Cadmium is located at the end of the second row of transition elements with atomic number 48, atomic weight 112.4, density 8.65 g cm^{-3}, melting point 320.9 °C, and boiling point 765 °C (Wuana and Okieimen 2011). It is similar to Zn in some

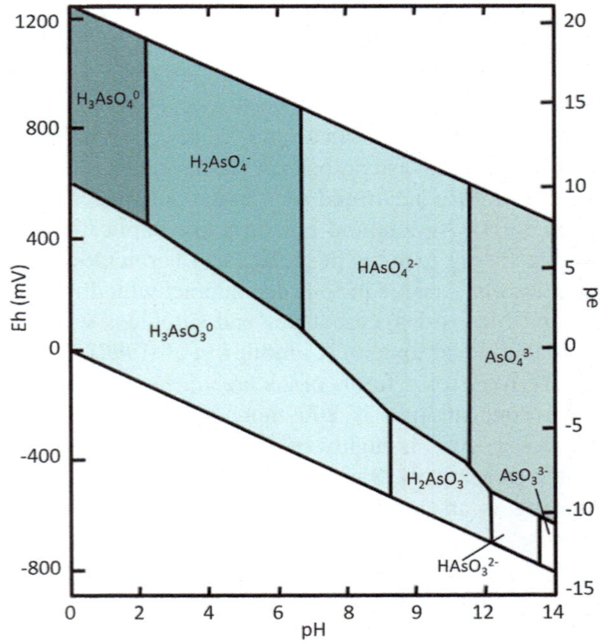

Fig. 8.7 Effect of pH and redox conditions (Eh) on the chemistry of As

respects but forms more complex compounds. Cadmium is soluble in acids but not in alkalis. Cadmium occurs naturally in the form of cadmium sulphide or carbonate (CdS or $CdCO_3$). It is recovered as a by-product from the mining of sulphide ores of Pb, Zn and Cu. Sources of Cd contamination include plating operations and the disposal of Cd-containing wastes (Smith et al. 1995). The most common forms of Cd include Cd^{2+}, Cd-cyanide complexes and $Cd(OH)_2$ (Cynthia and David 1997).

The chemistry of Cd in soil to a great extent is controlled by pH. Cadmium solubility increases at decreasing pH. At pH values greater than 6, Cd is adsorbed by the soil solid phase or is precipitated. Cadmium forms soluble complexes with inorganic and organic ligands which enhances the mobility of Cd in soils. Hydroxides and carbonates of Cd dominate at higher pH. Sorption is also influenced by the cation exchange capacity (CEC) of clays, carbonate minerals, and organic matter present in soils and sediments. In the presence of phosphate, arsenate, chromate and other anions, Cd precipitates. Cadmium is relatively mobile in aqueous systems and exists primarily as hydrated ions or as complexes with humic acids and other organic ligands (Cynthia and David 1997). Under reducing conditions, precipitation as CdS controls the mobility of cadmium (Smith et al. 1995). Cadmium salts with low solubility include Cd phosphate ($Cd_3(PO_4)_2$), sulphide (CdS), hydroxide (Cd $(OH)_2$ and carbonate ($CdCO_3$). The solubilities of these salts increase with decreasing pH.

Copper

Copper is a transition metal which belongs to period 4 and group IB of the periodic table with atomic number 29, atomic weight 63.5, density 8.96 g cm^{-3}, melting point 1083 °C and boiling point 2595 °C (Wuana and Okieimen 2011). It is mined as a primary ore product from CuS and oxide ores. Mining activities are the major source of Cu contamination. Copper has low chemical reactivity. In moist air Cu forms a greenish surface coating which protects the metal from further corrosive attack. Solution and soil chemistry have a strong influence on the speciation of Cu. In alkaline aerobic systems, $CuCO_3$ is the dominant Cu species. The cupric ion, Cu^{2+}, and Cu hydroxides, $CuOH^+$ and $Cu(OH)_2$, are also commonly present. Copper in soil over a wide pH range is strongly bound to SOM and soil minerals. As a result of this, it is hardly leached to the groundwater (Cynthia and David 1997). The affinity of Cu to SOM increases as pH increases. If S_2 is present in anaerobic environments, CuS is formed. The cupric ion (Cu^{2+}) is the most toxic species of Cu. Toxicity has also been demonstrated for $CuOH^+$ and $Cu_2(OH)_2{}^{2+}$ (LaGrega et al. 1994).

Chromium

Chromium is a first-row *d*-block transition metal of group VIB in the periodic table with the following properties: atomic number 24, atomic mass 52, density 7.19 g cm^{-3}, melting point 1875 °C, and boiling point 2665 °C. Chromium in nature does not occur in elemental form but only in compounds. It is mined as a primary ore product in the form of the mineral chromite, $FeCr_2O_4$. In contact with O_2, Cr immediately produces a thin oxide layer that is impermeable to oxygen and protects the metal below. Chromium in the environment exists in two stable oxidation states, Cr^{6+}

(commonly found at contaminated sites) and Cr^{3+}, which have different toxicity and transport characteristics (Cynthia and David 1997).

Hexavalent chromium (Cr^{6+}) is commonly present as CrO_4^{2-} (chromate) and $Cr_2O_7^{2-}$ (dichromate) and precipitates easily in the presence of Ba^{2+}, Pb^{2+}, and Ag^+. Chromate and dichromate also adsorb on soil minerals, particularly on Fe and Al oxides. Hexavalent Cr at high soil pH has a high solubility and tends to be mobile in the environment. It is acutely toxic, highly bioavailable and carcinogenic (Bianchi and Levis 1987). In contrast, trivalent Cr (Cr^{3+}), is the dominant form of Cr at pH <4. Trivalent Cr has a limited hydroxide solubility and forms strong complexes with ions such as Cl^-, CN^-, F^-, NH_4^+ and SO_4^{2-} making it relatively immobile and less bioavailable.

Mercury

Mercury belongs to same group of the periodic table with Zn and Cd. It is the only liquid metal at normal temperatures. It has atomic number 80, atomic weight 200.6, density 13.6 g cm^{-3}, melting point −13.6 °C, and boiling point 357 °C and is usually recovered as a byproduct of ore processing (Wuana and Okieimen 2011). Release of Hg from coal combustion and from manometers at pressure measuring stations along gas/oil pipelines as well as extraction of gold using mercury are major sources of Hg contamination. After release to the environment, Hg usually exists in mercuric (Hg^{2+}), mercurous (Hg_2^{2+}), elemental (Hg0), or alkylated forms (methyl/ethyl mercury). The redox potential and pH of the system determine the forms and stability of Hg. Mercurous and mercuric Hg are more stable under aerobic conditions. Under slightly reducing conditions Hg may be reduced to elemental Hg, which may then be converted to alkylated forms by biotic or abiotic processes. Mercury is most toxic in its alkylated forms which are soluble in water and volatile in air (Smith et al. 1995). Bivalent Hg forms strong complexes with a variety of both inorganic and organic ligands, making it very soluble in oxidized aquatic systems (Bodek et al. 1988). Sorption to soils, sediments, and humic materials is an important mechanism for the removal of Hg from soil solution. Sorption of Hg increases as pH increases. Under anaerobic conditions, both organic and inorganic forms of Hg may be converted to alkylated forms by sulfur reducing bacteria. Elemental Hg may also be formed under anaerobic conditions by demethylation of methyl Hg, or by reduction of bivalent Hg. Acidic conditions (pH <4) also favor the formation of highly toxic methyl mercury, whereas higher pH values favor precipitation of HgS (Smith et al. 1995).

Mercury is one of the most volatile metals known and once it gets evaporated, it becomes a colourless, odourless gas. In the atmosphere, it can be transformed into its various forms through abiotic and biogeochemical processes (Gochfeld 2003). Mercury has different common species which can have their own unique impact on the environment. In the atmosphere, mercury can be present in a gaseous phase, incorporated with atmospheric precipitation or associated with air borne particulate matter. In the gaseous phase Hg has been operationally divided into gaseous elemental Hg and gaseous oxidized Hg.

Nickel

Nickel is a transition element with atomic number 28 and atomic weight 58.69 (Wuana and Okieimen 2011). Nickel commonly exists in oxidation states between −1 and +4. The most important oxidation state of Ni is +2 and under acid conditions Ni^{2+} is very common (Pourbaix 1974). In neutral to alkaline solutions, it precipitates as nickelous hydroxide, $Ni(OH)_2$, which is a stable compound. This precipitate readily dissolves in acid solutions forming Ni(III). Under very alkaline conditions it forms nickelite ($HNiO_2$) which is water-soluble. In very oxidizing and alkaline conditions, nickel exists as stable Ni_3O_4 which is soluble in acid solutions. Other nickel oxides such as nickelic oxide (Ni_2O_3) and nickel peroxide (NiO_2) are unstable in alkaline solutions.

Nickel is resistant to corrosion by air and water under ambient conditions and combines readily with other metals including Cr, Cu, Fe and Zn to form alloys (Kabata-Pendias and Mukherjee 2007). Nickel is found in the environment combined primarily with oxygen or sulphur as oxides or sulphides.

Lead

Lead is a metal belonging to group IV and period 6 of the periodic table with atomic number 82, atomic mass 207.2, density 11.4 g cm^{-3}, melting point 327.4 °C, and boiling point 1725 °C (Wuana and Okieimen 2011). It is a naturally occurring, bluishgray metal usually found as a mineral combined with other elements, such as sulphur (i.e., PbS, $PbSO_4$), or oxygen ($PbCO_3$). Lead released to groundwater, surface water and land is usually in the form of elemental Pb, lead oxides and hydroxides, and lead metal oxyanion complexes (Smith et al. 1995). Lead occurs most commonly with an oxidation state of 0 or +2. Lead as Pb^{2+} is reactive and forms mononuclear and polynuclear oxides and hydroxides. Low solubility compounds are formed by complexation with inorganic anions (Cl^-, $CO_3{}^{2-}$, $SO_4{}^{2-}$, $PO_4{}^{3-}$) and organic ligands (humic and fulvic acids, EDTA, amino acids) (Cynthia and David 1997). Lead carbonate is formed above pH 6 in the presence of Ca^{2+}. Lead as PbS occurs when high sulphide is present under reducing conditions. The primary processes influencing the fate of Pb in soil include adsorption, precipitation and complexation with organic molecules. These processes limit the amount of Pb that can be leached into surface water or groundwater. The relatively volatile organo lead compound tetramethyl lead may occur in anaerobic sediments as a result of alkylation by microorganisms (Smith et al. 1995). The amount of dissolved Pb in surface water and groundwater depends on pH, the concentration of dissolved salts and the types of mineral surfaces present. In aquatic systems, a significant fraction of Pb is undissolved and occurs as precipitates ($PbCO_3$, Pb_2O, $Pb(OH)_2$, $PbSO_4$), sorbed ions or surface coatings on minerals, or as suspended organic matter.

Zinc

Zinc is a transition metal with the following characteristics: period 4, group IIB, atomic number 30, atomic mass 65.4, density 7.14 g cm^{-3}, melting point 419.5 °C, and boiling point 906 °C (Wuana and Okieimen 2011). Zinc does not occur naturally in elemental form and is usually extracted from mineral ores to form

ZnO. Zinc usually occurs in the +II oxidation state and forms complexes with a number of anions, amino acids and organic acids. Zn may precipitate as Zn $(OH)_2(s)$, $ZnCO_3(s)$, ZnS(s), or $Zn(CN)_2(s)$. Zinc is one of the most mobile potentially toxic elements in surface waters and groundwater because it is present as soluble compounds at neutral and acidic conditions. At higher pH values, Zn can form carbonate and hydroxide complexes which control its solubility. Zinc readily precipitates under reducing conditions and in highly polluted systems when it is present at very high concentrations, and may co-precipitate with hydrous oxides of iron or manganese (Cynthia and David 1997). Sorption to sediments or suspended solids, including hydrous iron and manganese oxides, clay minerals, and organic matter, is the primary fate of Zn in aquatic environments.

8.4 Pathways of Toxic Elements

Toxic elements follow definite pathways in the ecosystem. The main pathways by which humans and animals can ingest toxic elements occur from the soil to soil-water to the plants which are consumed. Humans and animals can also ingest contaminants directly through drinking water. Another pathway could be via contaminated water that runs off to the river and the sea. The water contaminates the ecosystem of sea foods and fish which may be consumed by humans. A simplified scheme representing the pathways in the food chain is given in Fig. 8.8.

Elevated concentrations of toxic elements in soils not only pose significant risk to flora and fauna but also to humans. Toxic elements negatively affect crop growth because of their interference with metabolic processes which sometimes may lead

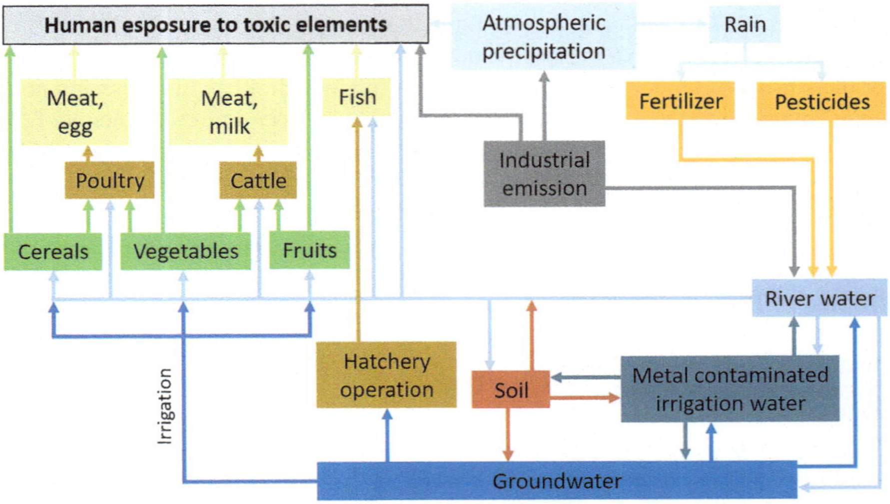

Fig. 8.8 Pathways of toxic elements in the food chain and human exposure

to plant death (Pal and Rai 2010). Humans are at risk through consumption of food grown on polluted soil. In view of the health risks posed by metals entering the food chain, the elements Ag, Cr, Sn and Ti may pose little risk because owing to their low solubility in soil, uptake and translocation by plants may be negligible (McLaughlin et al. 1999). Elevated concentrations of these elements in foods usually indicate direct contamination by soil or dust. The elements As, Hg and Pb are strongly sorbed by soil colloids. While they may be absorbed by plant roots, they are not readily translocated to aboveground plant tissues and therefore pose risks to human health only when root vegetables are grown on contaminated sites. In contrast, elements such as Cd, Cu, Mn, Ni and Zn are readily taken up by plants (McLaughlin et al. 1999).

8.4.1 Toxic Elements in Soil, Water, Air, Plants and Food Products

Arsenic in Soil

The mean concentration of As in the upper earth crust may be around 6 mg As kg^{-1} (Bissen and Frimmel 2003). These concentrations vary largely between different types of soil-forming rocks (Table 8.3). The As content of natural soils averages 5–6 ppm, but may range from 0.2 to 40 ppm (Jones 2007). Contents of As in soils are not only dependent on the geological background but also on anthropogenic activities. Table 8.9 shows ranges of As contents in soils and sediments found in different countries. The mobility of different As species depends on a number of factors (see 8.3.1 ff).

The highest values were found in China, followed by India, Japan and Italy. Mining is one of the main sources of high As accumulation in soils. Extremely high As contents were recorded in mining areas in Thailand, Ghana, Zimbabwe, England, Greece and Canada, probably because As is a by-product of mining and smelting of Cu, Pb, Co and Au ores (Bissen and Frimmel 2003). Near smelting

Table 8.9 Concentrations of arsenic in soils and sediments

Country	As concentrations (mg kg^{-1})
Argentina	0.8–22
Bangladesh	9.0–28
China	0.01–626
Germany	2.5–4.6
India	10–196
Italy	1.8–60
Japan	0.4–70
Mexico	2–40
South Africa	3.2–3.7
USA	1.6–72

Adapted from Arain et al. (2009)

operations and around older orchards where arsenical pesticides were used, soil levels of 100–2500 ppm As have been found (WHO 2000). Although it is estimated that about 80% of the total amount of anthropogenic (or man-made) As released into the environment resides in soil, most As compounds remain in particulate form and adsorb to soil particles being transported via leaching only short distances in the soil.

Arsenic in Water

In natural waters, As concentrations may vary between a few μg L^{-1} to hundreds of mg L^{-1} (Arain et al. 2009). Contamination of groundwater by As has been reported from various countries of the world. The occurrence of As in groundwater of Mexico (La Comarca Lagunera) is a result of the mineral weathering in rocks and soils producing high levels of As in well water (Parga et al. 2005). About 200,000 people were estimated to have been poisoned in India (West Bengal). Other cases were reported from Bangladesh, where geothermal activities in the Taupo volcanic zone increased As concentration in lakes and rivers (Aggette and Spell 1978). Groundwater As contamination has also been reported from Nepal (Shresther et al. 2003) and Vietnam, where several million people after consuming untreated groundwater run a considerable risk of chronic As poisoning (Berg et al. 2001). Arsenicosis (arsenic poisoning) might have become one of world's biggest environmental health disasters. Some heavy incidents of well water contamination with As are listed in Table 8.10.

Arsenic in Air

Arsenic concentrations of <1–3 ng m^{-3} have been found in air in remote areas, whereas 20–30 ng m^{-3} has been detected in urban areas without substantial industrial emissions, yielding estimated daily As intakes of 20–200 and 400–600 ng, respectively (WHO 2001).

Table 8.10 Heavy incidents of well water contamination

Country/location	Year	As concentrations (μg L^{-1})
Canada/Ontario	1937	100–410
Hungary	1941–1981	60–4000
Argentina/Cordoba (M. Quemado)	1955	100
North Mexico/Lagunera Region	1963–1983	8–624
USA/Oregon	1962	50–1700
Canada/Nova Scotia	1976	>3000
China/Xing-Jiang	1980	Up to 850
India/West Bengal	1983	Up to 3800
Bangladesh	1995	Up to 4730
Nepal	2001	Up to 2620
Vietnam	2001	Up to 3050

Adapted from Rahman et al. (2001)

Arsenic in Plants and Food Products

It has long been known that soils with elevated As levels (i.e., $\geq$20 ppm) produce plants with increased As levels (Jones 2007). However, it should be recognized that many of these studies were conducted with soils containing >500 ppm, whereas most soils contain $\leq$10 ppm As (Jones 2007). Concentrations of As may be 10–1000 times greater in soil than in plants growing on that soil. Moreover, the distribution of As among various plant parts is highly variable, with seeds and fruits having lower As concentration than leaves, stems, or roots. Roots and tubers generally have the highest As concentrations, with the skin having higher concentrations than the inner flesh (Peryea 2001). The edible parts of vegetables seldom accumulate high concentrations of As, because most plants will be killed or severely stunted long before the As concentration in their tissues reaches concentrations that pose a health risk (Jones 2007; Ontario Ministry of the Environment 2001). Compared to food crops and animal products, much higher concentrations of As were found in seafood (Table 8.11).

Cadmium in Soil

Cadmium in soil may exist in soluble form in soil water, or in insoluble complexes with inorganic and organic soil constituents. Cadmium in soil tends to be more available when the soil pH is low. Background cadmium levels in surface soils

Table 8.11 Contents of As in common foods

Food product	As concentrations (μg g^{-1})
Food crops	
Apples	0.04–1.72
Grapes	0.75–1.20
Lettuce	0.01–3.78
Orange juice	0.008–0.12
Oats	<0.1–2.28
Potatoes	0.0076–1.25
Rice grains	<0.07–3.53
Soybeans	0.05–1.22
Tomatoes	0.01–2.95
Wheat flour	0.01–0.09
Animal products	
Beef	0.008
Chicken	0.02
Pork	0.22–0.32
Eggs	0.005
Milk	0.0005–0.7
Seafood	
Crab	27.0–52.5
Oysters	0.3–3.7
Tuna	0.71–4.6

Adapted from Jones (2007) and National Academy of Sciences (2001)

range from 0.01 to 2.7 mg kg^{-1}, though values up to 1781 mg Cd kg^{-1} soil have been reported from very contaminated sites (Kabata-Pendias and Pendias 2001). In Europe, the mean cadmium concentration in cultivated soils is 0.5 mg kg^{-1} (Davister 1996).

Elevated Cd concentrations in soils (compared to background values) have been reported following the application of inorganic P fertilizers, sewage sludge and farmyard manure (EC 2001; Bergkvist et al. 2003). Since cadmium is taken up by plants, an increased soil concentration can result in increased levels in food and feeds. However, the concentration of cadmium in soils is not the primary determinant of cadmium in plants. Some studies have concluded that soil pH is the major factor influencing plant uptake of cadmium from soils (Smith 1994). Soil type also affects uptake of cadmium by plants. For soils with the same total cadmium content, Cd has been found to be more more plant-available in sandy soil than in clay soil (He and Singh 1994). Similarly, Cd mobility and bioavailability are higher in noncalcareous than in calcareous soils (Thornton 1992).

Cadmium in Water

Cadmium in drinking water does not contribute much to the total human intake of cadmium. The cadmium concentration in ground water increases with decreasing pH. In surface water and groundwater, cadmium can exist as free ion, or as ionic complexes with other inorganic or organic substances. In seawater, the most common forms are chlorine ion complexes, and in freshwater the free hydrated or carbonated ions (depending on the pH) are the most frequent forms (EFSA 2009). Concentrations of Cd up to 5 mg kg^{-1} have been reported in sediments from rivers and lakes, and from 0.03 to 1 mg kg^{-1} in marine sediments. The average cadmium content of seawater is about 5–20 ng L^{-1} in open seas, but increased concentrations of 80–250 ng L^{-1} have been reported in French and Norwegian coastal zones. Concentrations measured in European rivers generally vary between 10 and 100 ng L^{-1} (OSPAR 2002).

Cadmium in Air

Cadmium concentrations in ambient air are generally low. Industrial activities are the main sources of cadmium release to the air. In Europe and North America, emissions of metallic elements have shown a decreasing trend over the last decades, as a consequence of the reduction in coal consumption, the improvement of industrial manufacturing processes and the tightening of environmental legislation. In urban areas of the EU, Cd concentrations in air are in the range between 1 and 10 ng m^{-3} (EFSA 2009). Atmospheric Cd is in the form of particulate matter, which may consist of very small particles (<10 μm) if it is produced by combustion processes. The principal chemical species in air is Cd oxide, although some Cd salts, such as Cd chloride, can enter the air, especially during incineration (IARC 1993). These are stable compounds that do not undergo significant chemical transformation. Cadmium pollutants present in the air may be transported from a hundred to a few thousand kilometers and have a typical atmospheric residence time of about l–10 days before deposition occurs by wet or dry processes (ATSDR 1999).

Table 8.12 Concentrations of Cd (means) in selected food products

Food product	Mean Cd concentration (μg g^{-1})
Food crops	
Cereals and cereal products	0.023
Vegetables, nuts and pulses	0.067
Starchy roots or potatoes	0.021
Fruits	0.0039
Fruit and vegetable juices	0.0016
Coffee, tea, cocoa (expressed as liquid)	0.0041
Animal products	
Meat and meat products	0.0974
Eggs	0.0030
Milk and dairy products	0.0046
Seafood	
Fish and seafood	0.0923

Data from 18 EU member states, Iceland and Australia; adapted from EFSA (2009)

Cadmium in Plants and Food Products

Cadmium uptake has been reported for food crops, grasses, poultry, cattle, and wildlife (ATSDR 1999). Generally, Cd accumulates particularly in the leaves of plants and, therefore, is more of a risk in leafy vegetables grown in contaminated soil than in seed or root crops. He and Singh (1994) reported that, for plants grown in the same soil, accumulation of Cd decreased in the order leafy vegetables > root vegetables > grain crops. In virtually all foods Cd is present, but the concentrations vary to a great extent, depending on the type of food. Thus there are great individual variations in the dietary Cd intake due to differences in dietary habits. The amounts of Cd ingested daily with food in most countries are in the range of 10–20 μg day^{-1} (Bernard 2008). Table 8.12 shows data of aggregated food categories from the Concise European Food Consumption Database of EFSA (2009).

Cadmium accumulates in freshwater and marine animals to concentrations hundreds to thousands of times higher than in the water (ATSDR 1999). Bioconcentration factors range from 113 to 18,000 for invertebrates, from 3 to 4190 for fresh water aquatic organisms, and from 5 to 3160 for saltwater aquatic organisms (EFSA 2009).

Copper in Soil

Most Cu deposited in soil from the atmosphere, agricultural use, and solid waste and sludge disposal will be strongly adsorbed and remain in the upper few centimeters of soil. Copper's movement in soil is determined by a host of physical and chemical interactions between Cu and soil components. In general, Cu will adsorb to organic matter, carbonate minerals, clay minerals, or hydrous iron and manganese oxides. As a result of this, even at high concentrations, Cu hardly ever enters groundwater (Cynthia and David 1997). Acute inhalation of Cu dust or fumes at concentration of 0.075–0.12 mg m^{-3} may cause metal fume fever with symptoms such as cough, chills and muscle ache (Yasir et al. 2009).

Copper in Water

Copper is found in surface water, groundwater, seawater and drinking water, and is primarily present in complexes or as particulate matter (ATSDR 2002). Concentrations of Cu in surface waters ranged from 0.0005 to 1.0 mg L^{-1} in several studies in the USA; the median value was 0.01 mg L^{-1} (ATSDR 2002). In the UK, the mean Cu concentration in the River Stour was 0.006 mg L^{-1} (range 0.003–0.019 mg L^{-1}). Background levels in the upper catchment control site were 0.001 mg L^{-1}. Concentrations of Cu were four-fold increased downstream of a sewage treatment plant (IPCS 1998). In an unpolluted zone of the River Periyar in India, Cu concentrations ranged from 0.0008 to 0.010 mg L^{-1} (IPCS 1998).

Concentrations of Cu in drinking water vary widely as a result of variations in water characteristics, such as pH, hardness and copper availability in the distribution system. A number of studies in Europe, Canada and the USA indicate that Cu levels in drinking water can range from $\leq$0.005 to >30 mg L^{-1}, with the primary source most often being the corrosion of interior copper plumbing (WHO 2004). Drinking water may contribute 0.1–1 mg of the daily Cu uptake in most situations. Consumption of standing or partially flushed water from a system that includes copper pipes may considerably increase the daily copper exposure, especially for infants fed formula reconstituted with tap water (WHO 2004).

Copper in Air

In the atmosphere, Cu is present from wind dispersion of particulate geological materials and particulate matter from smokestack emissions (WHO 2004). Alltogether, these sources may account for only 0.4% of the copper released into the environment (Barceloux 1999a). In a study by the US EPA (1987) for the years 1977–1983, the range of copper concentrations in 23,814 air samples from the USA was 0.003–7.32 μg m^{-3}. Concentrations of copper determined in over 3800 samples of ambient air from 29 sites in Canada during 1984–1993 averaged 0.014 μg m^{-3} (IPCS 1998).

Copper in Plants and Food Products

Food is the primary sources of copper exposure in developed countries. Major sources of Cu include liver and other organ meats, seafood, nuts and seeds (IOM 2001). In the USA about 40% of dietary Cu originates from yeast breads, potatoes, tomatoes, cereals, dried beans, lentils and beef (Subar et al. 1998). The average copper requirements are 12.5 μg kg^{-1} of body weight per day for adults and about 50 μg kg^{-1} of body weight per day for infants (WHO 1996b). In general, dietary Cu intakes for adults range from 1 to 3 mg day^{-1} (IPCS 1998; IOM 2001). Together with the uptake with water, daily copper intakes for adults usually range from 1 to 5 mg day^{-1}. The IOM (2001) recommended 10 mg day^{-1} as a tolerable upper intake level for adults from foods and supplements. Levels of Cu in different foods are reported in Chap. 7.

Chromium in Soil

Levels of Cr in soil vary according to area and the degree of contamination from anthropogenic Cr sources. Concentrations may range from 1 to 1000 mg Cr kg^{-1}

soil, with an average concentration ranging from 14 to about 70 mg kg^{-1} (US EPA 1984). Surface runoff from soil can transport both soluble and bulk precipitate of chromium to surface water. Soluble hexavalent and trivalent Cr complexes in soil may leach into groundwater. Lower soil pH may facilitate leaching of acid-soluble Cr^{6+} compounds. Chromium has a relatively low mobility for translocation from roots to aboveground parts of plants.

Chromium in Water

The Cr input from the atmosphere to surface waters is commonly low because average concentration of Cr in rainwater ranges between 0.2 and 1.0 μg L^{-1} (WHO 2003). Most surface waters contain between 1 and 10 μg of Cr L^{-1}. Chromium is mainly introduced into natural waters due to the discharge of a variety of industrial wastewaters. In surface waters of industrialized areas of the USA, levels up to 84 μg Cr L^{-1} have been found (WHO 2003). The groundwater concentration of Cr in the USA is generally low, with measurements in the range of 2–10 μg L^{-1} in shallow groundwater (Deverel and Millard 1988), however, in some supplies levels as high as 50 L^{-1} have been reported (WHO 2003). In shallow groundwater in California, median levels of 2–10 μg L^{-1} have been found. In the Rhine, Cr levels were below 10 μg L^{-1} (RIWA 1989). In 50% of the natural stream waters in India the Cr concentration was below 2 μg L^{-1} (Handa 1988). The Cr concentration in most groundwaters does not exceed 1 μg L^{-1}. In the Netherlands, a mean concentration of 0.7 μg L^{-1} has been measured, with a maximum of 5 μg L^{-1}, and in India, 50% of 1473 water samples from dug wells contained less than 2 μg L^{-1} (WHO 2003; Handa 1988).

Chromium in Air

As chromium is almost ubiquitous in nature, in the air it may originate from wind erosion of shales, clay and other kinds of soil materials. Both trivalent and hexavalent Cr are released into the air. In countries where chromite is mined, production processes may constitute a major source of airborne Cr. Chromium in air is present in the form of aerosols. It is commonly removed from the atmosphere by wet and dry deposition. The residence time of Cr in the atmosphere is expected to be <10 days (Nriagu and Pacyna 1988). Most of the Cr in lakes and rivers will ultimately be deposited in the sediments. Data on Cr speciation in ambient air are rarely available, but the proportion present as hexavalent Cr may range between 0.01% and 30% (WHO 2003). In arctic air, Cr concentrations of 5–70 pg m^{-3} have been measured. Ambient air concentrations in the USA were generally below 300 ng m^{-3}, and median levels less than 20 ng m^{-3}. In non-industrialized areas, concentrations above 10 ng m^{-3} are uncommon (National Academy of Sciences 1980). Concentrations in urban areas are two to four times higher than regional background concentrations (Nriagu and Nieboer 1988). Ranges of Cr levels in member states of the European community were given as 0–3 ng m^{-3} for remote areas, 4–70 ng m^{-3} for urban areas, and 5–200 ng m^{-3} for industrial areas (Lahmann et al. 1986).

Table 8.13 Concentrations of Cr in selected food products

Food product	Cr concentrations (μg g^{-1})
Food crops	
Vegetables	0.03–0.23
Fruits	0.02–0.19
Grains and cereals	0.04–0.22
Animal products	
Meat	0.11–0.23
Dairy products	0.10
Fish and seafood	
Fish	0.05–0.16
Seafoods	0.12–0.47

Adapted from US EPA (1984)

Chromium in Plants and Food Products

The intake of Cr from typical North American diets was found to be 60–90 μg day^{-1} (US EPA 1984) and may be generally in the range 50–200 μg day^{-1}. Levels of Cr of different foods are reported in Table 8.13.

Depending on the diet, the Cr intake from food and water may range between 52 and 943 μg day^{-1} (WHO 2003).

Mercury in Soil

Mercury is a global environmental concern. Centuries of human activities have been mobilizing increasing amounts of Hg in the atmosphere, ocean, and terrestrial systems. Soil represents the largest store of Hg in terrestrial ecosystems. Most part of the Hg on the global land surface has been deposited as bivalent Hg(II) from the atmosphere. A portion of the deposited mercury will revolatilize to the atmosphere and the remainder will be incorporated into the soil, mainly into SOM where it binds strongly to reduced sulfur groups (Skyllberg et al. 2003) and, to a lesser extent, via sorption on clays. Figure 8.9 shows the main pathways in the global Hg cycle.

Mercury is naturally present in soils at concentrations ranging between 0.003 and 4.6 μg g^{-1} soil (Steinnes 1997). In contaminated sites, concentrations of up to 11,500 have been observed (e.g. Gray et al. 2002; Neculita et al. 2005). In these sites Hg accumulation is particularly due to surface spills, wastewater discharge, and/or by condensation of atmospheric Hg (Biester et al. 2002). Maximum sorption onto soil organic surfaces occurs in the range of pH 3–5, whereas as pH increases, sorption decreases, mainly because of the increase in dissolved organic matter complexed with Hg (Yin et al. 1996).

Mercury in Water

Contamination of freshwater systems with Hg is widespread due to both point source and diffuse (from the atmosphere and through surface runoff from watersheds) deposition. Wet and dry deposition to surface waters is predominantly as bivalent Hg which can reduce to Hg(0) and may then volatilize to the atmosphere. A small portion of bivalent Hg biologically converted to the more toxic form of

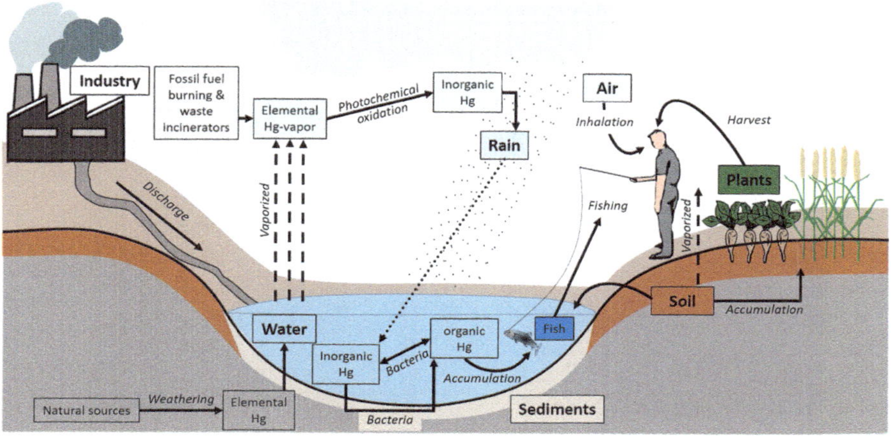

Fig. 8.9 The global Hg cycle including humans

methyl mercury (Selin 2009). Wetlands and lake sediments are important environments where methylation occurs. Methyl mercury can accumulate in living organisms and further concentrate up the food chain. This process of bioaccumulation means that methyl mercury concentrations in predatory fish can be elevated relative to water by a factor of $\geq 10^6$ (Engstrom 2007).

Mercury in Air

Mercury enters into the atmosphere through volcanic and geological activity in its elemental form (Hg(0)). Anthropogenic sources can mobilize Hg in two different forms, bivalent mercury (Hg^{2+}, ${Hg_2}^{2+}$) and Hg associated with particulate matter (Hg(P)). The most abundant form of Hg in the atmosphere is Hg(0), with a global mean concentration of about 1.6 ng m^{-3} in surface air (Selin 2009). With an atmospheric lifetime of about 0.5–1 year, Hg(0) is well distributed over the global atmosphere (Fitzgerald and Lamborg 2005). Because bivalent Hg and Hg(P) are more soluble in water than Hg(0), they are the dominant forms of Hg deposited to ecosystems through wet and dry deposition. Bivalent Hg and Hg(P) have a short lifetime in the atmosphere (days to weeks). Their surface concentrations range from 1 to 100 pg m^{-3} (Selin 2009). Quantification of Hg deposition to ecosystems up to now is very limited because measurement networks exist only in a few regions.

Mercury in Plants and Food Products

Although some individuals are primarily exposed to Hg sources other than food (e.g. workplace), for most people the main route of exposure to these toxic elements is through the diet. It is well established that there are notable differences in both food consumption and food contamination by Hg. An overview of Hg contents in food products is given in Table 8.14.

Crop levels of Hg were generally low (Table 8.15) and the corresponding soil contents of Hg were in the range of those commonly found, with a median value of

Table 8.14 Contents of Hg in food products reported in studies from countries in Europe, Canada and the USA

Food product	Hg concentrations ($\mu g\ g^{-1}$)
Food crops	
Lettuce	0.0005–0.0028
Tomato	0.0001–0.008
Cucumber	0.0001–0.002
Potato	<0.0001–0.017
Wheat	<0.0001–0.029
Barley	0.001–0.030
Oats	<0.0001–0.020
Apple	0.0004–0.003
Animal products	
Meat (different animal species)	0.001–0.005
Liver	0.003–0.045
Kidney	0.007–0.058
Fish and seafood	
Tilefish	0.65–3.37
Swordfish	0.10–3.33
King mackerel	0.30–1.67
Shark	0.05–4.54
Tuna	n.d.[a]–1.30
Crab	0.02–0.24
Lobster	0.05–1.31

Compiled from Wiersman et al. (1986), Vos et al. (1988) and Counter and Buchanan (2004)
[a]*n.d.* not detectable

0.07 μg Hg g^{-1}. At elevated soil contents of Hg, e.g. due to long-term application of heavy metal-contaminated sludge, crop Hg contents may be increased. Cappon (1981) for different crops found significantly increased Hg concentrations on a sludged vs. a control soil (e.g. for broccoli 0.017 vs. 0.003 μg Hg g^{-1}; cabbage: 0.013 vs. 0.0029 μg Hg g^{-1}; lettuce: 0.032 vs. 0.0029 μg Hg g^{-1}; onion: 0.017 vs. 0.0067 μg Hg g^{-1}). In case of lettuce the Hg concentration was elevated by a factor of 10. The percentage of highly toxic methyl Hg in crops was higher on the sludged compared to the control soil and reached up to 33.1% in broccoli.

The meat, liver and kidney values of Hg were not related to the age of the investigated animals and may be regarded as low (Table 8.15). In contrast, Hg may be highly accumulated in some fish species such as tilefish, swordfish, king mackerel and shark Counter and Buchanan 2004). In fish-consuming populations (e.g. in Catalonia, Spain) about one third to one half of the total daily Hg intake (16.7 to μg 18.6 day^{-1}) may be through the consumption of fish (Llobet et al. 2003). Mercury poisoning of humans is commonly the result of consumption of Hg-contaminated food, particularly fish. Microorganisms convert inorganic Hg in aquatic environments to methyl Hg. These organisms are consumed by progressively larger

Table 8.15 Contents of Ni in food products

Food product	Ni concentrations (μg g^{-1})
Food crops	
Salad	<1.4
Carrots	<0.16
Spinach	0.02–2.99
Cabbage	0.01–0.63
Tomatoes	0.01–0.25
Potatoes	0.44
Apples	<0.03
Canned fruits	0.02–1.36
Wheat flour	0.03–0.3
Oatmeal	0.80–4.7
Rice	0.08–0.45
Animal products	
Beef	0.01–0.03
Pork	<0.02
Chicken	0.02–0.24
Liver, kidney (different animal species)	<0.94
Milk	<0.13
Eggs	0.01–0.35
Fish and seafood	
Fish	0.005–0.30

Compiled from Flyvholm et al. (1984)

species. In this way, methyl Hg is biomagnified throughout the food chain reaching its most toxic concentrations in the larger and long-lived fish species which are consumed by humans. The serious health consequences of methyl Hg exposure was dramatically illustrated in 1953, when an epidemic of methyl Hg occurred around Minamata Bay in Japan ("Minamata disease"), from the consumption of fish (Tsubaki and Irukayama 1977).

Nickel in Soil

Nickel is normally distributed uniformly through the soil profile but accumulates at the surface from deposition by industrial and agricultural activities (Cempel and Nikel 2006). Nickel may present a major problem in peri-urban areas, in industrial areas, or even in agricultural land receiving sewage sludge. Its content in soil varies in a wide range from 3 to 1000 mg kg^{-1} (Bencko 1983; Scott-Fordsmand 1997). A survey of soils in England and Wales by McGrath and Loveland (1992) reported a geometric mean concentration of 20 mg kg^{-1}. Urban UK soils were found to contain Ni concentrations in the range 7.07–102 mg kg^{-1}, with a mean value of 28.5 mg kg^{-1} (Environment Agency 2007).

Nickel can exist in soils as inorganic crystalline minerals or precipitates, as chelated metal complexes, as exchangeable Ni or in soil solution (Garrett 2000; Cempel and Nikel 2006). The solubility and mobility of Ni increases with decreasing pH (Kabata-Pendias and Mukherjee 2007; McGrath 1995). Many nickel compounds are soluble at a pH less than 6.5. Compared with other heavy metals such as Cd and Zn, Ni is rather weakly sorbed to clay and iron oxides, and likely to be more mobile (McGrath 1995).

Nickel in Water

Nickel concentrations in drinking water rarely exceed 10 μg L^{-1} (Cempel and Nikel 2006). However, Ni may be leached from nickel-containing plumbing fittings, and levels of up to 500 μg L^{-1} have been recorded in water left overnight in such fittings (Andersen et al. 1983). In drinking water of areas with Ni ore mining, levels of up to 200 μg Ni L^{-1} have been recorded. In the USA, the average level of Ni in drinking-water in public water supply systems was 4.8 μg L^{-1} in 1969. Assuming a concentration of 5–10 μg L^{-1}, a daily consumption of 2 l of drinking-water would result in a daily Ni intake of 10–20 μg. In summary, the incidence of health impairments due to higher intakes of Ni in drinking water is not frequent (WHO 1991).

Nickel in Air

Concentration of Ni in ambient air may show considerable variation and the highest values have been reported from industrialized areas. Typical average concentrations of airborne nickel are 0.00001–0.003 μg m^{-3} in remote areas, 0.003–0.03 μg m^{-3} in urban areas having no metallurgical industry and 0.07–0.77 μg m^{-3} in Ni processing areas (Cempel and Nikel 2006). Concentrations of 0.018–0.042 μg m^{-3} were recorded in eight United States cities (Saltzman et al. 1985). Ranges of 0.010–0.060 μg m^{-3} have been reported in European cities and values of 110–180 μg m^{-3} have been reported from heavily industrialized areas (Bennett 1994). Nickel from man-made sources is probably represented mostly by oxides and sulfates of rather small particle size (median diameter about 1 μm). Assuming a daily respiratory rate of 20 m^3, the amount of airborne Ni entering the respiratory tract may range between 0.1 and 0.8 μg day^{-1} when concentrations are 0.005–0.040 μg m^{-3} in ambient air.

Nickel in Plants and Food Products

Food intake is the major route of Ni exposure for the general population, while inhalation from air, drinking water, oral and dermal routes could serve as secondary sources. Rather little is known about the actual chemical forms of Ni in various foods or whether dietary Ni has distinct "organic" forms with enhanced bioavailability analogous to those of Fe and Cr. Some foods may have obtained Ni during the manufacturing process but in most it apparently occurred naturally (Solomons et al. 1982). Food processing methods apparently add to the Ni levels already present in foodstuffs via contamination from Ni-containing alloys in food processing equipment made from stainless steel, the milling of flour and catalytic hydrogenation of fats and oils by use of Ni catalysts (Clarkson 1988). Levels of Ni in most foodstuffs range from <0.1 to 0.5 μg g^{-1}. Rich food sources of Ni include oat meal, dried beans and peas, nuts, dark chocolate and soya products, and consumption of these products in larger amounts may increase the Ni intake to

900 μg per person day^{-1} or more (Flyvholm et al. 1984). A requirement for Ni has not been conclusively demonstrated in humans. Scattered studies indicate a highly variable dietary intake of Ni but typical daily intake from food ranges from 100 to 300 μg day^{-1} in most countries. Recovery studies indicate an absorption rate of less than 15% from the gastrointestinal tract (Sundermann et al. 1989). Table 8.15 summarizes Ni contents in different food products.

Lead in Soil

Lead is present naturally in soil, though in most regions at relatively low levels. However, soils have been contaminated with Pb worldwide. Soil can be an important source of exposure to Pb particularly for children. The lead content in uncontaminated top soils of remote areas is generally within the range of 10–30 mg Pb kg^{-1}. Due to atmospheric deposition from atmospheric sources, the concentration of Pb in topsoils may vary considerably (US ATSDR 2007). For example, in Europe Pb concentrations in soils are heterogeneously distributed and range between <10 and $>>70$ mg Pb kg^{-1} soil. Among various anthropogenic sources of Pb contamination, industrial emissions and previously used leaded petrol may have the greatest environmental impact (Kabata-Pendias and Mukherjee 2007). Levels of Pb in soils beside roads and in towns were reported to amount to several thousands mg Pb kg^{-1}, and soils adjacent to smelters and battery factories were reported to exhibit up to 60,000 mg Pb kg^{-1} (UNEP 2008). Metal and coal mining as well contribute to Pb releases to land.

Mobility of Pb in soils is very limited. The movement of inorganic Pb and Pb salts to groundwater by leaching is therefore very slow under most natural conditions. Content and quality of SOM, presence of dissolved organic matter and pH influence the mobility of Pb in soils (UNEP 2008). Lead is strongly adsorbed to SOM and clays. There is also a strong association between Pb hydroxide and hydroxides of Fe and Mn (Kabata-Pendias and Mukherjee 2007). Acidic conditions increase Pb solubility. Lead adsorbed in a soil matrix may enter surface waters particularly as a result of erosion of Pb-containing soil particles (US ATSDR 2007).

Lead in Water

Lead in the aquatic environment can occur in ionic form (highly mobile and bio-available), in organic complexes with dissolved humic materials or attached to colloidal particles such as iron oxide or clay (strongly bound) (OECD 1993). The speciation of Pb in water is controlled by a number of factors. In acidic aquatic environments, Pb is typically present as PbSO4, PbCl2, ionic lead and/or cationic forms of Pb hydroxide (US ATSDR 2007). At higher pH ($\geq$7.5), lead forms hydroxide complexes ($PbOH^+$, $Pb(OH)_2$, $Pb(OH)_3^-$, $Pb(OH)_4^{2-}$) and insoluble PbCO3. Tetraalkyl-lead compounds in water, such as tetraethyl-Pb and tetramethyl-Pb undergo photolysis and volatilisation. Degradation proceeds from trialkyl species to dialkyl species, and eventually to inorganic Pb oxides (UNEP 2008).

The mean Pb concentration in worldwide river waters has been reported to be 0.08 μg L^{-1} (Millot et al. 2004). Mean values for Pb concentration in waters for the period 1999–2001 from nine European countries ranged from 0.02 to 14 μg L^{-1}

(EFSA 2010). From 1990 to 2002, sites in the Czech Republic showed a clear decline in lead (from 1.6 μg L^{-1} in 1990 to 0.2 μg L^{-1} in 2002), due to a decrease in deposition, while sites in Germany and Latvia showed no clear trend (EFSA 2010). The values for German sites during these years were in the range from <1 to 2 μg L^{-1} with some instances of higher concentrations, whilst lead concentrations for Latvian sites varied from <0.1 to 0.4 μg L^{-1} (Skjelkvale and Traaen 2003).

In tap water Pb is present to some extent as a result of its dissolution from natural sources, but partly also from household plumbing systems in which the pipes, solder, fittings or service connections to homes contain Pb. In public water distribution systems in the USA, Pb levels averaged 2.8 μg L^{-1} and the median level of lead in drinking-water samples collected in five Canadian cities was 2.0 μg L^{-1} (WHO 2011). In the UK in 1975–1976, there was virtually no Pb in the drinking-water in two thirds of households, but levels were above 50 μg L^{-1} in 10% of homes in England and 33% in Scotland (Quinn and Sherlock 1990). If a concentration of 5 μg L^{-1} in drinking-water is assumed, the total intake of lead from this source can be calculated to range from 3.8 μg day^{-1} for an infant to 10 μg day^{-1} for an adult.

Lead in Air

Lead is emitted into the atmosphere mainly from anthropogenic sources, such as metal production, manufacturing industries and power plants. Levels of Pb in ambient air range from 7.6×10^{-5} μg m^{-3} in remote areas such as Antarctica, to >10 μg m^{-3} near smelters (US ATSDR 2007). The nearly worldwide elimination of leaded gasoline has drastically changed the sources and fluxes of Pb. For example, concentrations of Pb in air, averaged over a number stations in Europe during the period 1990–2003, decreased from about 0.020 to about 0.005 μg m^{-3} (EMEP 2001). Airborne lead in the USA has decreased by a factor of 20 or more since 1980 (US EPA 2003). Airborne Pb in Shanghai dropped from about 0.500 to 0.200 μg m^{-3} after the phaseout of leaded Pb there (Tan et al. 2006). The annual average lead level in air according to WHO (2000) should not exceed 0.5 μg m^{-3}. Lead in the atmosphere primarily exists in particle/aerosol form as $PbSO_4$ and $PbCO_3$ (US ATSDR 2007) but the chemical forms emitted into the atmosphere depend on the source. Coal combustion causes emissions of $PbCl_2$, PbO, PbS and insoluble mineral particles (Wadge and Hutton 1987), whereas oil combustion produces mainly PbO (Kabata-Pendias and Mukherjee 2007). In the atmosphere, Pb particles/aerosols are transformed by chemical and physical processes and are ultimately removed from the atmosphere by dry and wet deposition in terrestrial or aquatic ecosystems (Mishra et al. 2004).

Lead in Plants and Food Products

Food is the major source of human exposure to Pb. In 19 European countries, daily dietary exposure of Pb ranged from 0.36 to 1.24 μg kg^{-1} body weight for adults with average exposure and from 0.73 to 2.43 μg kg^{-1} for high consumers (EFSA 2010). Cereals and cereal products, potatoes, leafy vegetables, meat products, seafood and tap water were identified as the major contributors to lead exposure. Considerable variation exists between countries in the contribution of different food categories. For example, per capita-based Pb intakes were estimated to be 27 μg day^{-1}

Table 8.16 Contents of Pb in food products in Europe

Food product	Pb concentrations ($\mu g\ g^{-1}$)
Food crops	
Lettuce	0.034–0.042
Endive	0.066–0.081
Tomato	0.008–0.016
Cucumber	0.008–0.013
Potato	0.018–0.023
Carrots	0.018–0.021
Broccoli	0.007–0.012
Wheat	0.028–0.035
Rye	0.027–0.037
Oats	0.040–0.047
Corn	0.027–0.032
Rice	0.020–0.032
Apple	0.010–0.015
Animal products	
Livestock meat	0.002–0.024
Poultry	0.002–0.041
Liver (different animal species)	0.014–0.051
Kidney (different animal species)	0.017–0.136
Fish and seafood	
Tuna	0.009–0.025
Mackeral	0.021–0.039
Halibut	0.009–0.015
Carp	0.016–0.026
Sole	0.028–0.048
Crab	0.031–0.042
Lobster	0.025–0.034

Compiled from EFSA (2012)

in Sweden, 66 $\mu g\ day^{-1}$ in Finland, 90 $\mu g\ day^{-1}$ in Belgium and 177 $\mu g\ day^{-1}$ in Mexico (WHO 2011). An overview of Pb contents in different food products is given in Table 8.16.

Zinc in Soil

Zinc is an important component in living cells and participates in biochemical processes. On the other hand, it becomes highly toxic upon its excessive amount. Total concentration of Zn in soils was reported to be in the range of 10–300 mg Zn kg^{-1} (Kiekens 1995). However, only a small proportion of the total Zn content of a non-contaminated soil (<1 mg Zn kg^{-1}) may be present in the soil solution to govern Zn supply to plants. Many soils were contaminated with Zn due to the operation of smelters with outdated pyrometallurgical equipment that emitted a lot of dust and smoke enriched in Zn. In the vicinity of a lead and zinc smelter in Canada, the content of Zn in soils reached 1390 mg kg^{-1} against the background equal to 50–75 mg kg^{-1} (Ladonin 2002). Industrial air dust, Zn fertilizers, and sewage sludge are the main sources of anthropogenic Zn entering soils (Robson 1993). Compared with Zn toxicity, Zn deficiency is also a risk in many soils (Chap. 7). Low total Zn contents are found primarily in sandy soils (developed on sandstones or sandy drift) and in strongly leached tropical soils (e.g. Ferralsols) developed on highly weathered parent materials. Sandy-textured soils have at least 65% sand-sized grains and less than 18% clay in the top 100 cm of the soil profile. In the World Reference Base for Soil Resources (IUSS Working Group WRB 2007), they occur in the soil units: Arenosols, Regosols, Leptosols, and sandy variants of Fluvisols and other soil groups.

Zinc in Water

The concentration of Zn usually does not exceed 10 μg L^{-1} in natural surface waters, and is in the range of 10–40 μg L^{-1} in groundwaters (Elinder 1986). In tapwater, the Zn concentration can be much higher as a result of Zn release from piping and fittings (Nriagu 1980). Low pH, high carbon dioxide content, and low mineral salts content enhance corrosion and release of Zn. In a Finnish survey of 67% of public water supplies, the median zinc content in water samples taken upstream and downstream of the waterworks did not exceed 20 μg L^{-1}, whereas concentrations in tapwater were up to 1.1 mg L^{-1} (WHO 1996a). In summary, drinking water usually makes a negligible contribution to human Zn intake unless high concentrations of Zn occur as a result of corrosion of pipings and fittings.

Zinc in Air

Sources of Zn in the air include wind-blown particles of sea spray, soil and rock. Zinc is also emitted to the atmosphere through human activities such as burning of coal and oil, waste incineration, industrial processes (including non-ferrous metal smelting) and general urban/industrial emissions. Emissions of Zn to the atmosphere can result in highly variable amounts deposited on soils. This mainly affects agricultural land in or near to areas of industrial activity and high urban populations. Long-distance transportation can distribute aerosol-sized particles containing Zn over hundreds or thousands of kilometres (Alloway 2008). Examples include contamination of soils in southern Norway from sources in the UK, deposition in Canada from sources in the North Eastern USA and deposition of soil particles in Northern Canada from sources in China (Alloway 2008).

In rural areas, atmospheric zinc concentrations are typically between 10 and 100 ng m^{-3}, whereas levels in urban areas commonly are within the range 100–500 ng m^{-3} (Nriagu 1980). Evaluations for amounts of Zn deposited in ten European countries give an average value of 217 g ha^{-1} $year^{-1}$ (Nicholson et al. 2003). In these countries, Zn deposition rates range from low values of 20 g ha^{-1} $year^{-1}$ in Finland and 68 g ha^{-1} $year^{-1}$ in Norway to high values of 540 g ha^{-1} $year^{-1}$ in both Germany and Poland. Zn showed the highest amount of deposition of any of the trace elements monitored (As, Cd, Cr, Cu, Hg, Ni, Pb and Zn) (Nicholson et al. 2003). The levels of Zn deposition from the atmosphere in many industrialized countries were probably higher in the past because the introduction of strict emissions limits has brought about a reduction in amounts of pollutants emitted to the atmosphere.

Zinc in Plants and Food Products

Zinc is toxic to humans when it is taken up in high concentrations. On the other hand, Zn is known to be involved in the activity of >300 enzymes in most major metabolic pathways and as an integral component of proteins that regulate DNA transcription (Cousins and McMahoh 2000). Values of 5–22 mg Zn have been reported in studies on the mean daily human intake of Zn in different regions (Elinder 1986). In North America, the average daily intake of Zn from foodstuffs varies between 10 and 15 mg day^{-1} (Solomons 1986). The zinc content of typical diets in Finland was calculated to be calculated to be 16 mg day^{-1} (Varo and Koivistoinen 1980). The recommended dietary allowance for adult men is set at 15 mg day^{-1}, for adult women 12 mg day^{-1} and for children 10 mg day^{-1} (Cousins and Hempe 1990). Grains, vegetables, except for some legumes, and fruit are relatively low in Zn, whereas meat and some seafood contain high concentrations of Zn, and are therefore good sources of dietary Zn. Table 8.17 summarizes Zn contents in different food products.

Table 8.17 Contents of Zn (means) in food products

Food product	Mean Zn concentrations (μg g^{-1})
Food crops	
Ground nuts (flour)	28.0
Kidney beans	15.0
Cow peas	10.0
Chinese cabbage	7.0
Sweet potato	2.0
Cassava	3.0
Maize flour	22.0
Sorghum flour	14.0
Rice	18.0
Animal products	
Beef	51.0
Pork	24.0
Chicken	26.0
Beef liver	52.0
Milk	4.0
Eggs	11.0
Fish and seafood	
Flatfish	6.0
Salmon	5.0
Crustaceans	76.0
Oyster	908.0

Compiled from EFSA (2012)

8.5 Human Exposure, Clinical Effects and Therapy Associated with Toxic Elements

As levels of toxic elements rise in soil, air and water, they may also increase in human bodies, contributing to chronic diseases, cancer, disorders, dementia, and premature aging. For example, heavy metals poison humans by disrupting cellular enzymes (Lindquivist 1995). Trace elements such as Co and Zn are required by the human body in small quantities to perform different functions but when their concentrations exceed a certain limit, they become toxic. Metals like Cd, Pb and Hg have no physiological functions in the human body and are accumulated to a level they become toxic to living organisms. Studies have shown that As, Cd, Ni and Hg are human-toxic carcinogens and that there is no identifiable threshold below which these substances do not pose a risk to human health (Brown and Welton 2008).

The consumption of food contaminated by lead, mercury, arsenic, cadmium and other metals can seriously deplete the body stores of iron, Vitamin C and other essential nutrients leading to decreased immunological defenses, intra uterine growth retardation, impaired psycho-social faculties and disabilities associated with malnutrition (Iyengar and Nair 2000). The intake of arsenic and some heavy metals may cause chromosomal aberration, genetic abnormalities and can lead to cancer. There are many studies indicating the link between heavy metals and cancer. For example, the high concentration of lead in fruits and vegetables in the Van region of Eastern Turkey were found to be related to a high upper gastrointestinal cancer rates (Turkdogan et al. 2003). In Bangladesh, where there is a high concentration of arsenic in groundwater and some foods, various diseases ranging from melanosis, leukomelanosis, keratosis, hyperkeratosis, non-pitting edema, gangrene, and skin cancer are prevalent (Hindmarsh et al. 2002). Table 8.18 gives an overview of major toxicological effects of potentially toxic elements.

8.5.1 Arsenic

Human Exposure

Human exposure to inorganic As can occur through consumption of drinking water, and to a lesser extent, through meats and crop products. Water sources contain mostly inorganic arsenate, though in some locations arsenite may be prevalent (WHO 2001). Children may have additional exposures from ingestion of contaminated soils (e.g., mine tailings) and dust. Smelter workers can have exposure to airborbne As trioxide. Smoking as well is a source of inorganic As. Inorganic As is well absorbed from the gastrointestinal tract and absorbed to a lesser degree through inhalation, but is poorly absorbed dermally (WHO 2001).

Table 8.18 Major toxicological effects of potentially toxic elements

Element	Toxicological effects
As	After several years of low arsenic exposure various skin lesions appear: hyperpigmentation (dark spots), hypopigmentation (white spots) and keratoses of hands and feet. After a dozen or more years skin cancers are expected
	Longer exposures may lead to malfunctioning and cancers of the lungs, kidneys, liver, bladder and prostate, and may cause neurological symptoms of toxicity
Cd	Cadmium exposure causes bone degeneration (osteoporosis), neurological disorder, irritation of the lungs and gastrointestinal tract, kidney damage, abnormalities of the skeletal system and cancer of lungs and prostate
Cr	Chromium exposure causes allergic contact dermatitis, possibly carcinogenic in some cases of workers being exposed, mostly because of Cr^{+6}
Cu	Chronic effects of copper exposure can damage the liver and kidneys. Its excessive concentration can cause vomiting, hematemesis, gastrointestinal distress and hemolytic anemia
Ni	Components like $Ni(CO)_4$, Ni_3S_2, NiO and Ni_2O_3 lead to pneumonitis with adrenal cortical insufficiency, pulmonary oedema, and hepatic degeneration, cancer of the respiratory tract due to chronic inhalation of nickel oxide, pulmonary eosinophilia, asthma, nasal and sinus problems. It can also cause skin rash
Pb	Lead poisoning includes general fatigue, tremors, headache, vomiting and seizures. Also it interferes with hemoglobin synthesis and damages kidney functions. It can damage internal organs, the brain and nervous system. Chronic exposure to Pb induces peripheral neuropathy accompanied by abdominal pain, constipation and microcytic anemia
	Symptoms for children are often reduction in intellectual quotient, hyperactivity and hearing loss
	Symptoms for adults include particularly increased blood pressure, liver, kidney and fertility damage
Zn	High Zn intake may affect cholesterol metabolism. Zinc chloride fumes have caused injury to mucous membranes and pale grey cyanosis, metal fume fever, anemia, pancreas damage and lower levels HDL. Inhalation causes throat dryness, cough, aching, chills, fever, nausea and vomiting

Adapted from Bradl (2005), Brown and Welton (2008), Davydova (2005), Fergusson (1990), Pierzynsky et al. (2005) and Wang et al. (2009)

Clinical Effects

Once in the body, As is widely distributed and reduced to arsenite (oxidation state +3) which is then oxidatively methylated to the monomethylarsonic acid (MMA) and dimethylarsinic acid (DMA). Direct uptake of DMA and MMA may result from exposure to the pesticides cacodylic acid and monosodium methylarsenate. Both methylated forms are excreted particularly with urine (NRC 2001). Inorganic arsenic and its metabolites have elimination half-lives of roughly 2–4 days (Lauwerys and Hoet 2001). Variation in the degree of methylation among persons is related to the susceptibility of As-induced disease and may involve consideration of genetic polymorphisms, dose level, age, selenium, and folate status (Chen et al. 2007; Tseng 2007; WHO 2001). Inorganic forms of As are of high acute toxicity, with arsenite (AS^{3+}) being more toxic than arsenate (AS^{5+}) (WHO 2001).

The reduced form of MMA (oxidation state +3) shows greater toxicity than arsenite itself (Cohen et al. 2006) and thioarsenic metabolites may also be as toxic (Raml et al. 2007). Arsenic in the human body leads to inhibition of numerous enzymes, substitution in phosphate metabolism, interference in signal transduction pathways and alteration of gene expression. These actions may decrease production of energy and increase oxidative stress, cytotoxicity and endothelial injury (Kumagai and Sumi 2007). The presence of arsenite in the body inhibits pyruvate dehydrogenase by binding to the sulfhydryl groups of dihydrolipoamide, and also inhibits succinate dehydrogenase, leading to a decrease in adenosine triphosphate (ATP) generation. Glucose uptake by cells, gluconeogenesis, fatty acid oxidation, and generation of glutathione may be also affected. Although arsenate is reduced in the body to arsenite, it may have its separate toxic action by substituting for phosphate in glycolysis and other pathways, and by uncoupling oxidative phosphorylation (NRC 2001; WHO 2001).

Symptoms resulting from the uptake of large amounts of trivalent As include hemorrhagic gastritis with nausea, vomiting, and diarrhea, which may cause dehydration and shock. Cardiac arrhythmias, hepatotoxicity, renal failure, and peripheral neuropathy may also occur with large doses or after surviving an acute overdose. Chronic human intake of arsenic at less than acutely toxic doses, including drinking water sources with elevated arsenic levels (e.g., Bangladesh, Vietnam), may cause peripheral sensorimotor neuropathies, peripheral vascular disease, noncirrhotic portal hypertension, hematocytopenias, hyperkeratosis, and hyperpigmentation of the skin (NRC 2001; WHO 2001). With chronic exposure, some of these effects may take several years to develop. Chronic elevated arsenic intakes have been associated with diabetes, hypertension, and childhood neurodevelopmental effects (Kapaj et al. 2006; WHO 2001). Chronic arsenic exposure in humans is considered to be a cause of skin, lung, and bladder cancer (NRC 2001). The risk of lung cancer appears more pronounced when exposure occurs through smoking or when large environmental exposures start in childhood. In animal studies, arsenic has been fetotoxic and teratogenic, but generally only at maternally toxic doses (WHO 2001).

It is known that during pregnancy As crosses the barrier of the placenta easily thus exposing the fetus to arsenic contamination. There is thus a possibility of the transfer of As from the mother's body to the milk of the mammary glands. However, As accumulation in breast milk appears to be low, even under high exposure. In Argentine, indigeneous women that were exposed to about 200 μg As L^{-1} in drinking water showed very low concentrations in breast milk of about three (Concha et al. 1998). Fängström et al. (2008) found that mothers that were exposed to high As concentration had a low As concentration in their breast milk. However, in Japan, 100 deaths were recorded among infants in 1955 due to poisoning from powdered milk with an arsenic concentration of 4–7 mg L^{-1} (Dakeishi et al. 2006). A dosis of 0.01–0.05 g As_2O_3 is toxic and 0.3 g is lethal for humans (Reichl and Ritter 2011).

Therapy
Dimercaptopropanol and derivates such as dimercaptosuccinic acid (DMSA) and dimercaptopropanesulfonate (DMPS) are suitable antidotes in humans for poisonings with As. For example, DMPS with As forms a stable, five-membered ring compound which can be more easily voided with the urine. Experiments with animals have shown that administration of colestyramine interrupts the enterohepatic cycling of the As-DMPS complex which enhances As elimination in the feces.

8.5.2 *Cadmium*

Human Exposure
Cadmium is taken up via inhalation and ingestion. Inhalation of cigarette smoke is a predominant source of exposure in smokers. The gastrointestinal absorption of dietary Cd is about 5% in adult men and 10% or higher in women (Horiguchi et al. 2004). Iron deficiency may increase Cd absorption which contributes to higher absorption by women (Diamond et al. 2003).

Clinical Effects
Cough, headache and fever are the first symptoms when Cd compounds are taken up by inhalation (Reichl and Ritter 2011). Chronic exposure causes Cd accumulation in liver and kidneys where it is bound to the metal-binding protein metallothionein. Up to one half of the total body burden accumulates in the kidney tissues (Nordberg and Nordberg 2001) and kidneys show the earliest sign of Cd toxicity. The estimated half-life of Cd in the kidneys ranges between one to four decades (Diamond et al. 2003). High dose chronic exposure may result in renal tubular and glomerular damage, manifested by irreversible proteinuria and progressive reduction in the glomerular filtration rate (Roels et al. 1999). Low level environmental exposure to Cd resulted in associations between higher urine or blood Cd levels and an increased prevalence of various biomarkers of renal tubular effects (Alfven et al. 2002; Olsson et al. 2002). In Japan, during the 1950s and the 1960s, a condition of painful osteoporosis known as "itai itai"affected postmenopausal women living in a Cd-polluted region. Kidney dysfunction that led to osteoporosis was also associated with very high urine Cd levels in residents of an area of China with extensive environmental Cd pollution (Jin et al. 2004; James and Meliker 2013). Advanced renal tubular damage led to increased urinary excretion of calcium and phosphorus and decreased hydroxylation of vitamin D metabolites. Older adults and postmenopausal women with increased urine Cd levels may have an increased risk for bone fracture due to diminished bone mineral density (Alfven et al. 2002).

Heavy exposure to airborne dusts and fumes has resulted in severe pneumonitis (Fernandez et al. 1996). Chronic inhalation exposure to Cd particulates was associated with changes in pulmonary function and chest radiographs that were consistent with emphysema (Davidson et al. 1988). Workplace exposure to airborne Cd particulates was associated with decreases in olfactory function (Mascagni et al. 2003). An inverse relationship was found between Cd in cord blood, maternal blood and birth weight (Nishijo et al. 2002). Cadmium has also been associated with lung cancer in humans.

Therapy
The treatment of Cd poisoning by inhalation or by oral uptake of Cd compounds is mainly symptomatic (Reichl and Ritter 2011). In case of inhalation of Cd vapors, airways should be kept free and fresh air should be applied immediately. After a latency period of several hours, toxic pulmonary edema may develop which cause damage to the alveolar structures. As a first aid, inhalation of high doses of glucocorticoid is indicated. In case of acute intoxication by inhalation, Administration of the British antilewisite (BAL, dimercaptopropanol) is recommended as an antidote. Ingestion of Cd calls for induction of vomiting or gastric lavage. Use of activated charcoal may be helpful. Administration of BAL is not recommended in case of chronic intoxification following inhalation or ingestion because Cd may be redistributed from the tissue to kidneys and may thus damage kidneys.

8.5.3 Copper

Human Exposure
Copper is an essential micronutrient for humans and is required in trace level in order to help haemoglobin formation and carbohydrate metabolism. Both Cu deficiency and Cu overload cause disease symptoms in humans. Symptoms of Cu deficiency are discussed in Chap. 7. Copper is mainly taken up by ingestion but can also be inhalated in form of dust or fumes. Up to 40% of the ingested Cu is absorbed in the gastrointestinal tract.

Clinical Effects
Acute inhalation of Cu may cause metal fume fever with symptoms such as cough, chills and muscle ache (Yasir et al. 2009). Acute Cu toxicity is characterized by lethargy, vomiting, and icterus (Reichl and Ritter 2011). In the blood Cu is mainly bound to albumin and transported to the liver where it is linked to metallothionein. The latter transfers Cu to ceruloplasmin which distributes the Cu to the tissues. Chronic Cu toxicity causes Wilson's disease, an autosomal recessive genetic disease primarily in brain and liver (Brewer and Askari 2005). Patients with liver disease may present with a hepatitis or recurrent hepatitis picture, cirrhosis, or liver failure (Brewer 2001). Neurological symptoms are present as a movement disorder, often with dysarthria, dysphagia, incoordination, tremor, and dystonia occurring in any combination. Often patients present with behavioral abnormalities before developing neurologic symptoms (Brewer 2001). These include depression, loss of emotional control, inability to focus on tasks, loss of inhibitions, and occasionally, bizarre behavior.

Therapy
Anticopper drugs include Penicillamine which is a reductive chelator, and acts to mobilize copper from hepatic and other stores, and cause its excretion in the urine (Brewer and Askari 2005). The usual dose is 1.0 g day^{-1}, given as 500 mg twice daily or 250 mg four times daily. As treatment proceeds the freely available copper pool is lessened. Pyridoxine in a dose of 25 mg day^{-1} must be taken by patients on penicillamine therapy to avoid pyridoxine deficiency.

Triethylene tetramine (TRIEN) is also a chelator and acts similarly to penicillamine, although in a somewhat gentler manner. The dose and precautions about food intake are identical to penicillamine. However, the amount of urine copper is much less, perhaps in the 1.0–3.0 mg day^{-1} range for newly treated patients, rapidly tailing off to 1.0 mg day^{-1} or less. TRIEN is also fully effective in Wilson's disease as long as the patient complies with therapy. Compliance and the copper status of the patient are monitored identically to penicillamine, with non-ceruloplasmin plasma copper being a better monitoring tool than 24 h urine copper.

8.5.4 Chromium

Human Exposure
Chromium as Cr^{3+} ion is regarded as essential for humans and has an important role in the maintenance of the carbohydrate, lipid and protein metabolism. Excessive exposure has shown that chromium can act as an acute irritant, as a carcinogen and as an allergen to humans (Dayan and Paine 2001). Depending on the compound, Cr can be taken up via inhalation, through the gastrointestinal tract and through the skin.

Clinical Effects
The Cr^{6+} ion has been accepted as the principal cause of toxic responses, and Cr^{3+} compounds have been regarded as irritants, but not as carcinogens or allergens. Dermal exposure to chromium containing materials is the cause for chronic ulcers of the skin and acute irritative dermatitis (WHO 1990). The strong acid and powerful oxidising properties of soluble Cr ions are regarded as the primary causes of its irritant action on epithelia.

After human exposure to Cr^{3+} by inhalation, urinary concentrations of chromium were found to be increased indicating respiratory absorption. Inhalation of Cr^{6+} compounds causes marked irritation of the respiratory tract (Dayan and Paine 2001; Haines and Nieboer 1988). Exposure to air containing more than 2 μg Cr m^{-3} for several hours may lead to ulceration and perforation of the nasal septum, bronchitis, pneumoconiosis, rhinorrhea and bronchial asthma (Reichl and Ritter 2011).

Oral uptake of chromic chloride (Cr^{3+}) by humans resulted in 99% of the dose being recovered in feces and only about 0.5% was excreted in urine26 indicating poor absorption of Cr^{3+} following oral ingestion. When Cr^{3+} enters the blood stream, it is taken up selectively by erythrocytes, reduced, and bound to hemoglobin, while Cr^{3+} is bound to plasma proteins, such as transferrin (Aaseth et al. 1982). Reduction of Cr^{6+} occurs during transport in the blood which is consistent with the finding that only $Cr^{3+}$is present in the urine (Mertz 1969). The lethal oral dose for adult humans is considered to be 50–70 mg soluble chromates per kilogram body weight (Dayan and Paine 2001). The clinical features of acute poisoning are vomiting, diarrhoea, hemorrhage and blood loss into the gastrointestinal tract, causing cardiovascular shock (Sharma et al. 1978). If the patient survives for more than about 8 days, the major effects resulting from oral ingestion of toxic doses of chromium are liver and kidney necrosis (WHO 1990). Chronic exposure to hexavalent Cr is reported to induce renal failure, anemia, hemolysis and liver failure. Allergic reactions and hypersensitivity have also been described following Cr exposure (Dayan and Paine 2001).

Therapy
In case of skin contact with soluble Cr compounds, the contaminated part of the skin should be immediately washed with water and $CaNa_2$-EDTA solution in polyglycol should be applied. After ingestion of Cr the chromium should be provoked and subsequently plenty of water should be consumed. Ascorbic acid helps to helps to transform Cr^{6+} into less toxic Cr^{3+}. Dimercaptopropanesulfonate is an effective antidote in case of Cr poisoning.

8.5.5 Mercury

Human Exposure
Elemental Hg is absorbed mainly by inhaling volatilized vapor (Hg^0), and is distributed to most tissues, with the highest concentrations in the central nervous system and in the kidneys (Barregard et al. 1999). After absorption of elemental Hg, it is oxidized to mercurous and mercuric inorganic forms.

Clinical Effects
Blood concentrations decline initially with a rapid halflife of approximately 1–3 days followed by a slower half-life of approximately 1–3 weeks (Barregard et al. 1992). Excretion of Hg occurs mainly through the kidneys (Sandborg-Englund et al. 1998). Peak levels in urine can lag behind peak blood levels by days to several weeks (Barregard et al. 1992).

Only up to 15% of inorganic Hg is absorbed from the gastrointestinal tract (Rahola et al. 1973). In blood the half-life of inorganic Hg is similar to the slow-phase half-life of Hg after inhalation of elemental Hg. Excretion of Hg occurs by renal and fecal pathways. The portion of methyl Hg absorbed from the gastrointestinal tract may amount up to 90% (Miettinen et al. 1971; Reichl and Ritter 2011). Methyl Hg enters the brain and other tissues (Vahter et al. 1994) and subsequently undergoes dealkylation to inorganic Hg. Methyl Hg declines in blood and in the body with a half-life of around 50 days. Most elimination occurs through fecal excretion. Methyl mercury is accumulated by growing hair (Cernichiari et al. 1995), a useful marker of exposure. Transplacental Hg transport has been observed in animals (Kajiwara et al. 1996). In the cord blood Hg levels were observed to be higher than in the mother's blood (Stern and Smith 2003). However, the newborne's levels may decline gradually during a couple of weeks (Bjornberg et al. 2005). Inorganic Hg and methyl Hg is accumulated in human breast milk in relatively low concentrations (Oskarsson et al. 1996) and the Hg levels in breast milk decline during the weeks after birth (Bjornberg et al. 2005).

Acute, high-dose exposure to elemental mercury vapor may cause severe pneumonitis. At levels below those that cause acute lung injury, symptoms of chronic inhalation may include tremor, gingivitis, and neurocognitive and behavioral disturbances, particularly irritability, depression, short-term memory loss, fatigue, anorexia, and sleep disturbance (Smith et al. 1970, 1983). Inorganic Hg is mostly taken up by ingestion. Large amounts may cause irritant effects on the gastrointestinal system (Sanchez-Sicilia et al. 1963). The most prominent effect of absorbed Hg is on kidneys where it accumulates and may cause renal tubular necrosis.

Methyl Hg particularly affects the central nervous system, causing parasthesias, ataxia, dysarthria, hearing impairment, and progressive constriction of the visual fields. Excessive prenatal exposure may result in mental retardation, cerebellar ataxia, dysarthria, limb deformities, altered physical growth, sensory impairments, and cerebral palsy (NRC 2000). Recent investigations have suggested a possible link between chronic ingestion of methyl Hg and an increased risk for cardiovascular disease (Vupputuri et al. 2005).

Therapy

In case of poisoning with inorganic Hg compounds, Administration of the British antilewisite (BAL, dimercaptopropanol) is recommended as an antidote. For poisoning with Hg^0 and organic mercury compounds, BAL is not recommended because the Hg accumulates in the central nervous system (Reichl and Ritter 2011). Dimercaptosuccinic acid (DMSA) and dimercaptopropanesulfonate (DMPS) have proved to be effective antidotes in humans for intoxication with Hg^0, Hg^+ and Hg^{2+}.

8.5.6 Nickel

Human Exposure

Although Ni is vital for the function of many organisms, concentrations in some areas from both anthropogenic release and naturally varying levels may be toxic to humans. Human Ni exposure to air and from drinking water are generally less important than dietary intake and ingestion (Cempel and Nikel 2006). The absorption of Ni is dependent on its physicochemical form, with water-soluble forms (chloride, nitrate, sulphate) being more readily absorbed compared to other forms. The way in which Ni is consumed may greatly affect its bioavailability (Barceloux 1999a, b; Haber et al. 2000). Absorption of Ni is influenced by the amount of food, the acidity of the gut and the presence of dietary constituents, possibly phosphate, phytate, fibres and similar metal ion binding components, which may make Ni much less available for absorption than nickel dissolved in water and ingested on an empty stomach (Cempel and Nikel 2006).

Clinical Effects

Due to its slow uptake from the gastrointestinal tract, ingested Ni compounds are considered to be relatively non-toxic, with the primary symptom being irritation. However, when taken orally in large doses (>0.5 g), some forms of Ni may be acutely toxic to humans. The major target organs for Ni-induced toxicity are the lungs and the upper respiratory tract for inhalation exposure and the kidney for oral exposure. Other target organs include the cardiovascular system, the immune system and blood (Coogan et al. 1989; Nielssen et al. 1999).

The toxic effects of Ni result from its ability to replace other metal ions (mainly Fe^{2+}, Mn^{2+}, Ca^{2+}, Zn^{2+}, Cu^{2+} and Mg^{2+}) in enzymes, proteins or bind to cellular compounds (Cempel and Nikel 2006). The diverse clinical manifestations of nickel toxicology include acute pneumonitis from inhalation of Ni tetracarbonyl ($Ni(CO)_4$), chronic rhinitis and sinusitis from inhalation of nickel aerosols, cancers of nasal cavities and lungs, and dermatitis and other hypersensitive reactions from cutaneous and parenteral exposures to nickel alloys. Early symptoms of Ni intoxication are observed when Ni concentrations of the urine exceed about 100 μg L^{-1} (Reichl and Ritter 2011).

Ingestion of Ni can cause allergic skin reactions in previously sensitised individuals. Nickel is a constituent of jewelry. Direct contact of Ni leads to skin rash which is the most common type of reaction to Ni exposure (Environmental Agency 2010). Recent studies have demonstrated enhanced lipid peroxidation in the liver, kidney, lung, bone marrow and serum, and dose-effect relationships for lipid peroxidation in some organs were observed (Coogan et al. 1989; Chen et al. 1999; Sundermann et al. 1985).

Therapy

Triethylene tetramine (TRIEN) is an effective antidote for Ni intoxication, similar as for copper poisoning (Reichl and Ritter 2011). Co-application of cyclohexylsulfamic acid (calcium cyclamate) enhances Ni elimination. In case of poisoning with Ni tetracarbonyl, Na diethyldithiocarbamate trihydrate (DDTC) is effective. For Ni sensitized individuals, a Ni-deficient diet may help alleviate the symptoms. Among the foods rich in Ni are cereals, legumes, mushrooms, tomatoes, asparagus tea and cacao.

8.5.7 Lead

Human Exposure

Lead absorption by the body occurs through inhalation of fine particulates of fumes or ingestion of soluble Pb compounds via food products, water and soil.

Clinical Effects

In the blood Pb is bound to erythrocytes and then is distributed to tissues and bones which act as a storage depot of Pb. About 40–70% of Pb in blood originates from the skeleton in Pb-exposed adults (Smith et al. 1996). Lead can be transported through the placenta and accumulate in the developing fetal brain. Lead may also be transferred to babies via mother's milk. Ettinger et al. (2004) found a correlation between the Pb concentration in the breast milk of lactating mothers and the corresponding blood Pb of infants. Lead is removed from the blood and soft tissues with a half-life of 1–2 months. It is lost much more slowly from bone tissue, with a half-life of years to decades. About 70% of Pb excretion occurs via the urine, with lesser amounts eliminated via the feces. Much smaller amounts are lost through sweat, hair and nails (Leggett 1993).

The toxic effects of lead result from its interference with the physiologic functions of Fe^{2+}, Ca^{2+} and Zn^{2+}, thus inhibiting enzymes. For example, Pb exposure inhibits the δ-aminolevulinic acid dehydrase (δ-ALA-D), an enzyme that catalyzes the conversion of δ-aminolevulinic acid (δ-ALA) to porphobilinogen (Reichl and Ritter 2011). The δ-ALA levels then rise which is an important diagnostic parameter. Other mechanisms include alterings in gene expression (ATSDR 2007).Some diseases caused by Pb intoxication exhibit clinical symptoms whereas others are sub-clinical, i.e. without any recognizable symptoms. Lead mobilized continuously from the bone tissue into the blood may cause health hazards including mental deficiency. This symptom is more common in children because the barrier between their blood and the brain compared to adults is less developed (Pueschel et al. 1996). The brain of infants is thus most vulnerable to Pb exposure. Chronic Pb exposure can also lead to brain dysfunction in adults (Carton 1988; Ling et al. 2006). A high percentage of children from mothers exposed to Pb contamination suffer seizures and show symptoms of mental retardation and anemia (Carton 1988; Pueschel et al. 1996). Lead can also lead to movement disorders, kidney dysfunction and abnormal perception during fetal growth and development. Prenatal and postnatal Pb exposure also predisposes children to dental caries (Larkin 1997). In children, neurodevelopmental effects may occur at lead levels <10 μg Pb dL^{-1} (Canfield et al. 2003). In adults, neurocognitive effects have been reported at blood concentrations as low as 20–30 μg Pb dL^{-1} (Schwartz et al. 2001). Overt encephalopathy, seizures, and peripheral neuropathy occurred at levels higher than 200–300 μg Pb dL^{-1}. Low level Pb exposure may be associated with small decrements in renal function (Muntner et al. 2003). High dose occupational lead exposure, with blood concentrations >40 μg Pb dL^{-1} may alter sperm morphology, reduce sperm count, and thus decrease fertility (Telisman et al. 2000). In women low Pb exposure may be associated with hypertension during pregnancy, premature delivery, and spontaneous abortion (Bellinger 2005; Borja-Aburto et al. 1999).

Therapy

Effective antidotes for Pb intoxication include chelators such as calcium disodium ethylenediaminetetraacetatic acid ($CaNa_2$ EDTA), D-penicillamine, British antilewisite (BAL, dimercaptopropanol) and dimercaptosuccinic acid (DMSA) (Reichl and Ritter 2011).

8.5.8 Zinc

Human Exposure

Zinc absorption by the human body occurs through inhalation, skin contact, or by ingestion (Plum et al. 2010). Each exposure type affects specific parts of the body and allows the uptake of different amounts of Zn.

Clinical Effects

The so-called metal fume fever, which is mainly caused by inhalation of zinc oxide, is the most commonly known effect of inhaling Zn-containing smoke. This acute syndrome is an industrial disease which mostly occurs by inhalation of metal fumes with a particle size <1 μm (Vogelmeier et al. 1987). Symptoms beginning few hours after acute exposure include fever, muscle soreness, nausea, fatigue, and respiratory effects like chest pain, cough, and dyspnea (Rohrs 1957). The respiratory symptoms have been shown to be accompanied by an increase in bronchiolar leukocytes. Metal fume fever is generally not life-threatening and the respiratory effects disappear within 1–4 days (Brown 1988).

Dermal absorption of zinc occurs, but the number of studies is limited and the mechanism is still not clearly defined (Plum et al. 2010). The pH of the skin, the amount of zinc applied, and its chemical speciation influence the absorption of zinc (Ågren 1990). Among Zn compounds, zinc chloride is the strongest irritant but the irritation does not necessarily indicate a toxic effect of zinc. It should be noted that Zn is a supplement for treatment of wounds and several dermatological conditions (Lansdown 1993). Based on the existing data, it can be concluded that dermal exposure to zinc does not constitute a noteworthy toxicological risk.

The gastrointestinal tract is directly affected by ingested Zn before it is distributed through the body. Zinc salts tend to be corrosive and ingestion can result in injury to the mouth, throat and stomach. Initial symptoms include burning of the mouth and pharynx with vomiting and may be accompanied by erosive pharyngitis, esophagitis, and gastritis.

Therapy

The symptoms of Zn exposure may persist for several days and thereafter, clinical effects, including the the corrosive injury of the gastrointestinal tract, commonly shows regression (Plum et al. 2010). Besides avoiding (further) Zn absorption, no therapy is therefore recommended.

8.6 Measures to Reduce Human Exposure to Toxic Elements

Contamination of soils with toxic elements affects large areas worldwide. Hot spots of contamination are located around industrial sites, close to large cities and in the vicinity of mining and smelting plants. In response to a growing need to address environmental pollution, remediation technologies have been developed to treat contaminated soils mainly using mechanically, physico-chemically or biologically based remediation methods (Yao et al. 2012). Agriculture in these areas faces major problems due to heavy metal transfer into crops and subsequently into the food chain (Puschenreiter et al. 2005).

8.6.1 Approaches to Reduce Toxic Element Uptake by Plants

On soils with low to medium toxic element levels, toxic element concentration in edible plants may be not high enough to cause acute toxicity, but in the long term it may cause chronic toxicity. In regions where remediation technologies are not available, other approaches may help to reduce toxic element accumulation in food crops. Different crops vary in the potential of taking up toxic elements. Table 8.19 gives an overview of the relative uptake of toxic elements by some important food crops.

Due to it's high toxicity and bioavailability, a number of studies focused on the accumulation Cd. The Cd uptake by crop species decreases in the order leaf vegetables > root vegetables > grain crops (Bingham 1979; Page et al. 1987). There exist large differences in the transfer of metals from soil to different plant parts. In order to investigate the relationship between the heavy-metal content of soils and corresponding crops, the transfer factor, defined as the ratio of toxic element concentration in plant to that in soil, is commonly used to assess plants' capability to absorb toxic elements from the soil (Kabata-Pendias and Pendias 1992). Variation in toxic element accumulation was also observed between different plant parts. Except for leaves, the highest concentrations were found in roots, whereas the lowest are commonly observed in seeds (Machelett et al. 1993). Generative organs are less concerned by toxic element accumulation compared to vegetative organs. In a field study Puschenreiter and Horak (2000) found that the transfer factor of Cd from soil to plant was twice as high for straw compared to grain.

In Table 8.20 transfer factors of toxic elements from soil to plant parts of different crop species are indicated. It has been shown that crop selection is effective to reduce the transfer into human food chain. The lowest transfer factors

Table 8.19 Relative uptake of toxic elements by important food crops

Relative uptake level			
Very low	Low	Medium	High
Pea	Maize	Potato	Spinach
Bean	Cauliflower	Cabbage	Mangold
Tomato	Brussels sprouts	Radish	Lettuce
Melon	Broccoli	Red beet	Endive
Pepper	Celery	Turnip	Carrot

Adapted from Kloke et al. (1984)

Table 8.20 Transfer factors (ratio of element concentration in plant ($mg\ kg^{-1}$) to that in soil ($mg\ kg^{-1}$)) of toxic elements from soil to different plant parts

Crop species	Plant part	Cd	Cu	Ni	Pb	Zn
Bean	Seeds	0.08	0.14	0.28	0.04	0.25
Winter rye	Grain	0.16	0.12	0.10	0.01	0.61
Maize	Cob	0.30	0.11	0.15	0.01	0.68
Maize	Straw	1.09	0.10	0.06	0.09	1.53
Potato	Tuber	0.33	0.18	0.14	0.05	0.21
Tomato	Fruit	0.38	0.18	0.15	0.03	0.21
Onion	Tuber	0.47	0.06	0.09	0.01	0.54
Fodder beet	Leaves	5.55	0.25	0.52	0.07	6.04
Fodder beet	Beet body	0.84	0.23	0.28	0.02	1.18
Spinach	Leaves	5.00	0.22	0.25	0.13	1.27
Celery	Leaves	2.82	0.15	0.15	0.04	1.22
Celery	Root	2.09	0.32	0.18	0.04	0.74
Lucerne	Shoot	1.73	0.18	0.60	0.02	1.66

Adapted from Machelett et al. (1993)

for Cd were observed for bean seeds, winter rye grain and maize cob. The highest Cd values were found in leaves of fodder beet, spinach and celery, as well as in roots of several plants. In contrast to Cd, differences in transfer factors for Cu, Ni and Pb were less pronounced. Among the crop species, spinach and other leaf vegetables had generally very high transfer factors. This means that these species should not be cultivated on polluted sites. Crops excluding toxic elements such as beans, cereals or potatoes or cereals are less critical.

The cultivation of industrial plants has been considered as an option for agricultural use of toxic element polluted soils. Among these are fibre plants such as cotton, flax and hemp (Angelova et al. 2004) and energy crops, such as *Salix* trees and reed canary grass (Börjesson 1999).

8.6.2 Incorporation of Amendments for Toxic Element Immobilization

Cation exchange, adsorption, complexation and precipitation are general phenomena that can reveal the primary mechanism in immobilization of toxic elements. Different elements have distinct mobilities and it is not easy to find uniform fixation agents to reduce the bioavailability of toxic elements.

8.6.2.1 Organic Amendments

Organic amendments such as farmyard manure, compost, bio-solids or crop residues may effectively reduce the bio-availability of distinct heavy metals in soils by accelerating the immobilizing processes such as adsorption, precipitation and

Table 8.21 Organic amendments suitable for toxic element immobilization

Amendment	Source	Toxic element	Reference
Straw	Cotton, rice, maize	Cd, Cr, Pb	Suran et al. (1998)
Sewage sludge	Municipal origin	Cd	John and Laerhoven (1976)
Sugar cane bagasse	Sugar cane	Pb	Janusa et al. (1998)
Xylogen	Paper industry	Zn, Pb, Hg	Rei (2000)
Bark sawdust	Timber industry	Cd, Pb, Hg, Cu	Suran et al. (1998)
Cattle manure	Cattle farm	Cd	Bolan et al. (2003)
Poultry manure	Poultry farm	Cu, Zn, Pb, Cd	Ihnat and Fernandez (1996)

complexation (Adriano et al. 2004). An additional advantage of this technique is that most amendments are inexpensive and available in larger quantities (Guo et al. 2006). High organic matter contents in the soil may retain Cd against both leaching and crop uptake (Jones and Johnston 1989). Organic amendments which have been used successfully include straw materials, sewage sludge, sugarcane bagasse, xylogen, bark sawdust and manures (Table 8.21). The use of organic materials is thus an option to reclaim contaminated soils while effectively diverting materials from the waste stream and reusing them (Gadepalle et al. 2007).

8.6.2.2 Inorganic Amendments

Among the numerous inorganic treatment agents, particularly limes, phosphates, and industrial co-products are typical fixation agents (Table 8.22). Liming is a common practice to overcome some of the problems associated with acidic soil conditions. Liming materials such as CaO, $Ca(OH)_2$, $CaMgCO_3$, and $CaCO_3$, and phosphate induced substances including $CaHPO_3$, $Ca(H_2PO_3)_2$, K_2HPO_4, H_3PO_4, and $(NH_4)HPO_4$ are common amendments used for the in situ immobilization in order to decrease the metal concentration in soil solution and extractability and to reduce the metal uptake by plants. Dermatas and Meng (1996) considered that compared to other liming materials CaO was a more effective in immobilizing toxic elements because of its high reactivity and the marked pH effect. Other inorganic amendments suitable for toxic element immobilization include fly ash and slag from thermal power plants. Minerals which have been used successfully include Ca-montmorillonite, bentonite, bauxite, ettringite and synthetic zeolithes.

8.6.3 *Technologies for Remediation of Heavy Metal Contaminated Soils*

Remediation technologies have been developed to treat contaminated soil in face of a growing need to address environmental contamination. Most of the technologies are mechanically or physio-chemically based. The major disadvantages are that

Table 8.22 Inorganic amendments suitable for toxic element immobilization

Amendment	Source	Toxic element	Reference
Liming materials	Lime factory	Cd, Cr, Cu, Hg, Ni, Pb, Zn	Dermatas and Meng (1996) and Bolan et al. (2003)
Rock phosphate	Phosphorus deposit	Pb, Zn, Cd	Basta et al. (2001)
Hydroxiapatite	Biogenic sources	Zn, Pb, Cu, Cd	Ma et al. (1993)
Fly ash	Thermal power plant	Cd, Pb, Cu, Zn, Cr	Ciccu et al. (2003)
Slag	Thermal power plant	Cd, Pb, Zn, Cr	Deja (2002)
Ca-montmorillonite	Mineral deposit	Zn, Pb	Auboiroux et al. (1996)
Bentonite	Mineral deposit	Pb	Geebelen et al. (2002)
Bauxite residue	Bauxite deposit	Cd, Pb	Lombi et al. (2002)
Ettringite	Bauxite deposit	Cd, Cu, Pb, Zn, Cr	Albino et al. (1996)
Synthetic zeolithes	Industrial product	Cd, Cu, Ni, Pb, Zn	Mueller and Pluquet (1997)

Table 8.23 Costs of some remediation technologies

Technology	Costs in US$ Mg^{-1}
Soil replacement/land filling	100–500
Chemical treatments	100–500
Vitrification	75–425
Electrokinetics	20–200
Phytoextraction	5–40

Adapted from Glass (1999)

they are expensive and soil disturbing, sometimes rendering the land useless as a medium for plant growth. Phytoremediation makes use of the naturally occurring processes by which plants and their microbial rhizosphere organisms sequester, degrade or immobilize pollutants for remediating soils contaminated with toxic elements (Marques et al. 2009). The costs of the different remediation technologies vary considerably (Table 8.23).

8.6.3.1 Mechanically or Physio-chemically Based Methods

Technologies mostly used for remediation of toxic element polluted soils are given in Table 8.24.

Table 8.24 Technologies for remediation of toxic element contaminated soils

Technology	Processes involved
Soil replacement	Replacement contaminated soil with clean soil
Soil washing with fresh water	Separation of coarse soil (sand and gravel) from fine soil (silt and clay), where contaminants tend to bind and sorb. The fine fraction must be further treated
Soil washing with reagents	Metal ions are exchanged through inorganic extraction reagents or chelates and then removed from the leachate
Thermal desorption	Contaminated soil is excavated and heated to temperatures such that the boiling point of the contaminants is reached, and they are released from the soil. The vaporized contaminants are to be collected and further treated
Electrochemical remediation	Voltage is applied at two locations of a contaminated soil forming an electric field gradient. The pollutant moves to the poles via electromigration, electroosmotic flow or electrophoresis and then needs to be treated further
Solidification	An inert matrix is used at low temperatures to fix the activity into a form where it cannot readily be dispersed
Vitrification	A powerful source of energy is used to "melt" soil at extremely high temperatures (1600–2000 °C), immobilizing most inorganics into a chemically inert, stable glass product and destroying organic pollutants by pyrolysis
Encapsulation	Physical isolation and containment of the contaminated material. The impacted soils are isolated by low permeability caps or walls to limit the infiltration of precipitation

Compiled from Marques et al. (2009), Peng et al. (2009) and Yao et al. (2012)

Soil Replacement

This is an in situ method by which contaminated soil is replaced by uncontaminated soil. The removed soil needs to be further treated or disposed. Due to the high labour costs, this is suitable for severely polluted soils with small area (Zhou et al. 2004).

Soil Washing

Soil washing is a relatively simple and useful ex situ remediation method, which involves washing water by that toxic elements can be transferred from the contaminated soil to wash solution. Various additives can be employed to enhance the performance of sediment washing. Among these are H_2SO_4 and HNO_3, chelating agents such as ethylenediaminetetraacetic acid (EDTA), diethylenetriamine pentaacetic acid (DTPA) and ethylenediaminedisuccinic acid (EDDS) or surfactants like rhamnolipid (Peng et al. 2009). These additives can enhance solubilization, dispersal and desorption of metals from contaminated soils. This technology is most appropriate for weaker bound metals in form of exchangeable ions, hydroxides, carbonates and reducible oxides. Residual fractions are not affected by the washing process (Ortega et al. 2008). EDTA can effectively remove Cd, Cu, Pb and Zn with the removal efficiencies ranging between 65% and 86% (Polettini et al. 2006). However, due to possible adverse environmental effects, such as enhanced leaching of toxic elements to the groundwater, EDTA is currently under scrutiny (Peng et al. 2009). Similar problems may arise when EDTA or other chelators are used in phytoremediation technologies.

Thermal Desorption
The thermal desorption is on the basis of pollutant's volatility and heat the contaminated soil using steam, microwave, infrared radiation to make the pollutant (e.g. Hg, As) volatile. The volatile elements are then collected using the vacuum negative pressure or carrier gas and achieve the aim of removing the heavy metals (Li et al. 2010). The traditional thermal desorption can be classified into high temperature desorption (320–560 °C) and low temperature desorption (90–320 °C). This technology has advantages of simple process, devices with mobility and the remediated soil being reused. However, limiting factors such as the expensive devices and long desorption time limit its application in the soil remediation (Peng et al. 2009).

Electrochemical Remediation
Electrochemical remediation involves applying a low potential gradient to electrodes that are inserted into the soil and encompass the contaminated zone. When electric fields are applied migration of charged ions occurs. Positive ions are attracted to the negatively charged cathode, and negative ions move to the positively charged anode. For example, under an induced electric potential, the anionic Cr(VI) migrated towards the anode, while the cationic Cr(III), Ni(II) and Cd (II) migrated towards the cathode (Peng et al. 2009). Once the remediation process is over, the contaminants accumulated at the electrodes are to be extracted by methods such as electroplating, precipitation/co-precipitation, pumping water near the electrodes, or complexing with ion-exchange resins (Krishna et al. 2001). Because the electric conductivity is the highest in the fine particles, this method is well suited for fine-textured soils.

Solidification
Solidification is a suitable method for treatment of severely contaminated sites by involving an inert matrix. Examples of such matrices include cements, polymer modified cements, organic polymers and inorganic fixing agents such as solid silica polymers. Since these processes normally increase the volume of waste requiring disposal, they are of less value where there are large volumes of material requiring treatment.

Vitrification
Vitrification is the high temperature equivalent of solidification which involves heating the soil to 1600–2000 °C at which all organic materials are oxidized. The melt after cooling forms a vitreous rock in that the toxic elements are trapped. For ex situ remediation, the energy can be supplied from fossil fuel or heating electrodes. For in-situ remediation, the heat can be applied through electrodes inserted into the contaminated soil. In summary, this technology can immobilize contaminants in an efficient way. However, as vitrification is an energy-intensive process it is more suitable for low volume contaminated material (Fu 2008).

8.6.3.2 Phytoremediation

The term phytoremediation, from the Greek "phyto" ("plant") and the Latin "remedium" ("cure") refers to the use of plants to extract, sequester, and/or detoxify pollutants and has been reported to be an effective, non-intrusive, inexpensive, aesthetically pleasing, socially accepted technology to remediate contaminated soils (Alkorta et al. 2004). Within the field of phytoremediation of toxic element contaminated sites, different categories have been defined such as, among others, phytoextraction, phytostabilization and phytovolatilization.

Phytoextraction

Phytoextraction, also referred as phytoaccumulation, is a suitable approach to remove the contamination from the soil and isolate it without destroying the soil structure and fertility. This technology is best suited for the remediation of diffusely contaminated areas, where pollutants occur surface-near and at relatively low concentrations (Rulkens et al. 1998). Phytoextraction is based on the use of so-called hyperaccumulator plants with exctremely high capacity for metal (As, Cd, Co, Cu, Mn, Ni, Pb, Sb, Se, Tl, Zn) accumulation. Compared to nonaccumulator plants, in shoots of hyperaccumulators, metal levels can be more than 100-fold higher. Thus, a hyperaccumulator plant will concentrate >10 mg kg^{-1} Hg, >100 mg kg^{-1} Cd, >1000 mg kg^{-1} Co, Cr, Cu, and Pb, and $>10{,}000$ mg kg^{-1} Zn and Ni (Wuana and Okieimen 2011). Unil present, over 500 plant species from at least 100 plant families have been reported to hyperaccumulate metals (Sarma 2011). Among these are *Asterraceae*, *Brassicaceae*, *Convolvulaceae*, *Crassulaceae*, *Fabaceae*, *Euphorbiaceae*, *Lamiaceae*, *Poaceae*, *Rosaceae* and *Pteridaceae* (species from the fern family) (Marques et al. 2009; Lasat 2002; Salt et al. 1998; Sarma 2011). Examples of some hyperaccumulators and their accumulation potential are given in Table 8.25.

Hypertolerance to high levels of toxic elements seems to be the key to make hyperaccumulation possible. This tolerance results from the exclusion of toxic elements or their metabolic tolerance to specific elements. The major mechanism in tolerant plant species appears to be compartmentalization of metal ions (i.e., sequestration in the vacuolar compartment or cell walls), which excludes them from cellular sites where processes such as cell division and respiration occur, thus providing an effective protective mechanism (Chaney et al. 1997; Sarma 2011). During phytoextraction of toxic elements from soils, a sequence of processes are involved: (i) toxic elements are adsorbed at the surface of roots, (ii) bioavailable elements moves across the cellular membrane into root cells, (iii) a fraction of the elements taken up by the roots is immobilized in the vacuole, (iv) intracellular mobile elements cross cellular membranes into the root vascular tissue (xylem), and (v) the elements are translocated from the root to stems and leaves (Fig. 8.10). Once inside the target organ of the plant, most toxic elements are too insoluble to move freely in the vascular system so they commonly form carbonates, sulphates, or phosphates that are precipitated in apoplastic or symplastic compartments.

Table 8.25 Extremely high metal accumulation levels determined for the aboveground sections of some plant species growing on contaminated soils

Plant families	Species	Element	Accumulation mg kg^{-1}
Asteraceae	*Berkheya codii*	Ni	5500
Brassicaceae	*Iberis intermedia*	Ti	3070
	Thlaspi caerulescens	Zn	19,410
	Thlaspi caerulescens	Cd	80
	Arabis paniculata	Cd	1127
	Brassica juncea	Ni	3916
	Rorippa globosa	Cd	219
	Thlaspi praecox	Cd	>1000
	Thlaspi tatrense	Zn	20,100
	Streptanthus polygaloydes	Ni	14,800
Convolvulaceae	*Ipomoea alpine*	Cu	12,300
Crassulaceae	*Sedum alfredii*	Cd	2183
	Sedum alfredii	Zn	13,799
Dichapetalaceae	*Dichapetalum gelonioides*	Zn	30,000
Fabaceae	*Sesbania drummondi*	Cd	1687
	Astragalus racemosus	Se	14,920
Laminaceae	*Aeollanthus subacaulis*	Cu	13,700
Poaceae	*Agrostis tenuis*	Pb	13,400
	Zea mays	Cr	2538
	Sorghum sudanense	Cu	5330
	Phragmites australis	Cr	4825
Plumbaginaceae	*Armeria maritime*	Pb	1600
Potamogetonaceae	*Potamogeton crispus*	Cd	49
Pteridaceae	*Pteris vittata*	As	23,000
Rosaceae	*Potentilla griffithii*	Zn	19,600
Violaceae	*Viola calamanaria*	Zn	10,000

Adapted from Marques et al. (2009) and Sarma (2011)

In recent years, many efforts have been made to improve phytoextraction of heavy metals. Besides genetic improvements, plant responses to mycorrhizae, bacteria and chelators have been investigated in order to assess plant tolerance and metal uptake, and the most appropriate doses and ways of application of these means. Major results have been summarized by Vamerali et al. (2010), Wuana and Okieimen (2011), and Marques et al. (2007a, b).

Arbuscular mycorrhizal fungi, which are a direct link between soil and roots, have been shown to be very important for availability and uptake of toxic elements by plants (Leyval et al. 2002). When the host is exposed to metal stress, the role of mycorrhiza in the plant stress response is variable. Some studies indicate reduced metal concentrations in plants due to mycorrhizal colonization (e.g. Heggo et al. 1990; Jentschke et al. 1998). However, other reports indicate enhanced metal uptake and accumulation in plants due to mycorrhizal colonization (e.g. Jamal et al. 2002; Marques et al. 2007a). In summary, there seems to exist a species-

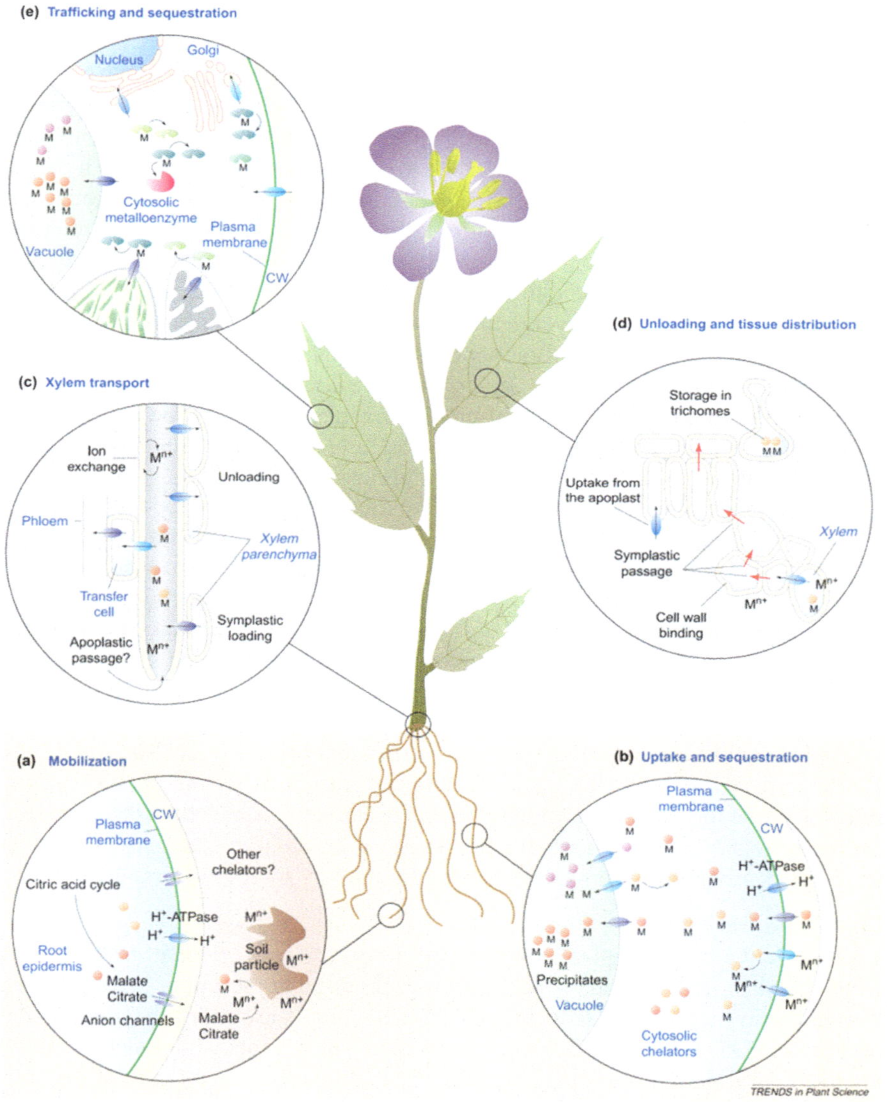

Fig. 8.10 Molecular mechanisms involved in metal accumulation by plants. (**a**) Metal ions in soil are mobilized by secretion of chelators and by acidification of the rhizosphere. (**b**) Uptake of hydrated metal ions or metal-chelate complexes by various uptake systems residing in the plasma membrane. Inside the cell, metals are chelated and excess metal is sequestered in the vacuole. (**c**) From the roots, metals are transported to the shoot via the xylem. Inside the xylem, metals are present as hydrated ions or as metal-chelate complexes. (**d**) After reaching the apoplast of the leaf, metals are differentially captured by different leaf cell types and move cell-to-cell through plasmodesmata. Storage occurs preferentially in trichomes. (**e**) Uptake into the leaf cells is catalyzed by various transporters (not indicated in (**e**)). Intracellular distribution of essential transition metals (=trafficking) is mediated by specific metallochaperones and transporters localized in endomembranes. *CW* cell wall, *M* metal; filled circles: chelators; filled ovals: transporters; bean-shaped structures: metallochaperones (Clemens et al. 2002, p. 310; with kind permission from Elsevier)

specific effect of arbuscular mycorrhizal fungi associations on plant metal uptake and accumulation.

Bacteria communities in the rhizosphere which promote plant growth can also play an important tool in the decontamination of contaminated soils. Plant growth can be enhanced through various mechanisms, including reduction of ethylene production, thus allowing plants to develop longer roots and better establish during early stages of growth, fixation of nitrogen by symbiotic and asymbiotic microorganisms, microbe-based enzyme activities (Khan 2005), supply of phosphorous and other trace elements for plant uptake, and production of phytohormones such as auxins, cytokinins, and gibberelins (Glick et al. 1995). These microorganisms can also produce antibiotic and other pathogen-depressing substances such as siderophores and chelating agents that protect plants from diseases (Kamnev and van der Lelie 2000). Different microorganisms may play assorted roles in plant growth and/or metal tolerance via different mechanisms, so it can be beneficial for the design of a phytoremediation plan to select appropriate multifunctional microbial combinations, which may include mycorrhizae and bacteria.

Studies concerning chelate-assisted phytoextraction have been reported, with the use of chelating agents. Nevertheless, the majority of the reports indicate ethylenediaminetetraacetic acid (EDTA) and/or ethylenediaminedissucinic acid (EDDS) as the main chelates applied in these studies, being those that more successfully improve heavy metal uptake by plants. As an example, Marques et al. (2007b) reported that the addition of EDTA to contaminated soils promoted an increase in the concentration of Zn accumulated by *Solanum nigrum* of up to 231% in the leaves, 93% in the stems, and 81% in the roots, while EDDS application enhanced the accumulation in leaves, stems, and roots up to 140%, 124%, and 104%, respectively, with the plants accumulating up to 8267 mg Zn kg^{-1} in the stems. Despite the possible usefulness of this technology, some concerns have been expressed regarding the potential inherent risk of leaching of metals to groundwater. Similar problems arise when chelators are used as additives for soil washing.

Phytostabilisation

This technique bases on plant roots' ability to reduce toxic element mobility and bioavalability in the soil which can occur through sorption, precipitation, complexaction, or valence reduction of the metal and is useful for the remediation of Pb, As, Cd, Cr, Cu, and Zn contaminated sites (Jadia and Fulekar 2009). Phytostabilisation improves the chemical and biological characteristics of contaminated soil by increasing the amount of organic matter, nutrient levels, cation exchange capacity and biological activity (Arienzo et al. 2004). Another important effect is the reduction of the amount of water percolating through the soil matrix, which may otherwise result in the formation of hazardous leachate and promote soil erosion that distributes toxic metals to other areas.

Application of soil amendments, enhances the reduction of metal bioavailability and thus contributes to preventing both plant uptake and leaching. A significant fraction of toxic elements can thus be stored in the rooting zone, especially under perennial species (Vamerali et al. 2010). According to Sutton and Dick (1987), soil acidity is the main constraint for the establishment of vegetation in these environments, although the application of organic residues and liming materials can attenuate the effects. The choice of plant species is an important task in a phytostabilisation-based technique (Rizzi et al. 2004). Plants must be able to develop extended and abundant root systems and keep the translocation of metals from roots to shoots as low as possible (Mendez and Maier 2008). The major disadvantage is that the contaminant remains in soil as it is, and therefore requires regular monitoring (Gosh and Singh 2005).

Phytovolatilization

This method involves the use of plants to take up contaminants from the soil, transforming them into volatile form and transpiring them into the atmosphere. Phytovolatilization occurs as growing trees and other plants that take up water together with the contaminant. Some of these contaminants can pass through the plants to the leaves and volatilise into the atmosphere at comparatively low concentrations (Mueller et al. 1999). Phytovolatilization has been primarily used for the removal of Hg. The mercuric ion in the plant is transformed into less toxic elemental Hg. The disadvantage is that Hg released into the atmosphere is likely to be recycled by precipitation and redeposition into the ecosystem (Henry 2000).

Phytovolatilization is also an important way of increasing the efficiency of Se removal from soils and thus decreasing Se ecotoxicity. Because of the chemical similarity of sulfur (S) and Se, plants are able to take up inorganic and organic forms of Se and metabolize them to volatile forms via the S assimilation pathway. Biological volatilization has the advantage of removing Se from a contaminated site in relatively non-toxic forms, such as dimethylselenide (DMSe), which is 500–700 times less toxic than SeO_4^{2-} or SeO_3^{2-} (Wilber 1980). Although the volatilized Se may eventually be redeposited in other areas, this should not be a problem in regions where much of the soil area is deficient in Se with respect to the nutrition of plants and animals, which require Se in low concentrations (LeDuc and Terry 2005).

References

Aaseth J, Alexander J, Norseth T (1982) Uptake of ^{51}Cr chromate by human erythrocytes – a role of glutathione. Acta Pharmacol Toxicol 50:310–315

Adriano DC (2003) Trace elements in terrestrial environments: biogeochemistry, bioavailability and risks of metals, 2nd edn. Springer, New York

Adriano DC, Wenzel WW, Vangronsveld J, Bolan NS (2004) Role of assisted natural remediation in environmental cleanup. Geoderma 122:121–142

Aggette J, Spell AC (1978) Arsenic from geothermal sources, New Zealand Research and Development report. No 35. Department of Agriculture, New Zealand

Ågren MS (1990) Percutaneous absorption of zinc from zinc oxide applied topically to intact skin in man. Dermatological 180:36–39

Albino V, Cioffi R, Marroccoli M, Santoro L (1996) Potential application of ettringite generating systems for hazardous waste stabilization. J Harzard Mat 51:241–252

Alfven T, Jarup L, Elinder CG (2002) Cadmium and lead in blood in relation to low bone mineral density and tubular proteinuria. Environ Health Perspect 110:699–702

Alkorta I, Hernandez-Allica J, Becerril JM, Amezaga I, Albizu I, Garbisu C (2004) Recent findings on the phytoremediation of soils contaminated with environmentally toxic heavy metals and metalloids such as zinc, cadmium, lead, and arsenic. Rev Environ Sci Biotechnol 3:71–90

Allaway WH (1968) Agronomic control over the environmental cycling of trace elements. Adv Agron 20:235–274

Alloway BJ (1995) Heavy metals in soils. Blackie Academic and Professional, London

Alloway BJ (2008) Zinc in soils and crop nutrition. International Zinc Association and International Fertilizer Industry Association, Brussels

Andersen KE, Nielsen GD, Flyvholm MA, Fregert S, Gruvberge B (1983) Nickel in tap water. Contact Dermatitis 9:140–143

Angelova V, Ivanova R, Delibaltova V, Ivanov K (2004) Bio-accumulation and distribution of heavy metals in fibre crops (flax, cotton and hemp). Ind Crop Prod 19(3):197–205

Angino EE, Magnuson LM, Waugh TC, Galle OK, Bredfeldt J (1970) Arsenic in detergents-possible danger and pollution hazard. Science 168:389–392

Arain MB, Kazi TG, Baig JA, Jamali MK, Afridi HI, Shah AQ, Jalbani N, Sarfraz RA (2009) Determination of arsenic levels in lake water, sediment, and foodstuff from selected area of Sindh, Pakistan: estimation of daily dietary intake. Food Chem Toxicol 47:242–248

Arienzo M, Adamo P, Cozzolino V (2004) The potential of Lolium perenne for revegetation of contaminated soils from a metallurgical site. Sci Total Environ 319:13–25

ATSDR (Agency for Toxic Substance and Disease Registry) (2007) Toxicological profile for lead. http://www.atsdr.cdc.gov/toxprofiles/tp.asp?id=96&tid=22. Accessed 20 July 2017

ATSDR (Agency for Toxic Substances and Disease Registry) (1999) Toxicological profile for cadmium (final report). NTIS Accession No. PB99-166621, p 434

ATSDR (Agency for Toxic Substances and Disease Registry) (2002) Toxicological profile for copper. Atlanta, GA, US Department of Health and Human Services, Public Health Service, Agency for Toxic Substances and Disease Registry (Subcontract No. ATSDR-205-1999-00024)

Auboiroux M, Baillif P, Touray JC, Bergaya F (1996) Fixation of Zn^{2+} and Pb^{2+} by a Ca-montmorillonite in brines and dilute solutions: preliminary results. Appl Clay Sci 11:117–126

Barceloux DG (1999a) Copper. Clin Toxicol 37:217–230

Barceloux DG (1999b) Nickel. Clin Toxicol 37:239–258

Barregard L, Sallsten G, Schutz A, Attewell R, Skerfving S, Jarvholm B (1992) Kinetics of mercury in blood and urine after brief occupational exposure. Arch Environ Health 7 (3):176–184

Barregard L, Sallsten G, Conradi N (1999) Tissue levels of mercury determined in a deceased worker after occupational exposure. Int Arch Occup Environ Health 72:169–173

Basta NT, Gradwohl R, Snethen KL, Schroder JL (2001) Chemical immobilisation of lead, zinc and cadmium in smelter contaminated soils using biosolids and rock phosphate. J Environ Qual 30:1222–1230

Basta NT, Ryan JA, Chaney RL (2005) Trace element chemistry in residual-treated soil: key concepts and metal bioavailability. J Environ Qual 34(1):49–63

Bech J, Tobias FJ, Roca N, Rustullet J (1998) Trace elements in some mediterranean red soils from the NE of Spain. Agrochimica XLII(1–2):26–40

Bellinger D (2005) Teratogen update: lead and pregnancy. Birth Defects Res Part A 73:409–420

Bencko V (1983) Nickel: a review of its occupational and environmental toxicology. J Hyg Epidemiol Micro Immun 27:237–247

Bennett BJ (1994) Environmental nickel pathways to man. In: Sunderman FW Jr (ed) Nickel in the human environment. Lyon International Agency for Research on Cancer, Lyon, pp 487–495

Berg M, Trans HC, Nguyeu TC, Pham MV, Scherteuleib R, Giger W (2001) Arsenic contamination of groundwater and drinking water in Vietnam: a human health threat. Environ Sci Technol 35:2621–2626

Bergkvist P, Jarvis N, Berggren D, Carlgren K (2003) Long-term effects of sewage sludge applications on soil properties, cadmium availability and distribution in arable soil. Agric Ecosyst Environ 97(1–3):167–179

Bernard A (2008) Cadmium and its adverse effects on human health. Indian J Med Res 128:557–564

Bianchi V, Levis AG (1987) Recent advances in chromium genotoxicity. Toxicol Environ Chem 15:1–24

Biester HG, Mueller G, Schoeler HF (2002) Binding and mobility of mercury in soils contaminated by emissions from chlor-alkali plants. Sci Total Environ 284(1–3):191–203

Bingham FT (1979) Bioavailability of Cd to food crops in relation to heavy metal content of sludge-amended soil. Environ Health Perspect 28:39–43

Bires J, Dianovsky J, Bartko P, Juhasova Z (1995) Effects on enzymes and the genetic apparatus of sheep after administration of samples from industrial emissions. Biol Met 8:53–58

Bissen M, Frimmel FH (2003) Arsenic – a review. Part I: occurrence, toxicity, speciation, mobility. Acta Hydrochim Hydrobiol 31:9–18

Bjornberg KA, Vahter M, Berglund B, Niklasson B, Biennow M, Sandborg-Englund G (2005) Transport of methyl mercury and inorganic mercury to the fetus and breast-fed infant. Environ Health Perspect 113(10):1381–1385

Bjuhr J (2007) Trace metals in soils irrigated with waste water in a periurban area downstream Hanoi City, Vietnam, Institutionen for markvetenskap, Sveriges lantbruksuniversitet (SLU), Uppsala, Sweden

Bodek I, Lyman WJ, Reehl WF, Rosenblatt DH (1988) Environmental inorganic chemistry: properties, processes and estimation methods. Pergamon Press, Elmsford

Bolan NS, Adriano DC, Duraisamy P, Mani A, Arulmozhiselvan K (2003) Immobilization and phytoavailability of cadmium in variable charge soils: I. Effect of phosphate addition. Plant Soil 250:83–94

Borja-Aburto VH, Hertz-Picciotto I, Rojas LM, Farias P, Rios C, Blanco J (1999) Blood lead levels measured prospectively and risk of spontaneous abortion. Am J Epidemiol 150 (6):590–597

Börjesson P (1999) Environmental effects of energy crop cultivation in Sweden – I: identification and quantification. Biomass Bioenergy 16:137–154

Bradl HB (2005) Heavy metals in the environment: origin, interaction and remediation. Elsevier, London

Brewer GJ (2001) Wilson's disease: a clinician's guide to recognition. Kluwer Academic, Boston

Brewer GJ, Askari FK (2005) Wilson's disease: clinical management and therapy. J Hepatol 42: S13–S21

Bronger A, Bruhn-Lobin N (1993) Paleopedology of mediterranean soils – case studies from NW Morocco. Abstracts of the 2nd International Meeting on Red Mediterranean Soils, Adana (Turquía), p 183

Brown JJ (1988) Zinc fume fever. Brit J Radiol 61:327–329

Brown SE, Welton WC (2008) Heavy metal pollution. Nova Science publishers, New York

Canfield RL, Henderson CR, Cory-Slechta DA, Cox C, Jusko TA, Lanphear BP (2003) Intellectual impairment in children with blood lead concentrations below 10 μg dL^{-1}. N Engl J Med 348:1517–1526

Cannon HL, Connally GG, Epstein JB, Parker JG, Thornton I, Wixson G (1978) Rocks: geological sources of most trace elements. In: Report to the workshop at south plantation Captiva Island, FL, US. Geochem Environ 3:17–31

Cappon CJ (1981) Mercury and selenium content and chemical form in vegetable crops grown on sludge-amended soil. Arch Environ Contam Toxicol 10(6):673–689

Carton JA (1988) Saturnismo. Med Clin 91:538–540

Castaldi S, Rutigliano FA, Virzo de Santo A (2004) Suitability of soil microbial parameters as indicators of heavy metal pollution. Water Air Soil Pollut 158:21–35

Cempel M, Nikel G (2006) Nickel: a review of its sources and environmental toxicology. Pol J Environ Stud 15(3):375–382

Cernichiari E, Brewer R, Myers GJ, Marsh DO, Lapham LW, Cox C, Shamlaye CF, Berlin M, Davidson PW, Clarkson TW (1995) Monitoring methylmercury during pregnancy: maternal hari predicts fetal brain exposure. Neurotoxicology 16(4):705–710

Chaney RL, Malik M, Li YM, Brown SL, Brewer EP, Angle JS, Baker AJM (1997) Phytoremediation of soil metals. Curr Opin Biotechnol 8:279–284

Chen CY, Sheu JY, Lin TH (1999) Oxidative effects of nickel on bone marrow and blood of rats. J Toxicol Environ Health A 58(8):475

Chen Y, Hall M, Graziano JH, Slavkovich V, van Geen A, Parvez F, Ahsan H (2007) A prospective study of blood selenium levels and the risk of arsenic-related premalignant skin lesions. Cancer Epidemiol Biomark Prev 16(2):207–213

Ciccu R, Ghiani M, Serci A, Fadda S, Peretti R, Zucca A (2003) Heavy metal immobilization in the mining-contaminated soils using various industrial wastes. Min Eng 16:187–192

Clarkson TW (1988) Biological monitoring of toxic metals. Plenum Press, New York, pp 265–282

Clemens S, Palmgren MG, Krämer U (2002) A long way ahead: understanding and engineering plant metal accumulation. Trends Plant Sci 7(7):309–315

Cohen SM, Arnold LL, Eldan M, Lewis AS, Beck BD (2006) Methylatedarsenicals: the implications of metabolism and carcinogenicity studies in rodents to human risk assessment. Crit Rev Toxicol 36(2):99–133

Concha G, Vogler G, Nermell B, Vahter M (1998) Low-level arsenic excretion inbreast milk of native Andean women exposed to high levels of arsenic in the drinking water. Int Arch Occup Environ Health 71(1):42–46

Coogan TP, Latta DM, Snow ET, Costa M (1989) Toxicity and carcinogenicity of nickel compounds. Crit Rev Toxicol 19(4):341–384

Counter SA, Buchanan LH (2004) Mercury exposure in children: a review. Toxicol Appl Pharmacol 198:209–230

Cousins RJ, Hempe JM (1990) Zinc. In: Brown ML (ed) Present knowledge in nutrition. International Life Sciences Institute, Washington, DC

Cousins RJ, McMahoh RJ (2000) Integrative aspects of zinc transporters. J Nutr 130:1384–1387

Cumming ED, Caccavo F, Fendor S, Rosenzweig FR (1999) Arsenic mobilization by the dissimilatory Fe(III)-reducing bacterium Shewanella alga BrY. Environ Sci Technol 33:723–729

Cynthia E, David D (1997) Remediation of metals-contaminated soil and groundwater. Technology Evaluation Report (TE-97-01) in Groundwater Remediation Technologies. Analysis Centre (GWRTAC) E series

D'Amore JJ, Al-Abed SR, Scheckel KG, Ryan JA (2005) Methods for speciation of metals in soils a review. J Environ Qual 34:1707–1745

Dakeishi M, Murata K, Grandjean P (2006) Long-term consequences of arsenic poisoning during infancy due to contaminated milk powder. Environ Health 5:31

Davidson AG, Fayers PM, Newman Taylor AJ, Venables KM, Darbyshire J, Pickering CAC, Chettle DR, Franklin D, Guthrie CJ, Scott MC, O'Malley D, Holden H, Mason HJ, Wright AL, Gompertz D, Taylor AJ (1988) Cadmium fume inhalation and emphysema. Lancet 1:663–667

Davister A (1996) Studies and research on processes for the elimination of cadmium from phosphoric acid. In: Proceedings of Fertilizers as a Source of Cadmium OECD/Inter-Organization Programme for the Sound Management of Chemicals (IMOC), Saltsjoebaden, Sweden, pp 21–30

Davydova S (2005) Heavy metals as toxicants in big cities. Microchem J 79:133–136

Dayan AD, Paine AJ (2001) Mechanisms of chromium toxicity, carcinogenicity and allergenicity: review of the literature from 1985 to 2000. Hum Exp Toxicol 20:439–451

Deja J (2002) Immobilization of Cr^{6+}, Cd^{2+}, Zn^{2+} and Pb^{2+} in alkali-activated slag binders. Cem Concr Res 32:1971–1979

Dermatas D, Meng XG (1996) Stabilization/Solidification (S/S) of heavy metal contaminated soils by means of a quicklime-based treatment approach. Stabilization and solidification of hazardous, radioactive, and mixed wastes, ASTM STP 1240, vol 1. American Society for Testing and Materials, Philadelphia, pp 449–513

Deverel SJ, Millard SP (1988) Distribution and mobility of selenium and other trace elements in shallow ground water of the Western San Joaquin Valley, California. Environ Sci Technol 22:697–702

Diamond GL, Thayer WC, Choudhury H (2003) Pharmacokinetic/pharmacodynamics (PK/PD) modeling of risks of kidney toxicity from exposure to cadmium: estimates of dietary risks in the U.S. population. J Toxicol Environ Health 66(Part A):2141–2164

Du Laing G, Rinklebe J, Vandecasteele B, Meers E, Tack FM (2009) Trace metal behaviour in estuarine and riverine floodplain soils and sediments: a review. Sci Total Environ 407:3972–3985

Dudka S, Adriano DC (1997) Environmental impacts of metal ore mining and processing: a review. J Environ Qual 26:590–602

EC (European Commission) (2001) DG enterprise: analysis and conclusions from member states' assessment of the risk to health and the environment from cadmium in fertilisers. Contract No. ETD/00/503201

EFSA (European Food Safety Authority) (2009) Cadmium in food. Scientific opinion of the Panel on Contaminants in the Food Chain. EFSA J 980:1–139

EFSA (European Food Safety Authority) (2010) Scientific opinion on lead in food. EFSA J 8 (4):1570

EFSA (European Food Safety Authority) (2012) Lead dietary exposure in the European population. EFSA J 10(7):2831

Elinder CG (1986) Zinc. In: Friberg L, Nordberg GF, Vouk VB (eds) Handbook on the toxicology of metals, 2nd edn. Elsevier, Amsterdam, pp 664–679

EMEP (European Monitoring and Evaluation Programme) (2001) Cyromazine summary report (2). Committee for veterinary medicinal products. European Agency for the Evaluation of medicinal products, veterinary medicines and inspections (EMEA). EMEA/MRL/770/00-FINAL, January 2001. http://www.emea.europa.eu/pdfs/vet/mrls/077000en.pdf. Accessed 20 July 2017

Engstrom DR (2007) Fish respond when the mercury rises. Proc Natl Acad Sci U S A 104:16394–16395

Environment Agency (2007) UK soil and herbage pollutant survey. Report No. 7: environmental concentrations of heavy metals in UK soil and herbage. Environment Agency, Bristol

Environmental Agency (2010) Contaminants in soil: updated collation of toxicological data and intake values for humans. Nickel. Science Report SC050021/SR TOX8

Ettinger AS, Téllez-Rojo MM, Amarasiriwardena C, Bellinger D, Peterson K, Schwartz J, Hu H, Hernández-Avila M (2004) Effect of breast milk lead on infant blood lead levels at 1 month of age. Environ Health Perspect 112(14):245–275

Ettler V, Johan Z, Kribek B, Sebek O, Mihaljevic M (2009) Mineralogy and environmental stability of slags from the Tsumeb smelter, Namibia. Appl Geochem 24:1–15

Fängström B, Moore S, Nermell B, Kuenstl L, Goessler W, Grandér M, Kabir I, Palm B, Arifeen S, Vahter M (2008) Breast-feeding protects against arsenic exposure in Bangladeshi infants. Environ Health Perspect 116(7):963–969

Feng L, Yuan-Ming Z, Ji-Zheng H (2009) Microbes influence the fractionation of arsenic in paddy soils with different fertilization regimes. Sci Total Environ 407:2631–2640

Fergusson JE (1990) The heavy elements: chemistry, environmental impact and health effects. Pergamon Press, Oxford

Fernandez MA, Sanz P, Palomar M, Serra J, Gadea E (1996) Fatal chemical pneumonitis due to cadmium fumes. Occup Med 46:372–374

Fitzgerald WF, Lamborg CH (2005) Geochemistry of mercury in the environment. In: Lollar BS (ed) Treatise on Geochemistry. Elsevier, New York, pp 107–148

Flyvholm MA, Nielsen GD, Andersen A (1984) Nickel content of food and estimation of dietary intake. Z Lebensm Unters Forsch 179:427–431

Fu JH (2008) The research status of soil remediation in China. Ann Meet Chin Soc Environ Sci 2008:1056–1060

Gadepalle VP, Ouki SK, Van Herwijnen R, Hutchings T (2007) Immobilization of heavy metals in soil using natural and waste materials for vegetation establishment on contaminated sites. Soil Sediment Contam 16:233–251

Garbarino JR, Hayes H, Roth D, Antweider R, Brinton TI, Taylor H (1995) Contaminants in the Mississippi river. U.S. Geological Survey Circular 1133, Virginia. http://www.pubs.usgs.gov/circ/circ1133/). Accessed 20 July 2017

Garrett RG (2000) Natural sources of metals to the environment. In: Centeno JA, Collery P, Fernet G, Finkelman RB, Gibb H, Etienne JC (eds) Metal ions in biology and medicine. John Libbey Eurotext, Paris, pp 508–510

Geebelen W, Vangronsveld J, Adriano DC, Carleer R, Clijsters H (2002) Amendment induced immobilization of lead in a lead-spiked soil: evidence from phytotoxicity studies. Water Air Soil Pollut 140(1–4):261–277

Ghosh M, Singh SP (2005) A review on phytoremediation of heavy metals and utilization of its byproducts. Appl Ecol Environ Res 3(1):1–18

Giller KE, Witter E, McGrath SP (1998) Toxicity of heavy metals to microorganisms and microbial processes in agricultural soils: a review. Soil Biol Biochem 30:1389–1414

Glass DJ (1999) U.S. and International Markets for Phytoremediation, 1999–2000. D. Glass Associates, Needham

Glick BR, Karaturovic DM, Newell PC (1995) A novel procedure for rapid isolation of plant-growth promoting pseudomonads. Can J Microbiol 41:533–536

Gochfeld M (2003) Cases of mercury exposure, bioavailability, and absorption. Ecotoxicol Environ Safe 56:174–179

Gray CW, Mc Laren RG, Roberts AHC, Condron LM (1998) Sorption and desorption of Cadmium from some New Zealand soils: effects of pH and contact time. Aust J Soil Res 36:199–216

Gray JE, Crock JG, Lasorsa BK (2002) Mercury methylation at mercury mines in the Humboldt River Basin, Nevada, USA. Geochem: Explor Environ Anal 2(2):143–149

Gregor M (2004) Metal availability, uptake, transport and accumulation in plants. In: Prasad MNV (ed) Heavy metal stress in plants – from biomolecules to ecosystems. Springer, Berlin, pp 1–27

Grigalavičienė I, Rutkovienė V, Marozas V (2005) The accumulation of heavy metals Pb, Cu and Cd at roadside forest soil. Pol J Environ Stud 14(1):109–115

Guo GL, Zhou QX, Ma LQ (2006) Availability and assessment of fixing additives for the in situ remediation of heavy metal contaminated soils: a review. Environ Monit Assess 116:513–528

GWRTAC (1997) Remediation of metals-contaminated soils and groundwater, Tech Rep TE-97-01, GWRTAC, Pittsburgh, Pa, USA, 1997, GWRTAC-E Series

Haber LT, Erdreicht L, Diamond GL, Maiera M, Ratney R, Zhao Q, Dourson ML (2000) Hazard identification and dose response of inhaled nickel-soluble salts. Regul Toxicol Pharmacol 31:210–230

Haines AT, Nieboer E (1988) Chromium hypersensitivity. In: Nriagu JO, Nieboer E (eds) Chromium in the natural and human environments. Wiley, New York, pp 497–532

Handa BK (1988) Occurrence and distribution of chromium in natural waters of India. Adv Environ Sci Technol 20:189–214

Harris RF, Karlen DL, Mulla DJ (1996) A conceptual framework for assessment and management of soil quality and health. In: Doran JW, Jones AJ (eds) Methods for assessing soil quality, SSSA Spec Publ, vol 49. Soil Sci Soc Am, Madison, pp 61–82

He QB, Singh BR (1994) Effect of organic matter on the distribution, extractability and uptake of cadmium in soils. Eur J Soil Sci 44(4):641–650

Heggo A, Angle A, Chaney RL (1990) Effects of vesicular arbuscular mycorrhizal fungi on heavy metal uptake of soybeans. Soil Biol Biochem 22:865–869

Henriques FS, Fernandez C (1991) Metal uptake and distribution in rush (Juncus conglomerates L.) plants growing in pyrites mine tailings at Lousal, Portugal. Sci Total Environ 102:253–260

Henry JR (2000) An overview of phytoremediation of lead and mercury. NNEMS Report. Washington, DC

Hindmarsh JT, Abernethy CO, Peters GR, McCurdy RF (2002) Environmental aspects of arsenic toxicity. In: Sarkar B (ed) Heavy metals in the environment. Marcel Dekker, New York, pp 217–229

Horiguchi H, Oguma E, Sasaki S, Miyamoto K, Ikeda Y, Machida M, Kayama F (2004) Comprehensive study of the effects of age, iron deficiency, diabetes mellitus and cadmium burden on dietary cadmium absorption in cadmium-exposed female Japanese farmers. Toxicol Appl Pharmacol 196:114–123

IARC (International Agency for Research on Cancer) (1993) Beryllium, Cadmium, mercury and exposures in the glass manufacturing industry. IARC Monographs on the Evaluation of Carcinogenic Risk of Chemicals to Humans, vol 58. Lyon, France, p 444 http://monographs.iarc.fr/ENG/Monographs/vol58/volume58.pdf. Accessed 20 July 2017

Ihnat M, Fernandes L (1996) Trace elemental characterization of composted poultry manure. Bioresour Technol 57:143–156

IOM (Institute of Medicine) (2001) Dietary reference intakes for vitamin A, vitamin K, arsenic, boron, chromium, copper, iodine, iron, manganese, molybdenum, nickel, silicon, vanadium and zinc, A report of the panel on micronutrients, subcommittees on upper reference levels of nutrients and of interpretation and use of dietary reference intakes, and the standing committee on the scientific evaluation of dietary reference intakes. Food and Nutrition Board, Institute of Medicine. National Academy Press, Washington, DC

IPCS (International Programme on Chemical Safety) (1998) Copper. World Health Organization Geneva, Switzerland

IUSS Working Group WRB (2007) World reference base for soil resources 2006, first update 2007, World Soil Resources Report No. 103. FAO, Rome

Iyengar V, Nair P (2000) Global outlook on nutrition and the environment meeting the challenges of the next millennium. Sci Total Environ 249:331–346

Jadia CD, Fulekar MH (2009) Phytoremediation of heavy metals: recent techniques. Afr J Biotechnol 8(6):921–928

Jaerup L, Hellström L, Alfvén T, Carlsson MD, Grubb A (2000) Low level exposure to cadmium and early kidney damage: the OSCAR study. Occup Environ Med 57:668–672

Jamal A, Ayub N, Usman M, Khan AG (2002) Arbuscular mycorrhizal fungi enhance Zn and nickel uptake from contaminated soil by soybean and lentil. Int J Phytoremediation 4:205–221

James AK, Meliker JR (2013) Environmental cadmium exposure and osteoporosis: a review. Int J Public Health 58:737–745

Janusa MA, Champagne CA, Fanguy JC (1998) Solidificationstabilization of lead with the aid of bagasse as an additive to Portland cement. Microchem J 65:255–259

Jentschke G, Marschner P, Vodnik D, Marth C, Bredemeier M, Rapp C, Fritz E, Gogala N, Godbold DL (1998) Lead uptake by Picea abies seedlings: effects of nitrogen source and mycorrhizaes. J Plant Physiol 153:97–104

Jin T, Nordberg G, Ye T, Bo M, Wang H, Zhu G, Kong Q, Bernard A (2004) Osteoporosis and renal dysfunction in a general population exposed to cadmium in China. Environ Res 96 (3):353–359

John MK, Van Laerhoven CJ (1976) Effects of sewage sludge composition, application rate, and lime regime on plant availability of heavy-metals. J Environ Qual 5:246–251

Jones FT (2007) A broad view of arsenic. Poult Sci 86:2–14

Jones KC, Johnston AE (1989) Cadmium in cereal grain and herbage from long-term experimental plots at Rothamsted, UK. Environ Pollut 57:199–216

Jones LHP, Jarvis SC (1981) The fate of heavy metals. In: Green DJ, Hayes MHB (eds) The chemistry of soil processes. Wiley, New York, p 593

Jung MC (2001) Heavy metal contamination of soils and waters in and around the Imcheon Au–Ag mine, Korea. Appl Geochem 16:1369–1375

Jung U, Sang W, Hyo T, Kyoung W, Jin S (2009) Enhancement of Arsenic mobility by indigenous bacteria from mine tailing as response to organic supply. Environ Int 35:496–501

Kaasalainen M, Yli-Halla M (2003) Use of sequential extraction to assess metal partitioning in soils. Environ Pollut 126(2):225–233

Kabata-Pendias A, Mukherjee AB (2007) Trace elements from soil to human. Spinger, Berlin

Kabata-Pendias A, Pendias H (1992) Trace elements in soils and plants, 2nd edn. CRC Press, Boca Raton

Kabata-Pendias A, Pendias H (2001) Trace metals in soils and plants, 2nd edn. CRC Press, Boca Raton

Kajiwara Y, Yasutake A, Adachi T, Hirayama K (1996) Methyl mercury transport across the placenta via neutral amino acid carrier. Arch Toxicol 70(5):310–314

Kamnev AA, van der Lelie D (2000) Chemical and biological parameters as tools to evaluate and improve heavy metal phytoremediation. Biosci Rep 20:239–258

Kapaj S, Peterson H, Liber K, Bhattacharya P (2006) Human health effects from chronic arsenic poisoning – a review. J Environ Sci Health Part A 41:2399–2428

Khan AG (2005) Role of soil microbes in the rhizospheres of plants growing on trace metal contaminated soils in phytoremediation. J Trace Elem Med Biol 18:355–364

Khan S, Cao Q, Zheng YM, Huang YZ, Zhu YG (2008) Health risks of heavy metals in contaminated soils and food crops irrigated with wastewater in Beijing, China. Environ Pollut 152(3):686–692

Kiekens L (1995) Zinc. In: Alloway BJ (ed) Heavy metals in soils, 2nd edn. Blackie Academic and Professional, London, pp 284–305

Kloke A, Sauerbeck DR, Vetter H (1984) The contamination of plants and soils with heavy metals and the transport of metals in terrestrial food chains. In: Nriagu JO (ed) Changing metal cycles and human health. Springer, Berlin, pp 113–141

Kribek B, Majer V, Veselovsky F, Nyambe I (2010) Discrimination of lithogenic and anthropogenic sources of metals and sulphur in soils of the central northern part of the Zambian Copper belt Mining District: a topsoil vs. subsurface soil concept. J Geochem Explor 104:69–86

Krishna RR, Xu CY, Supraja C (2001) Assessment of electrokinetic removal of heavy metals from soils by sequential extraction analysis. J Hazard Mater 84:279–296

Kristensen AKB, Thomsen JF, Mikkelsen S (2013) A review of mercury exposure among artisanal small-scale goldminers in developing countries. Int Arch Occup Environ Health. https://doi.org/10.1007/s00420-013-0902-9

Kumagai Y, Sumi D (2007) Arsenic: signal transduction, transcription factor, and biotransformation involved in cellular response and toxicity. Annu Rev Pharmacol Toxicol 47:243–262

Ladonin DV (2002) Heavy metal compounds in soils: problems and methods of study. Pochvovedenie 6:682–692

LaGrega MD, Buckingham PL, Evans JC (1994) Hazardous waste management. McGraw Hill, New York

Lahmann E, Munari S, Amicarelli V, Abbaticchio P, Gabellieri R (1986) Heavy metals: identification of air quality and environmental problems in the European Community, Vol 1 & 2

Lansdown AB (1993) Influence of zinc oxide in the closure of open skin wounds. Int J Cosmet Sci 15:83–85

Larkin M (1997) Lead in mothers' milk could lead to dental caries in children. Sci Med 35:789

Lasat MM (2002) Phytoextraction of toxic metals: a review of biological mechanism. J Environ Qual 31:109–120

Lauwerys RR, Hoet P (2001) Industrial chemical exposure. Guidelines for biological monitoring, 3rd edn. Lewis Publishers, Boca Raton

LeDuc DL, Terry N (2005) Phytoremediation of toxic trace elements in soil and water. J Ind Microbiol Biotechnol 32:514–520

Leggett RW (1993) Age-specific kinetic model of lead metal in humans. Environ Health Perspect 101(7):598–616

Levy DB, Barbarick KA, Siemer EG, Sommers LE (1992) Distribution and partitioning of trace metals in contaminated soils near Leadville, Colorado. J Environ Qual 21(2):185–195

Leyval C, Jones EJ, Del Val C, Haselwandter K (2002) Potential of arbuscular mycorrhizal fungi for bioremediation. In: Gianinazzi S, Schuepp H, Barea JM, Hasewandter K (eds) Mycorrhizal technology in agriculture. Birkhaeuser Verlag, Berlin, pp 175–186

Li J, Zhang GN, Li Y (2010) Review on the remediation technologies of POPs. Hebei Environ Sci 65–68

Lindquivist O (1995) Environmental impact of mercury and other heavy metals. J Power Sources 57:3–7

Ling C, Ching Y, Hung-Chang L, Hsing-Jasine C, Ming J, Bor-Cheng H (2006) Effect of mother's consumption of traditional Chinese herbs on estimated infant daily intake of lead from breast milk. Sci Total Environ 354:120–126

Llobet JM, Falcoa G, Casas CA, Teixidoa A, Domingo JL (2003) Concentrations of arsenic, cadmium, mercury and lead in common foods and estimated daily intake by children, adolescents, adults, and seniors of Catalonia, Spain. J Agric Food Chem 51:838–842

Lombi E, Zhao FJ, Zhang GY, Sun B, Fitza W, Zhang H, McGrath SP (2002) In situ fixation of metals in soils using bauxite residue: chemical assessment. Environ Pollut 118:435–443

Lone MI, He Z, Stoffella PJ, Yang X (2008) Phytoremediation of heavy metal polluted soils and water: progress and perspectives. J Zhejiang Univ Sci B 9(3):210–220

Lottermoser BG (2007) Mine wastes. Characterization, treatment and environmental impacts, 2nd edn. Springer, Berlin

Ma LQ, Tcaina SJ, Logan TJ (1993) In situ lead immobilization by apatite. Environ Sci Technol 27:214–220

Machelett B, Metz R, Bergmann H (1993) Schwermetalltransferuntersuchungen an landwirtschaftlichen und gärtnerischen Nutzpflanzen unter gleichen Anbaubedingungen. VDLUFA-Schriftenreihe 37:579–582

Magbagbeola NO, Oyeleke O (2003) Heavy metal poisoning and fish depopulation in Nigeria: a case study of the Lagos Lagoon. J Afr Fish Fish 8:192–193

Marques APGC, Oliveira RS, Samardjieva KA, Pissarra J, Rangel AOSS, Castro PML (2007a) Solanum nigrum in contaminated soil: effect of arbuscular mycorrhizal fungi on zinc accumulation and histolocalisation. Environ Pollut 145:691–699

Marques APGC, Oliveira RS, Samardjieva KA, Rangel AOSS, Pissarra J, Castro PML (2007b) EDDS and EDTA-enhanced zinc accumulation by Solanum nigrum inoculated with arbuscular mycorrhizal fungi grown in contaminated soil. Chemosphere 70:1002–1014

Marques APGC, Rangel AOSS, Castro PML (2009) Remediation of heavy metal contaminated soils: phytoremediation as a potentially promising clean-up technology. Crit Rev Environ Sci Technol 39:622–654

Mascagni P, Consonni D, Bregante G, Chiappino G, Toffoletto F (2003) Olfactory function in workers exposed to moderate airborne cadmium levels. Neurotoxicology 24:717–724

Maslin P, Maier RM (2000) Rhamnolipid-enhanced mineralizationof phenanthrene in organic-metal co-contaminated soils. Biorem J 4(4):295–308

Matschullat J, Ottenstein R, Reimann C (2000) Geochemical background – can we calculate it? Environ Geol 39(9):990–1000

Mc Grath SP, Zhao FJ, Lombi EL (2001) Plant and rhizosphere process involved in phytoremediation of metal contaminated soils. Plant Soil 232:207–214

McGrath SP (1995) Nickel. In: Alloway BJ (ed) Heavy metals in soils, 2nd edn. Blackie, London

McGrath SP, Loveeland PJ (1992) The soil geochemical atlas of England and Wales. Blackie, London
McLaren RG (2003) Micronutrients and toxic elements. In: Benbi DK, Nieder R (eds) Handbook of processes and modeling in soil-plant system. Haworth Press, New York, pp 589–625
McLaren RG, Backes CA, Rate AW, Swift RS (1998) Cadmium and cobalt desorption kinetics from soil clays: effects of sorption period. Soil Sci Soc Am J 62:332–337
McLaughlin MJ, Parker DR, Clarke JM (1999) Metals and micronutrients – food safety issues. Field Crop Res 60:143–163
McLaughlin MJ, Hamon RE, McLaren RG, Speir TW, Rogers SL (2000) Review: a bioavailability-based rationale for controlling metal and metalloid contamination of agricultural land in Australia and New Zealand. Aust J Soil Res 38(6):1037–1086
Mendez MO, Maier RM (2008) Phytostabilization of mine tailings in arid and semiarid environments – an emerging remediation technology. Environ Health Perspect 116:278–283
Mengel K, Kirkby EA (1978) Principles of plant nutrition. International Potash Institute, Worblaufen
Mertz W (1969) Chromium occurrence and function in biological systems. Physiol Rev 49:163–239
Miettinen JK, Rahola T, Hattula T, Rissanen K, Tillander M (1971) Elimination of ^{203}Hg-methyl mercury in man. Ann Clin Res 3(2):116–122
Millot R, Allegre CJ, Gaillardet J, Roy S (2004) Lead isotopic systematics of major river sediments: a new estimate of the Pb isotopic composition of the Upper Continental Crust. Chem Geol 203:75–90
Mishra VK, Kim KH, Kang CH, Choi KC (2004) Wintertime sources and distribution of airborne lead in Korea. Atmos Environ 38:2653–2664
Mitchell RL (1964) Trace elements in soils. In: Bear FE (ed) Chemistry of the soil. AACS Monograph Series, New York, pp 320–368
Mueller I, Pluquet E (1997) Einfluß einer Fe-(Oxid)gabe auf die Cd-Verfügbarkeit eines kontaminierten Auenbodens. Mitt Dt Bodenkundl Ges 85:311–314
Mueller B, Rock S, Gowswami D, Ensley D (1999) Phytoremediation decision tree. – Prepared by – Interstate Technology and Regulatory Cooperation Work Group, p 1–36
Muntner P, Vupputyuri S, Coresh J, Batuman V (2003) Blood lead and chronic kidney disease in the general United States population: results from NHANES III. Kidney Int 63:1044–1050
Nagajyoti PC, Lee KD, Sreekanth TVM (2010) Heavy metals, occurrence and toxicity for plants: a review. Environ Chem Lett 8:199–216
Naicker K, Cukrowska E, McCarthy TS (2003) Acid mine drainage arising from gold mining activity in Johannesburg, South Africa. Environ Pollut 22(1):29–40
National Academy of Sciences (1980) Drinking water and health, vol 3. National Academy Press, Washington, DC
National Academy of Sciences (2001) Dietary reference intakes for vitamin A, vitamin K, arsenic, boron, chromium, copper, iodine, iron, manganese, molybdenum, nickel, silicon, vanadium and zinc. National Academy Press, Washington, DC
Neculita CM, Zagury GJ, Deschênes L (2005) Mercury speciation in highly contaminated soils from chlor-alkali plants using chemical extractions. J Environ Qual 34(1):255–262
Newman KD, Ahmann D, Morel MF (1998) A brief review of microbial arsenate respiration. J Geom 15:255–268
Nicholson F, Smith SR, Alloway BJ, Chambers BJ (2003) An inventory of heavy metals inputs to agricultural soils in England and Wales. Sci Total Environ 311:205–220
Nieder R, Weber TKD, Paulmann I, Muwanga A, Owor M, Naramabuye FX, Gakwerere F, Biryabarema M, Biester H, Pohl W (2014) The geochemical signature of rare-metal pegmatites in the Central Africa Region: soils, plants, water and stream sediments in the Gatumba tin-tantalum mining district, Rwanda. J Geochem Explor 144:539–551

Nishijo M, Nakagawa H, Honda R, Tanebe K, Saito S, Teranishi H, Tawara K (2002) Effects of maternal exposure to cadmium on pregnancy outcome and breast milk. Occup Environ Med 59:394–397

Nordberg GF, Nordberg M (2001) Biological monitoring of cadmium. In: Clarkson TW, Friberg L, Nordberg GF, Sager PR (eds) Biological monitoring of toxic metals. Plenum Press, New York, pp 151–168

Norland MR, Veith DL (1995) Revegetation of coarse taconite iron ore tailing using municipal waste compost. J Hazard Mater 41:123–134

NRC (National Research Council) (2000) Toxicological effects of methylmercury. National Academy Press, Washington, DC

NRC (National Research Council) (2001) Arsenic in drinking water-2001 update. National Academy Press, Washington, DC

Nriagu JO (ed) (1980) Zinc in the environment. Part I, ecological cycling. Wiley, New York

Nriagu JO, Nieboer E (eds) (1988) Chromium in the natural and human environments. Wiley, New York

Nriagu JO, Pacyna JM (1988) Quantitative assessment of worldwide contamination of air water and soils by trace metals. Nature 333:134–139

Obrador A, Alvarez JM, Lopez-Valdivia LM, Gonzalez D, Novillo J, Rico MI (2007) Relationships of soil properties with Mn and Zn distribution in acidic soils and their uptake by a barely crop. Geoderma 137:432–443

OECD (Organization for Economic Cooperation and Development) (1993) Lead background and national experience with reducing risk. Risk reduction monograph no. 1, Paris, p 1–295

Olsson IM, Bensryd I, Lundh T, Ottosson H, Skerfving S, Oskarsson A (2002) Cadmium in blood and urine – impact of sex, age, dietary intake, iron status, and former smoking – association of renal effects. Environ Health Perspect 110:1185–1190

Ontario Ministry of the Environment (2001) Arsenic in the environment. http://www.ene.gov.on.ca/cons/3792e.htm. Accessed 20 July 2017

Oremland RS, Stolz JF (2003) The ecology of arsenic. Science 300:939–944

Ortega LM, Lebrun R, Blais JF, Hauslerd R, Drogui P (2008) Effectiveness of soil washing, nanofiltration and electrochemical treatment for the recovery of metal ions coming from a contaminated soil. Water Res 42:1943–1952

Oskarsson A, Schultz A, Skerfving S, Hallen IP, Ohlin B, Lagerkvist BJ (1996) Total and inorganic mercury in breast milk and blood in relation to fish consumption and amalgam fillings in lactating women. Arch Environ Health 51(3):234–241

OSPAR (The Convention for the Protection of the Marine Environment of the North-East Atlantic) (2002) Cadmium. Hazardous Substances Series 151. OSPAR Commission, p 58. http://www.ospar.org/v_publications/download.asp?v1=p00151. Accessed 20 July 2017

Ostermann A, Gao J, Welp G, Siemens J, Roelcke M, Heimann L, Nieder R, Xue QY, Lin XY, Sandhage-Hofmann A, Amelung W (2014) Identification of soil contamination hotspots with veterinary antibiotics using heavy metal concentrations and leaching data – a field study in China. Environ Monit Assess 186:7693–7707

Pacyna JM (1986) Atmospheric trace elements from natural and anthropogenic sources. In: Nriagu JO, Davidson CI (eds) Toxic metals in the atmosphere. Wiley, New York

Page AL, Logan TJ, Ryan JA (1987) Land application of sludge: food chain implications. Lewis Publications, Chelsea

Pal R, Rai JPN (2010) Phytochelatins: peptides involved in heavy metal detoxification. Appl Biochem Biotechnol 160:945–963

Parga JR, Cocke DL, Jesus LV, Gomes JA, Kesmez M, George I, Moreno H, Weir M (2005) Arsenic removal via lector coagulation of heavy metal contaminated ground water in La Comarca Lagurera, Mexico. J Hazard Mater 124:247–254

Peng JF, Song YH, Yuan P, Cui XY, Qiu GL (2009) The remediation of heavy metals contaminated sediment. J Hazard Mater 161:633–640

Peplow D (1999) Environmental impacts of mining in Eastern Washington. Center for Water and Watershed studies fact sheet. University of Washington, Seattle

Peryea FJ (2001) Gardening on lead and arsenic containing soils. Washington State Coop. Extension Bull No EB1884

Pierce ML, Moore CB (1982) Adsorption of arsenite and arsenate on amorphous iron hydroxide. Water Res 16:1247–1253

Pierzynsky GM, Sims JT, Vance GF (2005) Soils and environmental quality. CRC Press, New York

Plum LM, Rink L, Haase H (2010) The essential toxin: impact of zinc on human health. Int J Environ Res Public Health 7:1342–1365

Polettini A, Pomi R, Rolle E, Ceremigna D, Propris LD, Gabellini M, Tornato A (2006) A kinetic study of chelant-assisted remediation of contaminated dredged sediment. J Hazard Mater 137:1458–1465

Pourbaix M (1974) Atlas of electrochemical equilibria. Pergamon Press, New York

Prabhakaran KP, Cottenie A (1971) Parent material – soil relationship in trace elements – a quantitative estimation. Geoderma 5:81–97

Pueschel SM, Linakis JG, Anderson AC (1996) Lead poisoning in childhood. Paul Brooks, Baltimore

Puschenreiter M, Horak O (2000) Influence of different soil parameters on the transfer factor soil to plant of Cd, Cu and Zn for wheat and rye. Bodenkultur 51:3–10

Puschenreiter M, Horak O, Friesl W, Hartl W (2005) Low-cost agricultural measures to reduce heavy metal transfer into the food chain – a review. Plant Soil Environ 51(1):1–11

Putila JJ, Guo NL (2011) Association of arsenic exposure with lung cancer incidence rates in the United States. PLoS One 6:e25886

Quinn MJ, Sherlock JC (1990) The correspondence between U.K. action levels for lead in blood and in water. Food Addit Contam 7:387–424

Rahman MM, Chowdhury UK, Mukherjee SC, Mondal BK, Paul K, Lodh D, Chanda CR, Basu GK, Biswas BK, Saha KC, Roy S, Das R, Palit SK, Quamruzzamman Q, Chakraborti D (2001) Chronic arsenic toxicity in Bangladesh and West Bengal, India: a review and comentry. J Toxicol Clin Toxicol 39(7):683–700

Rahola T, Hattula T, Korolainen A, Miettinen JK (1973) Elimination of free and protein-bound ionic mercury ($^{203}Hg^{2+}$) in man. Ann Clin Res 5:214–219

Raml R, Rumpler A, Goessler W, Vahter M, Li L, Ochi T, Francesconi KA (2007) Thio-dimethylarsinate is a common metabolite in urine samples from arsenic-exposed women in Bangladesh. Toxicol Appl Pharmacol 222(3):374–380

Rei GY (2000) Research advances in ionic sorbent of heavy metals. Foreign Mineral Mill Running 137(10):2–6. (in Chinese)

Reichl FX, Ritter L (2011) Illustrated handbook of toxicology. Thieme, Stuttgart

RIWA (1989) De samenstelling van het Rijnwater in 1986 en 1987 (Composition of the water of the Rhine in 1986 and 1987). RIWA, Amsterdam

Rizzi L, Petruzelli G, Poggio G, Vigna Guidi G (2004) Soil physical changes and plant availability of Zn and Pb in a treatability test of phytostabilization. Chemosphere 57:1039–1046

Robson AD (ed) (1993) Zinc in soil and plants. Kluwer, Australia

Roels HA, Hoet P, Lison D (1999) Usefulness of biomarkers of exposure to inorganic mercury, lead, or cadmium in controlling occupational and environmental risks of nephrotoxicity. Ren Fail 21(3–4):251–262

Rohrs LC (1957) Metal-fume fever from inhaling zinc oxide. AMA Arch Ind Health 16:42–47

Ross SM (1994) Toxic metals in soil-plant systems. Wiley, Chichester

Rulkens WH, Tichy R, Grotenhuis JTC (1998) Remediation of polluted soil and sediment: perspectives and failures. Water Sci Technol 37:27–35

Sadiq M (1997) Arsenic chemistry in soils: an overview of thermodynamic predictions and field observations. Water Air Soil Pollut 93:117–136

Salt DE, Smith RD, Raskin I (1998) Phytoremediation. Annu Rev Plant Biol 49:643–668

Saltzman BE, Cholak J, Schafer LJ, Yaeger DW, Meiners BG, Svetlik J (1985) Concentrations of six metals in the air of eight cities. Environ Sci Technol 19:328–333

Sanchez-Sicilia L, Seto DS, Nakamoto S, Kolff WJ (1963) Acute mercury intoxication treated by hemodialysis. Ann Intern Med 59(5):692–706

Sandborg-Englund G, Elinder CG, Langworth S, Schutz A, Ekstrand J (1998) Mercury in biological fluids after amalgam removal. J Dent Res 77(4):615–624

Sarma H (2011) Metal hyperaccumulation in plants: a review focusing on phytoremediation technology. J Environ Sci Technol 4:118–138

Schwartz BS, Lee BK, Lee GS, Stewar WF, Lee SS, Hwang KY, Ahn KD, Kim YB, Bolla KI, Simon D, Parsons PJ, Todd AC (2001) Associations of blood lead, dimercaptosuccinic acid-chelatable lead, and tibia lead with neurobehavioral test scores in South Korean lead workers. Am J Epidemiol 153:453–464

Scott F, Matthew JL, Guangchao L (2004) Temporal changes in soil partitioning and bioaccessibility of As, Cr and Pb. J Environ Qual 33:2049–2055

Scott-Fordsmand JJ (1997) Toxicity of nickel to soil organisms in Denmark. Rev Environ Contam Toxicol 148:1–34

Seaward MRD, Richardson DHS (1990) Atmospheric sources of metal pollution and effects on vegetation. In: Shaw AJ (ed) Heavy metal tolerance in plants evolutionary aspects. CRC Press, Boca Raton

Selin NE (2009) Global biogeochemical cycling of mercury: a review. Annu Rev Environ Resour 34:43–63

Shaheen SM (2009) Sorption and lability of cadmium and lead in different soils from Egypt and Greece. Geoderma 153:61–68

Shaheen SS, Tsadilas CD, Rinklebe J (2013) A review of the distribution coefficients of trace elements in soils: influence of sorption system, element characteristics, and soil colloidal properties. Adv Colloid Interf Sci 201–202:43–56

Sharma BK, Singhal PC, Chugh KS (1978) Intravascular haemolysis and acute renal failure following potassium dichromate poisoning. Postgrad Med J 54:414–415

Shresther RR, Upadhyay NP, Pradhan R, Khadka R, Maskey A, Maharjan M, Tuladhar S, Dahal BM, Shresther K (2003) Groundwater arsenic contamination, its health impact and mitigation program in Nepal. J Environ Sci Health A A38(10):185–200

Silveira MLA, Alleoni LRF, Guilherme LRG (2003) Biosolids and heavy metals in soils. Sci Agric 60(4):64–111

Skjelkvale BL, Traaen TS (2003) Heavy metals in surface waters; results from ICP Waters. In: ICP waters Report, 73/2003. Convention on Long-range Transboundary Air Pollution. International Cooperative Programme on Assessment and Monitoring of Acidification of Rivers and Lakes. The 15-year report: Assessment and monitoring of surface waters in Europe and North America; acidification and recovery, dynamic modelling and heavy metals. Norwegian Institute for Water Research NIVA

Skyllberg U, Qian J, Frech W, Xia K, Bleam WF (2003) Distribution of mercury, methyl mercury and organic sulphur species in soil, soil solution and stream of a boreal forest catchment. Biogeochemistry 64:53–76

Smith SR (1994) Effect of soil pH on availability to crops of metals in sewage sludge-treated soils. II. Cadmium uptake by crops and implications for human dietary intake. Environ Pollut 86 (1):5–13

Smith KS, Huyck HLO (1999) An overview of the abundance, relative mobility, bioavailability, and human toxicity of metals. In: Plumlee GS, Logsdon MJ (eds) The environmental geochemistry of mineral deposits, Part A: process, techniques, and health issues. Society of Economic Geologists. Chelsea, Michigan, pp 29–70

Smith RG, Vorwald AJ, Pantil LS, Mooney TF (1970) Effects of exposure to mercury in the manufacture of chlorine. Am Ind Hyg Assoc J 31:687–700

Smith PJ, Langolf GD, Goldberg J (1983) Effects of occupational exposure to elemental mercury on short term memory. Br J Ind Med 40:413–419

Smith LA, Means JL, Chen A (1995) Remedial options for metals-contaminated sites. Lewis Publishers, Boca Raton

Smith DR, Osterlod JD, Flegal AR (1996) Use of endogenous, stable lead isotopes to determine release of lead from the skeleton. Environ Health Perspect 104(1):60–66

Solomons NW (1986) Competitive interaction of iron and Zn in the diet: consequences for human nutrition. J Nutr 116:927–935

Solomons NW, Viteri F, Shuler TR, Nielsen FH (1982) Bioavailabilty of nickel in man: effects of foods and chemically-defined dietary constituents on the absorption of inorganic nickel. J Nutr 112(1):39–50

Sposito G (1989) The chemistry of soils. Oxford University Press, New York

Sposito G, Page AL (1984) Metal ions in biological systems. In: Sigel H (ed) Circulation of metal ions in the environment. Marcel Dekker, New York

Steinnes E (1997) Mercury. In: Alloway BJ (ed) Heavy metals in soils, 2nd edn. Blackie Academics and Professional Press, London, pp 245–259

Stern AH, Smith AE (2003) An assessment of the cord blood: maternal blood methyl mercury ratio: implications for risk assessment. Environ Health Perspect 111(12):1465–1470

Stevenson FJ (1982) Humus chemistry: genesis, composition, reactions. Wiley, New York

Subar AF, Krebs-Smith SM, Cook A, Kahle LL (1998) Dietary sources of nutrients among US adults, 1989 to 1991. J Am Diet Assoc 98:537–547

Sumner ME (2000) Beneficial use of effluents, wastes, and biosolids. Commun Soil Sci Plant Anal 31(11–14):1701–1715

Sundermann FW Jr, Marzouk A, Hopfer SM, Zaharia O, Reid MC (1985) Increased lipid peroxidation in tissues of nickel chloride-treated rats. Ann Clin Lab Sci 15(3):229–236

Sundermann FW Jr, Hopfer SM, Sweney KR, Marcus AH, Most BM, Creason J (1989) Nickel absorption and kinetics in human volunteers. Proc Soc Exp Biol Med 191:5–11

Suran E, Beiley TJ, Olin R (1998) A review of potentially low-cost sorbents for heavy metals. Water Res 33(11):2469–2479

Sutton P, Dick WA (1987) Reclamation of acidic mined lands in humid areas. Adv Agron 41:377–406

Tan MG, Zhang GL, Li XL, Zhang YX, Yue WS, Chen JM, Wang YS, Li AG, Li Y, Zhang YM, Shan ZC (2006) Comprehensive study of lead pollution in Shanghai by multiple techniques. Anal Chem 78:8044–8050

Taylor RM, McKenzie RM (1966) The association of trace elements with manganese minerals in Australian soils. Aust J Soil Res 4:29–39

Telisman S, Cvitkovic P, Jurasovic J, Pizent A, Gavella M, Rocic B (2000) Semen quality and reproductive endocrine function in relation to biomarkers of lead, cadmium, zinc and copper in men. Environ Health Perspect 108(1):45–53

Telmer KH, Veiga MM (2009) World emissions of mercury from artisanal and mall scale gold mining. In: Pirrone N, Mason RP (eds) Mercury fate and transport in the global atmosphere: emissions, measurements and models. Springer, Dordrecht, pp 131–173

Thomas LDK, Hodgson S, Nieuwenhuijsen M, Jarup L (2009) Early kidney damage in a population exposed to cadmium and other heavy metals. Environ Health Perspect 117(2):181–184

Thornton I (1992) Sources and pathways of cadmium in the environment. In: Nordberg GF, Herber RFM, Alessio L (eds) Cadmium in the human environment: toxicity and carcinogenicity. IARC Scientific Publications, Lyon, pp 149–162

Tong S, von Schirnding YE, Prapamontol T (2000) Environmental lead exposure: a public health problem of global dimensions. Bull World Health Organ 78:1068–1077

Tseng CH (2007) Arsenic methylation, urinary arsenic metabolites andhuman diseases: current perspective. J Environ Carcinog Ecotoxicol Rev 25(1):1–22

Tsubaki T, Irukayama K (eds) (1977) Minamata disease: methyl mercury poisoning in Minamata and Niigata, Japan. Elsevier, New York

Turkdogan MK, Kulicel F, Kara K, Tuncer I, Uyang I (2003) Heavy metals in soils, vegetable and fruit in the endermic upper gastro intestinal cancer region of Turkey. Environ Toxicol Pharmacol 13(3):175–179

Underwood EJ (1981) The mineral nutrition of livestock, 2nd edn. Commonwealth Agricultural Bureau, Slough

UNEP (United Nations Environment Programme) (2008) Interim review of scientific information on lead. https://www.google.de/#q=UNEP+(United+Nations+Environment+Programme)+(2008)+Interim+review+of+scientific+information+on+lead.+Version+of+March+2008. Accessed 20 July 2017

US ATSDR (United States Agency for Toxic Substances and Disease Registry) (2007) Toxicological profile for lead. U.S. Department of Health and Human Services, pp 1–582

US EPA (Environmental Protection Agency) (2003) Latest findings on National Air Quality, 2002 Status and Trends, U.S. Environmental Protection Agency, http://www.epa.gov/air/airtrends/aqtrnd02/. Accessed 20 July 2017

US EPA (United States Environmental Protection Agency) (1984) Health assessment document for chromium. Research Triangle Park, Final report No. EPA600/8-83-014F, Washington, DC

US EPA (United States Environmental Protection Agency) (1987) Drinking water criteria document for copper. US Environmental Protection Agency, Office of Health and Environmental Assessment, Environmental Criteria and Assessment Office, Cincinnati

US EPA (United States Environmental Protection Agency) (1996) Report: recent developments for in situ treatment of metals contaminated soils. U.S. Environmental Protection Agency, Office of Solid Waste and Emergency Response, Washington, DC

Vahter H, Mottet NK, Friberg L, Lind B, Shen DD, Burbacher T (1994) Speciation of mercury in the primate blood and brain following long-term exposure to methyl mercury. Toxicol Appl Pharmacol 124:221–229

Vamerali T, Bandiera M, Mosca G (2010) Field crops for phytoremediation of metal-contaminated land. A review. Environ Chem Lett 8:1–17

Van den Enden E (1999) Arsenic poisoning. http://www.itg.be/evde/Teksten/sylabus/49_Arsenicism.doc. Accessed 17 Feb 2016

Varo P, Koivistoinen P (1980) Mineral element composition of Finnish foods. Acta Agric Scand 22:165–171

Verkleji JAS (1993) The effects of heavy metals stress on higher plants and their use as biomonitors. In: Markert B (ed) Plant as bioindicators: indicators of heavy metals in the terrestrial environment. VCH, New York, pp 415–424

Visioli G, Marmiroli N (2013) The proteomics of heavy metal hyperaccumulation by plants. J Proteome 79:133–145

Vogelmeier C, Koenig G, Bencze K, Fuhrmann G (1987) Pulmonary involvement in zinc fume fever. Chest 92:946–948

Vos G, Lammers H, van Delft W (1988) Arsenic, cadmium, lead and mercury in meat, livers and kidneys of sheep slaughtered in the Netherlands. Z Lebensm Unters Forsch 187:1–7

Vupputuri S, Longnecker MP, Daniels JL, Gou S, Sandler DP (2005) Blood mercury level and blood pressure accrocc US women: results from the National Health and Nutrition Examination Survey 1999–2000. Environ Res 97(2):195–200

Wadge A, Hutton M (1987) The leachability and chemical speciation of selected trace elements in fly ash from coal combustion and refuse incineration. Environ Pollut 48:85–99

Wang LK, Chen JP, Hung Y, Shammas NK (2009) Heavy metals in the environment. CRC Press, Boca Raton

WHO (World Health Organisation) (1990) Chromium, Environmental Health Criteria 61. International Programme on Chemical Safety, Geneva

WHO (World Health Organisation) (1991) Nickel. Environmental health criteria 108. Geneva International Programme on Chemical Safety, Geneva

WHO (World Health Organisation) (1996) Zinc in drinking water. Guidelines for drinking-water quality, Health criteria and other supporting information, vol 2, 2nd edn. World Health Organization, Geneva

WHO (World Health Organization) (1996) Trace elements in human nutrition and health. World Health Organization, Geneva

WHO (World Health Organization) (2000) Air quality guidelines for Europe, 2nd edn. WHO Regional Publications, European Series, No 91, Copenhagen

WHO (World Health Organization) (2001) Arsenic and arsenic compounds. 2nd ed. Environmental Health Criteria 224, Geneva. http://www.inchem.org/documents/ehc/ehc/ehc224.htm. Accessed 20 July 2017

WHO (World Health Organization) (2003). Chromium in drinking water. World Health Organization, Geneva. http://www.who.int/water_sanitation_health/dwq/chemicals/chromium.pdf. Accessed 20 July 2017

WHO (World Health Organization) (2004) Copper in drinking water. Background document for development of WHO (World Health Organization) guidelines for drinking-water quality. World Health Organization, Geneva

WHO (World Health Organization) (2011) Lead in drinking water. Background document for development of WHO Guidelines for Drinking-water Quality. http://www.who.int/water_sanitation_health/dwq/chemicals/lead.pdf. Accessed 20 July 2017

Wiersman D, van Goor BJ, van der Veen NG (1986) Cadmium, lead, mercury and arsenic concentrations in crops and corresponding soils in the Netherlands. J Agric Food Chem 34:1067–1074

Wilber CG (1980) Toxicology of selenium: a review. Clin Toxicol 17(2):171–230

Wuana RA, Okieimen FE (2011) Heavy metals in contaminated soils: a review of sources, chemistry, risks and best available strategies for remediation. International Scholarly Research Network ISRN Ecology, Volume 2011, Article ID 402647, doi:https://doi.org/10.5402/2011/402647

Yao ZT, Li JH, Xie HH, Yu CH (2012) Review on remediation technologies of soil contaminated by heavy metals. Procedia Environ Sci 16:722–729

Yasir F, Tufail M, Tayyeb JM, Chaudhry MM, Siddique N (2009) Road dust pollution of Cd, Cu, Ni, Pb and Zn along Islamabad Expressway, Pakistan. Microchem J 92:186–192

Yin Y, Allen HE, Huang CP, Sparks DL, Sanders PF (1996) Adsorption of mercury(II) by soil: effects of pH, chloride, and organic matter. J Environ Qual 25(4):837–844

Zhang XP, Deng W, Yang XM (2002) The background concentrations of 13 soil trace elements and their relationships to parent materials and vegetation in Xizang (Tibet), China. J Asian Earth Sci 21:167–174

Zhang S, Zhang F, Liu X, Wang Y, Zou S, He X (2005) Determination and analysis on main harmful composition in excrement of scale livestock and poultry feedlots. Plant Nutr Fert Sci 11:822–829. (in Chinese)

Zhang MK, Liu ZY, Wang H (2010) Use of single extraction methods to predict bioavailability of heavy metals in polluted soils to rice. Commun Soil Sci Plant Anal 41(7):820–831

Zhou DM, Hao XZ, Xue Y et al (2004) Advances in remediation technologies of contaminated soils. Ecol Environ Sci 13(2):234–242

Chapter 9
Health Risks Associated with Radionuclides in Soil Materials

Abstract Radionulides in soils from natural and man-made sources constitute a direct route of exposure to humans. The most significant part of the total exposure is due to natural radiation. Soil- or rock-borne radionuclides generate a significant component of the background radiation people are exposed to. Naturally occurring radionuclides with half-lives comparable with the age of the earth and their corresponding decay products existing in terrestrial material, such as thorium (^{232}Th), uranium (^{238}U, ^{235}U) and potassium (^{40}K), are of great importance. Their spatial distribution depends on geological parent materials and plays an important role for radiation protection. Another source of exposure to natural radiation is expressed through high energy cosmic ray particles in earth's atmosphere. Additional amounts of natural radionuclides are released into the environment through human activities such as mining and milling of mineral ores, processing and enrichment, nuclear fuel fabrications, and handling of the fuel cycle tail end products. Radionuclides produced by humans originate from nuclear industrial activities, nuclear reactor accidents, or military activities. The most important man-made radionuclide is cesium (^{137}Cs) with a half-life of 30.17 years, which is released from nuclear fission and activation processes. A large amount of ^{137}Cs was released into the atmosphere during the nuclear weapons tests in the 1950s and 1960s. Atmospheric deposition of this ^{137}C has made it a typical background component of northern hemisphere top soils. The most severe civil nuclear reactor accidents, which also released large quantities of ^{137}Cs, occurred at Chernobyl (April 26, 1986) and Fukushima Daiichi (March 11, 2011) power stations. During the years after those weapon tests and reactor accidents, the bioactivity and environmental mobility of ^{137}Cs declined markedly, resulting in large changes in contamination of soils, surface water and foodstuffs.

Radiation is harmful to life. Depending on the dose, it can cause cancer, genetic and organs damages, cell killing as well as rapid death. The potential damage from an absorbed dose depends on the type of radiation and the sensitivity of different tissues and organs. Radionuclides can generally be internalized through inhalation, ingestion, wound contamination and percutaneous absorption. If radionuclide contamination is likely, the first step is to remove sources of potential contamination. External decontamination procedures are vital in reducing the risk of additional internal contamination events. Isotope-specific pharmacological treatments can begin once thorough external decontamination is performed. For remediation of

R. Nieder et al., *Soil Components and Human Health*,
https://doi.org/10.1007/978-94-024-1222-2_9

radionuclide-contaminated sites a number or remediation techniques is available. The choice of the technique requires consideration of performance, reliability and maintenance requirements, cost, available supporting infrastructure, risk to workers and public during implementation, environmental impact, future land use and regulatory and community acceptance.

Keywords Alpha particles • Beta particles • Gamma particles • Cosmogenic radiation • Natural terrestrial radiation • Nuclear weapons tests • Nuclear fuel cycle • Nuclear accidents • Radionuclides in food and water • Radiation exposure • Clinical effects • Therapy • Remediation of radionuclide-contaminated sites

9.1 Types of Radiation

Unstable atoms are said to be radioactive. In order to reach stability, these atoms give off, or emit, the excess energy or mass. These emissions are called radiation. The three main types of ionizing radiation are called alpha (α), beta (β) and gamma (γ) radiation.

9.1.1 Alpha Particles

Alpha radiation is a particle, consisting of two protons and two neutrons, making it identical to a helium atom, but without the electrons. Compared to a β-particle the α-particle has a large size. It travels very fast and thus has a large amount of energy. When a positively charged α-particle passes near an atom, it excites its electrons and pulls an electron from the atom. This is the process of ionization. With each ionization, the α-particle loses some energy and slows down. At the end of its path it will take two electrons from other atoms and become a complete helium atom, which has no biological effect.

9.1.2 Beta Particles

Beta particle emission occurs when the ratio of neutrons to protons in the nucleus is too high. In this case, an excess neutron transforms into a proton and an electron. The proton stays in the nucleus and the electron is ejected energetically. This process increases the number of protons by one and decreases the number of neutrons by one. Since the number of protons in the nucleus of an atom determines the element, the conversion of a neutron to a proton changes the radionuclide to a different element. Beta particles have either a positive or a negative charge. Most β-particles are negatively charged. They are much lighter and more penetrating than α-particles.

9.1.3 *Gamma Particles*

Gamma radiation is very high-energy ionizing radiation. Gamma photons have about 10,000 times as much energy as the photons in the visible range of the electromagnetic spectrum. Different from α- and β-radiation, γ-photons are pure electromagnetic energy having no mass and no electrical charge. Because of their high energy, gamma photons travel at the speed of light and can cover hundreds to thousands of meters in air before spending their energy.

9.2 Determination of Radioactivity: Definition of the Units Used

Radioactivity can be quantified in different categories. The corresponding units are briefly described. Radioactivity is expressed with the unit Bequerel (Bq), Gray (Gy) describes the absorbed dose and Sievert (Sv) is representative for the effects on health of an individual which has been exposed to radionuclides (Table 9.1).

9.2.1 *Bequerel*

Becquerel is the unit of radioactivity measurement and one Becquerel is equivalent to one disintegration per second. One Curie (originally used unit) is equivalent to the radioactivity of 1 g of radium which corresponds to 37 billion disintegrations per second.

9.2.2 *Gray*

Gray represents the amount of radiation absorbed by any matter. One Gray corresponds to 1 J absorbed per kilogram of matter. Originally the Röntgen, and more recently, the Rad (Radiation Absorbed Dose) were common. One Rad is equivalent to 10^{-2} Gy.

Table 9.1 Current and former units for quantification of radioactivity

Current unit	Former unit	Equivalence	Quantity
Bequerel (Bq)	Curie (Ci)	1 Ci = 3.7×10^{10} Bq	Radioactivity
Gray (Gy)	Rad (rad)	1 rad = 10^{-2} Gy	Absorbed dose
Sievert (Sv)	Rem (rem)	1 rem = 10^{-2} Sv	Biological effect

Compiled from different sources

9.2.3 Sievert

Sievert is a unit of radiation weighted dose and is used to distinguish the effects induced on living tissues by the different types of radioactive particles and ionising radiation (alpha, beta, gamma, neutron), along with the radiosensitivity differences amongst different organs and tissues. The Sievert is a risk management unit used to assess the equivalent dose and the effective dose. The equivalent dose shows the effect of the different types of radioactive particles and ionising radiation on tissues. It corresponds to the absorbed dose, expressed in Gy, multiplied by a radiation weighting factors which depend on the energy delivered by the radiation. Originally, the Rem (Röntgen Equivalent Man) was used. One Rem is equivalent to 10^{-2} Sv. The effective dose, besides the different types of radioactive particles and ionising radiation on tissues, also describes the more or less radiosensitive nature of the exposed tissue or organ. It corresponds to the equivalent dose, expressed in Sv, multiplied by a tissue weighting factor which depends on the radiosensitivity of the tissue.

9.3 Naturally Occurring Radionuclides

Humans are exposed to ionizing radiation spontaneously emitted by naturally occurring atomic species since his existence on the earth. Three types of radiations, alpha, beta and gamma are emitted by different radioactive materials, which differ in their energy and penetrating power. Until recent times, life on earth was exposed to radiation only from natural sources. Sources of radiation can vary from place to place and there are areas in some part of the world in that the background radiation levels have been found to be abnormally high. Such areas are referred to as high background radiation areas. For example, the coastal regions of Espirito Santo and the Morro Do Forro in Brazil (Paschoa 2000), Ramsar and Mahallat in Iran (Ghiassi-nejad et al. 2002), the Southwest coast of India (Paul et al. 1998) and Yangjiang in China (Wei and Sugahara 2000) were identified as high background radiation areas. Monazite sands have been found to be the source of high radiation levels in parts of Brazil, China, Egypt and India (UNSCEAR 2000; Paschoa 2000; Ghiassi-nejad et al. 2002) while in parts of Southwest France, uranium minerals form the source of natural radiation (Delpoux et al. 1997), and in Ramsar, the very high amounts of ^{226}Ra and its decay products brought to the surface by hot springs (Ghiassi-nejad et al. 2002) have been found to be the source. Humans have created other sources of exposures such as radioactive waste and industrial, medical and agricultural use of radioisotopes, fallout from weapon tests and radioactive releases from nuclear reactor operations and accidents. However, the major contribution to the average annual background radiation still arises from natural sources.

9.3.1 Cosmogenic and Terrestrial Sources of Radionuclides

Exposure of human beings to naturally occurring radiation arises mainly from two different sources, the first coming directly from cosmic radiation from the outer space and the second from terrestrial radioactive materials. Exposure through naturally occurring radiation accounts for up to 85% of the total annual exposure dose received by the world population (World Nuclear Association 2011). The interactions of cosmic ray particles in the atmosphere can create a number of radioactive nuclei such as ^{3}H (half-life: 12.3 years), ^{7}Be (half-life: 53.2 day), ^{14}C (half-life: 5700 years) and ^{22}Na (half-life: 2.6 years) (National Council on Radiation Protection and Measurements 1975). Apart from the exposure from cosmic radiation, natural exposures originate mainly from the primordial radionuclides, which are spread widely and are present in almost all geological materials on earth (Wilson 1994). These radionuclides are known as Naturally Occurring Radioactive Material (NORM).

Only very long-lived nuclides, with half-lives comparable to the age of the earth, and their decay products, contribute to natural radiation background in significant quantities (United Nations Scientific Committee on the Effects of Atomic Radiation 2000). The majority of naturally occurring radionuclides belong to ^{238}U and ^{232}Th series and the single decay ^{40}K. Generally ^{235}U, ^{87}Rb and other trace elements are negligible. The decay chain of ^{238}U (^{232}Th) includes eight (6) alpha decays and six (4) beta decays respectively, which are often associated with gamma transitions (IAEA 2006). Radionuclides of the Th and U decay series and their half-lives are given in Fig. 9.1.

An overview of the radioisotopes produced by cosmic rays and terrestrial sources is given in Table 9.2. Radionuclides, which emit alpha or beta particles may be taken into the body by ingestion or inhalation and can give rise to internal exposures. Some of these nuclear species may also emit gamma rays following their

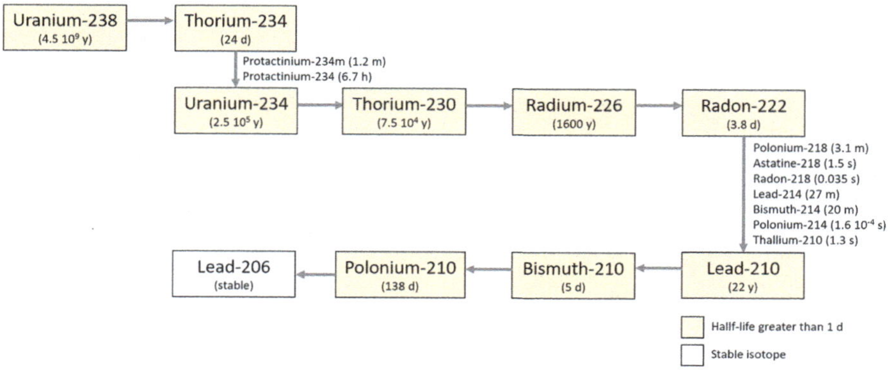

Fig. 9.1 Radionuclides of the thorium and uranium decay series

Table 9.2 Radioisotopes produced by cosmic rays and from terrestrial sources

Cosmogenic radioisotopes	Type of radiation	Terrestrial radioisotopes	Type of radiation
^{14}C	β	^{222}Rn (Radon)	α
^{32}Si	β	^{218}Po (RaA)	α
^{39}Ar	β	^{214}Pb (RaB)	β, γ
^{3}H	β	^{214}Bi (RaC)	α, β, γ
^{22}Na	β, γ	^{210}Pb (RaD)	β
^{35}S	β	^{210}Bi (RaE)	β
^{7}Be	γ	^{210}Po (RaF)	α
^{37}Ar	γ	^{220}Rn (Thoron)	α
^{33}P	β	^{216}Po (ThA)	α
^{32}P	β	^{212}Pb (ThB)	β, γ
^{24}Na	β, γ	^{212}Bi (ThC)	α, β, γ

Adapted from Ramachandran (2011)

Table 9.3 Concentrations of ^{238}U, ^{232}Th, and ^{40}K (ranges and averages) in rocks and soils

	^{40}K		^{232}Th		^{238}U	
Type of rock	K (%)	Bq kg^{-1}	μg g^{-1}	Bq kg^{-1}	μg g^{-1}	Bq kg^{-1}
Igneous rocks						
Basalt (crustal average)	0.8	300	3.0–4.0	10–15	0.5–1.0	7–10
Granite (crustal average)	>4.0	>1000	17	70	3.0	40
Sedimentary rocks						
Shale, sandstones	2.7	800	12	50	3.7	40
Arkose	2–3	600–900	2.0	<8.0	1.0–2.0	10–25
Carbonate rocks	0.3	70	2.0	8.0	2.0	25
Soils (global average)	1.5	400	9.0	37	1.8	22
Continental crust (average)	2.8	850	10.7	44	2.8	36

Compiled from National Council on Radiation Protection and Measurements (1988)

radioactive decay and represent the main sources of external (whole body) exposures to humans (Watson et al. 2005).

Levels of the radioactivity in soils are related to the mineral composition of the parent rock from which soils have developed and the soil forming processes. Igneous rocks, such as granite, generally exhibit higher radioactivity compared to sedimentary rocks (excluding some shales and phosphate rocks). Table 9.3 gives typical natural radioactivity concentrations in rocks and soils.

A number of studies focusing on naturally occurring radioactive materials in soil media provide information on levels of background radiation. Most soils contain ^{40}K and nuclides of the U and Th series, with broadly varying concentrations. For example, activity concentration levels arising from ^{238}U, ^{232}Th and ^{40}K in surface soils in Cyprus ranged from 0.01 to 39.3, 0.01 to 39.8, and 0.04 to 565.8 Bq kg^{-1}, respectively (Tzotzis et al. 2004). Mean natural radioactivity levels in the Firtina Valley (Turkey) of ^{238}U, ^{232}Th, ^{40}K, and ^{137}Cs were found to be 50, 42, 643 and

85 Bq kg^{-1} in soil samples, and 39, 38, 573 and 6 Bq kg^{-1} in river sediment samples (Kurnaz et al. 2007). Radioactivity levels in soils, developed from various geological parent materials in the northern Highlands of Jordan, were 42.5, 49.9, 26.7 and 291.1 Bq kg^{-1} for ^{226}Ra, ^{238}U, ^{232}Th, and ^{40}K, respectively (Al-Hamarneh and Awadallah 2009). Radioactivity levels in sediment samples taken along the Upper Egypt Nile River for ^{238}U, ^{232}Th, and ^{40}K ranged from 3.8 to 34.94, 2.8 to 30.10 and 112.31 to 312.98 Bq kg^{-1}, respectively (El-Gamal et al. 2007). In soils from Bahawalpur (Pakistan) mean activity levels of ^{226}Ra, ^{232}Th, ^{40}K, and ^{137}Cs were 32.9, 53.6, 647.4 and 1.5 Bq kg^{-1} (Matiullah et al. 2004).

In agricultural soils, application of fertilizers, particularly of inorganic P, has influenced radionuclide and trace element concentrations to a large extent (National Council on Radiation Protection and Measurements 1975). The application of phosphate fertilizers has substantially increased worldwide (Khater and Al-Sewaidan 2008). Fertilizers may also affect the chemical form of natural radionuclides in soils and thus their bioavailability (Klement 1982).

9.3.2 *Natural Sources Modified by Humans*

The term NORM is used more specifically for all naturally occurring radioactive materials where human activities have increased the potential for exposure compared with the unaltered situation. The large production of NORM in the last decades and the potential long-term radiological hazards represent an increasing level of concern. NORM are found as products, by-products and/or wastes of industrial activities, such as production of non-nuclear fuels (e.g. coal, oil and gas), mining and milling of metalliferous and non-metalliferous ores (e.g. aluminum, iron, copper, gold and mineral sand), industrial minerals (e.g. phosphate and clays), radioisotope extraction and processing, as well as water treatments (IAEA 2003).

9.3.2.1 Fossil Fuels

Oil and Gas Production

Radioactivity in oil, gas and coal originates from cosmogenic and primordial radionuclides which are released to the environment through burning of these fuels (Ramachandran 2011). Rocks that hold oil and gas also contain U and Th at the order of some mg kg^{-1}, corresponding to a total specific activity of some tens of Bq kg^{-1}. However, these are commonly not mobilized from the rock formations. Oil and gas reservoirs contain a natural water layer (formation water) that lies under the hydrocarbons in that U and Th do not go into solution. However, the formation water tends to reach a specific activity of the same order of the rock matrix (Metz et al. 2003) due to dissolution of radium isotopes. In the water co-produced during

Table 9.4 Radioactivity in oil and gas production

Radionuclide	Crude oil Bq g^{-1}	Natural gas Bq m^{-3}	Produced water Bq L^{-1}	Hard scale Bq g^{-1}	Sludge Bq g^{-1}
^{238}U	<0.01		<0.1	0.001–0.5	0.005–0.1
^{226}Ra	<0.04		0.002–1200	0.1–15,000	0.05–800
^{210}Po	<0.01	0.002–0.08		0.02–1.5	0.004–160
^{210}Pb		0.005–0.02	0.05–190	0.02–75	0.1–1300
^{222}Rn		5–200,000			
^{232}Th	<0.002		<0.01	0.001–0.002	0.002–0.1
^{228}Ra			0.3–180	0.05–2800	0.5–50
^{224}Ra			0.05–40		

Compiled from Jonkers et al. (1997)

oil and gas extraction, ^{226}Ra, ^{224}Ra, ^{228}Ra and ^{210}Pb are mobilized. These isotopes can precipitate out of solution, along with sulphate and carbonate deposits as scale or sludge in pipes. The immediate decay product of ^{226}Ra is ^{222}Rn which preferentially follows gas lines and is transformed through several rapid steps to ^{210}Pb which can precipitate as a thin film in gas extraction equipment. The level radioactivity varies significantly, depending on the radioactivity of the reservoir rock and the salinity of the water co-produced from the well (Table 9.4). The radioactivity of scales and sludges can vary considerably and is much higher than that of formation water. Fracking for gas production releases significant radioactivity in drill cuttings and water.

Exposure to radionuclides in the oil and gas industries poses a risk to workers during maintenance, waste transport, processing and decommissioning. In particular ^{210}Pb deposits and films, as a beta emitter, is a concern only when pipe internals become exposed. Internal exposures can be reduced through hygiene practices. External exposure is generally low enough and does not require protective measures.

Coal Production

Coal contains uranium and thorium, their decay products and ^{40}K. Total levels of individual radionuclides typically are not great and are generally about the same as in rocks near the coal. Enhanced radionuclide concentration in coal tends to be associated with the presence of other heavy metals and S. Characteristic values are given in Table 9.5.

Coals from Australia, the US, may contain up to 4 ppm U, those from Germany up to 13 ppm U and coals from Brazil and China <20 ppm U. Concentrations of Th are often three times higher than those of U. The biggest producers of coal are China, the US and India, producing together more than two thirds of the world total. Coal has reached a global consumption of about 4.8 Pg in 2001 (IEA 2001).

Table 9.5 Radioactivity in coal (in Bq kg^{-1}; adapted from IAEA 2003)

Country	^{238}U	^{226}Ra	^{210}Pb	^{210}Po	^{232}Th	^{228}Ra	^{40}K
Australia[a]	8.5–47	19–24	20–33	16–28	11–69	11–64	23–140
Germany[a]		10–145			10–63		100–700
Germany[b]		<58			<58		4–220
Greece[b]	117–390	44–206	59–205			9–41	
Hungary[a]	20–480					12–97	30–384
Poland[a]	<159					<123	<785
Romania[a]	<415	<557	<510	<580	<170		
UK[a]	7–19	8–22			7–19		55–314
USA[a]	6–73	9–59	12–78	3–52	4–21		

[a]Hard coal
[b]Lignite

According to the World Coal Association 42% of the world's electricity is currently generated by coal (WCA 2017).

Most coal is used for electricity generation in coal-fired power plants. During combustion of coal, radionuclides are concentrated in fly ash and bottom ash. The concentrations can vary widely depending on the mineral composition of the coal, which can differ between mining regions. Differences also exist in residues from hard coal and lignite. The volatile ^{210}Po shows a significant enrichment in fly ashes (Godoy et al. 2000). In older and smaller power plants, the absence or inadequacy of flue gas filtering and scrubbing may result in the atmospheric dispersal of fly ashes. In modern units, almost all of the fly ash is retained. While a lot of ash is deposited into surface impoundments and landfills, or is backfilled into the mines, some is also recycled together with gypsum for building construction. This may lead to exposure of the inhabitants to radiation. In 2003 the estimated worldwide fly ash production was 0.39 Pg (Mukherjee et al. 2008). In 2003 the US and EU fly ash production was about 0.06 Pg and 0.04 Pg, respectively. According to US GS (2002) the world coal consumption in 2035 is projected to reach about 9.4 Pg, corresponding to an estimated fly ash production of 0.75 Pg. In Table 9.6 some values for the radioactivity of coal ash and slag are given.

A special situation arises when coal is burnt domestically. Emissions of particles from coal-based heating and cooking systems can be considerable, particularly in less developed countries (Clarke 1992). As coal mining produces a significant amount of waste rock and drainage water, the mining process itself also contributes to increased levels of radioactivity. Underground coal mines are subject to increased levels of Rn, while elevated levels of radium and ^{40}K can be found in mining waste rocks and soil. Sediments discharged in waste water into the environment have been measured with activities as high as 55,000 Bq kg^{-1} of ^{226}Ra and 15,000 Bq kg^{-1} of ^{228}Ra (IAEA 2003).

Table 9.6 Radioactivity in coal ash and coal slag from power plants (in Bq kg^{-1}; adapted from IAEA 2003)

Country	^{238}U series	^{232}Th series	^{40}K
Germany[a]	6–166	3–120	125–742
Germany[b]	68–245	76–170	337–1240
Egypt[a]	16–41	9–11	
Egypt[b]	41–90	24–34	
Hungary[a, b]	200–2000	200–300	300–800
USA[a]	100–600	30–300	100–1200

[a]Fly ash or bottom ash
[b]Slag

9.3.2.2 Mining of Metal Ores

Mineral Sands

Mineral sands are sources of zircon, ilmenite, and rutile with xenomite and monazite which are mined in many countries. Mineral sand ore is important for the production of titanium, tin and zirconium bearing minerals and rare earth elements. The minerals have been concentrated by marine, alluvial and/or wind processes. These placer deposits can be found also in vein deposits, mostly disseminated in alkaline intrusions in hard rocks. The production amounts to millions of Mg per year of zirconium and titanium (from rutile and ilmenite). Thorium, tin and rare earth elements are associated. Depending on the placer geology, the radioactivity concentration of mineral sand can vary within a large range. The minerals in the sands are subject to gravity concentration and some of the concentrates are highly radioactive (Table 9.7).

In terms of NORM, monazite is highly relevant. Monazite is a rare earth phosphate containing a variety of rare earth minerals such as cerium, lanthanum, xenotime and yttrium phosphate with traces of uranium and thorium (World Nuclear Association 2014). Industrial use of zirconium mainly occurs in the form of zircon (zirconium silicate). This mineral occurs naturally and requires little processing. It is used chiefly in foundries, refractories manufacture and the ceramics industry. Zircons typically have activities of up to 10,000 Bq kg^{-1} of ^{238}U and ^{232}Th. No attempt is usually made to remove radionuclides from the zircon as this is not economical. Because zircon is used directly in the manufacture of refractory materials and glazes, the products will contain similar amounts of radioactivity. During mining and milling of zircon, care must be taken to keep dust levels down. When zircon is fused in refractories or ceramics manufacture, silica dust and fumes must be collected. This may contain the more volatile radionuclides, ^{210}Pb and ^{210}Po, and the collection of these gases means that filters become contaminated. The main radiological issue is occupational exposure to these radionuclides in airborne dusts in the processing plant. Waste produced during zirconia/zirconium production can be high in ^{226}Ra, which presents a gamma hazard, and waste must be stored in metal containers in special repositories. Powders from filters used during zirconia manufacture have been assayed as high as 200,000 Bq kg^{-1}

Table 9.7 Radioactivity in mineral sand products

	^{232}Th series		^{238}U series	
Mineral sand products	μg g^{-1}	Bq kg^{-1}	μg g^{-1}	Bq kg^{-1}
Ore	5–70	40–600	3–10	70–250
Mineral concentrate (primary separation)	80–800	600–6600	<10–70	<250–1700
Ilmenite	50–500	400–4100	<10–30	<250–750
Rutile	<50–350	<400–2900	<10–20	<250–500
Zircon	150–300	1200–2500	150–300	3700–7400
Xenomite	~15,000	~120	~4000	~100
Monazite	50,000–70,000	41,000–575,000	1000–3000	25,000–57,000
Tailings waste	200–6000	1500–50,000	10–1000	12,000–60,000

Adapted from IAEA (2003)

of ^{210}Pb and 600,000 Bq kg^{-1} ^{210}Po (World Nuclear Association 2014). Tin is sometimes a by-product of mineral sand production. Slag from smelting tin often contains high levels of niobium and tantalum and so may form the feedstock for their extraction. It also typically contains enhanced level of radionuclides (World Nuclear Association 2014).

Uranium Mining and Milling

Uranium mining and extraction of U from the ore have generated the largest volume of radioactive waste worldwide. Globally, the total estimated volume of mill tailing is 938×10^6 m^3 produced at roughly 4400 mines. The radioactivity of these tailings depends on the grade of ore mined and varies from less than 1000 Bq kg^{-1} to more than 100,000 Bq kg^{-1} (Abdelouas 2006). Kazakhstan has produced by far the largest volume of tailings (209×10^6 m^3), while the production in the USA is about 120×10^6 m^3. The radionuclides in uranium mill tailings include ^{238}U, ^{235}U, ^{234}U, ^{230}Th, ^{226}Ra, and ^{222}Rn. The isotopes ^{238}U and ^{230}Th are long-lived emitters, while ^{222}Rn has a short half-life (3.8 days). Radon concentrations persist in mill tailings because ^{222}Rn is a decay product of the longer-lived ^{226}Ra (half-live 1600 years), with the gamma radiation constituting the principal radiation risks from uranium tailings. In addition to radioactivity, uranium mill tailings are often associated with elevated concentrations of highly toxic heavy metals. Because of the oxidation of high sulfide content, uranium tailings generate acidic waters and accelerates the releases of radioactive and hazardous elements (Abdelouas 2006).

Bauxite Mining for Alumina Production

Bauxites are mined for alumina production. They generally contain concentrations of ^{232}Th and ^{238}U greater than the Earth's crustal average. Activities for ^{232}Th and

Table 9.8 Radioactivity in red mud from alumina processing

Country	Activities (Bq kg^{-1}) ^{226}Ra	^{232}Th	References
Greece	13–185	15–412	Papatheodorou et al. (2005)[a]
Jamaica	370–1047	328–350	Pinnock (1991)[a]
Hungary	225–568	219–392	Somlai et al. (2008)[a]
Turkey	128–285	342–357	Turhan et al. (2011)[a]
Germany	122 ± 18	183 ± 33	Philipsborn and Kühnast (1992)[b]
Australia	310 ± 20	1350 ± 40	Cooper et al. (1995)[b]

[a]Ranges indicating minima and maxima
[b]Means±standard deviation

^{238}U in bauxites were reported to be in a range of 400–600 Bq kg^{-1} for ^{238}U and 300–400 Bq kg^{-1} for ^{232}Th (UNSCEAR 2000). Parent rock composition significantly influences the ^{232}Th and ^{238}U abundances in bauxites. Those derived from acid igneous rocks show a concentration higher than those extracted from basic igneous rocks, whereas the bauxites mined from deposits of shales and carbonate rocks are characterized by intermediate concentrations. The process of lateritization during bauxite formation contributes to increase the Th: U ratio, which generally exceeds 4 (Adams and Richardson 1960). In 2009 the worldwide production of bauxite and alumina was 199 × 10^{6} Mg and 123 × 10^{6} Mg, respectively. Considering that the worldwide bauxite: alumina ratio is about 2.7, an amount of 1.7 kg of red mud is generated per kg of alumina (USGS 2002). During the process of extracting alumina from bauxite, over 70% of the radionuclides are concentrated in the red mud (Adams and Richardson 1960). Radionuclide activities in red mud reported in several studies are given in Table 9.8.

Tantalum, Niobium and Tin production

The main application of Ta is in the electronics industry as a major constituent of capacitors. Tantalum usually is associated with the chemically-similar niobium, often in tantalite and columbite. Tantalum ores, often derived from pegmatites, comprise a wide variety of more than a hundred minerals, some of which contain U and/or Th (Lehmann et al. 2013). Hence the mined ore and concentrate in some regions may contain both of these and their decay products in their crystal lattice. Concentration of the Ta minerals is generally by gravity methods (as with mineral sands), so the lattice-bound radioisotope impurities if present will report with the concentrate. While this has little radiological significance in the processing plant, concentrates shipped to customers sometimes exceed the Transport Code threshold of 10 kBq kg^{-1}, requiring declaration and some special documentation, labeling and handling procedures. Some concentrates reach 75 kBq kg^{-1} (World Nuclear Association 2014). Niobium (Nb) slags can reach radioactivity levels in excess of 100 kBq kg^{-1}. The largest producers of tantalum are Australia and Africa, most niobium comes from Brazil (World Nuclear Association 2014). In some situations

Ta mineralisation is associated with tin-bearing minerals. The primary tantalum/tin ore contains trace quantities of ^{238}U and ^{232}Th associated with the minerals. Activities of ^{238}U and ^{232}Th in the ore are less than 60 Bq kg^{-1} and less than 5 Bq kg^{-1}, respectively (Cooper 2005). After the initial dry and wet separation some ^{238}U and ^{232}Th may remain with the concentrates and are removed by acid leaching. Neutralisation of the leach solutions produces a solid tails containing ^{238}U and ^{232}Th. The levels of ^{238}U and ^{232}Th in the Ta products may reach radioactivity levels up to 75 kBq kg^{-1} (Cooper 2005).

Copper Mining

Copper is a metal of major importance with a range of uses, with the main application being for electrical installations and the electronics industry. Copper and gold are often associated with silver and uranium. After milling the ore, Cu is separated by flotation to produce a concentrate with a copper content of about 30%. The concentrates are smelted to remove volatile or less dense impurities. Further purification of the Cu melt from the smelter produces a primary form of the metal, known as blister copper. Electro-refining produces higher purity Cu. Most of the Au and U minerals are separated from the Cu concentrates during the flotation stage. Uranium and Th may be present in significant quantities in the Cu mineralization. Partitioning of ^{210}Pb and ^{210}Po from U occurs into the Cu concentrate during the smelting process. These radionuclides are vaporised at the smelting stage and may accumulate in dusts collected from off gases. Unless U is present in commercial quantities and separated during processing, the U will remain in the tailings from the flotation stage or will be present in the Cu concentrate and partition to the slag from the Cu smelter (Cooper 2005).

9.3.2.3 Phosphate Fertilizers

Phosphate fertilisers for agricultural use are derived from phosphate rock. Superphosphate contains approximately 20% available P. Higher P contents are present in triple superphosphate, mono/diammonium phosphate, and dicalcium phosphate. The main phosphate-rock (phosphorite) deposits are both of igneous and sedimentary origin and are part of the apatite group. They are commonly encountered as fluorapatite and francolite, respectively. Uranium is incorporated in sedimentary phosphates through ionic substitution into the carbonate-flourapatitic crystals or by adsorption. Igneous phosphorite contains less U, but more Th. High contents of P usually correspond to high contents of U (50–300 mg kg^{-1}; IAEA 2003). The world's largest producers in 2007 were the USA, Morocco and China, the global total was 156 Mt P in the same year (World Nuclear Association 2014). Concentrations of radionuclides in phosphate rocks are given in Table 9.9.

Normal superphosphate is produced by adding sulphuric acid to phosphate rock to form soluble mono-calcium phosphate. Phosphoric acid is used as the acidulating agent for higher grades of superphosphate and for ammonium phosphates.

Table 9.9 Radioactivity in phosphate rocks

Country	^{238}U (Bq kg^{-1})	^{232}Th (Bq kg^{-1})	^{226}Ra (Bq kg^{-1})	^{228}Ra (Bq kg^{-1})
USA	259–3700	3.7–22	1.540	–
Brazil	114–880	204–753	330–700	350–1550
Morocco	1500–1700	10–200	1500–1700	–
South Africa	100–200	483–564	–	–
Israel	1500–1700	–	–	–
Jordan	1300–1850	–	–	–
Australia	15–900	5–47	28–900	–

Adapted from IAEA (2003)

Table 9.10 Radioactivity in phosphate fertilizers

Product	^{238}U (Bq kg^{-1})	^{232}Th (Bq kg^{-1})	^{226}Ra (Bq kg^{-1})
Phosphoric acid	1200–1500	–	300
Superphosphate	520–1100	15–44	110–960
Triple superphosphate	800–2100	44–48	230–800
Ammonium phosphate	2000	63	20
Diammonium phosphate	2300	<15	210
Dicalcium phosphate	–	<37	740
PK fertilizers	410	<15	370
NP fertilizers	920	<30	310
NPK fertilizers	440–470	<15	210–270

Adapted from IAEA (2003)

Phosphoric acid itself is produced by treating rock phosphate with excess sulphuric acid. The major solid waste product from the phosphate industry is the large quantities of calcium sulphate (phosphogypsum) arising during phosphoric acid production. Unless the acid is to be used for fertiliser production, purification of the phosphoric acid is carried out by solvent extraction (Cooper 2005). In phosphogypsum 80–90% of the ^{226}Ra along with a high contents of ^{210}Pb and ^{210}Po (Carvalho 1995; Beddow et al. 2006) are found. About 80–85% of the ^{238}U (Beddow et al. 2006) and roughly 70% of ^{232}Th (Tayibi et al. 2009) concentrate in phosphoric acid. The radioactivity of P fertilizers is variable and depends on the radionuclide content of the phosphate rock and on the method of production. Radioactivity levels of different P products are given in Table 9.10.

9.3.2.4 Ceramics and Building Materials

Building materials may contain radionuclides due to their occurrence in raw materials (e.g. ^{238}U and ^{232}Th natural stones) or may be added with industrial products such as zircon sand or with industrial byproducts like coal ash, phosphogypsum and furnace slags. Generally, the radioactivity in the endproduct

Table 9.11 Radioactivity in different building materials

Material	^{226}Ra (Bq kg^{-1})	^{232}Th (Bq kg^{-1})	^{40}K (Bq kg^{-1})
Concrete	1–250	1–190	5–1570
Clay bricks (red)	9–2200	<220	180–1600
Sand-lime bricks/limestone	6–50	1–30	5–700
Natural building stones	1–500	1–310	1–4000
Natural gypsum	<70	<100	7–280
Cement	7–180	7–240	24–850
Tiles	30–200	20–200	160–1410
Phosphogypsum plasterboard	4–700	1–53	25–120
Blast furnace slagstone	30–120	30–220	–

Adapted from IAEA (2003)

will be lower compared to the original by-product because of the presence of other inert material in the building material. Typical levels of radioactivity in different materials are given in Table 9.11.

9.3.3 Anthropogenic Radionuclides

9.3.3.1 Release of Radionuclides by Nuclear Weapons Testing

The era of nuclear weapon testing began with the Trinity test of 1945. A total of 543 atmospheric weapon tests were carried during the period 1945–1980 giving a total yield of 440 Mt (comprising of 189 Mt due to fission and 251 Mt due to fusion) by all nuclear weapon countries all over the world. These were carried out at different locations on and above the earth's surface. Depending on the location of the detonation the radioactive debris entered the local, regional, or global environment. For tests conducted on the earth's surface, a portion of the radioactive debris is deposited at the site of the test (local fallout) and regionally up to several thousand kilometers down wind (intermediate fallout). Most radionuclides released in nuclear tests have a short half life. The most important fission products are given in Table 9.12.

Additionally, radionuclides such as ^{236}U, ^{237}Np, $^{238\text{-}242}$Pu and ^{241}Am are formed by neutron capture reactions. In terms of radioactivity, ^{3}H, ^{90}Sr, ^{137}Cs and plutonium isotopes are currently the radionuclides of great importance.

9.3.3.2 Release of Radionuclides by Reactors and Reprocessing Plants

Generation of electrical energy by nuclear means has grown steadily from the start of the industry in 1956. Currently, 439 reactors in 31 different countries are in operation (IAEA 2008). Nuclear power currently provides 16% of electricity in

Table 9.12 Radionuclides released to the atmosphere due to nuclear testing

Radionuclide	Estimated release (EBq)	Half-life (days or years)	Radionuclide	Estimated release (EBq)	Half-life (days or years)
^{3}H	186	12.3 years	^{125}Sb	0.74	2.7 years
^{14}C[a]	0.21	5730 years	^{131}I	675	8.0 days
^{54}Mn	3.98	312.3 years	^{137}Cs	0.95	30.0 years
^{55}Fe	1.53	2.7 years	^{140}Ba	759	12.8 days
^{89}Sr	117	50.5 days	^{141}Ce	263	32.5 days
^{90}Sr	0.62	28.8 years	^{144}Ce	30.7	284.9 days
^{91}Y	120	58.5 days	^{239}Pu	0.0065	24,110 years
^{95}Zr	148	64.0 days	^{240}Pu	0.0044	6560 years
^{103}Ru	247	39.3 days	Pu	0.14	14.4 years
^{106}Ru	12.2	1.0 years			

Adapted from UNSCEAR (2000)
[a]All ^{14}C was assumed to be due to fusion

Russia, 18% in the UK, 30% in Japan, 32% in Germany, 48% in Sweden and Ukraine, 78% in France, 19% in the USA and 30% for the entire European Union (Hu et al. 2010).

During the operation of a typical reactor, more than 200 radionuclides are produced most of which are relatively short-lived and decay to low levels within a few decades (Crowley 1997). A number of radionuclides are emitted from normal operation of NPP. For example, the annual discharge of gaseous ^{14}C to the atmosphere from pressured water reactors in Germany was 280 ± 20 GBq GWe^{-1} in 1999, on average 30% is thought to have emitted in the form of CO_2, the rest in other forms. In France, ^{14}C discharges were estimated to be 140 GBq $year^{-1}$ per unit of 900 MWe and 220 GBq $year^{-1}$ per unit of 1300 MWe (Roussel-Debet et al. 2006). Based on combined worldwide operable nuclear reactors of 3.72×10^5 MWe (World Nuclear Association 2007), the global annual discharge of ^{14}C may amount to 60 TBq $year^{-1}$. For comparison, cosmogenic natural production in the upper atmosphere is at a rate of approximately 1.54×10^3 TBq $year^{-1}$ and all atmospheric nuclear tests emitted about 2.13×10^5 TBq of ^{14}C. Monitoring of radionuclides in terrestrial and aquatic environments, including soil, plant and foodstuff samples, has been performed to assess the potential environmental contamination from normal operation of nuclear power plants (Hu et al. 2010).

9.3.3.3 Nuclear Waste

Countries using nuclear power must deal with high-level radioactive waste. High level waste and spent nuclear fuel are stored at some 125 sites in 39 states, with over 161 million people residing within 121 km of temporarily stored nuclear waste (Hu et al. 2010). Spent nuclear fuel remains highly radioactive for thousands of years. Isolating this high-level waste from people and the environment has been an important and challenging issue for all countries that use nuclear power. Although high-level waste makes up only about 3% of the world's total volume of radioactive waste, it contains roughly 95% of all the radioactivity in all kinds of radioactive wastes.

Countries with high-level waste and spent nuclear fuel plan to dispose of these materials in a permanent repository which is commonly a deep geologic disposal facility. The waste is destined for vitrification in a borosilicate glass before permanent disposal. For example, in France and Germany, more than 500 Mg of waste have been vitrified (Hu et al. 2010). However, up to now, not a single country has disposed of high level waste. Along with the USA, Belgium, Canada, China, Finland, France, Germany, Japan, Russia, Spain, Sweden, Switzerland and the United Kingdom have invested significant resources in their radioactive waste management programs because of their reliance on nuclear energy (Hu et al. 2010). Deep geologic disposal has been expected to be the best method for isolating highly radioactive, long-lived waste (Witherspoon and Bodvarsson 2001).

9.3.3.4 Releases and Environmental Impacts from the Chernobyl and Fukushima Accidents

Radioactive contamination of the environment has occurred as a result of nuclear accidents. The most notable accident is the one that destroyed Unit 4 of the Chernobyl nuclear complex in the Ukraine in April 1986. The accident at the Fukushima Daiichi nuclear power plant, following the earthquake and tsunami in March 2011, was one of the biggest nuclear disasters in recent years. In both accidents, most of the radioactivity released was due to volatile radionuclides, such as noble gases (^{85}Kr, ^{133}Xe), iodine (^{129}I, ^{131}I, ^{133}I), cesium (^{134}Cs, ^{136}Cs, ^{137}CS) and tellurium (^{129}Te, ^{132}Te) (reviewed by Steinhauser et al. 2014). However, the amount of refractory elements (including actinides) emitted in the course of the Chernobyl accident was approximately four orders of magnitude higher than during the Fukushima accident. While for Chernobyl, a total release of 5300 PBq (excluding noble gases) has been estimated, for Fukushima, a total source term of 520 (340–800) PBq was established (Steinhauser et al. 2014). In the course of the Fukushima accident, more than 80% of the radioactivity may have been deposited in the Pacific Ocean.

Initially, Chernobyl's "exclusion zone" covered a radius of 30 km, corresponding to 2800 km^2 around the nuclear complex. Fallout of hot particles in the vicinity of the reactor caused a considerable contamination of the soil surface, with ^{137}Cs up to 10^6 Bq m^{-2} (Hu et al. 2010). About 116,000 people within this zone were evacuated to less contaminated areas in the months following the accident. The evacuation started only 3–11 days after the accident, which was already critical for parts of the population (Pröhl et al. 2002). The exclusion zone was later extended and comprised 4300 km^2 in 1996 (EC/IAEA/WHO 1996). The Chernobyl catastrophe resulted in significant contamination also outside the evacuation zone, as well as in many European states, including Belarus, Russia, Sweden, Finland, Norway, Germany, Slovenia, Croatia, Austria, Greece, and others (Fig. 9.2).

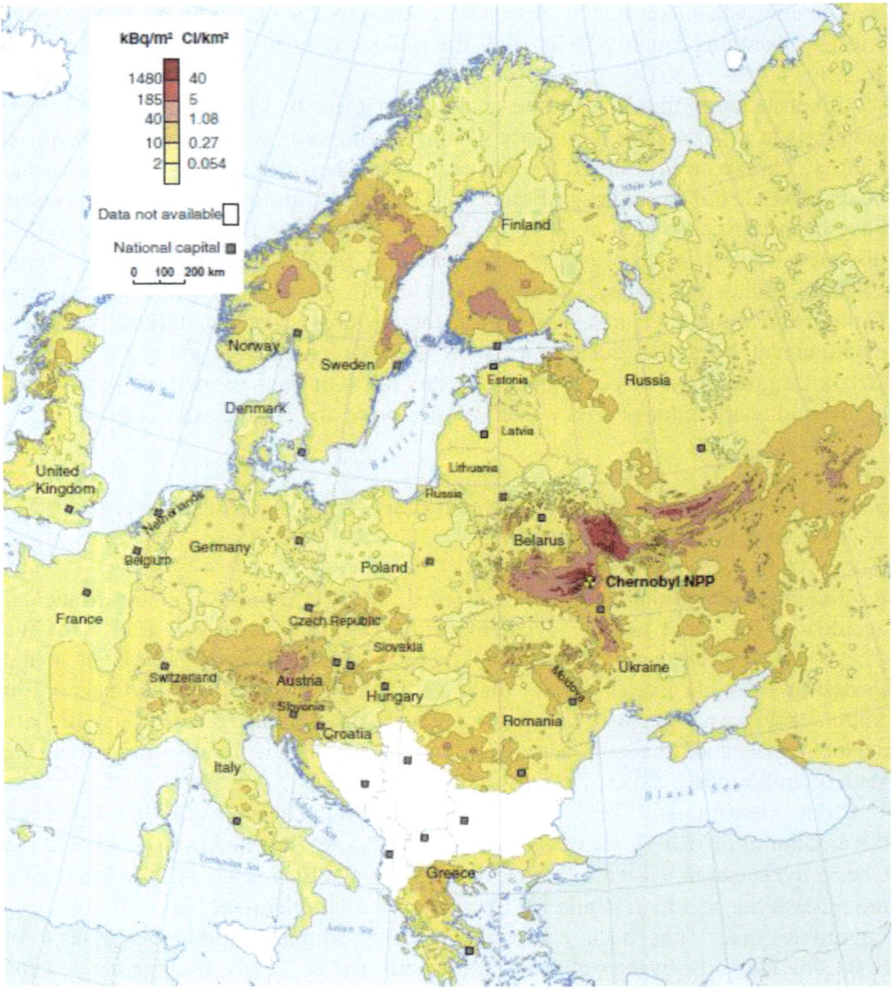

Fig. 9.2 Surface contamination with ^{137}Cs in Europe after the Chernobyl nuclear accident (Steinhauser et al. 2014, p. 804; with kind permission from Elsevier)

The contribution of ^{137}Cs from the Chernobyl plume was significant even at 2000 km distance. However, the total surface contamination was at least two orders of magnitude below the level within the 30 km exclusion zone. Overall, there were 28 deaths from the acute radiation syndrome as a result of the Chernobyl accident (Eisenbud and Gesell 1997). The exclusion zone around the Fukushima reactor encompassed an area of less than 600 km^2 (Yoshida and Takahashi 2012). Radionuclides were mainly deposited northwest of the reactor, causing the greatest contamination in a strip about 40 km in length (Hirose 2012). On 12 March 2011, a 20 km radius around the Fukushima reactor was designated as the "stay-away evacuation" zone. On 15 March 2011, the time of major releases of radionuclides, the surrounding area between 20 and 30 km was declared an "indoor evacuation" zone. On 16 March 2011, evacuees less than 40 years of age were instructed to leave the stay-away evacuation zone and to take pills or syrup of stable iodine (Hamada et al. 2012). Both the Chernobyl and Fukushima accidents caused radionuclide contamination of the atmosphere, hydrosphere, biosphere and pedosphere over the whole northern hemisphere. Not only the distance but also local weather conditions cause large variability in radionuclide concentrations in environmental media. After the Chernobyl catastrophe, maximum radioactivity in air resulting from the release of ^{131}I and ^{137}Cs was equivalent to 750,000 and 120,000 Bq m^{-3} at the reactor, 200 Bq m^{-3} and 9.9 Bq m^{-3} at 400 km distance, and < 12 Bq m^{-3} and < 10 Bq m^{-3} at distances greater than 1100 km from the reactor (Steinhauser et al. 2014). After the Fukushima accident, maximum values for ^{131}I were 5600 Bq m^{-3} at the reactor (data for ^{137}Cs are not available). Values for ^{131}I and ^{137}Cs at 190 km distance were 32 and 0.016 Bq m^{-3}, and for distances greater than 1100 km, they did not exceed 0.026 and 0.0032 Bq m^{-3}, respectively. Particularly the rainwater concentrations after the Chernobyl accident reflected the great influence of meteorological conditions on radionuclide concentrations. For example, in Göteborg, Sweden (1300 km distance to reactor) maximum values for ^{131}I and ^{137}Cs corresponded to 3000 and 950 Bq L^{-1} while in Munich, Germany with a similar distance to Chernobyl (1400 km), the corresponding values were 58,000 and 6500 Bq L^{-1}, respectively. The latter values were the highest that were determined for rainwater after the Chernobyl accident. Highest values after the Fukushima accident were determined at 200 km distance from the reactor (for ^{131}I: 6072 Bq L^{-1} and for ^{137}Cs: 752 Bq L^{-1}).

Maximum radionuclide concentrations in soils following the Chernobyl and Fukushima accidents are listed in Table 9.13. Part of the long-lived radionuclides (^{137}Cs, ^{90}Sr and $^{239+240}Pu$) can be present from previous accidents or nuclear weapon tests and may exhibit a significant background contamination that is difficult to distinguish from recent deposition. In summary, the impact of the Chernobyl accident was greater than of the Fukushima accident. The highly contaminated areas as well as the evacuated areas were smaller around Fukushima and the projected health effects in Japan were estimated to be lower than after the Chernobyl disaster. This is mainly due to the fact that evacuations and food safety campaigns worked quickly after the Fukushima accident.

9.3.3.5 Releases Through Medical Uses of Radiation

In medicine, radiation is used for both diagnostic and therapeutic purposes. As the benefits of procedures become more widely disseminated, medical uses of radiation increase from year to year (UNSCEAR 2000). The physicians, technicians, nurses and others involved constitute the largest group occupationally exposed to anthropogenic sources of radiation. Occupational doses received by staff varies according to the source of exposure, thereby characterizing different occupational groups.

Table 9.13 Activities of selected radionuclides (in Bq kg^{-1}) in soil after the Chernobyl and Fukushima accidents

Location	Distance to power plant (km)	^{131}I	^{137}Cs	^{90}Sr	$^{239+240}Pu$
Chernobyl					
Pripyat/Ukraine	4	n.d.	1,239,000	420,000	n.d.
Christogalovka/ Ukraine	5	n.d.	74,000	36,000	n.d.
Christinovka/ Ukraine	64	n.d.	15,000	130	17.8
Kupetsch/Ukraine	100	n.d.	3460	44	1.1
Vöcklabruck, Austria	1250	n.d.	506	n.d.	n.d.
Athens, Greece	1600	n.d.	22,000	n.d.	n.d.
Mumbai, India	5100	1.5	9.5	n.d.	n.d.
Fukushima					
Fukushima Daiichi	0.9	49,000	1,790,000	1700	~2200
Fukushima Daiichi	4.3	10,000	2,740,000	232	~0.5
Sendai/Japan	95	n.d.	5000	~3	n.d.
Kashiwa/Japan	195	n.d.	421,000	35	~0.5
Mekong Delta/ Vietnam	4700	n.d.	35	n.d.	n.d.
Vienna/Austria	9000	~0.4	21.4	n.d.	n.d.

Adapted from Steinhauser et al. (2014)
n.d. not detected or not reported

9.4 Behaviour of Important Radionuclides in Soil-Water Systems

Radionuclides of potential concern include long-lived fission products having relatively high to moderate waste inventories and high to moderate environmental mobilities (^{99}Tc, ^{129}I, ^{79}Se, ^{126}Sn), transuranic radionuclides of relatively high waste inventories and ingestion dose conversion factors ($^{239+240}$Pu, ^{241}Am, ^{242}Cm, ^{237}Np), long-lived neutron-activation products contained in irradiated reactor waste materials, such as spent fuel disassembly hardware, spent control rods, reactor internals, and spent primary coolant demineralizer resin (^{10}Be, ^{14}C, $^{59+63}$Ni, ^{94}Nb, ^{108}Ag) and naturally-occurring actinide radionuclides having relatively high waste inventories ($^{235+238}$U, ^{232}Th). The short-lived tritium (^{3}H) is considered a suitable water tracer. The isotopes ^{90}Sr and ^{137}Cs are the major fission products, yet they do not pose long-term risk because of their relatively short half-lives and strong sorption in the subsurface. In general, the mobility of actinides in aqueous systems is low. Of particular importance to the environment and risk assessment are radionuclides such as ^{99}Tc, ^{129}I, and ^{237}Np, because of their long half-lives (2.13×10^5, 1.57×10^7 and 2.14×10^7 years for ^{99}Tc, ^{129}I and ^{237}Np, respectively) and presumably high mobility (Hu and Smith 2004; Arnold et al. 2006). Activities conducted at European nuclear reprocessing facilities (mainly Sellafield and La Hague) have led to the increased radioactivity in the Arctic marine environment due to discharge of ^{99}Tc (Hu et al. 2010).

The most important processes governing radionuclide interactions with soils include adsorption/desorption (including ion exchange) and precipitation/dissolution. Adsorption occurs primarily in response to electrostatic attraction. The degree of adsorption of radionuclides is governed by the pH of the solution because the magnitude and polarity of the net surface charge of a solid changes with pH. Mineral surfaces become increasingly more negatively charged as pH increases. The sorption of radionuclides in soils is frequently quantified by the partitioning coefficient (Kd). The Kd is a factor related to the partitioning of a radionuclide between solid and aqueous phases and is defined as the ratio of the quantity of the adsorbate adsorbed per mass of solid to the amount of the adsorbate remaining in solution at equilibrium. Radionuclides that adsorb very strongly to soil have large Kd values (typically greater than 100 mL g^{-1}). Radionuclides that do not adsorb to soil and migrate essentially at the same rate as the water flow have Kd values near 0. The key geochemical processes affecting the mobility and bioavailability of selected radionuclides in soils are briefly described in the following paragraphs.

9.4.1 *Nickel-59,63 ($^{63,59}Ni$)*

In aqueous systems, the most important oxidation state of Ni is +2 (Baes and Mesmer 1976). At pH values less than 10, in aqueous systems the uncomplexed cation Ni^{2+} is the dominant Ni species. At pH values greater than 10, dissolved Ni is present in form of hydroxides. Nickel also forms aqueous complexes with ligands, such as dissolved chloride, sulfate and carbonate (Rai et al. 1984). The Ni concentrations in most soils are controlled by surface sorption processes. Nickel is known to be adsorbed by iron and manganese oxides and clays (Rai et al. 1984). Nickel is moderately to highly sorbed by soils with Kd values ranging from several tens to thousands milliliters per gram (Thibault et al. 1990). The adsorption of Ni^{2+} cations is greatest at pH values less than 11, decreases with deceasing pH, and minimal under acidic conditions. At pH values greater than 11, Ni adsorption to soil may decrease if the dominant Ni aqueous species is anionic, such as $Ni(OH)_3^-$. The sorption of Ni to clay minerals results in the formation of nickel hydroxide or nickel aluminum hydroxide surface precipitates (Scheckel and Sparks 2001). The formation of such surface precipitates strongly reduces Ni migration and remobilization in soil-water systems.

9.4.2 Selenium-79 (^{79}Se)

Selenium exists in the −2, 0, +4, and +6 oxidation states (Baes and Mesmer 1976). Dissolved Se is commonly present in the +6 oxidation state under oxidizing conditions as the dominant species $HSeO_4^-$ and SeO_4^{2-} at pH values less than and greater than 2, respectively. Under moderately oxidizing to reducing conditions, the Se(IV) species $H_2SeO_3^o$ (aq), $HSeO_3^-$, and $HSeO_3^{2-}$ may be dominant at pH values less than approximately 2.5, from 2.5 to 7, and greater than 7, respectively. The Se(−2) species H_2Se^o (aq) and HSe^- are the dominant aqueous Se species at pH values less than and greater than about 4, respectively, under highly reducing conditions. Dissolved Se in the −2, +4, and +6 oxidation states is present as anionic species at pH values greater than 4 under all redox conditions within the thermodynamic stability range of water. Due to the relatively high vapor pressure of these compounds, methylated forms of Se can be significant contributors to the mobility of Se. In some soil systems under moderately and highly reducing conditions, the concentration of dissolved Se may be controlled by the precipitation of solid forms of Se, such as Se^o. Elemental Se is relatively insoluble in soils over a wide range of pH under moderately reducing conditions. In highly reducing and organic-rich systems containing dissolved sulfide or bisulfide, Se-sulfide solids and metal selenides, such as ferroselite (FeSe2), are insoluble and would limit the concentration of dissolved Se and its mobility. Selenium can be reduced to its lower oxidation states by abiotic and biotic processes, which may result in the bioreduction of Se to insoluble forms, such as solid elemental Se (Se^o) or $S_{1-x}Se$. This will limit Se mobility and bioavailability in soils.

Because the dominant aqueous species of Se(IV) and Se(VI) are anionic over the pH range of most soils, the adsorption of Se to mineral surfaces would be expected to be low in most soils. However, hydrous oxides of Fe and Al and amorphous alumino-silicates have a high sorptive affinity for Se(VI) and Se(IV) (Rai et al. 1984). Values of Kd typically range from a few milliliters to several tens of milliliters per gram. The adsorption of Se also depends on pH with adsorption being strong under acidic conditions and decreasing with increasing pH. This pH dependency is consistent with that of other radionuclides and inorganic contaminants present primarily as anion.

9.4.3 Strontium-90 (^{90}Sr)

In aqueous solutions, dissolved Sr over the whole pH range up to pH 11 is present predominantly as uncomplexed Sr^{2+}. At pH values greater than 11, the neutral carbonate complex $SrCO_3^{\circ}$ (aq) is predicted to be the dominant aqueous complex (Robertson et al. 2003).

In alkaline soils, the precipitation of strontianite ($SrCO_3$) or coprecipitation in calcite may be important mechanisms for controlling the concentrations of dissolved Sr (Lefevre et al. 1993). Strontium as an alkaline-earth element can form similar solid phases as those with calcium. In certain soil systems, celestite ($SrSO_4$) and strontianite are potentially two important solubility controls for strontium, but most strontium minerals are highly soluble. Celestite may precipitate in acidic soil environments at elevated concentrations of total dissolved Sr and sulfate. In contrast, strontianite is only stable in highly alkaline soils. However, Sr does not commonly precipitate as a pure mineral, because the total concentrations of dissolved Sr^{2+} in most environmental systems are less than the solubility limits of Sr-containing minerals and much lower than the concentrations of dissolved Ca^{2+}. Because the ionic radii for Sr^{2+} (1.12 Å) and Ca^{2+} (0.99 Å) are similar, Sr^{2+} can substitute for Ca^{2+} in minerals to coprecipitate as a Sr-containing calcite ($Ca_{1-x}Sr_xCO_3$) (Veizer 1983). In most soils, Sr adsorption is controlled primarily by cation exchange. The important factors that affect adsorption and Kd values for Sr include CEC, pH, and concentrations of calcium and stable strontium naturally present in soil. Strontium Kd values increase with increasing CEC and pH. The correlation between strontium Kd values and pH is likely the result of H^+ ions competing with Sr^{2+} for exchange sites. Strontium Kd values of less than 1 mL g^{-1} to more than 30,000 mL g^{-1} have been reported (Sheppard and Thibault 1991). The adsorption of Sr^{2+} has also been found to decrease with increasing concentrations of competing cations, such as Ca^{2+}. Because Ca^{2+} concentrations in environmental systems are usually several orders of magnitude greater than stable strontium concentrations and many orders of magnitude greater than ^{90}Sr concentrations, the significantly greater mass of calcium increases the possibility that Ca^{2+} will outcompete Sr^{2+}, especially ^{90}Sr.

9.4.4 *Technetium-99 (^{99}Tc)*

Technetium exists in oxidation states from +7 to −1. Depending on the redox conditions, Tc exists in two stable oxidation states. It forms a reduced species [predominantly Tc(IV)] at Eh values below about 220 mV under neutral pH conditions. At higher Eh, it occurs as $Tc(VII)O_4^-$. Technetium(VII) can be reduced to Tc(IV) by abiotic and biotic processes. This reduction results in a decrease in the dissolved concentrations of Tc due to the precipitation of the weakly soluble, amorphous $TcO_2 \cdot 2\,H_2O$. In reduced iron-sulfide systems, Tc(VII) can be reduced to Tc(IV) by coprecipitation with FeS solid (mackinawite) (Wharton et al. 2000). Dissolved Tc is present under aerobic conditions as the aqueous Tc(VII) oxyanion species TcO_4^- over the complete pH range of natural waters. The TcO_4^- anion is essentially nonadsorptive. Due to its weak interaction with mineral surfaces, TcO_4^- is one of the most mobile radionuclides in the environment. In contrast, transport of Tc(IV) species ($TcO_2 \times nH_2O$) are expected to be strongly retarded because of sorption and/or precipitation (Eriksen et al. 1992). Lieser and Bauscher (1987) by varying the redox potential observed a change in the Kd value of about three orders of magnitude over a small range of Eh at 170 ± 60 mV and a pH of 7 ± 0.5. In carbonate-containing waters, Tc(IV) carbonate complexes, such as $TcCO_3(OH)_2^o$

(aq) and $TcCO_3(OH)_3^-$, may become important aqueous complexes of Tc (Eriksen et al. 1992).

Studies on the sorption of Tc on sediments, soils, pure minerals, oxide phases, and crushed rock materials consist primarily of measurements of Kd values for Tc (VII). The adsorption of Tc(VII) oxyanion TcO_4^- is very low to zero, i.e., Kd values of ≈ 0 mL g^{-1}, at near neutral and basic pH conditions and to increase when pH values decrease to less than 5. However, Kd values for Tc(VII) sorbed on sediments high in organic matter can be considerable (Thibault et al. 1990). The sorption of TcO_4^- is positively correlated to the soil organic matter content (Wildung et al. 2000). Adsorption of Tc(VII) in experiments conducted with organic material as well as with crushed rock and Fe(II)-containing minerals has been attributed to the reduction of Tc(VII) to Tc(IV). Technetium(IV) is considered to be immobile because it readily precipitates as hydrous oxides and forms complexes with surface sites on Fe and Al oxides and clay minerals.

9.4.5 Tin-126 (^{126}Sn)

Tin exists in compounds in several oxidation states from −4 to +4 but only the +2 and +4 states are important for Sn in natural systems (Baes and Mesmer 1976). However, Sn^{2+} and its hydrolysis products are limited to very reducing conditions at or below the stability boundary for the breakdown of water. According to Séby et al. (2001) information regarding the hydrolysis of Sn(IV) and speciation in general is limited due to the low solubility of SnO_2 (the mineral cassiterite) and its precipitation in laboratory studies of Sn(IV) speciation. Ashby and Craig (1988) found that Sn methylation may occur by abiotic and biotic process and thus may contribute to the mobility of Sn in the environment. As Sn(IV) will be present in the +4 oxidation state in most environmental systems, Sn is likely to be adsorbed in soils due to the low solubility of cassiterite, presupposed that no anionic aqueous complexes are formed.

9.4.6 Iodine-129 (^{129}I)

Similar to ^{99}Tc, ^{129}I has a complex chemistry in the environment which is caused by the multiple redox states. However, in contrast to ^{99}Tc, iodine has a minimal retardation under reducing conditions when I^- is the predominant form. The −1, +5, and molecular I2 oxidation states are those most relevant for iodine in environmental systems. Different iodine species (I^-,IO_3^-, and organic iodine species) are known to coexist in various aqueous systems (Hu et al. 2005). The iodide anion (I^-) is highly mobile. Under more oxidizing conditions, iodine may be present as the iodate anion (IO_3^-), which is more reactive than iodide.

Adsorption of iodine species in soils seems to be controlled mainly by soil organic matter and Fe and Al oxides which becomes increasingly important under more acidic conditions. Results from numerous Kd studies suggest that the adsorption of iodide increases with increasing soil organic matter content, but the majority of the reported Kd values for iodide are limited to soils containing less than 0.2% organic matter contents (reviewed by Robertson et al. 2003). Values of Kd for iodide have been reported in the range from 1 to 10 mL g^{-1} for the pH range from 4 to 10, but most of the reported Kd values are typically less than 3 mL g^{-1}. Because iodine is present as either the anions I^- or IO_3^- in most soils, their adsorption on soils and most minerals should be negligible at near neutral and alkaline pH conditions.

Iodine volatilization from soils to the atmosphere may be a result of both chemical and microbiological processes. The chemical processes generally result in molecular iodine or hydrogen iodide, and the microbiological processes yield methyl iodide which is not strongly retained by soil components and is only weakly soluble in water (Whitehead 1984).

9.4.7 Cesium-137 (^{137}Cs)

Cesium in the environment exists in the +1 oxidation state. Compared to the other contaminants the speciation of Cs in environmental systems is relatively simple. Cesium does not form any important aqueous complexes with ligands and organic matter found in natural systems and Cs containing solids are highly soluble in aqueous systems. Therefore, the precipitation and coprecipitation of Cs-containing solids are not important processes in controlling the concentration of dissolved Cs in environmental systems.

The behavior of Cs in soils is similar to K^+. Cesium due to strong sorption on most minerals has large Kd values (Robertson et al. 2003). The sorption of Cs occurs primarily by ion exchange in most soil systems except when mica-like minerals are present. For example, when mica-like clay minerals such as illite and vermiculite are present, selective fixation of Cs between structural layers of these minerals occurs. The extent to which Cs is fixed will depend on the concentration of mica-like clays in the soil, and the concentration of major cations, such as K^+, as K^+ can effectively compete with Cs^+ for ion exchange sites because its hydrated ionic radii are similar and smaller than those for the other alkali and alkaline earth ions. Cesium may also adsorb to iron oxides by complexation of cesium to surface mineral sites whose abundance is pH dependent (Schwertmann and Taylor 1989). In contrast, the sorption of Cs to humic substances is generally weak (Bovard et al. 1968).

9.4.8 Thorium-232 (^{232}Th)

In natural soil-water environments Th exists only in the +4 oxidation state. In the absence of dissolved ligands other than hydroxide, the uncomplexed ion Th^{4+} is the main aqueous species at pH values less than about 3.5. At pH values exceeding 3.5, the hydrolysis of Th is dominated by the aqueous species $Th(OH)_2^{2+}$ and $Th(OH)_4^{o}$ (aq) with increasing pH (Robertson et al. 2003). At pH values greater than 5 the neutral hydroxide complex $Th(OH)_4^{o}$ (aq) dominates the aqueous speciation of dissolved Th. Dissolved Th can form strong aqueous complexes with inorganic ligands, like dissolved carbonate, fluoride, phosphate, chloride, nitrate, and organic ligands. The complex ThF_2^{2+} may be dominant at pH values less than 5. Phosphate complexes may be important at acidic and near-neutral pH conditions. In addition, Th carbonate complexes $Th(OH)_3CO_3^{-}$ and $Th(CO_3)_5^{6-}$ may be of some importance for Th mobility.

Concentrations of Th in soils may also be controlled by adsorption processes. Oxides of Fe and Mn are expected to be important adsorbents of Th. Humic substances are considered particularly important in the adsorption of Th (Gascoyne 1982). Thibault et al. (1990) reviewed published Kd data for Th as a function of soil type and reported Kd values for Th that range from 207 to 1.3×10^{7} mL g^{-1}.

9.4.9 Uranium-235 (^{235}U)

Uranium in aqueous environments exists in different oxidation states (+3, +4, +5, +6). Uranium(VI), occurring as UO_2^{2+} (uranyl), and U(IV) in natural environments are the most common oxidation states of U. Under oxidizing to mildly reducing environments, U exists in the +6 oxidation state. Under reducing conditions, U (IV) is considered relatively immobile because U(IV) forms hardly soluble minerals, like uraninite (UO_2). Because the oxidation state of U has a significant effect on its mobility in waste streams and the natural environment, the reduction of U (VI) to U(IV) by abiotic and biotic processes has recently received considerable attention. These processes may be useful for certain remediation technologies, such as permeable barriers composed of zero-valent iron particles, or sodium-dithionite-reduced soils. Microbial reduction of U(VI) has also been suggested as a potential mechanism for removal of U from contaminated waters and soils (Lovley 1995). In carbonate-containing waters the aqueous speciation of U(VI) at near neutral and basic higher pH values is dominated by anionic aqueous carbonate complexes (e.g., $UO_2CO_3^{o}$ (aq), $UO_2(CO_3)_2^{2-}$, and $UO_2(CO_3)_3^{4-}$). The formation of anionic U (VI) carbonate complexes at pH values greater than 6 result in an increase in U (VI) solubility, decreased U(VI) adsorption, and thus increased U mobility because anions do not readily adsorb to mineral surfaces at basic pH conditions. Uranium can also form stable complexes with other naturally occurring inorganic and organic ligands. Complexes of UO_2^{2+} with phosphate ($UO_2HPO_4^{o}$ (aq) and $UO_2PO_4^{-}$) may be important in aqueous systems with a pH between 6 and 9 when the total concentration ratio PO_4(total)/CO_3(total) is greater than 0.1 (Sandino and Bruno 1992). Complexes with sulfate, fluoride, and possibly chloride may be important uranyl species where concentrations of these anions are high. However, their stability is considerably less than the carbonate and phosphate complexes (Grenthe et al. 1992). Organic complexes may also be important to U aqueous chemistry, thereby increasing their solubility and mobility. The uncomplexed uranyl ion has a greater tendency to form complexes with fulvic and humic acids than many other metals with a +2 valence (Kim 1986) which was attributed to the greater "effective charge" of the uranyl ion compared to other divalent metals.

In soils, U(VI) adsorbs onto a variety of minerals, including clays (Chisholm-Brause et al. 1994), oxides and silicates (Waite et al. 1994), and organic material (Read et al. 1993). Environmental parameters controlling U adsorption include pH, redox conditions, and concentrations of complexing ligands. The adsorption of uranium to humic substances may occur through ion exchange and complexation processes that result in the formation of stable U(VI) complexes involving the acidic functional groups (Borovec et al. 1979).

9.4.10 Neptunium-237 (^{237}Np)

Neptunium may exist in the +3, +4, +5, +6, and +7 valence states, but only the +4, +5, and possibly +6 states are relevant to natural environments. Neptunium in aqueous systems has a large stability range for Np(V) (Lieser and Mühlenweg 1988). The pentavalent NpO_2^+ species is dominant at pH values <8 whereas Np (V) carbonate complexes tend to dominate at higher pH values (Kaszuba and Runde 1999). Since Np(V) solid phases are relatively soluble and Np(V) aqueous species do not easily sorb onto common minerals, Np(V) is relatively mobile in the environment (Nakata et al. 2002). Under reducing conditions, Np(IV) is present as the low solubility Np(OH)4 (aq) species at pH values >5 (Kaszuba and Runde 1999).

Neptunium(V) species in soils to some extent adsorb to Fe oxides and clay minerals, but not on most common minerals (Nakata et al. 2002). Because NpO_2^+ does not compete favorably with dissolved Ca^{2+} and other divalent ions for adsorption sites on soils, the Kd values for Np(V) are relatively low (Kaplan and Serne 2000). Especially for iron oxides, the adsorption of Np(V) has a strong dependence on pH, (Kohler et al. 1999). Typically, the sorption of Np(V) on minerals is negligible at pH values less than pH 5, and increases at pH values between 5 to 7. This pH-dependency is expected for ions that are present in solution primarily as cations, such as NpO_2^+.

9.4.11 Plutonium-239 + 240 ($^{239+240}$Pu)

Under most environmental conditions Pu can occur in the +3, +4, +5, and +6 oxidation states. Under oxidizing conditions, Pu(IV), Pu(V), and Pu(VI) may exist, while Pu(III) and Pu(IV) would be present under reducing conditions (Allard and Rydberg 1983). In aqueous systems, Pu(III) species, such as Pu^{3+}, would be dominant up to pH values of approximately 8.5 under reducing conditions. The Pu (IV) species Pu(OH)4° (aq) is predicted to have a large stability range extending above near neutral pH values at moderately oxidizing conditions to pH values greater than 8 under reducing conditions (Robertson et al. 2003). Plutonium may form stronger complexes with dissolved carbonate, sulfate, phosphate, and fluoride, relative to those with ligands such as chloride and nitrate. Dissolved Pu may also form complexes with dissolved organic matter, such as fulvic and humic material. Associated with organic matter, Pu is mainly present in the +4 oxidation state (Nelson et al. 1987).

Dissolved Pu in the environment is commonly present at less than 10–15 mol L^{-1} which indicates that adsorption may be the most important process affecting the retardation of plutonium in soils. However, Kd values for Pu can extremely vary, depending on the properties of the substrate, pH, and the composition of solution. According to Thibault et al. (1990) they may range from 27 to 190,000 mL g^{-1}. If no complexing ligands are present, the adsorption of Pu commonly increases with increasing pH from about pH 5 to 9. At pH values exceeding 7, concentrations of dissolved carbonate and hydroxide will decrease Pu adsorption of and increase its mobility in soils as a result of the formation of strong mixed ligand complexes with Pu. According to Sanchez et al. (1985) increasing carbonate concentrations decreased the adsorption of Pu(IV) and Pu(V) on the surface of goethite [α-FeO (OH)]. At low pH in the presence of high concentrations of dissolved organic carbon, Pu-organic complexes may control Pu adsorption and mobility in soils (Robertson et al. 2003).

9.4.12 Americum-241 (^{241}Am)

Americium exists in several oxidation states (+3, +4, +5, and +6) but Am(III) is the most stable and important oxidation state in environmental systems. All of the higher oxidation states are strong oxidizing agents and stable only in systems containing no oxidizable substances (Ames and Rai 1978). Americium is present as Am(III) in all of the dominant species predicted to be stable for the Eh-pH region of environmental interest. The uncomplexed ion Am^{3+} is the dominant aqueous species at moderately to highly acidic conditions. At near neutral to alkaline pH conditions, Am(III) carbonate and hydroxyl complexes will dominate the aqueous species of Am(III). Sorption studies indicate that Am(III) readily sorbs to soils, pure minerals, and crushed rock materials and exhibits high Kd values that are often in the range of 1000 to greater than 100,000 mL g^{-1}. Americium(III) is, therefore, considered one of the most immobile actinide elements in the environment. The adsorption of Am(III) increases with increasing pH with peak adsorption occurring between pH 5 and 6. This pH dependence is because of the dominant aqueous species of Am(III) in the pH range of natural waters which are primarily Am^{3+} complexes at acidic and cationic carbonate and hydroxyl complexes at basic pH values. Americium(III) is more mobile at low to moderate pH values where the net surface charge on minerals becomes more positive. The tendency of Am(III) to strongly sorb to soil particles suggests that there is a potential for colloid-facilitated transport of Am(III). Colloids of clay and humic acids are potentially important substances for the transport of actinides in soil/water systems (Robertson et al. 2003).

9.4.13 *Curium-242 (^{242}Cm)*

Curium can exist in the +3 and +4 oxidation states. However, the +3 state is the dominant oxidation state in natural waters. Curium(IV) is not stable in solutions because of self-radiation reactions (Onishi et al. 1981). In natural waters, Cm(III) is predicted to be the dominant species at pH values less than 7 and may form complexes with inorganic ligands (Wimmer et al. 1992). The complexes $CmCO_3^+$ and $Cm(OH)_2^+$ dominate the aqueous speciation of Cm(III) in the pH range from 7 to 9 and at pH values greater than 9, respectively. Other aqueous complexes of Cm(III) include those with sulfate, fluoride, chloride, and humic substances (reviewed by Robertson et al. 2003). Sorption studies indicate that Cm(III) readily sorbs to minerals, crushed rock, and soil materials. Compared to other actinides, Cm(III)is considered to be immobile in soil environments and exhibits high Kd values. Adsorption of Cm(III) is strongly pH dependent and increases with increasing pH with peak adsorption occurring between pH values of 5 and 6 (Robertson et al. 2003). Similar to the environmental behaviour of Am(III), the tendency of Cm(III) to strongly adsorb to soil particles suggests that there is a potential for colloid-facilitated transport of Cm(III). The mobility of Cm(III) in soils may thus be enhanced by its migration in colloid form (Kaplan et al. 1994).

9.5 Routes of Exposure

There are two major routes of radionuclide exposure (Fig. 9.3). The first exposure route involving external contamination, originates from natural and anthropogenic sources of ionizing radiation. Part of the natural radiation is cosmic radiation from space and the other part is due to radioactive materials in soil and in building materials.

Higher levels of natural radioactive material are left in products or on the land due to human activities. Skin can be contaminated by contact with aerosols or radionuclide-contaminated surfaces. For aerosols that deposit on the skin, factors such as particle size may be important in that larger particles may deliver larger doses to the skin, especially for beta-emitting radionuclides. Contamination of skin with beta-emitting radionuclides can also cause serious burns.

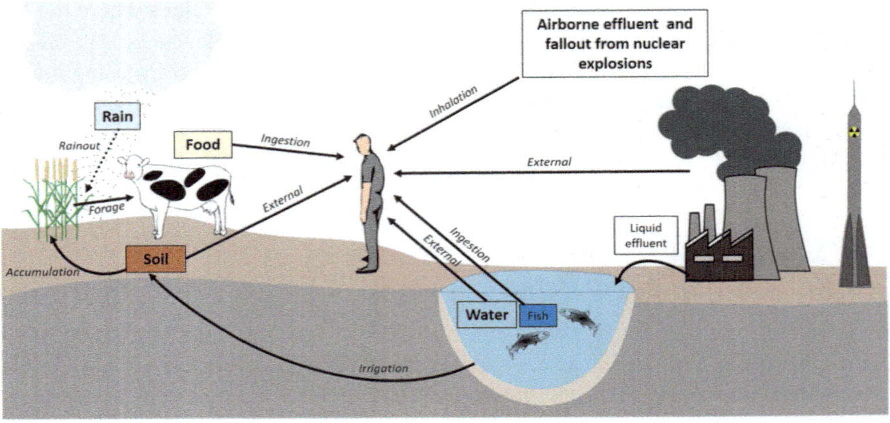

Fig. 9.3 Major exposure pathways of radionuclides

The second major exposure route is by internal contamination, which occurs when the radioactive source is incorporated into the human body through inhalation, ingestion or through uptake via skin or wounds. As with other routes of exposure, the nature and physiochemical properties of the radionuclide are of major importance when determining the effect of the internalized isotope. Inhalation is the primary route of exposure for internalized radionuclides. Their fate depends on the size of the inhaled substances and the solubility of the radionuclide. Approximately 25% of inhaled radionuclides are immediately exhaled. Of the remaining 75%, particles greater than 10 μm tend to remain in upper areas of the lung while those less than 5 μm in diameter can reach the alveolar space (Morrow et al. 1967). In the lung, radionuclides such as tritium, phosphorus, and cesium are rapidly solubilized and enter the circulatory system. Less soluble radionuclides, such as the oxides of plutonium, uranium, cobalt, and americium, may be removed through phagocytosis by alveolar macrophages. Until being removed, the radioactive particle will continue to irradiate surrounding tissues. In most cases, the internalized radionuclide will have both soluble and insoluble components (Harrison and Muirhead 2003). Larger particles which are unable to access the alveolar space, may be removed from the lung via mucocilliary clearance and thus enter the gastrointestinal tract.

In case of ingestion of radioactive material, the majority of radionuclides are poorly absorbed by the intestinal tract. Some exceptions include strontium, tritium, and cesium. The amount of damage inflicted will be determined by the transit time through the alimentary canal, with the greatest potential for damage occurring in the descending colon prior to the ingested radionuclide being excreted in the feces. Transit times are affected by a variety of factors, including diet, fluid intake levels, and physical activity, but generally range from 1 to 5 days (Eve 1966). The International Commission on Radiological Protection (ICRP 1999) has presented recommendations for the protection of the public in situations of prolonged exposure to radiation, including justification for remediation of contaminated sites. The ICRP dose limit for exposure of a member of the public from all relevant practices is 1 mSv year^{-1}. Examples of dose rates from sources of major concern for the public are given in the following sections.

9.5.1 Exposure from Cosmogenic Radiation

External exposure from cosmogenic radionuclides, except for ^{3}H, ^{7}Be, ^{14}C and ^{22} Na, contribute only to a small extent to radiation doses. UNSCEAR (2000) for these radionuclides gave estimates of effective dose rates (μSv year^{-1}) of 12 from ^{14}C, 0.03 from ^{7}Be, 0.15 from ^{22}Na and 0.01 from ^{3}H. Cosmic ray doses increase by a factor of about two for every 1500 m increase above the mean sea level (AMSL), resulting in 260 to 270 μSv year^{-1} at 0–150 m AMSL, 390 to 520 μSv year^{-1} at 1220 to 1830 m AMSL and 1070 μSv year^{-1} at 3400 m AMSL (Ramachandran 2011). However, contribution of this source compared to other sources (particularly terrestrial radiation) for most part of the global population is small.

9.5.2 Exposure from Natural Terrestrial Radiation

Outdoor values of exposure from terrestrial radiation for different countries range between 18 and 93 nGy h^{-1} (0.16–0.82 mGy year^{-1}) (Thorne 2003). The population weighted average may be around 59 nGy h^{-1} (0.52 mGy year^{-1}). Indoor rates are commonly larger than outdoor rates. UNSCEAR (2000) estimated a global average effective dose rate of 0.48 mSv year^{-1} (0.41 mSv year^{-1} indoors and 0.07 mSv year^{-1} outdoors). Results for individual countries commonly range between 0.3 and 0.6 mSv year^{-1}. External dose rates in regions of high natural background can be significantly increased. These include areas with monazite sand deposits, which have high levels of Th, regions with Andosols (volcanic soils) and precipitates of Ra associated with hot springs. For example, over themonazite sands in Kerala and Madras, India, absorbed dose rates in air of 200–4000 nGy h^{-1} (1.8–35 mGy year^{-1}) have been detected. On monazite sand beaches of Guarapari (Brazil) dose rates of up to 90,000 nGy h^{-1} (0.79 Gy year^{-1}) have been measured (UNSCEAR 2000).

9.5.3 Exposure from Nuclear Weapons Tests

UNSCEAR (2000) has published global means of effective dose rates from fallout radionuclides. Average effective dose rates for external radiation peaked in 1962 and

1963 at around 38 μSv $year^{-1}$, with ^{95}Zr and ^{95}Nb contributing about 50% of the total. In 1999, the global average was 2.90 μSv $year^{-1}$, almost entirely from ^{137}Cs.

Global average effective dose rates from ingested radionuclides peaked at 35 μSv $year^{-1}$ in 1962, particularly from ^{131}I (20.4 μSv $year^{-1}$), ^{137}Cs (10.3 μSv $year^{-1}$) and ^{90}Sr (3.1 μSv $year^{-1}$). Until 1999, the effective dose rate was reduced to 0.90 μSv $year^{-1}$, particularly from ^{137}Cs (0.35 μSv $year^{-1}$) and ^{90}Sr (0.56 μSv $year^{-1}$). The effective dose rate from ^{3}H and ^{14}C, with 12.7 μSv $year^{-1}$ reached its maximum in 1962, with 7.2 μSv $year^{-1}$ from ^{3}H and 5.5 μSv $year^{-1}$ from ^{14}C. Until 1999, the total effective dose rate decreased to 1.7 μSv a^{-1}, almost completely from ^{14}C. For inhalation, global means of effective dose rates with 36 μSv $year^{-1}$ reached their maxima in 1963, particularly from ^{144}Ce (15.3 μSv $year^{-1}$), ^{106}Ru (9.3 μSv $year^{-1}$), ^{90}Sr (1.9 μSv $year^{-1}$) as well as from radioisotopes of Pu and Am (7.7 μSv $year^{-1}$). In 1985, inhalation doses made only a small contribution to exposure from fallout (Thorne 2003).

The mean total effective dose rate from fallout in the Northern and the Southern Hemisphere in 1963 reached 125 μSv and 59 μSv $year^{-1}$, respectively. The global mean is thus dominated by the Northern Hemisphere with a larger population compared to the Southern Hemisphere. The global averages for the Northern and Southern Hemisphere and both hemispheres in 1999 were 5.87, 2.68 and 5.51 μSv $year^{-1}$, respectively (Thorne 2003).

9.5.4 Exposure from the Nuclear Fuel Cycle

The global average effective dose from the nuclear fuel cycle may generate a per caput effective dose of 0.1 μSv to the world population. This would represent only 0.005% of the average exposures to natural sources of radiation. The average percentage contribution may be 37.2 from reactor fabrication, 34.4% from mining, 14.3% from mile and mill tailings, 7.2% from fuel processing each 2.9% from milling and transportation, 1.15% from airborne reactor release and 0.07% from fuel fabrication. The contribution from mining, milling and mill tailings is thus almost equal to that from reactor and fuel processing.

9.5.5 Exposure from Nuclear Accidents

Radiation exposure from the soil and floral cover due to nuclear accidents will depend on a number of factors, particularly radionuclide composition of the fallout, the time of exposure after the fallout, the precipitation conditions at the time of deposition and afterwards, and upon the nature of the aerial biomass at the time of deposition. These factors probably account for the different external radiation dose rates per Becquerel per square meter implied in some of the post-Chernobyl reports. For example, in the Federal Republic of Germany "top level" ground deposits of 100,000 Bq m^{-2} representing an "absorbed dose" equivalent to 0.06 mSv $hr.^{-1}$

were indicated for persons "permanently on such ground". Respective data from Sweden, in contrast, implied that in the "regions of highest contamination" local levels of the order of 1 MBq m^{-2} (i.e., for ^{131}I and ^{137}Cs) were reached and involved external irradiation dose rates of the order of 10 mSv hr^{-1}, i.e., approximately 1 mSv hr^{-1} for a deposit of 100,000 Bq m^{-2} (IAEA 1986).

9.5.6 Contaminated Food and Water

Concentrations of natural radioactivity in food are commonly in the range of 40–600 Bq kg^{-1} food. The radioactivity from potassium alone may be 50 Bq kg^{-1} in milk, 420 Bq kg^{-1} in milk powder, 165 Bq kg^{-1} in potatoes, and 125 Bq kg^{-1} in beef (Shahbazi-Gahrouei et al. 2013). According to Ramachandran and Mishra (1989) the concentration of ^{40}K radioactivity in different foods varied from 45.9 to 649.0 Bq kg^{-1}, that of ^{226}Ra varied from 0.01 to 1.16 Bq kg^{-1}, and that of ^{228}Th from 0.02 to 1.26 Bq kg^{-1}. Food, water, and air also contain trace amounts of alpha emitters from the uranium, thorium, and actinium series. Some of the radon (^{222}Rn, and to a lesser extent ^{220}Rn and ^{219}Rn) gas diffuses into the food supply. For example, radon and its decay products, are deposited on the soil and the vegetation.

To derive the corresponding dose in mSv year^{-1}, it is necessary to take into account the energy and fraction deposited in the body, besides taking into account not only the radioactive lifetime but also the biological lifetime of the isotopes in the human body. The natural radioactivity from the ^{40}K isotope, which is a constant fraction (0.0117%) of the potassium content in food, varies significantly with potassium concentration in different food sources.

In Europe, following the accident of the Chernobyl atomic power station in 1986, high radioactivity levels of some wild-growing mushroom species were observed (Kalac 2001). Until 1985, activities of ^{137}Cs, from nuclear weapons testing, were commonly below 1 kBq kg^{-1} dry matter. After the Chernobyl accident, activities up to tens of kBq kg^{-1} dry matter of ^{137}Cs and to a lesser extent of ^{134}Cs were observed in the following years in edible mushrooms. The species most heavily contaminated included *Xerocomus badius*, *Xerocomus chrysenteron*, *Suillus variegatus*, *Rozites caperata* and *Hydnum repandum* (Kalac 2001). Wild-growing mushroom consumption contributed up to 0.2 mSv to the effective dose in individuals consuming about 10 kg (fresh weight) of heavily contaminated species per year. The radioactivity of cultivated mushrooms is negligible. Wild animals, consuming contaminated mushrooms, have elevated levels of radionuclides in their tissues.

9.5.7 Total Radiation Exposure

Excluding extreme scenarios such as atomic power plant accidents and releases of radionuclides through nuclear weapons tests, it can be summarized that on a global

Table 9.14 Radiation exposure and contribution from natural, modified and anthropogenic sources to the Indian population

Source	Dose (mSv $year^{-1}$)	Contribution (%)
All natural	2.299	96.07
Modified natural	0.00124	0.052
Nuclear weapons tests	0.045	1.88
Nuclear fuel cycle	0.00005	0.0021
Medical exposure	0.048	2.01
Total	2.393	100.0

Adapted from Ramachandran (2011)

scale, the natural sources, particularly soil materials, account for major share of total radiation exposure. An example is given in Table 9.14 for the Indian population. However, Ramachandran (2011) reported that this pattern is similar to the global situation. In future, these proportions may change, depending on the technological and industrial development, the energy policy and the trend in medical uses of radiation.

9.6 Clinical Effects of Radiation

Ionizing radiation is a form of energy that may be destructive in biological systems and can cause serious diseases such as cancer and mutations and prodromal symptoms such as anorexia, nausea, vomiting, diarrhea, nervousness, confusion and consciousness in humans (Sharma et al. 2010). The types of radiation include alpha particles, beta particles, gamma particles, neutrons and X-rays (see Sect. 9.1).

9.6.1 Alpha Radiation

Because of their large mass and charge, α-particles strongly ionize biological tissues. If the alpha particle is from radioactive material that is outside the human body, it will lose all its energy before passing through the dead outer layer of the human skin. This means that exposure to α-radiation only exists if α-particles are taken up by ingestion or inhalation. The α-particles can cause damage to tissues in the body.

9.6.2 Beta Radiation

Most β-particles (except for tritium) have enough energy to pass through the dead outer layer of a person's skin and irradiate the live tissue underneath. Human exposure is also possible through ingestion or inhalation of β-particles that lose their energy by exciting and ionizing atoms along their path. When all of the kinetic energy is spent, negative β-particles (negatrons) become ordinary electrons and have no more effect on the body. A positive β-particle (positron) collides with a nearby electron, and this electron-positron pair turns into a pair of gamma rays called annihilation radiation, which can interact with other molecules in the body.

9.6.3 Gamma Radiation

Gamma particles can pass through many kinds of materials, including human tissues. The gamma ray source can be relatively far away, like the radioactive materials in nearby construction materials, soil, and asphalt. A γ-ray may pass through the body without hitting anything, or it may hit an atom and give that atom all or part of its energy. This normally knocks an electron out of the atom (and ionizes the atom). This electron then uses the energy it received from the γ-ray to ionize other atoms by knocking electrons out of them as well. Since a γ-ray is pure energy, once it loses all its energy it no longer exists. Very dense materials, such as lead, are commonly used as shielding to slow or stop γ-photons.

9.7 Isotopes of Concern for Human Health

There are numerous natural and manufactured radioisotopes that could result in internal contamination. Of these, about 40 radionuclides are potentially hazardous to humans (Casarett 1968). Only a limited number including plutonium, cesium, strontium, radioiodine, and tritium provide most significant health hazards to humans (Prasad 1995). Isotopes of serious human health concern and their target organs are listed in Table 9.15.

9.8 Biological Significance of Radiation

Biological significance is a result of a combination of high decay energy, bioavailability and energy transfer to biological systems. The biological significance of radiation results from the enormous amount of energy contained in each emission. Visible light has an energy range of 1.77–4.13 electron volts. Most chemical changes occur within a range of 5–7 electron volts. Biologically significant radiation levels range from 18,610 electron volts for the weak beta emitting tritium (^{3}H) (half-life: 12.35 years) over 511,630 electron volts for ^{137}Cs (half-life: 30.174 years) to 5,155,400 electron volts for plutonium (^{239}Pu) (half-life: 24,131 years). These highly energetic emissions carry enough energy to tear electrons from neutral atoms and molecules. External exposure to ^{239}Pu poses

Table 9.15 Radiation exposure and contribution from natural, modified and anthropogenic sources to the Indian population

Element	Isotope of concern	Type of radiation	Major target organs
Americum	^{241}Am	α, γ	Lung, liver, bone
Cesium	^{137}Cs	β, γ	Whole body, particularly kidney
Cobalt	^{60}Co	β, γ	Whole body, particularly liver
Iodine	^{131}I	β, γ	Thyroid
Iridium	^{192}Ir	β, γ	Spleen
Phosphorus	^{32}P	β	Bone
Plutonium	^{238}Pu, ^{239}Pu	α, (γ)	Lung, liver, bone
Polonium	^{210}Po	α	Kidney, spleen
Radium	^{226}Ra	α, β, γ	Bone
Strontium	^{90}Sr	β, γ	Bone
Tritium	^{3}H	β	Whole body
Uranium	^{235}U, ^{238}U	α, β, γ	Kidney, bone

Adapted from Kalinich (2012)

very little health risk, since plutonium isotopes emit α-radiation, and almost no β- or γ-radiation. In contrast, internal exposure to ^{239}Pu is an extremely serious health hazard. It generally stays in the body for decades, exposing organs and tissues to radiation, and increasing the risk of cancer. Plutonium is also a toxic metal, and may cause damage to the kidneys. The radionuclide ^{137}Cs, a β-emitter with a γ-component, is biologically significant due to its high energy level, its long half-life, its ubiquitous production during the fission process, and its tendency to follow the potassium cycle in nature, giving a whole body dose to those who ingest it. The weaker β-radiation of ^{3}H is slightly more penetrating than α-radiation. Its biological significance comes from its ubiquitous production during the fission process, its tendency to follow the water cycle in nature, thus also contaminating soil materials, and its ability to become tissue bound in humans and the biotic environment.

9.9 Therapy

9.9.1 Determination of Radioactive Contamination

Contaminated persons could suffer acute symptoms of radiations injury and in the longer term could develop genetic damage or cancer. Early detection of the contaminants helps in early response for decontamination and decrease the potency of hazards. Information from the attack scene will provide the first information on the radioisotopes involved, but in many cases, the isotope and route of exposure are unknown. However, several simple assessments can determine whether the contaminant is a β or γ emitter, and may also indicate possible routes of exposure. A preliminary step is a body survey using a Geiger-Müller meter that is capable of detecting β and γ emitting isotopes, but not those emitting γ particles (Kalinich

2012). The first scan should be made with the shield of the Geiger-Müeller meter open to detect the presence and location of β and γ contamination. The next scan is to be conducted with the closed shield. Results from this scan will indicate what proportion of the contamination is due to γ emitting isotopes alone (Kalinich 2012). Radioactive contamination by α emitters should also be determined. A variety of instruments is available to measure α radiation. However, special training in use of these instruments is necessary for making accurate measurements.

9.9.2 External Decontamination

9.9.2.1 Skin Decontamination

External decontamination is essential to prevent the spread of radionuclides. Decontamination procedures decrease external radiation levels, allowing for a more accurate determination of internal radionuclide contamination. The primary objective is to remove the highest possible degree the radioactive contaminant from external body parts. The process requires removing the contaminants not just from the skin surface, but also from the clothing and protective equipments (Owens and Peacock 2004). The decontamination of skin involves different methods. The most commonly used one is simple washing with warm water and soap using face cloths and brushes. Apart from soap, the decontaminating agents include washing pastes, shampoos and various decontamination solutions (Turner 2007). To decontaminate radioactive material from eye or nose they should be washed with 0.9% saline solution, or if not available with tap water. Care should be taken that the water does not get swallowed and the solution for decontamination is not irritant to skin and other organs (Sharma et al. 2010).

9.9.2.2 Wound Decontamination

Washing and cleaning of the wound in most cases is sufficient (Kalinich 2012). Adding a chelating agent to the washing solution in some cases supports contaminant removal. For wound contamination with plutonium, americium, or curium, chelation therapy with Ca-DTPA and Zn-DTPA is recommended. In case of wounds containing embedded radioactive fragments it may be necessary to surgically remove the fragment. The latter should be placed in shielded containers.

9.9.3 Internal Decontamination

Radioactive contaminants in the body may cause significant health risks. The risks are long term and depend on type, nature and concentration of the radioactive material. Agents used for radionuclide decontamination can be categorized into

Table 9.16 Internal decontamination options

Target radionuclide	Compound	Application
^{241}Am	Calcium and zinc salts of DTPA[a]	Intravenous infusion
^{137}Cs	Prussian blue	Oral
^{60}Co	Calcium and zinc salts of DTPA[a]	Intravenous infusion
^{131}I	Potassium iodide	Oral
^{32}P	Phosphate, dibasic potassium and Na salts	Oral
^{238}Pu, ^{239}Pu	Calcium and zinc salts of DTPA[a]	Intravenous infusion
^{210}Po	2,3-Dimercaptopropanol	Intramuscular injection
^{226}Ra	Ammonium chloride, Ca carbonate, Ca gluconate, Na alginate	Oral
^{90}Sr	Al hydroxide, Al phosphate, ammonium chloride, Ca carbonate, Ca gluconate, Na alginate	Oral
^{235}U, ^{238}U	Sodium bicarbonate	Intravenous infusion or oral

Adapted from Kalinich (2012)
[a]*DTPA* Diethylenetriamine pentacetic acid

absorption-reducing agents, blocking and diluting agents, mobilizing agents, and chelating agents (Kalinich 2012). However, a single compound does not work for all contaminants, demonstrating the need for identification of the radioactive source. An overview of internal decontamination options is given in Table 9.16.

9.9.3.1 Absorption-Reducing Agents

Absorption can be reduced by lavage of stomach, emetics, purgatives, laxatives and ion exchangers. Stomach lavage is effective only if it is performed immediately after contamination and ingested dose is high. Commonly used emetic agents are apomorphine (5–10 mg, subcutaneous) and ipecac (1–2 g in capsule or 15 ml in syrup), which should be given concomitantly with 200–300 ml of water (Kalinich 2012). If the person is unconscious then emetics are contradicted and purgatives should not be used in individuals with abdominal pain. Certain non-absorbable binding resins are capable of preventing the uptake of various radioactive materials in the gut. For example, Prussian blue, which is a non-absorbable pigmented resin has been used orally to enhance the faecal excretion of cesium and thallium by means of ion exchange process. Several antacids (e.g. Al containing antacids, Al-hydroxide) are shown to reduce the absorption of radioactive strontium if given immediately after the exposure (Kalinich 2012).

9.9.3.2 Blocking and Diluting Agents

These agents are used to reduce the uptake of radionuclides from the blood tissues into target tissue (Torngren and Persson 1998). For example, potassium iodide is recommended to prevent radioactive iodine from being sequestered in the thyroid. Elements with chemical properties similar to the internalized radionuclide are also used as blocking agents. For example, calcium, and to a lesser extent phosphorus, can be applied to block the uptake of radioactive strontium. Diluting agents simply dilute the concentration of radionuclides in the body and decrease their absorption (Kalinich 2012). Water, for example, can be used to increase the excretion of tritium.

9.9.3.3 Mobilizing Agents

Mobilizing agents enhance the release of deposited radionuclides from the tissue by increasing their natural turnover process rate. Examples of mobilizing agents include diuretics, propylthiouracil, expectorants, parathyroid extract and corticosteroids, ammonium chloride and sodium bicarbonate (Dubois et al. 1994). Ammonium chloride results in acidification of the blood and increased elimination of internalized radiostrontium. Sodium bicarbonate is used to increase urinary pH which is useful for preventing the precipitation of uranium that passes through the kidney (Kalinich 2012).

9.9.3.4 Chelating or Complexing Agents

Chelating or complexing agents such as ethylenediamine tetraacetic acid (EDTA), diethylenetriamine pentacetate (DTPA), nitrilotriaceticacid (NTA) form complexes with certain radionuclides producing water soluble compounds which can be more rapidly eliminated from the body via excretion by the kidneys (Dalvi et al. 1980). Calcium and zinc salts of DTPA are approved as decorporating agent. DTPA forms water soluble, stable complexes with transuranium elements and increase their elimination from the body. Both Ca-DTPA and Zn-DTPA are safe for the treatment of plutonium, americium, or curium (Sharma et al. 2010). Ca-DTPA is administered as a single intravenous injection or inhaled immediately possible after contamination, and repeated doses of Zn-DTPA administered intravenously which may be given daily as maintenance therapy, as necessary. Uranium contamination has been treated with oral sodium bicarbonate, regulated to maintain an alkaline urine pH, and accompanied by diuretics (Sharma et al. 2010).

9.10 Measures for Remediation of Radionuclide Contaminated Sites

9.10.1 Classification of Radionuclide-Contaminated Sites

The size of radionuclide-contaminated land may stretch over different ranges, from small- and mid-size areas with diameters from several tens to hundreds meters to large-size areas with diameters of thousands of meters. For smaller areas, mechanically or physio-chemically based technologies may be convenient. Larger areas such as the landscapes contaminated by the Chernobyl or Fukushima Daiichi accidents for several reasons may be mainly suited for phytoremediation technologies. The classification of contaminated areas relates to the phenomenological description of the site and the processes most important for the fate of the contaminants, such as the basic chemistry, generation, internal transport, outflow, adsorption, etc. These processes are in turn strongly influenced by the characteristics of a given site, including the geochemical constituents. In terms of remediation, the major radioactive contaminants can be grouped in two classes, including (i) radionuclides originating either from original ore or from nuclear waste, such as ^{238}Pu, ^{239}Pu, ^{240}Pu, ^{241}Pu, ^{241}Am, ^{237}Np, ^{234}U, ^{235}U, ^{238}U, ^{230}Th, ^{226}Ra, ^{228}Ra, ^{222}Rn, ^{210}Pb, ^{231}Pa, ^{227}Ac, and (ii) radioisotopes of other elements, originating from low- or intermediate level radioactive waste, like ^{14}C, ^{36}Cl, ^{63}Ni, ^{90}Sr, ^{93}Zr, ^{94}Nb, ^{99}Tc, ^{107}Pd, ^{126}Sn, ^{129}I, ^{135}Cs and ^{137}Cs. This information is required for adequate site restoration.

9.10.2 Mechanically or Physio-chemically Based Technologies

The application of remediation techniques requires consideration of performance, reliability and maintenance requirements, cost, available supporting infrastructure, risk to workers and public during implementation, environmental impact, and regulatory and community acceptance. The choice of remediation technology for radioactively contaminated land also needs to consider future land use. Remediation technologies can include complete or partial remove, stabilization, immobilization (e.g. cement-based solidification, chemical immobilization) or isolation of the contamination (US EPA 1996; Mallett 2004). The most commonly used remediation techniques include (i) excavation of contaminated soil for on-site storage or offsite disposal, (ii) covering with inert materials for reducing or avoiding external exposure, creation of dust or infiltration by rainwater (and hence transfer of radioactivity to groundwater), and (iii) vertical and horizontal in-ground barriers to prevent contamination migration (IAEA 2002).

The suitability of restoration techniques for treating different radionuclides depends upon their chemical and physical properties as well as on site-specific

Table 9.17 Effectiveness of restoration techniques in treating radionuclide-contaminated sites

	Radionuclide						
Restoration technique	Co	Sr	Cs	Ra	Th	U	Pu
Excavation	x	x	x	x	x	x	x
Soil washing			x	x	x	x	x
Flotation				x		x	x
Filtration				x		x	x
Chemical solubilisation	x		x	x	x	x	x
Desorption		x		x		x	
Biosorption	x	x	x			x	x
Surface barriers	x	x	x	x	x	x	x
Sub-surface barriers	x	x	x	x	x	x	x
Solidification	x	x	x	x	x	x	x
Vitrification	x	x	x	x	x	x	x
Chemical immobilization	x	x	x	x	x	x	x

Compiled from US EPA (1996)

factors. Excavation of the contaminant, containment (surface and sub-surface barriers) and immobilization (solidification and chemical immobilization) are likely to be suited for most radionuclides (Table 9.17). However, while excavation removes contaminated soil from the site it creates large quantities of contaminated wastes. Excavated wastes necessitate long-term monitoring which limits the amounts of materials requiring disposal at landfill sites. Disposal of excavated materials is a problem worldwide. Potential alternative technologies for remediation of radionuclide contaminated soils include soil washing, desorption, electrochemical remediation and vitrification. Although very energy consuming, the latter method has been shown to be very efficient in immobilizing radionuclides in the long term. For example, on a test site in Australia, using an array of four electrodes, 3–6 Mg of soil per hour were melted to depths of 7–15 m. Retention of Am, Sr and Pu in the vitrified residue approached 100% while retention of Cs was >90% (IAEA 1999).

The above methods are also used for remediation of sites contaminated with toxic elements. They have been described in detail in Chap. 8. In the latter, the costs of different remediation technologies are indicated as well. However, for remediation of radionuclide-contaminated soils, additional high costs arise for the specific disposal of radioactive materials. These have been indicated with 2000–3000 EUR m^{-3} for disposal of radioactive soil, and radioactive waste from soil washing, flotation, filtration, desorption, chemical solubilisation and biosorption (Zeevaert and Bousher 1999). Indirect costs arising from monitoring and loss or gain of income and taxes, are site-specific and cannot be calculated from the characteristics of the remediation technologies.

Another important aspect in mechanically based site remediation is soil compaction caused by heavy equipment that is necessary to remove or transport soil. Compaction deteriorates physical conditions of the soil that have major impacts on chemical and microbiological processes which ultimately affect plant growth. Treating the soil with dispersing and chelating chemicals removes not only radionuclides but also soil nutrients which are necessary for microbial and plant growth. Dispersing chemicals often adversely affect not only soil chemical processes, but also soil physical processes. After the soil is replaced, establishment of plants on a physically, chemically and microbiologically compromised soil is a great challenge (Entry et al. 1996).

9.10.3 Phytoremediation

Radionuclides released into the environment are taken up by plants and redistributed throughout the ecosystem. A serious problem may arise when agro-ecosystems become contaminated. Soils contaminated with radionuclides provide a particular challenge to soil decontamination and hence a useful perspective on phytoremediation. The term phytoremediation has already been defined in Chap. 8, as this technology is being used for toxic element-contaminated sites as well (McGrath et al. 2002). Phytoextraction, phytostabilization, and phytovolatilization have been suggested as suitable remediation techniques for radionuclides (Beresford 2006). Besides the fact that phytoremediation is a non-intrusive technique, estimates from the USA suggest that this method is considerably more cost-effective than other remediation techniques (Fiore et al. 2000). However, major limiting factors for phytoextraction are low radionuclide availability in soils and limited translocation from roots to surface plant parts. For effective application of this technique the radionuclide uptake by plant roots needs to result in translocation to shoots. The major part of ^{137}Cs taken up by plants accumulates in roots (Clint and Dighton 1992). Phytoremediation studies have been performed with a wide range of radionuclides, including ^{3}H, ^{90}Sr, ^{95}Nb, ^{99}Tc, ^{106}Ru, ^{144}Ce, ^{226,228}Ra, ^{239,240}Pu, ^{241}Am, ^{228,230,232}Th, ^{244}Cm and ^{237}Np (Dushenkov 2003).

9.10.3.1 Phytoextraction

A wide array of plant species occupying different habitats have been shown to accumulate large amounts of radionuclides from contaminated soils. Some are given in Table 9.18. For effective remediation it is necessary to grow plants producing high amounts of biomass that should exceed 3 Mg dry matter ha^{-1} in a cropping season and accumulate >1000 mg radionuclide kg^{-1}. A variety of factors that may help to bring radionuclides into the soil solution and make it more available for plant uptake. These include manipulation of soil pH, addition of chelators, amending soil with chemicals stimulating radionuclide desorption, interaction with microorganisms and plant exudates. Bioavailability of radionuclides is strongly influenced by pH. For example, uranyl cations (UO_2^{2+}) that dominate at

Table 9.18 Plants with potential for phytoextraction of various radionuclides

Plant species	Radionuclide	References
Annual plants		
Sinapis ssp.	^{137}Cs	Dushenkov et al. (1999)
Amaranthus retroflexus	^{137}Cs, ^{90}Sr, U	Fuhrmann et al. (2002)
		Huang et al. (1998)
Brassica ssp.	U	Huang et al. (1998)
		Lasat et al. (1998)
Phaseolus acutifolius	^{137}Cs	Negri and Hinchman (2000)
Phalaris arundinaceae	^{137}Cs	Negri and Hinchman (2000)
Grasses		
Festuca ssp.	^{137}Cs	Dahlman et al. (1969)
Lolium perenne	^{137}Cs	Salt et al. (1992)
Trees		
Acer rubrum	^{244}Cm, ^{137}Cs, ^{238}Pu, ^{226}Ra and ^{90}Sr	Pinder et al. (1984)
Cocos nucifera	^{137}Cs	Robison and Stone (1992)
Pinus radiata	^{137}Cs, ^{90}Sr	Entry et al. (1993)
Eucalyptus tereticornis	^{137}Cs, ^{90}Sr	Entry et al. (1996)

pH 5 were more readily taken up and translocated in plants compared to hydroxyl (pH 6) and carbonate (pH 8) complexes of U. Chelators (see Chap. 8) have shown a tenfold increase in ^{137}Cs availability and an almost hundredfold increase in Pb and U availability (Huang et al. 1998). In greenhouse experiments, shoots U concentration in plants grown in a U-contaminated soil increased to more than 5000 mg kg^{-1} in citric acid treated soil compared to 5 mg kg^{-1} in control pots (Huang et al. 1998). In hydroponic experiments it was shown that shoots to roots ratio of ^{137}Cs was significantly higher in mycorrhizal plants of *Calluna vulgaris* L. (heather) compared to non-mycorrhizal plants (Clint and Dighton 1992).

In the Chernobyl exclusion zone mustard was used to noticeably reduce ^{137}Cs activity (Dushenkov et al. 1999). Using three subsequent mustard crops, ^{137}Cs activity was reduced from 2558 to 2239 (averages of the exclusion zone) Bq kg^{-1} soil. Within only one growth period, the portion of the area with ^{137}Cs levels exceeding 3000 Bq kg^{-1} soil could be reduced from 29.4% to 7.7%. At a contaminated field site within the Brookhaven National Laboratory (New York, USA) about 50% of ^{137}Cs and ^{90}Sr was estimated to be removed by twice-yearly cropping of *Amaranthus retroflexus* within 15 and 7 years, respectively (Fuhrmann et al. 2002). Trees have also been shown to take up substantial amounts of radionuclides. *Acer rubrum, Liriodendron tulipifera* and *Liquidambar stryaciflua* accumulated high quantities of a large spectrum of radionuclides (Pinder et al. 1984). *Cocos nucifera* took up substantial amounts of ^{137}Cs from soils contaminated by nuclear weapons testing on Bikini Atoll (Robison and Stone 1992). Entry et al. (1993) documented that *Pinus radiata* seedlings accumulated more ^{137}Cs and ^{90}Sr than *Pinus ponderosa*. Entry and Emmingham (1995) found that potted *Eucalyptus tereticornis* seedlings removed 31.0% of the ^{137}Cs and 11.3% of the ^{90}Sr in sphagnum peat soil after one month of exposure.

To increase plant uptake of radionuclides or toxic elements, genetic modification of plants has also been suggested (Entry et al. 1997; McGrath et al. 2002). However, Wolfe and Bjornstad (2002) warn that the development of genetically engineered plants for phytoremediation may not be accepted by the society. An alternative may be to exploit phylogenetic and ecological correlates of leaf mineral contents to select plant species with high radionuclide accumulation potentials (Broadley et al. 1999; Jansen et al. 2002).

9.10.3.2 Phytostabilisation

This technique is applied for radionuclide-contaminated sites mainly to hold contaminants in place to prevent secondary contamination and exposure. Capturing radionuclides in situ is often the best solution at sites with low contamination levels or vast contaminated areas where a large-scale removal action or other in situ remediation is not practicable. Phytostabilisation can result in a remarkable risk reduction, especially if radionuclides with relatively short half-lives are present. Fast growing and deep-rooted plants have gained popularity in their use for radionuclides stabilization in soil. Establishment of a vegetation cover prevents the formation of windblown dust, a major pathway for human exposure at radionuclide-contaminated sites (Berti and Cunningham 2000). The ability of plants to transpire high volumes of water prevents migration of leachate towards groundwater or surface waters. This may be useful in confining areas leaking radioactive materials. Phytostabilization may be particularly suitable for controlling tailings from strip and open pit uranium mines. It is important to keep in mind that while phytostabilization of radionuclides may reduce the environmental and human health risk it does not remove the source of radioactivity from the site (Dushenkov 2003). Thus, the potential risk of radiation exposure remains, which needs be considered when decision on the best remediation approach is made.

9.10.3.3 Phytovolatilization

This method exploits plants' ability to transpire high amounts of water and is currently used for tritium (^{3}H) remediation. Tritium is a radioactive isotope of hydrogen and is found in the environment typically as tritiated water. It decays to a stable helium with a half-life of about 12 years. As a weak β emitter, ^{3}H is easily shielded by air and skin and produces almost no external radiation exposure. Its incorporation in water and organic compounds, however, presents a health hazard when absorbed into the body. The most cost-effective approach for the minimization of ^{3}H risk is neither removal of ^{3}H from the effluent nor a typical isolation of the water, but rather a change in the path of ^{3}H exposure to the public (Dushenkov 2003). A reduction in dose by 40% can be achieved by releasing tritiated water as water vapor to the atmosphere, as opposed to allowing it to flow off site in surface water streams (Fulbright et al. 1996). For example, tritiated water applied as irrigation to the floor of forests either evaporates from the soil surface, is absorbed into the transpiration stream by plant roots or drains through the soil to the water table. Only a small portion of ^{3}H that was absorbed by plant roots remains in the plant tissues in the form of exchangeable hydroxyl ions or is incorporated into organic molecules through photosynthesis (Dushenkov 2003).

References

Abdelouas A (2006) Uranium mill tailings: geochemistry, mineralogy, and environmental impact. Elements 2(6):335–341

Adams JAS, Richardson KA (1960) Thorium, uranium and zirconium concentration in bauxite. Econ Geol 55:1653–1675

Al-Hamarneh IF, Awadallah MI (2009) Soil radioactivity levels and radiation hazard assessment in the Highlands of Northern Jordan. Radiat Meas 44:102–110

Allard B, Rydberg J (1983) Behaviour of plutonium in natural waters. In: Carnall W T, Choppin GR (eds) Plutonium chemistry. Am Chem Soc Symp Ser 216:275–295

Ames LL, Rai D (1978) Radionuclide interactions with soil and rock media. Vol 1: processes influencing radionuclide mobility and retention, element chemistry and geochemistry, and conclusions and evaluations. EPA 520/6-78-007-a. U.S. Environmental Protection Agency, Las Vegas

Arnold BW, Meijer A, Kalinina E, Robinson BA, Kelkar S, Jove-Colon C, Kuzio S.P, James S, Zhu M (2006) Impacts of reducing conditions in the saturated zone at Yucca Mountain. In: Proceedings of the 11th international high-level radioactive waste management conference (IHLRWM), Las Vegas, NV, pp 345–352

Ashby JR, Craig PJ (1988) Environmental methylation of tin: an assessment. Sci Total Environ 73:127–133

Baes CF Jr, Mesmer RE (1976) The hydrolysis of cations. Wiley, New York

Beddow H, Black S, Read D (2006) Naturally occurring radioactive material (NORM) from a former phosphoric acid processing plant. J Environ Radioact 86:289–312

Beresford NA (2006) Land contaminated by radioactive materials. Soil Use Manag 21:468–474

Berti WR, Cunningham SC (2000) Phytostabilization of metals. In: Raskin I, Ensley BD (eds) Phytoremediation of toxic metals: using plants to clean up the environment. Wiley, New York, pp 71–88

Borovec Z, Kribek B, Tolar V (1979) Sorption of uranyl by humic acids. Chem Geol 27:39–46

Bovard P, Grauby A, Saas A (1968) Chelating effect of organic matter and its influenceon the migration of fission products. In: Isotopes and radiation in soil organic matter studies, STI/PUB-190 (CONF-680725), proceedings series. International Atomic Energy Agency (IAEA), Vienna, Austria, pp 471–495

Broadley MR, Willey NJ, Mead A (1999) A method to assess taxonomic variation in shoot caesium concentration amongst flowering plants. Environ Pollut 106:341–349

Carvalho FP (1995) 210Pb and 210Po in sediments and suspended matter in the Tagus estuary, Portugal. Local enhancement of natural levels by wastes from phosphate ore processing industry. Sci Total Environ 159:201–214

Casarett AP (1968) Radiation biology-radiation detection and dosimetry, 1st edn. Prentice-Hall Inc, Englewood Cliffs

Chisholm-Brause C, Conradson SD, Buscher CT, Eller PG, Morris DE (1994) Speciation of uranyl sorbed at multiple binding sites on montmorillonite. Geochim Cosmochim Acta 58 (17):3,625–3,631

Clarke LB (1992) Applications for coal-use residues., Rep. IEA CR/50. IEA Coal Research, London

Clint GM, Dighton J (1992) Uptake and accumulation of radiocaesium by mycorrhizal and non-mycorrhizal heather plants. New Phytol 122:555–561

Cooper MB (2005) Naturally Occurring Radioactive Materials (NORM) in Australian industries – review of current inventories and future generation. EnviroRad Services Pty. Ltd. http://www.arpansa.gov.au/pubs/norm/cooper_norm.pdf. Accessed 29 Apr 2016

Cooper MB, Clarke PC, Robertson W, McPharlin IR, Jeffrey RC (1995) An investigation of radionuclide uptake into food crops grown in soils treated with Bauxite mining residues. J Radioanal Nucl Chem 194:379–387

Crowley KD (1997) Nuclear waste disposal: the technical challenges. Phys Today 50(6):32–39

Dahlman RC, Auerbach SI, Dunaway PB (1969) Environmental contamination by radioactive materials. International Atomic Energy Agency and World Health Organization, Vienna

Dalvi RR, McGowan C, Ademoyero A (1980) In vivo and in vitro effect of chelating agents on drug metabolizing enzymes of the rat. Toxicol Lett 6:25–28

Delpoux MA, Dulieu LA, Dalebroux M (1997) Experimental study on the genetic effects of high levels of natural radiation in south France. In: Proceedings of the fourth international conference on high levels of natural radiation: radiation doses and health effects, 1996, Beijing, China. Elsevier, Tokyo, 397–406

Dubois A, King GL, Livengood DR (1994) Radiation and the gastrointestinal tract, 1st edn. CRC Taylor and Francis Publishing Group, Boca Raton, pp 24–122

Dushenkov S (2003) Trends in phytoremediation of radionuclides. Plant Soil 249:167–175

Dushenkov S, Mikheev A, Prokhnevsky A, Ruchko M, Sorochinsky B (1999) Phytoremediation of radiocesium-contaminated soil in the vicinity of Chernobyl, Ukraine. Environ Sci Technol 33:469–475

EC (European Commission)/IAEA (International Atomic Energy Agency)/World Health Organization) (1996) Summary of the conference results "One Decade after Chernobyl – Summing up the consequences of the accident". Vienna 8–12 April 1996. IAEA Proceedings Series, Vienna

Eisenbud M, Gesell T (1997) Environmental radioactivity from natural, industrial, and military sources, 4th edn. Academic, London

El-Gamal A, Nasr S, El-Taher A (2007) Study of the spatial distribution of natural radioactivity in the Upper Egypt Nile River sediments. Radiat Meas 42:457–465

Entry JA, Emmingham WH (1995) Sequestration of ^{137}Cs and ^{90}Sr from soil by seedlings of *Eucalyptus tereticorinus*. Can J For Res 25:1044–1047

Entry JA, Rygiewicz PT, Emmingham WH (1993) Accumulation of ^{137}Cs and ^{90}Sr in *Pinus ponderosa* and *Pinus radiata* seedlings. J Environ Qual 22:742–746

Entry JA, Vance NC, Hamilton MA, Zabowski D, Watrud LS, Adriano DC (1996) Phytoremediation of soil contaminated with low concentrations of radionuclides. Water Air Soil Pollut 88:167–176

Entry JA, Lidia SW, Manasse RS, Vance NC (1997) Phytoremediation and reclamation of soils contaminated with radionuclides. Am Chem Soc Symp Ser 664:299–306

Eriksen TE, Ndalamba P, Bruno J, Caceci M (1992) The solubility of TcO_2 x nH_2O in neutral to alkaline solutions under constant pCO_2. Radiochim Acta 58(59):67–70

Eve IS (1966) A review of the physiology of the gastrointestinal tract in relation to radiation doses from radioactive materials. Health Phys 12:131–161

Fiore J, Rampertaap A, Greeves J, Mackinney J, Raguso M, Selstrom J (2000) Radioactive residues at nuclear sites in the United States of America. Restorations of environments with radioactive residues. International Atomic Energy Agency, Vienna, pp 25–47

Fuhrmann M, Lasat MM, Ebbs SD, Kochian LV, Cornish J (2002) Plant uptake of cesium-137 and strontium-90 from contaminated soil by three plant species; application to phytoremediation. J Environ Qual 31:904–909

Fulbright HH, Schwirian-Spann AL, Jerome KM, Looney BB, Brunt VV (1996) Status and practicality of detritiation and tritium reduction strategies for environmental remediation. Westinghouse Savannah River Company, Aiken, pp 1–10

Gascoyne M (1982) Geochemistry of the actinides and their daughters. In: Ivanovich M, Harmon RS (eds) Uranium series disequilibrium: applications to environmental problems. Clarendon Press, Oxford, pp 33–55

Ghiassi-nejad M, Mortazavi SMJ, Cameron R, Niroomand-rad A, Karam PA (2002) Very high background radiation areas of Ramsar, Iran: preliminary biological studies. Health Phys 82:87–93

Godoy MLDP, Roldao LA, Godoy JM, Conti, LFC (2000) Natural radionuclides and heavy metals in surface soils around a coal fired power plant in Brazil. Proc 5th Int. Conf on high levels of natural radiation and radon areas: radiation dose and health effects, Munich, 122

Grenthe I, Fuger J, Konings RJM, Lemire RJ, Muller AB, Nguyen T C, Wanner H (1992) Chemical thermodynamics 1: chemical thermodynamics of uranium. Elsevier Science Publishing Company, North-Holland

Hamada N, Ogino H, Fujimichi Y (2012) Safety regulations of food and water implementedin the first year following the Fukushima nuclear accident. J Radiat Res 53:641–671

Harrison JD, Muirhead CR (2003) Quantitative comparisons of cancer induction in humans by internally deposited radionuclides and external radiation. Int J Radiat Biol 79:1–13

Hirose K (2012) Fukushima Dai-ichi nuclear power plant accident: summary of regional radioactive deposition monitoring results. J Environ Radioact 111:13–17

Hu CH, Weng JQ, Wang JS (2010) Sources of anthropogenic radionuclides in the environment: a review. J Environ Radioact 101:426–437

Hu Q, Zhao P, Moran JE, Seaman JC (2005) Sorption and transport of iodine species in sediments from the Savannah River and Hanford Sites. J Contam Hydrol 78:185–205

Hu QH, Smith DK (2004) Field-scale migration of ^{99}Tc and ^{129}I at the Nevada test site. In: Hanchar JM, Stroes-Gascoyne S, Browning L (eds) Scientific basis for nuclear waste management XXVIII, Materials Research Society Symposium Proceedings, vol 824. Materials Research Society, Pittsburgh, pp 399–404

Huang JW, Blaylock MJ, Kapulnik Y, Ensley BD (1998) Phytoremediation of uranium contaminated soils: role of organic acids in triggering uranium hyperaccumulation in plants. Environ Sci Technol 32(13):2,004–2,008

IAEA (International Atomic Energy Agency) (1986) Collection of post-chernobyl reports from Austria, Belgium, Federal Republic of Germany, Hungary, Iran, Monaco, Poland, Switzerland, Turkey. Vienna

IAEA (International Atomic Energy Agency) (1999) Technologies for remediation of radioactively contaminated sites. Vienna

IAEA (International Atomic Energy Agency) (2002) Radiation legacy of the 20th Century: environmental restoration. IAEA-TECDOC-1280. Vienna

IAEA (International Atomic Energy Agency) (2003) Extent of environmental contamination by Naturally Occurring Radioactive Material (NORM) and technological options for mitigation, Technical Reports Series, vol 419. IAEA, Vienna

IAEA (International Atomic Energy Agency) (2006) Assessing the need for radiation protection measures in work involving minerals and raw materials, Safety Reports Series, vol 49. IAEA, Vienna

IAEA (International Atomic Energy Agency) (2008) Nuclear power reactors in the world. Reference data series No. 2. http://www-pub.iaea.org/MTCD/Publications/PDF/RDS2-28_web.pdf. Accessed 20 Jul 2017

ICRP (International Commission on Radiological Protection) (1999) Protection of the public in situations of prolonged exposure. ICRP Publication 82. Annals of the ICRP 29

IEA (2001) International energy agency, Coal Information 2001. OECD/IEA, Paris

Jansen S, Broadley MR, Robbrecht E, Smets E (2002) Aluminum hyperaccumulation in angiosperms: a review of its phylogenetic significance. Bot Rev 68:235–269

Jonkers G, Hartog FA, Knaepen AAI, Lancee PFJ (1997) Characterization of NORM in the oil and gas production (E&P) industry, Radiological problems with natural radioactivity in the non-nuclear industry, Proc Int Symp Amsterdam 1997. KEMA, Arnhem

Kalac P (2001) A review of edible mushroom radioactivity. Food Chem 75:29–35

Kalinich JF (2012) Treatment of internal radionuclide contamination. In: Lenhard MK (ed) Medical consequences of radiological and nuclear weapons. Office of The Surgeon General Department of the Army, United States of America and US Army Medical Department Center and School Fort Sam Houston, Houston, pp 73–81

Kaplan DI, Serne RJ (2000) Geochemical data package for the Hanford immobilized low-activity tank waste performance assessment (ILAW PA), PNNL-13037, Rev 1. Pacific Northwest National Laboratory, Richland

Kaplan DI, Bertsch PM, Adriano DC, Orlandini KA (1994) Actinide association with groundwater colloids in a coastal plain aquifer. Radiochim Acta 66(67):181–187

Kaszuba JP, Runde WH (1999) The aqueous geochemistry of neptunium: dynamic control of soluble concentrations with applications to nuclear waste disposal. Environ Sci Technol 33:4,427–4,433

Khater AEM, Al-Sewaidan HA (2008) Radiation exposure due to agricultural uses of phosphate fertilisers. Radiat Meas 43:1402–1407

Kim JJ (1986) Chemical behavior of transuranic elements in aquatic systems. In: Freeman AJ, Keller C (eds) Handbook on the physics and chemistry of the actinides. Elsevier Science Publishers, Amsterdam, pp 413–455

Klement AW (1982) CRC handbook of environmental radiation. CRC Press, Boca Raton

Kohler M, Honeyman BD, Leckie JO (1999) Neptunium(V) sorption on hematite (α-Fe_2O_3) in aqueous suspension: the effect of CO_2. Radiochim Acta 85:33–48

Kurnaz A, Kucukomeroglu B, Keser R, Okumusoglu NT, Kprkmaz F, Karahan G, Cevik U (2007) Determination of radioactivity levels and hazardsof soil and sediment samples in Firtina Valley (Rize, turkey). Appl Radiat Isot 65:1281–1289

Lasat MM, Fuhrmann M, Ebbs SD, Cornish JE, Kochian LV (1998) Phytoremediation of a radiocesium-contaminated soil: evaluation of cesium-137 bioaccumulation in the shoots of three plant species. J Environ Qual 27:165–169

Lefevre R, Sardin M, Schweich D (1993) Migration of strontium in clayey and calcareous sandy soil: precipitation and ion exchange. J Contam Hydrol 13:215–229

Lehmann B, Halder S, Ruzindana Munana J, Ngizimana JP, Biryabarema M (2013) The geochemical signature of rare-metal pegmatites in Central Africa: magmatic rocks in the Gatumba tin-tantalum mining district, Rwanda. J Geochem Explor 144:528–538

Lieser KH, Bauscher CH (1987) Technetium in the hydrosphere and in the geosphere. I. Chemistry of technetium and iron in natural waters and influence of the redox potential on the sorption of technetium. Radiochim Acta 42:205–213

Lieser KH, Mühlenweg U (1988) Neptunium in the hydrosphere and in the geosphere. I. Chemistry of neptunium in the hydrosphere and sorption of neptunium from groundwaters on sediments under aerobic and anaerobic conditions. Radiochim Acta 44(45):129–133

Lovley DR (1995) Bioremediation of organic and metal contaminants with dissimilatory metal reduction. J Indust Microbio 14:85–93

Mallett H (2004) Technical options for managing contaminated land. ENVIROS, Abingdon. http://www.ciria.org/safegrounds/pdf/technical_options_april04.pdf. Accessed 21 Jul 2017

Matiullah A, Ur-Rehman S, Ur-Rehman A, Faheem M (2004) Measurement of radioactivity in the soil of Behawalpur Division, Pakistan. Radiat Prot Dosim 112(3):443–447

McGrath SP, Zhao FJ, Lombi E (2002) Phytoremediation of metals, metalloids, and radionuclides. Adv Agron 75:1–56

Metz V, Kienzler B, Schüßler W (2003) Geochemical evaluation of different groundwater-host rock systems for radioactive waste disposal. J Contam Hydrol 61:265–279

Morrow PE, Gibb FR, Gazioglu KM (1967) A study of particulate clearance from the human lungs. Am Rev Respir Dis 96:1209–1221

Mukherjee AB, Zevenhoven R, Bhattacharya P, Sajwan KS, Kikuchi R (2008) Mercury flow via coal and coal utilization by-products: a global perspective. Resour Conserv Recycl 52:571–591

Nakata K, Nagasaki S, Tanaka S, Sakamoto Y, Tanaka T, Ogawa H (2002) Sorption and reduction of neptunium (V) on the surface of iron oxides. Radiochim Acta 90:665–669

National Council on Radiation Protection and Measurements (1975) Natural background radiation in the United States, NCRP Report No 45. NCRP, Washington, DC

National Council on Radiation Protection and Measurements (1988) Exposure of the population in the United States and Canada from natural background radiation, NCRP Report No 94. Bethesda

Negri CM, Hinchman RR (2000) The use of plants for the treatment of radionuclides. In: Raskin I (ed) Phytoremediation of toxic metals: using plants to clean up the environment. Wiley-Interscience, Wiley, New York, pp 107–132

Nelson DM, Larson RP, Penrose WR (1987) Chemical speciation of plutonium in natural waters. In: Pinder JE, Alberts JJ, McLeod KW, Schreckhise RG (eds) Environmental research on actinide elements, CONF-841142. Office of Scientific and Technical Information, US Department of Energy, Washington, DC, pp 27–48

Onishi Y, Serne RJ, Arnold EM, Cowan CE, Thompson FL (1981) Critical review: radionuclide transport, sediment transport, and water quality mathematical modeling; and radionuclide adsorption/desorption mechanisms, NUREG/CR-1322 (PNL-2901) prepared for the US Nuclear Regulatory Commission, Washington, DC. Pacific Northwest Laboratory, Richland

Owens A, Peacock A (2004) Compound semiconductor radiation detectors. Nucl Instr Meth A531:18–37

Papatheodorou G, Papaethymiou H, Maratou A, Ferentinos G (2005) Natural radionuclides in bauxitic tailings (red-mud) in the Gulf of Corinth. Greece Radioprotect 40:5549–5555

Paschoa AS (2000) More than forty years of studies of natural radioactivity in Brazil. Technology 7:193–212

Paul AC, Pillai PMB, Haridasan P, Radhakrishnan S, Krishnamony S (1998) Population exposure to airborne thorium at the high natural radiation areas in India. J Environ Radioact 40:251–259

Philipsborn HV, Kühnast E (1992) Gamma spectrometric characterization of industrially used African and Australian Bauxites and their red mud tailings. Radiat Prot Dosim 45:741–744

Pinder JE III, McLeod KW, Alberts JJ, Adriano DC, Corey JC (1984) Uptake of ^{244}Cm, ^{238}Pu and other radionucides by trees inhabiting a contaminated floodplain. Health Phys 47:375–384

Pinnock WR (1991) Measurements of radioactivity in Jamaican building materials and gamma dose equivalents in a prototype red mud house. Health Phys 61:647–651

Prasad KN (1995) Handbook of radiobiology, 2nd edn. CRC Taylor and Francis Publishing Group, Denver, pp 46–109

Pröhl G, Mück K, Likhtarev I, Kovgan L, Golikov V (2002) Reconstruction of the ingestion doses received by the population evacuated from the settlements in the 30-km zone around the Chernobyl reactor. Health Phys 82:173–181

Rai D, Zachara JM, Schwab AP, Schmidt RL, Girvin DC, Rogers JE (1984) Chemical attenuation rates, coefficients, and constants in leachate migration, A Critical Review. EA-3356, vol 1. Electric Power Research Institute, Palo Alto

Ramachandran TV (2011) Background radiation, people and the environment. Iran J Radiat Res 9 (2):63–76

Ramachandran TV, Mishra UC (1989) Measurement of natural radioactivity levels in Indian foodstuffs by gamma spectrometry. Appl Radiat Isot 40:723–726

Read D, Lawless TA, Sims RJ, Butter KR (1993) Uranium migration through intact sandstone cores. J Contam Hydrol 13:277–289

Robertson DE, Cataldo DA, Napier BA, Krupka AM, Sasser LB (2003) Literature review andassessment of plant and animal transfer factors used in performance assessment modeling, US Nuclear Regulatory Commission Office of Nuclear Regulatory Research Washington, DC 20555-0001. Pacific Northwest National Laboratory, Richland

Robison WL, Stone EL (1992) The effect of potassium on the uptake of 137Cs *in* food crops grown on coral soils: coconut at Bikini Atoll. Health Phys 62:496

Roussel-Debet S, Gontier G, Siclet F, Fournier M (2006) Distribution of carbon 14 in the terrestrial environment close to French nuclear power plants. J Environ Radioact 87:246–259

Salt CA, Mayes RW, Elston DA (1992) Effects of season, grazing intensity and diet composition on the radiocesium intake by sheep on a re-seeded hill pasture. J Appl Ecol 29:378

Sanchez AL, Murray JW, Sibley TH (1985) Adsorption of Pu (IV) and (V) on goethite. Geochim Cosmochim Acta 49:2,297–2,307

Sandino A, Bruno J (1992) The solubility of $(UO_2)_3(PO_4)_2 \cdot 4\ H_2O(s)$ and the formation of U (VI) phosphate complexes: their influence in uranium speciation in natural waters. Geochim Cosmochim Acta 56:4,135–4,145

Scheckel KG, Sparks DL (2001) Temperature effects on nickel sorption kinetics at the mineral–water interface. Soil Sci Soc Am J 65:719–728

Schwertmann U, Taylor RM (1989) Iron oxides. In: Dixon JB, Week SB (eds) Minerals in soil environments, 2nd edn. Soil Science Society of America, Madison, pp 379–438

Séby F, Potin-Gautier M, Giffaut E, Donard OFX (2001) A critical review of thermodynamic data for inorganic tin species. Geochim Cosmochim Acta 65(18):3041–3054

Shahbazi-Gahrouei D, Gholami M, Setayandeh S (2013) A review on natural background radiation. Adv Biomed Res 2(3):1–6

Sharma D, Sunil Kamboj S, Kamboj S, Nairl AB, Al J (2010) Nuclear and radiological agents: contamination and decontamination of human beings. Int J Pharm Sci Rev Res 5(3):95–101

Sheppard MI, Thibault DH (1991) A four-year mobility study of selected trace elements and heavy metals. J Environ Qual 20:101–114

Somlai J, Jobbágy V, Kovács J, Tarján S, Kovács T (2008) Radiological aspects of the usability of red mud as building material additive. J Hazard Mater 150(3):541–545

Steinhauser G, Brandl A, Johnson TE (2014) Comparison of the Chernobyl and Fukushima nuclear accidents: a review of the environmental impacts. Sci Total Environ 470–471:800–817

Tayibi H, Choura M, Lopez FA, Alguacil FJ, Lopez-Delgado A (2009) Environmental impact and management of phosphogypsum. J Envir Manage 90:2377–2386

Thibault DH, Sheppard MI, Smith PA (1990) A critical compilation and review of default soil solid/liquid partition coefficients, Kd, for use in environmental assessments, AECL-10125, Whiteshell Nuclear Research Establishment. Atomic Energy of Canada Limited, Pinawa

Thorne MC (2003) Background radiation: natural and man-made. J Radiol Prot 23:29–42

Torngren S, Persson SA (1998) Personal decontamination after exposure to simulated liquid phase contaminants. Functional assessment of a new unit. J Toxicol Clin Toxicol 36:567–573

Turhan S, Arikan IH, Demirel H, Güngör N (2011) Radiometric analysis of raw materials and end products in the Turkish ceramics industry. Radiat Phys Chem 80(5):620–625

Turner JE (2007) Atoms, radiation and radioactive protection – method of radiation detection, 3rd edn. Wiley VCH Verlag GmbH, Hoboken

Tzotzis M, Svoukis E, Tsertos H (2004) A comprehensive study of natural gamma radioactivity levels and associated dose rates from surface soils in Cyprus. Radiat Prot Dosim 109 (3):217–224

UNSCEAR (United Nations Scientific Committee on the Effects of Atomic Radiation) (2000) Sources and effects of ionizing radiation, UNSCEAR 2000 Report Vol.1 to the General Assembly, with scientific annexes, United Nations Sales Publication, United Nations, New York

US EPA (United States Environmental Protection Agency) (1996) Technology screening guide for radioactively contaminated sites, Report EPA 402-R-96-017. USEPA, Office of Air and Radiation, Washington, DC

USGS (US Geological Survey) (2002) Data series 140. In: Kelly T, Buckingham D, DiFrancesco C, Porter K, Goonan T, Sznopek J, Berry C, Crane M (eds). Historical statistics for mineral commodities in the United States. US Geological Survey open-file report 01-006, minerals.usgs.gov/minerals/pubs/of-01-006/

Veizer J (1983) Trace elements and isotopes in sedimentary carbonates. In: Reeder RJ (ed) Carbonates: mineralogy and chemistry, Reviews in Mineralogy, vol 11. Mineralogical Society of America, Washington, DC, pp 265–299

Waite TD, Davis JA, Payne TE, Waychunas GA, Xu N (1994) Uranium(VI) adsorption to ferrihydrite: application of a surface complexation model. Geochim Cosmochim Acta 58 (24):5465–5478

Watson SJ, Jones AL, Oatway WB, Hughes JS (2005) Ionising radiation exposure of the UK population: 2005 review. Health Protection Agency, Centre for Radiation, Chemical and Environmental Hazards, Radiation Protection Division, Chilton

WCA (World Coal Association) (2017) Coal and electricity. https://www.worldcoal.org/coal/uses-coal/coal-electricity. Accessed 20 Jul 2017

Wei L, Sugahara T (2000) An introductory overview of the epidemiological study on the population at the high background radiation areas in Yangjiang, China. J Radiat Res 41:1–7

Wharton MJ, Atkins B, Charnock JM, Livens FR, Pattrick RAD, Collison D (2000) An X-ray absorption spectroscopy study of the coprecipitation of Tc and Re with mackinawite (FeS). Appl Geochem 15:347–354

Whitehead DC (1984) The distribution and transformation of iodine in the environment. Environ Int 10:321–339

Wildung RE, Gorby YA, Krupka KM, Hess NJ, Li SW, Plymale AE, McKinley JP, Fredrickson JK (2000) Effect of electron donor and solution chemistry on the products of the dissimilatory reduction of technetium by *Shewanella putrefaciens*. Appl Environ Microbiol 66:2,452–2,460

Wilson WF (1994) A guide to naturally occurring radioactive material. Pennwell Books, Oklahoma

Wimmer H, Kim JI, Klenze R (1992) A direct speciation of Cm(III) in natural aquatic systems by time-resolved laser-induced fluorescence spectroscopy (TRLFS). Radiochim Acta 58 (59):165–171

Witherspoon PA, Bodvarsson GS (eds) (2001) Geological challenges in radioactive waste isolation. Lawrence Berkeley National Laboratory, University of California, Berkeley

Wolfe AK, Bjornstad DJ (2002) Why should anyone object? An exploration of social aspects of phytoremediation acceptability. Crit Rev Plant Sci 21:429–438

World Nuclear Association (2007) World nuclear power reactors 2006–07 and Uranium requirements. http://www.world-nuclear.org/info/reactors.html. Accessed 26 Apr 2016

World Nuclear Association (2011) Radiation and nuclear energy. http://www.world-nuclear.org/info/inf30.html. Accessed 29 Apr 2016

World Nuclear Association (2014) Naturally-Occurring Radioactive Materials (NORM). http://www.world-nuclear.org/info/Safety-and-Security/Radiation-and-Health/Naturally-Occurring-Radioactive-Materials-NORM/. Accessed 29 Apr 2016

Yoshida N, Takahashi Y (2012) Land-surface contamination by radionuclides from the Fukushima Daiichi nuclear power plant accident. Elements 8:201–206

Zeevaert T, Bousher A (1999) Restoration techniques: characteristics and performances. Westlakes Report 980132/01; SCK.CEN BLG-816. Mol, Belgium

Chapter 10
Health Risks Associated with Pesticides in Soils

Abstract The world population is expected to grow 50% until 2050 to nine billion people which combined with sophisticated diet demands may double the world food demand until the mid of this century. Increasing food demand requires the intensification of agriculture, as there is very limited scope for expanding the global agricultural land area. Intensifying agriculture involves increase of soil fertility, the use of improved crop varieties as well as the more efficient use of plant nutrients and water. It also goes along with the more intensive use of pesticides that have significantly contributed to the increase of agricultural productivity and food supply. However, they are a source of concern because of human and environmental health side effects. Pesticides themselves do not directly contribute to better crop yields but help to control the potential losses caused by weeds, plant pathogens (fungi, viruses and bacteria) and animal pests (e.g. insects, mites, nematodes and rodents). On a world scale an estimated 35% of potential crop production is lost to pests, pathogens and weeds each year.

Pesticides are a broad group of biologically active chemical compounds used for pest management. The term "pesticide" is a composite term that comprises chemicals used to control or kill pests. In agriculture, this term includes herbicides (weeds), insecticides (insects), fungicides (fungi), nematicides (nematodes), and rodenticides (vertebrate poisons). It is estimated that there are roughly 2000 pesticides in commercial use worldwide. Assuming the trend of the past several decades will continue, pesticide production worldwide will be 2.7 times higher in 2050 compared to 2000.

The application of pesticides is often not very precise. Unintended exposures occur to non-target organisms in the general area where pesticides are applied. Part of the pesticides applied remain in the soil as so-called "bound residues" while other part is lost via vaporization, surface runoff or leaching to the groundwater following precipitation or irrigation. These losses may cause large-scale, long-term damage to the environment. Ecosystem impacts of pesticides include soil microorganism response, effects on natural enemies of pests and crop pollinators, domestic and wild animal response, and effects on human health. The dispersal of pesticides in the environment on the one hand depends on the chemical and physical properties of the compounds, such as molecular structure, water pressure, solubility in water, stability and adsorption properties. On the other hand, pesticide dispersal

R. Nieder et al., *Soil Components and Human Health*,
https://doi.org/10.1007/978-94-024-1222-2_10

depends on conditions in the atmosphere (precipitation, wind, temperature, UV-radiation, moisture, particles), on soil (texture, structure, pH, adsorption capacity, biological activity, oxygen content, temperature, moisture, etc.) and water conditions (pH, biological activity, oxygen content, etc.).

Although the past decades have been characterized by efforts to reduce toxicity and to improve the effectiveness of pesticides through the use of new technologies and better information, this group of chemicals can severely affect animal and human health. Even low levels of exposure may have adverse health effects. Children are particularly vulnerable to the harmful effects of pesticides. Pesticide exposure can cause loss of coordination and memory, reduced speed of response to stimuli, reduced visual ability and altered mood. Some pesticides are known to disrupt hormones and reduce the ability to successful reproduction. Others have been associated with specific cancers or even death. The World Health Organization estimates that there are three million cases of pesticide poisoning each year and up to 220,000 deaths, primarily in developing countries.

A number of agronomic strategies and alternative control measures are available which in some cases can combat more than one weed, pest or disease. These measures can help reduce not only the current requirements for pesticides but also the buildup of resistance to pesticides. However, suitable decisions for the specific measures require deep knowledge of field conditions, crop rotations and pests.

Keywords Types of pesticides • Environmental fate • Pesticides in non-target organisms • Pesticides in food products • Human exposure • Clinical effects • Therapy • Measures for reducing pesticide dispersal in the environment • Technologies for remediation of pesticide-contaminated soils • Measures for minimizing pesticide use • Organic farming

10.1 History of Pesticide Use

Most crops are attacked by pathogens each adapted to certain conditions. For example, downy mildews and fruit rots prefer moist while rust fungi only need limited moisture. Besides leaf pathogens soil-borne pathogens exist as well. A number of historical examples of crop losses from disease had detrimental effects on food supply. A severe event was the appearance of late blight of potatoes in Ireland in the nineteenth century caused by *Phytophthora infestans* which led to mass starvation in the country. The impact of plant diseases and the most important plant pathogens can be differentiated by crop. Important diseases of major crops are summarized in Table 10.1.

Actual crop losses 2001–2003 due to weeds, animal pests and diseases on a worldwide base have been estimated to be 28.2% for wheat, 31.2% for maize and 28.8% for cotton, with a contribution of weeds of 7.7% (wheat), 10.5% (maize) and 8.6% (cotton) (Oerke 2006). The loss potential of diseases is higher in areas of high

Table 10.1 Important pest organisms and disease syndromes (in brackets) of major food and cash crops

Crop	Fungi	Bacteria	Insects and other animals
Wheat	*Blumeria graminis* (powdery mildew) *Tapesia yallundae* (eyespot) *Gaeumannomyces graminis* (take-all disease) *Septoria ssp.* (leaf and nodorum blotch) Different rust fungi		
Rice	*Magnaporthe grisea* (rice blast fungus) *Tanatephorus cucumeris* (bare patch) *Cochliobolus miyabeanus* (brown spot)		
Maize	*Peronosclerospora sorghi* (downy mildews)		Corn borers
Beans	*Thielaviopsis basicola* (black root rot) *Sclerotinia sclerotiorum* (sclerotinia rot) *Rhizoctonia* ssp. (rots, stem canker) *Isariopsis griseola* (angular leaf spot) *Uromyces appendiculatus* (bean rust)	*Pseudomonas syringae* (bacterial blight)	Diverse nematodes
Tomato	*Phytophtora infestans* (late blight) *Alternaria solani* (early blight) *Botrytis cinerea* (grey mould)		
Cabbage	*Peronospora parasitica* (downy mildew) *Leptoshaeria biglobosa* (black leg) *Alternaria brassicicola* (dark leaf spot)		
Pome fruits	*Venturia inaequalis* (scab) *Podosphaera leucotricha* (powdery mildew) *Phytophtora* ssp. (crown and collar rot) *Cytospora* ssp. (canker) *Botrytis cinerea* (gray mould) *Penicillium expansum* (blue mould)		

(continued)

Table 10.1 (continued)

Crop	Fungi	Bacteria	Insects and other animals
Stone fruits	*Monilinia fructicola* (brown rot)		
	Blumeriella jaapii (cherry leaf spot)		
	Taphrina cerasi (leaf curl)		
	Phytophtora ssp. (root and crown rot)		
	Podosphaera clandestine (powdery mildew)		
Bananas	*Mycosphaerella musicola* (Sigatoka disease)		
	Fusarium oxysporium (Panama disease)		
Citrus	*Phytophtora ssp.* (foot and root rot)		
	Elsinoe fawcettii (scab)		
	Guignardia citricarpa (black spot)		
	Diaporthe citri (melanose)		
	Penicillium ssp. (mould diseases)		
Cotton			Diverse insects

Compiled from Oerke et al. (1994) and Oerke (2006)

input cultivation than in other regions. The global potential loss due to pests is much higher and may vary from about 50% in wheat to more than 80% in cotton production (Oerke 2006). Crop losses due to harmful organisms may be reduced by crop protection measures. The most important measure of pest control is through the application of pesticides.

The history of pesticide use can be divided into three major periods (Zhang et al. 2011). In the time period before the 1870s (first period), natural pesticides such as sulfur were used to control pests. The time period 1870–1945 (second period) was the period of inorganic synthetic pesticides in that natural materials and inorganic compounds were mainly used. The third period (1945 to now) is the era of organic synthetic pesticides. Since 1945, man-made organic pesticides (chemical pesticides such as DDT, 2,4-D, γ-HCH, dieldrin) have terminated the era of inorganic and natural pesticides. In the earlier phase of the period after 1945, mainly insecticides (carbamates, organophosphorus and organochlorined insecticides) were produced. Soon after that, herbicides and fungicides achieved a considerable development as well. Pesticide consumption and expenditures in 2007 reached 2.37 billion kg (Table 10.2).

Since the 1960s, worldwide consumption structure of pesticides has undergone significant changes (Table 10.3). The proportion of herbicides in pesticide

Table 10.2 Global consumption of pesticides in 2007

Pesticides	Consumption 10^6 kg
Fungicides	235
Herbicides	955
Insecticides	411
Others	775
Total	2370

Adapted from US EPA (2011)

Table 10.3 Changes of pesticide consumption worldwide, expressed in sales

	Sales (10^6 US $)					
Pesticide category	1960	1970	1980	1990	2000	2007
Fungicides and bactericides	340	599	2181	5545	5306	9216
Herbicides	170	939	4756	11,625	12,885	15,512
Insecticides	310	1002	4025	7655	7559	11,158
Others	30	159	638	1575	1354	3557
Total	850	2700	11,600	26,400	27,104	39,443

Compiled from Zhang et al. (2011) and US EPA (2011)

Table 10.4 Changes of pesticide consumption worldwide

	Pesticide application (kg ha^{-1} agricultural land $year^{-1}$)		
Region	1989–1991	1994–1996	1998–2000
Developing countries	1.8	2.5	2.8
Developed countries	6.4	5.4	5.2
World average	4.0	3.8	3.9

Adapted from Sexton et al. (2007)

consumption increased rapidly, from 20% in 1960 to 39% in 2007. The proportion of consumption of insecticides and fungicides/bactericides declined despite their sales increased.

Consumption of pesticides has declined in several regions of the world, particularly in developed countries (Table 10.4). This trend is mainly due to growing ecological awareness, increasing pesticide resistance, a better understanding of the costs and benefits of pesticide applications, and the development of new technologies and pest management strategies. In contrast, it seems that in developing countries, pesticide application has increased. However, for example in Europe, the number of treatments has increased, with 66% of the cropped area using two or more herbicide types, and 80% using two or more insecticides during treatment (Bedos et al. 2002).

Pesticide use in Europe amounted to about 500×10^6 kg $year^{-1}$ towards the end of the 1990s (Candela 2003), with an average dose of 4.3 kg ha^{-1} (4.4 in France, 2.4 in Germany, 4.9 in the UK and about 14 kg ha^{-1} in the Netherlands and Italy).

10.2 Types of Pesticides

Agricultural pests include animals, insects, fungi, bacteria and plants that lead to reductions in crop yield or even crop shortfall. The use of chemicals has been the dominant paradigm for pest control in recent decades and has succeeded in reducing crop losses (Sexton et al. 2007). Agricultural productivity has improved greatly particularly since the beginning of the post-World War II period. While the global population more than doubled in the past 50 years, per capita food production increased. Currently, more than six billion people consume a daily average of 2700 kcals per person compared to the daily average of 2450 kcals consumed by each of the 2.5 billion people in 1950 (WHO/FAO 2003). The significant gains in productivity can be largely attributed to improved crop varieties, new irrigation and harvesting technologies, and to developments in chemical fertilizers and pesticides (Sexton et al. 2007). Pesticides represent a group of chemicals controlling organisms that compete with humans for food and fibre or cause injury to crops, livestock and man.

Worldwide, roughly 2000 chemicals are regularly used (Aherns 2008). Most pesticides are known by their trade names, of which there is a large variety. Of the commonly used pesticide compounds, herbicides, insecticides and fungicides each comprise approximately 25% of the total compound list (Table 10.5). Acaricides make up another 10%, and the remaining 15% are due to plant and animal hormones, molluscicides, nematicides, avicides and mammalian toxins.

Table 10.5 Pesticides, their target organisms and number of available compounds

Type of pesticide	Target organism	Number of available compounds
Acaricides	Mites and spiders	201
Algaecides	Algae	12
Antifeedants[a]		4
Avicides	Birds	5
Bactericides	Bacteria	19
Bird repellents	Birds	9
Chemosterilants[b]		17
Fungicides	Fungal diseases	409
Herbicides including herbicide safeners	Weed	551
Insecticides including insect attractants	Insects	498
Mammal repellents	Mammals	4
Mating disruptors	Insects	3
Molluscicides	Molluscs	17
Nematicides	Nematodes	44
Pesticide-miscellaneous and synergists	Different organisms	18
Plant activators and growth activators[c]		97
Rodendicides	Rodents	36
Viricides	Viruses	2
Total		1946

Adapted from Aherns (2008)
[a]Substance that inhibits normal feeding behaviour
[b]Chemical compound that causes reproductive sterility in an organism
[c]Promote plant establishment and stimulate growth

Pesticides can be grouped according to the chemical structure of the compound, the target pest, the action mode, or the degree of health hazard involved. The most important pesticides and the highest numbers of available compounds are represented through fungicides, herbicides and insecticides (Table 10.5). Table 10.6 shows a selection of these pesticides which are commonly used. The chemical structures of important pesticide groups are shown in Fig. 10.1.

Table 10.6 Commonly used fungicides, herbicides and insecticides, their chemical names and chemical classes

Pesticide group/ common name	Chemical name	Chemical class
Fungicides		
Carboxin	5,6-Dihydro-2-methyl-1,4-oxathiin-3-carboxanilide	Oxathiin
Chlorothalonil	Tetrachloroisophthalonitrile	Substituted benzene
Herbicides		
2,4-D	2,4-Dichlorophenoxy acetic acid	Phenoxyalkanoic acid
2,4-DP	Butoxyethyl ester of (±) 2-(2,4-dichlorophenoxy) propanoic acid	Phenoxyalkanoic acid
Alachlor	2-Chloro-*N*-(2,6-diethylphenyl)-*N*-(methoxymethyl) acetamide	Chloracetanilide
Atrazine	2-Chloro-4-ethylamino-6-isopropylamino-*S*-triazine	Triazine
Cyanazine	2[[4-Chloro-6(ethylamino)-*S*-triazin-2-yl]amino]-2-methylpropionitrile	Triazine
Propazine	6-Chloro-*N*,*N*_-*bis*(1-methylethyl)-1,3,5-triazine-2,4-diamine	Triazine
Simazine	2-Chloro-4,6-bis(ethylamino)-s-triazine	Triazine
Bromacil	5-Bromo-3-(*sec*-butyl)-6-methyluracil	Substituted uracil
Dalapon	2,2 Dichloropropionic acid (sodium salt)	Dichloropropionic acid
DCPA	Dimethyl 2,3,5,6-tetrachloro-1,4-benzenedicarboxylate	Dimethyl tetrachlorophtalate
Dicamba	2-Methoxy-3,6-dichlorobenzoic acid	Chlorophenoxy
Dinoseb	2-*sec*-Butyl-4,6-dinitrophenol	Dinitrophenol
Glyphosate	2-[(phosphonomethyl)amino]acetic acid	Organophosphate
Tebuthiuron	*N*-[5-(1,1-Dimethyl)-1,3,4-thiadizol-2-yl]-*N*,*N*-dimethylurea	Urea class herbicide
Trifluoralin	2,6-Dinitro-*N*,*N*-dipropyl-4-(trifluoromethyl) benzenemamine	Dinitroaniline
Insecticides		
Aldicarb	2-Methyl-2-(methylthio)-propionaldehyde *O*-(methylcarbamoyl)oxime	Carbamate
Carbaryl	1-Naphthyl-*N*-methylcarbamate	Carbamate
Carbofuran	2,3-Dihydro-2,2-dimethyl-7-benzofuranyl-*n*-methylcarbamate	Carbamate
Diazinon	*O*, *O*-Diethyl-*O*-(2-isopropyl-4-methyl-6-pyrimidinyl)phosphorothiote	Organophosphate

Adapted from Loague and Corwin (2005)

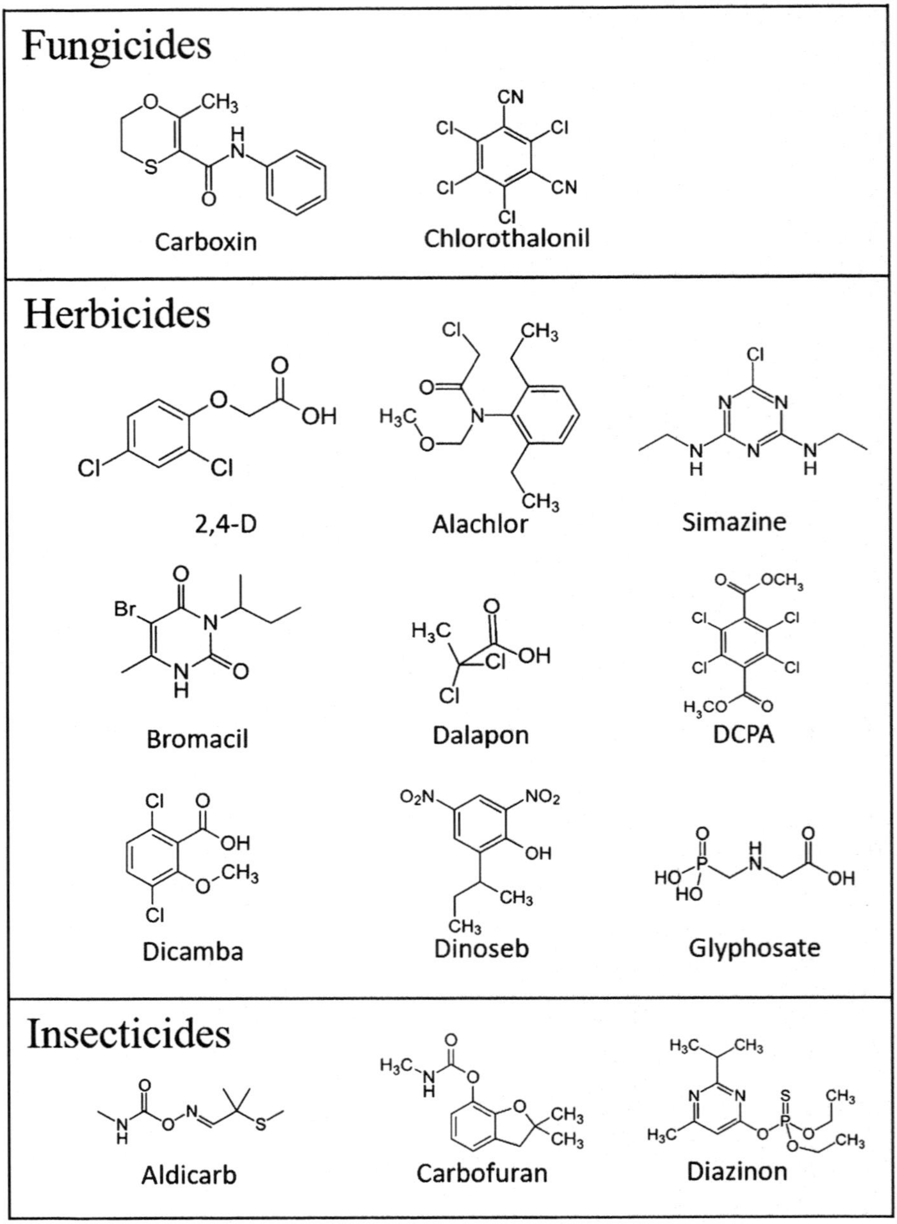

Fig. 10.1 Chemical structures of some important pesticides

10.3 Environmental Fate of Pesticides

Pesticides are found in the environment worldwide, both in areas where they are used and in areas where they never have been applied such as in the Arctic region. The large scale use of pesticides started during the 1950s and 1960s. The use was rather careless during that time and authorities and users were not aware of the dangers of using pesticides as regards to dispersal and side-effects in the environment. Since that period, we have learned much about the environmental fate of pesticides (Fig. 10.2).

Many countries have begun systematic educational programs for farmers to teach them about safer handling of pesticides. The environmental risk of a pesticide depends on the amount applied, the toxic properties of the compound, the type of formulation, method and time of application as well as its mobility and persistence. The potential risk involved when a compound is released to the environment depends on the type of formulation, method and time of application, its mobility and its persistence. The type of formulation has a great influence on the bioavailability and persistence of a compound. For example, compared to liquid forms, granules commonly increase the persistence of a compound. The way of application also has an effect and extensive treatments create more environmental problems compared to local treatments. The time of application is problematic when certain pests are to be eliminated at a time that is dangerous for beneficial species. Extensive and frequent applications may rise resistance phenomena (Navarro et al. 2007).

Climate and geomorphology also play an important role. The effects observed in a distinct region of the world are not necessarily transmittable to regions with other climatic conditions. Solar radiation may favor photodecomposition and the conversion of the original product into one or more of higher or lower toxicity. Facts about the toxicity of a pesticide and its metabolites will be of great use for evaluating the possible risks for humans, the end-users of agricultural products. Highly toxic pesticides may even have not harmful effects on the environment if they are applied in a way that they cause no damage to non-target organisms. Alternatively, compounds of low toxicity and persistency may have harmful effects if they are applied very frequently or in high doses. For an agricultural system to be sustainable,

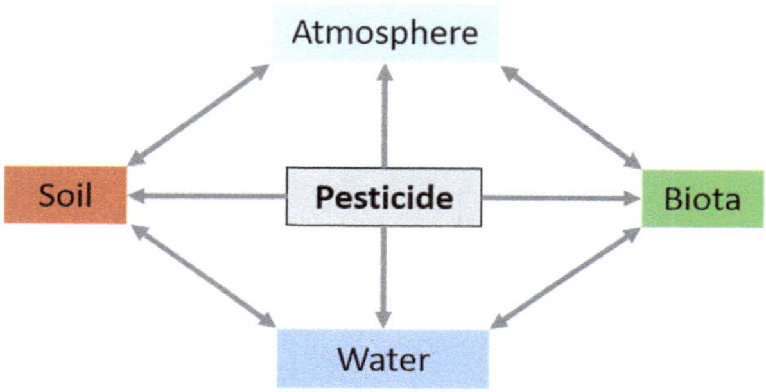

Fig. 10.2 Major pathways of pesticides in the environment

adverse environmental effects of agricultural production must be minimized. Degradation of surface and ground water quality through pesticides has been identified as an important concern with respect to the impact of agriculture on the environment. Pesticides may reach surface water and groundwater through runoff and leaching.

10.3.1 Soil-Pesticide Interactions

10.3.1.1 Adsorption and Desorption of Pesticides

Although the application target of pesticides may be weed or crop foliage and volatilization of part of the compound may occur, their fate is mostly decided in the soil (Helling et al. 1971). The remaining deposit may partly be infiltrated or translocated into the soil by living plants or with crop residues (Wauchope et al. 2002). Pesticides and other organic pollutants (Chap. 11) in the soil partition between the liquid and the solid phases of the pedosphere (Fig. 10.3).

The degree to that one phase is prefered will decide on the behaviour of a compound in soil, i.e., if a pesticide will be adsorbed or transported. In a mobile form, pesticides can be regarded as pollutants. Adsorption is the most important kind of interaction between soil and pesticides and controls the mobility of the latter (Gevao et al. 2000). The reverse process of adsorption is desorption. Desorption is

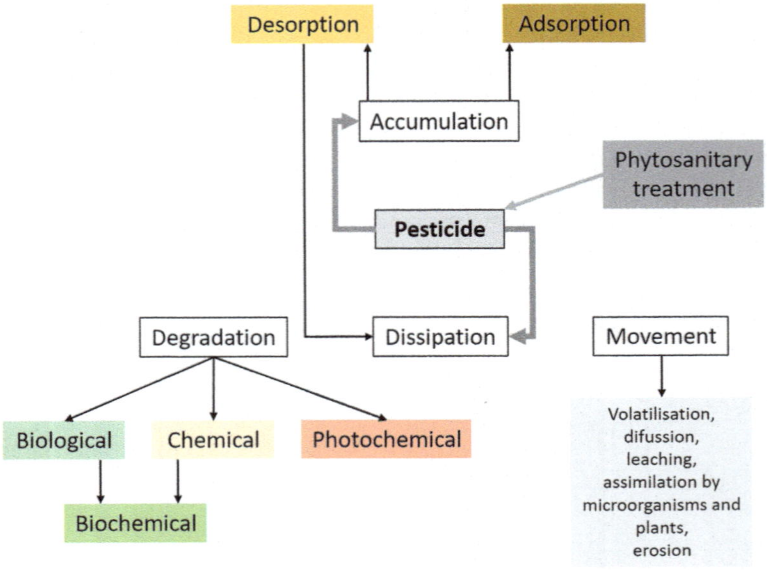

Fig. 10.3 Major soil-pesticide interactions (Adapted from Navarro et al. 2007)

inversely related to adsorption, being small when adsorption is great, and conversely. For example, the adsorption of atrazine is fully reversible (Celis et al. 1998) while that of glyphosate is not reversible (Mamy and Barriuso 2007).

Retention of pesticides is commonly determined with soil suspensions, known as batch experiments (OECD 2002). A defined volume of an aqueous solution of pesticide is added to a mass of dry sieved soil in glass centrifuge tubes. Soil suspensions are shaken mechanically for 24 h and then centrifuged. The amounts of adsorbed pesticide in the soil are calculated as the difference between initial pesticide concentration in solution and centrifuged supernatant concentration. This experiment is done at several initial pesticide concentrations to determine the adsorption isotherm of the pesticide. From this isotherm, distribution coefficients between soil and soil solution can be determined according to the Freundlich (Kf) model (Eq. 10.1):

$$\mathrm{Kf} = \mathrm{Qs}/\mathrm{Ce}^{\mathrm{nf}} \tag{10.1}$$

where Qs is the amount of adsorbed herbicide (mg kg^{-1} soil) at equilibrium concentration, Ce (mg L^{-1}) is pesticide concentration in supernatant solution, and nf is an empirical coefficient. The isotherm is linear when nf $=$ 1 and Kf $=$ Kd (L kg^{-1}). As organic carbon is a major adsorbent for pesticides, the K_{oc} (soil sorption constant: L kg^{-1}) is calculated as (Eq. 10.2):

$$K_{oc} = (\mathrm{Kd} \times 100)/C_{org} \tag{10.2}$$

where C_{org} is the percentage of organic carbon content in soil. For a given pesticide, the K_{oc} is commonly less variable than the Kd among different soils (Calvet 1989).

10.3.1.2 Soil Properties and Conditions that Determine the Behaviour of Pesticides

Minerals

Clays (particularly three-layer clay minerals), oxides and hydroxides are the most important mineral adsorbents involved in the adsorption of organic substances (Calvet 1989). Clay surfaces are mainly hydrophilic because of the presence of hydroxyl groups and exchangeable cations. The adsorption of pesticides on clay minerals is likely to occur on external surfaces of clay particles rather than in interlamellar space and increases with the specific surface of clays (Barriuso et al. 1994). For example, the adsorption of glyphosate increases in the sequence: kaolinite $<$ illite $<$ montmorillonite $<$ nontronite (McConnell and Hossner 1985). Oxides and hydroxides have a high surface activity and their charge strongly depends on the soil pH. The adsorption of glyphosate on iron and aluminium oxides and hydroxides is high at intermediate pH and driven by ionic bindings between the positive surface sites of the oxides and the negative acid groups of glyphosate (Morillo et al. 2000). Sorption is much lower at very acid or very alkaline pH because oxides will bear a similar charge as glyphosate.

Soil Organic Matter

Although SOM only represents only few percents of soil dry matter, it is a major sorbent of pesticides which is attributed to its high chemical reactivity towards both mineral surfaces and organic molecules. The sorption capacity of humic substances is not only controlled by their chemical composition, but also by their size, due to a greater number of sorptive sites related to a greater surface area (Benoit et al. 2008). In general, the adsorption of pesticides increases with SOM content, except for ionic molecules. The adsorption of pesticides thus decreases with soil depth because of a decrease in SOM content (Mamy and Barriuso 2007).

Soil pH

The pH is important particularly for the adsorption of ionic pesticides such as glyphosate (Mamy and Barriuso 2007). Depending on the charge of the pesticide, the adsorption will increase (or decrease) with pH. For example, the retention of glyphosate increases when the soil pH decreases because the number of negative charges of the molecule decreases, allowing the adsorption on negatively charged adsorbents like clay or SOM.

Soil Structure

The structure of a soil is characterized by the pore geometry and the bulk density that depend on agricultural practices and on climate. Pesticide movement through aggregated soils is mainly controlled by kinetic sorption and diffusion (Buckley et al. 2004). In static conditions, the rate of pesticide adsorption decreases when the density of soil aggregates increases (Chaplain et al. 2008). In dynamic conditions, retention depends on transport parameters such as pore water velocity and residence time (Pot et al. 2011). Compared to tilled soils, no-tilled or grassland soils are characterized by a higher presence of biopores (root channels and earthworm casts) and high SOM content in the surface layers. The retention of pesticides may therefore be higher in these soils (Larsbo et al. 2009). However, the increase in retention can be counter-balanced by increased preferential transport because no tillage leads to enhanced macropore connectivity.

Soil Water Content

The water content of a soil defines the specific exchange surface between liquid and solid phases. The adsorption of pesticides increases with water content as it facilitates pesticide diffusion to sorption sites. As water content increases, the organic matter also becomes more hydrophilic with greater sorption potential for hydrophilic pesticides (Roy et al. 2000). For hydrophobic pesticides like trifluralin, the adsorption decreases when the soil water content increases because the hydration of the surfaces of adsorbents decreases the accessibility to adsorption sites (Swann and Behrens 1972). Low soil moisture content might also favour access to the hydrophobic regions of humus by generating more hydrophobic surfaces, thereby increasing the sorption of hydrophobic substances (Roy et al. 2000).

Soil Temperature

The adsorption of most pesticides decreases when the temperature increases (Ten Hulscher and Cornelissen 1996). However, one should distinguish between fast sorption and slow sorption. While the fast sorption decreases with increasing temperature, the slow sorption is more rapid at higher temperature. This may explain why for some chemicals, overall sorption with short equilibration times is nearly independent on temperature. The slow sorption is generally due to diffusion of the pesticide through SOM and increasing temperature decreases the density thereby increasing diffusion (Ten Hulscher and Cornelissen 1996). For some pesticides that exhibit decreasing solubility at higher temperatures, an increase in the sorption with temperatures can be observed (Chiou et al. 1979).

10.3.1.3 Pesticide Properties That Determine Interactions with Soil

Factors such as molecular size, water solubility, lipophilicity, ionisability, polarisability and volatility determine the extent of adsorption (Navarro et al. 2007). The nature of adsorption may be physical, as with van der Waals forces, or chemical, as with electrostatic interactions. Pesticides can be grouped according to their chemical and physico-chemical properties (Table 10.7).

Chemical reactions between unaltered pesticides or their metabolites may result in the formation of stable chemical linkages which may enhance the persistence of a compound (Dec and Bollag 1997). Persistence and mobility of pesticides and their metabolites are determined by parameters such as water solubility, soil-sorption constant (K_{oc}) and half-life in soil (Andreu and Picó 2004). The tendency of a pesticide to contaminate groundwater through leaching is high if its water solubility is high, its sorption coefficient is low and its half-life is long. This is quite frequent because pesticides are increasingly polar, hydrosoluble and thermolabile to diminish their toxicity and to facilitate their disappearance from the environment.

Table 10.7 Pesticides grouped according to their chemical and physico-chemical properties

Non-ionic pesticides				Ionic pesticides			
Organo-chlorines	Organo-phosphorus	Dinitro-anilides	Carbanilates	Cationic	Basic	Acidic	Miscellaneous
DDT	Dimethoat	Benefin	Chlorpropham	Chlormequat	Ametryne	Dicamba	Bromacil
Endrin	Ethion	Trifluralin	Swep	Diquat	Atrazine	Ioxynil	Isocil
Lindane	Parathion	Oryzalin	Prophan	Morphamquat	Cyanacine	Mecoprop	Terbacil
Methoxychlor	Disulfoton	Nitralin	Barban	Paraquat	Prometone	2,4,5-T	MSMA
Endosulfan	Diazinon	Isopropalin			Metribuzine	Dichlorprop	DSMA
Heptachlor	Dursban	Dinitramine			Propazine	Fenac	
Dieldrin	Demeton				Simazine	MCPA	
Thiocarbamates	Anilides	Ureas	Methyl carbamates				
Cycloate	Alachlor	Norea	Carbaryl				
Metham	Propanil	Cycluron	Dichlormate				
Ferbam	Diphenamide	Linuron	Terbutol				
Nabam	Propachlor	Buturon	Zectran				
Butylate	Solan	Chloroxuron					
Diallate		Diuron					
Triallate		Buturon					

Adapted from Gevao et al. (2000)

Table 10.8 Pesticides identified by the Stockholm Convention

Pesticides	Currently listed	Under review (2009)
Aldrin	x	
α-Hexachlorocyclohexane (α-HCH)	x	
β-Hexachlorocyclohexane (β-HCH)	x	
γ -Hexachlorocyclohexane (γ-HCH; Lindane)	x	
Chlordane	x	
Chlordecone	x	
Dichlorodiphenyltrichloroethane (DDT)	x	
Dieldrin	x	
Endosulfan		x
Endrin	x	
Heptachlor	x	
Hexachlorobenzene (HCB)	x	
Mirex	x	
Toxaphene	x	

Adapted from US EPA (2010)

However, at the same time, they must persist long enough to enable acceptable pest control.

Together with several industrial substances (see Chap. 11) some pesticides, belonging to the organochlorines, were categorized as persistent organic pollutants (POPs) which have been described to be semi-volatile, bioaccumulative, persistent and toxic. Due to their possibly harmful effects on humans, international agreements have come into effect to reduce future environmental burdens and human exposure. One of such relevant agreements is the Stockholm Convention (2004 and 2009) on POPs, which focuses on a list including industrial POPs and pesticides. Pesticides identified by the Stockholm Convention are listed in Table 10.8.

Organochlorine pesticides are a major part of POPs, and are chemically stable under environmental conditions. They break down only very slowly and can remain in the environment for several years. Organochlorine pesticide residues have been detected in air, water, soil, sediment, fish, and birds worldwide even more than one decade after these pesticides were banned. The toxicological parameters and chemical properties of organochlorine pesticides can be drawn from Table 10.9.

Table 10.10 shows pesticides other than organochlorines and associated properties, their metabolites that have been found in soils in recent years and some characteristics and hazards for the environment and for humans. Numerous metabolites from a wide range of pesticides have been documented (e.g. Roberts 1998; Roberts and Hutson 1999). It is noteworthy that of the numerous pesticides up to now only a limited number has been monitored in soils. Some of the pesticides in Table 10.9 have been prohibited years ago but their metabolites continue persisting in soils.

Table 10.9 Toxicological and chemical parameters of pesticides listed by the Stockholm Convention

Pesticide	Formula	LD_{50} (mg kg^{-1})	Half-lifes (days)	Water solubility (mg L^{-1})	Log K_{oc}	Vapour pressure 25 °C (mm Hg)
Aldrin	$C_{12}H_8Cl_6$	39–64	53	0.01	7.67	1.2×10^{-4}
α-HCH	$C_6H_6Cl_6$	1000–4000	54–56	6.95	3.57	4.5×10^{-5}
β-HCH	$C_6H_6Cl_6$	<900	100–184	5.00	3.57	3.6×10^{-7} (20 °C)
γ-HCH (Lindane)	$C_6H_6Cl_6$	900–1000	1095–2190	0.01	6.08	1.09×10^{-5} (20 °C)
Chlordane	$C_{10}H_6Cl_8$	83–590	93–154	0.06	3.49–4.64	2.2×10^{-5}
Chlordecone	$C_{10}Cl_{10}O$	91–132	10	3.00	3.38–3.42	3.0×10^{-7}
DDT	$C_{14}H_9Cl_5$	45–63	22–365	0.03	5.18	1.6×10^{-7} (20 °C)
Dieldrin	$C_{12}H_8Cl_6O$	37–46	1825	0.05	6.67	6.0
Endosulfan	$C_9H_6Cl_6O_3S$	40–121	39–42	0.06–0.10	3.50	1.0×10^{-5}
Endrin	$C_{12}H_8Cl_6O$	7–43	5110	0.20	4.53	2.0×10^{-7}
Heptachlor	$C_{10}H_5Cl_7$	39–144	38–45	0.05	4.34	3.0×10^{-4}
Mirex	$C_{10}Cl_{12}$	365–740	62–107	17.0	3.57	4.2×10^{-5} (20 °C)
Toxaphene	$C_{10}H_{10}Cl_8$	80–293	10	0.60	3.76	3.0×10^{-7}

Adapted from US EPA (2010)

Table 10.10 Pesticides, their soil sorption constants (K_{oc}), half-lifes, metabolites that have been found in soils, main characteristics and hazards for the environment and for humans

Chemical class/pesticide name	K_{oc} (ml g^{-1} org.C)	Half-lifes (days)	Metabolites	Characteristics/features	Hazards
Carbamates					
Benomyl	1900	67	*N*,*N*-dibutylurea	Little persistent	Toxic
Carbofuran[a,b]	14–160	30–117	3-Hydroxicarbofuran	Hydrolysis product	Toxic
			3-Ketocarbofuran	Little persistent	
Chloroacetanilide					
Metolachlor	90	26	3-Ethyl-6-methylaniline	Detected often in groundwater	
			Metolachlor ethane sulfonic acid	Detected often in groundwater	
Alachlor[b]			2,6-Diethylaniline	Detected often in groundwater	
			2-Chloro-2′, 6′-diethylacetanilide	Detected often in groundwater	
			2-Hydroxy-2′, 6′-diethylacetanilide	Formed faster than metabolites from metolachlor	
Dinitroanilide					
Trifluoralin	8000–10,000	57–126	28 different metabolites	Strong binding to soil	Less toxic[c]
Organophosphorus					
Diazinon	1000	40	2-Isopropyl-6-methyl-4-pyrimidinol	Hydrolysis product	
Malathion	1800	1	Malaoxon	Oxidation product	More toxic[c]
Glyphosate	24,000	47	Aminimethylphosphoric acid		More toxic[c]

(continued)

Table 10.10 (continued)

Chemical class/pesticide name	K_{oc} (ml g^{-1} org.C)	Half-lifes (days)	Metabolites	Characteristics/features	Hazards
Phenoxialkanoic acid					
2,4-D	45	34–333	2,4-Dichlorophenol		More toxic[c]
MCPA	50–60	7–41	4-Chloro-2-methylphenol		More toxic[c]
Triazine herbicides					
Metribuzin	95	30–60	Deaminometribuzin	Microbial degradation	More toxic[c]
			Diketometribuzin	Microbial degradation	More toxic[c]
Simazine[b]	130	60	Monodeethylsimazine		More toxic[c]
Terbutylazine	500	30–90	Deethyltetbuthylazine		More toxic[c]
Atrazine[b]	100	60	Deethylatrazine		More toxic[c]
			Deisopropylatrazine		
			Deethyldeisopropylatrazine		
Urea herbicides	66	15–40	4,4-diisoprylazobenzene		Less toxic[c]
Isoproturon					

Adapted from Andreu and Picó (2004)
[a]Prohibited according to EC (2005)
[b]Prohibited according to US EPA (2015)
[c]Compared to original pesticide

10.3.1.4 Binding Mechanisms of Pesticides

Binding of pesticides to soil (particularly humus) leads to immobilization of the compound, thereby reducing its leaching and transport properties and a reduction in the toxicity (Gevao et al. 2000). The mechanisms responsible for the adsorption of pesticides to humic substances are described in the following paragraphs.

Ionic Bonding

Pesticides and their metabolites adsorbed by ionic bonding are present either in the cationic form in solution or can be protonated and become cationic. Ionic bonding involves ionised, or easily ionisable functional groups (carboxylic and phenolic hydroxyl groups) of humic substances. Bipyridilium pesticides (e.g. paraquat) bind to soil humic substances by ion exchange via their cationic group and form highly stable and unreactive bonds with the carboxyl groups of humic substances (Gevao et al. 2000). The effect of pH on binding has been reported for less basic pesticides such as the triazine herbicides (Weber et al. 1969). Ionic bonding can occur between a protonated secondary amino-group of the s-triazine and a carboxylate anion and, possibly, a phenolate group of the s-triazine (Senesi and Testini 1980). Maximum adsorption of basic compounds of s-triazine occurs at pH values close to their pKa value (Weber et al. 1969).

H Bonding

H-bonding can play an important role in the adsorption of several non-ionic polar pesticides, including substituted ureas and phenylcarbamates (Senesi and Testini 1983). Humic substances, with their functional groups form H-bonds with complimentary groups on pesticide molecules. Acidic and anionic pesticides, such as the phenoxyacetic acids (2,4-D and 2,4,5-T) and dicamba, can interact with soil organic matter by H-bonding at pH values below their pKa in non-ionised forms through their –COOH, –COOR and identical groups. Piccolo and Celano (1994) applied IR spectroscopy in the study of H-bonding using the complexes of the herbicide glyphosate with water-soluble humic acid. Following the formation of H-bonds between the glyphosate phosphono-group and the oxygen groups of humic acid, the IR spectra of the KBr sample showed two bands, one representing the P=O and another the P-O- stretch. The H-bonds were disrupted following the titration of the solution with NaOH.

Van der Waals Forces

Van der Waals forces consists of weak dipolar attractions that exist, in addition to stronger binding forces (Gevao et al. 2000). Since van der Waals forces are known to decay rapidly with distance, adsorption would be greatest for those ions which are in closest contact with the surface (Senesi 1992) and their contribution to adsorption increases with the size of the interacting molecule. The involvement of these binding forces has been observed for a large number of compounds, including cabaryl and parathion (Lenheer and Aldrichs 1971), benzonitrile and DDT (Pierce et al. 1971) and has been shown to be the major adsorption mechanism for picloram and 2,4-D (Kozak 1983).

Ligand Exchange

Adsorption by ligand exchange involves the replacement of relatively weak ligands, such as H_2O partially holding polyvalent cations associated with soil organic matter by suitable adsorbent molecules such as s-triazines and anionic pesticides (Senesi 1992; Gevao et al. 2000). The substitution may be facilitated by an entropy change, if a pesticide molecule succeeds in replacing several H_2O molecules associated with complexed metal ions.

Hydrophobic Adsorption

The hydrophobic attraction (due to the hydrophobic effect) can be regarded as a partitioning between a solvent and a non-specic surface. Hydrophobic adsorption by humic substances is suggested as an important mechanism for DDT and other organochlorine insecticides (Lenheer and Aldrichs 1971), methazole (Carringer et al. 1975), metolachlor (Kozak 1983), dicamba and 2,4-D (Gevao et al. 2000), and is considered a possible interaction mechanism for the s-triazine herbicides and polyureas (Khan and Mazurkevich 1974).

Covalent Bonding

Covalent bonding between pesticides and/or their metabolites and SOM, is often mediated by chemical, photochemical or enzymatic catalysts leading to incorporation into the soil. The pesticides which are most likely to bind covalently to SOM have properties similar to SOM (Bollag et al. 1992). Pesticides structurally resembling phenolic compounds can bind to SOM. Oxidative coupling is the process by which phenols, anilines and other compounds are linked together after oxidation by an enzyme or chemical agent which results in the formation of C–N and N–N between aromatic amines and C–C and C–O bonds between phenolic species (Sjoblad and Bollag 1977). Oxidative coupling reactions are mediated by a number of biotic and abiotic catalysts, including plant and microbial enzymes, inorganic chemicals, clay and soil extracts (Gevao et al. 2000). The incorporation of pesticides into SOM has been documented in tracer studies. For example, Wolf and Martin (1976) observed that ^{14}C-2,4-dichlorophenol and chlorpropham were incorporated into humic-like polymers. Compound classes that were observed to bind covalently to SOM without the intervention of microbial activity include acylanilides, phenylcarbamates, phenylureas, dinitroaniline herbicides, nitroaniline fungicides and organophosphate insecticides, such as parathion and methylparathion. They bind by mechanisms involving carbonyl, quinone and carboxyl groups of humic substances leading to hydrolysable and non-hydrolysable bound forms (Gevao et al. 2000).

10.3.2 Degradation of Pesticides and Other Organic Pollutants

10.3.2.1 Abiotic Degradation

Chemical processes such as hydrolysis, oxidation- reduction, and photolysis are responsible for degradation and transformation of many pesticides and other organic pollutants (Chap. 11) in soils. Abiotic degradation of pesticides in soil has mainly been studied through laboratory experiments performed under controlled conditions. To distinguish abiotic processes from biodegradation, soils are sterilized through autoclaving, adding a biocide (NaN_3, $HgCl_2$) or γ-irradiating with ^{60}Co.

Chemical Hydrolysis

Chemical hydrolysis is a major pesticide degradation process in the aqueous phase or on adsorbed forms on the solid-phase surfaces that often dominates among the transformation reactions of different groups of pesticides such as organophosphorus, carbamate, and sulfonylureas (Smith et al. 1988; Sarmah and Sabadie 2002). Hydrolysis is the cleavage of an intramolecular bond by water and the concomitant formation of a new bond involving the oxygen atom of water (Eq. 10.3).

$$RX + H_2O \rightarrow ROH + X^- + H^+ \tag{10.3}$$

Hydrolysis can occur in the presence of H_2O, H_3O^+ and OH^- and has been termed neutral, acid and alkaline hydrolysis, respectively. The process is temperature-dependent. For example, for chlorpyrifos Getzin (1981) reported a half-life of >20 days at 5 °C compared with only 1 day during incubation at 45 °C. Other factors that influence hydrolysis of organic compounds include SOM and clay minerals exhibiting larger surface area for facilitating hydrolytic degradation

(Yaron 1978). Hydrolysis may be favored by metal-ion catalysis, through direct polarization, where the metal coordinates the hydrolysable function and generation of a reactive metal hydroxo species. Surface-bound metals may also catalyze hydrolysis through electrostatic interactions that enhance the OH^- concentration in the aqueous phase (Smolen and Stone 1998). The most cited reactive metals are +II in solution (Cu(II), Pb(II), Mn(II), Fe(II)) whereas +III or +IV metals are surface-bound or solid particulate species (goethite α-FeOOH, γ-Al_2O_3, TiO_2). Clay minerals also act as hydrolysis catalyzers, mainly because of their surface acidic pH. The influence of soil pH on pesticide degradation depends on the nature of the chemical, e.g. some compounds are base-hydrolysed and others acid-hydrolysed. Sulfonylurea herbicides, for example, have the potential to persist for years in alkaline soils (Taylor et al. 1996), in contrast to their low persistence (weeks) in acidic to neutral soils (James et al. 1999).

Oxidation-Reduction

Oxidative processes in soils may be mediated by oxidative enzymes and abiotic catalysts such as metal oxides. Because of their reactivity and frequency in soils manganese oxides and hydroxides play an important role as oxidants (Li et al. 2003). Redox potentials of MnOOH and MnO_2 make those species able to oxidize organic contaminants with functionalities such as phenol (Lin et al. 2009), aniline (Laha and Luthy 1990) or triazine (Shin and Cheney 2004). To some extent Fe(III) oxides or Fe(III) adsorbed on smectite clays may also be involved in oxidation processes (Li et al. 2003). The oxidation process is initialized with the formation of a precursor complex between the chemical and the surface-bound metallic species, followed by electron transfer within the complex and release of an organic cationic radical, that will evolve to an oxidized form or bind to a vicinal reactive compound. Environmental factors may affect either the reaction kinetics of MnO_2-pollutant complex formation or the rate of electron transfer, directly or through action on the reactive surface sites (Zhang et al. 2008). The pH plays a role by affecting ionizable pollutants and reactive surface sites, changing the speciation of MnO_2 surface. Organic matter may affect the reaction by reducing reactive surface sites. Organic matter also reacts with the cationic radical and forms covalent binding, resulting in bound residues (Li et al. 2000). Another abundant oxidizing agent in soil upper layers is oxygen, both in gaseous form and dissolved in liquid surface films. As a marginal pathway in soils, oxygen causes autooxidation or weathering through a radicalar mechanism involving O_2^- production (Larson and Weber 1994).

Reduction of pesticides in soils occurs under suboxic and anoxic conditions in poorly drained or groundwater-fed soils, riparian zones, wetlands or flooded areas. Reductants include reduced metals, sulfide ions or organic compounds (Borch et al. 2010) as well as extracellular biochemicals such as metal chelated in porphyrin or corrinoid or as transition metal coenzymes (Kappler and Haderlein 2003). Studies using nitroaromatic pollutants (e.g. trifluralin) showed that they were predominantly reduced by Fe(II) associated with iron minerals surfaces.

Organic matter, particularly quinones may also be involved in reduction processes (Kappler and Haderlein 2003). It has also been observed that reducible organic contaminants compete with iron oxides for the electron flow generated by the microbially mediated oxidation of organic carbon and subsequent reduction of quinone functional groups associated with dissolved organic matter (Zhang and Weber 2009).

Photolysis

Exposure of pesticides present on the soil surface to solar radiation (mainly UV-A, with varying amounts of UV-B) may cause photolytic degradation. Photolysis can be direct, when the pesticide receives UV light within the spectrum of sunlight (<300 nm), or indirect, when the energy is absorbed by other compounds which subsequently transmit it to the pesticide molecule or give rise to different reactive species. Depending on the soil structure, light can penetrate in soils from 0.001 to 0.01 m. Solar radiation can therefore contribute significantly to the degradation of pesticides present on or close to the soil surface (Mansour and Korte 1986). The principal factors in the process are the presence of photochemical catalysts, the intensity and length of exposure to radiation, soil pH and aeration, chemical structure and physical state of the pesticide and degree of colloid adsorption.

Humic acids are capable of acting as sensitizers for the production of reactive intermediates such as singlet oxygen ($^{I}O_2$), hydroxyl radicals (·OH), hydrogen peroxide (H_2O_2), and peroxy radicals (ROO) (Mansour et al. 1989; Konstantinou et al. 2001). Metal oxides (ZnO, Fe_2O_3 and MnO_2) as well absorb radiation in the sunlight wavelength range and may accelerate pesticide degradation (Navarro et al. 2007). Some clay minerals are also known to catalyse oxidation reactions, as observed during degradation of parathion (Spencer et al. 1980). Significantly more rapid photolysis of triazine herbicides was observed in moist soil surfaces compared to dry soil surfaces (Burkhard and Guth 1979).

10.3.2.2 Biotic Degradation

Biological degradation, either complete mineralisation or partial decay to form metabolites, is the critical process controlling the ultimate fate of agro-chemicals in soil and water (Turco and Kladivko 1994). Microorganisms are the major agents involved in the degradation of many pesticides in the soil environment (Navarro et al. 2007). Soil microbial activities are strongly influenced by temperature, soil water content, soil pH, carbon and nutrient availability, size and species composition of the microbial population, and the nature of the pesticide itself (Bollag and Liu 1990). The microbial population capable of degrading the particular pesticide is more important than the total soil microbial biomass. The adaptation of relevant degrading bacteria is also of great importance because of previous exposure to a certain pesticide enhances the biodegradation process (Allan 1989). The half-lives of pesticides in these areas would probably be different compared to sites where repeated applications have not been carried out.

Some microorganisms are capable of using pesticides as a source of carbon and nitrogen, for example *Pseudomonas* (with 2,4-D and paraquat), *Nocardia* (with dalapon and propanyl) or *Aspergillus* (with trifluralin and picloram) (Higgins and Burns 1975). Microbial activity increased with atrazine pollution during long-term incubation (Moreno et al. 2007). In some cases, photochemical pre-treatment accompanied with microbial degradation lead to the complete degradation of pesticides, e.g. of atrazine (Chan et al. 2004).

The most important role of micobes in the transformation of pesticides is their ability to bring about detoxication (sometimes designated detoxification). Detoxication results in inactivation, with the toxicologically active ingredient being converted to an inactive product, because toxicological activity is associated with many chemical entities, substituents, and modes of action, detoxications similarly include a large array of different types of reactions some of which are summarized in Table 10.11.

In contrast, one of the most undesirable effects of microbial transformations in the soil is the formation of toxic substances. Some pesticides that may be harmless can be converted to products that possibly create risks to humans, animals, plants, and microorganisms. This conversion is known as activation which may involve a single reaction or a sequence in a cometabolic process (Navarro et al. 2007). However, toxic chemicals residing in soil become less toxic with time (Hatzinger and Alexander 1995). Alternatively, the harmful metabolite may be an intermediate in mineralization, yet it may persist long enough to create a pollution problem. The consequences of activation include the biosynthesis of carcinogens, mutagens, teratogens, neurotoxins, phytotoxins, and insecticidal and fungicidal agents (Alexander 1994).

Table 10.11 Biochemical transformations of organic substances in soils

Process	Original compound and transformation product
Oxidation	$RCH_3 \rightarrow RCH_2OH$
β-oxidation	$CH_3CH_2CH_2COOH \rightarrow CH_3COOH + CH_3COOH$
Oxidation of amino acids	$RNH_2 \rightarrow RNO_2$
Oxidation of compounds with S	$R_2S \rightarrow R_2SO$
Oxidative dealquilation	$ROCH_3 \rightarrow ROH + HCHO$
Decarboxylation	$RCOOH \rightarrow R\text{-}H + CO_2$
Aromatic hydroxylation	$Ar \rightarrow ArOH$
Ring rupture	$Ar(OH)_2 \rightarrow CHOCHCHCHCOHCOOH$
Hydrolytic dehalogenation	$RCHClCH_3 \rightarrow RCHOHCH_3 + Cl$
Dehydrohalogenation	$RCH_2CHClCH_3 \rightarrow RHC = CHCH_3$
Reductive dehalogenation	$RCCl_2R \rightarrow RCHClR + Cl$
Nitroreduction	$RNO_2 \rightarrow RNH_2$

Adapted from Navarro et al. (2007)

10.3.2.3 Kinetics of Pesticide Degradation

The degradation of pesticides and other organic substances is often studied using ^{13}C- or ^{14}C–labelled material. Mineralization rates of the pesticide (allowing the quantification of biodegradation) can be measured by trapping evolved $^{13}CO_2$ or $^{14}CO_2$, and a mass balance, including the formation of soil bound residues, can be established (OECD 2002). The knowledge of the kinetics of pesticide degradation is important for the evaluation of the persistence of chemicals and to assess exposure of plants, animals and humans. Many factors (e.g. temperature, pH, moisture of the soil, other C sources, etc.) influence the disappearance of pesticides from soil. However, most of the models proposed consider the pesticide concentration as the only dependent variable (reviewed by Navarro et al. 2007) which results in an exponential type of graph (Fig. 10.4), according to Eq. 10.4.

$$Rt = R_0\, e^{(-Kt)} \tag{10.4}$$

where Rt represents the residue at time t, R_0 is the residue at time zero, K is the rate constant for chemical disappearance and t is representative for the time elapsed since application.

First-order degradation is to be expected when microbial abundance and activity in the soil is limited, the microorganisms are not in abundance in the soil, possibly because of energy and nutrient limitation. First-order kinetics can be plotted logarithmically as a function of time. In this case, a straight line is obtained according to Eq. 10.5.

$$\ln Rt = \ln R_0 - K_t \tag{10.5}$$

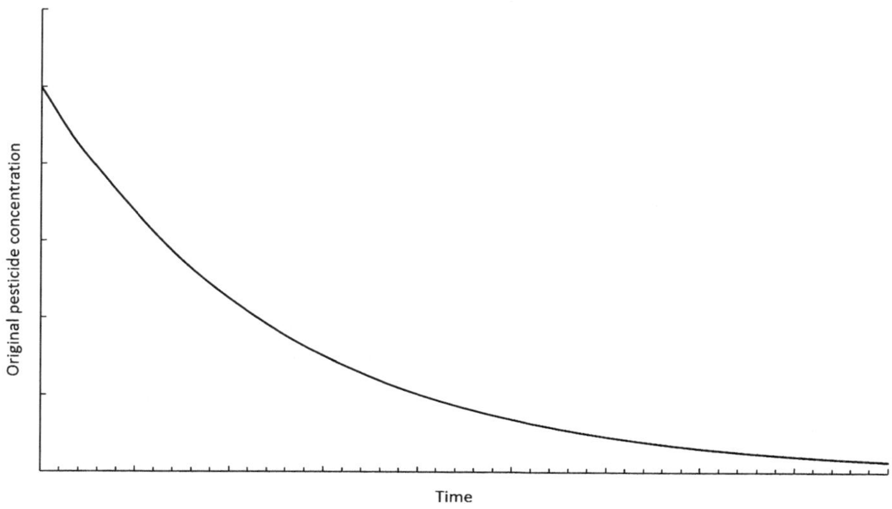

Fig. 10.4 Typical degradation curve of pesticides in soil (Adapted from Navarro et al. 2007)

Once the constant K has been calculated, the half life time (DT_{50}) can be calculated (Eq. 10.6).

$$t_{1/2} = \ln 2/K \tag{10.6}$$

With this expression the persistence in the soil of the different pesticides can be compared.

10.3.2.4 Bound Residues

The formation of bound residues in soil depends on a number of factors including climate, type and nature of soil, amount, type and nature of the pesticide applied and whether the soil has received only a single or multiple application. The ability of the soil to retain pesticides is attributed to adsorption phenomena and chemical reactions occurring on the active surfaces on minerals and SOM. The formation of bound residues on a percentage basis have been found to vary inversely with increased initial application rates (Gevao et al. 2000). For example, Gan et al. (1995) found that at application rates of 10 and 100 mg kg^{-1} soil of alachlor, 48% and 37% of bound residues were formed, respectively, in a clay loam soil, following a 40-week incubation period. Above 100 mg kg^{-1}, there was a very significant decrease in the formation of bound residues. The extent of the formation of bound residues also depends on type and nature of the chemical and whether the soil has received only a single or multiple application. Pesticides with high sorption coefficients and long half-lifes tend to form more bound residues than those with low sorption coefficients and short half-lifes (Table 10.6).

With multiple application, on a percentage basis, enhanced dissipation and decreased binding have been observed (Gevao et al. 2000). Repeated application of DDT and HCH in a tropical soil hindered bound residue formation, accelerated the rate of volatilisation and accelerated metabolite formation (Samuel and Pillai 1991). Suett and Jukes (1990), while studying the fate of mephosfolan in soil, found that more than 80% of the applied chemical was lost from pre-treated soil as compared to only 20–60% from previously untreated soils. Khan et al. (1989) found that repeated application of prometryn to soil already containing formerly bound residues resulted in the decline in the proportion of these residue formations, similar to the results of Smith and Aubin (1991) in their study with phenoxyalkanoic acid.

Increased contact time between soil and pesticides or ageing leads to the formation of a proportion of compounds being retained "permanently" in soil (Gevao et al. 2000). There is evidence that, with longer residence times in the soil, bound pesticide residues tend to lose their biological activity and become more resistant to degradation and extraction. Covalent interaction between pesticide compounds and SOM leads to the formation of very stable bonds. These interactions have been observed mainly with degradation products of chloroaniline degradation products of the urea and anilide herbicides as well as phenolic products from the phenoxy herbicides. This immobilization has been postulated to proceed through oxidative coupling reactions during humification (e.g. Bollag and Myers 1992; Senesi 1992).

10.3.3 Pesticide Losses to the Environment

10.3.3.1 Losses to the Atmosphere

The presence of pesticides in the atmosphere was first reported towards the end of the 1950s, when there was significant use of chlorinated pesticides such as DDT, lindane and dieldrin (Unsworth et al. 1999). Pesticides enter the atmosphere either by application drift, post-application volatilization or wind erosion of pesticide treated soil. The pesticides and their photodegradation products may be transported long distances (Cessna et al. 2006). At any point during transport, they are also subject to the removal processes of wet and dry deposition, both of which contaminate surface waters. In wet deposition, pesticides may be trapped in snow and hail or dissolved in rain. In a monitoring study of pesticides in rainfall in Europe, 48 pesticides and their isomers and metabolites have been detected with the most frequently monitored pesticides, being in descending order, lindane (γ HCH; HCH: hexachlorocyclohexane) and its main isomer (α-HCH), atrazine, MCPA (2-methyl-4-chlorophenoxyacetic acid), simazine, dichlorprop, isoproturon, mecoprop, DDT, terbuthylazine and Aldrin (Dubus et al. 2000). In dry deposition, pesticides are sorbed to particles of wind-eroded soil. Pesticides in the atmosphere are subjected to transport over distances which can range to thousands of kilometers leading to increasing levels of current use pesticides in regions largely free from direct inputs of industrial and agricultural chemicals, such as the Arctic (Hoferkamp et al. 2010). Dacthal, chlorothalonil, chlorpyrifos, metolachlor, terbufos and trifluralin have been detected in Arctic environmental samples (air, fog, water, snow) (Rice and Cherniak 1997; Garbarino et al. 2002). Other studies have identified pesticides to undergo short-range atmospheric transport (Muir et al. 2004) to ecologically sensitive regions such as the Chesapeake Bay (McConnell et al. 1997) and the Sierra Nevada mountains, USA (Le Noir et al. 1999).

Drift

During pesticide application, up to 30–50% of the amount applied can be lost to the air via drift loss (Van den Berg et al. 1999). Drift loss may be defined as the quantity of pesticide that is deflected out from the treated area by the action of climatic conditions during the application process (Gil and Sinfort 2005). Factors that influence droplets pesticide emission to air during application include equipment and application techniques, volatility and viscosity of the pesticide formulation, operator care, attitude and skill, horizontal wind speed and wind direction, temperature, relative humidity and stability of air at the application site (Gil and Sinfort 2005). The proportion of the total spray volume contained in droplet sizes below 150 μm can be used as an indicator of drift potential, because it is these small droplets that are most prone to movement under windy conditions.

Spray drift is one reason for atmospheric organic contamination. It can also damage nearby sensitive crops or contaminate crops ready to harvest and may be hazardous to people, domestic animals, or pollinating insects (Tiryaki and Temur 2010). Moreover, it can contaminate water in ponds, streams, ditches and harm aquatic plants and animals.

Volatilization

Volatilisation of a pesticide from the soil is determined by the inherent vapour pressure, soil conditions, climatic factors, as well as management factors (Taylor and Spencer 1990). Volatilisation of pesticides increases with temperature due to the increase in vapour density of the pesticide (Wienhold et al. 1993). Volatilisation also increases with an increase in soil water content (Walker and Bond 1977) due to the increase in pesticide vapour pressure as water displaces the chemical from adsorption sites. Due to the greater likelihood of stronger pesticide sorption to the soil, volatilisation losses are low from dry soil (Glotfelty et al. 1989). For example, for surface application of trifluralin, the volatilisation loss ranged from 25% in 50 h from a dry soil to 90% over 7 days from a moist soil (Glotfelty et al. 1984).

Water is also involved in the movement of pesticides to the soil surface for volatilisation. As water evaporates from the soil surface, soil water moves towards the surface by capillary rise, carrying with it dissolved pesticide residue. Hargrove and Merkle (1971) demonstrated that alachlor loss from the soil increased with rising temperature and relative humidity, with humidity having a stronger influence on volatility at higher temperatures. The pesticides dicamba, benomyl, and diuron generally have a low potential to volatilize from moist soil. In contrast, 1,3-dichloropropene and bensulide due to their high vapour pressure have a high potential to volatilize (Tiryaki and Temur 2010).

Wind Erosion

There is potential hazard of environmental transport of pesticides on windblown particles which may have implications for air and water quality. If dust containing pesticides is inhaled, there may be an impact on human health. Dust deposited in waterways may negatively affect water quality. Surface applied herbicides are adsorbed to soil particles in the very shallow surface layer which is the first to be removed by wind erosion. Hence the potential for removal of surface-applied chemicals by wind erosion is much higher than soil-incorporated pesticides. Larney et al. (1999) observed that overall wind erosion losses of the soil-incorporated herbicides were about three times lower than those of the surface-applied herbicides.

10.3.3.2 Losses to the Hydrosphere

Pesticides have been found in aquatic systems worldwide (Rovedatti et al. 2001; Chopra et al. 2011). For example, Agnihotri et al. (1994) reported that in Ganga river water near Farrukabad, India aldrin (organochlorine insecticide) residues often exceeded the World Health Organization (WHO) guideline value for drinking water. Matin et al. (1998) studied the organochlorine insecticide residues in surface and underground water from different regions of Bangladesh and found slight contamination of some of the water samples of both surface and underground sources with residues of DDT, heptachlor, lindane, and dieldrin. Qiu et al. (2005) studied the contribution of dicofol to the current DDT pollution in China and reported that the samples from Lake Taihu near Shanghai were contaminated with DDT and its derivatives. Sudo et al. (2002) studied the concentration of pesticide residues in Lake Biwa basin (Japan) and reported that the pesticides analyzed from six rivers flowing into the lake were the herbicides molinate, simetryn, oxadiazon and thiobencarb, the fungicide isoprothiolane, and the insecticides diazinon and fenitrothion. Cerejeira et al. (2003) studied pesticide concentrations in Portuguese surface and groundwater from three river basin from 1983 to 1999 and reported that various insecticides and herbicides were detected, particularly atrazine, chlorfenvinphos, α- and β-endosulfan, lindane and simazine. In groundwater collected from wells of seven agricultural areas from 1991 to 1998 various herbicides were detected including alachlor, atrazine, metolachlor, metribuzine, and simazine. Iyamu et al. (2007) studied the concentrations of residues from organochlorine pesticides in water and fish from some rivers in Edo State, Nigeria and reported that in all the water samples analyzed, organochlorine pesticides and their residues (lindane, aldrin, *p,p*-DDE, *o,p*-DDD, *p,p*-DDD, *o, p*-DDT, and *p,p*-DDT) were present.

The contamination of water bodies with pesticides can pose a significant threat to aquatic ecosystems and drinking water resources. Pesticides can enter water bodies via diffuse or via point sources. Diffuse-source pesticide inputs into water bodies are the inputs resulting from agricultural application on the field. Point-source inputs derive from a localized situation and enter a water body at a specific or restricted number of locations.

Surface Runoff

Pesticide runoff is the physical transport of pesticides over the soil surface by rainwater that does not soak into the soil. Pesticides move from fields while dissolved or suspended in runoff water or adsorbed (chemically attached) to eroded soil particles. Runoff may affect the quality of surface waters that include streams, rivers, lakes, reservoirs and oceans. Streams and reservoir supply approximately 50% of the drinking water in the world. Surface waters receive a portion of their water from snow melt or rainfall runoff. Pesticides susceptible to surface runoff are those within the runoff-soil interaction zone or the top 0.5–1 cm of soil (Tiryaki and Temur 2010). Several factors may affect the amount of pesticide present within this zone. These include type of pesticide application, soil type, physiochemical properties, formulation type and field half-life of the pesticide.

The presence of pesticides in surface water, even in very small amounts, compromises the life cycle of aquatic organisms. Surface water also can be contaminated directly by pesticide spray drift when the spray is applied close to surface water. Toxic pesticides can cause the death of fish and other aquatic organisms even at low concentrations.

Leaching

Leaching is the movement of pollutants through the soil by drainage water. The contamination of water bodies with pesticides can pose a significant threat to aquatic ecosystems and drinking water resources. Leaching of water and dissolved pesticides to groundwater occurs by matrix flow and preferential flow. Matrix flow is the slower transport process in which the simultaneous movement of pesticides with water is determined by soil structure, SOM and clay content and by the physico-chemical properties of the pesticide, including water solubility, vapour pressure and K_{oc} (Tiryaki and Temur 2010). Other factors affecting pesticide leaching are weather, application rate and season (Reichenberger et al. 2007). Preferential flow is the faster transport process and depends on the presence of cracks and macropores, including biopores (Fig. 10.5).

Leaching of pesticides occurs through mass transfer and molecular diffusion and is governed by mass flow and dispersion and is expected to be lower in soils high in SOM (Chesters et al. 1989). Except for diquat and paraquat (Helling 1970), SOM is more closely related to leaching than is clay, obviously due to SOM being the principal sorbent (Mueller and Banks 1991). Moreover, in fine-textured soils, macropores, which are principally root channels and wormholes, may contribute to the leaching of pesticides. Soil pH can also influence leaching indirectly by influencing sorption, as demonstrated for triazines and sulfonylurea herbicides. Studies on residues of various pesticides bound to SOM showed its significance for the accumulation, toxicity, and bioavailability of these bound residues, which greatly influence their leaching behaviour (Scheuner and Reuter 2000). There are many studies involving leaching experiments either at laboratory (e.g. Van Genuchten and Cleary 1979; Veeh et al. 1994) or field scale (Kookana et al. 1995; Flury 1996; Sarmah et al. 2000), which demonstrated that pesticides can move beyond the rooting zone. Understanding this process is therefore a prerequisite to quantify the groundwater contamination potential of a given pesticide.

Contamination of groundwater by pesticides is becoming an increasingly serious problem and monitoring of aquifers has become an essential aspect of environmental policy. This is related to the carcinogenic and health risks caused by the presence of pesticides in drinking water. Pesticides were found in groundwater in numerous instances, however, and it seems apparent that more instances will be discovered as more and more underground aquifers are sampled and tested for the presence of pesticides. In some cases, communities have had to use bottled water until other

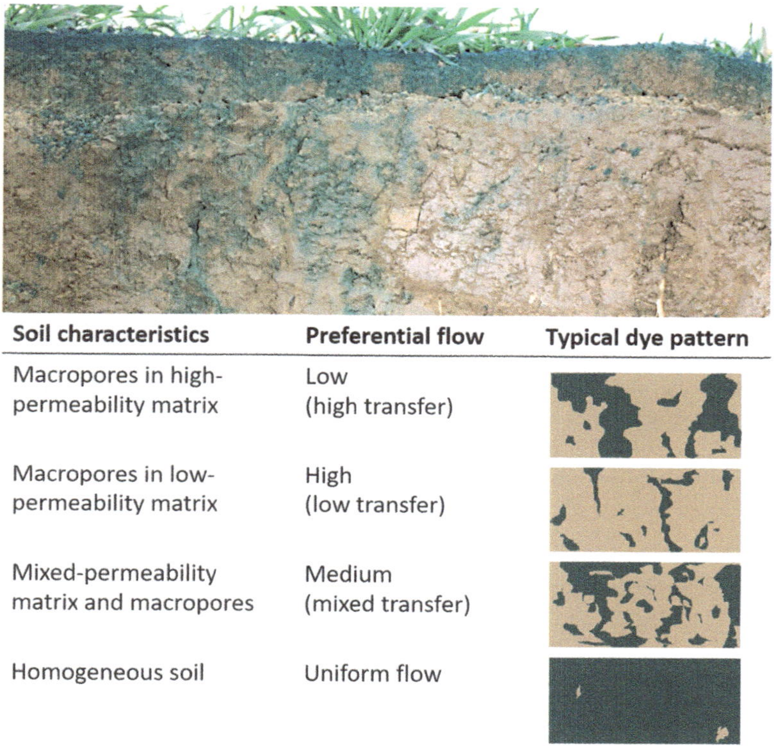

Fig. 10.5 Preferential flow of pesticides in soil visualized with the color tracer brilliant blue; (Photo in upper part: Otto Ehrmann, Bildarchiv Boden, with kind permission; panel in the lower part: adapted from Geyikçi 2011)

sources of drinking water were developed. In the USA, at least 143 pesticides and 21 of their transformation products were found in groundwater, from every major chemical class. Pesticides most frequently detected in ground water are triazine and acetanilide herbicides used extensively on corn and soybeans and the carbamate insecticide aldicarb (Tiryaki and Temur 2010). According to the EU drinking water standard, concentrations of a single pesticide may not exceed 0.1 μg L^{-1}. For a dose of 0.2 kg of a pesticide ha^{-1} and an annual recharge of 200 mm, this implies a maximum allowed leaching loss of only 0.1% of the applied amount.

10.3.4 *Pesticides in Non-target Organisms*

Pesticides are applied with the goal of maximizing productivity and economic returns. However, side effects on soil organisms are often neglected. Once present in the environment, pesticides can harm nontarget plants, soil organisms, aquatic organisms, insects, birds, and other wildlife.

10.3.4.1 Plants

Herbicides are designed to kill plants but they can also injure or kill desirable species if they are applied directly to such plants, or if they drift or volatilise onto them (Aktar et al. 2009). For example, volatilised esterformulation herbicides which have been transported off-site have been shown to cause severe damage to other plants (Straathoff 1986). In addition to killing non-target plants, pesticide exposure can cause sublethal effects on plants. Exposure to glyphosate has been shown to reduce seed quality (Locke et al. 1995) and can increase the susceptibility of plants to fungal disease (Brammall and Higgins 1988). This poses a special threat to endangered plant species. The US Fish and Wildlife Service has recognized 74 endangered plants that may be threatened by glyphosate alone (US EPA 1986). Exposure to the herbicide clopyralid can reduce potato yields (Lucas and Lobb 1987). Some insecticides and fungicides can also damage plants (Aktar et al. 2009). Plants can also suffer indirect consequences of pesticide applications when harm is done to soil microorganisms and beneficial insects.

10.3.4.2 Soil Organisms

Bünemann et al. (2006) in their review on the effect of pesticides on soil organisms found that herbicides generally had no major effects on soil organisms. However, butachlor was found to be an exception, because it was shown to be very toxic to earthworms at agricultural application rates (Panda and Sahu 2004). Phendimedipham induced avoidance behaviour in earthworms (Amorim et al. 2005) and collembola (Heupel 2002). These effects are expected to be relatively short-lived, as phendimedipham is degraded moderately rapidly (half-life: 25 days) in soil. Other effects of herbicides on soil organisms were mainly isolated changes in enzyme activities (Bünemann et al. 2006). Glyphosate, for example, was shown to suppress the phosphatase activity by up to 98% (Sannino and Gianfreda 2001) in a laboratory study. However, urease activity was stimulated by both glyphosate as well as atrazine.

Compared to herbicides, insecticides were shown to have a greater direct effect on soil organisms (Bünemann et al. 2006). Organophosphate insecticides (chlorpyrifos, quinalphos, dimethoate, diazinon, and malathion) had a range of effects including changes in bacterial and fungal numbers in soil (Pandey and Singh 2004), varied effects on soil enzymes (Singh and Singh 2005), as well as reductions in collembolan density (Endlweber et al. 2005) and earthworm reproduction (Panda and Sahu 1999). Carbamate insecticides (carbaryl, carbofuran, and methiocarb) had a range of effects on soil organsism, including a significant reduction of acetylcholinesterase activity in earthworms (Pandey and Singh 2004), mixed effects on soil enzymes (Sannino and Gianfreda 2001), and inhibition of nitrogenase in Azospirillum species (Kanungo et al. 1998). Persistent compounds including arsenic, DDT, and lindane caused long-term effects, including reduced microbial activity (Van Zwieten et al. 2003), reduced microbial biomass, and significant decreases in soil enzyme activities (Singh and Singh 2005).

Fungicides generally had even greater effects on soil organisms than herbicides or insecticides (Bünemann et al. 2006). As these chemicals are applied to control fungal diseases, they will also affect beneficial soil fungi and other soil organisms. Very significant negative effects were found for copperbased fungicides, which caused long-term reductions of earthworm populations in soil (Van Zwieten et al. 2004). Merrington et al. (2002) further demonstrated significant reductions in microbial biomass, while respiration rates were increased, and showed conclusively that copper residues resulted in stressed microbes. Other observed effects included the reduced degradation of DDT (Gaw et al. 2003). These effects are likely to persist for many years, as copper accumulates in surface soils and is not prone to biodegradation. Negative effects were also found for benomyl, which caused long-term reductions in mycorrhizal associations (Smith et al. 2000). Chlorothalonil and azoxystrobin, have recently been shown to affect on a biocontrol agent used for the control of Fusarium wilt (Fravel et al. 2005), illustrating potential incompatibilities of chemical and biological pesticides.

10.3.4.3 Aquatic Organisms

Pesticides enter aquatic habitats from direct application, surface runoff, leaching or wind-borne drift. Because there numerous pesticides in use around the world, studies have detected a variety of pesticides including the insecticides malathion, endosulfan and diazinon as well as the herbicides atrazine and glyphosate (Le Noir et al. 1999; Hayes et al. 2002; Kolpin et al. 2002; Thompson et al. 2004). Urea herbicides such as isoproturon and diuron often contaminate rivers, lakes, and groundwater. Most breakdown products of diuron are more toxic to microorganisms than diuron itself (Isenring 2010).

Many pesticides found in aquatic systems are not legally registered, but they appear (e.g. Thompson et al. 2004). The majority of the available pesticide studies regarding aquatic microorganisms describe effects on algae. Studies of herbicide effects dominate, in particular effects caused by atrazine. Far fewer pesticide studies exist for aquatic bacteria, fungi, and protozoa. DeLorenzo et al. (2001) published an extensive review on pesticide toxicity to aquatic microorganisms. The mechanism of pesticide action in microbial species may not be the same as for the target-organisms. In microorganisms, pesticides have been shown to interfere with respiration, photosynthesis, and biosynthetic reactions as well as cell division and growth, and molecular composition (DeLorenzo et al. 2001).

Algae and Cyanobacteria

Toxicity effects of the carbamate insecticides carbaryl and carbofuran to algae and cyanobacteria were tested by Peterson et al. (1994). Following carbaryl exposure to an expected environmental concentration, inhibition ranged from 35% to 86% with the cyanobacteria *Pseudoanabaena* and *Anabaena inaequalis*. In contrast, carbofuran had relatively low toxicity to cyanobacteria when applied at expected environmental concentration. Organophosphorus insecticides such as fenitrothion and pyridaphenthion cause growth inhibition in algae and cyanobacteria (Sabater and Carrasco 2001). Since herbicides are designed to kill plants, herbicide contamination of water could have devastating effects on photosyntetically active aquatic organisms. In a study with oxadiazon, algae growth was severely reduced (Ambrosi et al. 1978). Studies looking at the impacts of atrazine and alachlor on algae and diatoms in streams showed that even at fairly low levels, the chemicals damaged cells, blocked photosynthesis, and stunted growth in varying ways (Aktar et al. 2009). Little information is available about agricultural fungicide toxicity to aquatic microorganisms. However, it is commonly known that copper-based fungicides are highly toxic to aquatic organisms (Isenring 2010). Peterson et al. (1994) compared the sensitivity of ten algal species to the triazole derivative fungicide propiconazole. At expected environmental concentration, inhibition of ^{14}C uptake was <20% to a range of species of algae. Unexpectedly, propicanazole stimulated growth to some extent in cyanobacteria and diatoms.

Aquatic Invertebrates

Liess and Schulz (1999) and Schulz and Liess (1999) in a field-study in northern Germany investigated runoff-related insecticide input (parathion-ethyl, fenvalerate, and deltamethrin) on stream macroinvertebrate dynamics. Transient increase in discharge and insecticide contamination was observed in the stream subsequent to surface runoff from arable land. In the macroinvertebrate community, 8 of 11 abundant species disappeared, and the remaining 3 were reduced significantly in abundance following the 3 most highly contaminated runoff events (Liess and Schulz 1999; Schulz and Liess 1999). Following runoff events from cotton fields in Australia, the insecticide endosulfan caused negative effects on the macroinvertebrate community (Leonard et al. 1999). In this field study, five of the six dominant taxa were affected by elevated endosulfan concentrations. The herbicide trifluralin was found to be moderately to highly toxic to aquatic invertebrates, and highly toxic to estuarine and marine organisms like shrimps and mussels (US EPA 1996).

Fish

Pesticide toxicity to fish has been observed in several studies. The insecticide chlorpyriphos is highly toxic to fish, and has caused fish kills in waterways near treated fields (US EPA 2000). Surface water is frequently contaminated with insecticides through normal use at levels above those known to affect fish and aquatic invertebrates. Devastating fish deaths have been recorded in response to unintentional spreading of pesticides. In 1991 and 1995 huge amounts of eels

(*Anguilla anguilla*) died in Lake Balaton in Hungary, the largest freshwater lake in Europe (Bálint et al. 1997). The main reason was the use of deltamethrin and permethrin in the Lake Balaton area against mosquitoes. Herbicides can also be toxic to fish. Studies show that trifluralin is highly toxic to cold and warm water fish (US EPA 1996). In a series of different tests it was also shown to cause vertebral deformities in fish (Koyama 1996). The herbicides Ronstar and Roundup are also acutely toxic to fish (Shafiei and Costa 1990). In addition to direct acute toxicity, some herbicides may produce sublethal effects on fish that lessen their chances for survival and threaten the population as a whole. Glyphosate or glyphosate-containing products can cause sublethal effects such as erratic swimming and labored breathing, which increase the fish's chance of being easy prey (Liong et al. 1988). 2,4-D herbicides caused physiological stress responses in sockeye salmon (McBride et al. 1981) and reduced the food-gathering abilities of rainbow trout (Little 1990). 2,4-D or 2,4-D containing products have also been shown to be harmful to shellfish (Cheney et al. 1997). Elevated concentrations of ΣDDTs (sum of DDT-related compounds) were found in tuna samples from the western Pacific (>500 $\mu g\ g^{-1}$ fat), more specifically in the south China Sea and Sea of Japan (Ueno et al. 2006). ΣDDTs was also elevated in tuna of the Bay of Bengal region, likely contaminated from the use of DDT for agriculture and for malarial vector control.

Aquatic Mammals

Because of their high trophic level in the food chain and relatively low activities of drug-metabolising enzymes, aquatic mammals such as dolphins accumulate increased concentrations of persistent organic pollutants (Tanabe et al. 1988) and are thereby vulnerable to toxic effects from contaminant exposures. Dolphins inhabiting riverine and estuarine ecosystems are particularly vulnerable to the activities of humans because of the restricted confines of their habitat, which is in close proximity to point sources of pollution (Aktar et al. 2009). Exposure to great concentrations of DDT (1,1,1-trichloro-2,2-bis[*p*-chlorophenyl] ethane) and PCBs has been shown to elicit adverse effects on reproductive and immunological functions in aquatic mammals (Kannan et al. 1994; Colborn and Smolen 1996). Many species of marine mammals are highly contaminated by DDTs due to their top trophic positions and, in some animals, these chemicals have been linked with population declines. The highest levels of ΣDDTs were found in species inhabiting the midlatitudes of industrialized Asia, North America and Southern Europe. Very high concentrations in dolphin blubber were observed, sometimes exceeding 1000 $\mu g\ g^{-1}$ fat (Houde et al. 2005).

10.3.4.4 Wildlife, Birds and Insects

Of all agricultural intensification measures pesticide use has had the most negative effect on terrestrial species diversity. Insecticides, rodenticides, fungicides (for seed treatment) and herbicides threaten exposed wildlife through their toxicity. Pesticide poisoning through highly toxic carbamates and organophosphates can cause population declines which may threaten rare species. Pesticides accumulating in the food chain, particularly those which cause endocrine disruption, pose a long-term risk to animals. This is a particular concern to top predators such as mammals or raptors. Non-target predatory mammals (e.g. foxes) and raptors frequently suffer secondary poisoning by eating rats or mice poisoned by rodenticides some of which can bioaccumulate. In France, foxes were poisoned by residues of bromadiolone in prey tissue (Berny et al. 1997). In the UK, local wood mice, bank vole, and field vole populations declined significantly following rat control with rodenticides (Brakes and Smith 2005).

Increasing use of pesticides has been linked to periods of rapid bird decline. Organophosphate insecticides, including disulfoton, fenthion, and parathion are highly toxic to birds and have frequently poisoned raptors foraging in fields (Mineau et al. 1999). In the USA, some 50 pesticides are known to kill songbirds, gamebirds, raptors, seabirds and shorebirds (Isenring 2010). In the Argentine pampas, monocrotophos, an organophosphate, killed 6000 Swainson's hawks. Worldwide, over 100,000 bird deaths caused by this chemical have been documented (Isenring 2010). Besides lethal poisonings sublethal quantities of pesticides can affect the nervous system causing changes in behaviour. In an orchard, parent birds made fewer feeding trips after azinphos-methyl, an organophosphate, had been sprayed (Bishop et al. 2000).

Bees, spiders and beetles are threatened broad-spectrum insecticides (such as carbamates, organophosphates and pyrethroids) and thus cause serious population declines of insects part of which play a major role in the food web, as pollinators (e.g. bees) or as natural enemies of pest insects. Honey bees are under pressure from parasitic mites, viral diseases, habitat loss and pesticides. The carbamate bendiocarb, and the pyrethroids cypermethrin, deltamethrin and permethrin which are used in the UK are known to poison bee colonies (Isenring 2010). The neonicitinoid clothianidin, used to treat corn and sunflower seeds, in 2008 caused many bee poisonings and colony deaths in southern Germany. The pesticide has since been prohibited. Residues of the neonicotinoid imidacloprid in maize pollen grown from treated seed can also be a high risk to bees (Isenring 2010).

Pesticides can also have indirect effects by reducing the abundance of weeds and insects which are important food sources for many species. Herbicides can change habitats by altering vegetation structure, leading finally to decline in population.

10.4 Pesticides in Food Products

The widespread use of pesticides has led to concern over pesticide residues in food products. The major route through that most pesticides enter the food chain is due to their application as crop protection materials in the production of arable crops, vegetables and fruit crops and the treatment fruit, vegetables and grain products during storage. Pesticide residues may enter the human food chain directly or through the consumption of meat and milk products from animals that have consumed products containing pesticide residues. At the time when organo-chlorine insecticides were significant contaminants of milk as a result of their presence in the feed consumed by the animal. These materials were prohibitet in many world regions, including Europe and the USA, but are still used elsewhere and may occur in food products as a result of their long half-lifes in soil. Baker et al. (2002) compared pesticide residues in fruit and vegetables originating from conventional, integrated pest management and organic production systems in the USA. They found one or more residues in 23% of the organic samples, in 73% of the conventional and in 47% of the integrated pest management derived samples. About 40% of the residues in organic samples were derived from banned organo-chlorine compounds. However, pesticide presence in organic farming after removal of organo-chlorine pesticides is expected to decrease with time. Among the crops particularly prone to absorb organo-chlorines are carrots, potatoes and other root crops, cucurbits and leafy vegetables. Herbicides such as 2,4-D, 2,4,5-T and MCPA, which are still in use, have been found in milk after they had been used to treat pasture or food subsequently fed to cattle (Cowie and Swinburne 1977).

Current concerns have resulted in the EU, in the USA and many other countries setting up systems for the routine monitoring of food products. In the EU a coordinated community program was performed in 2010 including the 27 member states plus the EFTA countries Norway and Iceland to control pesticide contamination of food crops (EFSA 2013). The food commodities to be analyzed were apples, head cabbage, leek, lettuce, milk, peaches, pears, rye or oats, strawberries, swine meat and tomatoes. The program defined 157 pesticides to be analysed in food of plant origin and 34 pesticides in food of animal origin. A total number of 12,168 samples were analyzed. The results show that 197 (1.6%) of the 12,168 samples exceeded European legal limits, while 5802 (47.7%) of the samples had measurable residues below or at the European legal limit, and 6169 of the samples (50.7%) were free from measurable pesticide residues. The majority of total samples taken in 2010 were produced in one of the reporting countries (73%), while 23% of the samples originated from third countries.

Among different food commodities the highest percentage of samples exceeding European legal limit was identified for oats (5.3%), followed by lettuce (3.4%), strawberries (2.8%), peaches (1.8%), apples (1.3%), pears (1.3%), tomatoes (1.2%), leek (1.0%), head cabbage (0.9%) and rye (0.3%). In animal products (milk and swine meat) no legal limit exceedances were observed. Peaches had the highest percentage of samples with measurable pesticide residues below or at legal limits (71.2%), followed by 67.0% of the apple samples and 65.2% of strawberry samples. Samples of pears, swine meat or milk less frequently contained measurable residues at or below European legal limits (EFSA 2013). An overview of the pesticides most frequently found in selected food commodities of plant origin, swine meat and milk is given in Table 10.12.

The highest percentages of European legal limit exceedances were found for chlormequat in oats, where the limit was exceeded in 8.1% of all samples, followed

Table 10.12 Pesticides found in food products in the EU 27, Norway and Iceland in 2010

Food product	Number of samples	Number of pesticides found	Pesticides frequently found	Pesticides that exceeded European legal limits
Apples	2057	94	Dithiocarbamates, captan/folpet, diphenylamine, boscalid, chlorpyrifos, pyraclostrobin, thiacloprid pirimicarb, thiabendazole, carbendazim, benomyl	Thiacloprid, benomyl and carbendazim, pyrimethanil, propargite, tebufenpyrad, azinphos-methyl, dimethoate, dicofol, diazinon fenitrothion, phosalone, fenthion, oxydemethon-methyl, fenpropathrin, dichlorivos
Peaches	1200	79	Tebuconazole, dithiocarbamates, iprodione, spinosad, chlorpyrifos, triflumuron, etofenprox, cyprodinil, fenbuconazole	Captan, phosmet, dimethoate, carbendazim, benomyl
Tomatoes	1794	84	Bromide ion, dithiocarbamates	Ethephon, acetamiprid, pyraclostrobin,spiroxamine, oxadixyl, bifenthrin, procymidone, deltamethrin
Head cabbage	999	49	Dithiocarbamates	Dimethoate, dimethomorph, methoxyfenozide, oxamyl, cyproconazole, difenoconazole, ethion, procymidone
Leek	961	45	Dithiocarbamates, boscalid, tebuconazole, bromide ion	Bromopropylate, iprodione, indoxacarb, linuron, acrinathrin, triadimefon, thiabendazole, cypermethrin, cyprodinil
Lettuce	1568	68	Bromide ion, dithiocarbamates, iprodione, cyprodinil boscalid, proparmocarb, fludioxonil, imidacloprid	Bromide ion, dithiocarbamates, chlorothalonil, iprodione, chlorpyrifos, dimethoate,
Oats	246	20	Chlormequat, glyphosate, pirimiphos-methyl	Chlormequat
Rye	406	18	Chlormequat, bromide ion, mepiquat	Chlormequat
Swine meat	623	8	DDT, lindane, hexachlorobenzene, heptachlor, diazinon, pirimiphos-methyl, permethrin, cypermethrin	None
Milk	654	4	DDT, hexachlorobenzene, beta HCH, chlorpyrifos	None

Adapted from EFSA (2013)

by residues of ethephon in tomatoes (2.3%), amitraz in pears (1.3%) and bromide ion in lettuce (0.8%).

Many fresh products carry multiple pesticide residues which could have profound implications for the question of safety. Acceptable daily intakes as legal limits and safety levels are generally set for individual pesticides. They do not take into account combined effects of pesticides in foods. Studies in the USA have shown that combination of three pesticides at safe levels multiplied the toxicity by hundreds of times (Abou-Donia et al. 1996). Other researchers examining dietary exposures have found that mixtures (such as insecticides, herbicides and nitrate at low concentrations that are legally permitted in groundwater) can have effects on the reproductive, immune and nervous systems that individual chemicals do not have (Boyd et al. 1990; Porter et al. 1999).

Washing fresh food products with water has little effect on these residues as many are formulated to resist being washed off easily (for example, by rain). Tests with potatoes, apples and broccoli showed that between 50% and 93% of residues remained on the produce after washing with water (Nott 1997). While peeling, topping and tailing help, many pesticides are not only found on the surface of produce, but can also be taken up by a plant and contaminate its entire flesh. Peeling or washing is thus unable to remove very much of a systemic pesticide.

10.5 Human Health Risks Associated with Pesticides

Pesticides act selectively against certain organisms. However, absolute selectivity is difficult to achieve and most pesticides are a toxic risk to humans. Pesticides are an important method in self-poisoning, mainly in developing countries (Bolognesi 2003). Every year, about three million cases of pesticide poisoning, nearly 300,000 lethal, occur world-wide (Eddleston and Phillips 2004). While data on the acute toxicity of many of these pesticides is plentiful, knowledge on their chronic effects is much more limited. The potential carcinogenicity of a wide range of insecticides, fungicides, herbicides has been reviewed by IACR (2002). Fifty-six pesticides have been classified as carcinogenic to laboratory animals. Associations with cancer have been reported in human studies for chemicals such as phenoxy acid herbicides, 2,4,5-trichlorophenoxyacetic acid (2,4,5-T), lindane, methoxychlor, toxaphene and several organophosphates (IARC 2002). Meta-analyses showed that farmers were at risk for leukaemia (Zahm and Ward 1998) and multiple myeloma (Khuder and Mutgi 1997). Exposure to pesticides has also been the subject in view of congenital malformations (Bolognesi 2003). Exposure to pesticides of female workers in flower greenhouses may have reduced fertility (Abell et al. 2000). As various agrochemical ingredients possess mutagenic properties, pesticides have also been considered potential chemical mutagens (Bolognesi 2003).

10.5.1 Human Exposure

Humans may be exposed to pesticides through direct and indirect routes. Direct exposure occurs to individuals who personally apply pesticides in agricultural, occupational, or residential settings and is likely to result in the highest levels of exposure while indirect exposures occur through air, dust, drinking water and food. Pesticides can be taken up via oral ingestion, dermal absorption, and inhalation. Indirect exposures may occur more frequently than direct pesticide application (Alavanja et al. 2004). Pesticide exposure in occupational settings is influenced by both the pesticide application characteristics and personal behavior. Unintended events also contribute to an individual's pesticide exposure. Besides acute clinical effects of pesticide poisoning, chronic effects are also well-known.

10.5.2 Clinical Effects

There is globally scarce information on the magnitude of both intentional and unintentional poisoning through pesticides. However, it is well-known that the pesticides that cause most deaths in the world, are WHO Class I and II organophosphorus pesticides (Gunnell et al. 2007; WHO 2008). The vast majority of these deaths are intentional. Unintentional oral or dermal exposure to WHO Class I organophosphorus pesticides can cause severe poisoning but the doses are commonly smaller than with intentional poisoning which results in fewer deaths. WHO Class II organophosphorus pesticides are generally less toxic in unintentional poisoning. Where Class II organophosphorus pesticides are the most commonly used insecticides, unintentional poisoning is generally less likely to cause severe poisoning. In contrast, class II organophosphorus pesticides are highly toxic in intentional overdose (WHO 2008).

Other classes of pesticide that are common causes of significant poisoning include carbamate and organochlorine insecticides, the fumigant aluminium phosphide (a significant problem in north India), and the herbicide paraquat. Less common causes of significant poisoning include the herbicides chlorphenoxy acetic acid derivatives and propanil, some pyrethroid insecticides, avermectins, and amitraz (Eddleston 2000). The locally available pesticides will also determine how many poisoned people survive to hospital presentation (WHO 2008).

The acute features of poisoning generally develop within 1–2 h of exposure and can be grouped as those related to the muscarinic, nicotinic and central nervous system (Goel and Aggarwal 2007). Muscarinic or parasympathetic features include salivation, lacrimation, urination, defaecation, gastrointestinal cramps and emesis. Nicotinic or somatic motor and sympathetic features include fasciculations, muscle cramps, fatigue, paralysis, tachycardia, hypertension and rarely mydriasis. Neurological features include headache, tremors, restlessness, ataxia, weakness, emotional lability, confusion, slurring, coma and seizures.

In areas where highly toxic fast acting organophosphorus pesticides are used, the onset of poisoning can be so fast that many people die before they can be taken to hospital. By contrast, where slower acting pesticides are used, more patients will survive to reach hospital and medical care (Eddleston et al. 2008). The case fatality for different pesticides also varies markedly, from around 70% for both aluminium phosphide and paraquat, to close to 0% for many of the newer lower toxicity pesticides (Dawson and Buckley 2007; Eddleston 2000). More than half of global deaths from pesticide poisoning occur in China (Buckley et al. 2004; Phillips et al. 2002) where the organophosphorus pesticides are currently the major problem. More recently developed pesticides are commonly safer than the older pesticides (WHO 2008).

10.5.2.1 Cyclic Organochlorine Compounds

Organochlorine pesticides include the cyclodienes, hexachlorocyclohexane isomers, and DDT and its analogues (e.g., DDE, methoxyclor, dicofol). Organochlorine compounds have been banned in many countries due to their toxicity and propensity for accumulation in various body tissues. However, in low-income countries, the black marked industry still continues the sales of persistent and highly toxic pesticides (WHO 2008). DDT and related pesticides are completely absorbed in the gastrointestinal tract and through the skin. Exposure mainly results from ingestion with food, particularly lipid-rich foods such as milk and dairy products, meat fish and eggs (Reichl and Ritter 2011). The wide-spread occurrence of these substances has caused exposure of entire populations worldwide. Organochlorine compounds are lipophilic and thus distributed in fatty tissues from that they are mobilized slowly (Fig. 10.6).

Their half-lifes in the body are commonly long (about 1 year in case of DDT). Some organochlorine metabolites (e.g. DDE from DDT; heptachloroepoxide from heptachlor) are stored in fatty tissue like their parent substances (Reichl and Ritter 2011). When fat depots are consumed due to hunger or disease, concentrations increase rapidly in all other tissues. Fat deports are also broken down due to breastfeeding so that organochlorines may appear in the breast milk which may thus serve as a bioindicator. Some organochlorine pesticide metabolites are also monitored in urine. However, they are most commonly measusred as the intact pesticide and/or its metabolite in whole blood, serum, plasma, or other lipid-rich matrices (Barr and Nedham 2002).

When taken up in larger quantities, cyclic organochlorine compounds are neurotoxic. Typical poisoning symptoms include paresthesia in the tongue, lips, face and in the skin of limbs, and dizziness. After very high doses, symptoms such as agitation, seizures, vomiting, mydriasis, and unconsciousness have been observed (Reichl and Ritter 2011). The clinical picture may be complicated by additives (e.g. solvents) in commercial pesticides. The acute toxicity of dienes ($LD_{50\ oral}$: 0.03–0.9 g kg^{-1} body weight) is higher than that of HCH (0.15–1.0 g kg^{-1}), DDT (0.1–1.0 g kg^{-1}) and HCB (1.0–10 g kg^{-1}).

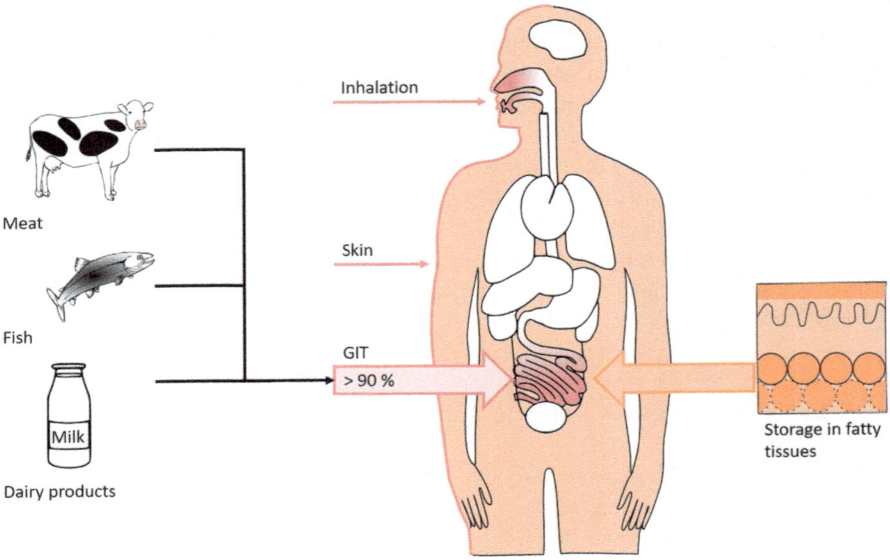

Fig. 10.6 Human exposure to organochlorines and reactions in the human body (Adapted from Reichl and Ritter 2011)

10.5.2.2 Organophosphates

This group of pesticides is comprised of a phosphate (or thio- or dithiophosphate) moiety and an organic moiety. Due to their high toxicity, the organophosphates methamidophos, methylparathion, parathion, monocrotophos and phoxim have been recently banned (Phillips et al. 2002). In most cases, the phosphate moiety is *O*,*O*-dialkyl substituted (Barr and Nedham 2002). These pesticides are potent cholinesterase inhibitors and can reversibly or irreversibly bind covalently with the serine residue in the active site of acetyl cholinesterase and prevent its natural function in catabolism of metabolites of neurotransmitters. Once in the human body, organophosphates are rapidly distributed in all organs and metabolized to the more reactive oxon form which may bind to cholinesterase. As a result, the organic part of the molecule is released. The cholinesterase-bound phosphate group may be "aged" by the loss of the *O*,*O*-dialkyl groups, or may be hydrolyzed to regenerate the active enzyme. These metabolites and hydrolysis products are then excreted in the urine. Alternatively, the intact pesticide may undergo hydrolysis prior to any conversion to the oxon form and the polar metabolites are excreted. The mechanism of action of organophosphates is shown in Fig. 10.7.

Acute toxicity is expressed through muscarinic effects (on parasympathetic nerve endings). These include nausea, anorexia, abdominal cramps, sweating, salivation and lacrimation (Reichl and Ritter 2011). Higher doses cause diarrhea, tenesmus, uncontrolled defecation and urination, pale skin, miosis, blurred vision,

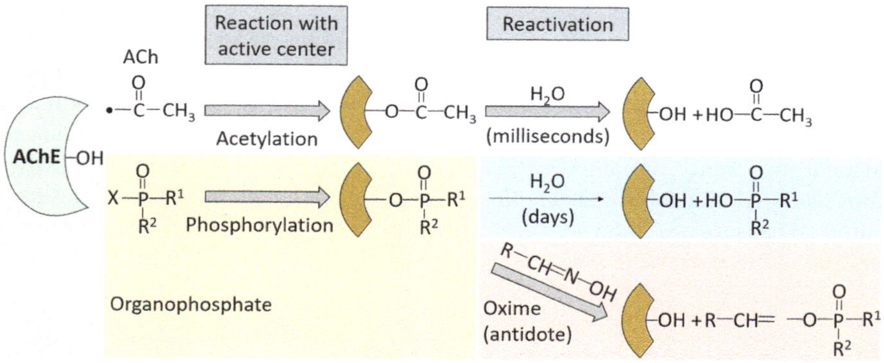

Fig. 10.7 Reactions of organophosphates in the human body (Adapted from Reichl and Ritter 2011)

pronounced bronchial secretion, asthma-like dyspnea and pulmonary edema. Next, nicotinic effects occur, including fibrillary twitching in the eylids, and the tongue, subsequently in the muscles of face, neck and eyes, and finally generalized twitches and muscle weakness. Effects on the central nervous system include malaise, restlesnenn, anxiety, headache and insomnia. More severe poisoning causes ataxia, tremor and in extreme cases coma, are flexia and cramps. The main risks are expressed through life-threatening respiratory dysfunction.

10.5.2.3 Carbamates

Aldicarb, benomyl, carbaryl, carbendazim, carbofuran, propuxur and triallate, are commonly used carbamates. These substances reversibly inhibit acetylcholinesterase and plasma pseudocholinesterase (Goel and Aggarwal 2007) and hydrolyse spontaneously from the enzymatic site within 48 h. They cause increased activity of acetylcholine at the nicotinic and muscarinic receptors during this transient period. The clinical features of carbamate ingestion are similar to those of organophosphate poisoning and the presenting symptoms include both muscarinic and nicotinic features. Central nervous system features are not very prominent in carbamate poisoning due to the poor permeability of these compounds across the blood–brain barrier. Measuring enzymatic activity to arrive at a diagnosis may be misleading due to a transient anticholinesterase effect of carbamates.

10.5.2.4 Dithiocarbamates

Dithiocarbamates are used in agriculture as fungicides, herbicides and insecticides. They are esters and salts of dithiocarbamic acid and classified as dialkyl (thiram, ziram, disulfiram) and alkylene dithio (maneb, Zineb) derivates (Reichl and Ritter

2011). After dermal uptake or ingestion, absorption is poor (~30%), particularly of alkylene dithiocarbamates. Absorbed substances are largely eliminated (mainly in the liver) within 1–3 days. Acute systemic poisoning through dithiocarbamates hardly occurs. Symptoms are expressed through local irritations of the skin and allergic contact dermatitis. Dialkyl derivates cause delayed ethanol degradation and alcohol intolerance by inhibiting alcohol dehydrogenase and acetaldehyde dehydrogenase. After alcohol consumption, people who absorbed dithiocarbamates may suffer of nausea, vomiting, an increase in respiratory rate and heart rate, and a rise or drop in blood pressure which may cause collapse of the circulation.

10.5.2.5 Chlorinated Phenoxicarbonic Acids

Chlorinated phenoxicarbonic acids are commonly known as hormonal weed killers. Besides dichloroprop, mecoprop and 2,4,5-T (trichlorophenoxyacetic acid) 2,4-D (2,4-dichlorophenoxyacetic acid) is the most commonly used agent. In Germany, since the 1980s, 2,4,5-T is no longer used because carcinogenic 2,3,7,8-TCDD is generated as a by-product during 2,4,5-T production (Reichl and Ritter 2011). Chlorinated phenoxicarbonic acids are well absorbed (>90%) after ingestion or dermal uptake. They are poorly metabolized. Because of their hydrophilic behavior, they are not stored in tissues and are rapidly (within hours or a few days) eliminated in the urine.

Acute intoxications occur after accidental or suicidal uptake of high doses. The symptoms are not characteristic. For example, ingestion of 50–60 mg kg^{-1} body weight of 2,4-D causes burning, nausea, vomiting, facial flushing and profuse sweating (Reichl and Ritter 2011). Ingestion of larger quantities causes headache, dizziness, muscle weakness, central nervous system depression, coma, rhabdomyolysis and respiratory distress. Renal injury produces oliguria and proteinuria (Broadberry et al. 2004). Extremely high doses may possibly lead to death.

10.5.2.6 Bipyridinium Compounds

Bipyridinium herbicides include paraquat and diquat which act on weeds and are inactivated upon contact with soil (Goel and Aggarwal 2007). Absorbed paraquat is sequestered in the lungs and causes release of hydrogen and superoxide anions which cause lipid damage in the cell membranes. A sixfold rise in plasma levels is observed 30 h after ingestion. Radicals (reactive oxygen species) are generated and are responsible of cell damage (Reichl and Ritter 2011).

Immediately after ingestion, patients complain of burning and ulceration of the throat, tongue and oesophagus. This is followed by toxic nephritis associated with renal dysfunction that commonly disappears after 5–10 days. The latter may be accompanied by liver disfunction (cell necrosis, icterus) and central nervous symptoms (headache, vertigo, rarely seizures and coma). After about 10 days, an acute alveolitis develops causing haemorrhagic pulmonary oedema or acute respiratory

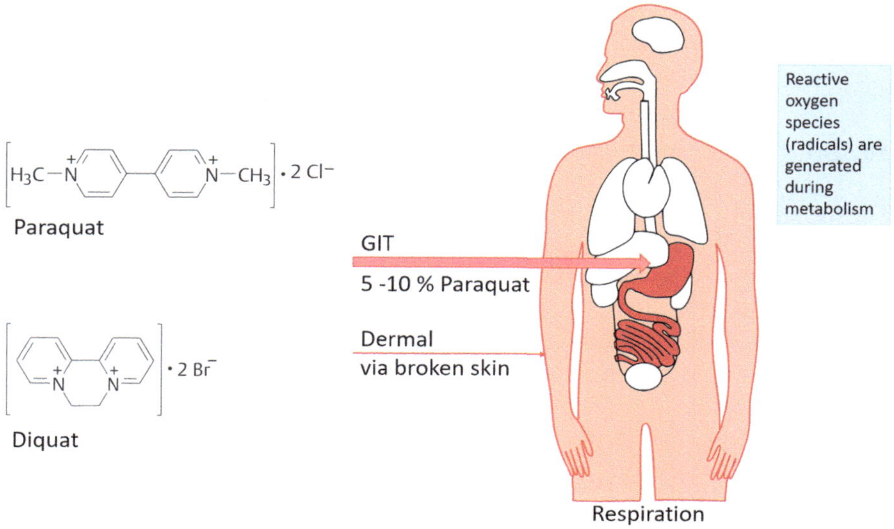

Fig. 10.8 Human exposure to bipyridinium compounds and reactions in the human body (Adapted from Reichl and Ritter 2011)

distress syndrome. Death after ingestion is due to hypoxaemia secondary to lung fibrosis. The LD_{50} for paraquat is 40–60 mg kg^{-1} body weight. The mechanism of action of bipyridinium compounds in the human body is shown in Fig. 10.8.

Systemically absorbed diquat is not selectively concentrated in the lung tissue, as is paraquat, and pulmonary injury due to diquat is less prominent. However, diquat has severe toxic effects on the central nervous system that are not typical of paraquat poisoning. These include nervousness, irritability, restlessness, combativeness, disorientation, nonsensical statements, inability to recognize friends or family members and diminished reflexes (Goel and Aggarwal 2007). Neurological effects may progress to coma accompanied by tonic–clonic seizures, and result in death. Other features include a corrosive effect on the gut leading to burning pain in the mouth, throat, chest and abdomen, intense nausea and vomiting, and diarrhoea. Renal and liver injury is common.

10.5.3 Therapy of Acute Toxicity

The treatment of pesticide poisoning is similar to other forms of poisoning, with gastric decontamination, supportive care and antidotes where available (Goel and Aggarwal 2007). Healthcare workers involved in decontamination must take adequate personal protection measures. Latex gloves give inadequate protection and rubber gloves should be used while decontaminating patients. The use of a full face mask with an organic vapor/high efficiency particulate filter has been recommended during skin decontamination (Simpson and Schuman 2002).

10.5.3.1 Cyclic Organochlorine Compounds

For therapy of organochlorine intoxication, no specific therapy is available. The general treatment includes gastric lavage and administration of charcoal and Na_2SO_4. In case of seizures, diazepam is recommended. Atripine is administered in case of bradycardia. In the diet, alcohol, fatty and oily substances need to be avoided because these substances enhance absorption (Reichl and Ritter 2011).

10.5.3.2 Organophosphates

Maintaining respiration (intubation, artificial ventilation; no mouth-to-mouth ventilation) and circulation are the first measures to be taken in case of organophosphate intoxication (Reichl and Ritter 2011). Respiratory failure is an important cause of death following the ingestion of organophosphorus poisoning and the complications of aspiration (Eddleston et al. 2006). Much aspiration results from poor initial care of the patient and/or unsafe gastric decontamination. Comatose patients should be intubated prior to gastric lavage to reduce the risk of aspiration.

Atropine is the most important antidote for pesticide poisoning, being effective in both organophosphorus and carbamate poisoning (Freeman and Epstein 1955). Atropine should be given in high doses (2–5 mg IV every 10 min) to antagonize the effects of acetylcholine (Reichl and Ritter 2011). The effectiveness of a second antidote for organophosphorus poisoning, an oxime for reactivation of cholinesterase intoxications, also varies markedly according to the organophosphorus and the ingested dose. Current recommendations are to give oximes to all organophosphorus poisoned patients requiring atropine (Johnson et al. 2000). However, many patients do not seem to benefit (Eddleston et al. 2005).

In case of ingestion, gastric lavage with an orogastric tube may be useful using 200–300 ml of tap water (5 ml kg^{-1} of normal saline in young children). Larger quantities of saline should be avoided since they push the gastric contents into the intestine, or may induce vomiting leading to aspiration. Charcoal is beneficial when given within 60 min of ingestion of the poison (WHO 2008). In case of absorption through the cutaneous route, skin decontamination is important and is done by washing the skin with large volumes of soap and water (Goel and Aggarwal 2007) or 5–10% $NaHCO_3$ solution (Reichl and Ritter 2011). Skin folds, areas under the fingernails, axillae and groins as well as other areas of the body that trap and retain chemicals should be carefully washed.

10.5.3.3 Carbamates

The treatment of carbamate intoxication is similar to that with organophosphate poisoning. The therapy is mainly supportive and includes high doses of atropine (Reichl and Ritter 2011). In case of carbamates, oximes should not be applied as they increase carbamate activity and increase damage due to its own toxicity.

10.5.3.4 Dithiocarbamates

The treatment of poisonings associated with dithiocarbamates is symptomatic.

10.5.3.5 Chlorinated Phenoxicarbonic Acids

The treatment is symptomatic. Urinary alkalinization enhances the elimination of 2,4-D. Haemodialysis may be considered (Goel and Aggarwal 2007).

10.5.3.6 Bipyridinium Compounds

Because of the high mortality, the treatment includes primary detoxification with gastric and intestinal lavage, activated charcoal, saline laxatives and secondary detoxification with forced diuresis, hemodialysis or hemoperfusion with application of charcoal (Reichl and Ritter 2011). Lowering the pH of urine via arginine-HCl infusion is helpful for elimination of the toxic substance by inhibiting renal reabsorption, as paraquat and diquat are strong bases. High partial pressure of oxygen is contraindicated early in the poisoning because of progressive oxygen toxicity to the lung tissue (increased fibrosis due to formation of peroxides). There may be therefore some advantage in placing the patient in a moderately hypoxic environment (e.g. 15–16% O_2; Goel and Aggarwal 2007). Renal insufficiency and hepatolysis are to be treated as required.

10.5.4 Chronic Toxicity Related to Long-Term Contact to Pesticides

Long-term contact to pesticides can harm human life and can disturb the function of different organs in the body, including nervous, endocrine, immune, reproductive, renal, cardiovascular, and respiratory systems. In this regard, there is mounting evidence on the link of pesticide's exposure with the incidence of human chronic diseases, including cancer, Alzheimer, Parkinson, multiple sclerosis, diabetes, aging, cardiovascular and chronic kidney disease (Mostafalou and Abdollahi

2013). The treatment of chronic diseases caused by contact to pesticides as described in the following sections is identical to the treatment of chronic diseases with other causes.

10.5.4.1 Cancer

Based on rising evidence by health studies associated with exposure to pesticides, different kinds of cancer have been reported, including cancers of breast, brain, lung, prostate, colorectum, testics, pancreas, esophagus, stomach, skin, and non-Hodgkin lymphoma (e.g. Alavanja and Bonner 2012; Weichenthal et al. 2010). For Hodgkin's disease meta-analyses showed a small but significant excess risk among farmers. This may be a result of exposure to infectious microorganisms or pesticides used on the farm (Alavanja et al. 2004). In a number of agricultural health studies, Lee et al. (2004a, b; 2007) found an association between exposure to pesticides and cancer incidence, particularly lymphohematopoietic cancers for alachlor, lung cancer for chlorpyrifos, and colorectal cancer for aldicarb. Chronic low-dose exposure to pesticides has been considered as one of the important risk factors for cancer expansion (Mostafalou and Abdollahi 2013). Pesticide formulations have also included carcinogenic solvents (Alavanja et al. 2004). Carcinogenicity tests are therefore required to detect carcinogenic potential of pesticides before allowing them to be marketed.

Factors that influence carcinogenic properties of pesticides include age, sex, individual susceptibility, amount and duration of exposure, and potential exposure to other cancer causing chemicals causing cancer. Carcinogenic mechanisms of pesticides can be examined in their potential to affect genetic material directly via structural or functional damage to chromosomes.

DNA, and Histone proteins, or indirectly disrupting the profile of gene expression through impairment of cellular organelles like mitochondria and endoplasmic reticulum, nuclear receptors, endocrine network, and other factors involved in maintenance of cell homeostasis (George and Shukla 2011; Rakitsky et al. 2000). An overview of the relation between exposure to individual pesticides and elevated incidence of cancer is given in Table 10.13. The widespread use of glyphosate (not listed in Table 10.13) has become increasingly controversial as studies have produced mixed results on the risk of causing cancer in humans.

10.5.4.2 Neurologic Diseases and Depression

Numerous case control studies have observed that pesticide exposure is associated with increased occurrence of Parkinson's disease (Alavanja et al. 2004) which is a motor progressive disorder of the central nerv characterized by degeneration of dopaminergic neurons in the substantia nigra. Only a few studies of pesticide exposure and Parkinson's disease risk have been able to implicate specific pesticides. Several studies found increased risk associated with exposure to either insecticides or herbicides (Gorell et al. 1998), and one study indicated that risk was elevated by exposure to organochlorines, organophosphates, or carbamates

Table 10.13 Pesticides associated with different kinds of cancer

Pesticides	Kind of cancer
Chlorpyrifos, diazinon, dicamba, dieldrin metolachlor, pendimethalin	Lung cancer
Heptachlor, chlorpyrifos, diazinon, fonofos, EPTC	Leukemia
Fonofos, methylbromide, butylate, chlordecone, butylate, DDT, lindane, simazine	Prostate cancer
EPTC, pendimethalin, DDT	Pancreatic cancer
Aldicarb, dicamba, EPTC, imazethapyr, trifluralin	Colon cancer
Chlordane, chlorpyrifos	Rectum cancer
Chlorpyrifos	Brain cancer
Imazethapyr	Bladder cancer
Carbaryl, toxaphene, parathion, maneb	Melanoma
Lindane, oxychlordane	Non-Hodgkin's lymphoma
Permithrin	Multiple myeloma

Adapted from Mostafalou and Abdollahi (2013)

(Seidler et al. 1996). Several case reports have described Parkinson's disease in individuals exposed to organophosphates (Bhatt et al. 1999), glyphosate (Blair et al. 1990), paraquat (Sanchez-Ramon et al. 1987), diquat (Sechi et al. 1992), maneb (Meco et al. 1994), and other ethylene bis-dithiocarbamates (Hoogenraad 1988). Higher concentrations of organochlorine residues, particularly dieldrin, have been found in postmortem brains of Parkinson's disease patients compared to patients with other neurological diseases (Corrigan et al. 2000). Several studies have suggested that risk of amyotrophic lateral sclerosis (ALS) is related to farming as an occupation although not necessarily to living in rural areas (Nelson 1995). Occupational pesticide exposure was also associated with Alzheimer's disease, vascular dementia and with risk of dementia among Parkinson's disease patients (Alavanja et al. 2004). High-intensity and cumulative pesticide exposure was also found to contribute to depression among pesticide applicators (Beseler et al. 2008).

10.5.4.3 Reproductive Disorders

Reproductive disorders are defined as conditions prejudicing the capacity of the reproductive system to reproduce (Mostafalou and Abdollahi 2013). Decreased fertility in both sex, demasculinization (antiandrogenic effects), elevated rate of miscarriage, altered sex ratio, and change in the pattern of maturity are among the most reported reproductive dysfunctions induced by chronic exposure to pesticides (Frazier 2007). These effects of pesticides deemed more important when their link to endocrinal disruption was explained. A number of pesticides, mostly the old organochlorine types like aldrin, chlordane, DDT, dieldrin, and endosulfan, the herbicide atrazine, and the fungicide vinclozolin have been identified as commonly believed endocrine disrupting chemicals (PAN 2009). Interfering with functions of the endocrine system has been implicated in most pesticides that caused reproductive toxicities (Tiemann 2008).

Birth defects, fecundability, fertility, altered fetal growth, fetal death, and mixed outcomes are the most important groups of reproductive outcomes. Specific birth defects include limb reductions, urogenital anomalies, central nervous system defects, orofacial clefts, heart defects and eye anomalies (Sanborn et al. 2007).

10.5.4.4 Genotoxicity

Genotoxicity is the ability of a pesticide to cause intracellular genetic damage resulting in DNA damage or chromosomal aberrations. Genetic damages are considered as a primary mechanism for chronic diseases within the context of carcinogenesis and teratogenesis (Mostafalou and Abdollahi 2013). They are studied in the field of genetic toxicology and can be detected by distinctive kinds of genotoxicity tests. Growing body of data concerning genetic toxicity of pesticides have been collected from epidemiological and experimental studies using different types of examinations, including chromosomal aberrations, micronucleus, sister chromatid exchanges and comet assay (Bolognesi 2003).

Genetic damages are expressed through premutagenic damages like DNA strand breaks, DNA adducts or unscheduled DNA synthesis, gene's mutation and chromosomal aberrations, including loss or gain of whole chromosome (aneuploidy), deletion or breaks (clastogenicity), and chromosomal segments or rearrangements. Premutagenic damages may be repaired prior to cell division while the damages through gene mutation and chromosomal are permanent and have the ability of transmission to daughter cells after cell division (Guy 2005).

10.5.4.5 Diabetes

Diabetes can be said that has become epidemic since 347 million people worldwide are appraised to be diabetic and diabetes deaths are expected to double between 2005 and 2030 (Mostafalou and Abdollahi 2013). Epidemiological studies published during the past few years indicated that exposure to pesticides can be a potential risk factor for developing diabetes (Everett and Matheson 2010; Montgomery et al. 2008). It has also been suggested that exposure to some pesticides can be a promoter for other risk factors of diabetes like obesity by distressing neural circuits that regulate feeding behavior or altering differentiation of adipocytes (Thayer et al. 2012).

10.5.4.6 Other Chronic Diseases

Other chronic diseases caused by pesticide exposure may include cardiovascular diseases, chronic nephorpathies, chronic respiratory disease and autoimmune diseases (Mostafalou and Abdollahi 2013). It was reported that chronic exposure to organophosphate pesticides may increase the risk of coronary artery disease presumably through diminished paraoxonase activity (Zamzila et al. 2011) and there have been a few evidences on the link between exposure to pesticides and atherosclerosis (e.g. Antov and Aianova 1980). Exposure to acetylcholinesterase inhibiting pesticides was associated with chronic renal failure (Peiris-John et al. 2006) and higher level of organochlorine pesticides was detected in chronic kidney disease patients along with a reduced glomerular filtration and increased oxidative stress (Siddharth et al. 2012). A significant relationship between chronic renal failure and environmental factors was found in farming areas in North Central Province of Sri Lanka (Wanigasuriya et al. 2007). Several reports have shown an increased rate of asthma in people occupationally exposed to pesticides (Hernandez et al. 2011). Exposure to pesticides may also cause autoimmune diseases like systemic lupus erythematous and rheumatoid arthritis (Parks et al. 2011).

10.6 Technical Measures for Reducing Pesticide Dispersal in the Environment

Technical measures aim at limiting the dispersion of pesticides in the environment and reduce pesticide use. Pesticide transfer depends on the properties of compounds, soil characteristics, climatic conditions and farming practices.

10.6.1 Storage, Mixing, Loading and Application of Pesticides

The first step in the chain of events involved with pesticide use should be checking storage facilities should. As containers are frequently opened in the storage area, a spill of a concentrated formulation can create serious soil and water pollution. Storage facilities should therefore have a concrete floor so that spilled concentrate can be easily cleaned up and disposed. Mixing and loading sites are areas where pesticides can be spilled on the ground. Repeated spills increase the concentration of the pesticide in the soil and increase the possibility of materials leaching through the soil to groundwater. Farmers who apply a lot of pesticide should construct a pit lined with clay or preferably concrete and filled with rock and soil. Mixing and loading can be carried out over the pit so that any spill is contained and the active ingredient is broken down without the possibility of leaching to groundwater. A reduction in losses during application, or immediately afterwards, can be achieved by improving the properties of commercial crop protection products and their conditions of application. Addition of adjuvants to active substances may have negative or contradictory effects on the risks of losses. For example, increasing the adhesion and wettability of foliage treatment products can reduce runoff from the leaves but may favour volatilization. However, it is possible to adjust inert substances or adjuvants to improve the properties of preparations, for example, by optimizing the size and density of drops to minimize losses due to droplet volatilization, increasing the rate of leaf penetration and increasing the resistance to abrasion of granules and seed coatings (Aubertot et al. 2005).

Pesticide applications should be prohibited during very windy conditions, very dry conditions, at extreme temperatures or when rain is forecast. For compounds which are highly volatile after application, it may also be possible to issue recommendations concerning the best time of day for application in order to limit the importance of losses, once the mechanisms involved in this process are fully understood. Periods of application must also take account of the type and state of the soil. Certain mobile substances can be transferred rapidly by surface runoff or by preferential flow in biopores or cracks of fissured clay soils. Compliance should be ensured with recommendations concerning incorporation of the substance into the soil, which is effective in reducing the volatilization of some compounds (Aubertot et al. 2005).

10.6.2 Reducing Flows of Applied Pesticides

Vegetated buffer strips along field edges and water bodies are a measure to reduce pesticide inputs into surface waters via runoff and erosion (Popov et al. 2005). Grassed waterways as well can substantially reduce pesticide runoff and erosion inputs (Asmussen et al. 1977). Other possibilities for mitigating pesticide runoff and erosion inputs into surface waters are common measures to reduce erosion from the field, such as conservation tillage including zero-tillage, mulching, cover crops, contour ploughing and planting. Also specific measures taken in vineyards to control erosion (e.g. vegetation between vine rows) belong to this category. There are also mitigation options concerning pesticide application, such as band spraying on row crops or, if feasible, reduction of the application rate (Reichenberger et al. 2007). Application as granules and incorporation of the pesticide into the soil are potential mitigation measures as well. The time passing between pesticide application and occurrence of a runoff event is also critical for runoff losses (Burgoa and Wauchope 1995). Hence, avoiding application in seasons with a high probability of occurrence of runoff events (due to high-intensity rainstorms or saturated soils) would be an effective mitigation measure. Irrigation is often inefficient and may thus enhance leaching of pesticides. Most irrigation water is not taken up by agricultural crops and thus can be considered lost commonly via drainage. A high percentage of irrigation water could be saved by using trickle or drip irrigation systems that send water directly above the individual plant's root system. This strategy helps keeping pesticides on the land, rather than leaching them to the groundwater.

Mitigation measures for reducing pesticide transport via leaching include application restrictions for vulnerable soils and/or wet climates, reducing the application rate, and shifting the application to an earlier or later date. To reduce pesticide leaching through the soil matrix, a possible mitigation measure is increasing the soil organic matter content by agronomic practices like incorporation of crop residues and manures, in order to increase sorption of nonionic pesticides (Reichenberger et al. 2007). Another option to reduce leaching by matrix flow would be switching to compounds with higher sorption and/or faster degradation (Flury 1996).

Mitigation measures for spray drift include the use of spray additives, drift-reducing nozzles, and windbreaks (Reichenberger et al. 2007). These measures should also have a side effect on volatilization and atmospheric deposition. Incorporating the pesticide into the soil to minimize volatilization is another mitigation measure. Pesticide input by wind erosion into surface waters can be mitigated by common measures for wind erosion control (e.g. windbreak hedges and ground cover).

10.7 Remediation of Pesticide-Contaminated Soils

Pesticide-contaminated soils pose potential risks to surface and groundwater quality, especially when concentrations are high due to discharges, spills or leaking storage tanks. Although the organic contaminants are generally biodegradable, high concentrations can inhibit microbial activity, resulting in slower degradation and enhanced mobility (Shea et al. 2004). For remediation of pesticide-polluted sites, physical, chemical and biological technologies are available.

10.7.1 Physical and Chemical Remediation Technologies

Conventional remediation techniques use physical and chemical processes that aim at eliminating, reducing, isolating or stabilizing a pesticide or multiple pesticides. The technology may either be applied at the place of the contamination (in situ) or the contaminated soil may be may be removed for ex situ treatment. Common in situ and ex situ technologies are listed in Table 10.14.

As each of these technologies has limitations and disadvantages, site-specific evaluations and decisions need to be made. Where possible, pesticide-contaminated soil is spread on agricultural land at labeled rates (Paulson 1998). Land spreading can be cost-effective if pesticide concentrations are low and suitable agricultural land is available, but may not be possible when the soil contains multiple pesticides or chemicals that are no longer approved for application. If multiple pesticides are involved, it may be useful to apply a combination of technologies to reduce pesticide concentrations to acceptable levels (Gavrilescu 2005). However, the majority of the above measures are extremely costly. Moreover, the deposition

Table 10.14 Common physical and chemical remediation technologies for treatment of pesticide-contaminated sites

Technology	In situ	Ex situ
Physical	Soil vapor extraction Containment systems and barriers	Land spreading Land filling Soil vapor extraction Thermal desorption Incineration
Chemical	Soil flushing Solidification Stabilization	Soil washing Solidification Stabilization Dehalogenation Solvent extraction Chemical reduction and oxidation

Adapted from Gavrilescu (2005)

(land spreading or land filling) or containment of contaminated soil may continue posing environmental and human health risks. Alternative low-cost and easy to implement treatments are therefore needed that will accelerate the degradation and natural attenuation of pesticides from multiple chemical classes.

A relatively new and effective chemical remediation technology involves treatment of soil with zerovalent iron (Fe^0) which promotes reductive dechlorination and nitro group reduction of a wide range of contaminants in soil and water (Comfort et al. 2001; Dombek et al. 2001; Shea et al. 2004). The rationale for using Fe^0 to treat pesticide contaminated soil is based on the fact that many pesticides contain moieties that can be reduced when coupled to the oxidation of iron metal, with Fe^0 and Fe(II) serving as reductants. For example, Fe^0 can enhance natural detoxification processes in soil, as demonstrated by increased mineralization of metolachlor [2-chloro-N-(2-ethyl-6-methylphenyl)-N-(2-methoxy-1-methylethyl)acetamide] (Comfort et al. 2001) and atrazine [6-chloro-N-ethyl-N0-(1-methylethyl)-1,3,5-triazine-2,4-diamine] (Singh et al. 1998) after treatment. Shea et al. (2004) employed Fe^0 for on-site treatment of soil containing >1000 mg metolachlor, >55 mg alachlor, >64 mg atrazine, >35 mg pendimethalin, and >10 mg chlorpyrifos kg^{-1}. Treatment with 5% (w/w) Fe^0 resulted in >60% destruction of the five pesticides within 90 days. Comfort et al. (2001) and Shea et al. (2004) demonstrated that the combined use of Fe^0 and $Al_2(SO_4)_3$ still enhanced degradation of several pesticides.

10.7.2 Bioremediation

To treat pesticide-contaminated sites, a number of biological remediation technologies have been developed. Compared to physical and chemical technologies, most

Table 10.15 Treatable media, treatment duration, cost ranges and contaminant removal efficiencies of conventional and bioremediation technologies for treatment of pesticide-contaminated sites

Technology	Treatable media	Treatment duration	Costs in US \$ m^{-3}	Removal efficiencies
Incineration	Soil, sludge, sediment	1 month	230–770	>99%
Thermal desorption	Soil, sludge, sediment	3 weeks	70–300	82–98%
Bioremediation	Soil, sludge, sediment, groundwater	3 months	6.4–150	>99%

Adapted from Gavrilescu (2005)

biological techniques are equally efficient but much less costly (particularly in situ) (Table 10.15). However, biological remediation procedures require more time.

Biological remediation procedures increase the rate of microbial decomposition of pesticides. The basic techniques which can be used include (i) stimulation of the activity of the indigenous microbial population by the addition of nutrients, regulation of redox conditions, optimizing pH conditions, (ii) inoculation of the sites with microbes of specific biodegradation abilities, (iii) application of immobilized enzymes, and iv) use of plants (phytoremediation) to remove, contain, or transform pollutants (Navarro et al. 2007).

In situ bioremediation techniques are commonly the most desirable options due to lower costs and limited soil disturbance since they enable the treatment in place. However, in situ treatment is limited by the depth of the soil that can be effectively treated. In many soils effective diffusion of O_2 for desirable rates of biodegradation range from several centimeters to about 30 cm, although in some soils, depths of more than 60 cm have been effectively remediated (Gavrilescu 2005). Ex situ bioremediation technologies include slurry phase biological treatment and composting. The first involves large tanks or bioreactor vessels in which contaminated soil is mixed with water, nutrients and other additives to form an aqueous slurry. Composting is an aerobic biological process by that organic materials are transformed into a stable, soil-like product. In order to achieve maximum efficiency, some conditions such as oxygen concentration, pH, moisture content, C/N ratio and particle size need to be optimized (Miller 1993). Composting can be performed using windrows, aerated static piles or composting vessels. The contained systems allow the treatment to be completed within less time compared to the windrow or aerated piles. However, the short remediation time when using composting vessels is offset by high initial costs of the system (Gavrilescu 2005).

10.8 Alternatives for Minimizing Pesticide Use

Alternatives for minimizing pesticide use include use of natural pesticides, biological pest controls (such as pheromones and microbial pesticides), methods of cultivation and use of plant genetic engineering. Release of other organisms that fight the pest is another example of an alternative to pesticide use. These organisms can include natural predators or parasites of the pests. Biological pesticides based on entomopathogenic fungi, bacteria and viruses cause disease in the pest species can also be used. Cultivation practices include polyculture (growing multiple types of plants), crop rotation, planting crops in areas where the pests that damage them do not live, timing planting according to when pests will be least problematic, and use of trap crops that attract pests away from the real crop. Another alternative to pesticides is the thermal treatment of soil through steam. Soil steaming kills pest and increases soil health. Some evidence shows that alternatives to pesticides can be equally effective as the use of chemicals. Major agricultural production systems that avoid or minimize synthetic pesticide use include organic farming and integrated pest management. Plant genetic engineering in some cases may also contribute to reduced pesticide use.

10.8.1 Organic Farming

Public concern about food quality has strongly increased in recent years. A series of food scares (e.g. Knowles and Moody 2007) and the controversy surrounding genetically modified crops have prompted heated debate about the safety and integrity of our food. Against this background, demand for organically produced food has been growing rapidly. Organic farming is a globally expanding economic sector and contributes to human health and prosperity, improves soil and water quality, protects birds and bees, and mitigates damage from global climate change. In organic farming, disease and weed control are managed with practices such as diversified crop rotation, weeding, mulching, intercropping, use of organic fertilizers, maintenance of natural predator populations and biological controls. In fields where crops are rotated regularly, pests, including weeds, insects, and pathogens, cannot adapt themselves to a single set of environmental conditions and therefore do not multiply as quickly. Predators in one crop are kept down by predators in the other. Intercropping (sometimes with trees and annual crops sharing the same field) still enhances these effects. Cultivation of legume cover crops builds soil organic matter, even when routine tillage is used for weed control. Increased soil organic matter enhances the soil's ability to increase carbon, nutrient and water storage capacities (Drinkwater and Snapp 2007). High soil organic matter levels also contribute to greater resilience under stresses including drought and flooding and produce crops with a greater ability to resist pests and diseases (Phelan et al. 1995; Liu et al. 2007). Table 10.16 summarizes key organic farming practices and their benefits.

While the use of chemical pesticides is severely restricted in organic farming, a small number of natural pesticides can be used. They tend to represent the final option for pest control when other methods have failed or are known to be ineffective. Organic farmers are allowed to use of nonsynthetic pesticides that have been approved on the basis of their origin, environmental impact and potential

Table 10.16 Organic farming practices and their benefits

Farming practice	Benefits
Diversified crop rotations	Increased soil health, carbon and nitrogen sequestration, disruption of insect, weed and disease life cycles
Application of manures, compost, green manures	Increased soil health, enhanced productivity, carbon and nitrogen sequestration
Use of cover crops	Increased soil health, reduction of soil erosion by wind and water, carbon and nitrogen sequestration
Introduction of habitat corridors, borders, and insectaries	Increased biodiversity, support of biological pest control
No use of synthetic fertilizers	Reduced surface and groundwater contamination, mitigation of salinization, reduction of costs
No use of synthetic pesticides	Increased soil health and biodiversity, improved water quality, prevention of pollinator disruption, reduction of costs

Table 10.17 Substances approved for pest control in organic farming in the EU

Substance origin	Approved substance	Approval pending
Plant origin	Hydrolized proteins	Azadirachtin
	Mint oil	Quassia
	Pyrethrins	
Microbial origin	Spinosad	
	Bacillus thuringensis	
Hormones	Pheromones	
Synthetic pyrethroid	Lambdacyhalothrin	
Gas	Ethylene	
Inorganic substances	Calcium hydroxide	
	Copper	
	Ferric phosphate	
	Potassium hydrogen carbonate	
	Sulphur	

Adapted from Hillocks (2012)

to persist as residues. Substances approved for pest control in organic farming in the EU are summarized in Table 10.17.

Microbial agents such as *Bacillus thuringiensis* are also permitted. For neem oil (azadirachtin) and quassia extract the approval is still pending. Substances used in organic farming are simpler than those used in conventional agriculture, tend to degrade quicker, and are typically only used on a non-routine basis following authorisation from the certifying body for a specific reason. However, copper is planned to be phased out of Europe-wide organic standards because of concerns about possible detrimental effects on earthworms and other soil life. Removal of copper would have a major impact on disease management in organic apple, grapes and potato production (Hillocks 2012).

Organically grown food is usually found to have no residues of synthetic pesticides. When residues are present, they are commonly of lower incidence and lower levels than residues in non-organically produced food (Reinhardt and Wolf 1986). Pesticide residues were found much less often in organically produced vegetables than typical non-organic levels (Woese et al. 1997). In contrast, tests on all fruits (organically and non-organically produced) in the UK in 1999 have shown that 48% contained detectable pesticide residues, as did 28.6% of all foods (including cereals, meat, dairy, fish and processed foods) tested (MAFF 2000).

10.8.2 Integrated Pest Management

Integrated pest management is an alternative pesticide control system that combines different crop protection practices with careful monitoring of pests and their natural enemies. In the EU, integrated pest management is defined through Directive 91/414/EEC "The rational application of a combination of biological, biotechnical, chemical, cultural or plant-breeding measures, where the use of plant protection products is limited to the strict minimum necessary to maintain the pest population at levels below those causing economically unacceptable damage or loss" (Hillocks 2012). The idea behind integrated pest management is that combining different practices together overcomes the shortcomings of individual practices (Table 10.18). The aim is not to eradicate pest populations but rather to manage them below levels that cause economic damage.

For outdoor crops, integrated pest management focuses mainly on targeted pesticide use, choice of cultivar and crop rotations, and mechanical weeding while biological control plays a central role in the production of many greenhouse crops. Some highly effective integrated pest management programmes are now in

Table 10.18 Measures proposed for integrated pest management

Category of practice	Individual measures
Cultivation	Crop rotation, intercropping, undersowing
	Mechanical weeding
Crop selection	Crops with partial or total pest resistance
Pesticide selection	Use of low-risk compounds
	Pesticides with high levels of selectivity
Alternative plant protection measures	Natural products (e.g. plant extracts)
	Biological control with natural enemies (e.g. predatory insects and mites, parasitoids, parasites, microbial pathogens, microbial antagonists of plant pathogens, microbial pathogens of weeds)
Decision support	No use of routine calendar spraying
	Calculation of economic action thresholds
	Optimized timing of pest activity
	Basic pest scouting

Adapted from Chandler et al. (2011)

place, based around the biocontrol of insect and mite pests using combinations of predators, parasitoids, parasitic nematodes and entomopathogens (Chandler et al. 2011). Short-persistence pesticides are used on an at-need basis if they are compatible with biological control. Pest management strategies are also determined through a close interaction between growers, consultants, biocontrol companies and retailers. Biological control requires considerable grower knowledge, but it has clear benefits in terms of reliable pest control, lack of phytotoxicity, a short harvest interval and better crop quality.

10.8.3 Plant Genetic Engineering

Introduction of genetically altered plants is another possible way to reduce pesticide use. Genetically modified crops with pest control attributes are widely grown throughout the world, particularly in the USA, China, Brazil, India and Australia (Hillocks 2012). The major crops involved are cotton, maize and soybeans which have been transformed to express genes for herbicide resistance or resistance to Lepidopteran insect pests or both traits. Herbicide resistant genetically modified crops were modified to resist mainly glyphosate, allowing the herbicide to be sprayed over the crop. Insect resistant crops are modified to produce insecticidal toxins from the bacterium *Bacillus thuringiensis* and are known as Bt crops. Insect resistant maize is the only genetically modified food crop approved for cultivation in the EU (Hillocks 2012). A number of European countries, Spain in particular, have been growing genetically modified maize for more than a decade. A genetically modified potato variety was approved in 2010 but only to produce industrial starch.

There are several groups opposed to genetic engineering due to several reasons. For example, it is widely feared that cultivation of genetically modified crops leads to increased pesticide use, based on increased use of herbicides on herbicide resistant crop varieties. Other criticisms include consequences such as locking farmers into new uses for chemical manufacturers' products. In fact, it is still not known whether the ultimate use of biotechnology and whether the benefits will go to chemical companies' sales rather than to reduce chemical use in the environment.

References

Abell A, Juul S, Bonde JP (2000) Time to pregnancy among female greenhouse workers. Scand J Work Environ Health 26:131–136

Abou-Donia MB, Wilmarth KR, Jensen KF, Oehme FW, Kurt TL (1996) Neurotoxicity resulting from coexposure to pyridostigmine bromide, DEET and permethrin: implications for gulf war chemical exposures. J Toxicol Environ Health 48:35–65

Agnihotri NP, Gajbhiye KM, Mohapatra SP (1994) Organochlorine insecticide residue in Ganga river water near Farrukhabad, India. Environ Monit Assess 30:105–112

Aherns M (2008) Review of organic chemicals of potential environmental concern in use in Auckland. Prepared by NIWA for Auckland Regional Council. Auckland Regional Council Technical Report 2008/028

Aktar MW, Sengupta D, Ashim Chowdhuri A (2009) Impact of pesticides use in agriculture: their benefits and hazards. Interdiscip Toxicol 2(1):1–12

Alavanja MCR, Bonner MR (2012) Occupational pesticide exposures and cancer risk: a review. J Toxicol Environ Health B Crit Rev 15(4):238–263

Alavanja MCR, Hoppin JA, Kamel F (2004) Health effects of chronic pesticide exposure: cancer and neurotoxicity. Annu Rev Public Health 25:155–197

Alexander M (1994) Biodegradation and bioremediation. Academic, NY

Allan ES (1989) Degradation, fate and persistence of phenoxyalkanoic acid herbicides in soil. Rev Weed Sci 4:1–24

Ambrosi D, Isensee A, Macchia J (1978) Distribution of oxadiazon and phoslone in an aquatic model ecosystem. Am Chem Soc 26(1):50–53

Amorim MJB, Rombke J, Soares AMVM (2005) Avoidance behavior of Enchytraeus albidus: effects of benomyl, carbendazim, phenmedipham and different soil types. Chemosphere 59:501–510

Andreu V, Picó Y (2004) Determination of pesticides and their degradation products in soil: critical review and comparison of methods. Trends Anal Chem 23(10–11):772–789

Antov G, Aianova A (1980) Effect of the pesticide, fundazol, on the myocardium of rats with experimental atherosclerosis. Probl Khig 5:58–67

Asmussen LE, White AW Jr, Hauser EW, Sheridan JM (1977) Reduction of 2,4-D load in surface runoff down a grassed waterway. J Environ Qual 6:159–162

Aubertot JN, Barbier JM, Carpentier A, Gril JJ, Guichard L, Lucas P, Savary S, Savini I, Voltz M (2005) Pesticides, agriculture et environnement. Réduire l'utilisation des pesticides et limiter leurs impacts environnementaux. Expertise scientifique collective, synthèse du rapport, INRA et Cemagref (France), 64 pp

Baker BP, Benbrook CM, Groth E, Benbrook KL (2002) Pesticide residues in conventional, integrated pest management (IPM) – grown and organic foods : insights from three US data sets. Food Addit Contam 19:427–446

Bálint T, Ferenczy J, Kátai F, Kiss I, Kráczer L, Kufcsák O, Láng G, Polyhos C, Szabó I, Szegletes T, Nemcsók J (1997) Similarities and differences between the massive eel (*Anguilla anguilla* L) devastations that occurred in Lake Balaton in 1991 and 1995. Ecotoxicol Environ Saf 37:17–23

Barr DB, Nedham LL (2002) Analytical methods for biological monitoring of exposure to pesticides: a review. J Chromatogr B 778:5–29

Barriuso E, Laird DA, Koskinen WC, Dowdy RH (1994) Atrazine desorption from smectites. Soil Sci Soc Am J 58(6):1632–1638

Bedos C, Cellier P, Calvet R, Barriuso E (2002) Occurrence of pesticides in the atmosphere in France. Agronomie 22:35–49

Benoit P, Madrigal I, Preston CM, Chenu C, Barriuso E (2008) Sorption and desorption of non-ionic herbicides onto particulate organic matter from surface soils under different land uses. Eur J Soil Sci 59(2):178–189

Berny PJ, Buronfosse T, Buronfosse F, Lamarque F, Lorgue G (1997) Field evidence of secondary poisoning of foxes (Vulpes vulpes) and buzzards (Buteo buteo) by bromadiolone, a 4-year survey. Chemosphere 35(8):1817–1829

Beseler C, Stallones L, Hoppin J, Alavanja M, Blair A, Keefe T, Kamel F (2008) Depression and pesticide exposures among private pesticide applicators enrolled in the agricultural health study. Environ Health Perspect 116(12):1713–1719

Bhatt MH, Elias MA, Mankodi AK (1999) Acute and reversible parkinsonism due to organophosphate pesticide intoxication: five cases. Neurology 52:1467–1471

Bishop CA, Ng P, Mineau P, Quinn JS, Struger J (2000) Effects of pesticide spraying on chick growth, behavior, and parental care in tree swallows (Tachycineta bicolor) nesting in an apple orchard in Ontario, Canada. Environ Toxicol Chem 19(9):2286–2297

Blair A, Axelson O, Franklin C (1990) Carcinogenic effects of pesticides. In: Baker SR, Wilkinson SF (eds) The effects of pesticides on human health. Princeton Sci., Princeton. Adv Mod EnvironToxicol XVIII:201–260

Bollag JM, Liu SY (1990) Biological transformation processes of pesticides. In: Cheng HH (ed) Pesticides in the soil environment: processes, impacts and modelling. SSSA, Madison, pp 169–211

Bollag J, Myers C (1992) Detoxification of aquatic and terrestrial sites through binding of pollutants to humic substances. Sci Total Environ 117(118):357–366

Bollag J, Myers CJ, Minard RD (1992) Biological and chemical interactions of pesticides with soils. Sci Total Environ 123(124):205–217

Bolognesi C (2003) Genotoxicity of pesticides: a review of human biomonitoring studies. Mutat Res 543:251–272

Borch T, Kretzschmar R, Kappler A, van Cappellen P, Ginder-Vogel M, Voegelin A, Campbell K (2010) Biogeochemical redox processes and their impact on contaminant dynamics. Environ Sci Technol 44(1):15–23

Boyd CA, Weiler MH, Porter WP (1990) Behavioural and neurochemical changes associated with chronic exposure to low-level concentrations of pesticide mixtures. J Toxicol Environ Health 30:209–221

Brakes CR, Smith RH (2005) Exposure of non-target small mammals to rodenticides: short-term effects, recovery and implications for secondary poisoning. J Appl Ecol 42(1):118–128

Brammall RA, Higgins VJ (1988) The effect of glyphosate on resistance of tomato to Fusarium crown and root rot disease and on the formation of host structural defensive barriers. Can J Bot 66:1547–1555

Broadberry SM, Proudfoot AT, Vale JA (2004) Poisoning due to chlorophenoxy herbicides. Toxicol Rev 23:65–73

Buckley NA, Karalliedde L, Dawson A, Senanayake N, Eddleston M (2004) Where is the evidence for the management of pesticide poisoning – is clinical toxicology fiddling while the developing world burns? J Toxicol Clin Toxicol 42:113–116

Bünemann EK, Schwenke GD, Van Zwieten L (2006) Impact of agricultural inputs on soil organisms – a review. Aust J Soil Res 2006(44):379–406

Burgoa B, Wauchope RD (1995) Pesticides in run-off and surface waters. In: Roberts TR, Kearney PC (eds) Environmental behavior of agrochemicals. Wiley, New York, pp 221–255

Burkhard N, Guth JA (1979) Chemical hydrolysis of 2-chloro-4,6-bis(alkylamino)-1,3,5-triazine herbicides and their breakdown in soil under the influence of adsorption. Pestic Sci 10:313–319

Calvet R (1989) Adsorption of organic chemicals in soils. Environ Health Perspect 83:145–177

Candela L (2003) Pesticide contamination in the EU. In: Proceedings of the XII symposium pesticide chemistry, Piacenza, p 767–780

Carringer RD, Weber JB, Monaco TJ (1975) Adsorption-desorption of selected pesticides by organic matter and montmorillonite. J Agric Food Chem 23:569–572

Celis R, Barriuso E, Houot S (1998) Effect of sewage sludge addition on atrazine sorption and desorption by soil. Chemosphere 37(6):1091–1107

Cerejeira MJ, Viana P, Batista S, Pereira T, Silva E, Valerio MJ, Silva A, Ferreira M, Silva-Fernandes AM (2003) Pesticides in Portuguese surface and ground waters. Water Res 37:1055–1063

Cessna AJ, Larney FJ, Kerr LA, Bullock MS (2006) Transport of trifluralin on wind-eroded sediment. Can J Soil Sci 86:545–554

Chan CY, Tao S, Dawson R, Wong PK (2004) Treatment of atrazine by integrative photocatalytic and biological processes. Environ Pollut 131:45–54

Chandler D, Bailey AS, Tatchell GM, Davidson G, Greaves J, Grant WP (2011) The development, regulation and use of biopesticides for integrated pest management. Philos Trans R Soc B 366:1,987–1,998

Chaplain V, Brault A, Tessier D, Défossez P (2008) Soil hydrophobicity: a contribution of diuron sorption experiments. Eur J Soil Sci 59(6):1,202–1,208

Cheney MA, Fiorillo R, Criddle RS (1997) Herbicide and estrogen eff ects on the metabolic activity of Elliptiocomplanata measured by calorespirometry. Comp Biochem Physiol 118C:159–164

Chesters G, Simsiman GV, Lavy J, Alhajjar BJ, Fatulla RN, Harkin JM (1989) Environmental fate of alachlor and metolachlor. Rev Environ Contam Toxicol 110:1–73

Chiou CT, Peters LJ, Freed VH (1979) A physical concept of soil-water equilibria for nonionic organic compounds. Science 206:831–832

Chopra AK, Sharma MK, Shamoli S (2011) Bioaccumulation of organochlorine pesticides in aquatic system – an overview. Environ Monit Assess 173:905–916

Colborn T, Smolen MJ (1996) Epidemiological analysis of persistent organochlorine contaminants in cetaceans. Rev Environ Contam Toxicol 146:91–172

Comfort SD, Shea PJ, Machacek TA, Gaber H, Oh BT (2001) Field-scale remediation of a metolachlor-contaminated spill site using zerovalent iron. J Environ Qual 30:1636–1643

Corrigan FM, Wienburg CL, Shore RF, Daniel SE, Mann D (2000) Organochlorine insecticides in substantia nigra in Parkinson's disease. J Toxicol Environ Health 59:229–234

Cowie AT, Swinburne JK (1977) Hormones, drugs, metals and pesticides in milk: a guide to the literature. Dairy Sci Abstr 39:391–402

Dawson A, Buckley NA (2007) Integrating approaches to paraquat poisoning. Ceylon Med J 52:45–47

Dec J, Bollag J (1997) Determination of covalent and non-covalent binding interactions between xenobiotic chemicals and soil. Soil Sci 162:858–874

DeLorenzo ME, Scott GI, Ross PE (2001) Toxicity of pesticides to aquatic microorganisms: a review. Environ Toxicol Chem 20:84–98

Dombek T, Dolan E, Schultz J, Klarup D (2001) Rapid reductive dechlorination of atrazine by zero-valent iron under acidic conditions. Environ Pollut 111:21–27

Drinkwater LE, Snapp SS (2007) Nutrients in agroecosystems: rethinking the management paradigm. Adv Agron 92:163–186

Dubus IG, Hollis JM, Brown CD (2000) Pesticides in rainfall in Europe. Environ Pollut 110:331–344

EC (European Commission) (2005) Health & Consumer Protection Directorate – General. Directive 79/117/EEC, Council Regulation 805/2004/EC, Directive 91/414/EEC and regulation (EC) of the European Parliament and of the Council No. 689/2008. http://ec.europa.eu/food/plant/protection/evaluation/exist_subs_rep_en.htm

Eddleston M (2000) Patterns and problems of deliberate self-poisoning in the developing world. Q J Med 93:715–731

Eddleston M, Phillips MR (2004) Self poisoning with pesticides. BMJ 328:42–44

Eddleston M, Eyer P, Worek F, Mohamed F, Senarathna L, von Meyer L, Juszczak E, Hittarage A, Azhar S, Dissanayake W, Sheriff MHR, Szinicz L, Dawson AH, Buckley NA (2005) Differences between organophosphorus insecticides in human self-poisoning: a prospective cohort study. Lancet 366:1452–1459

Eddleston M, Mohamed F, Davies JOJ, Eyer P, Worek F, Sheriff MHR, Buckley NA (2006) Respiratory failure in acute organophosphorus pesticide self-poisoning. Q J Med 99:513–522

Eddleston M, Buckley NA, Eyer P, Dawson AH (2008) Medical management of acute organophosphorus pesticide poisoning. Lancet 371:597–607

EFSA (European Food Safety Authority) (2013) The 2010 European Union Report on pesticide residues in food. EFSA J 11(3):3130

Endlweber K, Schadler M, Scheu S (2005) Effects of foliar and soil insecticide applications on the collembolan community of an early set-aside arable field. Appl Soil Ecol 31:136–146

Everett CJ, Matheson EM (2010) Biomarkers of pesticide exposure and diabetes in the 1999–2004 national health and nutrition examination survey. Environ Int 36(4):398–401

Flury M (1996) Experimental evidence of transport of pesticides through field soils – a review. J Environ Qual 25:25–45

Fravel DR, Deahl KL, Stommel JR (2005) Compatibility of the biocontrol fungus Fusarium oxysporum strain CS-20 with selected fungicides. Biol Control 34:165–169

Frazier LM (2007) Reproductive disorders associated with pesticide exposure. J Agromed 12 (1):27–37

Freeman G, Epstein MA (1955) Therapeutic factors in survival after lethal cholinesterase inhibition by phosphorus pesticides. New Engl J Med 253:266–271

Gan J, Koskinen WC, Becker RL, Buhler DD (1995) Effect of concentration on persistence of alochlor in soil. J Environ Qual 24:1162–1169

Garbarino JR, Snyder-Conn E, Leiker TJ, Hoffman GL (2002) Contaminants in Arctic snow collected over northwest Alaskan sea ice. Water Air Soil Pollut 139:183–214

Gavrilescu M (2005) Fate of pesticides in the environment and its bioremediation. Eng Life Sci 5 (6):497–526

Gaw SK, Palmer G, Kim ND, Wilkins AL (2003) Preliminary evidence that copper inhibits the degradation of DDT to DDE in pip and stonefruit orchard soils in the Auckland region, New Zealand. Environ Pollut 122:1–5

George J, Shukla Y (2011) Pesticides and cancer: insights into toxicoproteomic-based findings. J Proteome 74(12):2713–2722

Getzin LW (1981) Dissipation of chlorpyrifos from dry soil surfaces. J Econ Entomol 74:707–713

Gevao B, Semple KT, Jones KC (2000) Bound pesticide residues in soils: a review. Environ Pollut 108:3–14

Geyikçi F (2011) Pesticides and their movement in surface water and groundwater In: Stoytcheva M (ed) Pesticides in the modern world – risks and benefits. https://doi.org/10.5772/17301. Available from: https://www.intechopen.com/books/pesticides-in-the-modern-world-risks-and-benefits/pesticides-and-their-movement-surface-water-and-ground-water/

Gil Y, Sinfort C (2005) Emission of pesticides to the air during sprayer application: a bibliographic review. Atmos Environ 39:5183–5193

Glotfelty DE, Taylor AW, Turner BC, Zoller WH (1984) Volatilization of surface applied pesticides from fallow soil. J Agric Food Chem 32:638–643

Glotfelty DE, Leech MM, Jersy J, Taylor AW (1989) Volatilization and wind erosion of soil surface applied atrazine and toxaphene. J Agric Food Chem 37:546–551

Goel A, Aggarwal P (2007) Pesticide poisoning. Natl Med J India 20(4):182–191

Gorell JM, Johnson CC, Rybicki BA, Peterson EL, Richardson RJ (1998) The risk of Parkinson's disease with exposure to pesticides, farming, well water, and rural living. Neurology 50:1346–1350

Gunnell D, Eddleston M, Phillips MR, Konradsen F (2007) The global distribution of fatal pesticide self-poisoning: systematic review. BMC Public Health 7:357

Guy RC (2005) Toxicity testing, mutagenicity. In: Wexler P, Anderson BD, Peyster AD, et al (eds) Encyclopedia of toxicology. Elsevier, San Diego

Hargrove S, Merkle MG (1971) The loss of alachlor from soil. Weed Sci 19:652–654

Hatzinger PB, Alexander M (1995) Effect of ageing of chemicals in soil on their biodegradability and extractability. Environ Sci Technol 29:537–545

Hayes T, Haston K, Tsui M, Hoang A, Haeffele C, Vonk A (2002) Herbicides: feminization of male frogs in the wild. Nature 419:895–896

Helling CS (1970) Movement of s-triazine herbicides in soil. Resid Rev 32:175–210

Helling CS, Kearney PC, Alexander M (1971) Behaviour of pesticides in soils. Adv Agron 23:147–239

Hernandez AF, Parron T, Alarcon R (2011) Pesticides and asthma. Curr Opin Allergy Clin Immunol 11(2):90–96

Heupel K (2002) Avoidance response of different collembolan species to betanal. Eur J Soil Biol 38:273–276

Higgins IJ, Burns RG (1975) The chemistry and microbiology of pollution. Academic, New York, 248 pp

Hillocks RJ (2012) Farming with fewer pesticides: EU pesticide review and resulting challenges for UK agriculture. Crop Protect 31:85–93

Hoferkamp L, Hermanson MH, Muir DCG (2010) Current use pesticides in Arctic media; 2000–2007. Sci Total Environ 408:2985–2994

Hoogenraad T (1988) Dithiocarbamates and Parkinson's disease. Lancet 1:767

Houde M, Hoekstra PF, Solomon KR, Muir DC (2005) Organohalogen contaminants in delphinoid cetaceans. Rev Environ Contam Toxicol 184:1–57

IACR (International Agency for Cancer Research) (2002) IACR monographs on the evaluation of carcinogenic risk to humans, vol 1, 1971 to vol 82

Isenring R (2010) Pesticides reduce biodiversity. Pestic News 88:4–7

Iyamu IOK, Asia IO, Egwakhide F (2007) Concentrations of residues from organochlorine pesticide in water and fish from some rivers in Edo State Nigeria. Int J Phys Sci 2(9):237–241

James TK, Holland PT, Rahman A, Lu YR (1999) Degradation of the sulfonylurea herbicides chlorsulfuron and triasulfuron in a high organic matter volcanic soil. Weed Res 39:137–147

Johnson MK, Jacobsen D, Meredith TJ, Eyer P, Heath AJW, Ligtenstein DA, Marrs TC, Szinicz L, Vale JA, Haines JA (2000) Evaluation of antidotes for poisoning by organophosphorus pesticides. Emerg Med 12:22–37

Kannan K, Tanabe S, Tatsukawa R, Sinha RK (1994) Biodegradation capacity and residue pattern of organochlorines in Ganges river dolphins from India. Toxicol Environ Chem 42:249–261

Kanungo P, Ramakrishnan B, Rajaramamohan Rao V (1998) Nitrogenase activity of Azospirillum sp. isolated from rice as influenced by a combination of NH4+-N and an insecticide, carbofuran. Chemosphere 36:339–344

Kappler A, Haderlein SB (2003) Natural organic matter as reductant for chlorinated aliphatic pollutants. Environ Sci Technol 37(12):2714–2719

Khan SU, Mazurkevich R (1974) Adsorption of linuron on humic acids. Soil Sci 118:339–343

Khan SU, Behki RM, Dumrugs B (1989) Fate of bound ^{14}C residues in soil as affected by repeated treatment of prometryn. Chemosphere 18:2155–2160. 11/12:2155$\pm$2160

Khuder SA, Mutgi AB (1997) Meta-analyses of multiple myeloma and farming. Am J Ind Med 32:510–516

Knowles T, Moody R (2007) European food scares and their impact on EU food policy. Br Food J 109(1):43–67

Kolpin DW, Furlong ET, Meyer MT, Thurman EM, Zaugg SD, Barber LB, Buxton HT (2002) Pharmaceuticals, hormones, and other organic wastewater contaminants in US streams, 1999–2000: a national reconnaissance. Environ Sci Technol 36:1202–1211

Konstantinou IK, Zarkadis AK, Albanis TA (2001) Photodegradation of selected herbicides in various natural waters and soils under environmental conditions. J Environ Qual 30:121–130

Kookana RS, Di HG, Aylmore LAG (1995) A field study of degradation and leaching of nine pesticides in a sandy soil. Aust J Soil Res 33:1019–1030

Koyama J (1996) Vertebral deformity susceptibilities of marine fishes exposed to herbicide. Bull Environ Contam Toxicol 56:655–662

Kozak J (1983) Adsorption of prometryn and metholachlor by selected soil organic matter fractions. Soil Sci 136:94–101

Laha S, Luthy RG (1990) Oxidation of aniline and other primary aromatic amines by manganese dioxide. Environ Sci Technol 24(3):363–373

Larney FJ, Cessna AJ, Bullock MS (1999) Herbicide transport on wind-eroded sediment. J Environ Qual 28:1,412–1,421

Larsbo M, Stenström J, Etana A, Börjesson E, Jarvis NJ (2009) Herbicides sorption, degradation, and leaching in three Swedish soils under long-term conventional and reduced tillage. Soil Tillage Res 105(2):200–208

Larson RA, Weber EJ (1994) Reaction mechanisms in environmental organic chemistry. Lewis Publishers, Boca Raton

Le Noir JS, McConnell LL, Fellers GM, Cahill TM, Seiber JN (1999) Summertime transport of current use pesticides from California's central valley to the SierraNevada mountain range, USA. Environ Toxicol Chem 18:2715–2722

Lee WJ, Blair A, Hoppin JA, Lubin JH, Rusiecki JA, Sandler D, Dosemeci M, Alavanja MC (2004a) Cancer incidence among pesticide applicators exposed to chlorpyrifos in the Agricultural Health Study. J Natl Cancer Inst 96(23):1781–1789

Lee WJ, Hoppin JA, Blair A, Lubin JH, Dosemeci M, Sandler DP, Alavanja MC (2004b) Cancer incidence among pesticide applicators exposed to alachlor in the Agricultural Health Study. Am J Epidemiol 159(4):373–380

Lee WJ, Sandler DP, Blair A, Samanic C, Cross AJ, Alavanja MC (2007) Pesticide use and colorectal cancer risk in the Agricultural Health Study. Int J Cancer 121(2):339–346

Lenheer JA, Aldrichs J (1971) A kinetic and equilibrium study of the adsorption of cabaryl and parathion upon soil organic matter surfaces. Soil Sci Soc Am Proc 35:700–705

Leonard AW, Hyn RV, Lim RP, Chapman JC (1999) Effect of endosulfan runoff from cotton fields on macroinvertebrates in the Namoi River. Ecotoxicol Environ Saf 42:125–134

Li H, Lee LS, Jafvert CT, Graveel JG (2000) Effect of substitution on irreversible binding and transformation of aromatic amines with soils in aqueous systems. Environ Sci Technol 34 (17):3674–3680

Li H, Lee LS, Schulze DG, Guest CA (2003) Role of soil manganese in the oxidation of aromatic amines. Environ Sci Technol 37(12):2686–2693

Liess M, Schulz R (1999) Linking insecticide contamination and population response in an agricultural stream. Environ Toxicol Chem 18:1948–1955

Lin K, Liu W, Gan J (2009) Reaction of tetrabromobisphenol A (TBBPA) with manganese dioxide: kinetics, products, and pathways. Environ Sci Technol 43(12):4480–4486

Liong PC, Hamzah WP, Murugan V (1988) Toxicity of some pesticides towards freshwater fishes. Malays Agric J 54(3):147–156

Little EE (1990) Behavioral indicators of sublethal toxicity of rainbow trout. Arch Environ Contam Toxicol 19:380–385

Liu B, Tu C, Hu S, Gumpertz M, Ristaino JB (2007) Effect of organic, sustainable, and conventional management strategies in grower fields on soil physical, chemical, and biological factors and the incidence of Southern blight. Appl Soil Ecol 37:202–214

Loague K, Corwin DL (2005) Groundwater vulnerability to pesticides: an overview of approaches and methods of evaluation. Water Encycl 5:357–362

Locke D, Landivar JA, Moseley D (1995) The effects of rate and timing of glyphosate applications of defoliation efficiency, regrowth inhibition, lint yield, fiber quality and seed quality. In: Proceedings of the Beltwide Cotton Conference, National Cotton Council of America, pp 1088–1090

Lucas WJ, Lobb PG (1987) Response of potatoes, tomatoes and kumaras to foliar applications of MCPA, MCPB, 2,4-D, clopyralid, and amitrole. Proceedings of the 40th N.Z. weed and pest control conference, pp 59–63

MAFF (Ministry of Agriculture, Forestry and Fisheries) (2000) Annual report of the working party on pesticide residues 1999. Health and Safety executive, MAFF Publications, London

Mamy L, Barriuso E (2007) Desorption and time-dependent sorption of herbicides in soils. Eur J Soil 58(1):174–187

Mansour M, Korte F (1986) Abiotic degradation pathways of selected xenobiotic compounds in the environment. Studies on environmental science. In: Pawlowski L, Alaerts G (eds) Chemistry for protection of the environment, vol 29. Elsevier, Amsterdam, p 257

Mansour M, Feicht E, Meallier P (1989) improvement of the photostability of selected substances in aqueous medium. Toxicol Environ Chem 139:20–21

Matin MA, Malek MA, Amin MR, Rahman S, Khatoon J, Rahman M, Uddin MA, Miah AJ (1998) Organochlorine insecticide residues in surface and underground water from different regions of Bangladesh. Agric Ecosyst Environ 69(1):11–15
McBride JR, Dye HM, Donaldson EM (1981) Stress response of juvenile sockeye salmon (Oncorhynnchus nerka) to the butoxyethanol ester of 2,4-dichlorophenoxyacetic acid. Bull Environ Contam Toxicol 27:877–884
McConnell JS, Hossner LR (1985) pH-dependent adsorption isotherms of glyphosate. J Agric Food Chem 33(6):1075–1078
McConnell LL, Nelson E, Rice CP, Baker JE, Johnson WE, Harman JA, Bialek K (1997) Chlorpyrifos in the air and surface water of Chesapeake Bay: predictions of atmospheric deposition fluxes. Environ Sci Technol 31:1390–1398
Meco G, Bonifati V, Vanacore N, Fabrizio E (1994) Parkinsonism after chronic exposure to the fungicide maneb (manganese ethylene-bis-dithiocarbamate). Scand J Work Environ Health 20:301–305
Merrington G, Rogers SL, Zwieten LV (2002) The potential impact of long-term copper fungicide usage on soil microbial biomass and microbial activity in an avocado orchard. Aust J Soil Res 40:749–759
Miller FC (1993) Composting as a process based on the control of ecologically selective factors. In: Metting FB (ed) Soil microbial ecology. Marcel Dekker, New York, pp 515–544
Mineau P, Fletcher MR, Glaser LC, Thomas NJ, Brassard C, Wilson LK, Elliott JE, Lyon LA, Henny CJ, Bollinger T, Porter SL (1999) Poisoning of raptors with organophosphorus and carbamate pesticides with emphasis on Canada, U.S. and UK. J Raptor Res 33:1–37
Montgomery MP, Kamel F, Saldana TM, Alavanja MC, Sandler DP (2008) Incident diabetes and pesticide exposure among licensed pesticide applicators: agricultural health study, 1993–2003. Am J Epidemiol 167(10):1235–1246
Moreno JL, Aliaga A, Navarro S, Hernandez T, Garcia C (2007) Effects of atrazine on microbial activity in semiarid soil. Appl Soil Ecol 35:120–127
Morillo E, Undabeytia T, Maqueda C, Ramos A (2000) Glyphosate adsorption on soils of different characteristics. Influence of copper addition. Chemosphere 40(1):103–107
Mostafalou S, Abdollahi M (2013) Pesticides and human chronic diseases: evidences, mechanisms, and perspectives. Toxicol Appl Pharmacol 268:157–177
Mueller TC, Banks PA (1991) Furtamone adsorption and mobility in three Georgia soils. Weed Sci 39:275–279
Muir DCG, Teixeira C, Wania F (2004) Empirical and modeling evidence of regional atmospheric transport of currently used pesticides. Environ Toxicol Chem 23:2,421–2,432
Navarro S, Vela N, Navarro G (2007) Review. An overview on the environmental behaviour of pesticide residues in soils. Span J Agric Res 5(3):357–375
Nelson LM (1995–1996) Epidemiology of ALS. Clin Neurosci 3:327–331
Nott T (1997) Washing aid for fruit and vegetables. Pesticides News 35
OECD (2002) Guideline for the testing of chemicals. Aerobic and anaerobic transformation in soil. Organization for Economic Cooperation and Development, Paris
Oerke EC (2006) Crop losses to pests. J Agric Sci 144:31–41
Oerke EC, Dehne HW, Schönbeck F, Weber A (1994) Crop production and crop protection – estimated losses in major food and cash crops. Elsevier Science, Amsterdam
PAN (2009) List of lists: a catalogue of lists of pesticides identifying those associated with particularly harmful health or environmental impacts. P. A. Network
Panda S, Sahu SK (1999) Effects of malathion on the growth and reproduction of Drawida willsi (*Oligochaeta*) under laboratory conditions. Soil Biol Biochem 31:363–366
Panda S, Sahu SK (2004) Recovery of acetylcholine esterase activity of Drawida willsi (*Oligochaeta*) following application of three pesticides to soil. Chemosphere 55:283–290
Pandey S, Singh DK (2004) Total bacterial and fungal population after chlorpyrifos and quinalphos treatments in groundnut (Arachis hypogaea L.) soil. Chemosphere 55:197–205

Parks CG, Walitt BT, Pettinger M, Chen JC, de Roos AJ, Hunt J, Sarto G, Howard BV (2011) Insecticide use and risk of rheumatoid arthritis and systemic lupus erythematosus in the Women's Health Initiative Observational Study. Arthritis Care Res (Hoboken) 63(2):184–194

Paulson D (1998) Industrial aspects of remediation and environmental safety. In: Kearney PC, Roberts T (eds) Pesticide remediation in soils and water. Wiley, New York, pp 21–33

Peiris-John RJ, Wanigasuriya JK, Wickremasinghe AR, Dissanayake WP, Hittarage A (2006) Exposure to acetylcholinesterase-inhibiting pesticides and chronic renal failure. Ceylon Med J 51(1):42–43

Peterson HG, Boutin C, Martin PA, Freemark KE, Ruecker NJ, Moody MJ (1994) Aquatic phytotoxicity of 23 pesticides applied at expected environmental concentrations. Aquat Toxicol 28:275–292

Phelan PL, Mason JR, Stinner BR (1995) Soil fertility management and host preference by European corn borer, Ostrinia nubilalis (Hübner), on Zea mays L.: a comparison of organic and conventional chemical farming. Agric Ecosyst Environ 56:1–8

Phillips MR, Li X, Zhang Y (2002) Suicide rates in China, 1995–99. Lancet 359:835–840

Piccolo A, Celano G (1994) Hydrogen-bonding interactions between the herbicide glyphosate and water-soluble humic substances. Environ Toxicol Chem 13:1737–1741

Pierce RH, Olney CE, Felbeck GT (1971) Pesticide adsorption in soils and sediments. Environ Lett 1:157–172

Popov VH, Cornish PS, Sun H (2005) Vegetated biofilters: the relative importance of infiltration and adsorption in reducing loads of water-soluble herbicides in agricultural runoff. Agric Ecosyst Environ 114:351–359

Porter WP, Jaeger JW, Carlson IH (1999) Endocrine, immune, and behavioural effects of alicarb (carbamate), atrazine (triazine) and nitrate (fertiliser) mixtures at groundwater concentrations. Toxicol Ind Health 15:133–150

Pot V, Benoit P, Menn ML, Eklo OM, Sveistrup T, Kvaerner J (2011) Metribuzin transport in undisturbed soil cores under controlled water potential conditions: experiments and modelling to evaluate the risk of leaching in a sandy loam soil profile. Pest Manag Sci 67(4):397–407

Qiu X, Zhu T, Yao B, Hu J, Hu S (2005) Contribution of dicofol to the current DDT pollution in China. Environ Sci Technol 39:4385–4390

Rakitsky VN, Koblyakov VA, Turusov VS (2000) Nongenotoxic (epigenetic) carcinogens: pesticides as an example. A critical review. Teratog Carcinog Mutagen 20(4):229–240

Reichenberger S, Bach M, Skitschak A, Frede HG (2007) Mitigation strategies to reduce pesticide inputs into ground and surface water and their effectiveness; a review. Sci Total Environ 384:1–35

Reichl FX, Ritter L (2011) Illustrated handbook of toxicology. Thieme, Stuttgart

Reinhardt C, Wolf I (1986) Ruckstände an Pflanzenschutzmitteln bei alternativ und konventionell angebautem Obst und Gemüse. Bioland 2:14–17

Rice CP, Cherniak SM (1997) Marine arctic fog: an accumulator of currently used pesticide. Chemosphere 35:867–878

Roberts TR (1998) Metabolic pathway of agrochemicals. part I: herbicides and plant growth regulators. The Royal Society of Chemistry, Cambridge

Roberts TR, Hutson DH (1999) Metabolic pathway of agrochemicals. Part II: insecticides and fungicides. The Royal Society of Chemistry, Cambridge

Rovedatti MG, Castane PM, Topalian ML, Salibian A (2001) Monitoring of organochlorine and organophosphorus pesticides in the water of the Reconquista river (Buenos Aires, Argentina). Water Res 35:3457–3461

Roy C, Gaillardon P, Montfort F (2000) The effect of soil moisture content on the sorption of five sterol biosynthesis inhibiting fungicides as a function of their physicochemical properties. Pest Manag Sci 56(9):795–803

Sabater C, Carrasco JM (2001) Effects of pyridaphenthion on growth of five freshwater species of phytoplankton. A laboratory study. Chemosphere 44:1775–1781

Samuel T, Pillai MKK (1991) Impact of repeated application on the binding and persistence of [^{14}C]-DDT and [^{14C}]-HCH in a tropical soil. Environ Pollut 74:205–216
Sanborn M, Kerr KJ, Sanin LH, Cole CD, Bassil KL, Vakil C (2007) Non-cancer health effects of pesticides – Systematic review and implications for family doctors. Can Fam Physician 53:1712–1720
Sanchez-Ramon J, Hefti F, Weiner WI (1987) Paraquat and Parkinson disease. Neurology 37:728
Sannino F, Gianfreda L (2001) Pesticide influence on soil enzymatic activities. Chemosphere 45:417–425
Sarmah AK, Sabadie J (2002) Hydrolysis of sulfonylurea herbicides in soils and aqueous solutions: a review. J Agric Food Chem 50:6253–6265
Sarmah AK, Kookana RS, Alston AM (2000) Leaching and degradation of triasulfuron, metsulfuron-methyl, and chlorsulfuron in alkaline soil profiles under field conditions. Aust J Soil Res 38:617–631
Scheuner I, Reuter S (2000) Formation and release of residues of the ^{14}C-labelled herbicide isoproturon and its metabolites bound in model polymers and in soil. Environ Pollut 108:61–68
Schulz R, Liess M (1999) A field study of the effects of agriculturally derived insecticide input on stream macroinvertebrate dynamics. Aquat Toxicol 46:155–176
Sechi G, Agnetti V, Piredda M, Canu M, Deserra F, Omar HA, Rosati G (1992) Acute and persistent parkinsonism after use of diquat. Neurology 42:261–263
Seidler A, Hellenbrand W, Robra BP, Vieregge P, Nischan P, Joerg J, Oertel WH, Ulm G, Schneider E (1996) Possible environmental, occupational, and other etiologic factors for Parkinson's disease: a case-control study in Germany. Neurology 46:1275–1284
Senesi N (1992) Binding mechanisms of pesticides to soil humic substances. Sci Total Environ 123(124):63–76
Senesi N, Testini C (1980) Adsorption of some nitrogenated herbicides by soil humic acids. Soil Sci 10:314–320
Senesi N, Testini C (1983) The environmental fate of herbicides: the role of humic substances. Ecol Bull 35:477–490
Sexton SE, Lei Z, Zilberman D (2007) The economics of pesticides and pest control. Int Rev Environ Res Econ 1:271–326
Shafiei TM, Costa HH (1990) The susceptibility and resistance of fry and fingerlings of *Oreochromis mossambicus* Peters to some pesticides commonly used in Sri Lanka. J Appl Ichthyol 6:73–80
Shea PJ, Machacek TA, Comfort SD (2004) Accelerated remediation of pesticide-contaminated soil with zerovalent iron. Environ Pollut 132:183–188
Shin JY, Cheney MA (2004) Abiotic transformation of atrazine in aqueous suspension of four synthetic manganese oxides. Colloids Surf A: Physicochem Eng Asp 242(1–3):85–92
Siddharth M, Datta SK, Bansal S, Mustafa M, Banerjee BD, Kalra OP, Tripathi AK (2012) Study on organochlorine pesticide levels in chronic kidney disease patients: association with estimated glomerular filtration rate and oxidative stress. J Biochem Mol Toxicol 26(6):241–247
Simpson WM Jr, Schuman SH (2002) Recognition and management of acute pesticide poisoning. Am Fam Physician 65:1599–1604
Singh J, Singh DK (2005) Dehydrogenase and phosphomonoesterase activities in groundnut (*Arachis hypogaea* L.) field after diazinon, imidacloprid and lindane treatments. Chemosphere 60:32–42
Singh J, Shea PJ, Hundal LS, Comfort SD, Zhang TC, Hage DS (1998) Iron-enhanced remediation of water and soil containing atrazine. Weed Sci 46:381–388
Sjoblad RD, Bollag JM (1977) Oxidative coupling of aromatic pesticide intermediates by a fungal phenol oxidase. Appl Environ Microbiol 33:906–910
Smith AE, Aubin AI (1991) Effects of long-term 2,4-D and MCPA field applications on the soil breakdown of 2,4-D, MCPA, mecoprop, and 2,4,5-T. J Environ Qual 20:436–438
Smith JA, Witkowski PJ, Chiou CT (1988) Partition of nonionic organic compounds in aquatic systems. Rev Environ Contam Toxicol 103:127–151

Smith MD, Hartnett DC, Rice CW (2000) Effects of long-term fungicide applications on microbial properties in tallgrass prairie soil. Soil Biol Biochem 32:935–946

Smolen JM, Stone AT (1998) Metal (hydr)oxide surface-catalyzed hydrolysis of chlorpyrifos-methyl, chlorpyrifos-methyl oxon, and paraoxon. Soil Sci Soc Am J 62(3):636–643

Spencer WF, Adams JD, Shoup D, Spear RC (1980) Conversion of parathion to paraoxon on soil dusts and clay minerals as affected by ozone and UV light. J Agric Food Chem 28:366–371

Straathoff H (1986) Investigations on the phytotoxic relevance of volatilization of herbicides. Meded Faculteit Landbouwwetenschappen, Rijksuniversiteit Gent 51:433–438

Sudo M, Kunimatsu T, Okubo T (2002) Concentration and loading of pesticides residues in Lake Biwa basin (Japan). Water Res 36:315–329

Suett DL, Jukes AA (1990) Some factors influencing the accelerated degradation of mephosfolan in soils. Crop Protect 9:44–51

Swann CW, Behrens R (1972) Phytotoxicity of trifluralin vapors from soil. Weed Sci 20 (2):143–146

Tanabe S, Watanabe S, Kan H, Tatsukawa R (1988) Capacity and mode of PCB metabolism in small cetaceans. Mar Mamm Sci 4:103–124

Taylor AW, Spencer WF (1990) Volatilization and vapour transport processes. In: Cheng HH (ed) Pesticides in the soil environment: processes, impacts and modelling. SSSA, Madison, pp 213–269

Taylor JA, Skjemstad JO, Ladd JN (1996) Factors influencing the breakdown of sulfonylurea herbicides in solution and in soil. CSIRO Division of Soils, Adelaide, Divisional Report No. 127, pp 1–12

Ten Hulscher TEM, Cornelissen G (1996) Effect of temperature on sorption equilibrium and sorption kinetics of organic micropollutants – a review. Chemosphere 32(4):609–626

Thayer KA, Heindel JJ, Bucher JR, Gallo MA (2012) Role of environmental chemicals in diabetes and obesity: a national toxicology program workshop report. Environ Health Perspect 120 (6):779–789

Thompson DG, Wojtaszek BF, Staznik B, Chartrand DT, Stephenson GR (2004) Chemical and biomonitoring to assess potential acute effects of vision herbicide on native amphibian larvae in forest wetlands. Environ Toxicol Chem 23:843–849

Tiemann U (2008) In vivo and in vitro effects of the organochlorine pesticides DDT, TCPM, methoxychlor, and lindane on the female reproductive tract of mammals: a review. Reprod Toxicol 25(3):316–326

Tiryaki O, Temur C (2010) The fate of pesticides in the environment. J Biol Environ Sci 4 (10):29–38

Turco RK, Kladivko EJ (1994) Studies on pesticide mobility: laboratory vs. field. In: Honeycutt RC, Schabacker DJ (eds) Mechanisms of pesticide movement into groundwater. Lewis Publishers, Boca Raton, pp 63–80

Ueno D, Alaee M, Marvin C, Muir DCG, Macinnis G, Reiner E, Crozier P, Furdui VI, Subramanian A, Fillmann G, Lam PKS, Zheng GJ, Muchtar M, Razak H, Prudente M, Chung KH, Tanabe S (2006) Distribution and transportability of hexabromocyclododecane (HBCD) in the Asia-Pacific region using skipjack tuna as a bioindicator. Environ Pollut 144 (1):238–247

Unsworth JB, Wauchope RD, Klein AW, Dorn E, Zeeh B, Yeh SM, Akerblom M, Racke KD, Rubin B (1999) Significance of the long-range transport of pesticides in the atmosphere. Pure Appl Chem 71:1359–1383

US EPA (United States Environmental Protection Agency) (1986) Office of pesticides and toxic substances. Guidance for the reregistration of pesticide products containing glyphosate as the active ingredient. Washington, DC

US EPA (United States Environmental Protection Agency) (1996) Office of prevention, pesticides, and toxic substances. Reregistration eligibility decision (RED): trifluralin. Washington, DC

US EPA (United States Environmental Protection Agency) (2000) Reregistration eligibility science chapter for chlorpyrifos. Fate and environmental risk assessment chapter. http: // www.epa. gov/pesticides/op/chlorpyrifos/efedrra1.pdf./

US EPA (United States Environmental Protection Agency) (2010) Reference guide to non-combustion technologies for remediation of persistent organic pollutants in soil, 2nd ed. http://www.clu-in.org/contaminantfocus/default.focus/sec/Persistent_Organic_Pollutants_%28POPs%29/cat/Overview/

US EPA (United States Environmental Protection Agency) (2011) Pesticides and pesticide containers, regulation for acceptance and recommended procedures for disposal and treatment. Municipal Environmental Research Laboratory, Cincinnati. http://www.epa.ca

US EPA (United States Environmental Protection Agency) (2015) Restricted and canceled uses of pesticides. www.epa.gov/pesticides/regulating/restricted.htm#restricted

Van den Berg F, Kubiak R, Benjey WG (1999) Emission of pesticides into the air. Water Air Soil Pollut 115:195–218

Van Genuchten MT, Cleary RW (1979) Movement of solutes in soil. In: Vighi M, Funari E (eds) Soil chemistry. Part B. Physico-chemical models. CRC Lewis Publishers, London, pp 73–100

Van Zwieten L, Ayres MR, Morris SG (2003) Influence of arsenic co-contamination on DDT breakdown and microbial activity. Environ Pollut 124:331–339

Van Zwieten L, Rust J, Kingston T, Merrington G, Morris S (2004) Influence of copper fungicide residues on occurrence of earthworms in avocado orchard soils. Sci Total Environ 329:29–41

Veeh RH, Inskeep WE, Roe FL, Ferguson AH (1994) Transport of chlorsulfuron through soil columns. J Environ Qual 23:542–549

Walker A, Bond WJ (1977) Persistence of the herbicide AC92, 533, N-(lethylpropyl)-2, 6 dinitro-3, 4-xylidine in soils. Pestic Sci 8:359–365

Wanigasuriya KP, Peiris-John RJ, Wickremasinghe R, Hittarage A (2007) Chronic renal failure in north central province of Sri Lanka: an environmentally induced disease. Trans R Soc Trop Med Hyg 101(10):1,013–1,017

Wauchope RD, Yeh S, Linders JBHJ, Kloskowski R, Tanaka K, Rubin B, Katayama A, Kördel W, Gerstel Z, Lane M, Unsworth JB (2002) Pesticide soil sorption parameters: theory, measurement, uses, limitations and reliability. Pest Manag Sci 58:419–445

Weber JB, Weed SB, Ward TM (1969) Adsorption of s-triazines by soil organic matter. Weed Sci 17:417–421

Weichenthal S, Moase C, Chan P (2010) A review of pesticide exposure and cancer incidence in the Agricultural Health Study cohort. Environ Health Perspect 118(8):1117–1125

WHO/FAO (World Health Organization and Food and Agriculture Organization of the United Nations) (2003) Diet, nutrition and the prevention of chronic diseases. WHO technical report series, vol 916. World Health Organization, Geneva

WHO/FAO (World Health Organization and Food and Agriculture Organization of the United Nations) (2008) Clinical management of acute pesticide intoxication: prevention of suicidal behaviours. WHO Document Production Services, Geneva

Wienhold BJ, Sadeghi AM, Gish TJ (1993) Effect of starch encapsulated and temperature on volatilization of atrazine and alachlor. J Environ Qual 22:162–166

Woese K, Lange D, Boess C, Böel KW (1997) A comparison of organically and conventionally grown foods: results of a review of the relevant literature. J Sci Food Agric 74:281–293

Wolf DC, Martin JP (1976) Decomposition of fungal mycelia and humic-type polymers containing carbon-14 from ring and side-chain labelled 2,4-D and chlorpropham. Soil Sci Soc Am Proc 40:700–704

Yaron B (1978) Some aspects of surface interactions of clays with organophosphorus pesticides. Soil Sci 125:210–216

Zahm SH, Ward MH (1998) Pesticides and childhood cancer. Environ Health Perspect 106:893–908

Zamzila AN, Aminu I, Niza S, Razman MR, Hadi MA (2011) Chronic organophosphate pesticide exposure and coronary artery disease: finding a bridge. IIUM Research, Invention and Innovation Exhibition (IRIIE)
Zhang H, Weber EJ (2009) Elucidating the role of electron shuttles in reductive transformations in anaerobic sediments. Environ Sci Technol 43(4):1042–1048
Zhang H, Chen WR, Huang CH (2008) Kinetic modeling of oxidation of antibacterial agents by manganese oxide. Environ Sci Technol 42(15):5548–5554
Zhang WJ, Jiang FB, OU JF (2011) Global pesticide consumption and pollution: with China as a focus. Proc Int Acad Ecol Environ Sci 1(2):125–144

Chapter 11
Health Risks Associated with Organic Pollutants in Soils

Abstract The occurrence of organic pollutants at elevated levels has been of significant environmental and human health concern at numerous contaminated "hot spots" and their regional and global importance has received increasing attention in the last decade. Many different forms of organic pollutants exist. Among this group, persistent organic pollutants (POPs) play a key role. Importantly, POPs have the ability to enter the gas phase under environmental temperatures and may volatilize from soils, vegetation and water bodies into the atmosphere. Because of their resistance to breakdown reactions in the air they travel long distances before being re-deposited. The cycle of volatilization and deposition may be repeated several times, with the result that POPs could accumulate in an area far away from where they were used or emitted.

POPs are either intentionally produced for one or multiple purposes, or they are unintentionally formed as by-products in other processes. Minor quantities Polybrominated Diphenyl Ethersmay also result from natural processes. Several pesticides (Chap. 10), industrial chemicals and unintentionally formed substances are commonly mentioned in the context of POPs. This chapter focuses on industrial and unintentionally formed substances. The processes leading to unintentional production of POPs can be categorized as combustion and chemical-industrial processes. Some important examples are the emissions of dioxins (e.g. polychlorinated dibenzo-p-dioxins: PCDDs), furans (e.g. polychlorinated dibenzofurans: PCDFs) and polycyclic aromatic hydrocarbons (PAHs). Several POPs belong to various source categories, e.g. polychlorinated biphenyls (PCBs) that have been produced as an industrial chemical, but can also be unintentionally formed in combustion processes.

Persistent organic pollutants are toxic and resist to photolytic, biological and chemical degradation to varying degree. Persistent organic pollutants are also known to be semi-volatile, which permits these substances to occur either in the vapor phase or adsorbed on particles in the atmosphere. This enables their long-distance transport through the atmosphere. The persistence in combination with semi-volatility has resulted in worldwide, nonpoint source pollution of soils and ecosystems with organic pollutants. They occur even in remote regions including open oceans, deserts and poles where no local sources are present. Dumping of POPs in badly engineered and unsuitable landfill sites and dumps has often caused

R. Nieder et al., *Soil Components and Human Health*,
https://doi.org/10.1007/978-94-024-1222-2_11

point source pollution around these sites with widespread contamination of groundwater and surface water contamination particularly from hexachlorocyclohexane (HCH), hexachlorobenzene (HCB) and PCB.

Due to their bio-accumulating properties these substances build up in the food chain with exposure to animals and humans, possibly causing health impacts for current and future generations. Because of their harmful effects on man and wildlife, international agreements have come into effect to reduce future environmental risks. Persistent organic pollutants have been globally addressed by the Stockholm Convention which was enacted in 2004. This convention aims at protecting environmental and human health from adverse effects associated with exposure to POPs. Twelve of these POPs were originally listed but the number is increasing with nine new POPs added in 2009. While the original emphasis of the Stockholm Convention was on chlorinated compounds such as PCBs, HCBs, polychlorinated dibenzo-p-dioxins, dibenzofurans, POP pesticides and some others, fluorinated and brominated POPs were added to the convention in 2009. The latter include polybrominated biphenyls (PBBs) and polybrominated diphenyl ethers (PBDEs) which have been widely used as flame retardants for electronics, textiles, furniture, upholstery, insulation foam, etc. The first fluorinated additions to the Convention were perfluorooctane sulfonic acid (PFOS) together with its salts. PFOS and related compounds are increasingly entering waste streams in a wide range of products. Polychlorinated naphthalenes (PCNs) are under review by the Stockholm Convention as a candidate for POPs for their persistence, toxicity, bioaccumulation, and long-range atmospheric transport. Like other POPs, PCNs are globally distributed in air, sediments and biota. Short-chain chlorinated paraffins (SCCPs) are also found worldwide in the environment, and are bioaccumulative in wildlife and humans.

Organic pollutants are strongly linked to human health effects such as cancers, mesothelioma, skin disorders, respiratory diseases, eye disorders, asthma and endocrine disruption. Humans can be exposed to POPs through direct exposure, occupational accidents and the environment. Short-term exposures to high concentrations of POPs may result in illness and death. Chronic exposure may be associated with a wide range of adverse health and environmental effects.

Keywords Persistent organic pollutants (POPs) • Structure and properties of organic pollutants • Sources and emissions • Environmental fate • Organic pollutants in non-target organisms • Human exposure • Clinical effects • Therapy • Remediation of soils contaminated with organic pollutants

11.1 Organic Pollutants of Concern

Several groups of organic substances, particularly the so-called persistent organic pollutants (POPs) are likely to cause adverse environmental or human health effects at locations near and far from their sources. These pollutants are toxic, persistent, liable to bioaccumulate and are prone to long-range atmospheric transport and deposition (UN-ECE 1998).

Depending on their mobility, these substances can be of local, regional or global importance. Increasing concern about the potential effects of some man-made chemical substances on the environment and human health has prompted action from local to global level. The global extent of POP pollution became apparent with their detection in areas such as the Arctic, where they have never been used or produced. Persistent organic pollutants, including pesticides, industrial products and industrial by-products (e.g. polychlorinated dibenzo-p-dioxins (PCDDs) and dibenzofurans (PCDFs)) can persist in soils, sediments and waste deposits for periods ranging between decades and centuries. Historical PCDD/PCDF and dioxin-like polychlorinated biphenyl (PCB) contamination burdens are a result of the production of chlorine and of chlorinated organic chemicals. The production of chlorinated pesticides, PCBs and the related contaminated waste streams are responsible for wide distribution of toxic burdens (Weber et al. 2008). Along with such distributions, numerous PCDD/PCDF contaminated sites have resulted from recycling of wastes, application or improper disposal of polychlorinated biphenyls (PCBs) contaminated pesticides and other organochlorines. In some cases, PCDD/PCDF contaminated sites have been created by waste incinerators, secondary metal industries or recycling or deposition of specific waste (e.g. car shredder wastes or electronic waste) that often contain chlorinated or brominated organic pollutants. Contamination of fish with PCDD/PCDF and dioxin-like PCB in European rivers or the impact on other food resources demonstrate the relevance of these burdens for many human generations. Food contamination problems that have recently emerged in Europe show how PCDD/PCDF and dioxins like PCBs from historical sources can directly contaminate human and animal feedstuffs and demonstrate their relevance for presence and future. Sites contaminated with POPs will continue to represent an environmental issue for future generations and will need to be controlled, supervised and potentially remediated. Leachates and groundwater impacted by these sites will require continuous monitoring.

The international community has responded to the health concerns posed by organic pollutants through developing various treaties and organizations to address POPs chemicals and waste. A regional initiative on POPs was started by eastern and western European countries in community with Canada and the USA in 1992 with the establishment of a task force on POPs within the framework of the 1979 Convention on Long Range Transport of Air Pollution (Vallack et al. 1998). Starting from 107 substances, 16 priority substances were identified for initial inclusion in the list comprising 11 pesticides (aldrin, chlordane, chlordecone, DDT, dieldrin, endrin, heptachlor, hexachlorobenzene (HCB), hexachlorocyclohexane (HCH), mirex and toxaphene; see Chap. 10), industrial products such as hexabromobiphenyl and polychlorinated biphenyls (PCBs), and unintentional by-products of industrial processes and combustion including polycyclic aromatic hydrocarbons (PAHs), polychlorinated dibenzo-p-dioxins (PCDDs) and polychlorinated dibenzofurans (PCDFs).

Table 11.1 Persistent organic pollutants as industrial products or residues identified by the Stockholm Convention

Industrial products and by-products	Currently listed	Under review (2011; 2013)
Polychlorinated biphenyls (PCBs)	x	
Polybrominated diphenyl ethers (PBDEs) and other brominated flame retardants:		
Octabromodiphenyl ether	x	
Pentabromodiphenyl ether	x	
Hexabromobiphenyl (HBB)	x	
Hexabromocyclodecane (HBCD)		x
Polychlorinated dibenzo-p-dioxins (PCDDs)	x	
Polychlorinated dibenzofurans (PCDFs)	x	
Hexachlorobutadiene (HCBD)		x
Pentachlorobenzene (PeCB)	x	
Perfluorooctane sulfonate (PFOS)	x	
Polychlorinated naphtalenes (PCNs)		x
Short-chained chlorinated paraffins (SCCP)		x

Adapted from US EPA (2010), Liu and Zheng (2013) and Zhang et al. (2014)

In the Stockholm Convention on POPs, which came into operation in 2004, parties committed to reduce or eliminate the production, use, and release of the 12 POPs of greatest concern to the global community. In May 2009, another nine chemicals were added to the Stockholm Convention. Table 11.1 shows industrial products and by-products which are currently listed, or are under review, by the Stockholm Convention.

Another international convention regulating POPs is the Basel Convention (Basel, Switzerland) on the Control of Transboundary Movements of Hazardous Wastes and their Disposal. The Basel Convention was adopted on March 22, 1989 which was enacted in 1992 (US EPA 2010). In response to Stockholm Convention provisions requiring coordination with the Basel Convention on POPs waste issues, the Basel convention developed guidance on the environmentally sound management of POPs waste. In 2004, the Basel Convention invited signatories of the Stockholm Convention to consider its recommendations on environmentally sound management for POPs wastes (US EPA 2010).

Although polycyclic aromatic hydrocarbons (PAHs) are not listed by the Stockholm convention on POPs, they were included in the Convention on Long range Transboundary Air Pollution Protocol on Persistent Organic Pollutants (UN-ECE 2009). Their toxic effects on human and ecosystem health are well documented. PAHs can disperse regionally and globally through atmospheric long-range transport. For example, PAHs from East Asia are transported to the west coast of the USA under certain meteorological conditions and PAHs emitted from Russia influence the atmosphere PAH concentrations in the Arctic (e.g. Becker et al. 2006; Primbs et al. 2008).

11.2 Structure and Properties of Priority Organic Pollutants

Many different organic pollutants with different chemical structures and properties exist. The structures of important POPs (priority organic pollutants) are given in Fig. 11.1. Major toxicological and chemical parameters can be drawn from Table 11.2. An LD_{50} is a standard measurement of acute toxicity that is stated in milligrams (mg) of organic pollutant per kilogram (kg) of body weight. An LD_{50} represents the individual dose required to kill 50% of a population of test animals (e.g., rats, fish, mice, cockroaches). Because LD_{50} values are standard measurements, it is possible to compare relative toxicities among organic pollutants. The lower the LD_{50} dose, the more toxic is the substance. The persistence of a substance is characterized by its half-life (given in days), which is defined as the time necessary to degrade 50% of a substance. The organic carbon-water partitioning coefficient (K_{oc}) is the relation between chemical concentration sorbed to organic

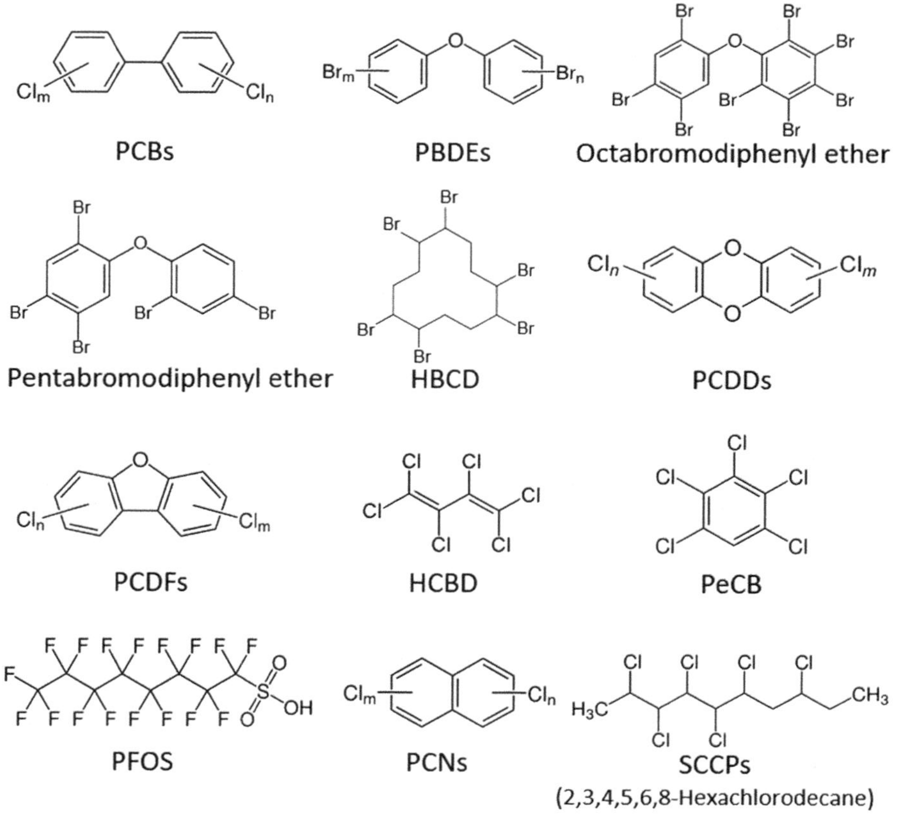

Fig. 11.1 Structures of selected chlorinated, brominated and fluorinated POPs

Table 11.2 Toxicological and chemical parameters of priority organic pollutants listed by the Stockholm Convention

Chemical	LD_{50} (mg kg^{-1})	Half-life (soil) (days)	Water solubility (mg L^{-1})	Log K_{oc}	Log K_{ow}	Vapor pressure 25 °C (Pa)
Polychlorinated biphenyls (PCBs)	1010–4250	2300	3.1×10^{-4}–0.1.6	n.a.	5.62–7.20	5×10^{-6}–1×10^{-2}
Polybrominated diphenyl ethers (PBDEs) and other brominated flame retardants:						
Octabromodiphenyl ether	65–149	76 days[a]	5×10^{-4}	n.a.	6.29	6.59×10^{-6} (21 °C)
Pentabromodiphenyl ether	65–149	150 days	1.3×10^{-2}	4.89–5.10	6.57	2.2×10^{-7}–5.5×10^{-7}
Hexabromobiphenyl (HBB)	65–149	>6 months	1.1×10^{-2}	3.33–3.87	6.39	5.2×10^{-8}
Hexabromocyclododecane (HBCD)	500–1000	66–101 days[b]	3.4×10^{-3}	n.a.	5.62	4.7×10^{-6}
Polychlorinated dibenzo-p-dioxins (PCDDs)	0.022–0.045	230–2300	8.4×10^{-3}–7.4×10^{-8}	n.a.	6.35–8.2	1.1×10^{-10}–1.0×10^{-4}
Polychlorinated dibenzofurans (PCDFs)	0.916	710–2300	1.16×10^{-6}–4.19×10^{-3}	n.a.	6.1–8.0	5.0×10^{-10}–2.0×10^{-6}
Hexachlorobutadiene (HCBD)	200–500	1.6 years	2.0–2.55	3.67	4.78–4.9	0.15
Pentachlorobenzene (PeCB)	33–330	260–7300 days	0.56	6.08	4.88–6.12	1.1–2.2×10^{-2}
Perfluorooctane sulfonate (PFOS)	199–318	>41 years[c]	519–680	2.57	n.a.	2.4×10^{-6}
Polychlorinated naphtalenes (PCNs)	530–710	180 days	8×10^{-5}-2.87	2.97–3.27	3.9–8.3	7.5×10^{-9}–1.2×10^{-4}
Short-chained chlorinated paraffins (SCCPs)	0.34	>1 year	6×10^{-6}-2×10^{-3}	n.a.	6.9–9.1	2.3×10^{-12}–2×10^{-3}

Adapted from Brändli et al. (2004) and US EPA (2010)
n.a. not available
[a]Half-life in air
[b]Half-life in sediments
[c]Half-life in water

carbon (mg g^{-1}) and chemical concentration in water (mg L^{-1}) which can be used to estimate the extent of sorption (see also Chap. 10). Bioaccumulation and lipophilicity of a substance can be characterized by the octanol-water partitioning coefficient (log K_{ow}). 1-octanol is used as a reference substance exhibiting similar properties to organic substances occurring in nature and organisms (e.g. cell membranes) (Fent 1998). Persistent substances with a log K_{ow} above three (e.g. PCDDs/PCDFs, PCBs, DDT) have a high potential for bioaccumulation.

Volatilization of substances depends on the Henry's law constant (H_c; unit: m^3 Pa mol^{-1} or USI; not given in Table 11.2). The higher the H_c constant, the higher is volatilization of a substance and hence the relative abundance in air. Naphthalene and low chlorinated PCBs are relatively volatile ($H_c \approx 10^{-2}$ USI). Pesticides (H_c 10^{-3} – 10^{-5} USI), substances like phthalates, PAHs and PCDD/PCDF (H_c 10^{-4} – 10^{-6} USI) volatilize to a lesser extent. Volatilization is mainly induced immediately after deposition of the compound and decreases with time. It depends not only on the Henry's law constant but also on the sorption which can be roughly estimated by log K_{ow}. Three classes of substances can be defined (Wild and Jones 1992) (i) high volatilization at $H_c > 10^{-4}$ USI and $H_c/K_{ow} > 10^{-9}$, (ii) medium volatilization at $H_c > 10^{-4}$ USI or $H_c/K_{ow} > 10^{-9}$ and (iii) low volatilization at $H_c < 10^{-4}$ and $H_c/K_{ow} < 10^{-9}$.

Solubility of organic pollutants in water is commonly low for organic pollutants. They are rather lipophilic and thus adsorbed onto the soil matrix (most important is organic matter or clay). Organic pollutants being often non polar compounds in soils are mainly adsorbed by organic matter exhibiting similar polarity. Affinity for sorption is characterized by the partition coefficient of a substance between organic carbon and the aqueous phase in the soil (K_{oc}) that is related to K_{ow} ($K_{oc} = 0.41^*$ K_{ow}; Karickhoff 1981). The octanol-air partitioning coefficient (K_{oa}; not given in Table 11.2) has become an important parameter in the understanding of pollutant sorption to surfaces of plant leaves. K_{oa} can be calculated from the compound's air-water partition coefficient (Henry's law constant) and K_{ow} (Paterson et al. 1994) or measured directly (Harner and Mackay 1995).

Polychlorinated Biphenyls (PCBs)

The chemical structure of PCBs consists of two benzene rings connected by a C-C bond which are substituted by one to ten chlorine atoms (Fig. 11.1). The number of possible congeners is 209. It is distinguished between ortho-substituted and coplanar PCBs. The latter represent 20 congeners that are highly toxic. PCBs are denominated by their number "Ballschmitter". Analyses of environmental samples commonly include congeners 28, 52, 101, 118, 153, 138 and 180. The degree of chlorination is responsible for physico-chemical properties of PCBs (Table 11.2). Their water solubility is low (PCB 28: 0.16 mg L^{-1}; PCB 180: 3.1×10^{-4} mg L^{-1}). Vapor pressure ranges between 0.01 and 5×10^{-6} Pa (at 25 °C). Similar to water solubility, volatilization decreases with increasing substitution with chlorine atoms. Increasing chlorination leads to slightly higher K_{ow} values (K_{ow} PCB 28: 5.62; K_{ow} PCB 180: 7.20). Similar to PAHs, PCBs are semivolatile compounds existing in the gas phase and associated to particles in the atmosphere. The lipophilicity of PCBs strongly reduces their bioavailability and hence their degradation in soils. The presence of humic substances enhances degradation under aerobic conditions due to higher activity of microorganisms (Fava and Piccolo 2002).

Polybrominated Diphenyl Ethers (PBDEs)

PBDEs consist of two connected phenyl rings which are substituted by one to ten bromine atoms (Fig. 11.1). They are structurally similar to PCBs and DDT. Depending on the number and positions of the bromine atoms on the two phenyl rings, PBDEs have a large number of congeners. The total number of possible congeners is 209, and the number of isomers for mono-, di-, tri-, tetra-, penta-, hexa-, hepta-, octa-, nona- and decabromodiphenyl ethers are 3, 12, 24, 42, 46, 42, 24, 12, 3 and 1, respectively (Rahman et al. 2001). PBDEs are denominated by IUPAC numbers (1–209). Analyses of environmental samples commonly include congeners 47, 85, 99, 100, 138, 153, 154 and 209. Most industrially manufactured PBDEs contain mixtures of brominated diphenyl ethers, their isomers and homologues. The commercial PBDEs consist predominantly of penta- (PeBDE), octa- (OBDE) and decabromodiphenylethers (DeBDE). PeBDE is mixed with tetrabromodiphenylethers (TeBDE) in which TeBDE is a major compound (Rahman et al. 2001). Other important brominated flame retardants are tetrabromobisphenol A (TBBPA) and hexabromocyclododecane (HBCD). Technical mixtures of the latter contain three diastereoisomers (α, β, γ) existing in proportions of approximately 6, 8 and 80%, respectively (Budakowski and Tomy 2003). Due to their similar chemical properties, persistence and distribution of PBDEs, PCBs and DDT in the environment follow similar patterns.

Similar to PCBs physico-chemical properties of PBDEs depend on the number of bromine atoms. Water solubility is between 9.3×10^{-3} mg L^{-1} (BDE 47) and 1.3×10^{-8} mg L^{-1} (BDE 209) (Palm et al. 2002). Vapor pressure ranges from 8.2×10^{-5} to 5.5×10^{-11} Pa (at 25 °C) for BDE 47 and 209, respectively. Similar to water solubility, volatilization decreases with increasing number of bromine atoms. Increasing number of bromine atoms goes also along with increasing K_{ow} values (K_{ow} BDE 47: 6.67; K_{ow} BDE 209: 11.15). Water solubility for TBBPA and HBCD are 4.2 mg L^{-1} and 3.4×10^{-3} mg L^{-1}, respectively. Values for vapor pressure are below 100 and 6.3×10^{-5} Pa (at 20 °C) and K_{ow} are 4.5 and 5.6 for TBBPA and HBCD, respectively (Sellström et al. 1999). Half-lives reported for BDE 47 are 10 days in air and 150 days in soil. Corresponding values for BDE 209 are 318 and 150 days respectively (Palm et al. 2002). Values of the priority PBDEs for water solubility, K_{ow}, vapour pressure and half-lives are near these numbers (Table 11.2).

Polychlorinated Dibenzodioxins and Dibenzofurans (PCDD/PCDF)

Polychlorinated Dibenzodioxins and Dibenzofurans (PCDDs/PCDFs) are ubiquitous compounds in the environment. They represent a group of tricyclic, aromatic ethers (Fig. 11.1). The molecules are substituted by one to eight chlorine atoms. There are 75 congeners for PCDDs and 135 for PCDFs. Analyses of environmental samples are commonly limited to tetra- to octahomologues. Water solubility and vapor pressures of PCDDs and PCDFs are low but they are highly lipophilic (Table 11.2) (Fiedler et al. 1994).

Hexachlorobutadiene (HCBD)
Hexachlorobutadiene (HCBD) is a halogenated aliphatic compound (Fig. 11.1), mainly created as a by-product in the manufacture of chlorinated aliphatic compounds (most likely tri- and tetrachloroethene and tetrachloromethane). The compound is poorly soluble in water, but miscible with ether and ethanol. The substance is lipophilic based on a log K_{ow} ranging between 4.78 and 4.9 and has a high vapour pressure (Table 11.2). The poor solubility in water of HCBD, its high vapour pressure, its high log K_{oc} and log K_{ow} values determine its behaviour and fate in environmental media. It can volatilize from moist soil and water. According IPCS (1994) HCBD has a turpentine-like odor.

Pentachlorobenzene (PeCB)
Pentachlorobenzene belongs to the group of chlorobenzenes, which are characterized by a benzene ring in which the hydrogen atoms are substituted by one or more chlorines (Fig. 11.1). Chlorobenzenes are neutral, thermally stable compounds. Stability, melting and boiling points increase with increasing chlorine substitution. Pentachlorobenzene has a low solubility in water. Vapor pressure varies between 0.11 and 0.22 Pa (Table 11.2). The log K_{ow} values vary between 4.88 and 6.12 (Mackay et al. 2006).

Perfluoroctane Sulfonate (PFOS)
Perfluoroctane sulfonate (PFOS) is a completely fluorinated anion, which is usually used as a salt or incorporated into larger polymers (Fig. 11.1). PFOS are members of the group of perfluoroalkyl sulphonate substances (PFAS). Perfluorinated substances with long carbon chains, including PFOS, are both lipid-repellent and water-repellent. PFOS-related substances are therefore used as surface-active agents in different applications. The extreme persistence of these substances makes them suitable for high temperature applications and for applications in contact with strong acids or bases. The persistence of perfluorinated substances is caused by the strong carbon-fluorine bindings.

Polychlorinated Naphtalenes (PCNs)
Polychlorinated naphtalenes (PCNs) represent a group of 75 compounds which are structurally similar to the PCBs (Stevens et al. 2002). They are based on a naphthalene ring system in which one or more hydrogen atoms have been replaced by chlorine (Fig. 11.1). PCNs have physical and chemical properties comparable with higher chlorinated PCBs (Stevens et al. 2002). Solubility of PCNs in water ranges between 8×10^{-5} and 2.87 mg L^{-1}. They exhibit low vapor pressure (7.5×10^{-9}–1.2×10^{-4} Pa) and K_{ow} ranges between 3.9 and 8.3 (Table 11.2). As for PCBs, physico-chemical properties of PCNs depend on the number of chlorine atoms. Solubility and vapor pressure decrease and K_{ow} increases with increasing number of chlorine atoms.

Short-Chained Chlorinated Paraffins (SCCPs)
Chlorinated paraffins are complex mixtures of straight chain chlorinated hydrocarbon molecules. They exhibit different chain lengths, ranging from short- (C10–13) over medium- (C14–17) to long-chained chlorinated paraffins (C18–30). The degree of chlorination on a weight basis varies between 40% and 70% (Fig. 11.1). The short-chained chlorinated paraffins (SCCPs) are considered as POPs. Data on physico-chemical properties of straight chain chlorinated hydrocarbons are sparse (Alcock et al. 1999). Solubility in water and vapour pressure are low (Table 11.2). K_{ow} varies between 6.9 and 9.1. Except for the vapor pressure which is lower for polychlorinated n-Alkanes, these values are comparable to those of higher chlorinated PCBs. The physico-chemical properties strongly depend on the chain length and the number of chlorine atoms. Solubility and vapor pressure decrease while K_{ow} increases with increasing molecular weight.

Polycyclic Aromatic Hydrocarbons (PAHs)
Polycyclic aromatic hydrocarbons consist of polynuclear aromatic cycles (Fig. 11.2). Although some hundreds of compounds have been detected, the priority is commonly on the 16 EPA compounds (US EPA 2001) which are classified according to their number of cycles (Table 11.3).

The physico-chemical properties of PAHs depend on the number of cycles. For example, PAHs with two cycles such as NAP are highly water soluble and volatile. Water solubility and volatility decrease with increasing number of cycles. NAP (2-ring PAH) exhibits a solubility in water of 31 mg L^{-1} and vapor pressure of 10.4 Pa (at 25 °C). The corresponding values for BGP (6-ring PAH) are 2.6×10^{-3}

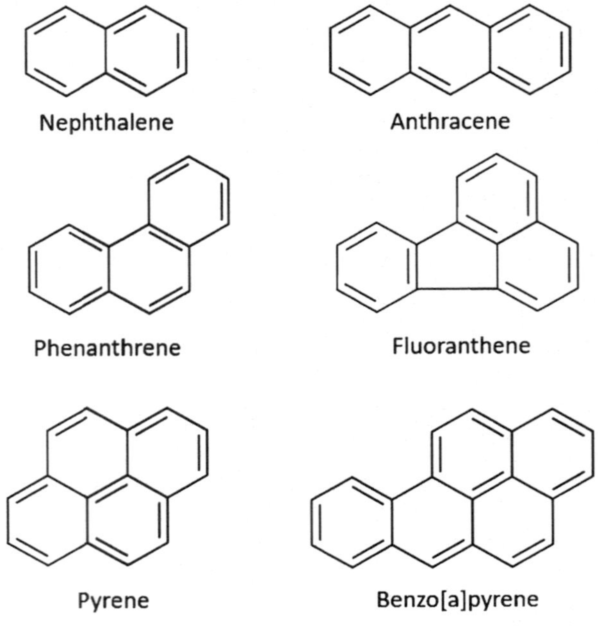

Fig. 11.2 Chemical structures of selected polycyclic aromatic hydrocarbons

Table 11.3 Classification of priority polycyclic aromatic hydrocarbons

PAHs	Number of cycles
Naphtalene (NAP)	2
Acenaphthylene (ACY), Acenaphthene (ACE), Fluorine (FLU), Phenanthrene (PHE), Anthracene (ANT)	3
Fluoranthene (FLT), Pyrene (PYR), Benz[a]anthracene (BaA), Chrysene (CHR)	4
Benzo[b]fluoranthene (BbF), Benzo[k]fluoranthene (BkF), Benzo[a]pyrene (BaP), Dibenz[a,h]anthracene (DBA)	5
Indeno[1,2,3-cd]pyrene (IND), Benzo[ghi]perylene (BGP)	6

mg L^{-1} and 1.39×10^{-10} Pa (at 25 °C) respectively. Increasing number of cycles goes along with increasing sorption capacity on particles (K_{ow} NAP: 3.37; K_{ow} BGP: 7.10) and the potential for accumulation with decreasing biodegradability. In the atmosphere sorption to aerosols is important particularly for 4- to 6-ring PAHs (Smith et al. 2001).

11.3 Sources and Emissions of Organic Pollutants

The knowledge about sources and emissions of POPs into the environment are essential to understand, quantify, and predict the source-receptor relationships and to make a plan to reduce the burden of POPs in the environment. Organic pollutants can be released directly to the environment during usage (e.g. spraying of pesticides; volatilization of PBDEs and other flame retardants from consumer products) or production (e.g., combustion releasing PAHs and PCDD/PCDFs). These pathways comprise the primary emission. When previously-deposited chemicals re-volatilize into the atmosphere from environmental media, like snow, soil, vegetation and water, often as a result of seasonal and diurnal changes in temperature, this is considered a secondary emission. Chemicals such as PCBs can also be released indirectly through biomass burning (e.g. forest fires and straw burning), and be transported to the remote Arctic (Hung et al. 2010).

Emitted POPs will deposit into environmental reservoirs, such as soils and water bodies, and re-emit to air through soil-air and water-air exchange, forming secondary emissions. For example, it has been found that reduction of ice cover in the Arctic may allow chemicals previously deposited into the Arctic Ocean, e.g. α-HCH, to volatilize back into the atmosphere (Jantunen et al. 2008). It is therefore important to take into account historical accumulations of POPs in environmental media and their possible re-emissions. Secondary emissions may become stronger with time, and may eventually exceed the primary emissions. Although secondary emissions may become increasingly important, they are commonly not an issue of emission inventories.

The efforts to develop global emission inventories can be traced back to the end of the 1990s and the beginning of the 2000s. There are many different kinds of emission inventories for POPs, as reviewed by Breivik et al. (2007). Examples of global inventories are for selected PCB congeners (Breivik et al. 2002, 2007), PAHs (Zhang and Tao 2009) and HCBs (Bailey 2001). Breivik et al. (2002) presented a global historical emission estimate for selected PCB congeners. Zhang and Tao (2009) compiled global emission inventories for PAHs and a global PFOS emission inventory has been compiled by Paul et al. (2009). A toolkit for compiling national inventories for dioxines and furans on a global scale has been published by UNEP (1999). However, uncertainties of emission inventories exist on different spatial scales. They are mainly related to the assumed emission factors and activities in principal not quantifiable. For most emission inventories, no information exists about completeness of the covered sources on global as well as on regional scales.

Polychlorinated Biphenyls

Polychlorinated biphenyls (PCBs) were produced in high amounts by the chemical industry and have been used in a large variety of applications. They share some physico-chemical properties with polybrominated diphenyl ethers (PBDEs), for example, their tendency for being readily emitted to the atmosphere (Breivik et al. 2004).

The USA, Germany and France were the first countries to produce PCBs in the 1930s (de Voogt and Brinkman 1989). The worldwide production ceased when manufacturing in Russia ended in 1993 (AMAP 2000). Estimates of global production since around 1930 range between about 1.2 million Mg (Holoubek 2001) and 2 million Mg (Fiedler 2001). The production of PCBs reached its maximum in 1970, with an estimated 76,000 Mg globally (Breivik et al. 2002). There currently seems to be no known industrial production of PCBs in the world. In contrast, PBDEs are still produced in significant amounts. In the early 1990s the global production of PBDEs was about 40,000 Mg (WHO 1994a).

The production of PCBs involved the chlorination of biphenyl in the presence of a catalyst. The degree of chlorination varied between 21% and 68% chlorine on a weight-by-weight basis (e.g. Ahlborg et al. 1992). A large number of technical mixtures with different chemical compositions were produced worldwide (Breivik et al. 2004). The production of PBDEs is similar to the production of PCBs, involving the bromination of diphenyl ethers in the presence of a catalyst.

Polychlorinated biphenyls were utilised in numerous open, nominally closed and closed systems (Breivik et al. 2004). Open use included the addition of PCBs in plasticisers, carbonless copy paper, lubricating oils, inks, laminating and impregnating agents, paints, adhesives, waxes, additives in cement and plaster, casting agents, dedusting agents, sealing liquids, fire retardants, immersion oils and pesticide extenders (de Voogt and Brinkman 1989). Nominally closed and closed usage included hydraulic and heat transfer fluids, small capacitors in cars and household electrical appliances, vacuum pumps as well as transformers and large capacitors. Globally, electrical companies were the largest users of PCBs (Breivik et al. 2004). The strong temperature dependence of vapour pressure, and therefore of the

Table 11.4 Emissions of PCBs to air from different industrial sectors in EU 27

Industrial sector	Total PCB emissions (kg year^{-1})				
	2000	2010	2020	2030	2050
Electrical products	20,300	5487	5487	0	0
Fuel combustion	1819	1144	529	161	30
Industrial and residential boilers	1071	745	426	142	28
Iron and steel production	1131	884	685	212	42
Waste incineration	194	84	37	8	4
Non-ferrous metal production	9	7	4	2	0
Sum of other sources	153	115	63	21	4
Total	24,676	8466	7230	545	110

Adapted from Strzelecka-Jastrzab et al. (2007)

evaporation rate from various environmental surfaces, has been shown to affect the atmospheric concentrations and congener distribution of PCBs (Breivik et al. 2004). This temperature dependency was observed to increase with the degree of chlorination.

A global PCB emission inventory for ΣPCB_{22} congeners from the start of their production in 1930 until 2005 has been developed by Breivik et al. (2007). Atmospheric emissions peaked in 1970 with more than 3000 Mg of ΣPCB22 being emitted that year on a global scale. Still, by 2009 about 250 Mg of ΣPCB22 were emitted into the atmosphere, which in brief is a combined effect of the persistence of these compounds and the long lifetime of various products and applications containing PCBs. However, as remaining uses of PCBs over time are phased-out, disposed or destroyed, emissions are expected to continue to decline into the future. An estimate of total emissions of PCBs from different sources to air in Europe (27 states) including a projection until 2050 is given in Table 11.4.

Polybrominated Diphenyl Ethers

The use of polybrominated diphenyl ethers (PBDEs) has enabled significant reduction of fire hazards during the last decades. Due to high volumes used, they are nowadays a world-wide pollution problem reaching even remote areas. PBDEs are still extensively used as additive flame retardants. The fact that BDEs are not chemically bound to the material it should protect has potential implications for emissions, possibly allowing PBDEs to evaporate from the material to the air and to affect the environment (Rahman et al. 2001). Bromine acts by reaction with the free radicals in the gas phase and thus slows down the ignition and combustion process. PBDEs have been used in a wide range of products e.g. housing and wiring of TV sets, computers and mobile phones as well as in electrical kitchen appliances, upholstery and textiles, building materials and many plastic products. Cars and airplanes also contain many of the above products (electronics, plastics, and upholstery) and have a great need for fire protection (Frederiksen et al. 2009). Thus, exposure to brominated flame retardants may occur in many situations in daily life. PBDEs have been sold as penta-, octa-, and deca-BDE mixtures which are named according to the dominating homologue group. Penta- and octa-BDE were banned from all products in the EU by 2004 as well as in ten states of the USA. These bans were followed by a voluntary cease of production of the penta- and octa-mixtures in the USA in 2004. However, deca-BDE is still produced and applied worldwide (56,100 t y^{-1} in 2001). Only Sweden has banned the use of deca-BDE (Frederiksen et al. 2009).

Polychlorinated Dibenzo-p-Dioxins and Dibenzofurans
Polychlorinated dibenzo-p-dioxins (PCDDs) and dibenzofurans (PCDFs) are key classes of unintentional by-products from various combustion and industrial processes. They usually occur as a mixture of congeners. Their presence in the incinerator fly ash samples was firstly reported in 1977 (Olie et al. 1977). However, dioxins had come to worldwide attention in 1982 after an explosion at ICMESA factory in Seveso, Italy, distributed dioxins over an area of 2.8 km^2 (Wilson 1982). Only 7 of the 75 possible PCDD congeners, and 10 of the 135 possible PCDF congeners, those with chlorine substitution in the 2,3,7,8 positions, have dioxin-like toxicity. Likewise, there are 209 possible PCB congeners, 12 of which have dioxin-like toxicity (US EPA 1994). These dioxin-like PCB congeners have four or more chlorine atoms. Physical and chemical properties of each congener vary according to the degree and position of chlorine substitution.

The largest source of PCDDs and PCDFs today is open burning of household waste, municipal waste, medical waste, landfill fires, and agricultural and forest fires (Kulkarni et al. 2008). Dioxin and furan compounds exhibit little potential for significant leaching or volatilization once sorbed to particulate matter and are extremely stable compounds under most environmental conditions. The only environmentally significant transformation process for these congeners is believed to be photodegradation of nonsorbed species in the gaseous phase, at the soil–air or water–air interface (Tysklind et al. 1993). PCDDs/PCDFs entering the atmosphere are removed either by photodegradation or by deposition. Burial in-place, resuspension back into the air, or erosion of soil to water bodies appears to be the predominant fate of PCDDs/PCDFs sorbed to soil. The ultimate environmental sink of PCDDs/PCDFs is believed to be aquatic sediments. Levels of PCDDs/PCDFs in fish and invertebrates have been found to be higher than those in the water column, suggesting bioaccumulation (Atkinson 1991).

There is considerable uncertainty about the global emissions of dioxins and furans. The most important source worldwide is the incineration of municipal, hospital and hazardous wastes. However, in Europe fuel combustion is still an important source of PCDDs and PCDFs as well as the production of steel and non-ferrous metals and vehicle operation. Due to the high number of single compounds PCDD/PCDF concentration is commonly characterized by International Toxicity Equivalents (I-TEQ) which is a parameter obtained by adding up the toxicities relative to 2,3,7,8-TCDD. In 1995, the PCDD/PCDF air emissions in Europe (17 countries) was estimated to have been 6500 g TEQ (International Toxic Equivalents) per year (Quaß and Fermann 1997). This would correspond to an average estimated deposition rate of about 5 TEQ m^{-2} day^{-1} which compares well

Table 11.5 Emissions of PCDDs and PCDFs to air from different industrial sectors in EU 27

	Total dioxin emissions (TEQ year^{-1})				
Industrial sector	2000	2010	2020	2030	2050
Industrial and residential boilers	1111	821	601	200	40
Iron and steel production	788	608	512	159	32
Waste incineration	471	47	11	2	1
Fuel combustion	160	105	47	14	3
Non-ferrous metal production	138	108	88	29	6
Transport	50	40	30	10	2
Cement production	32	24	17	6	1
Sum of other sources	63	54	36	12	2
Total	2812	1807	1340	432	87

Adapted from Strzelecka-Jastrzab et al. (2007)

with the lower limit of deposition rates for rural areas in Germany (Vallack et al. 1998). An estimate of the total emissions of PCDDs and PCDFs to air from different sources to air in Europe (27 states) including a projection until 2050 is given in Table 11.5.

Hexachlorobutadiene

The intentional production of hexachlorobutadiene (HCBD) in the UN-ECE region, in the USA and in Canada ended in the late 1970s (Lecloux 2004). Data about intentional production outside these regions are not available. However, monitoring data published by Li et al. (2008) suggest that in China production has continued at least until recently. Worldwide intentional production of HCBD was estimated at 10,000 Mg in 1982, but HCBD generated as waste by-product was as high as 14,000 Mg (1982) in the USA alone (Lecloux 2004).

HCBD is unintentionally generated during the production of chlorinated hydrocarbons, particularly of perchloroethylene, trichloroethylene and carbon tetrachloride (Lecloux 2004). It can also be formed during the production of vinyl chloride, allyl chloride and epichlorohydrin although a dossier prepared for the European chloralkali industry considers this extremely unlikely from a technological point of view (Lecloux 2004). In the USA in 1998 the majority of releases came from the chemical sector (1300 kg) and from the electrical, gas and sanitary services sectors (460 kg); in 1999 and 2000, total releases to air of 2635 kg and 1936 kg, respectively, were reported. The load to the atmosphere, however, does not include all possible releases from every type of industrial facility (ATSDR 1994). Some 15,000 Mg of HCBD appear to have been produced in 2000 as by-product, but recycled in production processes or treated and destroyed as waste, mainly by incineration on-site or in publicly-owned treatment works. Only 1835 kg of HCBD was released in waste. In the US, most of the disposal waste from chlorinated hydrocarbon manufacturing processes is incinerated.

In Europe, from recent surveys by Euro Chlor (Lecloux 2003) at 76 European chlor-alkali production sites, HCBD emissions to water were reported to have decreased from 100 kg year^{-1} in 1997 to 2.4 kg year^{-1} in 2002. Emissions to air decreased from 2 kg year^{-1} to almost zero over the same period. In 1990, HCBD emissions in Germany from the use of perchloroethylene as a solvent were estimated to be <0.56 and 620 kg year^{-1} to air and water, respectively, and discharges into Rhine and Elbe rivers were estimated at 70 and 150 kg year^{-1}, respectively (BUA 1991).

Pentachlorobenzene

Pentachlorobenzene (PeCB) is not known to be manufactured for any commercial uses at present (Bailey et al. 2009). In the past, PeCB was one component of a chlorobenzene mixture used to reduce the viscosity of PCB products employed for heat transfer (King et al. 2003). PeCB was also used in a chlorobenzenes mixture with PCBs in electrical equipment. PCBs are still in use in old electrical equipment throughout the world so there is a small potential for release of PeCB from this source (AMAP 2000). PCBs are being taken out of service and incinerated in mostcountries of the world so that any related PeCB emissions are expected to decrease with time. Global emissions of PCBs were estimated to be approximately 40 mt in the year 2000, compared to approximately 800 mt in 1970 (Breivik et al. 2002).

Current emissions of PeCB to the environment are estimated to be about 121 Mg year^{-1}. The major sources appear to be combustion of solid wastes (32 Mg year^{-1}) biomass burning (45 Mg year^{-1}) with degradation of an agricultural fungicide, quintozene, contributing 26 Mg year^{-1} (Bailey et al. 2009). Industrial releases are less important. PeCB has been measured in many environmental media over the past 35 years. Low but detectable concentrations of PeCB have been reported in the atmosphere, sediments and biota in remote areas of the world.

Perfluoroctane Sulfonate

Production of perfluorochemicals began in in 1949 (Paul et al. 2009). From 1966 to the 1990s, production and use grew due to their unique chemical attributes, including chemical stability and surface tension/leveling properties. The uses included inks, varnishes, waxes, firefighting foams, metal plating and cleaning, coating formulations, lubricants, water and oil repellents for leather, paper and textiles. Historical global production of perfluorooctane sulfonyl fluoride (POSF), a precursor of perfluoroctane sulfonate PFOS, was estimated to be 96,000 Mg (or 122,500 Mg, including unusable wastes) between 1970 and 2002, with an estimated global release of 45,250 Mg to air and water between 1970 and 2012 from direct (manufacture, use, and consumer products) and indirect (PFOS precursors and/or impurities) sources (Paul et al. 2009). Estimates indicate that direct emissions from POSF-derived products are the major source of PFOS to the environment resulting in releases of 450–2700 Mg PFOS into wastewater streams, primarily through losses from stain repellent treated carpets, waterproof apparel, and aqueous firefighting foams. Large uncertainties surround indirect sources and have not yet been estimated due to limited information on environmental degradation, although it can be assumed that some PFOS-derived chemicals will degrade to PFOS over time. The properties of PFOS (high water solubility, negligible vapor pressure, and relatively limited sorption to particles) imply it will reside in surface waters, predominantly in oceans. Measured oceanic data suggest ~235–1770 Mg of PFOS currently reside in ocean surface waters, similar to the estimated PFOS releases. Environmental monitoring from the 1970s onward shows strong upward trends in biota. A reduction in some compartments has been observed since cessation of POSF production in 2002 (Paul et al. 2009).

Polychlorinated Naphtalenes

Polychlorinated naphthalenes (PCNs) are a group of industrial chemicals which were produced in several countries under the tradenames Halowax, Nibren, Clonaicre and Seekay waxes and Cerifal Materials (Falandysz 1998). PCNs were first patented as flame retardants and dielectric fluids for capacitors in the early 1900s and found use in a variety of industrial applications, some of which were dye-making, fungicides in the wood, textile and paper industries, plasticizers, oil additives, casting materials for alloys and lubricants for graphite electrodes (Bidleman et al. 2010). PCNs also occur as trace contaminants in commercial polychlorinated biphenyl (PCB) mixtures (Taniyasu et al. 2003a; Bidleman et al. 2010). Production figures for PCNs are not well known, but have been estimated to be ~150,000 Mg, ~10% of the global PCB production (Crooks and Howe 1993; Falandysz 1998). Manufacture of PCNs is thought to have ended today, although illegal importation of PCN-containing products into Japan was reported after 2000 (Yamashita et al. 2003).

Sources of PCNs to the environment include evaporation and release during combustion (Helm and Bidleman 2003). Evaporation sources include emission of PCNs by volatilization from in-use or disposed products and from contaminated soil. Combustion release of PCNs occurs from various industrial and waste incineration processes and includes de novo synthesis as well as release of PCNs contained within incinerated waste. To a lesser extent, PCNs are also released during combustion of coal and wood (Lee et al. 2005). Studies have identified elevated levels of PCNs in some European cities, which could be a possible source to the Arctic. The PCNs averaged 85 pg m^{-3} in Chilton and 110 pg m^{-3} in Hazelrigg, UK in 2001, while a lower level of 15 pg m^{-3} was found at Mace Head, a remote station on the west coast of Ireland, in 2000 (Lee et al. 2005). Jaward et al. (2004a) reported 140–220 pg m^{-3} in London, Moscow and urban-industrialized regions of Poland by summing only 13 congeners, which may have underestimated actual PCN levels.

Short-Chained Chlorinated Paraffins

Short-chained chlorinated paraffins (SCCPs) belong to the group of polychlorinated n-Alkanes. Production of polychlorinated n-Alkanes used as pressure additives began in the 1930s. More than 200 commercial mixtures with different physical and chemical properties are produced for a wide range of applications such as secondary plasticizers in PVC and other plastics (C10–13 and C14–17), extreme pressure additives (C14–17), flame retardants (C10–13), sealants (C10–13 and C14–17) and paints (Alcock et al. 1999). About 70% of the production volume is used as secondary plasticizers (i.e. PVC), in adhesives and paints, 30% as flame retardants and additives for lubricating oils in the engineering industry (Kollotzek et al. 1998). The worldwide production volume was estimated at 300,000 Mg in 1985. The amount synthesized in Western Europe was 140,000 Mg in 1993. Within the OSPAR convention, production of short-chain PCAs was banned in 1999. Thus, a decline of the production volume can be expected (Kollotzek et al. 1998). Significant release into the atmosphere is expected for short-chain PCAs used as softeners but not for those used as flame retardants as the latter are bound to the matrix. Emissions to the environment are expected to occur from point sources (industry) and diffuse sources.

Polycyclic Aromatic Hydrocarbons (PAHs)

The majority of polycyclic aromatic hydrocarbons (PAHs) in the environment are unintentional by-products originating from incomplete combustion of carbonaceous materials during energy and industrial production processes (Zhang and Tao 2009). Natural processes, such as forest fires and volcanic eruptions, also produce PAHs (Xu et al. 2006). The major anthropogenic atmospheric emission sources of PAHs include biomass burning, coal and petroleum combustion, and coke and metal production (Zhang and Tao 2008, 2009). Fossil energy consumption and emissions of PAHs are closely correlated. Accumulation of PAHs in Greenland icesheet has thus been constant since the beginning of the industrial period (Masclet et al. 1995). Due to rapid population growth and the associated energy demand, PAH emissions from developing countries have increased significantly while in developed countries PAH emissions have decreased significantly in the past decades as the efficiency of energy utilization has improved (Pacyna et al. 2003).

Zhang and Tao (2009) estimated the emissions of the 16 PAHs listed as the US EPA priority pollutants for the reference year 2004 using reported emission activity and emission factor data. The sixteen PAHs listed included naphthalene (NAP), acenaphthylene (ACY), acenaphthene (ACE), fluorene (FLO), phenanthrene (PHE), anthracene (ANT), fluoranthene (FLA), pyrene (PYR), benz(a)anthracene (BaA), chrysene (CHR), benzo(b)fluoranthene (BbF), benzo(k)- fluoranthene (BkF), benzo(a)pyrene (BaP), dibenz(a,h)anthracene (DahA), indeno(l,2,3-cd) pyrene (IcdP) and benzo(g,h,i)perylene (BghiP). Total global atmospheric emission of these PAHs in 2004 was 520×10^3 Mg year^{-1} with biofuel (56.7%), wildfire (17.0%) and consumer product usage (6.9%) being the most important sources. China (114×10^3 Mg year^{-1}), India (90×10^3 Mg year^{-1}) and the USA (32×10^3 Mg year^{-1}) were the three countries with the highest PAH emission rates. The PAH sources in the individual countries varied significantly. In India, biofuel burning was the dominant PAH source, while wildfire emissions were the dominant PAH source in Brazil and consumer products were the major PAH emission source in the USA. In China, in addition to biomass combustion, coke ovens were a significant source of PAHs. The PAH emission density varied from 0.0013 kg km^{-2} year^{-1} in the Falkland Islands to 360 kg km^{-2} year^{-1} in Singapore with a global mean value of 3.98 kg km^{-2} year^{-1} (Zhang and Tao 2009).

11.3.1 Composts, Digestates and Sewage Sludge as Sources of Diverse Organic Pollutants

Composts, digestates and sewage sludge are materials suitable for recycling in agriculture and in horticulture. These materials being rich in organic matter as well as nitrogen and phosphorous are becoming appropriate material for use in agriculture as fertilizers or soil conditioners. Composts are produced using a conventional aerobic composting process in which temperature, pH and nutrient availability are constantly changing. It passes through several microbiological phases identified by temperature (Potter et al. 1999). Since microbial activities are crucial during the composting process, it is essential to achieve appropriate water and oxygen supply by possibly irrigating or covering the windrow with a semipermeable foil and turning it on a regular basis. Composting reduces the original mass of input material by about 40%, mainly through mineralization of the organic compounds and evolution of carbon dioxide (CO_2) and release of water through evaporation (Brändli et al. 2004).

Organic wastes for digestion after a pre-treatment are transferred into a tank where they pass through characteristic stages of fermentation. After the establishment of the anaerobic conditions macromolecules such as proteins, sugars and fats are degraded to amino acids, monosaccharide and fatty acids. In the following acidic phase organic acids such as acidic acid, butyric acid, lactic acid and other compounds such as alcohol, CO_2 and water are produced. In the subsequent methanogenic phase these acids are degraded by microorganisms to methane (CH_4) and CO_2. The CH_4 produced during fermentation is used for energy production and the digestate is commonly spread on agricultural land as fertilizer.

Numerous organic pollutants have been described as ubiquitous in the environment and thus have been detected even in remote areas indicating that organic pollutants are likely to be present in compost or digestate. Occurrence of PAHs, PCBs, PCDDs, PCDFs, phthalates, detergents and pesticides have been well documented (e.g. Houot et al. 2002; McGowin et al. 2001; Vanni et al. 2000; Zethner et al. 2000). During composting and digestion, degradation of organic pollutants occurs (Taube et al. 2002) but might be incomplete (Vorkamp et al. 2002; Hartlieb and Klein 2001) so that part of the pollutants might still be present as original substances or metabolites.

Sewage sludge, also called "biosolids", is a product that is left behind after water is cleaned in waste water treatment plants. Currently, the most widely available options for sewage sludge in the EU are agriculture use, waste disposal, land reclamation and restoration, and incineration (Martinez et al. 2007). Organic pollutants enter the wastewater stream via aerial deposition and runoff into urban drains, household domestic wastewater or industrial effluents Because of their lipophilic properties, many organic pollutants tend to adsorb on the solid particles and finally pass to the sludge during the wastewater treatment process (Stevens et al. 2001). Sludges with high levels of contamination with toxic substances have to be incinerated or landfilled since they would not be suitable for application to agricultural land.

11.3.2 Dumps and Landfills as Sources of Diverse Organic Pollutants

Dumps and landfills are regarded important disposal pathways for various wastes containing POPs, particularly before regulations on handling "POPs waste" came into effect. The historic dumping of POPs was often in badly engineered and unsuitably located landfill sites and dumps. POPs will persist in landfills for many decades and possibly centuries. Over these extended time periods engineered landfill systems and their liners are likely to degrade, thus posing a contemporary and future risk of releasing large contaminant loads to the environment (Weber et al. 2011). It's a common feature that groundwater and river systems are being contaminated by these old dumps (Weber et al. 2008) at a time when water resources are under increasing threat.

The amount of PCBs in the environment of Europe-25 annually directed to landfills (inert waste and hazardous waste landfills including underground disposal) is estimated to be 600 Mg (Chrysikou et al. 2008). Although the concentration of PCBs in municipal solid waste is low (<0.4 mg kg^{-1}), a yearly PCBs discharge of 87 Mg arises from municipal solid waste and about 66 Mg of them are annually disposed of at non-hazardous waste landfills. Landfills may thus be considered a continuous reservoir and emission source of PCBs to the atmosphere, via direct volatilization from contaminated surfaces, release through landfill vents. Landfill fires may be important emission sources of POPs like PCDD/PCDFs, PCBs and PAHs which are either non-combusted waste contaminants or are formed during incomplete combustion conditions (Capuano et al. 2005). Landfill fires generate a smoke plume consisting of inhalable particles. The potential for pollutant deposition from the smoke plume may be considerable, however, dispersion of pollutants will be complex, principally as the large range of physico-chemical properties of individual pollutants within the smoke plume will result in differential spatial and temporal distributions within the receiving environment.

11.4 Environmental Fate of Priority Organic Pollutants

According to their physico-chemical properties, POPs partition to air, water, soil, sediment, and other environmental compartments which creates obvious challenges for understanding and predicting their environmental fate and transport. Organic pollutants enter the environment in different ways (Fig. 11.3).

Fig. 11.3 Pathways of organic pollutants in the environment (Adapted from Galiulin et al. 2002)

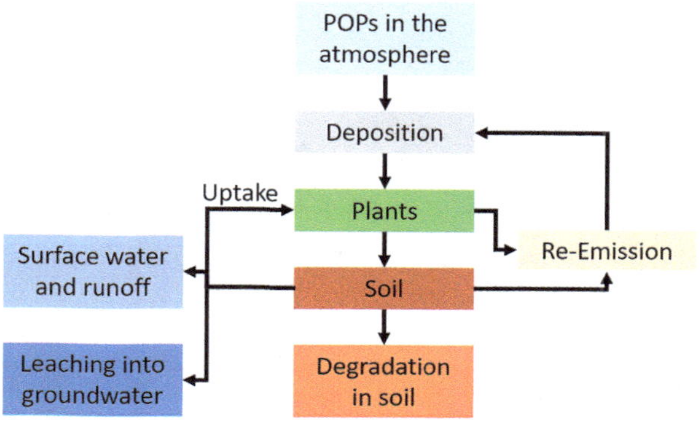

Unlike other classes of pollutants that have a natural background and have been cycling on the earth indefinitely, most persistent organic pollutants have only anthropogenic origins, with the first POPs manufactured approximately 80 years ago. Due to high production volumes, persistence and long range transport, POPs are ubiquitous in the environment. During transport, they might undergo transformation. After deposition, POPs may be bound or buried in soils, sediments, and ice or incorporated into the deep oceans. Subsequently, they might be re-emitted and distributed in the environment either by water or air.

11.4.1 Major Principles of Transportation Routes

In the atmosphere, the residence time of organic pollutants depends on their distributions among vapor, particle, and droplet phases. This partitioning is governed by the water solubilities and the vapor pressures of the substances, and by the concentrations and size distributions of particles and droplets in the atmosphere (Poster and Baker 1996). These factors determine whether substances undergo short- or long-range transport, i.e., their potential to reach remote areas such as the Arctic (Fig. 11.4).

Based on the partitioning coefficients octanol-air (K_{oa}), octanol-water (K_{ow}) and air-water (K_{aw}) of the individual POPs, Wania (2006) has classified POPs as "fliers", "multi hoppers", "single hoppers" and "swimmers". Most legacy POPs, e.g. lower molecular weight PCBs, highly chlorinated chlorobenzenes, HCHs, lower molecular weight PCDD/PCDFs, PAHs and many organochlorine pesticides, are "multi-hoppers". These are 'multimedia' chemicals with partitioning properties that allow for efficient exchange between air and terrestrial or aquatic surfaces. Multimedia partitioning enables the long-range transport of these chemicals through the atmosphere by means of repeated cycles of deposition and re-evaporation, driven by temperature changes along the path. Less volatile POPs, e.g. higher molecular weight PCDD/PCDFs, PAHs and decabrominated

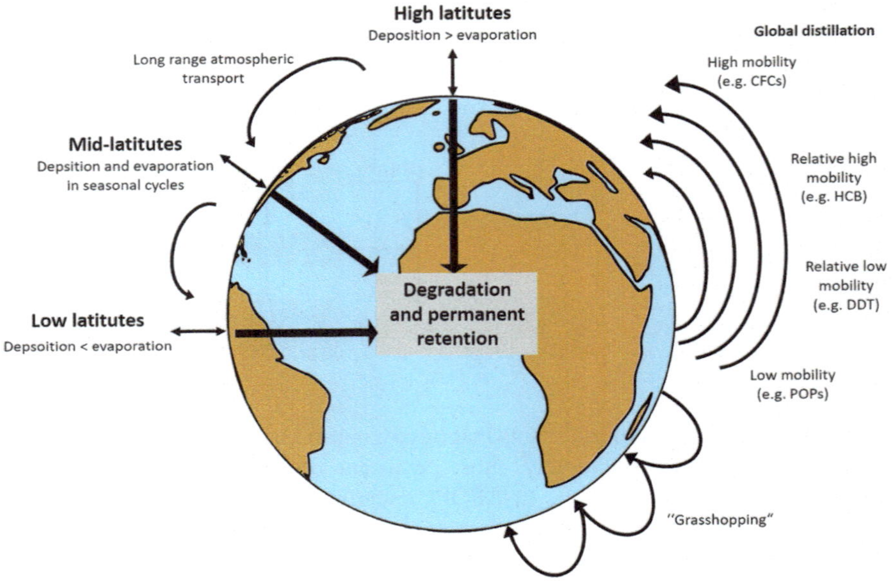

Fig. 11.4 Global transport patterns of organic pollutants (Adapted from Wania and Mackay 1996)

diphenyl ether, are "single hoppers". In the bulk atmosphere, they tend to associate with particles and so their ability to undergo long range transport is controlled by the transport of the atmospheric particles to which they sorb. After deposition to the soil surface they do not volatilize effectively and therefore must reach the receptor region, e.g. the Arctic, in one single hop without deposition along the path. However, climate variability can set these single-hoppers free to take a second hop through extreme events like dust storms, forest fires or floods.

While the atmosphere has been identified as the most rapid and major transport route for most legacy POPs, some new and emerging POPs, e.g. PFOS and perfluorinated carboxylates (PFCA) are swimmers. In some cases, chemicals may be both swimmers and multi-hoppers, with air-water exchange and varying degrees of transport in both air and water, playing key roles in their long range transport.

11.4.2 Concentrations of Organic Pollutants in the Atmosphere

Many organic pollutants are (semi)volatile and are thus released from soils or other sources in gaseous form or they can be transferred to the atmosphere bound to particles (aerosols). Processes for removal of substances from the atmosphere include photodegradation, OH-radical reaction and deposition (i.e. wet and dry deposition) (Sweetman and Jones 2000). Wet deposition includes wash-out of soluble volatile organic compounds as well as scavenging of particle organic matter. Atmospheric gases are dissolved in droplets within clouds and in falling raindrops. The rate is given by the mass transfer, the accommodation characteristics and the solubility of gases determined by the Henry's law constant. In contrast, particle scavenging results from physical processes controlled by cloud microphysics, meteorological conditions and the solubility, number density and size of ambient particles. Scavenging of atmospheric particles by precipitation is strongly size-dependent (Poster and Baker 1996).

Dry deposition includes deposition of particles as well as gases. Dry deposition of particles comprises (i) the gravitational sedimentation of particles not bound to precipitation, and (ii) interception, which is the filtering effect of the vegetation through horizontal impact of aerosols. Dry deposition of gases means that compounds in the gaseous phase are deposited by adsorption on the surface of plants, materials and soil and by stomata uptake into plants. After deposition organic pollutants may be irreversibly bound or buried in soils, sediments and ice or accumulated in deep oceans (Sweetman and Jones 2000). POPs can also be re-emitted which represents a secondary source.

Polychlorinated Biphenyls

In air samples of the USA, mean PCB concentration as a sum of 61 congeners was 427 pg m^{-3} (Poster and Baker 1996). Air samples collected during early spring in Ontario (Canada) exhibited PCBs concentrations as a sum of 47 congeners in a range from 96 to 950 pg m^{-3}, with a dominance of lower chlorinated (tri- to tetra-) congeners. Air samples collected in the Netherlands in 2000 and 2001 exhibited concentrations between 0.1 and 5.2 pg m^{-3} for PCBs congeners 8, 20, 28, 35, 52, 101, 118, 153, 138 and 180, with higher levels for the lower chlorinated congeners (Duyzer and Vonk 2003). At a semirural site in the UK, air samples have been collected from March to October and in December 1994. Concentrations for PCB congeners 28, 52, 101, 153, 138, 180 concentrations ranged between 54 and 375 pg m^{-3} (Lee and Jones 1999). Levels of PCBs (28, 52, 101, 118, 153, 138, 180) in London and Manchester were in the range of 200–2000 pg m^{-3}, with concentrations in London generally two to three times higher compared to Manchester (Coleman et al. 1997).

Polybrominated Diphenyl Ethers

On the basis of several studies in the 1990s, concentrations of PBDEs have been reported in the following ranges: PBDE 47: 0.7–670 pg m^{-3}, PBDE 99: 0.35–25 pg m^{-3}, PBDE 100: 0.07–5.4 pg m^{-3}, PBDE 209: not detected-0.34 pg m^{-3} (Palm et al. 2002). Jaward et al. (2004b) in samples from Norway and the UK collected in 2000–2002 found background concentrations (ΣPBDEs, excluding 209) ranging between 0.8–1.6 pg m^{-3} and 1.1–2.5 pg m^{-3}, respectively. In Canada (2002/2003), ΣPBDEs (excluding 209) ranged between 0.1 and 4.4 pg m^{-3} (Wilford et al. 2004). Outdoor air is commonly dominated by the congeners BDE-47, 99,100, and 209 (Strandberg et al. 2001). BDE 47 levels in air in North America were found to range between 3.7 and 50 pg m^{-3} (Alcock et al. 2003). In Kyoto (Japan) total

PBDE levels in the air (2000/2001) ranged from 6.5 to 80 pg m^{-3} (Watanabe and Sakai 2003), with BDEs 47, 99, 153, 183 and 209 as the main congeners. The levels of total PBDEs in atmospheric air samples of the Osaka area in Japan in 2001 ranged from 104 to 347 pg m^{-3}. BDE 209 formed 96% of the total concentration. The presence of PBDEs in in various products used every day may lead to additional exposure in the home environment, where dust seems to be an important source (Fredriksen et al. 2009). The PBDE exposure through dust is significant for toddlers who ingest more dust than adults. In aerosols, contents of PBDE 209 up to 3060 pg m^{-3} have been measured (Brändli et al. 2004). In the atmosphere, PBDEs may undergo photochemical decomposition. Break-down of PBDEs may result in the formation of lower brominated congeners (de Wit 2002).

Polychlorinated Dibenzodioxins and Dibenzofurans
In the atmosphere, PCDD/PCDF compounds in the gas phase might undergo photolysis in contrast to the particle bound ones which are rather persistent (Fiedler et al. 1994). PCDDs/PCDFs which are not removed by photodegradation may undergo deposition. PCDD/PCDF concentrations in air in two cities of the UK (Manchester and London) ranged between 0.05 and 0.2 pg I-TEQ m^{-3}. In London the concentration was slightly higher compared to Manchester (Coleman et al. 1997). Similar PCDD/PCDF concentrations were reported for Switzerland by Dettwiler et al. (1997). Compared to rural regions, in urban areas levels of PCDDs/PCDFs may be higher by a factor of 2–5 (Dettwiler et al. 1997). Fiedler et al. (1994) for the city of Hamburg reported a background deposition rate of 12 pg I-TEQ m^{-2} and a deposition rate of 30 pg I-TEQ m^{-2} in 1990. In a comprehensive literature review Lohmann and Jones (1998) found PCDD/PCDF concentrations in a range between 0.5 and 20 pg I-TEQ m^{-3} with a general gradient increasing from remote to rural to urban/industrial areas. This trend is consistent with expectations, given that combustion sources and chemical usage are the principal sources of PCDD/PCDF to the atmosphere.

Hexachlorobutadiene
Class and Ballschmiter (1987) from 18 locations sampled from 1982 to 1986 calculated an average concentration of 2 ng m^{-3} HCBD for the troposphere of the Northern Hemisphere. A study by the Swedish EPA indicated that the HCBD concentration in air varied from 2 to 5 pg m^{-3} in three locations which were regarded as "background levels" (Lecloux 2004). In a number of cities HCBD levels ranged from 0.02 to 0.12 μg m^{-3} (Lecloux 2004; Singh et al. 1982). Up to 0.41 μg m^{-3} were found in the basement air of homes near industrial and chemical waste disposal sites (Lecloux 2004). In a study of nine chemical plants, the highest levels of HCBD in air were found near those producing trichloroethene and tetrachloroethene (maximum 462 μg m^{-3} (IARC 1979). Concentrations of HCBD were reported at a level of 540 μg m^{-3} in Columbus, Ohio, and up to 1000 μg m^{-3} in Cincinnati, Ohio, from 1989 to 1991. However, a subsequent study conducted in the 1990s at six sampling locations in Columbus, Ohio, failed to detect HCBD in air (Spicer et al. 1996).

Pentachlorobenzene

Concentrations of PeCB are relatively constant in the atmosphere because the compound is released from a variety of processes and has a long lifetime combined with a wide distribution (Bailey et al. 2009). In a monitoring study in Canada 1988–1989, atmospheric PeCB was detected in 133 out of 143 samples (Hoff et al. 1992). In North America, PeCB concentrations were in a range of 17–136 pg m^{-3}, with an annual mean value of 45 pg m^{-3} (Shen et al. 2005). Atmospheric concentrations of PeCB in Southern Sweden in 2003 were reported to average 33 pg m^{-3} (Kaj and Palm 2004). In Germany, PeCB was found to be at significantly higher concentrations (5.7–28.6 pg m^{-3}) near industrial or urban areas compared to a rural site (0.31 pg m^{-3}) (Wenzel et al. 2006).

Perfluoroctane Sulfonate

Data on PFOS concentration in air are scarce. However, air sampling studies in temperate regions have confirmed that the atmospheric lifetime of perfluoroalkyl contaminants is sufficient for complete mixing in the northern hemisphere (e.g. Ellis et al. 2003; Martin et al. 2006). In the atmosphere, PFOS due to its low vapor pressure is hardly found in the gas phase but to some extent it is bound to particles (Stock et al. 2007). Concentrations of PFOS detected in the particulate phase above Japan (ranging from nondetect values to 21.8 pg m^{-3}; Sasaki et al. 2003) and Lake Ontario (concentrations up to 8.1 pg m^{-3}; Boulanger et al. 2005) were similar to mean values reported by Stock et al. (2007) for Cornwallis Island (Canadian Arctic). Concentrations of PFOA observed in samples from Cornwallis Island were several orders of magnitude less than the concentrations of PFOA (0.1–3.84 μg m^{-3}) observed on filter samples collected outside of a manufacturing facility (Barton et al. 2006).

Polychlorinated Naphtalenes

Like other POPs, PCNs are globally distributed in air, soils, water and biota. Atmospheric half-lives of PCN congeners range between 2 days (mono-CNs) and 343 days (octa-CN) (Puzyn et al. 2008). The Stockholm Convention (Annex D) recognizes an atmospheric half-life >2 days as a criterion for long range transport potential (UNEP 2001). Rural and semirural sites exhibited lower air concentrations than urban areas. Mean atmospheric PCN concentrations at an urban site (Chicago, USA) and a semiurban site (Toronto, Canada) were 68 and 17 pg m^{-3}, respectively. For urban air, approximately 40% was identified as 1,4,6-trichloronaphthalene (Harner and Bidleman 1997). Similarly, Dorr et al. (1996) reported PCN concentrations of 60 pg m^{-3} for urban air (Augsburg, Germany) and 24 pg m^{-3} for a rural area. Concentrations in urban air ranging up to 98 pg m^{-3} were reported by Helm et al. (2000), whereas concentrations at the Great Lakes (Canada/USA) ranged from 3 to 27 pg m^{-3}. More than 85% of the PCNs in the air samples were tri- and tetra-isomers. Lee et al. (2000) reported a mean PCN concentration of 152 pg m^{-3} for a semirural site at Lancaster, UK. They found that tri- and tetrachloronaphthalenes contributed >95% of the total. Harner et al. (2000a) in the air of urban areas in the UK found ΣPCN concentrations in a range of 60–150 pg m^{-3}. PCN levels of between 25 and 2900 ng m^{-3} were measured near a PCN manufacturing site in the USA, the congeners detected being mainly mono- (27%), di- (31%), and trichloronaphthalenes (37%), but other congeners were also detected (US EPA 1977; Erickson et al. 1978). Compared to urban or industrial areas, atmospheric burdens of PCN are much lower in remote areas such as the Arctic. Bidleman et al. (2010) reported ΣPCN air concentrations in arctic and subarctic regions ranging from 0.16 to 40 pg m^{-3}. Air concentrations of ΣPCN were found to be much higher at most European arctic–subarctic locations (up to 40 pg m^{-3} in the Barents sea area) than at sites in Siberia (0.7 pg m^{-3}), Iceland (0.9 pg m^{-3}), Alaska (2.3 pg m^{-3}) and the Canadian Arctic (up to 7.4 pg m^{-3}) (Bidleman et al. 2010).

Short-Chained Chlorinated Paraffins

SCCPs released to the atmosphere are expected to exist in the vapor and particulate phase in the ambient atmosphere. Vapor phase constituents are degraded in air by reaction with hydroxyl radicals with half-lives of less than one to slightly greater than 10 days. A half-life greater than about 2 days in the atmosphere can be a significant factor in facilitating long-range transport of chemicals. Barber et al. (2005) found concentrations of SCCPs in the UK atmosphere ranging between 185 and 3430 pg m^{-3}. Values were higher than 1997 levels at the same site. For the UK atmosphere, they calculated an average concentration of 600 pg m^{-3} of SCCPs. Borgen et al. (2000) measured SCCPs in Arctic air samples taken at Mt. Zeppelin, Svalbard, Norway in 1999. Concentrations ranged from 9.0 to 57 pg m^{-3} of SCCPs. The SCCPs were also detected in four individual samples of air collected in 1992 at Alert at the northern tip of Ellesmere Island in the high Arctic of Canada (Tomy et al. 1997a). Concentrations ranged from non-detectable to 8.5 pg m^{-3} in gas phase samples. Tomy et al. (1998) had previously analyzed air samples collected at Egbert, Ontario, in 1990 for SCCPs and found concentrations ranging between 65 and 924 pg m^{-3}, with chlorodecanes (C10) and chloroundecanes (C11) predominating. For Japan and China, SCCP concentrations in air were reported to be greater than in North America and Europe (Li et al. 2012). They ranged between 0.28 and 14.2 ng m^{-3} (Japan and China) and 13.5 and 517 ng m^{-3} (North America and Europe).

Polycyclic Aromatic Hydrocarbons

Concentrations of single PAH compounds measured in air samples in the USA ranged between 0.01 and 2 ng m^{-3} (Poster and Baker 1996). The dominating compound was PHE (~2 ng m^{-3}). The 3-ring PAHs (FLU, PHE, ANT) were mainly in the gaseous phase at concentrations of ~0.2–2 ng m^{-3}, whereas the 5-ring PAHs were particle bound at concentrations of ~0.01–0.04 ng m^{-3}. In Munich (Germany), concentrations of 16 EPA PAHs were between 1.9 and 5.0 ng m^{-3} and 0.8–2.9 ng m^{-3} at a major urban traffic junction and a residential site, respectively (Schauer et al. 2003). At a semirural site in UK concentrations of seven PAH (ACE, ANT, FLU, PHE, FLT, PYR, BaA) were between 1.4 and 40 ng m^{-3} (Lee and Jones 1999). Air samples collected in the Netherlands in 2000 and 2001 have yielded concentrations between 2 and 12 ng m^{-3} for 2- and 3-ring PAHs (highest value for NAP) whereas concentrations of the other compounds did not exceed 1 ng m^{-3} (Duyzer and Vonk 2003).

11.4.3 Concentrations of Organic Pollutants in Soils

Soil represents a major sink for organic pollutants. Agricultural soils as well as soils of public and private green areas are of particular interest as their fertility might be endangered by the accumulation of organic pollutants. As a consequence, its function for supply of healthy food for humans might be limited. Organic pollutants might enter soils via (i) atmospheric deposition (substances from incineration such as dioxins and ubiquitous compounds such as phthalates and plant protecting agents including organochlorine pesticides), (ii) road spray drift and deposition (mostly particle-bound), (iii) agricultural supplies such as fertilizers (mineral fertilizers, farmyard manure, sewage sludge, compost and other waste products), soil improvers and mulching agents (Brändli et al. 2004).

In soil, organic pollutants undergo several processes including volatilization, solubilization in water, sorption (adsorption and absorption) including formation of bound residues, physical, chemical or biological degradation, and bioaccumulation (uptake by soil organisms and plants). These processes are governed by the physico-chemical properties of the substances and the properties and conditions of soils (soil minerals, soil organic matter, pH, soil structure, water content, temperature, biological activity). In Chap. 10, the major principles of interactions of organic pollutants with soil and of degradation are described in detail. The following sections are exclusively focused on pollutant concentrations in soil.

Polychlorinated Biphenyls

The principal elimination pathway of PCBs from soils is volatilization. However, this process affects only the lightest congeners. This may explain why PCB concentrations in soils are still relatively high, although emissions of PCBs have decreased significantly during the last few decades. Meijer et al. (2003) estimated a global total soil PCB burden of 21,000 Mg. Basing on 191 global background surface (0–5 cm) soils they found differences of up to 4 orders of magnitude in concentrations of PCBs between sites. The lowest and highest PCB concentrations (26 and 97,000 pg g^{-1} soil) were found in samples from Greenland and mainland Europe (France, Germany, Poland), respectively. Background soil PCB concentrations were strongly influenced by proximity to source region and SOM content. A clear gradient in PCB concentrations is commonly found from remote over urban, suburban to industrial sites. Motelay-Massei et al. (2004) analysed soils of the Seine basin (France) and found PCB (Σ7 PCBs) concentrations in a range between 0.09 (remote sites) and 150 μg kg^{-1} (industrial areas). In Romania, Covaci et al. (2001) found mean concentrations of 4.0, 57.3 and 722 μg kg^{-1} (Σ9 congeners) for rural, urban and industrial sites, respectively. Notarianni et al. (1998) reported an average value of 0.8 μg kg^{-1} for remote sites and more than 20 μg kg^{-1} for urban sites. In Germany (Mannheim/Heidelberg region), PCB levels (Σ6 congeners) were higher in urban soils than in rural soils (median of 34 and 14 μg kg^{-1} respectively; Langenkamp and Part 2001). Soil samples from private gardens (30 μg kg^{-1}), parks (80 μg kg^{-1}) and industrial/traffic sites (48 μg kg^{-1}) exhibited higher concentrations than samples from arable land, grassland and forest (10–18 μg kg^{-1}).

Polybrominated Diphenyl Ethers

In surface soils (0–5 cm) from remote/rural woodland (coniferous and deciduous) and grassland locations on a latitudinal transect through the United Kingdom and Norway, concentrations as a sum of all PBDEs ranged between 65 and 12,000 ng kg^{-1} soil (Hassanin et al. 2004). PBDE-47, -99, -100, -153, and -154 as the major constituents of the penta-BDE technical products dominated the average congener pattern of the soils. In Danish soils PBDE concentrations were below 1 μg kg^{-1} for Σ4 PBDEs (BDE 47, 99, 100, 153), except for two soils (40.1 and 11.1 μg kg^{-1}, respectively) with had received high amounts of sludge (Vikelsøe et al. 2002). Alcock et al. (2003) reported values between 0.1 and 31.6 μg kg^{-1} for BDE 47 in soils from North America. On industrial sites in Taiwan and Japan concentrations up to 300 μg kg^{-1} for some congeners have been detected. Higher brominated PBDEs exhibited higher concentrations (Palm et al. 2002).

Polychlorinated Dibenzodioxins and Dibenzofurans

Major studies of different countries have indicated that PCDDs and PCDFs are ubiquitous in rural, urban and industrial areas and can persist in soils over timeframes ranging from decades to centuries or even longer (Weber et al. 2008). Background levels are about 1 pg (picogram) I-TEQ g^{-1} soil, whereas contaminated soils may show up to several ng (nanogram) I-TEQ g^{-1} (Eljarrat et al. 2001). Dettwiler et al. (1997) reported values in some arable soils of 2–5 ng I-TEQ kg^{-1}. In forest soils values were slightly higher probably due to interception. In urban areas concentrations of PCDDs/PCDFs can reach 10 ng I-TEQ kg^{-1} and on sites near point sources significantly higher values have been found (Brändli et al. 2004).

Hexachlorobutadiene

Soil plays an important role as a reservoir or a sink for HCBD, which was found to be toxic to soil microflora and earthworms (Neuhauser et al. 1985). However, in soils in Canada, HCBD could not be detected (detection limit: 0.05 μg g^{-1} soil) in 24 agricultural soils from across the country and in six samples from areas that had repeatedly received heavy applications of pesticides (Webber and Wang 1995). In contrast, soil samples from different industrial plants throughout the USA were found to have detectable levels of HCBD, and the concentrations were found to be associated with the types of production among the plants (Zhang et al. 2014). In China, Zhang et al. (2014) monitored HCBD levels in soils within and around a chemical plant. High concentrations of HCBD (27.9 ng g^{-1} soil) were found in the soil within the plant, which were most likely due to unintentional production as byproducts during the manufacturing of chlorinated hydrocarbons and other chlorinated chemicals. Rapid decreasing concentrations in soils were observed with increasing distance from the plant, indicating that the manufacturing plant is the primary source of these contaminants in this location. The concentrations in soil samples around the chemical plant ranged from 0.04 to 3.33 ng g^{-1} with a median concentration of 0.23 ng g^{-1} soil.

Pentachlorobenzene

Several analyses for different purposes yielded PeCB concentrations in soils ranging from non-detect to 1.7 ng g^{-1} soil (dry matter) (Ding et al. 1992; Gawlik et al. 2000). PeCB concentrations in background samples from woodland and grassland surface soils along an existing latitudinal UK-Norway transect were determined by Halse et al. (2015) on a soil organic matter (SOM) basis. The average concentration of PeCB in UK soils was about twice the average concentration for the Norwegian soils with 2 ng g^{-1} SOM and 1 ng g^{-1} SOM, respectively. The higher concentration of PeCB in UK soils may be due to proximity to past or ongoing source regions (Bailey et al. 2009). The average concentration of PeCB in grassland soils was more or less at the same level as woodland soils, which indicates that PeCB is not affected by the forest filter effect. There was a slightly stronger correlation between PeCB and SOM in Norwegian soils ($r = 0.80$) compared to UK soils ($r = 0.71$) which suggests that SOM may be somehow more important in controlling the occurrence in background soils in more remote regions of this transect.

Sediments and flood plain soils of the River Elbe watershed showed increased levels of PeCB, with concentrations up to 71 μg kg^{-1} (Witter et al. 1998; Schwarzbauer et al. 2001). PeCB concentrations in sediments from the Elbe River downstream from Hamburg ranged from 0.47 to 4.4 ng g^{-1} (Eder et al. 1987). Most soil and sediment samples collected in 1993 and 1994 in the Berlin area along the Spree and Havel rivers contained <10 μg g^{-1}, with maximum PeCB concentrations of 17 and 76 μg g^{-1}, respectively (Schwarzbauer et al. 2001).

Perfluoroctane Sulfonate

Illegal application of waste, accidents and spills, and the application of biosolids to agricultural land represent the most important sources of PFOS in soils (Sepulvado et al. 2011). However, data regarding concentrations of PFCs in soils are sparse in the literature and reported concentrations are often smaller than the detection limits (Zareitalabad et al. 2013). Strynar et al. (2012), based on analyses of ten soil samples from each of the countries USA, China, Japan, Norway, Greece, and Mexico estimated a global median soil concentration of 0.472 ng g^{-1} for PFOS. Li et al. (2010) reported much higher concentrations of PFOS for soils from Shanghai, China, where agricultural soils were not always less polluted than soils from residential or industrial areas. Application of sludge highly contaminated with perfluorinated compounds to agricultural land in North Rhine-Westphalia (Germany) led to concentrations of PFOS of a similar magnitude as the samples from Shanghai (LANUV 2010; cited by Zareitalabad et al. 2013). Concentrations larger than 500 ng g^{-1} soil were mostly associated with a heavy contamination of one site in Brilon-Scharfenberg (Zareitalabad et al. 2013), being part of the Möhne catchment. This contamination of soils was associated with severe contamination of surface water.

Polychlorinated Naphtalenes

Levels of PCNs have been measured in the USA in soils near various manufacturing plants where PCNs have been used (US EPA 1977; Erickson et al. 1978). Near a PCN manufacturing plant, levels of 130–2300 ng kg^{-1} were measured, made up of mainly tri-, tetra-, and pentachloronaphthalenes. PCN levels of between not detected and 21 μg kg^{-1} and between not detected and 470 μg kg^{-1} were measured near two capacitor manufacturing facilities, and levels ranging from not detected to 34 μg kg^{-1} were measured near a paper manufacturing plant.

Harner et al. (2000b) analyzed PCNs in rural soils in the UK dating back to the 1940s. They found a peak level of 12 μg kg^{-1} dry weight in the 1960s, falling to 0.5–1 μg kg^{-1} in 1990. More detailed analysis revealed that tetra- and pentachloronaphthalenes reached a peak in the 1950s, whereas peak values for trichlorinated isomers were recorded during the 1970s. PCNs have also been detected in contaminated soils from areas in the Netherlands that have been used for municipal waste disposal. The distribution of congeners was the same as that for Halowax 1013, suggesting that this was the source of contamination and that the composition of the PCNs had not changed, despite being buried in the landfill for 10–15 years. PCN levels of 31–38 mg kg^{-1} dry soil and 1180–1290 mg kg^{-1} dry soil were measured in two soils; a third soil contained no PCNs (De Kok et al. 1983). Kannan et al. (1998) found PCN concentrations of 17.9 mg kg^{-1} dry weight in soil near a former chlor-alkali plant. Hexa- and heptachloronaphthalene congeners accounted for >70% of the total concentration. Wang et al. (2012) reported PCN concentration and distribution in soils of the Pearl River Delta, China. The average total concentration was 59.9 ± 86.7 pg g^{-1}. Tri-CNs was the dominant homologue group, and CN 24 was the most abundant congener. There was a gradient of PCN levels between more and less developed areas.

Short-Chained Chlorinated Paraffins (SCCPs)

Halse et al. (2015) determined SCCPs in background samples from woodland and grassland surface soil on a SOM basis, collected along an existing latitudinal UK-Norway transect. They found a steep decline in concentrations of SCCPs with increasing latitude indicating that their occurrence is dictated by proximity to source regions. The average concentration of SCCPs for the UK sites was more than twice the average concentration for the Norwegian sites, with 50 ng g^{-1} SOM and 22 ng g^{-1} SOM, respectively. A forest filter effect could not be observed as the average SCCP concentration in grassland soils was 59 ng g^{-1} SOM, which is more than three times higher than the average concentration in woodland soils (17 ng g^{-1} SOM). Halse et al. (2015) from their results concluded that no sites at higher latitudes (>62° N) experienced concentrations of SCCPs above MDL which suggests that SCCPs have a limited potential for long range transport. However, studies by Reth et al. (2006) and Tomy et al. (1999) showed that SCCPs have been found in biota and sediments in the Arctic.

Due to sewage sludge application, arable soils are potentially major reservoirs of SCCPs. Nicholls et al. (2001) found total CP (SCCP + MCCP) concentrations in digested sewage sludge ranged from 1.8 to 93.1 μg g^{-1} dry matter. Stevens et al. (2002) found SCCP concentrations ranging from 6.9 to 200 μg g^{-1} dry matter in sewage sludge from 14 wastewater treatment plants in the UK. The highest concentrations of SCCPs were in sludge from industrial catchments. Wastewater irrigation as well can lead to higher accumulation of SCCPs in farm soils. For example, in farm soils from a wastewater irrigated area in China, SCCPs were detected in all topsoil samples, with the sum of all congeners in the range of 159.9–1450 ng g^{-1} (Zeng et al. 2012). Soil vertical profiles showed that ΣSCCP concentrations below the plough layer decreased exponentially and had a significant positive relationship with total organic carbon. Soil vertical distributions also indicated that lower chlorinated (Cl(5–6)) and shorter chain (C(10–12)) congeners are more prone to migrate to deeper soil layers compared to highly chlorinated and longer chain congeners (Zeng et al. 2012).

Polycyclic Aromatic Hydrocarbons

PAHs have been detected around the world even at sites which are located far from industrial activity such as the polar regions, and in the tropics (Wania and Mackay 1996). The PAH pattern in soil is dominated by two main types, which are indicative of background conditions on the one side (i.e., by biological and diffuse PAHs) and a strong impact by atmospheric deposition of anthropogenic emissions on the other side. Wilcke (2007) evaluated concentrations of 20 PAHs in 225 topsoil samples from 12 geographic regions. The Σ20PAHs concentrations ranged between 4.8 and 186,000 μg kg^{-1}. The maximum values were observed on roadsides in Germany. Concentrations of PAHs are strongly linked to the land use of the site. Mean concentrations of PAHs (16 US EPA priority pollutants) in topsoil in studies conducted in Europe between 1989 and 2000 were 328 μg kg^{-1} for arable soils, 284 μg kg^{-1} for grassland, 904 μg kg^{-1} for forest soils and 4420 μg kg^{-1} for urban soils (Wilcke 2000). The higher values of forest soils compared to agricultural soils most probably are due to higher interception rates of forest trees. Maliszewska-Kordybach (1996) measured an average of 264 μg kg^{-1} (for 16 PAHs) in agricultural soils in Poland. Aamot et al. (1996) found low values (144 μg kg^{-1} on average) in forest soils of Norway. In urbanized areas of Estonia, levels were between 2200 and 12,300 μg kg^{-1} (Trapido 1999). Mielke et al. (2004) found concentrations of 3700 μg kg^{-1} in the urban centre of New Orleans (USA). Studies on different soils in the UK exhibited values for 16 EPA PAHs ranging between 200 and 1000 μg kg^{-1} soil (Wild et al. 1991a, b). In a study considering 20 PAHs in topsoils of urban soils and rural sites with different management in Germany, values ranged between 160 and 186,000 μg kg^{-1}, with the maximum values on roadsides (Krauss and Wilcke 2003). In private gardens and industrial sites, concentrations were up to 10,000 μg kg^{-1}. The PAH concentrations generally decreased from the central urban to the rural area. In Denmark, levels in soils with high long-term sewage sludge application ranged between ~2000 and 2500 μg kg^{-1} soil (Vikelsøe et al. 2002).

11.4.4 Concentrations of Organic Pollutants in Water

Polychlorinated Biphenyls

PCBs (sum of 61 congeners) measured in rain water exhibited concentrations ranging from 0.97 to 2.2 ng L^{-1} (Brändli et al. 2004). Most PCBs were predominantly dissolved. Concentrations, profiles and dissolved/particulate fractions were observed to be significantly different for different rain events (Poster and Baker 1996). Duyzer and Vonk (2003) found concentrations of PCBs (congeners 8, 20, 28, 35, 52, 101, 118, 153, 138, 180) in the range of 0.1–25 ng L^{-1} for single compounds. Dominating compounds were PCBs 20, 28, 52 and 101, each with concentrations above 1 ng L^{-1}. In Switzerland, rain water concentrations of PCBs (congeners 28, 52, 101, 153, 138, 180) have been found between 2 and 12.7 ng L^{-1} in 1982–1984 (De Alencastro 1995). Background concentrations for the sum of 13 PCBs in river water measured in Canada were between 0.19 and 0.55 ng L^{-1} (Pham et al. 1999). In a study on water quality of the Detroit river (USA) the sum of bioavailable PCBs was in the range of 0.5–3.0 ng L^{-1} (Gewurtz et al. 2003). In Lake Michigan (USA), mean total PCB concentrations between 1980 and 1991 declined from 1.2 to 0.47 ng L^{-1} (Pearson et al. 1996).

Polybrominated Diphenyl Ethers

In rain water, PBDEs have been detected in the low ng L^{-1} range (Peters 2003). In coastal water from the Netherlands, values of BDE 47, 99, 153 and 209 were in the low pg L^{-1} range (Watanabe and Sakai 2003). Similarly, BDE 47 in background water from the USA was in the low pg L^{-1} range (Alcock et al. 2003). Concentrations of PBDEs in whole effluents from publicity-owned wastewater treatment works are usually in the sub-μg L^{-1} range. However, high water volumes discharged can yield in significant PBDE amounts. In a wastewater treatment effluent serving a US plastics product manufacturer BDE-209 at 12 μg L^{-1} was observed, most of which was associated with the suspended solids released (La Guardia et al. 2004). Analyses of effluent particulates in The Netherlands supports this view, where BDE-209 exceeded 100 μg kg^{-1} dry matter (de Boer et al. 2003). Water releases from textile facilities in the Netherlands exhibited BDE-209 burdens up to 4600 μg kg^{-1} on suspended particulates of surface waters, decreasing with distance from the facilities (de Boer et al. 2003).

Polychlorinated Dibenzodioxins and Dibenzofurans (PCDD/PCDF)

Due to their high affinity for soil organic matter and low water solubility, PCDD/PCDFs are generally considered relatively non-mobile. At present, only limited published documentation exists on leaching of PCDDs/PCDFs from soils to groundwater and to rivers. In non-contaminated areas, concentrations are commonly in the low pg I-TEQ L^{-1} range (Fiedler et al. 1994). A comparison of PCDD/PCDF loads derived from pesticide application and their deposition in sediments of the Tokyo bay (Japan) showed that only a small portion of PCDD/PCDF agrochemical impurities have been deposited in the Tokyo Bay sediment during the past 45 years (Masunaga 2004). This indicates that a large part of the PCDD/PCDF load still exists in the terrestrial soils or river sediments of this area, which resulting in a significant future leaching/runoff potential. However, experiences from landfills receiving PCDDs/PCDFs have revealed that such compounds can definitely leach from landfills into groundwater in high amounts (Persson et al. 2008). Experiences from the former pesticide production site and related landfills in Hamburg revealed that PCDD/PCDF, pesticide residues and other chlorinated aromatics and aliphatics can be leached readily from such sites depending e.g. on the quality of the landfill, the geological conditions and other wastes co-deposited with the PCDDs/PCDFs and pesticide residues. The high concentrations of these POPs in leachates from these landfills yielded concentrations of 2,3,7,8-TCDD up to 75,000 ng kg^{-1} in oily leachates from one landfill (Schnittger 2001). This case highlights that leachates and ground water around landfills/dumps and hot spots of former pesticide production sites need to be monitored for contaminants including PCDDs/PCDFs.

Hexachlorobutadiene

A study of HCBD in 108 samples of seawater collected in 1983 and 1984 from the Dutch coast of the North Sea reported an average HCBD concentration of 0.28 ng L^{-1} (Van de Meent et al. 1986). Pearson and McConnell (1975) reported average concentrations of 4 ng L^{-1} in the Liverpool Bay (UK) with maximum levels of 30 ng L^{-1}. A statistical analysis of the monitoring data of the EU COMMPS database (1998), which contains more than 10,000 measured HCBD concentrations from rivers of six European countries (Belgium, Germany, Spain, Greece, UK, The Netherlands) over the period 1994–1997 showed a mean value of 6 ng L^{-1} (Govaerts et al. 2004). Only 13% of the measured values were above the detection limit. However, Goldbach et al. (1976) reported values of HCBD near the mouth of the river IJssel in The Netherlands of about 130 ng L^{-1}. In the USA, HCBD was detected at 0.9–1.9 μg L^{-1} in Mississippi River water (IARC 1979). In Canadian surface waters, the highest reported concentration of HCBD was 1.3 μg L^{-1}, which was measured in the St. Clair River (Ontario) in 1984 (Lecloux 2004). Levels have decreased substantially since 1984, and after 1990, concentrations of HCBD in surface water from southern Ontario generally did not exceed 0.001 μg L^{-1} (Environment Canada 2000). HCBD has been detected at 0.27 μg L^{-1} in European drinking water and at 6.4 μg L^{-1} in the effluent from a European chemical plant (IARC 1979).

Pentachlorobenzene

A study of the distribution of chloroorganics in the North Pacific Ocean, the Bering and Chukchi Seas 1993 found PeCB in all samples (Strachan et al. 2001), with an average of 16 pg L^{-1} in the dissolved phase and 0.38 ng L^{-1} in suspended solids. Strachan et al. calculated that the flow of water northward into the Arctic Ocean carried 0.31–0.52 Mg y^{-1} PeCB. Yangtse River water near Nanjing analyzed for organochlorines in 1998 yielded total PeCB concentrations of about 0.4 ng L^{-1} (Jiang et al. 2000). In the Niagara River in North America, concentrations of PeCB during 1981–1983 ranged from 0.34 to 6.4 ng L^{-1} with a mean of 1.3 ± 1.0 ng L^{-1} (Oliver and Nicol 1984). At the inflow of the Niagara River into Lake Ontario concentrations of PeCB during the period 1987–1997 showed annual mean total (dissolved plus particulate) concentrations dropping from 0.351 to 0.093 ng L^{-1} (Williams et al. 2000).

Perfluoroctane Sulfonate
PFOS was found in surface fresh water samples from the cities of Calgary and Vancouver (Canada) in concentrations ranging from <0.05 to 0.1 ng L^{-1} (Tanaka et al. 2006). According to a number of other studies carried out on surface water of the Great Lakes region (Sinclair et al. 2004, 2006; Kannan et al. 2005), the PFOS concentrations were in the range <0.8–13 ng L^{-1}. In Tennessee river water (40 locations) in 2000 average PFOS concentrations were 32 ± 11 and 114 ± 19 ng L^{-1}, respectively upstream and downstream of the discharge of a fluorochemical plant (Hansen et al. 2002). PFOS was measured in drinking water samples collected from 1999 to 2000 in the cities Decatur, Mobile, Columbus, and Pensacola in that perfluorinated compounds were either manufactured or industrially used (EFSA 2008). Cleveland and Port St. Lucie were studied as controls. Only in Columbus and Pensacola PFOS was detected in drinking water-related samples (rawater, treated water, and/or tap water) with levels of 59 ng L^{-1} (average of 10 data) and from non-detect to 45 ng L^{-1}, respectively (EFSA 2008). The treatment process seemed to have little influence on the concentrations of PFOS. For Japan and other Asian areas Harada et al. (2003), Saito et al. (2004), and Tanaka et al. (2006) reported PFOS in drinking water at levels of <0.05–12.0 ng L^{-1}. In the Tokyo area of Kinuta, the PFOS concentrations were higher (up to 50.9 ng L) probably due to contamination of the Tama river from which the Kinuta Waterworks took the fresh water supply to treat into drinking water (Harada et al. 2003). PFOS levels in surface fresh waters from several locations in Asia, mostly in Japan but also in China, Malaysia, Thailand, and Vietnam were reported in the range <0.01–12 ng L^{-1} (Harada et al. 2003; Taniyasu et al. 2003a; Tanaka et al. 2006; So et al. 2007). Concentrations up to 135 ng L^{-1} were reported for several Japanese rivers by Saito et al. (2003). A concentration as high as 157 ng L^{-1} was found in vicinity of industrial waste water discharges. Environmental data from Norway, The Netherlands, and other European locations provided PFOS concentrations in surface fresh water of <0.02–0.48, <10–56, and <2–26 ng L^{-1}, respectively (Kallenborn et al. 2004; de Voogt et al. 2006; Skutlarek et al. 2006; Weremiuk et al. 2006). In Lake Maggiore (Italy) surface waters, PFOS was detected in the range 7.2–8.6 ng L^{-1}, whereas in nearby Alpine rivers the level of the chemical was close to non-detect (0.1 ng L^{-1}) (Loos et al. 2007). Concentrations in European drinking water samples were in the range 0.4–9.7 ng L^{-1}. However, in the Ruhr area in North Rhine-Westphalia (Germany), also determined PFOS in drinking water was in the range <2–22 ng L^{-1} (Skutlarek et al. 2006), likely reflecting contamination from the area. PFOS was measured by the same authors at concentrations between <2 and 193 ng L^{-1} in surface water of the rivers Ruhr and Möhne (river Rhine hydrological system). In selected tributaries of the river Möhne, concentrations up to 5900 ng L^{-1} were detected. Water contamination most likely stemmed from inorganic and organic waste materials applied to agricultural land on the upper part of the Möhne catchment.

Polychlorinated Naphtalenes

In the USA, levels of PCNs in water have been measured at various manufacturing sites where PCN use was suspected (US EPA 1977; Erickson et al. 1978). PCN levels of 0.6 and 1.4 μg L^{-1} were measured in two water samples near a PCN manufacturing plant. Levels of not detected to 5.5 μg L^{-1} were measured in water near two capacitor manufacturing sites. Monochloronaphthalene at levels of 650–750 ng L^{-1} and dichloronaphthalene at levels of 150–260 ng L^{-1} have been detected in the River Besós and River Llobregat, Barcelona, Spain. Both rivers receive a wide spectrum of waste discharges, including domestic, industrial, and agricultural wastes (Gomez-Belinchon et al. 1991). In groundwater, total PCNs (expressed as Halowax 1099 equivalents) ranged from <0.5 ng L^{-1} to 79.1 μg L^{-1} for the Llobregat aquifer, with tri- and tetrachloronaphthalenes the major groups of congeners identified. The authors reported that the higher levels probably originated from the poor disposal of illegal landfills closed during the 1970s (Espadaler et al. 1997; Martí and Ventura 1997). Total PCN concentrations of 0.89 and 2.6 ng L^{-1} were reported for a PCB-polluted river and percolating water at a city dump site (Stockholm, Sweden), respectively (Järnberg et al. 1997). Levels of chloronaphthalene and dichloronaphthalene have been measured in two samples of Tsukuba (Japan) tap water after chlorination. The detection limit in the experiment was 0.003 ng L^{-1}, and the levels of both chloronaphthalene and dichloronaphthalene were below the detection limit in the raw water before chlorination. After chlorination, levels of 0.03–0.44 ng L^{-1} for chloronaphthalene and levels of not detected to 0.15 ng L^{-1} for dichloronaphthalene were measured (Shiraishi et al. 1985).

Short-Chained Chlorinated Paraffins

Water samples collected in mid-Lake Ontario in July 1999 and October 2000 showed levels of SCCPs, with highest concentrations of 1.8 ng L^{-1} and 0.8 ng L^{-1}, respectively (Muir et al. 2001). Concentrations of SCCPs of 30 ± 14 ng L^{-1} were measured in the Red River in Selkirk, Manitoba (USA), over a 6-month period in 1995 (Tomy 1997b). Tomy et al. (1999) attributed the SCCPs in the water to a local source, possibly a metal machining/recycling plant in a neighboring town. In Canada, Tomy et al. (1997a, b) reported SCCP (C10–C13) levels ranging from 0.02 to 0.05 mg l^{-1} in water from the Red River downstream of the city of Winnipeg, a manufacturing center. In the UK, Willis et al. (1994) reported C10–C30 PCA concentrations ranging from 0.6 to 4 mg l^{-1}. Ballschmiter (1994, cited by Muir et al. 2000) reported C10–C30 PCA levels ranging from 0.05 mg l^{-1} for the River Lech, Augsburg Germany, to 1.2 mg l^{-1} for the River Danube, both close to industrial centers. The Ministry of the Environment (2006) in Japan monitored SCCPs in six surface water samples from across the country and did not find any concentrations above the detection limits.

Polycyclic Aromatic Hydrocarbons

In several studies, PAHs were measured in rain water. Concentrations for single compounds commonly did not exceed ~10 ng L^{-1}, except for the study by Duyzer and Vonk (2003) who found higher concentrations for single compounds between

3 and 110 ng L^{-1}. The dominating compounds in this study were NAP, FLU, PHE and PYR with concentrations exceeding 40 ng L^{-1}. The profiles were dominated by 5-ring PAHs and PHE. Concentrations and profiles differed significantly for different rain events (Poster and Baker 1996).

Due to their hydrophobic nature, PAHs entering the aquatic environment exhibit a high affinity for suspended particulates in the water column. After entering water, the physiochemical properties of PAHs make them quickly become adsorbed to organic or inorganic compounds and are mostly deposited in bottom sediments. Once adsorbed they are much more stable than pure compounds and are resistant to oxidation reactions to which they would otherwise be quite sensitive due to photochemical processes (Catoggio 1991). Low solubility of PAHs and the hydrophobic sorptive capacity (K_{ow}) correspondingly high, coupled with low volatilities and chemical stability mean that PAHs are environmentally persistent compounds that are strongly held to solids, both suspended particles and bottom sediment. PAH concentrations in water are usually quite low relative to the concentrations in bottom sediments (Moore and Ramamoorthy 1984). River water concentrations have been reviewed by Manoli and Samara (1999). Concentrations for single compounds were mainly below 50 ng L^{-1}. Background concentrations of 14 PAHs measured in river water in Canada were between ~10 and 20 ng L^{-1} (Pham et al. 1999). In the plume of sewage treatment, concentrations ranged between ~10 and 50 ng L^{-1}.

11.5 Organic Pollutants in Non-target Organisms

11.5.1 Plants

Vegetation is the link between the soil and atmosphere on one hand and the human and animal food supply on the other. Bioaccumulation of organic pollutants in plants has been recognized to be a serious threat to environmental and human health. Organic pollutants may enter vegetation by uptake from contaminated soil to the roots and be translocated in the plant by the xylem (Simonich and Hites 1995). They may also enter plants from the atmosphere by gas and particle deposition onto the waxy cuticle of the leaves or by uptake through the stomata and be translocated by the phloem (Fig. 11.5).

The uptake via the atmosphere is a function of the vapor-particle partitioning (V/P), the octanol-air partition coefficient (K_{oa}), the plant surface area (SA), the plant lipid concentration (lipids). The uptake from the soil depends on the octanol-water partition coefficient (K_{ow}), the water solubility (SOL_W), Henry's law constant (H), the soil organic matter content (SOM) and the plant species. Uptake from soil through a plant's roots is the predominant pathway of accumulation for organic compounds that have high water solubilities, low Henry's law constants, and low K_{ow} values. Compounds with a log K_{ow} less than about 4, do not partition to a great extent in soils rich in SOM (Simonich and Hites 1995). These compounds

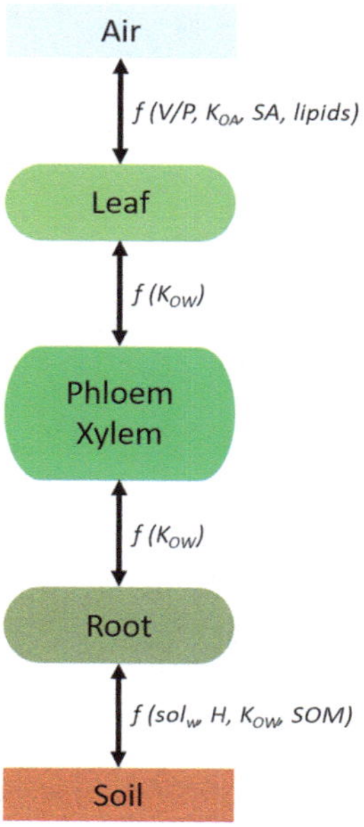

Fig. 11.5 Mechanism of pollutant uptake by plants (Adapted from Staci and Hites 1995)

move from the outer root to the inner root, are drawn into the plant by the xylem with short response times, and are distributed within the plant depending on their lipophilicity. In contrast, many lipophilic organic pollutants (log K_{ow} greater than about 4) such as PCDDs PCDFs, PCBs and PAHs partition to the epidermis of the root or to the soil particles and are not drawn into the inner root or xylem.

For lipophilic organic pollutants the main accumulation pathway is from the air to the leaf surface. The partitioning of lipophilic pollutants from the outer leaf to the inner leaf is slow. These compounds are thus rarely transported by the phloem. In general, gas-phase pollutants with a large K_{oa} are preferentially accumulated. The lipid concentration of the leaf and the leaf area index also influence the degree of accumulation (Simonich and Hites 1994). Lipid contents vary between species and within individual plants depending on the age of leaves and the season. However, it remains unclear which of the wild and cultivated plant species generally possess the largest potential for accumulation of organic pollutants, but it is widely accepted that conifer needles and kale (*Brassica oleraceae* var. acephala) are suitable species for the biomonitoring of organic air pollutants because of their high lipid contents. Partitioning of volatile organic pollutants between the atmosphere and vegetation is also dependent on the temperature. For example, Nakajima et al. (1995) found that at high ambient temperatures (summer), some PAHs volatilized back to the atmosphere and at low ambient temperatures in autumn and winter, PAH partition to vegetation. Higher lipid contents of plants in the winter might also favor accumulation of lipophilic air pollutants. While the PAH-profiles (i.e. "fingerprints", contribution of single compounds to PAH-sum) detected in different species are more

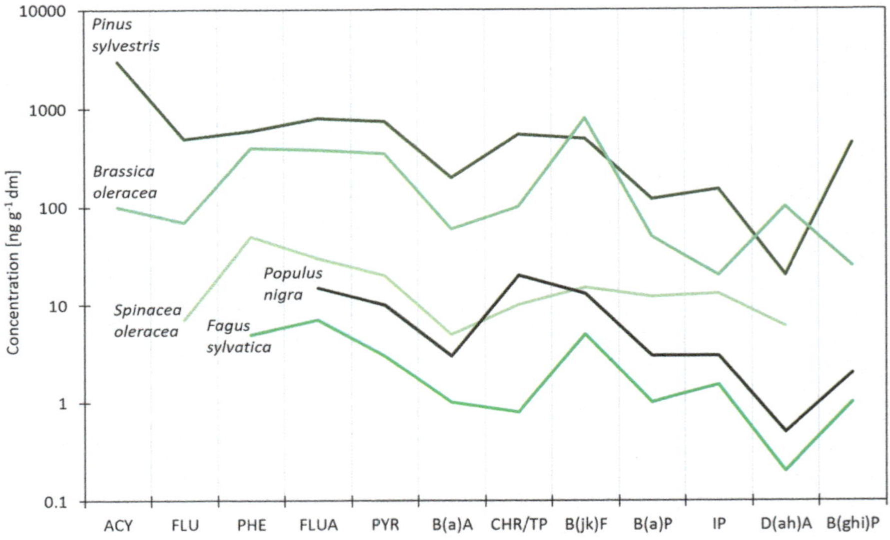

Fig. 11.6 PAH-concentrations in different plant species. The arrangement of PAHs is according to molecular weights (from left to right: *ACY* acenaphthylene, *FLU* fluorene, *PHE* phenanthrene, *FLUA* fluoranthene, *PYR* pyrene, *B(a)A* benzo(a)anthracene, *CHR/TP* chrysene/triphenylene, *B (jk)F* benzo(j,k)fluorene, *B(a)P* benzo(a)pyrene, *IP* indenopyrene, *D(ah)A* dibenzo(a,h)anthracene, *B(ghi)P* benzo(ghi)perylene) (Adapted from Franzaring and van der Eerden 2000)

or less identical, accumulation of organic air pollutants may vary some orders of magnitude (Fig. 11.6).

This particularly reflects different levels of atmospheric pollution, but morphology and physiology of different plant species might be important as well. Grasshopper effects (volatilisation in warmer climates and sorption in colder climates) may lead to the gradual accumulation of POPs in arctic ecosystems, boreal forests and mountain ecosystems (Franzaring and van der Eerden 2000). By scavenging large amounts of organic micropollutants from the atmosphere, plants may play a significant role in introducing organic pollutants into food webs. For example, Hülster et al. (1994) has shown that zucchini and pumpkins (both *Cucurbita pepo* L.) accumulate and translocate higher concentrations of PCDDs/PCDFs from contaminated soil than other fruits and vegetables, and this is the main contamination pathway for these species. Although the primary route of uptake for cucumber plants was found to be from the atmosphere, zucchini and pumpkin plants clearly showed uptake from PCDD/PCDF-contaminated soil. This may be due to root exudates, unique to these species that mobilize PCDDs/PCDFs from the soil particles and make these compounds available for uptake (Hülster et al. 1994).

11.5.2 Aquatic Food Chains

The concentrations of POPs in animals are often greater than the concentration of POPs in the animals nearby environment or the animal's food. The processes by which this occurs are called bioconcentration, bioaccumulation, and biomagnification (Gobas and Morrison 2000). Bioconcentration is the uptake of a chemical by an aquatic organism through respiration and dermal absorption, resulting in a greater concentration of the substance inside the organism than in the water. Bioaccumulation is the process by which the concentration of a chemical is greater in an aquatic organism than that in water, due to all exposure routes, including through diet, respiration, and dermal absorption. Through biomagnification, the concentration of the substance in the organism exceeds that of the organism's diet. A consequence of biomagnification is that the concentration of the substance increases at some or all steps in a food chain (Norstrom 2002).

Toxic and bioaccumulative pollutants are commonly found in only trace amounts in water, and often at elevated levels in sediments. Risk assessments based only on data derived from water analyses may thus be misleading. On the other hand, data from sediments may not be representative of pollutant concentrations in the overlying water column. Aquatic food chains are vulnerable to contamination by POPs as a result of the lipophilic characteristics of these chemicals and their resistance to breakdown. Organisms occupying high trophic levels are often exposed to high concentrations of such chemicals as a result of biomagnification (Fig. 11.7).

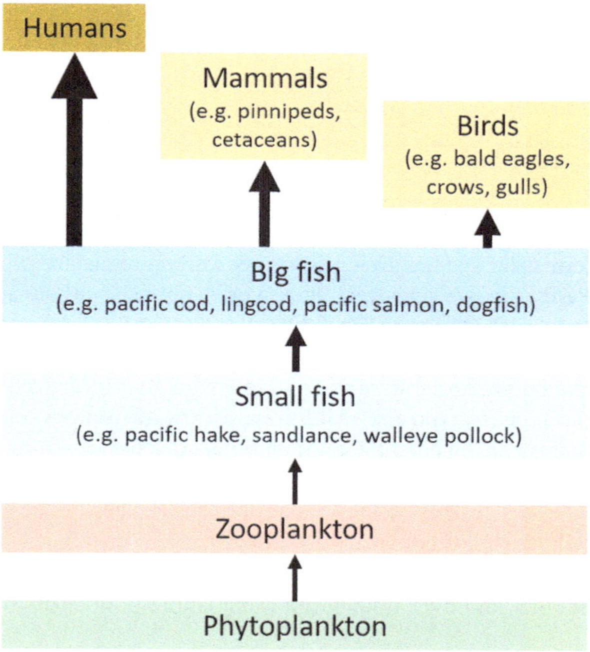

Fig. 11.7 Aquatic food chain (Adapted from Ross and Birnbaum 2003)

Since microalgae occupy the base of the food chain and are typically the most abundant life forms in aquatic environments, they are a particularly promising indicator species for organic pollutants. They play an important role in the dispersal, chemical transformation and bioaccumulation of many toxic compounds. Evidence points to a coupling between microalgae uptake and air–water organic pollutant concentration. Air–water exchange dynamics are influenced not only by physical parameters but also by phytoplankton biomass and growth rate. Pollutants with low octanol–water partition coefficients (log $K_{ow} < 5$) can be taken up and accumulated by aquatic organisms while more hydrophobic pollutants (log $K_{ow} > 5$) may partition in lipid membranes of cells leading to their biomagnification (Torres et al. 2008). Thus, microalgae may reduce pollutant exposure to organisms that do not consume them either directly or indirectly. Conversely, as a food source, microalgae may facilitate the uptake of contaminants into higher organisms, increasing the possibility of toxicity.

11.5.3 Wildlife, Fish, Aquatic Mammals

Many POPs have been detected in wildlife both near the primary emissions source and far away. The use of wildlife as early warning indicator has become an increasingly important issue in the area of human health. The exposure of wildlife to POPs in the environment will come from air, water, food, soil or sediment. These chemicals can be taken up by organisms from their diet, by inhalation or by absorbing them through the skin, where they travel in the blood to specific tissues and affect the endocrine system. Measureable concentrations of POPs tend to be found at the highest concentrations in animals at the top of the food web (e.g. humans, seals, polar bears, birds of prey, crocodilians) and in tissues and body fluids that are high in fat (e.g. blubber, mothers' milk, egg yolk). As top predators polar bears have some of the highest levels of PFOS, PCBs and other POPs of any species due to the long-range transport, deposition and food web biomagnification of these chemicals (Ramsey and Hobson 1991). While exposures of species living in remote locations are important to understand, aquatic and terrestrial species living in or near urban areas are often continuously exposed to POPs via sewage treatment works outfalls, urban and agricultural runoff, and industrial effluents (Fig. 11.8).

Once effluents are discharged to aquatic environments, POPs will be diluted in stream or river waters so that organisms living very close to the discharge will have the highest exposure. Additional exposure sources, especially in marine environments, are accidental or intentional discharges from oil tankers, ships and fuel extraction activities and oil spills. Water is the main route of exposure for wildlife to POPS. Fish will take them up through their gills, whereas birds and mammals will be exposed primarily through their drinking water. A diverse mixture of chemicals is present in surface waters and concentrations vary from one site to another and over time at the same site (e.g. Focazio et al. 2008). When EDCs are

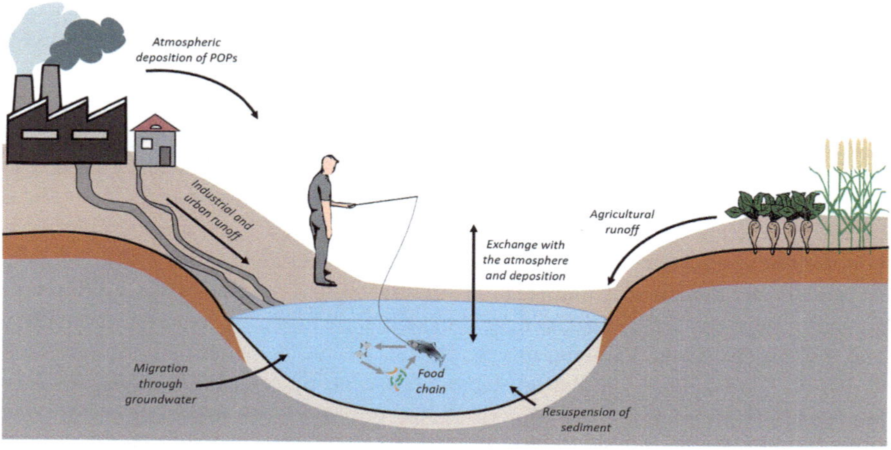

Fig. 11.8 Exposure of wildlife and fish in urban regions due to continuous release of POPs in effluents and to the atmosphere

released into the environment, some will bind to soils or to particles and sediments in rivers or other waterways. Organisms living in or on the soils or sediments (worms, snails, some insects) are exposed to these particle-bound POPs. Diet as well is an important source of POPs for wildlife.

Because of high affinities for fats and low solubility in water POPs are known to concentrate in organisms and through food webs, with higher levels in fish eating species than those that feed lower on the food web. The concentration of POPs in organisms depends on the type of diet it consumes. Diets high in fats will have the highest levels of fat-soluble chemicals like POPs. Because these chemicals are taken up into the body faster than they are lost, there is an accumulation of many of them as an organism grows and ages such that higher levels of POPs are found in the older, bigger animals. The most well studied POPs in wildlife tissues include PCBs, PCDDs/PCDFs and some dioxin-like PCBs, PBDEs, PFOS, PFCAs, PAHs and organochlorine pesticides such as DDT and DDE (for the latter see Chap. 10).

Polychlorinated Biphenyls

The global distribution of PCBs and chlorinated pesticides in skipjack tuna (*Katsuwonus pelamis*) has been reported by Ueno et al. (2003). The study regions included the Pacific Ocean, the Indian Ocean and the southwestern Atlantic. Elevated concentrations of ΣPCBs (>1000 ng g^{-1} fat) were found in tuna samples from the western Pacific, more specifically in the south China Sea and Sea of Japan. Relatively high levels of ΣPCBs were also found in tuna collected off the coast of Brazil (about 400 ng g^{-1} fat). In contrast, ΣPCBs were relatively low in samples from the Indian Ocean. In some animals, these chemicals have been linked with population declines. Dolphins (*Delphinidae*) are globally distributed and measurements exist from dead/stranded animals from around the world (Houde et al. 2005).

In general, the highest levels of ΣPCBs were found in species inhabiting the midlatitudes of Southern Europe (up to 1300 ng g^{-1} fat), industrialized Asia (>500 ng g^{-1} fat) and North America (>200 ng g^{-1} fat), reflecting the areas where these chemical compounds have been intensively used.

A wide ranging study of polar bears from Alaska, Canada, East Greenland and Svalbard sampled between 1996 and 2002 focused on geographical variations in chemical concentrations (Verreault et al. 2005). Concentrations of PCB + TeCB (1,2,3,4-tetrachorobenzene) were relatively uniform between these polar bear populations spread over about half of the Arctic with an average of 29.8 ± 9.5 ng g^{-1} lipid. PCB has also been detected in Antarctic biota. Corsolini et al. (2006) report concentrations of PCB to be 0.05 ng g^{-1} in krill, 0.09 ng g^{-1} in rockcod whole body and 0.68 ng g^{-1} in adelie penguin eggs.

Whole body bioaccumulation factors (BAFs) of pentachlorobenzene vary between 813 (mussel, *Mytilis edulis)* (Renberg et al. 1986) and 20,000 (rainbow trout, *Oncorhynchus mykiss)* (Oliver and Niimi 1983). A BAF of pentachlorobenzene of 401,000 was reported for earthworms *(Eisenia andrei)* (Belfroid et al. 1993).

Polybrominated Diphenyl Ethers

Contamination of wildlife by PBDEs has been documented worldwide over the past 10 years and reviews of spatial and temporal trends are available (e.g. Chen and Hale 2010; de Wit et al. 2010; Law et al. 2008). The major congeners detected were BDE-47, -99, -100, -153 and -154, resulting from the use of Penta- and Octa-BDE products. These congeners all exhibit significant trophic biomagnification, especially in aquatic food webs with fish and piscivorous birds as top predators. Skipjack tuna were also used to assess the global distribution of PBDEs and HBCDD (Ueno et al. 2006). Relatively high concentrations of ΣPBDEs and HBCDD were found in samples from the South China Sea and Sea of Japan, areas which are near large urban and industrial areas with use, recycling and manufacturing sites for BFRs. Relatively high levels of ΣPBDEs were also found in tuna collected off of the coast of Brazil. In contrast, ΣPBDEs and ΣHBCDs were relatively low in samples from the Indian Ocean. According to Chen and Hale (2010), North American birds (up to 23,600 ng g^{-1} fat in tissues or eggs) exhibited much higher concentrations of ΣPBDEs than those from Europe and Asia, each with concentrations up to about 2000 ng g^{-1} fat. This difference is in accord with the fact that the North American market has encompassed the bulk of the world's PentaBDE production. Major differences were also apparent between remote and urban regions. The Canadian and European Arctic (Svalbard) and Antarctica were much lower in concentrations compared to samples collected near urban areas. Chen and Hale (2010) also reviewed the available data on BDE-209 in birds around the world. Highest concentrations were observed in Chinese kestrels and USA peregrine falcons. The distinctive regional patterns observed with ΣPBDEs are not as apparent for BDE-209, with both North American (up to 490 ng g^{-1} fat in tissues or eggs) and East Asian birds (up to 1030 ng g^{-1} fat in tissues or eggs) having high exposures which may be explained by DecaBDE being the major PBDE product used in Asia, at least in the early 2000s (Chen and Hale 2010).

Time trend data in the Baltic related to PentaBDE products in guillemot eggs show significantly increasing concentrations of BDE-47 and BDE-99 and other congeners from the late 1960s until the early 1990s followed by decreasing values during 1998–2008 (Bignert et al. 2010). In juvenile polar bears from East Greenland ΣPBDEs achieved maximum concentrations in 2004–2005 (Dietz et al. 2012), which is about 10 years later than for top predators in the Baltic (Bignert et al. 2010) and in UK coastal waters (Law et al. 2010). This presumably reflects delayed exposure of the Greenland polar bears due to long-range atmospheric and oceanic transport.

Perfluorooctane Sulfonate

The presence of PFOS and PFOA in surface and tap water pose some environmental concerns because they could be taken up by animals that inhabit the surface water and bioaccumulated in the food chains. PFOS is the predominant PFC found in all species, tissues, and locations analyzed around the world (Houde et al. 2011). A survey conducted by Giesy and Kannan (2001) found PFOS to be distributed widely in animal tissues, even in such remote locations as the Arctic and North Pacific Oceans. The livers of some Alaskan polar bears were found to contain 180–680 ng g^{-1} PFOS, while the livers of some minks from the Midwestern USA, a more contaminated location, contained 970–3680 ng g^{-1} PFOS.

Taniyasu et al. (2003b) found the presence of PFOS in blood and liver samples of fish collected from several places in Japan at concentrations of 2–834 ng ml^{-1}. The highest mean PFOS concentration of 345 ng ml^{-1} was in the fish blood collected from Lake Biwa, followed by those collected from Tokyo bay (172 ng ml^{-1}), Osaka bay (100 ng ml-1), Seto inland sea (29 ng ml^{-1}), Ariake bay (28 ng ml^{-1}) and Okinawa (10 ng ml^{-1}). Although the PFOS concentration in Lake Biwa (a fresh water lake) was about four times less than that of Tokyo bay, as reported in the previous section, the PFOS concentration in the fish blood collected from Lake Biwa was the highest. This phenomenon could be due to the lake hydrodynamics and biological effects of the freshwater ecosystem which, in combination with other factors, probably contributed to the high PFOS accumulation in the fish blood.

In the USA, PFOS was found in all liver samples of fish and birds collected from New York State at concentrations of 9–315 and 11–882 ng g^{-1} (wet weight), respectively (Sinclair et al. 2006). Comparing with the PFOS concentrations in the water bodies, a bioconcentration factor (BCF) of PFOS in fish in this area was suggested as 8850. Based on the ratio of PFOS concentrations in fish-eating birds and in fish, a biomagnification factor (BMF) of PFOS in birds was estimated to be 8.9. Giesy and Kannan (2002) estimated a BMF of PFOS in fish-eating minks to be 22. From the PFOA data, Sinclair et al. (2006) estimated a BCF of PFOA in fish of the New York State to be 184, much less than that of PFOS. The BCFs of other POPs such as PAHs, PCDDs, PCDFs, DDT, and PCBs were reported to be 44–10,000; 135–31,600; 360–10,700; 37,200–69,200; and 1800–933,300, respectively (Lu et al. 2000), which are comparable with those of PFOS.

Polycyclic Aromatic Hydrocarbons

These chemicals have been widely measured in wildlife, often as part of monitoring responses to oil pollution events (Hellou 1996). In marine mammals, levels of PAHs in blubber are generally lower than POPs (Hellou 1996, Reijnders et al.). Lower molecular weight (two to four ring) PAHs predominate in most marine mammal blubber samples (Fair et al. 2010). Relatively high ΣPAH concentrations (on a fat weight basis) have been reported in blood. For example, sea otters from Alaska and the California coast had average Σ26PAH in blood serum ranging from 3.1 to 9.8 $\mu g\ g^{-1}$ fat (Jessup et al. 2010). Sea otter livers from the same regions had similar Σ16PAH (16 unsubstituted "priority" PAH) concentrations when expressed on a fat weight basis (Kannan et al. 2008). Highest Σ16PAH were from sea otters collected in Prince William Sound, the site of the Exxon Valdez oil spill. There has been limited study of PAHs in seabird tissues. Measurements of Σ16PAH in seabird livers from the Mediterranean and Eastern Atlantic found at low ng g^{-1} wet weight concentrations (Roscales et al. 2011). Seabird eggs have also been shown to generally have low PAH concentrations, with few distinctive differences between geographic areas (Pereira et al. 2009). Seabird fat from King George Island (Antarctica) had ΣPAHs (20 unsubstituted 2–6 rings + methyl naphthalenes) ranging from 1.5 to 5.7 $\mu g\ g^{-1}$ fat, with naphthalene and methyl naphthalenes predominating. Phenanthrene was the most abundant unsubstituted PAH in seabird eggs from the UK coast, while methylnaphthalenes predominated in most other locations (Pereira et al. 2009). Seabird blood has been used as a bioindicator of PAH exposure from oil spills (Pérez et al. 2008). Higher plasma concentrations of Σ16PAH were found in oil-exposed seabird colonies (Pérez et al. 2008).

11.6 Major Routes of Human Exposure to Organic Pollutants

The major exposure pathways for humans to many organic pollutants are via food and drinking water (Harrad 2001). In this context, soils play an important role because organic pollutants accumulated in surface or subsoils may be major sources for potential contamination of food chains. Figure 11.9 shows the major routes of human exposure to POPs.

Exposure to POPs can vary considerably and depends on individual habits (e.g. food choices), and the locations where people live. Humans eat many meats and animal products from species in which POPs have a tendency to bioaccumulate. Foods that can have especially high concentrations of POPs include fish, marine mammals, game, and milk. Fish represents important food for supplying essential trace elements and certain vitamins; moreover, the polyunsaturated *n*-3 fatty acids in fatty fish species are biologically important and have been associated with a decreased risk for cardiovascular disease (Svensson et al. 1995). Diets containing

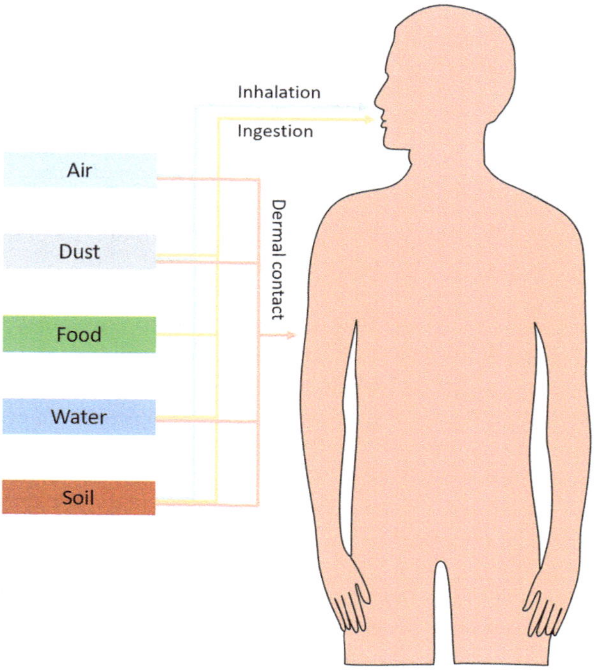

Fig. 11.9 Routes of human exposure to POPs

foods with elevated POPs concentrations generally result in higher body burdens of POPs.

Fish and wild game comprise a large fraction of the diet of many of the native peoples of northern Canada, Alaska, Greenland, and Scandinavia. This diet combined with the transport of POPs to the Arctic make human exposure to POPs in the Arctic an area of special concern. Several studies examining POPs in Inuit mothers indicated that the relatively high blood levels of dioxin-like PCBs in Inuit mothers are due to the traditional/country diet. In a comparison of breast milk from Inuit mothers in northern Quebec and Caucasian mothers in the same region, the concentrations of PCBs in Inuit breast milk were roughly five times those of the nearby Caucasian mothers and roughly double those of women in industrial areas further south (Dewailly et al. 1989; Harrad 2001). With almost no local sources of PCBs in the northern regions, it could be concluded that the Inuit diet was leading to the increased POPs concentrations. The concentrations of POPs in breast milk have also been examined in other studies.

Dioxins and Polychlorinated Biphenyls

Dioxins and PCBs are a major concern for food safety. In the general population, more than 95% of human exposure to PCDD/F and dioxin-like PCBs typically occurs via food (Weber et al. 2008). Dioxins as important contaminants of the food chain were first reported in the 1950s when thousands of chickens in the USA died due to dioxin intoxication (Firestone 1973). The problem was caused by the

Table 11.6 Levels of PCDDs/PCDFs/PCBs expressed as pg TEQ g^{-1} fresh weight in food from different countries

Food item	Sweden	Germany	UK	USA
Meats (general)	0.09	0.14	0.79	0.24
Milk	0.04	0.03	0.13	0.26
Cheese	n.i.	n.i.	0.18	0.33
Eggs	0.23	0.23	0.22	0.31
Fish	0.85	n.i.	0.57	0.55
Vegetables	n.i.	0.015	0.05	0.08

Compiled from Huwe (2002) and Darnerud et al. (2006)
n.i. not indicated

production of feed with fat scraped from cow hides treated with chlorophenols. Grazing farm animals, particularly sheep that spend most of their life outside in pastures ingest soil which in certain cases results in relatively high levels of PCDD/Fs and PCBs in animal food products (Hoogenboom et al. 2015). However, in most countries, human dioxin and PCB intake originates from various foods, including meat, meat products, milk and dairy products, and fish. Table 11.6 summarizes levels of analyzed dioxins and PCBs (PCDDs/PCDFs/PCBs) in food items of different countries.

Many surveys have shown relatively low background levels of dioxins (Table 11.6). However, occasionally highly contaminated samples were also found. Several dioxin contaminations have occurred in Europe. In 1998 during routine monitoring, dairy products were identified with dioxin levels two to four times higher than normal. The source of the contamination was traced to citrus pulp used as a cattle feed component (Malisch 2000). In another incident, PCB/PCDD/PCDF-contaminated oil was added to recycled fat used as an additive in feeds for Belgian poultry, dairy, and meat (Schepens et al. 2001). The contamination was discovered only after toxic effects were seen in chickens. In most Western countries dioxin and PCB intake originated from various foods (Table 11.6), whereas in Japan, Finland and Norway the intake was particularly due to consumption of fish and fish products (Arisawa et al. 2005). However, although Japanese people consume high quantities of fish, the intake of dioxins via food seems not to be higher compared to Western countries. An overview of mean dietary intake of dioxin-related compounds in different countries is given in Table 11.7.

Breast-fed babies have a much higher dioxin intake compared to adults and the values in Table 12.6 significantly exceed the tolerable daily intake set by the WHO of 4 pg TEQ kg^{-1} day^{-1} (Arisawa et al. 2005). Exposure of breastfed infants to PCDDs, PCDFs, and PCBs per unit body weight can be one to two orders of magnitude greater than that of adults (Harrad 2001). Nevertheless, it is not recommended to refrain from breast feeding, in view of the short duration of exposure and the merits of breast feeding, including balanced nutrition, immunologic protection ant the mother-to-infant bonding. It has been reported that in developed countries the content of dioxins and PCBs declined after the 1970s and the 1980s.

Table 11.7 Estimated dietary intakes of dioxins and PCBs (means and ranges) in humans (adults, children, breast-fed babies) in different countries

Compounds	Country	pg TEQ kg^{-1} day^{-1}	pg TEQ day^{-1}	Year of survey
PCDDs+PCDFs	Belgium	1,0[a]	65.3[a]	n.i.
	Germany	1.2[a]	85[a]	1984–1988
	"	2.0[a]	n.i.	n.i.
	"	149[b]	n.i.	n.i.
	"	1,6[c]	n.i.	1998
	Italy	n.i.	42[a]	1994–1996
	Spain	n.i.	63.8[a]	2002
	"	n.i.	210[a]	1996
	Japan	0.89[a]	44.7[a]	1999–2000
	New Zealand	0.18–0.44[a]	14.5–30.6	n.i.
PCDDs+PCDFs +PCBs	Belgium	2.0[a]	132.9[a]	2000–2001
	Netherlands	1.2 (median)[a]	n.i.	1998–1999
	Norway	n.i.	137–190[a]	1992–1994
	UK	n.i.	140[b]	1992
	"	2.4[a]	n.i.	1992
	"	39–170[b]	n.i.	1992
	Japan	2.25[a]	112.6[a]	1999–2000
	USA	1.2–3.6[a]	n.i.	1995
	"	2.2–2.4[a]	n.i.	1995
	"	42[b]	n.i.	1995
	New Zealand	0.3–0.8[a]	n.i.	n.i.

Adapted from Arisawa et al. (2005)
n.i. not indicated
[a]Adults
[b]Breast-fed infants
[c]Children

Polybrominated Diphenyl Ethers

Food items from high trophic levels like fish or lipid-rich oils contain relatively high concentrations of PBDEs, thus presenting an important exposure pathway to humans. According to Frederiksen et al. (2009), the mean reported ΣPBDE content on the basis of numerous samples in fish world-wide was estimated to be approximately 4200 pg g^{-1} wet weight (ww), while the median of the reported ΣPBDE was 1100 pg g^{-1} (ww). Concentrations will differ between tissues and organs of the fish. For example, it has been shown that in juvenile carps (*Cyprinus carpio*) the assimilation of PBDEs to the liver is higher than to the whole-body tissue (Stapleton et al. 2004). Variations in the same species from different regions have been observed. Wild salmon (*Oncorhynchus keta*) from Japan contained 703 pg g^{-1} (ww) ΣPBDE (Ohta et al. 2002), commercially caught salmon from the USA contained 433 pg g^{-1} (ww) ΣPBDE (Hayward et al. 2007), while Danish salmon from the Western Baltic Sea contained 1370–3760 pg g^{-1} (ww) ΣPBDE (Svendsen et al. 2007). However, none of these studies included BDE-209. Several studies have focused on the differences between wild and farmed salmon (Frederiksen et al. 2009). The general trend is that the levels of PBDEs are higher in the farmed salmon compared to wild salmon (Shaw et al. 2008). Fish from North America and Europe were not statistically different in ΣPBDE and the levels of ΣPBDE (excl. BDE-209) in Asia did not differ statistically from either North America or Europe (Frederiksen et al. 2009).

Levels of ΣPBDE in chicken eggs seem to be comparable for Europe (35–74 pg g^{-1} (ww)) and North America (85 pg g^{-1} (ww)) (summarized by Frederiksen et al. 2009). In contrast, the levels of ΣPBDE in meat from Europe, ranging from 1 to 102 pg g-1 (ww), and meat from North America, ranging from 154 to 283 pg g-1 (ww), were significantly different. PBDEs are most predominant in butter, fats and oils (119–569 pg g^{-1} (ww) ΣPBDE) (Frederiksen et al. 2009). Solid milk products like cheese also contain relatively high amounts of PBDEs (18–34 pg g-1 (ww) (Kiviranta et al. 2004; Voorspoels et al. 2007), while liquid milk products only contain very low concentrations of PBDEs (0.8 pg g-1 (ww) (Kiviranta et al. 2004). Due the higher water content and the low lipid content the level of PBDEs in vegetables is considerably lower than in the other food categories. Levels were below the limit of detection (LOD) for many European vegetables and those above LOD were in the range 1.3–17 pg g^{-1} (ww) (Bocio et al. 2003; Kiviranta et al. 2004). In contrast, Ohta et al. (2002) reported much higher concentrations in vegetables from Japan. Spinach contained large amounts (134 pg g^{-1} (ww)), greater than various meats from the same region (6.3–64 pg g^{-1} (ww)).

Mean levels of PBDEs (all congeners) in human milk in 2002–2005 were (in ng g^{-1} lipid): 3.7 in Europe, 1.57 in Japan, and 73.9 (range 6.2–419) in the USA (Costa et al. 2008). Levels of PBDEs in breast milk have been increasing in the past 20–30 years, along with serum levels in the general population (Costa and Giordano 2007), though a slight decline has started to emerge in the recent years. Internal human exposure to PBDEs has generally been found to be larger in North America than in Europe and Asia (Frederiksen et al. 2009). These differences cannot solely be explained by the dietary intake as levels of ΣPBDE in Europe and North America in most food groups seem to be similar. However, indoor air and dust concentrations have been found to be approximately one order of magnitude higher in North America than in Europe, possibly a result of different fire safety standards. Within Europe, higher PBDE concentrations in dust were found in the UK than in continental Europe.

Hexachlorobutadiene

Food may be contaminated with HCBD via environmental sources or by contact with contaminated water during food processing activity. Data on HCBD concentrations in food are limited. HCBD in food products sampled in the UK was found at concentrations of 0.00008 mg kg^{-1} in fresh milk, 0.002 mg kg^{-1} in butter, 0.0002 mg kg^{-1} in cooking oil, 0.0002 mg kg^{-1} in light ale, 0.0008 mg kg^{-1} in tomatoes, and 0.0037 mg kg^{-1} in black grapes (IARC 1979). Yip (1976) analyzed HCBD in food products within a 25-mile radius of tetrachloroethylene and trichloroethylene manufacturing plants that emit HCBD as a waste product. No HCBD was detected in 15 egg samples and 20 vegetable samples. One of 20 milk samples contained 1.32 mg kg^{-1} HCBD. Fish samples from the Mississippi River contained HCBD levels between 100 and 4700 ng g^{-1} (Laska et al. 1976). Hendricks et al. (1998) evaluated levels of HCBD in zebra mussel (*Dreissena polymorpha*) and eel (*Anguilla anguilla*) from approximately 30 locations in the Meuse-Rhine river catchment. In zebra mussel, levels of HCBD levels were 240 ng kg^{-1} at a background location and ranged from 950 to 14,000 ng kg wet weight within the study area. In eel, HCBD levels within the study area were found to range from 5000 to 55,000 ng kg^{-1} wet weight. Fish has relatively high concentrations of POPs and HCBD. An average estimate of adult general population exposure can be obtained by multiplying a mean concentration of 0.6 ng g^{-1} in fish (data from Kuehl et al. 1994) by a fish intake of 18 g day for the general population and dividing by a body weight of 70 kg. The resulting estimate is 0.15 ng kg^{-1} day^{-1}.

Pentachlorobenzene

There is only little information is available concerning the presence of chlorobenzenes in food. However, levels in various sea foods have been measured. All of the chlorobenzenes (mono- to penta-) were detected in trout from the Great Lakes of North America at levels ranging from 0.1 to 16 μg kg^{-1} whole-fish, wet weight (ww) (Oliver and Nicol 1982). Di- to pentachlorobenzenes were measured in samples of sprats from south-east Norway in the vicinity of an unspecified source of contamination. Levels of PeCB were highest, ranging from 0.01 to 3.7 mg kg^{-1}, while levels of trichloro- and tetrachlorobenzenes ranged between <0.01 and 0.5 mg kg^{-1} (Lunde and Ofstad 1976). In Slovenia, levels of total chlorobenzenes in the edible tissue of freshwater fish from highly polluted and industrial areas were 1.8 mg kg^{-1} lipid, whereas levels in fish from lightly polluted agricultural and woodland areas, and in marine fish, were 0.2 mg kg^{-1} and 0.4 mg kg^{-1}, respectively (Jan and Malnersic 1980). Studies investigating PeCB in human milk reported concentrations in the range of <1–5 μg kg^{-1} (WHO 1991).

Human intake from food was calculated on the basis of a total diet study conducted in the USA (Gunderson 1987). The total daily intake of pentachlorobenzene for age groups older than 6 months, was estimated to range from 0.0005 to 0.002 μg kg^{-1} bw day^{-1}. Based on the concentrations determined in breast milk, total intake for suckling infants aged 0–6 months was estimated to be up to several orders of magnitude greater, from 0.01 to 0.1 μg kg^{-1} bw day^{-1}.

Perfluorooctane Sulfonate

Noorlander et al. (2011) estimated levels of perfluorinated compounds in food and dietary intake of PFOS and PFOA in The Netherlands. The highest concentrations of PFOS were found in crustaceans (582 pg g^{-1} product) and in lean fish (308 pg g^{-1} product). Lower concentrations were found in beef, fatty fish, flour, butter, eggs, and cheese (concentrations between 29 and 82 pg g^{-1} product. Concentrations expressed as the sum of six perfluorinated compounds (PFOS were not detected) in milk, pork, bakery products, chicken, vegetable, and industrial oils were lower than 10 pg g^{-1} product. The median long-term intake for PFOS was 0.3 ng kg^{-1} bodyweight (bw) day^{-1} and for PFOA 0.2 ng kg^{-1} bw day^{-1}. The corresponding high level intakes were 0.6 and 0.5 ng kg^{-1} bw day^{-1}, respectively. These intakes were well below the tolerable daily intake values of both compounds (PFOS: 150 ng kg^{-1} bw day^{-1}; PFOA, 1500 ng kg^{-1} bw day^{-1}).

Fromme et al. (2009) compared the direct and the indirect exposure to PFOS/PFOA and their precursors of the population in Western countries. Direct PFOS and PFOA intakes on average were estimated at 1.6 ng kg^{-1} bw for PFOS and 2.9 ng kg^{-1} bw for PFOA. For an average human, the diet contributed to 96% (PFOS) and 99% (PFOA) of the direct exposure. House dust (50 mg day^{-1}; 37.8 ng g^{-1}) was responsible for 2% (PFOS) and 0.6% (PFOA) of the total intake, while air (indoor and outdoor together) was responsible for only 0.3% (PFOS) and 0.08% (PFOA), respectively. Drinking water (1.3 L day^{-1}; 1 ng L^{-1}) contributed to 1.5% (PFOS) and 0.8% (PFOA) of the daily intake. As with drinking water, the contribution of house dust to the total exposure of PFOS and PFOA may be limited in the average situation. However, in the case of high PFOS and PFOA dust concentrations, house dust may equal the diet as the route of exposure to these compounds. The contribution of PFOS/PFOA precursors to the PFOS exposure of the general population may be around 10% (Fromme et al. 2009). However, it should be noticed that PreFOS concentrations in different food products show considerable variation and may gain in importance as the source for PFOS and PFOA, in particular as the production of the latter compounds has been phased out.

Polychlorinated Naphtalenes

The probably most important route for the general population to PCNs is the diet (Falandysz 2003). However, the information concerning human exposure to PCNs through dietary intake up to now is very scarce. Domingo et al. (2003) examined quantitatively the levels of PCNs in various food groups and evaluated the dietary intake of PCNs by the general population in several cities of Catalonia, Spain. The highest concentration (wet weight) of total PCNs was found in oils and fats (447 pg g^{-1}), followed by cereals (71 pg g^{-1}), fish and shellfish (39 pg g^{-1}), and dairy products (36 pg g^{-1}). In contrast, milk (0.4 pg g^{-1}) and fruits (0.7 pg g^{-1}) were the groups showing the lowest concentrations of total PCNs. In general, tetraCNs was the predominant homologue in all food groups except for fruit and pulses, which had greater proportions of hexaCNs. Total dietary intake of PCNs by a male adult of 70 kg body weight was estimated 45.78 ng day^{-1}. This value is equivalent to 0.65 ng kg^{-1} body weight day^{-1}. The highest contribution to this intake corresponded to oils and fats with 40% of the total intake, followed by cereals with 32%. The lowest contributions in percentage corresponded to milk and pulses, while fish and shellfish contributed with approximately 8%, a similar percentage than those of meat and meat products, and dairy products.

Short-Chained Chlorinated Paraffins

Routes of potential human exposure to SCCPs include inhalation, dermal contact, and ingestion, primarily through contamination of foods. Because chlorinated paraffins are permitted in adhesives used in food packaging, the general population could be exposed to very low concentrations through ingestion of contaminated food products wrapped in these materials. SCCPs have also been found in food products contaminated through environmental exposure. For example, in Japan, SCCPs were found in foods containing high concentrations of lipids such as dairy products, vegetable oil, salad dressing, and mayonnaise, at a mean concentration of 140 ng g^{-1} (ww) (Bayen et al. 2006). Another contaminated food category in Japan was fish and shellfish, with SCCP concentrations of 16–18 ng g^{-1} (ww). The levels in Japanese foods would correspond to an average daily intake of 680 ng kg^{-1} of body weight for a 1-year-old female infant. In butter samples in Europe, SCCPs were measured at concentrations of 1.2 to 2.7 μg kg^{-1} of lipid content. Chlorinated paraffins have also been isolated from human tissues such as adipose tissue, liver and kidney at concentrations up to 1.5 mg kg^{-1} (ww) (Campbell and McConnell 1980), and from breast-milk at concentrations up to 0.8 mg kg^{-1} milk-fat (Thomas et al. 2006).

Polycyclic Aromatic Hydrocarbons

Human exposure to PAHs can occur in indoor or outdoor environments and by inhalation, ingestion and skin contact. PAHs formed by burning of fossil fuels, smoke from forest fires, and cigarette smoking are mainly available for inhalation, but they can also be ingested together with food (e.g., smoked foods, atmospheric deposition on vegetables, and coal used for cooking). For skin contact, the principal primary sources are exposure to tar, soot, and organic solvents (Burchiel and Luster 2001). Occupational exposure to PAHs may occur from workers breathing exhaust fumes (such as mechanics, street vendors, or motor vehicle drivers) and those involved in mining, metal working, or oil refining (Armstrong et al. 2004). The major route of exposure to PAHs in the general population is from eating food containing PAHs, breathing ambient (and indoor) air, smoking cigarettes, or breathing smoke from open fireplaces (Kim et al. 2013). A variety of PAHs from tobacco smoke are suspected human carcinogens (Lannerö et al. 2008).

For non-smokers and humans not occupationally exposed to PAHs the main route of exposure is through food. Processing (such as drying and smoking) and cooking of foods at high temperatures (grilling, roasting, and frying) are major sources of generating PAHs (Chen and Lin 2001). Some crops (such as wheat, rye, and lentils) may synthesize PAHs or absorb them via water, air, or, soil (Ciecierska and Obiedziński 2013). Intake of PAHs may also occur directly from contaminated soil via ingestion, inhalation, or dermal (skin) exposure and from inhalation of PAH vapors (Kim et al. 2013). PAH concentrations in crops are particularly high in urban areas and in vicinity of industrial activities (Table 11.8).

Concentrations of PAHs (Σ 8PAHs) outside strongly contaminated areas may range between about 20–50 ng g^{-1} dry weight in leafy vegetables, 0.6–9.0 ng g^{-1} in grains and 0.5–2.4 ng g^{-1} in fruits (Menzie et al. 1992). For the US population Menzie et al. (1992) estimated a median dietary intake of 3 μg total PAHs day^{-1} which represents around 96% of the total daily exposure for nonsmokers. In Italy, dietary exposure to PAHs (3 μg total PAHs day^{-1}) was estimated to be significantly higher than respiratory intake of PAHs from polluted urban air (370 ng total PAHs day^{-1}) (Lodovici et al. 1995). Tobacco smoke adds around 2–5 μg total PAHs day^{-1}

Table 11.8 Concentrations of PAHs in crops in urban areas and in vicinity of industrial activities

Country	Location	Number of PAHs	Crop	ΣPAH concentration in crop (ng g^{-1} dry wt)
Sweden	Highway	16	Lettuce	17–90
Canada	Highway	17	Onions[a], beet[a], tomatoes[a]	10–1900
Sweden	Al smelter	16	Lettuce	320–920
Sweden	Highway	16	Kale	500[b]; 5000[c]
Sweden	City street	16	Kale	5000[b]; 14,000[c]
Germany	Urban	13	Kale	1000–5000
UK	Urban	16	Grass	153
UK	Industrial	16	Grass	2400
Greece	Industrial	16	Various vegetables	25–239

Adapted from Srogi (2007)
[a]Washed leaves
[b]At 50 m distance from highway
[c]At 10 m distance from highway

for a one pack per day smoker, and up to 15 μg day^{-1} for a heavy (three packs per day) smoker (Menzie et al. 1992).

11.7 Clinical Effects

Evidence supporting a link between POPs exposure and a wide spectrum of diseases continues to grow. POPs share a number a common properties, including their long term persistence and diffusion in the environment, and their bioaccumulation through the food chain. For example, Canadian Inuit consume relatively large amounts of PCBs and PCDDs and bioaccumulate them in fatty tissues over years of exposure with the consequence that the intake of PCBs exceeds the 'tolerable daily intake' (TDI) for many Inuit consumers (Kinloch et al. 1992). The health effects due to low-level cumulative exposures are more difficult to determine than are the health effects due to acute exposures. The long timescales for effects, the low concentrations of contaminants, and the presence of confounding variables, such as other POPs, make conclusions about the health effects due to cumulative exposures to POPs much more difficult to reach than conclusions about acute exposures.

Dioxins and Polychlorinated Biphenyls

Toxic responses to dioxin/PCB exposure include dermal toxicity, immunotoxicity, carcinogenicity, and reproductive and developmental toxicity. Elevated body burden of dioxins and related compounds have also been associated with high blood pressure, elevated triglycerides and glucose intolerance (Porta et al. 2008).

Chloracne is the most widely recognised dermal effect of human exposure to the most toxic dioxin congener TCDD and has been observed in some workers after accidents at trichlorophenol (TCP) production facilities and among individuals involved in production of TCDD-contaminated products (Larsen 2006). Especially children experienced chloracne following the Seveso accident in 1976. Increases in serum levels of liver enzymes and of D-glucaric acid excretion in urine were also reported for children in Seveso and among TCP production workers with high serum TCDD levels. A change in sex ratio among newborns with an excess of females over males has been described in the period 1977–1984 for the most TCDD contaminated area in Seveso (Mocarelli et al. 1996). As an explanation, a possible role of hormonal disruption could not be excluded. Multiple, persistent effects were observed among children whose mothers had been exposed during pregnancy to high levels of PCB and PCDF from contaminated rice oil in the Yusho and Yucheng incidents in Japan and Taiwan. The effects included low birth weight, persistent developmental delays throughout childhood and behavioural disorders, hearing loss, and alterations in sexual development (Larsen 2006).

Weisglas-Kuperus et al. (1995) found that pre- and post-natal exposure of Dutch infants to PCBs and PCDDs was associated with certain immunological aberrations. Swedish investigations have reported that dietary intake of PCBs and PCDDs/PCDFs may be linked to reductions in the population of natural killer cells. These cells are believed to play a role in the body's defence against viruses and tumors (Vallack et al. 1998). High levels of PCBs in Greenlanders' adipose tissue and very high levels of PCBs in Inuit breast milk prompted concerns about their possible adverse effects on human (particularly infant) health, including increased susceptibility to infections (Mulvad et al. 1996). Canadian Inuit babies who accumulated high doses of organochlorines were significantly more likely to experience acute otitis (infection of the middle ear) and were harder to vaccinate since many failed to produce a primary antibody response to the usual vaccines (Vallack et al. 1998). Falck et al. (1992) reported higher levels of PCBs in mammary adipose tissue from women with malignant breast cancer compared with women having benign breast cancer. Dewailly et al. (1994) observed higher levels of plasma PCB in women with estrogen receptor-positive breast cancer. In a study of 866 infants in North Carolina (USA) signs of impaired neurological development in children, through exposure to PCBs in utero and in their mother's milk, were reported (Gladen et al. 1988). Early exposure of children to some organochlorine pollutants can also cause long-term intellectual impairment. At 7 months of age, in children exposed to higher PCB levels, there were signs of impaired cognitive function, and at 4 years of age they achieved lower scores in verbal and memory tests (Jacobson et al. 1990). At the age of 11 years these children were again tested, this time using a range of tests for intelligence quotient (IQ), arithmetic, spelling, reading and comprehension, and the results compared with their prenatal exposure to PCBs as measured in the placenta and mother's blood and milk (Jacobson and Jacobson 1996). Although much larger quantities of PCBs are transferred postnatally through lactation than in utero, intellectual impairment occurred only in relation to transplacental exposure. The authors concluded that these adverse effects were due to PCB exposure in utero, one possible mechanism being PCB-induced reduction in serum concentrations of thyroid hormones needed for stimulating neuronal and glial proliferation and differentiation.

In a dioxin/PCB breast milk study in the Netherlands, comprising 418 healthy mother/infant pairs, infants were examined for endocrine, immune and neurobehavioural development. Higher PCDD, PCDF and PCB levels in human milk (expressed as TEQs) correlated significantly with lower plasma levels of maternal total triiodothyronine and total thyroxine and higher plasma levels of thyroid stimulating hormone (TSH) levels in infants in the second week and third month after birth (Koopman-Esseboom et al. 1996). Infants exposed to higher than average human milk TEQ levels also had significantly lower plasma-free thyroxine and total thyroxine levels, in the second week after birth, than those exposed to lower than average human milk TEQs. It was concluded that elevated levels of PCBs, PCDDs and PCDFs can alter the human thyroid status. Infant neuro-developmental testing showed a small negative effect of prenatal PCB exposure on the psychomotor score at 3 months (Koopman-Esseboom et al. 1996). At 7 months of age both mental development and psychomotor scores were positively correlated with duration of breast-feeding although for breast-fed infants receiving higher cumulative TEQs the positive effect of breast-feeding on psychomotor outcome was diminished.

In most of the human studies examining the carcinogenicity of dioxins, people were exposed to mixtures of PCDD and TCDD. These epidemiological studies have involved subjects with the highest recorded exposures to TCDD. In a large study involving 12 industrial plants in the USA mortality from all cancers was slightly but significantly elevated Larsen (2006). Similarly, all-cancer mortality was significantly increased in several studies of workers from The Netherlands and Germany. Excesses at specific sites were reported for urinary bladder, kidney, the digestive system, lung, buccal cavity and pharynx, breast in females, as well as lymphatic and haematopoietic neoplasms and non-Hodgkin lymphoma (Larsen 2006).

Polybrominated Diphenyl Ethers

In contrast with the large volume of data on human body burden (levels of PBDEs in serum, adipose tissue, breast milk), there is only little information on possible developmental adverse effects in humans from PBDE exposure. Most potential adverse health effects on humans thus need to be extrapolated from animal data. In Scandinavia, human milk PBDE levels were associated with an increased incidence of cryptorchidism in newborn boys (Main et al. 2007). In another study in Taiwan, elevated PBDE levels in human breast milk were correlated with lower birth weight and length, lower head and chest circumference, and decreased body mass index (Chao et al. 2007). PBDEs have endocrine disrupting effects, as they have been shown to interact as antagonists or agonists at androgen, progesterone, and estrogen receptors (Legler and Brouwer 2003). PBDEs may exert direct neurotoxic effects in neuronal and glial cells. Few studies have investigated biochemical/molecular changes occurring in the central nervous system of animals following in vivo developmental exposure to PBDEs (Costa and Giordano 2007). For example, Viberg et al. (2003) reported changes in cholinergic nicotinic receptors in the hippocampus of mice upon exposure to BDE-99 and -153. Several studies (summarized by Costa et al. 2008) reported PBDEs to cause translocation of protein kinase C, stimulation of arachidonic acid release, and inhibition of calcium uptake in cerebellar granule neurons of rats.

Recent reports indicate that PBDEs cause oxidative stress in vitro. The chemicals DE-71 and BDE-47 were shown to cause oxidative stress in human neutrophil granulocytes (Reistad et al. 2005), an effect shared by other brominated fire retardants (Costa and Giordano 2007). BDE-47 was reported to induce oxidative stress in human neuroblastoma cells (He et al. 2008a), in rat hippocampal neurons (He et al. 2008b), and in rat fetal liver hematopoietic cells (Shao et al. 2008). Similar results were obtained in human hepatoma cells (Hu et al. 2007).

Hexachlorobutadiene

Up to now, only limited information is available on the human health effects associated with exposure to HCBD which is particularly true for the general population. German (1986) conducted two cytogenetic studies with workers of an HCBD production facility. The exposure levels reported by the manufacturer were between 1.6 and 16.9 mg m^{-3}. An increased frequency of chromosomal aberrations in the peripheral lymphocytes of exposed workers was found. However, the frequency of aberrations was not associated with duration of employment in the HCBD manufacturing facility (WHO 1994b), suggesting that factors other than HCBD might also have contributed to the observed effects. Krasniuk et al. (1969) evaluated health effects in 153 farm workers intermittently exposed over a period of 4 years to soil and grape fumigants containing HCBD. When compared to a control population of 52 unexposed workers, HCBD- exposed workers exhibited increased incidence of arterial hypotension, myocardial dystrophy, chest pains, upper respiratory tract changes, liver effects, sleep disorders, hand trembling, nausea, and disordered olfactory functions (US EPA 1991). Burkatskaya et al. (1982) reported adverse health effects in vineyard workers exposed to fumigants containing HCBD. However, the role of HCBD could not be evaluated because the workers were simultaneously exposed to other agrochemicals (WHO 1994b).

The primary target organ for HCBD is the kidney. Individuals with preexisting kidney damage may represent a potentially sensitive subpopulation for HCBD health effects. Studies in animals showed that the young rats and mice were more sensitive to the acute effects of oral HCBD than adults (Hook et al. 1983; Lock et al. 1984). Those data may suggest that infants may potentially be more susceptible to HCBD toxicity. HCBD has been classified as a Group C (possible human) carcinogen (US EPA 1986).

For humans, a no observed adverse effect level of 0.05 mg HCBD kg^{-1} body weight day^{-1} has been derived by IPCS (1994). The dose that corresponds to a lifetime excess cancer is 0.2 μg kg^{-1} day^{-1}. Under these conditions, it is not very likely that HCBD will occur with a frequency or at concentrations that are of concern for the general population.

Pentachlorobenzene

Although there are no clinical studies on humans available, there is sufficient evidence that PeCB has a high toxic potential and can adversely effect human health. Umegaki et al. (1993) studied the kinetics of PeCB in blood and tissues of rats given a single oral dose by gavage of either 15 mg or 20 mg. PeCB was observed in the blood, liver, kidney, brain, and fat tissue as well as in the feces (4.8% of the dose). In female Sherman rats ingesting diets containing >37.5 mg kg^{-1} bw day^{-1} PeCB for 100 days, there was an increase in liver weight and hypertrophy of hepatic cells, and at exposure levels >8.3 mg kg^{-1} bw day^{-1}, there was an increase in kidney weights and renal hyaline droplet formation in males (Linder et al. 1980). At about 80 mg kg^{-1} bw day^{-1} PeCB, the effects observed were an increase in adrenal weight and focal areas of renal tubular atrophy and interstitial lymphocytic infiltration in males, an increase in kidney weight in females a decrease in haemoglobin and an increase in white blood cells in both sexes, and decreases in red blood cells and haematocrit in males. LD50s for PeCB were reported to range between 940 and 1125 mg kg^{-1} bw in adult and weanling rats and 1175 and 1370 mg kg^{-1} bw in Swiss Webster mice (Linder et al. 1980).

Perfluorooctane Sulfonate

In contrast to the classical more lipophilic POPs like dioxins, furans, or PCBs, PFOS and PFOA do not typically accumulate in lipids but rather in body compartments with high protein content (Jones et al. 2003). Regarding the human health risk of PFOS and PFOA, the persistent nature of these compounds in the human body and the long-term exposure to these compounds via food, drinking water, air, and house dust lead to their accumulation in the body.

Not much information about acute and chronic effects with respect to PFOS and PFOA on humans is currently available. Alexander et al. (2003) suspected death of some workers exposed to PFOS to be caused by liver and bladder cancer. Gilliland and Mandel (1993) suggested a risk of mortality from prostate cancer in occupational workers exposed to PFOA. According to Austin et al. (2003), in rats exposed to PFOS, there was decreased food intake and body weight in a dose-dependent manner, including neuroendocrine effect. However, some epidemiological studies with occupationally exposed humans indicated no significant clinical hepato–toxicity at reported PFOA levels (Gilliland and Mandel 1996). The half-lives of PFOS and PFOA in the human body were reported to be 8.67 and 1–3.5 years, respectively (Hekster et al. 2003). Based on toxicity studies with rats, a dietary concentration of approximately 15 mg g^{-1} of PFOS was considered a safe level (Giesy and Kannan 2002). This concentration is much higher than those found in the tap water and animal tissues (Giesy and Kannan 2001).

Polychlorinated Naphtalenes

The major mechanism of action for the toxicity of PCNs is related to their ability to bind to and activate the aryl hydrocarbon receptor (AhR), which is a cytosolic, ligand-activated transcription factor (Blankenship et al. 2000). Toxic effects mediated through the AhR are species-, sex-, and tissue-specific, and include pleiotropic effects a characteristic wasting syndrome, thymic atrophy, immunosuppression, liver enlargement and necrosis, hyperplasia, chloracne, numerous biochemical effects, carcinogenesis, teratogenesis, and death (Poland and Knutson 1982). Exposure to PCNs has long been known to be associated with chloracne and lethality in occupationally exposed men (Hayward 1998). Experimental exposures of animals to PCN mixtures have resulted in chloracne in humans, "X-disease" in cattle, induction of cytochrome P450 enzyme activities and mortalities in chickens and eider ducklings, P450 induction in immature male wistar rats, three-spined stickleback, and rainbow trout sac fry (Blankenship et al. 2000).

Short-Chained Chlorinated Paraffins

According to UNEP (2009), acute toxicity of SCCPs is very low. SCCPs may cause skin and eye irritation upon repeated application, but do not appear to induce skin sensitization (EC 1999). There was no evidence of developmental effects in prenatal developmental toxicity studies in rats and rabbits (UNEP 2009). The liver, kidney and thyroid were major target organs in repeated-dose studies with rats and mice. When administered by gavage, chlorinated paraffins (C12, 60% chlorine) are carcinogenic in rats and mice of both sexes. The underlying mechanisms for the carcinogenicity of SCCP in rats and mice are not clearly known (US EPA 2009). SCCPs were not mutagenic to bacteria with or without metabolic activation. There are some indications that the SCCP-induced kidney tumors in male rats are associated with alpha 2 μ-globin accumulation in hyaline droplets, a mode of action unique to male rats and irrelevant to humans, but this could not be confirmed by immunocytochemical techniques. There are also data showing that the liver toxicity/carcinogenicity induced by SCCP is associated with peroxisome proliferation, whereas the thyroid effects are correlated to altered thyroid hormone homeostasis. However, humans are expected to be much less sensitive to thyroid hormone perturbation than rats and mice. There is no experimental evidence using human data that demonstrates carcinogenicity of SCCPs (US EPA 2009).

Polycyclic Aromatic Hydrocarbons

In the human body, since PAHs are lipophilic substances, they are readily dissolved and transported by cell membrane lipoproteins. The absorption rate depends on the specific PAH (Boström et al. 2002). In general, they are distributed throughout the body and found in any internal organ or tissue, particularly in lipid-rich tissues and the gastrointestinal tract, through the reabsorption of the product of hepatobiliary excretion. The harmful effects of PAHs depend on the mechanism of exposure (Kim et al. 2013). Unfortunately there is almost no study dealing with detailed information of human health effects following oral exposure to PAHs. In the majority of studies humans have been occupationally exposed to PAHs via inhalation, while only in a few studies the exposure has been via dermal contact. There is little information on human exposure to individual PAHs except for some accidental contact with naphthalene. In addition some data from controlled short-term studies of volunteers are available, but the conclusions are not transferable to the human exposure to PAHs via food (Kim et al. 2013). Many reports are on exposure to mixtures of PAHs, which also contained other potentially carcinogenic chemicals either in occupational or environmental situations, making it difficult to evaluate the effect of the PAHs alone.

Short-term or acute effects of PAHs on human health will depend mainly on the extent of exposure (e.g., length of time), the concentration of PAHs during exposure, the toxicity of the PAHs, and the route of exposure, e.g., via ingestion, inhalation, or skin contact (Kim et al. 2013).

Pre-existing health conditions and age also affect health effects. Short-term exposure to PAHs also has been reported to cause impaired lung function in asthmatics and thrombotic effects in people affected by coronary heart disease (Kim et al. 2013). However, it is not known which components of the mixture were responsible for these effects. Currently there is not a full understanding of the ability of PAHs at ambient concentrations to induce human health effects in the short-term. In contrast, occupational exposures to high levels of pollutant mixtures containing PAHs are known to result in symptoms such as eye irritation, nausea, vomiting and diarrhea, while mixtures of PAHs are also known to cause skin irritation and inflammation (Unwin et al. 2006). Anthracene, benzo(a)pyrene, and naphthalene are direct skin irritants. Anthracene and benzo(a)pyrene were reported to be a cause of an allergic skin response in animals and humans (Kim et al. 2013).

Long-term or chronic health problems (e.g., an increased risk of skin, lung, bladder, and gastrointestinal cancers) have been reported for workers exposed to mixtures of PAHs and other chemicals (Bach et al. 2003; Diggs et al. 2011; Olsson et al. 2010). Long-term exposure to low levels of some PAHs (e.g., pyrene and BaP) has been identified as the cause of cancer in laboratory animals (Diggs et al. 2012). Reactive metabolites (e.g., epoxides and dihydrodiols) of some PAHs have become one of the major health concerns because of their potential to bind to cellular proteins and DNA with toxic effects, despite the presence of some unmetabolized PAHs (Armstrong et al. 2004). The resulting biochemical disruption and cell damage can lead to mutations, developmental malformations, tumors, and cancer (Bach et al. 2003). Mixtures of PAHs may be more carcinogenic to humans than individual PAHs. Exposure to PAHs during pregnancy may be related to adverse birth outcomes including low birth weight, premature delivery, and delayed child development (Perera et al. 2005). High prenatal exposure to PAHs is also associated with low IQ at age three, increased behavioral problems at ages six to eight, and childhood asthma (e.g. Edwards et al. 2010; Perera and Herbstman 2011). Studies on lymphocytes from workers exposed to PAHs (including BaP) have identified DNA adducts of BaP (mainly the diol epoxide). In one of the previous studies on iron foundry workers, elevated levels of mutations at the hprt locus in lymphocytes were shown approximately to correlate with the levels of DNA adducts (WHO 1998). Genotoxicity plays an important role in the carcinogenicity process and may be in some forms of developmental toxicity as well. Figure 11.10 gives an overview of health effects due to short-term and long-term exposure to PAHs.

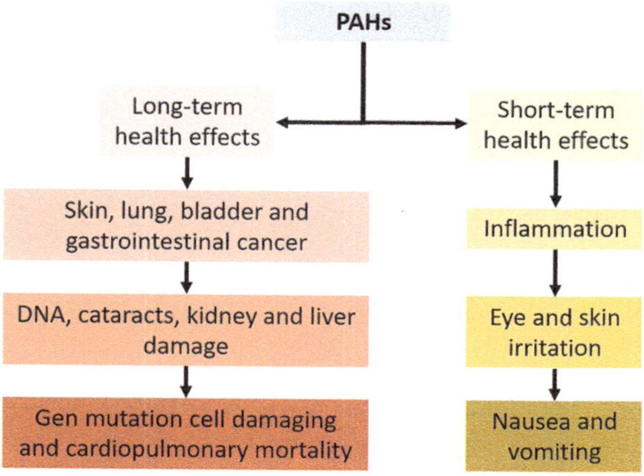

Fig. 11.10 Health effects as a consequence of short- and long-term exposure to PAHs (Adapted from Kim et al. 2013)

11.8 Therapy

POPs for the general population are not considered so harmful for their acute toxicity (also because of their low concentrations in the environment) but for long term, sub-lethal effects such as reproductive, developmental and immunological disturbances, and for cancer. Most health risks associated with POPs, except for occupation exposures or industrial accidents, are chronic, low-level exposures. Daily intakes of POPs in diets in most world regions are generally on the order of tens of micrograms per day, or less. These daily exposures are much smaller than these substances' acutely toxic doses (Ritter et al. 1996). There are three major routes to reduce/eliminate the body burden of organic pollutants including precautionary avoidance of exposures (Sect. 11.8.1), endogenous excretion (Sect. 11.8.2) and interventions to enhance elimination (Sect. 11.8.3).

11.8.1 Reduction of Exposure to Organic Pollutants

Preventative health care through reduction of exposure to organic pollutants, particularly of POPs, is primarily accomplished through dietary means. Generally, the mean contamination by POPs increases in the order of vegetables < dairy < meat < fish. Depending on the intake rates of these food categories, their relative importance will vary. If fish is a substantial part of the diet, it will be the main contributor to the total dietary exposure. The largest exposures to POPs occur in those who consume the most fish and game. An organic and largely plant-based diet may be the most effective for reducing POPs exposure (Pizzorno and Katzinger 2013). Special attention must be drawn to children because they eat and drink more per unit body weight than adults, and often have unique dietary patterns (e.g. consume more of a particular food group) that can increase their exposure to POPs (WHO 2010).

Diet is not the only source of exposure to many POPs. For example, approaches of action for reduction of exposure also include avoidance of contact with personal care products that contain phthalates, food containers that exhibit bisphenol A and in food packing materials which contain chlorinated paraffins. Over the past decades it has become clear that humans, in particular small children, are also exposed to POPs via dust and particles in indoor environments like homes, schools and childcare centers, and offices. These POPs, which are additives in electronics and electric products, textiles and furniture, are released from materials and goods in homes and at work. For example, humans can be exposed to PBDEs, which are used as flame retardants, through many household products such as polyurethane foam in furniture cushions. In the USA, 58% of the human PBDE intake result from dust and air (mainly indoor), and 42% from food (Costa et al. 2008). Also handling of waste (e.g. e-waste) and recycling have been identified as sources of external exposure to POPs for humans. Adults can be exposed to POPs via inhalation but this is mostly related to occupational exposure. Dermal uptake of POPs and other persistent and bioaccumulative compounds is regarded as a minor pathway for non-occupationally exposed people. Exposure of humans to unintentionally produced POPs, such as dioxins and furans, can result from chemical fires or accidents and from waste incineration.

It's a matter of fact that although some toxic chemicals have been restricted or banned, many of today's population have already been harmed by bioaccumulated pollutants. An important aspect is thus minimization of further exposure which is only possible through means of personal education, notification (e.g., labelling) and actions (e.g., remediation of living spaces and changes in diet) as well as governmental and industrial regulation with enforcement. This is fundamental to any successful strategy to diminish the toxicant burden of individuals and populations. From a clinical perspective, it is useful for the health provider to perform a comprehensive inventory of potential exposures including ingestion, breathing, skin contact, olfactory transmission via smell, mother to fetus/infant transmission, penetration of body tissues through processes such as surgery, dentistry, injection, or vector routes (Sears and Genius 2012). By identifying exposures and apprising individual patients regarding where and how they are being contaminated, patients are empowered to avoid further chemical contamination. With ongoing exposures minimized, the human organism is able to devote resources and energies of detoxification physiology to metabolizing and excreting retained compounds, with less devoted to ongoing exposures (Sears and Genius 2012).

11.8.2 Therapeutic Measures to Facilitate Excretion of Organic Pollutants

The risk associated with bioaccumulation of organic pollutants can generally be diminished through removal of such agents which can be achieved by endogenous processing of the chemical into a non-toxic metabolite or by enhancing the rate of excretion from the body. Research is currently focusing on effective techniques to diminish the body burden of such toxicants (Genius 2011). With the knowledge that perspiration acts a means to excrete organic pollutants, some reports suggest that induced sweating may be useful in diminishing toxicant bioaccumulation. Several studies have confirmed that induced perspiration e.g. through sauna therapy can reduce the body burden of assorted bioaccumulated toxicants including PCBs, polybrominated biphenyls, chlorinated pesticides, HCB, and various other pollutants in exposed individuals (Genius 2011).

With the knowledge that caloric restriction appears to greatly enhance lipolysis with release of assorted toxicants, restrictive dietary measures have also been proposed as a means to diminish the body burden of toxic compounds such as PCBs (Imamura and Tung 1984). In addition, vigorous exercise also appears to induce lipolysis and to enhance excretion of various toxicants (Genius 2011).

11.8.3 Therapeutic Measures to Enhance Elimination of Organic Pollutants

Organic compounds that remain lipophilic and their lipophilic metabolites tend to be excreted primarily through fecal excretion, while hydrophilic compounds and their metabolites are generally eliminated in the urine. Some bile-excreted organochlorines that are surface active, or which possess structural or behavioral properties similar to endogenous bile acids, however, may have slower elimination kinetics and remain persistent in the body due to enterohepatic recirculation (Jandacek and Tso 2007). The latter is a physiological re-absorption mechanism for recycling bile acids and acts as a means to conserve required biological compounds. Re-absorption results in prolongation of the half-life of the involved toxic compounds and in increase of the potential health risk. For example, PCFs remain persistent in the human body, in large part due to enterohepatic recirculation (Genius 2011).

Measures to circumvent reabsorption within the enterohepatic recirculation offer a chance to diminish the body burden of persistent toxicants by facilitating excretion. Several types of pharmaco-therapeutic interventions have been discussed as potentially useful. For example, therapies that preclude fat absorption have been found to enhance the excretion of fecal fat and lipophilic toxicants. These types of compounds, such as sucrose polyester (Olestra), have been effective in enhancing fecal elimination of a variety of organic pollutants such as PCDDs, PCDFs, PCBs, and hexachlorobenzene (Moser and McLachlan 1999).

Another mechanism involves the use of adsorbent compounds that accumulate toxic agents from within the gastrointestinal tract. For example, activated carbon such as charcoal and other compounds including bentonites, certain clays, and zeolites appear to have adsorbent action. Although animal studies have demonstrated promise, insufficient human research work has been done so far to conclusively determine the clinical value of these compounds for pollutant removal (Genius 2011). A further approach to circumvent enterohepatic recycling involves the binding of toxicants in the intestine via bile-acid sequestrants. The most studied of these is cholestyramine which is a strongly basic non-absorbable resin that effectively binds various agents through anion exchange, interrupts enterohepatic recycling, and prevents intestinal re-absorption. Bile-acid sequestrants have been shown to bind various bacterial endotoxins, mycotoxins, enterotoxins and PCBs (Genius 2011). Other non-absorbable compounds such as colestimide have also been effective at eliminating certain organic pollutants, including PCBs (Sakurai et al. 2006). In order to preclude potential adverse effects of these pharmaceutical compounds, particularly in children, current research is also focused on the use of herbal agents such as saponin compounds (originating from soy or the yucca plant) which are expected to have a similar mechanism of action (Genius et al. 2011). In case of acute poisoning after ingestion, e.g. of PCDDs/PCDFs, PCBs and chlorobenzenes, effective detoxification is required.

11.9 Remediation of Soils Contaminated with Organic Pollutants

Despite the fact that most POPs are banned or restricted for use in many countries, the contaminated soils and obsolete stock piles continue to have a significant impact on a number of ecosystems worldwide and pose a serious risk to humans, livestock, and other ecosystems (Abhilash and Yunus 2011). In times of global warming, the remediation and management of soils contaminated with organic pollutants is a challenging task as the increasing atmospheric temperature will accelerate the volatility and long-range transport of volatile POPs (Miraglia et al. 2009). Remediation of polluted sites has become a crucial issue because of the increasing global awareness of pollution, overexploitation of natural resources, and adverse effects of pollutants on human health and environment. Currently available remediation technologies can be classified as "established" and "emerging" in order to differentiate between those which have demonstrated their full-scale application and those which have proved their efficiency on a pilot level only. Established technologies are already discussed in the chapter on pesticides (Chap. 10). These include physical, chemical and biological techniques which are generally also suitable for the remediation of sites contaminated with organic pollutants discussed in this chapter. However, some "emerging", i.e. innovative technologies promise to be efficient for the treatment of sites/materials polluted with industrial chemicals and unintentionally formed substances. These technologies can be categorized as emerging physico-chemical and emerging thermal techniques, and have been reviewed by Lodolo et al. (2001). The principles are discussed in the following sections. All of the technologies are ex-situ.

11.9.1 Emerging Physico-chemical Technologies

11.9.1.1 Base Catalyzed Dechlorination

The base catalysed dechlorination technique is applicable for treatment of waste that contains up to 100,000 mg kg^{-1} of halogenated aliphatic or aromatic organic compounds such as PCBs. The formation of salt within the treated mixture may limit the concentration of halogenated material able to be treated. Rogers (1991) reported a reduction of chlorinated organics to less than 2 mg kg^{-1}. The process can involve direct dehalogenation or can be linked to a pre-treatment method such as thermal desorption. The procedure mainly involves chlorine stripping. When treating chlorinated aromatic hydrocarbons the removal of chlorine atoms causes an increased concentration of lower chlorinated species (Lodolo et al. 2001). This does not generally represent a problem for PCBs treatment, but with components such as dioxins, the lower congeners (e.g. TCDD) can be more toxic than the highly chlorinated congeners (e.g. OCDD). The process must be therefore well monitored to ensure that the reaction continues to completion.

11.9.1.2 Combined Solvent Extraction-Chemical Dehalogenation-Irradiation

This process reduces the volume of a pollutant that is to be decomposed. An extracting chemical is used to dissolve target contaminants from soils in a final solution suitable for treatment with recovery of the solvent. The technology can be applied to soils contaminated with volatile and semi-volatile organic compounds and other higher boiling point complex organics such as PAHs, petroleum hydrocarbons, PCBs, dioxins, pentachlorophenol (PCP) and pesticides. Approaches have been developed to combine solvent extraction with other techniques like chemical dehalogenation with immobilized reagents (CDP) and gamma-ray irradiation (Lodolo et al. 2001).

11.9.1.3 Solvated Electron

Sodium metal is dissolved in liquid anhydrous ammonia to produce a solution of solvated electrons. The latter acts as dehalogenating agents. Solvated electron solutions are rapidly formed when alkali or alkaline earth metals are dissolved in ammonia or in some amines, forming solutions containing the metal cation and free electrons. Halogens can be separate from organic halides to yield a fully substituted parent hydrocarbon and a metal halide. Halogens can be separated from organic halides to yield a fully substituted parent hydrocarbon and a metal halide. The method is claimed to be applicable to treat halogenated hydrocarbons, pesticides, The method is applicable to treat halogenated hydrocarbons, dioxins, PCBs, CFCs, pesticides and chemical warfare agents. Wastes have been successfully treated in bulk pure material, soils, sludge, sediments, porous and non-porous surfaces, oils, contaminated vessels, hardware, and contaminated clothing (Lodolo et al. 2001).

11.9.1.4 Gas Phase Chemical Reduction

This method is an alternative to incineration technologies and is based on gas phase thermo-chemical reaction of hydrogen with organic compounds (Lodolo et al. 2001). Hydrogen combines with organic compounds at 850 °C or higher, in a reductive reaction to form lighter hydrocarbons (mainly methane). For chlorinated organic compounds, such as PCBs, the products are methane and hydrogen chloride. The reaction is carried out with water that functions as a reducing agent and generates hydrogen. The technology is a hydrogenation process and adds hydrogen atoms to any incompletely hydrogenated organic molecule, dechlorinating molecules and breaking down aromatic rings, therefore is non selective in its treatment of organic substances. The process can quantitatively convert PCBs, PAHs, chlorophenols, dioxins, and chlorobenzenes to methane.

11.9.2 Emerging Thermal Technologies

11.9.2.1 Combined Thermal Desorption-Catalyzed Dehalogenation

This technology combines thermal desorption with base catalyzed dechlorination. The system uses an indirectly heated thermal desorber to split organic compounds from contaminated media. The system is designed to achieve feed material temperatures of up to 510 ° C allowing an effective treatment of soils and sludge polluted with a wide range of low and high boiling point compounds (Lodolo et al. 2001). The system is applicable for hydrocarbons, PCBs, coal byproducts, wood treating compounds, dioxins, furans and some pesticides.

11.9.2.2 Combined Thermal Desorption-Pyrolysis

This technology is based on thermal desorption combined with the flash pyrolysis technique, and followed by combustion. The operating temperature in the reactor ranges from 450 to 800 °C. The process applications include solid hazardous waste, PCB-contaminated soil, mercury- contaminated soil, hospital waste, municipal solid waste, sewage sludge and coal. In addition, the technology can treat a full range of chlorinated hydrocarbons, organochlorine pesticides, organic and/or inorganic materials combined with contamination of organics and halogens (Lodolo et al. 2001).

11.9.2.3 Plasma ARC Systems

This technology uses high temperature (around 10,000 ° C) for pyrolysis, which results from the discharge of a large electric current in an inert gas, to convert hazardous chemicals such as PCBs, pesticides and CFCs into innocuous and safe-emitted end products (Lodolo et al. 2001). The destructive process is made possible by the conversion of the hazardous compound by the superheated cloud of gas or plasma into atomic elements and subsequent treatment converts the atomic forms into innocuous substances.

References

Aamot E, Steinnes E, Schmid R (1996) Polycyclic aromatic hydrocarbons in Norwegian forest soils: impact of long range atmospheric transport. Environ Pollut 92:275–280

Abhilash PC, Yunus M (2011) Can we use biomass produced from phytoremediation? Biomass Bioenergy 35:1371–1372

Ahlborg UG, Hanberg A, Kenne K (1992) Risk assessment of polychlorinated biphenyls (PCBs). Nord 1992:26. Nordic Council of Ministers, Copenhagen. 99 pp

Alcock RE, Gemmill R, Jones KC (1999) Improvements to the UK PCDD/F and PCB atmospheric emissions inventory following an emission measurement programm. Chemosphere 38:759–770

Alcock RE, Sweetman AJ, Prevedouros K, Jones KC (2003) Understanding levels and trends of BDE-47 in the UK and North America: an assessment of principal reservoirs and source inputs. Environ Int 29(6):691–698

Alexander BH, Olsen GW, Burris JM, Mandel JH, Mandel JS (2003) Mortality of employees of a perfluorooctanesulphonyl fluoride manufacturing facility. Occup Environ Med 60:722–729

AMAP (Arctic Monitoring and Assessment Programme) (2000) PCB in the Russian Federation: inventory and proposals for priority remedial actions. AMAP Report 2000:3, Oslo, Norway, 27 pp

Arisawa K, Takeda H, Mikasa H (2005) Background exposure to PCDDs/PCDFs/PCBs and its potential health effects: a review of epidemiologic studies. J Med Investig 52:10–21

Armstrong BG, Hutchinson E, Unwin J, Fletcher T (2004) Lung cancer risk after exposure to polycyclic aromatic hydrocarbons: a review and meta-analysis. Environ Health Perspect 112 (9):970–978

Atkinson R (1991) Atmospheric lifetimes of dibenzo-para-dioxins and dibenzofurans. Sci Total Environ 104:17–33

ATSDR (Agency for Toxic Substances and Disease Registry) (1994) Toxicological profile for hexachlorobutadiene. Public Health Service, U.S. Department of Health and Human Services, Atlanta. Publication No. TP-93/08

Austin ME, Kasturi BS, Barber M, Kannan K, Mohan Kumar P, Mohan Kumar SMJ (2003) Neuroendocrine effects of perfluorooctane sulfonate in rats. Environ Health Perspect 111:1485–1489

Bach PB, Kelley MJ, Tate RC, McCrory DC (2003) Screening for lung cancer: a review of the current literature. Chest 123:72–82

Bailey RE (2001) Global hexachlorobenzene emissions. Chemosphere 43:167–182

Bailey RE, van Wijk D, Thomas PC (2009) Sources and prevalence of pentachlorobenzene in the environment. Chemosphere 75:555–564

Barber JL, Sweetman AJ, Thomas GO, Braekevelt E, Stern GA, Jones KC (2005) Spatial and temporal variability in air concentrations of short-chain (C10-C13) and medium-chain (C14-C17) chlorinated n-alkanes measured in the U.K. atmosphere. Environ Sci Technol 39:4407–4415

Barton CA, Butler LE, Zarzecki CJ, Flaherty J, Kaiser M (2006) Characterizing perfluorooctanoate in ambient air near the fence line of a manufacturing facility: comparing modeled and monitored values. J Air Waste Manag Assoc 56:48–55

Bayen S, Obbard JP, Thomas GO (2006) Chlorinated paraffins: a review of analysis and environmental occurrence. Environ Int 32:915e929

Becker S, Halsall CJ, Tych W, Hung H, Attewell S, Blanchard P, Li H, Fellin P, Stern G, Billeck B, Friesen S (2006) Resolving the long-term trends of polycyclic aromatic hydrocarbons in the Canadian Arctic atmosphere. Environ Sci Technol 40:3217–3222

Belfroid A, van Wezel A, Sikkenk M, van Gestel K, Semen W, Hermens J (1993) The toxicokinetic behavior of chlorobenzenes in earthworms *(Eisenia andrei):* experiments in water. Ecotoxicol Environ Safety 25:154–165

Bidleman TF, Helm PA, Braune BM, Gabrielsen GW (2010) Polychlorinated naphthalenes in polar environments – a review. Sci Total Environ 408:2919–2935

Bignert A, Danielsson S, Nyberg E, Asplund L, Nylund K, Berger U, Haglund P (2010) Comments concerning the national Swedish contaminant monitoring programme in marine biota, 2010. Swedish Museum of Natural History, Stockholm

Blankenship AL, Kannan K, Villalobos SA, Villeneuve DL, Falandysz J, Imagawa T, Jakobsson E, Giesy J (2000) Relative potencies of individual polychlorinated naphthalenes and halowax mixtures to induce Ah receptor-mediated responses. Environ Sci Technol 34:3153–3158

Bocio A, Llobet JM, Domingo JL, Corbella J, Teixido A, Casas C (2003) Polybrominated diphenyl ethers (PBDEs) in foodstuffs: human exposure through the diet. J Agric Food Chem 51:3191–3195

Borgen AR, Schlabach M, Gundersen H (2000) Polychlorinated alkanes in arctic air. Organohalogen Compd 47:272–274

Boström CE, Gerde P, Hanberg A, Jernström B, Johansson C, Kyrklund T, Rannug A, Törnqvist M, Victorin K, Roger Westerholm R (2002) Cancer risk assessment, indicators, and guidelines for polycyclic aromatic hydrocarbons in the ambient air. Environ Health Perspect 110:451–488

Boulanger B, Peck AM, Schoor JL, Hornbuckle CK (2005) Mass budget of perfluorooctane surfactants in Lake Ontario. Environ Sci Technol 39:74–79

Brändli R, Kupper T, Bucheli T, Mayer J, Stadelmann FX, Tarradellas J (2004) Occurrence and relevance of organic pollutants in compost, digestate and organic residues. Literature review. Final report in English with abstract in German and French of Module 1 Project Organic pollutants in compost and digestate in Switzerland. EPF Lausanne – ENAC – ISTE – CECOTOX/Agroscope FAL Reckenholz, 193 pp

Breivik K, Sweetman A, Pacyna JM, Jones KC (2002) Towards a global historical emission inventory for selected PCB congeners – a mass balance approach. 2. Emissions. Sci Total Environ 290(1–3):199–224
Breivik K, Alcock R, Li YF, Bailey RE, Fiedler H, Paczyna JM (2004) Primary sources of selected POPs: regional and global scale emission inventories. Environ Pollut 128:3–16
Breivik K, Sweetman A, Pacyna JM, Jones KC (2007) Towards a global historical emission inventory for selected PCB congeners – a mass balance approach: 3. An update. Sci Total Environ 377(2–3):296–307
BUA (Beratergremium für Umweltrelevante Altstoffe) (1991) In: Gesellschaft Deutscher Chemiker (ed) Hexachlorbutadien, BUA-Stoffbericht 62, August 1991. Hirzel Verlag, Stuttgart
Budakowski W, Tomy GT (2003) Congener-specific analysis of hexabromocyclododecane by high- performance liquid chromatography/electrospray tandem mass spectrometry. Rapid Commun Mass Spectrom 17(13):1399–1404
Burchiel SW, Luster MI (2001) Signaling by environmental polycyclic aromatic hydrocarbons in human lymphocytes. Clin Immunol 98:2–10
Burkatskaya EN, Viter VF, Ivanova ZV, Kaskevitch LM, Gorskaya NZ, Kolpakov IE, Deineka KA (1982) Clinico-hygienic data on working conditions during use of hexachlorobutadiene in vineyards. Vrach Delo 11:99–102. (in Russian)
Campbell I, McConnell G (1980) Chlorinated paraffins and the environment: environmental occurrences. Environ Sci Technol 14(10):1209–1214
Capuano F, Cavalchi B, Martinelli G, Pecchini G, Renna E, Scaroni I, Bertacchi M, Bigliardi G (2005) Environmental prospection for PCDD/PCDF, PAH, PCB and heavy metals around the incinerator power plant of Reggio Emilia town (Northern Italy) and surrounding main roads. Chemosphere 58(11):1563–1569
Catoggio JA (1991) Other organic toxic substances. In: Guidelines of lake management. Toxic Subst Manag Lakes Reserv 4:113–126
Chao HR, Wang SL, Lee WJ, Wang YF, Papke O (2007) Levels of polybrominated diphenyl ethers (PBDEs) in breast milk from central Taiwan and their relation to infant birth outcome and maternal menstruation effects. Environ Int 33:239–245
Chen D, Hale RC (2010) A global review of polybrominated diphenyl ether flame retardant contamination in birds. Environ Int 36(7):800–811
Chen BH, Lin YS (2001) Formation of polycyclic aromatic hydrocarbons in the smoke from heated model lipids and food lipids. J Agric Food Chem 49:5238–5243
Chrysikou L, Gemenetzis P, Kouras A, Manoli E, Terzi E, Samara C (2008) Distribution of persistent organic pollutants, polycyclic aromatic hydrocarbons and trace elements in soil and vegetation following a large scale landfill fire in northern Greece. Environ Int 34:210–225
Ciecierska M, Obiedziński MW (2013) Polycyclic aromatic hydrocarbons in the bakery chain. Food Chem 141:1–9
Class T, Ballschmiter K (1987) Global baseline pollution studies. Fresenius Z Anal Chem 327:198–204
Coleman PJ, Lee RGM, Alcock RE, Jones KC (1997) Observations on PAH, PCB, and PCDD/F. Trends in U.K. urban air, 1991–1995. Environ Sci Technol 31(2):120–2124
Corsolini S, Covaci A, Ademollo N, Focardi S, Schepens P (2006) Occurrence of organochlorine pesticides (OCPs) and their enantiomeric signatures, and concentrations of polybrominated diphenyl ethers (PBDEs) in the Adélie penguin food web, Antarctica. Environ Pollut 140:371–382
Costa LG, Giordano G (2007) Developmental neurotoxicity of polybrominated diphenyl ether (PBDE) flame retardants. Neurotoxicology 28:1047–1067
Costa LG, Giordano G, Tagliaferri S, Caglieri A, Mutti A (2008) Polybrominated diphenyl ether (PBDE) flame retardants: environmental contamination, human body burden and potential adverse health effects. Acta Biomed 79:172–183

Covaci A, Hura C, Schepens P (2001) Selected persistent organochlorine pollutants in Romania. Sci Total Environ 280:143–152

Crooks MJ, Howe PD (1993) Environmental hazard assessment: halogenated naphthalenes, TSD/13. Building Research Establishment, Garston

Darnerud PO, Atuma S, Aune M, Bjerselius R, Glynn A, Petersson Grawe´ K, Becker W (2006) Dietary intake estimations of organohalogen contaminants (dioxins, PCB, PBDE and chlorinated pesticides, e.g. DDT) based on Swedish market basket data. Food Chem Toxicol 44:1597–1606

De Alencastro LF (1995) Les polychlorobiphényles dans les systèmes de désapprovisionnement des eaux usées. PhD Thesis no.1328. Swiss Federal Institute of Technology, Lausanne

de Boer J, Wester PG, van der Horst A, Leonards PEG (2003) Polybrominated diphenyl ethers in influents, suspended particulate matter, sediments, sewage treatment plant and effluents and biota from the Netherlands. Environ Pollut 122(1):63–74

De Kok A, Geerdink RB, Brinkman UAT (1983) The determination of polychlorinated naphthalenes in soil samples by means of various gas and liquid chromatographic methods. Anal Chem Symp Ser 13:203–216

de Voogt P, Brinkman UA (1989) Production, properties and usage of polychlorinated biphenyls. In: Kimbrough RD, Jensen AA (eds) Halogenated biphenyls, terphenyls, napthalenes, dibenzodioxins and related products. Elsevier, Amsterdam, pp 3–45

de Voogt P, Berger U, de Coen W, de Wolf W, Heimstad E, McLachlan M, van Leeuwen S, van Roon A (2006) Perfluorinated organic compounds in the European environment (Perforce), Report to the EU. University of Amsterdam, Amsterdam, pp 1–126

de Wit CA (2002) An overview of brominated flame retardants in the environment. Chemosphere 46:583–624

de Wit CA, Herzke D, Vorkamp K (2010) Brominated flame retardants in the Arctic environment – trends and new candidates. Sci Total Environ 408(15):2885–2918

Dettwiler J, Karlaganis G, Studer C, Joss S, Stettler A, Chambaz D (1997) Dioxine und Furane – Standortbestimmung Beurteilungsgrundlagen, Massnahmen, Schriftenreihe Umwelt Nr. 290. Bundesamt für Umwelt, Wald und Landschaft (BUWAL), Bern

Dewailly E, Nantel A, Weber JP, Meyer F (1989) High levels of PCBs in breast milk of Inuit women from Arctic Quebec. Bull Environ Contam Toxicol 43:641–646

Dewailly E, Dodin S, Verreault R, Ayotte P, Sauvé L, Morin J, Brisson J (1994) High organochlorine body burden in women with estrogen receptor-positive breast cancer. J Natl Cancer Inst 86:232–234

Dietz R, Rigét FF, Sonne C, Born EW, Bechshøft T, McKinney M, Muir DCG, Letcher RJ (2012) Three decades (1984–2010) of legacy contaminant and flame retardant trends in East Greenland polar bears (*Ursus maritimus*). Environ Int 59:485–493

Diggs DL, Huderson AC, Harris KL, Myers JN, Banks LD, Rekhadevi PV, Niaz MS, Aramandla Ramesh A (2011) Polycyclic aromatic hydrocarbons and digestive tract cancers: a perspective. J Environ Sci Health C Environ Carcinog Ecotoxicol Rev 29(4):324–357

Diggs DL, Harris KL, Rekhadevi PV, Ramesh A (2012) Tumor microsomal metabolism of the food toxicant, benzo(a)pyrene, in ApcMin mouse model of colon cancer. Tumor Biol 33 (4):1255–1260

Ding WH, Aldous KM, Briggs RG, Valente H, Hilker DR, Connor S, Eadon GA (1992) Application of multivariate statistical analysis to evaluate local sources of chlorobenzene congeners in soil samples. Chemosphere 25:675–690

Domingo JL, Falco G, Llobet JM, Casas C, Teixido A, Müller L (2003) Polycyclic naphthalenes in foods: estimated dietary intake by the population of Catalonia, Spain. Environ Sci Technol 37:2332–2335

Dorr G, Hippelein M, Hutzinger O (1996) Baseline contamination assessment for a new resource recovery facility in Germany. Part V: analysis and seasonal/regional variability of ambient air concentrations of polychlorinated naphthalenes (PCN). Chemosphere 33(8):1563–1568

Duyzer JH, Vonk AW (2003) Atmospheric deposition of pesticides, PAHs and PCB in the Netherlands. TNO Environment, Energy and Process Innovation, Apeldoorn

EC (European Commission) (1999) European union risk assessment report alkanes, C10-13, chloro, CAS No. 85535-84-8, EUR 19010EN, vol 4. Office for Official Publications of the European Communities, Luxembourg

Eder G, Sturm R, Ernst W (1987) Chlorinated hydrocarbons in sediments of the Elbe River and Elbe Estuary. Chemosphere 16:2487–2496

Edwards SC, Jedrychowski W, Butscher M, Camann D, Agnieszka Kieltyka A, Mroz E, Elzbieta Flak E, Li Z, Shuang Wang S, Rauh V, Frederica Perera F (2010) Prenatal exposure to airborne polycyclic aromatic hydrocarbons and children's intelligence at 5 years of age in a prospective cohort study in Poland. Environ Health Perspect 118(9):1326–1331

EFSA (European Food Safety Authority) (2008) Perfluorooctane sulfonate (PFOS), perfluorooctanoic acid (PFOA) and their salts: scientific opinion of the panel on contaminants in the food chain. EFSA J 653:1–131

Eljarrat E, Caixach J, Rivera J (2001) Levels of polychlorinated dibenzo-*p*-dioxins and dibenzofurans in soil samples from Spain. Chemosphere 44:1383–1387

Ellis DA, Martin JW, Mabury SA, Hurley MD, Sulbeck Andersen MD, Wallington TJ (2003) Atmospheric lifetime of fluorotelomer alcohols. Environ Sci Technol 37:3816–3820

Environment Canada (2000) Canadian environmental protection act, 1999, priority substances list assessment report on hexachlorobutadiene, Environment Canada, Health Canada, November 2000. Co-published by Health Canada, ISBN 0-662-29297-9

Erickson MD, Michael LC, Zweidinger RA, Pellizzari ED (1978) Sampling and analysis for polychlorinated naphthalenes in the environment. J Assoc Off Anal Chem 61:1335–1346

Espadaler I, Eljarrat E, Caixach J, Rivera J, Marti I, Ventura F (1997) Assessments of polychlorinated naphthalenes in aquifer samples for drinking water purposes. Rapid Commun Mass Spectrom 11:410–414

Fair PA, Adams J, Mitchum G, Hulsey TC, Reif JS, Houde M, Muir D, Wirth E, Wetzel D, Zolman E, McFee W, Bossart GD (2010) Contaminant blubber burdens in Atlantic bottlenose dolphins (*Tursiops truncatus*) from two southeastern US estuarine areas: concentrations and patterns of PCBs, pesticides, PBDEs, PFCs, and PAHs. Sci Total Environ 408(7):1577–1597

Falandysz J (1998) Polychlorinated naphthalenes: an environmental update. Environ Pollut 101:77–90

Falandysz J (2003) Chloronaphthalenes as food-chain contaminants: a review. Food Addit Contam 20:995–1014

Falck F Jr, Ricci A Jr, Wolff MS, Godbold J, Deckers P (1992) Pesticides and polychlorinated biphenyl residues in human breast lipids and their relation to breast cancer. Arch Environ Health 47:143–146

Fava F, Piccolo A (2002) Effects of humic substances on the bioavailability and aerobic biodegradation of polychlorinated biphenyls in a model soil. Biotechnol Bioeng 77(2):204–211

Fent K (1998) Ökotoxikologie. Georg Thieme Verlag, Stuttgart

Fiedler H (2001) Global and local disposition of PCB. In: PCBs – recent advances in the environmental toxicology and health effects. University Press of Kentucky, pp 11–15

Fiedler H, Fricke H, Vogtmann H (1994) Bedeutung polychlorierter Dibenzo-p-dioxine und polychlorierter Dibenzofurane (PCDD/PCDF) in der Abfallwirtschaft, Organohalogen Compounds, vol 17. Ecoinforma Press, Germany

Firestone D (1973) Etiology of chick edema disease. Environ Health Perspect 5:59–66

Focazio MJ, Kolpin DW, Barnes KK, Furlong ET, Meyer MT, Zaugg SD, Barber LB, Thurman ME (2008) A national reconnaissance for pharmaceuticals and other organic wastewater contaminants in the United States – II. Untreated drinking water sources. Sci Total Environ 402:201–216

Franzaring J, van der Eerden LJM (2000) Accumulation of airborne persistent organic pollutants (POPs) in plants. Basic Appl Ecol 1:25–30

Frederiksen M, Vorkamp K, Thomsen M, Knudsen ML (2009) Human internal and external exposure to PBDEs – a review of levels and sources. Int J Hyg Environ Health 212:109–134

Fromme H, Tittlemier SA, Völkel W, Wilhelm M, Twardella D (2009) Perfluorinated compounds: exposure assessment for the general population in western countries. Int J Hyg Environ Health 212:239–270

Galiulin RV, Bashkin VN, Galiulina RA (2002) Review: behavior of persistent organic pollutants in the air-plant-soil system. Water Air Soil Pollut 137:179–191

Gawlik BM, Martens D, Schramm KW, Kettrup A, Lamberty A, Muntau H (2000) On the presence of PCDD/Fs and other chlorinated hydrocarbons in the second generation of the European reference soil set – the EUROSOILS. Fresen J Anal Chem 368:407–411

Genius SJ (2011) Elimination of persistent toxicants from the human body. Hum Exp Toxicol 30 (1):3–18

German IV (1986) Level of chromosome aberrations in workers coming in contact with hexachlorobutadiene during production. Gig Tr Prof Zabol 5:57–79. (in Russian)

Gewurtz SB, Lazar R, Haffner GD (2003) Biomonitoring of bioavailable PAH and PCB water concentrations in the Detroit River using the freshwater mussel, Elliptio Complanata. J Great Lakes Res 29(2):242–255

Giesy JP, Kannan K (2001) Global distribution of perfluorooctane sulfonate in wildlife. Environ Sci Technol 35(7):1339–1342

Giesy JP, Kannan K (2002) Peer reviewed: perfluorochemical surfactants in the environment. Environ Sci Technol 36(7):146A–152A

Gilliland FD, Mandel JS (1993) Mortality among employes of a perfluorooctanoic acid production plant. J Occup Med 35(9):950–954

Gilliland FD, Mandel JS (1996) Serum perfluorooctanoic acid and hepatic enzymes, lipoproteins, and cholesterol: a study of occupationally exposed men. Am J Ind Med 29:560–568

Gladen B, Rogan W, Hardy P, Thullen J, Tingelstad J, Tully M (1988) Development after exposure to polychlorinated biphenyls and dichlorodiphenyl dichloroethene transplacentally and through human milk. J Pediatr 113:991–995

Gobas FAPC, Morrison HA (2000) Bioconcentration and biomagnification in the aquatic environment. In: Boethling RS, Mackay D (eds) Handbook of property estimation methods for chemicals. CRC Press, Boca Raton, pp 189–231

Goldbach RW, Van Genderen H, Leeuwangh PL (1976) Hexachlorobutadiene residues in aquatic fauna from surface water fed by the River Rhine. Sci Total Environ 6(1):31–40

Gomez-Belinchon JI, Grimalt JO, Abaigés J (1991) Volatile organic compounds in two polluted rivers in Barcelona (Catalonia, Spain). Water Res 25:577–589

Govaerts B, Beck B, Lecoutre E, le Bailly C, Van den Eeckaut P (2004) From monitoring data to regional distributions: a practical methodology applied to water risk assessment. Environmetrics 15:1–19

Gunderson EL (1987) FDA Total Diet Study, April 1982–April 1986. Dietary intakes of pesticides, selected elements, and other chemicals. Association of Official Analytical Chemists, Arlington

Halse AK, Schlabach M, Schuster JK, Jones KC, Steinnes E, Breivik K (2015) Endosulfan, pentachlorobenzene and short-chain chlorinated paraffins in background soils from Western Europe. Environ Pollut 196:21–28

Hansen KJ, Johnson HO, Eldridge JS, Butenhoff JL, Dick LA (2002) Quantitative characterisation of trace levels of PFOS and PFOA in the Tennessee river. Environ Sci Technol 36:1681–1685

Harada K, Saito N, Sasaki K, Inoue K, Koizumi A (2003) Perfluorooctane sulfonate contamination of drinking water in the Tama River, Japan: estimated effects on resident serum levels. Bull Environ Contam Toxicol 71:31–36

Harner T, Bidleman TF (1997) Polychlorinated naphthalenes in urban air. Atmos Environ 31 (23):4009–4016

Harner T, Mackay D (1995) Measurements of octanol-air partition coefficients for chlorobenzenes, PCBs, and DDT. Environ Sci Technol 29:1599–1606

Harner T, Lee RGM, Jones KC (2000a) Polychlorinated naphthalenes in the atmosphere of the United Kingdom. Environ Sci Technol 34(15):3137–3142

Harner T, Meijer S, Halsall C, Johnston AE, Jones K (2000b) Polychlorinated naphthalenes in U.K. soils: time trends and equilibrium status. Organohalogen Compd 47:25–28

Harrad S (2001) Persistent organic pollutants: environmental behaviour and pathways of human exposure. Springer, Boston

Hartlieb N, Klein W (2001) Fate and behaviour of organic contaminants during composting of municipal biowaste. In: Rees RM, Ball BC, Campbell CD, Watson CA (eds) Sustainable management of soil organic matter. CABI Publishing CAB International, Oxon, pp 150–156

Hassanin A, Breivik K, Meijer SN, Steinnes E, Thomas GO, Jones KC (2004) PBDEs in European background soils: levels and factors controlling their distribution. Environ Sci Technol 38:738–745

Hayward D (1998) Identification of bioaccumulating polychlorinated naphthalenes and their toxicological significance. Environ Res 76(1):1–18

Hayward D, Wong J, Krynitsky AJ (2007) Polybrominated diphenyl ethers and polychlorinated biphenyls in commercially wild caught and farm-raised fish fillets in the United States. Environ Res 103:46–54

He W, He P, Wang A, Xia T, Xu B, Chen X (2008a) Effects of PBDE-47 on cytotoxicity and genotoxicity in human neuroblastoma cells in vitro. Mutat Res 649:62–70

He P, He W, Wang A, Xia T, Xu B, Zhang M, Chen X (2008b) PBDE-47-induced oxidative stress, DNA damage and apoptosis in primary cultured rat hippocampal neurons. Neurotoxicology 29:124–129

Hekster FM, Laane RWPM, de Voogt P (2003) Environmental and toxicity effects of perfluoroalkylated substances. Rev Environ Contam Toxicol 179:99–121

Hellou J (1996) Polycyclic aromatic hydrocarbons in marine mammals, finfish, and molluscs. In: Beyer WN, Heinz GH, Redmon-Norwood AWE (eds) Environmental contaminants in wildlife: interpreting tissues concentrations. Lewis Publishers, Boca Raton, pp 229–250

Helm PA, Bidleman TF (2003) Current combustion-related sources contribute to polychlorinated naphthalene and dioxin-like polychlorinated biphenyl levels and profiles in air in Toronto, Canada. Environ Sci Technol 37:1075–1082

Helm PA, Bidleman TF, Jantunen LMM, Ridal J (2000) Polychlorinated naphthalenes in Great Lakes air: sources and ambient air profiles. Organohalogen Compd 47:17–20

Hendricks AJ, Pieters H, De Boer J (1998) Accumulation of metals, polycyclic (halogenated) aromatic hydrocarbons, and biocides in zebra mussel and eel from the Rhine and Muese Rivers. Environ Toxicol Chem 17(10):1885–1898

Hoff RM, Muir DCG, Grift NP (1992) Annual cycle of polychlorinated biphenyls and organohalogen pesticides in air in Southern Ontario. 1. Air concentration data. Environ Sci Technol 26:266–275

Holoubek I (2001) Polychlorinated biphenyl (PCB) contaminated sites worldwide. In: PCBs – recent advances in the environmental toxicology and health effects. University Press of Kentucky, p 17–26

Hoogenboom R, Traag W, Fernandes A, Rose M (2015) European developments following incidents with dioxins and PCBs in the food and feed chain. Food Control 50:670–683

Hook JB, Ishmael J, Lock EA (1983) Nephrotoxicity of hexachloro-1,3-butadiene in the rat: the effect of age, sex, and strain. Toxicol Appl Pharmacol 67:122–131

Houde M, Hoekstra PF, Solomon KR, Muir DC (2005) Organohalogen contaminants in delphinoid cetaceans. Rev Environ Contam Toxicol 184:1–57

Houde M, De Silva A, Letcher RJ, Muir DCG (2011) Biological assessment and biomagnification of polyfluoroalkyl acids (PFAAs) in aquatic ecosystems: an updated review. Environ Sci Technol 45:7962–7973

Houot S, Clergeot D, Michelin J, Francou C, Bourgeois S, Caria G, Ciesielski H (2002) Agronomic value and environmental impacts of urban composts used in agriculture. Insam H, Riddech N, Klammer S Microbiology of composting, Springer, Berlin, 457–472

Hu XZ, Xu Y, Hu DC, Hui Y, Yang FX (2007) Apoptosis induction on human hepatoma cells Hep G2 of decabrominated diphenyl ether (PBDE-209). Toxicol Lett 171:19–28

Hülster A, Müller JF, Marschner H (1994) Soil–plant transfer of polychlorinated dibenzo-*p*-dioxins and dibenzofurans to vegetables of the cucumber family (Cucurbitaceae). Environ Sci Technol 28:1110–1115

Hung H, Kallenborn R, Breivik K, Su YS, Brorström-Lunden E, Olafsdottir K, Thorlacius JM, Leppänen S, Bossi R, Skoy H, Mano S, Patton G, Stern G, Sverko E, Fellin P (2010) Atmospheric monitoring of organic pollutants in the Arctic under the Arctic Monitoring and Assessment Programme (AMAP): 1993–2006. Sci Total Environ 408(15):2854–2873

Huwe JK (2002) Dioxins in food: a modern agricultural perspective. J Agric Food Chem 50:1739–1750

IARC (International Agency for Research on Cancer monographs on the evaluation of the carcinogenic risk of chemicals to humans) (1979) Hexachlorobutadiene, vol 20. World Health Organization, Lyon, pp 179–193

Imamura M, Tung TC (1984) A trial of fasting cure for PCB poisoned patients in Taiwan. Am J Ind Med 5:147–153

IPCS (International Programme on Chemical Safety) (1994) Environmental health criteria 156, Hexachlorobutadiene. World Health Organization, Geneva

Jacobson JL, Jacobson SW (1996) Intellectual impairment in children exposed to polychlorinated biphenyls in utero. New Engl J Med 335:783–789

Jacobson JL, Jacobson SW, Humphrey HEB (1990) Effects of in utero exposure to polychlorinated biphenyls and related contaminants on cognitive functioning in young children. J Pediatr 116:38–45

Jan J, Malnersic S (1980) Chlorinated benzene residues in fish in Slovenia (Yugoslavia). Bull Environ Contam Toxicol 24:824–827

Jandacek RJ, Tso P (2007) Enterohepatic circulation of organochlorine compounds: a site for nutritional intervention. J Nutr Biochem 18:163–167

Jantunen LM, Helm PA, Kylin H, Bidleman TF (2008) Hexachlorocyclohexanes (HCHs) in the Canadian Archipelago. 2. Air-water-gas exchange of α- and γ-HCH. Environ Sci Technol 42:465–470

Järnberg U, Asplund L, De Wit C, Egebäck AL, Wideqvist U, Jakobsson E (1997) Distribution of polychlorinated naphthalene congeners in environmental and source-related samples. Arch Environ Contam Toxicol 32(3):232–245

Jaward FM, Farrar NJ, Harner T, Sweetman A, Jones KC (2004a) Passive air sampling of PAHs and PCNs across Europe. Environ Toxicol Chem 23:1355–1364

Jaward FM, Meijer SN, Steinnes E, Thomas GO, Jones KC (2004b) Further studies on the latitudinal and temporal trends of persistent organic pollutants in Norwegian and UK background air. Environ Sci Technol 38:2523–2530

Jessup DA, Johnson CK, Estes J, Carlson-Bremer D, Jarman WM, Reese S, Dodd E, Tinker MT, Ziccardi MH (2010) Persistent organic pollutants in the blood of free-ranging sea otters (*Enhydra lutris* SSP.) in Alaska and California. J Wildl Dis 46(4):1214–1233

Jiang X, Martens D, Schramm KW, Kettrup A, Xu SF, Wang LS (2000) Polychlorinated organic compounds (PCOCs) in waters, suspended solids and sediments of the Yangtse River. Chemosphere 41:901–905

Jones PD, Hu W, De Coen W, Newsted JL, Giesy JP (2003) Binding of perfluorinated fatty acids to serum proteins. Environ Toxicol Chem 22:2639–2649

Kaj L, Palm A (2004) Screening av hexaklorbutadien (HCBD) i miljoen. IVI Rapport B1543. IVI Svenska Miljoeinstitutet AB

Kallenborn R, Berger U, Jarnberg U (2004) Perfluorinated alkylated substances (PFAS) in the Nordic environment. Norwegian Institute for Air Research, Oslo. 107 pp

Kannan K, Imagawa T, Blankenship AL, Giesy JP (1998) Isomerspecific analysis and toxic evaluation of polychlorinated naphthalenes in soil, sediment, and biota collected near the site of a former chlor-alkali plant. Environ Sci Technol 32(17):2507–2514

Kannan K, Tao L, Sinclair E, Pastva SD, Jude DJ, Giesy JP (2005) Perfluorinated compounds in aquatic organisms at various trophic levels in a Great Lakes food chain. Arch Environ Contam Toxicol 48:559–566

Kannan K, Moon HB, Yun SH, Agusa T, Thomas NJ, Tanabe S (2008) Chlorinated, brominated, and perfluorinated compounds, polycyclic aromatic hydrocarbons and trace elements in livers of sea otters from California, Washington, and Alaska (USA), and Kamchatka (Russia). J Environ Monit 10(4):552–558

Karickhoff SW (1981) Semiempirical estimation of sorpion of hydrophopic pollutants on natural sediments and soil. Chemosphere 10:833–846

Kim KH, Jahan SA, Kabir E, Brown RJC (2013) A review of airborne polycyclic aromatic hydrocarbons (PAHs) and their human health effects. Environ Int 60:71–80

King TL, Lee K, Alexander R (2003) Chlorobenzenes in snow crab (*Chionoecetes opilio*): time series monitoring following an accidental release. Bull Environ Contam Toxicol 71:543–550

Kinloch D, Kuhnlein H, Muir DCG (1992) Inuit foods and diet: a preliminary assessment of benefits and risks. Sci Total Environ 122:247–278

Kiviranta H, Ovaskainen ML, Vartiainen T (2004) Market basket study on dietary intake of PCDD/Fs, PCBs, and PBDEs in Finland. Environ Int 30:923–932

Kollotzek D, Hartmann E, Kassner W, Kurrle J, Lemmert-Schmitt E, Beck A (1998) Technische, analytische, organisatorische und rechtliche Massnahmen zur Verminderung der Klärschlammbelastung mit relevanten organischen Schadstoffen, Forschungsbericht 103 50 123 UBA-FB 98-037, vol 1. Umweltbundesamt, Berlin

Koopman-Esseboom C, Weisglas-Kuperus N, Deridder MAJ, Van der Paauw CG, Tuinstra LGMT, Sauer PJJ (1996) Effects of polychlorinated biphenyl dioxin exposure and feeding type on infants mental and psychomotor development. Pediatrics 97:700–706

Krasniuk EP, Ziritskaya LA, Bioko VG, Voitenko GA, Matokhniuk LA (1969) Health conditions of vine-growers contacting with fumigants hexachorobutadiene and polychlorbutan-80. Vrach Delo 7:111–115. (in Russian)

Krauss M, Wilcke W (2003) Polychlorinated naphthalenes in urban soils: analysis, concentrations, and relation to other persistent organic pollutants. Environ Pollut 122(1):75–89

Kuehl DW, Butterworth B, Marquis PJ (1994) A national study of chemical residues in fish. III. Study results. Chemosphere 29(3):523–535

Kulkarni PS, Crespo JG, Afonso CAM (2008) Dioxins sources and current remediation technologies – a review. Environ Int 34:139–153

La Guardia MJ, Hale RC, Harvey E (2004) Environmental debromination of decabrominated diphenyl ether. Presented at BFR 2004: the third international workshop on brominated flame retardants. Toronto, Canada

Langenkamp H, Part P (2001) Organic contaminants in sewage sludge for agricultural use. European Commission JRC

Lannerö E, Wickman M, van Hage M (2008) Exposure to environmental tobacco smoke and sensitisation in children. Thorax 63:172–176

Larsen JC (2006) Risk assessments of polychlorinated dibenzo*p*-dioxins, polychlorinated dibenzofurans, and dioxin-like polychlorinated biphenyls in food. Mol Nutr Food Res 50:885–896

Laska A, Bartell CK, Laseter JL (1976) Distribution of hexachlorobenzene and hexachlorobutadiene in water, soil, and selected aquatic organisms along the lower Mississippi River, Louisana. Bull Environ Contam Toxicol 15:535–542

Law RJ, Herzke D, Harrad S, Morris S, Bersuder P, Allchin CR (2008) Levels and trends of HBCD and BDEs in the European and Asian environments, with some information for other BFRs. Chemosphere 73(2):223–241

Law RJ, Barry J, Bersuder P, Barber JL, Deaville R, Reid RJ, Jepson PD (2010) Levels and trends of brominated diphenyl ethers in blubber of harbor porpoises (*Phocoena phocoena*) from the U.K., 1992–2008. Environ Sci Technol 44(12):4447–4451

Lecloux A (2003) Scientific activities of Euro Chlor in monitoring and assessing naturally and man-made organohalogens. Chemosphere 52:521–529

Lecloux A (2004) Hexachlorbutadiene – sources, environmental fate and risk characterization, Science Dossier, Euro Chlor representing the chlor-alkali industry, www.eurochlor.org, 43 pp

Lee RGM, Jones KC (1999) The influence of meteorology and air masses on daily atmospheric PCB and PAH concentrations at a UK location. Environ Sci Technol 33:705–712
Lee RGM, Burnett V, Harner T, Jones KC (2000) Short-term temperature-dependent air-surface exchange and atmospheric concentrations of polychlorinated naphthalenes and organochlorine pesticides. Environ Sci Technol 34:393–398
Lee RGM, Coleman P, Jones JL, Jones KC, Lohmann R (2005) Emission factors and importance of PCDD/Fs, PCBs, PCNs, PAHs and PM10 from the domestic burning of coal and wood in the U.K. Environ Sci Technol 39:1436–1447
Legler J, Brouwer A (2003) Are brominated flame retardants endocrine disruptors? Environ Int 29:879–885
Li MT, Hao LL, Sheng LX, Xu JB (2008) Identification and degradation characterization of hexachlorobutadiene degrading strain Serratia marcescens HL1. Bioresour Technol 99 (15):6878–6884
Li F, Zhang C, Qu Y, Chen J, Chen L, Liu Y, Zhou Q (2010) Quantitative characterization of short- and long-chain perfluorinated acids in solid matrices in Shanghai, China. Sci Total Environ 408:617–623
Li QL, Li J, Wang Y, Xu Y, Pan XH, Zhang G, Luo CL, Kobara Y, Nam JJ, Jones KC (2012) Atmospheric short-chain chlorinated paraffins in China, Japan, and South Korea. Environ Sci Technol 46:11948–11954
Linder R, Scotti T, Goldstein J, McElroy K, Walsh D (1980) Acute and subchronic toxicity of pentachlorobenzene. J Environ Pathol Toxicol 4(5–6):183–196
Liu G, Zheng M (2013) Perspective on the inclusion of polychlorinated naphtalenes as a candidate POP in Annex C of the Stockholm Convention. Environ Sci Technol 47:8093–8094
Lock EA, Ishmael J, Hook JB (1984) Nephrotoxicity of hexachloro-1,3-butadiene in the mouse: the effect of age, sex, strain, monooxygenase modifiers, and the role of glutathione. Toxicol Appl Pharmacol 72:484–494
Lodolo A, Gonzalez-Valencia E, Miertus S (2001) Overview of remediation technologies for persistent toxic substances. Arh Hig Rada Toksikol 52:253–280
Lodovici M, Dolara P, Casalini C, Ciappellano S, Testolin G (1995) Polycyclic aromatic hydrocarbon contamination in the Italian diet. Food Addit Contam 12:703–713
Lohmann R, Jones KC (1998) Dioxins and furans in air and deposition: a review of levels, behaviour and processes. Sci Total Environ 219:53–81
Loos R, Wollgast J, Huber T, Hanke G (2007) Polar herbicides, pharmaceutical products, perfluorooctanesulfonate (PFOS), perfluorooctanoate (PFOA), and nonylphenol and its carboxylates and ethoxylates in surface and tap waters around Lake Maggiore in Northern Italy. Anal Bioanal Chem 387:1469–1478
Lu X, Tao S, Hu H, Dawson RW (2000) Estimation of bioconcentration factors of nonionic organic compounds in fish by molecular connectivity indices and polarity correction factors. Chemosphere 41(10):1675–1688
Lunde G, Ofstat EB (1976) Determination of fat-soluble chlorinated compounds in fish. Z Anal Chem 282:395–399
Mackay D, Shiu WY, Ma KC, Lee SC (2006) Physical-chemical properties and environmental fate for organic chemicals, 2nd edn. CRC Press, Taylor and Francis Group, Boca Raton
Main KM, Kiviranta H, Virtanen HE, Sundqvist E, Tuomisto JT, Tuomisto J, Vartiainen T, Skakkebaek NE, Toppari J (2007) Flame retardants in placenta and breast milk and cryptorchidism in newborn boys. Environ Health Perspect 115:1519–1526
Malisch R (2000) Increase of the PCDD/F-contamination of milk, butter and meat samples by use of contaminated citrus pulp. Chemosphere 40:1041–1053
Maliszewska-Kordybach B (1996) Polycyclic aromatic hydrocarbons in agricultural soils in Poland: preliminary proposals for criteria to evaluate the level of soil contamination. Appl Geochem 11:121–127
Manoli E, Samara C (1999) Occurrence and mass balance of polycyclic aromatic hydrocarbons in the Thessaloniki sewage treatment plant. J Environ Qual 28:176–186
Martí I, Ventura F (1997) Polychlorinated naphthalenes in groundwater samples from the Llobregat aquifer (Spain). J Chromatogr 786A:135–144

Martin JW, Ellis DA, Mabury SA, Hurley MD, Wallington TJ (2006) Atmospheric chemistry of perfluoroalkanesulfonamides: kinetic and product studies of the OH radical and Cl atom initiated oxidation of N-ethyl perfluorobutanesulfonamide. Environ Sci Technol 40:864–872

Martínez K, Abad E, Palacios O, Caixach J, Rivera J (2007) Assessment of polychlorinated dibenzo-p-dioxins and dibenzofurans in sludges according to the European environmental policy. Environ Int 33:1040–1047

Masclet P, Hoyau V, Jaffrezo JL, Legrand M (1995) Evidence for the presence of polycyclic aromatic hydrocarbons in the polar atmosphere and in the polar ice of Greenland. Analusis 23:250–252

Masunaga S (2004) Trend and sources of dioxin pollution in Tokyo Bay, estimated based on the statistical analyses of congener specific data. China–Japan–Korea symposium on environmental analytical chemistry, October 18–21 2004, Beijing, China, p 127–131

McGowin AE, Adom KK, Obubuafo AK (2001) Screening of compost for PAHs and pesticides using static subcritical water extraction. Chemosphere 45(6–7):857–864

Meijer SN, Ockenden WA, Sweetman A, Breivik K, Grimalt JO, Jones KC (2003) Global distribution and budget of PCBs and HCB in background surface soils: implications for sources and environmental processes. Environ Sci Technol 37:667–672

Menzie CA, Potocki BB, Santodonato J (1992) Exposure to carcinogenic PAHs in the environment. Environ Sci Technol 26:1278–1284

Mielke HW, Wang GD, Gonzales CR, Powell ET, Le B, Quach VN (2004) PAHs and metals in the soils of inner-city and suburban New Orleans, Louisiana, USA. Environ Toxicol Pharmacol 18 (3):243–247

Ministry of the Environment (Japan) (2006) Chemicals in the environment, Report on Environmental Survey and Monitoring of Chemicals in FY 2005. Environmental Health Department, Ministry of the Environment, Japan

Miraglia M, Marvin HJ, Kleter GA, Battilani P, Brera C, Coni E (2009) Climate change and food safety: an emerging issue with special focus on Europe. Food Chem Toxicol 47:1009–1021

Mocarelli P, Brambilla P, Gerthoux PM, Patterson DG Jr, Needham LL (1996) Change in sex ratio with exposure to dioxin. Lancet 348:409

Moore JW Ramamoorthy S (1984) Aromatic hydrocarbons-polycyclics. In: Organic chemicals in natural waters: applied monitoring and impact assessment. Springer, New York, p 67–87

Moser GA, McLachlan MS (1999) A non-absorbable dietary fat substitute enhances elimination of persistent lipophilic contaminants in humans. Chemosphere 39:1513–1521

Motelay-Massei A, Ollivon D, Garban B, Teil M, Blanchard M, Chevreuil M (2004) Distribution and spatial trends of PAHs and PCBs in soils in the Seine River basin, France. Chemosphere 55:555–565

Muir DCG, Stern G, Tomy G (2000) Chlorinated paraffins. In: Paasivirta J (ed) The handbook of environmental chemistry, vol 3. Springer-Verlag, Berlin, Heidelberg

Muir DCG, Bennie D, Teixeira C, Fisk AT, Tomy GT, Stern GA, Whittle M (2001) Short chain chlorinated paraffins: are they persistent and bioaccumulative? In: Lipnick R, Jansson B, Mackay D, Patreas M (eds) Persistent, bioaccumulative and toxic substances, vol 2. ACS Books, Washington, DC, pp 184–202

Mulvad G, Pederson HS, Hansen JC, Dewailly E, Jul E, Pedersen MB, Bjerregaard P, Malcom GT, Deguchi Y, Middaugh JP (1996) Exposure of Greenlandic Inuit to organochlorines and heavy metals through the marine food-chain: an international study. Sci Total Environ 186:137–139

Nakajima D, Yoshida Y, Suzuki J, Suzuki S (1995) Seasonal changes in the concentration of polycyclic aromatic hydrocarbons in azalea leaves and relationship to atmospheric concentration. Chemosphere 30:409–418

Neuhauser EF, Loehr RC, Malecki MR, Milligan DL, Durkin PR (1985) The toxicity of selected organic chemicals to the earthworm Eisenia fetida. J Environ Qual 14:383–388

Nicholls CR, Allchin CR, Law RJ (2001) Levels of short and medium chain length polychlorinated n-alkanes in environmental samples from selected industrial areas in England and Wales. Environ Pollut 114:415–430

Noorlander CW, van Leeuwen SPJ, te Biesebeek JD, Mengelers MJB, Zeilmaker MJ (2011) Levels of perfluorinated compounds in food and dietary intake of PFOS and PFOA in The Netherlands. J Agric Food Chem 59:7496–7505

Norstrom RJ (2002) Understanding bioaccumulation of POPs in food webs: chemical, biological, ecological, and environmental considerations. Environ Sci Pollut Res 9:300–303

Notarianni V, Calliera M, Tremolada P, Finizio A, Vighi M (1998) PCB distribution in soil and vegetation from different areas in Northern Italy. Chemosphere 37(14–15):2839–2845

Ohta S, Ishizuka D, Nishimura H, Nakao T, Aozasa O, Shimidzu Y, Ochiai F, Kida T, Nishi M, Miyata H (2002) Comparison of polybrominated diphenyl ethers in fish, vegetables, and meats and levels in human milk of nursing women in Japan. Chemosphere 46:689–696

Olie K, Vermeulen PL, Hutzinger D (1977) Chlorobenzo-p-dioxins and chlorodibenzofurans are trace components of fly ash and flue gas of some municipal incinerators in the Netherlands. Chemosphere 6:455–459

Oliver BG, Nicol KD (1982) Chlorobenzenes in sediments, water, and selected fish from lakes Superior, Huron, Erie, and Ontario. Environ Sci Technol 16:532–536

Oliver BG, Nicol KD (1984) Chlorinated contaminants in the Niagara River, 1981–1983. Sci Total Environ 39:57–70

Oliver BG, Niimi AJ (1983) Bioconcentration of chlorobenzenes from water by rainbow trout: correlations with partition coefficients and environmental residues. Environ Sci Technol 17:287–291

Olsson AC, Fevotte J, Fletcher T, Cassidy A, ‘t Mannetje A, Zaridze D, Szeszenia-Dabrowska N, Rudnai P, Lissowska J, Fabianova E, Mates D, Bencko V, Foretova L, Janout V, Brennan P, Boffetta P (2010) Occupational exposure to polycyclic aromatic hydrocarbons and lung cancer risk: a multicenter study in Europe. Occup Environ Med 67:98–103

Pacyna JM, Breivik K, Munch J, Fudala J (2003) European atmospheric emissions of selected persistent organic pollutants, 1970–1995. Atmos Environ 37:119–131

Palm A, Cousins IT, Mackay D, Tysklind M, Metcalfe C, Alaee M (2002) Assessing the environmental fate of chemicals of emerging concern: a case study of the polybrominated diphenyl ethers. Environ Pollut 117(2):195–213

Paterson S, Mackay D, McFarlane C (1994) A model of organic chemical uptake by plants from soil and the atmosphere. Environ Sci Technol 28:2259–2266

Paul AG, Jones KC, Sweetman AJ (2009) A first global production, emission, and environmental inventory for perfluorooctane sulfonate. Environ Sci Technol 2009:386–392

Pearson CR, McConnell G (1975) Chlorinated C1 and C2 hydrocarbons in the marine environment. Proc R Soc London Ser B 189:305–332

Pearson RF, Hornbuckle KC, Eisenreich SJ, Swackhamer DL (1996) PCBs in Lake Michigan water revisited. Environ Sci Technol 30:1429–1436

Pereira MG, Walker LA, Wright J, Best J, Shore RF (2009) Concentrations of polycyclic aromatic hydrocarbons (PAHs) in the eggs of predatory birds in Britain. Environ Sci Technol 43 (23):9010–9015

Perera F, Herbstman J (2011) Prenatal environmental exposures, epigenetics, and disease. Reprod Toxicol 31(3):363–373

Perera F, Tang D, Whyatt R, Lederman SA, Jedrychowski W (2005) DNA damage from polycyclic aromatic hydrocarbons measured by benzo[a]pyrene-DNA adducts in mothers and newborns from Northern Manhattan, the World Trade Center Area, Poland, and China. Cancer Epidemiol Biomark Prev 14(3):709–714

Pérez C, Velando A, Munilla I, López-Alonso M, Daniel O (2008) Monitoring polycyclic aromatic hydrocarbon pollution in the marine environment after the Prestige oil spill by means of seabird blood analysis. Environ Sci Technol 42(3):707–713

Persson Y, Shchukarev A, Öberg L, Tysklind M (2008) Dioxins, chlorophenols and other chlorinated organic pollutants in colloidal and water fractions of groundwater from a contaminated sawmill site. Environ Sci Pollut Res 15(6):463–471

Peters RJB (2003) Hazardous chemicals in precipitation, TNO Report R 2003/198. AH. TNO Environment, Energy and Process Innovation, Apeldoorn

Pham TT, Proulx S, Brochu C, Moore S (1999) Composition of PCBs and PAHs in the Montreal urban community wastewater and in the surface water of the St. Lawrence River (Canada). Water Air Soil Pollut 111(1–4):251–270

Pizzorno JE, Katzinger JJ (2013) Clinical implications of persistent organic pollutants – epigenetic mechanisms. J Restor Med 2:4–13

Poland A, Knutson JC (1982) 2,3,7,8-Tetrachlorodibenzo-p-dioxin and related halogenated aromatic hydrocarbons: examination of the mechanism of toxicity. Ann Rev Pharmacol Toxicol 22:517–554

Porta M, Puigdomenech E, Ballester F, Selva J, Ribas-Fitó N, Llop S, Tomàs López T (2008) Monitoring concentrations of persistent organic pollutants in the general population: the international experience. Environ Int 34(4):546–561

Poster DL, Baker JE (1996) Influence of submicron particles on hydrophobic organic contaminants in precipitation. II. Concentrations and distribution of polyciclic aromatic hydrocarbons and chlorinated biphenyls in rain water. Environ Sci Technol 30(1):349–354

Potter CL, Glaser JA, Chang LW, Meier JR, Dosani MA, Herrmann RF (1999) Degradation of polynuclear aromatic hydrocarbons under bench- scale compost conditions. Environ Sci Technol 33(10):1717–1725

Primbs T, Wilson G, Schmedding D, Higginbotham C, Simonich SM (2008) Influence of Asian and Western United States agricultural areas and fires on the atmospheric transport of pesticides in the Western United States. Environ Sci Technol 42:6519–6525

Puzyn T, Mostrag A, Suzuki N, Falandysz J (2008) QSPR-based estimation of the atmospheric persistence for chloronaphthalene congeners. Atmos Environ 42:6627–6636

Quaß U, Fermann M (1997) Identification of relevant industrial sources of dioxins and furans in Europe, (The European Dioxin Inventory). Final Report, Materialen No. 43, published by North Rhine-Westphalia State Environment Agency (LUA NRW)), ISSN 0947-5206

Rahman F, Langford KH, Schrimshaw MD, Lester JN (2001) Polybrominated diphenyl ether (PBDE) flame retardants. Review. Sci Total Environ 275:1–17

Ramsay MA, Hobson KA (1991) Polar bears make little use of terrestrial food webs evidence from stable-carbon isotope analysis. Oecologia 86(4):598–600

Reijnders PJH, Aguilar A, Borrell A (2009) Pollution and marine mammals. In: William FP, Bernd W, Thewissen JGM (eds) Encyclopedia of marine mammals, 2nd edn. Academic, London, pp 890–898

Reistad T, Mariussen E, Fonnum F (2005) The effect of a brominated flame retardant, tetrabromobisphenol-A, on free radical formation in human neutrophil granulocytes: the involvement of the MAP kinase pathway and protein kinase C. Toxicol Sci 83:89–100

Renberg L, Tarkpea M, Sundstrom G (1986) The use of the bivalve *Mytilus edulis* as a test organism for bioconcentration studies. Ecotoxicol Environ Saf 11:361–372

Reth M, Ciric A, Christensen GN, Heimstad ES, Oehme M (2006) Short- and medium-chain chlorinated paraffins in biota from the European Arctic e differences in homologue group patterns. Sci Total Environ 367:252–260

Ritter L, Solomon KR, Forget J, Stemeroff M, O'Leary C (1996), Persistent organic pollutants: an assessment report on: DDT, aldrin, dieldrin, endrin, chlordane, heptachor, hexachlorobenzene, mirex, toxaphene, polychlorinated biphenyls, dioxins, and furans, inter-Organization Programme for the Sound Management of Chemicals. http://cdrwww.who.int/ipcs/assessment/en/pcs_95_39_2004_05_13.pdf. Accessed 19 May 2015

Rogers C (1991) Australian patent application No.74463/91 (PCT/US91/01112)

Roscales JL, González-Solís J, Calabuig P, Jiménez B (2011) Interspecies and spatial trends in polycyclic aromatic hydrocarbons (PAHs) in Atlantic and Mediterranean pelagic seabirds. Environ Pollut 159(10):2899–2905

Ross PS, Birnbaum LS (2003) Integrated human and ecological risk assessment: a case study of persistent organic pollutants (POPs) in humans and wildlife. Hum Ecol Risk Assess 9:303–324

Saito N, Sasaki K, Nakatome K, Harada K, Yoshinaga T, Koizumi A (2003) Perfluorooctane sulfonate concentrations in surface water in Japan. Arch Environ Contam Toxicol 45:149–158

Saito N, Harada K, Inoue K, Sasaki K, Yoshinaga T, Koizumi A (2004) Perfluorooctanoate and perfluorooctane sulfonate concentrations in surface water in Japan. J Occup Health 46:49–59
Sakurai K, Fukata H, Todaka E, Saito Y, Bujo H, Mori C (2006) Colestimide reduces blood polychlorinated biphenyl (PCB) levels. Intern Med 5:327–328
Sasaki K, Harada K, Saito N, Tsutsui T, Nakanishi S, Tsuzuki H, Koizumi A (2003) Impacts of air-borne perfluorooctane sulfonate on the human body burden and the ecological system. Bull Environ Contam Toxicol 71:408–413
Schauer C, Niessner R, Poschl U (2003) Polycyclic aromatic hydrocarbons in urban air particulate matter: decadal and seasonal trends, chemical degradation, and sampling artifacts. Environ Sci Technol 37:2861–2868
Schepens PJC, Covaci A, Jorens PG, Hens L, Scharpe S, van Larebeke N (2001) Surprising findings following a Belgian food contamination with polychlorobiphenyls and dioxins. Environ Health Perspect 109:101–103
Schnittger P (2001) Sanierung der Deponie Georgswerder in Hamburg. In: Handbuch der Altlastensanierung, Rd Nr. 7. CF Müller Verlag, Hüthig GmbH & Co. KG Heidelberg
Schwarzbauer J, Ricking M, Franke S, Francke W (2001) Organic compounds as contaminants of the Elbe River and its tributaries. Part 5. Halogenated organic contaminants in sediments of the Havel and Spree Rivers (Germany). Environ Sci Technol 35:4015–4025
Sears ME, Genuis SJ (2012) Environmental determinants of chronic disease and medical approaches: recognition, avoidance, supportive therapy, and detoxification. J Environ Pub Health 2012:356798., 15 pages. https://doi.org/10.1155/2012/356798
Sellström U, Kierkegaard, A, Alsberg T, Jonsson P, Wahlberg C, DE WIT CA (1999) Brominated flame retardants in sediments from European estuaries, the Baltic Sea and in sewage sludge. Organohalogen Compd 40:383–386
Sepulvado JG, Blaine AC, Hundal LS, Higgins CP (2011) Occurrence and fate of perfluorochemicals in soil following the land application of municipal biosolids. Environ Sci Technol 45:8106–8112
Shao J, White CC, Dabrowski MJ, Kavanagh TJ, Eckert ML, Gallagher EP (2008) The role of mitochondrial and oxidative injury in BDE 47 toxicity to human fetal liver hematopoietic stem cells. Toxicol Sci 101:81–90
Shaw SD, Berger ML, Brenner D, Carpenter DO, Tao L, Hong CS, Kannan K (2008) Polybrominated diphenyl ethers (PBDEs) in farmed and wild salmon marketed in the Northeastern United States. Chemosphere 71:1422–1431
Shen L, Wania F, Lei YD, Teixeira C, Muir DCG, Bidleman TF (2005) Atmospheric distribution and long-range transport behavior of organochlorine pesticides in North America. Environ Sci Technol 39:409–420
Shiraishi H, Pilkington NH, Otsuki A, Fuwa K (1985) Occurrence of chlorinated polynuclear aromatic compounds in tap water. Environ Sci Technol 19:585–590
Simonich SL, Hites RA (1994) Importance of vegetation in removing polycyclic aromatic hydrocarbons from the atmosphere. Nature 370:49–51
Simonich SL, Hites RA (1995) Organic pollutant accumulation in vegetation. Environ Sci Technol 29:2905–2914
Sinclair E, Taniyasu S, Yamashita N, Kannan K (2004) Perfluorooctanoic acid and perfluorooctane sulfonate in Michigan and New York Waters. Organohalogen Compd 66:4019–4023
Sinclair E, Mayack DT, Roblee K, Yamashita N, Kannan K (2006) Occurrence of perfluoroalkyl surfactants in water, fish, and birds from New York State. Arch Environ Contam Toxicol 50:398–410
Singh HB, Sales LJ, Stiles RE (1982) Distribution of selected gaseous organic mutagens and suspect carcinogens in ambient air. Environ Sci Technol 16:872–880
Skutlarek D, Exner M, Farber H (2006) Perfluorinated surfactants in surface and drinking waters. Environ Sci Pollut Res 13:299–307

Smith KEC, Thomas GO, Jones KC (2001) Seasonal and species differences in the air-pasture transfer of PAHs. Environ Sci Technol 35(11):2156–2165

So MK, Miyake Y, Yeung WY, Ho YM, Taniyasu S, Rostkowski P, Yamashita N, Zhou BS, Shi XJ, Wang JX, Giesy JP, Yu H, Lam PK (2007) Perfluorinated compounds in the Pearl River and Yangtze River of China. Chemosphere 68:2085–2095

Spicer CW, Buxton B, Holdren MW (1996) Variability of hazardous air pollutants in an urban area. Atmos Environ 30(20):3443–3456

Srogy K (2007) Monitoring of environmental exposure to polycyclic hydrocarbons: a review. Environ Chem Lett 5:169–195

Staci LS, Hites R (1995) Organic pollutant accumulation in vegetation. Environ Sci Technol 29(12):2905–2914

Stapleton HM, Letcher RJ, Li J, Baker JE (2004) Dietary accumulation and metabolism of polybrominated diphenyl ethers by juvenile carp(Cyprinus carpio). Environ Toxicol Chem 23:1939–1946

Stevens JL, Green NJL, Jones KC (2001) Survey of PCDD/Fs and non-ortho PCBs in UK sewage sludges. Chemosphere 44(6):1455–1462

Stevens JL, Northcott GL, Stern GA, Tomy GT, Jones KC (2002) PAHs, PCBs, PCNs, organochlorine pesticides, synthetic musks and polychlorinated n-alkanes in UK sewage sludge: survey results and implications. Environ Sci Technol 37:462–467

Stock NL, Furdui VI, Muir DCG, Maburi SA (2007) Perfluoroalkyl contaminants in the Canadian Arctic: evidence of atmospheric transport and local contamination. Environ Sci Technol 41:3529–3536

Strachan WMJ, Burniston DA, Williamson M, Bohdanowicz H (2001) Spatial differences in persistent organochlorine pollutant concentrations between the Bering and Chukchi seas (1993). Mar Pollut Bull 43:132–142

Strandberg B, Dodder NG, Basu I, Hites RA (2001) Concentrations and spatial variations of polybrominated diphenyl ethers and other organohalogen compounds in Great Lakes air. Environ Sci Technol 35:1078–1083

Strynar MJ, Lindstrom AB, Nakayama SF, Egeghy PP, Helfant LJ (2012) Pilot scale application of a method for the analysis of perfluorinated compounds in surface soils. Chemosphere 86:252–257

Strzelecka-Jastrzab E, Panasiuk D, Pacyna JM, Pacyna EG, Fudala J, Hlawiczka S, Cenowski M, Dyduch B, Glodek A (2007) Emission projections for the years 2010 and 2020 and assessment of the emission reduction scenario implementation costs, 21 pp, Norwegian Institute for Air Research (NILU). DROPS D1.3 Report. http://drops.nilu.no

Svendsen TC, Vorkamp K, Ronsholdt B, Frier JO (2007) Organochlorines and polybrominated diphenyl ethers in four geographically separated populations of Atlantic salmon (*Salmo salar*). J Environ Monit 9:1213–1219

Svensson BG, Nilsson A, Josson E, Schütz A, Åkesson B, Hagmar L (1995) Fish consumption and exposure to persistent organochlorine compounds, mercury, selenium and methylamines among Swedish fishermen. Scand J Work Environ Health 21:96–105

Sweetman AJ, Jones KC (2000) Declining PCB concentrations in the UK atmosphere: evidence and possible causes. Environ Sci Technol 34(5):863–869

Tanaka S, Fujii S, Lien NPH, Nozoe M, Fukagawa H, Wirojanagud W, Anton A, Lindstrom G (2006) A simple pre-treatment procedure in PFOS and PFOA water analysis and its' application in several countries. Organohalogen Compd 68:527–530

Taniyasu S, Kannan K, Holoubek I, Ansorgova A, Horii Y, Hanari N, Yamashita N, Aldous KM (2003a) Isomer-specific analysis of chlorinated biphenyls, naphthalenes and dibenzofurans in Delor: polychlorinated biphenyl preparations from the former Czechoslovakia. Environ Pollut 126:169–178

Taniyasu S, Kannan K, Horii Y, Hanari N, Yamashita N (2003b) A survey of perfluorooctane sulfonate and related perfluorinated organic compounds in water, fish, birds, and humans in Japan. Environ Sci Technol 37:2634–2639

Taube J, Vorkamp K, Forster M, Herrmann R (2002) Pesticide residues in biological waste. Chemosphere 49(10):1357–1365

Thomas GO, Farrar D, Braekevelt E, Stern GA, Kalantzi OI, Martin FL, Jones KC (2006) Short and medium chain length chlorinated paraffins in UK human milk fat. Environ Int 32(1):34–40

Tomy GT, Stern GA, Muir DCG, Fisk AT, Cymbalisty CD, Westmore JB (1997a) Quantifying C10-C13 polychloroalkanes in environmental samples by high-resolution gas chromatography/ electron capture negative ion high-resolution mass spectrometry. Anal Chem 69:2762–2771

Tomy GT, Stern GA, Muir DCG, Lockhart L, Westmore JB (1997b) Occurrence of polychloro- *n*-alkanes in Canadian mid-latitude and Arctic lake sediments. Organohalogen Compd 33:220–226

Tomy GT, Fisk AT, Westmore JB, Muir DCG (1998) Environmental chemistry and toxicology of polychlorinated n-alkanes. Rev Environ Contam Toxicol 158:53–128

Tomy GT, Stern GA, Lockhart WL, Muir DCG (1999) Occurrence of C-10-C-13 polychlorinated n-alkanes in Canadian midlatitude and arctic lake sediments. Environ Sci Technol 33:2858–2863

Torres MA, Barros MP, Campos SCG, Ernani Pinto E, Rajamani S, Sayre RT, Pio Colepicolo P (2008) Biochemical biomarkers in algae and marine pollution: a review. Ecotoxicol Environ Saf 71:1–15

Trapido M (1999) Polycyclic aromatic hydrocarbons in Estonian soil: contamination and profiles. Environ Pollut 105:67–74

Tysklind M, Fangmark I, Marklund S, Lindskog A, Thaning L, Rappe C (1993) Atmospheric transport and transformation of polychlorinated dibenzo-pdioxins and dibenzofurans. Environ Sci Technol 27:2190–2197

Ueno D, Takahashi S, Tanaka H, Subramanian AN, Fillmann G, Nakata H, Lam PKS, Zheng J, Muchtar M, Prudente M, Chung KH, Tanabe S (2003) Global pollution monitoring of PCBs and organochlorine pesticides using skipjack tuna as a bioindicator. Arch Environ Contam Toxicol 45(3):378–389

Ueno D, Alaee M, Marvin C, Muir DCG, Macinnis G, Reiner E, Crozier P, Furdui VI, Subramanian A, Fillmann G, Lam PKS, Zheng GJ, Muchtar M, Razak H, Prudente M, Chung KH, Tanabe S (2006) Distribution and transportability of hexabromocyclododecane (HBCD) in the Asia-Pacific region using skipjack tuna as a bioindicator. Environ Pollut 144 (1):238–247

Umegaki K, Ikegami S, Ichikawa T (1993) Effects of restricted feeding on the absorption, metabolism and accumulation of pentachlorobenzene in rats. J Nutr Sci Vitaminol 39:11–22

UN-ECE (United Nations Economic Commission for Europe) (1998) Draft protocol to the convention on long-range air pollution on persistent organic pollutants, EB.AIR:1998:2, Convention on Long-range Transboundary Air Pollution, United Nations Economic and Social Council, Economic Commission for Europe

UN-ECE (United Nations Economic Commission) (2009) Protocol on persistent organic pollutants (POPs). Amendments to the Protocol adopted on 18 December 2009, ECE/EB.AIR/104, Convention on Long-range Transboundary Air Pollution, United Nations Economic and Social Council, Economic Commission for Europe

UNEP (United Nations Environment Programme) (1999) Dioxin and furan inventories: national and regional emissions of PCDD/PCDF., 253 pp. UNEP Chemicals, Geneva

UNEP (United Nations Environment Programme) (2001) Final act of the plenipotentiaries on the Stockholm Convention on persistent organic pollutants. Annex D, Information requirements and screening criteria. United Nations environment program chemicals, Geneva

UNEP United Nations Environment Programme (2009) Stockholm convention on persistent organic pollutants (POPs). Persistent Organic Pollutants Review Committee. Revised draft risk profile: short-chained chlorinated paraffins. 9 July 2009. UNEP/POPS/POPRC.5/2 http://chm.pops.int/Convention/POPsReviewCommittee/hrPOPRCMeetings/POPRC5/POPRC5Documents/tabid/592/language/en-US/Default.aspx. Accessed 21 Aug 2015

Unwin J, Cocker J, Scobbie E, Chambers H (2006) An assessment of occupational exposure to polycyclic aromatic hydrocarbons in the UK. Ann Occup Hyg 50(4):395–403

US EPA (United States Environmental Protection Agency) (1977) Environmental monitoring near industrial sites: polychloronaphthalenes. Washington, DC (EPA 560/6–77-019)

US EPA (United States Environmental Protection Agency) (1986) Guidelines for carcinogen risk assessment. Fed Regist 51(185):33992–34003

US EPA (United States Environmental Protection Agency) (1991) Drinking water health advisories: hexachlorobutadiene. In: Volatile organic compounds. Office of Drinking Water. Ann Arbor, MI: Lewis Publishers, p 51–68

US EPA (United States Environmental Protection Agency) (1994) Health assessment document for 2, 3, 7, 8-tertachlorodibenzo-p-dioxin (TCDD) and related compounds. Washington, DC (EPA 600/Bp-92/001c)

US EPA (United States Environmental Protection Agency) (2001) Guidance for reporting toxic chemicals: polycyclic aromatic compounds category. Washington, DC (EPA 260-B-01-03)

US EPA (United States Environmental Protection Agency) (2009) Short-chain chlorinated paraffins (SCCPs) and other chlorinated paraffins action plan. Washington, DC

US EPA (United States Environmental Protection Agency) (2010) Reference guide to non-combustion technologies for remediation of persistent organic pollutants in soil, 2nd edn. Office of Research and Development, Washington, DC

Vallack HW, Bakker DJ, Brandt I, Broström-Lundén E, Brouwer A, Bull KR, Gough C, Guardans R, Holoubek I, Jansson B, Koch R, Kuylenstierna J, Lecloux A, Mackay D, McCutcheon P, Mocarelli P, Taalman RDF (1998) Controlling persistent organic pollutants–what next? Environ Toxicol Pharmacol 6:143–175

Van de Meent D, den Hollander HA, Pool WJ, Vredenbregt MJ, van Oers HAM, Greef E, Luijten JA (1986) Organic micropollutants in Dutch coastal waters. Water Sci Technol 18:73–81

Vanni A, Gamberini R, Calabria A, Pellegrino V (2000) Determination of presence of fungicides by their common metabolite, 3,5-Dca, in compost. Chemosphere 41(3):453–458

Verreault J, Muir D, Norstrom R, Stirling I, Fisk A, Garielsen G, Derocher A, Evans T, Dietz R, Sonne C, Sandala G, Gebbink W, Riget F, Born E, Taylor M, Nagy J, Letcher R (2005) Chlorinated hydrocarbon contaminants and metabolites in polar bears (Ursus Maritimus) from Alaska, Canada, East Greenland, and Svalbard: 1996–2002. Sci Total Environ 351–352:369–390

Viberg H, Fredriksson A, Eriksson P (2003) Neonatal exposure to polybrominated diphenyl ether (PBDE 153) disrupts spontaneous behavior, impairs learning and memory, and decreases hippocampal cholinergic receptors in adult mice. Toxicol Appl Pharmacol 192:95–106

Vikelsøe J, Thomsen M, Carlsen L, Johansen E (2002) Persistent organic pollutants in soil, sludge and sediment. A multianalytical field study of selected organic chlorinated and brominated compounds, NERI Technical Report No 402. National Environmental Research Institute, Roskilde

Voorspoels S, Covaci A, Neels H, Schepens P (2007) Dietary PBDE intake: a market-basket study in Belgium. Environ Int 33:93–97

Vorkamp K, Kellner E, Taube J, Moller KD, Herrmann R (2002) Fate of methidathion residues in biological waste during anaerobic digestion. Chemosphere 48(3):287–297

Wang Y, Cheng Z, Li J, Luo C, Xu Y, Li Q, Liu X, Zhang G (2012) Polychlorinated naphthalenes (PCNs) in the surface soils of the Pearl River Delta, South China: distribution, sources, and air-soil exchange. Environ Pollut 170:1–7

Wania F (2006) Potential of degradable organic chemicals for absolute and relative enrichment in the Arctic. Environ Sci Technol 40:569–577

Wania F, Mackay D (1996) Tracking the distribution of persistent organic pollutants. Environ Sci Technol 30:390A–396A

Watanabe I, Sakai S (2003) Environmental release and behavior of brominated flame retardants. Environ Int 29(6):665–682

Webber MD, Wang C (1995) Industrial organic compounds in selected Canadian soils. Can J Soil Sci 75(4):513–524

Weber R, Gaus C, Tysklind M, Johnston P, Forter M, Hollert H, Heinisch E, Holoubek I, Lloyd-Smith M, Masunaga S, Moccarelli P, Santillo D, Seike N, Symons R, Torres JPM, Verta M, Gerd Varbelow G, Vijgen J, Watson A, Costner P, Woelz J, Wycisk P, Zennegg M (2008) Dioxin- and POP-contaminated sites – contemporary and future relevance and challenges. Environ Sci Pollut Res 15:363–393

Weber R, Watson A, Forter M, Oliaei F (2011) Persistent organic pollutants and landfills – a review of past experiences and future challenges. Waste Manag Res 29(1):107–121

Weisglas-Kuperus N, Sas TCJ, Koopman-Esseboom C, van der Zwan CW, de Ridder MAJ, Keishuizen A, Hooijkaas H, Sauer PJJ (1995) Immunological effect of background prenatal and postnatal exposure to dioxins and polychlorinated biphenyls in Dutch infants. Pediatr Res 38:404–410

Wenzel KD, Hubert A, Weissflog L, Kühne R, Popp P, Kindler A, Schüürmann G (2006) Influence of different emission sources on atmospheric organochlorine patterns in Germany. Atmos Environ 40:943–957

Weremiuk AM, Gerstmann S, Hartmut F (2006) Quantitative determination of perfluorinated surfactants in water by LC-ESI-MS/MS. J Sep Sci 29:2251–2255

WHO (World Health Organization) (1991) International programme on chemical safety. Environmental Health Criteria (EHC) 128: chlorobenzenes other than hexachlorobenzene. United Nations Environment Programme. International Labour Organisation. World Health Organization, Geneva

WHO (World Health Organization) (1994a) Brominated diphenyl ethers, Environ Health Crit 162. World Health Organization, Geneva

WHO (World Health Organization) (1994b) Hexachlorobutadiene, Environ Health Crit 156. World Health Organization, Geneva

WHO (World Health Organization) (1998) International programme on chemical safety. Selected nonheterocyclic polycyclic aromatic hydrocarbons, Environmental Health Criteria 202. World Health Organization, Geneva

WHO (World Health Organization) (2010) Persistent organic pollutants: impact on child health. World Health Organization, Geneva

Wilcke W (2000) Polycyclic aromatic hydrocarbons (PAHs) in soil – a review. J Plant Nutr Soil Sci 163(3):229–248

Wilcke W (2007) Global patterns of polycyclic aromatic hydrocarbons (PAHs) in soil. Geoderma 141:157–166

Wild SR, Jones KC (1992) Organic chemicals entering agricultural soils in sewage sludges – screening for their potential to transfer to trop plants and livestock. Sci Total Environ 119:85–119

Wild SR, Berrow ML, Jones KC (1991a) The persistence of polynuclear aromatic hydrocarbons (PAHs) in sewage sludge amended agricultural soils. Environ Pollut 72(2):141–157

Wild SR, Obbard JP, Munn CI, Berrow ML, Jones KC (1991b) The long-term persistence of polynuclear aromatic hydrocarbons (PAHs) in an agricultural soil amended with metal-contaminated sewage sludges. Sci Total Environ 101(3):235–253

Wilford BH, Harner T, Zhu J, Shoeib M, Jones KC (2004) Passive sampling survey of polybrominated diphenyl ether flame retardants in indoor and outdoor air in Ottawa, Canada: implications for sources and exposure. Environ Sci Technol 38:5312–5318

Williams DJ, Neilson MA, Merriman J, L'Italien S, Painter S, Kuntz K, El-Shaarawi AH (2000) The Niagara River upstream/downstream program 1986/87–1996/97. Concentrations, loads, trends, Report No. EHD/ECB-OR/00-01/I. Environmental Conservation Branch/Ontario Region, Ecosystem Health Division, Environment Canada, Burlington

Willis B, Crookes MJ, Diment J, Dobson SD (1994) Environmental hazard assessment: chlorinated paraffins. Toxic Substances Division. Dept. of the Environment, London

Wilson DC (1982) Lessons from Seveso. Chem Br 18:499–504

Witter B, Francke W, Francke S, Knauth HD, Miehlich G (1998) Distribution and mobility of organic micropollutants in River Elbe floodplains. Chemosphere 37:63–78

Xu SS, Liu WX, Tao S (2006) Emission of polycyclic aromatic hydrocarbons in China. Environ Sci Technol 40:702–708

Yamashita N, Taniyasu S, Hanari N, Falandysz J (2003) Polychlorinated naphthalene contamination of some recently manufactured industrial products and commercial goods in Japan. J Environ Sci Health A 38:1745–1759

Yip G (1976) Survey of hexachloro-1,3,-butadiene in fish, eggs, milk, and vegetables. J Assoc Off Anal Chem 59:559–561

Zareitalabad P, Siemens J, Hamer M, Amelung W (2013) Perfluorooctanoic acid (PFOA) and perfluorooctanesulfonic acid (PFOS) in surface waters, sediments, soils and wastewater – a review on concentrations and distribution coefficients. Chemosphere 91:725–732

Zeng L, Wang T, Han W, Yuan B, Liu Q, Wang Y, Jiang G (2012) Spatial and vertical distribution of short chain chlorinated paraffins in soils from wastewater irrigated farmlands. Environ Sci Technol 45(6):2100–2106

Zethner G, Götz B, Amlinger F (2000) Qualität von Kompost aus der getrennten Sammlung. Monographien, vol 133. Federal Environment Agency, Wien

Zhang YX, Tao S (2008) Emission of polycyclic aromatic hydrocarbons (PAHs) from indoor straw burning and emission inventory updating in China. Ann N Y Acad Sci 1140:218–227

Zhang YX, Tao S (2009) Global atmospheric emission inventory of polycyclic aromatic hydrocarbons (PAHs) for 2004. Atmos Environ 43(4):812–819

Zhang H, Wang Y, Sun C, Yu M, Gao Y, Wang T, Liu J, Jiang G (2014) Levels and distributions of hexachlorobutadiene and three chlorobenzenes in biosolids from wastewater treatment plants and in soils within and surrounding a chemical plant in China. Environ Sci Technol 48:1525–1531

Chapter 12
Occurrence and Fate of Human and Veterinary Medicinal Products

Abstract Medicinal products are a class of emerging environmental contaminants that are increasingly being used in human and veterinary medicine. These products are designed to have a specific mode of action, and many of them for some persistence in the human body. Up to now, only little is known about ecotoxicological effects of medicinal products on aquatic and terrestrial organisms including wildlife. There is thus a need to focus on sources and long-term exposure assessment regarding specific modes of action of medicinal products to better judge their effects on the environment and on the human body. Contamination of the environment with pharmaceuticals has received increased attention in recent years. Unlike agrochemicals, which are applied to fields in pulsed events, pharmaceuticals enter the environment more or less continuously. Many different pharmaceuticals are used in human medicine, and antibiotics and anti-inflammatory drugs are used in veterinary medicine throughout the world. A number of pharmaceuticals have been detected in many environmental samples worldwide. Their occurrence has been reported in sewage treatment plant effluents, surface water, seawater, groundwater, soils, sediments plants and fish. In several countries of the EU, such as England, Germany, and Austria some pharmaceutical products are used in quantities of more than 100 Mg per year. Sources of human medicinal products include release from industrial production of pharmaceuticals, discharge of pharmaceuticals from wastewater treatment plants into rivers, field-application of sewage sludge as organic fertilizer, and use of treated wastewater for irrigation. Sources of veterinary pharmaceuticals include medical treatment of livestock and medicines from surface-applied liquid or farmyard manure. Pharmaceuticals used in animals raised on pastures are excreted directly to the grassland. Pharmaceuticals entering the terrestrial environment can reach surface water and groundwater.

Medicinal products are bioactive substances that are designed to target-specific metabolic and molecular pathways in humans and animals, but they can also have negative side effects. Pharmaceuticals and their residues may affect significantly the environment. Modes of action of pharmaceutical contaminants on lower organisms are little known and thus make toxicity prediction difficult. Pharmaceutical concentrations measured in surface waters are in a ng L^{-1} to lower µg L^{-1} range, and are generally well below concentrations that are known to cause acute toxicity to organisms. However, chronic exposure to pharmaceuticals has the potential for

R. Nieder et al., *Soil Components and Human Health*,
https://doi.org/10.1007/978-94-024-1222-2_12

numerous more subtle effects on non-target organisms, such as metabolic or reproductive changes. Despite numerous reports on environmental occurrence of pharmaceuticals the environmental significance is largely unknown. As an exception, the synthetic oestrogen ethinyl-estradiol is well-known for its potential for endocrine disruptive and reproductive effects. Several papers describe such effects after exposure to either natural or synthetic oestrogens in laboratory settings. For instance, exposure to low levels of ethinyl-estradiol delayed the embryonic development in zebrafish and caused vitellogenin induction in rainbow trout. Downstream sewage treatment plants in the UK, fish displayed intersex characteristics, leading to reduced reproductive success were found. These effects are likely to affect the population dynamics locally. Effects observed in the environment, however, are difficult to assign to a specific compound as mixtures with possible overlapping and/or interactive properties are likely to be present.

Although studies showing environmental effects up to now are very scarce, there is indirect evidence that exposure to antibiotics has brought about the resistance of bacteria in the environment. For example, the use of antibiotics in pig production coincided with the discovery of resistant *E. coli* where there previously was no resistance in the guts of pigs and in meat products. Later, the resistance had spread to the gut flora of pig farmers, their families, and citizens of the community. Sewers receiving effluent from a hospital displayed an increased prevalence of bacteria resistant to oxytetracyline, while sewers receiving effluent from a pharmaceutical plant showed an increased prevalence of bacteria resistant to multiple antibiotics. Fluoroquinolone use in poultry husbandry has promoted the evolution of fluoroquinolone-resistant *Campylobacter jejuni*, an important human pathogen (see Chap. 13). Antibiotic resistant genes, within microbes or as naked DNA bound to clay particles, were found to persist in soils, river sediments, dairy lagoons, wastewater effluent and water treatment plants. However, further understanding of how resistance is acquired and maintained in bacterial populations is needed before a link between the presence of antibiotics in the environment and antibiotic resistance can be made. Thus, there is an enormous public and scientific need to learn more about the extent of the occurrence of pharmaceutical residues, about their fate during sewage treatment and in the environment, and about natural and technological processes that are able to remove such residues from sewage or raw waters used for drinking water supply.

Keywords Human and veterinary pharmaceuticals • Physico-chemical properties • Environmental fate • Pharmaceuticals in water and edible plants • Pharmaceuticals in non-target organisms • Human exposure to pharmaceuticals • Human health threats • Antibiotic resistance • Options to reduce the release of pharmaceuticals into the environment

12.1 Sources of Medicinal Products

There are currently no data available about the total global use of medicinal products and the consumption may vary significantly between countries (Kümmerer 2009a). In industrial countries consumption of active pharmaceutical ingredients is estimated to be between 50 and 150 g per capita and year with less than 50 compounds making up 95% of the total amount of active pharmaceutical ingredient consumption (Alder et al. 2006). Up to 59% of the women in Europe of reproductive age take a contraceptive pill containing ethinyl-estradiol as the main active compound, compared to 16% in North America and 2.3% in Japan (United Nations 2004). In some countries medicinal products are sold over the counter without prescription, while in others they are only available via prescription. Antibiotics such as streptomycins are used in the growing of fruits (pomology). In the USA, the heavy use of streptomycins in the growing of fruits is being discussed as a possible reason for the high resistance of pathogenic bacteria against these pharmaceuticals (Kümmerer 2009a). In Germany, the use of streptomycins for this purpose has been banned. Antimicrobials are among the most widely used medicines in livestock production, particularly in lage-scale animal farming (Sarmah et al. 2006; Kümmerer 2009a). About 50,000 different drugs were registered in Germany in 2001, 2700 of which accounted for 90% of the total consumption and which, in turn, contained about 900 different active substances, corresponding to 38,000 Mg of active compounds (Greiner and Rönnefahrt 2003, cited by Kümmerer 2009a). A total of 6000–7000 Mg year of active substances are of potential environmental concern in Germany, which is approximately 0.45 kg per capita and year. According to Ongerth and Khan (2004), data for Australia are in a similar range. There are currently no legally regulated maximum permitted concentrations of pharmaceuticals in the environment, despite their unknown impact on the environment and human health. Based on the precaution principle, the European Union Water Framework Directive produces an updated list of priority substances every 4 years (2000/60/EC) and has identified compounds from pharmaceuticals as potential pollutants (Rivera-Utrilla et al. 2013).

12.1.1 Human Medicinal Products

In the European Union (EU) about 3000 different compounds are used in human medicine. The most common include analgetics and anti-inflammatory drugs, antibiotics, beta-blockers, contraceptives, antilipidemics and neuroactive compounds (Fent et al. 2006). Major categories of human pharmaceuticals and the most commonly used products are shown in Table 12.1.

Table 12.1 Major classes of human medicines and medicinal products

Class of medicine/ subclass	Medicinal product
Analgetics and anti-inflammatory drugs	Acetylsalicylic acid (Aspirin), diclofenac, ibuprofen; acetaminophen, metamizol, codeine, indometacine; naproxen, phenazone
Antibiotics/	
Aminoglycosides	Streptomycin
Aminopenicillines	Amoxycillin
Dihydrofolate reductase inhibitors	Trimetoprim
Fluoroquinolones	Ofloxacin, flumequine, ciprofloxacin
Lincodamides	Linomycin
Macrolides	Erytromycin, spiramycin
Sulfonamides	Sulfamethoxazole
Tetracyclines	Chlortetracycline, oxytetracycline
Antidepressant	Mianserin
Antiepileptic	Carbamazepine
Antilipidemics	Bezafibrate, gemfibrozil, clofibric acid, fenofibrate
Beta-blockers	Metoprolol, propranolol, nadolol, atenolol, sotalol, betaxolol
Cancer therapeutics	Cyclophosphamide, ifosphamide
Diuretic	Furosemide
Steroids and related hormones	17-β-estradiol, estrone, 17-α-ethinyl-estradiol, diethylstilbestrol, diethylstilbestrol acetate
Tranquilizers	Diazepam

Adapted from Nikolaou et al. (2007)

Among these compounds diclofenac, clofibric acid, acetaminophen, ibuprofen, aspirin, carbamazepine, artorvastatin, gemfibrozil, fluoxetine, 17-α-ethinyl-estradiol have become ubiquitous in surface water and waste water (Nikolaou et al. 2007).

The pattern of consumed medicinal products varies among different countries and some drugs may be forbidden or replaced by related drugs. In Australia, the UK and Germany the amounts for frequently used medicines are in the hundreds of Mg per year (Table 12.2). The data represent the annual consumptions and include mainly prescribed drugs, some include also sales over the counter, some a mixture of both, and internet sales are not included. Therefore, the real amount of applied drugs is uncertain, but probably significantly higher for some of the medicinal products reported compared to the values in Table 12.2.

12.1.2 Veterinary Medicinal Products

A large number of drugs and food additives are used in veterinary medicine. They fall into pharmacological categories such as anesthetic, antacid, anthelmintic,

Table 12.2 Consumption of different classes of human medicines in different countries

Class of medicine/ medicinal compound	Country/year of survey					
	Austria 1997	Denmark 1997	Germany 2001	Italy 2001	UK 2000	Australia 1998
	Mg year^{-1}					
Analgetics, antipyretics, anti-inflammatory/						
Acetylsalicilic acid	78.5	0.21	836.3	n.i.	n.i.	20.4
Salicylic acid	9.6	n.i.	71.7	n.i.	n.i.	n.i.
Paracetamol	38.1	0.24	621.7	n.i.	390.9	295.9
Naproxen	4.6	n.i.	n.i.	n.i.	35.1	22.8
Ibuprofen	6.7	0.03	344.9	1.9	162.2	14.2
Diclofenac	6.14	n.i.	85.8	n.i.	26.1	n.i.
Antilipidemics/						
Gemfibrazol	n.i.	n.i.	n.i.	n.i.	n.i.	20.0
Benzabibrate	4.5	n.i.	n.i.	7.6	n.i.	n.i.
Antiacidics/						
Ranitidine	n.i.	n.i.	85.8	26.7	36.3	33.7
Cimetidine	n.i.	n.i.	n.i.	n.i.	35.7	n.i.
Beta-blockers/						
Atenolol	n.i.	n.i.	n.i.	22.1	29.0	n.i.
Metoprolol	2.4	n.i.	93.0	n.i.	n.i.	n.i.
Diuretics/						
Furosemide	n.i.	3.7	n.i.	6.4	n.i.	n.i.
Sympatomimetica/						
Terbutalin	n.i.	0.5	n.i.	n.i.	n.i.	n.i.
Salbutamol	n.i.	0.2	n.i.	n.i.	n.i.	n.i.
Others/						
Metformin	26.4	n.i.	516.9	n.i.	205.8	90.9
Estradiol	n.i.	0.1	n.i.	n.i.	n.i.	n.i.

Adapted from Fent et al. (2006)
n.i. not indicated

antihistimine, anti-infective, steroidal and non-steroidal anti-inflammatory, antibacterial, antimicrobial, antiparasitic, antiseptic, astringent, bronchodilator, diuretic, emetic, emulsifier, estrus synchronization, growth promotant, nutritional supplement, sedative or tranquilizer (Sarmah et al. 2006). Of the drugs approved for livestock production, antibiotics are among the most widely administered for animal health and management.

The pharmaceuticals are delivered to the animals orally, through water and feed, by injection, implant, drench, paste, topically or pour on. The way how a medicinal product is delivered depends on the use and length of treatment and whether the drug is delivered to an individual animal, a herd or flock. Some of the important uses of veterinary pharmaceuticals are to treat and prevent infectious diseases (e.g. tetracycline, b-lactams, antibiotics and steroid anti-inflammatories), manage reproductive processes (e.g. steroids, oxytocin, ergonovine, GnRH, HCG and prostaglandins, progesterone, and FSH) and production (e.g. bovine somatotropin; hormonal growth implants; ionophores; sub-therapeutic antibiotics), control parasites (e.g. dewormers, insecticides), and control non-infectious diseases (e.g. nutritional supplements; Sarmah et al. 2006) Volume data for veterinary pharmaceuticals are generally not publicly available. However, Kools et al. (2008) estimated the use of important groups of veterinary pharmaceuticals in the European Union. Data were collected from different sources and used to calculate the average use, related with the total meat production in the respective countries. An extrapolation, based on food-stuff production data, yields a rough estimate of the use volume for European countries, largely based on 2004 data (Table 12.3).

Table 12.3 Use of different categories of veterinary pharmaceutics in the EU-25 states (in Mg year^{-1}) on the basis of reported data (roman) and extrapolation (*italics*) by meat production

Country	Meat production (× 1000 Mg)	Antibiotics 2004	Antiparasitics 2004	Hormones 2004
Austria	837	*113*	*4.1*	*0.1*
Belgium	1320	*178*	*6.4*	*0.15*
Cyprus	66	*9*	*0.32*	*0.008*
Czech Republic	755	*102*	*3.7*	*0.09*
Denmark	2149	111	0.24	0.03
Estonia	54	*7*	*0.26*	*0.006*
Finland	377	13.3	*1.8*	*0.0*
France	5869	1179	*28.5*	*0.7*
Germany	6612	668.8	46.3	0.67
Greece	485	*65*	*2.4*	*0.06*
Hungary	909	*123*	*4.4*	*0.11*
Ireland	981	*132*	*4.8*	*0.11*
Italy	3556	*479*	*17.2*	*0.41*
Latvia	73	*10*	*0.35*	*0.008*
Lithuania	195	*26*	*0.94*	*0.02*
Luxemburg	22	*3*	*0.11*	*0.003*
Malta	16	*2*	*0.08*	*0.002*
Netherlands	2321	453	*11.3*	*0.3*
Poland	3152	*425*	*15.3*	*0.36*
Portugal	693	*93*	*3.36*	*0.08*
Slovakia	291	*39*	*1.41*	*0.03*
Slovenia	127	*17*	*0.62*	*0.015*
Spain	5308	*715*	*26*	*0.61*
Sweden	536	16.1	3.86	0.28
UK	3329	414	10.84	0.48
Total EU-25	40,034	5393	194	4.63

Adapted from Kools et al. (2008)

An estimated use of 5393 Mg antibiotics, 194 Mg antiparasitics and 4.6 Mg hormones, sums up to 5592 Mg for the 25 EU countries. Antibiotics are the most widely used veterinary pharmaceuticals administered for animal health and management in agriculture. In the EU antibiotics used in the highest quantities are tetracyclines and beta-lactams together with cephalosporins, followed by other types of antibiotics (Kools et al. 2008). In the USA antibiotics are routinely used at therapeutic levels in livestock operations to treat disease and at sub-therapeutic levels (<0.2 g kg^{-1}) to increase feed efficiency and improve growth rate. Antibiotics used in animal feeding in the USA have increased from nearly 91.0 Mg in 1950 to 9.300 Mg in 1999 (AHI 2002). Of the 9.300 Mg of antibiotics used, about 8.000 Mg were used for treatment and prevention of disease and only 1.300 Mg were used for improving feed efficiency and enhancing growth. For Australia like Canada, there are no available data on the quantities of various antimicrobials used in animal production. In New Zealand, livestock production accounts for about 93.0 Mg $year^{-1}$ of antibiotics use (Kools et al. 2008). Data on the consumption of antibiotics in livestock production in African countries are lacking. Mitema et al. (2001) for the second half of the 1990s assessed consumption of antimicrobial substances to be approximately 14.6 Mg $year^{-1}$ in animal food production in Kenya, of which, tetracyclines and sulfonamides plus trimethoprim accounted for nearly 78% of the use.

12.2 Physico-Chemical Properties of Pharmaceuticals

Most medicinal products are complex, polar compounds with different physico-chemical and biological properties and specific functionalities. They have been detected in different environmental media and in plants. Their molecular weights range commonly from 200 to 1000 g Mol^{-1}, their half-life from a few days to more than 1000 days (Table 12.4).

Compounds belonging to the same therapeutic group may have very different physicochemical properties, such as log K_{ow} and pK_a values. One example is difference in sorption patterns of ofloxacin and sulfamethoxazole, both belonging to the group of antibiotics. This demonstrates the importance of a compound's individual properties to the analysis of its behavior in environmental matrices. The log K_{ow}, or log of the octanol-water partitioning coefficient, represents a compound's propensity to partition into either non-polar or polar mediums. Compounds with a high log K_{ow} are typically hydrophobic, with a low water solubility, while those with a low or negative log K_{ow} are typically hydrophilic with a higher water solubility. The pK_a, or acid dissociation constant, represents a compound's ionization state. This is important because chemicals with a pK_a lower than neutral (i.e., acidic pharmaceuticals), such as ibuprofen, diclofenac, and ketoprofen, will most likely be ions in the environment, making them more likely to dissolve into water

Table 12.4 Physico-chemical properties of frequently used pharmaceuticals, environmental media and plant parts in that they were detected

Class of medicine/medicinal compound	Molecular weight (g Mol^{-1})	Half-life (d)	pK_a	Log K_{ow}	Detecton in environmental media	Detection in plant parts
Analgesic/						
Diclofenac	296	20	4.18	4.06	SW, GW	Not significant
Indomethacin	357	n.i.	4.5	4.23	"	"
Antibiotic/						
Chlortetracycline	479	17–46	3.3	−0.53	Soil	Leaves
Erythromycin	734	30	8.8	3.06	"	n.i.
Ofloxacin	361	360–1.386	6.0–8.3	0.35	"	n.i.
Oxytetracycline	460	44	3.3	−0.9	"	Roots
Sulfadiazine	250	1.6	6.4	−0.09	SW, GW	"
Sulfadimethoxine	310	n.i.	2.1	1.17	"	"
Sulfamethazine	278	74–462	7.6	0.9	"	"
Sulfamethoxalole	253	n.i.	1.8	0.48	"	"
Sulfamonomethoxine	280	n.i.	2.6	0.2	"	Cotyledons, roots
Tetracycline	444	55–578	3.3	−1.3	Soil	Leaves
Trimethoprim	290	22–41	3.2	0.73	SW	Cotyledons, roots
Antiepileptic/						
Carbamazepine	236	75–495	7.0	2.45	Soil and water	Leaves
Anti-inflammatory/						
Ibuprofen	206	6	4.3	3.72	SW, GW	Not significant
Ketoprofen	254	n.i.	4.5	3.0	"	"
Naproxen	230	17	4.2	3.18	"	Leaves

Data from Carvalho et al. (2014), Drillia et al. (2005), Jones et al. (2006), Kang et al. (2013), Tanoue et al. (2012), Thiele-Bruhn (2003), Walters et al. (2010), and Wu et al. (2012)
SW surface water, *GW* groundwater, *n.i.* not indicated

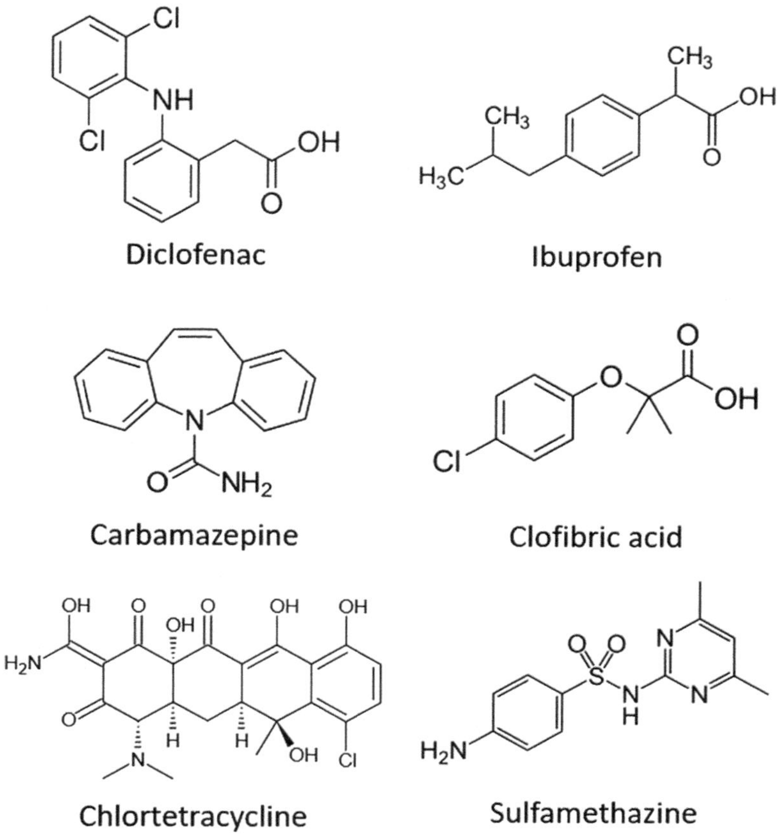

Fig. 12.1 Chemical structures of selected pharmaceuticals: antirheumatic diclofenac and anti-inflammatory ibuprofen; antiepileptic carbamazepine and antilipidemic clofibric acid; antibiotics chlortetracycline and sulfamethazine

than adsorb to solids (Fent et al. 2006). Examples of chemical structures of some important pharmaceuticals are given in Fig. 12.1.

Diclofenac, ibuprofen, carbamazepine and clofibric acid are applied in relatively high amounts and have become ubiquitous in waste water and surface water. However, they show different properties concerning biodegradation and mobility in the aqueous environment. Carbamazepine and clofibric acid both can bind to sediments and soils and particularly the latter shows a persistent behaviour. Antibiotics such as chlortetracycline and sulfamethazine are used for the therapy of infectious diseases in both human and veterinary medicine. As a result of high consumption and excretion, they are disseminated via excrements and enter soil and water environments in large areas. Due to their antibiotic effect, antibiotics can cause antimicrobial resistance in soil microbes. Moreover, resistant microorganisms can reach the soil with contaminated microbes. When pathogens are resistant, humans and animals are endangered to suffer from infectious diseases that cannot be treated successfully with antimicrobial therapy.

12.3 Environmental Fate of Pharmaceuticals

Increasing scientific attention has been devoted to the occurrence of pharmaceuticals in the environment which has resulted in increasing numbers of reports of medicinal products detected in a variety of environmental samples, such as soils, sediments, river water, seawater, and wastewater. Attention has focused on pharmaceuticals used in both veterinary and human medicine. However the environmental exposure scenarios are quite different for these modes of entry. The occurrence of medicinal products was first reported in 1976 by Garrison et al. (1976) who detected clofibric acid in treated wastewater in the USA at concentrations from 0.8 to 2 μg L^{-1}. The occurrence of pharmaceutical compounds in river waters in the UK was reported in 1981, and in 1986 ibuprofen and naproxen were detected in wastewaters in Canada (Nikolaou et al. 2007). However, information about the fate of pharmaceuticals in the environment is still rather limited. The low volatility of pharmaceutical products indicates that distribution in the environment will occur primarily by aqueous transport, but also through transport within the food chain. Although the persistence of pharmaceutical products is low, they are ubiquitous in the environment because rates of release are greater than their rates of transformation (Bendz et al. 2005). Figure 12.2 shows possible sources and pathways for the occurrence of pharmaceuticals and their metabolites in the environment.

Pharmaceuticals may be disseminated into the environment from both human and agricultural sources, including excretion, flushing of old and out-of-date prescriptions, medical waste, discharge from wastewater treatment facilities, leakage from septic systems and agricultural waste storage structures. Other pathways for dissemination are via land application of human and agricultural waste, surface runoff and unsaturated zone transport. Once in the environment, like any other organic chemicals, their efficacy depends on their physio-chemical properties, prevailing climatic conditions, soil types and variety of other environmental factors.

The amount of pharmaceuticals released into the environment depends on the quantity of drugs produced, dosage amount and frequency, metabolism, excretion efficiency of the parent compound and metabolites, ability of the pharmaceutical to sorb to solids, and the decomposition capability of subsequent sewage treatment. Consumption patterns may change depending on the season. During the winter season loads of macrolide antibiotics in sewage treatment works were twice as high as in the summer months (McArdell et al. 2003). Seasonal differences are due to lower biological activity in sewage systems during winter or because of higher input in winter. Monthly sales of macrolide antibiotics were twice as high in January/February as in summer because they are mainly used to cure infections

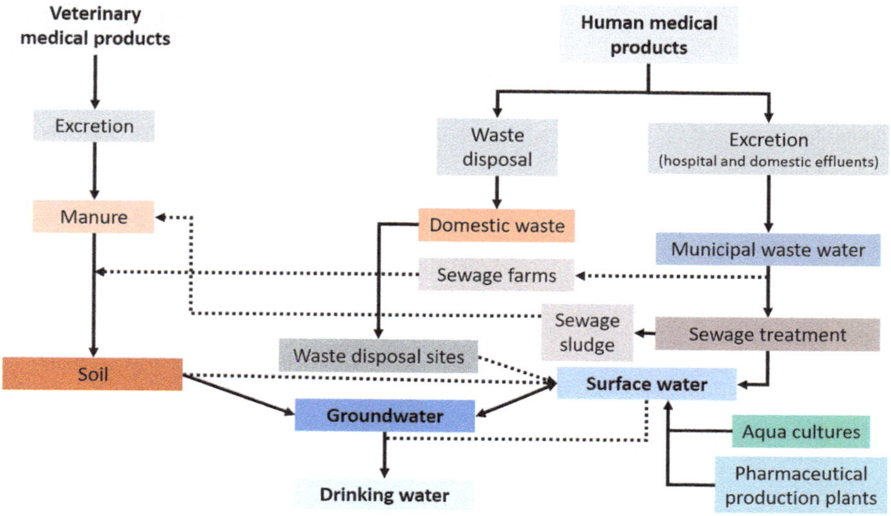

Fig. 12.2 Sources and fate of pharmaceutical compounds (Adapted from Heberer 2002)

of the respiratory tract which are more prevalent in the cold and wet winter season (McArdell et al. 2003).

12.3.1 Human Pharmaceuticals in the Waste Stream

12.3.1.1 Human Excretion

The majority of pharmaceuticals will enter the human body and will ultimately be excreted in urine or feces either unaltered or as metabolites that may or may not closely resemble the parent compound (Mückter 2006). Metabolism includes two major phases. In the first phase, oxidation, reduction and hydrolysis processes involve the formation of new or modified reactive, functional groups which commonly increase polarity compared to the original compound. In the second phase, the metabolites are conjugated with endogenous molecules to increase their water solubility. Urinary excretion rates for a number of commonly used pharmaceuticals are summarized in Table 12.5.

Unchanged pharmaceuticals that are sufficiently water-soluble will be primarily excreted in urine and only the less soluble pharmaceuticals are likely to be excreted in faeces. However, many pharmaceuticals are excreted as metabolites which are commonly water soluble and excreted via urine. Because of an incomplete elimination in wastewater-treatment plants, residues of many toxic organic compounds, including pharmaceutical products, are present in surface waters. Municipal and hospital wastewaters are the most important sources of human pharmaceutical

Table 12.5 Urinary excretion rates of unaltered human pharmaceuticals

Class of medicine/medicinal compound	Original compound excreted (%)	References
Analgesic/		
Ibuprofen	10	Bound and Voulvoulis (2005)
Antibiotic/		
Amoxycillin	60	Bound and Voulvoulis (2005)
Erythromycin	25	"
Sulfamethoxalole	15	"
Antiepileptic/		
Carbamazepine	3	Bound and Voulvoulis (2005)
Felbamate	40–50	"
Antihistamine/		
Cetirizine	50	Bound and Voulvoulis (2005)
Anti-inflammatory/		
Diclofenac	15	Alder et al. (2006)
Indometacin		"
Antipyretic/		
Paracetamol	4	Bound and Voulvoulis (2005)
Beta-blockers/		
Atenolol	90	Bound and Voulvoulis (2005)
Metoprolol	10	"
Propranolol	<1	Alder et al. (2006)
Lipid regulator/		
Benzafibrate	50	Bound and Voulvoulis (2005)

compounds, with contributions also from wastewater, manufacturers and landfill leachates, and from disposal of unused medicines into the environment.

12.3.1.2 Hospital Wastewater as a Source of Human Pharmaceuticals

Hospital wastewater is a major source of pharmaceuticals, especially antibiotics, anti-cancer agents, and iodinated contrast media containing individual pharmaceuticals at higher concentrations than household effluents due to the lower dilution in the wastewater (Alder et al. 2006). Most hospital sewers are directly connected to the municipal sewer system and no additional treatment is performed prior to disposal to public sewer. Table 12.6 shows concentrations of pharmaceutical products found in hospital wastewater.

Table 12.6 Concentrations of pharmaceutical products found in hospital wastewater

Class of medicine/medicinal compound	Concentration range (mean) (μg L^{-1})
Analgesic/	
Codeine	0.01–5.7 (0.9)
Ibuprofen	1.5–151 (19.8)
Antibiotic/	
Erythromycin	0.01–0.03 (0.02)
Metronidazole	1.8–9.4 (5.9)
Trimethoprim	0.01–0.03
Antiepileptic/	
Carbamazepine	0.03–0.07 (0.04)
Anti-inflammatory/	
Diclofenac	0.06–1.9 (1.4)
Ketoroloc	0.5–59.5 (4.2)
Beta-blockers/	
Atenolol	0.1–122 (3.4)
Propranolol	0.2–6.5 (1.4)
H_2 antagonist/	
Ranitidine	0.4–1.7 (1.0)
Cancer therapeutic/	
Cyclophosphamide	19 ng L^{-1} to 4.5 μgL^{-1}

Data from Gómez et al. (2006) and Steger-Hartmann et al. (1997)

12.3.1.3 Release of Human Pharmaceuticals from Sewer Systems

Leakages from house connection pipes and public sewer systems into the subsoil are a possible source for contamination of groundwater. The amount of leaking sewage is controlled by the size and geometry of the leak, the water level in the pipe, the chemical composition of the sewage and the existence and condition of a colmation layer. Outside of the pipe, the surrounding material and the seepage distance from the pipe to the groundwater is important in regard to degradation and chemical reactions. A number of potential marker substances were analysed, including several pharmaceutical residues, in an assessment of the sewer-groundwater interactions in the medium sized city Rastatt in Germany (Wolf et al. 2003). However, none were detected in the groundwater samples analysed, although significant loads were present in wastewater. On the other hand, the iodated x-ray contrast media amidotrizoic acid (66 ng l^{-1}) and iothalamic acid (72 ng l^{-1}) were measured, suggesting a percentage of wastewater in the groundwater in a range between 5% and 12% (Wolf et al. 2003). A number of studies have identified evidence for groundwater contamination as a result of sewer leakage on a citywide scale in the UK, with an estimated loss of 5% reported for the Greater London Region (Ellis 2001). Studies carried out in other countries have estimated exfiltration losses of 3–5% for pre-1960 sewer pipes (Ellis et al. 2004). Since this source results in a very diffuse input to the environment, the contribution to the concentration of pharmaceutical concentrations in water systems is difficult to estimate but is unlikely to be significant except perhaps in some very localized situations.

12.3.1.4 Human Pharmaceuticals in Wastewater Treatment Works

Sewage systems serving as the disposal path for most contaminants naturally excreted by humans constitute the main route of transfer of human pharmaceuticals. Centralized municipal wastewater-treatment plants are thus important sources of pharmaceuticals to soils, surface water and groundwater (Daughton and Ternes 1999). The ability of treatment plants to remove pharmaceuticals is variable and depends largely on the type of wastewater treatment and type of pharmaceutical. Most treatment works use activated sludge technology to treat wastewater (Drillia et al. 2005; Fig. 12.3).

During this procedure, solid matter is first separated from the waste stream during preliminary treatment. Subsequently, additional heavier particles are allowed to settle out by gravity in a sedimentation process referred to as primary treatment. Material removed at this phase is primary sludge and generally undergoes further treatment before being used as biosolids (UNEP DTIE 2002). Next, organic compounds are removed via microbial activity in secondary treatment, after which particles are allowed to settle out again to form secondary sludge. During the secondary treatment, activated sludge removed from secondary sedimentation may be added back into the waste stream to allow for microorganisms used in the secondary treatment to grow (UNEP DTIE 2002).

Pharmaceuticals most frequently found in water treatment effluents are antibiotics, antacids, steroids, antidepressants, analgesics, anti-inflammatories, antipyretics, beta-blockers, lipid-lowering drugs, tranquilizers, and stimulants. These pharmaceuticals have been detected in the surface and ground waters of Germany, The Netherlands, Switzerland, Italy, Spain, the, USA, Canada, China and Brazil (summarized by Rivera-Utrilla et al. 2013). The fate of pharmaceutical products and their metabolites in sewage treatment plants may be mineralization to carbon dioxide and water, adsorption on suspended solids (if the compound is lipophilic), or release in the effluent either as the original compound or as a degradation product. Mineralization of some pharmaceuticals and sorption of hydrophobic pharmaceuticals to sewage sludge may reduce concentrations present in the treated effluent.

Sorption is likely to be a particularly important process for pharmaceuticals that have low water solubility and lipophilic properties since the sewage sludge is comprised primarily of biosolids, which have a very high organic matter content. The concentration of a substance sorbed per liter of wastewater (C_{sorbed}) can be expressed as a simplified linear Eq. 12.1:

$$C_{sorbed} = Kd \times SS \times C_{dissolved} \tag{12.1}$$

where Kd is the sorption constant, defined as the partition coefficient of a compound between sludge and the water phase; SS is the concentration of suspended solids in the wastewater; and $C_{dissolved}$ is the dissolved concentration of the substance (Ternes et al. 2004a).

Sorption behaviour can be estimated using the sorption coefficient (Kd), which depends mainly on characteristics of the compound but is also influenced by the nature of the sludge. Ternes et al. (2005) found no correlation of the observed Kd values for a number of pharmaceuticals with the literature values for octanol water partitioning, K_{ow}, or partitioning to soil organic carbon, K_{oc}. Although there are a number of relationships that have been established between Kd and K_{ow}, they are compound type specific. Electrostatic interactions are also relevant for sorption of polar pharmaceuticals onto activated sludge. Additionally, for compounds containing functional groups which can be protonated and de-protonated, the pH of the sludge plays a crucial role (Ternes et al. 2004b).

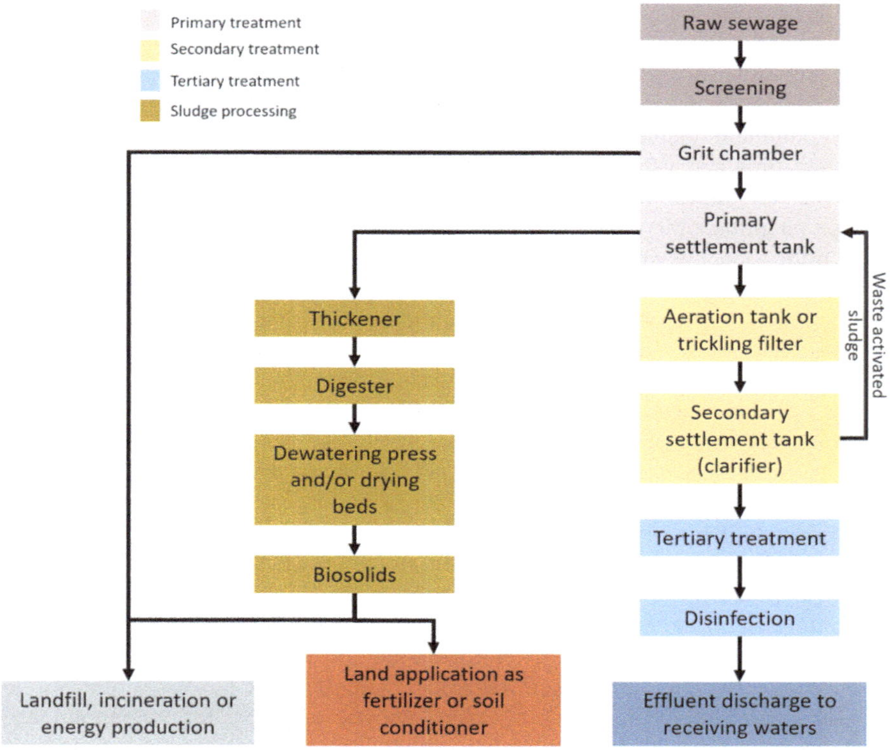

Fig. 12.3 Wastewater treatment processes and biosolids generation (Adapted from UNEP DTIE 2002)

Acidic pharmaceuticals such as acetylsalicylic acid, ibuprofen, fenoprofen, ketoprofen, naproxen, diclofenac and indomethacin have p*K*a values ranging from 4.9 to 4.1, for clofibric acid and bezafibrate the pKa is 3.6 (Fent et al. 2006). At neutral pH, these negatively charged pharmaceuticals therefore occur mainly in the dissolved phase in the wastewater. With lower pH, however, adsorption increases. For these compounds and the antitumor agent ifosfamide sorption by non-specific interactions seems not to be relevant (Kümmerer et al. 1997). Sorption of acidic pharmaceuticals to sludge is thus suggested to be not very important for the elimination of pharmaceuticals from wastewater. Therefore, levels of pharmaceuticals in digested sludge and sediments are suggested to be relatively low, as was demonstrated in several monitoring studies (e.g. Ternes et al. 2004a). In contrast, basic pharmaceuticals and zwitterions can adsorb to sludge to a significant extent, as has been shown for fluoroquinolone antibiotics (Golet et al. 2002). For the hydrophobic EE2 (17 alpha-ethinylestradiol: synthetic steroid hormone used mainly in oral contraceptives), exhibiting a log K_{ow} value of 4.0, sorption to sludge is likely to play a role in the removal from wastewater. Degradation in sludge seems not significant. As a consequence, EE2 occurs in digested sludge, where concentrations of 17 ng g^{-1} were reported (Ternes et al. 2002).

If a pharmaceutical is occurring mainly in the dissolved phase, biodegradation is suggested to be the most important elimination process in wastewater treatment. It can occur either in aerobic (and anaerobic) zones in activated sludge treatment, or anaerobically in sewage sludge digestion. In general, biological decomposition of micro-pollutants including pharmaceuticals increases with increase in hydraulic retention time and with age of the sludge in the activated sludge treatment. For example, diclofenac was shown to be significantly biodegraded only when the sludge retention time was at least 8 days (Kreuzinger et al. 2004). In contrast, data from Metcalfe et al. (2003) indicate that the neutral drug carbamazepine, which is hardly biodegradable, is only poorly eliminated (normally less than 10%), independent from hydraulic retention times. Pharmaceuticals are often excreted mainly as nonconjugated and conjugated polar metabolites. Conjugates can, however, be cleaved in sewage treatment plants, resulting in the release of active parent compound as for estradiol and EE2 (reviewed by Fent et al. 2006).

Elimination rates during the sewage treatment plant process are mainly calculated from influent and effluent concentrations. They vary according to the construction and treatment technology, hydraulic retention time, season and performance of the sewage treatment plant. Removal rates are variable even for the same pharmaceutical between different treatment plants. Elimination efficiencies of pharmaceuticals were shown to vary between 0% and 99% (Fent et al. 2006). According to a study conducted in Spain removal rates of 73 pharmaceuticals ranged from 40% to 99% at seven wastewater treatment plants in Spain (Gros et al. 2010). Average elimination for specific pharmaceuticals varied from 7% to 8% for carbamazepine up to 81% for acetylsalicylic acid, 96% for propranolol, and 99% for salicylic acid. Lowest average removal rates were found for diclofenac (26%), the removal of bezafibrate was 51%, but varied significantly between treatment plants, and high removal rates of 81% were found for naproxen. Very high total elimination of 94–100% of ibuprofen, naproxen, ketoprofen and diclofenac was found in three treatment plants in the USA (Thomas and Foster 2004). While in the primary treatment only 0–44% were removed, efficient removal took place mainly in the secondary treatment step (51–99% removal). To the contrary, X-ray contrast media (diatrizoate, iopamidol, iopromide, iomeprol) were not significantly eliminated (Ternes and Hirsch 2000). The anticancer drug tamoxifen (antiestrogen) was not eliminated as well (Roberts and Thomas 2006). The antineoplasic drug cyclophosphamide has been detected in sewage treatment plant influents and effluents, as well as in surface waters (Santos et al. 2010). In summary, the variation in elimination rates is not surprising, since pharmaceuticals form heterogeneous compounds with a wide diversity in chemical properties. Independent from the chemical characteristics of the compounds, the efficiencies of various sewage treatment plants also vary for the same compound due to their equipment and treatment steps but also to other factors such as temperature and weather. For instance, diclofenac showed largely different elimination rates between 17% and 100% (Fent et al. 2006). However, the concentration in ambient waters connected with sewage treatment works depends not only on the pharmaceutical loads of the effluents but also on the share of treated wastewater discharge to the receiving waters and therefore of the dilution of the wastewater that occurs.

12.3.1.5 Human Pharmaceuticals in Sewage Sludge

Prior to being applied to soil as a biosolid, sewage sludge is treated through means such as thickening, stabilization, conditioning, and de-watering (Fig. 12.3). The aim of these treatments is to reduce pollutants and pathogens of concern while optimizing the sludge's fertilizing and conditioning properties (Jones-Lepp and Stevens 2007). However, these practices do not eliminate all pharmaceuticals, and many compounds are present after sludge is applied to soils.

In 1993, the US EPA conducted a risk assessment for compounds found in sewage sludge, primarily metals and other priority pollutants, which demonstrated possible pathways for contaminants from sewage sludge to the environment (Jones-Lepp and Stevens 2007).

The composition of sewage sludge can vary based on inputs to the wastewater treatment works and differences in processes used to generate sludge. Concentrations of human pharmaceuticals in sewage sludge also vary widely. However, some compounds have been detected in some amount in the majority of studies published. The most frequently studied analgetics, antibiotics, antiepileptics and anti-inflammatory compounds detected in sewage sludge are listed in Table 12.7.

Pharmaceutical compounds contained in sewage sludge can adsorb to soil particles, leach into groundwater, run off in storm water to contaminate surface waters, or volatilize into the air. However, there currently exists no evidence that pharmaceuticals partition into the air in significant amounts.

Table 12.7 Concentrations of human pharmaceuticals products found in sewage sludge

Class of medicine/medicinal compound	Concentration range of human pharmaceuticals detected in sewage sludge (μg kg^{-1})
Analgesic/	
Diclofenac	1.1–425
Antibiotic/	
Ciprofloxazin	0.9–40,800
Clarithromycin	0.3–67,000
Erythromycin	30–111
Ofloxacin	0.1–58,100
Sulfamethazine	0.13–0.8
Sulfamethoxalole	0.6–68,000
Trimethoprim	0.6–133
Antiepileptic/	
Carbamazepine	4.7–1200
Anti-inflammatory/	
Ibuprofen	6.3–548
Ketoprofen	1.3–211
Naproxen	0.9–242

Data from Clarke and Smith (2011), Díaz-Cruz et al. (2009), Jelić et al. (2011), Jones et al. (2014), Jones-Lepp and Stevens (2007), Kinney et al. (2006), Martín et al. (2015), Nieto et al. (2010), and Yu and Wu (2012)

12.3.2 *Veterinary Pharmaceuticals in the Manure Waste Stream*

The most important routes of entry of veterinary pharmaceuticals into the environment are likely the excretion of substances in urine and feces of livestock animals, the wash-off of topical treatments from livestock animals and the direct discharge of aquaculture products. Contributions from the manufacturing process are likely low in the EU and in the USA and, where manufacture and formulation are subject to tight regulatory controls (Boxall et al. 2003).

12.3.2.1 Management of Animal Waste from Livestock Production

In most industrialized countries, until about the 1970s, livestock operations were usually part of integrated farming systems producing crops and animal products. Under these systems, waste and effluent from a modest number of animals was applied rotationally over different fields, effectively diluting nutrients and recycling waste for fertilizer use (Chee-Sandford et al. 2009). During the last 20–30 years, production in many countries has largely shifted from such integrated farming systems to concentrated animal feeding operations that may house thousands of animals. With the advent of concentrated animal feeding operations, large quantities of waste are concentrated in a single location and/or region, and producers may not own or access sufficient tracts of land suitable for disposal of manure through land application. Methods of waste storage vary among operations, but usually correspond to one of three primary types: (i) a slatted floor over a shallow pit with outdoor areas for slurry storage, (ii) a slatted floor over a deep concrete pit, and (iii) a slatted floor over a shallow pit with outdoor lagoon treatment. Additional land area is often required to house secondary waste storage systems (Chee-Sandford et al. 2009). The most common method to dispose of swine and feedlot cattle waste effluent following lagoon or pit storage is through land application (Sarmah et al. 2006). There are three primary methods used to apply effluent: (i) surface application, (ii) surface application followed by incorporation, and (iii) direct soil injection. One primary reason to incorporate surface-applied effluent is to strongly reduce gaseous losses of nitrogen compared to surface application alone (Rotz 2004). To use and dispose of the manure effluent, concentrated feeding animal operators often contract with neighboring growers to apply effluent to their land or apply it to land surrounding the facilities. However, because it is costly to transport liquid effluent over any greater distance, there is an incentive to apply effluent as close to the source as possible (Chee-Sandford et al. 2009).

Application of manure commonly occurs between crop cycles. For many locations, manure is stored for 6 months to 1 year before being applied to crop fields as fertilizer. Effluent differs from fresh manure in that it has a much greater water volume. Fresh swine waste contains approximately 10% solids, while deep pit and lagoon effluents are 4–8% solids and <0.5–1%, respectively (Chee-Sandford et al. 2009). O'Dell et al. (1995) found the solids content ranged from 4 to 10 g L^{-1} in 18 separate tank loads of swine effluent that had been agitated for 24 h before application, suggesting effluent application rates can be highly variable. The practice of stockpiling fresh manure and applying directly to fields is also used in the beef cattle industry (Dolliver and Gupta 2008). Poultry waste management differs somewhat from swine and cattle in that poultry litter is a dry mixture of excrement, bedding material, and feed, and the composition and disposal largely depends on the type of bird produced. Pit storage is often used in production of layer hens and for all types of poultry, direct land application of litter is the primary method of disposal, with a small percentage using composting. It is obvious that the amounts of manure generated by commercial livestock producers will further increase. Large animal facilities are increasingly a source of surface and groundwater contamination with veterinary pharmaceuticals, and elevated levels of antibiotic resistance in humans and animals have been linked to the practice of antimicrobial growth promotant use at poultry and swine farms (Gilchrist et al. 2007).

12.3.2.2 Urinary and Fecal Excretion of Veterinary Pharmaceuticals

Antibiotics are the most widely used veterinary pharmaceuticals, many of which are poorly adsorbed in the gut of the animal, resulting in a high percentage of the parent compound being excreted. Veterinary antibiotics commonly found in pig, cattle,

Table 12.8 Urinary and fecal excretion rates of unaltered veterinary antibiotics

Antibiotic compound	Original compound excreted (range)
Chlortetracycline	65–72%
Sulfamethazine	~90%
Tylosin	28–100%

Adapted from Kim et al. (2011)

and turkey manures included tetracyclines, tylosin, sulfamethazine, amprolium, monensin, virginiamycin, penicillin, and nicarbazine (reviewed by Kim et al. 2011). Excretion rates in feces and manures for widely used antibiotics were in a range between 28% and 100% (Table 12.8).

As a consequence of excretion, a significant portion of the administered antibiotics may be released to the environment. Antibiotic metabolites can also be bioactive in the manures and can be transformed back to the parent compound after excretion. For example, the excreted sulfamethazine metabolite, glucoronide of N-4-acetylated sulfamethazine, is converted back to the parent form in liquid manure (Berger et al. 1986). After administration, sulfamethazine undergoes conjugation with sugars present in the liver and thus inactivates the compound. After excretion, microbes can rapidly degrade the sugars, thereby allowing the compounds back to their bioactive forms (Renner 2002). It has been observed that due to their high water solubility up to 90% of one dose of antibiotics can be excreted in urine and up to 75% in animal feces (Halling-Sørensen 2001). It is therefore likely that when animal wastes are applied as fertilizer they can find their way into the environment and can be present either as metabolite or as the parent compound (Sarmah et al. 2006).

12.3.2.3 Veterinary Antibiotics in Manures

Concentrations of antibiotics in manures varied with both the excretion rate and the total amount fed to the animals. Chlortetracycline was a frequently detected compound in manures at mostly higher concentrations than other antibiotics. Examples are given in Table 12.9.

The high values of chlortetracycline, sulfamethazine and tylosin for poultry are because poultry was medicated with extremely high doses of antibiotics. According to a literature review by Chee-Sandford et al. (2009) antibiotic concentrations detected in manures from swine and poultry ranged from 2.5 to 240 μg kg^{-1} for linomycin, 0.1–1000 μg kg^{-1} for chlortetracycline, 25–410 μg kg^{-1} for tetracycline, 4.0–41.2 mg kg^{-1} for oxytetracycline and 2.5–380 μg kg^{-1} for sulfamethazine.

Tetracycline concentrations in some swine lagoons were as great as 1 mg L^{-1} (Campagnolo et al. 2002).

Antibiotics in manures are often concentrated in the solid phase because of sorption dynamics (Kolz et al. 2005). For a variety of antibiotics in manure, half-lives that have been reported to be less than the anticipated storage period of

Table 12.9 Concentrations of antibiotics in animal manures

Group of antibiotic/medicinal compound	Manure source: animal type	Antibiotic concentrations (μg kg^{-1}; range)
Tetracyclines/		
Chlortetracycline	Pig	46–880
	Cattle	11–208
	Poultry	57–11,900
Oxytetracycline	Pig	29–78
	Poultry	62
Tetracycline	Pig	23–81
	Poultry	69
Sulfonamides/		
Sulfadimidine	Pig	20–87
	Poultry	91–100
Sulfadimethoxine	Pig	85
	Poultry	103
Sulfamethazine	Pig	9990
	Poultry	10,800
Sulfamethoxazole	Pig	87
	Poultry	101
Macrolides/		
Tylosin	Pig	12.4
	Poultry	3700

Adapted from Kim et al. (2011)

manure classes (Boxall et al. 2004). This suggests that significant degradation of the parent compounds might occur before land application. Quinolones and tetracyclines were the most persistent with half-lives up to 100 days. Kolz et al. (2005) in an anerobic incubation experiment (22 °C) found that 90% of tylosin, tylosin B, and tylosin D that were added at the start of the experiment were not detected in the extractable fraction of the slurry mixture within 30–130 h. Aeration of the slurries reduced the time to achieve 90% loss of tylosin to 12–26 h. Although biodegradation and abiotic degradation occurred, the primary mechanism for tylosin loss was thought to be irreversible sorption to manure solids (Kolz et al. 2005). Residual tylosin and its breakdown product, dihydrodesmycosin, were also detected in the slurries after 8 months.

Gavalchin and Katz (1994) determined the persistence of seven antibiotics in a soil-feces matrix under laboratory conditions and found that the order of persistence was chlortetracycline > bacitracin > erythromycin > streptomcycin ≥ bambermycin ≥ tylosin ≥ penicillin with regard to their detection in the soil. The application of manure to agricultural fields also introduces breakdown products into the environment along with the parent compound. However, there is a lack of information on breakdown metabolites in natural environments which can largely be attributed to analytical difficulties and instability of suspected or unknown metabolites (O'Connor and Aga 2007).

12.3.3 Pharmaceuticals in Soils

Pharmaceuticals reach soils particularly through land application of sewage sludge from wastewater treatment plants (human pharmaceuticals) or animal manures (slurry or solid manure) from livestock production farms (veterinary pharmaceuticals). They can affect water quality, soil microbial activity, and plant growth. High percentages of sewage sludge are applied to soils as biosolids, with around 50% in the USA (Kinney et al. 2006), 53% in the European Union (Martín et al. 2015), 40% in Canada (Topp et al. 2008), and around 14% in Japan (Tanoue et al. 2012). This percentage is likely to increase as phosphorus supplies decrease and demand for food production increases (Martín et al. 2015). A model for the behavior of pharmaceuticals in soils is difficult to conceptualize due to the wide variations in pharmaceutical compounds, physicochemical differences in pharmaceuticals, differences in soil types, soil organic matter content, and climactic differences. However, a general scheme is presented in Fig. 12.4.

Pharmaceuticals interact with the soil solid phase in sorption and desorption reactions that control not only their mobility and uptake by plants but also their biotransformation and biological effects. For most antibiotics, sorption is not only a function of their polarity and water solubility, but it is particularly controlled by pH and its effect on the chemical speciation and charge of the compounds (Jechalke et al. 2014). Degradation of pharmaceutical compounds is an important mechanism of removal from the environment. Anaerobic processes, such as photodegradation and hydrolysis often break pharmaceuticals down into their component parts (Thiele-Bruhn 2003). Microbial biodegradation can also occur, in which pharmaceutical compounds are metabolized or broken down by soil microbes and removed from the environment. However, for many antibiotics, mineralization frequently accounts for less than 2% of the added compounds (Jechalke et al. 2014). Volatilization is not relevant for the fate of pharmaceuticals in soil because of their low vapour pressure. Surface runoff and particle-facilitated transport, however, may disperse all antibiotics in the environment (Jechalke et al. 2014).

Ibuprofen and naproxen have been shown to be more readily biodegraded than other pharmaceuticals. The antiepileptic carbamazepine, in particular, was most resistant to microbial biodegradation and thus was detected more frequently in leachate from soils (Gielen et al. 2009). However, a higher initial concentration of pharmaceutical compounds has shown to decrease microbial degradation (Xu et al. 2009). During decomposition, a part of the compound in soil may undergo mineralization (i.e., conversion to CO_2), which is viewed as complete detoxification. So far little information is available on mineralization of pharmaceuticals in soil due to the limited availability of ^{14}C-labeling compounds. ^{14}C-naproxen and ^{14}C-diclofenac were found to be mainly (up to 80%) mineralized to $^{14}CO_2$ in different soils (Dodgen et al. 2014), while only a small part of ^{14}C-carbamazepine (<1.2%) and ^{14}C-acetaminophen (17%) were mineralized (Li et al. 2013, 2014).

Carbamazepine is relatively resistant to biodegradation and is very commonly found in sewage sludge, making it more likely to accumulate in soils as it is

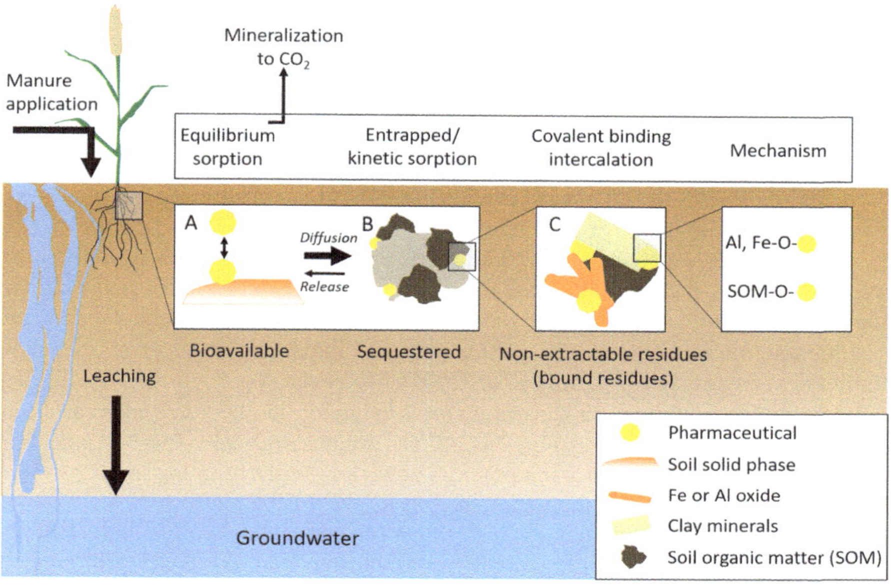

Fig. 12.4 Overall fate of pharmaceuticals in soils (Adapted from Jechalke et al. 2014)

reintroduced with each application (Kinney et al. 2006). One study found that carbamazepine's half-life in soils was around 495 days (Table 12.4), which was significantly higher than the predicted 75 days (Walters et al. 2010). This may be due to the compound's tendency to sorb more strongly to soil particles with high organic content, which was described by Gielen et al. (2009). Interestingly, in some studies, carbamazepine was detected in greater quantities in soils than in the sludge applied to those soils. It has been hypothesized that carbamazepine conjugates that form in the human body are cleaved in aerobic soils, releasing more of the parent compound (Gielen et al. 2009). This factor may also contribute to the concentration of carbamazepine being higher than expected in soils. Despite its apparent abundance in soils, one study found that carbamazepine was found in greater quantities in surface and groundwater than sulfamethoxazole or diclofenac. Carbamazepine has an average tendency to bind to soils compared to other pharmaceuticals and is found both in the soil and aqueous matrices (Drillia et al. 2005).

Some anti-inflammatory medicines, like ketoprofen, have been shown to adsorb more strongly to soils containing higher amounts of organic matter, but not those with lower organic content. Others, like diclofenac and ibuprofen, have been shown to adsorb weakly to soil particles in general, making them more likely to leach out of soils and into groundwater (Xu et al. 2009). Runoff into surface water is a significant fate for anti-inflammatory medicines, with ibuprofen and naproxen shown to partition rapidly into stormwater runoff after land application of biosolids (Topp et al. 2008). In a study by Gielen et al. (2009) ibuprofen and naproxen were not found in high concentrations in leachate from soils because they were likely degraded by soil microbes. Indeed, ibuprofen, naproxen, and diclofenac were shown to have soil half-lives of around 6, 17, and 20 days, respectively (Xu et al. 2009), far lower than the half-lives for antibiotics and anticonvulsants studied by Walters et al. (2010). With allowances made for differences in chemical structure, anti-inflammatory medicines seem less likely to concentrate in soils than antibiotics or antiepileptics, making them more of a concern in groundwater and surface water.

Antibiotics such as ofloxacin have a relatively high soil organic carbon-water partitioning coefficient (K_{oc}) (Thiele-Bruhn 2003), which is calculated from the K_{ow} and represents a compound's likelihood to bind to soil. Depending on the K_{oc}, a compound may preferentially bind to organic matter or soil particles, and a higher log K_{oc} generally indicates lower water solubility. The log K_{oc} which depends on soil organic matter content, clay content, pH and ionization state, varies substantially between soil types (Drillia et al. 2005). However, log K_{oc} remains a useful indicator when determining a compound's behavior in the environment. It is important to note that the log K_{ow} has a linear relationship with log K_{oc}, and can therefore be used to predict a compound's behavior similarly to the log K_{oc} when experimental data are unavailable for the log K_{oc} (US EPA 1996).

In spite of major difficulties in predicting a pharmaceutical compound's behavior in the environment, some general trends can be seen in scientific literature. Some types of antibiotics, particularly fluoroquinolones, sulfonamides, and tetracyclines, may be removed through biodegradation as well as photodegradation (Thiele-Bruhn 2003). However, quinolone antibiotics have been shown to have half-lives in soils of 1000 days or more due to their sorption to soil organic matter (Table 12.4). Tetracyclines also have longer half-lives than might be predicted, sometimes around 500 days (Walters et al. 2010). This could be attributable to other components in land applied sludge that may inhibit or interfere with microbial activity. Additionally, some biodegradation reactions may be reversible, and are often inhibited by the compound's binding to soil particles such as clay.

A pharmaceutical's tendency to move through soil is correlated particularly with its tendency to sorb to soil particles (Drillia et al. 2005). Leaching of pharmaceuticals into the groundwater mainly occurs in preferential flow paths and is restricted to a few pharmaceuticals while the largest fraction is usually retained in the surface soil (Jechalke et al. 2014). The propensity to sorb to soil particles is dependent on a compound's physicochemical properties (Table 12.4). For example, in one study, the antibiotic ofloxacin sorbed most strongly to soil particles, while diclofenac, carbamazepine, and sulfamethoxazole showed weaker sorption, respectively. This indicates that ofloxacin is more likely to remain tightly bound to soil particles and is less bioavailable, while the opposite is true for sulfamethoxazole, which is more likely to leach off into groundwater or surface water (Drillia et al. 2005). The history of fertilization and other solutes can also affect sorption. Sorption of a sulfamethoxazole was reduced in soils fertilized and irrigated with untreated wastewater for 14–100 years, most likely as a result of a saturation of high-affinity sorption sites, sorption competition with other solutes, and changes of soil organic matter properties, whereas sorption of ciprofloxacin was not affected (Dalkmann et al. 2014).

In regions with high livestock densities, the usage of antibiotics poses a particular risk in manured soils. However, until recently, information regarding the transport of antibiotics under field conditions has been limited. In a sandy soil that had repeated swine liquid manure applications, tetracycline and chlortetracycline were detected down to a depth of 30 cm (Hamscher et al. 2005). The highest tetracycline (198 μg kg^{-1}) and chlortetracyline (7.3 μg kg^{-1}) concentrations were detected at soil depths of 10–20 cm and 20–30 cm, respectively. Sulfamethazine was generally not detected in soil samples, but was detected in groundwater collected at a depth of 1.4 m. Oxytetracycline, sulfadiazine, sulfathiazole, sulfamerazine, sulfamethoxypyridazine, sulfamethoxazole, sulfadimethoxine, and tylosin were not detected in any soil or groundwater samples. While it appeared some of the tetracyclines could accumulate in soil, none of the antibiotics from the study were detected at soil depths >30 cm and only sulfamethazine was detected in groundwater suggesting limited transport, even in highly porous sandy soils. In a field study with clay loam soil that received swine manure spiked with the sulfonamide, sulfachlorpyridazine, the antibiotic was found to be mobile and readily entered the field drain, with a maximum concentration of 590 μg L^{-1} detected 7 day after manure application (Boxall et al. 2002). In the same study conducted with sandy loam field soil, sulfachlorpyridazine concentrations in soil pore water were significantly lower (maximum concentration 0.78 μg L^{-1}) than the field with clay loam, and contrasted with laboratory sorption studies that predicted larger soil water concentrations. In another soil transport study, sulfachlorpyridazine and oxytetracycline were detected in soil at concentrations up to 365 and 1691 μg kg^{-1}, respectively (Kay et al. 2004). Similar to other investigations, these compounds were not detected below a depth of about 37 cm. Sulfachlorpyridazine and oxytetracycline were detected in tile drainage at peak concentrations of 613 and 36 μg L^{-1}, respectively. A survey of the occurrence of various tetracyclines and sulfamethazine (sulfonamide group) in sandy soils fertilized with liquid manure was carried out in northwestern Germany by Pawelzick et al. (2004). The reported maximum concentrations for the compounds screened in this study were 27 μg kg^{-1} (oxytetracycline), 443 μg kg^{-1} (tetracycline), 93 μg kg^{-1} (chlortetracycline), and 4.5 μg kg^{-1} (sulfamethazine) in the top 0–30-cm soil. At least 3 of the 14 total agricultural fields used in this study had higher than EMEA (European Agency for the Evaluation of Medicinal products) trigger values of 100 μg kg^{-1} for tetracyclines (Pawelzick et al. 2004). Elsewhere in Germany, Winckler and Grafe (2000) also found tetracyclines to persist in agricultural soils at concentrations of 450–900 μg kg^{-1}.

In the peri-urban district of the Beijing Municipality (China), intensive animal farming results in extremely high livestock densities (e.g. Hou et al. 2012). Thus, the available arable land receives excessive amounts of veterinary pharmaceuticals (Ostermann et al. 2014). In this region, soil was sampled from agricultural fields and a dry riverbed that had received a constant influx of pig farm and biogas plant effluent for the past 20 years. The antibiotic concentrations reached 110 μg kg^{-1} sulfamethazine, 111 μg kg^{-1} chlortetracycline and 62 μg kg^{-1} enrofloxacin in the topsoil of agricultural fields. In the subsoils, sulfamethazine was found down to

≥2 m depth on agricultural sites and down to ≥4 m depth in the riverbed (Ostermann et al. 2014). Li et al. (2011) found similar maximum concentrations for sulfamethazine and chlortetracycline in Chinese vegetable fields, whereas their maximum enrofloxacin concentrations reached 1348 μg kg^{-1}. Compared to the findings by Ostermann et al. (2014) and Hu et al. (2010) in surface soil horizons detected distinctly higher tetracycline concentrations (1079 μg kg^{-1}), much lower sulfonamide concentrations, and fluoroquinolone concentrations in the same range. These differences probably resulted from regional variations concerning types and intensities of antibiotic applications. To scan for maximum pollution loads, Ostermann et al. (2014) analyzed sediment of the dry riverbed as a potential local contamination hotspot. As expected, this site exhibited high sulfamethazine (844 μg kg^{-1}) and chlortetracycline (2375 μg kg^{-1}) concentrations in 0–20 cm depth.

12.3.4 Concentrations of Human Pharmaceuticals in Water Matrices

Pharmaceutical compounds are called "micro-pollutants" because in the aquatic environment they are mostly found in the μg L^{-1} or ng L^{-1} range. Concentrations of pharmaceuticals in different EU surface waters are comparable to each other when considering the share of treated wastewater discharged into the river due to different dilution factors. For most EU countries wastewater is expected to be diluted between 10 and 100 times in the receiving waters. The dilution factor is thus a crucial parameter in order to be able to compare different studies and to predict environmental concentrations of pharmaceuticals from amounts used (Alder et al. 2006).

12.3.4.1 Surface Waters

In German municipal sewage treatment plant effluents and river waters, 32 pharmaceuticals from different medicinal classes have been detected (Ternes 1998). In the river Elbe diclofenac, ibuprofen, carbamazepine, a variety of antibiotics, and lipid regulators have been measured in a range between 20 and 140 ng L^{-1}) (Wiegel et al. 2004). In the Höje River, Sweden, the pharmaceuticals ibuprofen, ketoprofen, naproxen, diclofenac, atenolol, metoprolol, propranolol, trimetoprim, sulfametoxazole carbamazepine, and gemfibrozil have been found in maximum concentrations ranging from 0.12 to 2.2 μg L^{-1} (Bendz et al. 2005). In the rivers Po and Lambro, Italy, atenolol, bezafibrate, furosemide, ranitidine, clofibric acid, and diazepam have been detected (Calamari et al. 2003). Fifteen pharmaceuticals, including stimulants, antirheumatic, antiepileptic, analgesic, antimicrobial, and cytostatic agents were detected in the Somes River, Romania, with concentrations ranging from 30 ng L^{-1} to 10 μg L^{-1}. Pentaoxifilline, ibuprofen,

Table 12.10 Concentrations of selected human pharmaceuticals in surface waters found in Europe

	Median concentration (maximum) (ng L^{-1})			
Country	Benzafibrate	Carbamazepine	Diclofenac	Ibuprofen
Austria	20 (160)	75 (294)	20 (64)	Not detectable
Finland	5 (25)	70 (370)	15 (40)	10 (65)
France	102 (430)	78 (800)	18 (41)	23 (120)
Germany	350 (3.100)	250 (110)	150 (1.200)	70 (530)

Adapted from Ternes et al. (2005)

formylaminophenazone, p-chlorophenyl sulfone, N,N-bis(3,3-dimethyl-2-oxetanyl)-3,3-dimethyl-2-oxetanamine were measured in levels between 100 and 300 ng L^{-1}; while concentrations of aspirin, triclosan, carbamazepine, codeine, diazepam, and cyclophosphamide did not exceed 100 ng L^{-1} (Moldovan 2006).

In the UK in a monitoring study propranolol (median level 76 ng L^{-1}) was detected in all sewage treatment plant effluents whereas diclofenac (median 424 ng L^{-1}) was measured in 86%, ibuprofen (median 3086 ng L^{-1}) in 84%, mefenamic acid (median 133 ng L^{-1}) in 81%, dextropropoxyphene (median 195 ng L^{-1}) in 74%, and trimethoprim (median 70 ng L^{-1}) in 65% of samples (Ashton et al. 2004). Concentrations of pharmaceuticals (acetylsulfamethoxazole, clofibric acid, clotrimazole, dextropropoxyphene, diclofenac, erythromycin, ibuprofen, mefenamic acid, paracetamol, propranolol, sulfamethoxazole, tamoxifen, and trimethoprim compounds) in the Tyne estuary, UK, ranged from 4 to 2370 ng L^{-1} (Roberts and Thomas 2006). In the North Sea, maximum concentrations of clofibric acid and caffeine were 1.3, and 16 ng L^{-1}, respectively, while diclofenac and ibuprofen were detected only in the estuary of the river Elbe (6.2 and 0.6 ng L^{-1}, respectively) (Roberts and Thomas 2006). Table 12.10 shows concentration levels of important pharmaceuticals found in surface waters of four European countries.

Almost 100 micropollutants, including steroids, caffeine, triclosan (an antimicrobial compound), and antibiotics, were measured in samples from 139 streams downstream of urban areas and livestock production in the USA (Kolpin et al. 2002). In 44 rivers across the USA, carbamazepine has been detected (average 60 ng L^{-1} in water and 4.2 ng mg^{-1} in river sediment) (Thaker 2005). Ternes et al. (1999) reported monitoring data for oestrone, 17β-oestradiol and ethinyl-estradiol in 15 German rivers and streams. Oestrone was detected at very low levels of 0.7–1.6 ng l-1 in only three of the rivers. Other natural oestrogens and contraceptives (including 17β-oestradiol and EE) were not detected in any of the rivers, the reported limit of detection being 0.5 ng L^{-1}. Similarly, Belfroid et al. (1999) could only detect very low concentrations (generally below the level of detection up to about 5 ng L^{-1}) of 17β-oestradiol, oestrone and ethinyl-estradiol in surface waters in the Netherlands and the glucuronide conjugates were not detected. Oestrone was detected the most frequently (7 of 11 locations) whereas ethinyl-estradiol was detected only at 3 locations. Antidepressants such as fluoxetine, paroxetine, setraline, etc., are among the most commonly detected pharmaceuticals in both surface water and wastewater treatment effluents, reflecting their usage volumes in human medicine. They are generally present at concentrations in the ng L^{-1} to low μg L^{-1} range (Kolpin et al. 2002; Metcalfe et al. 2003).

12.3.4.2 Groundwater

Groundwater is an important water resource in many countries and regions and is difficult to remediate once contaminated with pharmaceuticals (Sui et al. 2015). In particular the presence of antibiotics in groundwater has attracted attention worldwide recently. A national reconnaissance carried out in the USA concerning pharmaceuticals in water resources reported the presence of antibiotics in a sampling network of 47 groundwater sites with the detection frequency exceeding 30% (Barnes et al. 2008). Among the various antibiotics, sulfonamides are the most extensively studied and have been found at high concentrations in several studies. Sulfonamides were reported in the groundwater were detected in the groundwater in studies conducted in Switzerland, Spain, the USA and China (Sui et al. 2015). Extremely high concentrations of sulfonamides (range: 10 μgL^{-1}–1 mg L^{-1}) were reported in the groundwater down gradient of a landfill site (Holm et al. 1995). Bartelt-Hunt et al. (2011) investigated the occurrence of veterinary pharmaceuticals in lagoons and adjacent groundwater at operating swine and beef cattle facilities, and sulfonamides like sulfamerazine, sulfamethazine, sulfamethazole and sulfathiazole, along with macrolides like erythromycin, lincomycin, monensin and tiamulin were detected in groundwater samples, with concentrations ranging from 29 ng L^{-1} to over 2000 ng L^{-1}. Zhou et al. (2012) analyzed water samples collected from a pig farm in Guangxi Province and found the presence of sulfadiazine, sulfamethazine and sulfamonomethoxine in groundwater with concentrations of 1.5, 130, and 19 ng L^{-1} respectively.

The most commonly detected anti-inflammatories and analgesics in groundwater include ibuprofen, diclofenac and paracetamol because of their large consumption in daily life (Sui et al. 2015). High peak concentrations of diclofenac and ibuprofen (120 and 250 ng L^{-1}, respectively) at a depth of about 0.5 m below the main trench sewer pipe in North East London provided evidence of contamination due to wastewater exfiltration to groundwater (Ellis et al. 2002). Rabiet et al. (2006) assessed the consequences of treated water recycling in surface and groundwater of a medium-sized Mediterranean catchment and found diclofenac and paracetamol were among the dominant pharmaceuticals in wells supplying drinking water, and the maximum concentration of paracetamol was measured as high as 211 ng L^{-1}, probably contaminated by wastewater. Carbamazepine was detected in 42% of the samples collected from 164 groundwater locations in 23 European countries, with a maximum concentration of 390 ng L^{-1} (Loos et al. 2010). In Leipzig and Halle (Germany), carbamazepine was detected in the groundwater at concentrations of 2–75 and 2–51 ng L^{-1}, respectively (Musolff et al. 2009; Osenbrück et al. 2007). In a city of Montana, USA, 12 of 38 well water samples analyzed were found to

Table 12.11 Concentrations of human pharmaceuticals in drinking water

Class of medicine/medicinal compound	Maximum concentration (ng L^{-1})	Country
Analgesic and anti-pyretic/		
Diclofenac	6	Germany
Ibuprofen	3	"
Phenazone	250, 400	", 2 studies
Prophylphenazone	80, 120	", 2 studies
Antiepileptic/		
Carbamazepine	24	Canada
"	258	USA
Antineoplastic/		
Bleomycin	13	UK
Lipid regulator/		
Benzafibrate	27	Germany
Gemfibrozil	70	Canada
Psychiatric drug/		
Diazepam	10	UK

Adapted from Jones et al. (2005)

contain carbamazepine, with the maximum concentration almost reaching 400 ng L^{-1} (Miller et al. 2006). Lopez-Serna et al. (2013) investigated two common beta-blockers, propranolol and metoprolol, in the urban groundwater underlying Barcelona, Spain. The latter one could be found in all the groundwater samples with the maximum concentration of 355 ng L^{-1} while the former one exhibited relatively low detection frequency (23%) and concentrations up to 9.38 ng L^{-1}.

12.3.4.3 Tap Water

Human pharmaceuticals have also been frequently detected in drinking water (tap water). Table 12.11 summarizes concentrations of drug compounds found in finished drinking water worldwide.

Even with advanced water treatment, all drugs obviously cannot be completely removed. Concentrations in drinking water are generally in the ng L^{-1} range (Jones et al. 2005). Heberer and Stan (1997) found clofibric acid and the drug metabolite N-(phenylsulfonyl)-sarcosine in the majority of drinking water samples collected from 14 waterworks in the Berlin area, with the maximum concentrations in drinking water samples being 270 ng L^{-1}. In Berlin drinking water the analgesic and antipyretic drugs phenazone and propylphenazone were also found in the ng L^{-1} level (Jones et al. 2005). The major source of most of the pharmaceutical pollution in Berlin drinking waters is thought to be from the use of groundwater contaminated with sewage as a water source. A highest maximal concentration of 1200 ng L^{-1} and 1250 ng L^{-1} has been detected in Germany, respectively for diatrizoate (Perez and Barcelo 2007) and the metabolite AMDOPH (Heberer et al. 2004; Reddersen et al. 2002).

Mompelat et al. (2009) in a literature review on the occurrence of pharmaceuticals in tap water found that 17 pharmaceutical products and 5 by products have been detected in concentrations ranging from 1.4 to 1250 ng L^{-1}. Non-steroidal anti-inflammatory drugs and to a small extent anticonvulsants are the mainly detected in Europe (Germany, France, Finland). Besides, iodinated contrast media are the most susceptible to be encountered in drinking water because of their very low lipophilicity (Perez and Barcelo 2007). Vulliet et al. (2009) investigated the occurrence of pharmaceuticals and hormones in French drinking waters treated from surface waters. In surface waters, 27 of the 51 target compounds were detected at least once. The highest concentration of 71 ng L^{-1} was observed for paracetamol but concentrations rarely exceed 50 ng L^{-1}. Salicylic acid, the main metabolite of aspirin, was present in drinking waters, up to 19 ng L^{-1}. Carbamazepine (maximum concentration 10.7 ng L^{-1}) and atenolol (maximal concentration 2 ng L^{-1}) were also present in more than 30% of the studied drinking waters. The antibiotics sulfamethoxazole and trimethoprim have been detected once, respectively at 0.8 and 1.0 ng L^{-1}. In treated waters, hormones have been detected at low concentrations, in the ng L^{-1} range of. Progestagens and androgens seem to be the more resistant to drinking water treatments. Levonorgestrel, progesterone and testosterone have been found in treated waters at concentrations ranging from 1 ng L^{-1} up to 10 ng L^{-1}. A comprehensive study of 28 sampling stations conducted by Associated Press in the USA (Donn et al. 2008) demonstrated the presence of pharmaceutical traces in the drinking water of 24 major cities, including Philadelphia, Washington, New York and San Francisco.

12.3.5 Pharmaceuticals in Edible Plants

Pharmaceuticals have been detected in the roots, stems, leaves, and seeds of plants grown in soil treated with sewage sludge (Prosser et al. 2014). The degree of plant uptake varies among pharmaceutical compounds. Membrane permeability plays an important role in the ability of a plant to take up a compound and store it in its tissue. More lipophilic compounds have a greater ability to pass through phospholipid membranes, but studies have shown that compounds that are somewhere between lipophilic and hydrophilic have a greater ability to partition into plant tissues (Tanoue et al. 2012). Soil conditions also have an effect on the ability of a plant to take up pharmaceutical compounds. For example, a soil's pH can affect the ionization state of compounds it contains, which can lead to those compounds becoming more or less able to cross biological membranes. Tanoue et al. (2012) found that pharmaceuticals are readily taken up when they are near their neutral ionization states. Additionally, higher soil organic matter can lead to compounds binding more tightly to soil and resisting uptake.

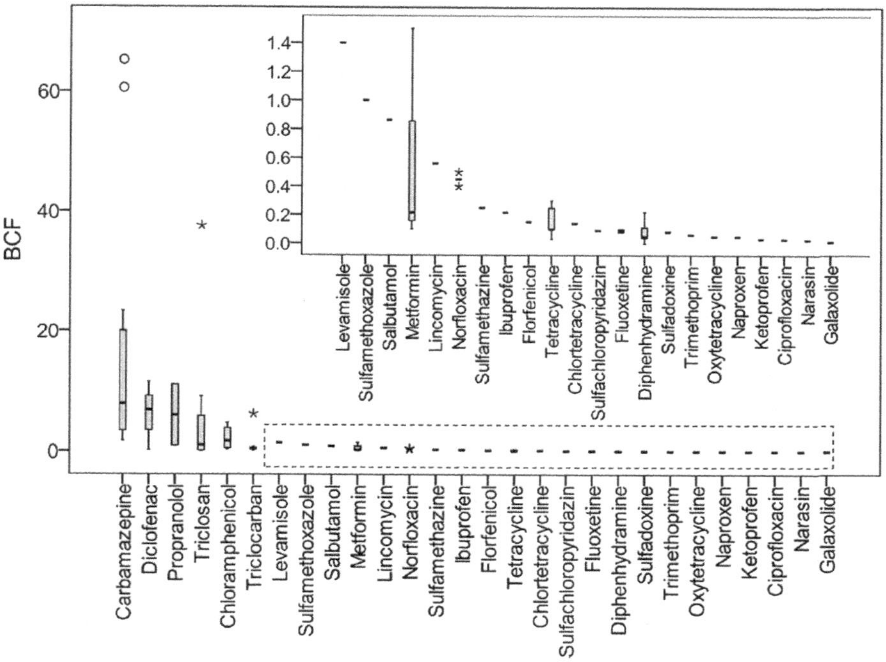

Fig. 12.5 Bioconcentration factors (BCF) between plants and soil. The BCF value was calculated as the ratio of the substance concentration in plant leaves/stems to its concentration in soil. The open circle represents the mild outlier and the star represents the extreme outlier (Wu et al. 2015; p. 662; with kind permission from Elsevier)

An increasing number of studies have considered the plant uptake of medicines from soils spiked with pharmaceutical standards, irrigated with pharmaceutical-contaminated water or treated wastewater, or amended with manure, sewage sludge, or biosolids (reviewed by Wu et al. 2015). The extent of plant uptake is commonly evaluated using the bioconcentration factor (BCF), which is the ratio of the substance concentration detected in the plant tissue to the spiked concentration in the growth medium. Compared to BCFs obtained from hydroponic studies, BCFs from soil studies were much lower, indicating that interactions between medicines and soil as well as degradation in soil significantly decreased the bioavailability of pharmaceuticals in soil. Bioconcentration factors found between plants and soil are shown in Fig. 12.5.

When treated wastewater or spiked water was used for irrigation, the antiepileptic carbamazepine appeared to be commonly detected in roots and other plant parts (Wu et al. 2015), suggesting that this substance has high bioavailability in soil and is relatively easy to transfer from soil to plants and within plants. On the other hand, when biosolids were used for soil amendment, triclocarban and triclosan are usually of concern because of their abundance in biosolids (accounting for up to 65% of the total pharmaceuticals in biosolids) (McClellan and Halden 2010).

Triclosan and triclocarban were found to be taken up by plant roots and subsequently translocated to stems, leaves, and even fruits (Wu et al. 2015). Some other pharmaceuticals have also been detected in plants grown in soils after application of biosolids as organic fertilizer, such as carbamazepine, salbutamol, and diphenhydramine, while sulfamethoxazole and trimethoprim were reported to have limited accumulation in plants grown in biosolids-amended soils (summarized by Wu et al. (2015)). Most studies on plant uptake of pharmaceuticals from soil were carried out in laboratory or greenhouse experiments. So far the information on accumulation of pharmaceuticals in crops receiving biosolids application or treated wastewater irrigation under field conditions is limited. Gottschall et al. (2012) investigated more than 20 pharmaceuticals in the grain of wheat grown in the field for about 1 year following a high single application of municipal biosolids, but pharmaceuticals could not be detected in grain.

Sabourin et al. (2012) examined the uptake of organic micropollutants including 118 pharmaceuticals by tomato, carrot, potato and sweet corn from field soils treated with municipal biosolids. The results suggested that the potential for micropollutants to enter edible parts of food crops was generally low under normal farming conditions. The detected pharmaceuticals included atenolol, cocaine, ciprofloxacin, metformin, minocycline, norfloxacin, naproxen, glyburide, sulfamerazine, penicillin G, triamterene, and trimethoprim. The concentrations ranged from 0.02 to 14 ng g^{-1} (dry weight). Although triclocarban and triclosan were the predominant pharmaceuticals existing in biosolids, they were not found in the plant tissue samples (Sabourin et al. 2012). In contrast, Prosser et al. (2014) detected triclosan and triclocarban in edible portions of green pepper, carrot, cucumber, tomato, radish, and lettuce plants grown in a field with biosolids application. Triclosan was detected in cucumber and radish up to 5.2 ng g^{-1} (dry weight), and triclocarban was detected in carrot, green pepper, tomato, and cucumber up to 5.7 ng g^{-1} (dry weight). Calderón-Preciado et al. (2011) reported the occurrence of hydrocinnamic acid, salicylic acid, caffeine, ibuprofen, methyl dihydrojasmonate, and galaxolide, in apple tree leaves and alfalfa irrigated with reclaimed wastewater, with concentrations of 0.016–16.9 ng g^{-1} (wet weight). In carrots and sweet potatoes Malchi et al. (2014) found that nonionic pharmaceuticals (carbamazepine, caffeine, and lamotrigine) were detected at significantly higher concentrations than ionic pharmaceuticals (metoprolol, bezafibrate, clofibric acid, diclofenac, gemfibrozil, ibuprofen, ketoprofen, naproxen, sulfamethoxazole, and sildenafil).

In contrast to carbamazepine and some antibiotics, anti-inflammatory drugs have a relatively low detection rate in plant tissues. In one study, neither ibuprofen nor diclofenac were detected in the leaves of lettuce or spinach, while naproxen was only detected in the leaves of spinach at the relatively low concentration of 0.04 μg kg^{-1}. For comparison, in the same study carbamazepine was detected at 28.7 μg kg^{-1} in lettuce and 2.9 μg kg^{-1} in spinach leaves (Wu et al. 2012). Ibuprofen and naproxen have been shown to be taken up into alfalfa tissue, but only in small amounts (Carvalho et al. 2014). One study detected neither diclofenac nor indomethacin in the stems or leaves of pea plants, and those compounds were found only at very low levels in roots. Ketoprofen was found in roots and shoots at comparatively low levels (Tanoue et al. 2012). The low detection of anti-inflammatory drugs could be attributable to their lower concentration in soils, biodegradation, and ionization states. Pan et al. (2014) investigated five antibiotics in five edible crops that were irrigated with either domestic wastewater (largely untreated) or fishpond water. Norfloxacin was consistently found at the highest concentrations (4.6–23.6 μg kg^{-1}) in crop tissues, followed by chloramphenicol (2.6–22.4 μg kg^{-1}) and tetracycline (4.0–10.1 μg kg^{-1}), while sulfamethazine and erythromycin were not detected in most of the vegetable crops.

Factors influencing plant uptake of pharmaceuticals from soil up to now are not completely understood. Goldstein et al. (2014) reported that crops grown in soils with low SOM and clay contents were at greater risk for uptake and accumulation of pharmaceuticals. They also found that the uptake of acidic pharmaceuticals by cucumber was inhibited, probably because of the interactions between the acidic pharmaceuticals and dissolved organic matter (DOM) present in treated wastewater. Therefore, the influence of factors such as SOM, DOM, soil pH, and sources of pharmaceuticals needs further investigations (Holling et al. 2012). There is also only little information on the plant uptake of pharmaceutical metabolites, which may be present at levels similar to or even greater than the parent compounds in wastewater effluents or in soils from biotic/abiotic transformations (Wu et al. 2015). Considering that some metabolites have similar biological activity to the parent compound, the uptake behaviors of these metabolites and their ecotoxicological and human health risks require further research.

12.3.6 Effects of Pharmaceuticals on Non-target Organisms

Veterinary pharmaceuticals enter the environment via excretion from animals following agricultural therapeutic and prophylactic use, and/or liquid manure application, or via direct application at fish farms. Human pharmaceuticals particularly enter via sewage treatment plant effluent discharge after therapeutic use. In the receiving environment, these contaminants may sorb to suspended particles, soil and sediments, remain in the aquatic phase or sequester to the organic-lipid compartment and enter the food chain. Acute effects attributed directly to pharmaceutical contamination of the environment have been largely limited to the feminization of some fish exposed to natural and xenobiotic endocrine disruptors. Chronic effects as a result of exposure of aquatic animals to these and other human pharmaceuticals have also been reported. However, chronic effects associated with pharmaceuticals are difficult to detect and manage under the current risk assessment process which is based on acute toxicity testing and high (acute) action limits. As pharmaceuticals are ubiquitous and globally distributed, specifically designed to alter biological functions; associated with a wide range of side effects in non-target organisms, can possibly cause chronic toxicity at environmentally relevant concentrations, they are of specific concern. For this assumption, data from acute and chronic ecotoxicity tests on species belonging to different trophic levels such as bacteria, algae, crustaceans and fish among others, is relevant to illustrate adverse effects that environmental exposure to measured concentrations of these contaminants can have.

12.3.6.1 Analgesics and Anti-inflammatory Drugs

Non-steroidal anti-inflammatory drugs act by inhibiting either reversibly or irreversibly one or both of the two isoforms of the cyclooxygenase enzyme (COX-1 and COX-2), which catalyze the synthesis of different prostaglandins via the oxidation of arachidonic acid (Vane and Botting 1998). Since non-steroidal anti-inflammatory drugs inhibit nonspecifically prostaglandin synthesis, most side effects are related to the physiological function of prostaglandins (reviewed by Fent et al. 2006). In the kidney, prostaglandins are involved in maintenance of the equilibrium between vasoconstriction and vasodilatation of the blood vessel that supply glomerular filtration. Renal damages and renal failure after chronic treatment seems to be triggered by the lack of prostaglandins in vasodilatation-induction. Gastric damages are thought to be caused by inhibition of both COX isoforms. In contrast, liver damages are apparently due to building of reactive metabolites (e.g. acyl glucuronides) rather than inhibition of prostaglandins synthesis (Bjorkman 1998). Prostaglandins are formed in a range of vertebrates and invertebrates. However, in lower invertebrates such as corals, their synthesis is independent of COX, involving other enzymes (Song and Brash 1991). In arthropods and molluscs, COX-like activity is apparently responsible for the formation of prostaglandins (Pedibhotla et al. 1995). In birds, prostaglandins play a role in the biosynthesis of egg shells and treatment with the COX-inhibitor indometacine resulted in egg shell thinning (Lundholm 1997). Diclofenac seems to be the compound having highest acute toxicity within the class of non-steroidal anti-inflammatory drugs (Fent et al. 2006). Short-term acute toxicity was analyzed in algae and invertebrates, phytoplankton was found to react more sensitive than zooplankton. Side effects of diclofenac have been observed in humans in the liver with degenerative and inflammatory alterations (Banks et al. 1995), in the lower gastrointestinal tract and in the esophagus (Bjorkman 1998).

In fish, prostaglandins were found in numerous cells and tissues, including red blood cells, macrophages, and oocytes, and their key roles include in reproduction, where they have a paracrine role in stimulating ovulation, eliciting female sexual behaviour through effects on the brain, and stimulating male sexual behaviour (Corcoran et al. 2010). Indomethacin has been shown to disrupt oocyte maturation and ovulation in zebrafish (*Danio rerio*) at a concentration of 100 mg L^{-1} (Lister and van der Kraak 2008) and ibuprofen has been shown to alter the pattern of spawning in Japanese medaka (*Oryzias latipes*) at the μg L^{-1} level (Flippin et al. 2007). Exposure of rainbow trout (*Oncorhynchus mykiss*) to ibuprofen at a concentration of 1 mg L^{-1} has been shown to impair ion regulation and so the hyposmoregulatory capacity in seawater (Gravel et al. 2009). It has been demonstrated that anti-inflammatory drugs can disrupt cortisol production in trout (Gravel and Vijayan 2007). Cortisol is known to play an important part in osmoregulation in fish, specifically the development and proliferation of chloride cells in the gills and stimulation of Na^+-, K^+-ATPase activity for seawater adaptation (McCormick 2001). In fish, as occurs in mammals, diclofenac hinders the stimulation of prostaglandin synthesis in the head kidney, and in brown trout (*Salmo trutta*) this has been shown to occur at environmentally relevant concentrations of 0.5–50 μg L^{-1} (Höger et al. 2005). In rainbow trout, harmful effects of diclofenac, including the induction of glomeruloneophritis, necrosis of endothelial cells, and hyaline droplet degeneration, have been shown in the kidney, and pillar cell necrosis, epithelial lifting, hyperplasia, and hypertrophy of epithelial chloride cells in gills have been shown to occur at exposure concentrations between 1 and 5 μg L^{-1} (Triebskorn et al. 2004). Diclofenac has been associated with a serious effect in wildlife, causing renal failure in particular species of exposed Asian vultures (e.g. *Gyps bengalensis*) that has caused widespread population crashes (Corcoran et al. 2010).

12.3.6.2 Antibiotics

Modes of action of antibiotics vary according to type. Penicillins impede synthesis of the bacterial cell wall; tetracyclines bind to ribosomes and impair protein manufacture; and sulfonamides competitively inhibit the bacterial enzyme dihydropteroate synthase. Despite the different modes of actions, the ultimate effect is the suppression of bacterial growth, and so they are used therapeutically to prevent and treat bacterial infections, as well as growth promoters in farming and aquaculture.

Antibiotics could be classified as extremely toxic to microorganisms (EC_{50} below 0.1 mg L^{-1}) and very toxic to algae (EC_{50} between 0.1 and 1 mg L^{-1}). Chronic toxicity tests performed on algae have shown high sensitivity to antibacterial agents as deduced from growth inhibition measurements (Halling-Sorensen 2000; Holten et al. 1999). The presence of antibiotics in the aquatic environment has more been investigated in terms of the development of bacterial resistance (Daughton and Ternes 1999), and knock on effects regarding human health by the transfer of resistance to human pathogens (Witte et al. 1999), rather than for any concern for possible toxicity to aquatic organisms. At high exposure concentrations (>100 mg L^{-1}), sulfonamides were found to be acutely toxic to fish medaka (Kim et al. 2006). There is also some evidence to suggest that tetracyclines can have a suppressive effect on the immune systems in fish, with effect concentrations overlapping with those sometimes occurring in the environment (0.1–50 μg L^{-1}; Grondel et al. 1985; Wishkovsky et al. 1987). It is most likely that antibiotics may affect microbial functions in both aquatic and terrestrial ecosystem in turn affecting processes such as denitrification, nitrogen fixation, and organic matter breakdown.

12.3.6.3 Antiepileptics and Antidepressants

Antiepileptic drugs act on the central nervous system by decreasing the overall neuronal activity. This can be achieved either by blocking voltage-dependent sodium channels of excitatory neurons (e.g. carbamazepine), or by enhancing of inhibitory effects of the GABA neurotransmitter by binding on a specific site in the gamma subunit of the corresponding receptor (e.g. diazepam) (Fent et al. 2006). Evidence of the occurrence of the GABA system in fish (*O. mykiss*) was found, whereas no studies have been found indicating the occurrence of sodium voltage dependent channels in fish or lower invertebrate. The antiepileptic carbamazepine is considered carcinogenic in rats but is not mutagenic in mammalian cells. Sublethal effects occurred in *Daphnia magna* at 92 μg L^{-1} and the lethal concentration in zebra fish was 43 μg L^{-1} (Thaker 2005). In a study with *Hydra vulgaris*, diazepam was shown to inhibit polyp regeneration at 10 μg L^{-1} (Pascoe et al. 2003).

Selective serotonin reuptake inhibitor antidepressants exert therapeutic effects by inhibiting monoamine transporters and thus inhibiting the reuptake of the neurotransmitter serotonin (5-hydroxytryptamine) at presynaptic neuronal membranes (Hiemke and Härtter 2000). This elevates the concentration of serotonin in the synaptic gap (Fent et al. 2006). As an important neurotransmitter, serotonin in fish has been implicated in several physiological functions, influencing behavior (aggression, appetite), endocrine, and reproductive parameters. Furthermore, serotonin is involved in social hierarchy and feeding rank in some species (Corcoran et al. 2010), with subordinate individuals having higher brain serotonergic activity. Serotonin reuptake inhibitor antidepressants have thus the potential to disrupt a wide range of processes in fish.

12.3.6.4 Antilipidemics

Lipid regulators, particularly fibrate drugs such as bezafibrate, gemfibrozil, and fenofibrate, are peroxisomal proliferators with the ultimate therapeutic effect being the lowering of blood plasma lipid levels. Acyl–coenzyme A oxidase, an enzyme which initiates peroxisomal β oxidation of fatty acids, is involved in lipid metabolism (Desvergne and Wahli 1999). This increased enzymatic activity, coupled with increased peroxisomal volume leads to the removal of fatty acids and cholesterol from the blood. One reported side effect of this is increased production of hydrogen peroxide (H_2O_2) in the cell, which may lead to oxidative stress and hepatocarcinogenesis (Gonzalez et al. 1998). Indeed, a strong correlation has been shown between exposure to fibrates and hepatocarcinogenesis in rodents (Fent et al. 2006). It has also been demonstrated that clofibrate and fenofibrate induce oxidative stress in rainbow trout (*Oncorhynchus mykiss*) hepatocytes (Laville et al. 2004), indicating fish may be susceptible to the effects of fibrates. Gemfibrozil was reported to bioconcentrate in goldfish with a bioconcentration factor of 113 after 14 days of exposure (Mimeault et al. 2005) and, as such, fish may be chronically exposed to high levels of these compounds.

12.3.6.5 Beta Blockers

Beta blockers, such as atenolol, propranolol, metoprolol, celiprolol, etc., are used in the treatment of high blood pressure (hypertension), and to treat patients after heart attack to prevent further attacks (Fent et al. 2006). They work as competitive β-adrenergic receptor antagonists on cardiac muscle to decrease heart rate and contractility (Owen et al. 2007). Beta blockers have the potential to impact on a range of physiological systems. Propranolol compared to other beta-blockers showed the highest acute toxicity to phytoplankton, benthos, zooplankton and fish (Fent et al. 2006). In medaka, this substance was shown to have a LC_{50} of 24.3 mg L^{-1} (48 h) and caused a decreased growth after 14 days' exposure to 0.5 mg L^{-1} (Huggett et al. 2002). Similarly, propranolol impaired growth in juvenile rainbow trout at 10 mg L^{-1} after 10 days (Owen et al. (2007). With reference to cardiovascular effects, propranolol has been shown to affect the heart rate in zebrafish (Fraysee et al. 2006) as well as blood flow through the gills (Payan and Girard 1977). There are limited reports of effects of beta blockers on reproductive and growth of fish. Egg production and hatching success were both reduced in medaka after 28 days of exposure to propranolol at 5 μg L^{-1} (Huggett et al. 2002). Atenolol has been shown to affect growth in fathead minnow (*Pimephales promelas*) embryolarvae, at relatively high concentration of 10 mg L^{-1}after a 28 days of exposure (Winter et al. 2006).

12.3.6.6 Steroids and Related Hormones

There exists a lot of information on the biological effects of the synthetic oestrogen ethinyl-estradiol which is extremely potent in fish. Environmentally relevant concentrations have been shown to induce feminisation in fish, including induction of the female yolk precursor vitellogenin in males, formation of a female reproductive duct in the testis and induction of intersex (reviewed by Corcoran et al. 2010). These effects have been documented for a wide range of fish species. Life time exposure to relatively low concentrations of ethinyl-estradiol in the water (5 ng L^{-1}) has been shown to cause reproductive failure in colonies of laboratory maintained zebrafish (Nash et al. 2004), and similarly the dosing of a lake in Canada with 4–6 ng L^{-1} ethinyl-estradiol resulted in complete failure of the fathead minnow (*Pimephales promelas*) fishery (Kidd et al. 2007). Exposure of roach (*Rutilus rutilus*), a species in which widespread sexual disruption has been shown in wild populations living in UK rivers downstream of wastewater treatment discharges, to 4 ng L^{-1} for a 3-year period resulted in complete feminization of the exposure populations (Lange et al. 2009). Other steroidal oestrogens used in formulations for contraception or in hormone replacement therapy include conjugated oestrone (oestrogen sulphate) and equine oestrogens. Oestrone sulphate, in its conjugated form, is biologically active in fish (Corcoran et al. 2010) and is also readily metabolised to oestrone, which is strongly oestrogenic in fish at environmentally relevant concentrations (Thorpe et al. 2003). Similarly, equilenin and dihydroequilenin are biologically active in fish, with effective concentration inducing vitellogenin synthesis in rainbow trout *(Oncorhynchus mykiss)* of 4.2 and 0.6 ng L^{-1}, respectively (Tyler et al. 2009).

12.3.6.7 Cancer Therapeutics

There are different modes of actions of anticancer drugs. For example, tamoxifen as an antiestrogenic drug is used for breast cancer treatment and acts by competitive inhibiting the estrogenic receptor at least in mammary gland (Fent et al. 2006). Methotrexate acts as a potent inhibitor of the folate dehydroreductase enzyme, which is responsible for the purine and pyrimidine synthesis (Schalhorn 1995). Doxorubicin is an intercalating substance inducing DNA-strand brakes (heart arrhythmia may be a side effect in humans). Anticancer drugs are finally designed to kill cells that are proliferating excessively such as those found in pathological cancer conditions. Therefore, a similar effect on any other growing eukaryotic organisms can be expected (Santos et al. 2010). They possess genotoxic, mutagenic, carcinogenic, teratogenic and fetotoxic properties and can constitute from 14% to 53% of the administered drug excreted in urine (Sanderson et al. 2004). Cyclophosphamide and ifosfamide ecotoxicity have yielded EC_{50} values of 8.2 and 70 mg L^{-1} for algae and fish respectively, whereas *Daphnia magna* registered a LC_{50} of 1795 mg L^{-1} (Sanderson et al. 2004). Toxicity tests performed on the algae *Pseudokirchneriella subcapitata* and the invertebrate *Daphnia magna* showed that cyclophosphamide slightly increased the growth of the former and reduced offspring number in the latter at all tested concentrations of the drug (10–100 mg L^{-1}) (Grung et al. 2008). Methotrexate revealed teratogenicity for fish embryos with an EC_{50} of 85 mg L^{-1} after 48 h of exposure and acute effects in the ciliate *Tetrahymena pyriformis* with an EC_{50} for 48 h of 4 5 mg L^{-1} (Henschel et al. 1997). Acute and chronic toxicity of tamoxifen and its photoproducts was studied by Della Greca et al. (2007), showing that both the active pharmaceutical and its photoproducts affected the rotifer *Brachionus calyciflorus* and the crustacean *Thamnocephalus platyurus* with LC_{50} values ranging from 0.95 to 1.31 mg L^{-1} and 0.40 to 1.59 mg L^{-1} respectively. In chronic toxicity tests, *Ceriodaphnia dubia* proved the most sensitive organism. An EC_{50} value of 0.81 μg L^{-1} for tamoxifen and EC_{50} values ranging from 0.41 to 2.8 μg L^{-1} for its photoproducts, relative to population growth inhibition, were found after a 7 days of experiment (Della Greca et al. 2007).

12.3.6.8 H_2-Antagonists

H_2-antagonists are used to treat gastric ulceration. Ranitidine and cimetidine act by inhibiting the histamine receptors type 2 in the gastric system, thus inhibiting the acid secretion (antacid). Since H_2-histamine receptors are found also in the brain, both drugs may elicit central nervous system reactions and side effects (Cannon et al. 2004). Peitsaro et al. (2000) demonstrated the presence of H_3-histamine receptors in central nervous system of zebrafish (*Danio rerio*), but the lack of histamine in the periphery of this fish was also reported. However, interspecies differences may occur; cod (*Gadus morhua*) and carp (*Cyprinus* ssp.) seem to have histamine and H_2-receptors in the periphery (Peitsaro et al. 2000).

12.3.6.9 X-Ray Contrast Media

Contrast media are used as diagnostic tools for capturing detailed X-ray images of soft tissues. Toxicity tests have shown that iopromide or its main metabolite do not have a toxic effect in luminescent bacteria, algae (*Scenedesmus subspicatus*), daphnids (*Daphnia rerio*) or fish (*Leuciscus idus*) even at concentrations as high as 1 g L^{-1} (Steger-Hartmann et al. 1999, 2002).

12.4 Major Routes of Human Exposure to Pharmaceuticals

More than 3000 different pharmaceuticals are currently in use legally in the European Union (Touraud et al. 2011). The concentration of pharmaceuticals in effluents, surface waters and groundwater are in the ng L^{-1} or low μg g^{-1} range. Little is known about the occurrence and fate of drug metabolites. In general, it is likely that metabolites will be less biologically active than the parent medicines. Humans can be exposed to very low concentrations of pharmaceuticals from two major sources, consumption of water or consumption of fishes which have accumulated pharmaceutical residues. There are no legal quality standards regarding pharmaceuticals in drinking water or drinking water sources. There are, however, signaling values, target values and provisional limit values. For example, The Danube, Meuse and Rhine MEMORANDUM 2008 (Wirtz 2009), which refers to the implementation process of the EU Water Framework Directive and its sub-directives has defined a signaling value of 1.0 μg L^{-1} for all anthropogenic substances. The memorandum (Wirtz 2009) also takes the drinking water interests in the Elbe river basin into account. This memorandum is intended to assist and guide politicians, authorities and decision-makers in industry and water management to make the necessary quality improvement of surface water bodies providing drinking water.

12.5 Possible Human Health Threats

12.5.1 Adverse Health Effects

The risk of adverse effects on humans through the ingestion of pharmaceuticals present in drinking water or in food products seems to be very limited (Kümmerer 2009a). Thus, the risks posed to humans from pharmaceuticals in the environment seem to concern environmental hygiene rather than toxicology or pharmacology. The maximum possible intake within a life-span (2 L of drinking water per day over 70 years) is far below the dosages used in therapy (Kümmerer 2009a; Kümmerer and Al-Ahmad 2010). However, this statement relies on the assumptions that (i) the effects and side effects during therapeutical use (short-term, high dosage) are the same in quality and quantity as for lifelong ingestion (long-term ingestion, low dosage), (ii) the effects are the same for fetuses, babies, children, healthy adults and elderly people; and (iii) the risk posed by a single compound is comparable to the one posed by a mixture. Concerning the latter, it has been found that elderly people who take several different medications at a time suffer more often from unwanted side effects of drugs during therapy. Up to now risk assessments have been undertaken for single substances only and not for mixtures.

12.5.2 Antibiotic Resistance

In contrast to the limited risk of adverse effects on humans through the ingestion of pharmaceuticals, antibiotic resistance in humans and animals is a growing public health concern worldwide. When a person or animal is infected with an antibiotic-resistant bacterium, treatment becomes more difficult because standard antibiotic therapies become less effective or may not work at all. According to Kümmerer (2009b) the most prominent medical examples are multi-resistant pseudomonads, vancomycin-resistant enterococci, and methicillin-resistant *Staphylococcus aureus* (MRSA), the latter being a virulent pathogen that is currently the most common cause of infections in hospitalized patients. Diseases range from relatively mild infections of the skin and soft tissue to life-threatening sepsis (Lowy 1998). Resistance to most β-lactam antibiotics, including the semisynthetic (β-lactamase-resistant) penicillins, such as methicillin and flucloxacillin, is due to the expression of the low-affinity penicillin binding protein PBP2a (Hartman and Tomasz 1984).

As prospective population-based surveillance studies are not frequently performed, the exact incidence of MRSA is difficult to assess. In Scandinavian countries, where data from the national surveillance of SAB are routinely collected, the annual incidence is approximately 26 per 100,000 population (Bentfield et al. 2007; Jacobsson et al. 2007). A similar incidence of 19.7 per 100,000 population was reported in a study in Canada (Laupland et al. 2008), while in countries with a greater burden of MRSA, such as the USA (State of Minnesota) and Australia, incidence rates are between 35 and 39 per 100,000 population (Collignon et al. 2005; El Atrouni et al. 2009). Even higher rates for the USA, approximately 50 per 100,000 population, were inferred from surveillance data by Morin and Hadler (2001). These large discrepancies probably reflect differences in health care systems, infection control practices, and the completeness of surveillance data.

The incidence of MRSA increases with advancing age, with the lowest rates observed in pediatric populations, at approximately 8.4 per 100,000 population per year (Fridkin et al. 2005). Similarly, younger adults have lower incidence rates than older adults (Bentfield et al. 2007). Other factors associated with higher incidences include male gender, ethnicity, community-onset MRSA, and specific patient subgroups that have frequent health care contact, including hemodialysis patients (Morin and Hadler 2001; Laupland et al. 2008). Tiemersma et al. (2004) have shown significant increases in MRSA between 1999 and 2002 in hospitals of European countries, particularly Belgium, Germany, Ireland, the Netherlands and the UK. MRSA prevalence varied widely, from $<1\%$ in northern Europe to $>40\%$ in southern and western Europe. Within countries, MRSA proportions varied between hospitals with highest variance in countries with a prevalence of 5–20%. The large differences between hospitals indicate that efforts may be most effective at hospital levels. As the prevalence of healthcare-associated infections caused by multidrug-resistant organisms continues to increase (Muto 2006), it seems essential to prevent MRSA transmission and reduce the number of MRSA infections. For healthcare workers it is important that MRSA rates should be controlled, as the average MRSA carrier rate in healthcare workers was found to be 4.6%, and that about 5.1% of these carriers had symptomatic MRSA infections (Albrich and Harbarth 2008).

Antibiotic resistance is an evolutionarily conserved natural process. Resistant bacteria can thus be found in natural environments (Davison 1999). However, naturally occurring resistance is only associated with some bacterial strains (Kümmerer 2009b), but its occurrence is increasing particularly in medical environments, while the tools for combatting it are decreasing in power and number (Finch and Hunter 2006). The transfer of resistant bacteria to humans could occur via water or food if plants are watered with surface water or sewage sludge, if manure is used as a fertilizer, or if resistant bacteria are present in meat or meat products (Perretin et al. 1997; Khachatourians 1998; Salyers 2002; Dolliver and Gupta 2008). To minimize the transfer of antibiotic resistance from animals to humans the antibiotic content of meat is monitored by authorities in many countries. Especially for consumers, when considering meat and meat products which are usually cooked or heated prior to use, bacteria including their resistance genes will probably be destroyed (Kümmerer 2009b).

An important source for the resistance material found in hospital effluents, municipal sewage and sewage treatment plants is the input of bacteria that have already become resistant through the use of antibiotics in medical treatment. The widespread use of triclosan and quaternary ammonium compounds used in hospitals and homes could select for antibiotic resistant bacteria (Hingst et al. 1995; Russel 2000). Triclosan has been shown to select for low level antibiotic resistance in Escherichia coli (McMurry et al. 1998), and high-level ciprofloxacin resistance in triclosansensitive *Pseudomonas aeruginosa* mutants (Chuanchuen et al. 2001). However, it has been observed that the presence of resistant bacteria and genetic material correlated with resistance do not correspond with the concentrations and activity spectrum of compounds found in the environment (Kümmerer 2009a). For example, β-lactams have been detected in the environment at only very low concentrations and they are easily hydrolyzed at ambient temperature (Längin et al. 2009), whereas resistant bacteria and genetic material encoding resistance against certain β-lactams have been found in sewage treatment plants. Moreover, resistance against vancomycin has been found in European sewage and waters, even though only small quantities of vancomycin are used in Europe (Kümmerer 2009b).

Another important source for resistance material results from the application of veterinary antibiotics to food animals which is supposed to enhance the selection for strains resistant to antibiotics used in human medicine (Kemper 2008). Since the agricultural use of antibiotics might act as an important source of resistance in bacteria affecting human health, the banning of antibiotic growth promoters in Europe in 2006 was administered as a precaution (EC 2003). Transmission of strains resistant to antibiotics may be performed via direct contact with animals or via the food chain (including water) to the consumers. In a study including 1600 pigs on 40 German farms, 70% of the farms carried strains resistant against MRSA (Köck et al. 2009). Other studies suggested transmission between pigs and humans causing MRSA colonization in 23–45% of pig-farmers and 4.6% of pig-care veterinarians, indicating that livestock might represent a relevant source for MRSA imports to hospitals (Köck et al. 2009). In summary, there is an urgent need to use antibiotics judiciously in both human and animal medicine to slow the development of resistance.

12.6 Mitigation Options to Reduce the Release of Pharmaceuticals to the Environment

Major strategies to reduce releases of pharmaceuticals to the environment include control of pharmaceuticals at the source, segregation of sources, treatment of waste products to remove pharmaceutical compounds, introduction of husbandry practices and improvement of disposal systems for out-of-date medicines and waste containers (Table 12.12).

Source controls include labelling, controlled disposal and urine separation. Segregating sources of pharmaceuticals, such as hospital wastewater, which is likely to be heavily contaminated with pharmaceuticals and antibiotic-resistance bacteria, should make it possible to focus treatment resources on the most contaminated waters (Boxall 2004). Pharmaceuticals can be removed when treated through biological, physical or chemical processes, such as biological degradation, sorption,

Table 12.12 Strategies for reduction of amounts of pharmaceuticals released to the environment

Strategy	Human pharmaceuticals	Veterinary pharmaceuticals
Source control	Medicine labels provide information on possible environmental impact allowing medical doctors and veterinary surgeons to consider potential environmental impact when prescribing	
Source separation	Separation of hospital waste from other wastes to allow targeted treatment; separation of solid material from urine	Separation of untreated from treated animals
Improvement of treatment and husbandry practices		Advice to farmers on usage and husbandry methods to reduce environmental exposure; for example, pasture animals should not be allowed near water bodies for *several* days after administration
Treatment of slurry and water	Wastewater treatment: Sorption removes fluoroquinolones; Biodegradation removes bezafibrate, sulphamethoxazole, ibuprofen, ethinylestradiol, and diclofenac; Ozonation removes estrone, beta blockers, antiphlogistics and antibiotics	Slurry and sludge storage; for example, tylosin degrades very rapidly during slurry storage which might be a mechanism to reduce amounts released to soils
Best farming practice	Sewage sludge and slurry should be applied to land according to good farming practice for reducing the potential for pharmaceuticals to contaminate surface water and groundwater	
Appropriate disposal	Advice on appropriate disposal of unused medicines and containers	

Adapted from Ternes et al. (2002, 2004a) and Daughton (2003a, b)

filtration or chemical reactions, for instance, through treatment with chlorine or ozone. Many of the treatment methods may produce transformation products that are more persistent and mobile than the parent compounds, some of which may also have similar or enhanced toxicity. Little work has been performed to assess the environmental impacts of these transformation products on the environment. In the following sections, the most common measures to reduce pharmaceutical concentrations in the environment are highlighted.

12.6.1 Preventive Measures as Mitigation Options

Reduction of human exposure to pharmaceuticals through drinking-water can be achieved through a combination of preventive measures, such as public guidance and consumer education to encourage the proper disposal of unwanted pharmaceuticals and minimize the introduction of pharmaceuticals into the environment, as well as take-back programmes and regulations.

12.6.1.1 Raising Physicians' Awareness

Over-prescription of pharmaceuticals due to over-diagnosis or misdiagnosis has been documented as part of the problem of the overuse of medications. Physicians with limited time to diagnose and treat patients may be quick to write prescriptions for patients who expect that their ailments can be cured with pills. Antibiotics are a class of medicines that are frequently over-prescribed based on over-diagnosis or misdiagnosis. Knowledge and information about the environmental relevance of pharmaceuticals raise the awareness of physicians in the consultation of patients (Kümmerer 2009a). In order to facilitate the integration of the problem into physicians' everyday practice, it has to be implemented during medical education and advanced training by policy makers in education and health care.

12.6.1.2 Raising Patients' Awareness

Inappropriate disposal practices, such discarding unwanted pharmaceuticals in household waste or flushing them down toilets are common and often a significant contributor of pharmaceuticals present in wastewater and other environmental media (e.g. surface waters and landfill leachate). A study from Germany's Management Strategies for Pharmaceutical Residues in Drinking Water research programme showed that consumers discarded 23% of liquid pharmaceuticals prescribed and 7% of tablets. While some went into household trash, the equivalent amount of pharmaceuticals that was flushed away in Germany is approximately 364 Mg every year (Lubick 2010). Another survey of households in the UK in 2003 found that 63% of unwanted pharmaceuticals were discarded in household waste and 11.5% were flushed down sinks or toilets (Bound and Voulvoulis 2005). Proper information about how to handle left over drugs will result in a reduction of the environmental burden of drugs.

12.6.1.3 Establishing Take-Back Programs for Excess Pharmaceuticals

In several countries, take-back programs have been established by government and private organizations to reduce the amount of drugs entering the environment (e.g. Daughton 2003a, 2004; Glassmeyer et al. 2009). A survey of households in the United Kingdom in 2003 showed that 22% of excess pharmaceuticals were returned to pharmacists (Bound and Voulvoulis 2005). These programs can be of different scales, ranging from small one-day collection events to regular and systematic regional collection, ongoing return of unused and excess medicines to participating pharmacies and mail-back programs where excess medicines are returned in prepaid packs to government-supervised mailboxes. Several household hazardous waste collection programs have also added pharmaceuticals to the list (Glassmeyer et al. 2009). In Australia, the Commonwealth Department of Health & Ageing Services provided funds to establish a system for the collection and disposal of unwanted medicines, known as the Return Unwanted Medicines (RUM) Project. Estimates from RUM showed that in 2010–2011, more than 34 Mg of unwanted medicines on average were collected monthly by community pharmacies across Australia and subsequently incinerated according to guidelines (RUM 2011).

In the USA, many scheduled pharmaceutical collection events facilitate disposal of unwanted medications at the regional level, such as the successful "Great Lakes Earth Day Challenge", which collected 4.5 million pills for safe disposal. The US EPA has also awarded grants to support take-back of non-controlled, unused medicines at pharmacies and mail-back of unused medicines with appropriate involvement of law enforcement (US EPA 2010). Europe has widespread standardized take-back programs. According to a report by the European Environment Agency (EEA 2010) most countries in Europe collect unused drugs separately from household waste, usually at pharmacies. The national systems are operated and funded by the pharmaceuticals industry, retail pharmacies or the public sector.

12.6.1.4 Raising Awareness About the Substances' Hazardous Properties

A good example of increased awareness about the substances' hazardous properties is the Swedish Environmental Classification and Information System (SECIS) for pharmaceuticals which is a national and voluntary classification system for pharmaceuticals (Gunnarson and Wennmalm 2008). So far, pharmaceutical companies in Sweden have shown a high degree of interest to participate in this initiative (Mattson 2007). In February 2010, 691 of the 1090 substances available on the Swedish market had been assessed, however environmental data is still missing for a substantial part of the substances: ecotoxicity data is missing for 59% of the substances, 47% lack data on the substances' persistency, and 40% lack bioaccumulation data (Ågerstrand and Rudén 2010). Within SECIS, the pharmaceutical companies provide environmental data and classify their products according to pre-defined criteria and a guidance document (Swedish Association of the Pharmaceutical Industry 2007). SECIS is based on an assessment of risk, i.e. combining information about the substances' hazardous properties with information about the estimated environmental concentrations. The system has four risk classification categories; insignificant, low, moderate, and high risk. A classification according to the SECIS guidance document furthermore includes an assessment of the pharmaceutical substance potential to persist in the environment and to bioaccumulate in organisms (Swedish Association of the Pharmaceutical Industry 2007).

12.6.2 *Pharmaceutical Removal from Wastewater Treatment Plant Effluent*

During the last few years, a number of studies have been conducted on the removal of pharmaceuticals from effluent at wastewater treatment plants. For example, graphene adsorption reactor technology removed carbamazepine, diclofenac, and ibuprofen by 96% (Rizzo et al. 2015). Membrane filtration enhanced with carbon nanotubules produced effluent with less than 10% of pharmaceutical compounds detected (Gethard et al. 2012). These new technologies can also reduce toxicity of effluent by a significant degree (Rizzo et al. 2015).

12.6.2.1 Pharmaceutical Removal from Sewage Sludge

Several studies have focused on removing pharmaceutical compounds from sewage sludge destined for land application. An interesting new technology focusing on fungi as an alternative to physical or chemical treatment has shown promise. White rot fungi, or *Trametes versicolor*, break down lignin in wood through an enzymatic process involving laccases and peroxidases. *T. versicolor* has a cytochrome P450 system similar to that of mammals that allows it to break down ingested compounds as well (Rodríguez-Rodríguez et al. 2011). These properties have led to the study of *T. versicolor* as a method of biodegradation and toxicity reduction of pharmaceutical compounds in sewage sludge.

In a recent study, pharmaceutical-spiked sterile sewage sludge was inoculated with *T. versicolor* in a solid-phase bioreactor to evaluate the potential for degradation and toxicity reduction. Out of 14 pharmaceuticals tested, 7 were completely removed, including the antibiotics clarithromycin and sulfamethoxazole. The anti-epileptics diazepam and carbamazepine were both removed by 43%, while the anti-inflammatory medicines ibuprofen, diclofenac, and mefenamic acid were removed by 75%, 64%, and 72%, respectively (Rodríguez-Rodríguez et al. 2011). This method proved to be far more effective for antibiotics than previous studies on anaerobic digestion and natural attenuation, where sulfamethazine was removed by 0–43% (Rodríguez-Rodríguez et al. 2011). Toxicity was also reduced by 56% in standard *Daphnia magna* tests, and completely eliminated in *Vibrio fischeri* tests (Rodríguez-Rodríguez et al. 2011). Rodríguez-Rodríguez et al. (2012) conducted a second study with *T. versicolor* focusing on non-sterile sewage sludge in which removal rates of up to 85% for clarithromycin, up to 9% for carbamazepine, up to 60% for diclofenac, and up to 61% for ibuprofen were observed. While *T. versicolor* survived well in the sewage sludge, it is likely that the naturally-occurring microbiota may have interfered with its ability to remove pharmaceutical compounds from sludge (Rodríguez-Rodríguez et al., 2012). This suggests that a preliminary sanitization step might be useful in the implementation of *T. versicolor* biodegradation treatment. Research into treatment with fungi is ongoing and may

Table 12.13 Removal of pharmaceuticals in drinking water treatment works by different methods

Method	Water source tested	Medicinal compound	Concentration in source water (ng L^{-1})	Removal (%)
Filtration[a] (aeration plus active clay and sand filters)	Highly contaminated groundwater, Germany	Phenazone	3950	90
		Propiphenazone	1230	90
		Dimethyl-aminophenazone	400	>95
		AMDOPH[h]	1200	25
Coagulation[b] (with 50 mg L^{-1} ferric sulphate)	Lake Roine (water samples spiked with pharmaceuticals)	Benzafibrate	30–40 μg L^{-1}	<5
		Carbamazepine	"	<5
		Diclofenac	"	30
		Ibuprofen	"	10
		Sulfamethoxalole	"	<5
Use of chlorine dioxide[c] (11.5 $mgClO_2\ L^{-1}$)	Drinking water, Germany	Benzafibrate	1000	0
		Carbamazepine	"	0
		Diazepam	"	0
		Diclofenac	"	100
		Gemifibrozil	"	41
		Ibuprofen	"	0
		Phenazone	"	100
		Propiphenazone	"	100
		Dimethyl-aminophenazone	"	100
Use of activated carbon[d] (granules)	Paldang lake, South Korea	Carbamazepine	4,8	100
		Ibuprofen	15	100
		Iopromide	143	100
Ozonation[e] (1.2 $mgO_3\ L^{-1}$)	Rhine water	Caffeine	500	100
		Carbamazepine	"	100
		Diazepam	"	65
		Ifosfamide	"	50
		Pentoxyfylline	"	100
		Phenazone	"	100
		Propiphenazone	"	100
		Dimethyl-aminophenazone	"	100
Reverse osmosis[f]	Teltowkanal, Germany	Carbamazepine	330	>99.7
		Clofibric acid	155	>99.4
		Diclofenac	329	>99.7
		Naproxen	38	>95.0
		Prophylenazone	170	>99.4

(continued)

Table 12.13 (continued)

Method	Water source tested	Medicinal compound	Concentration in source water (ng L^{-1})	Removal (%)
Coupled treatments[g]	Detroit River	Colifibric acid	103	100 (O_3 + coag + sed + Cl_2)
	"	Naproxen	63	"
	Mississippi River	"	64	0 (PAC + coag +sed)
	"	"	64	100 (Cl_2)

Data from
[a]Reddersen et al. (2002)
[b]Vieno et al. (2006)
[c]Huber et al. (2005)
[d]Kim et al. (2007)
[e]McDowell et al.(2005)
[f]Heberer et al. (2002)
[g]Boyd et al. (2003)
[h]*AMDOPH* 1-acetyl-1-methyl-2-dimethyl-oxamoyl-2-phenylhydrazide

lead to more effective removal of a greater amount of compounds with further refinement.

12.6.2.2 Pharmaceutical Removal in Drinking Water Treatment Systems

In drinking water treatment works a wide range of processes are used primarily to remove taste and odor causing compounds, but they also lead to removal of some pharmaceuticals. Filtration, sorption to particles removed by coagulation, use of chlorine, chlorine dioxide and/or of activated carbon may reduce concentrations present in the treated effluent for some substances. Table 12.13 shows examples of removal rates of pharmaceuticals due to application of processes frequently used.

Filtration

Filtration provides polishing of a potable water supply and follows sedimentation in a sedimentation basin if it is provided. Water moves through tanks that contain sand and other filter media. Fine solids that did not settle during the sedimentation procedure will be entrapped in the filter. Removal of the phenazone pharmaceuticals was very successful (Table 12.13). However removal of the phenazone metabolite, AMDOPH appeared to be poor.

Coagulation

Coagulation is the process of adding Fe or Al salts in a mixing tank to precipitate with other suspended particles to form larger, more readily settled particles (Droste 1997). Vieno et al. (2006) carried out a number of jar test experiments using spiked deionized water, lake water and commercial humic solutions using $Al_2(SO_4)_3$ (pH 6) and $Fe_2(SO_4)_3$ (pH 4). In deionised water, less than 10% of the pharmaceuticals were removed by coagulation with the exception of diclofenac, which was removed up to 66% with $Fe_2(SO_4)_3$. This was also the only pharmaceutical removed during the coagulation of lake water with ferric sulphate (Table 12.13). The coagulation of pharmaceuticals was impaired by the presence of soluble organic matter (OM). It is thought that lower molecular weight OM is responsible for this impairment and that high concentrations of high MW NOM may increase coagulation of pharmaceuticals by up to 50%. Neutral pharmaceuticals such as carbamazepine and sulfamethoxazole cannot be removed by coagulation.

Use of Chlorine and Chlorine Dioxide

Application of chlorine is a widespread treatment for disinfecting drinking waters (Rivera-Utrilla et al. 2013). Various studies on the chlorination of aromatic compounds demonstrated that the chlorine reaction rate can be strongly affected by the presence of different functional groups in the benzene ring. The reaction is usually rapid in pharmaceuticals containing amines, giving rise to chlorinated compounds (Pinkston and Sedlak 2004). Thus, metropolol and sulfamethoxazole give rise to chloramines as an oxidation product. Studies on the removal of acetaminophen, the active compound of paracetamol, showed that it reacts with chlorine to form numerous subproducts, which partly have been identified as toxic compounds (Glassmeyer and Shoemaker 2005; Bedner and MacCrehan 2006a). At least five subproducts are formed in diclofenac chlorination, although none of them are chloramines (Bedner and MacCrehan 2006b), and the degree of mineralization achieved is not acceptable. Chlorine dioxide (ClO_2) is a more potent oxidant than chlorine and can degrade numerous organic compounds by oxidation (Table 12.13). However, it does not react with ammonium or chlorinate organic substances, which is an advantage with respect to taste, smell, and the formation of organochlorinated toxic species. Research into the use of ClO_2 to oxidize persistent pharmaceutical products in the environment found that it is only effective for certain pharmaceuticals, including diclofenac, phenazone, propiphenazone and dimethylaminophenazone (Huber et al. 2005). ClO_2 reacts selectively with functional groups with high electron density, such as tertiary amines and phenoxides. In comparison to ozone, ClO_2 reacts slower and with fewer compounds; in comparison to chlorine, it reacts faster with sulfonamides, macrolides, and estrogens, but, unlike chlorine, it does not react with bezafibrate, carbamazepine, diazepam, or ibuprofen (Huber et al. 2005).

Use of Activated Carbon

Activated carbon generally demonstrates a high capacity to adsorb pharmaceuticals (Rivera-Utrilla et al. 2013). Dissolved, colloidal and particulate substances are attracted and attached to the surface of the carbon particles. Precipitation and other chemical reactions also occur on the carbon surface. A variety of carbon adsorbers can be designed, including batch and continuous flow units. The adsorption capacity of the carbon is eventually exhausted and the carbon is regenerated by heating, which oxidises and volatilises the accumulated substances. The activated carbon can be in form of granules or powder (Droste 1997). Kim et al. (2007) investigated the treatment efficiency of a number of pharmaceuticals, in drinking water processes using granular activated carbon. All compounds were reduced to below the analytical reporting limits in the finished drinking water (Table 12.13).

Another advantage of using activated carbon to remove pharmaceuticals is that it does not generate toxic or pharmacologically active products. Rivera-Utrilla et al. (2009) conducted an extensive study on the adsorption of antibiotics (nitroimidazoles) on different types of activated carbons, finding an increase in the adsorption rate with a decrease in the percentage of oxygen and an increase in the hydrophobicity of the carbon. Hence, in general, hydrophobic interactions appear to govern the adsorption kinetics. Nitroimidazole adsorption was largely determined by the chemical properties of the carbon. In the case of partial or coupled treatments, e.g.: oxidationadsorption, a specific oxidation treatment to remove persistent pharmaceuticals is followed by the adsorption of intermediate products on the activated carbon, diminishing their toxicity and pharmacological activity (Rivera-Utrilla et al. 2013).

Ozonation

Ozone (O_3) is a powerful oxidizing agent and a very effective biocide. Ozone reacts with most organic matter either by direct attack or indirectly through the formation of hydroxyl radicals (•OH) formed from O_3. McDowell et al. (2005) spiked several neutral pharmaceuticals into flocculated and sand filtered water taken from the River Rhine, Germany (Table 12.13). The neutral pharmaceuticals exhibit a high rate of oxidation from ozone because their structures contain fast reacting double bonds (caffeine, propyphenazone), and tertiary amino groups (phenazone, dimethylaminophenazone, pentoxifylline). The slower oxidation of ifosfamide, cyclophosphamide and diazepam is due to a lack of susceptible functional groups in these three compounds, which hinders a fast electrophilic attack by ozone, meaning that the oxidation occurs mainly through reactions with OH radicals (McDowell et al. 2005).

Reverse Osmosis

Reverse osmosis separates dissolved solids from the water by forcing the water through a membrane. Suspended solids must be removed to a low level before water is subjected to reverse osmosis to prevent fouling of the membrane. Heberer et al. (2002) tested mobile drinking water purification systems using water taken from the Teltowkanal in Berlin, Germany. This canal is highly polluted with municipal sewage effluents. Reverse osmosis proved effective at removing pharmaceuticals by more than 95% (Table 12.13). This was due to the size of the membrane pores that do not permit chemical species to pass through the membrane that are of a certain molecular size.

Coupled Treatments

There are a number of papers that summarize the removal of pharmaceuticals by single water treatment processes, while the number of studies dealing with coupled treatments is very limited. In the latter, different methods are combined to diminish concentration and toxicity of pollutants. In a study in the USA, conventional drinking water processes, i.e. coagulation, flocculation and sedimentation together with the continuous addition of activated carbon did not remove completely naproxen from Mississippi River waters. However, chlorination, ozonation and dual media filtration processes reduced the concentration of naproxen to below the limits of detection in both Mississippi and Detroit River waters tested (Boyd et al. 2003; Table 12.13).

References

Ågerstrand M, Rudén C (2010) Evaluation of the accuracy and consistency of the Swedish environmental classification and information system for pharmaceuticals. Sci Total Environ 408:2327–2339

AHI (Animal Health Institute) (2002) Available from: http://www.ahi.org/. Accessed 1 Aug 2017

Albrich WC, Harbarth S (2008) Health-care workers: source, vector, or victim of MRSA? Lancet Infect Dis 8:289–301

Alder A, Bruchest A, Carballa M, Clara M, Joss A, Loffler D, McArdell C, Miksck K, Omil F, Tukhanen T, Ternes T (2006) Consumption and occurrence. In: Ternes T, Joss A (eds) Human pharmaceuticals hormones and fragrances, the challenge of micropollutants in urban water management. IWA Publishing, London

Ashton D, Hilton M, Thomas KV (2004) Investigating the environmental transport of human pharmaceuticals to streams in the United Kingdom. Sci Total Environ 333:167–184

Banks AT, Zimmerman HJ, Ishak KG, Harter JG (1995) Diclofenac-associated hepatotoxicity: analysis of 180 cases reported to the food and drug administration as adverse reactions. Hepatology 22(3):820–827

Barnes KK, Kolpin DW, Furlong ET, Zaugg SD, Meyer MT, Barber LB (2008) A national reconnaissance of pharmaceuticals and other organic wastewater contaminants in the United States – I groundwater. Sci Total Environ 402:192–200

Bartelt-Hunt S, Snow DD, Damon-Powell T, Miesbach D (2011) Occurrence of steroid hormones and antibiotics in shallow groundwater impacted by livestock waste control facilities. J Contam Hydrol 123:94–103

Bedner M, MacCrehan WA (2006a) Reactions of the amine-containing drugs fluoxetine and metoprolol during chlorination and dechlorination processes used in wastewater treatment. Chemosphere 65:2130–2137

Bedner M, MacCrehan WA (2006b) Transformation of acetaminophen bychlorination produces the toxicants 1,4-benzoquinone and n-acetyl-pbenzoquinone imine. Environ Sci Technol 40:516–522

Belfroid AC, Van der Horst A, Vethaak AD, Schäfer AJ, Rijs GBJ, Wegener J, Cofino WP (1999) Analysis and occurrence of estrogenic hormones and their glucuronidesin surface water and waste water in The Netherlands. Sci Total Environ 225:101–108

Bendz D, Paxéus NA, Ginn TR, Loge FJ (2005) Occurrence and fate of pharmaceutically active compounds in the environment, a case study. J Hazard Mater 122:195–204

Benfield T, Espersen F, Frimodt-Møller N, Jensen AG, Larsen AR, Pallesen LV, Skov R, Westh H, Skinhøj P (2007) Increasing incidence but decreasing in-hospital mortality of adult *Staphylococcus aureus* bacteraemia between 1981 and 2000. Clin Microbiol Infect 13:257–263

Berger K, Peterson B, Buening-Pfaune H (1986) Persistence of drugs occurring in liquid manure in the food chain. Arch Leb 37:99–102

Bjorkman D (1998) Nonsteroidal anti-inflammatory drugassociated toxicity of the liver, lower gastrointestinal tract, and esophagus. Am J Med 105(5):17–21

Bound A, Voulvoulis N (2005) Household disposal of pharmaceuticals as a pathway for aquatic contamination in the United Kingdom. Environ Health Perspect 113:1705–1711

Boxall ABA (2004) The environmental side effects of medication. EMBO Rep 5(12):1110–1116

Boxall ABA, Blackwell P, Cavallo R, Kay P, Toll J (2002) The sorption and transport of a sulphonamide antibiotic in soil systems. Toxicol Lett 131:19–28

Boxall ABA, Kolpin DW, Halling-Sorensen B, Tolls J (2003) Are veterinary medicines causing environmental risks? Environ Sci Technol 37(15):286A–294A

Boxall ABA, Fogg LA, Blackwell PA, Kay P, Pemberton EJ, Croxford A (2004) Veterinary medicines in the environment. Rev Environ Contam Toxicol 180:1–91

Boyd GR, Reemsta H, Grimm DA, Mitra S (2003) Pharmaceuticals and personal care products (PPCPs) in surface water and treated waters of Louisiana, USA and Ontario, Canada. Sci Total Environ 311:135–149

Calamari D, Zuccato E, Castiglioni S, Bagnati R, Fanelli R (2003) Strategic survey of therapeutic drugs in the Rivers Po and Lambro in Northern Italy. Environ Sci Technol 37:1241–1248

Calderón-Preciado D, Jiménez-Cartagena C, Matamoros V, Bayona JM (2011) Screening of 47 organic microcontaminants in agricultural irrigation waters and their soil loading. Water Res 45:221–231

Campagnolo ER, Hohnson KR, Karpati A, Rubin CS, Kolpin DW, Meyer MT, Esteban JE, Currier RW, Smith K, Thu KM, McGeehin M (2002) Antimicrobial residues in animal water and water resources proximal to large-scale swine and poultry feeding operations. Sci Total Environ 299:89–95

Cannon KE, Fleck MW, Hough LB (2004) Effects of cimetidine-like drugs on recombinant GABA A receptors. Life Sci 75(21):2551–2558

Carvalho PN, Basto MCP, Almeida CMR, Brix H (2014) A review of plant-pharmaceutical interactions: from uptake and effects in crop plants to phytoremediation in constructed wetlands. Environ Sci Pollut Res 21:11729–11763

Chee-Sandford JC, Mackie RI, Koike S, Krapac IG, Lin YF, Yannarell AC, Maxwell S, Aminov RI (2009) Fate and transport of antibiotic residues and antibiotic resistance genes following land application of manure waste. J Environ Qual 38:1086–1108

Chuanchuen R, Beinlich K, Hoang TT, Becher A, Karkhoff-Schweizer R, Schweizer HP (2001) Cross-resistance between triclosan and antibiotics in Pseudomonas aeruginosa is mediated by multidrug efflux pumps: exposure of a susceptible mutant strain to triclosan selects nfxB mutants overexpressing MexCD-Oprj. J Antimicrob Agents Chemoth 45:428–432

Clarke BO, Smith SR (2011) Review of 'emerging' organic contaminants in biosolids and assessment of international research priorities for the agricultural use of biosolids. Environ Int 37:226–247

Collignon P, Nimmo GR, Gottlieb T, Gosbell IB (2005) *Staphylococcus aureus* bacteremia, Australia. Emerg Infect Dis 11:554–561

Corcoran J, Winter MJ, Tyler CR (2010) Pharmaceuticals in the aquatic environment: a critical review of the evidence for health effects in fish. Crit Rev Toxicol 40(4):287–304

Dalkmann P, Willaschek E, Schiedung H, Bornemann L, Siebe C, Siemens J (2014) Long-term wastewater irrigation reduces sulfamethoxazole sorption, but not ciprofloxacin binding, in Mexican soils. J Environ Qual 43:964–970

Daughton CG (2003a) Cradle-to-cradle stewardship of drugs for minimizing the deposition whilst promoting human health. I. Rationale for and avenues toward a green pharmacy. Environ Health Perspect 111:757–774

Daughton CG (2003b) Cradle-to-cradle stewardship of drugs for minimizing the deposition whilst promoting human health. II. Drug disposal, waste reduction and future directions. Environ Health Perspect 111:775–785

Daughton CG (2004) Non-regulated water contaminants: emerging research. Environ Impact Assess Rev 24(7):711–732

Daughton CG, Ternes TA (1999) Pharaceuticals and personal care products in the environment: agents of subtle change? Environ Health Perspect 107:907–938

Davison J (1999) Genetic exchange between bacteria in the environment. Plasmid 42:73–91

DellaGreca M, Iesce MR, Isidori M, Nardelli A, Previtera L, Rubino M (2007) Phototransformation products of tamoxifen by sunlight in water. Toxicity of the drug and its derivatives on aquatic organisms. Chemosphere 67:1933–1939

Desvergne B, Wahli W (1999) Peroxisome proliferator-activated receptors: nuclear control of metabolism. Endocr Rev 20:649–688

Díaz-Cruz MS, García-Galán MJ, Guerra P, Jelić A, Postigo C, Eljarrat E, Farré M, López de Alda MJ, Petrović M, Barceló D (2009) Analysis of selected emerging contaminants in sewage sludge. TRAC – Trend Anal Chem 28(11):1263–1275

Dodgen LK, Li J, Wu X, Lu Z, Gan JJ (2014) Transformation and removal pathways of four common PPCP/EDCs in soil. Environ Pollut 193:29–36

Dolliver HA, Gupta SC (2008) Antibiotic losses from unprotected manure stockpiles. J Environ Qual 37:1238–1244

Donn, J, Mendoza M, Pritchard J (2008) AP probe finds drugs in drinking water. Associated Press National Investigative Team. Syndicated nationally Mar 10, 2008

Drillia P, Stamatelatou K, Lyberatos G (2005) Fate and mobility of pharmaceuticals in solid matrices. Chemosphere 60:1034–1044

Droste RL (1997) Theory and practice of water and wastewater treatment. Wiley, New York

EC (European Commission) (2003) Regulation of the European parliament and of the council of 22 September 2003 on additives for use in animal nutrition. No. 1831/2003

EEA (2010) Pharmaceuticals in the environment: results of an EEA workshop. Copenhagen, European Environment Agency. EEA Technical Report No. 1. http://www.eea.europa.eu/publications/pharmaceuticals-in-the-environment-result-of-an-eea-workshop. Accessed 1 Aug 2017

El Atrouni WI, Knoll BM, Lahr BD, Eckel-Passow JE, Sia IG, Baddour LM (2009) Temporal trends in the incidence of Staphylococcus aureus bacteremia in Olmsted County, Minnesota, 1998 to 2005: a population-based study. Clin Infect Dis 49:130–138

Ellis JB (2001) Sewer infiltration/exfiltration and interactions with sewer flows andgroundwater quality. Interurba II, Lisbon, Portugal, 19–22 February 2001. http://www.insa-lyon.fr. Accessed 1 Aug 2017

Ellis JB, Revitt DM, Lister P, Willgress C, Buckley A (2002) Experimental studies of sewer exfiltration. Water Sci Technol 47:61–67

Ellis JB, Revitt DM, Blackwood DJ, Gilmou (2004) Leaky sewers: assessing the hydrology and impact of exfiltration in urban sewers. Hydrology: Science & Practice for the 21st Centruty: Vol II. Brit Hydrol Soc, pp 266–271

Fent K, Weston AA, Caminada D (2006) Ecotoxicology of human pharmaceuticals. Aquat Toxicol 76:122–159

Finch R, Hunter PA (2006) Antibiotic resistance-action to promote new technologies. J Antimicrob Chemother 58:3–22

Flippin JL, Huggett D, Foran CM (2007) Changes in the timing of reproduction following chronic exposure to ibuprofen in japanese medaka *(Oryzias latipes)*. Aquat Toxicol 81:73–83

Fraysse B, Mons R, Garric J (2006) Development of a zebrafish 4-day embryolarval bioassay to assess toxicity of chemicals. Ecotoxicol Environ Saf 63:253–267

Fridkin SK, Hageman JC, Morrison M, Thomson Sanza L, Como-Sabetti K, Jernigan JA, Harriman K, Harrison LH, Lynfield, Farley MM (2005) Methicillin-resistant *Staphylococcus aureus* disease in three communities. N Engl J Med 352:1436–1444

Garrison AW, Pope JD, Allen FR (1976) Analysis of organic compounds in domestic wastewater. In: Keith CH (ed) Identification and analysis of organic pollutants in water. Ann Arbor Science, Michigan, pp 517–566

Gavalchin J, Katz SE (1994) The persistence of fecal-borne antibiotics in soil. J AOAC 77:481–485

Gethard K, Sae-Khow O, Mitra S (2012) Carbon nanotube enhanced membrane distillation for simultaneous generation of pure water and concentrating pharmaceutical waste. Sep Purif Technol 90:239–245

Gielen GJHP, van den Heuvel MR, Clinton PW, Greenfield LG (2009) Factors impacting on pharmaceutical leaching following sewage application to land. Chemosphere 74:537–542

Gilchrist MJ, Greko C, Wallinga DB, Beran GW, Riley DG, Thorne PS (2007) The potential role of concentrated animal feeding operations in infectious disease epidemics and antibiotic resistance. Environ Health Perspect 115:313–316

Glassmeyer ST, Shoemaker JA (2005) Effects of chlorination on the persistence of pharmaceuticals in the environment. Bull Environ Contam Toxicol 74:24–31

Glassmeyer ST, Hinchey EK, Boehme SE, Daughton CG, Ruhoy IS, Conerly O, Daniels RL, Lauer L, McCarthy M, Nettesheim TG, Sykes K, Thompson VG (2009) Disposal practices for unwanted residential medicines in the U.S. Environ Int 35:566–572

Goldstein M, Shenker M, Chefetz B (2014) Insights into the uptake processes of wastewater- borne pharmaceuticals by vegetables. Environ Sci Technol 48:5593–5600

Golet EM, Alder AC, Giger W (2002) Environmental exposure and risk assessment of fluoroquinolone antibacterial agents in wastewater and river water of the Glatt Valley Watershed, Switzerland. Environ Sci Technol 36(17):3645–3651

Gómez MJ, Petrovic M, Fernandez-Alba AR, Barcelo D (2006) Determination of pharmaceuticals of various therapeutic classes by solid-phase extraction and liquid chromatography-tandem mass spectrometry analysis in hospital effluent wastewaters. J Chromatogr 1114:224–233

Gonzalez FJ, Peters JM, Cattley RC (1998) Mechanism of action of the nongenotoxic peroxisome proliferators; role of the peroxisome proliferatoractivated receptor α. J Natl Cancer Inst 90:1702–1709

Gottschall N, Topp E, Metcalfe C, Edwards M, Payne M, Kleywegt S, Russell P, Lapen DR (2012) Pharmaceutical and personal care products in groundwater, subsurface drainage, soil, and wheat grain, following a high single application of municipal biosolids to a field. Chemosphere 87:194–203

Gravel A, Vijayan MM (2007) Salicylate impacts the physiological responses to an acute handing disturbance in rainbow trout. Aquat Toxicol 85:87–95

Gravel A, Wilson JM, Pedro DFN, Vijayan MM (2009) Non-steroidal anti-inflammatory drugs disturb the osmoregulatory, metabolic and cortisol responses associated with seawater exposure in rainbow trout. Comp Biochem Physiol C 149:481–490

Greiner P, Rönnefahrt I (2003) Management of environmental risks in the life cycle of pharmaceuticals. European conference on human and veterinary pharmaceuticals in the environment. Lyon, April 14–16, 2003

Grondel JL, Gloudemans AG, Van Muiswinkle WB (1985) The influence of antibiotics on the immune system II. Modulation of fish leukocyte responses in culture. Vet Immunol Immunopathol 9:251–260

Gros M, Petrović M, Ginebreda A, Barceló D (2010) Removal of pharmaceuticals during wastewater treatment and environmental assessment using hazard indexes. Environ Int 36:15–26

Grung M, Kallqvist T, Sakshaug S, Skurtveit S, Thomas KV (2008) Environmental assessment of Norwegian priority pharmaceuticals based on the EMEA guideline. Ecotoxicol Environ Saf 71:328–340

Gunnarson B, Wennmalm A (2008) Mitigation of the pharmaceutical oulet into the environment – experiences from Sweden. In: Kümmerer K (ed) Pharmaceuticals in the environment. Sources, fate, effects and risk, 3rd edn. Springer, Berlin, pp 475–488

Halling-Sørensen B (2000) Algal toxicity of antibacterial agents used in intensive farming. Chemosphere 40:731–739

Halling-Sørensen B (2001) Inhibition of aerobic growth and nitrification of bacteria in sewage sludge by antibacterial agents. Arch Environ Contam Toxicol 40:451–460

Hamscher G, Pawelzick HT, Höper H, Nau H (2005) Different behavior of tetracyclines and sulfonamides in sandy soils after repeated fertilization with liquid manure. Environ Toxicol Chem 24:861–868

Hartman BJ, Tomasz A (1984) Low-affinity penicillin-binding protein associated with beta-lactam resistance in Staphylococcus aureus. J Bacteriol 158:513–516

Heberer T (2002) Occurrence, fate, and removal of pharmaceutical residues in the aquatic environment: a review of recent research data. Toxicol Lett 131:5–17

Heberer T, Stan HJ (1997) Occurrence of polar organic contaminants in Berlin drinking water (in German). Vom Wasser 86:19–31

Heberer T, Feldmann D, Redderson K, Altmann HJ, Zimmermann T (2002) Production of drinking water from highly contaminated surface waters: removal of organic, inorganic and microbial contaminants applying mobile membrane filtration units. Acta Hydrochim Hydrobiol 30(1):24–33

Heberer T, Mechlinski A, Fanck B, Knappe A, Massmann G, Peldeger A, Fritz B (2004) Field studies on the fate and transport of pharmaceutical residues in bank filtration. Ground Water Monit Remediat 24(2):70–77

Henschel KP, Wenzel A, Diedrich M, Fliedner A (1997) Environmental hazard assessment of pharmaceuticals. Regul Toxicol Pharmacol 25:220–225

Hiemke C, Härtter S (2000) Pharmacokinetics of selective serotonin reuptake inhibitors. Pharmacol Therapeut 85:11–28

Hingst V, Klippel KM, Sonntag H (1995) Epidemiology of microbial resistance to biocides. Zbl Hyg Umweltmed 197:232–251

Höger BA, Köllner B, Dietrich DR, Hitzfeld B (2005) Water-borne diclofenac affects kidney and gill integrity and selected immune parameters in brown trout (*Salmo trutta* f. fario). Aquat Toxicol 75:53–64

Holling CS, Bailey JL, Heuvel BV, Kinney CA (2012) Uptake of human pharmaceuticals and personal care products by cabbage (*Brassica campestris*) from fortified and biosolids-amended soils. J Environ Monit 14:3029–3036

Holm JV, Rügge K, Bjerg PL, Christensen TH (1995) Occurrence and distribution of pharmaceutical organic compounds in the groundwater downgradient of a landfill (Grindsted, Denmark). Environ Sci Technol 28:1415–1420

Lutzhoft HHC, Halling-Sorensen B, Jorgensen SE (1999) Algal toxicity of antibacterial agents applied in Danish farming. Arch Environ Contam Toxicol 36:1–6

Hou Y, Gao Z, Heimann L, Roelcke M, Ma WQ, Nieder R (2012) Nitrogen balances of smallholder farms in major cropping systems in a peri-urban area of Beijing, China. Nutr Cycl Agroecosyst 2012(92):347–361

Hu X, Zhou Q, Luo Y (2010) Occurrence and source analysis of typical veterinary antibiotics in manure, soil, vegetables and groundwater from organic vegetable bases, northern China. Environ Pollut 158:2992–2998

Huber MM, Korhonen S, Ternes TA, Von Gunten U (2005) Oxidation of pharmaceuticals during water treatment with chlorine dioxide. Water Res 39:3607–3617

Huggett DB, Brooks BW, Peterson B, Foran CW, Schlenk D (2002) Toxicity of select beta-adrenergic receptor blocking pharmaceuticals (β blockers) on aquatic organisms. Arch Environ Contam Toxicol 43:229–235

Jacobsson G, Dashti S, Wahlberg T, Andersson R (2007) The epidemiology of and risk factors for invasive *Staphylococcus aureus* infections in western Sweden. Scand J Infect Dis 39:6–13

Jechalke S, Heuer H, Siemens J, Amelung W, Smalla K (2014) Fate and effects of veterinary antibiotics in soil. Trends Microbiol 22:536–545

Jelić A, Gros M, Ginebreda A, Cespedes-Sánchez R, Ventura F, Petrović M, Barceló D (2011) Occurrence, partition, and removal of pharmaceuticals in sewage water and sludge during wastewater treatment. Water Res 45:1165–1176

Jones OAH, Lester JN, Voulvoulis N (2005) Pharmaceuticals: a threat to drinking water? Trends Biotechnol 23:163–167

Jones OAH, Voulvoulis N, Lester JN (2006) Partitioning behavior of five pharmaceutical compounds to activated sludge and river sediment. Arch Environ Contam Toxicol 50:297–305

Jones V, Gardner M, Ellor B (2014) Concentrations of trace substances in sewage sludge from 28 wastewater treatment works in the UK. Chemosphere 111:478–484

Jones-Lepp TL, Stevens R (2007) Pharmaceuticals and personal care products in biosolids/ sewage sludge: the interface between analytical chemistry and regulation. Anal Bioanal Chem 387:1173–1183

Kang DH, Gupta S, Rosen C, Fritz V, Singh A, Chander Y, Murray H, Rohwer C (2013) Antibiotic uptake by vegetable crops from manure-applied soils. J Agric Food Chem 61:9992–10001

Kay P, Blackwell PA, Boxall ABA (2004) Fate of veterinary antibiotics in a macroporous tile drained clay soil. Environ Toxicol Chem 23:1136–1144

Kemper (2008) Veterinary antibiotics in the aquatic and terrestrial Environment. Ecol Indic 8:1–13

Khachatourians G (1998) Agricultural use of antibiotics and the evolution and transfer of antibiotic-resistant bacteria. Can Med Assoc J 159:1129–1136

Kidd KA, Blanchfield PJ, Mills KH, Palace VP, Evans RE, Lazorchak JM, Flick RW (2007) Collapse of a fish population after exposure to a synthetic oestrogen. Natl Acad Sci 104:8897–8901

Kim Y, Choi K, Jung J, Park S, Kim PG, Park J (2006) Aquatic toxicity of acetaminophen, carbamazepine, cimetidine, diltiazem, and six major sulfonamides and their potential ecological risks in Korea. Environ Int 33:370–375

Kim SD, Cho J, Kim IS, Vanderford BJ, Snyder SA (2007) Occurrence and removal of pharmaceuticals and endocrine disruptors in South Korean surface,drinking, and waste waters. Water Res 41:1013–1021

Kim KR, Owens G, Kwon SI, So KH, Lee DB, Ok YS (2011) Occurrence and environmental fate of veterinary antibiotics in the terrestrial environment. Water Air Soil Pollut 214:163–174

Kinney CA, Furlong ET, Zaugg SD, Burkhardt MR, Werner SL, Cahill JD, Jorgensen GR (2006) Survey of organic wastewater contaminants in biosolids destined for land application. Environ Sci Technol 40:7207–7215

Köck R, Harlizius J, Bressan N, Laerberg R, Wieler LH, Witte W, Deurenberg RH, Voss A, Becker K, Friedrich AW (2009) Prevalence and molecular characteristics of methicillin-resistant Staphylococcus aureus (MRSA) among pigs on German farms and import of livestock-related MRSA into hospitals. Eur J Clin Microbiol Infect Dis 28:1375–1382

Kolpin DW, Furlong ET, Meyer MT, Thurman EM, Zaugg SD, Barber LR, Buxton HT (2002) Pharmaceuticals, hormones, and other organic wastewater contaminants in US streams, 1999–2000: a National Reconnaissance. Environ Sci Technol 36:1202–1211

Kolz AC, Moorman TB, Ong SK, Scoggin KD, Douglass EA (2005) Degradation and metabolite production of tylosin in anaerobic and aerobic swine-manure lagoons. Water Environ Res 77:49–56

Kools SAE, Moltmann JF, Knacker T (2008) Estimating the use of veterinary medicines in the European Union. Regul Toxicol Pharmacol 50:59–65

Kreuzinger N, Clara M, Strenn B, Kroiss H (2004) Relevance of the sludge retention time (SRT) as design criteria for wastewater treatment plants for the removal of endocrine disruptors and pharmaceuticals from wastewater. Water Sci Technol 50(5):149–156

Kümmerer K (2009a) The presence of pharmaceuticals in the environment due to human use – present knowledge and future challenges. J Environ Manag 90:2354–2366

Kümmerer K (2009b) Antibiotics in the aquatic environment – a review – part II. Chemosphere 75:435–441

Kümmerer K, Al-Ahmad A (2010) Estimation of the cancer risk to humans resulting from the presence of cyclophosphamide and ifosfamide in surface water. Environ Sci Pollut Res 17:486–496

Kümmerer K, Steger-Hartmann T, Meyer M (1997) Biodegradability of the anti-tumour agent ifosfamide and its occurrence in hospital effluents and communal sewage. Water Res 31 (11):2705–2710

Lange A, Paull GC, Coe TS, Katsu Y, Urushitani H, Iguchi T, Tyler CR (2009) Sexual reprogramming and estrogenic sensitisation in wild fish exposed to ethinylestradiol. Environ Sci Technol 43:1219–1225

Längin A, Alexy R, König A, Kümmerer K (2009) Deactivation and transformation products in biodegradability testing of b-lactams piperacillin and amoxicillin. Chemosphere 75 (3):347–354

Laupland KB, Ross T, Gregson DB (2008) *Staphylococcus aureus* bloodstream infections: risk factors, outcomes, and the influence of methicillin resistance in Calgary, Canada, 2000–2006. J Infect Dis 198:336–343

Laville N, Aït-Aïssa S, Gomez E, Casellas C, Porcher JM (2004) Effects of human pharmaceuticals on cytotoxicity, EROD activity and ROS production in fish hepatocytes. Toxicology 196:41–55

Li X, Wu X, Mo C, Tai Y, Huang X, Xiang L (2011) Investigation of sulfonamide, tetracycline, and quinolone antibiotics in vegetable farmland soil in the Pearls River Delta area, Southern China. J Agric Food Chem 59:7268–7276

Li J, Dodgen L, Ye Q, Gan J (2013) Degradation kinetics and metabolites of carbamazepine in soil. Environ Sci Technol 47:3678–3684

Li J, Ye Q, Gan J (2014) Degradation and transformation products of acetaminophen in soil. Water Res 49:44–52

Lister A, Van der Kraak G (2008) An investigation into the role of prostaglandins in zebrafish oocyte maturation and ovulation. Gen Comp Endocrinol 159:46–57

Loos R, Locoro G, Comero S, Contini S, Schwesig D, Werres F, Balsaa P, Gans O, Weiss S, Blaha L, Bolchi M, Gawlik BM (2010) Pan-European survey on the occurrence of selected polar organic persistent pollutants in ground water. Water Res 44:4115–4126

López-Serna R, Jurado A, Vázquez-Suñé E, Carrera J, Petrovic M, Barcelo D (2013) Occurrence of 95 pharmaceuticals and transformation products in urban groundwaters underlying the metropolis of Barcelona, Spain. Environ Pollut 174:305–315

Lowy FD (1998) *Staphylococcus aureus* infections. N Engl J Med 339:520–532

Lubick N (2010) Drugs in the environment: do pharmaceutical take-back programs make a difference? Environ Health Perspect 118(5):A211–A214

Lundholm CE (1997) DDE-induced eggshell thinning in birds: effects of *p,p'*-DDE on the calcium and prostaglandin metabolism of the eggshell gland. Comp Biochem Physiol C118(2):113–128

Malchi T, Maor Y, Tadmor G, Shenker M, Chefetz B (2014) Irrigation of root vegetables with treated wastewater: evaluating uptake of pharmaceuticals and the associated human health risks. Environ Sci Technol 48:9325–9333

Martín J, Santos JL, Aparicio I, Alonso E (2015) Pharmaceutically active compounds in sludge stabilization treatments: anaerobic and aerobic digestion, wastewater stabilization ponds, and composting. Sci Total Environ 503-504:97–104

Mattson B (2007) A *voluntary* environmental classification system for pharmaceutical substances. Drug Inf J 41(2):187–191

McArdell CS, Molnar E, Suter MJF, Giger W (2003) Occurrence and fate of macrolide antibiotics in wastewater treatment plants in the Glatt Valley watershed, Switzerland. Environ Sci Technol 37:5479–5486

McClellan K, Halden RU (2010) Pharmaceuticals and personal care products in archived US biosolids from the 2001 EPA national sewage sludge survey. Water Res 44:658–668

McCormick SD (2001) Endocrine control of osmoregulation in teleost fish. Am Zool 41:781–794

McDowell DC, Huber MM, Wagner M, von Gunten U, Ternes T (2005) Ozonation of carbamazepine in drinking water: identification and kinetic study of major oxidation products. Environ Sci Technol 39:8014–8022

McMurry LM, Oethinger M, Levy SB (1998) Overexpression of marA, soxS or acrAB produces resistance to triclosan in laboratory and clinical strains of Escherichia coli. FEMS Microbiol Lett 166:305–309

Metcalfe CD, Koenig BG, Bennie DT, Servos M, Ternes TA, Hirsch R (2003) Occurrence of neutral and acidic drugs in the effluents of Canadian sewage treatment plants. Environ Toxicol Chem 22(12):2872–2880

Miller KJ, Meek J, Valley H (2006) Ground water: pharmaceuticals, personal care products, endocrine disruptors (PPCPs) and microbial indicators of fecal contamination. Montana Department of Environmental Quality, Helena

Mimeault C, Woodhouse AJ, Miao X-S, Metcalfe CD, Moon TW, Trudeau VL (2005) The human lipid regulator, gemfibrozil, bioconcentrates and reduces testosterone in the goldfish *Carassius auratus*. Aquat Toxicol 73:44–54

Mitema ES, Kikuvi GM, Wegener HC, Stohr K (2001) An assessment of antimicrobial consumption in food producing animals in Kenya. J Vet Pharmacol Therap 24:385–390

Moldovan Z (2006) Occurrences of pharmaceutical and personal care products as micropollutants in rivers from Romania. Chemosphere 64(11):1808–1817

Mompelat S, Le Bot B, Thomas O (2009) Occurrence and fate of pharmaceutical products and by-products, from resource to drinking water. Environ Int 35:803–814

Morin CA, Hadler JL (2001) Population-based incidence and characteristics of community-onset *Staphylococcus aureus* infections with bacteremia in 4 metropolitan Connecticut areas, 1998. J Infect Dis 184:1029–1034

Mückter H (2006) Human and animal toxicology of some water-borne Pharmaceuticals. In: Ternes TA, Joss A (eds) Human pharmaceuticals, Hormones and Fragrances: the challenge of micropollutants in urban water management. IWA Publishing, London

Musolff A, Leschik S, Möder M, Strauch G, Reinstorf F, Schirmer M (2009) Temporal and spatial patterns of micropollutants in urban receiving waters. Environ Pollut 157:3069–3077

Muto CA (2006) Methicillin-resistant *Staphylococcus aureus* control: we didn't start the fire, but it's time to put it out. Infect Control Hosp Epidemiol 27:111–115

Nash JP, Kime DE, Van der Ven LTM, Wester PW, Brion WF, Maack G, Stahlschmidt-Allner P, Tyler CR (2004) Long-term exposure to environmental concentrations of the pharmaceutical ethynylestradiol causes reproductive failure in fish. Environ Health Perspect 112:1725–1733

Nieto A, Borrull F, Pocurull E, Marcé RM (2010) Pressurized liquid extraction of pharmaceuticals from sewage-sludge. J Sep Sci 30:979–984

Nikolaou A, Meric S, Fatta D (2007) Occurrence patterns of pharmaceuticals in water and wastewater environments. Anal Bioanal Chem 387:1225–1234

O'Connor S, Aga DS (2007) Analysis of tetracycline antibiotics in soil: advances in extraction, clean-up, and quantifycation. Trends Anal Chem 26:456–465

O'Dell JD, Essington ME, Howard DD (1995) Surface application of liquid swine manure: chemical variability. Commun Soil Sci Plant Anal 26:3113–3120

Ongerth JE, Khan S (2004) Drug residuals: how xenobiotics can affect water supply sources. J AWWA 96:94–101

Osenbrück K, Gläser HR, Knöller K, Weise SM, Möder M, Wennrich R, Schirmer M, Reinstorf F, Busch W, Strauch G (2007) Sources and transport of selected organic micropollutants in urban groundwater underlying the city of Halle (Saale), Germany. Water Res 41:3259–3270

Ostermann A, Gao J, Welp G, Siemens J, Roelcke M, Heimann L, Nieder R, Xue Q, Lin X, Sandhage-Hofmann A, Amelung W (2014) Identification of soil contamination hotspots with veterinary antibiotics using heavy metal concentrations and leaching data – a field study in China. Environ Monit Assess 186:7693–7707

Owen SF, Giltrow E, Huggett DB, Hutchinson TH, Saye J, Winter MJ, Sumpter JP (2007) Comparative physiology, pharmacology and toxicology of β-blockers: mammals versus fish. Aquat Toxicol 82:145–162

Pan M, Wong CK, Chu LM (2014) Distribution of antibiotics in wastewater-irrigated soils and their accumulation in vegetable crops in the Pearl River Delta, Southern China. J Agric Food Chem 62:11062–11069

Pascoe D, Karntanut W, Müller CT (2003) Do pharmaceuticals affect freshwater invertebrates? A study with the cnidarian *Hydra vulgaris*. Chemosphere 51(6):521–528

Pawelzick HT, Höper H, Nau H, Hamscher G (2004) A survey of the occurrence of various tetracyclines and sulfamethazine in sandy soils in northwestern Germany fertilized with liquid manure. SETAC Euro 14th Annual Meeting, April 2004, Prague, Czech Republic, pp 18–22

Payan P, Girard J-P (1977) Adrenergic receptors regulating patterns of blood flow through the gills of trout. Am J Physiol Heart Circ Physiol 232:18–23

Pedibhotla VK, Sarath G, Sauer JR, Stanleysamuelson DW (1995) Prostaglandin biosynthesis and subcellular-localization of prostaglandin-H synthase activity in the lone star tick, *Amblyomma americanum*. Insect Biochem 25(9):1027–1039

Peitsaro N, Anichtchik OV, Panula P (2000) Identification of a histamine H3-like receptor in the zebrafish (*Danio rerio*) brain. J Neurochem 75(2):718–724

Perez S, Barcelo D (2007) Fate and occurrence of X-ray contrast media in the environment. Anal Bioanal Chem 387:1235–1246

Perretin V, Schwarz F, Cresta L, Boeglin M, Dasen G, Teubner M (1997) Antibiotic resistance spread in food. Nature 389:801–802

Pinkston KE, Sedlak DL (2004) Transformation of aromatic ether- and aminecontaining pharmaceuticals during chlorine disinfection. Environ Sci Technol 38:4019–4025

Prosser RS, Trapp S, Sibley PK (2014) Modeling uptake of selected pharmaceuticals and personal care products into food crops from biosolids-amended soil. Environ Sci Technol 48:11397–11404

Rabiet M, Togola A, Brissaud F, Seidel JL, Budzinski H, Elbaz-Poulichet F (2006) Consequences of treated water recycling as regards pharmaceuticals and drugs in surface and ground waters of a medium-sized mediterranean catchment. Environ Sci Technol 40:5282–5288

Reddersen K, Heberer T, Dünnbier U (2002) Identification and significance of phenazone drugs and their metabolites in ground- and drinking water. Chemosphere 49:539–544

Renner R (2002) Do cattle growth hormones pose an environmental risk? Environ Sci Technol 36:194A–197A

Rivera-Utrilla J, Prados-Joya G, Sánchez-Polo M, Ferro-García MA, Bautista-Toledo I (2009) Removal of nitroimidazole antibiotics from aqueous solution by adsorption/bioadsorption on activated carbon. J Hazard Mater 170:298–305

Rivera-Utrilla J, Sánchez-Polo M, Ángeles Ferro-García M, Prados-Joya G, Ocampo-Pérez R (2013) Pharmaceuticals as emerging contaminants and their removal from water. A review Chemosphere 93:1268–1287

Rizzo L, Fiorentino A, Grassi M, Attanasio D, Guida M (2015) Advanced treatment of urban wastewater by sand filtration and graphene adsorption for wastewater reuse: Effect on a mixture of pharmaceuticals and toxicity. J Environ Chem Eng 3:122–128

Roberts PH, Thomas KV (2006) The occurrence of selected pharmaceuticals in wastewater effluent and surface waters of the lower Tyne catchment. Sci Total Environ 356(1–3):143–153

Rodríguez-Rodríguez CE, Jelić A, Llorca M, Farré M, Caminal G, Petrović M, Barceló D, Vicent T (2011) Solid-phase treatment with the fungus *Trametes versicolor* substantially reduces pharmaceutical concentrations and toxicity from sewage sludge. Bioresour Technol 102:5602–5608

Rodríguez-Rodríguez CE, Jelić A, Pereira MA, Sousa DZ, Petrović M, Alves MM, Barceló D, Caminal G, Vicent T (2012) Bioaugmentation of sewage sludge with *Trametes versicolor* in solid-phase biopiles produces degradation of pharmaceuticals and affects microbial communities. Environ Sci Technol 46:12012–12020

Rotz CA (2004) Management to reduce nitrogen losses in animal production. J Anim Sci 82:E119–E137

RUM (Returning Unwanted Medicine) (2011) The national return and disposal of unwanted medicine limited, Returning Unwanted Medicine Project. http://www.returnmed.com.au/collections. Accessed 1 Aug 2017

Russel AD (2000) Do biocides select for antibiotic resistance? J Pharm Pharmacol 52:227–233

Sabourin L, Duenk P, Bonte-Gelok S, Payne M, Lapen DR, Topp E (2012) Uptake of pharmaceuticals, hormones and parabens into vegetables grown in soil fertilized with municipal biosolids. Sci Total Environ 431:233–236

Salyers AA (2002) An overview of the genetic basis of antibiotic resistance in bacteria and its implications for agriculture. Anim Biotechnol 13:1–5

Sanderson H, Brain RA, Johnson DJ, Wilson CJ, Solomon KR (2004) Toxicity classification and evaluation of four pharmaceuticals classes: antibiotics, antineoplastics, cardiovascular, and sex hormones. Toxicology 203:27–40

Santos LHMLM, Araujoa AN, Fachinia A, Penab A, Delerue-Matosc D, Montenegro MCBSM (2010) Ecotoxicological aspects related to the presence of pharmaceuticals in the aquatic environment. J Hazard Mater 175:45–95

Sarmah AK, Meyer MT, Boxall ABA (2006) A global perspective on the use, sales, exposure pathways, occurrence, fate and effects of veterinary antibiotics (VAs). Chemosphere 65:725–759

Schalhorn A (1995) Medikamentöse Therapie maligner Erkrankungen. Gustav Fischer Verlag, Stuttgart

Song WC, Brash AR (1991) Purification of an allene oxide synthase and identification of the enzyme as a cytochrome-P-450. Science 253:781–784

Steger-Hartmann T, Kümmerer K, Hartmann A (1997) Biological degradation of cyclophosphamide and its occurrence in sewage water. Ecotoxicol Environ Saf 36:174–179

Steger-Hartmann T, Lange R, Schweinfurth H (1999) Environmental risk assessment for the widely used iodinated X-ray contrast agent iopromide (Ultravist). Ecotoxicol Environ Saf 42:274–281

Steger-Hartmann T, Lange R, Schweinfurth H, Tschampel M, Rehmann I (2002) Investigations into the environmental fate and effects of iopromide (ultravist), a widely used iodinated X-ray contrast medium. Water Res 36:266–274

Sui Q, Cao X, Lu S, Zhao W, Qiu Z, Yu G (2015) Occurrence, sources and fate of pharmaceuticals and personal care products in the groundwater: a review. Emerg Contam 1:14–24

Swedish Association of the Pharmaceutical Industry (2007) Environmental classification of pharmaceuticals in fass.se – guidance to pharmaceutical companies. http://www.fass.se/LIF/miljo/miljoinfo.jsp. Accessed 1 Aug 2017

Tanoue R, Sato Y, Motoyama M, Nakagawa S, Shinohara R, Nomiyama K (2012) Plant uptake of pharmaceutical chemicals detected in recycled organic manure and reclaimed wastewater. J Agric Food Chem 60:10203–10211

Ternes T (1998) Occurrence of drugs in German sewage treatment plants and rivers. Water Res 32:3245–3260

Ternes T, Hirsch R (2000) Occurrence and behavior of X-ray contrast media in sewage facilities and the aquatic environment. Environ Sci Technol 34:2741–2748

Ternes TA, Stumpf M, Mueller J, Haberer K, Wilken RD, Servos M (1999) Behavior and occurrence of estrogens in municipal sewage treatment plants: I. Investigations in Germany. Canada and Brazil Sci Total Environ 225(1–2):81–90

Ternes T, Meisenheimer M, McDowell D, Sacher F, Brauch HJ, Haist-Glude B, Preuss G, Wilme U, Zulei-Seibert N (2002) Removal of pharmaceuticals during drinking water treatment. Environ Sci Technol 36:3855–3863

Ternes TA, Herrmann N, Bonerz M, Knacker T, Siegrist H, Joss A (2004a) A rapid method to measure the solid-water distribution coefficient (Kd) for pharmaceuticals and musk fragrances in sewage sludge. Water Res 38:4075–4084

Ternes TA, Joss A, Siegrist H (2004b) Scrutinizing pharmaceuticals and personal care products in wastewater treatment. Environ Sci Technol 38(20):392–399

Ternes TA, Janex-Habibi M-L, Knacker T, Kreuzinger N, Siegrist H (2005) Assessment of technologies for the removal of pharmaceuticals and personal care products in sewage and drinking water to improve the indirect potable water reuse. POSEIDON project detailed report. EU Contract No.EVK1-CT-2000-00047

Thaker (2005) Pharmaceutical data elude researchers. Environ Sci Technol 139:193A–194A

Thiele-Bruhn S (2003) Pharmaceutical antibiotic compounds in soils – a review. J Plant Nutr Soil Sci 166:145–167

Thomas PM, Foster GD (2004) Determination of nonsteroidal anti-inflammatory drugs, caffeine, and triclosan inwastewater by gas chromatography–mass spectrometry. J Environ Sci Health A 39(8):1969–1978

Thorpe KL, Cummings RI, Hutchinson TH, Scholze M, Brighty G, Sumpter JP, Tyler CR (2003) Relative potencies and combination effects of steroidal estrogens in fish. Environ Sci Technol 37:1142–1149

Tiemersma EW, Bronzwaer SL, Lyytikäinen O, Degener JE, Schrijnemakers P, Bruinsma N, Monen J, Witte W, Grundmann H (2004) Methicillin-resistant Staphylococcus aureus in Europe, 1999–2002. Emerg Infect Dis 10:1627–1634

Topp E, Monteiro SC, Beck A, Coelho BB, Boxall ABA, Duenk PW, Kleywegt S, Lapen DR, Payne M, Sabourin L, Li H, Metcalfe CD (2008) Runoff of pharmaceuticals and personal care products following application of biosolids to an agricultural field. Sci Total Environ 396:52–59

Touraud E, Roig B, Sumpter JP, Coetsier C (2011) Drug residues and endocrine disruptors in drinking water: risk for humans? Int J Hyg Environ Health 214(6):437–441

Triebskorn R, Casper H, Heyd A, Eikemper R, Köhler H-R, Schwaiger J (2004) Toxic effects of the non-steroidal anti-inflammatory drug diclofenac. Part II. Cytological effects in liver kidney gills and intestine of rainbow trout (*Oncorhynchus mykiss*). Aquat Toxicol 68:151–166

Tyler CR, Filby AL, Bickley LK, Cumming RI, Gibson R, Labadie P, Katsu Y, Liney KE, Shears JA, Silva-Castro V, Urushitani H, Lange A, Winter MJ, Iguchi T, Hill EM (2009) Environmental health impacts of equine estrogens derived from hormone replacement therapy. Environ Sci Technol 43(10):3897–3904

UNEP DTIE (United Nations Environment Programme Division of Technology, Industry and Economics) (2002) Biosolids Management: an environmentally sound approach for managing sewage treatment plant sludge, an introductory guide to decision-makers, 1st edn. UNEP DTIE International Environmental Technology Centre, Osaka. http://www.unep.or.jp/ietc/publications/freshwater/fms1. Accessed 1 Aug 2017

United Nations (2004) World contraceptive use 2003. United Nations publication, Sales No. E. 04. XIII.2

US EPA (US Environmental Protection Agency) (1996) Soil screening guidance: technical background document. Part 5: Chemical-specific parameters. Available: http://www.epa.gov/superfund/health/conmedia/soil/pdfs/part_5.pdf. Accessed 1 Aug 2017

US EPA (US Environmental Protection Agency) (2010) Guidance document: best management practices for unused pharmaceuticals at health care facilities. Washington, DC, United States Environmental Protection Agency (EPA- 821-R-10-006). Available:http://water.epa.gov/scitech/wastetech/guide/upload/unuseddraft.pdf. Accessed 1 Aug 2017

Vane JR, Botting RM (1998) Mechanism of action of anti-inflammatory drugs. Int J Tissue React 20(1):3–15

Vieno N, Tuhkanen T, Kronberg L (2006) Removal of pharmaceuticals in drinking water treatment: effect of chemical coagulation. Environ Technol 27:183–192

Vulliet E, Cren-Olivé C, Grenier-Loustalot MF (2009) Occurrence of pharmaceuticals and hormones in drinking water treated from surface waters. Environ Chem Lett 9:103–114

Walters E, McClellan K, Halden RU (2010) Occurrence and loss over three years of 72 pharmaceuticals and personal care products from biosolids-soil mixtures in outdoor mesocosms. Water Res 44:6011–6020

Wiegel S, Aulinger A, Brockmeyer R, Harms H, Loffler J, Reincke H, Schmidt R, Stachel B, Von Tumpling W, Wanke A (2004) Pharmaceuticals in the river Elbe and its tributaries. Chemosphere 57:107–126

Winckler C, Grafe A (2000) Abschätzung des Stoffeintrags in Böden durch Tierarzneimittel und pharmakologisch wirksame Futterzusatzstoffe. UBA-Texte 44/00, Berlin

Winter MJ, Caunter JE, Glennon Y, Hutchinson TH (2006) Atenolol: 28 day assessment of survival and growth in fathead minnow (Pimephales promlas) embryo-larvae. AstraZeneca Brixham Environmental Laboratory, Report Number BL8269A

Wirtz F (2009) Danube, Meuse and Rhine MEMORANDUM 2008. Environ Sci Pollut Res 16: S112–S115

Wishkovsky A, Roberson BS, Hetrick FM (1987) In vitro suppression of the phagocytic response of fish macrophages by tetracyclines. J Fish Biol 31:61–65

Witte W, Klare I, Werner G (1999) Selective pressure by antibiotics as feed additives. Infection 27:35–38

Wolf L, Eiswirth M, Hötzl (2003) Assessing sewer-groundwater interaction at the city scale based on individual sewer defects. RMZ-Mater Geoenviron 50:423–426

Wu X, Conkle JL, Gan J (2012) Multi-residue determination of pharmaceutical and personal care products in vegetables. J Chromatogr A 1254:78–86

Wu X, Dodgen LK, Conkle JL, Gan J (2015) Plant uptake of pharmaceutical and personal care products from recycled water and biosolids: a review. Sci Total Environ 536:655–666

Xu J, Wu L, Chang AC (2009) Degradation and adsorption of selected pharmaceutical and personal care products (PPCPs) in agricultural soils. Chemosphere 77:1299–1305

Yu Y, Wu L (2012) Analysis of endocrine disrupting compounds, pharmaceuticals, and personal care products in sewage sludge by gas chromatography-mass spectrometry. Talanta 89:258–263

Zhou LJ, Ying GG, Liu S, Zhao JL, Chen F, Zhang RQ, Peng FQ, Zhang QQ (2012) Simultaneous determination of human and veterinary antibiotics in various environmental matrices by rapid resolution liquid chromatography-electrospray ionization tandem mass spectrometry. J Chromatogr A 1244:123–138

Chapter 13
Soil as a Transmitter of Human Pathogens

Abstract Soils are the habitat of about 25% of the Earth's species. The majority of these organisms are not of any threat to human health, but rather function to provide numerous ecosystem services which emerge through the multitude of complex interactions between the organisms and the soil itself. These ecosystem services range from those which are vital for maintaining life on Earth, such as the formation of soil, the cycling of carbon and nutrients with the result of maintaining the global cycles of C and N, soil fertility, the filtering of water, as well as provision of useful compounds such as antibiotics, the majority of which have been isolated from soil organisms (Chap. 12). However, soils also contain microorganisms which are capable of causing diseases in humans. A wide variety of soil-related infections need to be considered, particularly in the case of wound, respiratory tract, or gastrointestinal infections. Soil-borne microbes that are pathogenic for humans include protozoa, fungi, bacteria, and also viruses which require a host for their survival. Over 400 genera of bacteria have been identified with possibly as many as 10,000 species and, with the exception of viruses, they are in most cases more abundant than any other organism in soils. Helminths, belonging to the mesofauna size class, are also important as human pathogens. The number of bacteria that can be cultured in the laboratory is probably less than 1%. Their actual diversity is thus probably much greater. Of the approximately 100,000 species of fungi currently recognized, only about 300 may cause human disease.

Soil organism may potentially enter surface water and groundwater via the soil. Thus soil is often the origin of water-borne infections. However, as soils mostly provide an effective barrier against pathogens reaching the groundwater, and the die-off time of most pathogens in the sub-surface is short, the number of viable organisms reaching groundwater maybe normally low. Most cases of waterborne disease from groundwater consumption are caused by viruses and bacteria, as protozoa and helminths are too large to be transmitted far through the soil pore system. However, some occurrences of groundwater contamination by the protozoa *Giardia* and *Cryptosporidium* have been recorded. Most cases of waterborne disease from wells where there is a thick soil cover are due to the faulty construction of head works, or to the use of manure near sinkholes, abandoned wells or other features allowing surface water and contaminated material direct access to the groundwater. The main acute disease risk associated with drinking water in

R. Nieder et al., *Soil Components and Human Health*,
https://doi.org/10.1007/978-94-024-1222-2_13

developing and transition countries is due to well-known viruses, bacteria, and protozoa, which spread via the fecal-oral route.

Feces are commonly applied to fields in order to dispose of animal waste and to fertilize soils. Enteric pathogens can survive for prolonged periods of time in animal manures and may serve as potential inoculum onto plants in the field. Many studies have demonstrated the presence of foodborne pathogenic bacteria on crops grown in soil to which naturally or artificially contaminated manure was applied. Additionally, poor hygiene practices by field workers and a lack of on-site sanitation facilities may result in produce-associated outbreaks, particularly enteric illness such as shigellosis, which is easily contracted from human feces because of the low infectious dose of the causal agent, *Shigella*. Crop irrigation with contaminated water also is considered as primary sources of inoculum in the field. This is of particular concern for production of vegetables and fruits in areas where the supply of fresh water is scarce, and where water reclaimed from effluents increasingly serves for agricultural purposes.

Several soil-borne diseases are capable of transmission to the air (e.g. Q fever, aspergillosis, tularemia, sporotrichosis) and may be then transported by dust (Chap. 3). These diseases are therefore likely to be those most directly affected by land management practices and land use change. Any activity which is associated with increased wind erosion seems likely to increase the incidence of such diseases in the surrounding area. Related activities could include land use change, for example by converting grassland into arable land, ploughing or tilling soil that is too dry, etc. These conditions may also increasingly occur under climate change, as that can lead to enhanced incidence of drought periods.

Keywords Soil-borne human pathogens • Viruses • Bacteria • Fungi • Protozoa • Helminths • Life conditions in soils • Ecophysiology of pathogens • Transmission of infectious diseases • Human exposure • Symptoms and treatment • Strategies for control of zoonotic diseases • Public health measures to prevent infections

13.1 Global Impact of Diseases Caused by Soil Pathogens

Infectious diseases are a major cause of human suffering and mortality, and account for an estimated 13 million deaths worldwide each year (Bultman et al. 2005). This number is expected to increase. Soil-borne human pathogens are important contributors to those numbers. It is estimated that globally, more than 1.2 billion people are infected with at least one soil-transmitted helminth species (Bethony et al. 2006) and that more than half of the world's population is at risk of infection (Horton 2003). Ascariasis (roundworm) caused 60,000 deaths in 1993, schistosomiasis killed 200,000 people in 1993, and *Clostridium tetani* in the 1990s killed 450,000 newborns and about 50,000 mothers each year (WHO 1996). In 2006, 290,000 persons died of tetanus, of which 250,000 were neonatal deaths (Heymann 2008). Worldwide, enteric bacteria are estimated to cause between four and six million deaths each year, mostly in developing countries, and are the second most common cause of infant mortality globally (Jeffery and van der Putten 2011). *Salmonella* is a pathogen of major worldwide public health concern, accounting for 93.8 million foodborne illnesses and 155,000 deaths per year (Eng et al. 2015). Many deaths hereof are contracted from microbes introduced into the soil via fecal waste and then ingested. The easily preventable diarrheal diseases caused by lack of sanitation and hygiene contribute to about 6.0% of all health-related deaths (Schwarzenbach et al. 2010).

13.2 Life Conditions of Pathogens in Soils

Infectious organisms capable of causing disease come from five major phylogenetic groups including bacteria, fungi, viruses, protozoa and helminths (nematodes). The type of organisms in the soil are determined by the properties of soils, the local climate and vegetation. Soil microbial populations are generally more abundant in surface horizons than in deeper horizons but are not uniformly distributed laterally (Coyne 1999). For example, *Penicillium* is a fungus that is found in both warm and cold soils, while the fungus *Aspergillus* prefers warm soils (Bultman et al. 2005). Lower temperatures favor bacterial and viral survival. Maximum survival times of viruses, bacteria, protozoa and helminths in soil and on plant surfaces are given in Table 13.1.

In laboratory studies, as the temperature increased from 15 to 40 °C, the inactivation rate increased significantly for poliovirus type 1 (Straub et al. 1992). Human viruses almost certainly have no functional role within the soil system which means that they can all be considered to be soil transmitted pathogens, as opposed to euedaphic pathogens. Plant viruses rarely survive in soils for long

Table 13.1 Maximum survival times of pathogens in soils and on plant surfaces

	Soil		Plants	
Pathogen group	Absolute maximum	Common maximum	Absolute maximum	Common maximum
Viruses	6 months	3 months	2 months	1 month
Bacteria	1 year	2 months	6 months	1 month
Protozoa	10 days	2 days	5 days	2 days
Helminths	7 years	2 years	5 months	1 month

Adapted from Gerba and Smith (2005)

periods; however, some insect viruses remain infective for years (Coyne 1999). Viruses are also known to infect many soil helminths and microbes. Some viruses are capable of surviving within the soil system for extended periods of time and to be able to adsorb to soil particles and so resist elution (e.g. poliovirus). Others adsorb less strongly to the soil and so can become mobilised after rainfall and either washed deeper into the soil profile or to different areas depending on the soil hydrology (Landry et al. 1979). Factors which most affect the survival of human pathogenic viruses in the soil are pH level, moisture content, temperature, exposure to sunlight and the presence of soil organic matter (WHO 1979). Viruses generally survive longer in cooler, wetter, pH neutral soils with low microbial activity. Humic and fulvic organic material may cause reversible loss of infectivity, but, some other organic materials may complex with the virus and protect it from inactivation by preventing adsorption to soil particles (Sobsey and Shields 1987). Specific species of viruses have different survivability. Also, viruses tend to survive longer when clustered together. Ultraviolet light from the sun inactivates viruses on the surface of the soil but viruses in deeper layers are not affected (Gerba and Bitton 1984).

The fungus *Fusarium* does not thrive in soils with the clay mineral smectite (Paul and Clark 1996). The soil-borne fungus *Coccidioides*, the etiological agent of coccidioidomycosis, is mostly found in dry, alkaline soils with a soil texture that includes large percentages of silt and very fine sand. Clays favor the adsorption of microorganisms to soil particles and this further reduces their die-off rates. Many microbes have an affinity for clay minerals within soils. Clays have large surface areas, are chemically reactive, and have a net negative charge. They are a source of inorganic nutrients, such as potassium and ammonium (Nieder et al. 2011). Also, clays strongly adsorb water making it less available for microbes. Clays also protect bacterial cells, and possibly viral particles, by creating a barrier against microbial predators and parasites (Roper and Marshall 1978). Thus, the survival rates of microbes are lower in sandy soils with a low water-holding capacity. The pH affects the adsorption characteristics of cells, so inactivation rates in acidic soils are lower. Increases in cation concentrations also result in increased adsorption rates, consequently affecting microbial survival (Santamaria and Toranzos 2003).

In a given soil, the distribution of microbes on soil particle or soil aggregate surfaces is uneven or irregular. Microbial populations occur in localized concentrations associated with various favorable and unfavorable microenvironments throughout the soil profile (Bultman et al. 2005). Microbes are thus generally concentrated in clusters where conditions are favorable for growth and there may be relatively large distances between the clusters. The determining factors for the locations of these "hot spots" include the geometry of pores, the soil water potential, the presence of gases in the air-filled soil pore spaces, the local composition of soil minerals and the distribution of soil organic matter. Particularly soluble organics increase survival and, in the case of bacteria, may favor their regrowth

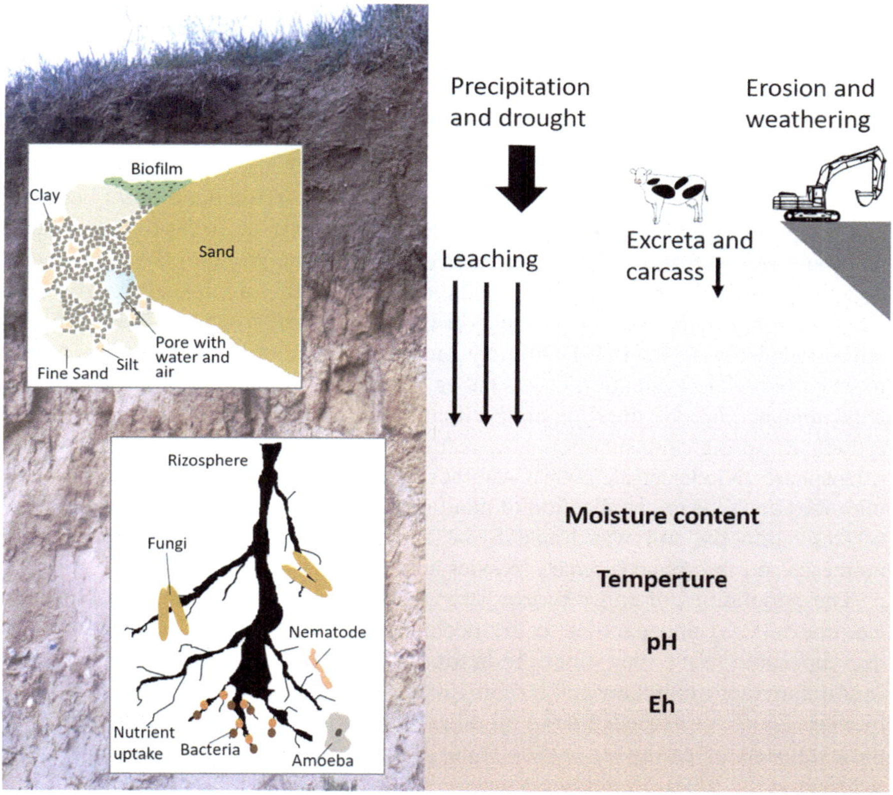

Fig. 13.1 Factors determining distribution of organisms in soil. Left panel: the upper viewframe represents soil particles, pore spaces, and bacteria in biofilm; the lower viewframe represents the rhizosphere and associated bacteria, fungi, protozoa and nematodes. The right panel represent climatic, physical, and geobiochemical factors (Adapted from Baumgardner 2012)

when degradable organic matter is present (Santamaria and Toranzos 2003). Competition, infection, predation, or all three by other bacteria, archaea, viruses, fungi, and larger organisms such as protozoa further select the predominant species, particularly in diverse, nutrient-rich zones such as the rhizosphere which is the soil area influenced by a plant roots (Fig. 13.1).

Biomass and activity of microbes strongly depend on the presence of plant roots (Nieder et al. 2008). The rhizosphere has chemical properties that differ from the bulk of the soil because of the uptake by the plant roots of moisture and nutrients and the exudations of soluble organic compounds by the roots. The stimulation of microbial activity in the rhizosphere results from the fact that plants secrete an array of low- and high-molecular weight organic molecules (e.g. amino and organic acids, sugars, proteins) into the soil as exudates, which may account for up to 40% of the dry matter produced by plants (Lynch and Whipps 1990). As free-living soil microorganisms are strongly carbon limited (Wardle 1992), a specialized microflora, typically consisting of fast growing bacteria, is triggered into activity by the carbon pulses provided as exudates (Semenov et al. 1999). Root-derived carbon leads to strongly increased levels of microbial biomass and activity around roots and channels energy to subsequent microfaunal grazers, where numbers of bacterial feeding protozoa and free-living nematodes may increase up to 30-fold compared with bulk soil (Griffiths 1990). Estimates of plant-below-ground investments vary widely, but even if the C transfer to exudation was 10–20% of total net fixed carbon, other microbial symbionts such as mycorrhizae (Smith and Read 1997) or N_2-fixing microorganisms (Ryle et al. 1979) may each consume another 10–20% of total net fixed carbon. The evaluation of a number of studies from agricultural ecosystems with annual crops of the temperate cimate zone indicated a seasonal microbial growth in spring and summer as a result of a stimulatory effect of an active rhizosphere (Nieder et al. 2008). Another factor that influences the dynamics of microbes in soil is the application of plant residues. For example, the incorporation of straw into the soil was found to be a management option that temporarily increases soil microbial biomass (Nieder and Benbi 2008).

The populations of soil protozoa strongly fluctuate through time (Janssen and Heijmans 1998) and parallel to the decline in protozoan numbers their rapidly decomposable tissue may enter the detrital food-web. The size of most protozoa in soil may range only between 10 and 100 μm in diameter. In most soils protozoan biomass equals or exceeds that of all other soil animal groups taken together, with the exclusion of earthworms (Sohlenius 1980; Schäfer and Schauermann 1990; Schröter et al. 2003). Roughly estimated, 70% and 15% of total respiration of soil animals might be attributed to protozoa and nematodes, respectively (Sohlenius 1980; Foissner 1987). High production rates of protozoa with 10–12 times their standing crop per year and minimum generation times of 2–4 h (Coleman 1994) suggest a strong grazing pressure on bacterial and algal biomass and subsequent significant effects on nutrient mineralization. Predation by protozoa has a significant effect in controlling bacterial populations in soil, and the degradation of bacteria undoubtedly contributes to the maintenance of soil fertility (Weekers et al. 1993). Likewise, protozoa play an integral part in the cycling of nutrients in aquatic food chains (Wright and Coffin 1984).

The distribution of helminths in soils fluctuates greatly with season, climate, and the amount of soil organic matter. Helminths typically prefer warm, moist soils with plentiful organic material. During favorable conditions, most helminths are found in the upper 10–15 cm of the soil profile. They may move vertically in the soil profile in response to seasonal weather changes (Bultman et al. 2005).

Life in soil strongly depends on the availability of water. Water that infiltrates the soil will surround soil particles, fill pore spaces, and may eventually move downward into the groundwater body. Microbial movement in soils is dependent on the water saturation state. Microorganisms move rapidly under saturated conditions, but only for a few centimeters, because microorganisms are in close contact with soil particles, promoting the adsorption of microorganisms onto the soil particles. When soil is water-saturated, all pores are filled with water, allowing microorganisms to pass through the soil. Thus, soil texture controls, in part, the movement of microorganisms, because fine-grained soils avoid movement while coarse-grained soils promote it (Abu-Ashour et al. 1994). The soil water potential is a quantitative measure of water availability. It is defined as the amount of work that must be done per unit quantity of water in order to transport reversibly and isothermally an infinitesimal quantity of water from a pool of pure water, at a specified elevation and at atmospheric pressure, to the soil water table (Bultman et al. 2005). More simply, the soil water potential is the amount of energy that is needed to extract water from soil. It may be expressed mathematically as (Eq. 13.1):

$$\psi_{soil} = \psi_g + \psi_m + \psi_s \tag{13.1}$$

The soil water potential ψ_{soil} is the sum of the gravitational potential ψ_g, the matric potential ψ_m and the solute (or osmotic) potential ψ_s. The water potential is expressed in mega pascals (MPa). One mega pascal equals 1×10^6 pascals. The matric potential is related to the attraction of water molecules to solid surfaces. In unsaturated soils it is always negative and becomes more negative as the surface area of the soil increases. Microbial growth rates in general are greatest for a matric potential (ψ_{soil}) near -0.01 MPa and decrease as soils become drier and have larger, negative water potentials. Microbial activity also decreases as soils become waterlogged or saturated, which results in ψ_{soil} values at or near zero.

Air-filled pores of the soil tend to be enriched in carbon dioxide (CO_2) and depleted in oxygen (O_2) due to biological activity (Bultman et al. 2005). Even in well-aerated soils, water may be blocking pore spaces limiting the diffusion of O_2 into and CO_2 out of the soil. The population of soil-dwelling microbes is also affected by human activities. Microbial populations are lower in tilled soils and compacted soils. Tilled soils are less moist than non-tilled and compacted soils have reduced pore space and aeration (Coyne 1999). Acid rain has changed the pH and mineralogy of soils and has also affected the microbial populations of these soils. Global warming is changing the characteristics of soils worldwide.

13.3 Classification of Pathenogenic Soil Organisms

Numerous infectious diseases may be considered to be soil-borne. However, several pathogens that occur in soils have also been classified as waterborne or foodborne. For example, pathogens categorized as waterborne are the result of animal or human fecal, urine, or other wastes introduced first into the soil environment and subsequently washed into surface water or transported to groundwater. Infection then follows by contact with, or consumption of, contaminated water. Similar circumstances occur with some food-borne diseases. The pathogen is again introduced into the soil via defecation or through contaminated waste material and then may be consumed on unwashed vegetables or raw fruit. A number of pathogens have complex life cycles that may involve hosts in which they live and reproduce. Many have biological (insects, animals) and physical vectors (wind, water) for transport, and different reservoirs with adverse environmental conditions during lifetime.

Bultman et al. (2005) introduced a sophisticated classification of soil-borne pathogens according to residence times and categorized pathogenic organisms as permanent, periodic, transient and incidental. Permanent pathogens can complete their entire life cycle within a soil environment. Examples are the bacteria *Clostridium botulinum*, *C. tetani* and *Listeria monocytogenes*. Also included are dimorphic organisms if one of their morphologic forms is capable of living and reproducing completely within the soil. Examples of permanent dimorphic soil pathogens are the fungi *Coccidioides* and *Histoplasma capsulatum*. Periodic pathogens require part of their life cycle to be completed within a soil environment on a regular, recurring basis. Examples are spores of *Bacillus anthracis* and the eggs laid in the soil by tick vectors that contain the bacterium *Rickettsia rickettsii*. Additional examples are eggs of the helminths *Ancylostoma duodenale* and *Necator americanus* (hookworms). Transient pathogens are organisms that may naturally occur in soil, but the soil environment is not necessary for the completion of the organism's life cycle. Examples are cysts of a protozoan parasite *G. lamblia* and viruses in the genus *Hantavirus* that are introduced into soil environments worldwide via urine and feces of rodent vectors. Also included are *Leptospira*, a bacterium shed in urine of animals on soil, skin, and water and spores of the bacterium *Coxiella burnetii*. Incidental pathogens are introduced into the soil via anthropogenic means such as in sewage sludge, waste water, septic systems, unsanitary living conditions, biologically toxic spills, dumping of biohazardous waste materials, and release of biological warfare agents. Examples of viruses are enterovirus Poliovirus, enterovirus Coxsackie A and B, and enterovirus Hepatitis A. Length of survival time and virulence can range from hours to years depending mainly on numerous physical and chemical factors of the soil and the effluent.

In order to clarify the origin of the etiological agent Santamaria and Toranzos (2003) have proposed the following classification scheme: (1) soil-associated diseases which are caused by opportunistic or emerging pathogens belonging to the normal soil microbiota (e.g. the very common fungus *Aspergillus fumigatus* that occurs in soils and can infect the lungs via inhalation of spores), (2) soil-related diseases, which result in intoxication from the ingestion of food contaminated with enterotoxins or neurotoxins (e.g. *Clostridium botulinum*, *C. perfrigens* and *Bacillus cereus*), (3) soil-based diseases caused by pathogens indigenous to soil (e.g. *C. tetani*, *B. anthracis*, *C. perfringens*) and (4) soil-borne diseases caused by enteric pathogens which get into soil by means of human or animal excreta. Enteric pathogens transmitted by the fecal–oral route are bacteria, viruses, protozoa and helminths.

Table 13.2 Examples of soil-borne pathogens and the corresponding infectious diseases and their categorization according to euedaphic and transmitted

Euedaphic pathogens	**Soil-transmitted pathogens**
Pathogen/infectious disease	*Pathogen*/infectious disease
Bacteria	**Viruses**
Actinomycetes/Actinomycetoma	*Hantavirus*/Hantavirus syndrome
Bacillus anthracis/Anthrax	*Enteroviruses*/Broad clinical spectrum
Bacillus cereus/Broad clinical spectrum	
Clostridium botulinium/Botulism	**Bacteria**
Campylobacter jejuni/Campylobacteriosis	*Coxiella burnetii*/Q fever
Leptospira interrogans/Leptospirosis	*Pseudomonas aeruginosa*/Broad clinical spectrum
Listeria monocytogenes/Listeriosis	*Escherichia coli*/Broad clinical spectrum
Clostridium tetani/Tetanus	*Salmonella enterica*/Salmonellosis
Francisella tularensis/Tularemia	**Helminths**
Clostridium perferingens/Gas gangrene	*Ascaris lumbricoides*/Ascariasis
Yersinia enterocolitica/Yersiniosis	*Ancylostoma duodenale*/Hookworm
Fungi	*Strongyloides stercoralis*/Strongyloidiasis
Aspergillus ssp./Aspergillosis	*Trichuris trichiura*/Trichuriasis
Blastomyces dermatitidis/Blastomycosis	*Echinococcus multicularis*/Echinococcosis
Coccidiodes immitis/Coccidioidomycosis	*Trichinella spiralis*/Trichinellosis
Histoplasma capsulatum/Histoplasmosis	**Protozoa**
Sporothrix schenckii/Sporotrichosis	*Entamoeba histolytica*/Amoebiasis
	Balantidium coli/Balantidiasis
	Cryptosporidium parvum/Cryptosporidiosis
	Cyclospora cayetanensis/Cyclosporiasis
	Giardia lambila/Giardiasis
	Isospora belli/Isosporiasis
	Toxoplasma gondii/Toxoplasmosis
	Shigella dyseneriae/Shigellosis

Adapted from Jeffery and van der Putten (2011)

Jeffery and van der Putten (2011) have introduced a simple classification scheme that categorizes soil-borne pathogens as (1) euedaphic pathogens and (2) soil-transmitted pathogens. Euedaphic pathogens are true soil organisms, while soil transmitted pathogens are not true soil organisms but may be able to survive in soil for extended periods of time (Table 13.2). The latter must infect a host in order to complete their life cycles.

It is highly probable that infection with euedaphic pathogens has occurred from the soil, but this is much less certain with transmitted pathogens. For example, salmonellosis can be a result of poor hygiene by an infected individual who is preparing food. Infectious organisms capable of causing disease come from five major phylogenetic groups including bacteria, fungi, viruses, protozoa and helminths (nematodes).

13.4 Gateways of Introducing Soil-borne Pathogens into Humans

To cause disease, soil-borne pathogens must come in physical contact with and establish itself in a human (Bultman et al. 2005). The most common gateway of introducing soil-borne pathogens into the human body is through ingestion which is commonly accidental. In some cultures, active soil uptake via geophagia is practiced which sometimes may lead to infectious diseases (see Chap. 2). Pathogen contamination in agricultural settings can originate from point or nonpoint sources. Point sources of fecal contamination include leaking septic and sewer systems, drains, manure piles, or wastewater lagoons. In contrast, nonpoint sources are more difficult to characterize because they involve spatial and/or temporal variability in pathogen sources, as well as spatial translocation in the environment external to a host. General releases of fecal contamination in agricultural settings appear to be largely from nonpoint sources (Bradford et al. 2013). In many countries, application of human or animal waste (manures, slurry) to agricultural fields is an efficient means of introducing many human pathogens into the soil. The concentration of viruses in feces can be very high, amounting to 10^6 virus particles enterovirus per gram of soil, and for hepatitis A and for rotavirus 10^9 virus particles per gram of soil (Sobsey and Shields 1987). In most industrialized countries, the application of human waste products in the form of treated sewage sludge is regulated and a high percentage of bacteria and viruses are removed by primary and secondary sewage treatment (Bertucci et al. 1987).

In countries where no such standards exist, pathogens from human waste have a much easier time entering soil and the environment. However, many bacteria are also known to survive the sewage treatment process. If pathogens are ingested that became soil-borne as human waste, the gateway of infection will be referred to as the fecal-soil-oral gateway. Other oral gateways for soil-borne pathogens require an intermediate host. For example, the trematode *Schistosoma mansoni* has a transient soil residency. In soil its eggs hatch into larvae, which can infect an intermediate host. This intermediate host releases infective larvae into water. Consumption of infected water results in the pathogen moving to the human host. The cestode *Taenia saginata* also has a temporary soil residency. Cattle ingest soil infected with this pathogen and develop cysts in their muscles. When humans consume incompletely cooked beef, they can become infected with the pathogen. Microbes rarely penetrate the intact skin. However, the bacterium *C. tetani* enters the body if contaminated soil makes contact with a break in the skin. The nematode *Strongyloides stercoralis* is capable of actively burrowing its way through healthy skin. A respiratory gateway for soil-borne pathogens occurs when soil-borne pathogens are inhaled as airborne dust (see Chap. 3). Every place on Earth receives dust from both local and distant sources and along with the minerals that make up the inorganic part of the dust ride fungi, bacteria, and viruses. Each year, several million Mg of airborne soil makes its way from Africa to the Americas, Europe, and the Middle East as dust. Asian dust crosses the Pacific Ocean and dust from the southwestern USA can make its way to Canada (Raloff 2001). African dust in the Caribbean has been shown to contain the fungus *Aspergillus sedowii*.

13.4.1 Land Application of Animal Manures

More than 150 microbial pathogens have been identified from all animal species that can be transmitted to humans by various routes (Gerba and Smith 2005). Pathogens can be transmitted from animals to humans when manure is used as a fertilizer for food crops eaten raw and by storm water runoff from manured soils to-surface waters or by its percolation to groundwater. Animal manures have been used as a source of nutrients since the beginning of agriculture. Approximately 10^{11} Mg of agricultural animal manure are currently produced worldwide each year (Yao et al. 2015). They are still widely used as fertilizers in developing countries, where chemical fertilizers are often not available or expensive. Animal manures are also still applied to soils in developed countries to reduce the amount of mineral fertilizers to be applied. They are also being re-introduced to agriculture as alternatives to chemical fertilizers through the increased popularity of organic farming. In intensive livestock production in particular, field application of manures is a disposal method for animal wastes, with partly deleterious consequences for the environment. Animal manures can contaminate groundwater with a variety of pathogens that can affect human health, either through the ingestion of unwashed crops, or through the consumption of polluted groundwater or surface water used as a source of drinking water. Animal manure can contain large numbers of bacteria, viruses, protozoa and helminths that can cause human disease, i.e. up to 10^6 pathogens per gram of manures (Gannon et al. 2004). However, not all of the species excreted by farm animals cause disease in humans. Pathogens from manure that may cause disease from consuming contaminated groundwater are given in Table 13.3.

Table 13.4 summarizes cases where serious disease outbreaks have occurred due to contact with pathogens from a manure source. One of the best documented outbreaks of disease caused by groundwater contamination by manure occurred in Walkerton, Canada, in May 2000 (Table 13.4). Six people died and more than 2300 people became seriously ill when pathogens from manure spread on a nearby farm was washed in surface runoff into a badly constructed and poorly monitored well used as a water supply for the town. The disease was caused by the virulent strain of *E. coli* O157:H7 and by *Campylobacter jejuni*.

In Quebec, Canada, human and pig enteroviruses were isolated from 70% of water samples collected from a river. The contamination source was attributed to a massive pig raising activity in the area (Payment 1989). To study the virus transport to groundwater and surface water resources, several experiments were carried out using vaccine poliovirus type 1 as the seed (Gerba 1987). Seeded viruses were detected up to 50 m from the source, indicating they could easily travel through silt loam; and they were also detected in a nearby lake at days 43 and 71 after seeding.

Table 13.3 Examples of human pathogens present in animal manures

Pathogen	Major source
Bacteria	
Brucella ssp.	Chicken excrements
Campylobacter ssp.	Cattle, pig and poultry excrements
Escherichia coli O157:H7	Cattle and sheep excrements, to a lesser extent goat, pig and chicken excrements
Leptospira species	Pig urine
Listeria moncytogenes	Excrements of different animals
Salmonella ssp.	Livestock (cattle, pig, sheep, poultry) and wild animal excrements
Yersinia enterocolitica	Pig excrements
Virus	
Hepatitis E virus	Pig, cattle and goat excrements
Protozoa	
Cryptosporidium parvum	Livestock excrements
Giardia lamblia	Livestock excrements

Compiled from Gerba and Smith (2005) and Appleyard and Schmoll (2006)

Table 13.4 Manure-related human outbreaks

Location	Year	Pathogen	Impact	Source
Cabool, Missouri, USA	1990	*E. coli* O157:H7	243 cases, 4 deaths	Water line breaks in farm community
Sakai City, Japan	1995	*E. coli* O157:H7	12,680 cases, 425 hospitalized, 3 deaths	Animal manure used in fields growing alfalfa sprouts
Maine, USA	1993	*E. coli* O157:H7	Several illnesses	Animal manure spread in apple orchard
Milwaukee, Wisconsin,	1993	*Cryptosporidium parvum*	400,000 cases, 87 deaths	Animal manure and/or human excrement
Bradford, UK	1994	*Cryptosporidium parvum*	125 cases	Storm runoff from farm fields
Swindon and Oxfordshire, UK	1989	*Cryptosporidium parvum*	516 excess cases	Runoff from farm fields
Carrollton, Georgia, USA	1989	*Cryptosporidium parvum*	13,000 cases	Manure runoff
New York, USA	1999	*E. coli* O157:H7 and *Campylobacter* spp.	2 deaths, 116 cases	Runoff at fairgrounds
Walkerton, Ontario, Canada	2000	*E. coli* O157:H7 and *Campylobacter* spp.	6 deaths, 2300 cases	Runoff from farm fields entering town's water supply

Adapted from Smith and Perdek (2003)

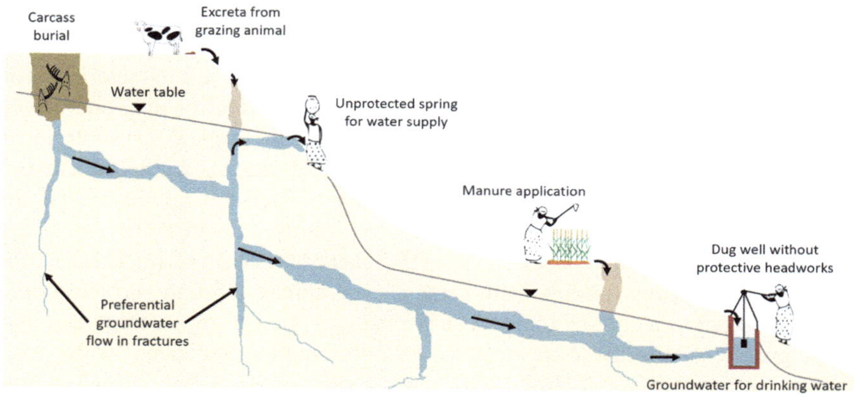

Fig. 13.2 Pathways for groundwater contamination by pathogens in agricultural areas (Adapted from Appleyard and Schmoll 2006)

In another field study in a farm that had received anaerobically digested sludge for 7 years, it was possible to detect viral nucleic acid sequences at points vertically and laterally displaced from sludge injections (Straub et al. 1995). However, it should be noted that the survival and transport of viruses in soil is highly dependent on the type of virus (Gerba 1987). Among the bacterial pathogens, numerous outbreaks were caused by *Escherichia coli* O157:H7, which is dangerous because of its low infective dose (as few as ten cells), high pathogenicity and ability to survive under frozen conditions (Tilden et al. 1996). Moreover, the virulence genes were found to be transferable to non-pathogenic *E. coli* strains (Herold et al. 2004).

Preferential water flow is a potentially important mechanism for pathogens to bypass portions of the soil matrix. It can occur as a result of funneling of water at textural interfaces, unstable flow behavior from spatially variable wettability or hysteresis, dynamic capillary properties, macropores, and fractured systems (Bradford et al. 2013). Contamination of groundwater by pathogens such as *E. coli* is often a significant issue in areas where there is a thin or no soil cover over fractured rock or karstic limestone (Fig. 13.2). In these terrains, pathogens can be rapidly carried through preferential flow paths into the groundwater with little or no attenuation.

Karstic aquifers may even allow larger organisms like protozoa to be transmitted through the aquifer to water supply wells. In a limestone terrain in Ireland studies of bacterial contamination in springs indicated contamination with coliform bacteria in spring water samples to be derived from dairy cattle (Thorn and Coxon 1992). Detected numbers of microorganisms varied greatly from 0 to 300 colony forming units (cfu: ability to multiply via binary fission under the controlled conditions) 100 mL^{-1} of water within a 2 h period. Contamination of groundwater by pathogens is also an issue in shallow fractured rock aquifers and in volcanic aquifers. Large cave systems may also be formed in some volcanic terrains as a result of the degassing of volatiles dissolved in molten lava flows, and the interface between individual lava flows may contain interconnected voids that may allow groundwater to transmit contaminants over long distances (Appleyard and Schmoll 2006). Therefore volcanic aquifers show some of the hydrogeological characteristics of karst aquifers, and share the same extremely high vulnerability to groundwater contamination. Karst-like features can also develop in tropical or subtropical regions with lateritic soils. Voids can form in lateritic duricrusts due to the erosion of soft, poorly consolidated clays in an otherwise cemented ferruginous matrix. In Brazil and in other tropical regions, voids in lateritic soils also appear to be caused by termite activity (Mendonça et al. 1994). The bacterial quality of water is usually at its worst after heavy rainfall. Pathogens can be spread through surface runoff from places where manure or biosolids have been applied or by leaching through the soil profile. Coliforms were detected in both shallow and deep wells, with bacterial contamination coinciding with the heaviest rainfalls (Gerba and Bitton 1984). Water from some springs was estimated to have travelled about 1 km from recharge areas within a 12–18 h period after heavy rain (Appleyard and Schmoll 2006).

Pathogens can also be rapidly mobilized and exported from soils by overland flow. Overland flow is initiated when the inflow (precipitation, snow melt, or runoff) rate exceeds the soil infiltration rate. Pathogen transport in agricultural settings is especially likely during extreme precipitation events that produce overland flow because raindrop impact can detach and consequently mobilize pathogens. The greater velocities in overland flow relative to subsurface flow can help translocate mobilized pathogens over long distances and can produce extremely high pathogen loads to surface water bodies (Dorner et al. 2006). Patterns of overland flow are complex and vary substantially over time (Thompson et al. 2010). Overland flow may occur as either sheet or rill flow. Sheet flow appears as a thin layer of water with deeper, faster flow near protruding objects. Rill flow is deeper and faster than sheet flow, and it occurs in small channels and gullies. Overland flow can range from laminar to fully turbulent, depending on the water depth, velocity, and raindrop impact (Bradford et al. 2013). Pathogen removal from overland flow depends strongly on the surface topography, vegetation, and soil texture and structure. Pathogen deposition will likely be enhanced at lower overland flow velocities, lower slopes and water depths, higher surface roughness, and higher vegetation density. Pathogens can readily associate with soil and sediment particles such as clays and organic/inorganic aggregates (Oliver et al. 2007). These pathogen-particle associations can thus be transported with overland flow. Pathogen removal from overland flow is expected to be enhanced when the microorganisms are associated with soil aggregates, but especially with denser inorganic particles because of increases in sedimentation (Muirhead et al. 2006). However, pathogens released from manure have been observed to often be in a free (unattached) state and higher vegetation density (Bradford et al. 2013). The presence of vegetation, including installation of vegetated buffer regions or filter strips for runoff management, has been observed to effectively trap and retain a variety of pathogens in overland flow, with reported pathogen removal quantities ranging from 25% to several logs (Bradford et al. 2013). Vegetation not only removes pathogens through direct attachment to plant surfaces, but also by increasing flow resistance and thus reducing overland flow velocities, ponding water, and enhancing infiltration (Fiener and Auerswald 2003).

13.4.2 Animal Feedlots

In many parts of the world, livestock densities on agricultural land have progressively increased due to an increasing demand for animal products. This has resulted in the development of animal feedlots, where animals are maintained in pens in a controlled environment. Feedlots are used for beef, pork and poultry (meat and eggs) production. They may either be completely enclosed within large buildings or exist as open-air facilities (Appleyard and Schmoll 2006). Dairies are similar to feedlots in that a large number of cows are gathered together for milking, although they may be allowed to run free range in between milking events. The large number of stock housed in animal feedlots generates great amount of wastes that can become a substantial point source of widespread groundwater pollution if not managed properly. For example, beef feedlots may contain 500 or more steers, and a typical 450 kg steer will produce up to 30 kg of solid and liquid wastes each day (Appleyard and Schmoll 2006).

The major sources of pathogens released from feedlots are manure, process wastewater (e.g. dairy wastes), feed, bedding materials and animal carcasses. Typically, in uncovered open-air facilities there is a much higher amount of waste generated than in enclosed feedlots with roofs. This is particularly the case if treatment and/or storage ponds do not have sufficient storage capacity to store water from intense rainfall events. Treated pond effluent is commonly used to irrigate pasture, but treatment ponds have to be designed to store water for the wettest period of the year when there is little opportunity to dispose of effluent through land irrigation (Appleyard and Schmoll 2006). Manure collected in feedlots will commonly be applied as fertilizer on cultivated fields and may become a diffuse source of groundwater pollution, particularly in areas where feedlots significantly increase stock densities in relation to application options for manure within economically viable distances (Appleyard and Schmoll 2006).

Generally, the risks of disposal of animal carcasses by burial on farms or in landfill sites are similar than those posed by the excessive use of animal manures, including contaminating groundwater with pathogens such as *E. coli*, *Campylobacter*, *Salmonella*, *Cryptosporidium* and *Giardia*. The risk of groundwater contamination are greatest when very large numbers of animals may be buried to control the spread of animal diseases in agricultural areas (Appleyard and Schmoll 2006). Of particular concern to public health are epidemics of animal disease that may also cause disease in humans, especially where the disease-causing agent is persistent in soil and groundwater. The mass burial of animal carcasses infected by such a disease-causing agent may pose a risk to nearby groundwater supplies. However, the risks may be reduced by burning of carcases prior to burial. One group of disease-causing agents of concern in this regard are prions, particularly the prion that causes the disease BSE (Mad Cow disease) in cattle (for more details see Chap. 14).

13.4.3 Land Application of Wastewater and Sewage Sludge

As with animal manures, there is also significant concern about a possible increase in soil-borne diseases in human populations given by the land disposal of wastewater and sewage sludges that result from wastewater treatment. These practices may favor the entry of considerable concentrations of enteric pathogens into soil, because large amounts of these materials are applied to lands or disposed of in landfills. Wastewater irrigation and spreading of sludges in agriculture, like use of manures, can cause serious pathogen contamination of groundwater, particularly in areas with high vulnerability (e.g. high water table, thin soil cover over fractured rock or karstic limestone) and features that allow rapid movement of pathogens in the subsurface. For example, Moore et al. (1981) found virus particles in groundwater up to 27 m below sites irrigated with sewage wastewater, and Jorgensen and Lund (1985) found enteroviruses 3 m below a forest site used for sludge application. As the infectious dose of some pathogens is low (e.g. enteric viruses, *Giardia*, *Cryptosporidium*), this could imply a high risk, especially in populations such as the immune-compromised and elderly people. There is also a possibility of regrowth of pathogenic bacteria, such as coliforms. The latter, being often used as an index of safety, do not predict the presence of other pathogens. Moreover, many diseases may be due to unknown agents and the methods for their detection have not yet been developed.

In developing countries, untreated domestic wastewater is an important source of enteric pathogens to soil because it is used in agricultural irrigation. This presents a high risk to farm workers and to consumers of food products irrigated with wastewater (Santamaria and Toranzos 2003). Other practices that favor the entry of considerable amounts of enteric pathogens into the soil environment are the use of human and animal excreta as manure. Feachem et al. (1983) reported survival times of some excreted pathogens in soil and on crop surfaces. Enteroviruses, thermotolerant coliforms and *Salmonella* spp. persist less than 20 days, *V. cholera* persists less than 10 days and helminth eggs may persist for several months. Blum and Feachem (1985) reviewed the epidemiological evidence of the agricultural use of excreta and they concluded that crop fertilization with untreated excreta causes significant infections with intestinal nematodes and bacteria in consumers and field workers.

13.4.4 *Municipal Solid Waste*

Municipal solid waste may be another source of enteric pathogens to soil because most landfill sites were constructed without a leachate collection system. This leachate may contain viruses and bacteria (Gerba 1996) which can percolate through soil and contaminate groundwater. In 1993, 62.4% of $205 \cdot 10^6$ Mg of municipal solid waste generated in the USA were sequestered in landfills and 50% of these sites were in the vicinity of water wells (Barlaz 1996). Enteric pathogens in municipal solid waste come from the excreta present in disposable diapers, pet feces, food waste and sewage sludge (Gerba 1996).

13.4.5 *Infections Caused by Consumption of Fruits and Vegetables*

Fresh produce can be contaminated with various human pathogens including bacteria, protozoa and viruses. Raw fruits and vegetables, especially fresh-cut leafy greens, are increasingly being recognized as the foremost transmitting vehicles of pathogens. Although, fresh produce can be contaminated at any point along the farm-to-consumption handling chain, the field application of raw manure or contaminated irrigation water in the primary production phase is a principal route of pathogen contamination. Hence, it is critical to prevent fresh produce from pre-harvest contamination in order to achieve the delivery of microbiologically safe produce to consumers (Fremaux et al. 2008). Figure 13.3 shows a schematic diagram of different sources of contamination that may introduce human pathogens into the field and contribute to the contamination cycle in agricultural areas. Manure is commonly applied to fields in order to fertilize soils. Enteric pathogens can survive for prolonged periods of time in animal feces (Hutchison et al. 2004) and may thus serve as potential inoculum onto plants in the field. Several studies have demonstrated the presence of foodborne pathogenic bacteria on crops grown in soil to which manure was applied (Brandl 2006). Thus, the use of improperly composted manure or the feces from free roaming domestic or wild animals in the fields, enhance the risk of microbial contamination.

Although bioaerosols have long been recognized as the primary mode of pathogen spread among livestock, the potential for air currents to disperse enteric pathogens from manure piles or contaminated soil, and to enable their immigration as viable cells onto crops remains to be investigated (Brandl 2006). It was found that *E. coli* was transmitted via insect vectors such as the fruit fly (*Ceratitis capitata*) (Janisiewicz et al. 1999) and the vinegar fly (*Drosophila melanogaster*) to apples (Sela et al. 2005), and for its excretion from inoculated houseflies (Sasaki et al. 2000). Plant pathogenic bacteria and their antagonists are effectively vectored by honeybees on and among apple and pear flowers in fruit orchards (Johnson et al. 2000), and the dissemination of epiphytic bacteria by insects on wet leaf surfaces

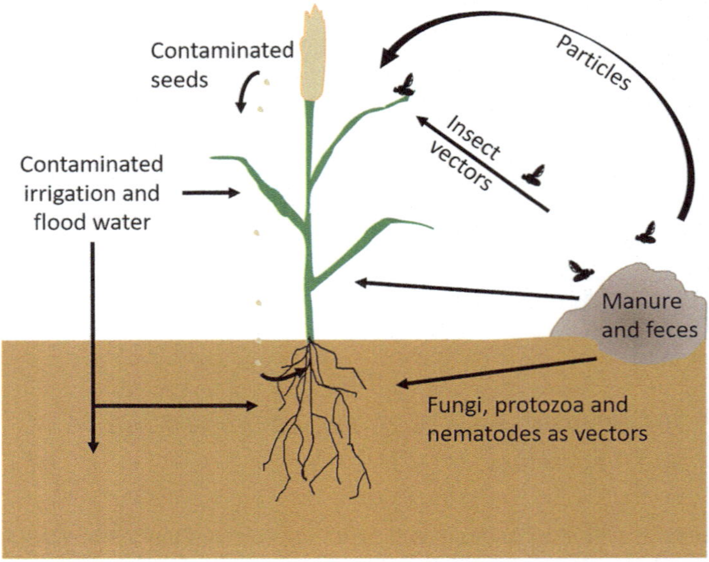

Fig. 13.3 Factors that can contribute to the contamination of fruit and vegetables with human enteric pathogens in the field (Adapted from Brandl 2006)

has been reported (Hirano and Upper 2000). The ubiquitous presence of insects on manure piles, in feedlots, and also in fruit and vegetable fields warrants examination of this type of transmission as a factor in the preharvest contamination of produce.

Crop irrigation and application of pesticides with contaminated water also are considered as primary sources of inoculum in the field. This is of particular concern for production of fruits and vegetables in areas where the supply of fresh water is scarce, and where water reclaimed from effluents increasingly serves for agricultural purposes. Because *E. coli* and *S. enterica* survive well in water sediments (Hendricks 1971), seasonal flooding of fields with overflowing stream water should be added to the risk factors of potential crop contamination.

13.4.5.1 Epidemics of Foodborne Infections Linked to Fresh Produce

Since the 1990s, awareness of the potential of fresh produce to cause foodborne disease has increased, and reported outbreaks associated with this commodity have grown steadily (Sivapalasingam et al. 2004). In the USA, produce caused the most cases (28,315 cases) of foodborne disease, and was the second most common single-food vehicle in the period of 1990–2003 (Brandl 2006). Lettuce, melon, seed sprouts, and fruit juice were the four most important single produce items implicated in epidemics of foodborne illness in the USA between 1973 and 1997 (Sivapalasingam et al. 2004). In England and Wales, salad, vegetables, and fruit caused 6.4% and 10.1% of all outbreaks with a known food vehicle in the periods of

Table 13.5 Factors contributing to the emergence of produce-linked outbreaks

Factors
Changes in consumer habits/
Increased consumption of fresh fruits and vegetables, and fresh fruit juices
Increased consumption of meals outside the home
Increased popularity of salad bars
Changes in the produce industry/
Wider distribution of produce over longer distances
Intensification and centralization of production
Introduction of minimally processed produce
Increased importation of fresh produce
Emerging pathogens with low infectious dose
Improved methods to identify and track pathogens
Enhanced epidemiological surveillance
Increased size of at-risk population (elderly, immune-compromised)

Adapted from Tauxe et al. (1997)

1993–1998 (Brandl 2006; WHO 2003) and 1999–2000 (Brandl 2006), respectively. Other fresh produce associated with outbreaks of bacterial enteric disease include tomato, cilantro, parsley, spinach, green onions, carrot, and cabbage (Brandl 2006). Factors that have likely contributed to the emergence of produce as a source of enteric illness are summarized in Table 13.5.

In addition to the major factors in the emergence of produce-associated outbreaks (Table 13.5), intensification of animal production raises the specter that the sources of contamination in the environment are themselves increasing. Several bacterial pathogens have caused fresh produce-associated epidemics of enteric illness, including *Salmonella enterica*, pathogenic *E. coli*, *Shigella* spp., *Campylobacter* spp., *Listeria monocytogenes*, *Staphylococcus aureus*, *Yersinia* spp., and *Bacillus cereus*. *S. enterica*, the most frequent etiologic agent of outbreaks from fresh produce, caused 48% of such outbreaks with a known etiology between 1973 and 1997 in the USA (Sivapalasingam et al. 2004). Of particular concern is the occurrence of outbreaks caused by multidrug-resistant strains of *S. enterica* serovar Typhimurium DT104 and linked to the consumption of lettuce (Brandl 2006). Pathogenic *E. coli* is the second most important causal agent of outbreaks from fresh produce (Sivapalasingam et al. 2004). Pathogenic strains of *E. coli* include the enterotoxigenic serotype O157:H7, a dangerous foodborne pathogen that can cause hemolytic uremic syndrome and lead to death, particularly in children, the elderly, and the immune-compromised. *E. coli* O157:H7 caused 21% of all produce-linked outbreaks in 1982–2002 in the USA (Rangel et al. 2005). In the USA, enteric pathogens other than *S. enterica* and *E. coli* have been implicated in relatively few produce-associated outbreaks (Table 13.6).

The data (Table 13.6) were categorized by causal agent and by plant part or entity that the produce represents. The leafy vegetables consisted mostly of lettuce and the pathogenic *E. coli* was predominantly serotype O157:H7. Mixed salads,

Table 13.6 Number of outbreaks linked to fresh produce in the USA in 1990–2004

Type of produce	*S. enterica*	Pathogenic *E. coli*	*Shigella* spp.	*Campylobacter* spp.
Leafy vegetables	8	13	3	3
Seed sprouds	9	6	0	0
Fruit	32	8	1	1
Total	49	27	4	4

Adapted from Brandl (2006)

which are associated with a high incidence of foodborne illness, were not included. Root vegetables, which have rarely been implicated in outbreaks, possibly because they are commonly cooked before consumption, were also not listed. During the 14-year period, 76% of the epidemics from contaminated fruit were caused by *S. enterica*, whereas pathogenic *E. coli* strains were the causal agent of the highest proportion associated with fresh leafy vegetables. The sources of the produce implicated in these different outbreaks varied widely in geographical location (Brandl 2006). A given agricultural area may thus harbor a certain type of pathogen and may recurrently cause epidemics via the same type of produce.

13.4.5.2 Life Conditions of Foodborne Pathogenic Bacteria on Plant Surfaces

Enteric pathogens encounter harsh physicochemical conditions and frequent changes on plant surfaces. Such changes, for example in temperature and osmotic conditions within short time periods, may not be experienced by enteric pathogens in the animal and human gut. Aerial plant surfaces are overall poor in nutrients, relatively aerobic, and exposed to UV (ultraviolet) radiation. In contrast, the intestinal environment is replete in nutrients, anaerobic, shielded from solar rays and supports vigorous bacterial growth, with *E. coli* alone reaching between 10^6 and 10^9 cells per g colon content (Brandl 2006). The spatial distribution of physicochemical conditions on leaf and root surfaces at the microscale is highly heterogeneous. For example, although sugars are present in small amounts on leaf surfaces and limit bacterial growth on leaves (Wilson and Lindow 1994), whole-cell bacterial biosensors for sucrose, fructose, and glucose have shown that these sugars are abundant in oases on leaves. Sucrose, fructose, and glucose are the dominant carbon sources in the phyllosphere of plant species examined so far (Mercier and Lindow 2000). Spatial patterns of sucrose, amino acids, and nitrate abundance have also been mapped in the rhizosphere (Jaeger et al. 1999). Similarly, large variations in water potential between microsites on a given leaf surface were reported, with only a subset of the cell population experiencing detectable water deprivation (Axtell and Beattie 2002). In addition to starvation and osmotic and matric stress, bacteria on aerial plant surfaces have to cope with the effects of solar radiation. Besides the genes involved in pigmentation and in the repair of UV-damaged DNA, evidence also indicates that physical avoidance of UV radiation in shaded sites on leaves may be an important strategy for survival of bacteria on plant surfaces (Jacobs et al. 2005).

13.4.5.3 Persistence of Foodborne Pathogenic Bacteria Attached to Plant Surfaces

Studies performed in the laboratory or in the field have demonstrated the ability of enteric pathogens to survive for long periods of time on vegetables and fruits (reviewed by Brandl 2006). Ercolani (1979) demonstrated that *S. enterica* Typhi was still detectable on mature lettuce plants that were inoculated at a young stage in the field. Field studies conducted with an avirulent mutant of *S. enterica* Typhimurium and a nontoxigenic mutant of *E. coli* O157:H7 revealed that both pathogens were able to persist in the lettuce and parsley phyllosphere for up to 3 and 6 months, respectively, after the seedlings were planted and exposed to the pathogens via contaminated manure or irrigation water (Islam et al. 2004a). In a similar field study, *S. enterica* Typhimurium also survived on field-grown radishes and carrots for 84 and 203 days, respectively, after the seeds were sown in soil contaminated with either compost or irrigation water (Islam et al. 2004b). *E. coli* O157:H7 survived in the soil longer in field plots with long-season crops than in those with short-season crops (Islam et al. 2004a). Low water activity is considered to be one of the main limitations of bacterial survival on plant surfaces. Enteric pathogens are exposed to osmotic stress in the animal and human gut and have developed a range of mechanisms to adapt to such conditions. Additionally, *S. enterica* has a high tolerance to long-term desiccation stress in non-host environments (Mitscherlich and Marth 1984). This may explain why *S. enterica* is the most common etiologic agent of foodborne illness linked to spices (Vibha et al. 2006) and the occurrence of two salmonellosis outbreaks associated with dry raw almonds (Brandl 2006). It was hypothesized that in at least one of these almond-linked outbreaks, *S. enterica* multiplied on the moist almond fruits on the orchard floor at harvest and survived on the dry kernels to which it had migrated (Uesugi and Harris 2006). Under conditions of low relative humidity imposed for 24 h, *S. enterica* declined at rates similar to those of epiphytic bacteria on the dry leaf surface of cilantro plants, and was capable of resuming growth at significant rates upon reoccurrence of wet conditions on the leaves (Brandl and Mandrell 2002). Thus, the tolerance of *S. enterica* to desiccation stress, combined with its potential to multiply under subsequent wet conditions may confer on this pathogen the ability to persist despite the high fluctuations in water availability on plant surfaces in the field. It implies also that even small populations of the pathogen that survive on crop plants before harvest may increase to infectious dose levels during growth-conducive conditions while the produce is stored or processed for consumption. Epiphytic bacteria survive desiccation stress on leaves at higher rates as aggregated cells than as solitary cells, which suggests an important role for aggregate formation in bacterial tolerance to low water availability on plants (Jacques et al. 2005).

13.5 Ecophysiology of Pathenogenic Soil Organisms

13.5.1 Viruses

Viruses are the smallest pathogens, most having maximum dimension of less than 30 nm (Coyne 1999). They are acellular organisms, have no cell membrane, and occur in many shapes including cubic, helical, and icosahedral. Viruses comprise a nucleic acid core of either DNA or RNA, surrounded by a protein coat. Their genetic material (the nucleic acid core) contains only 10–200 genes (Coyne 1999). They require living cells in order to replicate and generally have a very restricted host range. They are always host specific and infect animals, plants, fungi, protozoans, algae, and bacteria. Viruses do not multiply in foods or water, or in any other environmental sample, including soil. However, viruses can survive outside living cells and remain infectious. In a host, they attach to a cell, use enzymes to break through the cell wall, and inject their nucleic acid core into the cell. Once in the cell, the genetic material from the virus begins making three types of proteins. It replicates its own genetic material, builds protein coating, and assembles proteins that will help it get out of the cell (Bultman et al. 2005). Outside of the host, viruses are inert. They do not grow or reproduce. Human pathogenic viruses with protective coatings can remain infectious in the environment for up to 6 months (Gerba and Smith 2005).

13.5.1.1 Hantaviruses

Hantaviruses are single-strained RNA viruses that represent a separate genus in the *Bunyaviridae* family (Bi et al. 2008). Hantaviruses form predominantly spherical or irregular particles 80–120 nm in diameter. In contrast to other genera of the *Bunyaviridae*, hantavirus is transmitted to humans not by arthropod but from contact with persistently infected rodents (Lednicky 2003). Transmission commonly occurs through contact with infected excreta, i.e. saliva, urine and feces (Fig. 13.4). Though the aerosol route of infection is undoubtedly the most common means of transmission among rodents and to humans (Vapalahti et al. 2003), virus transmission by bite occurs among rodents (Hinson et al. 2004) and may also result in human infection. Like other enveloped viruses, hantaviruses are readily inactivated by heat, detergents, UV irradiation, organic solvents and hypochlorite solutions (Kraus et al. 2005). Hantavirus produces chronic persistent infection in the host rodent (Easterbrook et al. 2007). Transmission of these viruses across a species barrier may result in human infections. The virus type correlates closely with disease severity.

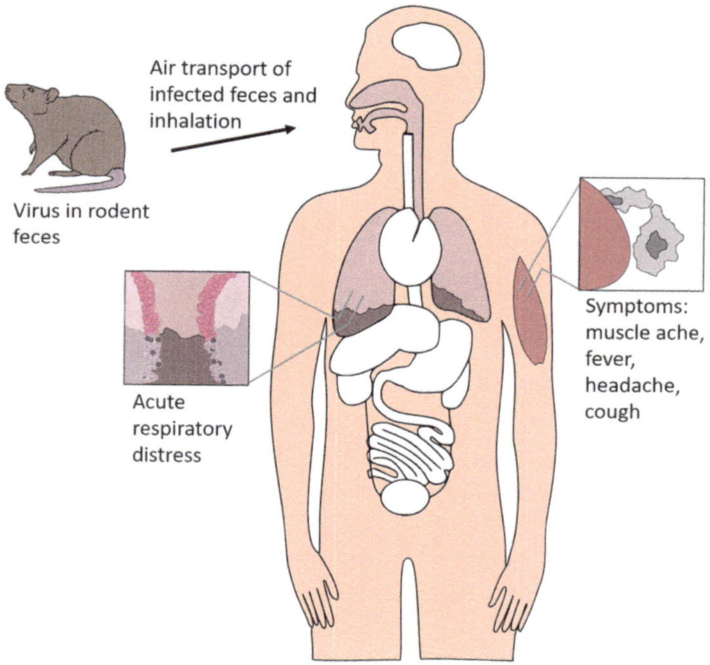

Fig. 13.4 Infection of humans by hantavirusses

13.5.1.2 Enteroviruses

Human enteric viruses (also known as enteroviruses) are a genus of picornaviruses which are small (~30 nm), nonenveloped viruses containing a single-stranded ribonucleic acid (RNA) (de Crom et al. 2016). They include polioviruses, coxsackieviruses and echoviruses. The *Picornaviridae* family is one of the largest RNA virus families and contains an array of pathogens that infect both humans and animals (de Crom et al. 2016). The family is classified into 29 genera including the genus enterovirus and parechovirus.

Land disposal of sewage sludge and wastewater has a potential of pollution of raw potable water sources or infiltration into the groundwater and contamination of crops (Rao et al. 1986). Fresh produce such as leafy green vegetables and berry fruit can be extensively handled during harvesting, especially strawberries and raspberries which are often picked by hand. Extensive handling post-harvest may also occur; for instance with green onions at least three workers can be required to peel, trim and bundle them (Dentinger et al. 2001). Such practices increase the chances of contamination by infected persons. Enteroviruses were found to be able to survive for extended periods in some soil environments with survival times of up to 170 days reported for virus particles in loamy and sandy loamy soils (WHO 1979). Enteric viruses have been recovered at considerable distances from their

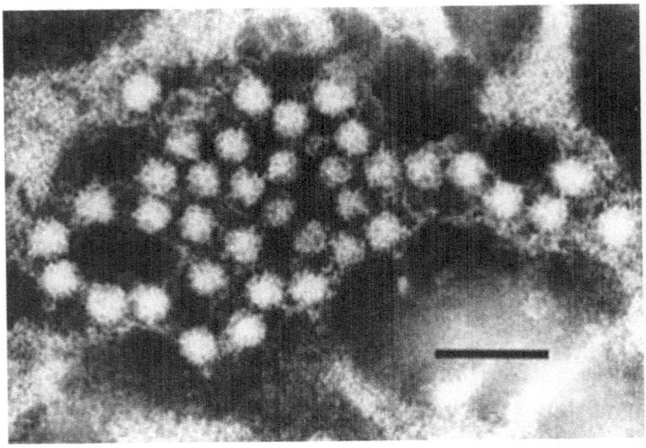

Fig. 13.5 Electron micrograph of negatively stained HEV virus particles in a bile sample from a chicken; bar = 100 nm (Haqshenas et al. 2001, p. 2451; with kind permission from Microbiology Society)

source owing to their generally relatively poor binding to soil particles and soil organic matter meaning that they are easily eluted and carried in rainwater. The exception to this is poliovirus which seems to adsorb relatively strongly to soil particles (Landry et al. 1979).

13.5.1.3 Hepatitis A and E Virus

Hepatitis A virus (HAV) belongs to the genus *Hepatovirus*, in which it is the only species. The genus itself lies within the family *Picornaviridae*. HAV particles are approximately 28 nm in diameter, possessing an icosahedral capsid enclosing the single-stranded positive-sense RNA genome. The HAV genome contains four structural and seven non-structural genes. The infectious dose of HAV particles may be very low, perhaps in some cases even single particles (Cliver 1985).

Hepatitis E virus (HEV) was designated in 2004 as the sole member of the genus *Hepevirus* in the family *Hepeviridae* (Emerson et al. 2005). It is a small, non-enveloped, single-stranded, positive-sense RNA virus (Haqshenas et al. 2001; Fig. 13.5), about 35 nm in diameter (Krawczynski 1993). The genome contains short stretches of untranslated regions at both ends (Tam et al. 1991). HEV variants most closely related to those infecting humans can be divided into at least four genotypes (Panda et al. 2007). Genotypes 1 and 2 only seem to affect humans. Genotype 1 viruses are predominantly isolated from outbreaks and sporadic cases in Asia and Africa, whereas genotype 2 strains mainly have been observed in outbreaks in Mexico and Africa. Genotypes 3 and 4 are zoonotic and are present in different animal species and sporadic human cases, worldwide for HEV genotype 3 and mainly in Asia for HEV genotype 4.

Foodborne transmission of HAV and HEV has become a serious concern (Cliver 1985; Colson et al. 2010), raising questions about the survival of these hepatic viruses in a range of different foods and in the environment. In contaminated soil and sediments, the persistence of infectious HAV may be relatively long. In samples of saturated sand, clay and organic soil saturated with HAV-contaminated primary sewage and stored at 25 °C, 3.2%, 0.8% and 0.4% of the original number of infectious units remained after 84 days (Sobsey et al. 1988). Examining the persistence of HAV in spiked soil samples (soil type not given, possibly taken from a riparian site), Parashar et al. (2011) monitored persistence of HAV RNA, and observed that it could be detected up to and including the 13th week in samples stored at 37 °C, 5 weeks after it could not be detected in samples placed in a location where they were exposed to environmental temperatures. This may indicate that temperature fluctuations may affect persistence of intact virus. However it should be noted that this study was conducted in Pune, India, where diurnal temperatures are higher than those normally encountered in Europe. Parashar et al. (2011) also examined the persistence of HEV genotype I under the same conditions as described above. Again, HEV RNA was detected for longer (week 10) at 37 °C than in samples exposed to environmental temperatures (week 9), although the difference was not pronounced.

Ingestion of HAV and HEV through contaminated food plants and contact with sewage sludge-polluted soils and waters are assumed to be the most important infection routes, the latter being used for irrigation or washing. Foods may also acquire viral contamination by contact with infected persons during harvesting or preparation, when viruses are transferred or shed onto foodstuffs via faecally contaminated hands. One milligram of faeces may contain up to 10^7 genome copies in an immunocompetent HAV-infected patient, so microscopic quantities of faeces could harbour sufficient virus particles to constitute a hazard (Pintó et al. 2012).

13.5.1.4 Other Viruses

Numerous other human viruses have either been shown or are thought to only survive relatively poorly in the soil system. However, studies on many viruses are currently still lacking. However, the rubella virus, mumps virus, rhinovirus and parainfluenza virus among others, have all been found to survive in the external environment, including in the soil, for only several hours and occasionally for a couple of days (Walther and Ewald 2004). Therefore, these diseases are not considered within this chapter.

13.5.2 Bacteria

According to modern taxonomy, bacteria form one of the three domains of life, along with archaea and eukaryota. They are a remarkably diverse group of micro-organisms. Over 400 genera of bacteria have been identified with possibly as many as 10,000 species and, with the exception of viruses, they are in most cases more abundant than any other organism in soils (Bultman et al. 2005). Bacteria are small, generally less than 50 μm in length and 4 μm in width. One bacterium weighs about

10^{-12} g. Their small size affords them a high surface area to volume ratio, which allows them to maximize nutrient uptake through diffusion. They also possess a very high metabolism and the ability to reproduce through binary fission. The number of bacteria that can be cultured in the laboratory is probably less than 1%; thus their actual diversity is probably much greater. Bacteria play a vital role in global functioning through the ecosystem services and functions that they provide. Bacteria which are pathogenic in humans include obligate pathogens, and opportunistic pathogens which can be divided into different groups. For example, enteric bacteria are rod shaped, gram negative bacteria which occur most commonly in the intestines of humans and other animals where they may cause disease, mainly diarrhea, in some instances. They include bacteria from the genera *Campylobacter*, *E. coli* and *Shigella* sp. Other groups of bacteria, such as the actinomycetes, are found more commonly in the external environment but there are some species which may be capable of causing disease in humans.

13.5.2.1 Infectious Actinomyces

Actinomycetes are a group of branching unicellular organisms, which reproduce either by fission or by means of special spores or conidia (Jeffery and van der Putten 2011). They are closely related to the true bacteria. Frequently, they are considered as higher, filamentous bacteria. Classified in *Actinomycetales* order, they also belonged to gram-positive bacteria with a high Guanine-plus-cytosine (G+C) content in their DNA.

Infectious actinomycete species, often *Actinomyces israelii*, cause actinomycosis which is a soil dwelling species that is found in decaying organic matter (Roque et al. 2010). Actinomycetes are generally soil inhabiting saprophytes although some species are capable of causing diseases in plants, animals or humans. They are anaerobic or facultative anaerobic organisms, with some genera such as *Frankia* forming symbiotic relationships with the roots of non-leguminous plants including trees. They play an important role in maintaining soil fertility and plant nutrition due to their ability to fix nitrogen. Many species of actinomycetes can also be found colonizing the gut, mouth and vagina of humans although the majority of these do not cause diseases.

13.5.2.2 *Bacillus anthracis*

The bacterium, causing the zoonotic disease anthrax, is gram positive, unmovable, rod shaped, aerobic and spore forming (Jernigan et al. 2001; Fig. 13.6). It forms spores, which can survive for years in the environment and is spread over many countries worldwide where it can affect livestock and wildlife. The nonuniform distribution of *B. anthracis* in soil may be largely determined by its preference for black soils, which are rich in organic matter and calcium carbonate that promotes spore viability (Hugh-Jones and Blackburn 2009).

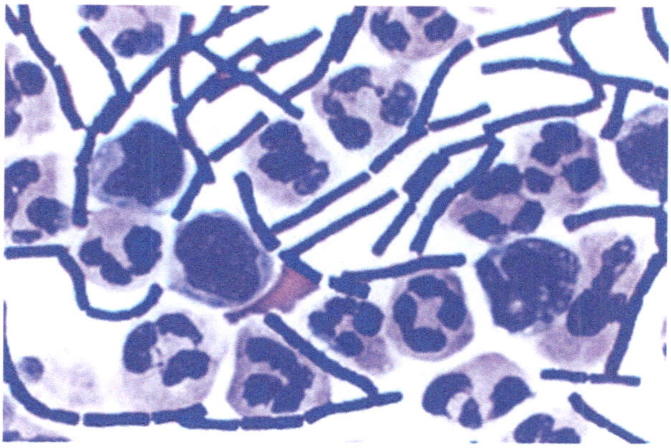

Fig. 13.6 Gram stain of cerebrospinal fluid of a patient with a diagnosis of meningitis showing *Bacillus anthracis* (Jernigan et al. 2001, p. 936; reproduced with kind permission from Emerging Infectious Diseases)

It also has been proposed that spores accumulate in low-lying areas during rainfall, followed by exposure of grazing animals during dry periods. Because of the concentration of spores needed, it seems that human acquisition of anthrax directly from soil is very unlikely. Rather, host animals are required for spore germination and anthrax is naturally acquired from infected meat and animal products such as goat hair (Baumgardner 2012; Blackburn et al. 2007). Animals generally get anthrax from grazing on soils which contain spores of the bacterium *B. anthracis*. Humans then may become infected through touching infected animals or animal products where the bacterial spores may enter into wounds if present, or be inhaled as is the case for "wool sorter's diseases". Often fatal infection results from the handling of infected wool (Van Ness 1971). Evidence exists that soil management practices such as tillage may increase the risk of human infection of anthrax due to inhalation of spores. The spore form is highly resistant to desiccation and other environmental stresses and can remain in this inactive form in soil for many years, only becoming active once environmental conditions become favourable again.

13.5.2.3 *Bacillus cereus*

Bacillus cereus is a gram positive, aerobic-to-facultative, spore-forming rod widely distributed environmentally and bearing close phenotypic and genetic (16S rRNA) relationships to several other *Bacillus* species, especially *B. anthracis* (Ash et al. 1991). The bacterium exists as a spore former and vegetative cell in nature and as a vegetative cell when colonizing the human body. This bacterium is found naturally in decaying organic matter, in and out of soil, fresh and salt water, plants, dusts, fomites, and the intestinal tract of invertebrates. Spores germinate within an insect or animal host or on contact with organic matter, entering the soil via the droppings of an animal host or upon the host's death. Saprophytic growth in soil, including transition from a single cell to a multicellular form, then ensues (Baumgardner

2012). Cells and spores may then contaminate plant material and enter food processing areas. *B. cereus* spores resist extreme environmental conditions including heat, freezing, drying, and radiation. The hydrophobic spore surface allows attachment to food and processing equipment, where biofilm formation may further protect forms of the organism (Bottone 2010).

13.5.2.4 *Clostridium botulinium*

Clostridium botulinum is as soil-borne pathogen and prefers to grow in decaying organic matter and is the causative agent of botulism (Valério et al. 2010). It is not a well-defined species of bacterium but rather refers to distinct groups of bacteria that are spore forming, gram positive and anaerobic. Their principal habitat is the soil, although their distribution can be highly regional. They multiply fast in an oxygen deficient, warm and moist environment (Desta et al. 2016). While the bacteria are soil-borne, infection generally occurs through eating contaminated food, although it can also be transmitted into wound infections directly from the soil.

13.5.2.5 *Campylobacter jejuni*

Bacteria from the genus *Campylobacter*, most commonly the species *Campylobacter jejuni,* are the causative agents of campylobacteriosis. It is an enteric bacterium and is one of the most common forms of bacterial infections in humans. The main route of transmission of *Campylobacter* sp. are the ingestion of contaminated food or water (particularly undercooked poultry or unpasteurised milk), the faecal-oral route, or person to person contact. However, *C. jejuni* has been demonstrated to survive in the soil for at least 25 days, with indications that it can survive considerably longer (Ross and Donnison 2006) meaning that the soil is also a possible route of transmission, particularly owing to the relatively small infectious dose need to cause disease.

13.5.2.6 *Leptospira interrogans*

Bacteria of the genus *Leptospira* cause leptospirosis, also known as Weil's diseases (Jeffery and van der Putten 2011). Leptospires are tightly coiled spirochetes, usually 0.1 μm by 6 to 0.1 by 20 μm, but occasional cultures may contain much longer cells (Fig. 13.7). The helical amplitude is approximately 0.1–0.15 μm, and the wavelength is approximately 0.5 μm (Levett 2001). The cells have pointed ends, either or both of which are usually bent into a distinctive hook. Two axial filaments (periplasmic flagella) with polar insertions are located in the periplasmic space (Swain 1957).

Aerobic spirochete *Leptospira* are found in the soil as well as in aquatic ecosystems (Henry and Johnson 1978). The genus *Leptospira* is most often divided

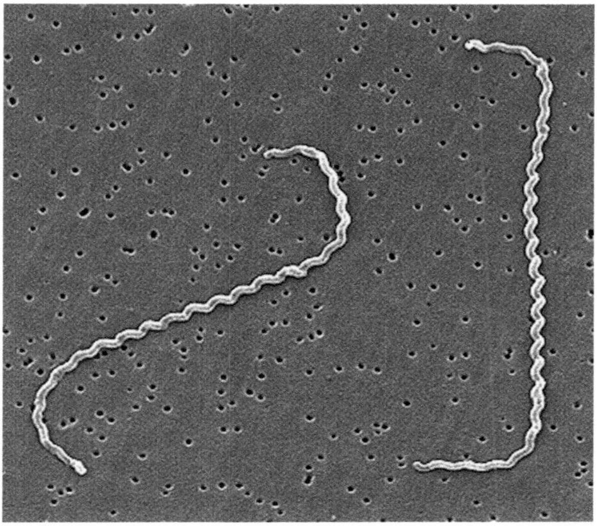

Fig. 13.7 Scanning electron micrograph of *Leptospira interrogans* serovar icterohaemorrhagic strain RGA bound in a 0.2 mm membrane filter (Levett 2001, p. 299; with kind permission from American Society for Microbiology)

into two complexes, the parasitic and the non-parasitic complex. The parasitic complex includes 13 named and 4 unnamed species, of which *L. interrogans* is the most well-known, and has over 200 known pathogenic serologic variants. Transmission of leptospirosis generally occurs through contact with soil, fresh water, or possibly vegetation which has been contaminated by the urine of infected animals. The bacteria are capable of surviving for at least 42 days in acidic (pH ~5.5) soil (Hellstrom and Marshall 1978) and up to 74 days in more neutral soils (Zaitsev et al. 1989). This demonstrates that while pathogenic species require an infected individual to transfer them to the soil environment through their urine, they are then capable of surviving a considerable time after the infected individual has left the area until a susceptible host comes along. The fact that the bacteria can survive for extended periods in water means that they can be transported relatively large distances in times of floods, or through being transported via overland flow and in waterways (Jeffery and van der Putten 2011). Increased risk of contracting the disease is associated with famers, vets, loggers, sewer workers and those that partake in water sports including fresh water swimming, canoeing and kayaking.

13.5.2.7 *Listeria monocytogenes*

Listeria are gram-positive, facultative anaerobic, non-spore-forming, rod-shaped bacteria with a low G+C content. The bacterium *Listeria monocytogenes* is the causative agent of listeriosis. Infection usually occurs in humans after eating food contaminated with the bacterium. While it is a generally considered to be a foodborne disease, the bacterium has been demonstrated to survive for extended periods in soil (Welshimer 1960) as well as being associated with sheep and cattle.

The *Listeria* species are tolerant to extreme conditions such as low pH, low temperature and high salt conditions (Sleator et al. 2003; Liu et al. 2005). Therefore they can be found in a variety of environments, including soil, sewage, silage, water, effluents and foods. The main factor affecting the survival of *Listeria monocytogenes* in soil was found to be soil moisture content with survival averaging up to approximately 67 days in soils where the moisture content was not controlled as compared to up to 295 days in soils which were protected from evaporation (Welshimer 1960). The ability to survive and grow in harsh conditions makes of this pathogen a high concern to the food industry.

13.5.2.8 *Clostridium tetani*

The bacterium is the causative agent of the tetanus disease. *C. tetani* has a worldwide distribution in soil and dust (where spores can persist for years), feces (which can reinfect soil), and other agents (including contaminated heroin) (Baumgardner 2012; Wilkins et al. 1988). The organism may be isolated from surface soils of school and hospital grounds, fields, roadsides, and along waterways (Ebisawa et al. 1986). *C. tetani* bacteria, as all other species of *Clostridium*, are mobile, spore forming, gram positive and obligate anaerobes. Climate and soil pH relate to the increased prevalence of tetanus in tropical zones (Brook 2008), and clusters of infections may occur in developing countries, particularly after natural disasters such as earthquakes and tsunamis (Afshar et al. 2011). In the USA, about half of tetanus cases followed known injuries, 45% result from infection of preexisting wounds, wounds of unknown cause, parenteral drug abuse, or animal-related injuries (Baumgardner 2012). The association that many people have with tetanus and rusty metals is somewhat misconstrued (Jeffery and van der Putten 2011). The bacterium does not have any propensity to grow on rusty metal, but rather owing to its highly resilient spores is capable of surviving on rusty metal. The rusty metal itself sometimes provides a means of infection through piercing the skin of individuals who cut themselves on the metal or stand on rusty nails, etc.

The spore of *C. tetani* is incompletely destroyed by boiling but eliminated by autoclaving at 1 atmosphere pressure and 120 °C for 15 min. It is rarely cultured, as the diagnosis of the disease is clinical. *C. tetani* produces its clinical effects via a powerful exotoxin. The role of the toxin within the organism is not known (Mellanby and Green 1981).

13.5.2.9 *Francisella tularensis*

The bacterium is a non-mobile gram-negative coccobacillus of which several serotypes exist, each with varying degrees of virulence. *Francisella tularensis* is known to survive for weeks at low temperatures in water, soil and animal carcasses, and for more than a year in mud (Parker et al. 1951). Evidence suggests that the main reservoir of the bacterium in the environment may be the cells of protozoa, particularly the protozoa *Acanthamoeba castellanii* which is ubiquitous in the environment (Abd et al. 2003).

The diseases caused by *Francisella tularensis* is known as rabbit fever, or deer fly fever owing to the fact that the primary vectors are deer flies or ticks, as well as other arthropods. However, more than 100 species of mammals have been noted to be infected with tularemia, includings rabbits, hares, muskrats, prairie dogs, skunks, raccoons, rats, voles, squirrels, sheep, cattle, and cats (Nada and Harik 2013).

13.5.2.10 *Clostridium perferingens*

Like other *Clostridia* species such as *Clostridium botulinium*, *C. perfringens* are gram positive, anaerobic and spore forming bacilli. It has been reported that *C. perfringens* is more widely spread than any other pathogenic bacterium (Matches et al. 1974). Pathogenic *Clostridium* are highly prevalent in soils (DeSpain Smith and Gardner 1949), as well as the intestinal tract of humans and animals (Bryan 1969). *Clostridium perferingens* is also found on vegetable products and in other raw and processed foods. The organism is found frequently in meats, generally through fecal contamination of carcasses (Brynestat and Granum 2002). Major contributing factors leading to food poisoning associated with *C. perfringens* include its ability to form heat-resistant spores that can survive commercial cooking operations, as well as the ability to germinate, outgrow, and multiply at a very rapid rate during postcook handling, primarily under conditions conducive to germination. Contamination also occurs through ingredients such as spices.

13.5.2.11 *Yersinia enterocolitica*

Yersinia enterocolitica is a species of gram negative coccobacillus bacterium which is part of the family Enterobacteriaceae, along with *Salmonella* and *E. Coli* (Rosner et al. 2010). The pathogen consists of a relatively diverse group of bacteria which can be divided into either 18 or 54 different serotypes depending on the criteria used (Jeffery and van der Putten 2011). The survival of *Yersinia enterocolitica* in soil is strongly affected by environmental conditions, particularly soil moisture, as the bacterium is not well adapted to desiccation. Air-drying soil over 10 days reduced the number of viable cells to only 0.1% of the original population (Chao et al. 1988), although different strains of the bacterium have different survival rates (Tashiro et al. 1991). *Y. enterocolitica* species can be isolated from a variety of domestic and wildlife animals, e.g., pigs, cattle, sheep, goats, dogs, cats, wild boars, and small rodents (Bottone 1999). Pigs are considered to be the main reservoir of human pathogenic strains, largely because of the high prevalence of these strains in pigs and the high genetic similarity between porcine and human isolates (Fredriksson-Ahomaa et al. 2006).

13.5.2.12 *Pseudomonas aeruginosa*

The genus *Pseudomonas* consists of more than 120 species that are ubiquitous in moist environments such as soil and water ecosystems and pathogenic to plants, animals and humans (Peix et al. 2009). The gram negative and rod shaped *Pseudomonas* species are easily detectable on agar due to the production of pigments such as pyoverdine which is a yellow-green, fluorescent pigment, and pyocanin that is a blue-green pigment (Meyer 2000). Many sources of environmental water could potentially be acting as a reservoir for potentially pathogenic strains of *P. aeruginosa*. Various studies have shown that water resources (including sewage treatment plants and river water) are highly polluted with pathogenic bacteria including *P. aeruginosa* (Streeter and Katouli 2016). Public recreational swimming pools have also shown *P. aeruginosa* contamination (Moore et al. 2002).

13.5.2.13 *Escherichia coli*

E. coli is a rod-shape, gram-negative, gammaproteobacterium in the family *Enterobactericeae*, and is a member of the fecal coliform group of bacteria. The primary habitat of *E.coli* is thought to be the lower intestine of warm-blooded animals, including humans. Greater than 10^6 *E. coli* cells are generally present in 1 g of colon material, and are often released into the environment through fecal deposition (Ishii and Sadowsky 2008). Until relatively recently, *E. coli* was believed to survive poorly in the environment, and not to grow in secondary habitats, such as soil, water and sediment (Winfield and Groisman 2003). Recent studies, however, have shown that *E. coli* can survive for long periods of time in the environment, and potentially replicate, in soils, in water, on algae, and in tropical, subtropical, and temperate environments (Ishii and Sadowsky 2008). Relatively high concentrations of nutrients and warm temperatures in tropical and subtropical environments are likely factors enabling *E. coli* to survive and grow outside of the host (Winfield and Groisman 2003). *E. coil* can survive in the soil for sufficient periods of time to lead to infection even when the infectious individual that was the source of the contamination has long left the area. The main factor affecting the length of time that *E. coli* can survive in soil appears to be soil moisture content, with cells surviving for 14 days in dry soils and longer in wet soils (Chandler and Craven 1980). However, Avery et al. (2004) demonstrated that the pathogenic strain O157 could survive on surface vegetation for up to 6 weeks, or in the underlying soil for 8 weeks.

The ability of *E. coli* to survive and grow in the environment is likely due to its versatility in energy acquisition. *E. coli* is a heterotrophic bacterium, requiring only simple carbon and nitrogen sources, plus phosphorus, sulfur, and other trace elements for their growth. This bacterium can also degrade various types of aromatic compounds such as phenylacetic acid and benzoic acid, to acquire energy (Díaz et al. 2001). In addition, *E. coli* can grow both under aerobic and anaerobic conditions, which they may face in a variety of fluctuating environments. Moreover, *E. coli* can grow over a broad range of temperatures (7.5–49 °C), with has a growth optimum of 37 °C (Jones et al. 2004). The long-term survival of *E. coli* under freezing temperature has also been reported (Ansay et al. 1999; Bollman et al. 2001). The ability of *E. coli* to grow and survive under various conditions likely allows them to become an integrated member of microbial communities in a large variety of environments (Ishii and Sadowsky 2008).

13.5.2.14 *Coxiella burnetii*

The bacterium *Coxiella burnetii* is the causative agent of Q (for query) fever. *C. burnetii* was originally named *Rickettsia burnetii* since it shares some characteristics with the Rickettsiae, such as being an obligate intracellular organism and having a tick reservoir (Parker et al. 2006). The pathogen is a small, gram negative obligate intracellular coccobacillus meaning that it has to infect the cells of a host in order to complete its life cycle (Heinzen et al. 1999; Fig. 13.8). It was discovered in 1937 in Brisbane, Australia, where the disease was originally described. Q fever is considered to be a zoonotic disease with a reservoir in cattle, goats and sheep. The bacterium is found in all parts of the world with the exception of New Zealand (Greenslade et al. 2003) and Antarctica (Jones et al. 2011). *Coxiella burnetii* is very resistant to heat, desiccation and many common disinfectants, and so is able to persist in the environment, including soil, for up to 150 days (Jones et al. 2011). Infection occurs from the inhalation of the bacteria in spore form, usually from farmyard dust. The presence of vegetation and increased soil moisture appear to reduce the transmission of *Coxiella* by reducing the amount of dust available for

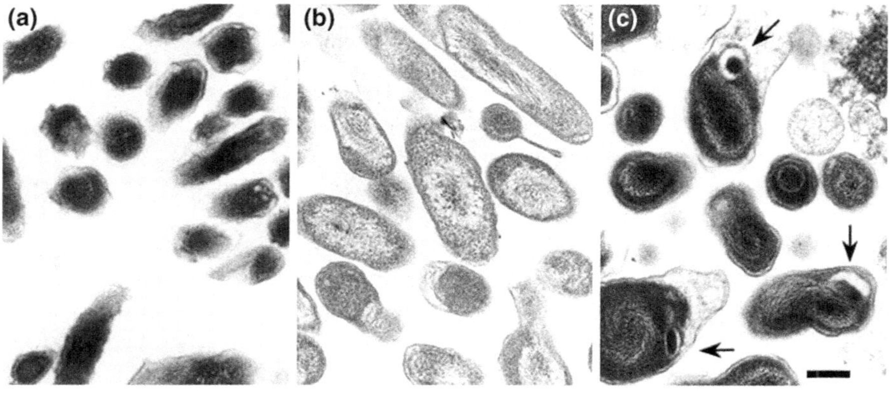

Fig. 13.8 Electron micrographs of *Coxiella burnetii* small cell variants SCVs and large cell variants (LCVs) containing spore-like particles (SLPs); (**a**) Electron micrograph of purified SCVs; (**b**) Electron micrograph of purified LCVs; (**c**) Electron micrograph of purified LCVs harboring SLPs (arrows); Scale bar: 0.2 μm (Heinzen et al. 1999, p. 151; with kind permission from Elsevier)

dispersion of the bacteria (van der Hoek et al. 2011). The bacterium can also be contracted through contact with urine, body fluids or milk of infected animals as well as through ticks.

13.5.2.15 *Salmonella enterica*

Salmonella is a rod-shaped, gram negative facultative anaerobe that belongs to the family Enterobacteriaceae (Barlow and Hall 2002) Within the genus *Salmonella*, around 2600 serotypes have been identified and most of these serotypes have the ability to adapt within a variety of animal hosts, including humans (Allerberger et al. 2003; Grassl and Finlay 2008). *Salmonella,* along with *Campylobacter* and *E. Coli*, are the most frequently isolated foodborne pathogens, and are predominantly found in poultry, eggs and dairy products (Silva et al. 2011). The slaughtering process of food animals at abattoirs is considered one of the important sources of organ and carcass contamination with *Salmonella* (Gillespie et al. 2005). Other food sources that are involved in the transmission of *Salmonella* include fresh fruits and vegetables (Pui et al. 2011). Although humans generally contract salmonellosis through eating contaminated food, one of the main routes in which pathogenic *Salmonella* come into contact with vegetables is from the soil (Islam et al. 2004b). *Salmonella* sp. along with *E. coli* and *Campylobacter*, are often introduced into the soil with liquid manure on agricultural land (Bech et al. 2010). Once introduced into the soil, pathogenic species of *Salmonella* have been demonstrated to be able to persist for up to 231 days and to be capable of contaminating vegetables grown in such soils (Islam et al. 2004b).

13.5.3 *Fungi*

Over 100,000 species of fungi exist of which about 300 are known to be pathogenic (Bultman et al. 2005). Diseases are caused by are caused by opportunistic pathogenic strains of fungi which pose the greatest risk to immuno-compromised patients. Like humans, the fungi are from the domain "eukaryota.". Fungal cells are thus more similar to human cells than they are to bacterial cells. This means that complications arise with regard to treatment of fungal diseases as the two types of cell (human and fungus) are relatively similar compared to bacterial cells (Jeffery and van der Putten 2011). The chemicals which can be used to combat fungal infection are thus relatively limited compared to the antibiotics used to fight bacterial infection. Fungi do not contain chlorophyll and are therefore not capable of photosynthesis. The cell walls are chitin-based and the cells can be grouped into molds and yeasts (Bultman et al. 2005). Molds are composed of branching filaments called hyphae that grow by elongation at their tips. Hyphae can be composed of one cell with continuous cytoplasm, called coenocytic hyphae, or are composed of cells separated by walls (septa) in which case they are called septate hyphae. The mass of hyphae of an individual organism is referred to as mycelium. Reproduction is through sexual or asexual spores and fragmentation of hyphae. Single cell nonfilamentous fungi are called yeasts which are generally spherical or ovoid in shape and reproduce by budding. Some fungi are dimorphic in that they can switch between filamentous or yeast growth.

The majority of human pathogenic fungi are soil inhabiting saprotrophs which means that they feed on dead organic matter within the soil (Jeffery and van der Putten 2011). Most molds are aerobic and cannot survive in saturated soils. They need to be able to extend hyphae into air-filled pores that contain oxygen. Many yeasts are facultative anaerobes and some yeasts are capable of surviving in anaerobic environments. In moist soils, the largest fraction of the microbial biomass is made up of fungi. Soil-borne fungi are more tolerant of acidic soils, grow best between 6 and 50 °C, and are usually found in the top 15 cm of the soil (Coyne 1999). When coming into the contact with immuno-compromised patients or other susceptible individuals, such as in an open wound or through the inhalation of spores, some fungi can become very aggressive forms of infection.

13.5.3.1 *Aspergillus* ssp.

Fungi of the genus *Aspergillus* (particularly A. *fumigatus* and *A. flavus*; less commonly: *A. terreus, A. nidulans, A. niger*) cause a disease named aspergillosis (Bultman et al. 2005). The pathogens are relatively wide-spread in soils and leaf litter all around the world including forest soils, wetland soils, cultivated soils and desert soils (Klich 2002). In the soil *Aspergillus* sp. function as saprophytes, and as with most other soil fungi they play an important role in the decomposition of organic matter. Despite being very wide spread globally, meaning that most people are frequently exposed to the fungus, infections only occur very rarely in people with normal immune systems (Zieve et al. 2010).

13.5.3.2 *Blastomyces* ssp.

Blastomycosis refers to fungal infections caused by the dimorphic fungi *Blastomyces dermatitidis*, which is endemic in parts of North America, and *B. brasiliensis* (also known as *Paracoccidioides brasiliensis*) which is endemic in some areas of South America (Kayser et al. 2005). Dimorphism refers to the ability of these fungi to undertake growth forms, either hyphal growth like a mould, or single cellular growth like a yeast (Baumgardner 2012; Fig. 13.9). The mycelial form is commonly found in the soil and the yeast form in the infected tissues (Yildiz et al. 2016).

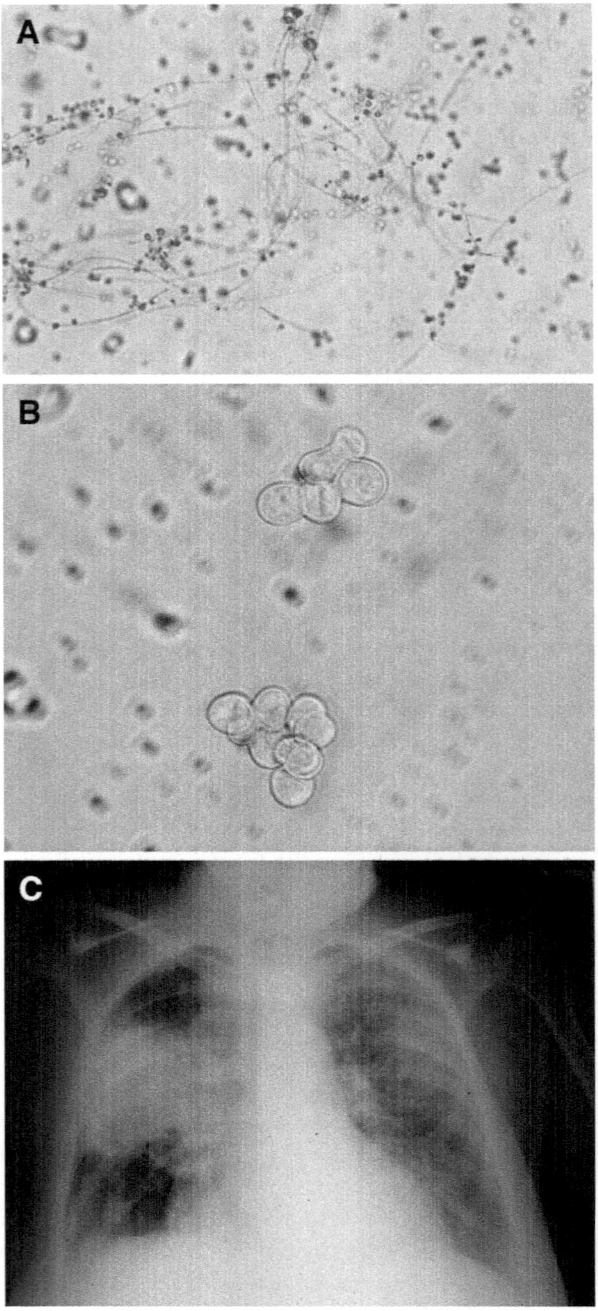

Fig. 13.9 *Blastomyces dermatitidis:* example of endemic dimorphic fungal pathogen causing primarily pulmonary infection. (**a**) Infectious mycelial (mold) forms of *B. dermatitidis* grown on Sabouraud dextrose agar at 20 °C (magnification 400). (**b**) Yeast forms of *B. dermatitidis* grown on brain-heart infusion agar at 37° (magnification 400). (**c**) Chest radiograph illustrating 2 of several nonspecific radiographic patterns of pulmonary blastomycosis. A dense opacity in the right mid-lung fields and patchy infiltrates in the left lung are seen in this adult woman (Baumgardner 2012, p. 740; reproduced with kind permission from the American Board of Family Medicine)

13.5.3.3 *Coccidiodes* ssp.

The fungi *Coccidiodes immitis* and *Coccidiodes posadasii* are the causative agents of coccidioidomycosis (also known as "San Joaquin Valley Fever" or "Desert Rheumatism") (Laniado-Laborín et al. 2012). Both species are morphologically identical but are distinct both genetically and epidemiologically. *Coccidioides* sp. are mainly found in alkaline, sandy soils from semi-desert and desert regions with hot summers, gentle winters, and annual rainfall between 100 and 500 mm. These fungi are usually found 10–30 cm beneath the surface (Brown et al. 2013). While *C. immitis* is endemic to the San Joaquin valley region of California, *C. posadasii* is endemic to desert areas of South and Central America, Northern Mexico and South-western parts of the USA (Oppenheimer et al. 2010). Cases of coccidioidomycosis have also been reported in Europe, however, these have been associated with patients who have travelled to the USA and other places where the fungus is endemic (Jeffery and van der Putten 2011).

13.5.3.4 *Histoplasma capsulatum*

The fungus *Histoplasma capsulatum* causes histoplasmosis. It is also known as Darling's disease (first described by Darling 1906 concerning a case that looked like disseminated tuberculosis), Ohio valley disease or Cave disease. The infection occurs due to the inhalation of spores disturbed from the soil or from bird or bat droppings. It is endemic to the Ohio, Missouri and Mississippi river valleys in the USA as well as other river valleys of North and Central America, in caves in southern and East Africa and eastern and southern Europe, eastern Asia and Australia (Fayyaz and Lessnau 2010). *Histoplasma capsulatum* is a dimorphic fungus which grows in mycelial form in the soil at ambient temperatures but which grows in a single celled yeast form at body temperature in mammals (Kauffman 2007; Fig. 13.10). The most infectious soil occurs in areas inhabited by birds and bats. However, birds cannot transmit the diseases, while bats can. Bird excrement, however, enriches the soil and provides favourable conditions for the fungus to grow. In contrast, bats can become infected with the fungus and can transmit it through their guano. Owing to the fact that *Histoplasma capsulatum* can survive well and even thrive within the soil, contaminated areas can remain infectious for long (Fayyaz and Lessnau 2010). Soil samples from sites where birds have roosted have remained contaminated for at least 10 years after the roost has been cleared, even in urban areas (Moquet et al. 2012).

13.5.3.5 *Sporothrix schenckii*

Sporotrichosis is a chronic fungal skin infection which is caused by *Sporothrix schenckii* which is a dimorphic fungus (Lima Barros et al. 2011). In nature, the

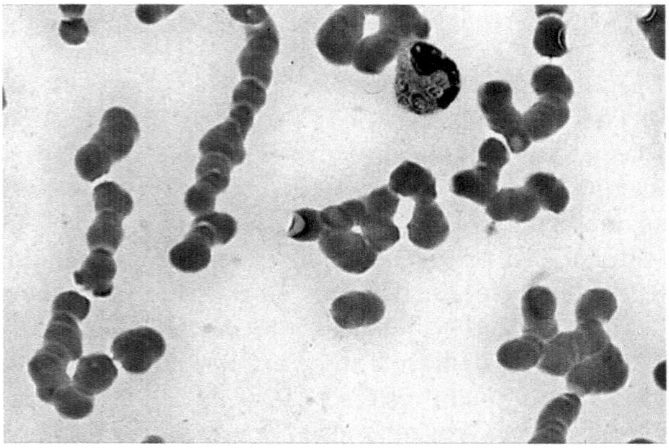

Fig. 13.10 Yeast forms of *Histoplasma capsulatum* found in a neutrophil on a peripheral blood smear (Kauffman 2007, p. 124; with kind permission from American Society for Microbiology)

fungus has been found to live as a saprophyte on living and decaying vegetation, animal excreta, and soil (Lima Barros et al. 2011). Organic material in soil is fundamental for mycelium development. The fungus thrives in soil plentiful in cellulose, with a pH range from 3.5 to 9.4 and a temperature of 31 °C (Noriega et al. 1993). In its saprophytic stage or when cultured at 25 °C, it assumes a filamentous form, composed of hyaline, septate hyphae 1–2 μm wide, with conidiogenous cells arising from undifferentiated hyphae forming conidia in groups on small, clustered denticles. The one-celled conidia are tear shaped to clavate and do not yield chains (Sigler et al. 1990). The transition from mold to yeast form in *S. schenckii* can be attained by culturing mycelia or conidia on rich culture media such as brain heart infusion agar at 35–37 °C (Morris-Jones 2002). This transition process also occurs after patients are infected with filamentous *S. schenckii*. The fungus is widely spread in the soil with a global distribution (Lima Barros et al. 2011) and may also be found on plant materials such as hay. Infection usually occurs via infection through broken skin while handling plant materials and as such is often associated with farmers, horticulturalist and gardeners. Disseminated sporotrichosis may also occur in immuno-compromised people who inhaled dust containing spores. Sporotrichosis has been sporadically associated with scratches or bites from animals such as mice, armadillos, squirrels, dogs, and cats (Kauffman 1999).

13.5.3.6 Fungi of the Order Mucorales

Ubiquitous filamentous fungi of the Mucorales order of the class of Zygomycetes cause an emerging angioinvasive infection called Mucormycosis. It is an opportunistic infection usually occurring in immuno-compromised patients such uncontrolled diabetes, hematological malignancies, renal failure, patients on chemotherapy, patients on long term steroids and AIDS (Fogarty et al. 2006). Nevertheless, there are few reported cases of mucormycosis occurring in immune-competent patients. Mucoralean fungi are thermotolerant and generally saprotrophic. They grow on organic matter such as dead plant or animal material within soil. However, some species of Mucorales are parasites or pathogens of plants or animals, including humans. Fungi from the genera *Rhizopus* are the most common causative agents of these diseases although fungi from the genera *Mucor and Absidi* are also frequently implicated in causing diseases (Ahamed and Thobaiti 2014).

13.5.4 Protozoa

Protozoa are a highly diverse group of single celled eukaryotic organisms. There are over 30,000 species of that only a small number are parasites of humans (Bultman et al. 2005). Protozoa commonly range in size from 0.01 to 0.1 mm. One of the most famous parasites is represented by *Plasmodium* sp., the causative agent of malaria. However, unlike the *Plasmodium* sp. which has mosquitoes as a vector, several groups can be soil-borne. Protozoa are distributed globally and are found in high abundances in soils where they feed on bacteria, algae, fungi and organic matter and in turn provide a food source for other soil invertebrates (Jeffery and van der Putten 2011). There are about 10,000–100,000 protozoa per gram of upper soil surface (Coyne 1999). The protozoan life cycle ranges from binary fission in a single host to many morphological transformations in a series of hosts. There are no eggs, larva, or adults.

13.5.4.1 *Entamoeba histolytica*

There are at least eight different amoebas (*E. histolytica*, *E. dispar*, *E. moshkovskii*, *E. coli*, *E. hartmanni*, *E. polecki*, *Iodamoeba butschlii*, and *Endolimax nana*) that live in the human intestinal lumen, however, these are usually accepted as commensals except for *E. histolytica* (Raza et al. 2013). The latter is the causative agent of amoebiasis which is distributed worldwide and is an anaerobic parasitic protozoan. It is most commonly found in tropical and subtropical regions. The highest incidences of amoebiasis are in developing countries, particularly in areas where there is inadequate sanitation and hygiene. Amoebiasis is second only to malaria with regard to deaths caused by protozoa globally (Jeffery and van der Putten 2011). Man is the definitive host. As with *G. lamblia, E. histolytica* exists in two forms: the active parasite (trophozoite) and the dormant parasite (cyst). The trophozoites, 10–60 μm in diameter, live in the intestine and feed on bacteria or on the wall of the intestine. They are expelled in feces and die rapidly. However, cysts (10–15 μm in diameter) expelled in the feces are very hardy and can survive weeks or months in soil. In water, cysts can live for up to 30 days. Nonetheless, they are rapidly killed by desiccation, and temperatures below 5 °C and above 40 °C (Raza et al. 2013). Mature cysts can infect hosts who later consume food or water contaminated with soil.

13.5.4.2 *Balantidium coli*

Balantidiasis is infection by *Balantidium coli*, the largest protozoan parasite (Bultman et al. 2005) and the only ciliate parasite known to infect humans. It is found worldwide but is most common in Latin America, Southeast Asia, The Philippines and Papua New Guinea (Jeffery and van der Putten 2011). The parasite inhabits a variety of hosts, especially primates (Nakauchi 1999). Humans are most commonly infected by contact with infected pigs. In some pig-raising areas of New Guinea, human infection rates are as high as 28% (Radford 1973). The life cycle and methods of transmission of the parasite are similar to *E. histolytica* (Bultman et al. 2005). The trophozoite of *B. coli* is 50–200 μm by 40–70 μm in diameter. Trophozoites are ovoid, with a cell membrane covered with uniform cilia. The cysts are spherical and vary from 45 to 75 μm in diameter. The latter can remain viable for several days in stool. In soil, *Balantidium coli* can survive for weeks to months in cyst form (Jeffery and van der Putten 2011). Humans are infected by ingesting cysts, the infective stage of *B. coli*, in contaminated water or food.

13.5.4.3 *Giardia lambila*

The flagellate protozoan *Giardia lambila* is the causative agent of the zoonotic disease giardiasis. It is the most common protozoan intestinal parasite worldwide which has been isolated from the stools of various animals including cats, dogs, sheep, cattle and various rodents. *G. lambila* trophozoites (the active stage of organism) live in the large intestine of infected humans or animals (Adam 2001; Fig. 13.11). At times they form cysts, millions of these (and trophozoites) are released in the feces and may enter the soil. *Giardia lambila* does not strongly bind to soil particles and in the event of rainfall leading to overland flow, the protozoan is easily washed away and can be transported for long distances (Dai and Boll 2006). Cysts of *G. lambila* are capable of surviving for up to 3 months in water at 4 °C. While data on survivability in soil is not currently available it seems likely that it will survive for a similar amount of time in soil in its resistant spore form, although this may well be dramatically reduced in dry soils (Jeffery and van der Putten 2011). The cysts can persist for some time (up to many months) in the environment, which includes soil, food, water, or surfaces that have been contaminated (Bultman et al. 2005). Infection results from ingestion of the cyst, usually in contaminated water or food.

13.5.4.4 *Toxoplasma gondii*

The protozoan *Toxoplasma gondii* is the causative agent of toxoplasmosis. *T. gondii* is very common and it is estimated that approximately one third of the world's population has been exposed to it (Jeffery and van der Putten 2011). It is an obligate

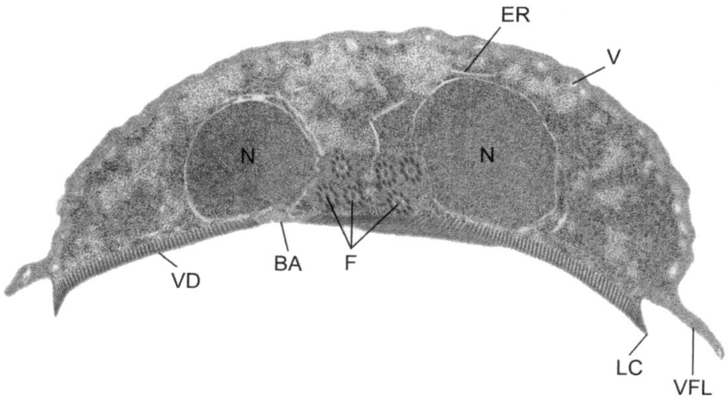

Fig. 13.11 *Giardia lamblia* trophozoite coronal section demonstrating the nuclei (*N*), endoplasmic reticulum (*ER*), flagella (*F*), and vacuoles (*V*). A mechanical suction is formed when the ventral disk (*VD*) attaches to an intestinal or glass surface. Components of the ventral disk include the bare area (*BA*), lateral crest (*LC*), and ventrolateral flange (*VLF*) (Adam 2001, p. 457; with kind permission from American Society for Microbiology)

intracellular parasite. Humans are only intermediate hosts of the protozoan, along with other mammals. Cats are the primary host for *T. gondii* who generally acquire the infection through consumption of infected rodents (Bultman et al. 2005). After a cat consumes the tissue containing cysts or oocysts, viable tachyzoites invade the small intestine. These eventually form oocysts that are excreted. *T. gondii* has been shown to be detectable in soils where infected cats defecate and so can be highly localised (Afonso et al. 2008). When in the soil, the oocysts of *T. gondii* have been found to be able to survive for at least 3 months (Lindsay et al. 2002) which means that soils can remain infectious for long periods of time after the originally infected individual has left.

There are three infective stages of *T. gondii*, a rapidly dividing invasive tachyzoite, a slowly dividing bradyzoite in tissue cysts, and an environmental stage, the sporozoite, protected inside an oocyst. These infective stages are crescent-shaped cells, approximately 5 μm long and 2 μm wide, with a pointed apical end and a rounded posterior end (Robert-Gangneux and Dardé 2012). The life cycle of *T. gondii* includes two phases called the intestinal (or enteroepithelial) and extraintestinal phases (Bultman et al. 2005). The intestinal phase occurs in cats only (wild as well as domesticated cats) and produces oocysts. The extraintestinal phase occurs in all infected animals (including cats) and produces tachyzoites and, eventually, bradyzoites or oocysts. Humans can become infected in several ways including the ingestion of cysts through contaminated food or soil or ingestion of undercooked meat (e.g., lamb, pork, or beef) infected with cysts (Bultman et al. 2005). Toxoplasma infection after soil contact or ingestion is particularly high for children.

13.5.4.5 *Cryptosporidium* ssp.

The coccidial, spore-forming parasites *Cryptosporidium hominis* and *Cryptosporidium parvum* in humans usually cause cryptosporidiosis (Abubakar et al. 2007). Cryptosporidiosis is generally a water-borne disease, but *Cryptosporidium* sp. only reach the environment through human and animal waste. *C. parvum* can survive in surface water for 6 months and in liquid manure tanks for many months (Bultman et al. 2005). *Cryptosporidia* is capable of surviving in soil for weeks or months in cyst form. The organism must travel through soils to become water-borne in many cases. Like other protozoa *Cryptosporidia in cyst form* are non-motile. However, they have been found to leach down through soil due to rainfall (Mawdsley et al. 1996) and as such may possibly migrate from the initial area of contamination, particularly if they enter a water source such as a river.

Many animal species, including man, act as a reservoir for *C. parvum*. Infected hosts excrete sporulated oocysts in the feces. In this fashion, oocysts may enter the soil as animal or human waste products. Infection begins when oocysts are ingested, most often through contaminated water and food or by direct fecal-oral transmission (Bultman et al. 2005). After ingestion, the oocysts mature to sporozoites and parasitize the epithelial cells of the gastrointestinal tract. These parasites then undergo asexual followed by sexual reproduction. As with the other protozoan diseases cryptosporidiosis is usually spread through the faecal-oral route and so is most prevalent in areas of poor sanitation.

13.5.4.6 *Cyclospora cayetanensis*

The coccidian protozoan *Cyclospora cayetanensis* causes cyclosporiasis in humans, a diarrheal disease found worldwide (Ghimire and Sherchan 2008). Little is known about possible animal reservoirs. *C. cayetanensis* is most commonly found in tropical and subtropical regions (Bultman et al. 2005). The disease has a life cycle similar to *Cryptosporidium parvum* with the exception that when passed in the feces, the oocyst is not infective. Freshly passed oocysts sporulate after spending days or weeks in the environment at temperatures of between 22 and 32 °C and become infective (Bultman et al. 2005). The protozoan is capable of surviving for extended periods in soil and as such contact with contaminated soil has been suggested to be an important mode of transmission (Chacín-Bonilla 2010). Humans are also infected when they ingest the sporulated oocysts by swallowing contaminated food or by drinking contaminated water (Ghimire and Sherchan 2008).

13.5.5 Helminths

All human pathogenic helminths are parasites and have man as their definitive host. Most inhabit the human intestines at some point in their life cycle (Bultman et al. 2005). Some are systemic in the lymph system or in other tissue. Helminths can be grouped into the nematodes (including hookworms, roundworms, whipworms, and pinworms), the trematodes (flukes), and the cestodes (tapeworms) (Bultman et al. 2005). Human diseases caused by cestodes and trematodes are called saprozoonoses which are zoonotic diseases where the transmission of the disease requires a non-animal development site or reservoir. In many cases this site is the soil.

Most human pathogenic helminths are nematodes of which there are approximately 10,000 species and approximately 1000 of these are found in soils (Bultman et al. 2005). Almost all soil-borne human pathogenic nematodes inhabit the intestines. The main soil-transmitted human pathogenic nemathodes, *Ascaris lumbricoides*, *Trichuris trichiura*, *Ancylostoma duodenale* and *Necator americanus* cause very common clinical disorders in humans (Bethony et al. 2006; Fig. 13.12).

Trematodes are soft-bodied invertebrate animals with bilateral symmetry that are also called flukes. They can cause parasitic infections in humans (Bultman et al. 2005). Trematodes have complex life cycles that always involve an intermediate host that is a mollusk. Trematodes and cestodes are members of a group of animals that are commonly called flatworms, because most species are flattened dorsoventrally. This shape is due to the fact that they must respire by diffusion and no cell can be too far from the surface. Trematodes and cestodes have only one opening to the gut, which must both take in food and expel waste. Several *Schistosoma* species are very important human pathogens.

Cestodes are often referred to as tapeworms because of the shape of their long ribbon-like body. They resemble a colony of animals in that their bodies are divided into a series of segments each with its own set of internal organs (Bultman et al. 2005). Adults of the species can reach 100 m in length. There are several parasitic tapeworm infections where man is the definitive host. Two examples of soil-borne tapeworms are *Taenia saginata* and *Taenia solium* both of which have a periodic soil residency. The bodies of cestodes can be divided into three regions: the scolex or head, the neck, and the strobila. The strobila is composed of a series of segments called proglottids and each proglottid has its own complete set of internal reproductive organs. As the organism grows, proglottids are added, and the result may be thousands in a mature animal. This pattern of growth forms a long, ribbon-like body referred to as a tapeworm (Bultman et al. 2005). Mature proglottids containing eggs which exit the body in feces and enter the environment.

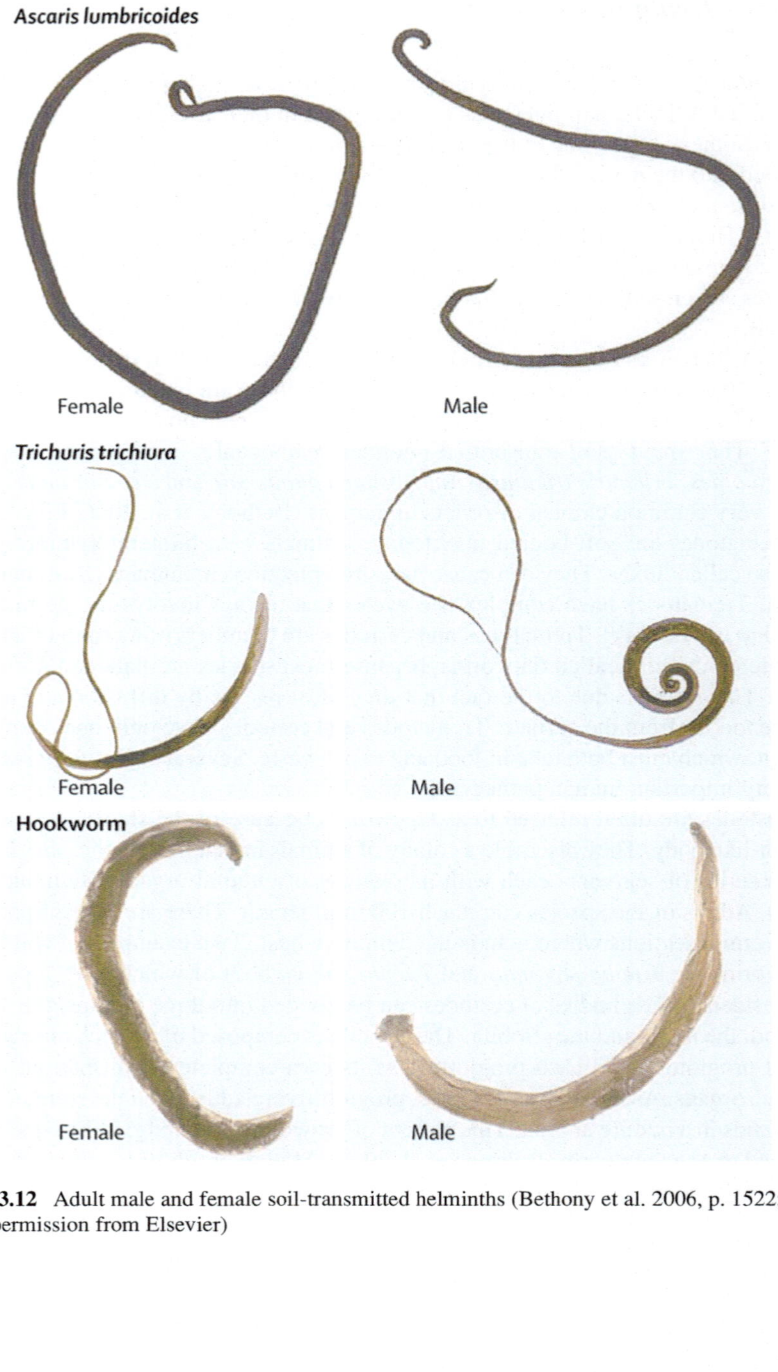

Fig. 13.12 Adult male and female soil-transmitted helminths (Bethony et al. 2006, p. 1522; with kind permission from Elsevier)

13.5.5.1 *Ascaris lumbricoides*

The pathogen is the causative agent of ascariasis is which is the largest intestinal nematode and the most common form of human helminth infection worldwide (Jeffery and van der Putten 2011). Ascariasis is most common in tropical and sub-tropical areas and is generally associated with poor sanitation. Adult *A. lumbricoides* are capable of reaching more than 30 cm in length, which is more than two orders of magnitude greater than most nematodes which are found in the soil. The life cycle of *Ascaris* involves eggs from an infected individual being excreted along with faeces which may contaminate the soil if a person defecates outside, or if the faeces of an infected individual are used as a fertilizer (Jeffery and van der Putten 2011). *Ascaris* eggs are more resistant to desiccation than hookworm or *Trichuris* eggs. Given favorable environmental conditions, they have been reported to survive for up to 15 years (O'Lorcain and Holland 2000). They are also very sticky, and attach easily to fruit, vegetables, soil and dust particles, children's toys, currency notes, flies and cockroaches (Scott 2008). Transmission of Ascaris eggs is typically associated with accidental ingestion of soil, but deliberate ingestion of soil and ingestion of contaminated vegetables, greens and fruit are also important (Scott 2008).

13.5.5.2 *Trichuris trichiura*

Trichuris trichiura, also known as whipworm, is the causative agent of trichuriasis which is a very common intestinal parasite with approximately one quarter of the world's population being though to carry the nematode (Jeffery and van der Putten 2011). The name "whipworm" is due to their shape as they have a wider posterior end with a longer, thinner anterior end which means that they can look like a whip with a handle. Eggs which are deposited onto the soil become embryonated after 10–14 days in the soil which is the infective stage of their life cycle. People become infected with *T. trichiura* by ingesting eggs from soil in areas of poor sanitation or via consumption of contaminated food (e.g., fruits and vegetables). Children are particularly vulnerable to infection (Bultman et al. 2005) because of their high exposure risk and because partial protective immunity is thought to develop with age. *T. trichiura* larvae develop within the small intestine before passing into the cecum where they penetrate the mucosa before going on to complete their development into adult worms in the large intestine. Adult worms bury their thin, threadlike anterior parts into the intestinal mucosa and feed on tissue secretions (not including blood). Females produce up to more than 5000 eggs per day (Bethony et al. 2006). The eggs are excreted with the feces and undergo a maturation phase in the environment. Approximately 3 months is needed for eggs to develop into mature worms which may live for between 1 and 5 years (Jeffery and van der Putten 2011). *T. trichiura* is often found in association with other helminth infections which inhabit similar conditions such as *A. lumbricoides.*

13.5.5.3 *Ancylostoma duodenale* and *Necator americanus*

Both species are parasitic nematodes which infect humans. They are also are known as hookworms and are found throughout the subtropics and tropics, with overlapping distributions, although *A. duodenale* is most common in North Africa, India and the Middle East, and formerly in southern Europe (Jeffery and van der Putten 2011). *N. americanus* on the other hand, is most common in the Americas, Sub-Saharan Africa, Indonesia, China and Southeast Asia. Hookworms are considerably smaller than *Ascaris lumbricoides*. Humans are the only known reservoir for hookworms. Soil-transmitted helminths do not reproduce within the host. Climate is an important determinant of transmission of these infections, with adequate moisture and warm temperature essential for larval development in the soil (Bethony et al. 2006). Infection rates can be as high as 80% in lesser developed countries in the most tropics, but is only 10–20% in drier climates, probably owing to the relatively poor survivability of the eggs and larvae in dry soils. Particle size distribution of soils and organic matter content have also been shown to play an important role in the survival of hookworm larvae in soils (Mabaso et al. 2004) possibly due to their impact on soil water retention. Infection occurs through skin contact with contaminated soil or through ingestion of eggs or larvae via the hand to mouth route after touching infected soil or water.

13.5.5.4 *Schistosoma* ssp.

Schistosoma mansoni, S. japonicum, and S. haematobium are the three species of trematodes that cause the most prevalent form of the disease. These trematodes are unusual in that they reproduce sexually and that they live in the mesenteric veins that lie outside of the liver (Bultman et al. 2005). The life expectancy of an adult is from 10 to 25 years. Adult schistosomes are 7–20 mm in length with a cylindrical body that features two terminal suckers, a complex tegument, a blind digestive tract, and reproductive organs (Gryseels et al. 2006). The females produce hundreds (African species) to thousands (oriental species) of eggs per day. Each ovum contains a ciliated miracidium larva which secretes proteolytic enzymes that help the eggs to migrate into the lumen of the bladder (*S. haematobium*) or the intestine (other species). The eggs are excreted in the urine or faeces and can stay viable for up to 7 days. On contact with water, the egg releases the miracidium. It searches for the intermediate host, freshwater snails, guided by light and chemical stimuli. After penetrating the snail, the miracidia multiply asexually into multicellular sporocysts and later into cercarial larvae with embryonic suckers and a characteristic bifurcated tail (Gryseels et al. 2006). The cercariae start leaving the snail 4–6 weeks after infection and spin around in the water for up to 72 h seeking the skin of a suitable definitive host. On finding a host, the cercariae penetrate the skin, migrate in the blood via the lungs to the liver, and transform into young worms or schistosomulae. These mature within 4–6 weeks in the portal vein, mate, and migrate to their perivesicular or mesenteric destination where the cycle starts again (Gryseels et al. 2006). The lifespan of an adult schistosome averages 3–5 years but can be as long as 30 years. The theoretical reproduction potential of one schistosome pair is up to 600 billion schistosomes.

13.5.5.5 *Taenia solium*

Taenia solium, the pork tapeworm, can cause both taeniasis and cysticercosis in humans (Bultman et al. 2005). The natural life cycle of *T. solium* includes humans as the only definitive hosts carrying the intestinal adult tapeworm, and pigs as the intermediate hosts infected with the metacestode larval stage (cysticercus), generally in the muscular tissue (Rabiela et al. 2000). *T. solium* has less than 1000 proglottids and is 2–7 m in length. Their proglottids each contain about 50,000 eggs. When the proglottids mature, eggs or proglottids are shed in the feces and into the soil. If swine ingest them, they mature to onchospheres that move to the muscles and grow into the larval form of *T. solium, Cysticercus cellulosae* (Bultman et al. 2005). Humans acquire a *T. solium* tapeworm infection by consumption of undercooked pork containing viable cysticerci. *T. solium* adults can live up to 25 years in the human intestine.

13.5.5.6 *Taenia saginata*

The adult *T. saginata* is found in man only and ranges from 5 to 15 m in length (Gebrie and Engdaw 2015). The life cycle of *T. saginata* is similar to that of *T. solium* except cattle are an intermediate host of *T. saginata*. The strobila of *T. saginata* is composed of 1000–2000 proglottids. The gravid proglottids, can contain more than 100,000 eggs which are either expelled from proglottid or released when it disintegrates (Gebrie and Engdaw 2015). Adult tapeworms have a life span of some years during which they produce millions of eggs which are intermittently released either free or as intact proglottids into the environment with the faeces (Allan et al. 1996). The eggs can survive in the environment for months or years. Cattle become infected by ingesting vegetation contaminated with eggs (or proglottids). The eggs develop in the intestines and release onchospheres that evaginate and invade the intestinal wall (Bultman et al. 2005). They then migrate to the striated muscles and develop into a cysticercus, which is capable of surviving for several years. Man becomes infected by ingesting raw or inadequately cooked meat. The cysticercus in man forms into adult tapeworm in the small intestine, starting the cycle all over again (Gebrie and Engdaw 2015).

13.6 Symptoms and Treatment of Diseases

13.6.1 Viral Diseases

13.6.1.1 Hantaviral Diseases

Hantaviruses have the potential to cause two different types of diseases in humans: hemorrhagic fever with renal syndrome (HFRS) and hantavirus pulmonary syndrome (HPS). The first outbreak of HFRS occurred during the Korean War and so HFRS was initially called Korean hemorrhagic fever (Bi et al. 2008). HFRS denotes a group of clinically similar illnesses that occur throughout the Eurasian landmass and adjoining areas, while HPS, recognized as a clinical entity since 1993, represents the prototype of emerging diseases occurring in the Western hemisphere (Vapalahti et al. 2003). HFRS manifests as mild, moderate, or severe disease, depending in part on the causative virus (Vapalahti et al. 2003). However, HFRS is the less fatal form of hantavirus infection with mortality rates in the region of 12% as compared to a 60% mortality rate with HPS (Jonsson et al. 2010). The clinical course of HPS is divided into three periods: the febrile prodrome, cardiopulmonary stage and convalescence. There is a 14–17 day incubation period after exposure followed by the prodrome phase typically lasting 3–6 days with myalgia, malaise, and fever of abrupt onset in the absence of cough and coryza. Other early symptoms include gastrointestinal disturbance, headache and chills (Bi et al. 2008).

Currently, there is no specific therapy available for both HFRS and HPS. The cornerstone of treatment thus includes supportive measures. The management should include early admission to an intensive care unit where blood and tissue oxygenation, cardiac output, central blood pressure and cerebral pressure can be monitored (Lázaro et al. 2007). Maintaining fluids balance is very important; it must be carefully monitored according to the patient's fluid status, amount of diuresis, and kidney function. Usually one or two haemodialysis sessions are needed for HFRS treatment, while mechanical ventilation when indicated and appropriate use of pressures are crucial to HPS patients (Vapalahti et al. 2003). Extracorporeal membrane oxygenation has been found useful as a rescue therapy in patients with severe HPS (Crowley et al. 1998). Although corticosteroids are not standard of care in the treatment of hantaviral infection, steroid was used to treat severe HFRS and HPS cases (Bi et al. 2008). Ribavirin was demonstrated to have antihantaviral effect both in vitro and in vivo by causing error catastrophe during hantavirus replication (Huggins et al. 1991).

13.6.1.2 Diseases Caused by Enteroviruses

Up to now, the mechanisms of enteric virus-induced disease have not been well-characterized (Palacios and Oberste 2005). However, much of the disease presumably results from tissue-specific cell destruction, but some disease manifestations, for example, enteric virus exanthemas and myocarditis are thought to result from the host immune response to the infection (de Crom et al. 2016). The primary replication sites of enteroviruses are the epithelial cells of the oropharyngeal and intestinal mucosa. Although some replication may occur in the nasopharynx with spread to upper respiratory tract lymphatics, most of the viruses are swallowed and transferred to the stomach and lower gastro-intestinal tract where they presumably bind to specific receptors on enterocytes (de Crom et al. 2016). The viruses cross the intestinal lining cells and reach the Peyer's patches in the lamina propria, where significant viral replication occurs. This is followed by a viremia that may lead to a secondary site of tissue infection (Palacios and Oberste 2005). Secondary infection of the central nervous system results in encephalitis or meningitis. Enteric viruses are the most important cause for viral meningitis, accounting for approximately 90% of all cases for which an etiological agent was identified (de Crom et al. 2016). Other tissue-specific infections can result in myocarditis or pleurodynia. Disseminated infection can lead to exanthemas, nonspecific myalgias, or severe multiple organ disease. After a primary infection with enteroviruses, there is still a possibility of viral shedding in the feces and respiratory system for several weeks (Kapusinszky et al. 2012).

Although enteric viruses bare a frequent cause of serious infection, particularly in children, there are limited tools available to fight these viruses. The standard therapy for aseptic meningitis caused by enterovirus infection continues to be intravenous immunoglobulin treatment. This passive immunization-based therapy might neutralize infectious virus circulating within the host, in addition to other non-specific inflammatory mechanisms (Ooi et al. 2010). In addition, novel antiviral therapies against enteric viruses continue to be developed (Wu et al. 2010). Vaccines are only available against Poliovirus (van der Linden et al. 2015). Advances have been made towards the development of a vaccine against the enterovirus serotype EV-71 (de Crom et al. 2016). However, developing vaccines against all members of the enterovirus genera is not feasible due to the large number of serotypes.

13.6.1.3 Hepatitis A

After ingestion, the hepatitis A virus (HAV) passes through the gastrointestinal tract and is carried to the liver (O'Connor 2000). The predominant site of replication is the lymphoid tissue of the gut, and the virus is therefore typically detectable in stool specimens. HAV replication in the liver causes damage to liver cells, which is known as hepatitis. In immune individuals, however, circulating antibodies prevent HAV from infecting the liver. There seems to be only one antigenic type of HAV, which elicits lifelong immunity. HAV tends to cause infections that are mild or without clinical symptoms in children. The incidence of infection is closely linked with hygiene and sanitation conditions, and most people in developing countries contract infections during early childhood. Typical clinical symptoms of infection are predominantly seen in adults. Although mortality is generally less than 1%, the disease may be quite severe and incapacitating. There may be substantial liver damage, and regeneration of the liver cells takes time (Zuckerman and Thomas 1993). Patients may feel ill and be confined to bed for up to 6 weeks or more. They commonly lack interest in foods that depend heavily on liver functions for digestion. The severity of illness and mortality may be dependent on underlying conditions such as immunodeficiencies and malnutrition, and on the general state of health. In adult populations of developing countries and communities, immunity to HAV may exceed 95%, in contrast to developed countries and communities where levels of immunity may be less than 50% (Iwarson 1992; Sathar et al. 1994; Tucker et al. 1996). People from developed countries who visit developing areas are therefore exposed to a high risk of infection. HAV typically occurs in all parts of the world and, beyond the link to standards of hygiene and sanitation, gives no indication of geographical preferences. HAV is not readily detectable by routine cell culture procedures. Many questions about the epidemiology of the virus, as well as about its occurrence and behaviour in the environment, therefore remain unanswered. However, there is little doubt that the virus is highly infectious and can cause explosive outbreaks when present in water or food. In addition, it is relatively resistant to unfavourable environmental conditions, including water treatment and disinfection processes.

The incubation period of hepatitis A may vary from 15 to 45 days, with a mean of 30 days (Reid and Dienstag 1997), i.e. some 10 days less than the incubation period of hepatitis E. Faecal excretion of HAV begins late in the incubation period, peaks just before onset of clinical symptoms of disease (usually the appearance of dark urine), and falls to barely detectable levels as the clinical illness evolves. The virus is present in blood in relatively low numbers for about 7–14 days, with a peak before the onset of clinical symptoms (Zuckerman and Thomas 1993).

Treatment for acute hepatitis caused by (HAV) is necessarily supportive in nature, because no antiviral therapy is available. Hospitalization is warranted for patients whose nausea and vomiting places them at risk for dehydration. Patients with acute liver failure require close monitoring to ensure they do not develop fulminant hepatic failure, which is defined as acute liver failure that is complicated by hepatic encephalopathy.

For people traveling to regions where HAV is endemic, vaccination is recommended. The inactivated HAV vaccines Havrix and Vaqta are administered as 1 mL (0.5 mL in children) intramuscular (IM) injections given more than 1 month before anticipated travel. This approach results in a better-than-90% likelihood of stimulating production of immunoglobulin G anti-HAV, with resulting immunity against HAV infection (Fiore et al. 2006). Passive postexposure immunization with hepatitis A immune globulin is an alternative to active immunization with HAV vaccine. Its effectiveness is highest when it is given within 48 h of exposure, but it may be helpful when given as far as 2 weeks into the incubation period.

13.6.1.4 Hepatitis E

Infection with hepatitis E virus (HEV) was initially referred to as enterically transmitted (or epidemic) non-A, non-B hepatitis. HEV had for many years been mistaken for HAV, because the two viruses share certain basic clinical and epidemiological properties (Purcell 1997). Both are transmitted primarily by the faecal-oral route, and are often associated with foodborne and waterborne outbreaks. However, viral hepatitis E tends to occur more often in young adults, many of whom are already immune to hepatitis A (Purcell 1997). In contrast to hepatitis A, which rarely causes complications, hepatitis E tends to give rise to more prominent cholestasis and the infection can present as acute fulminating hepatitis, particularly in pregnant women, for whom case fatality rates as high as 20–40% have been recorded. Hepatitis E has an incubation period of 14–16 days, with a mean of 40 days (Reid and Dienstag 1997), which is longer than that of hepatitis A. An exceptionally long viraemia is typical for HEV infection, generally lasting for as long as 6 weeks, and in some cases up to 16 weeks which is substantially longer than for HAV. Patients generally excrete HEV for 1–2 weeks. In one case, however, the virus was excreted for more than 7 weeks, well after clinical and biochemical recovery (Scharschmidt 1995; Purcell 1997).

There is no specific treatment capable of altering the course of acute hepatitis E. As the disease is usually self-limiting, hospitalization is generally not required. Hospitalization is required for people with fulminant hepatitis, however, and should also be considered for symptomatic pregnant women. Immunosuppressed people with chronic hepatitis E benefit from specific treatment using ribavirin, an antiviral drug. In some specific situations, interferon has also been used successfully.

13.6.2 Bacterial Diseases

13.6.2.1 Actinomycosis

Multiple different clinical features of actinomycosis have been described, as various anatomical sites (such as face, bone and joint, respiratory tract, genitourinary tract, digestive tract, central nervous system, skin, and soft tissue structures) can be affected. In any site, actinomycosis frequently mimics malignancy, tuberculosis, or nocardiosis, as it spreads continuously and progressively, and often forms a cold abscess (Smego and Foglia 1998; Wong et al. 2011).

Cervicofacial actinomycosis is the most frequent clinical form of actinomycosis, and "lumpy jaw syndrome", which is associated with odontogenic infection, is the most common clinical manifestation, representing approximately 60% of all reported cases (Smego and Foglia 1998; Wong et al. 2011; Oostman and Smego 2005). *Actinomyces* spp. could also be responsible for maxillary osteomyelitis in patients with odontogenic maxillary sinusitis. Although cervicofacial actinomycosis is the most frequent form of actinomycosis with bone involvement, *Actinomyces* spp. could also be involved in extrafacial bone and joint infection. Various clinical forms of extrafacial bone and joint actinomycosis have been described: (1) hematogenous spread of localized actinomycosis; (2) contiguous spread of pulmonary actinomycosis to the spine; and (3) polymicrobial bone and joint infection following bone exposition, especially in patients with paraplegia and osteomyelitis of the ischial tuberosity (Wong et al. 2011). Genitourinary tract actinomycosis is the second most frequent clinical form of actinomycosis. The main clinical feature of genitourinary tract actinomycosis is pelvic actinomycosis in women using an intrauterine device (Garner et al. 2007). However, other clinical presentations have been described, such as primary bladder actinomycosis and testicular actinomycosis (Bae et al. 2011).

Pulmonary actinomycosis is the third most common type of actinomycosis, after that occurring in cervicofacial and abdominopelvic locations. In children, pulmonary involvement is uncommon (Bates and Cruickshank 1957). The peak incidence is reported to be in the fourth and fifth decades of life (Apothéloz and Regamey 1996). And males are more often affected than women, with a 3:1 ratio. Pulmonary actinomycosis results mainly from aspiration of oropharyngeal or gastrointestinal secretions (Apothéloz and Regamey 1996). Consequently, individuals with poor oral hygiene, pre-existing dental disease, and alcoholism have an increased risk for developing pulmonary actinomycosis (Brown 1973; Cohen et al. 2007). Patients with chronic lung disease such as emphysema, chronic bronchitis, and bronchiectasis, and patients with pulmonary sequelae following tuberculosis, are considered to also be at risk for pulmonary actinomycosis (Bates and Cruickshank 1957).

Bacterial cultures and pathology are the cornerstones of diagnosis and require particular attention to prevent misdiagnosis. Prolonged bacterial cultures in anaerobic conditions are necessary for identification of the bacterium, and typical microscopic findings include necrosis with yellowish sulfur granules and filamentous Gram-positive fungal-like pathogen. Patients with actinomycosis require prolonged (6–12 month) high doses of penicillin G or amoxicillin, but the duration of antimicrobial therapy could likely be reduced (3 months) for patients in whom optimal surgical resection of infected tissues has been performed (Smego and Foglia 1998; Wong et al. 2011). Specific preventive measures (reduction of alcohol abuse, dental hygiene, change of IUD every 5 years) may limit the occurrence of actinomycosis.

13.6.2.2 Anthrax

Anthrax is an endemic disease in some countries in the world and has become a re-emerging disease in western countries with recent intentional outbreak. The main route of transmission is contact with or inhalation of *Bacillus anthracis* spores. Human cases may occur in an agricultural, an industrial environment or as a deliberately caused disease (Doganay and Metan 2009). Cutaneous, respiratory, and gastrointestinal forms result after exposure to anthrax spores. Cutaneous anthrax accounts for 95% of human cases globally. The respiratory form occurs most likely due to inhalation of the bacterial spores, whereas the gastrointestinal form happens after spores' ingestion. Data from preantibiotic and vaccine days indicated that 10–40% of untreated cutaneous anthrax cases might be expected to result in death (Doganay et al. 2010). With the treatment, <1% cases are fatal. The clinical picture varies from mild to severe form. Cutaneous anthrax can be self-limiting, and lesions resolve without complications or scarring in 80–90% of cases with treatment. Extensive edema and toxemic shock can be seen as a rare and potentially life-threatening complication of cutaneous anthrax (Doganay and Metan 2009).

Prophylactic, early diagnosis and proper treatment will reduce mortalities of anthrax (Doganay et al. 2010). The therapy may include volume replacement including fresh plasma and antibiotic administration. Corticosteroids and dopamine may be also given. Penicillin G is still the drug of choice, and doxycycline or ciprofloxacin are accepted as the best alternatives in the treatment of naturally-occurring anthrax. In the World Health Organization (WHO) Guidelines, intramuscular procaine penicillin, oral amoxicillin or penicillin V are recommended for the treatment of mild uncomplicated cases of cutaneous anthrax (Doganay et al. 2010). Intravenous penicillin G is recommended in cutaneous anthrax with extensive edema (Turnbull 2008). WHO Guidelines is suggesting to continue antimicrobial therapy for 3–7 days in uncomplicated cutaneous anthrax although there is no controlled clinical study about the duration of treatment in cutaneous anthrax (Doganay et al. 2010). Antibiotic treatment does not affect the progress of the lesion or other toxin-related systemic damage, and it does not alter the evolutionary stages. However, early treatment will limit the size of lesion. For this reason, early diagnosis of anthrax and early initiation of therapy are very important.

13.6.2.3 Disease Related to *Bacillus cereus*

Gastroenteritis related to *Bacillus cereus* is an important food-borne disease worldwide. Pathogenicity is aided by a variety of toxins. Emetic toxin (a peptide) induces nausea and vomiting a few hours after ingesting a meal contaminated with the toxin (Bottone 2010). A diarrheal syndrome results from protein enterotoxins produced in the small intestine. *B. cereus* also has been associated with pulmonary infections mimicking anthrax (perhaps by inhalation of contaminated dust), invasion of the oral cavity or upper respiratory tract in immunosuppressed patients, bone and soft tissue, and central nervous system infections, as well as endophthalmitis, often from a foreign body contaminated with soil or dust (Bottone 2010).

The clinical spectrum of *B. cereus* infections is multifaceted, and therapeutic options usually revolve around the antibiotic susceptibility pattern of the isolated strain. In general, most *B. cereus* isolates are resistant to penicillins and cephalosporins as a consequence of β-lactamase production. In the setting of a suspected *B. cereus* infection, empirical therapy may be necessary while awaiting the antibiotic susceptibility testing profile. Resistance of *B. cereus* to erythromycin, tetracycline, and carbapenem has been reported (Bottone 2010), which may complicate the selection of an empirical treatment choice. To address this issue, several investigators have undertaken in vitro susceptibility studies. For example, Luna et al. (2007) tested 42 *B. cereus* isolates that showed resistance to clindamycin, erythromycin, and trimethoprim-sulfa-methoxazole. Regarding newer antibiotics, Luna et al. (2007) extended their spectrum of therapeutic/prophylactic antimicrobials, which included gatifloxacin, levofloxacin, moxifloxacin, rifampin, daptomycin, and linezolid, to which all 42 isolates were 100% susceptible.

13.6.2.4 Botulism

Botulism is a severe neuroparalytic disease. *C. botulinium* produce a total of seven distinct toxins (called A to G), all of which have similar pharmacological actions (Smith 1979). Botulinum toxin is a dichain polypeptide which is the most lethal toxin and all seven types act in similar ways. Of these groups, only types A, B, E and more rarely F are capable of causing botulism in humans. Types C, D, E and G cause illness in other mammals, birds and fish (WHO 2002). It is an important disease in the world, particularly in the farm subject to periods of protein and phosphorous deficiency. It causes high mortality due to neurological disorder (Desta et al. 2016). Symptoms usually present within 12–36 h and are caused by toxins produced by the bacteria rather than by the organisms themselves. However, thorough cooking of contaminated foods (heating to >85 °C for 5 min or boiling for a few minutes) is sufficient to destroy the toxins as well as the bacteria (WHO 2002). Early symptoms usually include fatigue, weakness and vertigo, followed by a dry mouth and difficulty swallowing and speaking, and blurred vision. The disease can progress to weakness in the neck and arms after which respiratory muscles and muscles of the lower body become affected and this paralysis may make breathing difficult (WHO 2002).

Diagnosis is based on clinical sign and laboratory examination. The first critical therapeutic step that is given for botulism intoxicated individuals is polyvalent antitoxin which is effective against circulating toxin before reaching the neuromuscular junction. Equine antitoxin should be administered by infusion (Shapiro et al. 1998). Because of the risk of an allergic reaction to the equine serum, the patient should be asked about past history of asthma, hay fever or allergic reactions when in contact with horses. Antibiotic administration is indicated for inhalation pneumonia or wound infection (Desta et al. 2016). Although botulinum toxin is hazard to life, it is used as drug by using molecular technique. This toxin is a group of highly potent drugs with specific mechanism of action. Careful use of botulinum toxin and imparting knowledge about its various clinical applications to the physicians will ensure that it will be an important treatment option for improving quality of life of patients.

13.6.2.5 Campylobacteriosis

The incidence of *Campylobacter* infection varies throughout the world but appears to be declining in industrialized countries. The infectious dose of *C. jejuni* varies depending on the strain but may be as low as 500 organisms in milk (Robinson 1981). Campylobacters cause a nonspecific acute inflammatory enteritis involving the colon and small intestine. Edema of the infected area as well as an infiltrate composed of neutrophils and mononuclear cells is seen histologically (Young et al. 2007). After oral ingestion, the pathogen moves through the intestinal mucus layer via its flagellum and multiplies in the distal ileum and colon. Campylobacters cause diarrhea by damaging the gut epithelial cells either directly by invading the cells or indirectly by initiating an inflammatory response (Ketley 1997).

Symptoms of campylobacteriosis are usually present within 2–5 days. They include fever, which can reach 40 °C, headaches, and diarrhea which is classified as inflammatory diarrhea that may be bloody (also known as dysentery). *C. fetus* may also cause bacteraemia (i.e. infection of the blood), usually as an opportunistic disease in immuno-compromised hosts, where it may be associated with systemic illness, meningitis, vascular infections, abscesses and/or nonspecific abdominal pain (Jeffery and van der Putten 2011).

Campylobacter infection should be suspected in patients with fever and acute diarrhea, particularly those with visible blood and mucus in the stool, including international travelers. Because the clinical presentation is similar to that seen with other common enteric bacterial pathogens such as *Salmonella, Shigella, Yersinia, Clostridium difficile,* and *E. coli* O157:H7, a presumptive diagnosis based on clinical presentation cannot be made. Diagnosis is made by isolating campylobacters from stool samples. Most cases of campylobacteriosis are self-limiting in immune-competent patients without systemic signs of infection, requiring only supportive treatment with adequate hydration. Illness with severe diarrhea, abdominal pain, or high fever are considerations for hospital admission and fluid replacement (Pitkanen et al. 1983). Carriage of *Campylobacter* organisms following infection is usually less than 3 weeks even in untreated patients (Blaser et al. 1980). Use of antibiotics will favor patients with visible blood in the stool, fever, a large number of stools, and/or worsening of symptoms, as with other inflammatory diarrheas (Ruiz-Palacios 2007). Pregnant women and individuals with immunosuppressive medical conditions, including HIV/AIDS, should also receive antibiotics (Allos 2001). *C. jejuni* have been historically sensitive to macrolides, tetracyclines, fluoroquinolones, aminoglycosides, imipenem, and chloramphenicol but resistant to trimethoprim (Allos 2001). With the introduction of the fluoroquinolones, ciprofloxacin became the mainstay of empiric treatment for acute community-acquired bacterial diarrhea and for travelers' diarrhea (Adachi et al. 2002). However, rapid emergence of fluoroquinolone-resistant *Campylobacter* strains was noted in Europe in the 1980s which coincided with the introduction of quinolone use in poultry. Azithromycin should be used for travelers' diarrhea due to *Campylobacter* infection and empirically where quinolone resistance is anticipated (Guerrant et al. 2001). More severe, systemic disease can be treated with a variety of intravenous antibiotics, including cefotaxime, imipenem, ampicillin, and parenteral aminoglycosides, but antimicrobial sensitivities should always be checked.

13.6.2.6 Leptospirosis

Leptospirosis has been identified as one of the emerging infectious diseases, exemplified by large outbreaks in Nicaragua, Brazil, India, Southeast Asia and the USA (Levett 2001). Pathogenic *Leptospira* sp. can enter the human body through ingestion of contaminated food or water, through broken skin or through mucous membranes. The bacteria then enter the blood and can migrate around the body, particularly to the kidneys, and within 7–10 days the bacteria can be passed in urine. The spectrum of symptoms is extremely broad. The classical syndrome of Weil's disease represents only the most severe presentation (Levett 2001). The clinical presentation of leptospirosis is biphasic, with an acute or septicemic phase lasting about a week, followed by the immune phase, characterized by antibody production and excretion of leptospires in the urine (Edwards and Domm 1960; Turner 1967). Most of the complications of leptospirosis are associated with localization of leptospires within the tissues during the immune phase and thus occur during the second week of the illness. Complications include meningitis, renal failure and liver damage.

Treatment of leptospirosis differs depending on the severity and duration of symptoms at the time of presentation. Patients with mild, flu-like symptoms require only symptomatic treatment but should cautioned to seek further medical help if they develop jaundice (Levett 2001). Patients who present with more severe anicteric leptospirosis will require hospital admission and close observation. If the headache is particularly severe, a lumbar puncture usually produces a dramatic improvement. The management of icteric leptospirosis requires admission of the patient to the intensive care unit initially. Patients with prerenal azotemia can be rehydrated initially while their renal function is observed, but patients in acute renal failure require dialysis as a matter of urgency.

Antibiotic treatment was reported with mixed results (Levett 2001). A major difficulty in assessing the efficacy of antibiotic treatment results from the late presentation of many patients with severe disease, after the leptospires have localized in the tissues. Doxycycline (100 mg twice a day for 7 days) was shown to reduce the duration and severity of illness in anicteric leptospirosis by an average of 2 days (McClain et al. 1984). Doxycycline (200 mg orally, once weekly) has been shown to be effective for short-term prophylaxis in high-risk environments (Gonsalez et al. 1998).

13.6.2.7 Listeriosis

Listeria monocytogenes has been shown to be of world-wide prevalence and is associated with serious disease in a wide variety of animals, including humans. Infection usually occurs in humans after eating food contaminated with the bacterium. It mainly affects infants, the elderly and immuno-compromised patients (Jeffery and van der Putten 2011). Symptoms of listeriosis usually include fever, vomiting and myalgia which last for 7–10 days. Infection causes a spectrum of illness, ranging from febrile gastroenteritis to invasive disease, including bacteraemia, sepsis, and meningoencephalitis.

Listeriosis often requires antimicrobial therapy. However, in the cases of bacteraemia, it is important to start treatment as early as possible, because of the severity of the disease and the high associated mortality. The choice of treatment consist of a β-lactam antibiotic, normally ampicillin. Because the penicillins are bacteriostatic, some studies have attempted drug combinations. For example, the simultaneous use of ampicillin and an aminoglycoside (usually gentamicin) is one of the most useful methods, especially in patients over age 50 (Temple and Nahata 2000). The dose is important in the treatment of invasive disease, which requires a dose of 6 g or higher (Temple and Nahata 2000). The duration of treatment of bacteraemia is usually about 2 weeks, in accordance with the clinical evolution. But the appropriate duration of treatment is not clear. After 2 weeks of treatment, there have been reported recurrences in immune-compromised patients. Therefore, it seems appropriate to prolong the time of therapy in such cases depending on the clinical manifestations of the patients (Lorber 2010).

13.6.2.8 Tetanus

Once in anaerobic tissue, *Clostridium tetani* spores convert to the vegetative form, multiply, and produce the neurotoxin tetanospasmin, which migrates to the central nervous system via a peripheral nerve at the site of infection (Baumgardner 2012). There may be no apparent local infection. The incubation period ranges from 3 to 21 days; the farther the site of initial spore contact from the central nervous system, the longer the incubation period and the milder the disease (Brook 2008; Afshar et al. 2011). Diagnosis of tetanus is based on the presentation of symptoms as there are currently no blood tests which are capable of detecting tetanus. Furthermore, *C. tetani* bacteria are only recovered from the wound in approximately 30% of cases, as well as occasionally being isolated from wounds of individuals who do not have symptoms of tetanus which is testament to how widespread this species of bacteria is.

The classical presentation of tetanus seen in patients begins with trismus or 'locked jaw' due to spasms of the masseter. Rigidity then spreads down the arms and trunks over the next 1–2 days, progressing to generalized muscle rigidity, stiffness, reflex spasms, opisthotonus and dysphagia (Rodrigo et al. 2014). Even minute sensory stimulation can precipitate prolonged spasms. The generalized spasms are also accompanied by autonomic disturbances, such as swings in blood pressure, arrhythmias, hyperpyrexia and sweating. Exhaustion, autonomic disturbances, and complications from muscle spasms (for example, asphyxiation, pneumonia, rhabdomyolysis, pulmonary emboli) can contribute to the high fatality rates observed in severe tetanus (Rodrigo et al. 2014).

The traditional management strategy in tetanus involves sedation, neuromuscular paralysis and elective ventilation combined with wound debridement, antibiotic therapy and administration of human tetanus immunoglobulin to neutralize the toxin. Sedatives used vary from benzodiazepines such as midazolam and diazepam to anesthetic agents such as propofol (Firth et al. 2011). Intravenous magnesium sulfate reduces muscle spasms and autonomic dysfunction (Karanikolas et al. 2010) but it may not be suitable as sole therapy to relieve spasms in severe tetanus and has no proven mortality benefit. Intrathecal baclofen is an effective option to relieve spasms till recovery but its use is limited due to costs and the risks of introducing concurrent central nervous system infection (Müller et al. 1987).

Benefits reported with dantrolene (e.g. Checketts and White 1993), botulinum toxin for local forms of spasms (e.g. García-García et al. 2007) in reducing muscle spasms and clonidine to reduce autonomic dysfunction need to be further evaluated before being recommended as standard therapy. The beneficial role of intrathecal human tetanus immunoglobulin or equine antitetanus sera is not well established (Rodrigo et al. 2014). However, the majority of studies are in favor of intrathecal administration. The use of this mode of administration should be at the treating physician's discretion. The evidence base for the efficacy of antibiotics in tetanus is limited. Metronidazole and penicillin can be used as the bacterium is susceptible to both. There is a theoretical advantage of using metronidazole but its clinical correlations have not been well established by trials (Rodrigo et al. 2014).

13.6.2.9 Tularemia

Francisella tularensis causes a zoonotic infection called Tularemia. There are four distinct subspecies of *F. tularensis,* however, disease is mainly caused by *F. tularensis* subspecies *tularensis* (type A) and *F. tularensis* subspecies *holarctica* (type B) (Nada and Harik 2013). Type A is more virulent and is primarily found in North America, whereas type B is found throughout the Northern Hemisphere, mainly in Europe and Asia, and causes milder infection than type A (Nada and Harik 2013).

Tularemia can be transmitted to humans by the bite of an insect vector such as a tick, deer fly or flea. Insects become infected when they feed on an infected animal. Ticks can also become infected by transovarian passage (Hopla 1974). Besides bites from arthropods, the diseases can be contracted through direct contact between a break in the skin with an infected animal or its dead body (most often rabbits, hence the name rabbit fever (Schaffner 2007). The disease may also be contracted through inhalation or ingestion of infected contaminated dust or soil.

F. tuuarensis is highly contagious. Inoculation or inhalation of as few as ten cells of *F. tularensis* bacteria can be sufficient to cause disease in humans (Abd et al. 2003). The incubation period after infection is usually 3–5 days after exposure and the illness usually starts very suddenly and may continue for several weeks. Symptoms include chills, fever, headache, joint stiffness and muscle pain and red spots on the skin which grow to become ulcers. Tularemia is fatal in approximately 5% of untreated cases, but less than 1% of treated cases (Dugdale and Vyas 2011).

There are six major tularemia clinical syndromes each with different clinical presentations: ulceroglandular tularemia, glandular tularemia, oropharyngeal tularemia, oculoglandular tularemia, typhodial tularemia, and pneumonic tularemia (Nada and Harik 2013). The diagnosis of tularemia is usually made clinically, taking into account exposure history and clinical manifestations and confirmed by serologic testing. Antibiotic therapy should be initiated as soon as tularemia is suspected, rather than awaiting results of serologic testing. The illness may be prolonged, complications are more likely to occur, and treatment failure is more frequent if antibiotic therapy is delayed (Snowden and Stovall 2011; Penn and Kinasewitz 1987). The aminoglycosides streptomycin and gentamicin are the drugs of choice for the treatment of tularemia (Nada and Harik 2013). Alternative antibiotic therapies for tularemia include doxycycline and ciprofloxacin. However, relapse is possible even after appropriate antibiotic therapy.

13.6.2.10 Gas Gangrene

Gas gangrene or Clostridial myonecrosis is a necrotic infection of skin and soft tissue and it is characterized by the presence of gas under the skin which is produced by pathogenic *Clostridium* (Aggelidakis et al. 2011). It is a potentially lethal disease which spreads quickly in soft tissues of the body. Gas gangrene can occur anywhere on the body, but it most commonly affects the arms or legs. Tissue necrosis is due to production of exotoxins by gas producing bacteria in an environment of low oxygen. In 80–90% of cases he causative agent of gas gangrene is the bacterium *Clostridium perfringens* (previously known as *Clostridium welchii*). The other 10–20% of cases are caused by the bacteria *C. novio, C. septicum, C. histolyticum, C. bifermentans* and *C. fallax* (Jeffery and van der Putten 2011). In both humans and animals, *Clostridium perfringens* is an important cause of histotoxic infections and diseases originating in the intestines, such as enteritis and enterotoxemia. The virulence of this gram-positive, anaerobic bacterium is heavily dependent upon its prolific toxin-producing ability.

Gas gangrene is subclassified in two categories. Traumatic or postoperative is the most common form accounting for 70% of the cases followed by spontaneous or non-traumatic gangrene (Aggelidakis et al. 2011). Infection only occurs if the organisms are inoculated into a tissue where the oxygen tension is below 30%, such as in some deep cuts. Incubation is usually between 12 and 24 h, although this can be as short as 1 h in some instances, or may take several weeks (Jeffery and van der Putten 2011). Initial symptoms are usually the increasing of pain after surgery or a trauma, which is out of proportion to what would be expected from the surgery or wound, and which may have a sudden onset. Development of blisters containing foul smelling brownish liquid with gas bubbles, soft tissue induration and discoloration may also be present (Hoffman et al. 1971).

Severe pain, toxicity and high creatinine phosphokinase levels with or without radiographic findings are indications for surgery in order to achieve early debridement and obtain tissue for appropriate cultures. The mainstay of treatment is early aggressive surgical intervention, antibiotic therapy and intensive care support. In severe cases of gas gangrene, amputation of a limb may be necessary to prevent the infection from spreading to the rest of the body. Delay of the operation for more than twelve hours may be associated with higher overall morbidity (Sudarsky et al. 1987). Wide resection of all necrotic tissue is necessary. Only viable muscle that bleeds when cut or contracts upon stimulation with electrodiathermy should be left behind (Aggelidakis et al. 2011).

Initial empirical antibiotic treatment should cover Clostridia, gram positive cocci, aerobes and anaerobes. Ampicillin-sulbactam or piperacillin-tazobactam or ticarcillin-clavulate in combination with clindamycin or metronidazone are suggested empiric regimens, whereas antibiotic treatment should be tailored according to the susceptibility results (Bryant and Stevens 2010). Neutralization of clostridial or streptococcal circulating toxins by the use of intravenous immune globulin has shown promising results (Norrby-Teglund et al. 2005).

13.6.2.11 Yersiniosis

Yersinia enterocolitica is the causative agent of yersiniosis. It is a species of gram negative coccobacillus bacterium which is part of the family Enterobacteriaceae, along with *Salmonella* and *E. coli*. Infections are thought to be primarily transmitted to humans by food, in particular, raw or undercooked pork and pork products (Bottone 1999). The primary transmission route of human yersiniosis is proposed to be fecal-oral via contaminated food. In particular, pork and pork products have been implicated as the major source of human *Yersinia enterocolitica* infection, with some epidemiological studies linking consumption of uncooked or undercooked pork (Bottone 1999; Fredriksson-Ahomaa et al. 2000, 2006; Fredriksson-Ahomaa and Korkeala 2003). However, risk factors such as contaminated drinking water or pet animal contact, have also been reported (Rosner et al. 2010). Typically, symptoms disappear within 1–2 weeks after onset. Clinical symptoms of yersiniosis first appear after an incubation period of about 5 days (range 1–11 days) and include diarrhea, fever, vomiting, tenesma and abdominal pain. In older children and young adults, abdominal pain in the right lower abdomen can occur, which may be mistaken for appendicitis (pseudoappendicitis). Sometimes diarrhea may be bloody and fever may be high, especially in infants (Jones et al. 2003). Sequelae such as reactive arthritis or erythema nodosum may sometimes occur (Cover and Aber 1989).

Uncomplicated cases of diarrhea usually resolve on their own without antimicrobial treatment. In more severe infections, like systemic infection and bacteremia, antimicrobials may be useful. A number of antimicrobial agents are active in vitro against enteropathogenic *Yersinia* strains isolated from human and nonhuman sources. These include aminoglycosides, the third-generation cephalosporins, co-trimoxazole, tetracyclines, chloramphenicol, and fluoroquinolones (Baumgartner et al. 2007).

13.6.2.12 Disease Related to *Pseudomonas aeruginosa*

Within the *Pseudomonas* species, *P. aeruginosa* is most frequently associated with causing human infection (Streeter and Katouli 2016). The bacterium is an opportunistic pathogen which rarely causes disease in healthy individuals but can cause a wide range of diseases in immuno-compromised individuals (Brown et al. 2012). When normal physiological function is disrupted, the pathogen is capable of causing a wide-spectrum of infections. Such disruptions include damaged epithelial barriers, depleted neutrophil production, altered mucociliary clearance and the use of medical devices (Streeter and Katouli 2016). *P. aeruginosa* is rarely associated with causing chronic infections in previously healthy patients, although fatal cases of *P. aeruginosa* infections in previously healthy people have been reported (Hatchette et al. 2000). *P. aeruginosa* can cause a wide range of diseases in immuno-compromised individuals including respiratory tract infection, bacteraemia (infection of the blood), endocarditis (infection of the heart), urinary tract, gastrointestinal infection, as well as infecting bones, joints and the central nervous system. Symptoms usually include inflammation of the infected area and sepsis, and if critical organs become infected can be fatal (Balcht and Smith 1994).

P. aeruginosa is often a severe and life-threatening disease. Management of this infection represents a difficult therapeutic challenge for critical care physicians, as one characteristic of *Pseudomonas aeruginosa* is that this bacterium expresses a variety of factors that confer resistance to a broad array of antimicrobial agents. Empirical antibiotic therapy is often inadequate because cultures from initial specimens grow strains that are resistant to initial antibiotics. Although doripenem and biapenem are new carbapenems that have excellent activity against *P. aeruginosa*, they lack activity against strains that express resistance to the currently available carbapenems (El Sol and Alhajhusain 2009). The polymyxins remain the most consistently effective agents against multidrug-resistant *P. aeruginosa*. Strains that are panantibiotic-resistant are rare, but their incidence is increasing. The use of combination therapy is thought to minimize the emergence of resistance and to increase the likelihood of therapeutic success through antimicrobial synergy (Klastersky and Zinner 1982). Levofloxacin and imipenem might be an effective combination for preventing the emergence of resistance during treatment of *P. aeruginosa* infections (Lister and Wolter 2005). Experimental polypeptides may provide a new therapeutic approach, although the efficacy of these regimens has yet to be established in clinical studies (El Sol and Alhajhusain 2009).

13.6.2.13 Disease Related to *Escherichia coli*

Many strains of the bacterium *E. coli* are harmless and in fact are prevalent within the human gut and do not causes diseases while, some strains can cause human diseases (Ishii and Sadowsky 2008). Shiga toxin-producing *E. coli* (STEC), including enterohemorrhagic *E. coli* (EHEC), can cause bloody diarrhea as well as potentially fatal human diseases, such as hemolytic uremic syndrome and hemorrhagic colitis (Nataro and Kaper 1998). In addition to STEC and EHEC, at least five additional pathogroups of *E. coli* have been identified. Enteropathogenic *E. coli* (EPEC) are one of the major causes of watery diarrhea in infants, especially in developing countries. Enterotoxigenic *E. coli* (ETEC) are the main cause of traveler's diarrhea and enteroaggregative *E. coli* (EAEC) can cause persistent diarrhea, lasting for more than 2 weeks. Enteroinvasive E. coli (EIEC) are genetically, biochemically, and pathogenically closely related to *Shigella* (Nataro and Kaper 1998). Several, researchers consider *Shigella* as being a subgroup of *E. coli* (Pupo et al. 2000). While extraintestinal pathogenic *E. coli* (ExPEC), including uropathogenic and avian pathogenic strains, were reported to be harmless while when they are in the intestinal tracts, they can cause neonatal meningitis/sepsis and urinary tract infections if acquired by others (Ishii and Sadowsky 2008).

E. coli O157:H7 is among the most recognized serotypes of EHEC. The strain O104:H4 caused the outbreak in Europe in 2011. This is a particularly virulent strain of the bacterium which was identified Germany and caused the third largest outbreak of *E. coli* with about 2200 infected patients, as well as one of the most lethal with at least 22 people who died (Nettleman 2011). The infectious dose of only 10–100 bacteria needed to cause diseases, as opposed to over a million bacteria needed for most other pathogenic strains of *E. coli* demonstrates the relatively high virulence of this serotype. This new strain produces shiga toxin and is very similar to the 0157:H7 strain about which more is known. However, this strain also has the ability to attach to cells within the gastrointestinal tract in much the same way as Enteroaggregative (EAEC) strains (Nettleman 2011).

Despite the concerning trends in antimicrobial resistance among *E. coli* isolates worldwide, a growing armamentarium of antimicrobial agents provides multiple options for treating *E. coli* infections. As with other Enterobacteriaceae, antimicrobial testing of the infecting strain should direct therapy. In other situations, knowledge of recent local susceptibility patterns is useful for guiding treatment. In general, monotherapy with trimethoprim-sulfamethoxazole, aminoglycoside, cephalosporin, or a fluoroquinolones is recommended as the treatment of choice for most known infections with *E. coli*, although many broad spectrum agents (such as ß-lactam/ß-lactamase inhibitor combinations and the carbapenems) remain highly active.

However, the presence of extended-spectrum β-lactamases (ESBLs) and AmpC b-lactamases complicates antibiotic selection especially in patients with serious infections such as bacteraemia. ESBLs are most often present in *E. coli a*nd *K. pneumoniae and* have the ability to hydrolyse the penicillins, cephalosporins and monobactams, but not the cephamycins and carbapenems. ESBLs are inhibited by "classical" β-lactamase inhibitors such as clavulanic acid, sulbactam and tazobactam (Paterson and Bonomo 2005). Antibiotics that are regularly used for empiric therapy of serious community-onset infections, such as the third generation cephalosporins or fluoroquinolones are often not effective against ESBL and or AmpC-producing bacteria (Pitout 2013). This multiple drug resistance has major implications for selection of adequate empiric therapy regimens. Empiric therapy is prescribed at the time when an infection is clinically diagnosed while awaiting the results of cultures and anti-microbial susceptibility profiles. Multiple studies in a wide range of settings, clinical syndromes, and organisms have shown that failure or delay in adequate therapy results in an adverse mortality outcome. This is also true of infections caused by ESBL-producing bacteria (Schwaber and Carmeli 2007). A major challenge when selecting an empiric regimen is to choose an agent that has adequate activity against the infecting organism(s). Empirical antibiotic choices should be individualized based on institutional antibiograms.

The carbapenems are widely regarded as the drugs of choice for the empiric treatment of severe infections due to AmpC- and ESBL-producing *E. coli* (Pitout 2012). It is reasonable to suggest that ertapenem should be used for serious community-onset infections in cases where ESBL-producing isolates are suspected to be the source (Pitout 2010). This would include patients with urinary tract infections, underlying renal pathology, recent administration of previous antibiotics (including cephalosporins and fluoroquinolones), previous hospitalization, nursing home residents, older males, Diabetes Mellitus, underlying liver pathology and recent international travel to high risk areas (e.g. the Indian subcontinent) (Rodriguez-Bano and Pascual 2008). Imipenem or meropenem or doripenem would be more appropriate for the empiric treatment of serious hospital-onset infections in cases where ESBL-producing isolates are suspected to be the source (Pitout 2010). The existing data suggest that piperacillin-tazobactam may be a useful agent for the treatment of some infections with ESBL-producing pathogens (Retamar et al. 2013). However, this potential recommendation still must be interpreted cautiously, because it is based on a relatively small database of information. Definitive conclusions regarding the efficacy of piperacillin-tazobactam for the treatment of infections caused by *E. coli* that produce ESBLs must await large-scale, prospective, randomized clinical trials. Oral agents such as nitrofuratoin, and fosfomycin show good in-vitro activity against ESBL and AmpC-producing from different areas of the world and are adequate options for the empiric treatment of uncomplicated lower urinary tract infections (Pitout 2010). However, it is important for medical practitioners to know their local susceptibility rates for nitrofuratoin against these multi-resistant bacteria, since in certain areas high resistance rates had been reported.

13.6.2.14 Q Fever

Coxiella burnetii is highly virulent in humans with an infective dose of just one organism being necessary to cause disease in some instances (Tigertt et al. 1961). Symptoms of acute Q fever usually appear 2–3 weeks after infection, although as many as half of the people infected with *C. burnetii* do not show symptoms. If symptoms do present they can vary greatly from person to person and can include high fever, severe headache, nausea, vomiting, diarrhea, abdominal and/or chest pain, myalgia, chills and/or sweats and a non-productive cough (Parker et al. 2006). However, complications may occur including pneumonia, inflammation of the liver or heart or central nervous system complications. Q fever can also appear in a chronic form which occurs in <5% of infected patients, particularly in pregnant women, immuno-compromised patients and people with pre-existing heat valve defects, and usually presents as endocarditis.

Doxycycline, 100 mg twice daily for 14 days is recommended for acute illness (Parker et al. 2006). Antibiotic treatment lessens the time in which the patient has fever, and hastens recovery from pneumonia (Marrie 2003). Anti-inflammatory agents could be useful when symptoms do not respond to antibiotics. Co-trimoxazole is recommended for children younger than 8 years, and the newer macrolides might also prove useful (Maltezou and Raoult 2002).

13.6.2.15 Salmonellosis

Almost all strains of *Salmonella* are pathogenic as they have the ability to invade, replicate and survive in human host cells, resulting in potentially fatal disease (Eng et al. 2015). The severity of *Salmonella* infections in humans varies depending on the serotype involved and the health status of the human host. Children below the age of 5 years, elderly people and patients with immunosuppression are more susceptible to *Salmonella* infection than healthy individuals. In human infections, the four different clinical manifestations are enteric (typhoid) fever, gastroenteritis, bacteraemia and other extraintestinal complications, and chronic carrier state (Sheorey and Darby 2008).

The term "enteric fever" is used collectively for both typhoid and paratyphoid fevers, and both *S. typhi* and *S. paratyphi* are referred as typhoid *Salmonella* (Connor and Schwartz 2005). Enteric fever can present with nonspecific features such as diarrhea, vomiting or respiratory symptoms. Approximately 10–15% of patients develop severe diseases which presents with persistent and high fever which generally is accompanied by relative bradycardia (slower than normal heart rate) as well as rose spots on the back, arms and legs in 25% of cases (Eng et al. 2015). The tongue is usually coated, the abdomen tender and the liver and spleen may become inflamed.

Salmonella strains other than *S. typhi* and *S. paratyphi* are referred to as non-typhoid Salmonella (NTS), and are predominantly found in animal reservoirs.

NTS infections are characterized by gastroenteritis or "stomach flu", an inflammatory condition of the gastrointestinal tract which is accompanied by symptoms such as non-bloody diarrhoea, vomiting, nausea, headache, abdominal cramps and myalgias (Eng et al. 2015). Symptoms such as hepatomegaly and splenomegaly are less commonly observed in patients infected with NTS (Hohmann 2001). Compared to typhoid infections, NTS infections have a shorter incubation period (6–12 h) and the symptoms are usually self-limiting and last only for 10 days or less.

Salmonella bacteraemia is a condition whereby the bacteria enter the bloodstream after invading the intestinal barrier. Almost all the serotypes of *Salmonella* can cause bacteraemia, while *S. dublin* and *S. cholearaesuis* are two invasive strains that are highly associated with the manifestations of bacteraemia (Woods et al. 2008). Similar to enteric fever, high fever is the characteristic symptom of bacteraemia, but without the formation of rose spots as observed in patients with enteric fever. In severe conditions, the immune response triggered by bacteraemia can lead to septic shock, with a high mortality rate.

The status of chronic carrier is defined as the shedding of bacteria in stools for more than a year after the acute stage of *Salmonella* infection. Since humans are the only reservoir of typhoid *Salmonella*, carriers of *S. typhi* and *S. paratyphi* are responsible for the spreading of enteric fever in endemic regions, as the common transmission route is the ingestion of water or food contaminated with the faeces of chronic carriers (Bhan et al. 2005). About 4% of patients with enteric fever, predominantly infants, elderly people and women, may become chronic carriers (Gonzalez-Escobedo et al. 2011).

Salmonella infections that involve invasive serotypes are often life threatening and the development of multi-drug resistance in the serotypes of *Salmonella* has a significant impact on the antibiotic treatment of *Salmonella* infections. Quinolones and third generation cephalosporins have been the antibiotics of choice in treating infections with multi-drug resistant *Salmonella* (Karon et al. 2007). However, the emergence of *Salmonella* serotypes resistant to quinolones and cephalosporin poses a new challenge in treating infected patients, and the lack of an effective antibiotic therapy may lead to an increase in the morbidity and mortality rates.

13.6.3 Fungal Diseases

13.6.3.1 Aspergillosis

Aspergillus is ubiquitous in air, and the sources of spores include soil, decaying vegetation, and dust (Baumgardner 2012). Huge spore dispersal may follow disasters such as storms. The relative importance of soil as a source of infection compared with plants, flowers, building materials, water, and hospital environments is unclear (Hajjeh and Warnock 2001). Infections caused by *Aspergillus* sp. (excluding allergic bronchopulmonary aspergillosis) are usually in immuno-compromised patients and include invasive pulmonary aspergillosis (cough, dyspnea, possible fever, chest pain, hemoptysis, wheezing), pulmonary or sinus fungus balls, chronic pulmonary aspergillosis (cavitary, fibrosing, subacute), sinusitis, endocarditis; and other superficial or disseminated forms (Boucher and Patterson 2008). Most infections occur after inhalation of conidia. No specific virulence factors have been identified for *Aspergillus*.

Invasive aspergillosis is a major cause of morbidity and mortality in severely immuno-compromised patients. Risk factors for invasive aspergillosis include prolonged and severe neutropenia, hematopoietic stem cell and solid organ transplantation, advanced AIDS, and chronic granulomatous disease (Segal and Walsh 2006). Invasive aspergillosis most commonly involves the sinopulmonary tract reflecting inhalation as the principal portal of entry. Chest computed tomography scans and new non-culture–based assays such as antigen detection and polymerase chain reaction may facilitate the early diagnosis of invasive aspergillosis, but have limitations. In recent years, there has been a significant expansion in the antifungal armamentarium. The second-generation triazole, voriconazole, is superior to conventional amphotericin B as primary therapy for invasive aspergillosis, and is the new standard of care for this infection (Segal and Walsh 2006).

13.6.3.2 Blastomycosis

Infection with *B. dermatitidis* typically occurs when conidia is produced from the mycelial phase in soil or decaying organic matter are inhaled into the lungs (Brömel and Sykes 2005). Direct inoculation of the organism via skin puncture wounds is rarely seen. The increase in temperature within the body causes conversion from the spore phase to a large broad-based budding yeast cell (Yildiz et al. 2016). When yeasts reach into the lungs they cause pyogranulomatous pneumonia. Moreover, they cause granulomatous or pyogranulomatous inflammation in many other organs via vascular or lymphatic system (lymph nodes, eyes, bones, central nervous system, kidneys, liver, spleen, skin, genitourinary system, heart, adrenal glands). However, the mostly effected tissues are respiratory system, lymphatic tissues, eyes, skin and bones (Yildiz et al. 2016). Symptoms normally include low grade fevers, chest pain, a mild but persistent productive cough and haemoptysis (i.e. coughing up of blood) (Fang et al. 2007). Secondary infections may also be detected as lesions in the skin which may remain localised or, in some instances, may spread throughout the body leading to extensive ulceration. The most successful methods in the way to diagnose blastomycosis are cytologic samples from concerned tissues (Yildiz et al. 2016).

Currently several antifungal agents exist which have demonstrated efficacy in the treatment of human blastomycosis including amphotericin B, ketoconazole, itraconazole, posaconazole, isavuconazole, voriconazole and fluconazole. However, there have been no published therapeutic trials comparing different antifungal agents in blastomycosis, thus our understanding of what constitutes the most effective therapy for this disorder is largely based on small to moderate sized clinical trials, case reports, retrospective reviews and anecdotal experience. Historically, conventional (deoxycholate) amphotericin B (AmB) has been the mainstay of therapy for all forms of blastomycosis since its availability in 1958. Clinicians generally prefer a lipid formulation of AmB over the deoxycholate formulation due to less nephrotoxicity (Chapman et al. 2008; Cook 2001). The published efficacy of AmB in all forms of blastomycosis ranges between 66% and 93% depending on the dose of drug, duration of therapy, underlying illness and disease severity. For mild to moderate disease, efficacy generally exceeds 90%. Unfortunately, all formulations of AmB must be administered by the intravenous route and they have significant renal, electrolyte, hematologic, and infusion-associated toxicities. Consequently, orally administered and less toxic azole antifungal agents are an attractive alternative to AmB (Chapman et al. 2008).

13.6.3.3 Coccidioidomycosis

The number of cases of coccidioidomycosis in dryland areas have been reported to increase markedly in the later summer and early autumn (i.e. the dry season) where soil disturbances due to wind erosion, or other anthropogenic factors, such as agricultural practices, are thought to make the fungus airborne and thereby increases the risk of its inhalation (Oppenheimer et al. 2010). Coccidioidomycosis may cause a heterogeneous clinical spectrum and be classified into primary and secondary disease. Primary disease generally affects the lungs, which is by far the most common site involved. In this case, it is acquired by direct inhalation of arthroconidia. Primary involvement of the skin is quite uncommon. It is acquired by direct inoculation of the fungus by means of splinters and abrasions. Secondary disease can affect the lungs and become a chronic process. On the other hand, disseminated disease (usually resulting from hematogenous spread of a primary lung infection) can involve many organs, including the skin, bones, joints, nervous system and meninges (Arenas 2008).

Pulmonary coccidioidomycosis is often self-limited and treatment is often not indicated for immunocompetent patients (Laniado-Laborín et al. 2012). However, following the availability of triazoles, most clinicians favor treating all symptomatic patients with antifungal therapy and some of them claim that early treatment may reduce symptom duration and risk of dissemination, though there is no evidence for this assertion (Sharma and Thompson 2012). For primary pulmonary coccidioidomycosis, immuno-competent patients without risk factors for dissemination may be treated with fluconazole or itraconazole for 3–6 months. Patients with severe symptoms lasting more than 6 weeks, or with risk factors for dissemination (e.g. immunosuppression, AIDS, pregnant women and comorbidities) should also be treated. In pregnant women, the mainstay of treatment is amphotericin B (due to azoles teratogenicity). All other patients with primary coccidioidomycosis and risk factors for dissemination will be treated with fluconazole or itraconazole for 4–12 months (Galgiani et al. 2000; Sharma and Thompson 2012). In cases of chronic pulmonary coccidioidomycosis, fluconazole or itraconazole for 12–18 months or more are indicated. Severe or refractory infections may be treated with amphotericin B, deoxycholate and liposomal forms, while disseminated, meningeal forms of coccidioidomycosis require life-long treatment with drugs such as fluconazole or itraconazole (Galgiani et al. 2000; Sharma and Thompson 2012).

13.6.3.4 Histoplasmosis

Pulmonary infection is the primary manifestation of histoplasmosis, varying from mild pneumonitis to severe acute respiratory distress syndrome. The extent of disease depends on the number of conidia inhaled and the function of the host's cellular immune system. The majority of infected persons have either no symptoms or a very mild illness that is not recognized as being histoplasmosis. The usual case of acute pulmonary histoplasmosis is a self-limited illness occurring mostly in children exposed to the organism for the first time. Symptoms include fever, malaise, headache, and weakness (Goodwin et al. 1981). Acute self-limited pulmonary histoplasmosis is accompanied by rheumatologic and/or dermatologic manifestations in approximately 5% of patients. Erythema nodosum and erythema multiforme are the most common skin manifestations. They occur most frequently in young women and are thought to be associated with a hypersensitivity response to the antigens of *H. capsulatum* (Medeiros et al. 1966). Myalgias and arthralgias are common symptoms during acute infection. However, a minority of patients may develop self-limited, polyarticular, symmetrical arthritis (Rosenthal et al. 1983).

Patients with a strong suspicion of histoplasmosis with or without severe symptoms should be treated with intravenous amphotericin B or oral itraconazole. Although amphotericin B is fungicidal and has shown its efficacy in terms of survival, it is also nephrotoxic (Johnson et al. 2002). Itraconazole is fungistatic and is associated with drug interactions that may complicate patient care in the context of profound immuno-suppression. In a randomized clinical trial, intravenous liposomal amphotericin B was more effective than deoxycholate amphotericin B, with a quicker clinical response, a decreased toxicity, and a decreased mortality (Johnson et al. 2002). Thus, for the induction treatment of moderately severe and severe presentations of disseminated histoplasmosis, liposomal amphotericin B is the recommended strategy for 2 weeks or until clinical improvement. A relay with oral itraconazole must then be initiated for at least 1 year (Wheat et al. 2007). In severe forms of histoplasmosis, renal failure is often described, either associated with multiorgan failure, or secondary to amphotericin B nephrotoxicity. Thus, clinicians tend to switch to itraconazole in order to not aggravate renal failure. For non-severe cases, itraconazole is the first-line treatment. The response was positive in 85% of the cases (Wheat et al. 1995).

13.6.3.5 Sporotrichosis

Clinical presentations of sporotrichosis among other factors may vary according to the immunological status of the host, the load and depth of the inoculum, and the pathogenicity and thermal tolerance of the strain (Lima Barros et al. 2011). Continuous exposure to small amounts of conidia in an area of endemicity gradually conferred immunity (Rippon 1988). According to the location of the lesions, sporotrichosis can be classified into cutaneous, mucosal, and extracutaneous forms (Lima Barros et al. 2011). Symptoms usually include a red lump which develops at the site of infection and which eventually turns into an ulcer. As the fungus sometimes grows through the lymphatic channels in the body small ulcers may appear in lines on the skin passing up a leg or arm from the initial site of infection. Disseminated sporotrichosis may lead to lung and breathing problems as well as infection of bone or the central nervous system (Lima Barros et al. 2011).

Potassium iodide has been traditionally used in the treatment of sporotrichosis since the early twentieth century, with satisfactory results (De Beurmann and Gougerot 1912). However, the exact mechanism of action remains unknown. Due to adverse effects related to this medication, in the 1990s the azole compounds were introduced, and itraconazole is currently the first-choice treatment (Kauffman et al. 2007). Itraconazole has been used effectively and safely in most cases of sporotrichosis, with low toxicity and good tolerance, even in long-term treatments.

13.6.3.6 Mucormycosis

Most human infections result from inhalation of fungal sporangiospores that have been released in the air or direct inoculation of organisms into disrupted skin or mucosa (Ibrahim et al. 2004). Based on its clinical presentation and anatomic site, invasive mucormycosis is classified as one of the major clinical forms rhinocerebral, pulmonary, cutaneous, gastrointestinal, disseminated, and uncommon rare forms, such as endocarditis, osteomyelitis, peritonitis, and renal infection (Petrikkos et al. 2012). Any of the species of the Mucorales may cause infection at these sites. The most common reported sites of invasive mucormycosis have been the sinuses, lungs, and skin (Torres-Narbona et al. 2007). The skin and gut are affected more frequently in children than in adults.

Medical treatment with conventional antifungals and non-conventional therapeutics are corner stone for successful treatment of mucormycosi (Shetty and Punya 2008). Polyenes like amphotericin-deoxycholates and lipid complex are primary therapeutic agents for mucormycosis. There should be close monitoring of serum electrolytes, as polyenes are known to cause potassium imbalance (Prasad et al. 2012). Data from comparison of lipid complex Ampho-B and deoxycholate Ampho-B showed improved survival rates less side effects with lipid complex Ampho-B (Shetty and Punya 2008). Triazoles are not recommended as primary therapy for mucormycosis (Spellberg et al. 2009). These can be used as supportive treatment when patients are intolerant to amphotericin or in whom the use is limited by nephrotoxicity. Debridement of necrotic tissue in combination with medical therapy is mandatory for patient survival (Ahamed and Thobaiti 2014). In rhinocerebral disease, surgical care includes drainage of the sinuses and may require excision of the orbital contents and involved brain. Repeated surgery may be required, especially for rhinocerebral mucormycosis. Non-conventional therapeutic agents like anti diabetics, iron chelating agents, statins, granulocyte transfusions, cytokines, and hyperbaric oxygen have increased survival rates (Shetty and Punya 2008).

13.6.4 Diseases Caused by Protozoa

13.6.4.1 Amoebiasis

The primary target organ colonized by *Entamoeba histolytica* is the intestinal mucosa of the sigmoid and colon. The highly invasive strains of species can display their invasive potential by producing tissue damage. Therefore, 10% of the world's total population was infected by amoeba and only 1% of infected personages developed the aggressive form of disease (Raza et al. 2013). The annual mortality rate appraised at that time for invasive forms of amebiasis was around 100,000 deaths. The majority of deaths were a concern of severe complications associated with intestinal or extra-intestinal offensive disease (Clark and Diamond 1993). In most of *E. histolytica* infections, symptoms remain absent or very mild. The patients with non-invasive disease excrete cysts for a short period of time and get clear from infection within 12 months. Whereas, most frequent clinical manifestation are amoebic colitis and amoebic liver abscess (Stanley 2003). Colitis results when the trophozoites penetrate the mucand in layer lectin attaches to intestinal epithelium. The penetration depends on genetic makeup of parasite to produce proteolytic enzymes and resistance to complement mediated lyses. Throphozoites invasion is initiated by destroying epithelial cell lining and inflammatory cells (Seydel et al. 1998).

Nitromidazole derivatives like metronidazole, tinidazole and ornidazole are considered as foundation stone of the treatment for amoebiasis (Stanley 2003). Amoebiasis treatment by metronidazole is followed by luminal agents (Paromomycin, Iodoquinol) to eradicate colonization especially in amoebic colitis (Pehrson and Bengtsson 1984). Asymptomatic patients carrying *E. histolytica* should be treated with the luminal agents to eradicate infection which is suggested because of the known risk of development of disease in the carrier and the chances that this individual can shed the cysts of *E. histolytica* and pose a threat to public health (Haque et al. 2001). Surgically drainage of liver abscess is generally unnecessary and should be avoided as liver abscess in amoebiasis can be treated with single dose of metronidazole without drainage (Akgun et al. 1999).

13.6.4.2 Balantidiasis

After infection, *Balantidium coli* may inhabit the bowel lumen without invading tissue or provoking clinical symptoms. Parasites that invade tissue do so by mechanical action of the cilia and by lytic action, particularly in patients weakened by underlying factors such as malnutrition or immunosuppression (Vasilakopoulou et al. 2003). Balantidiasis can mimic intestinal amebiasis. The acute form of the disease is marked by rapid onset of diarrhea or dysentery, with 20 bowel movements or more per day (Castro et al. 1983). Other frequent complaints are abdominal colic, tenesmus, nausea, and vomiting. Chronic balantidiasis produces intermittent diarrheal episodes alternating with normal bowel movements or constipation. Patients may occasionally have headache, insomnia, anorexia, weight loss, or muscular weakness. *Balantidium coli* may cause appendicitis and lung involvement, and has been attributed to urinary tract disease (Koopowitz et al. 2010; Maino et al. 2010).

The drugs of choice for treatment of balantidiasis are tetracycline, iodoquinol or metronidazole. Other drugs such as ampicillin, carbasone, diodoquin, nitrimidazine, and paromomycin have been used with varying results (e.g. García-Laverde and de Bonilla 1975; García 1999; Wolfe 2000).

13.6.4.3 Giardiasis

For symptomatic disease the incubation period is usually 3–20 days, but can be much longer. Among healthy immunocompetent individuals, infection with *G. lamblia* is frequently self-limited (Adam 1991). The disease usually presents with a broad spectrum of symptoms with some people experiencing an abrupt onset of explosive watery diarrhea with abdominal cramps, vomiting, fever, malaise and foulflatus which last for 3–4 days before transition into the more common symptoms of giardiasis (Jeffery and van der Putten 2011). However, most people experience a slower onset of symptoms. These include stools which become malodorous, mushy and greasy which may alternate with watery diarrhea and constipation. Upper and mid abdominal cramps may occur, as well as sulphurous belching, substantial burning and acid indigestion as well as loss of appetite, fatigue and malaise. Prolonged infections with a high level of post-infectious fatigue and symptoms indistinguishable from those of irritable bowel syndrome have been reported (Morch et al. 2009).

Tinidazole is the first-line drug treatment of giardiasis, as it requires only a single dose to cure infection in most individuals (Petri 2005). The related drug metronidazole is as effective, but it requires 5–7 days of three times a day therapy. Nitazoxanide appears in limited studies to be as effective as tinidazole or metronidazole, and it does not have the bitter taste of nitroimidazoles. A good alternate for use during pregnancy is paromomycin (Petri 2005). Cure of infection varies between 60% and 100% with one course of treatment. Less effective and/or less well-tolerated drugs for the treatment of giardiasis include albendazole, quinacrine, and furazolidone. The use of these drugs should be reserved for giardiasis refractory to treatment with the first-line agents.

13.6.4.4 Toxoplasmosis

Toxoplasmosis is generally benign and often remains unnoticed in immunocompetent individuals (Jeffery and van der Putten 2011). However, non-specific flu-like symptoms, lymphadenopathy, and some rare complications might be associated with primary infection. Under some conditions, toxoplasmosis can cause serious pathology, including hepatitis, pneumonia, blindness, and severe neurological disorders (Bultman et al. 2005). The infection also has health-threatening and fatal implications in congenitally infected foetuses. While the primary infection is mostly asymptomatic in pregnant women, the parasite might cross the placenta, infect the foetus, and cause retinochoroiditis, hydrocephaly, mental retardation, seizures, or even foetal death (Remington et al. 2006).

Treatment of immunocompetent adults with lymphadenopathic toxoplasmosis is rarely indicated as this form of the disease is usually self-limited. If visceral disease is clinically evident or symptoms are severe or persistent, treatment may be indicated for 2–4 weeks. Treatment for ocular diseases should be based on a complete ophthalmologic evaluation. The decision to treat ocular disease is dependent on numerous parameters including acuteness of the lesion, degree if inflammation, visual acuity, and lesion size, location, and persistance. The "classic therapy" for ocular toxoplasmosis includes pyrimethamine plus sulfadiazine plus folinic acid (leucovorin). If the patient has a hypersensitivity reaction to sulfa drugs, pyrimethamine plus clindamycin can be used instead (de-la-Torre et al. 2011). The combination of trimethoprim with sulfamethoxazole has been used as an alternative, as well as other drugs such as atovaquone and pyrimethamine plus azithromycin, which have not been extensively studied (de-la-Torre et al. 2011). Management of maternal and fetal infection varies depending on the treatment center. In general, spiramycin is recommended (for the first and early second trimesters) or pyrimethamine/sulfadiazine and leucovorin (for late second and third trimesters) for women with acute *T. gondii* infection (Montoya and Liesenfeld 2004).

13.6.4.5 Cryptosporiosis

Infection with *Cryptosporidia* is often asymptomatic. However, the disease affects immunocompetent, particularly children under the age of 5 years, and immunocompromised individuals worldwide (Abubakar et al. 2007). It causes diarrhoea lasting about 1–2 weeks, extending up to 2.5 months among the immunocompetent and a more severe life-threatening illness among immunocompromised individuals

(Hunter and Nichols 2002). Nausea, low grade fever and abdominal cramps are also common symptoms.

Several antimicrobials have been proposed for the treatment of cryptosporidiosis. However, some evidence of effectiveness was only identified for nitazoxanide in a combined population of immunocompetent and -compromised individuals (Abubakar et al. 2007). The absence of effective therapy highlights the importance of preventive interventions particularly in this group of patients. Supportive management including rehydration therapy, electrolyte replacement and antimotility agents will remain the main treatment strategies until better drugs emerge (Abubakar et al. 2007).

13.6.4.6 Cyclosporiasis

Infection with *Cyclospora cayetanensis* usually presents with diarrhea which may be explosive and accompanied with abdominal cramps, fatigue, malaise and weight loss and which may be interspersed with periods of remission. The diarrhea may persist for weeks to months if left untreated (Jeffery and van der Putten 2011).

The drug of choice is trimethoprim–sulfamethoxazole (Ghimire and Sherchan 2008). Combination of these two antibiotics inhibits two sequential steps in bacterial folate synthesis and has a wide spectrum of this drug. It is usually administered orally, but can be administered intravenously if the patient can't tolerate per oral medications because of nausea, vomiting or underlying gastrointestinal problems (Ghimire and Sherchan 2008).

13.6.5 Diseases Caused by Helminths

13.6.5.1 Ascariasis

Ascaris is transmitted by the ingestion of eggs (Scott 2008). These eggs hatch and larvae penetrate through the intestine and migrate through the portal vessels to the liver and lungs where they are coughed up and swallowed, a process that takes several weeks. After the worms return to the intestine, they mature as adult male and female worms, typically measuring about 20 and 30 cm in length, respectively (Bethony et al. 2006; Fig. 13.13). The period from ingestion of eggs to their detection in feces ranges from 10 to 11 weeks, and adult worms live for 1–2 years. During this time, adults mate and Ascaris eggs are passed in the feces.

The migratory phase is responsible for inflammatory and hypersensitivity reactions in the lung, including pneumonitis and pulmonary eosinophilia. Pathology induced by the adult worms includes malabsorption, intestinal obstruction, and invasion of the bile duct or appendix, leading to acute pancreatitis and appendicitis. Ascaris has also been associated with impaired cognitive function (Ezeamama et al. 2005).

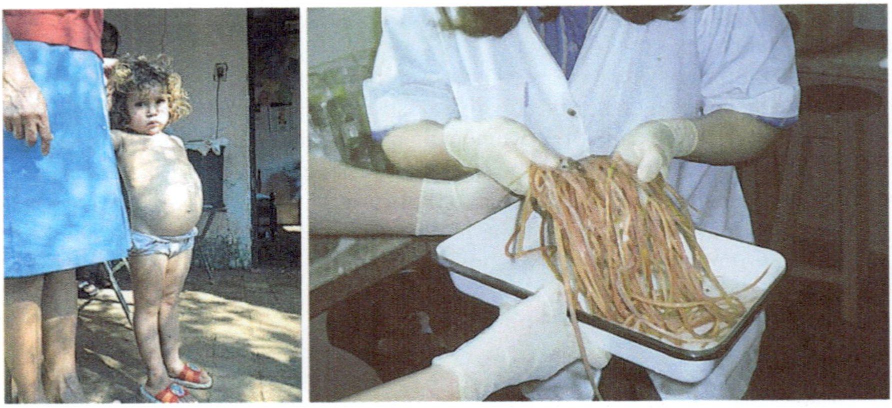

Fig. 13.13 Girl from Paraguay with heavy ascaris infection before deworming and worms extracted (Bethony et al. 2006, p. 1527; with kind permission from Elsevier)

13.6.5.2 Trichuriasis

Most infections with *Trichuris trichiura* are asymptomatic. Clinical symptoms are more frequent with moderate to heavy infections. Heavy chronic infection with *T. trichiura* can cause inflammation and colitis. Stools can be loose and often contain mucus and/or blood. Chronic disease signs include diarrhea, impaired growth, anemia, and finger clubbing. The most serious manifestation of heavy *T. trichiura* infection is chronic dysentery and rectal prolapse (Bundy and Cooper 1989). This occurs primarily in the setting of heavy infection, and embedded worms may be directly visualized in the mucosa of the inflamed rectum. Children who are heavily infected may have impaired growth and/or cognition (Nokes et al. 1992). However, it can be difficult to quantify the role of trichuriasis in isolation from comorbidities and other social factors.

13.6.5.3 Ancylostomiasis

Ancylostomiasis is a disease associated with hookworm (*Ancylostoma duodenale* or *Necator americanus*) infection. After oral uptake or skin penetration, *A. duodenale* and *N. americanus* larvae enter subcutaneous venules and lymphatic vessels to access the host's circulation (Bethony et al. 2006). Ultimately, the larvae become trapped in pulmonary capillaries, enter the lungs, pass over the epiglottis, and migrate into the gastrointestinal tract (Hotez et al. 2004). About 5–9 weeks are needed from skin penetration until development of egg-laying adults.

Several cutaneous syndromes result from skin-penetrating larvae of *N. americanus* and *A. duodenale*. Repeated exposure to hookworm third-stage larvae results in ground itch, a local erythematous and papular rash accompanied by pruritus on the hands and feet (Hotez et al. 2004). After skin invasion, hookworm third-stage larvae travel through the vasculature and enter the lungs, although the resulting pneumonitis is not as great as in ascaris infection (Hotez et al. 2004). Oral ingestion of *A. duodenale* larvae can result in Wakana syndrome, which is characterised by nausea, vomiting, pharyngeal irritation, cough, dyspnoea, and hoarseness (Hotez et al. 2004). Soil-transmitted helminth infections of moderate and high intensity produce clinical manifestations in the gastrointestinal tract, with the highest-intensity infections most common in children (Chan et al. 1994). Hookworm also contributes to moderate and severe anaemia in children (Bethony et al. 2006).

13.6.5.4 Schistosomiasis

Schistosomiasis or bilharzia is a tropical parasitic disease caused by blood-dwelling fluke worms of the genus *Schistosoma*. The disease is spread by bathing or wading in infected rivers, lakes, and irrigation systems (WHO 1996). People who are repeatedly infected can face liver, intestinal, lung, and bladder damage (Bultman et al. 2005). Acute schistosomiasis is a systemic hypersensitivity reaction against the migrating schistosomulae, occurring a few weeks to months after a primary infection (Gryseels et al. 2006). The disease starts suddenly with fever, fatigue, myalgia, malaise, non-productive cough, eosinophilia, and patchy infiltrates on chest radiography. Abdominal symptoms can develop later, caused by the migration and positioning of the mature worms. Most patients recover spontaneously after 2–10 weeks, but some develop persistent and more serious disease with weight loss, dyspnoea, diarrhoea, diffuse abdominal pain, toxaemia, hepatosplenomegaly and widespread rash (Gryseels et al. 2006). The main lesions in established and chronic infection are due not to the adult worms but to eggs that are trapped in the tissues. The eggs secrete proteolytic enzymes that provoke typical eosinophilic inflammatory and granulomatous reactions, which are progressively replaced by fibrotic deposits. Immunopathological reactions against schistosome eggs trapped in the tissues lead to inflammatory and obstructive disease in the urinary system (*S. haematobium*) or intestinal disease, hepatosplenic inflammation, and liver fibrosis (*S. mansoni*, *S. japonicum*) (Cheever et al. 1978; Gryseels et al. 2006). The severity of the symptoms is related both to the intensity of infection and to individual immune responses.

13.6.5.5 Taeniasis and Cysticercosis

Taeniasis is the intestinal infection of human with the adult stage of the tapeworm of the genus *Taenia*. There are two kinds of taeniasis, one caused by *T. saginata* and the other by *T. solium*. In the case of *T. solium*, clinical signs are generally insignificant in humans with adult tapeworms (Gebrie and Engdaw 2015). However, when humans are infected by cysticerci through the ingestion of *T. solium* egg or proglottid, rupture within the host intestine can cause larvae to be transported by the circulatory system and dispersed throughout the body producing cysts (García and Del Brutto 2000). Cysticerci may be found in every organ of the body in humans but most commonly in the subcutaneous tissue, eye and brain (Bourke and Petana 1994). The central nervous system is the most frequent localization of cysts which cause neurocysticercosis. This is considered to be the most common parasitic infection of the human nervous system and the most frequent preventable cause of epilepsy in the developing world (Willingham and Engels 2006). A wide spectrum of symptoms may be expressed, including headaches, dizziness and occasionally seizures (Gebrie and Engdaw 2015). In most severe case, dementia or hyper tension can occur due to perturbation of the normal circulation.

13.6.5.6 Treatment of Soil-Transmitted Helminth Infections

The treatment goal for soil-transmitted helminth infections is to remove adult worms from the gastrointestinal tract. The drugs most commonly used for the removal of soil-transmitted helminth infections are mebendazole and albendazole. These benzimidazole drugs bind to nematode β-tubulin and inhibit parasite microtubule polymerisation (Lacey 1990), which causes death of adult worms through a process that can take several days. Although both albendazole and mebendazole are deemed broad-spectrum anthelmintic agents, important therapeutic differences affect their use in clinical practice. Both agents are effective against ascaris in a single dose. However, in hookworm, a single dose of mebendazole has a low cure rate and albendazole is more effective (Bennett and Guyatt 2000). Conversely, a single dose of albendazole is not effective in many cases of trichuriasis (Adams et al. 2004). For both trichuriasis and hookworm infection, several doses of benzimidazole anthelmintic drugs are commonly needed. Another important difference between the two drugs is that mebendazole is poorly absorbed from the gastrointestinal tract so its therapeutic activity is largely confined to adult worms. Albendazole is better absorbed, especially when ingested with fatty meals, and the drug is metabolised in the liver to a sulphoxide derivative, which has a high volume of distribution in the tissues (Dayan 2003). Systemic toxic effects, such as those on the liver and bone marrow, are rare for the benzimidazole anthelmintic drugs in the doses used to treat soil-transmitted helminth infections. However, transient abdominal pain, diarrhoea, nausea, dizziness, and headache may occur.

Praziquantel is the drug of choice for treatment of schistosomiasis (Gryseels et al. 2006). It is an acylated quinoline-pyrazine that is active against all schistosome species. The drug acts within 1 h of ingestion by paralysing the worms and damaging the tegument. Side-effects are mild and include nausea, vomiting, malaise, and abdominal pain. In heavy infections, acute colic with bloody diarrhoea can occur shortly after treatment, probably provoked by massive worm shifts and antigen release (Stelma et al. 1995). Praziquantel has very low toxicity in animals, and no important long-term safety difficulties have been documented in people (Dayan 2003). It is also judged safe for treatment of young children and pregnant women. Vaccines are not yet available.

Both species of adult Taenia (*T. solium* and *T. saginata*) are treated similarly (Gebrie and Engdaw 2015). Patients with *T. solium* taeniasis should be treated immediately after diagnosis. Therapy of cysticercosis in swine is generally not indicated. *T.solium* can be treated with niclosamide or praziquantel (Opens membrane calcium channels, causing paralysis of the worm, thus aiding the body in expelling the parasite through peristalsis), the latter is also effective against cerebral cyst of *T. solium* in man (Gebrie and Engdaw 2015).

13.7 Examples of Strategies for Control of Zoonotic Diseases Transmission and Infection

Interruption of the epidemiologic chain at the level of host (human) and of intermediate hosts (animals) is one of the important strategies to control spread of zoonotic disease. For example, control and prevention of *Taenia saginata* and *Taenia solium* is achieved by protection cattle/pig from grazing on feces or sewage polluted grass, not using untreated human feces as fertilizer for pasture land which may contain segments and ova, avoiding eating raw or insufficiently cooked meat/ pork which may contain infective larva, inspecting meat for larva and treating infected person (Gebrie and Engdaw 2015).

Toxoplasma gondii is a ubiquitous parasite using almost all warm-blooded vertebrates as hosts, including humans. Humans can become infected by ingestion of oocysts, excreted by cats, via contaminated soil or water, or foods. A second means of transmission is the ingestion of tissue cysts present in raw or undercooked meat or meat products from so-called intermediate hosts such as sheep and cattle (Saadatnia and Golkar 2012). In the absence of an effective vaccine in humans, prevention of zoonotic transmission might be the best way to approach the problem of toxoplasmosis, and must be done by limiting exposure to oocysts or tissue cysts. Recommendations for accomplishing this include practicing good hygiene (e.g. hand washing after soil contact, washing fruits and vegetables that are eaten raw), freezing meat at −12 °C for 24 h (Jones et al. 2007) and/or cooking meat until an internal temperature of 66 °C is reached, and not drinking untreated water (Dubey and Jones 2008). When women become infected during pregnancy, vertical transmission from mother to child can also occur. However, cats are the only definitive hosts for toxoplasma, responsible for shedding oocysts into the environment. It is therefore recommended to keep cats indoors, feed them commercially prepared diets, and clean their litter boxes daily, because it takes at least 1 day for the organisms to sporulate and become infectious after being shed (Vollaire et al. 2005).

Measures to prevent soil-transmitted helminth infections are the provision of improved sanitary infrastructure, safe water, abstaining from night-soil use for farming (commonly practiced in Southeast and East Asia), and adequate personal and food hygiene as well as wearing shoes (Ziegelbauer et al. 2012). Individuals can also avoid helminth infections by peeling or cooking their food, boiling drinking water, and by consistent use of soap for hand washing. Travelers should also be alerted to the risk of soil-transmitted helminth infections and the simple prevention measures. Hookworm infections can be prevented by reducing skin contact with potentially infected soil, by wearing closed shoes or gloves, and by placing infants on mats.

Effective sanitation and use of water from protected sources are the most useful measures of protection from balantiasis (Schuster and Ramirez-Avila 2008). Amoebiasis as well can be controlled by adopting proper sanitation and hygienic measures (Abd-Alla and Ravdin 2002). Improved sanitation, strict personal hygiene, and hand washing may also help prevent transmission of hepatitis A and E. The virus is inactivated by household bleach or by heating (Seymour and Appleton 2001). In addition, travelers to endemic areas should not drink untreated water or ingest raw seafood or shellfish. Fruits and vegetables should not be eaten unless they are cooked or can be peeled.

The most effective way of controlling hantaviral diseases is to reduce human exposure to infected rodents and their excrement. Monitoring of hantavirus prevalence in rodent populations may give some warning of expected increase in the numbers of human cases. Rodent-proofing of homes, reduction of rodent cover around houses, minimization of food available for rodents, trapping in and around dwellings, and the careful disposal of dead rodents are recommended (Bi et al. 2008). Similarly, workplaces and conditions in agriculture, forestry and military activities should be modified when possible to reduce human-rodent exposure. *Coxiella burnetii* infection is common in dairy cattle herds. To control Q fever by *Coxiella burnetii* infection, animals should come from seronegative herds. As animals may become infected by direct contact with infected animals and contaminated environments, transmission can be reduced with good hygiene and other management practices that reduce environmental load, such as immediately removing and disposing of aborted fetuses, dead newborns, and placentas (Porter et al. 2011). Pregnant women, the immunocompromised, and those with known heart valve defects should be excluded from high risk situations, unless they are immune.

Food has been suggested to be the main source of yersiniosis and diseases caused by other bacteria. Fresh produce may become contaminated with pathogenic *Yersinia* during irrigation, harvesting, packing, shipping, and processing. Preventing the access of wild animals to irrigation water and fields could reduce the risk of contamination of surface water and soil (Jalava et al. 2006). Furthermore, the access by rodents and birds to storage facilities should be prevented, and the cleaning of processing equipment should be adequate (Jalava et al. 2006).

Common strategies for tularemia prevention include avoiding dead or sick animals and areas that are tick-infested (Nada and Harik 2013). For hunters or abbatoir workers, animals should be skinned using gloves. All wild game should be cooked thoroughly before eating. Patients should be counseled to not drink untreated water. When engaging in outdoor activities, to prevent bites from tick and deer flies, protective clothing should be worn (Nada and Harik 2013). Insect repellents provide protection against ticks but need to be reapplied frequently in tularemia-endemic areas. Ticks should be removed as soon as possible using tweezers, not fingers, by grabbing the tick as close to the skin surface as possible, then pulling straight up. Hands should be washed immediately after removing a tick.

Control measures related to microbial diseases take advantage of the microorganism's limitations of growth with respect to oxygen, pH, curing salts, organic acids, natural inhibitors and effective heating-cooling schedules (e.g. Riha and Solberg 1975; Sabah et al. 2003, 2004). The best way to prevent infections via dermal or wound contact is to practice proper hygiene. In case of injury, it should be made sure to clean the skin thoroughly and to cover the wound with a bandage. At the first signs of infection, a physician should be contacted immediately. Signs of infection include redness, swelling, pain, and discharge. It is important to take any prescribed antibiotics according to the physician's instructions.

Gas gangrene, caused by subspecies of *Clostridium,* is a necrotic infection of soft tissue associated with high mortality, often necessitating amputation in order to control the infection. Prompt initiation of antimicrobial treatment covering aerobic and anaerobic organism is therefore critical (Aggelidakis et al. 2011).

13.8 Public Health Measures to Prevent Infections with Pathogens

Many countries have regulations that help protect the general public from infectious diseases. Public health measures typically involve eliminating the pathogen from its reservoir or from its route of transmission. Measures include vaccination programs, ensuring safe water supply, mitigation measures in agricultural practice, protecting of the public from foodborne infections, as well as information and education.

13.8.1 Vaccination

Vaccination is a common strategy to control, eliminate, eradicate, or contain disease (i.e., mass immunization strategy). A vaccine is either a killed or weakened (attenuated) strain of a particular pathogen, or a solution containing critical antigens from the pathogen. The body's immune system will respond to these vaccines as if they contain the actual pathogen, even though the vaccine is not capable of causing the disease. The aim of vaccination is to protect individuals who are at risk of a disease. The children, the elderly, immune-compromised individuals, people living with chronic diseases, and people living in disease-endemic areas are those most commonly at risk (Lahariya 2016). Since the discovery of smallpox vaccine, a number of effective vaccines have become available. The vaccine efficacy is quite high against measles (90–95%), mumps (72–88%) and rubella (95–98%) (Reinert et al. 2003) and vaccination was probably the most successful intervention against soil-borne *Clostridium tetani* in history. In 2010, the World Health Organization estimated a 93% reduction in newborns dying from tetanus worldwide compared to the situation in the late 1980s (Rodrigo et al. 2014). Most countries now boast more than 90% coverage of infants in tetanus immunization programs. However, it is difficult to eradicate the disease due to the abundance of tetanus bacterial spores in the environment (Rodrigo et al. 2014). There is also an effective vaccine for hepatitis A and it has been suggested that food handlers should be vaccinated (Cliver 1997). Several kinds of hantavirus vaccines based on inactivated viruses have been demonstrated safe and effective in China (Chen et al. 2002). For Q fever, an effective whole-cell vaccine was licensed in Australia. Live and acellular vaccines have also been studied, but are not currently licensed (Parker et al. 2006).

13.8.2 Safe Water Supply

13.8.2.1 Physical Separation and Elimination Systems

As many waterborne pathogens are spread primarily via feces-contaminated water, a clear separation between wastewater and drinking water systems is key to successful water management (reviewed by Schwarzenbach et al. 2010). A number of conventional treatment methods, including feasible options for low-income countries are available (Nelson and Murray 2008). Most of these methods rely on physical elimination of the pathogens by coagulation, sedimentation, and filtration, typically eliminating pathogens by 1–3 log units. Excreta treated by freezing or high temperature seem to have lower concentrations of pathogens, although many of them can survive these treatments, most likely as a result of the high concentration of organic material around them serving as an insulator. For example, *Salmonella* cells are not inactivated by freezing and are relatively resistant to drying (Plym-Forshell and Ekesbo 1993), while helminth egg concentrations are reduced by these treatments but are not completely eliminated. Taking into account that the cyst infectious dose is low, composting practices do not completely eliminate the risk of infection (Santamaria and Toranzos 2003).

In industrialized countries, connectivity to municipal wastewater treatment plants may be in the range of 50–95%, whereas more than 80% of the municipal wastewater in low-income countries is discharged without any treatment, polluting rivers, lakes, and coastal areas of the seas (Schwarzenbach et al. 2010). The water efficiency of industrial wastewater treatment (i.e., the product revenues per treated volume of process water) is highly variable, ranging from approximately US$ 140 m^{-3} in Denmark to only US$ 10 m^{-3} in the USA and even less in low-income countries (Schwarzenbach et al. 2010). These numbers depend on the type of industrial activity.

13.8.2.2 Inactivation by UVC Irradiation or Chemicals

In several countries, disinfection of treated wastewater by UVC irradiation or chemicals (UVC, chlorination, ozone) is performed. Chlorination is still the most widely used technique for disinfecting drinking water because it is effective and economical, and it maintains a disinfectant residual concentration during distribution as additional security measure. The formation of chlorinated disinfection by-products is considered insignificant when compared to the health benefits from the inactivation of pathogens (Nelson and Murray 2008).

13.8.2.3 Subsurface Multiple Barriers

Also disinfection of the raw wastewater is practiced occasionally. One of the main ways of producing safe drinking water is by the removal and/or inactivation of pathogenic microbes through multiple barriers (Schwarzenbach et al. 2010). These barriers include filtration by soil aquifer treatment, riverbank filtration, sand filtration, or membrane systems. Constructed wetlands are systems that utilize natural processes for wastewater treatment. They are recognized as a viable technology to treat domestic wastewaters, treatment plant effluents, stormwater, and agricultural wastewaters (Bradford et al. 2013). Similar to natural wetlands, constructed wetlands consist of porous media and biota adapted to flooding or water clogging. Various configurations of constructed wetlands have been developed, including surface flow systems, horizontal subsurface flow systems, and vertical subsurface flow systems (Bradford et al. 2013). Surface flow systems often have standing water above the substrate up to 0.6 m deep, whereas subsurface flow systems have no visible surface water. While surface flow constructed wetlands historically have been popular in North America and Australia, horizontal subsurface systems are more prevalent in Europe (Vymazal 2011). Nonetheless, a recent analysis showed that these two types of wetlands perform similarly (Kadlec 2009). Both biotic and abiotic mechanisms can contribute to removal of microorganisms, including predation, die-off, release of antibiotics by plants and other microbes, sedimentation, filtration, adsorption, chemical oxidation, and UV irradiation in the case of surface flow systems (Werker et al. 2002). A wide range of microorganism removal efficiencies has been reported. For example, Decamp and Warren (2000) reported a removal efficiency of 96.6–98.9% for *E. coli* in a pilot-scale horizontal subsurface flow system. Morgan et al. (2008) found that after passing through a series of anaerobic, aerobic, and clarifier reactors and wetland cells, coliform and *E. coli* concentrations were decreased by at least 99% in dairy wastewater. In four surface flow constructed wetlands treating agricultural irrigation return flow in California, removal efficiencies were observed to be 66–91% for *E. coli*, and 86–94% for *enterococci* (Díaz et al., 2010). The removal of cysts of *Cryptosporidium parvum* and *Giarda lamblia* was reported to be two log in two pilot subsurface systems for treating sewage wastewater with respective 10,000 and 300 population equivalents in Germany (Redder et al. 2010).

13.8.2.4 Membrane Technologies

In urban areas that are under water stress (e.g. California, Australia, Singapore), direct or indirect potable reuse is practiced on large scales. These systems mostly rely on membrane technologies (microfiltration followed by reverse osmosis) to treat secondary wastewater effluent and remove pathogens efficiently (Schwarzenbach et al. 2010). In particular membrane based processes became cost-effective for their application in municipal water treatment and are increasingly used as polishing steps to remove microbes and viruses from pretreated water (Peter-Varbanets et al. 2009). Gravity driven low-flow ultrafiltration may become a valid option for producing drinking water directly from low-quality source water (Peter-Varbanets et al. 2009).

13.8.2.5 Water Treatment at the Household Level

Drinking water is often microbially contaminated and, in many cases people may treat their water at the household level to make it safer for consumption. Particularly in low-income countries, water treatment at the household level is required not only in rural areas but also in cities with existing centralized systems. The reliability of such treatments is of primary importance because even occasional consumption of unsafe water results in an increased health risks, particularly for children (Hunter et al. 2009). According to WHO (2014), an estimated 1.1 billion people report treating water in the household, with the practice being particularly common in the Western Pacific region (66.8%) and South East Asia region (45.4%). Boiling is the most commonly used method and it is known to be very efficient in reducing pathogens (Clasen et al. 2008). Filtration is fairly commonly reported in South East Asia and Western Pacific regions, while chlorination is more common in Latin America, Caribbean and African countries (WHO (2014)).

13.8.3 Mitigation Measures in Agricultural Practice

13.8.3.1 Reduction of the Number of Pathogens in Biosolids

Common measures of reducing spread of pathogens with biosolids include effective waste management to reduce pathogen loading and proper application of waste to lands to minimize pathogen movement to adjacent waters. To date, however, it is still a challenge to prevent transmission of pathogens applied with manures. Multiple steps are needed to reduce the amount of viable pathogens entering pathways and to mitigate the transport of pathogens in agricultural settings (Topp et al. 2009). The number of viable pathogens in manure can be greatly reduced by storing manure before use. Survival times of disease-causing bacteria and protozoa are greatly affected by ambient temperature. Viruses may become dormant and can persist for long periods in manure. For example, the infectious avian influenza virus can survive in water for 207 days at 17 °C) and rotaviruses are stable in manure for 7–9 months (Appleyard and Schmoll 2006). However, the longevity of some viruses can be reduced by the presence of predatory bacteria. The lifetime of pathogens can be further reduced by either aerobic composting or drying of manure, and most pathogens die within a week if manure is treated in this manner when the temperature in the compost pile reaches at least 55 °C (Appleyard and Schmoll 2006). Equally, the use of sludges which were treated either by composting or by other disinfection methods generally pose a lower risk of groundwater contamination due to greatly reduced numbers of pathogens. Risks to groundwater will depend on both good agricultural practices in use of sludge (e.g. maintaining buffer zones between areas used for application and water supply wells) and good sanitation practices in sludge treatment.

13.8.3.2 Vegetative Treatment Areas

Vegetative treatment areas are designed land areas with vegetation (e.g., perennial grass) that are widely used to control the transport of soil sediment, nutrients, and microorganisms from agricultural sources (Koelsch et al. 2006). Vegetative treatment areas could be an effective technique to reduce the loadings of coliforms (Tate et al. 2006) or protozoa (Tate et al. 2004) to adjacent surface or subsurface waters if properly implemented. It is generally believed that the efficacy of vegetative treatment areas is related to length, slope, soil type and structure, and vegetation cover (Bradford et al. 2013). Concentrated flow at the soil surface and preferential or macropore flows in the subsurface could reduce the effectiveness of vegetative treatment areas (Faulkner et al. 2011). Therefore, the development of optimal design criteria is urgently needed for better utilization of this technique as a barrier to pathogen movement.

13.8.3.3 Riparian Buffer Strips

Riparian buffer strips function similarly to vegetative treatment areas, but are more critical because they are the last control point before the pathogens enter streams (Bradford et al. 2013). Field and modeling studies suggest that riparian buffer strips could effectively hinder pathogen delivery to streams (Collins and Rutherford 2004). However, complications arise from groundwater-surface interactions and bank erosion that may release significant numbers of pathogens to streams. In this case, the efficiency of riparian buffer strips remains uncertain, and more research is needed to optimize performance of these systems (Bradford et al. 2013; Collins and Rutherford 2004).

13.8.4 *Protection of the Public from Foodborne Infections*

The presence of various pathogens in different foods poses a health hazard and rise concerns about the safety of these food products. Safety measures for vegetables, fruits and animal products have received much attention because an increasing proportion of foodborne disease outbreaks has been associated with these products. Good agricultural practices are recommended to prevent contamination from various sources.

13.8.4.1 Measures to Be Taken by Plant Producers

Fruits and vegetables in the field can be contaminated from a number of sources (Beuchat and Ryo 1997), including feces from man and animals, contaminated manure, irrigation water, water used for pesticide application, insects and seeds. The relative importance of these sources is not known, but it would seem that conditions resulting in direct contact with feces or feces-contaminated water are among the most important. The increase in foodborne disease from plant products may be due to factors such as an increased consumption of raw plants (fruits, fruit juices, lettuce, sprouts and other vegetables), import of products that have been produced under suboptimal safety conditions, packaging methods that extend shelf-life but do not improve safety, application of contaminated manure and water, and marginal acidity of fruit juices. The potential for contamination must be considered significant, and both pre- and postharvest safety actions are needed. When dealing with plants, the primary preharvest food safety concern is with produce. Crops commonly grow close to the ground, with increased risk of contamination, and are often eaten raw. Grains and other crops are also exposed to foodborne pathogens, but most of these crops undergo some processing before being eaten and therefore pose a minor risk. According to Delazari et al. (2006), good agricultural practices in plant production include:

- Provision and use of toilet facilities and facilities for hand-washing or hand disinfection for field workers.
- Decontamination of manures used as fertilizer by composting, long-term (several weeks) stacking or drying.
- Use of sewage or slurry for the irrigation of fruits and vegetables that are eaten raw should be prohibited for a period of 10 months preceding harvest.
- Use of clean (disinfected) water for irrigation and pesticide application.
- Control of wildlife (e.g. through fencing).
- Control of insects that serve as vectors for foodborne pathogens.
- Treatment of seeds (disinfection) in the production of salad sprouts.

Most vegetables have little or no killing effect on foodborne pathogens. In contrast, organic acids in fruits have a detrimental effect on pathogens. However, the acidity is not always sufficient, as evidenced by outbreaks of *E. coli* O157:H7 infections caused by unpasteurized apple juice (Tauxe 1997). Dipping vegetables (alfalfa sprouts) in various disinfectants may result in a five to seven log cycle reduction of *Salmonella*, while dipping of fruit (cantaloupe) resulted in less than one log cycle reduction (Beuchat and Ryo 1997).

13.8.4.2 Measures to Be Taken by Animal Producers

Modern food production requires a risk-based approach in which several critical control points are monitored. Parasite-free farming implies strict indoor housing of animals including pest control, proper feed preparation and storage, and general hygienic measures. Such an approach is effective to prevent *Trichinella* infections in swine and might also be effective to prevent other food-borne infections such as Toxoplasma. However, the current trend is for more extensive (outdoor) livestock rearing to comply with increasing public demand for management practices to improve animal welfare. Such free-ranging livestock, in close contact with both the environment and wildlife, constitutes an emerging risk for parasitic infections (Van der Giessen et al. 2007). Generally, methods for the purpose of eradication of foodborne disease agents from animal population are the same as those developed over the years to eradicate livestock diseases. Some of these are also effective to reduce pathogens in manures which are used as fertilizers. New methods rather merely increased the number of combinations of measures available for control of infections in animals. According to Delazari et al. (2006), methods include the following:

- Farm hygiene, including disinfection after slaughter of infected animals is a potentially important step in the control of enteric foodborne agents in farm animals.
- Much effort has been spent to develop effective vaccines against foodborne disease agents that colonize farm animals. Some *Salmonella* vaccines have shown promise, but none have so far been shown to prevent infection of herds/ flocks. Experimental vaccination of calves against *Cryptosporidium parvum* has shown some beneficial effect, but failed to prevent infection (Harp and Goff 1998).
- Slaughter of infected and exposed animals has been successfully used in control of tuberculosis and brucellosis, and is also being used in control of *Salmonella* infections and paratuberculosis, especially in breeding flocks/herds.
- Mass screening based on laboratory tests or direct tests on animals is used to assure freedom from defined infectious agents; it is based on laboratory testing or direct tests on animals, and played an indispensable role in the eradication of brucellosis and tuberculosis.

- Quarantine is used to some extent by individual farmers to control *Salmonella* infections, for example in dairy herds, where replacements are kept in isolation until there is evidence that they are not infected. Quarantine is not effective against diseases with long incubation periods and no reliable laboratory diagnostic procedures, such as bovine spongiform encephalopathy (BSE) and paratuberculosis.

Combinations of the methods mentioned above have been used to eradicate some foodborne zoonoses that have a fairly narrow reservoir and also cause animal losses. Brucellosis and tuberculosis in cattle have been eradicated in many countries by the slaughter of infected animals, quarantine, mass screening, farm hygiene and, to some extent, vaccination (Delazari et al. 2006).

13.8.4.3 Measures to Be Taken in the Slaughterhouse

Measures Against Pathogenic Microbes

Meat is a highly perishable food and constitutes often the means through which foodborne illnesses may spread. Therefore, strict application of good hygiene practices during slaughtering process is of great importance for public health preservation and quality assurance (Milios et al. 2014). Microbiological parameters that have been used as indicators in slaughterhouses include total viable count, total coliforms, *Enterobacteriaceae*, *Escherichia coli*, fecal streptococci and aeromonads, while *Listeria* sp., *enterococci* and *bifidobacteria* have also been suggested for this purpose (Milios et al. 2014). Meat pathogens that have been related in the past with food borne diseases are *Salmonella*, *E. coli* O157:H7, non O157 STEC *E. coli*, *Listeria*, *Campylobacter*, *Clostridium perfrigens* and *Yersinia*. The most important are E. coli O157:H7, non O157 STEC E. coli and *Salmonella* (Koohmaraie et al., 2005), mainly found in ruminants' meat. Methicillin-resistant *Staphylococcus aureus* in pork meat is also a major global public health concern and could be a safety issue (Lasok and Tenhagen 2013).

The current legislation in the EU requires microbiological examinations for certain indicator organisms (hygiene criteria for the slaughtering process), such as total viable count, *Enterobacteriaceae* and *Salmonella* (Milios et al. 2014). A three-class system, classifying microbiological results from carcasses (for each animal species) into satisfactory, acceptable and unsatisfactory is used to determine the hygiene performance of the operator. However, no proven correlation has been found between indicator organisms and prevalence/levels of pathogens. Therefore, the microbiological data, based on the indicators, should be interpreted only to assess general trends in the hygiene process of the operator. Detected values above the defined criteria require an improvement in slaughter hygiene and the review of process control (Barco et al. 2015). However, microbiological results used alone may be insufficient. The results are obtained only after dressing and before chilling, at the end of the slaughtering process and therefore, do not provide information on the cause of the problem (Buncic 2006). Therefore, 'process-based' microbiological criteria which are based on values measured at various stages of the process, including final carcass values, have been recently proposed (Milios et al. 2014). Finally, in order to implement an adequate monitoring system, non-destructive techniques of carcass sampling could be used instead of excision. The microbial recovery may be lower, but it is proportional to the excision recovery and therefore, non-destructive techniques, like swabbing with sponges, could be a practical sampling method for the estimation of indicators during the slaughtering procedure and hygiene evaluation (Milios et al. 2014).

Measures Against Parasites

Among the major food-borne parasites are *Toxoplasma gondii*, *Sarcocystis* spp., *Taenia* spp. and *Trichinella* spp. Humans get infected by eating raw or undercooked meat infected with cyst stages of these parasites. In most countries measures are taken to prevent humans from becoming infected with meatborne helminths by inspecting the meat in the slaughterhouse or laboratory (Dorny et al. 2009). For toxoplasmosis and sarcocystosis no specific meat inspection is done. Meat inspection for cysticercosis has a low sensitivity, resulting in a high number of infected carcasses entering the food chain (Dorny et al. 2009). In addition, in developing countries a large proportion of the carcasses escape meat inspection because it is not practised or because the animals are not slaughtered in abattoirs. Cooking is effective in killing the parasites if the appropriate temperature is reached in the core of the meat product (Dorny et al. 2009). Freezing and other meat processing techniques such as drying, smoking, curing etc. are other effective ways to reduce the risk of infection by consuming contaminated meat, except for some species of *Trichinella*, which show remarkable resistance.

13.8.4.4 Measures by Food Industry and Food Handlers

Food industry management has the responsibility to provide food of a high degree of safety (Delazari et al. 2006). The industry also has to provide consumers with information that enables them to protect their food from contamination or spoilage. This is done through clear and detailed label instructions that tell the consumers how to store, handle and prepare the food correctly. Labeling should include a complete listing of food constituents, including allergenic substances. Companies that transport food have the duty to keep and maintain food environment conditions that are specified by the producers and processors, from the point of origin to the final destination. Additionally, it is imperative that the vehicles are used only for the transport of harmless products.

Food handlers, retailers and consumers have the duty to follow label instructions and keep food under the environmental conditions specified by producers and processors (Delazari et al. 2006). Consumers and food handlers also have to apply good handling practices and hygienic procedures when preparing food for serving. Human contact with food is generally greater at the service and preparation levels than in food production and processing, and mandates good personal hygiene.

13.8.5 Information and Education

A constant dialogue between public health, veterinary and food safety experts, with multidisciplinary skills, is essential in order to signal new threats, to monitor changing trends in well-recognised diseases, to detect emerging pathogens, to understand transmission routes, to develop control effective strategies and to ensure the priority of food hygiene during production and processing.

The education of farmers, veterinary and public health professionals and consumers is essential to improve knowledge and awareness of the risks to humans caused by zoonotic diseases circulating in livestock and the measures needed to reduce spread of all pathogens to products of the food chain. In case of contaminated food products, prevention can be achieved by simple measures including the strict personal hygiene. Finally, consumers should be aware of the risks of eating raw food and the possible routes of getting food-borne infections.

References

Abd H, Johansson T, Golovliov I, Sandström G, Forsman M (2003) Survival and growth of *Francisella tularensis* in *Acanthamoeba castellanii*. Appl Environ Microbiol 69:600–606

Abd-Alla MD, Ravdin JI (2002) Diagnosis of amoebic colitis by antigen capture ELISA in patients presenting with acute diarrhoea in Cairo. Egypt Trop Med Int Health 7:365–370

Abu-Ashour J, Joy DM, Lee H, Whiteley HR (1994) Transport of microorganisms through soil. Water Air Soil Pollut 75:141–158

Abubakar I, Aliyu SH, Arumugam C, Usman NK, Hunter PR (2007) Treatment of cryptosporidiosis in immunocompromised individuals: systematic review and meta-analysis. Br J Clin Pharmacol 63(4):387–393

Adachi JA, Mathewson JJ, Jiang ZD, Ericsson CD, DuPont HL (2002) Enteric pathogens in Mexican sauces of popular restaurants in Guadalajara, Mexico, and Houston, Texas. Ann Intern Med 136:884–887

Adam RD (1991) The biology of *Giardia* spp. Microbiol Rev 5(4):706–732

Adam RD (2001) Biology of *Giardia lamblia*. Clin Microbiol Rev 14(3):447–475

Adams VJ, Lombard CJ, Dhansay MA, Markus MB, Fincham JE (2004) Efficacy of albendazole against the whipworm *Trichuris trichiura*: a randomised, controlled trial. S Afr Med J 94:972–976

Afonso E, Lemoine M, Poulle ML, Ravat MC, Romand S, Thuillez P, Villena I, Aubert D, Rabilloud M, Riche B, Gilot-Fromont E (2008) Spatial distribution of soil contaminated by *Toxoplasma gondii* in relation to cat defecation behaviour in an urban area. Int J Parasitol 38:1017–1023

Afshar M, Raju M, Ansell D, Bleck TP (2011) Narrative review: tetanus–a health threat after natural disasters in developing countries. Ann Intern Med 154:329–335

Aggelidakis J, Lasithiotakis K, Topalidou A, Koutroumpas J, Kouvidis G, Katonis P (2011) Limb salvage after gas gangrene: a case report and review of the literature. World J Emerg Surg 6:28

Ahamed SK, Thobaiti YA (2014) Mucormycosis: a challenge for diagnosis and treatment – case reports and review of literature. OHDM 13(3):703–706

Akgun Y, Tacyildiz IH, Celik Y (1999) Amebic liver abscess: changing trends over 20 years. World J Surg 23:102–106

Allan JC, Velasquez-Tohom M, Torres-Alvarez R, Yurrita P, Garcia-Noval J (1996) Field trial of the coproantigen-based diagnosis of Taenia solium taeniasis by enzyme-linked immunosorbent assay. Am J Trop Med Hyg 54:352–356

Allerberger F, Liesegang A, Grif K, Khaschabi D, Prager R, Danzl J, Hock F, Ottl J, Dierich MP, Berghold C, Neckstaller I, Tschäpe H, Fisher I (2003) Occurrence of Salmonella enterica serovar Dublin in Austria. Wien Med Wochenschr 153:148–152

Allos BM (2001) *Campylobacter jejuni* infections: update on emerging issues and trends. Clin Infect Dis 32:1201–1206

Ansay SE, Darling KA, Kaspar CW (1999) Survival of *Escherichia coli* O157:H7 in ground-beef patties during storage at 2, −2, 15 and then −2°C, and −20°C. J Food Prot 62:1243–1247

Apothéloz C, Regamey C (1996) Disseminated infection due to *Actinomyces meyeri*: case report and review. Clin Infect Dis 22(4):621–625

Appleyard S, Schmoll O (2006) Agriculture: potential hazards and information needs. In: Schmoll O, Howard G, Chilton J (eds) Protecting groundwater for health: managing the quality of drinking-water sources. I Chorus. IWA Publishing, London, pp 243–273

Arenas R (2008) Micología médica ilustrada, 3rd edn. McGraw Hill, Mexico

Ash C, Farrow JA, Dorsch M, Stackenbrandt E, Collins MD (1991) Comparative analysis of *Bacillus anthracis*, *Bacillus cereus* and related species on the basis of reverse transriptase of 16S rRNA. Int J Syst Bacteriol 41:343–346

Avery LM, Hill P, Killham K, Jones DL (2004) *Escherichia coli* O157 survival following the surface and sub-surface application of human pathogen contaminated organic waste to soil. Soil Biol Biochem 36:2101–2103

Axtell CA, Beattie GA (2002) Construction and characterization of a *proU-gfp* transcriptional fusion that measures water availability in a microbial habitat. Appl Environ Microbiol 68:4604–4612

Bae JH, Song R, Lee A, Park JS, Kim MR (2011) Computed tomography for the preoperative diagnosis of pelvic actinomycosis. J Obstet Gynaecol Res 37(4):300–304

Balcht AL, Smith RP (1994) Pseudomonas aeruginosa: infections and treatment. Marcel Dekker, New York

Barco L, Belluco S, Roccato A, Ricci A (2015) A systematic review of studies on Escherichia coli and Enterobacteriaceae on beef carcasses at the slaughterhouse. Int J Food Microbiol 207:30–39

Barlaz MA (1996) Microbiology of solid waste landfills. In: Palmisano AC, Barlaz MA (eds) Microbiology of solid waste. CRC Press, New York

Barlow M, Hall BG (2002) Origin and evolution of the AmpC beta-lactamases of Citrobacter freundii. Antimicrob Agents Chemother 46:1190–1198

Bates M, Cruickshank G (1957) Thoracic actinomycosis. Thorax 12(2):99–124

Baumgardner DJ (2012) Soil-related bacterial and fungal infections. J Am Board Fam Med 25 (5):734–744

Baumgartner A, Kuffer M, Suter D, Jemmi T, Rohner P (2007) Antimicrobial resistance of *Yersinia enterocolitica* strains from human patients, pigs and retail pork in Switzerland. Int J Food Microbiol 115(1):110–114

Bech TB, Johnsen K, Dalsgaard A, Laegdsmand M, Jacobsen OH, Jacobsen CS (2010) Transport and distribution of *Salmonella enterica* serovar typhimurium in loamy and sandy soil monoliths with applied liquid manure. Appl Environ Microbiol 76:710–714

Bennett A, Guyatt H (2000) Reducing intestinal nematode infection: efficacy of albendazole and mebendazole. Parasitol Today 16:71–74

Bertucci JJ, Sedita SJ, Lue-Hing C (1987) Viral aspects of applying sludges to land. In: Rao VC, Melnick JL (eds) Human viruses in sediments, sludges, and soils. CRC Press, Boca Raton

Bethony J, Brooker S, Albonico M, Geiger SM, Loukas A, Diemert D, Hotez PJ (2006) Soil transmitted helminth infections: ascariasis, trichuriasis, and hookworm. Lancet 367:1521–1532

Beuchat LR, Ryo JH (1997) Produce handling and processing practices. Emerg Infect Dis 3:459–465

Bhan MK, Bahl R, Bhatnagar S (2005) Typhoid and paratyphoid fever. Lancet 366(2):749–762

Bi ZQ, Formenty PBH, Roth CE (2008) Hantavirus infection: a review and global update. J Infect Dev Countries 2(1):3–23

Blackburn JK, McNyset KM, Curtis A, Hugh-Jones ME (2007) Modeling the geographic distribution of *Bacillus anthracis*, the causative agent of anthrax disease, for the contiguous United States using predictive ecologic niche modeling. Am J Trop Med Hyg 77:1103–1110

Blaser MJ, LaForce FM, Wilson NA, Wang WL (1980) Reservoirs for human campylobacteriosis. J Infect Dis 141:665–669

Blum D, Feachem RG (1985) Health aspects of nightsoil and sludge use in agriculture and aquaculture. Part III. An epidemiological perspective. IRCWD, International Reference Centre for Waste Disposal, Duebendorf

Bollman J, Ismond A, Blank G (2001) Survival of *Escherichia coli* O157:H7 in frozen foods: impact of the cold shock response. Int J Food Microbiol 64:127–138

Bottone EJ (1999) *Yersinia enterocolitica*: overview and epidemiologic correlates. Microbes Infect 1:323–333

Bottone EJ (2010) Bacillus cereus, a volatile human pathogen. Clin Microbiol Rev 23:382–398

Boucher HW, Patterson TF (2008) Aspergillosis. In: Hospenthal DR, Rinaldi MG (eds) Diagnosis and treatment of human mycoses. Humana Press, Totowa, pp 181–199

Bourke GJ, Petana WB (1994) Human *Taenia cysticercosis*: a bizarre mode of transmission. Trans R Soc Trop Med Hyg 88:680

Bradford SA, Morales VL, Zhang W, Harvey RW, Packman AI, Mohanram A, Welty C (2013) Transport and fate of microbial pathogens in agricultural settings. Crit Rev Environ Sci Technol 43:775–893

Brandl MT (2006) Fitness of human enteric pathogens on plants and implications for food safety. Annu Rev Phytopathol 44:367–392

Brandl MT, Mandrell RE (2002) Fitness of *Salmonella enterica* serovar Thompson in the cilantro phyllosphere. Appl Environ Microbiol 68:3614–3621

Brömel C, Sykes JE (2005) Epidemiology, diagnosis, and treatment of blastomycosis in dogs and cats. Clin Tech Small Anim Pract 20:233–239

Brook I (2008) Current concepts in the management of *Clostridium tetani* infection. Expert Rev Anti-Infect Ther 6:327–336

Brown JR (1973) Human actinomycosis. A study of 181 subjects. Hum Pathol 4(3):319–330

Brown SP, Cornfort DM, Mideo N (2012) Evolution of virulence in opportunistic pathogens: generalism, plasticity and control. Trends Microbiol 20(7):336–342

Brown J, Benedict K, Park BJ, Thompson GR (2013) Coccidioidomycosis: epidemiology. Clin Epidemiol 5:185–197

Bryan FL (1969) What the sanitarium should know about *Clostridium perfringens* foodborne illness. J Milk Food Technol 32:381–389

Bryant AE, Stevens DL (2010) Clostridial myonecrosis: new insights in pathogenesis and management. Curr Infect Dis Rep 12(5):383–391

Brynestat S, Granum PE (2002) *Clostridium perfringens* and foodborne infections. Int J Food Microbiol 74:195–202

Bultman MW, Fisher FS, Pappagianis D (2005) The ecology of soil-borne human pathogens. In: Selenius O, Alloway B, Centeno JA, Finkelman RB, Fuge R, Lindhu U, Smedley P (eds) Essentials of medical geology: impacts of the natural environment on public health. Elsevier-Academic Press, Burlington, pp 481–512

Buncic S (2006) Integrated food safety and veterinary public health. CABI Publishing, Wallingford

Bundy DAP, Cooper ES (1989) Trichuris and trichuriasis in humans. Adv Parasitol 28:107–173

Castro J, Vazquez-Iglesias JL, Arnal-Monreal F (1983) Dysentery caused by *Balantidium coli*-report of two cases. Endoscopy 15:272–274

Chacín-Bonilla L (2010) Epidemiology of *Cyclospora cayetanensis*: a review focusing in endemic areas. Acta Trop 115:181–193

Chan MS, Medley GF, Jamison D, Bundy DA (1994) The evaluation of potential global morbidity attributable to intestinal nematode infections. Parasitology 109:373–387

Chandler DS, Craven JA (1980) Relationship of soil moisture to survival of *Escherichia coli* and *Salmonella typimurium* in soils. Crop Pasture Sci 31:547–555

Chao WL, Ding RJ, Chen RS (1988) Survival of *Yersinia entercolitica* in the environment. Can J Microbiol 34:753–756

Chapman SW, Dismukes WE, Proia LA, Bradsher RW, Pappas PG, Threlkeld MG, Kauffman CA (2008) Clinical practice guidelines for the management of blastomycosis: update by the Infectious Diseases Society of America. Clin Infect Dis 46(12):1801–1812

Checketts MR, White RJ (1993) Avoidance of intermittent positive pressure ventilation in tetanus with dantrolene therapy. Anaesthesia 48:969–971

Cheever AW, Kamel IA, Elwi AM, Mosimann JE, Danner R, Sippel JE (1978) *Schistosoma mansoni* and *S haematobium* infections in Egypt, III: extrahepatic pathology. Am J Trop Med Hyg 27:55–75

Chen HX, Luo ZZ, Zhang JJ, Hantavirus Vaccine Efficacy Evaluation Working Group (2002) Large scale field evaluation on vaccines of hemorrhagic fever with renal syndrome in China. Chin J Epidemiol 23:145–147

Clark CG, Diamond LS (1993) Entamoeba histolytica: a method for isolate identification. Exp Parasitol 77:450–455

Clasen TF, Do HT, Boisson S, Shipin O (2008) Microbiological effectiveness and cost of boiling to disinfect drinking water in rural Vietnam. Environ Sci Technol 42(12):4.255–4.260

Cliver DO (1985) Vehicular transmission of hepatitis A. Public Health Rev 13:235–292

Cliver DO (1997) Virus transmission via food. Food Technol 51:71–78

Cohen RD, Bowie WR, Enns R, Flint J, Fitzgerald JM (2007) Pulmonary actinomycosis complicating infliximab therapy for Crohn's disease. Thorax 62(11):1013–1014

Coleman DC (1994) The microbial loop concept as used in terrestrial soil ecology studies. Microb Ecol 28:245–250

Collins R, Rutherford K (2004) Modelling bacterial water quality in streams draining pastoral land. Water Res 38:700–712

Colson P, Borentain P, Queyriaux B, Kaba M, Moal V, Gallian P, Heyries L, Raoult D, Gerolami R (2010) Pig liver sausage as a source of hepatitis E virus transmission to humans. J Infect Dis 202:825–834

Connor BA, Schwartz E (2005) Typhoid and paratyphoid fever in travellers. Lancet Infect Dis 5:623–628

Cook PP (2001) Amphotericin B lipid complex for the treatment of recurrent blastomycosis of the brain in a patient previously treated with itraconazole. S Med J 94(5):548–549

Cover TL, Aber RC (1989) Yersinia enterocolitica. N Engl J Med 321:16–24

Coyne MS (1999) Soil microbiology: an exploratory approach. Delmar Publishers, Albany

de Crom SCM, Rossen JWA, van Furth AM, Obihara CC (2016) Enterovirus and parechovirus infection in children: a brief overview. Eur J Pediatr 175:1023–1029

Crowley MR, Katz RW, Kessler R, Simpson SQ, Levy H, Hallin GW, Cappon J, Krahling JB, Wernly J (1998) Successful treatment of adults with severe Hantavirus pulmonary syndrome with extracorporeal membrane oxygenation. Crit Care Med 26:409–441

Dai X, Boll J (2006) Settling velocity of *Cryptosporidium parvum* and *Giardia lamblia*. Water Res 40:1321–1325

Darling ST (1906) A protozoon general infection producing pseudotubercles in the lungs and focal necrosis in the liver, spleen and lymphnodes. JAMA 46:1283

Dayan AD (2003) Albendazole, mebendazole and praziquantel. Review of non-clinical toxicity and pharmacokinetics. Acta Trop 86:141–159

De Beurmann L, Gougerot H (1912) Les sporotrichoses. Librarie Félix Alcan, Paris

Decamp O, Warren A (2000) Investigation of *Escherichia coli* removal in various designs of subsurface flow wetlands used for wastewater treatment. Ecol Eng 14:293–299

Delazari I, Riemann HP, Hajmeer M (2006) Food safety. In: Riemann HP, Cliver DO (eds) Foodborne infections and intoxications, 3rd edn. Academic Press (Elsevier), London, pp 833–884

Dentinger CM, Bower WA, Nainan OV, Cotter SM, Myers G, Dubusky LM, Fowler S, Salehi ED, Bell BP (2001) An outbreak of hepatitis A associated with green onions. J Infect Dis 183:1273–1276

DeSpain Smith L, Gardner MV (1949) The occurrence of vegetative cells of *Clostridium perfringens* in soil. J Bacteriol 58:407–408

Desta S, Melaku M, Abdela N (2016) Botulinum toxin and its biological significance: a review. Austin J Vet Sci Anim Husb 3(1):1021

Díaz E, Ferrandez A, Prieto MA, Garcia JL (2001) Biodegradation of aromatic compounds by Escherichia coli. Microbiol Mol Biol Rev 65:523–569

Díaz FJ, O'Geen AT, Dahlgren RA (2010) Efficacy of constructed wetlands for removal of bacterial contamination from agricultural return flows. Agric Water Manag 97:1813–1821

Doganay M, Metan G (2009) Human anthrax in Turkey from 1990 to 2007. Vector Borne Zoonot Dis 9(2):131–140

Doganay M, Metan G, Alp E (2010) A review of cutaneous anthrax and its outcome. J Infect Publ Health 3:98–105

Dorner SM, Anderson WB, Slawson R, Kouwen MN, Huck PM (2006) Hydrologic modeling of pathogen fate and transport. Environ Sci Technol 40:4746–4753

Dorny P, Praet N, Deckers N, Gabriel S (2009) Emerging food-borne parasites. Vet Parasitol 163:196–206

Dubey JP, Jones JL (2008) Toxoplasma gondii infection in humans and animals in the United States. Int J Parasitol 38:1257–1278

Dugdale DC, Vyas JM (2011) Tularemia. URL http://www.nlm.nih.gov/medlineplus/ency/article/000856.htm. Accessed 6 Mar 2017

Easterbrook JD, Zink MC, Klein SL (2007) Regulatory T cells enhance persistence of the zoonotic pathogen Seoul virus in its reservoir host. Proc Natl Acad Sci U S A 104:15502–15507

Ebisawa I, Takayangi M, Kurata M, Kigawa M (1986) Density and distribution of *Clostridium tetani* in the soil. Jpn J Exp Med 56:69–74

Edwards GA, Domm BM (1960) Human leptospirosis. Medicine 39:117–156

El Sol AA, Alhajhusain A (2009) Update on the treatment of *Pseudomonas aeruginosa* pneumonia. J Antimicrob Chemother 64(2):229–238

Emerson SU, Arankalle VA, Purcell RH (2005) Thermal stability of hepatitis E virus. J Infect Dis 192:930–993

Eng SK, Pusparajah P, Mutalib NSA, Ser HL, Chan KG, Lee LH (2015) *Salmonella*: a review of pathogenesis, epidemiology and antibiotic resistance. Front Life Sci 8:284–293

Ercolani GL (1979) Differential survival of *Salmonella typhi*, *Escherichia coli*, and *Enterobacter aerogenes* on lettuce in the field. Zentralbl Bakteriol Naturwiss 134:402–411

Ezeamama AE, Freidman JF, Acosta LP, Bellinger DC, Langdon GC, Manalo DL, Olveda RM, Kurtis JD, McGarvey ST (2005) Helminth infection and cognitive impairment among Filipino children. Am J Trop Med Hyg 72:540–548

Fang W, Washington L, Kumar N (2007) Imaging manifestations of blastomycosis: a pulmonary infection with potential dissemination. Radiographics 27:641–655

Faulkner JW, Zhang W, Geohring LD, Steenhuis TS (2011) Tracer movement through paired vegetative treatment areas receiving silage bunker runoff. J Soil Water Conserv 66:18–28

Fayyaz J, Lessnau KD (2010) Histoplasmosis clinical presentation. URL http://emedicine.medscape.com/article/299054-clinical. Accessed 19 Mar 2017

Feachem GG, Bradley DJ, Garelick H, Mara DD (1983) Sanitation and disease—health aspects of excreta and wastewater management. World bank studies in water supply and sanitation 3. Wiley, Chichester

Fiener P, Auerswald K (2003) Effectiveness of grassed waterways in reducing runoff and sediment delivery from agricultural watersheds. J Environ Qual 32:927–936

Fiore AE, Wasley A, Bell BP (2006) Prevention of hepatitis A through active or passive immunization: recommendations of the Advisory Committee of Immunization Practices. MMWR Recomm Rep 19:1–23

Firth PG, Solomon JB, Roberts LL, Gleeson TD (2011) Airway management of tetanus after the Haitian earthquake: new aspects of old observations. Anesth Analg 113:545–547

Fogarty C, Regennitter F, Viozzi CF (2006) Invasive fungal infection of the maxilla following dental extractions in a patient with chronic obstructive pulmonary disease. J Can Dent Assoc 72:149–152

Foissner W (1987) Soil protozoa: fundamental problems, ecological significance, adaptations in ciliates and testaceans, bioindicators, and guide to the literature. Prog Protistol 2:69–212

Fredriksson-Ahomaa M, Korkeala H (2003) Low occurrence of pathogenic *Yersinia enterocolitica* in clinical, food and environmental samples: a methodological problem. Clin Microbiol Rev 16:220–229

Fredriksson-Ahomaa M, Bjorkroth J, Hielm S, Korkeala H (2000) Prevalence and characterization of pathogenic *Yersinia enterocolitica* in pig tonsils from different slaughterhouses. Food Microbiol 17:93–101

Fredriksson-Ahomaa M, Stolle A, Siitonen A, Korkeala H (2006) Sporadic human *Yersinia enterocolitica* infections caused by bioserotype 4/O:3 originate mainly from pigs. J Med Microbiol 55:747–749

Fremaux B, Prigent-Combaret C, Vernozy-Rozand C (2008) Long-term survival of Shiga toxin-producing Escherichia coli in cattle effluents and environment: an updated review. Vet Microbiol 132:1–18

Galgiani JN, Catanzaro A, Cloud GA, Johnson RH, Williams PL, Mirels LF, Nassar F, Lutz JE, Stevens DA, Sharky PK, Singh VR, Larsen RA, Delgado KL, Flanigan C, Rinaldi MG (2000) Comparison of oral fluconazole and itraconazole for progressive, nonmeningeal coccidioidomycosis: a randomized, double-blind trial. Mycoses Study Group. Ann Intern Med 133:676–686

Gannon VPJ, Humenik F, Rice M, Cicmanec JL, Smith JE, Carr R (2004) Control of zoonotic pathogens in animal wastes. In: Cotruvo JA, Dufour A, Rees G, Bartram J, Carr R, Clover DO, Craun GF, Frayer R, Gannon VPJ (eds) Waterborne zoonoses: identification, causes and control. IWA Publishing, London, pp 409–425

García LS (1999) Flagellates and ciliates. Clin Lab Med 19:621–638

García HH, Del Brutto OH (2000) Taenia solium cysticercosis. Infect Dis Clin N Am 14:97–119

García-García A, Gandara-Rey JM, Crespo-Abelleira A, Jorge-Barreiro J (2007) Botulinum toxin for treating muscular contractures in cephalic tetanus. Br J Oral Maxillofac Surg 45:573–575

García-Laverde A, de Bonilla L (1975) Clinical trials with metronidazole in human balantidiasis. Am J Trop Med Hyg 24:781–783

Garner JP, Macdonald M, Kumar PK (2007) Abdominal actinomycosis. Int J Surg 5(6):441–448

Gebrie M, Engdaw TA (2015) Review on taeniasis and its zoonotic importance. Eur J Appl Sci 7 (4):182–191

Gerba CP (1987) Transport and fate of viruses in soil: field studies. In: Rao VC, Melnick JL (eds) Human viruses in sediments, sludges and soils. CRC Press, Boca Raton

Gerba CP (1996) Microbial pathogens in municipal solid waste. In: Palmisano AC, Barlaz M (eds) Microbiology of solid waste. CRC Press, New York, pp 155–117

Gerba CP, Bitton G (1984) Microbial pollutants: their survival and transport pattern to groundwater. In: Bitton G, Gerba CP (eds) Groundwater pollution microbiology. Wiley, New York, pp 39–54

Gerba CP, Smith JE (2005) Sources of pathogenic microorganisms and their fate during land application of wastes. J Environ Qual 34:42–48

Ghimire TR, Sherchan JB (2008) Human infection of *Cyclospora cayetanensis*: a review on its medico-biological and epidemiological pattern in global scenario. J Nepal Health Res Counc 4 (2):25–40

Gillespie IA, O'Brien SJ, Adak GK, Ward LR, Smith HR (2005) Foodborne general outbreaks of *Salmonella* Enteritidis phage type 4 infection, England and Wales, 1992–2002: where are the risks? Epidemiol Infect 133:759–801

Gonsalez CR, Casseb J, Monteiro FG, Paula-Neto JB, Fernandez RB, Silva MB, Camargo ED, Mairinque JM, Tavares LC (1998) Use of doxycycline for leptospirosis after high-risk exposure in Sao Paulo, Brazil. Rev Inst Med Trop Sao Paulo 41:59–61

Gonzalez-Escobedo G, Marshall JM, Gunn JS (2011) Chronic and acute infection of the gall bladder by *Salmonella typhi*: understanding the carrier state. Nat Rev Microbiol 9:9–14

Goodwin RA, Loyd JE, des Prez RM (1981) Histoplasmosis in normal hosts. Medicine (Baltimore) 60:231–266

Grassl GA, Finlay BB (2008) Pathogenesis of enteric salmonella infections. Curr Opin Gastroenterol 24:22–26

Greenslade E, Beasley R, Jennings L, Woodward A, Weinstein P (2003) Has *Coxiella burnetii* (Q-fever) been introduced into New Zealand? Emerg Infect Dis 9:138–140

Griffiths BS (1990) A comparison of microbial-feeding nematodes and protozoa in the rhizosphere of different plants. Biol Fertil Soils 9:83–88

Gryseels B, Polman K, Clerinx J, Kestens L (2006) Human schistosomiasis. Lancet 368:1106–1118

Guerrant RL, Van Gilder T, Steiner TS, Thielman NM, Slutsker L, Tauxe RV, Hennessy T, Griffin PM, DuPont H, Sack RB, Infectious Diseases Society of America (2001) Practice guidelines for the management of infectious diarrhea. Clin Infect Dis 32:331–351

Hajjeh RA, Warnock DW (2001) Counterpoint: invasive aspergillosis and the environment—rethinking our approach to prevention. Clin Infect Dis 33(9):1549–1552

Haque R, Ali IM, Sack RB, Farr BM, Ramakrishnan G, Petri WA (2001) Amebiasis and mucosal IgA antibody against the *Entamoeba histolytica* adherence lectin in Bangladeshi children. J Infect Dis 183:1787–1793

Haqshenas G, Shivaprasad HL, Woolcock PR, Read DH, Meng XJ (2001) Genetic identification and characterization of a novel virus related to human hepatitis E virus from chickens with hepatitis-splenomegaly syndrome in the United States. J Gen Virol 82:2449–2462

Harp JA, Goff TP (1998) Strategies for the control of *Cryptosporidium parvum* infection in calves. J Dairy Sci 81:289–294

Hatchette TF, Gupta R, Marrie TJ (2000) *Pseudomonas aeruginosa* community acquired pneumonia in previously healthy adult: case report and review of the literature. Clin Infect Dis 31 (6):1349–1356

Heinzen RA, Hackstadt T, Samuel JE (1999) Developmental biology of Coxiella Burnetii. Trends Microbiol 7(4):149–154

Hellstrom JS, Marshall RB (1978) Survival of *Leptospira interrogans* serovar pomona in an acidic soil under simulated New Zealand field conditions. Res Vet Sci 25:29–33

Hendricks CW (1971) Increased recovery rate of salmonellae from stream bottom sediments versus surface waters. Appl Microbiol 21:379–380

Henry RA, Johnson RC (1978) Distribution of the genus *Leptospira* in soil and water. Appl Environ Microbiol 35:492–499

Herold S, Karch H, Schmidt H (2004) Shiga toxin-encoding bacteriophages–genomes in motion. Int J Med Microbiol 294:115–121

Heymann DL (ed) (2008) Control of communicable diseases manual, 19th edn. American Public Health Association, Washington, DC

Hinson ER, Shone SM, Zink MC, Glass GE, Klein SL (2004) Wounding: the primary mode of Seoul virus transmission among male Norway rats. Am J Trop Med Hyg 70:310–317

Hirano SS, Upper CD (2000) Bacteria in the leaf ecosystem with emphasis on *Pseudomonas syringae* – a pathogen, ice nucleus, and epiphyte. Microbiol Mol Biol Rev 64:624–653

van der Hoek W, Hunink J, Vellema P, Droogers P (2011) Q fever in The Netherlands: the role of local environmental conditions. Int J Environ Health Res 11:1–11

Hoffman S, Katz JF, Jacobson JH (1971) Salvage of a lower limb after gas gangrene. Bull N Y Acad Med 47:40–49

Hohmann EL (2001) Nontyphoidal salmonellosis. Clin Infect Dis 15(32):263–269

Hopla CE (1974) The ecology of tularemia. Adv Vet Sci Comp Med 18:25–53

Horton J (2003) Global anthelmintic chemotherapy programs: learning from history. Trends Parasitol 19:405–409

Hotez PJ, Brooker S, Bethony JM, Bottazzi ME, Loukas A, Xiao S (2004) Hookworm infection. N Engl J Med 351:799–807

Huggins JW, Hsiang CM, Cosgriff TM, Guang MY, Smith JI, Wu ZO, LeDuc JW, Zheng ZM, Meegan JM, Wang QN (1991) Prospective, double-blind, concurrent, placebo-controlled clinical trial of intravenous Ribavirin therapy of hemorrhagic fever with renal syndrome. J Infect Dis 164:1119–1127

Hugh-Jones M, Blackburn J (2009) The ecology of Bacillus anthracis. Mol Asp Med 30:356–367

Hunter PR, Nichols G (2002) Epidemiology and clinical features of Cryptosporidium infection in immunocompromised patients. Clin Microbiol Rev 15:145–154

Hunter PR, Zmirou-Navier D, Hartemann P (2009) Estimating the impact on health of poor reliability of drinking water interventions in developing countries. Sci Total Environ 407:2621–2624

Hutchison ML, Walters LD, Moore A, Crookes KM, Avery SM (2004) Effect of length of time before incorporation on survival of pathogenic bacteria present in livestock wastes applied to agricultural soil. Appl Environ Microbiol 70:5111–5118

Ibrahim A, Edwards JE, Filler SG (eds) (2004) Mucormycosis. Harcourt Brace, Philadelphia

Ishii S, Sadowsky J (2008) *Escherichia coli* in the environment: implications for water quality and human health. Microbes Environ 23(2):101–108

Islam M, Doyle MP, Phatak SC, Millner P, Jiang X (2004a) Persistence of *E. coli* O157:H7 in soil and on leaf lettuce and parsley grown in fields treated with contaminated manure composts or irrigation water. J Food Prot 67:1365–1370

Islam M, Morgan J, Doyle MP, Phatak SC, Millner P, Jiang X (2004b) Fate of *Salmonella enterica* serovar typhimurium on carrots and radishes grown in fields treated with contaminated manure composts or irrigation water. Appl Environ Microbiol 70:2497–2502

Iwarson S (1992) New vaccines against hepatitis A enter the market-but who should be vaccinated? Infection 20:192–193

Jacobs JL, Carroll TL, Sundin GW (2005) The role of pigmentation, UV radiation tolerance, and leaf colonization strategies in the epiphytic survival of phyllosphere bacteria. Microb Ecol 49:104–113

Jacques MA, Josi K, Darrasse A, Samson R (2005) *Xanthomonas axonopodis* pv. *phaseoli* var. fuscans is aggregated in stable biofilm population sizes in the phyllosphere of field-grown beans. Appl Environ Microbiol 71:2008–2015

Jaeger CH, Lindow SE, Miller S, Clark E, Firestone MK (1999) Mapping of sugar and amino acid availability in soil around roots with bacterial sensors of sucrose and tryptophan. Appl Environ Microbiol 65:2685–2690

Jalava K, Hakkinen M, Valkonen M, Nakari UM, Palo T, Hallanvuo S, Ollgren J, Siitonen A, Nuorti P (2006) An outbreak of gastrointestinal illness and erythema nodosum from grated carrots contaminated with *Y. pseudotuberculosis*. J Infect Dis 194:1209–1216

Janisiewicz WJ, Conway WS, Brown MW, Sapers GM, Fratamico P, Buchanan RL (1999) Fate of *Escherichia coli* O157:H7 on fresh-cut apple tissue and its potential for transmission by fruit flies. Appl Environ Microbiol 65:1–5

Janssen MPM, Heijmans GJSM (1998) Dynamics and stratification of protozoa in the organic layer of a Scots pine forest. Biol Fertil Soils 26:285–292

Jeffery S, van der Putten WH (2011) Soil borne human diseases. JCR Scientific and Technical Reports, no. 65787. Publications Office of the European Union. https://doi.org/10.2788/36703

Jernigan JA, Stephens DS, Ashford DA, Omenaca C, Topiel MS, Galbraith M, Tapper M, Fisk TL, Zaki S, Popovic T, Meyer RF, Quinn CP, Harper SA, Fridkin SC, Sejvar JJ, Shepard CW, McConnell M, Guarner J, Shieh WJ, Malecki JM, Gerberding JL, Hughes JM, Perkins BA (2001) Bioterrorism-related inhalational anthrax: the first 10 cases reported in the United States. Emerg Infect Dis 7(6):933–944

Johnson KB, Stockwell VO, Sawyer TL, Sugar D (2000) Assessment of environmental factors influencing growth and spread of *Pantoea agglomerans* on and among blossoms of pear and apple. Phytopathology 90:1285–1294

Johnson PC, Wheat LJ, Cloud GA, Goldman M, Lancaster D, Bamberger DM, Powderly WG, Hafner R, Kauffman CA, Dismukes WE (2002) Safety and efficacy of liposomal amphotericin B compared with conventional amphotericin B for induction therapy of histoplasmosis in patients with AIDS. Ann Intern Med 137(2):105–109

Jones TF, Buckingham SC, Bopp CA, Ribot E, Schaffner W (2003) From pig to pacifier: chitterling-associated yersiniosis outbreak among black infants. Emerg Infect Dis 9:1007–1009

Jones T, Gill CO, McMullen LM (2004) The behaviour of log phase *Escherichia coli* at temperatures that fluctuate about the minimum for growth. Lett Appl Microbiol 39:296–300

Jones JL, Kruszon-Moran D, Wilson M, McQuillan G, Navin T, McAuley JB (2007) Toxoplasma gondii infection in the United States, 1999–2004, decline from the prior decade. Am J Trop Med Hyg 77:405–410

Jones RM, Hertwig S, Pitman J, Vipond R, Aspän A, Bälske GR, McCaughey C, McKenna JP, van Rotterdam BJ, de Bruin A, Ruuls R, Buijs R, Roest HJ, Sawyer J (2011) Interlaboratory comparison of real-time polymerase chain reaction methods to detect *Coxiella Burnetii*, the causative agent of Q fever. J Vet Diagn Invest 23:108–111

Jonsson CB, Figueiredo LTM, Vapalahti O (2010) A global perspective on hantavirus – ecology, epidemiology, and disease. Clin Microbiol Rev 23:412–441

Jorgensen PH, Lund E (1985) Detection and stability of enteric viruses and indicator bacteria in sludge, soil and groundwater. Water Sci Technol 17:185–195

Kadlec RH (2009) Comparison of free water and horizontal subsurface treatment wetlands. Ecol Eng 35:159–174

Kapusinszky B, Minor P, Delwart E (2012) Nearly constant shedding of diverse enteric viruses by two healthy infants. J Clin Microbiol 50(11):3427–3434

Karanikolas M, Velissaris D, Marangos M, Karamouzos V, Fligou F, Filos KS (2010) Prolonged high-dose intravenous magnesium therapy for severe tetanus in the intensive care unit: a case series. J Med Case Rep 4:100

Karon AE, Archer JR, Sotir MJ, Monson TA, Kazmierczak JJ (2007) Human multidrug-resistant *Salmonella* newport infections, Wisconsin, 2003–2005. Emerg Infect Dis 13:1777–1780

Kauffman CA (1999) Sporotrichosis. Clin Infect Dis 29:231–237

Kauffman CA (2007) Histoplasmosis: a clinical and laboratory update. Clin Microbiol Rev 20 (1):115–132

Kauffman CA, Bustamante B, Chapman SW, Pappas PG (2007) Clinical practice guidelines for the management of sporotrichosis: 2007 update by the Infectious Diseases Society of America. Clin Infect Dis 45:1255–1265
Kayser FH, Bienz KA, Eckert J, Zinkernagel RM (2005) Medical microbiology. Thieme, Stuttgart
Ketley JM (1997) Pathogenesis of enteric infection by *Campylobacter*. Microbiology 143:5–21
Klastersky J, Zinner SH (1982) Synergistic combinations of antibiotics in Gram-negative bacillary infections. Rev Infect Dis 4:294–301
Klich MA (2002) Biogeography of *Aspergillus* species in soil and litter. Mycologia 94:21–27
Koelsch RK, Lorimor JC, Mankin KR (2006) Vegetative treatment systems for management of open lot runoff: review of literature. Appl Eng Agric 22:141–153
Koohmaraie M, Arthur TM, Bosilevac JM, Guerini M, Shackelford SD, Wheeler TL (2005) Post-harvest interventions to reduce-eliminate pathogens in beef. Meat Sci 71:79–91
Koopowitz A, Smith P, van Rensburg N, Rudman A (2010) *Balantidium coli*-induced pulmonary haemorrhage with iron deficiency. S Afr Med J 100:534–536
Kraus AA, Priemer C, Heider H, Kruger DH, Ulrich R (2005) Inactivation of Hantaan virus-containing samples for subsequent investigations outside biosafety level 3 facilities. Intervirology 48:255–261
Krawczynski K (1993) Hepatitis E. Hepatology 17:932–941
Lacey E (1990) Mode of action of benzimidazoles. Parasitol Today 6:112–115
Lahariya C (2016) Vaccine epidemiology: a review. J Fam Med Prim Care 5(1):7–15
Landry EF, Vaughn JM, McHarrell TZ, Beckwith CA (1979) Adsorption of enteroviruses to soil cores and their subsequent elution by artificial rainwater. Appl Environ Microbiol 38:680–687
Laniado-Laborín R, Alcandar-Schramm JM, Cazares-Adame R (2012) Coccidioidomycosis: an update. Curr Fungal Infect Rep 6:113–120
Lasok B, Tenhagen BA (2013) From pig to pork: methicillin-resistant *Staphylococcus aureus* in the pork production chain. J Food Prot 6:1095–1108
de-la-Torre A, Stanford M, Curi A, Jaffe GJ, Gomez-Marin JE (2011) Therapy for ocular toxoplasmosis. Ocul Immunol Inflamm 19:314–320
Lázaro ME, Cantoni GE, Calanni LM, Resa AJ, Herrero ER, Iacono MA, Enria DA, González Cappa SM (2007) Clusters of hantavirus infection, southern Argentina. Emerg Infect Dis 13:104–110
Lednicky JA (2003) Hantavirus: a short review. Arch Pathol Lab Med 127:30–35
Levett PN (2001) Leptospirosis. Clin Microbiol Rev 14(2):296–326
Lima Barros MB, de Almeida Paes R, Schubach AO (2011) Sporothrix schenckii and Sporotrichosis. Clin Microbiol Rev 24(4):633–654
van der Linden L, Wolthers KC, van Kuppeveld FJ (2015) Replication and inhibitors of enteroviruses and parechoviruses. Virus 7(8):4529–4562
Lindsay DS, Blagburn BL, Dubey JP (2002) Survival of nonsporulated *Toxoplasma gondii* oocysts under refrigerator conditions. Vet Parasitol 103:309–313
Lister P, Wolter D (2005) Levofloxacin and imipenem combination prevents the emergence of resistance among clinical isolates of *Pseudomonas aeruginosa*. Clin Infect Dis 40:105–114
Liu D, Lawrence M, Austin FW, Ainsworth AJ (2005) Comparative assessment of acid, alkali and salt tolerance in Listeria monocytogenes virulent and avirulent strains. FEMS Microbiol Lett 243:373–378
Lorber B (2010) Listeria monocytogenes. In: Mandell GL, Bennett JE, Dolin R (eds) Principles and practice of infectious diseases, 7th edn. Churchill Livingstone, Philadelphia
Luna VA, King DS, Gulledge JD, Cannons AC, Amuso PT, Cattani J (2007) Susceptibility of Bacillus anthracis, Bacillus cereus, Bacillus mycoides, Bacillus pseudomycoides and Bacillus thuringiensis to 24 antimicrobials using Sensititre automated microbial dilution and Etest agar gradient diffusion methods. J Antimicrob Chemother 60:555–567
Lynch JM, Whipps JM (1990) Substrate flow in the rhizosphere. Plant Soil 129:1–10

Mabaso ML, Appleton CC, Hughes JC, Gouws E (2004) Hookworm (*Necator americanus*) transmission in inland areas of sandy soils in Kwa Zulu-Natal, South Africa. Tropical Med Int Health 9:471–476

Maino A, Garigali G, Grande R, Messa P, Fogazzi GB (2010) Urinary balantidiasis: diagnosis at a glance by urine sediment examination. J Nephrol 23:732–737

Maltezou HC, Raoult D (2002) Q fever in children. Lancet Infect Dis 2:686–691

Marrie TJ (2003) *Coxiella burnetii* pneumonia. Eur Respir J 21:713–719

Matches JR, Liston J, Curran D (1974) *Clostridium perfringens* in the environment. Appl Microbiol 28:655–660

Mawdsley JL, Brooks AE, Merry RJ (1996) Movement of the protozoan pathogen *Cryptosporidium parvum* through three contrasting soil types. Biol Fertil Soils 21:30–36

McClain JBL, Ballou WR, Harrison SM, Steinweg DL (1984) Doxycycline therapy for leptospirosis. Ann Intern Med 100:696–698

Medeiros AA, Marty SD, Tosh FE, Chin TDY (1966) *Erythema nodosum* and *Erythema multiforme* as clinical manifestations of histoplasmosis in a community outbreak. N Engl J Med 274:415–420

Mellanby J, Green J (1981) How does tetanus toxin work? Neuroscience 6:281–300

Mendonça AF, Pires ACB, Barros JGC (1994) Pseudosinkhole occurrences in Brasilia, Brazil. Environ Geol 23:36–40

Mercier J, Lindow SE (2000) Role of leaf surface sugars in colonization of plants by bacterial epiphytes. Appl Environ Microbiol 66:369–374

Meyer JM (2000) Pyoverdines: pigments, siderophores and potential taxonomic markers of fluorescent *Pseudomonas* species. Arch Microbiol 174(3):135–142

Milios KT, Drosinos EH, Zoiopoulos PE (2014) Food safety management system validation and verification in meat industry: carcass sampling methods for microbiological hygiene criteria. A review. Food Control 43:74–81

Mitscherlich E, Marth EH (1984) Microbial survival in the environment. Springer Verlag, Berlin

Montoya JG, Liesenfeld O (2004) Toxoplasmosis. Lancet 363:1965–1976

Moore BE, Sagik BP, Sorber CA (1981) Viral transport to ground water at a wastewater land application site. J Water Pollut Contr Found 53:1492–1502

Moore JE, Heaney N, Millar BC, Crowe M, Elborn JS (2002) Incidence of *Pseudomonas aeruginosa* in recreational and hydrotherapy pools. Commun Dis Public Health 5(1):23–26

Moquet O, Blanchet D, Simon S, Veron V, Michel M, Aznar C (2012) Histoplasma capsulatum in Cayenne. Mycopathologia, French Guiana

Morch K, Hanevik K, Rortveit G, Wensaas KA, Langeland N (2009) High rate of fatigue and abdominal symptoms 2 years after an outbreak of giardiasis. Trans R Soc Trop Med Hyg 103 (5):530–532

Morgan JA, Hoet AE, Wittum TE, Monahan CM, Martin JF (2008) Reduction of pathogen indicator organisms in dairy wastewater using an ecological treatment system. J Environ Qual 37:272–279

Morris-Jones R (2002) Sporotrichosis. Clin Exp Dermatol 27:427–431

Muirhead RW, Collins RP, Bremer PJ (2006) Interaction of *Escherichia coli* and soil particles in runoff. Appl Environ Microbiol 72:3406–3411

Müller H, Borner U, Zierski J, Hempelmann G (1987) Intrathecal baclofen for treatment of tetanus-induced spasticity. Anesthesiology 66:76–79

Nada S, Harik MD (2013) Tularemia: epidemiology, diagnosis, and treatment. Pediatr Ann 42 (7):288–292

Nakauchi K (1999) The prevalence of *Balantidium coli* infection in fifty-six mammalian species. J Vet Med Sci 61:63–65

Nataro JP, Kaper JB (1998) Diarrheagenic *Escherichia coli*. Clin Microbiol Rev 11:142–201

Nelson KL, Murray A (2008) Sanitation for unserved populations: technologies, implementation challenges, and opportunities. Annu Rev Environ Resour 33:119–151

Nettleman MD (2011) *Escherichia coli* 0157:H7 (E. coli 0157:H7) overview. URL http://www.emedicinehealth.com/e_coli_escherichia_coli_0157h7_e_coli_0157h7/page11_em.htm. Accessed 11 Mar 2017

Nieder R, Benbi DK (2008) Carbon and nitrogen in the terrestrial environment. Springer Verlag, Heidelberg

Nieder R, Harden T, Martens R, Benbi DK (2008) Microbial biomass in arable soils of Germany during the growth period of annual crops. J Plant Nutr Soil Sci 171(6):878–885

Nieder R, Benbi DK, Scherer H (2011) Fixation and defixation of ammonium in soils: a review. Biol Fertil Soils 47:1–14

Nokes C, Grantham-McGregor SM, Sawyer AW, Cooper ES, Robinson BA, Bundy DA (1992) Moderate to heavy infections of *Trichuris trichiura* affect cognitive function in Jamaican school children. Parasitology 104(3):539

Noriega CT, Garay RR, Sabanero G, Basurto RT, Sabanero-Lopez M (1993) *Sporothrix schenckii*: culturas en diferentes suelos. Rev Latinoam Micol 35:191–194

Norrby-Teglund A, Muller MP, McGeer A, Gan BS, Guru V, Bohnen J, Thulin P, Low DE (2005) Successful management of severe group A streptococcal soft tissue infections using an aggressive medical regimen including intravenous polyspecific immunoglobulin together with a conservative surgical approach. Scand J Infect Dis 37:166–172

O'Lorcain P, Holland CV (2000) The public health importance of *Ascaris lumbricoides*. Parasitology 121:S51–S71

O'Connor JA (2000) Acute and chronic viral hepatitis. Adolesc Med 11:279–292

Oliver DM, Clegg CD, Heathwaite AL, Haygarth PM (2007) Preferential attachment of *Escherichia coli* to different particle size fractions of an agricultural grassland soil. Water Air Soil Pollut 185:369–375

Ooi MH, Wong SC, Lewthwaite P, Cardosa MJ, Solomon T (2010) Clinical features, diagnosis, and management of enterovirus 71. Lancet Neurol 9:1097–1105

Oostman O, Smego RA (2005) Cervicofacial actinomycosis diagnosis and management. Curr Infect Dis Rep 7(3):170–174

Oppenheimer AP, Arsura EL, Hospenthal DR (2010) Coccidioidomycosis. URL http://emedicine.medscape.com/article/215978-overview. Accessed 19 Mar 2017

Palacios G, Oberste MS (2005) Enteroviruses as agents of emerging infectious diseases. J Neurovirol 11:424

Panda SK, Thakral D, Rehman S (2007) Hepatitis E virus. Rev Med Virol 17:151–180

Parashar D, Khalkar P, Arankalle VA (2011) Survival of hepatitis A and E viruses in soil samples. Clin Microbiol Infect 17:E1–E4

Parker RR, Steinhaus EA, Kohls GM, Jellison WL (1951) Contamination of natural waters and mud with *Pasteurella tularensis* and tularemia in beavers and muskrats in the northwestern United States. Bull Natl Inst Health 193:1–161

Parker NR, Barralet JH, Bell AM (2006) Q fever. Lancet 367:679–688

Paterson DL, Bonomo RA (2005) Extended-spectrum beta-lactamases: a clinical update. Clin Microbiol Rev 18:657–686

Paul EA, Clark FE (1996) Soil microbiology and biochemistry. Academic, New York

Payment P (1989) Presence of human and animal viruses in surface and ground water. Water Sci Technol 21:283–285

Pehrson PO, Bengtsson EA (1984) Long-term follow up study of amoebiasis treated with metronidazole. Scand J Infect Dis 16:195–198

Peix A, Ramirez-Bahena MH, Velazquez E (2009) Historical evolution and current status of the taxonomy of genus *Pseudomonas*. Infect Genet Evol 9(6):1132–1147

Penn RL, Kinasewitz GT (1987) Factors associated with a poor outcome in tularemia. Arch Intern Med 147:265–268

Peter-Varbanets M, Zurbrugg C, Swartz C, Pronk W (2009) Decentralized systems for potable water and the potential of membrane technology. Water Res 43:245–265

Petri WA (2005) Treatment of giardiasis. Curr Treat Options Gastroenterol 8:13–17

Petrikkos G, Skiada A, Lortholary O, Roilides E, Walsh TJ, Kontoyiannis DP (2012) Epidemiology and clinical manifestations of mucormycosis. Clin Infect Dis 54(1):S23–S34
Pintó RM, D'Andrea L, Perez-Rodriguez FJ, Costafreda MI, Ribes E, Guix S, Bosch A (2012) Hepatitis A virus evolution and the potential emergence of new variants escaping the presently available vaccines. Future Microbiol 7:331–346
Pitkanen T, Ponka A, Pettersson T, Kosunen TU (1983) *Campylobacter* enteritis in 188 hospitalized patients. Arch Intern Med 143:215–219
Pitout JD (2010) Infections with extended-spectrum beta-lactamase-producing enterobacteriaceae: changing epidemiology and drug treatment choices. Drugs 70:313–333
Pitout JD (2012) Extraintestinal pathogenic *Escherichia coli*: an update on antimicrobial resistance, laboratory diagnosis and treatment. Expert Rev Anti-Infect Ther 10:1165–1176
Pitout JD (2013) Enterobacteriaceae that produce extended-spectrum beta-lactamases and AmpC beta-lactamases in the community: the tip of the iceberg? Curr Pharm Des 19:257–263
Plym-Forshell L, Ekesbo I (1993) Survival of Salmonella in composted and not composted animal manure. J Veterinary Med Ser B 40:654–658
Porter SR, Czaplicki G, Mainil J, Guattéo R, Saegerman C (2011) Q fever: current state of knowledge and perspectives of research of a neglected zoonosis. Int J Microbiol 2011:248418
Prasad KI, Lalitha RM, Reddy EK, Ranganath K, Srinivas DR, Singh J (2012) Role of early diagnosis and multimodal treatment in rhinocerebral mucormycosis: experience of 4 cases. Int J Oral Maxillofac Surg 70:354–362
Pui CF, Wong WC, Chai LC, Nillian E, Ghazali FM, Cheah YK, Nakaguchi Y, Nishibuchi M, Radu S (2011) Simultaneous detection of *Salmonella* spp., *Salmonella typhi* and *Salmonella typhimurium* in sliced fruits using multiplex PCR. Food Control 22:337–342
Pupo GM, Lan R, Reeves PR (2000) Multiple independent origins of *Shigella* clones of Escherichia coli and convergent evolution of any of their characteristics. Proc Natl Acad Sci U S A 97:10567–10572
Purcell RH (1997) Hepatitis E virus. In: Fields BN, Knipe DM, Howley PM (eds) Fields virology, 3rd edn. Lippincott-Raven, Philadelphia, pp 2831–2843
Rabiela MT, Hornelas Y, Garcia-Allan C, Rodriguez-del-Rosal E, Flisser A (2000) Evagination of *Taenia solium* cysticerci: a histologic and electron microscopy study. Arch Med Res 31 (6):605–607
Radford AJ (1973) Balantidiasis in Papua New Guinea. Med J Aust 1:238–241
Raloff J (2001) Dust storms ferry toxic agents between countries and even continents. Sci News 160:218–220
Rangel JM, Sparling PH, Crowe C, Griffin PM, Swerdlow DL (2005) Epidemiology of *Escherichia coli* O157:H7 outbreaks, United States, 1982–2002. Emerg Infect Dis 11:603–609
Rao VC, Metcalf TG, Melnick JL (1986) Human viruses in sediments, sludges, and soils. Bull WHO 64(1):1–14
Raza A, Iqbal Z, Ghulam M, Muhammad G, Khan MA, Hanif K (2013) Amoebiasis as a major risk to human health: a review. Int J Mol Med Sci 3(3):13–24
Redder A, Dürr M, Daeschlein G, Baeder-Bederski O, Koch C, Müller R, Exner M, Borneff-Lipp M (2010) Constructed wetlands: are they safe in reducing protozoan parasites? Int J Hyg Environ Health 213:72–77
Reid AE, Dienstag JL (1997) Viral hepatitis. In: Richman DD, Whitley RJ, Hayden FG (eds) Clinical virology. Churchill Livingstone, New York, pp 69–86
Reinert P, Soubeyrand B, Gauchoux R (2003) 35-year measles, mumps, rubella vaccination assessment in France. Arch Pediatr 10:948–954
Remington JS, McLeod R, Thulliez P, Desmonts G (2006) Toxoplasmosis. In: Remington JS, Klein JO, Wilson CB, Baker CJ (eds) Infectious diseases of fetus and newborn infant, 6th edn. Elsevier-Saunders, Philadelphia, pp 947–1091
Retamar P, Lopez-Cerero L, Muniain MA, Pascual A, Rodriguez-Bano J, Group ERG (2013) Impact of the MIC of piperacillin-tazobactam on the outcome of patients with bacteremia due

to extended-spectrum-beta-lactamase-producing Escherichia coli. Antimicrob Agents Chemother 57:3402–3404

Riha WE, Solberg M (1975) *Clostridium perfringens* inhibition by sodium nitrite as a function of pH, inoculum size, and heat. J Food Sci 40:439–442

Rippon J (1988) Sporotrichosis. In: Rippon J (ed) Medical mycology-the pathogenic fungi and the pathogenic actinomycetes, 3rd edn. WB Saunders Company, Philadelphia, pp 325–352

Robert-Gangneux F, Dardé ML (2012) Epidemiology of and diagnostic strategies for toxoplasmosis. Clin Microbiol Rev 25(2):264–296

Robinson DA (1981) Infective dose of *Campylobacter jejuni* in milk. Br Med J 282:1584

Rodrigo C, Fernando D, Rajapakse S (2014) Pharmacological management of tetanus: an evidence-based review. Crit Care 18:217

Rodriguez-Bano J, Pascual A (2008) Clinical significance of extended-spectrum beta-lactamases. Exp Rev Anti-Infect Ther 6:671–683

Roper MM, Marshall KC (1978) Effects of a clay mineral on microbial predation and parasitism on *Escherichia coli*. Microb Ecol 4:279–289

Roque MR, Roque BL, Foster CS (2010) Actinomycosis in ophthalmology. URL http://emedicine.medscape.com/article/1203061-overview. Accessed 27 Feb 2017

Rosenthal J, Brandt KD, Wheat LJ, Slama TG (1983) Rheumatologic manifestations of histoplasmosis in the recent Indianapolis epidemic. Arthritis Rheum 26:1065–1070

Rosner B, Stark K, Werber D (2010) Epidemiology of reported *Yersinia enterocolitica* infections in Germany, 2001–2008. BMC Public Health 10:337

Ross CM, Donnison AM (2006) *Campylobacter jejuni* inactivation in New Zealand soils. J Appl Microbiol 101:1188–1197

Ruiz-Palacios GM (2007) The health burden of *Campylobacter* infection and the impact of antimicrobial resistance: playing chicken. Clin Infect Dis 44:701–703

Ryle GJA, Powell CE, Gordon AJ (1979) Respiratory costs of nitrogen-fixation in soybean, cowpea, and white clover. 1. Nitrogen-fixation and the respiration of the nodulated root. J Exp Bot 30:135–144

Saadatnia G, Golkar M (2012) A review on human toxoplasmosis. Scand J Infect Dis 44 (11):805–814

Sabah JR, Thippareddi H, Marsden JL, Fung DYC (2003) Use of organic acids for the control of *Clostridium perfringens* in cooked vacuum-packaged restructured roast beef during an alternative cooling procedure. J Food Prot 66:1408–1412

Sabah JR, Juneja VK, Fung DYC (2004) Effect of spices and organic acids on the growth of *Clostridium perfringens* during cooling of cooked ground beef. J Food Prot 67:1840–1847

Santamaria J, Toranzos GA (2003) Enteric pathogens and soil: a short review. Int Microbiol 6:5–9

Sasaki T, Kobayashi M, Agui N (2000) Epidemiological potential of excretion and regurgitation by *Musca domestica* (Diptera: Muscidae) in the dissemination of *Escherichia coli* O157:H7 to food. J Med Entomol 37:945–949

Sathar MA, Soni PN, Fernandes-Costa FJTD, Wittenberg DF, Simjee AE (1994) Racial differences in the seroprevalence of hepatitis A virus infection in Natal/KwaZulu, South Africa. J Med Virol 44:9–12

Schäfer M, Schauermann J (1990) The soil fauna of beech forests: comparison between a mull and a moder soil. Pedobiology 34:299–314

Schaffner W (2007) Tularemia and other *Francisella* infections. In: Goldman L, Ausiello D (eds) Cecil medicine. Elsevier, Philadelphia, pp 2251–2253

Scharschmidt BF (1995) Hepatitis E: a virus in waiting. Lancet 346:519–520

Schröter D, Wolters V, De Ruiter PC (2003) C and N mineralization in the decomposer food webs of a European forest transect. Oikos 102:294–308

Schuster FL, Ramirez-Avila L (2008) Current world status of *Balantidium coli*. Clin Microbiol Rev 21:626–638

Schwaber MJ, Carmeli Y (2007) Mortality and delay in effective therapy associated with extended-spectrum beta-lactamase production in Enterobacteriaceae bacteraemia: a systematic review and meta-analysis. J Antimicrob Chemother 60:913–920
Schwarzenbach RP, Egli T, Hofstetter TB, von Gunten U, Wehrli B (2010) Global water pollution and human health. Annu Rev Environ Resour 35:109–136
Scott ME (2008) Ascaris lumbricoides: a review of its epidemiology and relationship to other infections. Ann Nestlé 66:7–22
Segal BH, Walsh TJ (2006) Current approaches to diagnosis and treatment of invasive aspergillosis. Am J Respir Crit Care Med Vol 173:707–717
Sela S, Nestel D, Pinto R, Nemny-Lavy E, Bar-Joseph M (2005) Mediterranean fruit fly as a potential vector of bacterial pathogens. Appl Environ Microbiol 71:4052–4056
Semenov AM, van Bruggen AHC, Zelenev VV (1999) Moving waves of bacterial populations and total organic carbon along roots of wheat. Microb Ecol 37:116–128
Seydel KB, Li E, Zhang Z, Stanley SL (1998) Epithelial cell-mediated inflammation plays a crucial role in early tissue damage in amebic infection of human intestine. Gastroenterology 115:1446–1453
Seymour IJ, Appleton (2001) Foodborne viruses and fresh produce. J Appl Microbiol 91:759–773
Shapiro RL, Hatheway C, Swerdlow DL (1998) Botulism in the United States: a clinical and epidemiologic review. Ann Intern Med 129(3):221–228
Sharma S, Thompson GR III (2012) How I treat Coccidioidomycosis. Curr Fungal Infect Rep 7:29–35
Sheorey H, Darby J (2008) Searching for *Salmonella*. Aust Fam Phys 37:806–810
Shetty SR, Punya VA (2008) Palatal mucormycosis: a rare clinical dilemma. Oral Surg 1:145–148
Sigler L, Harris JL, Dixon DM, Flis AL, Salkin IF, Kemna M, Ducan RA (1990) Microbiology and potential virulence of *Sporothrix cyanescens*: a fungus rarely isolated from blood and skin. J Clin Microbiol 28:1009–1015
Silva J, Leite D, Fernandes M, Mena C, Gibbs PA, Teixeira P (2011) *Campylobacter* spp. as a foodborne pathogen: a review. Front Microbiol 2:200
Sivapalasingam S, Friedman CR, Cohen L, Tauxe RV (2004) Fresh produce: a growing cause of outbreaks of foodborne illness in the United States, 1973 through 1997. J Food Prot 67:2342–2353
Sleator RD, Gahan CGM, Hill C (2003) A postgenomic appraisal of osmotolerance in *Listeria monocytogenes*. Appl Environ Microbiol 69:1–9
Smego RA, Foglia G (1998) Actinomycosis. Clin Infect Dis 26(6):1255–1261
Smith LDS (1979) *Clostridium botulinum*: characteristics and occurrence. Clin Infect Dis 1:637–641
Smith JE, Perdek JM (2003) Assessment and management of watershed microbial contaminants. Crit Rev Environ Sci Technol 33:1–27
Smith SE, Read DJ (1997) Mycorrhizal symbiosis. Academic, London
Snowden J, Stovall S (2011) Tularemia: retrospective review of 10 years' experience in Arkansas. Clin Pediatr 50:64–68
Sobsey MD, Shields PA (1987) Survival and transport of viruses in soils: model studies. In: Rao VC, Melnick JL (eds) Human viruses in sediments, sludges, and soils. CRC Press, Boca Raton, pp 155–177
Sobsey MD, Shields PA, Hauchman FS, Davies AL, Rullman VA, Bosch A (1988) Survival and persistence of hepatitis A virus in environmental samples. In: Zuckerman AJ (ed) Viral hepatitis and liver disease. Alan R. Liss Inc, New York, pp 121–124
Sohlenius B (1980) Abundance, biomass and contribution to energy flow by nematodes in terrestrial ecosystems. Oikos 34:186–194
Spellberg B, Walsh TJ, Konntoyiannis DP, Edward JJ, Ibrahim AS (2009) Recent advances in the management of mucormycosis: from bench to beside. Clin Infect Dis 48:1743–1751
Stanley SL (2003) Amoebiasis. Lancet 361:1025–1034

Stelma FF, Talla I, Sow S, Kongs A, Niang M, Polman K, Deelder AM, Gryseels B (1995) Efficacy and side effects of praziquantel in an epidemic focus of *Schistosoma mansoni*. Am J Trop Med Hyg 53:167–170

Straub TM, Pepper IL, Gerba CP (1992) Persistence of viruses in desert soils amended with anaerobically digested sewage sludge. Appl Environ Microbiol 58:636–641

Straub TM, Pepper IL, Gerba CP (1995) Comparison of PCR and cell culture for detection of enteroviruses in sludge-amended field soils and determination of their transport. Appl Environ Microbiol 61:2066–2068

Streeter K, Katouli M (2016) *Pseudomonas aeruginosa*: a review of their pathogenesis and prevalence in clinical settings and the environment. Infect Epidemiol Med 2(1):25–32

Sudarsky LA, Laschinger JC, Coppa GF, Spencer FC (1987) Improved results from a standardized approach in treating patients with necrotizing fasciitis. Ann Surg 206:661–665

Swain RHA (1957) The electron-microscopical anatomy of Leptospira canicola. J Pathol Bacteriol 73:155–158

Tam AW, Smith MM, Guerra ME, Huang CC, Bradley DW, Fry KE, Reyes GR (1991) Hepatitis E virus (HEV): molecular cloning and sequencing of the full-length viral genome. Virology 185:120–131

Tashiro K, Kubokura Y, Kato Y, Kaneko Y, Ogawa M (1991) Survival of *Yersinia enterocolitica* in soil and water. J Vet Med Sci 53:23–27

Tate KW, Pereira MDGC, Atwill ER (2004) Efficacy of vegetated buffer strips for retaining *Cryptosporidium parvum*. J Environ Qual 33:2243–2251

Tate KW, Atwill ER, Bartolome JW, Nader G (2006) Significant *Escherichia coli* attenuation by vegetative buffers on annual grasslands. J Environ Qual 35:795–805

Tauxe RV (1997) Emerging foodborne diseases: an evolving public health challenge. Emerg Infect Dis 3(4):425–434

Tauxe RV, Kruse H, Hedberg C, Potter M, Madden J, Wachsmuth K (1997) Microbial hazards and emerging issues associated with produce; a preliminary report to the national advisory committee on microbiologic criteria for foods. J Food Prot 60:1400–1408

Temple ME, Nahata MC (2000) Treatment of listeriosis. Ann Pharmacother 34(5):656–661

Thompson SE, Katul GG, Porporato A (2010) Role of microtopography in rainfall-runoff partitioning: an analysis using idealized geometry. Water Resour Res 46:W07520

Thorn RH, Coxon CE (1992) Hydrogeological aspects of bacterial contamination of some Western Ireland karstic limestone aquifers. Environ Geol Water Sci 20(1):65–72

Tigertt WD, Benenson AS, Gochenour WS (1961) Airborne Q fever. Microbiol Mol Biol Rev 25:285–293

Tilden J, Young W, McNamara AM, Custer C, Boesel B, Lambert Fair M, Majkowski J, Vugia D, Werner SB, Hollingsworth J, Morris JG (1996) A new route of transmission for Escherichia coli: infection from dry fermented salami. Am J Public Health 86:1142–1145

Topp E, Scott A, Lapen DR, Lyautey E, Duriez P (2009) Livestock waste treatment systems for reducing environmental exposure to hazardous enteric pathogens: some considerations. Bioresour Technol 100:5395–5398

Torres-Narbona M, Guinea J, Martinez-Alarcon J, Munoz P, Gadea I, Bouza E (2007) Impact of zygomycosis on microbiology workload: a survey study in Spain. J Clin Microbiol 45:2051–2053

Tucker TJ, Kirsch RE, Louw SJ, Isaacs S, Kannemeyer J, Robson SC (1996) Hepatitis E in South Africa: evidence for sporadic spread and increased seroprevalence in rural areas. J Med Virol 50:117–119

Turnbull PC (2008) Anthrax in humans and animals, 4th edn. World Health Organization, Geneva

Turner LH (1967) Leptospirosis I. Trans R Soc Trop Med Hyg 61:842–855

Uesugi AR, Harris LJ (2006) Growth of *Salmonella* Enteritidis phage Type 30 in almond hull and shell slurries and survival in drying almonds hulls. J Food Prot 69:712–718

Valério E, Chaves S, Tenreiro R (2010) Diversity and impact of prokaryotic toxins on aquatic environments: a review. Toxins 2:2359–2410

Van der Giessen J, Fonville M, Bouwknegt M, Langelaar M, Vollema A (2007) Seroprevalence of Trichinella spiralis and Toxoplasma gondii in pigs from different housing systems in The Netherlands. Vet Parasitol 148:371–374

Van Ness GB (1971) The ecology of Anthrax. Science 172:1303–1307

Vapalahti O, Mustonen J, Lundkvist A, Henttonen H, Plyusnin A, Vaheri A (2003) Hantavirus infections in Europe. Lancet Infect Dis 3:653–752

Vasilakopoulou A, Dimarongona K, Samakovli A, Papadimitris K, Avlami A (2003) *Balantidium coli* pneumonia in an immunocompromised patient. Scand J Infect Dis 35:144–146

Vibha V, Ailes E, Wolyniak C, Angulo F, Klontz K (2006) Recalls of spices due to bacterial contamination monitored by the U.S. Food and Drug Administration: the predominance of Salmonellae. J Food Prot 69:233–237

Vollaire MR, Radecki SC, Lappin MR (2005) Seroprevalence of *Toxoplasma gondii* antibodies in clinically ill cats in the United States. Am J Vet Res 66(5):874–877

Vymazal J (2011) Constructed wetlands for wastewater treatment: five decades of experience. Environ Sci Technol 45:61–69

Walther BA, Ewald PW (2004) Pathogen survival in the external environment and the evolution of virulence. Biol Rev 79:849–869

Wardle DA (1992) A comparative assessment of factors which influence microbial biomass carbon and nitrogen levels in soil. Biol Rev 67:321–358

Weekers PHH, Bodelier PLE, Wijen JPH, Vogels GD (1993) Effects of grazing by the free-living soil amoebae *Acantbamoeba castellanii, Acantbamoeba pobphaga* and *Hartmannella vermiformis* on various bacteria. Appl Environ Microbiol 59:2317–2319

Welshimer HJ (1960) Survival of *Listeria monocytogenes* in soil. J Bacteriol 80:316–320

Werker AG, Dougherty JM, McHenry JL, Van Loon WA (2002) Treatment variability for wetland wastewater treatment design in cold climates. Ecol Eng 19:1–11

Wheat J, Hafner R, Korzun AH, Limjoco MT, Spencer P, Larsen RA, Hecht FM, Powderly W (1995) Itraconazole treatment of disseminated histoplasmosis in patients with the acquired immunodeficiency syndrome. AIDS Clin Trial Group Am J Med 98(4):336–342

Wheat LJ, Freifeld AG, Kleiman MB, Baddley JW, McKinsey DS, Loyd JE, Kauffman CA (2007) Clinical practice guidelines for the management of patients with histoplasmosis: 2007 update by the Infectious Diseases Society of America. Clin Infect Dis 45(7):807–825

WHO (World Health Organization and Food and Agriculture Organization of the United Nations) (2003) Eighth report of WHO surveillance program for control of foodborne infections and intoxications in Europe, Collab Cent Res Train Food Hygiene Zoonoses. World Health Organization, Geneva

WHO (World Health Organization and Food and Agriculture Organization of the United Nations) (2014) Preventing diarrhoea through better water, sanitation and hygiene: exposures and impacts in low- and middle-income countries. World Health Organization, Geneva

WHO (World Health Organization) (1979) Human viruses in water, wastewater and soil. In: WHO technical report series 639. World Health Organisation, Geneva

WHO (World Health Organization) (1996) Foodborne, waterborne, and soilborne diseases, world health report, fighting disease fostering development. World Health Organization, Geneva

WHO (World Health Organization) (2002) Botulism; factsheet no 270. WHO Media Centre, Geneva

Wilkins CA, Richter MB, Hobbs WB, Whitcomb M, Bergh N, Carstens J (1988) Occurrence of *Clostridium tetani* in soil and horses. S Afr Med J 73:718–720

Willingham AL, Engels D (2006) Control of *Taenia solium* cysticercosis/taeniosis. Adv Parasitol 61:509–566

Wilson M, Lindow SE (1994) Coexistence among epiphytic bacterial populations mediated through nutritional resource partitioning. Appl Environ Microbiol 60:4468–4477

Winfield MD, Groisman EA (2003) Role of nonhost environments in the lifestyles of *Salmonella* and *Escherichia coli*. Appl Environ Microbiol 69:3687–3694

Wolfe MS (2000) Miscellaneous intestinal protozoa. In: Strickland GT (ed) Hunter's tropical medicine and emerging infectious diseases, 8th edn. WB Saunders, Philadelphia, pp 603–606

Wong VK, Turmezei TD, Weston VC (2011) Actinomycosis. BMJ 343:d6099

Woods DF, Reen FJ, Gilroy D, Buckley J, Frye JG, Boyd EF (2008) Rapid multiplex PCR and real-time TaqMan PCR assays for detection of *Salmonella enterica* and the highly virulent serovars Choleraesuis and Paratyphi C. J Clin Microbiol 46:4018–4022

Wright RT, Coffin RB (1984) Measuring microzooplankton grazing by its impact on bacterial production. Microb Ecol 10:917–925

Wu KX, Ng MM, Chu JJ (2010) Developments towards antiviral therapies against enterovirus 71. Drug Discov Today 15:1041–1051

Yao ZY, Yang L, Wang HZ, Wu JJ, Xu JM (2015) Fate of Escherichia coli O157: H7 in agricultural soils amended with different organic fertilizers. J Hazard Mater 296:30–36

Yildiz K, Dokuzeylul B, Ulgen S, Or ME (2016) Blastomycosis: a systematical review. RRJVS 2:84–88

Young KT, Davis LM, Dirita VJ (2007) *Campylobacter jejuni* molecular biology and pathogenesis. Nat Rev Microbiol 5:665–679

Zaitsev SV, Chernukha IG, Evdokimova OA, Belov AS (1989) Survival rate of *Leptospira pomona* in the soil at a natural leptospirosis focus. Zh Mikrobiol Epidemiol Immunobiol 2:64–68

Ziegelbauer K, Speich B, Mäusezahl D, Bos R, Keiser J, Utzinger J (2012) Effect of sanitation on soil-transmitted helminth infection: systematic review and metaanalysis. PLoS Med 9: e1001162

Zieve DZ, Eltz DR, Dugdale DC (2010) Aspergillosis. URL http://www.nlm.nih.gov/medlineplus/ency/article/001326.htm. Accessed 18 Mar 2017

Zuckerman AJ, Thomas HC (1993) Viral hepatitis: scientific basis and clinical management. Churchill Livingstone, London

Chapter 14
Soil as an Environmental Reservoir of Prion Diseases

Abstract Prions are recognized as misfolded, pathologic isoforms of the normal mammalian prion protein, which can uniquely cause infectious inherited or spontaneous disease. They are agents of transmissible spongiform encephalopathies (TSEs). The normal, benign, host-encoded forms of PrP are denoted PrP^{C}, and the infectious disease-associated, misfolded conformers are designated PrP^{TSE}. Earlier it was hypothesized that TSEs were caused by a new type of "slow virus" that was too small to purify but had virus-like phenotypes such as transmissibility and heritability. This was the predominantly held theory until 1982, when Stanley Prusiner proposed that the causative agent was exclusively a protein. He solidified the "protein-only" hypothesis which had been developing and termed the agent ***pro***teinacious ***i***nfectious ***on***ly, or prion. Prion diseases are a family of inevitably fatal neurodegenerative disorders affecting a variety of mammalian species, including human diseases such as Creutzfeldt-Jakob disease (CJD), variant CJD (vCJD), Kuru, Fatal Familial Insomnia (FFI) and Gerstmann-Sträussler-Scheinker Syndrom (GSSS). Animal prion diseases include chronic wasting disease (CWD) in North American deer, elk and moose; scrapie in sheep and goats, bovine spongiform encephalopathy (BSE: "mad cow" disease) in cattle, and transmissible mink encephalopathy (TME) in mink. These diseases are characterized by long incubation periods, spongiform degeneration of the brain and accumulation of an abnormally folded isoform of the prion protein, designated PrP^{Sc}, in brain tissue. The unusual nature of prions has created a formidable challenge for detection and study of the agent. Studies have shown prions to adsorb strongly to soil components, remain infectious and persist for years. Indirect transmission most likely occurs through incidental and geophagic ingestion of soil or other contaminated fomites, as well as deer sign-post behavior such as scraping and marking overhanging branches.

Scrapie has been known in sheep since the early eighteenth century. Clinically infected sheep exhibit the obvious feature of excessive rubbing and scratching of the skin. The terminus "scrapie" goes back to this symptom. The origin of the scrapie agent is unknown, but a familial pattern exists in natural sheep scrapie suggesting that genetics and, possibly, vertical transmission are important. Scrapie has a world-wide distribution and has been documented wherever sheep are raised, with the exception of Australia and New Zealand. In the 1920s a similar

R. Nieder et al., *Soil Components and Human Health*,
https://doi.org/10.1007/978-94-024-1222-2_14

neurodegenerative disease in humans was recognized which was later named Creutzfeldt-Jakob Disease (CJD) after the discovering physicians, Hans Creutzfeldt and Alfons Jakob. In the 1960s, Kuru, a devastating disease of the Fore people in Papua New Guinea, was also identified as a TSE, with an epidemiology suggesting transmission through cannibalistic traditions. CWD was identified in the late 1960s and recognized as a TSE in 1980.

Clinical manifestations of prion diseases generally include progressive neurologic deterioration resulting in ataxia, dementia, and behavioral changes. The diagnostic hallmark of a TSE disease is the presence of extracellular plaques composed of prion aggregates in neurologic or lymphatic tissues. Prion plaque deposits are generally associated with spongiform destruction to brain tissue and elevated levels of astrogliosis in the central nervous system (CNS). Although there are examples of prion transmission through iatrogenic means or infection through blood or tissue grafts, the primary and most natural route is through oral-nasal exposure. This is true for Kuru, BSE, vCJD, transmissible mink encephalopathy (TME), scrapie and CWD. Upon ingestion, studies suggest that the prion infects the gut associated lymphatic tissue, which includes M cells, Peyer's patches (PP) and the follicle-associated epithelium. In challenge studies the PPs appeared to be the first infected tissue, with detectable prions associated with lymphatic tissue in as early as 1 week of oral inoculation. M cells sample the prions from the intestinal lumen then traffic the prion to the PP. Once inside the PP the prion is trafficked by macrophages and DCs to germinal centers in lymph tissues including mesenteric lymph node, retropharyngeal lymph node and spleens, where they are transferred to cells of the immune system (follicular dendritic cells: FDCs). Studies suggest that the process of prion replication and retrograde neuroinvasion is dependent on FDCs and their proximity to the enteric nervous system.

The general nomenclature for the infectious agent is PrP^{Sc}, for the pathologic isoform associated with scrapie, or generically PrP^{RES} since the isoform is generally resistant to protease degradation. Other biochemical hallmarks include conversion of the normal α helical rich PrP^{C} conformation to a β sheet rich PrP^{RES} form. This conformational change allows for the formation of insoluble amyloidogenic aggregates. However, the exact tertiary structure of the prion protein in either conformation up to now is not completely clear.

Prions and their resulting TSE diseases, were hardly known to the general public until the occurrence of BSE and the epidemiologically linking it to vCJD. The possibility of prion transmission to humans triggered a large research effort into BSE, CJD, vCJD, scrapie, CWD and other prion diseases. This extensive research on prions has led to the recognized commonalities between prion diseases and other protein misfolding diseases. The "prion like" or prionoid term is now applied to diseases that include Alzheimer's disease, Parkinson's disease, ALS, Huntington's disease and more. While most recognized protein misfolding diseases are not considered transmissible at the host level, the cell to cell and tissue to tissue transmission that occurs in the host is very similar to prion misfolding, replication and transmission. These diseases, similar to prion diseases, can result from inherited or acquired mutations, spontaneous misfolding events, or environmental or stress factors not yet identified.

Transmission of prion diseases orally is well-known which raises concern about interspecies transmission of animal TSEs to humans. For example, interspecies transmission of BSE to sheep, felines and ungulates has occurred which is regarded to be responsible for the emergence of vCJD in humans. The primary infection pathway for BSE that emerged in the UK in the mid-1980s has probably been via dietary exposure to industrial animal feed. However, risks of transmission of PrP^{BSE} from cattle to humans are reduced by preventing tissues, including brain, spinal cord, dorsal root ganglia, trigeminal ganglia, eyes and tonsils for ruminants 30 months or older and the distal ileum of all ruminants, designated as specified risk material, from entering the food chain. Removal and management of SRM brings up the obvious problem of safe disposal. The disposal of specified risk material wastes infected with PrP^{TSE} is challenging, as the infectivity of these materials exhibits resistance to inactivation by a wide range of physicochemical methods, which are commonly used for disinfection. Currently approved decontamination procedures for specified risk material include thermal or alkaline hydrolysis, incineration, gasification and combustion with extreme temperature and pH requirements. Although these procedures are highly successful in destroying all infectious materials, they are difficult to implement, particularly when a large volume of specified risk material requires decontamination. As a consequence, specified risk materials are still being rendered and deposited in landfills in several countries. However, this procedure does not eliminate risk, because PrP^{TSE} may still be present in these materials, and the potential for leakage from landfills may result in contamination of the environment.

Keywords Prion diseases • Creutzfeld-Jakob disease • Gerstmann-Sträussler-Scheinker syndrome • Fatal familial insomnia • Kuru • Prions in the environment • Prion transmission • Pathogenesis • Therapy • Public health management • Treatment of environmental material infected with prions • Treatment of wastewater

14.1 Overview of Prion Diseases

Transmissible spongiform encephalopathies (TSEs) comprise a class of inevitably fatal neurodegenerative diseases afflicting a number of mammalian species. Prusiner (1982) was the first to confirm the "protein-only" hypothesis which termed the agent ***pr****oteinacious* ***i****nfectious* ***on****ly*, or prion. Besides the "protein-only" hypothesis other hypotheses concerning the nature of the scrapie agent have been disproved, for example, the virino hypothesis and the hypothesis that stoichiometric transformation of PrP^{C} to PrP^{Sc} in vitro requires specific RNA molecules. The "protein only" hypothesis is currently the most widely accepted model, even though data and scientific opinions do not absolutely conform to this idea (Aguzzi and Heikenwalder 2003). As already outlined in general terms by Griffith (1967), characterized in detailed form by Prusiner (1982), and refined by Weissmann (1991), it suggests that the infectious agent causing TSE is devoid of nucleic acid and is identical to a posttranslationally modified form (PrP^{Sc}) of a host protein (PrP^{C}). Probably it differs from the latter only in the conformational state (Cohen et al. 1994). The prion is a pathogen apparently lacking nucleic acid and is composed of an abnormally folded isoform of the prion protein, designated PrP^{TSE}. In mammals, prions reproduce by recruiting normal cellular prion protein (PrP^{C}) and stimulating its conversion to the disease-causing (scrapie) isoform (PrP^{Sc}). A major feature that distinguishes prions from viruses is that PrP^{Sc} is encoded by a chromosomal gene (Prusiner 2001).

PrP^{C} is encoded by *PRNP*, a small, single-copy, housekeeping gene on chromosome 20, which is expressed at highest levels in neurons (Oesch et al. 1985). The gene has only three exons and the entire open reading frame is in one exon. The human PrP^{C} protein is synthesised as a 253 aminoacid polypeptide chain from which the first 22 aminoacids (signal peptide) are cleaved shortly after translation commences. Limited proteolysis of PrP^{Sc} produces a smaller, protease-resistant molecule of approximately 142 amino acids, designated PrP^{C}, which polymerizes into amyloid (McKinley et al. 1991). The polypeptide chains of PrP^{C} and PrP^{Sc} are identical in composition but differ in their three-dimensional, folded structures. PrP^{C} is rich in α-helixes (spiral-like formations of amino acids) and has little β-sheet (flattened strands of amino ac ids), whereas PrP^{Sc} is less rich in α -helixes and has much more β -sheet (Pan et al. 1993; Fig. 14.1). The structural transition from α-helixes to β-sheet in PrP is the fundamental event underlying prion diseases (Prusiner 2001; DeArmond and Bouzamondo 2002). Extension of the β-sheet region confers insolubility to detergents, resistance to protease digestion, and the propensity to aggregate.

PrP^{C} is a highly conserved protein in mammals, and paralogs are present in turtles and possibly even in amphibians (Aguzzi et al. 2008). *PRNP* was identified in 1986 (Basler et al. 1986). The broad, diverse, developmentally regulated expression pattern of PrP^{C} in skeletal muscle, kidney, heart, secondary lymphoid organs, and the CNS suggests a conserved and broad function (Aguzzi and Polymenidou 2004). Many different functions have been attributed to PrP^{C}, including immunoregulation, signal transduction, copper binding, synaptic transmission, induction of apoptosis or protection against apoptotic stimuli, and many others (Aguzzi and Polymenidou 2004). Within the CNS, high PrP^{C} expression levels can be detected in synaptic membranes of neurons. In the periphery, PrP^{C} expression is reported on lymphocytes and at high levels on follicular dendritic cells (Aguzzi and Polymenidou 2004; Aguzzi et al. 2008).

14.1.1 Prion Diseases in Humans

Sporadic CJD makes up around 85% of all recognised human prion disease, about 10–15% of cases are familial, 1% iatrogenic, and variant CJD is a regional disease limited largely to the UK and France (Johnson 2005). The most important prion

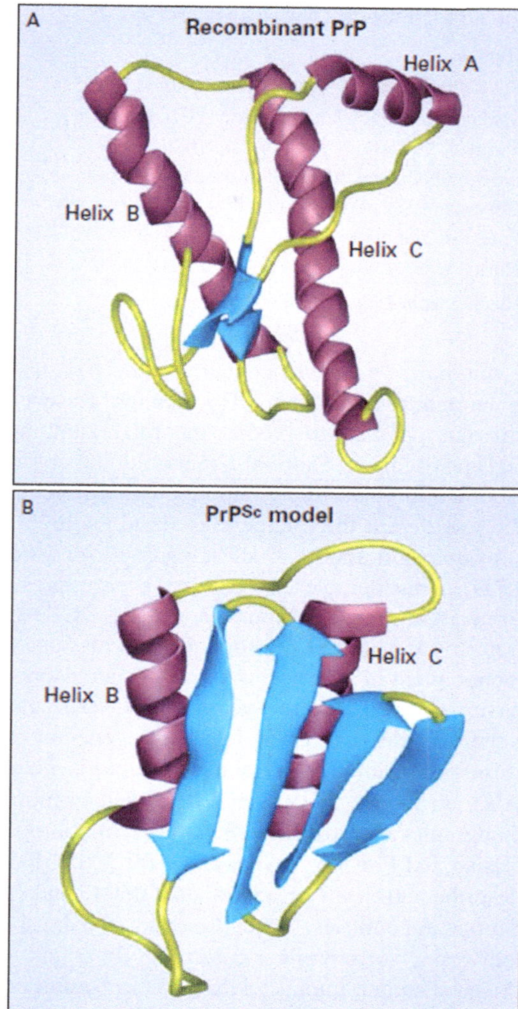

Fig. 14.1 Structures of Prion Protein (PrP) isoforms (Prusiner 2001, p. 1517) (with kind permission from Massachusetts Medical Society). Panel **a**: PrP^{C} with three α-helixes and two short β-strands. Panel **b**: PrP^{Sc} with only two a-helixes and more β-strands

diseases in humans including the year of first report is summarized in Table 14.1. The historical evolution of prion diseases has been summarized by Dalsgaard (2002).

Sporadic Creutzfeldt-Jakob disease (sCJD) occurs ubiquitously with an incidence of 0.5–2 per 1,000,000 per year and is known to have a similar incidence and prevalence among all races and in all climatic zones (Dalsgaard 2002). The disease is reported in all age groups, but the median age at onset is the seventh decade, and it very rarely occurs in persons younger than 40 years. Men and women are equally affected (Ravilochan and Tyler 1993). Variant CJD (vCJD) has well defined clinical features that distinguish it from classic CJD. Infection is believed to occur from bovine prions transmitted through consumption of contaminated beef.

Table 14.1 Important human TSEs

Humans	Year of first report (country)
Sporadic Creutzfeld-Jakob disease (CJD)	1920 (Germany)
Variant Creutzfeld-Jakob disease (vCJD)	1996 (UK)
Familial Creutzfeld-Jakob disease (fCJD)	1924 (Germany)
Iatrogenic Creutzfeld-Jakob disease (iCJD)	1974
Gerstmann-Sträussler-Scheinker syndrome (GSSS)	1928 (Austria)
Fatal familial insomnia	1986 (Italy)
Kuru	1957 (New Guinea)

Adapted from Dalsgaard (2002)

A number of estimates concerning the size of the anticipated epidemic have been generated, ranging from a few hundred to several hundreds of thousands, but it is still not possible to predict the full extent with any certainty (Collinge 1999; Dalsgaard 2002). Familial Creutzfeldt-Jakob disease (fCJD) for the first time was recorded in 1924. Many families with different mutations have been found around the world with three large clusters in Chile, Slovakia, and among Israeli Jews of Libyan origin, the latter all being due to the same point mutation (Gambetti et al. 1999). Considerable concern has been expressed that blood and blood products from asymptomatic donors incubating vCJD may pose a risk for the iatrogenic transmission of vCJD. Iatrogenic transmission of CJD was first suggested in 1974 in the recipient of a corneal transplant from a donor who died of CJD, although it may have occurred several years earlier (Dalsgaard 2002). Gerstmann-Sträussler-Scheinker disease (GSS) is a heterogeneous group of diseases, occurring in five forms that can each be related to a point mutation. GSS was first reported in an Austrian family in 1928 and described in more detail in 1936 (Dalsgaard 2002). Geographically it has been reported in Europe, in the USA, Canada and Japan (Hsiao and Prusiner 1990; Dalsgaard 2002). Familial fatal insomnia (fFI) was first described in 1986 in an Italian family, and has since been reported in several European countries, North America, and Japan (Dalsgaard 2002). Kuru (meaning 'shivering' – from either cold, fear, fever, agitation or exultation) was an extraordinary disorder amongst a tribe in the South Fore region of the almost inaccessible Eastern Highlands in central Papua New Guinea, where the disease was once endemic (Dalsgaard 2002). The disease occurred chiefly in children and women, who practiced an isolated and odd form of (endo)cannibalism by consuming deceased family members, including their brains.

14.1.2 Prion Diseases in Animals

In animals prion diseases include scrapie, bovine spongiform encephalopathy, chronic wasting disease, transmissible mink encephalopathy, feline spongioform encephalopathy and Exotic ungulate encephalopathy (Table 14.2).

Table 14.2 Important TSEs in animals

Disease	Affected species	Year of first report (country)
Scrapie	Sheep and goats	1732 (worldwide)
Bovine spongioform encephalopathy (BSE)	Cattle	1986 (UK)
Chronic wasting disease (CWD)	Wild cervids, elk, captive mule deer	1980 (USA)
Transmissible mink encephalopathy (TME)	Domestic mink	1947 (USA)
Feline spongioform encephalopathy (FSE)	Domestic cats	1990
Exotic ungulate encephalopathy (EUE)	Antelopes in zoological gardens	1986

Adapted from Dalsgaard (2002)

Scrapie, which is a progressive and fatal neurological disease of sheep (and rarely of goats), has been present in Europe for more than two centuries, and has been recorded in various countries and continents (Schreuder 1994). Its descriptive name originates on the phenomenon that in some cases the infected animals scrape off their wool presumably due to intense itching. The first case of bovine spongiform encephalopathy (BSE or 'mad cow disease') occurred in 1985 and was diagnosed in 1986. Since then an epizootic of considerable dimensions has developed, with more than 180,000 affected cattle having been registered to date in the UK. A few thousand cases occurred in other European countries and a few cases were registered in Japan (Dalsgaard 2002). Chronic wasting disease (CWD) of captive mule deer in Colorado, USA, was first reported in 1980, and in 1982 it was found in captive Rocky Mountain elk (Dalsgaard 2002). During recent years it has also been found in free-ranging animals all along the Rocky Mountains from Saskatchewan in Canada to Oklahoma in the USA. Transmissible mink encephalopathy (TME) is a rare disease in ranch-ranged mink with clinico-pathological features similar to those of scrapie. TME has occurred as sporadic outbreaks, and was first reported in 1947 in Wisconsin, USA. Feline spongiform encephalopathy (FSE) has been reported in captive members of the cat family such as puma, cheetah, ocelot and tiger. Since the first case of FSE in a domestic cat in 1990 in the UK many other cases have appeared (Nathanson et al. 1999). Exotic ungulate encephalopathy has affected captive antelope, greater kudu, gemsbok, eland, nyala, Arabian oryx, scimitar-horned oryx, bison (Dalsgaard 2002).

14.2 Prions in the Environment

Prions are capable of horizontal transmission between animals and indirect transmission from contaminated environments. For reasons that remain unclear, indirect environmental transmission of prions appears to concentrate on scrapie (PrP^{Sc}) and

CWD prions (PrP^{CWD}), and does not appear to be an ecological component of BSE or other TSEs. Hence, this section focuses mainly on scrapie and CWD prions. An accurate in vitro model for evaluation of prion fate in the environment up to now has not been established. Most studies to date have used rodent models (hamster and murine brain homogenates infected with various prion strains) or recombinant PrP models (Saunders et al. 2008). These models, however, lack important features of PrP^{Sc} from natural prion diseases. While recombinant PrP displays the same amino acid composition of PrP^{Sc}, it lacks the post-translational modifications and does not share the unique properties of PrP^{Sc}, which include increased stability upon treatment with proteases and denaturing agents, and importantly it does not possess infectivity (Maddison et al. 2010). Furthermore, rodent-passaged PrP^{Sc}, while displaying the broad biochemical properties of PrP^{Sc}, can display altered molecular traits and pathogenesis compared to the 'natural' prion disease prior to rodent adaptation. Studies with authentic prion strains are therefore still required (Maddison et al. 2010).

Animals infected with PrP^{TSE} can cause deposition of infectious particles in the surrounding environment. The resistance of PrP^{TSE} to physiochemical degradation allows the infectious agent to persist in the environment up to many years. A number of studies show that PrP^{TSE} can survive in the environment and poses a long-term hazard outside of the host. The variety of ways prions can be deposited in the environment are shown in Fig. 14.2.

Prion diseases can be transmitted via horizontal, indirect environmental routes, and recent studies suggest that soil may play a key role in natural prion transmission (reviewed by Saunders et al. 2012). Prion shedding can occur many months prior to clinical manifestation of the disease (Gough and Maddison 2010). Prions are shed from diseased hosts in urine, feces, blood, saliva, skin, milk, placenta, and nasal mucus (Gough and Maddison 2010). They also enter the environment after decomposition of diseased animal carcasses, as prions are present near-ubiquitously throughout a diseased host (Saunders et al. 2012). Uptake of prions by animals or humans can occur via ingestion or inhalation of contaminated material (Kincaid and Bartz 2007). However, significant routes of natural exposure up to now have remained uncertain (Saunders et al. 2012). Direct contact between infected and uninfected animals may result in disease transmission via contaminated saliva, blood, milk, skin, or nasal secretions, but the efficacy of direct contact transmission for sustaining endemic prion incidence is not clear. In contrast, indirect transmission has been implicated in multiple CWD and scrapie outbreaks (Georgsson et al. 2006; Miller et al. 2006). Environmental transmission of PrP^{Sc} and PrP^{CWD} has been demonstrated experimentally in multiple studies (Saunders et al. 2012).

Epidemiological records indicate numerous instances of scrapie recurrence on farms previously exposed to scrapie upon reintroduction of animals following 1–19 intervening fallow years (Sigurdarson 1991; Georgsson et al. 2006). For example, PrP^{CWD} deposited in a paddock by infected mule deer has been shown to remain infectious for at least 2 years (Miller et al. 2004). Sheep housed in a facility 16 years after it held a flock clinically diagnosed with PrP^{Sc} developed spongiform

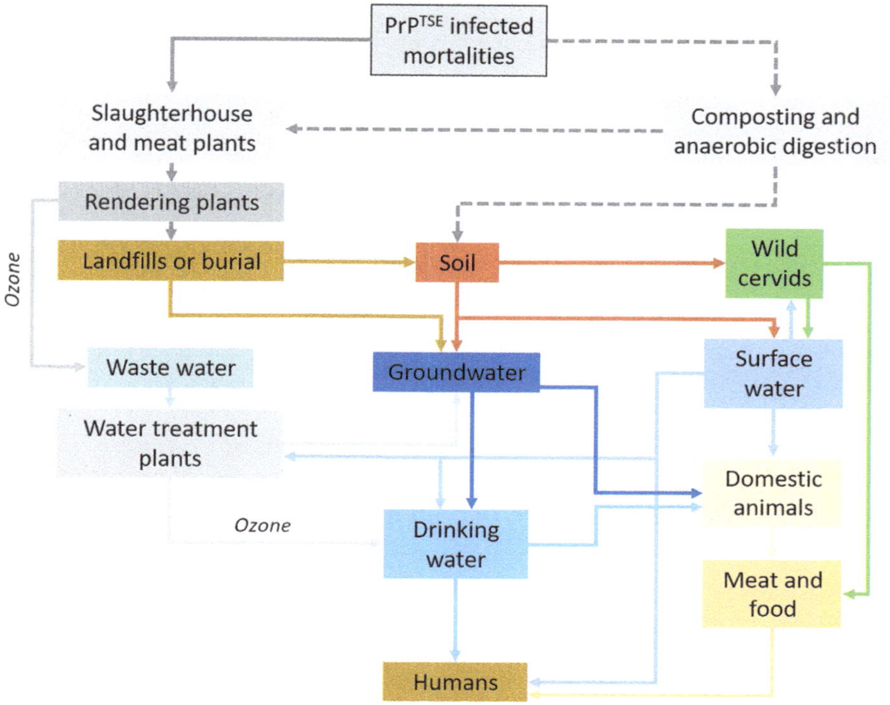

Fig. 14.2 Possible pathways of prions in the environment (Adapted from Xu et al. 2014)

encephalopathies (Georgsson et al. 2006). PrP^{BSE} has been shown to be more resistant to proteolysis by proteinase K compared to PrP^{CWD} and PrP^{Sc} (Breyer et al. 2012).

Pastures can retain infectious CWD prions at least 2 years post-exposure (Miller et al. 2004). Environmental transmission of BSE, Kuru and CJD up to now has not been thoroughly investigated. The relatively low number of vCJD cases worldwide (>220) compared to the high number of confirmed BSE cases (>190,000) shows the limited ability of PrP^{BSE} to cross the bovine-human species barrier (reviewed by Xu et al. 2014). Since the introduction of feed bans in Europe (1994), the USA (1997) and Canada (1997), the number of BSE cases has declined worldwide (Xu et al. 2014).

Wild animals usually ingest about 100 g of soil per day either incidentally, by feeding and/or grooming (Abrahams and Steigmajer 2003), or deliberately in order to complement grass and other forage plants (Freer and Dove 2002) for the ~22 essential chemical elements that are necessary to sustain life and to allow reproduction of the species (Underwood and Suttle 1999). Thus, inadvertent soil ingestion or even dust inhalation could be a pathway for parallel nutrient uptake and scrapie infection. CWD was further reported to jump among species, first to white tail deer and last year to moose.

Table 14.3 Overview of factors potentially affecting spatial variance of prion disease incidence

Environmental factors	Host population factors
Climate	Host/habitat concentrations
Vegetation	Genetics
Vector populations	Demographics
	Predator prevalence
	Host movement
Soil factors	Human factors
Texture	Surveillance
Mineral composition	Game management
SOM content	Hunting
Moisture content	Farm management
Ionic strength	
Organic chemicals	
Trace elements	

Adapted from Saunders et al. (2012)

14.2.1 Major Factors Affecting Spatial Distribution of Prions

Prion diseases such as CJD in humans, CWD, and scrapie exhibit significant geographic variance. A large number of factors may influence spatial distribution of these diseases, including animal movement patterns and habitat prevalence, population genetics, predator prevalence, and human impacts (Table 14.3). Climate, vegetation, the presence of potential vectors, as well as soil characteristics may also influence prion disease incidence for a given area by affecting the prion along its transmission pathway.

The link of prion diseases of domestic animals to environmental factors is difficult to demonstrate due to the fact that animals may be born in one place, raised in another, fattened up in the third, and slaughtered in the fourth (Charlet et al. 2008). Scrapie prevalence across Europe is thus not uniform (Table 14.4).

Outbreaks are most common in Cyprus, Slovenia, Greece, Italy, UK, and France and rates are far from uniform within one country. For instance, hotspots are clustered in France in Basque Country and SW Massif Central and in Italy in Tuscany, Sardinia, Sicily, and Emilia-Romagna regions. In Iceland they were first reported at the beginning of the twentieth century in the mid-north part of the country and by the middle of the century it had spread to both the NW and NE coasts.

14.2.2 Prions in Soils

Soil is a natural sink of prion proteins and may thus facilitate scrapie and CWD transmission by serving as an infectivity reservoir. Once introduced into the soil a protein may be associated with preformed organic and inorganic soil components.

Table 14.4 Occurence of scrapie (cases per 10,000) in goats and sheep in Europe in 2005

State	Sheep	Goats
Cyprus	3596.0	2631
Slovenia	462.2	67.7
Greece	389.2	76.3
Finland	n.d.	48.2
Italy	121.2	3.2
Uk	93.9	15.3
France	69.8	2.1
Slovakia	34.3	n.d.
Spain	34.3	2.5
Netherlands	32.0	n.d.
Ireland	22.4	n.d.
Czech Republic	22.3	n.d.
Belgium	13.5	n.d.
Germany	9.5	n.d.
Portugal	7.0	n.d.
Finland	7.5	n.d.
Sweden	3.1	

Adapted from Charlet et al. (2008)
n.d. Not detected

Understanding the retention and/or the dissemination of prion proteins in soils may be important for dealing with the dissemination of TSE infectivity. Prions can come into contact with soil as a consequence of accidental contamination or intentional disposal of infectious material. The latter can occur via dispersion from storage plants of meat and bone meal, burial of carcasses and natural contamination of pasture soils by grazing herds (Pucci et al. 2012).

Basic soil properties may affect prion transmission pathways. Imrie et al. (2009) found possible correlations between soil pH and soil organic matter (SOM) content and scrapie incidence in the UK. A study on CWD in northern Colorado has suggested a correlation between soil texture and CWD incidence in free-living cervids where the soil clay content of a deer's home range appeared to be positively correlated with the risk of PrP^{CWD} infection (Walter et al. 2011). Specifically montmorillonite, the most commonly occurring smectite clay, has been implicated in the adsorption of prions in the environment and it has been hypothesized that like other proteins, prions may enter the interlayer area of smectite clay. Montmorillonite is prevalent throughout the US Rocky Mountains, including CWD-endemic areas (Violante et al. 1995; Walter et al. 2011). Models suggest that the prevalence of montmorillonite at a landscape level may explain and predict CWD prevalence, which can exceed 45% in free-ranging cervids (Miller et al. 2008). Results for other deer habitat factors such as distance to riparian habitat, location near wintering concentration areas, and landownership (private/public), were less conclusive.

Environmental manganese (Mn) may also be a possible risk factor for prion diseases. Davies and Brown (2009) have shown that exposure to Mn in a soil matrix caused a dramatic increase in prion protein survival (tenfold) over a 2 year period. Manganese also increased infectivity of mouse passaged scrapie to culture cells by two logs. These results clearly verify that manganese is a risk factor for both the survival of the infectious agent in the environment and its transmissibility.

14.2.2.1 Prion Adsorption in Soils

Prions are bound strongly to soil components by adsorption with or entrapment by clays, or participate in the formation of organic and organo-mineral soil complexes which may be responsible for their longevity in the environment. With increasing clay content, soils increase in cation exchange capacity due to increasing negative charge. Electrostatic and hydrophobic interactions between the prion protein and clay are thought to mediate this non-specific adsorption activity. Structural alterations and conformational changes of protein molecules associated to soil colloids may take place with variations of the protein functionality (Rao et al. 2007). Structural molecular alterations under some environmental conditions may also result from protein release from soil aggregates.

Leita et al. (2006) investigated the interaction of a scrapie-specific isoform of the prion protein or of a brain homogenate from hamster-adapted scrapie 263 K strain with mineral soil constituents and whole soils, respectively. Revault et al. (2005) and Rigou et al. (2006) investigated the adsorption/desorption of an ovine recombinant PrP onto montmorillonite and some sandy soils. In all of the above studies it was suggested that the N-terminal part of the ovine recombinant PrP rich in positive amino acid residues and conferring to the protein a net positive charge played a main role in the adsorption/desorption processes. According to Rigou et al. (2006) at least 95% of prion protein was retained by Wyoming montmorillonite for protein/clay ratios varying from 500:1 to 1:1. Comparable maximal adsorbed amounts were measured onto mica sheets by Vasina et al. (2005) who identified the electrostatic attraction between the protein and highly negatively charged mica surfaces as the main interaction mechanism. The sorption capacity of a silty clay loam soil was at least three times higher than a sandy loam soil (or 400 times higher at initial equilibrium) and 2000 times higher than fine quartz sand (Saunders et al. 2009). Sorption of purified PrP^{Sc} to montmorillonite was at least 100 times greater than fine quartz sand (Johnson et al. 2006). Thus, prions contacting clay soils could be rapidly immobilized on the soil surface, forming potent reservoirs for efficient transmission (Saunders et al. 2009). In contrast, prions contacting sandy soils may be more readily transported below the surface and diluted by surface or groundwater.

If active proteins are involved in the formation of SOM, protein-organic and protein-organo-mineral complexes with different structural and functional properties may form. However, the specific sorption properties of SOM with respect to PrPs up to now have been poorly investigated. Studies on interactions of prion proteins with SOM are carried out mainly using chemical extraction methods which may modify both organic substances and mineral phases resulting in possible artifacts when sorption properties of the purified fractions are investigated (Pucci et al. 2008). Rao et al. (2007) demonstrated an irreversible interaction with both synthetic humic-like and humic–mineral complexes (catechol polymers and birnessite-catechol polymers mixtures) and recombinant PrP. However, the highest level of complexity was obtained when PrP was copolymerized within the humic-like complexes. It has also been observed that the interaction of proteins with the prevalently hydrophobic surfaces of SOM might cause structural conformational changes in the immobilized protein (Rao et al. 2007). In the case of the prion protein, such conformational changes may result in the transformation from a non-pathogenic to a pathogenic form of the protein. On the other hand, the results suggest that PrPs should be strongly retained in soils rich in SOM, with very low risks of release and subsequent dissemination away from the initial application site. However, prions adsorbed to soil particles remain generally infectious through oral consumption (Johnson et al. 2007).

To date, there is very limited data on the interaction with soil of natural prion strains of environmental importance such as ovine scrapie, cervid CWD, and BSE of cattle. Maddison et al. (2010) investigated the interaction of natural sheep scrapie and cattle BSE with a wide range of soil types for up to 18 months. The results demonstrated that all detectable ovine scrapie and bovine BSE PrP^{Sc} bound to a range of soil types within 24 h. This highly efficient binding of prions to soils was characterized by truncation of desorbed PrP^{Sc} in a soil-dependent way, with clay-rich soils resulting in N-terminal truncation of the PrP^{Sc} and sand-rich soils yielding full length PrP^{Sc} species. Persistence of PrP^{Sc} within soil and their interaction with soil particles of distinct sizes was determined by both the soil type and the source of the prion, with ovine scrapie being apparently more persistent in some soils than cattle BSE (Maddison et al. 2010). These data indicate that natural ruminant prion strains are stable in the soil environment for at least 18 months and that PrP^{Sc}-soil interaction is dictated by both the soil properties and the strain/host species of PrP^{Sc}.

14.2.2.2 Replication and Infectivity of Soil-Bound Prions

The replication efficiency of soil-bound prions depends significantly on soil properties. For example, the replication ability of PrP^{Sc} is significantly reduced when bound to clay, clay soil, and SOM compared to unbound and sand-bound PrP^{Sc} (Saunders et al. 2011a). In accord with this, higher levels of cell culture infectivity were observed with prions adsorbed to pure sand compared to prions adsorbed to a sandy loam soil (Genovesi et al. 2007). Oral bioassay data also suggest differences in soil-bound prion infectivity among soil types (Johnson et al. 2007). Although variance in soil-bound replication efficiency has been observed independent of soil particle size (Saunders et al. 2011a), soil particle size may be a key determinant of soil-bound oral prion infectivity. The uptake of soil-bound prions by the gut or by nasal cavity may be required prior to replication, potentially favoring smaller, clay-sized soil particle sizes. Johnson et al. (2007) have reported an increase in oral infectivity of purified PrP^{Sc} when bound to two different clays and silicon dioxide powder (particles with similar sizes but distinct mineralogy) compared to unbound controls (Johnson et al. 2007). The infectivity of larger soil particles has not been evaluated so far. It is still unknown if prions must desorb from soil particles to initiate infection in the host. Soil-bound prion uptake and infection initiation thus represent strong research needs.

14.2.2.3 Prion Inactivation in Soils

It is known that prions are long-lived in the environment and unusually resistant to most inactivation treatments for conventional pathogens but the resistance of soil-bound prions to environmental or host degradation up to now remains poorly understood (Saunders et al. 2008). crapie and CWD can remain infectious after years in the environment. Several studies suggest the potential for environmental degradation of PrP^{Sc}. Scrapie-infected sheep tissues placed in a compost pile contained no detectable PrP^{Sc} after 108–148 days (Huang et al. 2007). Similarly, PrP^{Sc} was shown to be degraded in soil below lamb carcasses in a field burial site (Rapp et al. 2006). The degradation was correlated with an increase in proteolytic activity and substrate induced respiration. Scrapie agent buried in garden soil remained infectious for 3 years (Brown and Gajdusek 1991).

Laboratory studies have shown that prions bound to soil with high SOM content may degrade more rapidly compared to prions bound to clay minerals and sand (Saunders et al. 2011b, 2012). An increase of survival of clay-bound PrP was reported in the presence of manganese (Davies and Brown 2009). Enzymatic digestion of soil-bound prions under environmentally relevant conditions was observed to be effective across a number of soil types, although prions bound to SOM may be more susceptible than those bound to other surfaces (Saunders et al. 2012). Use of sterilized and unsterilized soils yielded comparable PrP sorption, suggesting that active soil biota does not significantly alter PrP sorption (Saunders et al. 2012). Degradation of PrP^{Sc} by select lichen extracts has been shown (Johnson et al. 2011) and treatment with Mn oxide under acidic conditions also caused PrP^{Sc} degradation (Russo et al. 2009). So far, the occurrence of certain microorganisms, oxidants, or lichens in soils could reduce prion survival in these soils but the resistance of soil-bound PrP^{Sc} to these degradants has not been evaluated. Preliminary results indicate that binding to soil decreases prion resistance to heat desiccation irrespective of soil-type. This suggests that soils retaining moisture could favor prion persistence (Saunders et al. 2012).

14.2.2.4 Prion Mobility in Soils

The mobility of prions in soil has important implications for the disposal of potentially infected materials, delivery of PrP^{TSE} to surface waters, management of TSE-contaminated pastures, and the bioaccessibility of TSE agents in natural environments (Smith et al. 2011). The potential for partially purified PrP^{TSE} to migrate through soils and porous landfill materials has been investigated in saturated column experiments, (e.g. Jacobson et al. 2009, 2010). The migration of PrP^{TSE} through fine quartz sand, shredded municipal solid waste (fresh and aged), and daily cover materials (natural soils and green waste residual) was examined using landfill leachate as the eluent (Jacobson et al. 2009). All detectable PrP^{TSE} was retained near the point of introduction in columns packed with soil or fine quartz sand, but 0.3–28% of the added PrP^{TSE} eluted from columns containing green residual waste (i.e., composted materials) and municipal solid waste. A similar study using protease-digested brain homogenate as the prion source assessed the potential for PrP^{TSE} to migrate through natural soils (Jacobson et al. 2010). All detectable PrP^{TSE} was retained in columns of five different soils of high sand or silt content when artificial rainwater was used as the eluant. A study using recombinant PrP also reported limited transport in column experiments monitored for 6–9 months under conditions of normal and reduced microbial activity and varying water content (simulating a fluctuating water table) (Cooke and Shaw 2007). Overall, limited prion mobility in soil columns was consistent with the strong attachment of PrP^{TSE} to soil particles. Dissolved organic carbon may contribute to PrP^{TSE} mobility in soil and therefore warrants further investigation. Studies conducted to date suggest that in the absence of preferential flow paths or facilitated transport, prion mobility in most soils and subsurface environments may be quite limited. This suggests maintenance of TSE agents near the soil surface where they would be more accessible to grazing animals. Prions at the soil surface would also be more available for entrainment into overland flow and delivery to water bodies (Nichols et al. 2009).

14.2.3 Prions in Wastewater Treatment Systems

Prions are highly resistant to degradation and to many disinfection procedures. Reports of urinary excretion of PrP^{Sc} by CJD patients and TSE-infected animals (Shaked et al. 2001) have led to concern that CJD agent may enter wastewater treatment plants and be subsequently applied to agricultural fields in treated biosolids. If prions enter wastewater treatment systems through sewers and/or septic systems (e.g., from slaughterhouses, necropsy laboratories, rural meat processors, private game dressing) or through leachate from landfills that have received TSE-contaminated material, they may survive conventional wastewater treatment (Hinckley et al. 2008). There is little knowledge about the degree to which conventional wastewater treatment processes inactivate TSE agents and on the possible transmission of prion disease via contaminated sewage sludge (Yamamoto et al. 2006). Most conventional wastewater treatment facilities rely on aerobic biological treatment processes to remove biodegradable organic matter and on anaerobic digestion to reduce the mass of activated sludge biomass produced. Some bacteria may be able to degrade PrP^{TSE} (Scherebel et al. 2006) and proteolytic enzymes produced by activated sludge and/or anaerobic digester sludge consortia may be capable of inactivating prions.

In a study on prion stability in wastewater treatment processes, anaerobic digester sludge was spiked with PrP^{TSE} and incubated under mesophilic (37 °C) or thermophilic (55 °C) conditions (Kirchmayr et al. 2006). None of the conditions degraded all added PrP^{TSE} which suggests that some prion infectivity could remain in biosolids. Hinckley et al. (2008) examinined the partitioning and persistence of PrP^{TSE} during simulated wastewater treatment processes including activated and mesophilic anaerobic sludge digestion. They found that incubation with activated sludge did not result in significant PrP^{TSE} degradation. PrP^{TSE} and prion infectivity partitioned strongly to activated sludge solids and are expected to enter biosolids treatment processes. A large fraction of PrP^{TSE} survived simulated mesophilic anaerobic sludge digestion. The small reduction in recoverable PrP^{TSE} after 20-days anaerobic sludge digestion appeared attributable to a combination of declining extractability with time and microbial degradation. In summary, the studies by Kirchmayr et al. (2006) and Hinkley et al. (2008) suggest that if prions were to enter municipal wastewater treatment systems, most would partition to activated sludge solids, survive mesophilic anaerobic digestion, and be present in treated biosolids that may be applied to soils as fertilizers.

14.3 Prion Transmission

Herbivores ingest soil during grazing both deliberately and incidentally (Beyer et al. 1994). Locations where infected carcasses, gut piles or placentas have been deposited may represent foci of TSE transmission via ingestion of prions associated with soils. The frequency that cervids and sheep visit such sites and the amount of soil taken up during visits has not been determined so far. In cervids (mule deer, white-tailed deer, elk and moose) ingestion of soil occurs at mineral licks, artificial salt licks and scrapes (Atwood and Weeks 2002). Since minerals from licks are persistent and may draw deer to their location for many years, they may serve as points of prion accumulation from deposited feces, urine and saliva. Scrapes are created by heavy pawing of the soil to remove surface detritus and subsequently marked with deposits of feces, urine or glandular secretions (Miller et al. 1987). Particularly during the rut male deer create these chemical signpoints to communicate with other deer. Multiple males visit each scrape with little indication of revisitation (Alexy et al. 2001). It is not known if females take up soil at scrapes. Soil is estimated to comprise a minimum of 2% of a deer's diet on a dry matter basis (Weeks and Kirkpatrick 1976). The average soil uptake by sheep pastured all year was estimated at 4.5% of the dry matter intake (Fries 1996).

Prion diseases can be categorized as those that are readily transmitted between susceptible individuals resulting in high disease penetrance within a population, such as scrapie and CWD, and those that appear to show limited transmissibility between individuals such as cattle BSE and CJD, and that are transmitted almost exclusively through iatrogenic or foodborne carriage (Gough and Madison 2010). The BSE epidemic that emerged in the mid-1980s was fueled by the feeding of BSE prion-contaminated bone and meat meal to cattle (Kimberlin and Wilesmith 1994). The kuru epidemic that developed in the first half of the twentieth century in Papua New Guinea, caused by ritualistic cannibalism is believed to have originated from a case of sporadic CJD (Weissmann et al. 2002). vCJD is thought to come about by ingestion of BSE prion-contaminated foodstuff. Mice, sheep, calves, and nonhuman primates can be experimentally infected with the BSE agent by the oral route (Weissmann et al. 2002). The occurrence of vCJD in predominantly young persons is probably due to infection by contaminated foodstuff through wounds resulting from teething and tooth loss between early infancy and adolescence.

Transmission of prions can be more efficient within the same than between different species. This phenomenon defines the so-called species barrier. However, the interpretation of the latter has been complicated by the finding that although mice inoculated with prions from another species failed to develop disease and thus appeared to be resistant. Nonetheless, they accumulate PrP^{Sc} and infectivity, albeit only very late after inoculation (Hill et al. 2000; Race et al. 1998, 2001). At present it is unknown which routes of transmission facilitate disease spread in scrapie- and CWD-affected populations, but there have been considerable advancements in the understanding of prion excretion and maintenance within the environment in recent years.

14.3.1 In Vivo Dissemination

The most likely infection route of the acquired prion diseases is via oral intake. An accumulation and amplification of prion infectivity follows in lymphoid tissues associated with the gut. For example, scrapie and CWD display considerable in vivo dissemination, with PrP^{Sc} and infectivity being found in a range of peripheral tissues. This in vivo dissemination facilitates the excretion of prions through multiple routes such as from skin, feces, urine, milk, nasal secretions, saliva and placenta. Excreted scrapie and CWD agent was detected within environmental samples such as water and soil and on the surfaces of inanimate objects (Gough and Maddison 2010).

14.3.2 Excretion/Secretion of Prions

14.3.2.1 Excretion with Urine

Gregori et al. (2008) have shown prion infectivity in urine from clinical animals and also demonstrated the presence of scrapie infectivity within urinary bladder and kidney tissues. Deposition of prion has also been detected within renal tissues of both scrapie-infected sheep 39 and CWD-infected deer, indicating that analogous secretion may be occurring with prion infections of natural hosts. Indeed, deer with clinical CWD and mild to moderate nephritis produced urine containing PrP^{Sc} and CWD-infectivity (Gough and Maddison 2010).

14.3.2.2 Excretion with Fecal Matter

Sheep, naturally infected with PrP^{Sc} excreted prions in their feces during both the preclinical and clinical phases of disease (Gough and Maddison 2010). Safar et al. (2008) used a rodent-scrapie model to demonstrate the presence of prions within feces. Prions were found in feces from hamsters throughout the incubation of the disease and from terminally sick animals. White-tailed deer orally inoculated with a mixture of urine and feces from CWD-infected deer were shown to be sub-clinically infected with CWD, demonstrating that such excreta contain biologically relevant amounts of prion infectivity (Haley et al. 2009).

14.3.2.3 Secretion with Milk

Milk has been proven to contain biologically relevant levels of disease agent with scrapie being transmitted readily within this matrix from ewe to lamb (Konold et al. 2008). The milk harbored infectivity and lambs accumulated PrPSc within their lymphoreticular tissues. Lacroux et al. (2008) used ovinized mice to measure prion infectivity in immunoprecipitates from ovine milk and found infectivity in milk fractions such as cell pellet, casein whey and cream fractions from both colostrum and milk. It was estimated that a milliliter of whole milk could contain prion infectivity equivalent to that found in 6 μg of posterior brain stem from an animal with clinical scrapie. The route through which prions are secreted into milk is still under discussion. It has been speculated that PrP^{Sc} is likely to be transported via exocytosis from epithelial cells and via apocrine secretion of milk fat globules (Didier et al. 2008).

14.3.2.4 Nasal or Saliva Secretions

Tissue homogenates of nasal mucosa from scrapie-affected small ruminants were shown to harbor infectivity (Hadlow et al. 1982). Prions were also present within the papillae of the tongue in sheep with natural scrapie (Casalone et al. 2005). Vascellari et al. (2007) demonstrated that PrP^{Sc} is present within both major (parotid and mandibular) and minor (labial, buccal and palatine) salivary glands; including within epithelial cells and within the lumina of salivary ducts draining saliva into the oral cavity. It was also shown that with both CJD in humans and ovine scrapie PrP^{Sc} is present within olfactory tissues (Gough and Maddison 2010). Saliva from CWD-infected deer was shown to transmit disease to other susceptible naïve deer when harvested from the animals in both the clinical and preclinical stages of infection, although within relatively large volumes (50 ml) of saliva (Mathiason et al. 2006, 2009).

14.3.2.5 Secretion with Parturient Tissues

Placental materials can harbor prion infectivity and it seems likely that such infectivity will be disseminated during and after parturition when lambs are exposed to placental tissues (Gough and Maddison 2010). When placental accumulation of PrP^{Sc} occurs, it has been shown that this can be with ewes in the clinical and preclinical stages of disease (Caplazi et al. 2004). It therefore appears that in ovine scrapie the placenta harbors very high levels of prion but in utero transmission is not occurring (Gough and Maddison 2010). However, following parturition placental tissue is a likely source of infective agent in the transmission of disease from ewes to lambs. In the human placenta, the fetal epithelium is bathed within maternal blood and it is known that vCJD can be transmitted through blood transfusions (Llewelyn et al. 2004). This has led to speculation that human prion diseases may be transmitted through the placenta. However, to date there is no evidence of vertical transmission in human vCJD (Gough and Maddison 2010).

14.3.2.6 Skin as a Source of Prion Infections

As mammalian skin is composed of different strata and cell types that are innervated and carry blood vessels, there is the potential for skin to harbor prions and facilitate its excretion into the environment. Procedures of prion shedding from skin may include natural sloughing of skin, abrasions of the skin and, in the case of sheep, skin cuts that are introduced during shearing (Gough and Maddison 2010). Cervine CWD prions have been found within antler velvet (Angers et al. 2009), a skin layer covering the developing antler of male deer, which is shed after the ossification of antlers. Cunningham et al. (2004) demonstrated the presence of prion infectivity in the skin of BSE-affected kudu. Thomzig et al. (2007) used a rodent model to show that skin-associated prions could be identified from late preclinical stages of disease for both scrapie and BSE, estimating that there is likely to be 5000–10,000 times less prion in the skin of hamsters challenged with scrapie than in the brains of those same animals. This study also demonstrated the presence of prions in the skin of sheep during the late stages of naturally acquired scrapie where PrP^{Sc} was associated with small nerve fibers within the skin. Notari et al. (2010) found PrP^{Sc} within the skin of a vCJD patient.

14.4 Clinical Features of Human TSEs

The human phenotypic range of TSEs includes Creutzfeldt-Jakob disease and its variant form, kuru, Gerstmann-Sträussler-Scheinker syndrome, and fatal familial insomnia (Collins et al. 2004). Notwithstanding the generally low incidence of TSE and their limited infectiousness, major epidemics such as BSE and kuru arise in situations where intraspecies recycling of the abnormal protein is sustained. Moreover, evidence of chronic subclinical infection in animals offers insights into pathogenesis and prompts re-evaluation of the notion of species barriers and present infection control measures.

14.4.1 Spread of Prions Within Organisms

Under natural conditions, prion disease is mainly acquired through oral infection (Weissmann 2004). Orally ingested prions are intestinally absorbed and transported to the blood and lymphoid fluids. Prions are replicated and accumulated in the lymphoreticular system. Transport of prions from the peripheral entry site, particularly the digestive tract, to the lymphoreticular system is attributed to myeloid dendritic cells (Huang et al. 2002). After peripheral replication in the spleen, appendix, tonsils or other lymphoid tissues, prions are transported to the CNS primarily by peripheral nerves (Fig. 14.3).

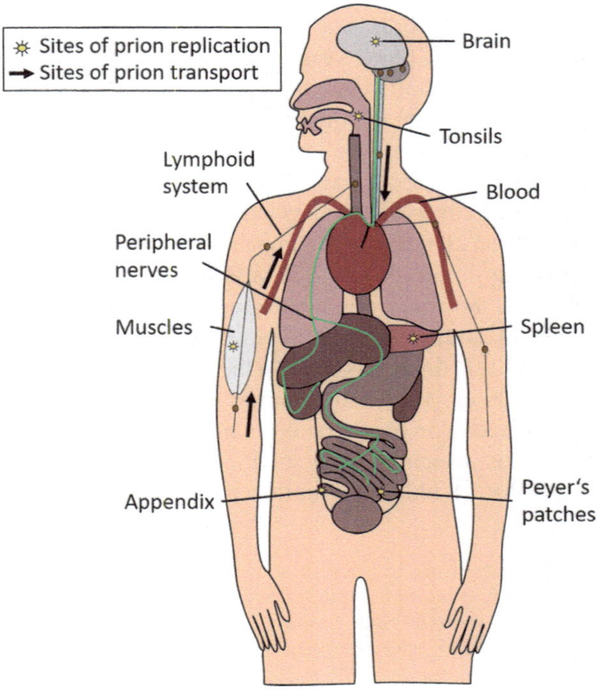

Fig. 14.3 Human tissues and blood involved in replication and transport of prions (Adapted from Aguzzi and Heikenwalder 2006)

14.4.2 Pathogenesis

Following oral infection, an early increase in prion infectivity was observed in the distal ileum. In some experimental rodent scrapie models, as in natural sheep scrapie, infectivity was first detectable in the spleen and other lymphoreticular tissues (Fraser et al. 1992). Spleen titres rise to a plateau early in the incubation period, long before neuroinvasion is detectable. After amplification in the spleen, prions are transferred to the CNS, presumably via the sympathetic nervous system, which provides the main innervation of lymphoid organs. CNS prion replication then rises to high levels and the clinical phase follows next. The most important pathological changes due to prion infection are found in the CNS, with vacuolation, neuronal cell death and astrocytosis being the most common clinical pictures (Weissmann 2004). Histological features of prion-affected brain are shown in Fig. 14.4, morphological features are presented in Fig. 14.5.

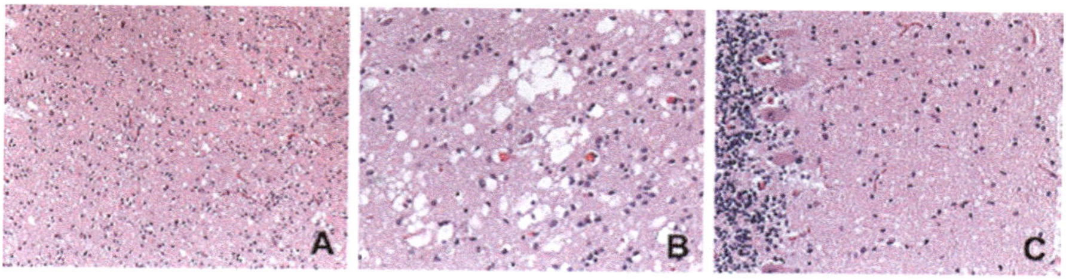

Fig. 14.4 Histological findings in a case with sporadic CJD. (**a**) Cerebral cortex with typical spongioform change accompanied by slight neuronal loss and gliosis. (**b**) Focal area with large confluent vacuoles in the cortex of the same patient. (**c**) Cerebellar cortex with spongioform change in the molecular layer (Unterberger et al. 2005, Figure 1 A–C, p. 34; with kind permission from Springer)

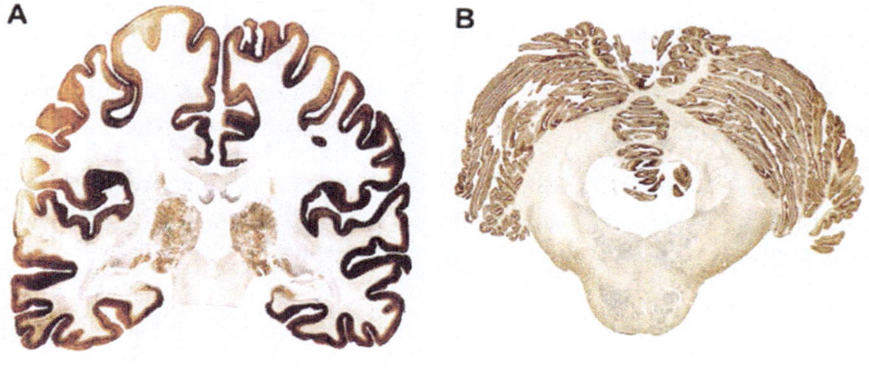

Fig. 14.5 Bihemispheric sections immunostained for PrPSc. Cerebral (**a**) and cerebellar (**b**) cortex of a patient suffering from sporadic CJD with prominent diffuse cortical PrP deposition, but without significant brain atrophy and hippocampal involvement (Unterberger et al. 2005, Figure 2 A, B, p. 35; with kind permission from Springer)

14.4.2.1 Sporadic Creutzfeld-Jakob Disease (CJD)

Sporadic CJD occurs in all countries with an apparently random distribution and annual incidence of 1–2 per million (Collinge 2005). Hypothesised causes include spontaneous production of PrP^{Sc} via rare stochastic events, somatic mutation of PRNP, or unidentified environmental prion exposure. Sporadic CJD typically presents as a rapidly progressive dementia, often accompanied by cerebellar ataxia and myoclonus, with death in an akinetic-mute state after a median of 4–5 months. Around 90% of patients die within 12 months, although survival for more than 2 years is recognized (Brown et al. 1994). The mean age at onset is about 60 years, with little difference in age-adjusted sex incidence. The incidence of sporadic CJD declines after age 70 years (Collins et al. 2004). In up to 40% of patients, sleep disturbance, anxiety and weight loss occurred in the days to weeks immediately before clearcut features of CNS disturbance (Collins et al. 2004). Very abrupt presentations over days were also recognised in up to 20% of patients. Cognitive decline and behavioural disturbance were invariable features, with myoclonus and cerebellar ataxia in 70–80% of cases (Brown et al. 1994). Other extrapyramidal features, such as rigidity, developed in up to 50% of patients. Pyramidal dysfunction and brainstem disturbance with diplopia are frequent. Less common presentations include isolated disturbed vision, and non-myoclonic involuntary movements such athetosis, as chorea, and hemiballismus (Collins et al. 2004).

14.4.2.2 Variant Creutzfeld-Jakob Disease (vCJD)

Contrasting to sporadic CJD, patients with variant disease are much younger (median age at death 29 years), and about 60% present with psychiatric symptoms including anxiety, insomnia, or withdrawal (Spencer et al. 2002). Neurological features were evident in about 35% of individuals at presentation, with unpleasant or painful sensory experiences the most common symptom (Spencer et al. 2002). Nearly 60% of patients reported neurological symptoms by 2 months, but it is generally more than 4 months before clearcut neurological signs such as gait disturbance, slurred speech, and tremor were evident, and longer than 6 months before involuntary movements (dystonia, chorea, or myoclonus), cognitive impairment, and ataxia were manifest (Spencer et al. 2002). Illness duration is usually longer than in sporadic CJD, with a median of 14 months. Death in an akinetic-mute state is a typical outcome. Neuropathologically, brains of patients with variant CJD harbour high burdens of widespread PrP plaques, some of which are encircled by vacuoles, prompting the designation florid plaques.

14.4.2.3 Familial Creutzfeld-Jakob Disease (fCJD)

Familial CJD cases show autosomal dominant inheritance of mutations in *PRNP*. Over 50 different mutations in *PRNP* have been found in kindreds with fCJD

(Johnson 2005). In general, fCJD has earlier age of onset and longer clinical course than sporadic CJD.

14.4.2.4 Iatrogenic Creutzfeld-Jakob Disease (iCJD)

Outbreaks of iCJD have occurred after distribution of contaminated dural graft material and human growth hormone. Since 1985, more than 100 cases of iCJD have occurred 16 months to 18 years after surgical use of human cadaveric dura mater (Brown et al. 2000). In 1985, four patients under 40 years of age developed CJD, all of whom had previously received human growth hormone made from pooled human cadaveric pituitary glands (Brown et al. 1985). Over 8000 children and adolescents in the USA had received this preparation. The product was withdrawn in most countries, and a recombinant human growth hormone was quickly licensed. Since then, however, over 130 young adults have developed CJD 5–30 years after discontinuing injections (Brown et al. 2000). The extremely long incubation period after growth hormone injections presumably reflects the peripheral route of inoculation in contrast to intracerebral placement of contaminated dura mater.

14.4.2.5 Gerstmann-Sträussler-Scheinker Syndrome (GSSS)

Gerstmann-Sträussler-Scheinker disease is characterized by onset at age 20–40 years with progressive cerebellar ataxia and, in many patients, spastic paraparesis (Johnson 2005). In some families myoclonus is not prominent, and dementia develops late. In contrast to sporadic CJD the course of GSSS may last 5–11 years. The pathological changes are also unique with amyloid plaques throughout the brain (Masters et al. 1981).

14.4.2.6 Fatal Familial Insomnia

Fatal familial insomnia is a very strange phenotype of familial prion diseases. The clinical course is dominated by progressive insomnia, autonomic dysfunction and dementia. Polysomnography shows little sleep, loss of sleep spindles, and near absence of rapid-eye-movement sleep. The neuropathological changes are localized largely to neuronal loss in the thalamus. Particularly the anterior ventral and mediodorsal nuclei, and the olivary nuclei of the brainstem are concerned and there is little vacuolization (Johnson 2005).

14.4.2.7 Kuru

Kuru, transmitted by ritual endocannibalism in a remote area of New Guinea, was a progressive cerebellar ataxia leaving victims helpless within a few months. Cognitive changes developed only in advanced stages of disease (Hornabrook 1968).

Remission or survival was never recorded. The adult ratio of women to men with the disease was about ten to one; children over 5 years of age were affected at an intermediate rate with boys and girls being infected equally. Hadlow (1959) described the similarities in epidemiology, clinical signs, and pathological findings between kuru and scrapie.

14.4.3 *Therapeutic Strategies*

Effective treatment of neurodegenerative disease is one of the major challenges facing biomedical research (Collinge 2005). Prion diseases are invariably fatal following a relentlessly progressive course. Although up to now no proven treatment for human or non-human TSEs exists, there have been significant recent advances in understanding prion propagation and neurotoxicity suggesting effective therapeutics for human disease is realistic (Mallucci and Collinge 2005).While attempts at therapeutics in experimental models have focussed on drugs that target PrP^{Sc} formation, the cause of cell death in prion neurodegeneration remains unclear. A key area of controversy is whether PrP^{Sc} is itself directly neurotoxic. There are prion diseases in which PrP^{Sc} levels in brain are very low and, conversely, subclinical prion infection occurs with high levels of PrP^{Sc} and no clinical symptoms (Hill and Collinge 2003). Also, PrP^{Sc} itself is not directly toxic to neurons that do not express PrP^{C}.

Considerable progress has been made in immunotherapeutic approaches. Antibodies against several PrP epitopes inhibited PrP^{Sc} propagation in cell culture (Collinge 2005). These antibodies had little or no affinity for native PrP^{Sc}, and might act by binding cell surface PrP^{C} and reducing its availability for incorporation into propagating prions. Wild type mice that had been peripherally infected with prions were passively immunised with anti-PrP monoclonal antibodies. PrP^{Sc} levels and prion infectivity in the spleens of scrapie infected mice were markedly reduced even when antibodies were administered at the point of near maximal PrPSc accumulation (Collinge 2005). Treated animals remained healthy more than 300 days after untreated animals had succumbed to the disease (White et al. 2003). As antibodies do not readily cross the bloodbrain barrier, there was no protective effect in intracerebrally infected mice. Nevertheless, humanised anti-PrP monoclonal antibodies might be used for post-exposure prophylaxis of particular risk groups (Collinge 2005). For established clinical disease, such antibodies could in principle be given by intracerebroventricular infusion, although effective tissue penetration throughout the CNS would be a problem.

14.5 Public Health Management

Up to present, significant challenges in studies of TSEs in the environment include the lack of sufficiently sensitive and quantitative detection methods and difficulties in extracting TSE agents from environmental matrices. Current detection procedures are based on immunochemical methods (commonly immunoblotting and enzyme-linked immunosorbent essays) that suffer of a limited linear response range and high detection limits (Pedersen et al. 2006).

14.5.1 Avoidance of Iatrogenic Transmission of Prions

Iatrogenic routes of TSE transmission include the use of inadequately sterilized intracerebral electrodes, dura mater and corneal grafting, and from the use of human cadaveric pituitary derived growth hormone or gonadotrophin. There is also evidence that vCJD prion infection is transmissible by blood transfusion (Collinge 2005). In the UK policy for some time has been to leucodeplete all blood plasma products from outside the UK. Another possible route of transmission of TSEs is via contaminated surgical and medical instruments. As prions resist conventional sterilization methods, neurosurgical instruments are known to be able to act as a vector for prion transmission (Bernoulli et al. 1977; Blattler 2002). Several cases of iatrogenic transmission of sporadic CJD prions via neurosurgical instruments were documented (Collinge 2005). The latter TSE may also be transmitted by other surgical procedures. Prions resist conventional sterilization and decontamination methods (Taylor 2000), particularly on stainless steel surfaces (Zobeley et al. 1999). Mild detergents, alcohols, hydrogen peroxide, chlorine dioxide, potassium permanganate, aldehydes, ultraviolet irradiation, and ethylene oxide are ineffective (Collins et al. 2004). Autoclaving at 134 °C for at least 18 min in a porous load device or for about 1 h at standard autoclave temperatures in a gravity displacement sterilizer, or soaking instruments in 1 mol L^{-1} sodium hydroxide or concentrated sodium hypochlorite (more than 5000 ppm available chlorine) for 1 h, are the recommended methods for reducing infectivity (Rutala and Weber 2001). Enzymatic proteolytic inactivation methods (alone or in combination with detergents) are under development (Collins et al. 2004).

Occupational groups are at particular risk of exposure to human prions, for example, neurosurgeons and other operating staff, pathologists and morticians, histology technicians, as well as laboratory workers (Collinge 2005). Because of the prolonged incubation periods to prions following administration to sites other than the CNS, which is associated with clinically silent prion replication in the lymphoreticular tissue, treatments inhibiting prion replication in lymphoid organs may represent a viable strategy for rational secondary prophylaxis following accidental exposure.

14.5.2 Minimizing Risks of BSE Transmission from Cattle to Humans

The relatively low number of vCJD cases worldwide (>220) compared to the high number of confirmed BSE cases (>180,000) is a testament to the limited ability of PrP^{BSE} to cross the bovine–human species barrier (Collinge et al. 1996; Xu et al.

2014). The number of BSE cases has declined worldwide since the introduction of feed bans in Europe (1994), in the USA (1997) and in Canada (1997), which prevented the feeding of animal proteins, including specified risk material to livestock (Xu et al. 2014). Additionally, an enhanced feed ban was introduced in these countries about 10 years later to require that particularly specified risk material be removed completely from the food chain (Xu et al. 2014). Currently, risks of transmission of PrP^{BSE} from cattle to humans are reduced by preventing specific risk materials, including brain, spinal cord, dorsal root ganglia, trigeminal ganglia, eyes and tonsils for ruminants 30 months or older and the distal ileum of all ruminants from entering the food chain (Xu et al. 2014).

14.5.3 Depopulation of Herds Affected by TSE-Infected Animals

Efforts to control transmission of TSEs involve depopulation of herds associated with infected animals. For example, when the first BSE case was discovered in the USA in 2003, 2000 Mg of potentially infected beef products were discarded (Pedersen et al. 2006). By these measures, large amounts of infected waste are generated, creating the need for safe disposal for carcasses and other materials. Potentially viable disposal methods include landfilling and different methods for treatment of infected material.

14.5.4 Landfilling

Landfilling may represent a low-cost disposal option. However, the risks associated with landfilling TSE-contaminated waste materials up to now are not sufficiently known (Pedersen et al. 2006). For example, there are only limited data available on prion fate and transport in porous media (soils, solid wastes). The extraordinary recalcitrance of prions may imply that they may persist within a landfill for many years and may also pose a risk on other environmental compartments, including the hydrosphere. Several of the currently available methods for treatment of specific risk material may be sufficient to reduce PrP^{TSE} to levels that would allow release of treated material into the environment.

14.6 Methods for Treatment of Environmental Material Infected with TSEs

The inactivation of environment-related prions is problematic, as no other pathogens have the unique composition and extreme recalcitrance of PrP^{TSE}. The fact that many PrP^{TSE} detection methods are not well suited for environmental samples add to these challenges.

14.6.1 Aerobic Treatment

Composting is primarily an aerobic biological process and compost is well known for its ability to kill a variety of pathogens (Xu et al. 2014), as well as to produce a product that has valuable properties as a soil amendment. Compost contains a wide variety of both bacterial and fungal consortia that produce numerous proteases, some of which may have the capacity to degrade PrP^{TSE}. A number of bacteria that produce proteases with the ability to degrade PrP^{TSE} have been isolated from compost, including *Bacillus spp.*, *Streptomyces spp.* and *Thermus spp.* (Ryckeboer et al. 2003). Currently, composting of animal mortalities and risk material using static piles, bins, vessels or windrows, has been shown to be an effective means of disposing of sheep, cattle and deer mortalities (Xu et al. 2014). Composting requires minimal infrastructure and can be used on site, thereby eliminating the need to transport risk material to disposal sites. Compost is commonly alkaline (pH 8–10), with temperatures commonly exceeding 55 °C for weeks or even months (Xu 2012). These conditions may further enhance the degradation of PrP^{TSE}. Huang et al. (2007) reported that PrP^{Sc} was near or below the detection limit by directly testing scrapie-infected tissues after 108–148 days of composting. Xu et al. (2014) further observed a 1–2 $\log_{10}$ degradation of PrP^{CWD} and PrP^{BSE} after 28 days of composting in a laboratory composting system. These results indicate that composting can result in significant degradation of PrP^{TSE} and a reduction in the risk of composted risk material or animal mortalities contributing to prion transmission.

14.6.2 Anaerobic Treatment

For anaerobic digestion a wide range of organic substrates have been used, including farm, industrial and municipal wastes (Xu et al. 2014). However, there exist only few studies related to PrP^{TSE} inactivation during anaerobic digestion. The high protein and fat content of slaughterhouse wastes (Mcllwain and Bachelard 1985) makes them a viable substrate for renewable energy production by way of anaerobic digestion. Anaerobic digestion of specific risk material can offer economic returns in the form of biogas production and reduction of rendering costs. Gilroyed et al. (2010) observed that inclusion of risk material as a substrate at the ratios of 10% or 25% (wet weight basis) during 90 days of anaerobic digestion of cattle manure enhanced methane production under both mesophilic (35 °C) and thermophilic (55 °C) conditions. Moreover, the diversion of slaughterhouse or mortality wastes from landfills to biodigesters also offers environmental advantages. Additional benefit of anaerobic digestion of risk material may lie in the inactivation of PrP^{TSE}. Under mesophilic conditions, anaerobic digestion was incapable of reducing or eliminating PrP^{BSE} in sludge after 12 days (Kirchmayr et al. 2006). In contrast, a reduction of PrP^{BSE} was observed under thermophilic conditions but without quantifying the reduction level (Kirchmayr et al. 2006). Hinckley et al. (2008) in a similar manner observed a minor reduction in hamster-adapted PrP^{TME} after 20 days of anaerobic digestion under mesophilic conditions. Miles et al. (2011) reported 2.4 and 3.4 $\log_{10}$ reduction in the infectivity of mouse-adapted PrP^{Sc} as assessed by a cell culture assay after 2 weeks of mesophilic and thermophilic anaerobic digestion, respectively. In summary, these studies suggest that anaerobic biodigestion results in some degree of inactivation of PrP^{TSE}, but at present, it is doubtful whether this inactivation is sufficient to allow biodigestion sludge to be applied to land as a fertilizer.

14.6.3 Thermal and Chemical Procedures

There are several methods available for safe disposal of specific risk material possibly infected with TSEs (Table 14.5). Incineration is currently the primary method of SRM disposal in Europe (Paisley and Hostrup-Pedersen 2005). It is considered the most effective method, as combustion at 1000 °C destroys the infectivity of PrP^{TSE}. Residual infectivity may still remain after treatment at 600 °C (Brown et al. 2004). Gasification is a process through which solid and liquid carbonaceous materials are converted to gaseous products. These gases can be combusted to provide energy or be used for a variety of industrial applications, but the process is not competitive with fossil fuels as dry specific risk material only contains about 60–75% of the energy of coal (Fedorowicz et al. 2007).

Thermal hydrolysis involves exposure of risk material to 180 °C at the pressure of 12 atmospheres for a period of no less than 40 min (Schieder et al. 2000). Alkaline hydrolysis requires processing risk material at 150 °C and 4 atmospheres in a solution with 15% sodium hydroxide or 19% potassium hydroxide added for at

Table 14.5 Procedures for safe disposal of specific risk material possibly infected with TSEs and their costs

Procedure	Conditions	Cost in US $
Alkaline hydrolysis	150 °C and 4 atmospheres for minimum 180 min; 15% NaOH or 19% KOH (wet weight basis) added	Capital cost ~140,000 to 180,000
		410–550 per Mg of risk material
Thermal hydrolysis	180 °C and 12 atmospheres for minimum 40 min	Capital cost ~25 million
Combustion	1000 °C, reduced to ash	Not indicated
Gasification	1000 °C, reduced to ash	Capital cost: 4.4 million
		260–625 per Mg of risk material
Incineration	1000 °C, reduced to ash	Capital cost ~110,000
		200–520 per Mg

Adapted from Xu et al. (2014)

least 180 min to achieve hydrolysis of risk material into small peptides, amino acids and sugars (Murphy et al. 2009). However, these approved methods may not be practical or economical for the disposal of the large amount of potential PrP^{TSE}-contaminated wastes generated from slaughterhouses, rendering plants and waste-water treatment facilities (Xu et al. 2014). In order to deal with this material safely, a successful procedure must effectively inactivate PrP^{TSE}. The production of large amounts of risk material is a financial burden on the agriculture industry, and there is urgent need for new approaches to disposal.

14.7 Methods for Treatment of Wastewater

There is currently a lack of effective approaches for the treatment of PrP^{TSE} in wastewater (Xu et al. 2014). Chlorine and ultraviolet (UV) irradiation are usually used to disinfect microorganisms in wastewater treatment plants. However, PrP^{TSE} are resistant to UV irradiation (Alper et al. 1967), and although chlorine in the form of sodium hypochlorite has been reported to inactivate PrP^{TSE} (Taylor 1999), the concentrations must be extremely high (>5000 mg L^{-1}), enabling the formation of the carcinogenic trihalomethanes.

Ozone (O_3) has a short half-life and thus forms fewer and less toxic disinfection by-products as compared to other oxidants such as chlorine and has a high oxidation capacity. It is widely used to degrade a wide range of organic contaminants and pathogens in both potable water and wastewater (Li et al. 2001; Zuma et al. 2009), raising the possibility that O_3 may have the capacity to degrade PrP^{TSE}. Ding et al. (2012, 2013) quantified at least a 4 $\log_{10}$ inactivation in PrP^{Sc} after 5 s of exposure to 13 mg L^{-1} of ozone at pH 4.4 and 20 °C. Ding et al. (2014) further assessed the O_3 inactivation of PrP^{Sc} in raw, gravity-separated and dissolved air flotation treated rendering plant wastewater and in the municipal final effluent. It was found that the presence of organics reduced the ability of O_3 to inactivate PrP^{Sc}, whereas the use of dissolved air flotation largely overcame this deterrent (Ding et al. 2014). At 44.6 mg L^{-1} of O_3, at least a 4 $\log_{10}$ inactivation of PrP^{Sc} was achieved after 5 min of exposure in dissolved air flotation-treated rendering plant wastewater (Ding et al. 2014). Similar reduction was observed when O_3 was applied at 22×5 mg L^{-1} in municipal final effluent (Ding et al. 2014). These results indicate that O_3 could serve as a final barrier for prion inactivation in primary and/or secondary treated wastewater.

References

Abrahams PW, Steigmajer J (2003) Soil ingestion by sheep grazing the metal enriched floodplain soils of mid-Wales. Environ Geochem Health 25(1):17–24

Aguzzi A, Heikenwalder M (2003) Prion diseases: cannibals and garbage piles. Nature 423:127–129

Aguzzi A, Polymenidou M (2004) Mammalian prion biology. One century of evolving concepts. Cell 116:313–327

Aguzzi A, Heikenwalder M (2006) Pathogenesis of prion diseases: current status and future outlook. Nat Rev Microbiol 4:765–775

Aguzzi A, Sigurdson C, Heikenwaelder M (2008) Molecular mechanisms of prion pathogenesis. Annu Rev Pathol Mech Dis 3:11–40

Alexy KJ, Gassett JW, Osborn DA, Miller KV (2001) Remote monitoring of scraping behaviors of a wild population of white-tailed deer. Wildl Soc Bull 29(3):873–878

Alper T, Cramp WA, Haig DA, Clarke MC (1967) Does the agent of scrapie replicate without nucleic acid? Nature 214(5090):764–766

Angers RC, Seward TS, Napier D, Green M, Hoover E, Spraker T, O'Rourke K, Balachandran A, Telling GC (2009) Chronic wasting disease prions in elk antler velvet. Emerg Infect Dis 15:696–703

Atwood TC, Weeks HP (2002) Sex- and age-specific patterns of mineral lick use by white-tailed deer (*Odocoileus virginianus*). Am Midl Nat 148:289–296

Basler K, Oesch B, Scott M, Westaway D, Walchli M, Groth DF, McKinley MP, Prusiner SB, Weissmann C (1986) Scrapie and cellular PrP isoforms are encoded by the same chromosomal gene. Cell 46:417–428

Bernoulli C, Siegfried J, Baumgartner G, Regli F, Rabinowicz T, Gajdusek DC, Gibbs CJ (1977) Danger of accidental personto-person transmission of Creutzfeldt-Jakob disease by surgery. Lancet 1:478–479

Beyer WN, Connor EE, Gerould S (1994) Estimates of soil ingestion by wildlife. J Wildl Manag 58:375–382

Blattler T (2002) Implications of prion diseases for neurosurgery. Neurosurg Rev 25:195–203

Breyer J, Wemheuer WM, Wrede A, Graham C, Benestad SL, Brenig B, Richt JA, Schulz-Schaeffer WJ (2012) Detergents modify proteinase K resistance of PrPSc in different transmissible spongiform encephalopathies (TSEs). Vet Microbiol 157:23–31

Brown P, Gajdusek DC (1991) Survival of scrapie virus after 3 year's interment. Lancet 337:269–270

Brown P, Gajdusek DC, Gibbs CJ Jr, Asher DM (1985) Potential epidemic of Creutzfeldt-Jakob disease from human growth hormone therapy. N Engl J Med 313:728–731

Brown P, Gibbs CJ Jr, Rodgers-Johnson P, Asher DM, Sulima MP, Bacote A, Goldfarb LG, Gajdusek DC (1994) Human spongiform encephalopathy: the National Institutes of Health Series of 300 cases of experimentally transmitted disease. Ann Neurol 35:513–529

Brown P, Preece M, Brandel JP, Sato T, McShane L, Zerr I, Fletcher A, Will RG, Pocchiari M, Cashman NR, d'Aignaux JH, Cervenákova L, Fradkin J, Schonberger LB, Collins SJ (2000) Iatrogenic Creutzfeldt-Jakob disease at the millennium. Neurology 55:1075–1081

Brown P, Rau EH, Lemieux P, Johnson BK, Bacote A, Gajdusek DC (2004) Infectivity studies of both ash and air emissions from simulated incineration of scrapie-contaminated tissues. Environ Sci Technol 38:6155–6160

Caplazi PA, O'Rourke KI, Baszler TV (2004) Resistance to scrapie in PrP ARR/ARQ heterozygous sheep is not caused by preferential allelic use. J Clin Pathol 57:647–650

Casalone C, Corona C, Crescio MI, Martucci F, Mazza M, Ru G, Bozzetta E, Acutis PL, Caramelli M (2005) Pathological prion protein in the tongues of sheep infected with naturally occurring scrapie. J Virol 79:5847–5849

Charlet L, Chapron Y, Roman-Ross G, Hureau C, Hawkins DP, Ragnarsdottir KV (2008) Prions, metals and soils. In: Barnett MO, Kent DB (eds) Adsorption of metals by Geomedia II: variables, mechanisms, and model applications. Elsevier, Amsterdam, The Netherlands, pp 125–152

Cohen FE, Pan KM, Huang Z, Baldwin M, Fletterick RJ, Prusiner SB (1994) Structural clues to prion replication. Science 264:530–531

Collinge J (1999) Variant Creutzfeldt-Jakob disease. Lancet 354:317–323

Collinge J (2005) Molecular neurology of prion disease. J Neurol Neurosurg Psychiatry 76:906–919

Collinge J, Sidle KC, Meads J, Ironside J, Hill AF (1996) Molecular analysis of prion strain variation and the aetiology of 'new variant' CJD. Nature 383:685–690

Collins SJ, Lawson VA, Masters VL (2004) Transmissible spongiform encephalopathies. Lancet 363:51–61

Cooke CM, Shaw G (2007) Fate of prions in soil: longevity and migration of recPrP in soil columns. Soil Biol Biochem 39:1181–1191

Cunningham AA, Kirkwood JK, Dawson M, Spencer YI, Green RB, Wells GAH (2004) Distribution of bovine spongiform encephalopathy in Greater Kudu (*Tragelaphus strepsiceros*). Emerg Infect Dis 10:1044–1049

Dalsgaard NJ (2002) Prion diseases. An overview. APMIS 110:3–13

Davies P, Brown DR (2009) Manganese enhances prion protein survival in model soils and increases prion infectivity to cells. PLoS One 4:e7518

DeArmond SJ, Bouzamondo E (2002) Fundamentals of prion biology and diseases. Toxicology 181–182:9–16

Didier A, Gebert R, Dietrich R, Schweiger M, Gareis M, Märtlbauer E (2008) Cellular prion protein in mammary gland and milk fractions of domestic ruminants. Biochem Biophys Res Commun 369:841–844

Ding N, Neumann NF, Price LM, Braithwaite SL, Balachandran A, Belosevic M, El-Din MG (2012) Inactivation of template-directed misfolding of infectious prion protein by ozone. Appl Environ Microbiol 78(3):613–620

Ding N, Neumann NF, Price LM, Braithwaite SL, Balachandran A, Mitchell G, Belosevic M, El-Din MG (2013) Kinetics of ozone inactivation of infectious prion protein. Appl Environ Microbiol 79(8):2721–2730

Ding N, Neumann NF, Price LM, Braithwaite SL, Balachandran A, Belosevic M, El-Din MG (2014) Ozone inactivation of infectious prions in rendering plant and municipal wastewaters. Sci Total Environ 470–471:717–725

Fedorowicz EM, Miller SF, Miller BG (2007) Biomass gasification as a means of carcass and specified risk materials disposal and energy production in the beef rendering and meatpacking industries. Energy Fuel 21:3225–3232

Fraser H, Bruce ME, Davies D, Farguhar CF, McBride PA (1992) The lymphoreticular system in the pathogenesis of scrapie. In: Prusiner SB, Collinge J, Powell J, Anderton B (eds) Prion diseases of humans and animals. Ellis Horwood, London, pp 308–317

Freer M, Dove H (eds) (2002) Sheep nutrition. CABI, Canberra

Fries GF (1996) Ingestion of sludge applied organic chemicals by animals. Sci Total Environ 185:93–108

Gambetti P, Petersen RB, Parchi P, Chen SG, Capellari S, Goldfarb L, Gabizon R, Montagna P, Lugaresi E, Piccardo P, Ghetti B (1999) Inherited prion diseases. In: Prusiner SB (ed) Prion biology and diseases. Cold Spring Harbor Laboratory Press, New York, USA, pp 509–583

Genovesi S, Leita L, Sequi P, Andrighetto I, Sorgato MC, Bertoli A (2007) Direct detection of soil-bound prions. PLoS One 2:e1069

Georgsson G, Sigurdarson S, Brown P (2006) Infectious agent of sheep scrapie may persist in the environment for at least 16 years. J Gen Virol 87:3737–3740

Gilroyed BH, Reuter T, Chu A, Hao X, Xu W, McAllister T (2010) Anaerobic digestion of specified risk materials with cattle manure for biogas production. Bioresour Technol 101 (15):5780–5785

Gough KC, Maddison BC (2010) Prion transmission: prion excretion and occurrence in the environment. Prion 4:275–282

Gregori L, Kovacs GG, Alexeeva I, Budka H, Rohwer RG (2008) Excretion of transmissible spongiform encephalopathy infectivity in urine. Emerg Infect Dis 14:1406–1412

Griffith JS (1967) Self-replication and scrapie. Nature 215:1043–1044

Hadlow WJ (1959) Scrapie and kuru. Lancet 2:289–290

Hadlow WJ, Kennedy RC, Race RE (1982) Natural infection of Suffolk sheep with scrapie virus. J Infect Dis 146:657–664

Haley NJ, Mathiason CK, Zabel MD, Telling GC, Hoover EA (2009) Detection of sub-clinical CWD infection in conventional test-negative deer long after oral exposure to urine and feces from CWD+ deer. PLoS One 4:7990

Hill AF, Collinge J (2003) Subclinical prion infection. Trends Microbiol 11:578–584

Hill AF, Joiner S, Linehan J, Desbruslais M, Lantos PL, Collinge J (2000) Species barrier-independent prion replication in apparently resistant species. Proc Natl Acad Sci U S A 97:10248–10253

Hinckley G, Johnson CJ, Jacobson KH, Bartholomay C, McMahon KD, McKenzie D, Aiken JM, Pedersen JA (2008) Persistence of pathogenic prion protein during simulated wastewater treatment processes. Environ Sci Technol 42:5254–5259

Hornabrook RW (1968) Kuru-a subacute cerebellar degeneration: the natural history and clinical features. Brain 91:53–74

Hsiao K, Prusiner SB (1990) Inherited human prion diseases. Neurology 40:1820–1827

Huang FP, Farquhar CF, Mabbott NA, Bruce ME, MacPherson GG (2002) Migrating intestinal dendritic cells transport PrPSc from the gut. J Gen Virol 83:267–271

Huang H, Spencer JL, Soutryine A, Guan J, Rendulich J, Balachandran A (2007) Evidence of degradation of abnormal prion protein in tissues from sheep with scrapie during composting. Can J Vet Res 71:34–40

Imrie CE, Korre A, Munoz-Melendez G (2009) Spatial correlation between the prevalence of transmissible spongiform diseases and British soil geochemistry. Environ Geochem Health 31:133–145

Jacobson KH, Lee S, McKenzie D, Benson CH, Pedersen JA (2009) Transport of the pathogenic prion protein through landfill materials. Environ Sci Technol 43:2022–2028

Jacobson KH, Lee S, Somerville RA, McKenzie D, Benson CH, Pedersen JA (2010) Transport of pathogenic prion protein through soils. J Environ Qual 39:1145–1152

Johnson RT (2005) Prion diseases. Lancet Neurol 4:635–642

Johnson CJ, Phillips KE, Schramm PT, McKenzie D, Aiken JM, Pedersen JA (2006) Prions adhere to soil minerals and remain infectious. PLoS Pathog 2:296–302

Johnson CJ, Pedersen JA, Chappell RJ, McKenzie D, Aiken JM (2007) Oral transmissibility of prion disease is enhanced by binding of soil particles. PLoS Pathog 3:e93

Johnson CJ, Bennett JP, Biro SM, Duque-Velasquez JC, Rodriguez CM, Bessen RA, Roke TE (2011) Degradation of the disease-associated prion protein by a serine protease from lichens. PLoS One 6:e19836

Kimberlin RH, Wilesmith JW (1994) Bovine spongiform encephalopathy. Epidemiology, low dose exposure and risks. Ann N Y Acad Sci 724:210–220

Kincaid AE, Bartz JC (2007) The nasal cavity is a route for prion infection in hamsters. J Virol 81:4482–4491

Kirchmayr R, Reichl HE, Schildorfer H, Braun R, Somerville RA (2006) Prion protein: detection in 'spiked' anaerobic sludge and degradation experiments under anaerobic conditions. Water Sci Technol 53:91–98

Konold T, Moore SJ, Bellworthy SJ, Simmons HA (2008) Evidence of scrapie transmission via milk. BMC Vet Res 4:14

Lacroux C, Simon S, Benestad SL, Maillet S, Mathey J, Lugan S, Corbière F, Cassard H, Costes P, Bergonier D, Weisbecker JL, Moldal T, Simmons H, Lantier F, Feraudet-Tarisse C, Morel N, Schelcher F, Grassi J, Andréoletti O (2008) Prions in milk from ewes incubating natural scrapie. PLoS Pathog 4:1000238

Leita L, Fornasier F, De Nobili M, Bertoli A, Genovesi S, Sequi P (2006) Interactions of prion proteins with soil. Soil Biol Biochem 38:1638–1644

Li H, Gyurek LL, Finch GR, Smith DW, Belosevic M (2001) Effect of temperature on ozone inactivation of Cryptosporidium parvum in oxidant demand-free phosphate buffer. J Environ Eng 127(5):456–467

Llewelyn CA, Hewitt PE, Knight RS, Amar K, Cousens S, Mackenzie J, Will RG (2004) Possible transmission of variant Creutzfeldt-Jakob disease by blood transfusion. Lancet 363:417–421

Maddison BC, Owen JP, Bishop K, Shaw G, Rees HC, Gough KC (2010) The interaction of ruminant PrPSc with soils is influenced by prion source and soil type. Environ Sci Technol 44:8503–8508

Mallucci G, Collinge J (2005) Rational targeting for prion therapeutics. Nat Rev Neurosci 6:23–34

Masters CL, Gajdusek DC, Gibbs CJ Jr (1981) Creutzfeldt-Jakob disease virus isolations from the Gerstmann-Sträussler syndrome with an analysis of the various forms of amyloid plaque deposition in the virus-induced spongiform encephalopathies. Brain 104:559–588

Mathiason CK, Powers JG, Dahmes SJ, Osborn DA, Miller KV, Warren RJ, Mason GL, Hays SA, Hayes-Klug J, Seelig DM, Wild MA, Wolfe LL, Spraker TR, Miller MW, Sigurdson CJ, Telling GC, Hoover EA (2006) Infectious prions in the saliva and blood of deer with chronic wasting disease. Science 314:133–136

Mathiason CK, Hays SA, Powers J, Hayes-Klug J, Langenberg J, Dahmes SJ, Osborn DA, Miller KV, Warren RJ, Mason GL, Hoover EA (2009) Infectious prions in pre-clinical deer and transmission of chronic wasting disease solely by environmental exposure. PLoS One 4:5916

McKinley MP, Meyer RK, Kenaga L, Rahbar F, Cotter R, Serban A, Prusiner SB (1991) Scrapie prion rod formation in vitro requires both detergent extraction and limited proteolysis. J Virol 65:1340–1351

McIlwain H, Bachelard HS (eds) (1985) Biochemistry and the central nervous system. Churchill Livingstone, Edinburgh

Miles SL, Takizawa K, Gerba CP, Pepper IL (2011) Survival of infectious prions in Class B biosolids. J Environ Sci Health A 46(4):364–337

Miller KV, Marchinton RL, Forand KJ, Johansen KL (1987) Dominance, testosterone levels, and scraping activity in a captive herd of white-tailed deer. J Mammol 68:812–817

Miller MW, Williams ES, Hobbs NT, Wolfe LL (2004) Environmental sources of prion transmission in mule deer. Emerg Infect Dis 10:1003–1006

Miller MW, Hobbs NT, Tavener SJ (2006) Dynamics of prion disease transmission in mule deer. Ecol Appl 16:2208–2214

Miller MW, Swanson HM, Wolfe LL, Quartarone FG, Huwer SL, Southwick CH, Lukacs PM (2008) Lions and Prions and Deer Demise. PLoS One 3(12):e4019

Murphy RGL, Scanga JA, Powers BE, Pilon JL, VerCauteren KC, Nash PB, Smith GC, Belk KE (2009) Alkaline hydrolysis of mouse-adapted scrapie for inactivation and disposal prion-positive material. J Anim Sci 87:1787–1793

Nathanson N, Wilesmith J, Wells GA, Griot C (1999) Bovine spongiform encephalopathy and related diseases. In: Prusiner SB (ed) Prion biology and diseases. Cold Spring Harbor Laboratory Press, New York, USA

Nichols TA, Pulford B, Wyckoff AC, Meyerett C, Michel B, Gertig K, Hoover EA, Jewell JE, Telling GC, Zabel MD (2009) Detection of protease-resistant cervid prion protein in water from a CWD-endemic area. Prion 3:171–183

Notari S, Moreles FJ, Hunter SB, Belay ED, Schonberger LB, Cali I, Parchi P, Shieh WJ, Brown P, Zaki S, Zou WQ, Gambetti P (2010) Multiorgan detection and characterization of protease-resistant prion protein in a case of variant CJD examined in the United States. PLoS One 5:8765

Oesch B, Westaway D, Wälchli M, McKinley MP, Kent SBH, Aebersold R, Barry RA, Tempst P, Teplow DP, Hood LE, Prusiner SB, Weissmann C (1985) A cellular gene encodes scrapie PrP 27–30 protein. Cell 40:735–746

Paisley LG, Hostrup-Pedersen J (2005) A quantitative assessment of the BSE risk associated with fly ash and slag from the incineration of meat and bone meal in a gas-fired power plant in Denmark. Prevent Vet Med 68:263–275

Pan KM, Baldwin M, Nguyen J, Gasset M, Serban A, Groth D, Mehlhorn I, Huang Z, Fletterick RJ, Cohen FE (1993) Conversion of α-helices into β-sheets features in the formation of the scrapie prion proteins. Proc Natl Acad Sci U S A 90:10962–10966

Pedersen JA, McMahon KD, Benson CH (2006) Prions: novel pathogens of environmental concern? J Environ Eng 132(9):967–969

Prusiner SB (1982) Novel proteinaceous infectious particle causes scrapie. Science 216:136–144

Prusiner SB (2001) Shattuck lecture – neurodegenerative diseases and prions. N Engl J Med 344:1516–1526
Pucci A, D'Acqui LP, Calamai L (2008) Fate of prions in soil: interactions of RecPrP with organic matter of soil aggregates as revealed by LTA-PAS. Environ Sci Technol 42:728–733
Pucci A, Russo F, Rao MA, Gianfreda L, Calamai L, D'Acqui LP (2012) Location and stability of a recombinant ovine prion protein in synthetic humic-like mineral complexes. Biol Fertil Soils 48:443–451
Race R, Jenny A, Sutton D (1998) Scrapie infectivity and proteinase K–resistant prion protein in sheep placenta, brain, spleen, and lymph node: implications for transmission and antemortem diagnosis. J Infect Dis 178:949–953
Race R, Raines A, Raymond GJ, Caughey B, Chesebro B (2001) Long-term subclinical carrier state precedes scrapie replication and adaptation in a resistant species: analogies to bovine spongiform encephalopathy and variant Creutzfeldt-Jakob disease in humans. J Virol 75:10106–10112
Rao MA, Russo F, Granata V, Rita Berisio R, Zagari A, Gianfreda L (2007) Fate of prions in soil: Interaction of a recombinant ovine prion protein with synthetic humic-like mineral complexes. Soil Biol Biochem 39:493–504
Rapp D, Potier P, Jocteur-Monrozier L, Richaume A (2006) Prion degradation in soil: possible role of microbial enzymes stimulated by the decomposition of buried carcasses. Environ Sci Technol 40:6324–6329
Ravilochan K, Tyler KL (1993) Human transmissible neurodegenerative diseases (prion diseases). Semin Neurol 12:178–192
Revault M, Quiquampoix H, Baron MH, Noinville S (2005) Fate of prions in soil: trapped conformation of full-length ovine prion protein induced by adsorption on clays. Biochim Biophys Acta 1724:367–374
Rigou P, Rezaei H, Grosclaude J, Staunton S, Quiquampoix H (2006) Fate of prions in soil: adsorption and extraction by electroelution of recombinant ovine prion protein from montmorillonite and natural soils. Environ Sci Technol 40:1497–1503
Russo F, Johnson CJ, Johnson CJ, McKenzie D, Aiken JM, Pedersen JA (2009) Pathogenic prion protein is degraded by a manganese oxide mineral found in soils. J Gen Viro 90:275–280
Rutala WA, Weber DJ (2001) Creutzfeldt-Jakob disease: recommendations for disinfection and sterilization. Clin Infect Dis 32:1348–1356
Ryckeboer J, Mergaert J, Vaes K, Klammer S, De Clercq D, Coosemans J, Insam H, Swings J (2003) A survey of bacteria and fungi occurring during composting and self-heating processes. Ann Microbiol 53(4):349–410
Safar JG, Lessard P, Tamgüney G, Freyman Y, Deering C, Letessier F, DeArmond SJ, Prusiner SB (2008) Transmission and detection of prions in feces. J Infect Dis 198:81–89
Saunders SE, Bartz JC, Telling GC, Bartelt-Hunt SL (2008) Environmentally-relevant forms of the prion protein. Environ Sci Technol 42:6573–6579
Saunders SE, Bartz JC, Bartelt-Hunt SL (2009) Prion protein adsorption to soil in a competitive matrix is slow and reduced. Environ Sci Technol 43:7728–7733
Saunders SE, Shikiya RA, Langenfeld KA, Bartelt-Hunt SL, Bartz JC (2011a) Replication efficiency of soil-bound prions varies with soil type. J Virol 85:5476–5482
Saunders SE, Yuan Q, Bartz JC, Bartelt-Hunt SL (2011b) Effects of solution chemistry and aging time on prion protein adsorption and replication of soil-bound prions. PLoS One 6:e18752
Saunders SE, Bartz JC, Bartelt-Hunt SL (2012) Soil-mediated prion transmission: Is local soil-type a key determinant of prion disease incidence? Chemosphere 87:661–667
Scherebel C, Pichner R, Groschup MH, Müller-Hellwig S, Scherer S, Dietrich R, Märtlbauer E, Gareis M (2006) Degradation of scrapie associated prion protein (PrP^{Sc}) by the gastrointestinal microbiota of cattle. Vet Res 37:695–703
Schieder D, Schneider R, Bischof F (2000) Thermal hydrolysis (TDH) as a pretreatment method for the digestion of organic waste. Water Sci Technol 41(3):128–187
Schreuder BEC (1994) Animal spongiform encephalopathy – an update. Part 1. Scrapie and lesser known animal spongiform encephalopathies. Vet Q 16:174–181

Shaked GM, Shaked Y, Kariv-Inbal Z, Halimi M, Avraham I, Gabizon R (2001) A protease-resistant prion protein isoform is present in urine of animals and humans affected with prion diseases. J Biol Chem 276:31479–31482

Sigurdarson S (1991) Epidemiology of scrapie in Iceland. In: Bradley R, Savey M, Marchant B (eds) Sub-acute spongiform encephalopathies. Kluwer Academic Publishers, Dordrecht, pp 233–242

Smith CB, Booth CJ, Pedersen JA (2011) Fate of prions in soil: a review. J Environ Qual 40:449–461

Spencer MD, Knight RS, Will RG (2002) First hundred cases of variant Creutzfeldt-Jakob disease: retrospective case note review of early psychiatric and neurological features. BMJ 324:1479–1482

Taylor DM (1999) Inactivation of prions by physical and chemical means. J Hosp Infect 43:S69–S76

Taylor DM (2000) Inactivation of transmissible degenerative encephalopathy agents: a review. Vet J 159:10–17

Thomzig A, Schulz-Schaeffer W, Wrede A, Wemheuer W, Brenig B, Kratzel C, Lemmer K, Beekes M (2007) Accumulation of pathological prion protein PrPSc in the skin of animals with experimental and natural scrapie. PLoS Pathog 3:66

Underwood EJ, Suttle NF (1999) The mineral nutrition of livestock, 3rd edn. CABI Publishing, New York, p 614

Unterberger U, Voigtländer T, Budka H (2005) Pathogenesis of prion diseases. Acta Neuropathol 109:32–48

Vascellari M, Nonno R, Mutinelli F, Bigolaro M, Di Bari MA, Melchiotti E, Marcon S, D'Agostino C, Vaccari G, Conte M, De Grossi L, Rosone F, Giordani F, Agrimi U (2007) PrPSc in salivary glands of scrapie-affected sheep. J Virol 81:4872–4876

Vasina EN, Déjardin P, Rezaei H, Grosclaude J, Quiquampoix H (2005) Fate of prions in soil: adsorption kinetics of recombinant unglycosylated ovine prion protein onto mica in laminar flow conditions and subsequent desorption. Biomacromolecules 6:3425–3432

Violante A, DeCristofaro A, Rao M, Gianfreda L (1995) Physicochemical properties of protein-smectite and protein-Al(OH)(x)-smectite complexes. Clay Miner 30:325–336

Walter WD, Walsh DP, Farnsworth ML, Winkelman DL, Miller MW (2011) Soil clay content underlies prion infection odds. Nat Commun 2:200

Weeks HP, Kirkpatrick CM (1976) Adaptations of white-tailed deer to naturally occurring sodium deficiencies. J Wildl Manag 40:610–625

Weissmann C (1991) A 'unified theory' of prion propagation. Nature 352:679–683

Weissmann C (2004) The state of the prion. Nature Rev 2:861–871

Weissmann C, Enari M, Klöhn PC, Rossi D, Flechsig E (2002) Transmission of prions. J Infect Dis 186(2):S157–S165

White AR, Enever P, Tayebi M, Mushens R, Linehan J, Sebastian Brandner S, Anstee D, Collinge J, Hawke S (2003) Monoclonal antibodies inhibit prion replication and delay the development of prion disease. Nature 422:80–83

Xu S (2012) Composting as a method for disposal of specified risk material and degradation of prions. PhD thesis, University of Alberta, Edmonton, AB, USA

Xu S, Rasmussen J, Ding N, Neumann NF, El-Din MG, Belosevic M, McAllister T (2014) Inactivation of infectious prions in the environment: a mini-review. J Environ Eng Sci 9 (2):125–136

Yamamoto T, Kobayashi S, Nishiguchi A, Nonaka T, Tsutsui T (2006) Evaluation of bovine spongiform encephalopathy (BSE) infection risk of cattle via sewage sludge from wastewater treatment facilities in slaughterhouses in Japan. J Vet Med Sci 68:137–142

Zobeley E, Flechsig E, Cozzio A, Enari M, Weissmann C (1999) Infectivity of scrapie prions bound to a stainless steel surface. Mol Med 5:240–243

Zuma F, Lin J, Jonnalagadda SB (2009) Ozone-initiated disinfection kinetics of Escherichia coli in water. J Environ Sci Health A 44(1):48–56

Index

R. Nieder et al., *Soil Components and Human Health*,
https://doi.org/10.1007/978-94-024-1222-2

D

E

I

K

L

M

P

R

S

T

U

Printed by Printforce, the Netherlands